EHRLICH'S
Geomicrobiology
Sixth Edition

EHRLICH'S
Geomicrobiology
Sixth Edition

Edited by Henry Lutz Ehrlich,
Dianne K. Newman, and Andreas Kappler

CRC Press
Taylor & Francis Group
Boca Raton London New York

CRC Press is an imprint of the
Taylor & Francis Group, an **informa** business

CRC Press
Taylor & Francis Group
6000 Broken Sound Parkway NW, Suite 300
Boca Raton, FL 33487-2742

First issued in paperback 2020

ISBN-13: 978-1-4665-9240-7 (hbk)
ISBN-13: 978-0-367-65872-4 (pbk)

Visit the Taylor & Francis Web site at
http://www.taylorandfrancis.com

and the CRC Press Web site at
http://www.crcpress.com

CONTENTS

Preface / vii
Acknowledgments / ix
Contributors / xi

1 • INTRODUCTION / 1
Henry Lutz Ehrlich, Dianne K.
Newman, and Andreas Kappler

2 • EARTH AS A MICROBIAL HABITAT / 7
Henry Lutz Ehrlich

3 • EMERGENCE OF LIFE AND ITS EARLY
HISTORY / 19
Michael Russell

4 • UPPERMOST LITHOSPHERE AS A
MICROBIAL HABITAT / 55
Henry Lutz Ehrlich

5 • TERRESTRIAL SUBSURFACE ECOSYSTEM / 69
Michael Wilkins and James K. Fredrickson

6 • HYDROSPHERE AS MICROBIAL HABITAT / 97
Mak Saito

7 • GEOMICROBIAL PROCESSES:
A PHYSIOLOGICAL AND BIOCHEMICAL
OVERVIEW / 129
Henry Lutz Ehrlich

8 • CULTIVATION, *IN SITU* MEASUREMENTS,
AND GEOCHEMICAL TECHNIQUES FOR
GEOMICROBIOLOGICAL STUDIES / 157
Greg Druschel and Victoria J. Orphan

9 • MOLECULAR METHODS IN
GEOMICROBIOLOGY / 187
Maureen L. Coleman and Dianne K. Newman

10 • MICROBIAL FORMATION AND
DEGRADATION OF CARBONATES / 209
Tanja Bosak, Jaroslav Stolarski, and
Anders Meiborn

11 • GEOMICROBIAL INTERACTIONS WITH
SILICON / 237
Kurt Konhauser

12 • GEOMICROBIOLOGY OF ALUMINUM:
MICROBES AND BAUXITE / 257
Henry Lutz Ehrlich

13 • GEOMICROBIAL INTERACTIONS WITH
PHOSPHORUS / 265
Bernhard Schink and Diliana
D. Simeonova

14 • GEOMICROBIOLOGY OF NITROGEN / 281
Christopher A. Francis and Karen L. Casciotti

15 • GEOMICROBIAL INTERACTIONS WITH
ARSENIC AND ANTIMONY / 297
Ronald Oremland

16 • GEOMICROBIOLOGY OF MERCURY / 323
Robert Mason

17 • GEOMICROBIOLOGY OF IRON / 343
Andreas Kappler, David Emerson,
Jeffrey A. Gralnick, Eric E. Roden,
and E. Marie Muehe

18 • GEOMICROBIOLOGY OF MANGANESE / 401
Colleen M. Hansel and Deric R. Learman

ACKNOWLEDGMENTS

We, the editors, express sincere appreciation to each of the individuals who contributed to the revising and updating of different chapters and who are named under the respective chapter headings. In addition, we thank the following individuals, listed in alphabetical order, who reviewed individual revised and updated or rewritten chapters: Gregory Dick, Woody Fischer, William C. Ghiorse, Lisa Gieg, Brian Glazer, Jeff Gralnick, Tim Hollibaugh, Nick Lane, Lynne Macaskie, Ronald Oremland, Alexandre Poulain, Chad Saltikov, Allison Santoro, Gordon Southam, Brad Tebo, Dominique Tobler, and Nathan Yee. Thanks to Grant Ferris for reviewing Chapter 11 of the fifth edition and recommending that it be left unmodified for the sixth edition because it still reflected the state of knowledge on the topic at the time of preparation of the latter edition.

Kathryn Everett and Chuck Crumly of CRC Press, Taylor & Francis Group, were most helpful in their advice during the preparation of this edition.

Responsibility for the presentation and interpretation of the subject matter in this edition rests with the contributors and the editors of this edition.

HENRY LUTZ EHRLICH
DIANNE K. NEWMAN
ANDREAS KAPPLER

COLLEEN M. HANSEL
Department of Marine Chemistry and
    Geochemistry
Woods Hole Oceanographic Institution
Woods Hole, Massachusetts

IAN M. HEAD
School of Civil Engineering and Geosciences
Newcastle University
Newcastle upon Tyne, United Kingdom

ANDREAS KAPPLER
Department of Geomicrobiology
Center for Applied Geoscience (ZAG)
Eberhard-Karls-University of Tuebingen
Tuebingen, Germany

KURT KONHAUSER
Department Earth and Atmospheric Sciences
University of Alberta
Edmonton, Alberta, Canada

DERIC R. LEARMAN
Department of Biology
and
Institute for Great Lakes Research
Central Michigan University
Mt. Pleasant, Michigan

WILLIAM D. LEAVITT
Department of Earth and Planetary Sciences
Washington University in St. Louis
St. Louis, Missouri

JONATHAN R. LLOYD
School of Earth, Atmospheric and Environmental
    Sciences
The University of Manchester
Manchester, United Kingdom

ROBERT MASON
Department of Marine Sciences
University of Connecticut
Groton, Connecticut

ANDERS MEIBORN
Laboratory for Biological Geochemistry
School of Architecture, Civil and Environmental
    Engineering
Swiss Federal Institute of Technology in Lausanne
and
Center for Advanced Surface Analysis
Institute of Earth Sciences
University of Lausanne
Lausanne, Switzerland

E. MARIE MUEHE
Department of Geomicrobiology
Center for Applied Geoscience (ZAG)
Eberhard-Karls-University of Tuebingen
Tuebingen, Germany

DIANNE K. NEWMAN
Division of Biology and Biological Engineering
and
Division of Geological and Planetary Sciences
California Institute of Technology/HHMI
Pasadena, California

RONALD OREMLAND
National Research Program
Water Resources Discipline
United States Geological Survey
Menlo Park, California

VICTORIA J. ORPHAN
Division of Geological and Planetary Sciences
California Institute of Technology
Pasadena, California

ERIC E. RODEN
Department of Geoscience
University of Wisconsin–Madison
Madison, Wisconsin

MICHAEL RUSSELL
Jet Propulsion Laboratory
California Institute of Technology
Pasadena, California

MAK SAITO
Department of Microbiology
and
BioTechnology Institute
Woods Hole Oceanographic Institution
Woods Hole, Massachusetts

BERNHARD SCHINK
Fachbereich Biologie
Universitaet Konstanz
Konstanz, Germany

DILIANA D. SIMEONOVA
Fachbereich Biologic
Universitaet Konstanz
Konstanz, Germany

JAROSLAV STOLARSKI
Institute of Paleobiology
Polish Academy of Science
Warsaw, Poland

JOHN STOLZ
Department of Biological Sciences
and
Center for Environmental Research and Education
Duquesne University
Pittsburgh, Pennsylvania

MATTHEW P. WATTS
School of Earth Sciences
University of Melbourne
Melbourne, Victoria, Australia

MICHAEL WILKINS
School of Earth Sciences and Department of
    Microbiology
The Ohio State University
Columbus, Ohio

ADAM J. WILLIAMSON
Department of Plant and Microbial Biology
University of California, Berkeley
Berkeley, California

CHAPTER ONE

Introduction

Henry Lutz Ehrlich, Dianne K. Newman, and Andreas Kappler

CONTENTS

1.1   Scope / 1
1.2   History / 1
1.3   Comment on Chapter Authorship / 3
References / 3

1.1 SCOPE

Geomicrobiology deals with the role that microbes are playing on Earth in a number of fundamental geologic processes in the present and have played in the past since the beginning of life. These processes include the cycling of organic and some forms of inorganic matter at the surface and in the subsurface of Earth; the weathering of rocks; soil and sediment formation and transformation; the genesis, transformation, and degradation of a variety of minerals; and the genesis, preservation, and degradation of fossil fuels.

Geomicrobiology should not be equated with microbial ecology or microbial biogeochemistry. *Microbial ecology* is the study of interrelationships between different microorganisms; among microorganisms, plants, and animals; and between microorganisms and their environment. *Microbial biogeochemistry* is the study of microbially influenced geochemical reactions, enzymatically catalyzed or not, and their kinetics. These reactions are often studied in the context of cycling of inorganic and organic matter with emphasis on environmental mass transfer and energy flow. These subjects do overlap to some degree, as shown in Figure 1.1.

It is unclear when the word "geomicrobiology" was first introduced into the scientific vocabulary. It obviously derived from the term "geological microbiology." Beerstecher (1954) defined geomicrobiology as "the study of the relationship between the history of Earth and microbial life upon it." Kuznetsov et al. (1963) defined it as "the study of microbial processes currently taking place in the modern sediments of various bodies of water, in ground waters circulating through sedimentary and igneous rocks, and in weathered Earth crust [and also] the physiology of specific microorganisms taking part in presently occurring geochemical processes." Neither of the authors traced the history of the word, but they pointed to the important roles that scientists such as S. Winogradsky, S.A. Waksman, and C.E. ZoBell played in the development of the field.

1.2 HISTORY

Geomicrobiology is not a new scientific discipline, although until the 1980s, it did not receive much specialized attention. A unified concept of geomicrobiology and the biosphere can be said to have been pioneered in Russia under the leadership of V.I. Vernadsky (1863–1945) (see Ivanov (1967), Lapo (1987), Bailes (1990), and Vernadsky (1998) for insights into and discussion of early Russian geomicrobiology and its practitioners).

Certain early investigators in soil and aquatic microbiology may not have thought of themselves as geomicrobiologists, but they nevertheless exerted an important influence on the subject. One of the first contributors to geomicrobiology was Ehrenberg (1836, 1838), who in the second quarter

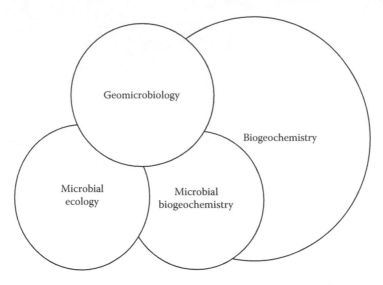

Figure 1.1. Interrelationships among geomicrobiology, microbial ecology, microbial biogeochemistry, and biogeochemistry.

of the nineteenth century discovered the association of *Gallionella ferruginea* with ochreous deposits of bog iron. He believed that the organism, which he classified as an infusorian (protozoan), but which we now recognize as a stalked bacterium (see Chapter 17), played a role in the formation of such deposits. Another important early contributor to geomicrobiology was S. Winogradsky, who discovered that *Beggiatoa*, a filamentous bacterium (see Chapter 20), could oxidize H_2S to elemental sulfur (1887) and that *Leptothrix ochracea*, a sheathed bacterium (see Chapter 17), promoted oxidation of $FeCO_3$ to ferric oxide (1888). He believed that each of the organisms gained energy from the corresponding process. Still other important early contributors to geomicrobiology were Harder (1919), a researcher trained as a geologist and microbiologist, who studied the significance of microbial iron oxidation and precipitation in relation to the formation of sedimentary iron deposits, and Stutzer (1912) and others, whose studies led to recognition of the significance of microbial oxidation of H_2S to elemental sulfur in the formation of sedimentary sulfur deposits. Our early understanding of the role of bacteria in sulfur deposition in nature received a further boost from the discovery of bacterial sulfate reduction by Beijerinck (1895) and van Delden (1903).

Starting with the Russian investigator Nadson (1903) (see also Nadson, 1928) at the end of the nineteenth century and continuing with such investigators as Bavendamm (1932), the important role of microbes in some forms of $CaCO_3$ precipitation began to be noted. Microbial participation in manganese oxidation and precipitation in nature was first recognized by Beijerinck (1913), Soehngen (1914), Lieske (1919), and Thiel (1925). Zappfe (1931) later related the activity to the formation of sedimentary manganese ore (see Chapter 17). A microbial role in methane formation (methanogenesis) became apparent through the observations and studies of Béchamp (1868), Tappeiner (1882), Popoff (1875), Hoppe-Seyler (1886), Omeliansky (1906), and Soehngen (1906); see also Barker (1956). A role of bacteria in rock weathering was first suggested by Muentz (1890) and Merrill (1895). Later, involvement of acid-producing microorganisms such as nitrifiers and of crustose lichens and fungi in such weathering was suggested (see Waksman, 1932). Thus, by the beginning of the twentieth century, many important areas of study of geomicrobial processes had begun to receive serious attention from microbiologists. In general, it may be said that most of the early geomicrobially important discoveries were made through physiological studies in the laboratory, which revealed the capacity of specific organisms to promote geomicrobially important transformations, causing later workers to study the extent of the occurrence of such processes in nature.

Geomicrobiology in the United States can be said to have begun with the work of Harder (1919) on iron-depositing bacteria. Other early

American investigators of geomicrobial phenomena include J. Lipman, S.A. Waksman, R.L. Starkey, and H.O. Halvorson, all prominent in soil microbiology, and G.A. Thiel, C. Zappfe, and C.E. ZoBell, all prominent in aquatic microbiology. ZoBell was a pioneer in marine microbiology (see Ehrlich, 2000).

Very fundamental discoveries in geomicrobiology continue to be made, some having been made as the twentieth century progressed and others very recently. For instance, the concept of environmental limits of pH and E_h for microbes in natural habitats was first introduced by Baas-Becking et al. (1960) (see Chapter 6). The pH limits as these authors defined them have since been extended at both the acid and the alkaline end of the pH range (pH 0 and pH 13) as a result of new observations. Life at high temperature was systematically studied for the first time in the 1970s by T.D. Brock and associates in Yellowstone Park in the United States (Brock, 1978). A specific acidophilic, iron-oxidizing bacterium, originally named *Thiobacillus ferrooxidans* and later renamed *Acidithiobacillus ferrooxidans*, was discovered by Colmer et al. (1950) in acid coal mine drainage in the late 1940s and thought by these investigators and others to be directly involved in its formation by promoting oxidation of pyrite occurring as inclusions in bituminous coal seams (see also Chapters 16 and 20). The subsequent demonstration of the presence of *A. ferrooxidans* in acid mine drainage from an ore body with sulfidic copper as chief constituent, located in Utah, Unites States (Bingham Canyon open pit mine), and the experimental finding that *A. ferrooxidans* can promote the leaching (mobilization by dissolution) of metals from various metal sulfide ores (Bryner et al., 1954) led to the first industrial application of a geomicrobially active organism to ore extraction (Zimmerley et al., 1958) (see also Ehrlich, 2001, 2004). Since these pioneering studies on microbial participation of *A. ferrooxidans* in the formation of acid mine drainage, other organisms with iron-oxidizing capacity have been discovered in acid mine drainage from different sources and implicated in its formation, as have other microorganisms associated in consortia with the iron oxidizers (see review by Ehrlich, 2004).

Paleomicrobiology had its beginnings with the discovery of Precambrian prokaryotic fossils by Tyler and Barghoorn (1954) and by Schopf et al. (1965) and by Barghoorn and Schopf (1965). These and subsequent discoveries (e.g., Perri and Tucker, 2007; Ivarsson et al., 2008; Sweetlove, 2011; Cosmidis et al., 2013) have had a profound influence on current theories on the origin and evolution of life on Earth (e.g., Schopf, 1983; Westall, 2004). Despite the time that has elapsed since the 1950s, paleomicrobiology remains a very immature subject.

Other important areas of geomicrobiology had their beginnings only relatively recently and remain to be fully explored or developed further. For instance, microbial life in and around hydrothermal vents on the seafloor at about 2500 m depth of the sea was unrecognized until reports such as those of Jannasch and Mottl (1985) and Tunnicliff (1992). Similarly, the existence of microbial life at depth of 100s and even 1000s of meters below Earth's terrestrial surface was not recognized until reports such as those by Ghiorse and Wilson (1988), Sinclair and Ghiorse (1989), Fredrickson et al. (1989), and Pedersen (1993), as was that of microbial life below the seafloor (Parkes et al., 1994; Mayer, 2012) (*see new chapter by Wilkins and Fredrickson*). These and subsequent observations have led to an awareness that Earth's biosphere encompasses a vastly greater space at and below Earth's surface than was thought possible previously (e.g., Onstott et al., 2009; Edwards et al., 2012). They have also had a major impact on the thinking in the field of astrobiology.

1.3 COMMENT ON CHAPTER AUTHORSHIP

Each of the following chapters, except for Chapters 12 and 21, is a revision and updating or an extensive rewriting of the corresponding chapter in the fifth edition of *Geomicrobiology* by the individual(s) named in the chapter heading, or in the case of Chapter 5, a new contribution to this book.

REFERENCES

Baas-Becking LGM, Kaplan IR, Moore D. 1960. Limits of the environment in terms of pH and oxidation and reduction potentials. *J Geol* 68:243–284.

Bailes KE. 1990. *Science and Russian Culture in the Age of Revolution: V.I. Vernadsky and His Scientific School, 1863–1945.* Bloomington, IN: Indiana University Press.

Barghoorn ES, Schopf JW. 1965. Microorganisms from the late Precambrian of Central Australia. *Science* 150:337–339.

CHAPTER TWO

Earth as a Microbial Habitat

Henry Lutz Ehrlich

CONTENTS

2.1   Geologically Important Features / 7
2.2   Biosphere / 12
2.3   Summary / 14
References / 15

2.1 GEOLOGICALLY IMPORTANT FEATURES

The interior of the planet Earth consists of three successive concentric regions (Figure 2.1), the innermost being the *core*. It is surrounded by the mantle, which in turn, is surrounded by the outermost region, the *crust*. The crust is surrounded by a gaseous envelope, the *atmosphere*.

The core, whose radius is estimated to be about 3450 km, is believed to be consisting of Fe–Ni alloy with an admixture of small amounts of the siderophile elements such as cobalt, rhenium, and osmium, very probably some sulfur and phosphorus, and perhaps even hydrogen (Mercy, 1972; Anderson, 1992; Wood, 1997; Alfe et al., 2000). The inner portion of the core, which has an estimated radius of about 1250 km, is solid, having a density of 13 g cm^{-3} and being subjected to a pressure of 3.7×10^{12} dyn cm^{-2}. The outer portion of the core has a thickness of about 2200 km and is molten, owing to a higher temperature but lower pressure than at the central core (1.3–3.2×10^{12} dyn cm^{-2}). The density of this portion is 9.7–12.5 g cm^{-3}.

The mantle, which has a thickness of about 2865 km, has a very different composition from the core and is separated from it by the Weichert–Gutenberg discontinuity (Madon, 1992; Jackson, 1998). Seismic measurements of the mantle regions have revealed distinctive layers called the upper mantle (365 km thick), the asthenosphere or transition zone (270 km thick), and lower mantle (1230 km thick) (Madon, 1992).

The mantle rock is dominated by the elements O, Mg, and Si with lesser amounts of Fe, Al, Ca, and Na (Mercy, 1972). The consistency of the rock in the upper mantle, although not truly molten, is thought to be plastic, especially in the region called the *asthenosphere*, situated 100–220 km below the Earth's surface (Madon, 1992). Upper mantle rock penetrates the crust on rare occasions and may be recognized as an outcropping, as in the case of some ultramafic rock on the bottom of the western Indian Ocean (Bonnatti and Hamlyn, 1978).

The crust is separated from the mantle by the Mohorovicic discontinuity. The thickness of the crust varies from as little as 5 km under ocean basins to as great as 70 km under continental mountain ranges. The average crustal thickness is 45 km (Madon, 1992; Skinner et al., 1999). The rock of the crust is dominated by O, Si, Al, Fe, Mg, Na, and K (Mercy, 1972). These elements make up 98.03% of the weight of the crust (Skinner et al., 1999) and occur predominantly in the rocks and sediments. The bedrock under the oceans is generally basaltic, whereas that of the continents is granitic to an average crustal depth of 25 km. Below this depth, it is basaltic to the Mohorovicic discontinuity (Ronov and Yaroshevsky, 1972, p. 243). Sediment covers most of the bedrock under the oceans. It ranges in thickness from 0 to 4 km. Sedimentary rock and sediment (soil in a nonaquatic context) cover the bedrock of the as a result of the movement of the subducting oceanic plate in the direction of the continental plate (Gurnis, 1992; Van Andel, 1992).

Oceanic plates grow along oceanic ridges, the sites of *crustal divergence*. Two examples of such divergence are represented by the Mid-Atlantic Ridge and the East Pacific Rise (Figure 2.3). The older portions of growing oceanic plates are destroyed through subduction with the formation of deep-sea trenches, such as the Marianas, Kurile, and Philippine trenches in the Pacific Ocean and the Puerto Rico Trench in the Atlantic Ocean. Growth of the oceanic plates at the mid-ocean ridges is the result of submarine volcanic eruptions of *magma* (molten rock from the deep crust or upper mantle). This magma becomes added to opposing plate margins along a mid-ocean ridge, causing adjacent parts of the plates to be pushed away from the ridge in opposite directions (Figure 2.4). The oldest portions of the interacting oceanic plates are consumed by subduction more or less in proportion to the formation of new oceanic plate at the mid-ocean ridges, thereby maintaining a fairly constant plate size.

Volcanism occurs not only at mid-ocean ridges but also in the regions of subduction where the sinking crustal rock undergoes melting as it descends toward the upper mantle. The molten rock may then erupt through fissures in the crust and contribute to mountain building at the continental margins (*orogeny*). It is plate collision and volcanic activity associated with subduction at continental margins that explain the existence of coastal mountain ranges. The origin of the Rocky Mountains and the Andes on the North and South American continents, respectively, is associated with subduction activity, whereas Himalayas are the result of collision of the plate bearing the Indian subcontinent with that bearing the Asian continent.

Volcanic activity may also occur away from crustal plate margins, at so-called hot spots. In the Pacific Ocean, one such hot spot is represented by the island of Hawaii with its active volcanoes. The remainder of the Hawaiian island chain had its origin at the same spot where the island of Hawaii is presently located. Crustal movement of the Pacific Ocean plate westward caused the remaining islands to be moved away from the hot spot so that they are no longer volcanically active.

The continents as they exist today are thought to have derived from a single continental mass, "Pangaea," which broke apart less than 200 million years ago as a result of crustal movement. Initially, this separation gave rise to "Laurasia" (which included present-day North America, Europe, and most of Asia) and "Gondwana" (which included present-day Africa, South America, Australia, Antarctica, and the Indian subcontinent). These continents separated subsequently into the continents we know today, except for the Indian subcontinent, which did not join the Asian continent until some time after this breakup (Figure 2.5) (Dietz and Holden, 1970; Fooden, 1972; Matthews, 1973; Palmer, 1974; Hoffman, 1991; Smith, 1992). The continents that evolved became modified by accretion of small landmasses through collision with plates bearing them. Pangaea itself is thought to have originated 250–260 million years ago from an aggregation of crustal plates bearing continental landmasses including Baltica (consisting of Russia west of the Ural Mountains, Scandinavia, Poland, and Northern Germany), China, Gondwana, Kazakhstania (consisting of present-day Kazakhstan), Laurentia (consisting of most of North America, Greenland, Scotland, and the Chukotsky Peninsula of eastern Russia), and Siberia (Bambach et al., 1980). Mobile continental plates are believed to have existed as long as 3.5 billion years ago (Kroener and Layer, 1992). The Earth seems to have had a crust as early as 4.35–4.4 eons ago, the age of the Earth being 4.65 eons (Amelin, 2005; Harrison et al., 2005; Watson and Harrison, 2005; Wilde et al., 2001).

The evidence for the origin and movement of the present-day continents has been obtained from at least three kinds of studies: (1) paleomagnetic and seismic examinations of the Earth's crust, (2) comparative sedimentary analyses of deep-ocean cores obtained from drillings by the *Glomar Challenger*, an ocean-going research vessel, and (3) paleoclimatic studies (Vine, 1970; Nierenberg, 1978; Bambach et al., 1980; Ritter, 1999). Although the separation of the present-day continents with the breakup of Pangaea had probably had no significant effect on the evolution of prokaryotes (they had pretty much evolved to their present complexity by this time), it did have a profound effect on the evolution of metaphytes and metazoans (McKenna, 1972; Raven and Axelrod, 1972). Flowering plants, birds, and mammals, for example, had yet to establish themselves.

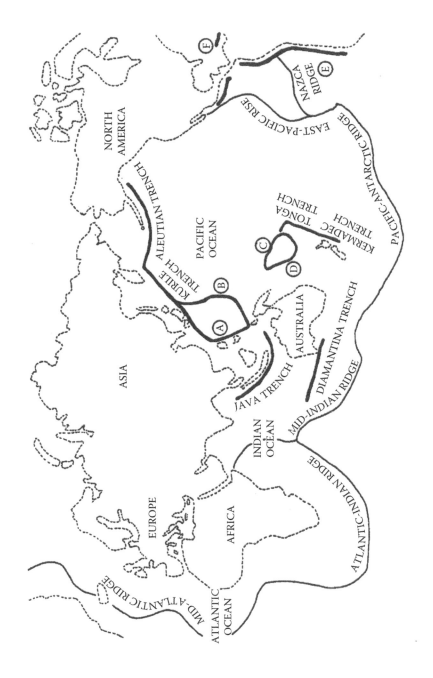

Figure 2.3. Major mid-ocean rift systems (thin continuous lines) and ocean trenches (heavy continuous lines). A, Philippine Trench; B, Marianas Trench; C, Vityaz Trench; D, New Hebrides Trench; E, Peru–Chile Trench; and F, Puerto Rico Trench. The East Pacific Ridge is also known as the East Pacific Rise.

CHAPTER THREE

Emergence of Life and Its Early History

Michael Russell

CONTENTS

3.1 Beginning / 19
    3.1.1 Origin of Life on Earth: Panspermia / 20
    3.1.2 Origin of Life on Earth: *De Novo* Appearance / 21
    3.1.3 Life from Abiotically Formed Organic Molecules in Aqueous Solution
        (Organic Soup Theory) / 21
    3.1.4 "Pyrite-Pulled" Surface Metabolism Theory / 22
    3.1.5 Origin of Life at a Submarine Alkaline Hydrothermal Mound / 23
    3.1.6 What Happened to the RNA World? / 28
3.2 Escape from the Mound / 29
    3.2.1 Early Evolution by Retrodiction / 29
    3.2.2 Emergence of the Methanogens / 29
3.3 Evolution of Life through the Precambrian: Geological and Biochemical Benchmarks / 30
    3.3.1 First Stromatolites and Accompanying Microbes / 31
    3.3.2 Origin of Oxygenic Photosynthesis / 33
3.4 Evidence / 36
3.5 Unresolved Issues / 39
3.6 Summary / 40
Acknowledgment / 41
References / 41

3.1 BEGINNING

Our Sun condensed from an accretion disk that itself resulted from gravitational collapse of interstellar matter derived from nearby supernova explosions (Bizzarro et al., 2007). Other components in the disk subsequently clumped together to form planetesimals of various sizes. And these in turn accreted ~4.6×10^9 years ago to form our Earth and the other three inner planets of our solar system, namely, Mercury, Venus, and Mars (Wood et al., 2006; Jacobsen et al., 2008; Wood and Halliday, 2010; Jackson and Jellinek, 2013). All four of these planets are rocky though the gas giants also probably have rocky interiors, as have some of their moons (Wilson and Militzer, 2012). Hydroxylated olivine particles and carbonaceous chondrites were the major building blocks, bringing water as well as magnesium iron silicates to the planets and their moons very early in their formation (Alexander et al., 2012; Marty, 2012; Vattuone et al., 2013; Sarafian et al., 2014). A collision of proto-Earth with Theia, a putative Mars-sized planet, produced a mostly molten Earth and Moon, accreted just beyond the Roche limit, perhaps around a distance of four Earth radii from our planet (Canup, 2012). There would have been many more days *and* months in the year in such an early conformation of the Earth–Moon system. Radiogenic, impact, and gravitational heating kept the Earth hot, resulting in the separation of silicates from iron, the latter differentiating and gravitating to the core, leaving a relatively oxidized from abiotically formed nucleotides (Cech, 1986; Gilbert, 1986; Joyce, 1991). As this self-reproducing RNA evolved, it acquired new functions through mutations and recombinations, with the result that an RNA world emerged. In time, a form of RNA (template RNA) arose that assumed a direct role in the assembly of proteins from constituent amino acids. Many of the proteins were enzymes (biocatalysts), and among these proteins were some that assumed a catalytic role in RNA synthesis. The protein catalysts were more efficient than RNA catalysts (Gilbert, 1986). Still later, deoxyribonucleic acid (DNA) acquired the information stored in RNA related to protein structure and function through reverse transcription, a process in which information stored in RNA was transcribed into DNA (Gilbert, 1986).

This speculative scenario has been proposed as a result of studies in the past two to three decades in which some RNAs were discovered in living cells that can modify themselves by self-splicing through catalysis of phosphoester cleavage and phosphoester transfer reactions (ribozyme activity) (Kruger et al., 1982; Guerrier-Takada et al., 1983; Cech, 1986; Doudna and Szostak, 1989). However, it must be said that these ideas take no account of thermodynamic drives and stabilities, electron transport, the synthesis of the phosphate anhydride bond in adenosine triphosphate (ATP) or pyrophosphate, and the fact that autotrophs occupy both the roots of the evolutionary tree and the base of the food chain, i.e., life is sourced from carbon dioxide which, to borrow a phrase from Bernal (1960), "produce(s) a small but ever-renewed stock of organic molecules." From calculations of the relative Gibbs free energies of organic molecule generation on mixing of Hadean hydrothermal solution with seawater, Amend and McCollom (2009) show that, in contrast to the amino and fatty acids, the nucleotides require energy for their formation and are therefore the output of metabolic process and not its basis. However, it is true that Powner et al. (2009) have generated nucleotides in the laboratory though the various conditions imposed vary from one step to the next and tend to result in complex mixtures.

Lane et al. (2010) subject this heterotrophic hypothesis to a rigorous critique, pointing out that soup offers no disequilibria with respect to pH and redox potential and therefore, would be incapable of energy transduction. In particular, there is no facility for driving chemiosmosis, the process vital to carbon and energy metabolism in the biosphere, and apparently applying to organisms at the base of the evolutionary tree and therefore, presumably required to get life started (Say and Fuchs 2010; Fuchs 2011). We turn therefore, to autotrophic theories for the emergence of life.

3.1.4 "Pyrite-Pulled" Surface Metabolism Theory

One alternative approach to how life may have originated on the Earth is the SMT formulated by Wächtershäuser (1988, 1992) and Huber et al. (2012). According to it, building blocks were synthesized and polymerized starting with key inorganic constituents (carbon dioxide, carbon monoxide, carbonyl sulfide, methane thiol, cyanide, pyrophosphate, hydrogen sulfide, hydrogen, nitrogen, and ammonia) on the surface of minerals with a positive (anodic) surface charge. Under this view, iron pyrites (FeS_2) formation is the driver. It is axiomatic for Wächtershäuser (1988) that polymerizations of surface-bound molecular building blocks are thermodynamically favorable, whereas polymerizations of the same molecular building blocks in solution in an organic soup scenario are thermodynamically unfavorable, i.e., any supposed polymers formed in the soup would be subject to hydrolytic cleavage.

In Wächtershäuser's surface metabolism scenario, building-block molecules were synthesized autocatalytically on a pyrite surface and, because of their ability to self-replicate, constituted the first life forms as 2D surface metabolists in Wächtershäuser's terminology. With the emergence of isoprenoid lipid synthesis, membranes with amphoteric properties formed that detached themselves from the mineral surface to form half-cells through a process he calls lipophilization. In time, these membranes completely enclosed the putative surface metabolists, thus becoming the first cells. The cells featured a membrane-enclosed cytosol in which the vital chemistry still occurred on the surface of a pyrite grain. However, with the passage of time and the appearance of some critical molecules, the vital chemistry became progressively independent of the mineral grain and assumed a distinct existence in the aqueous phase of the cytosol.

Thus Wächtershäuser (1988, 1992) considers the formation of pyrite to have provided the early reducing power in the original cells. The heteropolymers DNA and RNA, which eventually became key components of the genetic apparatus and assumed firm control of the cell's metabolic behavior and its perpetuation, originated independently through other surface metabolic processes in the precellular stage and did not exert control over them. The evolution of the genetic apparatus after the cellular stage had been attained involved, among other processes, the encoding in DNA via RNA of structural information of specific proteins and the development of a mechanism for deciphering this information to enable the synthesis of proteins, most of which had the ability to serve as enzymes (see Fang and Jing, 2010). After cellular metabolism became independent of the mineral grain, these enzymes took over as catalysts of the metabolic processes in the cytosol. The expression of the genetic determinants of the enzymes in the DNA regulated their formation and function as well as the timing of their appearance in the cell.

The appearance of enzymes in the cytosol made possible for the first time the utilization of accumulated building blocks through salvaging of surface-detached organic substances in the cytosol. Evolution of a membrane-bound respiratory chain is assumed to have followed the formation of the cell membrane. According to Wächtershäuser (1988, 1992), the respiratory chain, once it had arisen, liberated organisms that had developed it from having to rely on reactions involving ferrous iron and H_2S as a source of reducing power by enabling them to use reactions in which H_2 reduces elemental sulfur (S^0) or sulfate to H_2S (see Chapter 19, "Geomicrobiology of Sulfur" in the fifth edition). The surface metabolist theory also proposes that substrate-level phosphorylation (see Chapter 6, "Geomicrobial Processes: Physiological and Biochemical Overview" in the fifth edition) originated with autotrophic catabolism as an alternative means of "conserving" (i.e., converting free) energy. *Heterotrophy*, in which free energy needed by the cell is generated in the oxidation of reduced carbon compounds, is believed, by Wächtershäuser (1988), to have evolved from autotrophic catabolism in some cell lines.

In summary, the driving energy and reducing power for converting inorganic into organic carbon in autotrophic metabolism came from an interaction of ferrous iron and hydrogen sulfide in "a place deep down where a pyrite-forming autocatalyst once gave, and still is giving, birth to life." As both ferrous iron and H_2S should have been plentiful on the primitive Earth, the driving energy and reducing

power is supposed to have derived from iron pyrite (FeS_2) formation, i.e., in what Wächtershäuser (1988) termed a "pyrite-pulled reaction." Against this, the thermodynamic calculations of Schoonen et al. (1999) suggest pyrite formation is not sufficiently energetic to initiate prebiotic carbon fixation. And because of this lack of energetic drive, Pascal et al. (2013) argue that the process does not comply with the condition of irreversibility required of self-organizing systems. Nevertheless, using iron–nickel sulfides, Wächtershäuser and his collaborators have catalyzed the assembly of acetyl methane thiol (cf. acetyl coenzyme A [CoA]) from carbon monoxide and methane thiol, the amination of certain carboxylic acids to amino acids, and their subsequent polymerization to dipeptides, all vital reactions if we are to assume an autotrophic origin (Huber and Wächtershäuser, 1997, 1998; Huber et al., 2003; Fuchs, 2011).

3.1.5 Origin of Life at a Submarine Alkaline Hydrothermal Mound

In an alternative "metabolism-first" proposal by Russell et al. (1989, 1994, 2013), life is considered to have emerged in a long-lived alkaline submarine hydrothermal mound. Recent experimental support for the hypothesis has been offered by Yamaguchi et al. (2014) and Herschy et al. (2014). The sources of carbon in this autogenic origin of life theory are also the simplest of carbon-bearing molecules, atmospheric carbon dioxide, and hydrothermal methane. Under this view, what we may think of as the hatchery of life consisted of compartments mainly bounded by iron monosulfide (FeS) and double-layered hydroxides (e.g., ~$Fe_2(OH)_5$) that precipitated spontaneously and continually at the surface of a submarine hydrothermal mound, so separating the alkaline hydrothermal fluid from the carbonic ocean (Figure 3.1). These inorganic compartments were generated on the Hadean Ocean floor at conditions very far from equilibrium with the recipient ocean. That is, they are thought to have been formed at the interface between a hot ($\leq130°C$), alkaline (pH 10–11), extremely reduced, and hydrogen-, methane-, ammonia-, and bisulfide-bearing solution issuing from the ocean floor with a cool (~$10°C$) somewhat acidic (pH 5–6) iron-bearing carbonic Hadean Ocean shortly after it first rained out from the original steamy carbonic atmosphere, perhaps around 4.4–4.3

We may surmise that Pi's (~0.48 nm across) in the cytosol are pumped by protons into the interior of the funnel (Yamagata et al., 1991; Hagan et al., 2007). Once there, they are also condensed to the diphosphate anhydride bond. The protein undergoes a conformational transition that has the effect of opening the pump to the cytoplasm where the diphosphate now finds itself in strong disequilibrium with respect to Pi. A mineral that could potentially fill the same role and operate in a comparable way is the ferrous/ferric oxyhydroxide green rust, now known as fougèrite (~$[(Fe^{II},Mg)_2Fe^{III}(OH^-)_5 \cdot CO_3^{2-}]$) (Arrhenius, 2003; Feder et al., 2005; Génin et al., 2006; Trolard and Bourrié, 2012). Fougèrite is known to be flexible and has interlayer channels. Oxidation of the ferrous hydroxide component of the fougèrite comprising the exterior of the hydrothermal membrane precipitates leads to the opening of the interlayers through Fe^{III} to Fe^{III} charge repulsion to a gallery height at the edge of ~1.0 nm, just slightly shallower than the funnel entrance width of the H^+-pyrophosphatase. We might imagine the phosphate ions, perhaps along with the more soluble phosphite, dissolved in the mildly acidic ocean being driven by the steep proton gradient to intercalate the confined spaces between the layers where water activity is negligible (Yamagata et al., 1991; Hagan et al., 2007; Pasek et al., 2013). Here, any phosphite would be oxidized to phosphate by nitric oxide, and these and the original phosphates would be forced to condense to pyrophosphate ($HP_2O_7^{3-}$) and thence be driven to the interior to play its part—as the energy currency of the cell—in further condensations (see Russell et al., 2013 for details).

To be any use, such a putative fougèrite–pyrophosphate synthetase requires carboxylic and amino acids as substrates to condense to polymers that could take over the operations of the cell from their mineral precursors. Where might these be found if, as has been asserted, there was no source soup for these molecules? The main carbon-bearing molecules available to the mound are CO_2 in the ocean and CH_4 in the hydrothermal solution. But to produce organic molecules from both sources requires free energy (as well as contributions from hydrothermal ammonia and hydrogen sulfide as HS^-). This is because these first redox reactions—the reduction of CO_2 and the oxidation of CH_4—are endergonic (Maden, 2000). Life constructs and uses redox bifurcating enzymes to undertake these tasks whereby a strongly exergonic reaction is tightly coupled to a lesser endergonic reaction (Herrmann et al., 2008; Bertsch et al., 2013). In redox bifurcation at its simplest, one electron cannot resist being attracted exergonically to a high potential electron acceptor, while the other does the uphill endergonic job (Nitschke and Russell, 2009, 2013; Branscomb and Russell, 2013; cf. Schoonen et al., 1999; Schut and Adams, 2009; McInerney et al., 2010; Ramos et al., 2012; Wang et al., 2013). The proton gradient may be responsible for the contrasting electron energies (cf. Schuchmann and Müller, 2013) (Figures 3.1 and 3.2). Typically, quinones and flavins take these roles but molybdopterins also can (Schoepp-Cothenet et al., 2012). Molybdenum, soluble in alkaline-reducing hydrothermal fluids, would be contributed to the sulfide and oxyhydroxide membranes produced at off-ridge hydrothermal springs (Helz et al., 2014).

How might the possible exergonic reactions driving the first steps along a metabolic pathway, i.e., the reduction of CO_2 to CO and the conversion of CH_4 to a methyl group, involve the molybdenum atom? It has been assumed that Mo^{IV} would not be able to resist losing an electron to the high potential electron acceptors nitrate and nitrite in the ocean (derived from atmospheric NO) (Mancinelli and McKay, 1988; Martin et al., 2007; Ducluzeau et al., 2009; Nitschke and Russell, 2013; cf. Ettwig et al., 2010). At the same time, the leftover second electron potentially has the free energy to force the respective reductions in a catalytic environment where the degrees of freedom are limited to one (Branscomb and Russell, 2013; Nitschke and Russell, 2013). The ultimate source of the two bifurcating electrons—produced at an Fe–Ni hydrogenase—was hydrothermal hydrogen (Kelley et al., 2001). Iron sulfides comprising the inner wall of inorganic compartments at the hydrothermal mound could have acted in the same guise, i.e., as a protohydrogenase (Nakamura et al., 2010; Nitschke et al., 2013; Russell et al., 2014).

The immediate result of the supposed reduction of CO_2 and the oxidation of methane was to produce what is known as activated acetate or acetyl methane thiol (~CH_3COSCH_3), probably on a nickel–iron mineral cluster surface occupying the inorganic precipitate membrane (cf. Huber and Wächtershäuser, 1997)—the first stages in the putative denitrifying methanotrophic acetogenic pathway (Figure 3.2) (Nitschke and Russell, 2013).

Phosphorylations, hydrogenations, and carboxylations produced high-carbon-number carboxylic acids along the incomplete reversed tricarboxylic acid cycle terminating at succinate and glyoxylate (Fuchs and Stupperich, 1978; Morowitz et al., 2000; Wang et al., 2011). Some of these carboxylic acids could be aminated to amino acids with either hydrothermal ammonia or through ammonia produced by the abiotic reduction of nitrate on green rust (Hansen and Koch, 1998; Huber and Wächtershäuser, 2003; Novikov and Copley, 2013). Such amino acids can also be condensed to short peptides in the same environment with the same mineral catalysts as those precipitated in, and as, the membrane (cf. Huber et al., 2003). We may assume that peptides, possibly in the form of amyloid, took over the roles of both cell membrane and cell wall from the original inorganic compartments, so improving the various protobiochemical processes, including informational function (Zhang et al., 1993; Chernoff, 2004; Carny and Gazit, 2005; Milner-White and Russell, 2005; Maury, 2009). An attractive aspect of this hypothesis is that the peptides that would have been available to make up such a cellular boundary have the same structure as those that form natural nests chelating the thiolated sulfides and phosphates in manners comparable to those of modern proteins (Milner-White and Russell,

2005, 2011; Childers et al., 2009; Bianchi et al., 2012; Hong Enriquez and Do, 2012) (Figure 3.3).

As primitive cells evolved, peptides are envisaged to have taken over the roles of mineral free energy converters within their membranes (Milner-White and Russell, 2011; Branscomb and Russell, 2013; Russell et al., 2013). These peptides not only exerted positive or negative control over the passage of specific substances into and out of a cell but, for example, as pyrophosphatase synthetase, acted as energy transducers (cf. Lin et al., 2012; Tsai et al., 2014). And mineral clusters, sequestered in peptides, could conduct electrons toward the externally available terminal electron acceptors such as nitrite, nitrate, and Fe^{3+}, thereby reducing CO_2, ever more efficiently (Russell and Hall, 1997; Ducluzeau et al., 2009).

Of all the theories for the emergence of metabolism, it must be admitted that the submarine hydrothermal theory is the most complex or, at least, the most detailed. However, if only we could see it from the point of view of electrons as—inhibited at first—they sought out the most efficient paths from the rich source of transferable electrons constituting minerals in the early Earth's interior toward the electron-poor but proton-rich volatispheric exterior, all might become clear (cf. Szent-Györgyi, 1968a,b; Russell et al., 2014)!

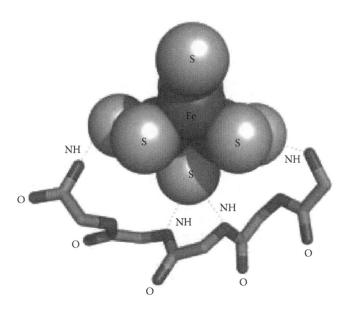

Figure 3.3. A five-mer peptide nest binding a thiolated Fe_3S_4 entity comparable to the binding loop of the ferredoxin from the archaebacterium *Sulfolobus* sp. (From Fujii, T. et al., *Nat. Struct. Biol.*, 3, 834, 1996; Milner-White, E. J. and Russell, M. J. *Orig. Life Evol. Biosph.*, 35, 19, 2005.)

3.1.6 What Happened to the RNA World?

The RNA world theory may be thought of as the orphan child of the organic soup hypothesis. From the point of view of metabolism first theory, RNA is a guidance molecule that emerged once metabolic pathways had emerged (Nitschke and Russell, 2013). This idea is supported by the calculations of Amend and McCollom (2009) who show that the synthesis of RNA is particularly endergonic. Thus metabolism is required to supply not only the building blocks of the nucleotides but also the free energy and the coupling devices or protoenzymes to drive their synthesis. But once metabolism was a going concern, RNA could direct a metabolic pathway in a direction appropriate to changing environments. However, in the welter of interactions taking place in the compartments and cells of first metabolists, how was RNA first assembled, and from what? While those working from the heterotrophic theory assume that the ribonucleotides were produced by the condensation of hydrogen cyanide and formaldehyde—molecules hard to come by on the early water world—from an autotrophic viewpoint, the components of the ribonucleotides that life now uses are so simple as to invite the possibility that the building materials have always been similar, if not the same (Berg et al., 2006). These components include phosphorylated sugars, amino acids, formyl and carbamoyl phosphate, and carbon dioxide—all potentially available or supplied through the envisaged protometabolic pathway discussed in Section 3.1.5. Once generated, RNAs could replicate and, acting like retroviruses, infect the local milieu. Some of these RNAs may subsequently have developed an ability to serve as templates in protein synthesis, making synthesis of specific proteins more orderly (Koonin and Martin, 2005). Other RNAs may have evolved into reactants (transfer RNAs) in the protein-assembly reactions in which amino acids are linked to each other in a specific sequence by peptide bonds, making the polymerization more efficient. As template RNA became more diverse through mutation and recombination, the diversity of catalytic proteins increased. This resulted in controlled accelerated synthesis of the building blocks (amino acids, fatty acids, sugars, nucleotides, etc.) from which vital polymers (proteins, lipids, polysaccharides, nucleic acids, etc.) could be synthesized by other newly evolved catalytic proteins. Enzyme-catalyzed synthesis was much more efficient than abiotic synthesis.

Yarus (2010) offers an idea of how the so-called Darwinian threshold might have been crossed (Woese, 2002). He suggests that phosphorylated aromatic structures, involving nitrogen, brought more sophistication to anabolism and proton transfer in the form of diphosphates such as nicotine amide dinucleotide and flavin adenine dinucleotide. Some of these structures may have even coded for each other (Yarus, 2010), the first step into the RNA world. There may have been a drive for such RNAs to polymerize as a further branch in an evolving metabolic network, but other functions of those primordial RNA molecules are also likely. For example, they could have acted as chelant, mobilizers, or enhancers of metal ion reactivity or even have been reproduced as the first (coding) parasites (Koonin and Martin, 2005). Another possibility is the initial RNA oligomers could have acted as secondary long-term energy storage entities. For example, as energy-rich polymers, they could have been efficient in preserving the chemical (osmotic) potentials, maintaining the system both in homeostasis and far from equilibrium and/or buffering the system against the vagaries of energy supply.

Whatever the details, it is clear that the LUCA had developed biosynthetic pathways leading to amino acid and nucleotide synthesis and was well equipped with most other aspects of the modern prokaryotic cell: electron transfer agents, metalloenzymes, the universal genetic code, tRNA and ribosomal proteins, RNA polymerases, DNA and DNA polymerases, translation factors, F_1-F_0 type, and ATPases (Poole and Logan, 2005; Martin and Russell, 2007). However, there was an as yet unexplained bifurcation of the protoprokaryotes with LUCA herself. The two offspring leading to the two domains of life were the archaea and the bacteria, and some of their differences are fundamental (Williams et al., 2013). For example, although their ribosomes and RNA polymerase are homologous, their cell membranes and walls are very different, and these differentiated and evolved independently after their split from the LUCA (Werner and Grohmann, 2011). But the first representatives of both made their way into the world at large, and we turn now to how they left the nest—always a transition fraught with danger and difficulty.

3.2 ESCAPE FROM THE MOUND

Compared to the putative hydrothermal hatchery, succored as it was with all the wherewithal required by life—mineral surfaces, materials, and fuels—the earliest ocean would have been a relative desert where any prokaryotes unlucky enough to be expelled from the mound would have been hard put to make a living. A more likely way to expand from the hatchery was through cell-by-cell budding and division and thereby slow propagation along the ocean floor and down into the ocean crust—the latter process perhaps inaugurating the deep biosphere. The supplies of electron donors and acceptors—some solid, some fluid—and of nutrients, while much diminished, nevertheless were of the same ilk as those at the periphery of the hydrothermal mound, just the turnover would have been so much slower (Whitman et al., 1998; Roussel et al., 2008; Morono et al., 2011; Hoehler and Jørgensen, 2013). Microbes gravitating to greatest depths—deprived of nitrate and ferric iron—may have evolved reductases enabling respiration to native sulfur, sulfite, and eventually sulfate (Nitschke et al., 2013). Recalling that all life lives off electron-rich substrates and that respiration requires these same electrons to escape to acceptors, minerals such as mackinawite and fougèrite might have acted as electron wires, allowing access to more distal acceptors or even donors (e.g., Ferris et al., 1992; Christiansen et al., 2009). An open question is whether early prokaryotes developed extracellular quinones, heme cytochromes or some other electron conductors to do the same job (Newman and Kolter, 2000; El-Naggar et al., 2010; Edwards et al., 2012; Pfeffer et al., 2012). Only when the deep biosphere was abducted or otherwise lifted to the ocean surface did fresh opportunities appeared for alternative metabolic pathways driven by different forms of free energy. We can think of evolution as the search engine for appropriate free energies and nutrients broadly comparable to those that drove life into being in the first place, but where only the "most fitting" are selected and survive. Let's see what new horizons offered and when.

3.2.1 Early Evolution by Retrodiction

If we accept the ability of cells to self-replicate as the most basic definition of life, then the appearance of coupled interactions between proteins and nucleic acids within reproducing cells with this ability marked the beginning of life and the generation of the LUCA, perhaps in a submarine hydrothermal mound. DNA must have rapidly evolved to become the repository for structural information of the different proteins that resided in the RNA templates. Many of these proteins were the catalytic molecular motors needed for the synthesis of monomers and for assembly of polymers from monomers (see DNA Learning Center Animations). As the primitive cells evolved, they must have developed the traits that we associate with modern *prokaryotic cells*, as suggested by micropaleontological evidence. This implies that they possessed a cell envelope or wall surrounding a plasma membrane enclosing an interior featuring a large DNA strand (repository of genetic information), nucleoprotein granules (ribosomes), and other proteins and smaller polymers and monomers. At least in the early beginnings, cell multiplication is likely to have been by binary fission, a process involving replication of all vital cell components and their equal partitioning between daughter cells. At this juncture, a *cytoskeleton* ensured equal partitioning and also assisted in the maintenance of cell shape (Gitai, 2005; Møller-Jensen and Löwe, 2005).

The plasma membrane of the cell had mechanisms for transporting externally available organic and inorganic molecules across it. These first prokaryotes were probably anaerobic *autotrophs* that, while fed mainly from hydrogen and methane, added to the drawdown of the thick, mainly CO_2 atmosphere that enveloped the Earth in the Hadean, a process augmenting the carbonation of the oceanic crust. Even in the hatchery mound, it is likely that heterotrophs would have differentiated from the autotrophs to take advantage of the organic waste products—part acetate effluent and part organic detritus. Indeed, these heterotrophs would have acted to clean the nest and so keep the chemical tensions at a maximum.

3.2.2 Emergence of the Methanogens

A reversal of methanotrophy by chemosynthetic autotrophs, though not via all the same enzymes, may have led to the evolution of the methanogens at least by the Eoarchean (Chistoserdova et al., 2004). However, it must be remarked that others have considered methanogenesis to be the first, or at least one of the very earliest, of the archaeal metabolisms (Sousa et al., 2013; Boyd et al., 2014; and see Ueno et al., 2006). Analyses by the techniques of molecular biology have certainly shown that methanogens have an ancient origin (Fox et al., 1977; Woese and Fox, 1977; Cox et al., 2008; Kelly et al., 2011) though others demur (Nitschke and Russell, 2013). Shen et al. (2009) have demonstrated from sulfur isotope analyses that sulfate-reducing microbes had emerged by 3.4 Ga. It is notable that their electron transport complexes contain heterodisulfide reductases similar to those occurring in the sulfate-reducing archaea (Hedderich et al., 2005; Pereira, 2008; Thauer et al., 2008; Ramos et al., 2012). However, whether the methanogens evolved from early sulfate-reducing archaea or vice versa is still moot (Sousa et al., 2013). Supposing methanogens did evolve from the sulfate reducers, what could have driven this evolutionary split? One might speculate that, as the respiring occupants of the deepest biosphere were deprived of the electron acceptors, notably sulfate, then the only mechanism left to rid autotrophic cells of their electrons was through the loss of volatile methane. In concert with this view, it is notable that methanogens are strict anaerobes, and methanogenesis is favored by low sulfate concentrations (Habicht et al., 2002).

Methane is 25 times more potent a greenhouse gas than carbon dioxide so, as CO_2 was drawn down and partly replaced by the CH_4 waste of the wide and deeply spread methanogens, hothouse conditions may have ensued (Kiehl and Dickinson, 1987; Kasting, 2005). A number of extant methanogens live off hydrothermal hydrogen at relatively high temperatures (+60°C to +90°C), and it has been argued by reference to oxygen isotope ratios in ancient cherts that these temperatures prevailed on the Earth throughout the Archean (Kasting and Siefert, 2002; Knauth and Lowe, 2003; Kasting, 2004; Robert and Chaussidon, 2006; Schwartzman et al., 2008). Under this view, as the temperatures rose toward 100°C, the Earth became unbearable even for methanogens, and thus a runaway greenhouse was prevented— perhaps the first example of a Gaian feedback to regulate climate (cf. Baas Becking, 1931; Margulis and Lovelock, 1974). It has been argued that only toward the end of the Archean did the Earth cool again sufficiently for oxygenic photosynthetic autotrophs to displace the methanogens as primary producers and controllers of the atmosphere, perhaps through a combination of wide continental littorals and lagoons, the extraordinary rapid burial of organic carbon, and thereby the development of the Snowball Earth, processes leading to the great oxidation event (Schidlowski et al., 1975; Knoll, 1979; Baker and Fallick, 1989; Kirschvink, 1992; Holland, 2002; Aharon, 2005; Anbar et al., 2007; Sessions et al., 2009). However, there are also cogent arguments suggesting the Archean climate was clement and that oxygenic photosynthesis evolved long before the great oxidation event but did not impact the atmosphere until around 2.5 Ga (Shields and Kasting, 2007; Zahnle et al., 2007; Thomazo et al., 2011; Crowe et al., 2013).

3.3 EVOLUTION OF LIFE THROUGH THE PRECAMBRIAN: GEOLOGICAL AND BIOCHEMICAL BENCHMARKS

The geologic timescale since the origin of the Earth is divided into the Precambrian, which extends from ~4.5 to 0.54 billion years BP (Ga), and the Phanerozoic, which extends from 0.54 Ga to the present. The Precambrian may be divided into three eras: Hadean (4.5–3.85 Ga), the Archean (3.85–2.5 Ga), and the Proterozoic (2.5–0.54 Ga). The Archean may be further subdivided into four periods: the Eoarchean (3.85–3.6 Ga), the Paleoarchean (3.6–3.2 Ga), the Mesoarchean (3.2–2.8 Ga), and the Neoarchean (2.8–2.5 Ga). The Proterozoic may be subdivided into three periods: the Paleoproterozoic (2.5–1.6 Ga), the Mesoproterozoic (1.6–1.0 Ga), and the Neoproterozoic (1.0–0.54 Ga) (Walker and Geissman, 2009).

As we have seen, the Earth seems to have had a partially differentiated crust as early as 4.35–4.4 Ga, and life may have first appeared shortly thereafter (Mojzsis et al., 2001, and see Chapter 2). The ocean floor consisted mainly of mafic and ultramafic lavas and lava plateaux with developments here and there of thick granodiorite and andesite that flows perhaps constituting the first protocontinents. However, as there was around twice as much water enveloping the Earth than now, at best, any dry landmasses would have been tiny and ephemeral (Bounama et al., 2001; Elkins-Tanton, 2008). Recent thinking based on geochemical evidence is that the first life, that is, the first living entities, whatever their form, had originated well before 3.7 Ga (Schopf et al., 1983; Rosing 1999; Mojzsis et al., 1996; Holland,

1997; Russell and Hall, 1997; Nutman et al., 2010). Mojzsis et al. (1996) found carbon isotopic evidence in carbonaceous inclusions (graphitized carbon) within grains of apatite (basic calcium phosphate) from the oldest known sediment sequences that supports the existence of biotic activity (Rosing, 1999). These metasediments are associated with the ~3.8-billion-year-old banded iron formation of the Isua supracrustal belt of West Greenland and a similar ~3.85-billion-year-old sedimentary formation nearby Akilia Island. Although arguments favoring a microbiological origin for the graphite concentrations there have been heavily contested (e.g., Naraoka et al., 1996; Fedo and Whitehouse, 2002), Ohtomo et al. (2014) have now shown that the high concentrations of graphite in the metasediments, in contradistinction to that in the veins, is ^{13}C depleted and best explained as microbial in origin (see Chapter 6 for an explanation of isotope enrichment and its significance). However, as the Isua rocks have suffered deformation and amphibolite facies metamorphism, differing interpretations as to the nature of the then biosphere have been entertained. It is only when we examine the Pilbara Craton, Western Australia, with rocks spanning 3.6–2.8 Ga is evidence for life absolutely beyond controversy.

3.3.1 First Stromatolites and Accompanying Microbes

Remnants of a *stromatolite* (in this instance a fossilized mat of filamentous microorganisms) have been found in the cherts of the Warrawoona Group in the Pilbara Block and in the Strelley Pool Chert in the same group (Figures 3.4a through c and 3.5) (Awramik et al., 1983; Allwood et al., 2006, 2009). Its age has been determined to be ~3.43 Ga (Lowe, 1980; Walter et al., 1980; Van Kranendonk et al., 2007, 2008). It may seem tempting to ascribe these well-structured microbialites to oxygenic photosynthesis through the operations of cyanobacteria. In support of such a notion, Hoashi et al. (2009) point to an extensive jasperite formation containing what appear to be hematite crystals in the same succession at 3.46 Ga. Against this and on the basis of extreme heavy Fe isotope enrichment, Li et al. (2013) suggest that bacterial photosynthesis at ~3.46 Ga was anoxygenic and relied on hydrogen as the electron donor along with the photosynthetic

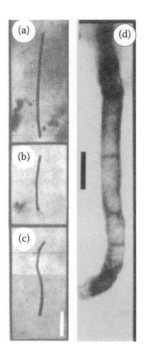

Figure 3.4. Rod-shaped, threadlike, juvenile forms of apparently nonseptate bacteria in petrographic thin section of stromatolitic black chert from the 3.5-billion-year-old Warrawoona Group (Pilbara Supergroup) of Western Australia. Scale mark in panel (c) is 5 μm and also applies to panels (a) and (b). (d) *Gunflintia* sp.: a septate filament with unusually elongated cells, preserved in dark-brown organic matter from Duck Creek Dolomite, Mount Stuart, Western Australia, ~2.02 billion years old. Scale mark on the left is 10 μm. (Reproduced from Schopf, J. W., *Earth's Earliest Biosphere: Its Origin and Evolution,* Princeton University Press, Princeton, NJ, 1983, photo 9-4C, D, and E, and 14-3J. With permission.)

oxidation of Fe^{II}, perhaps explaining the hematite (cf. Widdel et al., 1993; Kappler et al., 2005, 2010). Using a dispersion/reaction model of iron isotope fractionation—and with reference to extremely low uranium concentrations—Li et al. (2013) consider the ocean then to be essentially anoxic. Moreover, morphological cell types shown to have been sulfur metabolizing, yet resembling cyanobacteria, have also been found in the Strelley Pool Cherts (Wacey et al. 2011a) (cf. Schopf and Packer, 1987).

Of course, if the Archean atmosphere was controlled by a high partial pressure of methane as discussed in Section 3.2.2, then the temperature of the still extensive and rapidly circulating ocean may have been too high for oxygenic photosynthesizing cyanobacteria. Even today, having had vein, Olson (1970) had suggested that a Mn^{III}–cytochrome could have been oxidized by nitrite or nitric oxide. We have already explored the possibility that nitrate and nitrate were the first electron acceptors (Ducluzeau et al., 2009). Another possibility appeals to the cooption of a photolytic manganate such as ranciéite ($CaMn_4O_9 \cdot 3H_2O$) fortuitously adsorbed as a defense against hard UV (Russell et al., 2003, 2008). Anbar and Holland (1992) have demonstrated the photooxidation of a manganese complex to birnessite, a manganese homolog of ranciéite lacking calcium. Manganese, as Mn^{III} and Mn^{II}, reduced from Mn^{IV} in ambient minerals by organic molecules and sequestered into the membrane (Parker et al., 2004), would have adventitiously lent protection against hard photons (e.g., Daly et al., 2004). And it would have detoxified side products of this radiation, i.e., various reactive oxygen species, through Mn-dependent superoxide dismutase as it does in the cyanobacterium Chroococcidiopsis (Billi et al., 2000). Manganese could protect cells from hard UV injury because of its proneness to short-wavelength photooxidation and its variable valence states. At the same time, hydrogen and/or electrons and protons could have been released to the cell and entrained as further supply of free energy. Once adsorbed at carboxyl and hydroxyl sites in a prototype reaction center on the surface of an anoxygenic photosynthesizer, four manganese ions were likely to have organized themselves, as they would in the mineral hollandite, in one of two possible 3Mn open cubane structures ligated to a fourth, "dangling" Mn (Sauer and Yachandra, 2004). The open cubane was closed with Ca. Fortuitously, the Ca^{2+} and the distal Mn^{2+} were hydrated, as they can be in ranciéite. Photons gathered from antennae drove the initially hydrated ($Mn_3O_4Ca \cdot H_2O \cdot Mn \cdot H_2O$) complex to higher redox states in five stages. Two oxygen atoms from water were retained, while two protons and two electrons and two activated hydrogen atoms were released to the cell for biosynthesis. The oxygen was then released as O_2 waste, hence the descripter—"the OEC" (Messinger, 2004). The OEC is still housed at the outward-facing domain of photosystem (PSII) employed by all cyanobacteria and plants. Oxygen that was photosynthetically evolved initially probably reacted with oxidizable inorganic matter such as iron (Fe^{II}), forming iron oxides such as magnetite (Fe_3O_4) and hematite (Fe_2O_3) (see Chapter 16).

A few modern cyanobacteria are known that have the capacity to carry out not only oxygenic photosynthesis but also anoxygenic photosynthesis that requires anaerobic conditions and the presence of H_2S. In the latter instance, they photosynthesize like green bacteria, using the H_2S rather than H_2O to reduce CO_2 (Cohen et al., 1975; Shahak et al., 1999). Buick (1992) inferred from stromatolites in the Tumbiana Formation of the late Archean located in Western Australia that oxygenic photosynthesis occurred 2.7 billion years ago. This was the time that the first large continental littorals formed that might have supported protected nurseries for the emergence of cyanobacteria (Knoll, 1979; cf. the Cambrian explosion as described by Peters and Gaines [2012]).

Recent research strongly suggests that the cyanobacteria emerged from freshwater (Buick, 1992; Blank and Sánchez-Baracaldo, 2010; Blank, 2013; Dagan et al., 2013). Species of the genus Gloeobacter, a bacterium lacking a thylakoid membrane, that emerged preplastid (i.e., before the emergence of the eukaryotes) still live in wet freshwater, rock surfaces, or subaerial conditions well protected from harsh UV light (Rippka et al., 1974; Mareš et al., 2013; Saw et al., 2013). Given the likely high partial pressure of carbon dioxide toward the end of the Archean, it is notable that several species of freshwater cyanobacteria can also live in a 100% CO_2 atmosphere (Thomas et al., 2005).

Stromatolites became much more common in the Proterozoic (2.5–0.57 billion years BP) as the cyanobacteria achieved dominance as carbon-fixing and oxygen-evolving microorganisms. These structures formed as a result of the aggregation of some filamentous forms of cyanobacteria into mats that trapped siliceous and carbonaceous sediment, which in many instances contributed to their ultimate preservation by silicification and transformation into stromatolites. Environmental conditions at this time seemed to favor the mat-forming growth habit of cyanobacteria: continental emergence, development of shallow seas, and climatic and atmospheric changes resulting from oxygenic photosynthesis probably exerted selective pressure favoring this growth habit (Knoll and Awramik, 1983). Most fossil finds representing this period have been stromatolites, possibly because they are among the more easily recognized, if somewhat controversial (Lowe, 1994, Brasier et al., 2002, 2006). Unicellular microfossils would be much harder to find and identify, so one should not draw the conclusion that mat-forming cyanobacteria were necessarily the only common form of life at that time.

Because the biochemical oxygen reduction process leads to the formation of rather toxic superoxide radicals (O_2^-) (Fridovich, 1977), the oxygen-utilizing organisms protect themselves against them through superoxide dismutase and catalase, metalloenzymes that together catalyze the reduction of superoxide to water and oxygen. Superoxide dismutase catalyzes the reaction $2O_2^- + 2H^+ \rightarrow H_2O_2 + O_2$ and catalase catalyzes the reaction $2H_2O_2 \rightarrow 2H_2O + O_2$. Peroxidase may replace catalase as the enzyme for disposing of the toxic hydrogen peroxide $2RH + H_2O_2 \rightarrow 2H_2O + 2R$, where RH represents an oxidizable organic molecule (Katona et al., 2007).

Schopf et al. (1983) suggested that the first oxygen-utilizing prokaryotes were *amphiaerobic*, that is, they retained the ability to live anaerobically although they had acquired the ability to metabolize aerobically. (Present-day amphiaerobes are called *facultative* organisms.) From them, according to Schopf et al. (1983), obligate aerobes evolved more than 2 billion years ago. Towe (1990), however, has proposed that aerobes could have first appeared in the early Archean; a view supported by Hoashi et al. (2009), based on primary hematites in hydrothermal jasper precipitated around 3.46 Ga. In any case, aerobes probably became well established only 2 billion years ago. The evolutionary sequence leading from anaerobes or amphiaerobes to aerobes was probably complex because present-day facultative aerobes include some that have a respiratory system that can use nitrate, ferric iron, and manganese (IV) oxides.

Amphiaerobes are best defined as organisms that have the ability to respire aerobically in the presence of nitric oxides, dimethyl sulfoxide (DMSO), tetramethylene sulfoxide (TMSO), trimethylamine-N-oxide (TMAO), as well as trace oxygen. If we accept the view that the LUCA used nitrate as an electron acceptor, then even the earliest microbes could be so defined (Nitschke and Russell, 2013; Ducluzeau et al., 2014). On the other hand, the sulfate reducers appear to have emerged post-LUCA (Klein et al., 2001), and the methanogens, strict anaerobes, perhaps derived from sulfate reducers, would have appeared even later. As a group, the methanogens, like their modern counterparts, must have respired anaerobically using CO_2 as the terminal electron acceptor, although a few could

have produced methane by fermenting acetate, as some modern methanogens are able to do. A counter view holds that the LUCA could carry out acetogenesis *and* methanogenesis (Sousa et al., 2013).

Some strict anaerobes probably also evolved subsequent to the appearance of oxygen in the atmosphere. These were organisms with a capacity to respire anaerobically using oxidized inorganic molecules and anions as terminal electron acceptors that became available in sufficient quantity only after the appearance of O_2 in the atmosphere. Examples of such molecules are sulfate, DMSO, TMSO, TMAO, and arsenate (Duval et al., 2008). Sulfate reducers in the domain bacteria have been viewed as having made their first appearance in the early Archean (Shen et al., 2009). They have been important ever since their first appearance in the reductive segment of the sulfur cycle in the mesophilic temperature range (~15°C–40°C) (Schidlowski et al., 1983). At ambient temperature and pressure, bacterial sulfate reduction is the only process whereby sulfate can be reduced to hydrogen sulfide.

A group of extremely thermophilic, archaeal sulfate reducers have been isolated from marine hydrothermal systems in Italy (Stetter et al., 1987), which in the laboratory grew in a temperature range of 64°C–92°C with an optimum near 83°C. They reduce sulfate, thiosulfate, and sulfite with H_2, formate, formamide, lactate, and pyruvate. Stetter et al. (1987) suggested that these types of bacteria may have inhabited early Archean hydrothermal systems containing significant amounts of sulfate of magmatic origin.

With the appearance of oxygen-producing cyanobacteria and aerobic heterotrophs, the stage was set for extensive cellular compartmentalization of such vital processes as photosynthesis and respiration. New types of photosynthetic cells evolved, where photosynthesis was carried on in special organelles, the *chloroplasts*, and new types of respiring cells evolved in which respiration was carried on in other special organelles, the *mitochondria*. Based on molecular biological evidence, it is now generally accepted that these organelles arose by endosymbiosis, a process in which primitive cells that were incapable of photosynthesis or respiration were invaded by cyanobacterial-like organisms to evolve into chloroplasts and aerobically respiring bacteria to evolve into mitochondria (Martin and Müller, 1998; Martin et al., 2002; Raven and Allen, 2003;

stromatolites are confined to very special locations such as saline lagoons and isolated lakes (Zedef et al., 2000; Papineau et al., 2005).

No microfossils of the earliest life forms have been found, nor are they ever likely to be found on the Earth because of the weathering or diagenetic processes to which sedimentary rock on our planet has been subjected from the start and to which it is continuing to be subjected. This weathering was originally a physicochemical process involving water and various reactive substances in the planet's atmosphere. With the emergence of prokaryotic life, microbes also became important agents of weathering. Their present-day weathering activities are discussed in Chapters 4 and 9. Of course, there is always the teasing possibility that fossils biotic molecules or even structures might be found in the very earliest unmetamorphosed sediments on Mars that date from the early Noachian around 4.1 Ga.

Geochemical studies of Precambrian rocks can also tell us something about early life. For instance, measurements of stable isotope ratios of major elements important to life, namely, C, H, N, and S (Schidlowski et al., 1983) and Fe (Beard et al., 1999, but see also Mandernack et al., 1999, and Anbar et al., 2000; Li et al., 2013), can give an indication of whether a biological agent was involved in their formation or transformation. Such interpretation rests on observations that some present-day organisms can discriminate among stable isotopes of an element by metabolizing the lighter species faster than the heavier species. They attack ^{12}C more readily than ^{13}C, hydrogen more readily than deuterium, ^{14}N more readily than ^{15}N, ^{32}S more readily than ^{34}S, or ^{54}Fe more readily than ^{56}Fe. Abiotic reactions do not discriminate among stable isotopes to this extent though there is some overlap with Fischer–Tropsch reactions (McCollom, 2011). As a result, products of microbial isotope fractionation reactions will show enrichments with respect to the lighter isotope. The fractionation is most noticeable in the initial stages of a reaction in a closed system or in an open system with a low rate of substrate consumption. Residual substrates will show an enrichment in the heavier isotope in a closed system (Chapter 6). Carbon isotope studies of many sediment samples of Archean age indicate that life played a dominant role in the carbon cycle as far back as 3.5 billion years ago (Schidlowski et al., 1983). Sulfur isotopic data for early Archean sediments indicate the likely activity of photosynthetic bacteria. For instance, barites ($BaSO_4$) of this time were only slightly enriched in ^{34}S compared to sulfides from the same sequence (Schidlowski et al., 1983). They also lend support to the notion that sulfate respiration by prokaryotes was also operating 3.5 billion years ago (Ohmoto et al., 1993; Mojzsis et al., 2003; Ueno et al., 2008; Shen et al., 2009).

Organic geochemistry provides another approach to seeking clues to early life on the Earth. Organic matter trapped in sediment subjected to abiological transformation due to heat and pressure may be changed into products that are stable in situ over geologic time. Organic matter that underwent this kind of transformation is likely to have been in a form that was not rapidly degraded by biological means as would, for instance, carbohydrates, nucleic acids, and nucleotides and most proteins. Nevertheless, amino acids, fatty acids, porphyrins, n-alkanes, and isoprenoid hydrocarbons have been identified in sediments of Archean age (Kvenvolden et al., 1969; Schopf, 1977; Hodgson and Whiteley, 1980; Waldbauer et al., 2009; Hallmann et al., 2011). While some of these organic molecules may be found in carbonaceous chondrites, the latter are dominated by insoluble macromolecules, while the most of the amino acids are nonbiogenic (Sephton, 2002).

If the source compound of any stable organic product identified in an ancient sample of sedimentary rock is known, the latter can be used as an indicator or *biological marker* of the source compound. If the source compound such as porphyrin is a key compound in a particular process, it indicates that a process such as photosynthesis or respiration or both was occurring when the source material became trapped in sediment. *Kerogen* is an example of a stabilized substance formed from ancient organic matter. Its presence in an ancient sedimentary rock suggests the existence of life contemporaneous with the age of that rock.

Studies in molecular biology have revealed that the proportion and sequence of certain monomers in some bioheteropolymers such as ribosomal RNAs are highly conserved in various organisms, that is, they have not become significantly modified over very long times due to extremely slow mutation rates. Such conserved sequences can be used to study the degree of relatedness among different groups of organisms (see Fox et al., 1980; Woese, 1987; Olsen et al., 1994) and can, to a limited degree, also be used to estimate the geologic time at which they first appeared. Such studies have led to the conclusion that archaea and bacteria—the two primary domains of life—had diverged at least by the early in Archean times from a common prokaryotic ancestor, or possibly even from the LUCA, and have evolved ever since along independent parallel lines (Sousa et al., 2013; Williams et al., 2013). The fact that certain physiological processes such as protein synthesis, energy conversion by chemiosmosis, and some biodegradative as well as biosynthetic pathways are held in common by the Bacterial and Archaean domains, although differing in some details, suggests that these pathways may have existed in a common ancestor but became modified during *divergent* evolution, though, in some cases, the issue is clouded by convergent evolution.

Combining several lines of paleontological evidence can lend strong support to a model of an ancient biological process or microbe responsible for it (e.g., Williford et al., 2011). Summons and Powell (1986) found in a certain Canadian petroleum deposit of Silurian age (~400 million years ago) (1) the presence of characteristic biological markers indicating an ancient presence of aromatic carotenoids from green sulfur bacteria (*Chlorobiaceae*) and (2) an enrichment in ^{12}C in these markers relative to the saturated oils. Relating these findings to the paleoenvironmental setting of the oil deposit, the investigators deduced that microbial communities that included *Chlorobiaceae* must have existed in the ancient restricted sea in which the source material from which the oil derived was emplaced.

3.5 UNRESOLVED ISSUES

A lesson learned from cosmology is that as soon as a particular disequilibrium applies in a certain environment, the universe is quick to "invent" an engine to dissipate it (Cottrell, 1979; Russell et al., 2013). We have taken the general case to assume life began as soon as the ocean was cold enough to support open-system hydrothermal convection, i.e., at ~20°C around 4.4 Ga. Why then did it apparently take more than another billion years for oxygenic photosynthesis to emerge, or at least make itself felt, at around 3 Ga (Crowe et al., 2013; Li et al., 2013)? Schwartzman et al. (2008) have

argued that high temperatures (60°C–80°C) due to a high CH_4 greenhouse atmosphere precluded an earlier emergence though, as discussed in Section 3.2.2, this seems less likely. On the other hand, we are used to the apparent emergences of free energy–converting systems getting older as evidence accumulates. One point to make is that prior to 2.7 Ga, there were not the large littorals to the continents that could have supported high turnover rates for the first cyanobacteria or their antecedents, perhaps in extensive freshwater lagoons (Knoll, 1979). Moreover, given the attractiveness of oxygen as a high potential electron acceptor, such photosynthesizers may well have been accompanied by symbionts, mopping up much of the oxygen as soon as it was emitted not to mention the load of ferrous iron as an oxygen sink in the early oceans.

A comparable puzzle is the apparent late emergence of the eukaryotes at ~1.9 Ga, again around a billion years since the likely emergence of cyanobacteria (Knoll et al., 2006; Parfrey et al., 2011). This transition was marked by a unique endosymbiotic event between a bacterial symbiont—an alpha proteobacteria already tuned to a relatively high concentration of free oxygen—and an archaeal host. However, notwithstanding evidence for a "whiff of oxygen" at around 3 Ga (Anbar et al., 2007; Crowe et al., 2013), the bacterial mitochondrial symbiont to the archaeal host akin to rhodobacter (Martin and Müeller, 1998; Allen, 2005; Embley and Martin, 2006) is, by this evidence, unlikely to have emerged early in the history of the planet. However, the dilemma remains and turns on the description of Sugitani et al. (2013) of a flanged lenticular carbonaceous microfossils up to 100 μm in length in the 3.43 Ga Strelley Pool Chert similar to acritarchs occurring in much younger rocks. It has been argued that at least some acritarchs are protists, free living, or colonial eukaryotes. If so, could the eukaryotes date back well into the Archean after all, or are the organic structures merely colonies of prokaryotes—giant bacteria like *Thiomargarita* (Bailey et al., 2007; Cunningham et al., 2012)? However, as we have remarked, geological discoveries normally push back the age of any phenomena.

Another question remains—why did an endosymbiosis between representatives of the two domains of life, the archaea and bacteria, occur at all? Lane and Martin (2010) proposed that the mitochondria afforded a remarkable

200,000-fold expansion per haploid copy of each gene expressed in a cell assuming respiration across the plasma membrane, though genome size only increased by three to four orders of magnitude (and see Lane, 2011). Thus, not only could cells get bigger, the number of proteins expressed was also vastly increased. The metabolic power of a eukaryotic cell approaches 5000 times that of a prokaryotic cell, and the options for complexification and evolution are thereby greatly enhanced. This is because as the surface area of the plasma membrane increases, so does ATP synthesis, whereas protein synthesis scales with the cell volume.

So, given the power of evolution to fill new niches rapidly, we are left to ask if it was the environment that held back the appearance of oxygenic photosynthesis and of the eukaryotes. Or is it our failure to find the evidence in the intermittent and geologically and geochemically compromised ancient rock record for earlier signs? Perhaps, the fact that nitrite-respiring prokaryotes might have sprung from the LUCA explains such a microbe offering its services to a methanogen (Müller et al., 2012).

One final microbial puzzle: it appears "everything is everywhere, but, the environment selects" (a phrase first coined by Baas Becking [1934], as "alles is overal: maar het milieu selecteert") (De Wit and Bouvier, 2006). While Baas Becking invoked passive transport by the wind, this seems to fall short in the explanation for the occupants of the deep biosphere. Instead, perhaps the convective and advective motions of the more viscous hydrosphere and lithosphere over the last 4 Ga played the main role (e.g., Russell and Hall, 2006).

3.6 SUMMARY

The formation of Earth ~4.567 billion years ago and the condensation of a deep ocean shortly thereafter from a mainly CO_2-dominated atmosphere set the stage for an early onset of life (Elkins-Tanton, 2008). Realizing that the biosphere today is autotrophic, i.e., at base, it processes the simplest of nutrients, mainly carbon dioxide, suggests that life emerged to resolve the disequilibrium between this atmospheric gas and the reduced "electron-rich" interior. That carbon dioxide—reduced to the CO molecule and donated to acetyl methane thiol and later acetyl CoA—was also a major

provider of organic carbon, and being a significant electron acceptor along with nitrate and or ferric iron, at the origin of life, may be also gleaned from the nature of the roots to the evolutionary tree. At these roots, Fuchs and his coworkers show, for example, that gluconeogenesis predated glycolysis, i.e., that organic molecules were likely generated from the so-called C1 molecules on the early Earth—mainly CO_2, but also methane and formate (Say and Fuchs, 2010; Fuchs, 2011). Catalysts and protoenzymes were available as metal sulfide and oxyhydroxide clusters. These same clusters were likely sequestered by short peptides, the first fruits of emerging metabolism (Nitschke et al., 2013). Other fuels, nutrients, and electron acceptors were of course hydrogen as well as ammonia, phosphate, and hydrogen sulfide, while nitrate and ferric iron were electron acceptors. Once autotrophic microorganisms emerged, organic detritus and effluents offered further disequilibria to living system and the heterotrophs followed.

Noting the role of the biosphere at large is to hydrogenate carbon dioxide, it is not so surprising to see life using the same old catalysts, though often in new associations in its proteins as it evolves (Baymann et al., 2003; Nitschke and Russell, 2013). At some unknown time before around 3 billion years ago, life discovered a much more powerful way to reduce carbon dioxide, through photosynthesis, especially oxygenic photosynthesis through the cyanobacteria. The main fuel remained the same, i.e., hydrogen, as did its source, i.e., water. But instead of H_2 being produced geochemically and at some remove largely through serpentinization, it was generated in situ by the first bacteria to master the breakdown of water to hydrogen or, more accurately, to four protons and four electrons. And the effluent of this process of course was oxygen, a most effective electron acceptor. The effect on the planet was ultimately so remarkable as to constitute almost another "origin of life." Perhaps we can rehabilitate Darwin's warm little pond as a site for this "second origin"!

Oxygenic photosynthesis generates about 20 times the entropy as does chemosynthesis and moreover sets the stage for the third qualitative evolutionary event, the emergence of the eukaryotes and, through them, complex life (Lane and Martin, 2010). This was facilitated through endosymbiosis. The intracellular symbionts were mitochondria that gain energy using mainly glucose as fuel and chloroplasts that gain their energy, as do cyanobacteria from light, involving anaerobic prokaryotes as host cells multiplying bacterial power, and photosynthetic and anaerobically respiring prokaryotes as the symbionts that have since lost most of their genes, so saving on overheads. Knowing this, we realize that evolution is life's search engine for free energies commensurate with its origin and appropriate for metabolic processing within the bounds of the stability of water—an engine that took care of the colonization of the rest of the planet's deep biosphere and surfaces.

ACKNOWLEDGMENT

The research described in this chapter was carried out at the Jet Propulsion Laboratory, California Institute of Technology, under a contract with the National Aeronautics and Space Administration (NASA), with support by the NASA Astrobiology Institute (Icy Worlds).

REFERENCES

Aharon, P. (2005). Redox stratification and anoxia of the early Precambrian oceans: Implications for carbon isotope excursions and oxidation events. *Precambrian Research*, 137, 207–222.

Alexander, C. O. D., Bowden, R., Fogel, M. L., Howard, K. T., Herd, C. D. K., and Nittler, L. R. (2012). The provenances of asteroids, and their contributions to the volatile inventories of the terrestrial planets. *Science*, 337, 721–723.

Allen, J. F. (2005). A redox switch hypothesis for the origin of two light reactions in photosynthesis. *FEBS Letters*, 579, 963–968.

Allwood, A. and Anderson, M. (2009). 3.45 billion year old stromatolite reef of Western Australia: A rich, large-scale record of early biota, strategies and habitats. *Origins of Life and Evolution of Biospheres*, 39, 188–189.

Allwood, A. C., Walter, M. R., Kamber, B. S., Marshall, C. P., and Burch, I. W. (2006). Stromatolite reef from the Early Archaean era of Australia. *Nature*, 441, 714–718.

Allwood, A. C., Grotzinger, J. P., Knoll, A. H., Burch, I. W., Anderson, M. S., Coleman, M. L., and Kanik, I. (2009). Controls on development and diversity of Early Archean stromatolites. *Proceedings of the National Academy of Sciences of the United States of America*, 106(24), 9548–9555.

Amend, J. P. and McCollom, T. M. (2009). Energetics of biomolecule synthesis on early Earth. In: *Chemical Evolution II: From the Origins of Life to Modern Society*, Zaikowski, L., Friedrich, J. M., and Seidel, S. R. (eds.), pp. 63–94. ACS Symposium Series. Washington, DC: American Chemical Society.

Anbar, A. D., Duan, Y., Lyons, T. W., Arnold, G. L., Kendall, B., Creaser, R. A., Kauffman, A. J., and Buick, R. (2007). A whiff of oxygen before the great oxidation event? *Science*, 317, 1903–1906.

Anbar, A. D. and Holland, H. D. (1992). The photochemistry of manganese and the origin of banded iron formations. *Geochimica et Cosmochimica Acta*, 56(7), 2595–2603.

Anbar, A. D., Roe, J. E., Barling, J., and Nealson, K. H. (2000). Nonbiological fractionation of isotopes. *Science*, 288, 126–128.

Archibald, J. M. (2011). Origin of eukaryotic cells: 40 years on. *Symbiosis*, 54, 69–86.

Arrhenius, G. O. (2003). Crystals and life. *Helvetica Chimica Acta*, 86, 1569–1586.

Atri, D., Hariharan, B., and Griessmeier, J. M. (2013). Galactic cosmic ray–induced radiation dose on terrestrial exoplanets. *Astrobiology*, 13, 910–919.

Atri, D. and Melott, A. L. (2012). Cosmic rays and terrestrial life: A brief review. *arXiv Preprint*, arXiv, 1211.3962.

Awramik, S. M., Schopf, J. W., and Walter, M. R. (1983). Filamentous fossil bacteria form the Archean of Western Australia. *Precambrian Research*, 20, 357–374.

Baas Becking, L. G. M. (1931). *Gaia of leven en aarde*. Oratie gewoon hoogleraar Rijksuniv. Leiden, Mart. Nijhoff's Gravenhage.

Baas Becking, L. G. M. (1934). *Geobiologie of inleiding tot de milieukunde*. The Hague, the Netherlands: W.P. Van Stockum & Zoon.

Bada, J. L. (2004). How life began on Earth: A status report. *Earth and Planetary Science Letters*, 226, 1–15.

Bada, J. L., Bingham, C., and Miller, S. L. (1994). Impact melting of frozen oceans on the early Earth: Implications for the origin of life. *Proceedings of the National Academy of Sciences of the United States of America*, 91, 1248–1250.

Bada, J. L. and Lazcano, A. (2003). Prebiotic soup—Revisiting the Miller experiment. *Science*, 300, 745–746.

Bailey, J. V., Joye, S. B., Kalanetra, K. M., Flood, B. E., and Corsetti, F. (2007). Evidence of giant sulphur bacteria in Neoproterozoic phosphorites. *Nature*, 445, 198–201.

Baker, A. J. and Fallick, A. E. (1989). Evidence from Lewisian limestones for isotopically heavy carbon in two thousand million year old sea water. *Nature*, 337, 352–354.

Baltscheffsky, M., Schultz, A., and Baltscheffsky, H. (1999) H⁺-PPases: A tightly membrane-bound family. *FEBS Letters*, 457, 527–533.

Baymann, F., Brugna, M., Mühlenhoff, U., and Nitschke, W. (2001). Daddy, where did (PS) I come from? *Biochimica et Biophysica Acta (BBA)—Bioenergetics*, 1507(1), 291–310.

Baymann, F., Lebrun, E., Brugna, M., Schoepp-Cothenet, B., Giudici-Orticoni, M. T., and Nitschke, W. (2003). The redox protein construction kit: Pre last universal common ancestor evolution of energy-conserving enzymes. *Philosophical Transactions of the Royal Society of London B: Biological Sciences*, 358, 267–274.

Beard, B. L., Johnson, C. M., Cox, L., Sun, H., Nealson, K. H., and Anguillar, C. (1999). Iron isotope biosignatures. *Science*, 285, 1889–1892.

Bédard, J. H. (2006) A catalytic delamination-driven model for coupled genesis of Archaean crust and sub-continental lithospheric mantle. *Geochimica et Cosmochimica Acta*, 70, 1188–1214.

Bekker, A., Holland, H. D., Wang, P.-L., Rumble, D. III, Stein, H. J., Hannah, J. L., Coetze, L. L., and Beukes, N. J. (2004). Dating the rise of atmospheric oxygen. *Nature*, 427, 117–120.

Benner, S. (2013). Goldschmidt Conference 2013, Florence, Italy, August 25–30. http://www.goldschmidt.info/2013/.

Berg, J. M., Tymoczko, J. L., and Stryer, L. (2006). *Biochemistry*, International edn. New York: WH Freeman & Co.

Bernal, J. D. (1960). The problem of stages in biopoesis. In: *Aspects of the Origin of Life*, pp. 30–45.

Bernard, D. G., Netz, D. J. A., Lagny, T. J., Pierik, A. J., and Balk, J. (2013). Requirements of the cytosolic iron–sulfur cluster assembly pathway in Arabidopsis. *Philosophical Transactions of the Royal Society B*, 368, 20120259.

Bertsch, J., Parthasarathy, A., Buckel, W., and Müller, V. (2013). An electron-bifurcating Caffeyl-CoA reductase. *Journal of Biological Chemistry*, 288(16), 11304–11311.

Beveridge, T. J., Meloche, J. D., Fyfe, W. S., and Murray, R. G. E. (1983). Diagenesis of metals chemically complexed to bacteria: Laboratory formulation of metal phosphates, sulfides and organic condensates in artificial sediments. *Applied and Environmental Microbiology*, 45, 1094–1108.

Bianchi, A., Giorgi, C., Ruzza, P., Toniolo, C., and Milner-White, E. J. (2012). A synthetic hexapeptide designed to resemble a proteinaceous P-loop nest is shown to bind inorganic phosphate. *Proteins*, 80, 1418–1424.

Billi, D., Friedmann, E. I., Hofer, K. G., Caiola, M. G., and Ocampo-Friedmann, R. (2000). Ionizing-radiation resistance in the desiccation-tolerant cyanobacterium Chroococcidiopsis. *Applied and Environmental Microbiology*, 66(4), 1489–1492.

Bizzarro, M., Ulfbeck, D., Trinquier, A., Thrane, K., Connelly, J. N., and Meyer, B. S. (2007). Evidence for a late supernova injection of ⁶⁰Fe into the protoplanetary disk. *Science*, 316, 1178–1181.

Blank, C. E. (2013). Phylogenetic distribution of compatible solute synthesis genes support a freshwater origin for cyanobacteria. *Journal of Phycology*, 49(5), 880–895.

Blank, C. E. and Sánchez-Baracaldo, P. (2010). Timing of morphological and ecological innovations in the cyanobacteria—A key to understanding the rise in atmospheric oxygen. *Geobiology*, 8, 1–23.

Blankenship, R. E. (2008). *Molecular Mechanisms of Photosynthesis*. Oxford, U.K.: Blackwell Science.

Blankenship, R. E. and Hartman, H. (1998). The origin and evolution of oxygenic photosynthesis. *Trends in Biochemical Sciences*, 23(3), 94–97.

Bounama, C., Franck, S., and von Bloh, W. (2001) The fate of the Earth's ocean. *Hydrology and Earth System Sciences*, 5, 569–575.

Boussau, B., Szöllősi, G. J., Duret, L., Gouy, M., Tannier, E., and Daubin, V. (2013). Genome-scale coestimation of species and gene trees. *Genome Research*, 23, 323–330.

Boyd, E. S., Schut, G. J., Adams, M. W. W., and Peters, J. W. (2014). Hydrogen metabolism and the evolution of biological respiration. *Microbe*, 9, 361–367.

Boyer, P. D. (1975). A model for conformational coupling of membrane potential and proton translocation to ATP synthesis and to active transport. *FEBS Letters*, 58, 1–6.

Boyer, P. D. (1997). The ATP synthase—A splendid molecular machine. *Annual Review of Biochemistry*, 66, 717–749.

Branscomb, E. and Russell, M. J. (2013) Turnstiles and bifurcators: The disequilibrium converting engines that put metabolism on the road. *Biochimica et Biophysica Acta (BBA)—Bioenergetics*, 1827, 62–78.

Brasier, M., McLoughlin, N., Green, O., and Wacey, D. (2006). A fresh look at the fossil evidence for early Archaean cellular life. *Philosophical Transactions of the Royal Society B: Biological Sciences*, 361, 887–902.

Brasier, M. D., Green, O. R., Jephcoat, A. P., Kleppe, A. K., Van Krankendonk, M. J., Lindsay, J. F., Steele, A., and Grassineau, N. V. (2002). Questioning the evidence for Earth's oldest fossils. *Nature*, 416, 76–81.

Brown, G. (1999). *The Energy of Life: The Science of What Makes Our Minds and Bodies Work*. New York: Free Press.

Brown, J. R., Douady, C. J., Italia, M. J., Marshall, W. E., and Stanhope, M. J. (2001). Universal trees based on large combined protein sequence data sets. *Nature Genetics*, 28, 281–285.

Buick, R. (1992). The antiquity of oxygenic photosynthesis: Evidence from stromatolites in sulfate-deficient Archean lakes. *Science*, 255, 74–77.

Cairns-Smith, A. G. and Hartman, H. (eds.) (1986). *Clay Minerals and the Origin of Life*. Cambridge, U.K.: Cambridge University Press.

Canup, R. M. (2012). Forming a Moon with an Earth-like composition via a giant impact. *Science*, 338, 1052–1055.

Canuto, V. M., Levine, J. S., Augustsson, T. R., and Imhoff, C. L. (1982). UV radiation from the young Sun and oxygen and ozone levels in the prebiological palaeoatmosphere. *Nature*, 296, 816–820.

Carny, O. and Gazit, E. (2005). A model for the role of short self-assembled peptides in the very early stages of the origin of life. *FASEB Journal*, 19, 1051–1055.

Catling, D. C., Zhanle, K. J., and McKay, C. P. (2001). Biogenic methane, hydrogen escape, and the irreversible oxidation of early Earth. *Science*, 293, 839–843.

Cavalier-Smith, T. (2001). What are fungi? In: *Systematics and Evolution*, pp. 3–37. Berlin, Germany: Springer.

Cech, T. R. (1986). A model for the RNA-catalyzed replication of RNA. *Proceedings of the National Academy of Sciences of the United States of America*, 83, 4360–4363.

Chang, S., Des Marais, D., Mack, R., Miller, S. L., and Strathearn, G. E. (1983). Prebiotic organic syntheses and the origin of life. In: *Earth's Earliest Biosphere. Its Origin and Evolution*, Schopf, J. W. (ed.), pp. 53–92. Princeton, NJ: Princeton University Press.

Chernikova, D., Motamedi, S., Csürös, M., Koonin, E. V., and Rogozin, I. B. (2011). A late origin of the extant eukaryotic diversity: Divergence time estimates using rare genomic changes. *Biology Direct*, 6(1), 26.

Chernoff, Y. O. (2004). Amyloidogenic domains, prions and structural inheritance: Rudiments of early life or recent acquisition? *Current Opinion in Chemical Biology*, 8, 665–671.

Childers, W. S., Ni, R., Mehta, A. K., and Lynn, D. G. (2009). Peptide membranes in chemical evolution. *Current Opinion in Chemical Biology*, 13, 652–659.

Chistoserdova, L., Jenkins, C., Kalyuzhnaya, M. G., Marx, C. J., Lapidus, A., Vorholt, J. A., Staley, J. T., and Lidstrom, M. E. (2004). The enigmatic planctomycetes may hold a key to the origins of methanogenesis and methylotrophy. *Molecular Biology and Evolution*, 21(7), 1234–1241.

Christiansen, B. C., Balic-Zunic, T., Petit, P. O., Frandsen, C., Mørup, S., Geckeis, H., Katerinopoulou, A., and Stipp, S. L. (2009). Composition and structure of an iron-bearing, layered double hydroxide (LDH)–Green rust sodium sulphate. *Geochimica et Cosmochimica Acta*, 73, 3579–3592.

Claire, M. W. (2005). Modeling the rise of atmospheric oxygen. Earth System Processes 2, Abstract, Paper no. 16-5, *The Geological Society of America Meeting*, August 8–11, 2005.

Cleaves, H. J., Scott, A. M., Hill, F. C., Leszczynski, J., Sahai, N., and Hazen, R. M. (2012). Mineral-organic interfacial processes: Potential roles in the origins of life. *Chemical Society Reviews*, 41, 5365–5568.

Cockell, C. S. (2006). The origin and emergence of life under impact bombardment. *Philosophical Transactions of the Royal Society B*, 361, 1845–1856.

Cohen, J. (1995). Getting all turned around over the origins of life on Earth. *Science*, 267, 1265–1266.

Cohen, Y., Padan, E., and Shilo, M. (1975). Facultative anoxygenic photosynthesis in the cyanobacterium *Oscillatoria limnetica*. *Journal of Bacteriology*, 123, 855–861.

Cottrell, A. (1979). The natural philosophy of engines. *Contemporary Physics*, 20, 1–10.

Cox, C. J., Foster, P. G., Hirt, R. P., Harris, S. R., and Embley, T. M. (2008). The archaebacterial origin of eukaryotes. *Proceedings of the National Academy of Sciences of the United States of America*, 105, 20356–20361.

Crowe, S. A., Dossing, L. N., Beukes, N. J., Bau, M., Kruger, S. J., Frei, R., and Canfield, D. E. (2013). Atmospheric oxygenation three billion years ago. *Nature*, 501, 535–538.

Cunningham, J. A, Thomas, C.-W., Bengtson, S., Marone, F., Stampanoni, M., Turner, F. R., Bailey, J. V et al. (2012). Experimental taphonomy of giant sulphur bacteria: Implications for the interpretation of the embryo-like Ediacaran Doushantuo fossils. *Proceedings of the Royal Society: Biological Sciences*, 279(1734), 1857–1864.

Dagan, T. and Martin, W. (2006). The tree of one percent. *Genome Biology*, 7, 118.

Dagan, T., Roettger, M., Stucken, K., Landan, G., Koch, R., Major, P., Gould, S. V. et al. (2013). Genomes of stigonematalean cyanobacteria (Subsection V) and the evolution of oxygenic photosynthesis from prokaryotes to plastids. *Genome Biology and Evolution*, 5(1), 31–44.

Daly, M. J., Gaidamakova, E. K., Matrosova, V. Y., Vasilenko, A., Zhai, M., Venkateswaran, A., Hess, M. et al. (2004). Accumulation of Mn (II) in *Deinococcus radiodurans* facilitates gamma-radiation resistance. *Science*, 306(5698), 1025–1028.

Darwin, F. (ed.) 1888. *The Life and Letters of Charles Darwin*, Vol. 3. London, U.K.: John Murray.

Hodgson, G. W. and Whiteley, C. G. (1980). The universe of porphyrins. In: *Biogeochemistry of Ancient and Modern Environments*, Trudinger, P. A., Walter, M. R., and Ralph, B. J. (eds.), pp. 35–44. Canberra, Australian Capital Territory, Australia: Australian Academy of Science.

Hoehler, T. M. and Jørgensen, B. B. (2013). Microbial life under extreme energy limitation. *Nature Reviews Microbiology*, 11, 83–94.

Holland, H. D. (1997). Evidence for life on Earth more than 3850 million years ago. *Science*, 275, 38–39.

Holland, H. D. (2002). Volcanic gases, black smokers, and the Great Oxidation Event. *Geochimica et Cosmochimica Acta*, 66, 3811–3826.

Hong Enriquez, R. P. and Do, T. N. (2012). Interactions of iron-sulfur clusters with small peptides: Insights into early evolution. *Computational Biology and Chemistry*, 41, 58–61.

Horneck, G., Stöffler, D., Eschweiler, U., and Hornemann, U. (2001). Bacterial spores survive simulated meteorite impact. *Icarus*, 149(1), 285–290.

Huber, C., Eisenreich, W., Hecht, S., and Wächtershäuser, G. (2003). A possible primordial peptide cycle. *Science*, 301, 938–940.

Huber, C., Kraus, F., Hanzlik, M., Eisenreich, W., and Wächtershäuser, G. (2012). Elements of metabolic evolution. *Chemistry—A European Journal*, 18(7), 2063–2080.

Huber, C. and Wächtershäuser, G. (1997). Activated acetic acid by carbon fixation on (Fe,Ni)S under primordial conditions. *Science*, 276, 245–247.

Huber, C. and Wächtershäuser, G. (1998). Peptides by activation of amino acids with CO on (Ni, Fe) S surfaces: Implications for the origin of life. *Science*, 281(5377), 670–672.

Huber, C. and Wächtershäuser, G. (2003). Primordial reductive amination revisited. *Tetrahedron Letters*, 44, 1695–1697.

Jackson, M. G. and Jellinek, A. M. (2013). Major and trace element composition of the high $^3He/^4He$ mantle: Implications for the composition of a non-chondritic Earth. *Geochemistry, Geophysics, Geosystems*, 14, 2954–2976.

Jacobsen, S. B., Ranen, M. C., Petaev, M. I., Remo, J. L., O'Connell, R. J., and Sasselov, D. D. (2008). Isotopes as clues to the origin and earliest differentiation history of the Earth. *Philosophical Transactions of the Royal Society A: Mathematical, Physical and Engineering Sciences*, 366(1883), 4129–4162.

Javaux, E. J., Marshall, C. P., and Bekker, A. (2010). Organic-walled microfossils in 3.2-billion-year-old shallow-marine siliciclastic deposits. *Nature*, 463, 934–938.

Joyce, G. F. (1991). The rise and fall of the RNA world. *New Biologist*, 3, 399–407.

Kappler, A., Johnson, C. M., Crosby, H. A., Beard, B. L., and Newman, D. K. (2010). Evidence for equilibrium iron isotope fractionation by nitrate-reducing iron(II)-oxidizing bacteria. *Geochimica et Cosmochimica Acta*, 74, 2826–2842.

Kappler, A., Pasquero, C., Konhauser, K. O., and Newman, D. K. (2005). Deposition of banded iron formations by anoxygenic phototrophic Fe(II)-oxidizing bacteria. *Geology*, 33, 865–868.

Kasting, J. F. (1990). Bolide impacts and the oxidation state of carbon in the Earth's early atmosphere. *Origins of Life and Evolution of Biospheres*, 20, 199–231.

Kasting, J. F. (2001). The rise of atmospheric oxygen. *Science*, 293, 819–820.

Kasting, J. F. (2004). Today methane-producing microbes are confined to oxygen-free settings, such as the guts of cows, but Earth's distant past, they ruled the world: When methane made climate. *Scientific American*, 291, 78–81, 83–85.

Kasting, J. F. (2005). Methane and climate during the Precambrian era. *Precambrian Research*, 137, 119–129.

Kasting, J. F. and Siefert, J. L. (2002). Life and evolution of Earth's atmosphere. *Science*, 296, 1066–1068.

Katona, G., Carpentier, P., Nivière, V., Amara, P., Adam, V., Ohana, J., Tsanov, N., and Bourgeois, D. (2007) Raman-assisted crystallography reveals end-on peroxide intermediates in a nonheme iron enzyme. *Science*, 316, 449–453.

Kelley, D. S., Karson, J. A., Blackman, D. K., Früh-Green, G. L., Butterfield, D. A., Lilley, M. D., Olson, E. J. et al. (2001). An off-axis hydrothermal vent field near the Mid-Atlantic Ridge at 30°N. *Nature*, 412, 145–149.

Kelly, S., Wickstead, B., and Gull, K. (2011). Archaeal phylogenomics provides evidence in support of a methanogenic origin of the Archaea and a thaumarchaeal origin for the eukaryotes. *Proceedings of the Royal Society B*, 278, 1009–1018.

Kiehl, J. T. and Dickinson, R. E. (1987). A study of the radiative effects of enhanced atmospheric CO_2 and CH_4 on early Earth surface temperatures. *Journal of Geophysical Research*, 92(D3), 2991–2998.

Kirschvink, J. L. (1992). Late Proterozoic low-latitude global glaciation: The snowball earth. In: *The Proterozoic Biosphere*, Schopf, J. W. and Klein, C. (eds.), pp. 51–52. Cambridge, U.K.: Cambridge University Press.

Klales, A., Duncan, J., Nett, E. J., and Kane, S. A. (2012). Biophysical model of prokaryotic diversity in geothermal hot springs. *Physical Review E*, 85(2), 021911.

Klein, M., Friedrich, M., Roger, A. J., Hugenholtz, P., Fishbain, S., Abicht, H., Blackall, L. L., Stahl, D. A., and Wagner, M. (2001). Multiple lateral transfers of dissimilatory sulfite reductase genes between major lineages of sulfate-reducing prokaryotes. *Journal of Bacteriology*, 183(20), 6028–6035.

Knauth, L. P. and Lowe, D. R. (2003). High Archean climatic temperature inferred from oxygen isotope geochemistry of cherts in the 3.5 Ga Swaziland Supergroup, South Africa. *Geological Society of America Bulletin*, 115, 566–580.

Knoll, A. H. (1979). Archean photoautotrophy: Some alternatives and limits. *Origins Life*, 9, 313–327.

Knoll, A. H., Javaux, E. J., Hewitt, D., and Cohen, P. (2006). Eukaryotic organisms in Proterozoic oceans. *Philosophical Transactions of the Royal Society B*, 361, 1023–1038.

Knoll, A. N. and Awramik, S. M. (1983). Ancient microbial ecosystems. In: *Microbial Geochemistry*, Krumbein, W. E. (ed.), pp. 287–315. Oxford, U.K.: Blackwell Scientific.

Koonin, E. V. and Martin, W. (2005). On the origin of genomes and cells within inorganic compartments. *Trends in Genetics*, 21, 647–654.

Kruger, K., Grabowski, P. J., Zaug, A. J., Sands, J., Gottschling, D. E., and Cech, T. R. (1982). Self-splicing RNA: Autoexcision and autocatalyzation of the ribosomal RNA intervening sequence of *Tetrahymena*. *Cell*, 31, 147–157.

Krylov, I. N. and Semikhatov, M. A. (1976). Appendix II. Table of time ranges of the principal groups of Precambrian stromatolites. In: *Stromatolites*, Walter, M. R. (ed.), pp. 693–694. Amsterdam, the Netherlands: Elsevier.

Kump, L. R., Barley, M. E., and Kasting, J. F. 2001. Rise of atmospheric oxygen and the "upside-down" Archean mantle. *Geochemistry, Geophysics, Geosystems*, 2(2). American Geophysical Union. Online computer file: http://g-cubed.org/gs2001/2000GC000114/article2000GC000114.pdf.

Kvenvolden, K. A., Peterson, E., and Pollock, G. E. (1969). Optical configuration of amino-acids in Pre-Cambrian Fig Tree chert. *Nature*, 221, 141–143.

Lane, N. (2010). Why are cells powered by proton gradients? *Nature Education*, 3(9), 18.

Lane, N. (2011). Energetics and genetics across the prokaryote-eukaryote divide. *Biology Direct*, 6, 35.

Lane, N., Allen, J. F., and Martin, W. (2010). How did LUCA make a living? Chemiosmosis in the origin of life. *Bioessays*, 32, 271–280.

Lane, N. and Martin, W. (2010). The energetics of genome complexity. *Nature*, 467(7318), 929–934.

Larkum, A. W., Lockhart, P. J., and Howe, C. J. (2007). Shopping for plastids. *Trends in Plant Science*, 12(5), 189–195.

Li, W., Czaja, A. D., Van Kranendonk, M. J., Beard, B. L., Roden, E. E., and Johnson, C. M. (2013). An anoxic, Fe (II)-rich, U-poor ocean 3.46 billion years ago. *Geochimica et Cosmochimica Acta*, 120, 65–79.

Lin, S. M., Tsai, J. Y., Hsiao, C. D., Huang, Y. T., Chiu, C. L., Liu, M. H., Tung, J. Y., Liu, T. H., Pan, R. L., and Sun, Y. J. (2012). Crystal structure of a membrane-embedded H$^+$-translocating pyrophosphatase. *Nature*, 484(7394), 399–403.

Lowe, D. R. (1980). Stromatolites 3,400 Myr old from the Archean of Western Australia. *Nature (London)*, 284, 441–443.

Lowe, D. R. (1994). Abiological origin of described stromatolites older than 3.2 Ga. *Geology*, 22, 387–390.

Maden, B. E. H. (2000). Tetrahydrofolate and tetrahydromethanopterin compared: Functionally distinct carriers in C1 metabolism. *Biochemical Journal*, 350, 609–629.

Mancinelli, R. L. and McKay, C. P. (1988). The evolution of nitrogen cycling. *Origins of Life and Evolution of the Biospheres*, 18(4), 311–325.

Mandernack, K. W., Bazylinski, D. A., Shanks, W. C. III, and Bullen, T. D. (1999). Oxygen and iron isotope studies of magnetite produced by magnetotactic bacteria. *Science*, 285, 1892–1896.

Mareš, J., Hrouzek, P., Kaňa, R., Ventura, S., Strunecký, O., and Komárek, J. (2013). The primitive thylakoid-less cyanobacterium gloeobacter is a common rock-dwelling organism. *PLoS ONE*, 8(6), e66323.

Margulis, L. and Lovelock, J. E. (1974). Biological modulation of the Earth's atmosphere. *Icarus*, 21(4), 471–489.

Martin, R. S., Mather, T. A., and Pyle, D. M. (2007). Volcanic emissions and the early Earth atmosphere. *Geochimica et Cosmochimica Acta*, 71, 3673–3685.

Martin, W. and Müller, M. (1998). The hydrogen hypothesis for the first eukaryote. *Nature*, 392, 37–41.

Martin, W., Rotte, C., Hoffmeister, M., Theissen, U., Gelius-Dietrich, G., Ahr, S., and Henze, K. (2003). Early cell evolution, eukaryotes, anoxia, sulfide, oxygen, fungi first (?), and a tree of genomes revisited. *IUBMB Life*, 55(4-5), 193–204.

Martin, W., Rujan, T., Richly, E., Hansen, A., Cornelsen, S., Lins, T., Leister, D. et al. (2002). Evolutionary analysis of *Arabidopsis*, cyanobacterial, and chloroplast genomes reveals plastid phylogeny and thousands of cyanobacterial genes in the nucleus. *Proceedings of the National Academy of Sciences of the United States of America*, 99, 12246–12251.

Pascal, R., Pross, A., and Sutherland, J. D. (2013). Towards an evolutionary theory of the origin of life based on kinetics and thermodynamics. *Open Biology*, 3(11), 130156.

Pasek, M. A., Harnmeijer, J. P., Buick, R., Gull, M., and Atlas, Z. (2013). Evidence for reactive reduced phosphorus species in the early Archean ocean. *Proceedings of the National Academy of Sciences of the United States of America*, 110, 10089–10094.

Pereira, I. A. C. (2008). Membrane complexes in *Desulfovibrio*. In: *Microbial Sulfur Metabolism*, Friedrich, C. and Dahl, C. (eds.), pp. 24–35. Berlin, Germany: Springer-Verlag.

Peters, S. E. and Gaines, R. R. (2012). Formation of the 'Great Unconformity' as a trigger for the Cambrian explosion. *Nature*, 484, 363–366.

Pfeffer, C., Larsen, S., Song, J., Dong, M., Besenbacher, F., Meyer, R. L., Kjeldsen, K. U. et al., (2012). Filamentous bacteria transport electrons over centimetre distances. *Nature*, 491(7423), 218–221.

Pizzarello, S. and Shock, E. (2010). The organic composition of carbonaceous meteorites: The evolutionary story ahead of biochemistry. *Cold Spring Harbor Perspectives in Biology*, 2(3), a002105.

Poole, A. M. and Logan, D. T. (2005). Modern mRNA proofreading and repair: Clues that the Last Universal Common Ancestor (LUCA) possessed an RNA genome? *Molecular Biology and Evolution*, 22, 1444–1455.

Powner, M. W., Gerland, B., and Sutherland, J. D. (2009). Synthesis of activated pyrimidine ribonucleotides in prebiotically plausible conditions. *Nature*, 459, 239–242.

Puthiyaveetil, S., Ibrahim, I. M., and Allen, J. F. (2013). Evolutionary rewiring: A modified prokaryotic gene-regulatory pathway in chloroplasts. *Philosophical Transactions of the Royal Society B*, 368, 20120260.

Ramos, A. R., Keller, K. L., Wall, J. D., and Pereira, I. A. C. (2012). The membrane QmoABC complex interacts directly with the dissimilatory adenosine 5′-phosphosulfate reductase in sulfate reducing bacteria. *Frontiers in Microbiology*, 3, 137.

Raven, J. A. and Allen, J. F. (2003). Genomics and chloroplast evolution: What did cyanobacteria do for plants? *Genome Biology*, 4, 209.

Rippka, R., Waterbury, J., and Cohen-Bazire, G. (1974). Cyanobacterium which lacks thylakoids. *Archives of Microbiology*, 100, 419–436.

Rivera, M. C. and Lake, J. A. (2004). The ring of life provides evidence for a genome fusion origin of eukaryotes. *Nature*, 431, 152–155.

Robert, F. and Chaussidon, M. (2006). A palaeotemperature curve for the Precambrian oceans based on silicon isotopes in cherts. *Nature*, 443, 969–972.

Rosing, M. T. (1999). ^{13}C-depleted carbon microparticles in >3700-Ma sea-floor sedimentary rocks from West Greenland. *Science*, 283, 674–676.

Roussel, E. G., Bonavita, M. A. C., Querellou, J., Cragg, B. A., Webster, G., Prieur, D., and Parkes, R. J. (2008). Extending the sub-sea-floor biosphere. *Science*, 320, 1046–1046.

Russell, M. J., Allen, J. F., and Milner-White, E. J. (2008). Inorganic complexes enabled the onset of life and oxygenic photosynthesis. In: *Energy from the Sun: 14th International Congress on Photosynthesis*, Allen, J. F., Gantt, E., Golbeck, J. H., and Osmond, B. (eds.), pp. 1193–1198. Berlin, Germany: Springer.

Russell, M. J., Barge, L. M., Bhartia, R., Bocanegra, D., Bracher, P. J., Branscomb, E., Kidd, R. et al. (2014). The drive to life on wet and icy worlds. *Astrobiology*, 14(4), 303–343.

Russell, M. J., Daniel, R. M., Hall, A. J., and Sherringham, J. (1994). A hydrothermally precipitated catalytic iron sulphide membrane as a first step toward life. *Journal of Molecular Evolution*, 39, 231–243.

Russell, M. J. and Hall, A. J. (1997). The emergence of life from iron monosulfide bubbles at a submarine hydrothermal redox and pH front. *Journal of the Geological Society of London*, 154, 377–402.

Russell, M. J. and Hall, A. J. (2006). The onset and early evolution of life. In: *Evolution of Early Earth's Atmosphere, Hydrosphere, and Biosphere—Constraints from Ore Deposits*, Kesler, S. E. and Ohmoto, H. (eds.), pp. 1–32. Boulder, CO: Geological Society of America, Memoir 198.

Russell, M. J., Hall, A. J., and Mellersh, A. R. (2003). On the dissipation of thermal and chemical energies on the early Earth: The onsets of hydrothermal convection, chemiosmosis, genetically regulated metabolism and oxygenic photosynthesis. In: *Natural and Laboratory-Simulated Thermal Geochemical Processes*, Ikan, R. (ed.), pp. 325–388. Dordrecht, the Netherlands: Kluwer Academic Publishers.

Russell, M. J., Hall, A. J., and Turner, D. (1989). In vitro growth of iron sulphide chimneys: Possible culture chambers for origin-of-life experiments. *Terra Nova*, 1, 238–241.

Russell, M. J., Nitschke, W., and Branscomb, E. (2013). The inevitable journey to being. *Philosophical Transactions of the Royal Society of London B: Biological Sciences B*, 368, 20120254.

Rutherford, A. W. (1985). Orientation of EPR signals arising from components in photosystem II membranes. *Biochimica et Biophysica Acta (BBA)—Bioenergetics*, 807(2), 189–201.

Sarafian, A. R., Nielsen, S. G., Marschall, H. R., McCubbin, F. M., and Monteleone, B. D. (2014). Early accretion of water in the inner solar system from a carbonaceous chondrite-like source. *Science*, 346, 623–626.

Sasaki, S., (1990). The primary solar-type atmosphere surrounding the accreting Earth: H_2O-induced high surface temperature. In: *Origin of the Earth*, Newsom, H. E. and Jones, J. H. (eds.), pp. 195–209. Oxford University Press: New York.

Sauer, K. and Yachandra, V. K. (2004). The water-oxidation complex in photosynthesis. *Biochimica et Biophysica Acta (BBA)—Bioenergetics*, 1655, 140–148.

Saw, J. H., Schatz, M., Brown, M. V., Kunkel, D. D., Foster, J. S., Shick, H., Christensen, S., Hou, S., Wan, X., and Donachie, S. P. (2013). Cultivation and complete genome sequencing of *Gloeobacter kilaueensis* sp. nov., from a Lava Cave in Kīlauea Caldera, Hawai'i. *PLoS ONE*, 8(10), e76376.

Say, R. F. and Fuchs, G. (2010). Fructose 1, 6-bisphosphate aldolase/phosphatase may be an ancestral gluconeogenic enzyme. *Nature*, 464, 1077–1081.

Schidlowski, M., Eichmann, R., and Junge, C. E. (1975). Precambrian sedimentary carbonates: Carbon and oxygen isotope geochemistry and implications for the terrestrial oxygen budget. *Precambrian Research*, 2, 1–69.

Schidlowski, M., Hayes, J. M., and Kaplan, I. R. (1983). Isotopic inferences of ancient biochemistries: Carbon, sulfur, hydrogen, and nitrogen. In: *Earth's Earliest Biosphere: Its Origin and Evolution*, Schopf, J. W. (ed.), pp. 149–186. Princeton, NJ: Princeton University Press.

Schneider, D. A., Bickford, M. E., Cannon, W. F., Schulz, K. J., and Hamilton, M. A. (2002). Age of volcanic rocks and syndepositional iron formations, Marquette Range Supergroup: Implications for the tectonic setting of Paleoproterozoic iron formations of the Lake Superior. *Canadian Journal of Earth Sciences*, 39, 999–1012.

Schoepp-Cothenet, B., van Lis, R., Atteia, A., Baymann, F., Capowiez, L., Ducluzeau, A.-L., Duval, S., ten Brink, F., Russell, M. J., and Nitschke, W. (2013). On the universal core of bioenergetics. *Biochimica et Biophysica Acta (BBA)—Bioenergetics*, 1827, 79–93.

Schoepp-Cothenet, B., van Lis, R., Philippot, P., Magalon, A., Russell, M. J., and Nitschke, W. (2012). The ineluctable requirement for the trans-iron

elements molybdenum and/or tungsten in the origin of life. *Nature Scientific Reports*, 2, 263.

Schoonen, M. A., Xu, Y., and Bebie, J. (1999). Energetics and kinetics of the prebiotic synthesis of simple organic acids and amino acids with the FeS-H_2S/FeS_2 redox couple as reductant. *Origins of Life and Evolution of the Biosphere*, 29, 5–32.

Schopf, J. W. (1977). Evidences of Archean life. In: *Chemical Evolution of the Early Precambrian*, Ponnamperuma, C. (ed.), pp. 101–105. New York: Academic Press.

Schopf, J. W. (ed.) (1983). *Earth's Earliest Biosphere: Its Origin and Evolution*. Princeton, NJ: Princeton University Press.

Schopf, J. W., Hayes, J. M., and Walter, M. R. (1983). Evolution of Earth's earliest ecosystem: Recent progress and unsolved problems. In: *Earth's Earliest Biosphere: Its Origin and Evolution*, Schopf, J. W. (ed.), pp. 360–384. Princeton, NJ: Princeton University Press.

Schopf, J. W. and Klein, C. (1992). *The Proterozoic Biosphere, a Multidisciplinary Study*. New York: Cambridge University Press.

Schopf, J. W., Kudryavtsev, A. B., Sugitani, K., and Walter, M. R. (2010). Precambrian microbe-like pseudofossils: A promising solution to the problem. *Precambrian Research*, 179(1), 191–205.

Schopf, J. W. and Packer, B. M. (1987). Early Archean (3.3-billion-year-old) microfossils from Warrawoona Group, Australia. *Science*, 237, 70–73.

Schopf, J. W. and Walter, M. R. (1983). Archean microfossils: New evidence of ancient microbes. In: *Earth's Earliest Biosphere: Its Origin and Evolution*, Schopf, J. W. (ed.), pp. 214–239. Princeton, NJ: Princeton University Press.

Schuchmann, K. and Müller, V. (2013). Direct and reversible hydrogenation of CO_2 to formate by a bacterial carbon dioxide reductase. *Science*, 342(6164), 1382–1385.

Schut, G. J. and Adams, M. W. (2009). The iron-hydrogenase of *Thermotoga maritima* utilizes ferredoxin and NADH synergistically: A new perspective on anaerobic hydrogen production. *Journal of Bacteriology*, 191, 4451–4457.

Schwartzman, D., Caldeira, K., and Pavlov, A. (2008). Cyanobacterial emergence at 2.8 Gya and greenhouse feedbacks. *Astrobiology*, 8, 187–203.

Sephton, M. A. (2002). Organic compounds in carbonaceous meteorites. *Natural Product Reports*, 19(3), 292–311.

Sessions, A. L., Doughty, D. M., Welander, P. V., Summons, R. E., and Newman, D. K. (2009). The continuing puzzle of the great oxidation event. *Current Biology*, 19(14), R567–R574.

Williford, K. H., Grice, K., Logan, G. A., Chen, J., and Huston, D. (2011). The molecular and isotopic effects of hydrothermal alteration of organic matter in the Paleoproterozoic McArthur River Pb/Zn/Ag ore deposit. *Earth and Planetary Science Letters*, 301(1), 382–392.

Wilson, H. F. and Militzer, B. (2012). Rocky core solubility in Jupiter and giant exoplanets. *Physical Review Letters*, 108(11), 111101.

Woese, C. R. (1987). Bacterial evolution. *Microbiological Reviews*, 51, 221–271.

Woese, C. R. (2002). On the evolution of cells. *Proceedings of the National Academy of Sciences of the United States of America*, 99, 8742–8747.

Woese, C. R. and Fox, G. E. (1977). Phylogenetic structure of the prokaryotic domain: The primary kingdoms. *Proceedings of the National Academy of Sciences of the United States of America*, 74, 5088–5090.

Wood, B. J. and Halliday, A. N. (2010). The lead isotopic age of the Earth can be explained by core formation alone. *Nature*, 465, 767–770.

Wood, B. J., Walter, M. J., and Wade, J. (2006). Accretion of the Earth and segregation of its core. *Nature*, 441, 825–833.

Worth, R. J., Sigurdsson, S., and House, C. H. (2013). Seeding life on the moons of the outer planets via lithopanspermia. *Astrobiology*, 13, 1155–1165.

Yamagata, Y. and Inomata, K. (1997). Condensation of glycylglycine to oligoglycines with trimetaphosphate in aqueous solution II: Catalytic effect of magnesium ion. *Origins of Life and Evolution of Biospheres*, 27, 339–344.

Yamagata, Y., Watanabe, H., Saitoh, M., and Namba, T. (1991). Volcanic production of polyphosphates and its relevance to prebiotic evolution. *Nature*, 352, 516–519.

Yamaguchi, A., Yamamoto, M., Takai, K., Ishii, T., Hashimoto, K., and Nakamura, R. (2014). Electrochemical CO_2 reduction by Ni-containing iron sulfides: How is CO_2 electrochemically reduced at bisulfide-bearing deep-sea hydrothermal precipitates? *Electrochimica Acta*, 141, 311–318.

Yang, W., Holland, H. D., and Rye, R. (2002). Evidence for low or no oxygen in the late Archean atmosphere from the 2.76 Ga Mt. Roe #2 paleosol, Western Australia: Part 3. *Geochimica et Cosmochimica Acta*, 66, 3707–3718.

Yarus, M. (2010). Getting past the RNA world: The initial Darwinian ancestor. In: *RNA World IV*, Vol. 2, Cech, T., Gesteland, R., and Atkins, J. (eds.), p. a003590. CSH Lab Press.

Yoshida, M., Muneyuki, E., and Hisabori, T. (2001). ATP synthase—A marvellous rotary engine of the cell. *Nature Reviews Molecular Cell Biology*, 2, 669–677.

Yung, P. T., Shafaat, H. S., Connon, S. A., and Ponce, A. (2007). Quantification of viable endospores from a Greenland ice core. *FEMS Microbiology Ecology*, 59(2), 300–306.

Zahnle, K., Arndt, N., Cockell, C., Halliday, A., Nisbet, E., Selsis, F., and Sleep, N. H. (2007). Emergence of a habitable planet. *Space Science Reviews*, 129, 35–78.

Zedef, V., Russell, M. J., Fallick, A. E., and Hall, A. J. (2000). Genesis of vein stockwork and sedimentary magnesite and hydromagnesite deposits in the ultramafic terranes of southwestern Turkey: A stable isotope study. *Economic Geology*, 95(2), 429–445.

Zhang, S., Holmes, T., Lockshin, C., and Rich, A. (1993). Spontaneous assembly of a self-complimentary oligopeptide to form a stable macroscopic membrane. *Proceedings of the National Academy of Sciences of the United States of America*, 90, 3334–3338.

CHAPTER FOUR

Uppermost Lithosphere as a Microbial Habitat

Henry Lutz Ehrlich

CONTENTS

4.1 Rock and Minerals / 55
4.2 Mineral Soil / 57
    4.2.1 Origin of Mineral Soil / 57
    4.2.2 Some Structural Features of Mineral Soil / 58
    4.2.3 Effects of Plants and Animals on Soil Evolution / 59
    4.2.4 Effects of Microbes on Soil Evolution / 59
    4.2.5 Effects of Water in Soil Erosion / 60
    4.2.6 Water Distribution in Mineral Soil / 60
    4.2.7 Nutrient Availability in Mineral Soil / 61
    4.2.8 Some Major Soil Types / 62
    4.2.9 Types of Microbes and Their Distribution in Mineral Soil / 64
4.3 Organic Soils / 66
4.4 Summary / 66
References / 67

4.1 ROCK AND MINERALS

To understand how the lithosphere supports the existence of microbes on and in it and how microbes influence the formation and transformation of some of its constituent rocks and minerals, we must review some of the general chemical and physical features of the lithosphere components.

Geologically, the term "rock" refers to massive, solid, inorganic matter consisting usually of two or more intergrown minerals. Rock may be igneous in origin, that is, it may arise by cooling of "magma" (molten rock material) from the interior of the Earth (crust and/or asthenosphere). The cooling may be a slow or a fast process. In slow cooling, different minerals begin to crystallize at different times, owing to their different melting points, leading to the formation of rock with a visually distinguishable mixture of intergrown crystals of which granite is a typical example (Figure 4.1a). In fast cooling, rapid crystallization occurs, leading to the formation of

rock containing only tiny crystals that are not visible to the naked eye. Basalt is an example of rock formed in this way (Figure 4.1b).

Rock may also be *sedimentary* in origin, that is, it may arise through the accumulation and compaction of sediment that consists mainly of mineral matter derived from breakdown of other rock. In other instances, sedimentary rock may arise as a result of cementation of accumulated inorganic sediment by carbonate, silicate, aluminum oxide, ferric oxide, or a combination thereof. The cementing substance may result from microbial activity. These transformations of loose sediment into sedimentary rock are termed "lithification." Sedimentary rock facies often exhibit a layered structure in vertical section, reflecting changes in composition as the sediment is accumulated. Analysis of the different layers may tell something of the environmental conditions during which they accumulated. Examples of sedimentary rock are limestone, shale, and sandstone.

form cavities in limestone rock that they occupy by causing dissolution of the $CaCO_3$ (Golubic et al., 1975). In other cases, opportunistic microorganisms invade preformed cavities in rock (chasmolithic organisms) (Friedmann, 1982). Invertebrates, snails in particular, may feed on boring organisms (Golubic and Schneider, 1979; Shachak et al., 1987) or chasmolithic microorganisms by grinding away the superficial rock to expose them and consume them. The rock debris that the snails generate becomes part of a soil (Shachak et al., 1987; Jones and Shachak, 1990).

Microbes dissolve rock minerals through the corrosive action of metabolic products such as NH_3, HNO_3, and CO_2 (forming H_2CO_3 in water), and oxalic, citric, and gluconic acids they excrete. Organic compounds formed by microorganisms such as lichens have been shown in studies using scanning electron microscopy to cause distinct weathering (Jones et al., 1981). Waksman and Starkey as long ago as 1931 cited the following reactions as examples of how microbes can affect weathering of minerals:

$$2KAlSi_3O_8 + 2H_2O + CO_2 \rightarrow H_4Al_2Si_2O_9 + K_2CO_3$$
Orthoclase $\qquad\qquad\qquad + 4SiO_2$
$$\text{Kaolinite} \qquad (4.1)$$

$$12MgFeSiO_4 + 26H_2O + 3O_2 \rightarrow 4H_4Mg_3Si_2O_9 + 4SiO_2$$
Olivine $\qquad\qquad\qquad\qquad + 6Fe_2O_3{\cdot}3H_2O$
$$\text{Serpentine} \qquad (4.2)$$

Reaction 4.1 is promoted by CO_2 production in the metabolism of heterotrophic microorganisms, and Reaction 4.2 is promoted by O_2 production in oxygenic photosynthesis by cyanobacteria, algae, and lichens inhabiting the surface of rocks. Further investigations have extended these observations. In recent studies, reactions were examined in which organic acids that are excreted by microorganisms promote weathering of primary minerals such as feldspars and secondary minerals such as clays (e.g., Browne and Driscoll, 1992; Hiebert and Bennett, 1992; Lucas et al., 1993; Welch and Ullman, 1993; Brady and Carroll, 1994; Oelkers et al., 1994; Barker and Banfield, 1996, 1998; Bennett et al., 1996; Ullman et al., 1996). Some current weathering models favor protonation as a means of displacing cationic components from the crystal lattice followed by cleaving of Si–O and Al–O bonds (Berner et

Holdren, 1977; Chou and Wollast, 1984). Others favor complexation, for instance, of Al and/or Si in aluminosilicates, as a primary mechanism of dissolution (Wieland and Stumm, 1992; Welch and Vandevivere, 1995).

Mineral soil may derive from aquatic sediment or *alluvium* left behind after the water that carried it from its place of origin to its final site of deposition has receded. Mineral soil can also form in place as a result of progressive weathering of parent rock and subsequent differentiation of weathering products. Soils originating by either mechanism undergo eluviation (removal of some products by washing out) and/or alluviation (addition of new material by water transport). Any soil, once formed, undergoes further gradual transformation due to the biological activity it supports (Buol et al., 1980).

4.2.2 Some Structural Features of Mineral Soil

Mineral soil will vary in composition, depending on the source of the parent material, the extent of weathering, the amount of organic matter introduced into or generated in the soil, and the amount of moisture it holds. Its texture is affected by the particle sizes of its inorganic constituents (stones, >2 mm; sand grains, 0.05–2 mm; silt, 0.002–0.05 mm; clay particles, <0.002 mm), which determine its porosity and thus, its permeability to water and gases.

Many, but not all, mineral soils tend to be more or less obviously stratified. As many as three or four major strata or *horizons* may be recognizable in agricultural and forest soil *profiles*. A soil profile is a vertical section through soil (Figure 4.2). The strata are labeled O, A, B, and C horizons. The O horizon represents the litter zone, consisting of much undecomposed and partially decomposed organic matter. Some soil profiles may lack an O horizon. The A and B horizons represent the true soil. The C horizon represents the parent material from which the soil was formed. It may be bedrock or an earlier soil. The A and B horizons are often further subdivided, although these divisions are somewhat arbitrary. The A horizon is the biologically most active zone, containing most of the root systems of plants growing on it and the microbes and other life forms that inhabit soil. As is to be expected, the carbon content in this horizon is also greater. The biological

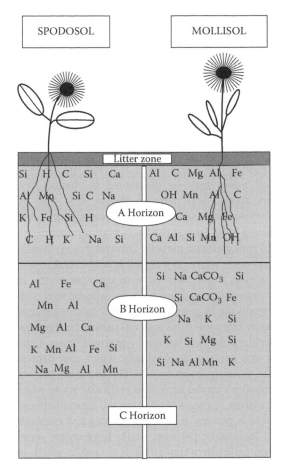

SPODOSOL	MOLLISOL

Litter zone

Si H C Si Ca   Al C Mg Al Fe
Al Mn   Si C Na   OH Mn Al C

A Horizon

K Fe Si H   Ca Mg Fe
C H K Na Si   Ca Al Si Mn OH

Al   Fe   Ca   Si Na CaCO$_3$ Si
                Si CaCO$_3$ Fe
Mn Al   **B Horizon**
                Na K Si
Mg Al Ca
                K Si Mg Si
K Mn Al Fe Si
                Si Na Al Mn K
Na Mg Al Mn

C Horizon

Figure 4.2. Schematic representation of the major soil horizons of spodosol and mollisol. The litter zone is also called the O horizon. The A and B horizons may be further subdivided on the basis of soil chemistry.

activity in the A horizon may cause solubilization of organic and inorganic matter, some or all of which, especially the inorganic matter, is carried by soil water into the B horizon. The A horizon is therefore, known at times as the *leached layer*, and the B horizon is at times known as the *enriched layer*. Both biological and abiological factors play a role in soil profile formation.

4.2.3 Effects of Plants and Animals on Soil Evolution

Plants assist in soil evolution by contributing organic matter through excretions from their root systems and as dead organic matter. The plant excretions may react directly with some soil mineral constituents, or they may first be modified together with dead plant matter by microbes resulting in

products that then react with soil mineral constituents. During their lifetime, plants remove some minerals from soil and contribute to water movement through the soil by water absorption via their roots and transpiration from their leaves. Their root system may also help prevent destruction of the soil through wind and water erosion by anchoring it.

Burrowing invertebrates, from small mites to large earthworms, help to break up soil, keep it porous, and redistribute organic matter. The habitat of some of these invertebrates is restricted to specific regions in the soil profile.

4.2.4 Effects of Microbes on Soil Evolution

Microbes contribute to soil evolution by mineralizing some or all of any added organic matter during the decay process. Some of the metabolic products from this decay, such as organic and inorganic acids, CO_2 and NH_3, interact slowly with soil minerals and cause their alteration or solution, an important step in soil profile formation (Berthelin, 1977; Welch and Ullman, 1993; Barker and Banfield, 1996, 1998; Ullman et al., 1996). For instance, the mineral chlorite has been reported to be bacterially altered in this manner through loss of Fe and Mg and an increase in Si. The mineral vermiculite has been reported to be bacterially altered through mobilization by dissolution of Si, Al, Fe, and Mg, thereby forming montmorillonite (Berthelin and Boymond, 1978). Certain microbes may interact directly (i.e., enzymatically) with certain inorganic soil minerals by oxidizing or reducing them or constituents in them (see Chapters 14, 15, and 18 through 21) (Ehrlich, 2001), resulting in their mobilization by dissolution or in the formation of new minerals (Berthelin, 1977). Microbes may also play an important role in "humus" formation.

Humus is an important constituent of soil, consisting of humic and fulvic acids, humins and amino acids, lignin, amino sugars, and other compounds of biological origin (Stevenson, 1994; Paul and Clark, 1996, pp. 148–152). Humic and fulvic acids are dispersible in solutions of NaOH or sodium pyrophosphate; humin is not. Humic acids are precipitated at acid pH whereas fulvic acids are not. The humus constituents humins and humic and fulvic acids represent components of soil organic matter that are only slowly decomposed. They are mostly formed by

Clay particles are especially important in ionic binding of organic or inorganic cationic solutes (those having a positive charge). Such particles exhibit mostly negative charges except at their edges, where positive charges may appear. Their capacity for ion exchange depends on their crystal structure. The partitioning of solutes between soil solution and mineral surfaces often results in the greater concentration of solutes on mineral surfaces than in the soil solution, and as a result, the mineral surfaces may be the preferred habitat of soil microbes that require these solutes in more concentrated form. On the other hand, ionically bound solutes on clay or other soil particles may be less available to soil microbes because the microbes may not be able to dislodge them from the particle surface. In that instance, soil solution may be the preferred habitat for microbes that have a requirement for such solutes. Ionic binding to soil particles may be beneficial if a solute subject to such binding is toxic and not readily dislodged (see Chapter 11).

4.2.8 Some Major Soil Types

Distinctive soil types may be identified by and correlated with climatic conditions and with the vegetation they support (Bunting, 1967; Buol et al., 1980). Climatic conditions determine the kind of vegetation that may develop. Thus, in the high northern latitudes, *tundra soil*, a type of *inceptisol*, prevails, which in that cold climate is often frozen and therefore, supports only limited plant and microbial development. It has a poorly developed profile. It may be slightly alkaline. Examples of tundra soil are arctic brown soil and bog soil. In the cool (i.e., temperate), humid zones at midlatitudes, "spodosols" (Figures 4.2 and 4.4) prevail, which support extensive forests, particularly of the coniferous type. Spodosols tend to be acidic, having a strongly leached, grayish A horizon depleted in colloids and compounds of iron, and aluminum and a brown B horizon enriched in colloids and compounds of iron and aluminum leached from the A horizon. In regions of moderate rainfall in temperate climates at midlatitudes, "mollisols" (Figures 4.2 and 4.5) prevail. These are soils that support grasslands (i.e., they are prairie soils). They exhibit rich black topsoil and show lime accumulation in the B horizon because they have neutral to alkaline pH. "Oxisols" are found at low latitudes in tropical, humid climates. They are poorly zonated, highly weathered, jungle soils with a B horizon rich in sesquioxides or clays. Owing to the hot, humid climate conditions under which they exist, these soils are intensely active microbiologically and require constant replenishment of organic matter by the vegetation growing on them and from animal excretions and remains to stay fertile. The neutral to alkaline pH conditions of oxisols promote leaching of silicate and precipitation of iron and aluminum as sesquioxides. When oxisols are denuded of the arboreal vegetation, as in slash-and-burn agriculture, they quickly lose their fertility as a result of

Figure 4.4. Soil profile: spodosol (podzol). (Courtesy of U.S. Department of Agriculture (USDA), Soil Conservation Service, Washington, D.C.)

EHRLICH'S GEOMICROBIOLOGY

Figure 4.5. Soil profile: mollisol (chernozem). (Courtesy of U.S. Department of Agriculture (USDA), Soil Conservation Service, Washington, D.C.)

intense microbial activity, which rapidly destroys soil organic matter. Because little organic matter is returned to the soil in its agricultural exploitation, conditions favor laterization, a process in which iron and aluminum oxides, silica, and carbonates are precipitated that cement the soil particles together and greatly reduce the porosity and water-holding capacity of the soil and make it generally unfavorable for plant growth.

Aridisols and *entisols* are desert soils that occur mostly in hot, arid climates at low latitudes. Aridisols feature an ochreous surface soil and may show one or more subsurface horizons as follows: argillic horizon (a layer with silica and clay minerals dominating), cambic horizon (an altered, light-colored layer, low in organic matter, with carbonates usually present), natric horizon (dominant presence of sodium in exchangeable cation fraction), salic horizon (enriched in water-soluble salts), calcic horizon (secondarily enriched in $CaCO_3$), gypsic horizon (secondarily enriched in $CaSO_4 \cdot 2H_2O$), and duripan horizon (primarily cemented by silica and secondarily by iron oxides and carbonates) (Fuller, 1974; Buol et al., 1980). Entisols are poorly developed immature desert soils without subsurface development. They may arise from recent alluvial deposits or from rock erosion (Fuller, 1974; Buol et al., 1980).

Desert soils are not fertile. It is primarily the lack of sufficient moisture that prevents the development of lush vegetation. However, insufficient nitrogen as major nutrient and zinc, iron, and sometimes copper, molybdenum, or manganese as minor nutrients may also limit plant growth. Desert soils support a specially adapted macroflora and fauna that cope with the stressful conditions in such an environment. They also harbor a characteristic microflora of bacteria, fungi, algae, and lichens. Actinomycetes and lichens may sometimes be dominant. Cyanobacteria seem to be more important in nitrogen fixation in desert soils than other bacteria. Desert soil can sometimes be converted to productive agricultural soils by irrigation. Such watering often results in extensive solubilization of salts from the sub-horizons where they have accumulated during on such counts, the largest number of organisms has been found characteristically in the upper A horizon and the smallest in the B horizon. Aerobic bacteria generally have been found more numerous than anaerobic bacteria, actinomycetes, fungi, or algae. In enumerations like those in Table 4.3, anaerobes were found to decrease with depth to about the same degree as aerobes. This seems contradictory but may reflect the fact that most of the anaerobes that were enumerated by the methodology then in use (1915) were facultative. Special techniques for cultivating anaerobes were not developed until much later (Levett, 1992; Chung and Bryant, 1997).

Determination of microbial distribution in soil, when done by culturing as in the study summarized in Table 4.3, never yields an absolute estimate because no universal culture medium exists on which all living microbes can grow. A somewhat better estimate can be obtained through direct counts using fluorescence microscopy with soil preparations treated with special fluorescent reagents (see Chapter 8), provided viable cells can be distinguished from dead cells.

4.3 ORGANIC SOILS

In some special locations, *organic soils* or *histosols* are found. They form from rapid accumulation and slow decomposition of organic matter, especially plant matter, as a result of displacement of air by water, which prevents rapid and extensive microbial decomposition of the organic matter. These soils are thus, sedimentary in origin and never the result of rock weathering. They consist of 20% or more of organic matter (Lawton, 1955; Buol et al., 1989; Atlas and Bartha, 1997). Their formation is associated with the evolution of swamps, tidal marshes, bogs, and even shallow lakes. An organic soil such as peat may have an ash content of 2%–50% and contain cellulose, hemicellulose, lignin, and derivatives, heterogeneous complexes, fats, waxes, resins, and water-soluble substances such as polysaccharides, sugars, amino acids, and humus (Lawton, 1955). The pH of organic soils may range from 3 to 8.5. Examples of such soils are peat and "mucks." They accumulate to depths ranging from less than a meter to more than 8 m (Lawton, 1955) and are not stratified like mineral soils. They are rare in occurrence. Some are agriculturally very productive.

4.4 SUMMARY

The lithosphere of the Earth consists of rock, which may be igneous, metamorphic, or sedimentary. Rock is composed of intergrown minerals. The rock surface and, in the case of porous rock, the interior of rock may be habitats for microbes. Rock may be broken down by weathering, which may ultimately lead to formation of mineral soil. Some of the rock minerals become chemically altered in the process. Weathering may be biological, especially microbiological, as well as chemical and physical.

Progress of mineral soil development is recognizable in a soil profile. A vertical section through mineral soil may reveal more or less well developed horizons. Typical horizons of spodosols and mollisols include the litter zone (O horizon), a leached layer (A horizon), an enriched layer (B horizon), and the parent material (C horizon). The aspect of the horizons varies with soil type. Climate is one of several important determinants of soil type. The horizons are the result of intense biological activity in the litter zone and A horizon. Much of the organic matter in the litter zone is microbially solubilized and at least partly degraded. Soluble components are washed into the A horizon or transported there by some invertebrates, where they may be further metabolized and where they contribute directly or indirectly to transformation of some of the mineral matter. Soluble products, especially inorganic ones, formed in the A horizon may be washed into the B horizon. The more refractory organic matter in the soil accumulates as humus, which contributes to the soil's texture, water-holding capacity, and general fertility. Mineral soil may be 50% solid matter and 50% pore space. The pore space is occupied by gases such N_2, CO_2, and O_2 and by water. Water also surrounds soil particles to varying degrees. Microbes, including bacteria, fungi, protozoa, and algae may inhabit the soil pores or live on the surface of soil particles. They are most numerous in the upper layer of soil.

Not all soils can be classified as mineral soils. A few are organic and have a different origin. They arise from the slow decomposition of organic matter, mainly plant residues, which accumulates by sedimentation as in swamps, marshes, and shallow lakes. They are not stratified and usually have low mineral content.

Soil is not the only important microbial habitat of the lithosphere. Microbes have been detected in the deep subsurface of the lithosphere, at depths in excess of 3500 m. Although aerobic bacteria, fungi, protozoa, and algae are found at shallower depths, anaerobic bacteria predominate in deeper zones. The organisms can be found in permeable strata formed by sediments, sedimentary rock, and cracked or fissured igneous rock. They inhabit the pore water in these strata and also the exposed mineral surfaces of mineral particles or rock, on which they may form microcolonies or biofilms.

The bacteria in the lithosphere exhibit great diversity morphologically and physiologically. Their average in situ metabolic rates in the deep subsurface appear to be very low owing to limitations in major or essential minor nutrients. Much remains to be discovered about life in the deep subsurface of the lithosphere.

REFERENCES

Alexander M. 1977. Introduction to Soil Microbiology, 2nd edn. New York: Wiley.

Atlas RM, Bartha R. 1997. Microbial Ecology. Fundamentals and Applications, 4th edn. Menlo Park, CA: Addison Wesley Longman.

Barker WW, Banfield JF. 1996. Biological versus inorganically mediated weathering reactions: Relationships between minerals and extracellular microbial polymers in lithobiontic communities. Chem Geol 132:55–69.

Barker WW, Banfield JF. 1998. Zones of chemical and physical interaction at interfaces between microbial communities and minerals: A model. Geomicrobiol J 15: 223–224.

Bennett PC, Hiebert FK, Choi WJ. 1996. Microbial colonization and weathering of silicates in petroleum-contaminated groundwater. Chem Geol 132:45–53.

Berner RA, Holdren GR Jr. 1977. Mechanism of feldspar weathering: Some observational evidence. Geology 5:369–372.

Berthelin J. 1977. Quelques aspects des mécanismes de transformation des minéraux des sols par les micro-organismes hétérotrophes. Sci Bull AFES 1:13–24.

Berthelin J, Boymond D. 1978. Some aspects of the role of heterotrophic microorganisms in the degradation of waterlogged soils. In: Krumbein WE, ed., Environmental Biogeochemistry and Geomicrobiology. Ann Arbor, MI: Ann Arbor Science Publishers, pp. 659–673.

Brady PV, Carroll SA. 1994. Direct effects of CO_2 and temperature on silicate weathering. Possible implications for climate control. Geochim Cosmochim Acta 58:1853–1856.

Brock TD. 1975. Effect of water potential on growth and iron oxidation by Thiobacillus ferrooxidans. Appl Microbiol 29:495–501.

Brock TD, Smith DW, Madigan MT. 1984. Biology of Microorganisms, 4th edn. Englewood Cliffs, NJ: Prentice-Hall.

Brown AD. 1976. Microbial water stress. Bacteriol Rev 40:803–846.

Browne BA, Driscoll CT. 1992. Soluble aluminum silicates: Stoichiometry, stability, and implications for environmental geochemistry. Science 256:1667–1670.

Bunting BT. 1967. The Geography of Soil. London, U.K.: Hutchinson University Library.

Buol SW, Hole FD, McCracken RJ. 1980. Soil Genesis and Classification, 2nd edn. Ames, IA: Iowa State University Press.

Buol SW, Hole FD, McCracken RJ. 1989. Soil Genesis and Classification, 3rd edn. Ames, IA: Iowa State University Press.

Campbell NER, Lees H. 1967. The nitrogen cycle. In: McLaren AD, Petersen GH, eds., Soil Biochemistry, Vol. 1. New York: Marcel Dekker, pp. 194–215.

Casida EL. 1965. Abundant microorganisms in soil. Appl Microbiol 13:327–334.

Chou L., Wollast R. 1984. Study of weathering of albite at room temperature and pressure with a fluidized bed reactor. Geochim Cosmochim Acta 48:2205–2218.

Chung K-T, Bryant MP. 1997. Robert E. Hungate: Pioneer of anaerobic microbial ecology. Anaerobe 3:213–217.

Costerton JW, Lewandowski Z, deBeer D, Caldwell D, Korber D, James G. 1994. Biofilms, the customized microniche. J Bacteriol 176:2137–2142.

Dommergues Y, Mangenot F. 1970. Ecologie Microbienne du Sol. Paris, France: Masson.

Ehrlich HL. 2001. Interactions between microorganisms and minerals under anaerobic conditions. In: Huang PM, Bollag J-M, Senesi N, eds., Interactions Between Soil Particles and Microorganisms. Impact on the Terrestrial Environment. IUPAC Series on Analytical Physical Chemistry of Environmental Systems, Vol. 8. New York: Wiley, Chapter 11, pp. 459–494.

Flemming H-C, Neu TR, Wozniak DJ. 2007. The EPS matrix: The "house of biofilm cells." J Bacteriol 189:7945–7947.

Friedmann EI. 1982. Endolithic microorganisms in the Antarctic cold desert. Science 215:1045–1053.

associated with relatively shallow unconsolidated aquifer sediments to fractures in bedrock formations that are more than a kilometer deep, where extreme lithostatic pressures and temperatures are encountered. While these different environments contain varying physical and chemical conditions, the absence of light is a constant. Despite this, diverse physiologies and metabolisms enable microorganisms to harness energy and carbon for growth in water-filled pore spaces and fractures. Carbon and other element cycles are driven by microbial activity, which has implications for both natural processes and human activities in the subsurface, e.g., bacteria play key roles in both hydrocarbon formation and degradation. Hydrocarbons are a major focus for human utilization of the subsurface, via oil and gas extraction and potential geologic CO_2 sequestration. The subsurface is also utilized or being considered for sequestered storage of high-level radioactive waste from nuclear power generation and residual waste from past production of weapons grade nuclear materials. While our understanding of the subsurface is continually improving, it is clear that only a small fraction of microbial habitats have been sampled and studied. In this chapter, we will discuss these studies in the context of the distribution of microbial life in the subsurface, the stresses that microorganisms must overcome to survive in these environments, and the metabolic strategies that are employed to harness energy in a region

of the planet far removed from sunlight. Finally, we will consider both beneficial and deleterious effects of microbial activity in the subsurface on human activities.

5.2 TERRESTRIAL SUBSURFACE BIOSPHERE ENVIRONMENTS

Previous analyses of the terrestrial subsurface biosphere have divided such environments into a wide range of categories, including sedimentary environments, permafrost, ancient salt deposits, caves, ice sheets, and bedrock environments (Heim, 2011). Indigenous microorganisms have been identified within all these environments: beneath the East and West Antarctic ice sheets in both accreted ice and the water column of relatively oligotrophic subglacial lakes (Karl et al., 1999; Christner et al., 2014), within fluid inclusions in deeply buried halite deposits (Schubert et al., 2009), in highly saline fluids within permafrost (Gilichinsky et al., 2003), and in cave systems (Sarbu et al., 1996). However, for the purposes of this chapter, we will primarily consider both shallow and deep microbial habitats in sedimentary and bedrock formations (Figure 5.1). Across all these different environments, a constant requirement for life is water. Although this resource is present in huge volumes in the terrestrial subsurface—some estimates put the volume at 10^{22} mL (McMahon and Parnell, 2014)—the properties of

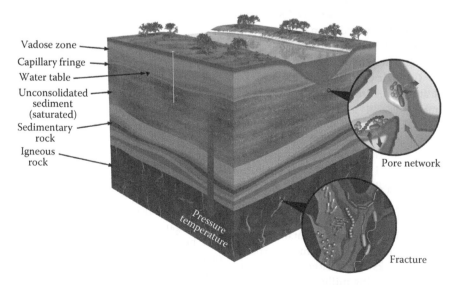

Vadose zone
Capillary fringe
Water table
Unconsolidated sediment (saturated)
Sedimentary rock
Igneous rock

Pressure temperature

Pore network

Fracture

Figure 5.1. Schematic representation of primary terrestrial subsurface biosphere environments. Water is a critical element for microbial life, and microorganisms are present predominantly in water-filled pore and fracture networks.

EHRLICH'S GEOMICROBIOLOGY

water important for microbial life are linked to both porosity and permeability in rock and sediment matrices.

5.2.1 Sedimentary Systems

A range of different environments fall under the category of sedimentary systems. Sedimentary environments are formed by the accumulation of particulate mineral and organic matter, primarily via the action of wind and water. The depositional environment plays a key role in the development of the physical sediment characteristics, with coarse materials (e.g., sands and gravels) deposited by flowing water leading to formation of permeable aquifer deposits, while accumulation of fine silts and clays on lake beds generates relatively impermeable layers. The physical and geochemical characteristics of these shallow environments may therefore, vary significantly with permeability and porosity linked to the relative fractions of coarse- and fine-grained particles (Kamann et al., 2007). In marine environments, deeper unconsolidated sedimentary deposits are generated by the transport and deposition of terrestrial sediment particles into seas and oceans, resulting in extremely thick sediment layers in some nearshore systems where sedimentation rates can be greater than 100 m myr^{-1} (Kallmeyer et al., 2012). Such rates can be contrasted with those in mid-ocean environments, such as the North and South Pacific gyres, where sedimentation occurs at less than 1 m myr^{-1} (Røy et al., 2012).

Generally, groundwater flows through unconsolidated sediment via networks of interconnected pore spaces, with the presence of organic carbon and water offering ideal microbial habitats (Krumholz et al., 1997), even in silt- and clay-dominated materials (Lin et al., 2012). At greater depths in sedimentary systems, diagenetic processes including compaction, mineral dissolution and precipitation, and cementation occur, resulting in a more competent formation with reduced porosity and permeability, relative to unconsolidated materials. Despite these physical changes, microbial habitats are still present in sedimentary rocks, as evidenced by the detection of microorganisms across a range of depths and formation types. Samples recovered from 100-million-year-old Cretaceous shale and sandstone formations in New Mexico revealed limited microbial activity in shales where pore sizes were restrictive for growth but greater activity at shale–sandstone interfaces where organic matter could diffuse from shale into the more porous sandstone (Fredrickson et al., 1997; Krumholz et al., 1997). Although sedimentary rocks undergo diagenetic changes over time, in the absence of significant subduction, no extremes of pressure or temperature are encountered during these processes that might destroy any microorganisms present. It is therefore likely that at least a fraction of the microorganisms present in deep sedimentary environments are direct descendants of microbes that were deposited along with mineral grains and organic matter millions of years earlier (Fredrickson et al., 1997). Extensive evaporite deposits may also be considered sedimentary systems. Such regions are typically hundreds of meters thick and underlay approximately one-quarter of the terrestrial land mass. The evaporation of seawater is primarily responsible for these deposits and is catalyzed by a range of physical and climatic conditions including restricted water flow, low rainfall, high temperatures, low humidity, and high wind speed (McGenity et al., 2000). On average, evaporation of seawater over a depth of approximately 1000 m will generate 14 m of evaporites, the majority of which are halite (Schreiber, 1986). The long-term survival of microorganisms within ancient halite has been a relatively controversial topic. How microorganisms survive within formations that may be more than 250 myr old is a key question, although the potential for environmental and laboratory contamination remains a concern (Hebsgaard et al., 2005). More recent data however have supported observations that halophilic bacteria and archaea are able to survive in fluid inclusions within halite (Schubert et al., 2009).

5.2.2 Igneous and Metamorphic Rocks

Hydrology in sedimentary systems can be contrasted with groundwater flow in igneous and metamorphic rocks that exhibit far lower porosity and permeability. Metamorphic rocks form from the transformation of existing rock types, while igneous rocks are generated from the cooling of either lava or magma. Both of these formation processes require extremes of heat or pressure that can destroy any indigenous microorganisms and transform and degrade organic matter that might allow microbial survival to depths of 10 km. In this environment, microorganisms have been detected in samples from depths between 4 and 5 km beneath the surface (Moser et al., 2005; Lin et al., 2006), suggesting that further drilling and sampling is necessary to probe beyond the lower limit of the deep terrestrial biosphere in this crustal region. Other drilling efforts appear to have reached the temperature limit for life; temperatures of 265°C were measured at 9100 m depth in Cretaceous–Tertiary boundary boreholes drilled by the German Continental Deep Drilling Program (Huber et al., 1994). Tests for microbial activity in samples recovered at this borehole depth were negative.

5.3.2 Energetics

With depth, higher temperatures lead to faster rates of decay for biomolecules. For all life forms, this decay of biomolecules exerts a constant demand for replacement to ensure cell survival. The ratio of biomolecule decay rate to metabolic energy flux must therefore, be at least equal to ensure survival of a cell. Maintaining this supply of biomolecules may be especially problematic for microbial cells in the deep subsurface, where essential organic and inorganic compounds are frequently limiting. The racemization (conversion between L- and D-forms) of amino acids and depurination (the loss of purine bases from the deoxyribose sugar) of nucleic acids likely determines lower limits for rates of molecular repair and resynthesis (Hoehler and Jorgensen, 2013) and may be the principal limit on microbial abundances and activity in hot, deep subsurface environments (Onstott et al., 2013). Where such elevated temperatures are encountered in terrestrial systems, cellular turnover times may be relatively short, on the order of 1–2 years at 54°C at 3 km depth (Onstott et al., 2013), to ensure that rapidly degrading biomolecules are replenished. This of course is reliant on the availability of sufficient carbon and energy resources. Under ambient temperature conditions in other deep biosphere systems (e.g., deep marine sediments or subpermafrost), maintaining a basal power requirement is key to the persistence of microbial life, with cell division only occurring occasionally when environmental conditions are suitable (Nealson et al., 2005). Such a basal power requirement in the deep subsurface is likely much lower than calculated

maintenance energies for cells, given that many nongrowth functions associated with maintenance energy are not requirements for cell survival, e.g., cell motility (Hoehler and Jorgensen, 2013). Estimates of cell doubling times can range from hundreds to even thousands of years in the deep subsurface (Jørgensen and D'Hondt, 2006; Lomstein et al., 2012). At these energetic limits for life, some cells may produce endospores as a survival mechanism, to enhance the chance that a fraction of the microbial community will survive current extreme conditions. Endospores are dormant, nonreproductive structures that may be highly abundant in some deep subsurface environments (Lomstein et al., 2012). In such environments in which conditions are relatively stable, this growth strategy may not necessarily favor spore-forming species. Energy flux in these systems may decrease slowly and steadily over time, and the energetic cost associated with transition from dormancy and active states may be deleterious. Indeed, such an effect has been observed in permafrost where slow metabolic activity and DNA repair appear to be superior survival strategies relative to dormancy (Johnson et al., 2007). Further evidence supports the presence of an active, viable deep biosphere. As reviewed by Hoehler and Jørgensen (2013), DNA and RNA structures in these environments are sufficiently intact to hybridize with primers and probes (Schippers et al., 2005; Biddle et al., 2006), while biomass is able to assimilate substrates during experimentation to estimate uptake rates (Morono et al., 2011). Both ^{13}C- and ^{14}C-labeled compounds have been used to identify active carbon assimilation in deep subsurface granite (Pedersen and Ekendahl, 1992) and sediments (Morono et al., 2011), with carbon assimilation rates between $14–67 \times 10^{-18}$ mol/cell per day observed for a range of substrates applied to 460,000 years old lower Pleistocene sediments (Morono et al., 2011).

5.3.3 Electron Acceptors and Donors

To derive energy in the subsurface, respiratory organisms need a source of electron donor (reductant) and electron acceptor (oxidant). A wide range of electron acceptors are present in both the shallow and deep subsurface, including oxygen, nitrate, oxidized iron and manganese minerals, and sulfate. Thermodynamic calculations of standard Gibbs free energies of reaction for each electron acceptor

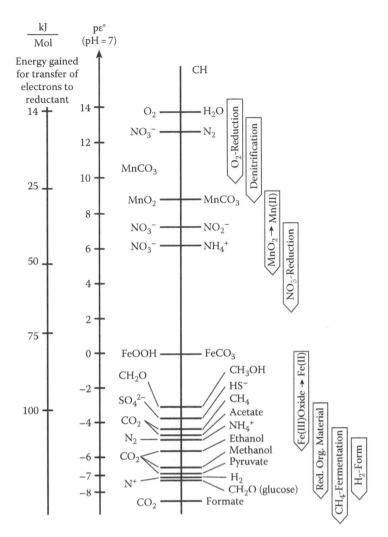

Figure 5.2. In closed aqueous systems such as the subsurface, the oxidation of organic matter is catalyzed predominantly by microorganisms. This oxidation is coupled to the reduction of electron acceptors in the order of decreasing pε. (From Fredrickson, JK and Onstott, TC: Biogeochemical and geological significance of subsurface microbiology, in: *Subsurface Microbiology and Biogeochemistry.* Fredrickson, JK and Fletcher, M, eds. 3–38. 2001. Wiley-LISS, Inc., New York. Copyright Wiley-VCH Verlag GmbH & Co. KGaA. Reproduced with permission; Originally adapted from Stumm, W and Morgan, JJ: *Aquatic Chemistry.* 3rd edn. 1996. John Wiley & Sons, New York. Copyright Wiley-VCH Verlag GmbH & Co. KGaA. Reproduced with permission.)

(Figure 5.2) suggest that redox zonation should occur with increasing depth in the subsurface. However, these calculations were made assuming 1 bar pressure at 25°C and are based on thermodynamic properties under standard state conditions, rather than environmental conditions. These caveats help explain why idealized redox zonation does not frequently occur in subsurface systems (Bethke et al., 2011). Moreover, oxygen and nitrate are generally rapidly consumed by microbial activity and are rarely present in deep systems.

Iron(III) reduction, sulfate reduction, methanogenesis, and homoacetogenesis are all

detected in the shallow and deep subsurface and are sensitive to variations in pH (Figure 5.3). The reduction of ferric iron may be encountered more frequently where fluctuating redox conditions drive the reoxidation of reduced ferrous iron phases and ensure a supply of poorly crystalline ferric oxides that are in general easier for microbes to reduce compared to more crystalline forms. At greater depths under strongly reducing conditions, enzymatic and/or chemical reduction of ferric iron in the absence of reoxidizing processes limits Fe(III) as an important electron acceptor. While more crystalline ferric

Fe(III)-reducing microorganisms. Methanogens play a foundational role in granitic groundwater communities, and can dominate some subsurface microbial populations. At Lidy Hot Springs, ID, United States, researchers estimated that between 95% and 99% of microbial cells in 200 m deep hydrothermal fluids were autotrophic methanogens and therefore, completely independent of photosynthesis (Chapelle et al., 2002). In addition to H_2, lithoautotrophic growth in the subsurface can also be supported by other sources of reducing power, including reduced iron, manganese, and sulfur species.

5.3.5 Physiological Adaptations

In addition to the metabolic strategies described earlier, some subsurface microorganisms, similar to their surface-dwelling relatives, exhibit physiological adaptations to tolerate extreme physical and geochemical conditions in the deep subsurface. Much of the water in the deep subsurface has had extremely long rock contact times, resulting in high concentrations of dissolved ions. As groundwater moves along flow paths from zones of recharge to discharge, its chemistry is altered by a range of geochemical and biogeochemical processes including the weathering and dissolution of rocks and minerals and by the products of microbial oxidation of organic matter and reduction of a range of electron acceptors. In response to increased levels of groundwater salinity, halotolerance has been observed in many bacterial strains isolated from deep terrestrial environments, including oil reservoirs (Tardy-Jacquenod et al., 1998), siltstone formations (Boone et al., 1995), sandstones (Dong et al., 2014), and granitic groundwater (Kotelnikova et al., 1998). Halophilic growth strategies are encountered in regions characterized by higher salinity concentrations, including ancient deep evaporite deposits (Schubert et al., 2009) and cryopegs in permafrost (Gilichinsky et al., 2003). The single-celled alga *Dunaliella* is an important primary producer in hypersaline environments and generates high intracellular concentrations of glycerol for osmoregulation. It has been hypothesized that leakage of this glycerol from active and dead (lysed) cells can support a network of heterotrophic bacteria and archaea in evaporite fluid inclusions (Lowenstein et al., 2011). Increased levels of salinity encountered in evaporites may decrease rates of DNA depurination (Lindahl, 1993), reducing energy requirements for cell maintenance. In addition, cell-rounding and cell-size reduction has also been observed in microbial cells trapped in halite for extended periods of geologic time (>10 ka). Depending on the nature of the host rock, dissolution can also result in high alkalinity to which subsurface organisms such as *Alkaliphilus transvaalensis* are well adapted with a pH range of 8.5–12.5 and a growth optimum of 10.0 (Takai et al., 2001). With the geothermal gradient ranging between 8°C and 30°C/1000 m planet wide, thermophily is a common physiological trait of deep subsurface microbial populations (Slobodkin and Slobodkina, 2014). Where energy fluxes are low, adaptations that limit the loss of hard-gained transmembrane ion gradients are beneficial. These might include membrane structures that are less prone to leakage or the preferential use of sodium ions that have lower transmembrane diffusion rates relative to protons, in the generation of membrane potential (Van De Vossenberg et al., 1995). Archaeal cell membranes are particularly good at limiting ion loss and may help explain the high abundances of these microorganisms across many "extreme" environments where energy fluxes are low (Van De Vossenberg et al., 1995; Hoehler and Jorgensen, 2013). It has also been suggested that high levels of metabolic regulation in the subsurface may be both energetically expensive and unnecessary in environments where geochemical conditions are relatively stable over extended time periods. Under such conditions, relatively simple metabolic strategies that require little regulation may be the most efficient mode of survival. However, under low energy flux conditions in the deep subsurface, the accumulation of beneficial mutations within a population may be hindered by slow rates of biomass turnover and the inability of surviving microorganisms to access favorable growth conditions (Hoehler and Jorgensen, 2013).

5.3.6 Viruses and Eukaryotes

Although generally less abundant than Bacteria and Archaea, microeukaryotes are common inhabitants of some terrestrial subsurface environments. In the deep Atlantic coastal plain sediments of South Carolina, viable protozoa (amoebae and flagellates) were present in core samples harboring the highest populations of viable bacteria, and fungi were cultivated from a large number of samples (Sinclair and Ghiorse, 1989). Protozoan predation of subsurface bacterial blooms as a result stimulation with acetate to promote the in situ immobilization of uranium was identified as a potentially important factor impacting remediation efforts (Holmes et al., 2013). Yeasts related to *Rhodotorula minuta* and *Cryptococcus* sp. were isolated from groundwater associated with the Äspö Hard Rock Laboratory, Sweden, between 201 and 444 m beneath the surface (Ekendahl et al., 2003). These isolates exhibited physiological properties consistent with the ability to grow in this particular deep subsurface environment. Groundwater ecosystems exhibit a remarkable ability to recover from hydrocarbon contamination including the support of a variety of eukaryotes including alveolates (anaerobic and predatory ciliates), stramenopiles, fungi, and even small metazoan flatworms (Yagi et al., 2010). Subsurface nematodes, including the entirely new species *Halicephalobus mephisto*, were detected in 0.9–3.5 km deep fracture water in deep mines of South Africa (Borgonie et al., 2011). These nematodes tolerated high temperatures and preferentially fed upon subsurface bacteria. Collectively, these results indicate that some deep subsurface ecosystems have more complex food webs than previously believed.

Investigations into the abundances and potential roles of viruses in the subsurface have revealed new controls on microbial community structure and function in such environments. Viruses are abundant in shallow unconsolidated sediments with relatively high rates of microbial growth and may play key roles in carbon cycling through cell lysis (Pan et al., 2014) and the efficiency of bioremediation efforts (Holmes et al., 2014). While viruses are highly abundant in some deep marine sediments, with virus-to-cell ratios as high as 225 in the most oligotrophic system sampled (South Pacific Gyre) (Engelhardt et al., 2014), insufficient data currently exist for the terrestrial deep subsurface to establish the relative importance of viruses. However, some evidence indicates the existence of viral activities that affect microbial community dynamics even in terrestrial hard rock systems. The genome of *Candidatus Desulforudis audaxviator*, a member of the *Firmicutes* detected across a range of deep subsurface environments, contains two clustered regularly interspaced short palindromic repeat regions that are used for viral defense, suggesting that interactions with viruses occur in such systems (Chivian et al., 2008). Despite the paucity of suitable habitats for microbial life in deep igneous formations that exhibit large solid surface area-to-water volume ratios, virus-like particles have been detected in granitic groundwater at the Äspö Hard Rock Laboratory in Sweden at abundances between 10^5 and 10^7 mL^{-1} (Kyle et al., 2008). Viruses have been implicated in the infection of metabolically active microorganisms (Kyle et al., 2008), indeed, isolated lytic phages from this environment were shown to infect the indigenous bacterium *Desulfovibrio aespoeensis*. Interestingly, these phage were unable to infect other *Desulfovibrio* species with up to 99.9% 16S rRNA gene sequence identity to *D. aespoeensis*, suggesting very specific host ranges (Eydal et al., 2009). Lysogenic lifestyles are expected to be a common strategy in the deep subsurface, where lytic particles may struggle to find hosts in sparse low biomass environments. In addition, the ability of lysogenic phage to repress host metabolic genes may have important beneficial implications for cell survival under energy-limited conditions. Anderson et al. (2013) suggested that under such conditions, the relationship between virus and host may be more representative of a mutualistic symbiosis than the traditional view of parasitic behavior.

5.4 PROBING THE SUBSURFACE

5.4.1 Drilling and Coring

Recovering representative samples from environments hundreds or thousands of meters beneath the land surface that are typically low in biomass is technically challenging and costly. Limited sample accessibility remains as a primary reason we still know relatively little about the diversity of organisms and habitats in the deep terrestrial subsurface. Microbiological exploration of the subsurface via caves (Sarbu et al., 1996), mines (Onstott et al., 2003), and deep caverns constructed for scientific purposes or as waste repositories (Kotelnikova and Pedersen, 1998) have been extensive, and much has been learned from such investigations.

Drilling and coring are common approaches to access subsurface materials for microbiological characterization, but contamination must be limited and assessed upon sample recovery. For shallow samples, techniques such as rotary or resonant sonic drilling have been effectively used to recover sediment material, without the use of drilling isolated from other microbial community members and photosynthetically derived carbon substrates. Genes were detected for sulfate reduction, carbon fixation, ammonium uptake, and nitrogen fixation, and all amino acid biosynthesis pathways (Chivian et al., 2008) increased analytical sensitivity and may enable biomass limitations to be overcome for future deep biosphere analyses. Single cell genomics offers an alternative method for revealing microbial physiology and metabolism through genomic reconstructions in low biomass systems. This technology relies on cell sorting to isolate individual microbial cells (biomass abundance is therefore, not such a concern) that are then subjected to lysis, multiple displacement DNA amplification, and DNA sequencing. This technique has recently been used to target uncultivated microorganisms and identify ecological roles for such members in the subsurface (Rinke et al., 2013). Finally, high-throughput functional microarrays have also been used to probe changes in functional gene diversity during changes in redox status at the Rifle site (Liang et al., 2012).

5.4.4 Ecological Analyses

High-throughput gene sequencing and microarray-based techniques have enabled understanding and quantification of the ecological processes governing the spatial structure of subsurface communities (Stegen et al., 2012, 2013; Zhou et al., 2014). These approaches are beginning to provide insights into the general rules that govern the relative influences of stochastic and deterministic processes structuring subsurface communities. For example, Stegen et al. (2013) found that drift consistently governed ~25% of the spatial turnover in community composition associated with an unconfined aquifer in SE Washington State. In deeper, fine-grained sediments selection was relatively strong and accounted for approximately 60% of the species turnover whereas in shallower, coarser-textured sediments selection was considerably weaker, accounting for only 30% of the turnover. Similar tools were applied to sediments that had been perturbed via the addition of carbon sources to stimulate microbial activity. Handley et al. (2014) demonstrated that neither neutral processes nor founding community structures influenced microbial community succession once excess carbon was added to the system.

5.5 SUBSURFACE–HUMAN INTERACTIONS

Humans interact with the terrestrial subsurface via a range of processes linked to energy recovery, groundwater extraction, waste disposal, and inadvertent contamination. Oil and gas extracted from the subsurface environment comprises a significant fraction of energy consumed in the United States. Meanwhile, CO_2 generated from the combustion of fossil fuels at power plants may be injected into subsurface reservoirs in attempts to reduce anthropogenic greenhouse gas (GHG) emissions to the atmosphere. Microbial activity during these processes can have undesirable outcomes, including oil field souring (the production of H_2S by sulfate-reducing bacteria), the corrosion of wells and pipelines, and pore clogging due to biomass accumulation and biogenic mineral precipitation around wells (Morozova et al., 2010). The economic costs of these deleterious processes runs into billions of dollars.

5.5.1 Gas and Hydrocarbon Production

Shale has become increasingly exploited as a source of natural gas in the United States (and Canada) to the point that in 2013, there were over 40,000 shale gas production wells across 20 states. Horizontal or directional drilling has allowed the tracking of relatively thin but productive shale layers that in combination with hydraulic fracturing ("hydrofracking" or simply "fracking") has made shale gas an energy "game changer" (Rivard et al., 2014). The number of unconventional wells in the Marcellus Formation in Pennsylvania alone increased from 8 in 2005 to >7200 in 2014 (Brantley et al., 2014). Risks associated with the development of unconventional gas and oil fields include contamination of overlying shallow aquifers with gas and other hydrocarbons, brines containing high concentrations of Na, Ca, Cl, Br, Sr, Ba, etc., and hydraulic fracturing fluids (Vengosh et al., 2013; Brantley et al., 2014). While the influence of such contaminants on the microbial populations in shallow aquifers is unknown, there exists potential for significant shifts in community structure and function given that methane and other hydrocarbons are excellent carbon and energy sources for microorganisms, especially when introduced into water where electron acceptors are available. This can be advantageous in terms of subsurface microorganisms serving as agents of natural attenuation of contaminants (see in the following text). Also unknown is the effect of the process of hydraulic fracturing on microbial populations and processes native to shale formations. In one such investigation, microbial community dynamics were tracked over a 328-day period in water samples from three hydraulically fractured wells in the Marcellus formation (Cluff et al., 2014). The authors reported an overall reduction in microbial richness and diversity after fracturing, with the majority of the postfracking community related to halotolerant taxa associated with fermentation, hydrocarbon oxidation, sulfur cycling, and methanogenesis. The hydrofracturing itself increases permeability by inducing artificial fractures via the high-pressure injection of large volumes of fracturing fluids, 10%–70% of which resurface along with formation brines as flowback fluids. Organic components of shale have previously been shown to stimulate the activity and growth of anaerobic organisms including sulfate-reducing bacteria and acetogens (Krumholz et al., 2002). The opening of fractures and increasing microbial access to previously inaccessible regions of shale has potential to simulate microbial processes although the long-term effect on the populations and associated processes in the deep subsurface from such extensive perturbations remain unknown.

Studies of deep oil reservoirs have revealed active microbial populations that can survive and grow in these environments, utilizing metabolic strategies including sulfate reduction, methanogenesis, and fermentation (Rueter et al., 1994; Jones et al., 2008). Microorganisms living in these locations are generally adapted to elevated temperatures and salinity and utilize a range of electron donors including H_2, volatile fatty acids, petroleum hydrocarbons, and inorganic compounds. Electron acceptors are most commonly sulfate and carbonate minerals in such systems, although some microorganisms have been shown to reduce Fe(III) (Lovley et al., 1989). There is a common sequence of removal different classes of compounds during crude oil degradation, starting with easily degradable straight-chain n-alkanes, followed by more recalcitrant branched acyclic and monocyclic hydrocarbons, and finishing with polycyclic steroidal and triterpenoid hydrocarbons and some aromatic hydrocarbons (Wenger et al., 2002). It has been estimated that degradation of light compounds within a 100 m oil column will occur over 1–2 myr, with n-alkane removal occurring over 5–15 myr (Head et al., 2003). The residue from these degradation processes is known as "heavy oil." Methanogenic oil degradation proceeds at a low rate (Jones et al., 2008) but is most likely responsible for a large fraction of the world's deposits of heavy oil (Head et al., 2003). Recent studies have demonstrated that the initial composition of crude oil plays a key role in the extent of degradation, with volatile hydrocarbons (e.g., benzene, toluene) playing an inhibitory role in oil biodegradation (Sherry et al., 2014).

Microbial metabolism can cause deleterious effects during oil extraction efforts. The flooding of reservoirs with brine or seawater to stimulate oil production introduces sulfate, nitrogen, and phosphorous sources, lowers the in situ reservoir temperature, and enhances conditions suitable for growth of SRB that subsequently generate sulfide. The microbial production of sulfide can increase rates of corrosion in pipelines and other equipment, plug pore spaces via mineral precipitation, and cause health effects associated with exposure of workers to toxic H_2S. Attempts to inhibit oilfield souring have focused on the injection of biocides (e.g., bronopol, formaldehyde), the removal of sulfate from injection waters, and addition of oxidized nitrogen species that can be used by heterotrophic nitrate-reducing bacteria and nitrate-dependent sulfide oxidizers. Nitrate-reducing bacteria are able to effectively outcompete SRB for electron donors owing to greater energy yields associated with nitrate reduction relative to sulfate reduction. A number of trials have demonstrated effective inhibition of SRB activity via this approach (Hubert and Voordouw, 2007; Hubert, 2010). Perchlorate is also being investigated as an alternative to nitrate/nitrite amendment, with specific toxicity and biocompetitive exclusion of SRB identified as possible inhibitory mechanisms (Engelbrektson et al., 2014).

Microbial metabolism involving the production of acids, solvents, gases, biosurfactants, biopolymers, and emulsifiers may also be beneficial to oil recovery in marginal wells from which recovery is less than 1.6 m^3 of oil per day. The removal of unwanted compounds and deposits from the wellbore and production equipment can be performed via the introduction of isolated microorganisms with characterized hydrocarbon-degrading

Nearby planets such as Mars share some of Earth's characteristics that potentially enabled the development of life in the subsurface. Although the contemporary surface environment of Mars is considered likely inhospitable due to the absence of liquid water, low temperatures, high UV radiation, and extremely low organic carbon concentrations, the physical and chemical characteristics of the subsurface are thought to be suitable for refuge from extreme surface and near-surface conditions associated with comet and asteroid bombardment (Sleep and Zahnle, 1998). Some of the same strategies that support microbial life in the deep biosphere on Earth may be relevant to subsurface niches on Mars, where CO_2 could potentially support autotrophic growth in hydrothermally warmed environments (Boston et al., 1992). A range of electron donors in such systems could be utilized for energy generation, including volcanically derived CH_4, H_2S, and H_2 from rock–water interactions (e.g., radiolysis, serpentinization) described earlier in this chapter (Schulte et al., 2006; Sherwood Lollar et al., 2007). Indeed, in analogous deep subsurface environments on Earth (e.g., Precambrian shields of Canada and Finland), some of the highest levels of dissolved H_2 ever reported have been identified, emphasizing the role that such reactions play in sustaining energy-rich environmental conditions that could support sulfate-reducing bacteria and methanogens (Sherwood Lollar et al., 2007). Silicates detected in the Martian crust (e.g., olivine) would support H_2 generation via serpentinization reactions. Furthermore, many of the trace elements and other compounds required for cell maintenance and growth are also detected in Mars basaltic minerals (Fisk and Giovannoni, 1999). Subpermafrost environments, which potentially contain water beneath near-surface permafrost layers, have also been suggested as suitable terrestrial analogues to Martian habitats. Microbial biogeochemistry studies beneath a 540 m thick permafrost layer in Canada identified 10^3 cells mL^{-1} planktonic biomass concentrations and evidence for extensive sulfur cycling (Onstott et al., 2009). Moving further into the solar system, Jupiter's moon Europa has been suggested as another location for possible unicellular life, because of an inferred subsurface "ocean" of liquid water beneath a frozen shell ~10 km thick. Methanogenesis at subsurface hydrothermal vents and oxidation of organic compounds (e.g., formaldehyde) generated from ion bombardment of water and CO_2 have been suggested as possible mechanisms for life to survive on Europa (Chyba and Phillips, 2001).

5.7 SUMMARY

The terrestrial subsurface is an aphotic environment consisting of a range of habitats from the vadose zone to near-surface saturated sediments to deep fractured bedrock that can extend multiple kilometers beneath the Earth's surface. These habitats cover a wide range of physical (pressure, temperature) and chemical (pH, salinity, nutrient availability) conditions to which the extant microbiota are well adapted. Collectively, the terrestrial subsurface has been estimated to harbor up to 19% of the total of Earth's biomass. It is dominated by microbial life, and the diversity of this life is expansive, including Bacteria, Archaea, viruses, and eukaryotes. A common characteristic of many subsurface habitats is the low availability of energy and nutrients that results in extremely slow rates of metabolism and cell growth. In some hot regions of the subsurface, the rate of biomolecule degradation may necessitate relatively rapid carbon turnover. Many subsurface microorganisms exhibit the remarkable ability to survive under a state of constant starvation or as dormant cells but with the ability to take advantage of a wide range of energy resources from ancient detrital organic carbon to gases and hydrocarbons generated from abiotic processes. The ability to survive under such conditions in an environment that is buffered from changes that can impact life at the Earth's surface have led scientists to consider the subsurface of planets and moons in our own solar system as possible refuges for microbial life. In spite of low rates of metabolic activity, subsurface microbiota have a major impact upon the geosphere contributing to the geochemical evolution of groundwater along hydrologic flow paths, the turnover of organic carbon, and the weathering of rocks and minerals.

A major challenge to exploration of the terrestrial subsurface is access. Other than caves and mines, accessibility of subsurface geological materials is limited largely to drilling and coring that can subject samples to contamination if proper precautions are not taken and extensive attention given during retrieval and processing of samples to ensure they are representative. Samples of subsurface waters from wells and fractures in mines and caves have provided critical knowledge to the understanding of microbial life in the subsurface. Wells have the advantage in that they can be sampled repeatedly. The geomicrobiological exploration of the subsurface has benefited greatly from the application of cultivation-independent molecular approaches that have allowed inference of diversity and metabolic potential in a wide range of habitats.

Interactions with the terrestrial subsurface have been increasing as humans have become more dependent upon its resources. These resources include fresh water for domestic purposes, agriculture, and industry and natural gas and hydrocarbons for energy and rocks and minerals as a source of precious metals. The deep subsurface is also being used, or under consideration for use, for the storage of various types of radioactive wastes and for the sequestration of CO_2 from the combustion of fossil fuels. As a result of industrial processes, processing of nuclear materials for weapons production, storage of hydrocarbons in tanks that have leaked, and through fossil fuel exploration and development, many aquifers have been contaminated worldwide. Subsurface microorganisms, either through natural processes or by stimulation of their activities via nutrient addition, are able to degrade many contaminant to harmless or less toxic products or greatly reduce their solubility and hence mobility. Future development of new analytical technologies coupled to continue sampling of subsurface environments will enhance our understanding of this key ecosystem.

REFERENCES

Adams J, Riediger C, Fowler M, Larter S. 2006. Thermal controls on biodegradation around the Peace River tar sands: Paleo-pasteurization to the west. J Geochem Explor 89:1–4.

Anderson C, Johnsson A, Moll H, Pedersen K. 2011. Radionuclide geomicrobiology of the deep biosphere. Geomicrobiol J 28:540–561.

Anderson RE, Brazelton WJ, and Baross JA. 2013. The Deep Viriosphere: Assessing the viral impact on microbial community dynamics in the deep subsurface. Rev Mineral Geochem 75:649–675.

Anderson RT, Chapelle FH, and Lovley DR. 1998a. Evidence against hydrogen-based microbial ecosystems in basalt aquifers. Science 281:976–977.

Anderson RT, Rooney-Varga J, Gaw CV, Lovley DR. 1998b. Anaerobic benzene oxidation in the Fe(III)-reduction zone of petroleum-contaminated aquifers. Environ Sci Technol 32:1222–1229.

Anderson RT, Vrionis HA, Ortiz-Bernad I, Resch CT, Long PE, Dayvault R, Karp K et al. 2003. Stimulating the in situ activity of Geobacter species to remove uranium from the groundwater of a uranium-contaminated aquifer. Appl Environ Microbiol 69:5884–5891.

Baldwin BR, Peacock AD, Park M, Ogles DM, Istok JD, McKinley JP, Resch CT, White DC. 2008. Multilevel samplers as microcosms to assess microbial response to biostimulation. Ground Water 46:295–304.

Balkwill DL, Murphy EM, Fair DM, Ringelberg DB, White DC. 1998. Microbial communities in high and low recharge environments: Implications for microbial transport in the vadose zone. Microb Ecol 35:156–171.

Barton HA, Taylor MR, Pace NR. 2004. Molecular phylogenetic analysis of a bacterial community in an oligotrophic cave environment. Geomicrobiol J 21:11–20.

Beeman RE, Suflita JM. 1989. Evaluation of deep subsurface sampling procedures using serendipitous microbial contaminants as tracer organisms. Geomicrobiol J 7:223–233.

Bekele E, Toze S, Patterson B, Higginson S. 2011. Managed aquifer recharge of treated wastewater: Water quality changes resulting from infiltration through the vadose zone. Water Res 45:5764–5772.

Belyaev SS, Borzenkov IA, Nazina TN, Rozanova EP, Glumov IF, Ibatullin RR, Ivanov MV. 2004. Use of microorganisms in the biotechnology for the enhancement of oil recovery. Microbiology 73:590–598.

Bethke CM, Sanford RA, Kirk MF, Jin Q, Flynn TM. 2011. The thermodynamic ladder in geomicrobiology. Am J Sci 311:183–210.

Biddle JF, Lipp JS, Lever MA, Lloyd KG, Sorensen KB, Anderson R, Fredricks HF et al. 2006. Heterotrophic Archaea dominate sedimentary subsurface ecosystems off Peru. Proc Natl Acad Sci USA 103:3846–3851.

Boone DR, Liu YY, Zhao Z-J, Balkwill DL, Drake GR, Stevens TO, Aldrich HC. 1995. Bacillus infernus sp. nov., an Fe(III)- and Mn(IV)-reducing anaerobe from the deep terrestrial subsurface. Int J Syst Bacteriol 45:441–448.

Borgonie G, Garcia-Moyano A, Litthauer D, Bert W, Bester A, Van Heerden E, Moeller C, Erasmus M, Onstott TC. 2011. Nematoda from the terrestrial deep subsurface of South Africa. Nature 474:79–82.

Horn JM, Masterson BA, Rivera A, Miranda A, Davis MA, Martin S. 2004. Bacterial growth dynamics, limiting factors, and community diversity in a proposed geological nuclear waste repository environment. *Geomicrobiol J* 21:273–286.

Huber H, Huber R, Ludemann H-D, Stetter KO. 1994. Search for hyperthermophilic microorganisms in fluids obtained from the KTB pump test. *Sci Drill* 4:127–129.

Hubert C. 2010. Microbial ecology of oil reservoir souring and its control by nitrate injection, in *Handbook of Hydrocarbon and Lipid Microbiology*, ed. K. Timmis. Berlin, Germany: Springer, pp. 2753–2766.

Hubert C, Voordouw G. 2007. Oil field souring control by nitrate-reducing *Sulfurospirillum* spp. that outcompete sulfate-reducing bacteria for organic electron donors. *Appl Environ Microbiol* 73:2644–2652.

Islam FS, Gault AG, Boothman C, Polya DA, Charnock JM, Chatterjee D, Lloyd JR. 2004. Role of metal-reducing bacteria in arsenic release from Bengal delta sediments. *Nature* 430:68–71.

Itävaara M, Nyyssönen M, Kapanen A, Nousiainen A, Ahonen L, Kukkonen I. 2011. Characterization of bacterial diversity to a depth of 1500 m in the Outokumpu deep borehole, Fennoscandian Shield. *FEMS Microbiol Ecol* 77:295–309.

Johnson SS, Hebsgaard MB, Christensen TR, Mastepanov M, Nielsen R, Munch K, Brand T et al. 2007. Ancient bacteria show evidence of DNA repair. *Proc Natl Acad Sci USA* 104:14401–14405.

Jones DM, Head IM, Gray ND, Adams JJ, Rowan AK, Aitken CM, Bennett B et al. 2008. Crude-oil biodegradation via methanogenesis in subsurface petroleum reservoirs. *Nature* 451:176–180.

Jørgensen BB, D'Hondt S. 2006. A starving majority deep beneath the seafloor. *Science* 314:932–934.

Kallmeyer J, Pockalny R, Adhikari RR, Smith DC, D'hondt S. 2012. Global distribution of microbial abundance and biomass in subseafloor sediment. *Proc Natl Acad Sci USA* 109:16213–16216.

Kamann PJ, Ritzi RW, Dominic DF, Conrad CM. 2007. Porosity and permeability in sediment mixtures. *Ground Water* 45:429–438.

Karl DM, Bird DF, Björkman K, Houlihan T, Shackelford R, Tupas L. 1999. Microorganisms in the Accreted Ice of Lake Vostok, Antarctica. *Science* 286:2144–2147.

Kashefi K, Lovley DR. 2003. Extending the upper temperature limit for life. *Science* 301:934.

Kato C, Nogi Y, Arakawa S. 2008. Isolation, cultivation, and diversity of deep-sea piezophiles, in *High-Pressure Microbiology*, eds. C. Michiels, D.H. Bartlett, and A. Aertsen. Washington, DC: ASM Press, pp. 203–217.

Kieft T, Fredrickson J, Mckinley J, Bjornstad B, Rawson S, Phelps T, Brockman F, Pfiffner S. 1995. Microbiological comparisons within and across contiguous lacustrine, paleosol, and fluvial subsurface sediments. *Appl Environ Microbiol* 61:749–757.

Kieft TL. 2010. Sampling the deep sub-surface using drilling and coring techniques, in *Handbook of Hydrocarbon and Lipid Microbiology*, ed. K. Timmis. Berlin, Germany: Springer, pp. 3427–3441.

Kieft TL, Brockman FJ. 2001. Vadose zone microbiology, in *Subsurface Microbiology and Biogeochemistry*, eds. J.K. Fredrickson and M. Fletcher. New York: Wiley-LISS, Inc.

Kieft TL, Kovacik WP, Ringelberg DB, White DC, Haldeman DL, Amy PS, Hersman LE. 1997. Factors limiting microbial growth and activity at a proposed high-level nuclear repository, Yucca Mountain, Nevada. *Appl Environ Microbiol* 63:3128–3133.

Kieft TL, Phelps TJ. 1997. Life in the slow lane: Activities of microorganisms in the subsurface, in *The Microbiology of the Terrestrial Deep Subsurface*, eds. P.S. Amy and D.L. Haldeman. Boca Raton, FL: CRC Press, pp. 137–164.

Kieft TL, Phelps TJ, Fredrickson JK. 2007. Drilling, coring, and sampling subsurface environments, in *Manual of Environmental Microbiology*, 3rd edn, ed. C. J. Hurst. ASM Press: Washington, D.C., pp. 799–817.

Kotelnikova S. 2002. Microbial production and oxidation of methane in deep subsurface. *Earth-Sci Rev* 58:367–395.

Kotelnikova S, Macario AJL, Pedersen K. 1998. Methanobacterium subterraneum sp. nov., a new alkaliphilic, eurythermic and halotolerant methanogen isolated from deep granitic groundwater. *Int J Syst Bacteriol* 48:357–367.

Kotelnikova S, Pedersen K. 1998. Distribution and activity of methanogens and homoacetogens in deep granitic aquifers at Äspö Hard Rock Laboratory, Sweden. *FEMS Microbiol Ecol* 26:121–134.

Krumholz LR, Harris SH, Suflita JM. 2002. Anaerobic microbial growth from components of cretaceous shales. *Geomicrobiol J* 19:593–602.

Krumholz LR, Harris SH, Tay ST, Suflita JM. 1999. Characterization of two subsurface H_2-utilizing bacteria, *Desulfomicrobium hypogeium* sp. nov. and *Acetobacterium psammolithicum* sp. nov., and their ecological roles. *Appl Environ Microbiol* 65:2300–2306.

Krumholz LR, McKinley JP, Ulrich GA, Suflita JM. 1997. Confined subsurface microbial communities in Cretaceous rock. *Nature* 386:64–66.

Kyle JE, Eydal HSC, Ferris FG, Pedersen K. 2008. Viruses in granitic groundwater from 69 to 450 m depth of the Aspo hard rock laboratory, Sweden. ISME J 2:571–574.

Lehman RM, Colwell FS, Ringelberg D, White DC. 1995. Combined microbial community-level analyses for quality assurance of terrestrial subsurface cores. J Microbiol Methods 22:263–281.

Liang YT, Van Nostrand JD, N'Guessan LA, Peacock AD, Deng Y, Long PE, Resch CT et al. 2012. Microbial functional gene diversity with a shift of subsurface redox conditions during in situ uranium reduction. Appl Environ Microbiol 78:2966–2972.

Lin L-H, Hall J, Lippmann-Pipke J, Ward JA, Sherwood Lollar B, Deflaun M, Rothmel R et al. 2005. Radiolytic H_2 in continental crust: Nuclear power for deep subsurface microbial communities. Geochem Geophy Geosyst 6:Q07003.

Lin LH, Wang PL, Rumble D, Lippmann-Pipke J, Boice E, Pratt LM, Lollar BS et al. 2006. Long-term sustainability of a high-energy, low-diversity crustal biome. Science 314:479–482.

Lin X, Kennedy D, Peacock A, McKinley J, Resch CT, Fredrickson J, Konopka A. 2012. Distribution of microbial biomass and potential for anaerobic respiration in Hanford Site 300 Area subsurface sediment. Appl Environ Microbiol 78:759–767.

Lindahl T. 1993. Instability and decay of the primary structure of DNA. Nature 362:709–715.

Lloyd JR, Sole VA, Van Praagh CV, Lovley DR. 2000. Direct and Fe(II)-mediated reduction of technetium by Fe(III)-reducing bacteria. Appl Environ Microbiol 66:3743–3749.

Lomstein BA, Langerhuus AT, D'Hondt S, Jørgensen BB, Spivack AJ. 2012. Endospore abundance, microbial growth and necromass turnover in deep subseafloor sediment. Nature 484:101–104.

Lovley DR, Baedecker MJ, Lonergan DJ, Cozzarelli IM, Phillips EJP, Siegel DI. 1989. Oxidation of aromatic contaminants coupled to microbial iron reduction. Nature 339:297–299.

Lovley DR, Phillips EJP, Gorby YA, Landa E. 1991. Microbial reduction of uranium. Nature 350:413–416.

Lowenstein TK, Schubert BA, Timofeeff MN. 2011. Microbial communities in fluid inclusions and long-term survival in halite. GSA Today 21:4–9.

McGenity TJ, Gemmell RT, Grant WD, Stan-Lotter H. 2000. Origins of halophilic microorganisms in ancient salt deposits. Environ Microbiol 2:243–250.

McMahon S, Parnell J. 2014. Weighing the deep continental biosphere. FEMS Microbiol Ecol 87:113–120.

Mitchell AC, Dideriksen K, Spangler LH, Cunningham AB, Gerlach R. 2010. Microbially enhanced carbon capture and storage by mineral-trapping and solubility-trapping. Environ Sci Technol 44:5270–5276.

Mitchell AC, Phillips AJ, Hamilton MA, Gerlach R, Hollis WK, Kaszuba JP, Cunningham AB. 2008. Resilience of planktonic and biofilm cultures to supercritical CO_2. J Supercrit Fluid 47:318–325.

Mitchell AC, Phillips AJ, Hiebert R, Gerlach R, Spangler LH, Cunningham AB. 2009. Biofilm enhanced geologic sequestration of supercritical CO_2. Int J Greenh Gas Contr 3:90–99.

Morono Y, Terada T, Nishizawa M, Ito M, Hillion F, Takahata N, Sano Y, Inagaki F. 2011. Carbon and nitrogen assimilation in deep subseafloor microbial cells. Proc Natl Acad Sci USA 108:18295–18300.

Morozova D, Wandrey M, Alawi M, Zimmer M, Vieth A, Zettlitzer M, Würdemann H. 2010. Monitoring of the microbial community composition in saline aquifers during CO_2 storage by fluorescence in situ hybridisation. Int J Greenh Gas Contr 4:981–989.

Moser DP, Gihring TM, Brockman FJ, Fredrickson JK, Balkwill DL, Dollhopf ME, Lollar BS et al. 2005. Desulfotomaculum and Methanobacterium spp. dominate a 4-to 5-kilometer-deep fault. Appl Environ Microbiol 71:8773–8783.

Moser DP, Onstott TC, Fredrickson JK, Brockman FJ, Balkwill DL, Drake GR, Pfiffner SM, White DC. 2003. Temporal shifts in the geochemistry and microbial community structure of an ultradeep mine borehole following isolation. Geomicrobiol J 20:517–548.

Mu N, Boreham C, Leong HX, Haese RR, Moreau JW. 2014. Changes in the deep subsurface microbial biosphere resulting from a field-scale CO_2 geosequestration experiment. Front Microbiol Terrestrial Microbiol 5:209.

Nealson KH, Inagaki F, Takai K. 2005. Hydrogen-driven subsurface lithoautotrophic microbial ecosystems (SLiMEs): Do they exist and why should we care? Trends Microbiol 13:405–410.

Onstott TC, Magnabosco C, Aubrey AD, Burton AS, Dworkin JP, Elsila JE, Grunsfeld S et al. 2013. Does aspartic acid racemization constrain the depth limit of the subsurface biosphere? Geobiology 12:1–19.

Onstott TC, McGown D, Bakermans C, Ruskeeniemi T, Ahonen L, Telling J, Soffientino B et al. 2009. Microbial communities in subpermafrost saline fracture water at the Lupin Au Mine, Nunavut, Canada. Microb Ecol 58:786–807.

Urrutia MM, Roden EE, Zachara JM. 1999. Influence of aqueous and solid-phase Fe(II) complexants on microbial reduction of crystalline iron(III) oxides. *Environ Sci Technol* 33:4022–4028.

Van De Vossenberg JLCM, Ubbink-Kok T, Elferink MGL, Driessen AJM, Konings WN. 1995. Ion permeability of the cytoplasmic membrane limits the maximum growth temperature of bacteria and archaea. *Mol Microbiol* 18:925–932.

Vargas M, Kashefi K, Blunt-Harris EL, Lovley DR. 1998. Microbiological evidence for Fe(III) reduction on early Earth. *Nature* 395:65–67.

Vengosh A, Warner N, Jackson R, Darrah T. 2013. The effects of shale gas exploration and hydraulic fracturing on the quality of water resources in the United States. *Proc Earth Planet Sci* 7:863–866.

Vreeland RH, Piselli AF, McDonnough S, Meyers SS. 1998. Distribution and diversity of halophilic bacteria in a subsurface salt formation. *Extremophiles* 2:321–331.

Wenger LM, Davis CL, Isaksen GH. 2002. Multiple controls on petroleum biodegradation and impact on oil quality. *SPE Reserv Eval Eng* 5:375–383.

Whicker FW, Hinton TG, Macdonell MM, Pinder JE, Habegger LJ. 2004. Avoiding destructive remediation at DOE sites. *Science* 303:1615–1616.

Whitman WB, Coleman DC, Wiebe WJ. 1998. Prokaryotes: The unseen majority. *Proc Natl Acad Sci USA* 95:6578–6583.

Wilkins MJ, Daly R, Mouser PJ, Trexler R, Sharma S, Cole DR, Wrighton KC et al. 2014a. Trends and future challenges in sampling the deep terrestrial biosphere. *Front Microbiol Front Extreme Microbiol* 5:481.

Wilkins MJ, Hoyt DW, Marshall MJ, Alderson PA, Plymale AE, Markilli LM, Tucker AE et al. 2014b. CO_2 exposure at pressure impacts metabolism and stress responses in the model sulfate-reducing bacterium *Desulfovibrio vulgaris* strain Hildenborough. *Front Microbiol* 5:507.

Wilkins MJ, Livens FR, Vaughan DJ, Beadle I, Lloyd JR. 2007. The potential of microbial redox cycling to control the mobility of key radionuclides in the subsurface at a low-level radioactive waste storage site. *Geobiology* 5:293–301.

Wilkins MJ, VerBerkmoes NC, Williams KH, Callister SJ, Mouser PJ, Elifantz H, N'Guessan AL et al. 2009. Proteogenomic monitoring of *Geobacter* physiology during stimulated uranium bioremediation. *Appl Environ Microbiol* 75:6591–6599.

Wilkins MJ, Wrighton KC, Nicora CD, Williams KH, McCue LA, Handley KM, Miller CS et al. 2013. Fluctuations in species-level protein expression occur during element and nutrient cycling in the subsurface. *PLoS ONE* 8:e57819.

Williams KH, Long PE, Davis JA, Wilkins MJ, N'Guessan AL, Steefel CI, Yang L et al. 2011. Acetate availability and its influence on sustainable bioremediation of uranium-contaminated groundwater. *Geomicrobiol J* 28:519–539.

Williams KH, Wilkins MJ, N'Guessan AL, Arey BW, Dodova E, Dohnalkova A, Holmes DE, Lovley DR, Long PE. 2013. Field evidence of selenium bioreduction in a uranium-contaminated aquifer. *Environ Microbiol Rep* 5:444–452.

Wood WW, Ehrlich GG. 1979. Use of bakers yeast to trace microbial movement in ground water. *Ground Water* 16:398–402.

Wrighton KC, Castelle CJ, Wilkins MJ, Hug LA, Sharon I, Thomas BC, Handley KM et al. 2013. Metabolic interdependencies between phylogenetically novel fermenters and respiratory organisms in an unconfined aquifer. *ISME J* 8:1452–1463.

Wrighton KC, Thomas BC, Sharon I, Miller CS, Castelle CJ, VerBerkmoes NC, Wilkins MJ et al. 2012. Fermentation, hydrogen, and sulfur metabolism in multiple uncultivated bacterial phyla. *Science* 337:1661–1665.

Wu B, Shao H, Wang Z, Hu Y, Tang YJ, Jun Y-S. 2010. Viability and metal reduction of *Shewanella oneidensis* MR-1 under CO_2 stress: Implications for ecological effects of CO_2 leakage from geologic CO_2 sequestration. *Environ Sci Technol* 44:9213–9218.

Wu W-M, Carley J, Fienen M, Mehlhorn T, Lowe K, Nyman J, Luo J et al. 2006b. Pilot-scale in situ bioremediation of uranium in a highly contaminated aquifer. 1. Conditioning of a treatment zone. *Environ Sci Technol* 40:3978–3985.

Wu W-M, Carley J, Gentry T, Ginder-Vogel MA, Fienen M, Mehlhorn T, Yan H et al. 2006a. Pilot-scale in situ bioremediation of uranium in a highly contaminated aquifer. 2. Reduction of U(VI) and geochemical control of U(VI) bioavailability. *Environ Sci Technol* 40:3986–3995.

Yagi JM, Neuhauser EF, Ripp JA, Mauro DM, Madsen EL. 2010. Subsurface ecosystem resilience: Long-term attenuation of subsurface contaminants supports a dynamic microbial community. *ISME J* 4:131–143.

Yelton AP, Williams KH, Fournelle J, Wrighton KC, Handley KM, Banfield JF. 2013. Vanadate and acetate biostimulation of contaminated sediments decreases diversity, selects for specific taxa, and decreases aqueous V^{5+} concentration. *Environ Sci Technol* 47:6500–6509.

Yoshida H, Metcalfe R, Yamamoto K, Murakami Y, Hoshii D, Kanekiyo A, Naganuma T, Hayashi T. 2008. Redox front formation in an uplifting sedimentary rock sequence: An analogue for redox-controlling processes in the geosphere around deep geological repositories for radioactive waste. *Appl Geochem* 23:2364–2381.

Youssef N, Elshahed MS, McInerney MJ. 2009. Microbial processes in oil fields: Culprits, problems, and opportunities, in *Advances in Applied Microbiology*, eds. A.I. Laskin, S. Sariaslani, and G.M. Gadd. Burlington, VT: Academic Press, pp. 141–251.

Zeng X, Birrien J-L, Fouquet Y, Cherkashov G, Jebbar M, Querellou J, Oger P, Cambon-Bonavita M-A, Xiao X, Prieur D. 2009. *Pyrococcus* CH1, an obligate piezophilic hyperthermophile: Extending the upper pressure-temperature limits for life. *ISME J* 3:873–876.

Zhou J, Deng Y, Zhang P, Xue K, Liang Y, Van Nostrand JD, Yang Y et al. 2014. Stochasticity, succession, and environmental perturbations in a fluidic ecosystem. *Proc Natl Acad Sci USA* 111:E836–E845.

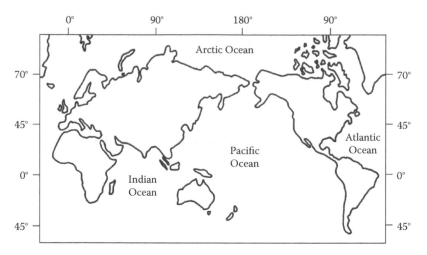

Figure 6.1. Oceans of the world.

Arctic Ocean, 2.8% of the Earth's surface. The average depth of all oceans is 3795 m. The average depth of the Atlantic Ocean is 3296 m, that of the Pacific Ocean is 4282 m, that of the Indian Ocean is 3693 m, and that of the Arctic Ocean is 1205 m. The greatest depths in the oceans occur in the *ocean trenches*. For instance, in the Pacific Ocean, the water depth of the Marianas Trench is 10,500 m; in the Atlantic Ocean, the water depth of the Puerto Rico Trench is 8650 m; and in the Indian Ocean, the water depth of the Java Trench is 7450 m. Shallow ocean depths are encountered in the marginal seas along the coasts of continents. These are usually <2000 m and frequently <1000 m deep. Figure 6.1 shows the oceans of the world (Williams, 1962; Bowden, 1975).

An ocean includes a basin with several structural features. Its walls are formed by the *continental margin*. Projecting from each continental shore is the *continental shelf*; all shelves together encompass ~7.5% of the ocean area. Each shelf slopes gently downward in the direction of the ocean to a water depth of ~130 m at an average angle of 7°. The average shelf width is ~65 km, but may range from 0 to 1290 km, the greatest width being represented by the shelf projecting from the coast of Siberia into the North Polar Sea. The waters over the continental shelves are a biologically important part of the oceans. They are sites of high biological productivity because they receive a significant contribution of nutrients in general runoff from the adjacent land, particularly from rivers emptying into the waters over a shelf.

At the edge of the continental shelf, the ocean floor drops sharply at an average angle of 4° (range 1°–10°) to abyssal depths of ~3000 m. This is the region of the *continental slope* and constitutes ~12% of the ocean area. In some places, slopes are cut by deep canyons, which often occur at the mouths of large rivers (e.g., Hudson River, Amazon River). Many canyons were formed by *turbidity currents* over geologic time. Such currents consist of strong water movements that carry high sediment load, picked up as a river flows oceanward. As the river meets the sea, the sediment is dropped, and when settling, it abrades the slope. Marine canyons may also be cut by slumping of an unstable sediment deposit on a portion of the continental slope and the consequent abrasion of the slope. Occasionally, the continental slope may be interrupted by a terraced region, as in the case of Blake Plateau off the southern Atlantic coast of the United States. This particular shelf is ~302 km wide and drops gradually from a depth of 732–1100 m over this distance. It was gouged out of the continental slope by the northward flowing Gulf Stream.

At the foot of the continental slope lies the *continental rise*, consisting of accumulations of sediment carried downslope by turbidity currents. Such deposits may extend for 100 km or more from the foot of the continental slope. The continental rise may form fanlike structures in some places and wedges in others. An idealized profile of a continental margin is shown in Figure 6.2.

The *ocean basin* takes up 80% of the ocean area. Its floor, far from being a flat expanse, as once believed by some, often exhibits a rugged

EHRLICH'S GEOMICROBIOLOGY

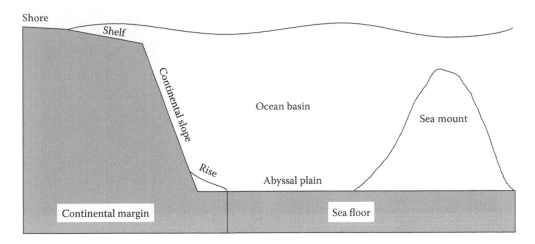

Figure 6.2. Schematic representation of a profile of an ocean basin.

topography. Submarine mountain ranges cut by fracture zones and rift valleys stretch over thousands of kilometers as the midocean ridge systems where the new ocean floor is created (see Chapter 2). Elsewhere, somewhat more isolated submarine mountains, some of which are active and others dormant volcanoes, dot the ocean floor. Some of the seamounts have flattened tops and have been given the special name of *guyots*. Some of the flattened tops of seamounts, especially in the Pacific Ocean, reach surface waters at depths of 50–100 m where the temperature is ~21°C. In these positions, the flat tops may serve as substratum for colonization by corals (coelenterates) and coralline algae, which then form atolls and reefs.

Covering the ocean floor almost everywhere are *sediments*. They range in thickness from 0 to 4 km, with an average thickness of 300 m. Their rate of accumulation varies, being slowest in midocean (<1 cm/10^3 years) and fastest on continental shelves (10 cm/10^3 years). These rates may be even greater in some inland seas and gulfs (e.g., 1 cm/10–15 years in the Gulf of California and 1 cm in 50 years in the Black Sea). In some regions of the deep ocean, the sediments consist mainly of deposits of siliceous and calcareous remains of marine organisms. The siliceous remains are derived from the frustules of diatoms (algae) and the support skeleton and spines of radiolarians (protozoa). The calcareous remains are derived from the tests of foraminifera (protozoa), carbonate platelets from the walls of coccolithophores (algae), and shells from pteropods (mollusks). *Diatomaceous oozes*

predominate in colder waters (e.g., in the North Pacific between 40° and 70° north latitude and 140° west to 145° east longitude; Horn et al., 1972). *Radiolarian oozes* predominate in warmer waters (e.g., in the North Pacific between 5° and 20° north latitude and 90° and 180° west longitude; Horn et al., 1972). *Calcareous oozes* are found mainly in warmer waters on ocean bottoms no deeper than 4550–5000 m (e.g., in the North Pacific between 0° and 10° north latitude and 80° and 180° west longitude; Horn et al., 1972). At greater depths, the combination of increased pressure and accumulating CO_2 from degrading sinking organic matter is high enough to cause dissolution of carbonate.

Other vast areas of the ocean floor are covered by clays (*red clay* or *brown mud*), which are probably of terrigenous origin and washed into the sea by rivers and general runoff from continents and islands and carried into the ocean basins by ocean currents, mudflows, and turbidity currents. At high latitudes in both hemispheres, particularly on and near continental shelves, ice-rafted sediments are found. They were dropped into the ocean by melting icebergs, which had previously separated from glacier fronts that had picked up terrigenous debris during glacial progression. Except for ice-rafted detritus, only the fine portion of terrigenous debris (clays and silts) is carried out to sea. The clay particles are defined as having a diameter <0.004 mm and the silt particles as having a size range of 0.004–0.1 mm in diameter. Figure 6.3 shows the appearance of some Pacific Ocean sediments under the microscope.

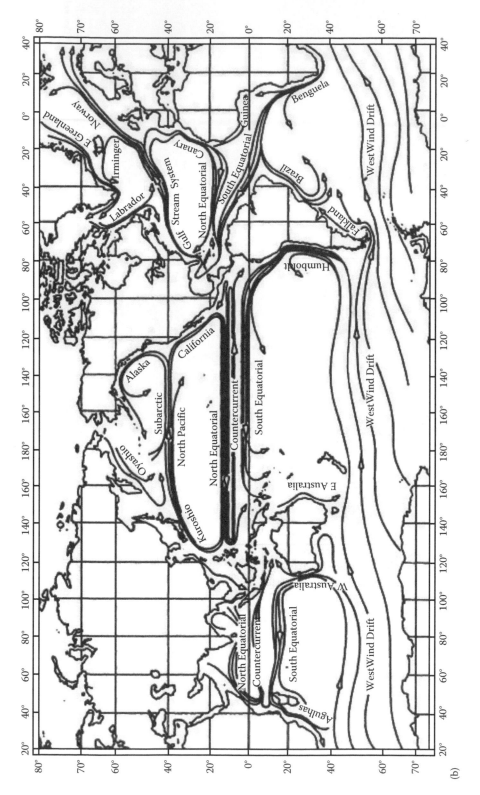

Figure 6.4. (*Continued*) Oceanic surface currents. (b) Average surface currents of the world's oceans. (From Williams, J, *Oceanogr*, Little, Brown, Boston, MA, 1962. With permission. [Jerome Williams is deceased.])

Meanders in the Gulf Stream in the Atlantic and the Kuroshio current in the Pacific Ocean may give rise to the so-called rings—small closed current systems that may measure as much as 300 km in diameter and may have a depth as great as 2 km. Such rings may move 5–10 km/day. The chemical, physical, and biological characteristics of the water enclosed in a ring may be significantly different from those of the surrounding water. A slow exchange of solutes and biota as well as heat transfer may take place across the boundary of a ring, while waters can be upwelled within the rings providing nutrient-rich waters to the euphotic zone (McGillicuddy et al., 1998). Rings thus constitute the means of nutrient transport from ocean currents. The eddies may ultimately rejoin the current that spawned them (Ring Group, 1981; Gross, 1982; Richardson, 1993). *Anticyclones* can also arise from the Gulf Stream in addition to the rings and *meddies* from the north of the Strait of Gibraltar (Richardson, 1993). Although rings have a cold water core surrounded by a warm water layer and counterclockwise rotation, anticyclones have a warm water core surrounded by colder water and clockwise rotation. Meddies have a core that is more saline than the surrounding ocean water and a clockwise rotation. Collectively, these formations are known as *ocean eddies*.

The deep ocean is also in motion. There is a large-scale movement of seawater connecting the ocean basins known as *thermohaline circulation* (or *meridional overturning circulation*), induced by the sinking of cooler and saltier waters (denser) primarily in the North Atlantic Ocean (Wunsch, 2003; Talley et al., 2011). This deepwater can return to the surface in a process called *upwelling*. This results from the moving apart of two surface water masses, causing the deepwater to rise to take the place of the divergent waters. The moving apart of the water masses is called *divergence* (Williams, 1962). Upwelling of deepwater may also result when winds blow large surface water masses away from coastal regions (Smith, 1968). This entire process of meridional overturning circulation can take more than a thousand years on its journey from the formation of deepwaters in the North Atlantic, around the Antarctic continent, traveling through the Indian and Pacific Ocean depths, prior to upwelling in the North Pacific Ocean.

Deepwater is rich in mineral nutrients, including nitrate, phosphate, silicic acid, and certain micronutrients, and thus upwelling is of great ecological significance because it replenishes biologically depleted nutrients in the surface waters. The major upwelling regions are therefore very fertile and economically important fisheries and occur at the eastern boundaries of the Pacific and Atlantic Ocean, such as the Peru, California, and Benguela upwelling regions. Upwelling regions are often prone to a mosaic of iron limitation due to its gradual scavenging and depletion in upwelled waters (Hutchins et al., 1998). A disturbance in the surface water circulation in the southern Pacific can result in failure of upwelling in this region (El Niño) and can spell temporary disaster for the fisheries of the area. In contrast to deepwaters, when dense surface waters sink creating deepwaters, they carry the chemical signature of their exposure to the surface. This signature includes oxygen and carbon dioxide from equilibration with the atmosphere as well as depleted nutrients due to phytoplankton growth. These nutrients then gradually accumulate in deepwaters during its circulation through the degradation of biological material that sinks from the surface ocean and is degraded into its chemical components.

6.1.3 Chemical and Physical Properties of Seawater

Seawater is saline. Some important chemical components of seawater, listed in decreasing order of concentration, are presented in Table 6.1 (Bruland and Lohan, 2003). Of these components, chloride (55.2%), sodium (30.4%), sulfate (7.7%), magnesium (3.7%), calcium (1.16%), potassium (1.1%), bromide (0.1%), strontium (0.04%), and borate (0.07%) account for 99.5% of the total salts in solution (percent, in weight per volume). Because these components generally occur in constant proportions relative to one another in true ocean waters, it has been possible to estimate salt concentrations in seawater samples by merely measuring chloride concentration. The chloride concentration in g kg^{-1} (chlorinity, Cl) is related to the total salt concentration (salinity, S) in g kg^{-1} by the relationship

$$S(‰) = 0.030 + 1.8050Cl\ (‰)* \tag{6.1}$$

---

* The symbol "‰" represents parts per thousand or g kg^{-1}. Equation 6.1 was amended to S(‰) = 0.030 + 1.80655Cl (‰) by UNESCO in 1969.

water with iron (Coale et al., 1996; Church et al., 2000; Arrieta et al., 2004). Growth stimulation of phytoplankton in the ocean by iron fertilization might offer a means of significant enhancement of CO_2 sequestration in the ocean and, thus, have a positive effect on climate control. These blooms tend to enhance the bioproduction of silica by growth stimulation of marine diatoms, an important food source for some plankton feeders, for instance, in the seas around Antarctica (Southern Ocean). However, theoretical considerations and results from field experiments examining these possibilities suggest that the extent of sequestration of carbon and its transfer to the ocean floor as particulate organic carbon is not great enough to have a significant impact on climate control (Bakker, 2002; Buessler and Boyd, 2003; Arrieta et al., 2004; Buessler et al., 2004; Brzezinski et al., 2005; Liss et al., 2005; Zeebe and Archer, 2005).

The salts dissolved in seawater impart a special osmotic property to it. The *osmotic pressure* of seawater is of the order of magnitude of the internal pressure of bacterial cells or the cell sap of eukaryotic cells. At a salinity of 35‰ and a temperature of 0°C, seawater has an osmotic pressure of 23.37 bar (23.07 atm), whereas at the same salinity but at 20°C, it has an osmotic pressure of 25.01 bar (24.96 atm). Clearly then, the osmotic pressure of seawater is not deleterious to living cells.

With increasing depth in the water column, HP becomes a significant factor in the life of microbes and other forms of life in the sea. On average, the HP in the open ocean increases ~1 atm (1.013 bar) for every 10 m of depth. Related to the weight of overlying water at a given depth, HP in the oceans ranges from 0 to more than 1000 atm (1013 bar). Thus, the highest pressures occur in the deep ocean trenches. Among the marine fauna, some members are adapted to live only in surface waters, others at intermediate depths, and still others at abyssal depths. Generally, none are known to live over the entire depth range of the open ocean. Although microorganisms such as bacteria appear to be more adaptable to changes in HP, facultative (pressure tolerant) and obligately barophilic (pressure-requiring) bacteria are known (see also Section 6.1.5).

Salinity and temperature affect the *density* of seawater. At 0°C, seawater having a salinity of 30‰–37‰ has a corresponding density range of 1.024–1.030 g/cm³. A variation in seawater density due to variation in salinity is one cause of water movement in the ocean, because denser water will sink below lighter water (convergence), or conversely, lighter water will rise above denser water (upwelling). The following processes may cause changes in salinity and, therefore, density: (1) dilution of seawater by runoff or groundwater inputs of less-saline water; (2) dilution by rain or snow; (3) concentration through surface evaporation; (4) freezing, which excludes salts from ice and thus leaves any residual, unfrozen water more saline; or (5) thawing of ice, which dilutes the already existent saline water.

As already stated, variation in salinity of seawater is not the sole cause of variation in density. The other important cause of density variation of seawater is temperature. Unlike freshwater, whose density is greatest at 4°C (Figure 6.6b), seawater with a salinity of 24.7‰ or greater has maximum density at its freezing point, that is, 0°C (Figure 6.6a). A body of freshwater thus freezes from its surface downward because freshwater at its freezing point is lighter than at a temperature of 4°C. Ocean water in the Arctic or Antarctic seas also freezes from the surface downward, but in this instance, ice, which excludes salts as it forms from seawater, is lighter than the seawater and will thus float on it.

The temperature of seawater ranges from about −2°C (the freezing point at 36‰ salinity) to +30°C, in contrast to the temperature of air over the ocean, which can range from −65°C to +65°C. The narrower temperature range for seawater can be related to (1) its heat capacity, (2) its latent heat of evaporation, and (3) heat transfer from lower to higher latitudes by surface currents in both hemispheres. The major source of heat in the ocean is solar radiation. More than half the surface waters of the ocean are at 15°C–30°C. Only 27‰ of the surface waters are below 10°C. From ~50° north latitude to 50° south latitude, the ocean is thermally stratified. In this range of latitudes, the seawater temperature below the depth of ~1000 m is below 4°C (deepwater). At depths from ~300 to 1000 m, the temperature drops rapidly with increasing depth. The zone of this rapid temperature change is called the *thermocline*. Its thickness and position vary with geographic location and season of the year. Above the thermocline lies the warm surface water, the *mixed layer*, which is extensively agitated by wind and water currents and thus exhibits relatively little temperature change with increasing depth.

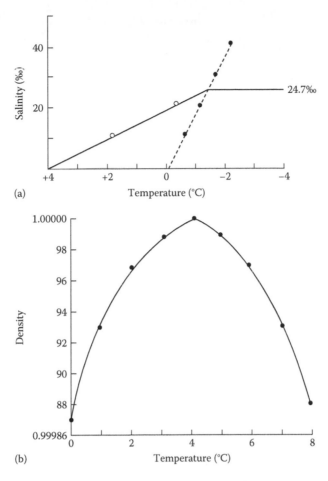

Figure 6.6. Density relationships in seawater and freshwater. (a) Relationship of seawater salinity to freezing point. Symbols: open circles, temperature of maximum density at a given salinity; closed circles, freezing-point temperature at a given salinity. Note that above a salinity of 24.7‰, seawater freezes at its maximum density as its temperature at maximum density cannot be lower than its freezing point. (b) Relationship of freshwater density (in g/cc) to temperature. Data points from chemically pure water are shown. Note that in the case of freshwater, its density at its freezing point is lower than its density at 4°C.

At latitudes higher than 50°N and 50°S, seawater is not thermally stratified. The waters around Antarctica, being cold (−1.9°C) and saline (34.82‰) owing to ice formation, are dense and thus sink below warmer, less dense water to the north and flow northward along the bottom of the ocean basin. This is an example of convergence. Similarly, Atlantic waters from the subarctic region, having a temperature in the range of 2.8°C–3.3°C and a salinity in the range of 34.9‰–34.96‰, sink and flow southward at near-bottom or bottom levels of the ocean. Because the Arctic Ocean bottom is separated from the other oceans by barriers such as the shallow Bering Strait in the case of the Pacific Ocean and a shallow ridge in the case of the Atlantic Ocean, it does not influence the water masses of the Pacific and Atlantic Oceans directly. Other convergences occur in the world's oceans in both hemispheres because of the interaction of waters of different densities. In these instances, the heavier waters sink to lesser depths because they have lower densities than the heavier waters at high latitudes.

The water convergences, and the resultant thermohaline circulation of the oceans described earlier helps to explain why ocean water is generally oxygenated at all depths (Figure 6.7; Kester, 1975). Of all ocean waters, only some coastal and near-coastal waters (e.g., estuarine waters, Cariaco Trench) may, as a result of intense biological activity, be devoid of oxygen at depth. At some coastal sites, intense biological activity is sometimes the direct result of

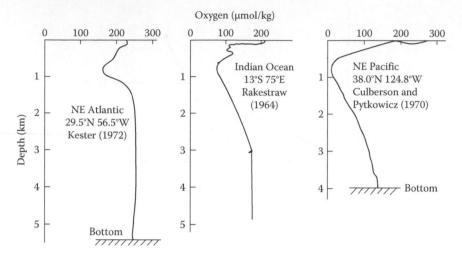

Figure 6.7. Vertical distribution of oxygen in the ocean. Profiles from three ocean basins. (Reprinted from *Chemical Oceanography*, Kester, DR, Academic Press, New York. Copyright 1975, with permission from Elsevier.)

pollution caused by human beings. Surface waters of the open ocean tend to be saturated or supersaturated with oxygen because of oxygenation by the atmosphere and, equally important, by the photosynthetic activity of the phytoplankton. Oxygenation by phytoplankton can occur to depths as great as 200 m, where light penetration is deeper in less productive waters due to fewer light-absorbing phytoplankton. Seawater salinity of 34.352‰ is saturated at 5.86 mL or 8.40 mg of oxygen per liter at 760 mm Hg and 15°C. The higher the salinity and the higher the temperature, the lower the solubility of oxygen in seawater.

Starting at the top of the water column, the oxygen concentration in seawater will at first decrease with depth, typically due to a combination of oxygen consumption by the respiration of living organisms and limited mixing of oxygenated surface waters with subsurface waters (Figure 6.7). Because many life forms in the oceans tend to be concentrated in the upper waters, oxygen concentration will fall to a minimum at ~600–900 m of depth, where respiration of sinking organic matter (oxygen consumption) by bacterioplankton and zooplankton occurs but not photosynthesis (oxygen production) by phytoplankton. Below this depth, because of rapidly decreasing biological activity, the oxygen concentration may at first increase once more and then slowly decrease again toward the bottom. Bottom water, however, may still be half-saturated with oxygen relative to surface water. This oxygen is not supplied by in situ photosynthesis, which cannot occur in

the absence of light at these depths, nor is it the result of significant oxygen diffusion from the atmosphere to these depths. As previously indicated, the oxygenated waters at depth derive from the Antarctic and sub-Arctic convergences. The oxygen-carrying waters from the Antarctic convergences flow northward along the bottom and at intermediate depths of the ocean basins, whereas the waters from subarctic convergence in the Atlantic flow southward at more intermediate depths. The oxygen content of these waters is only slowly depleted because of the low numbers of oxygen-consuming organisms in these deep regions of the oceans and the low rate of oxygen consumption in the upper sediments.

Photosynthetic activity of the phytoplankton is dependent on the penetration of sunlight into the water column because phytoplankton derive their energy almost exclusively from sunlight. It has been shown that light absorption by pure water in the visible range between 400 and 700 nm increases greatly toward the red end of the spectrum. It has also been shown that 60% of the light that penetrates transparent water is absorbed at a depth of 1 m. And 80% and 99% of the same light is absorbed at depths of 10 and 140 m, respectively. In less transparent coastal water, 95% of the light may have been absorbed at 10 m. Although the photosynthetic process of phytoplankters can use light over the entire visible spectrum, action spectra show peaks in the red and blue ends of the spectrum, where chlorophylls absorb optimally. Accessory pigments, such as carotenoids, absorb light at intermediate visible wavelengths. Clearly, light penetration limits the depth at which phytoplankton can grow. This depth is ~80–100 m on average (200 m maximally) and often much less in less transparent waters. Two exceptions have been noted, however. One was seen off the northern border of San Salvador Island in the Bahamas, where crustose coralline algae (Rhodophyta) were growing attached to rock at a depth of 268 m, observed from a submersible. At this location, the light intensity was only ~0.0005% of that at the surface (Littler et al., 1985). The other exception was noted in the Black Sea. Here, the photosynthetic sulfur bacterium *Chlorobium phaeobacteroides* was found to grow in a chemocline at a depth of 80 m, where light transmission from surface irradiance has been calculated to be 0.0005% (Overmann et al., 1992), as at the station at San Salvador Island.

The water layer from the ocean surface to the depth below which photosynthesis cannot take place constitutes the *euphotic zone*. Zooplankton and bacteria, except for cyanobacteria, may abound to lower depths than phytoplankton, being scavengers and able to feed on decaying phytoplankton and zooplankton fecal pellets as they sink through the water column. In addition, bacteria and archaea in the subsurface ocean can be chemolithotrophic, for example, oxidizing ammonia or nitrite with oxygen for energy and fixing carbon from dissolved CO_2.

6.1.4 Microbial Distribution in Water Column and Sediments

Microbial distribution in the open oceans is not uniform throughout the water column (Figure 6.8). Factors affecting this distribution are energy, carbon, nitrogen, and phosphorus limitations (Wu et al., 2000) and also temperature, HP, and salinity. Accessory growth factors, such as vitamins and trace metals, may also be limiting to those microbes, the latter of which cannot synthesize them by themselves (Moore et al., 2013). Phytoplankton distribution is limited to the euphotic zone of the water column primarily by available sunlight—the energy source for the organisms. However, phytoplankton distribution

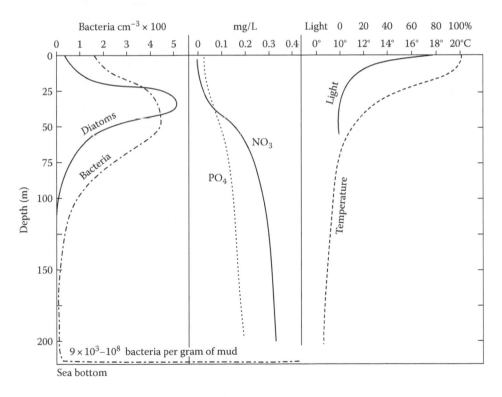

Figure 6.8. Vertical distribution of bacteria (number per cubic centimeter of water), diatoms (number per liter of water), PO_4 and NO_3 (milligrams per liter), light, and temperature in the sea based on average results at several different stations off the coast of Southern California. (From ZoBell, CE, Bacteria in the marine world, *Sci Mon*, 55, 320, 1942. Reprinted with permission from American Association for the Advancement of Science.)

in the euphotic zone is also controlled by the availability of nutrients and micronutrients. These nutrients diffuse upward into the euphotic zone where they are rapidly consumed by phytoplankton. Continual grazing by zooplankton and lysis by viruses can release nutrients as well as cause them to form larger particles that sink out of the euphotic zone. This process of continual recycling of nutrients and organic matter in the euphotic zone is known as the *microbial loop* (Azam et al. 1983; Azam, 1998; Pomeroy et al., 2007).

Nonphotosynthetic microorganisms are abundant in the photic zone and are found in lesser abundances throughout the ocean water column. The nonphotosynthetic microorganisms include predators (zooplankton), scavengers (zooplankton, fungi), chemolithotrophic bacteria and archaea, and decomposers (bacteria and fungi and, possibly to a small extent, zooplankton). Zooplankters therefore dominate the euphotic zone, where they can feed optimally on phytoplankton, zooplankton, and bacteria. Bacteria and fungi are also prevalent here, because they find sufficient sources of nutrients produced by the phytoplankton and zooplankton in secretions and excretions and in dead remains.

Abundant bacterial populations occur at the air–water interface of the ocean as a result of concentrations of organic carbon in the surface film, which may exceed concentrations in the waters below by three orders of magnitude. The bacterial population in this film is known as *bacterioneuston* (Sieburth, 1976; Sieburth et al., 1976; Wangersky, 1976). A comparison of 16S rRNA clone libraries constructed from DNA in samples from a surface microlayer of the North Sea off the coast of Northumbria with 16S rRNA clone libraries from the underlying pelagic seawater indicated a much more limited bacterial diversity in the surface microlayer than in the underlying pelagic seawater (Franklin et al., 2005). The bacterioneuston was dominated by *Vibrio* spp. (57%) and *Pseudoalteromonas* (32%), whereas in the pelagic water, these two groups represented only 13% and 8%, respectively, suggesting that the bacterial community in the bacterioneuston is specialized.

Marine sediments contain significant numbers of living bacteria, fungi, and other benthic microorganisms as well as higher forms of life. In 1940, Rittenberg reported the recovery of viable bacteria from 350 cm below the sediment surface in the Pacific Ocean (Rittenberg, 1940), a finding that at that time must have seemed remarkable and that ZoBell considered very significant (ZoBell, 1946; Ehrlich, 2000a). In recent years, living bacterial cells have been confirmed to exist in the deep biosphere as deep as 400 m sediment depth, which correspond to deposition more than 16 million years ago (Schippers et al., 2005).

Bacteria, protozoa, fungi, and metazoa are found associated with sediments of shallow as well as abyssal water depths. The chief function of the microbes is to aid in scavenging and decomposing organic matter that settled undecomposed or partially decomposed from the overlying regions of biological productivity in the water column. Most of the organic matter from the euphotic zone settles to the bottom in the form of fecal pellets from metazoa. It should be pointed out that not all settled organic matter in deep-sea sediments is utilizable by microbes, for reasons that are not yet clearly understood. The unutilizable organic matter constitutes a significant part of the *sedimentary humus*.

The metabolic activity of the free-living bacteria of deep-sea sediment has been observed to be at least 50 times lower than that of microorganisms in shallow waters or on sediments at shallow depths (Jannasch et al., 1971; Jannasch and Wirsen, 1973; Wirsen and Jannasch, 1974). Environmental factors contributing to this slow rate of bacterial metabolism seem to be low temperature (<5°C) and, especially, elevated HP. Turner (1973) observed that pieces of wood left for 104 days on sediment at a station in the Atlantic Ocean at a depth of 1830 m was rapidly attacked by two species of wood borers (mollusks). This observation led to the suggestion that the primary attackers of organic matter in the deep sea, including sediment, are metazoans. Bacteria and other microbes harbored in the digestive tract of these metazoans decompose this organic matter only after ingestion by the metazoa (see, for instance, Jannasch, 1979). The intestinal bacteria appear essential to the digestion of cellulose to enable these metazoans to assimilate it.

Schwartz and Colwell (1976) were the first to report that the bacterial flora of the intestines of amphipods (crustaceans) that they collected in the Pacific Ocean at a depth of 7050 m was able to grow and metabolize nearly as rapidly at 780 atm and 3°C as at 1 atm and 3°C in laboratory experiments. Their study suggested that these types of gut bacteria behave very differently in

EHRLICH'S GEOMICROBIOLOGY

response to HP from free-living bacteria from the same depths. These findings have been extended by other observations on amphipod microflora. Deming and Collwell (1981) reported finding barophilic and barotolerant bacteria in the intestinal flora of amphipods living at depths of 5200–5900 m. Yayanos et al. (1979) isolated a barophile from a decomposing amphipod that grew optimally at ~500 bar and 2°C–4°C and poorly at atmospheric pressure in this temperature range. The same workers (Yayanos et al., 1981) isolated an obligately barophilic bacterium from an amphipod recovered from 10,746 m in the Marianas Trench.

The generally low rate of metabolism of the biological community (benthos) on deep-sea sediments is also reflected by respiratory measurements carried out at 1850 m. The measurements revealed a rate of oxygen consumption that was orders of magnitude less than in sediments at shallow shelf depths (Smith and Teal, 1973).

Archaea have recently been found to be a major component of the microbial community in the ocean water column, particularly at intermediate depths (Karner et al., 2001; Francis et al., 2005). These crenarchaeota have been found to be chemolithotrophic, oxidizing ammonia with oxygen or conducting urea-based nitrification (Santoro and Casciotti, 2011; Alonso-Sáez et al., 2012).

Growth and reproduction of bacteria and fungi in ocean water also occur on surfaces of some living organisms and on the surface of suspended organic and inorganic particles (epiphytes), because at these sites essential nutrients may be very concentrated (Sieburth, 1975, 1976; Hermansson and Marshall, 1985). The microorganisms may form microcolonies or a biofilm on these surfaces. Detritus, although not a nutrient by itself, usually has adsorptive capacity, which helps to concentrate nutrients on its surface and thus makes for a preferred microbial habitat. The beneficial effect that the buildup of nutrients by adsorption to particle surfaces has on microbial growth is great, because the concentration of the nutrients in solution in seawater is very low (0.35–0.7 mg/L) (Menzel and Rhyter, 1970). ZoBell (1946) long ago showed a significant increase in the bacterial population in natural seawater during 24 h of storage in an Erlenmeyer flask. He attributed this to the adsorption of essential nutrients in the seawater to the walls of the flask, where the bacteria actually grew.

6.1.5 Effects of Temperature, Hydrostatic Pressure, and Salinity on Microbial Distribution in Oceans

Temperature and pressure may have a profound influence on where a given microbe may live in the ocean. Some will grow only in the temperature range of 15°C–45°C (mesophiles) with an optimum near 30°C, others only in the range from 0°C or slightly below to 20°C with an optimum at 15°C or below (psychrophiles), and still others in the range of 0°C–30°C or even higher (37°C) (Ehrlich, 1983) with an optimum near 25°C (psychrotrophs) (Morita, 1975). The mesophiles would be expected to grow only in waters of the mixed zone and near active hydrothermal vents, whereas psychrophiles would grow only below the thermocline and in polar seas. Psychrotrophs would be expected to grow above, in, and below the thermocline and the polar seas, although they might do better in and above the thermocline. Mesophiles can be recovered from cold waters and deep sediments, where they are able to survive but cannot grow (i.e., they are psychrotolerant). Even the most abundant of marine photosynthetic cyanobacterium, Prochlorococcus, has been shown to exist as distinct ecotypes organized by large-scale oceanic temperature gradients (Johnson et al., 2006).

Many bacteria that normally grow at atmospheric pressure are not inhibited by HPs up to ~200 atm (202.6 bar), but their growth is retarded at 300 atm (303.9 bar), and they will not grow above 400 atm (405.2 bar). One effect that increased HP has on these bacteria is on cell morphology and structure (Kaletunç et al., 2004). Many bacteria isolated from waters at 500 atm (506.5 bar) and 600 atm (607.8 bar) were found to grow better at these pressures under laboratory conditions than at atmospheric pressure. Such organisms are called barophiles. Some organisms, described in a pioneering study by ZoBell and Morita (1957), which had been recovered from extreme depths (10,000 m), were suspected of having been obligate barophiles. Since that time, an obligately barophilic bacterium has actually been isolated from an amphipod taken at 10,476 m in the Marianas Trench and studied (Yayanos et al., 1981). It exhibited an optimal growth rate (generation time of 25 h) at 2°C and 690 bar (681 atm) of HP. As already mentioned, Yayanos et al. (1979) also isolated a facultatively barophilic spirillum from 5700 m depth that grows fastest at ~500 bar (493.5 atm) and 2°C–4°C, with a generation time of 4–13 h.

diatom and picoeukaryote prymnesiophyte hosts, respectively (Foster et al., 2011; Thompson et al., 2012). The cyanobacteria in these associations provide a nitrogen source for the host, while the host provides a protective environment from grazing and nutritional components. Nonsymbiotic heterotrophic, mixotrophic, or autotrophic bacteria in the marine environment may be free living or attached to living organisms or inert particles suspended in the seawater column (e.g., inorganic matter such as clay particles and fecal pellets; for more information on bacterial attachment to fecal pellets, see Turner and Ferrante [1979]). They can also be detected in the sediment column in all parts of the ocean.

Marine bacteria are usually defined as types that will not grow in media prepared in freshwater because they need one or more of the salt constituents of seawater. Bacteria that do not require seawater for growth can, however, be readily isolated from seawater and marine sediment samples far from shore. Many of these organisms can grow readily in media prepared in seawater. They may represent terrestrial forms. The active bacteria in the marine environment are important as decomposers and in special marine niches as primary producers. Some may also play a role in mineral formation and mineral diagenesis (e.g., ferromanganese deposits, manganese oxides; see Chapter 18; for further discussion, see Ehrlich [1975, 2000b], Sieburth [1979], Jannasch [1984], and Kulm et al. [1986]).

6.2 FRESHWATER LAKES

Freshwater amounts to <3% of the total water on Earth. Like the oceans, it furnishes important habitats for certain life forms. Among these habitats are lakes, which are part of the *lentic* environments—the standing waters. Other lentic environments are ponds and swamps. Lakes represent only 0.009% of the total water in land areas (van der Leeden et al., 1990), most of the freshwater being tied up in ice caps and glaciers (2.14%) and in groundwater (0.61%) (van der Leeden et al., 1990).

Lakes have arisen in various ways. Some resulted from past glacial action. An advancing glacier gouged out a basin that, when the glacier retreated, was filled with water from the melting ice and was later kept filled by runoff from the surrounding *watershed*. Other basins resulted from

landslides that obstructed valleys and blocked the outflow from their watershed. Still others resulted from crustal up-and-down movement (dip-slip faulting) that formed dammed basins for the collection of runoff water. Some resulted from the solution of underlying rock, especially limestone, which led to the formation of basins in which water collected. Lakes have also been formed by the collection of water from glacier melts in craters of extinct volcanoes and by the obstruction of river flow or changes in river channels (Welch, 1952; Strahler and Strahler, 1974; Skinner et al., 1999).

Lakes vary greatly in size. The combined Great Lakes in the United States cover an area of 328,000 km², an unusually great expanse. More commonly, lakes cover areas of 26–520 km², but many are smaller. Most lakes are <30 m deep. However, the deepest lake in the world, Lake Baikal in southern Siberia (Russia), has a depth of 1700 m. The average depth of the Great Lakes is 700 m, and that of Lake Tahoe on the California–Nevada border is 487 m. The elevation of lakes ranges from below sea level (e.g., the Dead Sea at the mouth of the Jordan River) to as high as 3600 m (Lake Titicaca in the Andes on the border between Bolivia and Peru).

6.2.1 Some Physical and Chemical Features of Lakes

Some of the water of lakes may be in motion, at least intermittently. Most prevalent are horizontal currents, which result from wind action and the deflecting action of shorelines. Vertical movements are rare in lakes of average or small size. They may result from thermal, morphological, or hydrostatic influences. Thermal influence can result in changes of water density such that heavier (denser) water sinks below lighter water. Morphological influence can result from rugged bottom topography, which may deflect horizontal water flow downward or upward. Hydrostatic influence can result from springs at the lake bottom that force water upward into the lake. Besides horizontal and vertical movement, return currents may occur as a result of water being forced against a shore by wind and piling up. Depending on the type of lake and the season of the year, only a portion of the total water mass of a lake, or all of it, may be circulated by the wind. (For further discussion, see Welch [1952] and Strahler and Strahler [1974].)

The waters of lakes may vary in composition from very low salt content (e.g., Lake Baikal) to a very high salt content (e.g., Dead Sea between Israel and Jordan and Lake Natron in Africa) and from low organic content to high organic content. Salt accumulation in lakes is the result of input from runoff from the watershed, including stream flow, very gradual solution of sediment components and rock minerals in the lake bed, and evaporation.

The waters of lakes may or may not be thermally stratified, depending on various factors: geographic location, the season of the year, and lake depth and size. Thermal stratification, when it occurs, may or may not be permanent. Absence of stratification may be the result of complete mixing or turnover as a result of wind action. When waters are thermally stratified into a warmer layer in the upper portion of a lake (epilimnion) and a cooler layer in the lower portion (hypolimnion), complete mixing does not occur because of a density difference between the two layers. A thermocline forms between the epilimnion and the hypolimnion, which is a relatively thin layer of water in which a temperature gradient exists ranging from the temperature corresponding to the bottom of the epilimnion to that of the top of the hypolimnion. Lakes may be classified according to whether and when they turn over (Reid, 1961). The categories can be defined as follows:

Amictic lakes are bodies of water that never turn over, being permanently covered by ice. Such lakes are found in Antarctica and at high altitudes in mountains.

Cold monomictic lakes are bodies of water that contain waters never exceeding 4°C, which turn over once during summer, being thermally stratified the rest of the year.

Dimictic lakes turn over twice a year, in spring and fall. They are thermally stratified at other times. These are typically found in temperate climates and at higher altitudes in subtropical regions.

Warm monomictic lakes have water that is never colder than 4°C. They turn over once a year in winter and are thermally stratified the rest of the year.

Oligomictic lakes contain water that is significantly warmer than 4°C and turn over irregularly. Such lakes are found mostly in tropical zones.

Polymictic lakes have water just over 4°C and turn over continually. Such lakes occur at high altitude in equatorial regions.

Meromictic lakes are deep, narrow lakes whose bottom waters never mix with the waters above. The bottom waters generally have a relatively high concentration of dissolved salts, which makes them dense and separates them from the overlying waters by a chemocline. The upper waters in temperate climates may be thermally stratified in summer and winter and may undergo turnover in spring and fall.

A dimictic lake in a temperate zone during spring thaw accumulates water near 0°C, which, because of its lower density, floats on the remaining denser water, which is near 4°C. As the season progresses, the colder surface water is slowly warmed by the sun to near 4°C. At this point, all water has a more or less uniform temperature and thus uniform density. This allows the water to be completely mixed or turned over by wind agitation if the lake is not excessively deep like a meromictic lake. As the surface water undergoes further warming by the sun, segregation of water masses occurs as warmer, lighter water comes to lie over the colder, denser water. A thermocline is established between the two water masses, separating them into epilimnion and hypolimnion. The temperature of the water in the epilimnion may be higher than 10°C and vary little with depth (perhaps 1°C/m). However, the water in the thermocline will show a rapid drop in temperature with depth. This drop may be as drastic as 18°C/m but is more usually ~8°C/m. The thickness of the thermocline varies with position in the lake and between different lakes—an average thickness being ~1 m. The water in the hypolimnion will have a temperature well below that in the epilimnion and show a small drop in temperature with depth, usually <1°C/m.

The water in the epilimnion but not in the hypolimnion is subject to wind agitation and is thus fairly well mixed at all times. It is the greater density of the water in the hypolimnion that prevents it from getting mixed by the wind action. Continual warming by the sun and mixing by the wind produces horizontal currents and, in larger lakes, return currents over the thermocline, resulting in some exchange with water of cannot be completely decomposed because of an insufficiency of oxygen and alternative terminal electron acceptors such as ferric iron, nitrate, or sulfate. They are also typically deficient in assimilable phosphorus, in contrast to coastal oceans that are often nitrogen limited. The waters of such lakes are turbid and often acid. The origin of dystrophic conditions may be encroachment of the shoreline by plants, including reeds, shrubs, and trees.

6.2.4 Lake Evolution

Lakes have an evolutionary history. Once fully matured, they age progressively. Their basin slowly fills with sediment, partly contributed by feed streams, by the surrounding land through erosion and runoff, and partly by the biological activity in the lake. The size of the contribution that each process makes depends on the fertility of the lake. Changes in climate may also contribute to lake evolution (e.g., through lessened rainfall, which can cause a drop in water level, or through warming of the climate, which can cause more rapid water evaporation). These effects usually make themselves felt slowly. Ultimately, a lake may change into a swamp.

6.2.5 Microbial Populations in Lakes

The microbial population in eutrophic lakes tends to be orders of magnitude greater than in the seas. Numbers of culturable bacteria may range from 10^2 to 10^5 mL^{-1} of lake water and be of the order of 10^6 g^{-1} of lake sediment. The size of the bacterial population may be affected by runoff, which contributes soil bacteria. The culturable bacterial population of lakes consists predominantly of gram-negative rods (Wood, 1965, p. 36), although gram-positive spore-forming bacteria and actinomycetes can be readily isolated, especially from sediments. The freshwater bacteria have been found to have evolved from marine bacteria long ago (Logares et al., 2009), and recent surveys have found the *Actinobacteria* to represent the majority of DNA sequences found, despite having relatively few representatives group in culture (Newton et al., 2011).

Like marine bodies of water, bodies of freshwater, such as lakes and ponds, feature surface films at the air–water interface that harbor a number of microorganisms that are greater than

in the underlying water (reviewed by Wotton and Preston, 2005). The organisms in such surface films include bacteria, protozoa, and even invertebrate larvae (e.g., mosquito larvae). Some of the bacteria may occur in microcolonies, being embedded in an exopolymer (see figure 4 in Wotton and Preston, 2005). Accumulations of organic matter in the surface films help to sustain the heterotrophic microbial population in these films.

Few, if any, of the types of bacteria found in lakes seem to be exclusively limnetic organisms. The main activity of the bacteria, other than the cyanobacteria, is that of decomposers. The cyanobacteria along with algae serve as the primary producers. Fungi and protozoa are also found. Important functions of the former are as scavengers and decomposers and of the latter as predators.

Cyanobacteria and algae are abundant in eutrophic lakes. The algae include green forms as well as diatoms and pyrrhophytes. *Cyanobacterial and algal blooms* may occur at certain times when one species suddenly multiplies explosively and becomes the dominant phytoplankton member temporarily, often forming a carpetlike layer or mat on the water surface. After having reached a population peak, most of the cells in the phytoplankton bloom die off and are attacked by scavengers and decomposers. Especially favorable growth conditions appear to be the stimulus for such blooms.

6.3 RIVERS

Rivers, which account for only 0.0001% of the total water in the world (van der Leeden et al., 1990), are part of the *lotic* environment in which waters move in channels on the land surface. Such flowing water may start as a brook, then widen into a stream, and ultimately become a river. The source of this water is surface runoff and groundwater reaching the surface through springs or, more important, through general seepage. A riverbed is shaped and reshaped by the flowing water that scours the bottom and sides with the help of suspended particles ranging in size from clay particles to small stones. Young rivers may feature rapids and steep valley slopes. Mature rivers lack rapids and feature more uniform stream flow, owing to a smoothly graded river bottom and an ever-widening riverbed. Old rivers may develop meanders in wide, flat floodplains. The flow of the water is caused by gravity because the head of a river always lies above its mouth. Average flow of rivers range from 0 to 9 m/s. However when viewed in cross section, the flow of water in a river is not uniform. The water in some portions in such a cross section flows much faster than in other portions. This can be attributed to frictional effects related to the riverbed topography as well as to density differences of different parts of the water mass. Density variation may arise from temperature differences and from solute concentration differences between parts of the river. Portions of river water may exhibit strong turbulence caused in part by certain features of the river topography. Water velocity, turbulence, and terrain determine the size of particles a river may sweep along (see Strahler and Strahler, 1974; Stanley, 1985; Skinner et al., 1999).

Most river water is ultimately discharged into an ocean, but exceptions exist. The Jordan River, whose headwaters originate in the mountains of Syria and Lebanon, empties into a lake called the Dead Sea, which has no connection with any ocean. The Dead Sea does not overflow because it loses its water by evaporation, which accounts, in part, for its high salt accumulation. (Nowadays, actual shrinkage in the size of the Dead Sea is observed that is attributable to commercial exploitation for recovery of minerals from its waters and to a decrease in water inflow from the Jordan River as a result of water diversion as a source of potable and irrigation water.) The waters in the Dead Sea are nearly saturated with salts, which makes life impossible except for specially adapted organisms.

When river water is discharged into an ocean, an estuary is frequently formed where the less dense river water will flow over the denser saline water from the sea with incomplete mixing. Tidal effects of the sea may alter the water level of the discharging river, sometimes to a considerable distance upstream. Estuaries form special habitats for microbes and higher forms of life, which must cope with periodic changes in salinity, water temperature, nutrients, oxygen availability, and so forth, engendered by tidal movement.

Because of the relatively constant water movement, the water temperature of rivers tends to be rather uniform (i.e., rivers generally are not thermally stratified the way lakes are, when examined in cross section). Only where a tributary with a different water temperature enters a river there may be local temperature stratification. Different segments of a river may, however, differ in temperature. The pH of river water can range from very acidic (e.g., pH 3), for instance, in streams receiving acid mine drainage, which is the result of microbial activity (see Chapter 21), to alkaline (e.g., pH 8.6) (Welch, 1952, p. 413). Unless heavily polluted by human activity, rivers are generally well aerated. It has been thought that in unpolluted rivers most organic and inorganic nutrients supporting microbial as well as higher forms of life are largely introduced by runoff (allochthonous). It is now believed that a significant portion of fixed carbon in such streams and rivers may be contributed by photosynthetic autotrophs, mainly algae growing in parts of a river with quiet waters (autochthonous) (Minshall, 1978). Pollution may cause overloading, which, because of excessive oxygen demand, will result in anoxic conditions with the consequent elimination of many micro- as well as macroorganisms.

Planktonic organisms tend to be found in greater numbers in the more stagnant or slower-flowing waters of a river than in fast-flowing portions. The plankters include algae such as diatoms, cyanobacteria, green algae, protozoa, and rotifers. The proportions depend on the environmental conditions of a particular river and its sections. Sessile plants or algae tend to develop to significant extents only in sluggish streams or in the backwaters of otherwise rapidly flowing streams. Bacteria are present in significant numbers where physical and chemical conditions favor them. Rheinheimer (1980) reported total bacterial numbers, estimated by culture methods, in the River Elbe in Germany to range from 4.7×10^9 to 6.9×10^9 L^{-1} and bacterial biomass to range from 0.55 to 0.71 mg/L in an unspecified year. As in lakes, no unique microflora occurs in unpolluted rivers.

6.4 GROUNDWATERS

Water that collects below the land surface in soil, sediment, and permeable rock strata is called *groundwater*. It represents 0.61% of all water in the world (van der Leeden et al., 1990). Groundwater derives mainly from *surface water* whose origin is meteoritic precipitation such as rain and melted snow. Surface water includes the water of rivers and lakes (Figure 6.12). A minor amount of groundwater derives from *connate water, water of dehydration*, or *juvenile water*. Connate water, often

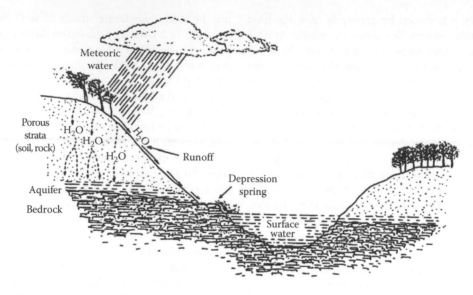

Figure 6.12. Interrelation of meteoritic, surface, and groundwaters.

of marine origin and therefore saline, is water that got trapped in rock strata in the geologic past by up- or downwarping or faulting and as a consequence became isolated as a stagnant reservoir. Its salt composition frequently has become highly altered from that of the original water from which it is derived as a result of long-term interaction with the enclosing rock. Connate waters are often associated with oil formations. Waters of dehydration are derived from waters of crystallization, which are part of the structure of certain crystalline minerals. They are released as a result of the action of heat and pressure in the lithosphere. Juvenile waters are associated with magmatism that causes them to escape from the interior of the Earth. They are termed juvenile because they are assumed to have never reached the Earth's surface before.

Surface water slowly infiltrates permeable ground as long as the ground is not already saturated. It passes through a zone of aeration or unsaturation called the *vadose zone* to the zone of saturation or *aquifer*, which lies over an impermeable stratum (Strahler and Strahler, 1974; Chapelle, 1993; Skinner et al., 1999). The vadose zone may include the soil, an intermediate zone, and the capillary fringe and can range in thickness from a few centimeters to 100 m or more. The resident water in the vadose zone is pore water (pellicular water; see Chapter 4) that is under less than atmospheric pressure and held there by capillarity. In soil, this water supports the soil micro- and macroflora and

fauna and plant growth. The water in the aquifer can be extracted by human intervention, but that in the vadose zone cannot (Hackett, 1972). The rate of infiltration of permeable strata depends not only on the surface water supply but also on the porosity of the permeable strata. Similarly, the water holding capacity of the aquifer depends on the porosity of the matrix rock. The cause of water movement below ground level is not only gravity but also intermolecular attraction between water molecules, capillary action, and hydrostatic head (see Chapter 4).

At a given location, two or more aquifers may occur, one over the other, separated by one or more impervious strata. An example of such multiple aquifers is found in the Upper Atlantic Coastal Plain Province in South Carolina, United States (Sargent and Fliermans, 1989). In such a sequence of aquifers, the uppermost may be directly rechargeable with surface water from above and is then called an *unconfined aquifer*, or it may be separated from a larger aquifer just below by a thin lens of material with low or no porosity, called an *aquiclude*, in which case the uppermost aquifer is called a *perched aquifer*. The material composing an aquiclude may be clay or shale. Aquifers underlying a perched aquifer are called *confined aquifers* and are rechargeable only where the aquiclude is absent. The matrix of aquifers may consist of sand and gravel, fractured limestone and other soluble rock, basalt and fractured volcanic rocks, sandstone and conglomerate, crystalline and

EHRLICH'S GEOMICROBIOLOGY

metamorphic rocks, or other porous but poorly permeable materials (Hackett, 1972).

Groundwater may escape to the surface or into the atmosphere through *springs* or by evaporation with or without the mediation of plants (transpiration). Some water will be accumulated by the vegetation itself. Depending on the relative rates of water infiltration and water loss, the level of the *water table* in the ground may rise, fall, or remain constant. Groundwater that reaches the surface through springs may do so under the influence of gravity, which may create sufficient head to force the water to the surface through a channel, as in an *artesian spring*. Groundwater may also reach the surface as a result of an intersection of the water table with a sloping land surface, as in a *depression spring*. Finally, groundwater may reach the surface in springs under the influence of thermal energy applied to reservoirs of water deep underground. Such *hot springs* in their most spectacular form are *geysers* from which hot water spurts forth intermittently. Some hot springs emit not only water derived from infiltration of surface water but also juvenile water.

As it infiltrates permeable soil and rock strata, surface water will undergo changes in the composition of dissolved and suspended organic and inorganic matter. These changes are the result of adsorption and ion exchange by surfaces of soil and rock particles. They are also the result of biochemical action of microbes, including bacteria, fungi, and protozoa, which exist mainly in biofilms on the surface of many of the rock particles and metabolize the adsorbed organic and (to a limited extent) inorganic matter (Cullimore, 1992; Costerton et al., 1994; Flemming et al., 2007). Plant roots and microbes associated with them in the rhizosphere affect the solute composition through absorption or excretion and metabolism. Polluted water infiltrating the ground may become thoroughly purified, provided that it moves through a sufficient depth and does not encounter major cracks and fissures, which, because of reduced surface area, would exert only limited filtering action. Under some circumstances, the groundwater may also become highly mineralized during filtration or after reaching the water table. If such mineralized water reaches the land surface, it may leave extensive deposits of calcium carbonate, iron oxide, or other material as it evaporates.

Systematic studies of microbes in groundwater have been undertaken. As many as 10^6 bacteria per gram have been recovered by viable and direct counts from the vadose zone and some shallow water table aquifers. They included gram-positive and gram-negative types, the former in apparently greater numbers (Wilson et al., 1983). Evidence of fungal spores and yeast cells was also seen in one instance (Ghiorse and Balkwill, 1983), but eukaryotic microbes were generally thought to be absent from subsurface samples associated with groundwater. Sinclair and Ghiorse (1987) showed that the number of protozoa, mostly flagellates and amoebae, decreased sharply to 28 g^{-1} (dry weight) with increasing depth in the vadose zone at the bottom of a clay loam subsoil material taken from a site in Lula, Oklahoma, United States. They were absent from the saturated zone except in gravelly, loamy sand matrix at a depth of 7.5 m, which also contained significant numbers of bacteria.

Special drilling methods that minimize the possibility of microbial contamination have been developed (see Russell et al., 1992; Fredrickson et al., 1993) for studying the microbiology of vadose zones and aquifers in the deep subsurface. Fredrickson et al. (1991) reported as many as 10^4–10^6 colony-forming heterotrophic bacteria per gram of Middendorf (~366–416 m deep) and Cape Fear (~457–470 m deep) sediments of Cretaceous age, which were obtained by drilling into the Atlantic Coastal Plain at Savannah River Plant, South Carolina. The isolates from the two sediments were physiologically distinct. Contrary to what they expected from their study of a shallow aquifer in Lula, Oklahoma (see earlier), Sinclair and Ghiorse (1989) found significant numbers of fungi and protozoa in samples from deep aquifer material from the Upper Atlantic Coastal Plain Province. The numbers of these organisms were highest where the prokaryotic population was high. The bacteria in the samples included members of the bacteria and archaea. Diverse physiological groups were represented, including autotrophs such as sulfur oxidizers, nitrifiers, and methanogens (Fredrickson et al., 1989; Jones et al., 1989) as well as heterotrophs such as aerobic and anaerobic mineralizers. The latter included denitrifiers, sulfur reducers, and iron(III) reducers (Balkwill, 1989; Francis et al., 1989; Hicks and Fredrickson, 1989; Madsen and Bollag, 1989; Phelps et al., 1989; Chapelle and Lovley, 1992). Hydrogen- and acetate-oxidizing methanogens were also found (Jones et al., 1989). Rates of metabolism in deep aquifers of the Atlantic coastal plain have been estimated to range from 10^{-4} to 10^{-6} mmol of CO_2 per liter per year, using a method of geochemical modeling of groundwater chemistry (see also Chapter 4) (Chapelle and Lovley, 1990).

Significant numbers of autotrophic methanogens and homoacetogens have been detected in groundwater from the deep granite aquifers (112–446 m) at the Äspö Hard Rock Laboratory in Sweden (Kotelnikova and Pedersen, 1998). According to the investigators, these organisms may be the primary producers in this aquifer, using available hydrogen as their energy source. However, only $0-4.5 \times 10^{-1}$ autotrophic methanogens and $0-2.2 \times 10^{1}$ homoacetogens per milliliter were found in groundwater samples from deep igneous rock aquifers in Finland (Haveman et al., 1999). These samples did contain 0 to $>1.6 \times 10^{4}$ sulfate reducers and 7.0×10^{0} to 1.6×10^{4} iron(III) reducers. A subsurface paleosol sample from a depth of 188 m in the Yakima borehole at the U.S. Department of Energy's Hanford Site in the state of Washington yielded a variety of bacteria, including members of the bacteria and archaea. The members of the bacteria included relatives of *Pseudomonas*, *Bacillus*, *Micrococcus*, *Clavibacter*, *Nocardioides*, *Burkholderia*, *Comamonas*, and *Erythromicrobium* in addition to six novel types with some affinity to the Chloroflexaceae. The members of the archaea showed an affinity to the Crenarchaeota branch (Chandler et al., 1998). The paleosol in which these bacteria were detected is part of a sedimentary formation overlying the Columbia River Basalt. It has been proposed that methanogens and homoacetogens along with sulfate and iron(III) reducers, present in groundwater from aquifers in the Columbia River Basalt, form a biocenosis that uses H_2 generated in an interaction between water and ferromagnesian silicates in basalt as its energy source (Stevens and McKinley, 1995; Stevens, 1997). However, the occurrence of a reaction between ferromagnesian silicates in basalt and water has been questioned (Lovley and Chapelle, 1996; Madsen et al., 1996; Stevens and McKinley, 1996; Anderson et al., 1998). Indeed, although it can be shown on a thermodynamic basis that H_2 can be formed in a reaction between Fe^{2+} and water at pH 7.0,

$$3Fe^{2+} + 4H_2O \rightarrow H_2 + Fe_3O_4 + 6H^+$$

$$\left(\Delta G_r^{0'} = -50.33 \text{ kJ} \right) \qquad (6.4)$$

it cannot be formed in a reaction between ferrous silicate and water at pH 7.0, for example,

$$3FeSiO_3 + H_2O \rightarrow H_2 + Fe_3O_4 + SiO_2$$

$$\left(\Delta G_r^{0'} = +52.63 \text{ kJ} \right) \qquad (6.5)$$

The hydrogen in the Columbia River Basalt may come from other sources, as pointed out by Madsen (1996) and Lovley and Chapelle (1996).

Much remains to be learned about the microbial populations of pristine aquifers and their associated vadose zones and the response of this population to environmental stresses (pollution).

6.5 SUMMARY

The hydrosphere is mainly marine. It occupies more than 70% of the Earth's surface. The world's oceans reside in basins whose walls arise from the margins of continental landmasses that project into the sea by way of the continental shelf, the continental slope, and the continental rise, bottoming out at the ocean floor. The ocean floor is traversed by mountain ranges cut by fracture zones and rift valleys—the midocean spreading centers—where new ocean floor is being formed. The ocean floor is also cut by deep trenches representing zones of subduction, where the margin of an oceanic crustal plate slips beneath a continental crustal plate. Parts of the ocean floor also feature isolated mountains that are live or extinct volcanoes and may project above sea level as islands. The average world ocean depth is 3975 m; the greatest depth is ~11,000 m in the Marianas Trench.

Most of the ocean floor is covered by sediment of 300 m average thickness, accumulating at rates <1 to >10 cm per 1000 years. Ocean sediments may consist of sand, silt, and clays of terrigenous origin and of oozes of biogenic origin, such as diatomaceous, radiolarian, or calcareous oozes.

Different parts of the ocean are in motion at all times, driven by wind stress, the Earth's rotation, density variations, and gravitational effects exerted by the sun and moon. Surface, subsurface, and bottom currents have been found in various geographic locations.

Where water masses diverge, upwelling occurs, which replenishes nutrients for plankton in the surface waters. Where the water masses converge, surface water sinks and carries oxygen to deeper levels of the ocean, ensuring some degree of oxygenation at all levels.

Seawater is saline (average salinity ~35%) owing to the presence of chloride, sodium, sulfate, magnesium, calcium, potassium, and some other ions. Variations in total salt concentrations affect the density of seawater, as does variation in water temperature. The ocean is thermally stratified between 50° north and 50° south latitude into a mixed layer (to a depth of ~300 m), with water at more or less uniform temperature between 15°C and 30°C depending on latitude, a thermocline (from a depth of ~300 to ~1000 m) in which the temperature drops to ~4°C with depth, and the deepwater (from the thermocline to the bottom), where the temperature is more or less uniform, that is, between <0°C and 4°C. HP in the water column increases by ~1 atm (1.013 bar) for every 10 m of increase in depth. Light penetrates to an average depth of ~100 m, which restricts phytoplankton to shallow depths. Zooplankton and bacterioplankton can exist at all depths but are found in greatest numbers at the seawater–air interface, near where the phytoplankton abounds, and on the ocean sediment. Intermediate depths are at most sparsely inhabited because of limited nutrient supply.

Marine phytoplankton consists of algal groups, including diatoms, dinoflagellates, and coccolithophorids, whereas marine zooplankton consists mainly of flagellated and amoeboid protozoa as well as some small invertebrates. Bacterioplankton is composed of members of bacteria, chiefly heterotrophs, and of members of archaea, whose traits we know little so far. Phytoplankton are the primary producers, zooplankton the predators and scavengers, and the heterotrophic bacteria the decomposers. The metabolic rate of free-living microorganisms decreases markedly with depth, probably as a result of the effects of high HP and low temperature. Different life forms in the ocean show different tolerances to salinity, temperature, and HP.

Freshwater is found in lakes and streams aboveground and in saturated and unsaturated strata belowground. Lakes are standing bodies of water, usually of low salinity, which may be thermally stratified into epilimnion, thermocline, and hypolimnion. The degrees of stratification may vary with the season of the year. Water

below the thermocline (i.e., the hypolimnion) may develop anoxia because the thermocline is an effective barrier to diffusion of oxygen into it. Only after the disappearance of the thermocline do these waters become reoxygenated due to total mixing by wind agitation. Lakes vary in nutrient quality. Phosphorus is usually the most limiting element to lake life. Phytoplankton, zooplankton, and bacterioplankton are important life forms in lakes. Phytoplankton is restricted to the epilimnion, whereas zooplankton and bacterioplankton together with fungi may be distributed throughout the water column and in the sediment.

Rivers constitute moving freshwater. They are generally not thermally stratified. Abundant life forms, such as phytoplankton, zooplankton, and bacterioplankton, are concentrated mainly in the quieter potions of streams, especially those that are not polluted.

Groundwaters are derived from surface waters that seep into the ground and accumulate in aquifers above impervious rock strata. Water from an aquifer may come to the surface again by way of springs or seepage. In passing through the ground, water is purified. Microorganisms as well as organic and inorganic chemicals are removed by adsorption to rock and soil particles. Organic matter may be mineralized by the microbial decomposers. Sediment samples from shallow as well as deep aquifers have revealed the presence of a significant microbial population with very diverse physiological potentials.

REFERENCES

Alonso-Sáez L, Waller AS, Mende DR, Bakker K, Farnelid H, Yager PL, Lovejoy C et al. 2012. Role for urea in nitrification by polar marine Archaea. *Proc Natl Acad Sci USA* 109(44):17989–17994.

Anderson RT, Chapelle FN, Lovley DR. 1998. Evidence against hydrogen-based microbial ecosystems in basalt aquifers. *Science* 281:976–977.

Arrieta JM, Weinbauer MG, Lute C, Herndl GJ. 2004. Response of bacterioplankton to iron fertilization in the Southern Ocean. *Limnol Oceanogr* 49:799–808.

Azam F. 1998. Microbial control of oceanic carbon flux: The plot thickens. *Science* 280:694–696.

Azam F, Fenchel T, Field JG, Gray JS, Meyer-Reil LA, Thingstad F. 1983. The ecological role of water-column microbes in the sea. *Mar Ecol Prog Ser* 10:257–263.

Kaletunç G, Lee J, Alpas H, Bozoglu F. 2004. Evaluation of structural changes induced by hydrostatic pressure in Leuconostoc mesenteroides. Appl Environ Microbiol 70:1116–1122.

Karner MB, DeLong EF, Karl DM. 2001. Archaeal dominance in the mesopelagic zone of the Pacific Ocean. Nature 409:507–510.

Kaye JZ, Baross JA. 2004. Synchronous effects of temperature, hydrostatic pressure, and salinity on growth, phospholipids profiles, and protein patterns of four Halomonas species isolated from deep-sea hydrothermal vent and sea surface environments. Appl Environ Microbiol 70:6220–6229.

Kester DR. 1975. Dissolved gases other than CO_2. In: Riley JP, Skirrow G, eds. Chemical Oceanography, Vol. 1, 2nd edn. New York: Academic Press, pp. 497–547.

Kotelnikova S, Pedersen K. 1998. Distribution and activity of methanogens and homoacetogens in deep granitic aquifers at Äspö Hard Rock Laboratory, Sweden. FEMS Microbiol Ecol 26:121–134.

Kulm LD, Suess E, Moore JC, Carson B, Lewis BT, Ritger SD, Kado DC et al. 1986. Oregon subduction zone: Venting, fauna, and carbonates. Science 231:561–566.

LaRock PA. 1969. The bacterial oxidation of manganese in a fresh water lake. PhD Thesis. Rensselaer Polytechnic Institute, Troy, NY.

Liss P, Chuck A, Bakker D, Turner S. 2005. Ocean fertilization with iron: Effect on climate and air quality. Tellus B Chem Phys Meteorol 57B:269–270.

Littler MM, Littler DS, Blair SM, Norris JN. 1985. Deepest known plant life discovered on an uncharted seamount. Science 227:57–59.

Logares R, Bråte J, Bertilsson S, Clasen JL, Shalchian-Tabrizi K, Rengefors K. 2009. Infrequent marine–freshwater transitions in the microbial world. Trends Microbiol 17:414–422.

Lovley DR, Chapelle FH. 1996. Technical comment. Hydrogen-based microbial ecosystems in the Earth. Science 272:896.

Mackenzie FT, Stoffym M, Wollast R. 1978. Aluminum in seawater: Control by biological activity. Science 199:680–682.

MacLeod RA. 1965. The question of the existence of specific marine bacteria. Bacteriol Rev 29:9–23.

Madsen EL, Bollag JM. 1989. Aerobic and anaerobic microbial activity in deep subsurface sediments from the Savannah River Plant. Geomicrobiol J 7:93–101.

Madsen EL, Stevens T, Lovley DR, Chapelle H, McKinley JP. 1996. Technical comment. Hydrogen-based microbial ecosystems in the Earth. Science 272:896–897.

Marine Chemistry. 1971. A Report to the Marine Chemistry Panel of the Committee of Oceanography. Washington, DC: National Academy of Sciences.

Marquis RE. 1982. Microbial barobiology. BioScience 32:267–271.

Martinez JS, Zhang GP, Holt PD, Jung H-T, Carrano CJ, Haygood MG, Butler A. 2000. Self-assembling amphiphilic siderophores from marine bacteria. Science 287:1245–1247.

Menzel DW, Rhyter JH. 1970. Distribution and cycling of organic matter in the ocean. In: Wood DE, ed. Organic Matter in Natural Waters. Institute of Marine Science Occasional Publ No. 1. Fairbanks, AK: University of Alaska, pp. 31–54.

McGillicuddy DJ, Robinson AR, Siegel DA, Jannasch HW, Johnson R, Dickey TD, McNeil J, Michaels AF, Knap AH. 1998. Influence of mesoscale eddies on new production in the Sargasso Sea. Nature 394:263–266.

Minshall GW. 1978. Autotrophy in stream ecosystems. BioScience 28:767–771.

Moore CM, Mills MM, Arrigo KR, Berman-Frank I, Bopp L, Boyd PW, Galbraith ED et al. 2013. Processes and patterns of oceanic nutrient limitation. Nat Geosci 6:701–710.

Morita RY. 1967. Effects of hydrostatic pressure on marine bacteria. Oceanogr Marine Biol Annu Rev 5:187–203.

Morita RY. 1975. Psychrophilic bacteria. Bacteriol Rev 39:144–167.

Morita RY. 1980. Microbial life in the deep sea. Can J Microbiol 26:1375–1385.

Newton RJ, Jones SE, Eiler A, McMahon KD, Bertilsson S. 2011. A guide to the natural history of freshwater lake bacteria. Microbiol Mol Biol Rev 75(1):14–49.

Nissenbaum A. 1979. Life in the Dead Sea—Fables, allegories, and scientific research. BioScience 29:153–157.

Overmann J, Cypionka H, Pfennig N. 1992. An extremely low light-adapted phototrophic sulfur bacterium from the Black Sea. Limnol Oceanogr 37:150–155.

Park PK. 1968. Seawater hydrogen-ion concentration: Vertical distribution. Science 162:357–358.

Paull CK, Hecker B, Commeau R, Freeman-Lynde RP, Neumann C, Corso WP, Golubic S, Hook JE, Sikes E, Curray J. 1984. Biological communities at the Florida Escarpment resemble hydrothermal vent taxa. Science 226:965–967.

Phelps TJ, Raione EG, White DC, Fliermans CB. 1989. Microbial activities in deep subsurface environments. Geomicrobiol J 7:79–91.

Pomeroy LR, Williams PJI, Azam F, Hobbie JE. 2007. The microbial loop. Oceanography 20(2):28.

Pope DH, Berger LR. 1973. Inhibition of metabolism by hydrostatic pressure: What limits microbial growth? *Arch Mikrobiol* 93:367–370.

Rappe MS, Connon SA, Vergin KL, Giovannoni SJ. 2002. Cultivation of the ubiquitous SAR11 marine bacterioplankton clade. *Science* 418:630–633.

Reid GK. 1961. *Ecology of Inland Waters and Estuaries*. New York: Reinhold.

Rheinheimer G. 1980. *Aquatic Microbiology*, 2nd edn. Chichester, U.K.: Wiley.

Richardson PL. 1993. Tracking ocean eddies. *Am Sci* 81:261–271.

Ring Group. 1981. Gulf-Stream cold-core rings: Their physics, chemistry, and biology. *Science* 212:1091–1100.

Rittenberg SC. 1940. Bacteriological analysis of some long cores of marine sediments. *J Mar Res* 3:191–201.

Russell BF, Phelps TJ, Griffin WT, Sargent KA. 1992. Procedures for sampling deep subsurface microbial communities in unconsolidated sediments. *Ground Water Monitor Rev* 12:96–104.

Saito MA, Noble AE, Tagliabue A, Goepfert TJ, Lamborg CH, Jenkins WJ. 2013. Slow-spreading submarine ridges in the South Atlantic as a significant oceanic iron source. *Nat Geosci* 6:775–779.

Santoro AE, Casciotti KL. 2011. Enrichment and characterization of ammonia-oxidizing archaea from the open ocean: Phylogeny, physiology and stable isotope fractionation. *ISME J* 5:1796–1808.

Sargent KA, Fliermans CB. 1989. Geology and hydrology of the deep subsurface microbiology sampling site at the Savannah River Plant, South Carolina. *Geomicrobiol J* 7:3–13.

Schippers A, Neretin LN, Kallmeyer J, Ferdelman TG, Cragg BA, John Parkes R, Jorgensen BB. 2005. Prokaryotic cells of the deep sub-seafloor biosphere identified as living bacteria. *Nature* 433:861–864.

Schwartz JR, Colwell RR. 1976. Microbial activities under deep-ocean conditions. *Dev Indust Microbiol* 17:299–310.

Seyfried WE Jr., Janecky DR. 1985. Heavy metal and sulfur transport during subcritical hydrothermal alteration of basalt: Influence of fluid pressure and basalt composition and crystallinity. *Geochim Cosmochim Acta* 49:2545–2560.

Shanks WC III, Bischoff JL, Rosenbauer RJ. 1981. Seawater sulfate reduction and sulfur isotope fractionation in basaltic systems: Interaction of seawater with fayalite and magnetite at 200–350°C. *Geochim Cosmochim Acta* 45:1977–1995.

Sieburth JMcN. 1975. *Microbial Seascape. A Pictorial Essay on Marine Microorganisms and Their Environments*. Baltimore, MD: University Park Press.

Sieburth JMcN. 1976. Bacterial substrates and productivity in marine ecosystems. *Annu Rev Ecol Syst* 7:259–285.

Sieburth JMcN. 1979. *Sea Microbes*. New York: Oxford University Press.

Sieburth JMcN, Willis P-J, Johnson KM, Burney CM, Lavoie DM, Hinga KR, Caron DA, French FW III, Johnson PW, Davis PG. 1976. Dissolved organic matter and heterotrophic microneuston in the surface microlayers of the North Atlantic. *Science* 194:1415–1418.

Sinclair JL, Ghiorse WC. 1987. Distribution of protozoa in subsurface sediments of a pristine groundwater study site in Oklahoma. *Appl Environ Microbiol* 53:1157–1163.

Sinclair JL, Ghiorse WC. 1989. Distribution of aerobic bacteria, protozoa, algae, and fungi in deep subsurface sediments. *Geomicrobiol J* 7:15–31.

Skinner BJ, Porter SC, Botkin DB. 1999. *The Blue Planet. An Introduction to Earth System Science*, 2nd edn. New York: Wiley.

Smith FGW, ed. 1974. *Handbook of Marine Science*, Vol. 1. Cleveland, OH: CRC Press, p. 617.

Smith KL, Teal JM. 1973. Deep-sea benthic community respiration: An in situ study at 1850 meters. *Science* 179:282–283.

Smith RL. 1968. Upwelling. *Oceanogr Marine Biol Annu Rev* 6:11–46.

Smith W, Pope D, Landau JV. 1975. Role of bacterial ribosome subunits in barotolerance. *J Bacteriol* 124:582–584.

Stanley SM. 1985. *Earth and Life through Time*. New York: WH Freeman.

Stein JL. 1984. Subtidal gastropods consume sulfur-oxidizing bacteria: Evidence from coastal hydrothermal vents. *Science* 223:696–698.

Stevens TO. 1997. Lithoautotrophy in the subsurface. *FEMS Microbiol Rev* 20:327–337.

Stevens TO, McKinley JP. 1995. Lithoautotrophic microbial ecosystems in deep basalt aquifers. *Science* 270:450–454.

Stevens TO, McKinley JP. 1996. Response to technical comment. Hydrogen-based microbial ecosystems in the Earth. *Science* 272:896–897.

Strahler AN, Strahler AH. 1974. *Introduction to Environmental Science*. Santa Barbara, CA: Hamilton.

Talley LD, Pickard GL, Emery WJ, Swift JH. 2011. *Descriptive Physical Oceanography: An Introduction*. Amsterdam, the Netherlands: Academic Press.

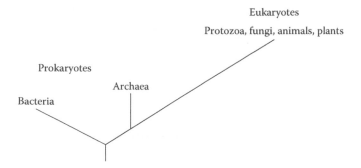

Figure 7.1. Phylogenetic relationships of the prokaryotes, domains of the bacteria and archaea, and the eukaryotes, based on the proposal by Woese et al. (1990).

acid (DNA). This structure is often called the bacterial chromosome, but unlike the chromosomes of eukaryotic cells, it does not contain structural protein such as histone, nor is it surrounded by a nuclear membrane. The molecular size of a prokaryotic chromosome measures on the order of 10^9 daltons (Da). Some genetic information in prokaryotes may also be located in one or more extrachromosomal circularized DNA molecules, called *plasmids*. The exact molecular size of different plasmids varies, depending on the amount of genetic information they carry, but generally ranges around 10^7 Da.

Another reason for classifying bacteria and archaea as prokaryotes is that they lack *mitochondria*, organelles that carry out respiration in eukaryotic cells, and *chloroplasts*, organelles that carry out photosynthesis in eukaryotic cells. In prokaryotes, respiratory activity is carried out in the plasma membrane, and in some gram-negative, anaerobic, and facultative bacteria, the activity may also involve the periplasm and outer membrane. Photosynthetic activity in members of the domain bacteria is carried out either by internal membranes derived from the plasma membrane (purple bacteria, cyanobacteria) or by special internal membranes (green bacteria). Similar photosynthetic activity is so far unknown in the domain archaea.

Archaea are not distinguishable from bacteria when seen as intact cells with a light microscope. At a submicroscopic level, however, archaea exhibit distinct differences from bacteria in the structure and composition of their cell envelope and plasma membrane and in the structure of their ribosomes, which are the sites of protein synthesis in bacteria. They also differ in key enzymes involved in nucleic acid and protein synthesis (Brock and Madigan, 1988; Atlas, 1997; Schaechter et al., 2006).

Eukaryotic microorganisms include algae, fungi, protozoa, and slime molds. They differ from prokaryotic microorganisms in possessing a true nucleus, which is an organelle enclosed in a double membrane in which the chromosomes, the bearers of genetic information, and the nucleolus, the center for ribonucleic acid synthesis, are located. Eukaryotic cells also feature mitochondria, chloroplasts, and vacuoles, all of which are membrane-bound organelles. The structure and mode of operation of their *flagella*, organs of locomotion, if present, also differ from those of the flagella of prokaryotic cells, when they have them. In eukaryotic cells, some key metabolic processes are highly compartmentalized, unlike those in prokaryotes. Figure 7.1 shows the phylogenetic interrelationship of the prokaryotic domains of bacteria and archaea and the Eukaryotes.

Examples of geomicrobially important archaea include *methanogens* (methane-forming bacteria), oxidizers of reduced forms of sulfur, extreme *halophiles*, and *thermoacidophiles*. Examples of geomicrobially important bacteria include some aerobic and anaerobic hydrogen-metabolizing bacteria, iron-oxidizing and iron-reducing bacteria, manganese-oxidizing and manganese-reducing bacteria, nitrifying and denitrifying bacteria, sulfate-reducing bacteria, sulfur-oxidizing and sulfur-reducing bacteria, anaerobic photosynthetic sulfur bacteria, and oxygen-producing cyanobacteria.

Examples of geomicrobially important eukaryotes include fungi that can attack silicate, carbonate, and phosphate minerals, among others. They are also important in initiating degradation of somewhat recalcitrant natural organic polymers such as lignin, cellulose, and chitin, as in the

O and A horizons of soil (see Chapter 4) or on and in surface sediments. Other geomicrobially important eukaryotes are algae, which together with cyanobacteria (prokaryotes) are a major source of oxygen in the atmosphere. Some algae promote calcium carbonate precipitation or dissolution, and others precipitate silica as frustules. Still other geomicrobially important eukaryotes include protozoa, some of which lay down silica, calcium carbonate, strontium sulfate, or manganese oxide tests, and others may accumulate preformed iron oxide on their cell surface.

7.2 GEOMICROBIALLY IMPORTANT PHYSIOLOGICAL GROUPS OF PROKARYOTES

Prokaryotes can be divided into various physiological groups such as chemolithoautotrophs, photolithoautotrophs, mixotrophs, photoheterotrophs, and heterotrophs (Figure 7.2). Each of these groups includes some geomicrobially important organisms. *Chemolithoautotrophs* (chemosynthetic autotrophs) include members of both the bacteria and the archaea. They are

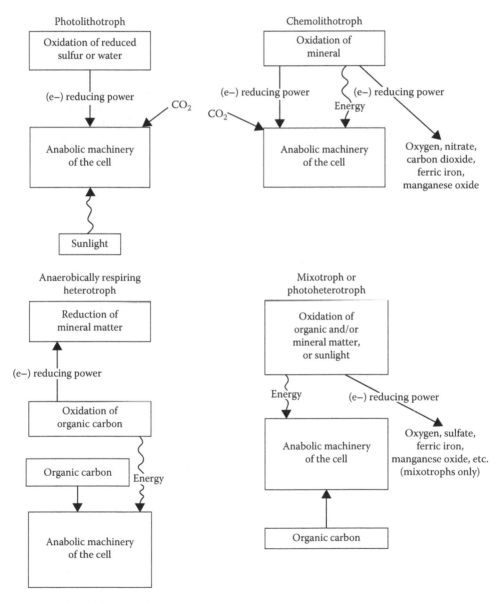

Figure 7.2. Geomicrobially important physiological groups among prokaryotes.

microorganisms that derive energy for doing metabolic work from the oxidation of inorganic compounds and that assimilate carbon as CO_2, HCO_3^-, or CO_3^{2-} (see, e.g., Wood, 1988). *Photolithoautotrophs* (photosynthetic autotrophs) include a variety of the bacteria, which use mostly forms of bacterial chlorophyll as their photosynthetic pigment, but include no known archaea. They are microorganisms that derive energy for doing metabolic work by converting radiant energy from the sun into chemical energy and use it, in part, in the assimilation of carbon as CO_2, HCO_3^-, or CO_3^{2-} as their carbon source (photosynthesis). Some of these microbes are anoxygenic (do not produce oxygen from photosynthesis), whereas others are oxygenic (produce oxygen from photosynthesis). *Mixotrophs* include some members of the bacteria and the archaea. They may derive energy simultaneously from the oxidation of reduced carbon compounds and oxidizable inorganic compounds, or they may derive their carbon simultaneously from organic carbon and CO_2, or they may derive their energy totally from the oxidation of an inorganic compound but their carbon from organic compounds. *Photoheterotrophs* include mostly bacteria but also a few archaea (extreme halophiles). They derive all or part of their energy from sunlight, but they derive their carbon by assimilating organic compounds. Unlike the bacteria, the archaea use proteorhodopsin in their cell membrane, in which they use as a hydrogen pump (Furutani et al., 2006; Ikeda et al., 2007), in conserving energy from sunlight (Béjà et al., 2000, Frigaard et al., 2006). Oxygen is not evolved in this process. *Heterotrophs* include members of both the bacteria and the archaea. They derive all their energy from the oxidation of organic compounds and most or all of their carbon from the assimilation of organic compounds. They may respire (oxidize their energy source) aerobically or anaerobically, or they may ferment their energy source by disproportionation (see the later discussion of catabolic reactions in Section 7.5).

7.3 ROLE OF MICROBES IN INORGANIC CONVERSIONS IN THE LITHOSPHERE AND HYDROSPHERE

A number of microbes in the biosphere can be considered to be *geologic agents*. They may serve as agents of concentration, dispersion, or fractionation of geologically important matter. As *agents of concentration*, they cause localized accumulation of inorganic matter by (1) depositing inorganic products of metabolism in or on special parts of the cell, (2) passive accumulations involving surface adsorption or ion exchange, or (3) promoting precipitation of insoluble compounds external to the cell (Ehrlich, 1999). An example of mineral accumulation of an inorganic metabolic product inside a cell is the deposition of polyphosphate (volutin) in the cytoplasm of bacteria such as *Spirillum volutans*, lactobacilli, and rhizobia. An example of metabolic product accumulation in the bacterial cell envelope is the deposition of elemental sulfur granules in the periplasm (region between the plasma membrane and outer membrane of gram-negative bacteria) by *Beggiatoa* and *Thiothrix* (Strohl et al., 1981; Smith and Strohl, 1991). An example of metabolic product accumulation at the cell surface is the formation of silica frustules by diatoms (algae), the frustules being their cell walls (de Vrind de Jong and de Vrind, 1997) (see also Chapter 11).

Examples of passive accumulation of inorganic matter by adsorption or ion exchange are the binding of specific metallic cations by carboxyl groups of peptidoglycan or phosphate groups of teichoic or teichuronic acids in the cell wall of gram-positive bacteria (e.g., *Bacillus subtilis*) or by the lipopolysaccharide phosphoryl groups of outer membranes of gram-negative bacteria (e.g., *Escherichia coli*). The bound cations may subsequently react with certain anions, such as sulfide, carbonate, or phosphate, and form insoluble salts that may serve as nuclei in the formation of corresponding minerals (Beveridge et al., 1983; Macaskie et al., 1987, 1992; Beveridge, 1989; Beveridge and Doyle, 1989; Doyle, 1989; Ferris, 1989; Geesey and Jang, 1989).

An example of extracellular inorganic accumulation is the precipitation of metal cations in the cellular surround (bulk phase) by sulfide produced in sulfate reduction by sulfate-reducing bacteria. Many such sulfides are very insoluble and fairly stable in the absence of oxygen (anoxic condition) (see Chapter 21).

As agents of dispersion, microbes promote dissolution of insoluble mineral matter, for example, in the dissolution of $CaCO_3$ by respiratory CO_2 (see Chapter 10) or in the biochemical reduction of insoluble ferric oxide or manganese(IV) oxide to corresponding soluble compounds (see Chapters 17 and 18).

As *agents of fractionation*, microbes may act on a mixture of insoluble inorganic compounds (minerals) by promoting selective mobilization involving one or a few compounds in the mixture. One example is the oxidation of arsenopyrite (FeAsS) in pyritic gold ore by *Acidithiobacillus* (formerly *Thiobacillus*) *ferrooxidans* (Ehrlich, 1964) (see also Chapter 15). In this process, some of the iron solubilized by oxidation reacts with arsenic simultaneously mobilized from the mineral to precipitate in the bulk phase as a new compound, ferric arsenate. Another example is the preferential solubilization by reduction of Mn(IV) over Fe(III) contained in ferromanganese nodules by bacteria (Ehrlich et al., 1973; Ehrlich, 2000) (see also Chapter 18).

Microbes may also cause fractionation by preferentially attacking the light isotope in a mixture of stable heavy and light isotopes of an element in a compound in preference to the heavier isotope(s). Examples are the reduction of $^{32}SO_4^{2-}$ in preference to $^{34}SO_4^{2-}$ by some sulfate-reducing bacteria and the assimilation of $^{12}CO_2$ in preference to $^{13}CO_2$ by some autotrophs, in either instance under conditions of slow growth (see discussion by Doetsch and Cook, 1973). Other isotope mixtures that may be fractionated by microbes include hydrogen/deuterium (Estep and Hoering, 1980), $^6Li/^7Li$ (Sakaguchi and Tomita, 2000), $^{14}N/^{15}N$ (Wada and Hattori, 1978), $^{16}O/^{18}O$ (Duplessy et al., 1981), $^{28}Si/^{30}Si$ (De La Rocha et al., 1997), and $^{54}Fe/^{56}Fe$ (Beard et al., 1999). In the laboratory, the magnitude of these fractionations may be relatively large and may involve significant changes in isotopic ratios in a relatively short time. In some natural settings, corresponding microbial isotope fractionations are also readily detectable but may be of somewhat smaller magnitude. Studies so far lead to the impression that only a few, mostly unrelated organisms have the capacity to fractionate stable isotope mixtures.

7.4 TYPES OF MICROBIAL ACTIVITIES INFLUENCING GEOLOGICAL PROCESSES

A number of geological processes at the Earth's surface and in the uppermost crust (deep subsurface) are under the influence of microbes. *Lithification* is a type of geological process in which microbes may produce the cementing substance that binds inorganic sedimentary particles together to form sedimentary rock. The microbially produced cementing substance may be calcium carbonate, iron or aluminum oxide, or silicate.

Some types of mineral formation may be the result of microbial activity. Iron sulfides such as pyrite (FeS_2), iron oxides such as magnetite (Fe_3O_4) or goethite (FeOOH), manganese oxides such as vernadite (MnO_2) or psilomelane (Ba, $Mn^{2+}Mn^{4+}O_{16}(OH)_4$), calcium carbonates such as calcite and aragonite ($CaCO_3$), and silica (SiO_2) may be generated *authigenically* by microbes (for a more extensive survey, see Lowenstamm, 1981).

In some instances, microbes may be responsible for mineral *diagenesis*, in which microbes may cause alteration of rock structure and transformation of primary into secondary minerals, as in the conversion of orthoclase to kaolinite (Chapter 4).

Rock weathering may be promoted by microbes through production and excretion of metabolic products, which attack the rock and cause solubilization or diagenesis of some mineral constituents of the rock. Rock weathering may also involve direct enzymatic attack by microbes of certain oxidizable or reducible rock minerals, thereby causing their solubilization or their diagenesis.

Microbes may contribute to *sediment accumulation* in the form of calcium carbonate tests like those from coccolithophores or foraminifera, silica frustules from diatoms, or silica tests from radiolaria or actinopods in oceans and lakes. The aging of lakes may be influenced by microbes through their rock weathering activity and/or their generation of organic debris from incomplete decomposition of organic matter (see Chapter 6).

Geological processes that are not influenced by microbes include *magmatic activity* or *volcanism*, *rock metamorphism* resulting from heat and pressure, *tectonic activity* related to crustal formation and transformation, and the allied processes of *orogeny* or mountain building. *Wind* and *water erosion* should also be included, although these processes may be facilitated by prior or concurrent microbial weathering activity. Even though microbes do not influence these geological processes, microbes may be influenced by them because these processes may create new environments that may be more or less favorable for microbial growth and activities than before their occurrence.

7.5 MICROBES AS CATALYSTS OF GEOCHEMICAL PROCESSES

Most of the influence that microbes exert on geological processes is physiological. They may act as *catalysts* in some geochemical processes, or they may act as producers or consumers of certain geochemically active substances and thereby influence the rate of a geochemical reaction in which such substances are reactants or products (see Ehrlich, 1996). In either case, the microbes act through their *metabolism*, which has two components. One of these components is *catabolism*, which provides the cell with needed energy through *energy conservation* and which may also yield to the cell some compounds that can serve as building blocks for polymers. A key reaction in energy conservation is the oxidation of a suitable nutrient or *metabolite* (a compound metabolically derived from a nutrient). The other component of metabolism is *anabolism*. It deals with assimilation (synthesis, polymerization) and leads to the formation of organic polymers such as nucleic acids, proteins, polysaccharides, and lipids. It also deals with the synthesis of "inorganic polymers" such as the polysilicates in diatom frustules and radiolarian tests and the polyphosphate granules that are formed by some bacteria and yeasts as energy storage compounds within their cells. Anabolism, by contributing to an increase in cellular mass and duplication of vital molecules, makes growth and reproduction possible. Catabolism and anabolism are linked to each other in that catabolism provides the energy and some or all of the building blocks that make anabolism, which overall is an energy-consuming process, possible. Both catabolism and anabolism may play a geomicrobial role. Catabolism is involved, for instance, in large-scale oxidation that brings about transformation of inorganic substances and degradation of organic molecules, whereas anabolism is involved, for instance, in the synthesis of organic compounds from which fossil fuels (peat, coal, and petroleum) are generated. Anabolism is also the process by which the diatom frustules and radiolarian tests that accumulate in siliceous oozes are formed.

Catabolism may take the form of aerobic or anaerobic respiration, both of which are oxidation processes, or fermentation. Catabolism may thus be carried on in the presence or absence of oxygen in air. The role of oxygen is that of terminal electron acceptor. Indeed, microorganisms can be grouped as *aerobes* (oxygen-requiring organisms), *anaerobes* (oxygen-shunning organisms), *microaerophilic organisms* (requiring low concentrations of oxygen), or *facultative organisms*, which can adapt their catabolism to operate in the presence or absence of oxygen in air. Facultative organisms use oxygen as terminal electron acceptor when it is available. When oxygen is not available, they use a reducible inorganic (e.g., nitrate or ferric iron) or an organic (e.g., fumarate) compound as a substitute terminal electron acceptor, or they ferment.

7.5.1 Catabolic Reactions: Aerobic Respiration

In *aerobic respiration*, hydrogen atoms or electrons are removed in the oxidation of organic compounds and electrons in the oxidation of inorganic entities by a variety of biochemical reactions and conveyed by an *electron transport system* (ETS) to oxygen to form water. Among these biochemical reactions, an important reaction sequence in which reducing power (hydrogen atoms, electrons) is generated as a part of aerobic respiration is the Krebs tricarboxylic acid cycle (Figure 7.3). By this reaction sequence, organic substances are completely oxidized to CO_2 and H_2O (Stryer, 1995; Schaechter et al., 2006). The reaction sequence is initiated when acetyl~SCoA, produced enzymatically in the oxidative degradation of a large variety of organic nutrients, is enzymatically combined with oxaloacetate to form citrate with the release of CoASH. The citrate is converted stepwise to isocitrate, α-ketoglutarate, succinate, fumarate, malate, and back to oxaloacetate. One turn of this cycle produces four hydrogen pairs and two CO_2 as well as one adenosine 5′-triphosphate (ATP) by *substrate-level phosphorylation*. The hydrogen pairs are the source of reducing power that is fed into the ETS and transported to oxygen as part of *aerobic respiration* to form water. In the transfer of the reducing power via the ETS, some of the energy that is liberated is conserved in special phosphate anhydride bonds of ATP by a chemiosmotic process called *oxidative phosphorylation* (see later). Upon hydrolysis, these bonds (Figure 7.4) yield 7.3 kcal mol^{-1} (30.5 kJ mol^{-1}) of free energy at pH 7.0 and 25°C (Stryer, 1995), as opposed to ordinary phosphate ester bonds, which release only about 2 kcal mol^{-1} (8.4 kJ mol^{-1}) of energy upon hydrolysis under these conditions. The energy in

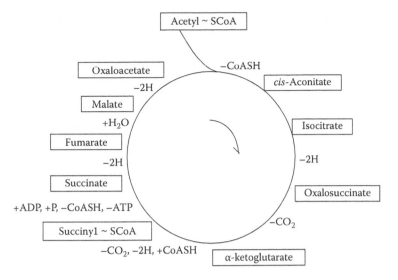

Figure 7.3. Krebs tricarboxylic acid cycle. One turn of the cycle converts one molecule of acetate to two molecules of CO_2 and four hydrogen pairs (2H), with formation of one molecule of ATP by substrate phosphorylation. An additional 11 ATP can be formed when the four hydrogen pairs are oxidized to H_2O with oxygen as terminal electron acceptor.

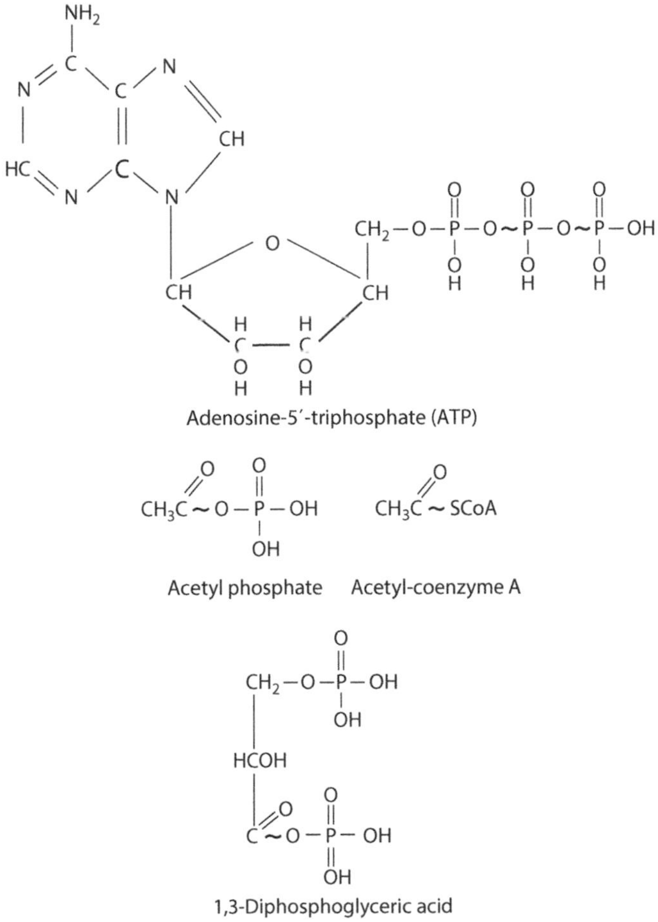

Adenosine-5′-triphosphate (ATP)

Acetyl phosphate   Acetyl-coenzyme A

1,3-Diphosphoglyceric acid

Figure 7.4. Examples of compounds containing one or more high-energy phosphate bonds (~).

consumed anaerobically is oxidized completely, the tricarboxylic acid cycle may be involved, but other pathways may be used instead. Among the best characterized of these anaerobic respiratory systems are those in which sulfate and nitrate are reduced as terminal electron acceptors.

7.5.3 Catabolic Reactions: Respiration Involving Insoluble Inorganic Substrates as Electron Donors or Acceptors

It is important to recognize that in prokaryotic cells the ETS is located in the plasma membrane (Figure 7.7). Sometimes, parts of it are located in the cell envelope (Figure 7.8a and b). By contrast, in eukaryotic cells, the ETS is located internally, in special organelles called mitochondria (Figure 7.9). As a result, those prokaryotes endowed with appropriate oxidoreductases (enzymes that

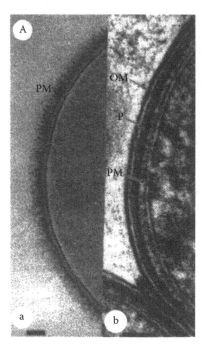

Figure 7.7. Location of electron transport system in typical prokaryotes. (a) Thin sections of the gram-positive cell wall of *Bacillus subtilis* and (b) the gram-negative cell wall of *Escherichia coli*. Both sections were prepared by freeze substitution. OM, outer membrane; PM, plasma membrane; P, periplasmic gel containing peptidoglycan located between the outer and plasma membranes. In both types of cells, the electron transport system is located in the plasma membrane. The bars in (a) and (b) equal 25 nm. (From Beveridge, TJ and Doyle, RJ: *Metal Ions and Bacteria.* 1989. Wiley, New York. Copyright Wiley-VCH Verlag GmbH & Co. KGaA. Reproduced with permission.)

transfer hydrogen atoms or electrons) in their cell surface are able to oxidize or reduce insoluble inorganic substrates when these are in physical contact with the cell surface (Figure 7.8a and b); in other words, these organisms can use *insoluble*, oxidizable, or reducible inorganic substrates (e.g., minerals) as electron donors or terminal electron acceptors, respectively, in their respiration by importing or exporting electrons, respectively. In some documented instances, bacteria such as *Geobacter sulfurreducens* and *Shewanella oneidensis* MR-1 have been shown to be able to facilitate electron transfer to the surface of Fe(III) oxides via special pili, filaments projecting from the cell surface (Reguera et al., 2005; Gorby et al., 2006), which eliminate the need for direct contact between the cells and the surface of an appropriate solid for electron transfer. Oxidoreductases in the cell envelope of some prokaryotes also enable these organisms to oxidize electron donors or acceptors without first taking them into the cytosol. This avoids any possible intracellular toxic effects from these dissolved electron donors and acceptors or from the intracellular accumulation of insoluble products resulting from the oxidation or reduction of these electron donors or acceptors that cannot be readily expelled from the cell. Examples of insoluble inorganic substrates that cannot enter a cell but can still serve as electron donors are elemental sulfur and iron sulfide (pyrite). Examples of inorganic compounds that cannot enter a cell but can nevertheless act as electron acceptors are iron(III) oxide and manganese(IV) oxide.

In gram-negative bacteria, enzymes and/or electron carriers in the periplasmic space of the cell envelope participate in the transfer of electrons in the appropriate direction between catalytic sites in the outer membrane and the ETS in the plasma membrane. The details of the mechanism of electron transfer in gram-positive bacteria and archaea that oxidize or reduce electron donors or acceptors at their cell surface remain to be elucidated (e.g., Ehrlich, 2008). Pham et al. (2008) found that the gram-positive *Brevibacillus* sp. strain PTH1 in culture was able to export electrons to the anode of a microbial fuel cell using acetate as electron donor in the presence of purified phenazine-1-carboxamide (PCN) from *Pseudomonas* sp. CMR12a and rhamnolipids as biosurfactants. The PCN appeared to serve as an electron shuttle. The *Brevibacillus* also appeared to be able to reduce goethite (FeOOH) under these conditions, but only

EHRLICH'S GEOMICROBIOLOGY

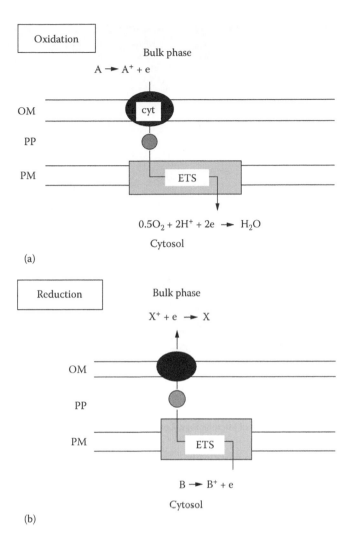

Figure 7.8. (a and b) Generalized diagrams of the electron transport system of gram-negative prokaryotes capable of oxidizing or reducing oxidizable or reducible constituents in insoluble minerals or dissolved electron donors or acceptors at their cell-surface/bulk phase interface by electron import or export, respectively. (a) Electron import (oxidation). (b) Electron export (reduction). OM, outer membrane; PP, periplasm; PM, plasma membrane; ETS, electron transport system in plasma membrane.

to a very limited extent. The ability of phenazines to act as an electron shuttle in electron export in gram-negative *Pseudomonas chlororaphis* PCL1391 and some other gram-negative bacteria was first shown by Hernandez et al. (2004).

Although most experimental evidence in support of electron transfer across the cell envelope has so far been gathered from studies of *gram-negative bacteria*, such as *S. oneidensis* MR-1 and *Geobacter* spp., which are able to respire anaerobically using ferric oxide or Mn(IV) oxide as terminal electron acceptors (e.g., Myers and Myers, 1992; Lovley, 2000; review by Ehrlich, 2002), evidence for the presence of c-type cytochrome Cyc2 in the outer membrane of iron-grown cells of *A. ferrooxidans* strain 33020 indicates that during Fe(II) oxidation by this organism, electrons are transferred from Fe(II) via this outer membrane cytochrome and rusticyanin and another cytochrome in the periplasmic space to the ETS in the plasma membrane (Appia-Ayme et al., 1999; Yarzábal et al., 2002a,b, 2004; see also Chapter 20). This electron transport mechanism probably operates as well when these organisms oxidize insoluble metal sulfide minerals like chalcocite (Cu$_2$S) and covellite (CuS) (see discussion in Chapter 21).

membrane and the inner mitochondrial membrane are impermeable to protons except at the sites where a protein complex, F_1F_0-ATP synthase/ATPase, is located. The F_1F_0-ATP synthase/ATPase, which has been shown to behave like a nanomotor (see discussion in Weber and Senior, 2003), is anchored in the plasma membrane and projects into the cytoplasm in bacteria and the matrix in mitochondria. It permits the reentry of protons into the cytosol of a prokaryotic cell or the matrix of the mitochondrion in a eukaryotic cell through a proton channel (Figure 7.5). It couples this proton reentry with ATP synthesis (Reaction 7.1). Proton reentry via F_1F_0-ATPase is facilitated in aerobes by the consumption of protons in the reduction of O_2 to water catalyzed by cytochrome oxidase on the inside of the plasma membrane or inner mitochondrial membrane. In anaerobically respiring bacteria, protons may be consumed in the reduction of an electron acceptor that replaces oxygen, the reduction being catalyzed by an enzyme other than cytochrome oxidase. The energy that drives Reaction 6.1 comes from the proton gradient and from the membrane potential

$$PMF = \Delta\psi - 2.3RT(\Delta pH/F) \qquad (7.2)$$

in which PMF is the proton motive force, $\Delta\psi$ is the transmembrane potential, ΔpH is the pH gradient across the membrane, R is the universal gas constant, T is the absolute temperature, and F is the Faraday constant. The overall process by which energy is generated and conserved in aerobic and anaerobic respiration is called chemiosmosis (see Hinkle and McCarty (1978), Stryer (1995), Weber and Senior (2003), and Schaechter et al. (2006) for further discussion of the process). As many as three molecules of ATP may be formed per electron pair transferred from donor to terminal electron acceptor in aerobic respiration and a probable maximum of two in anaerobic respiration.

Methanogens, which are anaerobic respirers, were once thought to conserve energy by a specialized energy-yielding process. But they have now been shown to conserve energy chemiosmotically, like other anaerobic respirers, in their catabolism of a limited range of substrates: acetate, CO_2 and H_2 or formate, and methanol, methylamines, and dimethylsulfide. Depending on the type of substrate, they employ different

versions of an electron transport chain to conserve energy. The charge-separation mechanism employed by methanogens for energy conservation (ATP synthesis) most frequently involves H^+ (Mountford, 1978; Butsch and Bachofen, 1984; Blaut and Gottschalk, 1984; Sprott et al., 1985; Müller et al., 1988; Ruppert et al., 1998; Kulkarni et al. 2009; Schlegel and Müller, 2011; Wang et al., 2011), but sometimes involves Na^+ instead as in some homoacetogens (Müller, 2003). A few hydrogen-utilizing methanogens have been shown to be able to use Fe(III) oxide, soluble anthraquinone-2,6-disulphonate, and humic acids as terminal electron acceptors (Bond and Lovley, 2002). The charge-separation mechanism used by methanogens for generating ATP by oxidative phosphorylation is probably illustrative of the earliest chemiosmotic mechanisms from which the more elaborate systems utilizing a variety of different membrane-bound electron carriers and enzymes found in modern aerobic and anaerobic respirers evolved.

In fermentation, useful energy is conserved by substrate-level phosphorylation, a process in which a high-energy bond, which traps some of the total energy released during oxidation, is formed on the substrate molecule (metabolite) that is being oxidized. An example is the oxidation of glyceraldehydes 3-phosphate to 1,3-diphosphoglycerate in glucose fermentation illustrated in Figure 7.11. Substrate-level phosphorylation may also occur during aerobic and anaerobic respiration, but it contributes only a small portion of the total energy conserved in high-energy bonds by cells. Clearly, aerobic and anaerobic respirations are much more efficient energy-yielding processes in a cell than fermentation. It takes less substrate to satisfy a fixed energy demand by a cell if a substrate is oxidized by aerobic or anaerobic respiration than by fermentation. If the energy-yielding substrate is organic, the greater efficiency may also result from the fact that respirers may oxidize a substrate completely to CO_2 and H_2O, whereas fermenters cannot.

Although many of the microbes that oxidize inorganic substrates to obtain energy are aerobes, a few are not. All autotrophically growing methanogens (domain archaea) oxidize hydrogen gas (H_2) by transferring electron from H_2 to CO_2 to form methane (CH_4), generating ATP by oxidative phosphorylation in the process. *Homoacetogens*

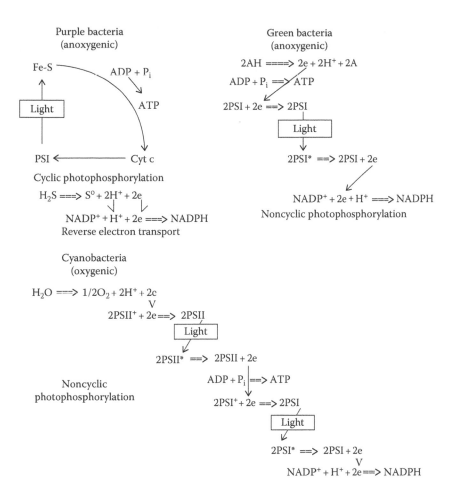

Figure 7.11. Diagrammatic representation of the mechanisms of photophosphorylation and generation of reducing power (NADPH) in purple and green photosynthetic bacteria and in cyanobacteria. PSI, photosystem I; PSII, photosystem II. (Adapted from Stanier, RY et al., *The Microbial World*, 5th edn., Prentice-Hall, Englewood Cliffs, NJ, 1986.)

(domain bacteria) carry out a similar reduction of CO_2 by hydrogen but form acetate instead of methane (Eden and Fuchs, 1983):

$$4H_2 + 2CO_2 \rightarrow CH_3COOH + 2H_2O$$
$$(\Delta G°, -25 \text{ kcal or } -104.8 \text{ kJ}) \qquad (7.3)$$

Some oxidizers of sulfur compounds can transfer electrons from a reduced sulfur substrate, such as thiosulfate or elemental sulfur, to nitrate in the absence of oxygen. In the presence of oxygen, these sulfur-oxidizing organisms transfer the electrons from the reduced sulfur compounds to oxygen. The maximum ATP yield in methane formation from H_2 reduction of CO_2 and in the oxidation of reduced sulfur by nitrate, two examples of anaerobic respiration, has not yet been established.

7.5.6 How Chemolithoautotrophic bacteria (Chemosynthetic Autotrophs) Generate Reducing Power for Assimilating CO_2 and Converting It to Organic Carbon

Unlike chemoheterotrophs, most chemolithoautotrophs when reducing CO_2 with NADPH have a special problem in generating NADPH. These chemolithoautotrophs, which possess an electron transport chain containing Fe–S proteins, quinones, and cytochromes in their plasma membrane whether they are aerobic or anaerobic respirers, depend on reverse electron transport to reduce $NADP^+$ to NADPH. In reverse electron transport, electrons must travel against a redox gradient with the expenditure of energy contained in high-energy bonds of ATP. The source of the electrons, which is also the source of

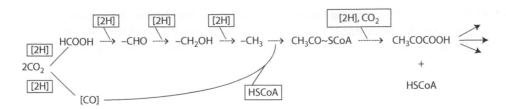

Figure 7.12. Pathway for carbon assimilation in methanogens (the activated acetate (CH₃CO~SCoA) pathway). Pyruvate (CH₃COCOOH) is a key intermediate for forming various building blocks for the cell, including sugars, amino acids, and fatty acids.

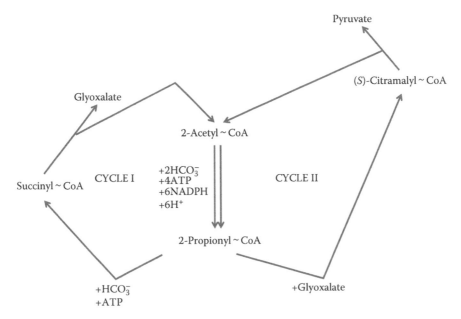

Figure 7.13. Hydroxypropionate cycles I and II by which CO_2 is assimilated by *Chloroflexus aurantiacus*. (Adapted from Zarzycki, J et al., *Proc Natl Acad Sci USA*, 106, 21317, 2009. By permission.)

1993; Eisenreich et al., 1993; Herter et al., 2001, 2002; Friedmann et al., 2006; Zarzycki et al., 2009), the most recent version of which is shown in Figure 7.13. In this version by Zarzycki et al. (2009), CO_2 as bicarbonate is incorporated into acetyl~SCoA, which is transformed into 2-propionyl~SCoA via 3-hydroxy-propionate, all with ATP consumption. Some of the 2-propionyl~SCoA is used by the cell to generate glyoxalate and regenerate acetyl~SCoA with further bicarbonate and ATP consumption. The glyoxalate is then combined with remaining 2-propionayl~SCoA and transformed into pyruvate and acetyl~SCoA. Pyruvate is the intermediate in the conversion of the assimilated CO_2 into cellular carbon.

Green sulfur bacteria of the genus *Chlorobium* fix CO_2 and reduce it by a reverse tricarboxylic acid cycle (Figure 7.14). In this process, CO_2 is combined with pyruvate in an ATP-consuming process to form oxalate, which is then converted via

malate, fumarate, and succinate to 2-ketoglutarate, the last step requiring consumption of ATP. The 2-ketoglutarate is a key precursor in amino acid synthesis as well as a precursor for citrate synthesis. Formation of citrate involves fixation of another CO_2. The citrate is cleaved to form oxaloacetate and acetate. The acetate serves as a precursor in the synthesis of pyruvate by CO_2 fixation and ATP consumption, thus completing the reverse tricarboxylic acid cycle. The pyruvate is a key precursor for the synthesis of other biochemical building blocks. NADH and NADPH needed in the operation of this cycle are generated by a noncyclic photoreduction mechanism (see previous). Although once thought unique to *Chlorobium*, the reverse tricarboxylic acid cycle has since been found to operate as a mechanism of CO_2 assimilation in other autotrophs, e.g., *Aquifex pyrophilus*, a chemolithoautotrophic, H_2-oxidizing

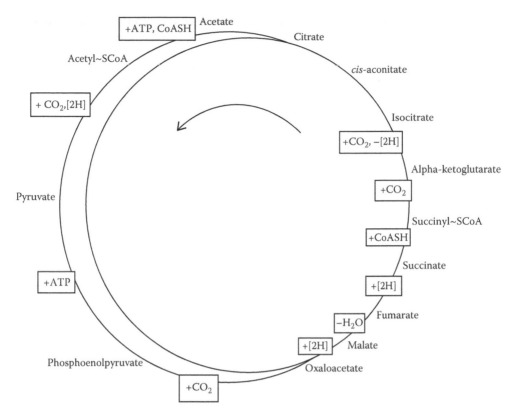

Figure 7.14. The reverse tricarboxylic acid cycle used by green sulfur bacteria for carbon assimilation. (Modified from Stanier, RY et al., *The Microbial World*, 5th edn., Prentice-Hall, Englewood Cliffs, NJ, 1986; scheme originally presented by Evans, MCW et al., *Proc Natl Acad Sci USA*, 55, 928, 1966.)

member of the domain bacteria, and *Thermoproteus neutrophilus*, a chemolithoautotrophic, thermophilic, H_2-oxidizing, and S^o-reducing archaeon (Beh et al., 1993). Most recently, the reverse tricarboxylic acid cycle has been reported in the ε-proteobacteria (e.g., *Thiomicrospira denitrificans*, *Candidatus Arcobacter sulfidicus*) (Hügler et al., 2005).

Oxygenic photolithoautotrophs fix CO_2 and reduce it to organic carbon by a reaction sequence similar to Reactions 7.9 through 7.11, also called the Calvin–Benson cycle. They produce NADPH for this process via noncyclic photophosphorylation and form ATP by both noncyclic and cyclic photophosphorylation.

7.5.9 Carbon Assimilation by Mixotrophs, Photoheterotrophs, and Heterotrophs

Because they fashion some monomeric building blocks by catabolism and acquire others preformed from the environment external to the cell, heterotrophs use much of the ATP, which they generate catabolically by respiration and/or fermentation, for polymerization reactions, as in the formation of proteins, polysaccharides, nucleotides and nucleic acids, lipids, and others. Mixotrophs and photoheterotrophs perform anabolic reactions that are similar to those performed by chemolithoautotrophs, photolithoautotrophs, and/or heterotrophs.

So far, nothing is known of detailed biochemical steps used by algae or protozoa in forming inorganic polymers such as polysilicate. Such polymerization is expected to involve the consumption of ATP or its equivalent.

7.6 MICROBIAL MINERALIZATION OF ORGANIC MATTER

Microbes play a major role in the transformation of organic matter in the upper lithosphere (soils, sediments, deep subsurface) and in the hydrosphere (oceans and bodies of freshwater). Because biological availability of carbon as well as of other nutritionally vital inorganic elements in the biosphere is limited, it is essential that these structural components of cells serving a particular function require somewhat different compositions and structures for stability and activity at different temperature intervals within the overall temperature range in which life exists. No organisms are known that are genetically endowed to produce respective components to cover all these different temperature intervals. The heat-stability range of the enzymes and critical cell structures, including cell membranes, of a microorganism reflect the temperature range in which it is able to grow. In other words, key molecules in organisms with different temperature requirements have different heat labilities (Brock, 1967; Tansey and Brok, 1972; Morgan-Kiss et al., 2006). Psychrophiles grow in a range from slightly below 0°C to about 20°C, with an optimum at 15°C or lower (Morita, 1975). Psychrotrophs grow over a wider temperature range than do psychrophiles (e.g., 0°C–30°C), with an optimum near 25°C. Mesophiles are microbes that grow in the range of 10°C–45°C, with an optimum range for some of about 25°C–30°C and for others of about 37°C–40°C. Thermophiles are microbes that live in a temperature range of 42°C–121°C, but the range for any given thermophile is considerably narrower. The temperature optimum for any one thermophilic organism depends on its identity and usually corresponds to the predominant temperature of its normal habitat. Extreme or hyperthermophiles, those growing optimally above 60°C, seem to be mostly archaea. Generally, thermophilic photosynthetic prokaryotes cannot grow at temperatures higher than 73°C. In contrast, thermophilic eukaryotic algae cannot grow at temperatures higher than 56°C (Brock, 1967, 1974, 1978). Thermophilic fungi generally exhibit temperature maxima around 60°C, and thermophilic protozoa, around 50°C. Only nonphotosynthetic, thermophilic prokaryotes exhibit temperature maxima that may be as high as 121°C (Kashefi and Lovley, 2003). For growth at temperatures at and above the boiling point of water, elevated hydrostatic pressure has to prevail to keep the water liquid. Liquid water is a requirement for life.

The parameters of pH and E_h also exert important influences on geomicrobial activity, as they do on biological activity in general. Each enzyme has its characteristic pH optimum, and E_h optimum in the case of redox enzymes, at which it catalyzes most efficiently. That is not to say that in a cell or, in the case of extracellular enzymes, outside a cell, an enzyme necessarily operates at its optimal pH and E_h. The interior of living cells tends to have a pH around neutrality and an E_h that may be lower or higher than its external environment. Enzymes with higher or lower pH optima will operate at less than optimal efficiency. This helps a cell to modulate and integrate individual enzyme reactions in a sequence so that no shortage or unneeded buildup of metabolic intermediates occurs in such a sequence. Changes in external pH that are within the physiological range of a microorganism do not affect its internal pH because of its plasma membrane barrier and its ability to control internal pH. Extreme changes will, however, have adverse effects.

Environmental pH and E_h control the range of distribution of microorganisms (see, however, Ehrlich (1993) for environmental significance of E_h), as recognized by Baas Becking et al. as long ago as 1960. As shown in Figure 7.15, Baas Becking et al. (1960) gave recognition to the prevalence of iron-oxidizing bacteria and, to some extent, thiobacteria in environments of relatively reduced potential and elevated pH. More recent studies have extended the environmental pH limits. For instance, iron-oxidizing Ferroplasma acidiphilum has been found to grow in a pH range of 1.3–2.2 (Golyshina et al., 2000) and iron-oxidizing Ferroplasma acidarmanus in a pH range of 0–2.5 (Edwards et al., 2000) (see also Chapter 17).

As mentioned in Chapter 6, hydrostatic pressure in excess of 400 atm (405 bar) at a fixed physiologically permissive temperature below the boiling point of water generally prevent the growth of nonbarophilic microbes. Pressure between 200 and 400 atm (203 and 405 bar) at such a temperature tends to interfere reversibly with cell division of bacteria (ZoBell and Oppenheimer, 1950). Barophilic organisms can grow at pressures above 400 atm (405 bar) at physiologically permissive temperatures. Facultative barophiles grow progressively more slowly with increasing pressure, whereas obligate barophiles grow best at or near the pressure and temperature of the native environment and grow progressively more slowly with decreasing pressure and usually not at all at atmospheric pressure at the same temperature (Yayanos et al., 1982). The growth-inhibiting effect of hydrostatic pressure is attributable to its effect on protein synthesis (Schwarz and Landau, 1972a,b; Pope et al., 1975; Smith et al., 1975). Many other biochemical reactions are much less pressure sensitive (Pope and Berger, 1973) (see also Chapter 6).

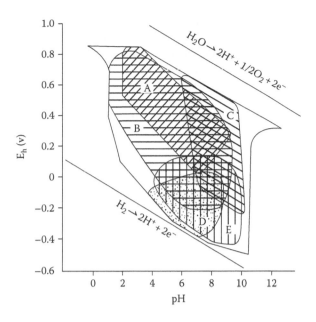

Figure 7.15. Environmental limits of E_h and pH for some bacteria. A, "iron bacteria"; B, thiobacteria; C, denitrifying bacteria; D, facultative and anaerobic heterotrophic bacteria and methanogens; E, sulfate-reducing bacteria. (Adapted from Baas Becking et al., *J Geol*, 68, 243, 1960, Copyright 1960, by permission from the University of Chicago Press.)

7.9 SUMMARY

Microbes may make a geologically significant contribution to lithification, mineral formation, mineral diagenesis, and sedimentation, but not to volcanism, tectonic activity, orogeny, or wind and water erosion. They may act as agents of concentration, dispersion, or fractionation of mineral matter. Their influence may be direct, through action of their enzymes, or indirect, through chemical action of their metabolic products, through passive concentration of insoluble substances on their cell surface, and through alteration of pH and E_h conditions in their environment. Their metabolic influence may involve anabolism or catabolism under aerobic and anaerobic conditions. Respiratory activity of prokaryotes may cause oxidation or reduction of certain inorganic compounds, resulting in their precipitation, often as minerals, or in their solubilization. Chemolithoautotrophic and some mixotrophic bacteria can obtain useful energy from the oxidation of some inorganic substances, such as H_2, Fe(II), Mn(II), H_2S, and $S°$. Photolithoautotrophic bacteria can use H_2S as a source of reducing power in the assimilation of CO_2 and, in the process, deposit sulfur. Anaerobically respiring organisms, which use any of various oxidized inorganic substances as terminal electron acceptors, are important in the mineralization of organic matter in environments devoid of atmospheric oxygen. Mineralization under aerobic conditions of organic matter by microbes in soil, freshwater, and marine environments leads to the formation of CO_2, H_2O, NO_3^-, SO_4^{2-}, PO_4^{3-}, and so on and under anaerobic conditions to the formation of CH_4, CO_2, NH_3, H_2S, PO_4^{3-}, and so on. Under some special conditions in the marine environment, CH_4 may be oxidized anaerobically to CO_2 and H_2O by collaboration of certain specific methanogens and sulfate reducers. Some organic matter is refractory to mineralization under anaerobic conditions and is microbially converted to humus. All microbial activities are greatly influenced by temperature, pH, and E_h conditions of an environment.

REFERENCES

Anderson DQ. 1940. Distribution of organic matter in marine sediments and its availability to further decomposition. *J Mar Res* 2:225–235.

Appia-Ayme C. Guilliani N, Ratouchniak J, Bonnefoy V. 1999. Characterization of an operon encoding two *c*-type cytochromes, an *aa*3-type cytochrome oxidase, and rusticyanin in *Thiobacillus ferrooxidans* ATCC 33020. *Appl Environ Microbiol* 65:4781–4787.

Jackson BE, McInerney MJ. 2000. Thiosulfate disproportionation by Desulfotomaculum thermobenzoicum. Appl Environ Microbiol 66:3650–3653.

Jackson TA. 1975. Humic matter in natural waters and sediments. Soil Sci 119:56–64.

Jannasch HW, Wirsen CO. 1973. Deep-sea microorganisms: In situ response to nutrient enrichment. Science 180:641–643.

Janssen PH, Schuhmann A, Bak F, Liesack W. 1996. Disproportionation of inorganic sulfur compounds by the sulfate-reducing bacterium Desulfocapsa thiozymogenes gen. nov., spec. nov. Arch Microbiol 166:184–192.

Kashefi K, Lovley DR. 2003. Extending the upper temperature limit for life. Science 301:934.

Kulkarni G, Kridelbaugh DM, Guss AM, Metcalf WW. 2009. Hydrogen is a preferred intermediate in the energy-conserving electron transport chain of Methanosarcina barkeri. Proc Natl Acad Sci USA 106:15915–15920.

Lovley DR. 2000. Fe(III) and Mn(IV) reduction. In: Lovley DR, ed., Environmental Microbe-Metal Interactions. Washington, DC: ASM Press, pp. 3–30.

Lowenstamm HA. 1981. Minerals formed by microorganisms. Science 211:1126–1131.

Macaskie LE, Dean ACR, Cheetham AK, Jakeman RJB, Skarnulis AJ. 1987. Cadmium accumulation by a Citrobacter sp.: The chemical nature of the accumulated metal precipitate and its location in the bacterial cells. J Gen Microbiol 133:539–544.

Macaskie LE, Empson RM, Cheetham AK, Grey CP, Skarnulis AJ. 1992. Uranium bioaccumulation by a Citrobacter sp. as a result of enzymatically mediated growth of polycrystalline HUO_2PO_4. Science 257:782–784.

Moore LR. 1969. Geomicrobiology and geomicrobial attack on sediment organic matter. In: Eglinton G, Murphy MTJ, eds., Organic Geochemistry: Methods and Results. New York: Springer-Verlag, pp. 264–303.

Morgan-Kiss RM, Priscu JC, Pocock T, Gudynaite-Savitch L, Huner NPA. 2006. Adaptation and acclimation of photosynthetic microorganisms to permanently cold environments. Microbiol Molec Biol Rev 70:222–252.

Morita RY. 1975. Psychrophilic bacteria. Bacteriol Rev 39:144–167.

Mountford DO. 1978. Evidence for ATP synthesis driven by a proton gradient in Methanobacterium barkeri. Biochem Biophys Res Commun 85:1346–1351.

Müller V. 2003. Energy conservation in acetogenic bacteria. Appl Env Microbiol 69:6345–6353.

Müller V, Winner C, Gottschalk G. 1988. Electon transport-driven sodium extrusion during methanogenesis from formaldehyde + H_2 by Methanosarcina barkeri. Eur J Biochem 178:519–525.

Myers CR, Myers JM. 1992. Localization of cytochromes to the outer membrane of anaerobically grown Shewanella putrefaciens MR-1. J Bacteriol 174:3429–3438.

Onishi T, Meinhardt SW, Yagi T, Oshima T. 1987. Comparative studies on the NADH-Q oxidoreductase segment of the bacterial respiratory chain. In: Kim CH, Teschi H, Diwan JJ, Salerno JC, eds., Advances in Membrane Biochemistry and Bioenergetics. New York: Plenum Press, pp. 237–248.

Palacas JG, Swanson VE, Moore GW. 1966. Organic geochemistry of three North Pacific deep sea sediment samples. US Geological Survey Professional Paper, Vol. 550C, pp. C102–C107.

Payne WE, Yang X, Trumpower BL. 1987. Biochemical and genetic approaches to elucidating the mechanism of respiration and energy transduction in Paracoccus denitrificans. In: Kim CH, Tedeschi H, Diwan JJ, Salerno JC, eds., Advances in Membrane Biochemistry and Bioenergetics. New York: Plenum Press, pp. 273–284.

Pezacka E, Wood HG. 1984. The synthesis of acetyl-CoA by Clostridium thermoaceticum from carbon dioxide, hydrogen, coenzyme A and methyltetrahydrofolate. Arch Microbiol 137:63–69.

Pham TH, Boon N, Aelterman P, Clauwaert P, De Schamphelaire L, Vanhaecke L, De Mayer K, Höfte M, Verstraete W, Rabaey K. 2008. Metabolites produced by Pseudomonas sp. enable a Gram-positive bacterium to achieve extracellular electron transfer. Appl Microbiol Biotechnol 77:1119–1129.

Pope DH, Berger LR. 1973. Inhibition of metabolism by hydrostatic pressure: What limits microbial growth? Arch Mikrobiol 93:367–370.

Pope DH, Smith WP, Swartz RW, Landau JV. 1975. Role of bacterial ribosomes in barotolerance. J Bacteriol 121:664–669.

Reguera G, McCarthy KD, Mehta T, Nicoll JS, Tuominen MT, Lovley DR. 2005. Extracellular electron transfer via microbial nanowires. Nature 435:1098–1101.

Ruppert C, Sönke W, Lemker T, Müller V. 1998. The A_1A_o ATPase from Methanosarcina mazei: Cloning of the 5′ end of the aha operon encoding the membrane domain and expression of the proteolipid in a membrane-bound form in Escherichia coli. J Bacteriol 180:3448–3452.

Sakaguchi T, Tomita O. 2000. Bioseparation of lithium isotopes by using microorganisms. Resource Environ Biotechnol 3:173–182.

Schaechter M, Ingraham JL, Neidhardt FC. 2006. *Microbe*. Washington, DC: ASM Press.

Schlegel K, Müller V. 2011. Sodium ion translocation and ATP synthesis in methanogens. *Methods in Enzymology* 494:233–255.

Schwarz JR, Landau JV. 1972a. Hydrostatic pressure effects on *Escherichia coli*: Site of inhibition of protein synthesis. *J Bacteriol* 109:945–948.

Schwarz JR, Landau JV. 1972b. Inhibition of cell-free protein synthesis by hydrostatic pressure. *J Bacteriol* 112:1222–1227.

Smith DW, Strohl WR. 1991. Sulfur oxidizing bacteria. In: Shively JM, Barton LL, eds., *Variations in Autotrophic Life*. London, U.K.: Academic Press, pp. 121–146.

Smith W, Pope D, Landau JV. 1975. Role of bacterial ribosome subunits in barotolerance. *J Bacteriol* 124:582–584.

Sprott GD, Bird SE, McDonald IJ. 1985. Proton motive force as a function of cell pH at which *Methanobacterium bryantii* is grown. *Can J Microbiol* 31:1031–1034.

Stanier RY, Ingraham JL, Wheelis ML, Painter PR. 1986. *The Microbial World*, 5th edn. Englewood Cliffs, NJ: Prentice-Hall.

Strauss G, Fuchs G. 1993. Enzymes of a novel autotrophic CO_2 fixation pathway in the phototrophic bacterium *Chloroflexus aurantiacus*, the 3-hydroxypropionate cycle. *Eur J Biochem* 215:633–643.

Strohl WR, Geffers I, Larkin JM. 1981. Structure of the sulfur inclusion envelopes from four Beggiatoas. *Curr Microbiol* 6:75–79.

Stryer L. 1995. *Biochemistry*, 4th edn. New York: WH Freeman.

Tansey MR, Brock TD. 1972. The upper temperature limit for eukaryotic organisms. *Proc Natl Acad Sci USA* 69:2426–2428.

Wada E, Hattori A. 1978. Nitrogen isotope effects in the assimilation of inorganic nitrogen compounds by marine diatoms. *Geomicrobiol J* 1:85–101.

Waksman SA. 1933. On the distribution of organic matter in the sea and chemical nature and origin of marine humus. *Soil Sci* 36:125–147.

Waksman SA, Hotchkiss M. 1937. On the oxidation of organic matter in marine sediments by bacteria. *J Mar Sci* 36:101–118.

Wang M, Tomb J-F, Ferry JG. 2011. Electron transport in acetate-grown *Methanosarcina acetivorans*. *BMC Microbiology* 11:165. http:/www.biomedcentral.com/1471-2180/11/165.

Weber J, Senior AE. 2003. ATP synthesis driven by proton transport in F_1F_0-ATP synthase. *FEBS Lett* 545:61–70.

Wirsen CO, Jannasch HW. 1975. Activity of marine psychrophilic bacteria at elevated hydrostatic pressure and low temperature. *Mar Biol* 31:201–208.

Woese CR, Kandler O, Wheelis ML. 1990. Towards a natural system of organisms: Proposal for the domains Archaea, Bacteria, and Eucarya. *Proc Natl Acad Sci USA* 87:4576–4579.

Wood P. 1988. Chemolithotrophy. In: Anthony C, ed., *Bacterial Energy Transduction*. London, U.K.: Academic Press, pp. 183–230.

Xavier KB, Da Costa MS, Santos H. 2000. Demonstration of a novel glycolytic pathway in the hyperthermophilic archaeon *Thermococcus zilligii* by ^{13}C-labeling experiments and nuclear magnetic resonance analysis. *J Bacteriol* 182:4632–4636.

Yarzábal A, Appia-Ayme C, Ratouchiak J, Bonnefoy V. 2004. Regulation of the expression of the *Acidithiobacillus ferrooxidans rus* operon encoding two cytochromes c, a cytochrome oxidase and rusticyanin. *Microbiology* 150:2113–2123.

Yarzábal A, Brasseur G, Bonnefoy V. 2002a. Cytochromes c of *Acidithiobacillus ferrooxidans*. *FEMS Microbiol Lett* 209:189–195.

Yarzábal A, Brasseur G, Ratouchniak J, Lund K, Lemesle-Meunier D, DeMoss JA, Bonnefoy V. 2002b. The high-molecular-weight cytochrome c Cyc2 of *Acidithiobacillus ferrooxidans* is an outer membrane protein. *J Bacteriol* 184:313–317.

Yayanos AA, Dietz AS, Van Boxtel R. 1982. Dependence of reproduction rate on pressure as a hallmark of deep-sea bacteria. *Appl Environ Microbiol* 44:1356–1361.

Zarzycki J, Brecht V, Müller M, Fuchs G. 2009. Identifying the missing steps of the autotrophic 3-hydroxypropionate CO_2 fixation cycle in *Chloroflexus aurantiacus*. *Proc Natl Acad Sci USA* 106:21317–21322.

ZoBell CE, Oppenheimer CH. 1950. Some effects of hydrostatic pressure on the multiplication and morphology of marine bacteria. *J Bacteriol* 60:771–781.

Cultivation, *In Situ* Measurements, and Geochemical Techniques for Geomicrobiological Studies

Greg Druschel and Victoria J. Orphan

CONTENTS

8.1   Introduction / 157
8.2   Considerations for Geomicrobiological Sampling / 158
    8.2.1   Terrestrial Surface/Subsurface Sampling / 159
    8.2.2   Aquatic Sampling / 159
    8.2.3   Sample Storage / 161
8.3   Detection, Isolation, and Identification of Geomicrobially Active Organisms / 163
8.4   In Situ Observation of Geomicrobial Agents / 163
8.5   In Situ Study of Ongoing Geomicrobial Activity / 166
8.6   Study of Solid Reaction Products of Geomicrobial Transformation / 169
8.7   Culture Isolation and Characterization of Active Agents from Environmental Samples / 171
8.8   Laboratory Reconstruction of Geomicrobial Processes in Nature / 171
8.9   Quantitative Study of Growth on Surfaces / 175
8.10 Test for Distinguishing between Enzymatic and Nonenzymatic Geomicrobial Activity / 177
8.11 In Situ Study of Past Geomicrobial Activity / 177
8.12 Summary / 178
References / 179

8.1 INTRODUCTION

Geomicrobial phenomena can be studied in the field (in *situ*) and in the laboratory (in *vitro*), in microcosms or as isolated reactions. Field study of a given geomicrobial phenomenon should ideally involve identification and enumeration of the active microorganisms and in *situ* measurements of their metabolic activities and growth rate. It should also involve chemical and physical identification of the substrates and nutrients, that is, the reactants (e.g., minerals [insoluble] and dissolved inorganic or organic substances) and the products that are formed in the processes surrounding metabolism and growth. Furthermore, it should involve measurement of the overall rate at which the process occurs and assessment of the impact of different environmental factors on it. In practice, however, it may happen that a suspected geomicrobial process is no longer operating at a given site but took place in the geologic past. In such instances, the role of the microorganisms in the process has to be reconstructed from microscopic observations (e.g., searching for biosignatures or microfossils associated with the starting materials, if still present, and especially the products of the process). It may also be reconstructed from geochemical observations such as biomarker (fingerprint) compounds in sedimentary rock, which indicate the past existence of an organism or a group of organisms that could have been the geochemical agents responsible. If applicable, evidence of isotopic fractionation of a key element relevant to the geomicrobial process should be sought.

In situ observations of an ongoing geomicrobial process should include the study of the setting in which it occurs in nature. In a terrestrial environment, the kinds of rocks, soil, or sediment, whichever are involved, and their constituent minerals ought to be characterized, and the prevailing temperature, pH, oxidation–reduction chemistry, sunlight intensity, seasonal cycles, and the source and availability of moisture, oxygen or other terminal electron acceptors, and nutrients ought to be identified. In an aqueous environment, water depth, availability of oxygen or other terminal electron acceptors, turbidity, light penetration, thermal stratification, pH, oxidation–reduction chemistry, chemical composition of the solutes in the water, nature of the sediment if part of the habitat, and nutrient sources and availability should be examined.

In the laboratory, a geochemical process can be studied and manipulated using microcosms. For this, a large sample of water, soil, sediment, or rock on or in which the process is occurring is collected. This sample is then transferred to a suitable vessel, which may be a flow-through chamber, a glass or plastic column, anaerobic bottles, or another kind of suitable vessel. Filter-sterilized water from the site at which a solid sample was collected or a synthetic media of a composition that approximates the nutrient supply available at the sampling site is added intermittently or continuously. The added nutrient solution should displace an equivalent volume of spent solution from the culture vessel. The experimental setup may be placed in the same environment from which the sample was taken, or it may be incubated in the laboratory at the temperature with the illumination and access to air, or lack of it, and humidity to which the sample was exposed at the sampling site. Measurement of the concentration of nutrients and products in the influent and effluent critical to the process under study will give a measure of the process rate. Solid products that are not recoverable in the effluent can be identified and measured in representative samples taken from the microcosm. Continuous or intermittent measurement of temperature, pH, oxidation–reduction chemistry, and oxygen availability in the microcosm will give information about any changes in these parameters, some of which may be a result of microbial activity in the microcosm.

The microcosm will probably contain a mixed population of bacteria, not all of which are likely to play a role in the geochemical process of interest. Manipulation of the microcosm through changes in nutrient supply, adjustment of pH or temperature, or a combination of these factors may cause selective increases of the organisms directly responsible for the geomicrobial process of interest and intensify the process. Because of enrichment biases, it is possible that a minor member of the indigenous population may grow to dominate the microcosm, and therefore, care must be taken before concluding that this member is the causative agent of the geomicrobial process in situ.

In vitro laboratory study of a geomicrobial process may be done by isolating the responsible microorganism(s) in pure culture, if possible, from a representative sample from the geomicrobially active site. The process originally observed in the field is then recreated with the isolate(s) in batch or continuous culture. Characterization of the process mechanism will involve qualitative and quantitative measurements of the biogeochemical transformation(s). It may include genetic and biochemical studies (see Chapter 9 for more details), where appropriate, as well as an assessment of environmental effects on the in vitro process. In vitro laboratory study may be important in lending support to field interpretation of geomicrobial processes that are occurring at present or have occurred in the past.

8.2 CONSIDERATIONS FOR GEOMICROBIOLOGICAL SAMPLING

Sampling of microbes or microbial biomarkers in the environment needs to be done with awareness of possible contamination, ideally coupled with sampling of minerals and water for chemical analysis of metabolites, nutrients, and other key physicochemical components of the geomicrobiological system under investigation. As natural systems vary in space and time, the timing and appropriate scale of the sampling must also be considered. Additionally, the investigator must also be aware that neither microbial activity nor geochemical reactions necessarily stop when a sample is collected and preservation of the sample is an additionally critical consideration in designing appropriate sampling procedures and may be further complicated considering atmospheric oxygen contamination.

Samples, whatever their nature, brought to the laboratory must be obtained under conditions as aseptic as possible, that is, with very little or preferably no contamination. Working surfaces of sampling tools should be thoroughly washed and alcohol flamed before use even if previously sterilized using an autoclave or UV sterilizer. Water, sediment, or mineral/rock samples should be collected in sterile containers, and if the collection is done manually, sterile surgical gloves should be worn and changed often. In the case of sampling solid materials, where the microorganisms of interest may be contained within sealed pore spaces, fractures, or fluid inclusions, careful sterilization of the exterior of a rock or core material that has been exposed to potential outside contamination is required. Tracers may be applied to outer surfaces as a control prior to processing for sampling inner areas. Contamination control assurances and assessment of the degree to which the sterilization methods are effective are necessary to firmly establish that the organisms present in the sample are not contamination artifacts (Weirich and Schweisfurth, 1985; Vreeland et al., 2000; Hebsgaard et al., 2005; Sankaranarayanan et al., 2011).

8.2.1 Terrestrial Surface/Subsurface Sampling

To sample the terrestrial subsurface down to 3000–4000 m or more, drilling methods that depend on the use of special drilling equipment have been devised, which causes minimal if any contamination during the collection of samples (cores) (Phelps and Russell, 1990; Pedersen, 1993; Griffin et al., 1997; Stroes-Gascoyne et al., 2007). One method uses modified wireline coring tools, with cores collected in Lexan- or polyvinyl chloride-lined barrels. The drill rig, rods, and tools are steam cleaned. The drilling fluid system includes a recirculation tank, the drilling fluid being chlorinated water. Tracers such as potassium bromide, the dye rhodamine, microbe-sized fluorescent beads, and perfluorocarbons are added to the drilling fluid and aid in determining to what extent, if any, the recovered core contaminated during drilling. The assessment is made by measuring the extent of tracer present in core samples (House et al., 2003; Onstott et al., 2003; Stroes-Gascoyne et al., 2007). The extent of bacterial contamination can be determined by quantitative enumeration of bacteria not expected to be part of the indigenous microbial assemblage (Beeman and Suflita, 1989). Analysis of the microbial diversity present in the core interior, exterior, and drill fluid using molecular techniques may also be used to assess potential contaminants (Onstott et al., 2003). If culturing of anaerobes as well as aerobes is sought from subsurface samples, the cores should be minimally exposed to air and/or processed in an oxygen-free atmosphere and stored/transported using anaerobic techniques that may include foil-lined sample bags and oxygen-consuming packs. Subsamples may then be tested for aerobes and anaerobes by appropriate culture techniques.

Soil and sediment samples from shallow depths may be collected manually with an auger or other coring device under aseptic conditions. Cores should be subdivided aseptically for sampling at different depths. If the cores cannot be obtained with a sterilized sampling device, they should be subsampled so as to obtain the least contaminated sample (e.g., the center portion of the core).

Sampling through, or within, ice is a necessary and technically challenging aspect of sampling in frozen settings once thought to be devoid of microbial life, but now known to offer many niches in which microorganisms have been shown to thrive (Gaidos et al., 2008; Murray et al., 2012; Christner et al., 2014). A critical part of this sampling is the sterility of operations to drill through ice and collect subglacial samples (see, e.g., Priscu et al., 2013), similar, in concept if not in practice, to concerns associated with geobiological sampling through rock strata as discussed earlier.

8.2.2 Aquatic Sampling

Water samples at any given depth below the surface, including deepwater samples in marine or freshwater bodies, may be obtained with specialized samplers designed to collect and isolate water samples from specific depth intervals or the use of a pump and tubing where the inlet position is controlled. Samplers include Van Dorn (horizontal tube position) and Niskin (vertical tube position) samplers, consisting of hollow plastic tubes with ends that can be triggered to seal at a specific depth by means of a brass messenger striking a trigger (or an electronic trigger) to seal the water sample for retrieval (Figure 8.1). Collecting particles from much larger water volumes (≥1000 L)

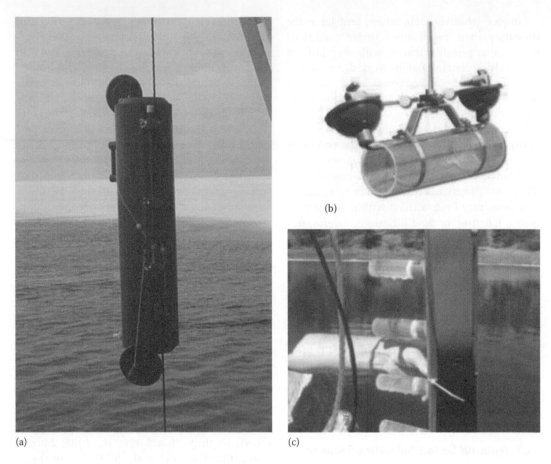

(a)

(b)

(c)

Figure 8.1. Examples of three water sampling devices. (a) Niskin bottle, note open end caps that are triggered to close using a messenger device at specified depths. (b) Van Dorn or horizontal water sampler with similar configuration compared to b with open end caps triggered with a messenger. (c) Syringe sampling device with syringe barrels that draw or equilibrate at specified depth intervals (example from Mahoney Lake sampling a very thin plate of purple sulfur bacteria). (Photo courtesy of Dr. Bill Gilhooly.)

can be achieved using tangential flow filtration (Giovannoni et al., 1990). This method is based on directing the flow of water tangentially across the filter rather than passing through it, and, when coupled with a submersible pump (e.g., McClain pump), large volumes of water can be processed in situ (e.g., Jones et al. 2008). Automated or manually manipulated syringe systems can also be used to sample water at finer spatial intervals, with sets of syringes at different positions set to draw samples by means of actuating the syringe plunger(s) at a specific time.

Aquatic sediment samples, including sediment porewater, may be obtained with dredging or coring devices. Dredging devices such as Ekman or Peterson (Clesceri et al., 1989, pp. 10–100) dredges can be used to collect large volumes of surface sediment. Ocean surface

sediment, rocks, concretions, and nodules may also be collected by dragging a bucket dredge or chain dredge over a desired area of the ocean floor. Such a sample will, however, consist of combined, mixed surface sediment encompassing the total surface area sampled. Coring should be used if different depths of sediment, porewater, and, in some cases, bottom water are needed. There are many different types of coring devices (e.g., box core, gravity core, piston core, multicore, vibracore, freeze core, and push core), and the type of equipment needed will depend on the scientific objective and research platform available. Researchers should consider factors such as the required sediment depth (10 cm to m), size of sample, preservation of the sediment–water interface, and precision of sampling at the seabed.

EHRLICH'S GEOMICROBIOLOGY

Hand cores can be collected by scuba divers from relatively shallow water (note that responsible scientific divers follow guidelines and standards established by the American Association of Underwater Scientists, http://www.aaus.org/) or using push coring devices deployed by manned or robotic submersibles operated from a research vessel. These cores (typically clear polycarbonate) enable precise sampling of target locations on the ocean floor or lake bed and collect surface sediment to a depth of ~30 cm while preserving the sediment–water interface. Core top water is also retained with push coring devices that can be later sampled for analysis. Syringe cores, porewater sampling wells, and long syringe needles have also been successfully used by divers in nearshore environments for collecting sediment and porewater samples (e.g., Fogaren et al., 2013). Freeze cores (utilizing aluminum tubing and a slurry of dry ice and ethanol inside to freeze sediment, porewater, and bottom water to the tubing) can also be used to collect even less disturbed samples than hand or push cores (Shapiro, 1958). Box cores are designed to collect a large sediment sample with minimum disturbance of the sediment–water interface to ~0.5 m penetration depth. The box corer is a large, typically square, box of rigid material that is lowered on a wire from the research vessel until it impacts and penetrates the sediment. After penetration, sediments are recovered by triggering a large plate to slide beneath the corer, sealing the bottom of the box before it is hoisted with a large mechanical winch onto the research vessel. Sediment samples deeper than the top meter cannot be sampled using box, freeze, or hand cores, but can be sampled using a multicore device, which extends ~1.5 m and also preserves the sediment–water interface (Figure 8.2). Gravity corers can also be used to collect deeper samples (typically 3–6 m) when preserving the sediment–water interface is not required (Crusius and Anderson, 1991). Gravity corers are long cylindrical tubes that are outfitted with a heavy lead weight and stabilizing fins at the top and a core catcher at the base to prevent sediment loss upon recovery. The core is deployed on a wire from a research vessel and gravity is used to penetrate the sediment. Alternatively, for very deep samples, piston coring can be used. Piston coring utilizes long sections of cylindrical piping that is driven deep into the sediment with a piston device and extracted

via a winch (Nesje, 1992). Once sediment cores are retrieved, samples are typically sectioned into spatially resolved sections and preserved for analysis. Alternatively, the core can be sliced lengthwise to reveal a cross section for analysis and subsampling. As with other sample types, care should be taken to minimize contamination.

Deepwater and sediment samples are also collected by use of manned submersibles or remotely operated vehicles, utilizing tools that can be manipulated by these vehicles (Luther et al., 2001b, 2008). Direct observation of sampling localities in the expanse of ocean floor largely untouched by human investigation to date is made possible by submersible vehicles and has driven many new and exciting discoveries in geomicrobiology (Orcutt et al., 2011). Sampling strategies utilizing these vehicles has some commonality to subsurface sampling described earlier, but much innovation has been necessary to both adapt to the high-pressure environments of the deep marine subsurface and mate these sampling strategies to the function of the submersibles. Additionally, there are deep subsurface boreholes in the ocean where new sampling procedures (e.g., subseafloor observatories or CORK's) have been developed to recover appropriate samples of the environments underneath the ocean floor (Wheat et al., 2010, 2011; Edwards et al., 2011).

8.2.3 Sample Storage

If samples cannot be examined immediately after collection, they should be stored so as to minimize microbial growth, loss of viability, and contamination. Cooling a sample is usually the best way to preserve it temporarily in its native state, but the extent of cooling may be critical. Freezing may be destructive to at least some archaea, bacteria, and microeukaryotes. However, icing may not prevent growth of psychrophiles or psychrotrophs. Vitrification of samples is possible and a powerful method to preserve both the microbial communities in natural samples and any associated nanoparticles for later analysis, including cryo-transmission electron microscopy (cryo-TEM) and cryo-scanning electron microscopy (cryo-SEM) (Comolli et al., 2012; Burrows and Penn, 2013). Fast freezing in cryogenic liquids is key to this technique, and the freezing must be fast enough that the water forms a glass instead of ice crystals, preserving microbial cells and

Figure 8.2. Examples of sediment coring devices. (a) Push core sampler (example of remotely operated vehicle Tiburon pushing a core tube into deep-sea sediment, retrieving sediment, and overlying water). (b) Gravity corer, note clear polycarbonate core barrel is filled half with sediment and half with overlying water (plunger at the top maintains seal to keep material in barrel until collection at surface). (c) Nesje percussion piston corer used to collect longer sediment cores (deeper sediment recovery) where core barrel is pounded into sediment with long rods and then brought to the surface. (d) Freeze core where a dry ice–ethanol slurry is used inside a plastic or metal sleeve to freeze material in place (example is sampling a cyanobacterial mat at Deer Lake, BC). (Photos courtesy of Dr. Victoria Orphan [a]; Dr. Broxton Bird [b and c]; and Dr. Greg Druschel [d].)

even the positions of nanoparticles associated with those cells. Preservation of geomicrobiological samples for molecular analysis is commonly accomplished through rapid freezing with dry ice or liquid nitrogen followed by storage at −80°C. If cryogenic agents are not available in the field, chemical stabilization agents can also be used (e.g., RNAlater, LifeTechnologies, Ussler et al., 2013; or Lifeguard [MOBIO], Desai et al., 2013; Lay et al., 2013). For preservation of samples for cell counts, fluorescence in situ hybridization (FISH), or electron microscopy (EM) samples should be fixed with a chemical agent (e.g., formaldehyde, paraformaldehyde, glutaraldehyde, or ethanol). The type of fixative, its concentration, and time of preservation is substrate and sample dependent and often needs to be tested to assess which preservation method will work best as well as to quantify any nontarget effects that may confound analyses. In general, samples preserved in paraformaldehyde at concentrations ranging from 2% to 4% work well for cell counting and FISH using epifluorescence microscopy. Glutaraldehyde can cause autofluorescence and is typically not suitable for FISH, but is the preferred fixative for preservation of cell ultrastructure if analyzed by TEM. The duration of storage before examination should not be longer than absolutely necessary, but fixed geomicrobiological samples for microscopy can be stored long term in −80°C sample storage freezers if transferred into ethanol to prevent freezing.

8.3 DETECTION, ISOLATION, AND IDENTIFICATION OF GEOMICROBIALLY ACTIVE ORGANISMS

A geomicrobial process may be the result of a single microbial species or an association of two or more microorganisms. Spatially well-structured interdependent associations between microorganisms are referred to as consortia (Overmann and Schubert, 2002). The basis for the association may be synergism, in which one type of organism is not capable of carrying out the complex process but in which each member of the consortium carries out part of the process in a sequential set of interactions (McInerney et al., 2009; Morris et al., 2013). It is also possible that not all members of a microbial association contribute directly to the observed geomicrobial process but instead carry out reactions that create environmental conditions relating, for instance, to pH or Eh that facilitate the process under consideration. Even if a geomicrobial process is the result of the action of a single microorganism, that organism rarely occurs in isolation in nature. While these co-associated microorganisms may not play a direct role in the geomicrobial process, they may influence the microorganisms involved in the geomicrobial process through competition for space and nutrients and through the production of metabolites that may have a stimulatory or inhibitory effect.

Two different classes of microorganisms may be recovered in a geomicrobial sample collected from the field: (1) indigenous organisms, who naturally reside in the habitat under investigation and should include the geomicrobially active organism(s), and (2) exogeneous organisms that were introduced into the habitat either by chance or through environmental manipulation (e.g., drilling) and that may or may not grow in the new environment but do survive in it. Distinctions among these groups are frequently difficult to make experimentally, but are critical to our interpretation and understanding of the geomicrobial processes occurring in situ.

A criterion for identifying indigenous organisms may be their frequency of occurrence in a given habitat and in similar habitats at different sites. Exogenous organisms may be identified by their inability to grow successfully in the in situ conditions of the habitat under study and their lower frequency of occurrence. Neither of these criteria is absolute, however, and identification of a contaminant may simply be based on the knowledge about the microorganism that would make its indigenous existence in the habitat under study unlikely. For example, the recovery of thermophilic anaerobes (e.g., Thermococcus sp., Methanothermobacter sp.) from deep high-temperature petroleum reservoirs with physiological properties that matched the in situ physicochemical conditions in the reservoir (temp ~80°C, anaerobic and sulfidic) was used as criterion for arguing an indigenous origin for these organisms relative to recovered 16S ribosomal RNA (rRNA) sequences from aerobic mesophilic bacteria (Pseudomonas sp.) that were likely contaminants associated with the well head (Stetter et al., 1993; Orphan et al., 2000). Another recent example includes the description of thermophilic endospores from sulfate-reducing Desulfotomaculum recovered from cold marine sediments—suggestive of the potential for widespread dispersal of subseafloor microorganisms originating from vents or petroleum reservoirs into nonnative habitats (Hubert et al., 2009; Muller et al., 2014)

8.4 IN SITU OBSERVATION OF GEOMICROBIAL AGENTS

Since the advent of the first homemade microscope by Antonie van Leeuwenhoek over 300 years ago, there have been exponential advances in microscopy that have enhanced our ability to

Pramer, 1966; Casida, 1971; Huber et al., 1985; Muyzer et al., 1987; Delong et al., 1989; Kepner and Pratt, 1994; Edwards et al., 1999).

Another approach is the use of SEM (LaRock and Ehrlich, 1975; Sieburth, 1975; Jannasch and Wirsen, 1981; Edwards et al., 1999) or thin-sectioned samples by TEM (Figure 8.3b), which can be used to examine extant microorganisms and their ultrastructure as well as identify putative fossilized microbes (Jannasch and Wirsen, 1981; Schopf, 1983; Jannasch and Mottl, 1985; Barker and Banfield, 1998; Wanner et al., 2008). Laser Raman spectroscopy can also be used for microbial visualization and chemical or isotopic characterization (see Wagner, 2009) and has also been used to generate 3D images of suspected microfossils in live position (Schopf and Kudryavtsev, 2005). Recently, synchrotron facilities have been used for collecting x-ray absorption near-edge structure (XANES) spectra of iron and carbon distribution within solid-phase marine and microbial samples (e.g., Prange et al., 2002, and Chan et al., 2011; more details on this are given in Section 8.6).

8.5 *IN SITU* STUDY OF ONGOING GEOMICROBIAL ACTIVITY

Microbes exist in a variety of potential states—growing, active, dormant, and deceased—geomicrobial "activity" is an assessment of the degree to which metabolism is affecting the levels of metabolites, nutrients, or cofactors in any environment (Blazewicz et al., 2013). Microbial activity can be estimated by characterizing and quantifying rRNA (assumed to be associated only with growing or active cells where ribosomal activity is tied to protein production [Kemp et al., 1993], but see Blazewicz et al., 2013, for a recent review of the topic), tracking the incorporation of a substrate (e.g., isotopic or fluorescently labeled) into cells (e.g., Wagner, 2009) or macromolecules (DNA, RNA, lipids, proteins; Radajewski et al., 2000) or by direct measurement of specific compounds associated with metabolism or growth (Luther et al., 2001b; Ferris et al., 2003; Jin and Bethke, 2005).

Ongoing geomicrobial activity may be quantifiable through measurement of metabolic compounds; these measurements are most accurate when done in situ or as quickly as possible because the highly reactive nature of these compounds that are always out of equilibrium (metabolism is only possible with redox compounds out of equilibrium

to gain energy) and often quite labile (Amend et al., 2003; Diaz et al., 2013). Methods for measurement of these compounds include radiolabeling of substrates or metabolites, electrochemical and spectroscopic measurements that can be done in the field or in laboratory culture in situ or immediately on sampling, or sampling with preservation of key compounds through separation, freezing, or derivatization for later analysis. Some compounds may be involved in more than one kind of reaction, making interpretation of rates of change observed more complicated as there can be multiple sources and sinks associated with specific compounds (Jorgensen, 2006; Boyd and Druschel, 2013).

Geomicrobial activity may be followed by the use of radioisotopes, a technique that can be powerful when chemical analysis alone is either not sensitive enough or not possible on relevant timescales, to investigate metabolic pathways. As isotopic fractionation occurs during metabolic transformations, partitions can be measured by separation of the substrate and metabolite followed by analysis using liquid scintillation, a technique that is extremely sensitive and permits analysis of chemical changes too small to be determined by other methods (Kallmeyer et al., 2004; Jorgenson, 2006). One advantage of using radioisotopes in the quantitative assessment of a specific geomicrobial transformation in nature is that their detection is extremely sensitive so that only minute amounts of radiolabeled substrate, which do not significantly change the naturally occurring concentration of the substrate, need to be added. Another advantage is that in cases where the rate of transformation is very slow although the natural substrate concentration is high, spiking the reaction with radiolabeled substrate allows analysis after a relatively brief reaction time because of the sensitivity of radioisotope detection.

For example, bacterial sulfate-reducing activity may be determined by adding a small quantity of $Na^{35}SO_4$ to water, soil, or sediment sample of known sulfate content in a closed vessel. After incubation under in situ conditions, the sample is analyzed for loss of $^{35}SO_4^{2-}$ and buildup of $^{35}S^{2-}$ by separating these two entities and measuring their quantity in terms of their radioactivity. In the case of a water or sediment sample, incubation of the reaction mixture in a closed vessel in the water column or sediment may be conducted at the depth from which the sample was taken, or within temperature-controlled pressure vessels.

An example of a direct application of this method includes Elsgaard and Jorgensen (1992) and Greef et al. (1998). It allows the estimation of the rate of sulfate reduction in the sample without having any knowledge of the number of physiologically active organisms present in it. A modified method is that of Sand et al. (1975). Their method allows an estimation of the sulfate-reducing activity in terms of the number of physiologically active bacteria in the sample as distinct from an estimation 'of the sum of physiologically active and inactive bacteria. The assay for the estimation of active bacteria can be set up either to measure percentage of sulfate reduced in a fixed amount of time that is proportional to the logarithm of the concentration of active cells or to measure the length of time required to reduce a fixed amount of sulfate to sulfide, which is related to the concentration of physiologically active sulfate-reducing bacteria in the sample.

Electrochemical probes are particularly well suited to determination of geomicrobial activity because they can be used to measure a number of key redox substrates and metabolites with fine spatial and temporal resolution and can be done in the presence of both minerals and microbes in situ and essentially in real time. Three major types of electrochemical analysis have proven very useful in the study of geomicrobial activity—potentiometry (where the potential of an analyte is determined in the absence of a current), amperometry (where current is measured at a constant potential), and voltammetry (where current is measured as a function of varied potential) (see Taillefert and Rozan, 2002; Moore et al., 2009, for a more detailed review of many common techniques and applications). The term microelectrode is used for many of these electrochemical sensors, a term that corresponds, at least contextually, to both the size of the physical sensor and how it minimally perturbs an environment it is inserted in and to some of the details of how the electrode surface itself behaves with respect to diffusion of analyte molecules to the sensing surface (Buffle and Tercier-Waeber, 2000). There is not general agreement on what size the term microelectrode really refers to, and Buffle and Tercier-Waeber (2000) suggests true microelectrodes (whose response would be independent of hydrodynamic conditions like stirring) correspond to sensors less than 10 μm, whereas macroelectrodes are typically greater than 100 μm, while Revsbech and Jorgensen (1986) suggest sensors at less than

200 μm can be called microsensors. That electrode response affected by hydrodynamic conditions that can vary aspects of the diffusion layer is an important consideration in the application of electrodes to geomicrobiological systems; this can be minimized by the use of smaller electrodes and faster measurement times, but as in any analytical technique, this possible cause of analytical variability should be considered and assessed in application of these devices in natural settings.

Potentiometric electrodes include pH electrodes, oxidation–reduction potential (ORP) electrodes, P_{CO_2} electrodes, Ag/AgS electrodes, and ion-specific electrodes. ORP electrodes can be used to investigate a number of redox couples via application of the Nernst equation and a measured electropotential, but caution must be exercised that not all species react at the Pt electrode surface. Not all reactions are reversible at this surface, and since redox chemistry is never fully at equilibrium in a geomicrobiological system, the hypothesis that one potential can fully describe any geobiological system is false (Lindberg and Runnels, 1984). Many sensors (pH, ORP, P_{CO_2}, Ag/AgS, NH^{4+}, NO^{3-}, or voltammetric sensors) have also been developed as microsensors that can be utilized even in underwater landers for analysis of waters and sediment porewaters (de Beer, 2002; Tercier-Waeber et al., 2005; Luther et al., 2008).

Amperometric electrodes include Clark-type oxygen electrodes and H_2S amperometric electrodes (Revsbech et al., 1980; Jeroschewski et al., 1996). These electrodes typically incorporate a membrane or other material for selection of a specific set of analytes; Clark-type O_2 sensors, for example, incorporate an O_2-permeable membrane that can be made of silicone, Teflon, polyethylene, or a double membrane of collodion and polystyrene (Revsbech et al., 1980; Revsbech, 1989; Taillefert et al., 2000). Amperometric sensors have been produced with sensing tips as thin as a few microns to minimize disturbance of the environment in which they carry out their measurements (Revsbech, 1989; de Beer and Kuhl, 2000). Switchable trace oxygen sensors are a relatively recent development to make a more sensitive O_2 sensor by the addition of a front guard cathode to the traditional Clark-type amperometric electrode, lowering effective detective limits a 1000-fold, to a few nM (Revsbech et al., 2011).

Voltammetric electrodes are different from other types of electrodes generally used in that they are capable of measuring multiple analytes, through manipulation of the potential (set relative to a reference electrode) between a working electrode and a counter electrode and measurement of current at each potential step (Buffle and Tercier-Waeber, 2000; Luther et al., 2008). Voltammetric working electrodes can be made of a wide range of materials that determine the working window in which the potential can be varied in a fluid and affect what analytes are *electroactive* at that working electrode surface via how reactions occur between the electrode material and specific analytes. Gold–amalgam (Au–Hg) electrodes are commonly used in natural waters and other geomicrobiological systems including sediment porewaters, open ocean water columns, deep-sea hydrothermal vents and the deep subsurface biosphere, and biofilms to measure a number of key O, Fe, Mn, and S geomicrobial species (Brendel and Luther, 1995; Glazer et al., 2002, 2006; Druschel et al., 2008, Cowen et al. 2012).

To fully appreciate *in situ* geomicrobial activity, knowledge of chemical, pH, and redox gradients over relatively narrow depth intervals (e.g., millimeters or less) is important. The application of specific microelectrodes and other techniques has made determination of redox gradients with micron to millimeter resolution in geobiologically active settings possible, as the following examples, a few among many, show.

Au–Hg voltammetric microelectrodes made it possible to measure simultaneously, and with a spatial resolution in millimeters, the vertical 3D distribution of O_2, Mn^{2+}, Fe^{2+}, HS^-, and I^- in porewater of undisturbed sediment cores from the Canadian continental shelf and slope and in the sediment surrounding an actively irrigated worm burrow in a mesocosm (Luther et al., 1998). Solid-state gold amalgam voltammetric microelectrodes have also made the monitoring of sulfur speciation *in situ* in sediments, microbial mats, and hydrothermal vent waters possible (Luther et al., 2001a) and the determination of limits governing neutrophilic iron-oxidizing bacterial use of Fe^{2+} (Druschel et al., 2008; Krepski et al., 2013). Use of a fiber-optic scalar irradiance microsensor and oxygen microelectrode spaced 120 μm apart made it possible to measure scalar irradiance and oxygenic photosynthesis with 100 μm spatial resolution in marine microbial mats dominated by cyanobacteria with a surface layer of pennate diatoms on sandy sediment along the coast of Limfjorden, Denmark (Lassen et al., 1992). Optodes combine chemiluminescent

reactions with microscopy to image oxygen gradient in biofilms (Staal et al., 2011) or aquatic sediments (König et al., 2005). Microelectrodes have also been employed in measuring O_2, H_2S, and pH microgradients at depth increments of 50 μm in Beggiatoa mats from marine sediments and *Thiovulum* films above the sediment to determine the microbial response to O_2 and H_2S (Jørgensen and Revsbech, 1983). Freshly prepared nitrate-selective microelectrodes with a liquid ion exchanger have been employed in determining nitrate gradients in sediments from a mesotrophic lake in the Netherlands (De Beer and Swearts, 1989).

Once redox gradient concentrations are defined, these profiles can be utilized to identify spatially resolved geochemical niches that may support specific organisms (Revsbech et al., 1983; Luther et al., 1998, 2001a) and to determine the flux of redox species produced or consumed by different geobiological processes (Froehlich et al., 1979; Berg et al., 1998; DeFlandre and Duchene, 2010). Water columns and sediment porewaters often exhibit a predictable sequence of redox species observable with electrodes, defining oxic, suboxic, and anoxic regions where steep, opposing gradients of key redox species control the energy available for microbial metabolisms (Figure 8.5)

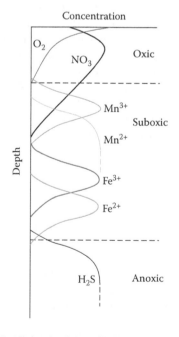

Figure 8.5. Idealized redox profile for sediment biogeochemistry with Mn(III) and Fe(III) major parts of the cycles. (From Madison, AS et al., *Science*, 341, 875, 2013.)

(Froehlich et al., 1979; Revsbech and Jorgensen, 1986; Madison et al., 2013). The discovery of key components of Mn cycling, for example, was determined by application of voltammetric electrodes to identify areas where Mn(III) is a significant intermediate species (Trouwborst et al., 2006). Combining a series of microbial and geochemical approaches with this spatial data has redefined our view of the sedimentary redox system and the reactions involving abiotic and biotic Mn reactions to recognize the critical importance of Mn(III) (Tebo et al., 2006; Madison et al., 2013). Once spatial information for redox species is obtained and if advection can be ignored or quantified, the production/consumption rates can be estimated using diffusion calculations (assuming steady-state conditions govern the observed spatial patterns) (Froehlich et al., 1979; Berg et al., 1998; DeFlandre and Duchene, 2010).

8.6 STUDY OF SOLID REACTION PRODUCTS OF GEOMICROBIAL TRANSFORMATION

Studies of geomicrobial phenomena require ingenuity in the application of standard microbiological, chemical, and physical techniques and often require collaboration among microbiologists, biochemists, geochemists, mineralogists, and other specialists to unravel a problem. Ideally, the products of geomicrobial transformation, if they are precipitates, should be studied not only with respect to chemical composition but also with respect to mineralogical properties through one or more of the following techniques: SEM or high-resolution transmission electron microscopy (HRTEM, including energy-dispersive x-ray measurements), microprobe examination (wavelength-specific x-ray spectroscopy), x-ray photoelectron spectroscopy, infrared spectroscopy (including attenuated total reflectance geometry), Raman spectroscopy, x-ray diffraction, small-angle x-ray scattering, scanning transmission x-ray microscopy (STXM), x-ray photoelectron emission spectromicroscopy (X-PEEM), extended x-ray absorption fine structure (EXAFS), and XANES spectroscopy, secondary ion mass spectrometry (SIMS), and stereoscopic or polarizing light microscopy. Ideally, the suite of techniques characterizes the mineral products in terms of the chemistry, structure, size, morphology, and surface properties of the solid material in question, which is often nanometers in size. Similar studies should ideally be undertaken on the substrate if it is an insoluble mineral complex—which can enable the detection of mineralogical changes occurring as a result of biological activity or temporal changes occurring during geomicrobial transformation.

For example, a key biomineral in many environments is a suite of iron (hydro)oxide minerals, associated often with the oxidation of iron(II) at near-neutral conditions. The distinct morphology of several organisms associated with this metabolism led to the early identification of these minerals as distinctly biological in origin (such as those associated with the stalk-forming *Gallionella* spp. and members of the Zetaproteobacteria), but much has been learned about the nature of this mineralization through application of modern microscopic and spectroscopic tools. These minerals are associated with extracellular polymeric material (Ghiorse, 1984, Chan et al., 2004), and imaging of these minerals with a number of techniques, including HRTEM, X-PEEM, STXM, and XANES, has illustrated both incredible detail of this association and newly discovered details on the structure of these minerals as affected by these materials (Figure 8.6; Chan et al., 2004, 2011). HRTEM and STXM imaging provides a detailed image of the minerals, including information about the internal structure of the atoms in these minerals that help identify specific mineral forms such as akaganeite (HRTEM) and the detailed chemistry of both the minerals and the associated EPS materials (STXM and X-PEEM combined with micro-X-ray fluorescence). Many of these nondestructive analytical methods can be used in concert to provide additional information about microorganisms and associated minerals. For example, in a study of neutrophilic iron-oxidizing bacteria, Chan et al. (2011) used a combination of STXM for high-resolution elemental mapping of the distribution of C, N, O, and Fe in the cell and its mineralized stalk combined with NEXAFS (near-edge X-ray absorption fine structure) spectroscopy to more specifically identify classes of biomolecules and oxidation state of the stalk-associated iron.

SIMS can also be used to characterize the isotopic and elemental composition of minerals, microbial cells, and select captured metabolites.

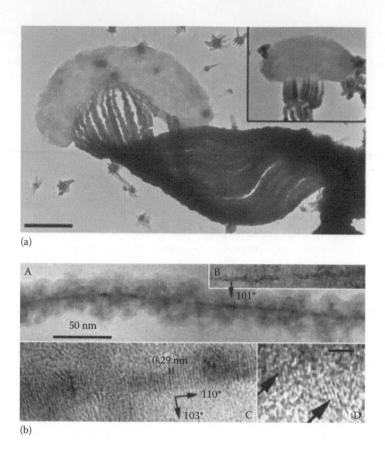

(a)

(b)

Figure 8.6. Transmission electron microscopy (TEM) micrographs of iron oxyhydroxide biominerals associated with neutrophilic iron-oxidizing microorganisms. (a) TEM images of *Mariprofundus ferrooxydans* cell attached to stalk, which is composed of individual filaments. Inset: smaller cell and stalk, displayed at the same scale showing that smaller cells produce narrower stalks with fewer filaments. Scale bar = 500 nm. (b) High-resolution transmission electron microscopy (HRTEM) images of natural FeOOH-mineralized filaments from the biofilm. (A) A filament showing overall structure of thin akaganeite core surrounded by amorphous and finely crystalline iron oxyhydroxide. (B) Akaganeite at the filament core. The 0.75 nm (101) lattice fringes are parallel to the filament length. (C) (−331) zone axis HRTEM image of akaganeite showing slight misorientations between nanocrystal segments. (D) Ferrihydrite nanoparticles (diameter ~2 nm) surrounding the akaganeite core. Scale bar, 1.8 nm. (From [a] Chan, CS et al., *ISME J*, 5, 717; [b] Chan, CS et al., *Science*, 303, 1656, 2004.)

SIMS is based on secondary ion detection after ejection from the surface of the sample using a focused primary beam of ions (Ireland, 1995). These highly specialized instruments enable the quantification of stable isotope (e.g., ^{13}C, ^{15}N, ^{34}S, and D/H) and elemental variations in small samples—biological and geological—with newer generation instruments (e.g., CAMECA nanoSIMS 50L) capable of simultaneous detection of several target masses and ion mapping of spatial distributions at micrometer to submicrometer resolution. Examples of applications of SIMS and nano-SIMS include measurements of isotope ratios in single cells and microfossils recovered directly from the environment (House et al., 2000; Orphan et al., 2001; Lepot et al., 2013) or after sample incubation with an isotopically labeled substrate (e.g., ^{15}N or ^{13}C labeled) (see reviews by Orphan and House, 2009, and Wagner, 2009) (Figure 8.7). SIMS analyses of microscale patterns in isotopic variation between putative biogenic minerals and metabolites (e.g., sulfides) have also been used to detect geobiological activity of microorganisms in both extant and paleoenvironments (Fike et al., 2009; Bontognali et al., 2012; Lever et al., 2013). For additional discussion of the application of SIMS in combination with molecular detection techniques for microorganisms (e.g., FISH), see Chapter 9.

EHRLICH'S GEOMICROBIOLOGY

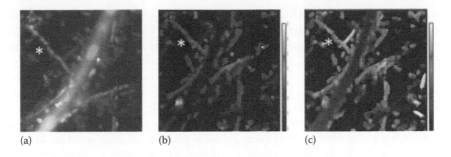

(a)   (b)   (c)

Figure 8.7. **(See color insert.)** Stable isotope probing and single-cell activity measurements of environmental microorganisms using fluorescence *in situ* hybridization–nano-secondary ion mass spectrometry (nano-SIMS). Environmental sample of white microbial mat collected from a sulfidic intertidal area after incubation with ^{13}C-labeled acetate and ^{15}N-labeled ammonium. (a) Fluorescence *in situ* hybridization image of putative sulfur-oxidizing gammaproteobacterial filaments, in green, and other bacteria (in blue and red). (b) Corresponding nano-SIMS image of ^{13}C/^{12}C ratios, showing active uptake of ^{13}C-acetate into segmented gammaproteobacterial filaments (color scale for ^{13}C/^{12}C ratios ranges from natural abundance values, 0.0112–0.635). (c) Secondary ion image of ^{15}N/^{14}N ratios of the same field of view demonstrating active microbial assimilation of ^{15}N-ammonium during the incubation (scale between 0.0036 and 0.120). Asterisks denote the same gammaproteobacterial filament across all three panels, raster image 20 μm × 20 μm. (Images courtesy of Kat Dawson and V Orphan.)

8.7 CULTURE ISOLATION AND CHARACTERIZATION OF ACTIVE AGENTS FROM ENVIRONMENTAL SAMPLES

To study microorganisms or consortia that are active in a geomicrobial process of interest, enrichment and pure culture isolations should be attempted as far as possible and isolated organisms should be independently tested for their ability to perform, or contribute to, the geomicrobial activity. Many geomicrobial processes are the collective result of the activity of a community or microbial consortium. Pure culture isolates may have the potential to carry out or participate in the specific geomicrobial activity of interest, but may require the involvement of other microorganisms and various combinations of isolates should be tested under a variety of coculture conditions. For example, many strains of cyanobacteria show enhanced growth in coculture with heterotrophic bacteria (Morris et al., 2008). Examples of geomicrobial metabolic cooperation among microorganisms include (1) manganese oxidation by the bacterium *Metallogenium symbioticum* in association with the fungus *Coniothyrium carpaticum* (Zavarzin, 1961; Dubinina, 1970, but see also Schweisfurth, 1969), (2) the bacteria *Corynebacterium* sp. and *Chromobacterium* sp. (Bromfield and Skerman, 1950), and (3) two strains of *Pseudomonas* (Zavarzin, 1962). Other examples include the process of sulfate-coupled anaerobic oxidation of methane that was shown to be catalyzed by a consortium consisting

of a methanotrophic archaeon (related to methanogens) and sulfur-metabolizing bacteria (e.g., Boetius et al., 2000; Orphan et al., 2001; Michaelis et al., 2002) (for more details, see Chapter 23).

Enrichment of and isolation from a mixed culture require selective conditions. Subsequently, it's estimated that less than 1% of microbes found in nature have been successfully cultivated. If a microbial agent with a specific geomicrobial attribute is sought, the selective culture medium should have ingredients incorporated that favor the geomicrobial activity of interest. Apart from special nutrients, pH, salinity, redox, temperature, and specific substrates' and electron acceptors' conditions may also have to be chosen to favor selective growth of the geomicrobial agents.

Isolation and characterization of pure cultures from enrichments should follow standard bacteriological technique, including determination of the physiology and molecular phylogeny of the isolates (for details, see, for instance, Gerhardt et al., 1981, 1993; Hurst et al., 1997; and Skerman, 1967).

8.8 LABORATORY RECONSTRUCTION OF GEOMICROBIAL PROCESSES IN NATURE

It is often important to reconstruct a naturally occurring geomicrobial process in the laboratory to investigate the mechanism whereby the process operates. Laboratory reconstruction can permit optimization of a process through the application of more favorable conditions than in nature. Examples are the use of a pure culture or a purified consortium to eliminate interference by competing microorganisms and the optimization of substrate availability, temperature, pH, Eh, and oxygen and carbon dioxide supply.

The activity of organisms growing on the surface of solid substrates such as soils, sediments, rocks, and ore may be investigated in batch culture, in air-lift columns, in percolation columns, or in a chemostat. A batch culture represents a closed system in which an experiment is started with a finite amount of substrate that is continually depleted during the growth of the culture. Cell population and metabolic product buildup and changes in pH and Eh are likely to occur. Conditions within the culture are thus continually changing and becoming progressively less like the starting conditions that were most favorable for the targeted organism or process of interest. Batch experiments may be least representative of a natural process, which usually occurs in an open system with continual or intermittent replenishment of substrate and removal of at least some of the metabolic wastes. A culture in an air-lift column (Barker and Wargen, 1981; Kumar and Das, 2012; Figure 8.8) is a partially open system, where the microbes grow and carry on their biogeochemical activity on a mineral charge in the column. They are continually fed with recirculated nutrient solution from which nutrients are depleted by the organisms. The recirculation removes metabolic products from the solid substrate charge in the column. Percolation columns (Figure 8.9) and downflow hanging sponge reactors (Figure 8.10) are even more open systems than air-lift columns. In them, microbes grow on the solid substrate charge in the columns, but they are fed with nutrient solution that is not recirculated. This fresh nutrient solution is added continually or at intervals, and wastes are removed at the same time in the effluent without recirculation, while pH, Eh, and temperature are held constant or nearly so. The retention of microbial biomass in anaerobic downflow hanging sponge reactors has been recently shown to be an effective method for enriching novel and slow-growing

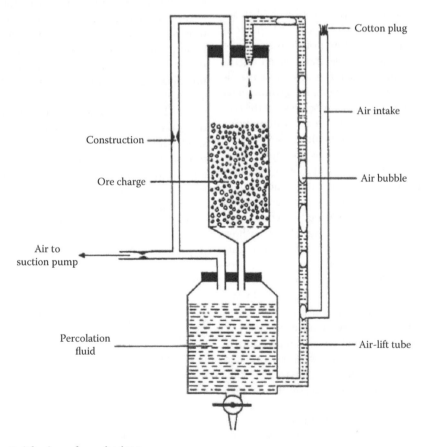

Figure 8.8. Air-lift column for ore leaching.

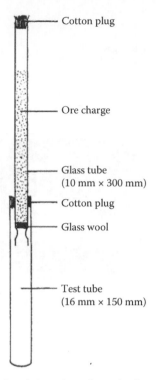

Figure 8.9. Percolation column for ore leaching.

microorganisms from environmental samples including deep subseafloor sediments (Imachi et al, 2011).

Steady-state conditions such as in a chemostat idealize the open culture system. They do not imitate nature because conditions are too constant. In open systems in nature, some fluctuation in various environmental parameters occurs over time. The chemostat is a liquid culture system of constant volume. The continuous introduction of fresh nutrient solution at a constant rate keeps the actively growing cell population and the concentration of accumulating metabolic products constant in the culture vessel through medium displacement to maintain constant volume. In other words, steady-state conditions are maintained. This can be expressed mathematically as

$$\frac{dx}{dt} = \mu x - Dx \qquad (8.1)$$

where

    dx/dt is the rate of cell population change in the chemostat

    μ is the instantaneous growth rate constant

    D is the dilution rate

    x is the cell concentration or cell number in the chemostat

The dilution rate is defined as the flow rate (f) of the influent feed or the effluent waste divided by the liquid volume (V) of the culture in the chemostat. Under steady-state conditions, $dx/dt = 0$, and therefore, $\mu x = Dx$ (i.e., instantaneous growth rate equals dilution rate, $\mu = D$). Under conditions where $D > \mu$, the cell population in the chemostat will decrease over time and may ultimately be washed out. Conversely, if $\mu > D$, the cell population in the chemostat will increase until a new steady state is reached, which is determined by the growth-limiting concentration of an essential substrate.

The steady state in the chemostat can also be expressed in terms of the rate change of growth-limiting substrate concentration (ds/dt). This is based on the principle that the rate of change in substrate concentration is dependent on the rate of substrate addition to the chemostat, the rate of washout from the chemostat, and the rate of substrate consumption by the growing organism:

$$\frac{ds}{dt} = D(S_{\text{inflow}} - S_{\text{outflow}}) - \mu(S_{\text{inflow}} - S_{\text{outflow}}) \qquad (8.2)$$

where

    D is the dilution rate

    S_{inflow} is the substrate concentration entering the chemostat

    S_{outflow} is the concentration of unconsumed substrate

    μ is the instantaneous growth rate constant

At steady state, $ds/dt = 0$. The substrate consumed, that is, $S_{\text{inflow}} - S_{\text{outflow}}$, is related to the cell mass produced (x) according to the relationship

$$x = y(S_{\text{inflow}} - S_{\text{outflow}}) \qquad (8.3)$$

where y is the growth yield constant (mass of cells produced per mass of substrate consumed). These relationships require modification if a solid substrate is included in the chemostat (see, for instance, Section 8.6).

The chemostat can be used, for example, to determine limiting substrate concentrations for the growth of bacteria under simulated natural conditions. Thus, the limiting concentrations of lactate, glycerol, and glucose required for growth at different relative growth rates (D/μm) of *Achromobacter aquamarinus* (strain 208) and *Spirillum lunatum* (strain 102) in seawater have been determined by this

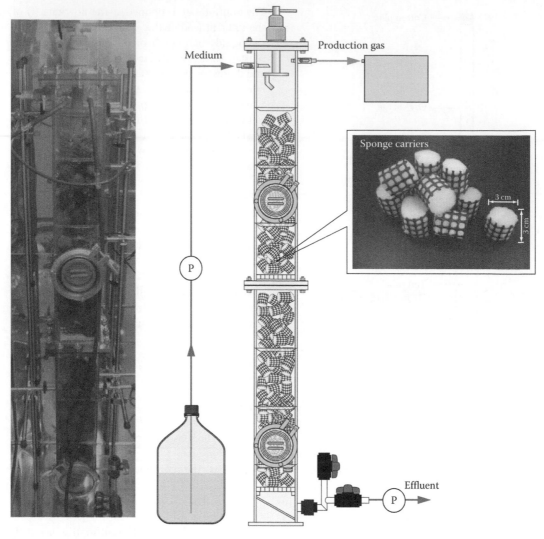

Figure 8.10. Photograph and schematic diagram of the down-flow hanging sponge (DHS) bioreactor. The DHS reactor was originally developed for the treatment of municipal wastewater. A distinctive feature of the DHS reactor is the use of polyurethane sponges that provide increased surface area for microbial colonization and extended residence time within the sponge pore space. In this system, the sponge carriers are not submerged in liquid; rather, the influent medium slowly drips down onto the sponges by gravity, permitting exchange of medium throughout the interior of the sponge.

method (Jannasch, 1967). The chemostat principle can also be applied to a study of growth rates of microbes in their natural environment by laboratory simulation (Jannasch, 1969) or directly in the natural habitat. For instance, the size of the algal population in an algal mat of a hot spring in Yellowstone National Park, Montana, was found to be relatively constant, implying that the algal growth rate equaled its washout rate from the spring pool. When a portion of the algal mat was darkened by blocking access of sunlight, thereby stopping photosynthesis and thus algal growth and multiplication in that part of the mat, the algal cells were washed out from it at a constant rate after a short lag (Figure 8.11). The washout rate under these conditions equaled the growth rate in the illuminated part of the mat. This follows from Equation 8.2 when $dx/dt = 0$ (Brock and Brock, 1968). Similarly, the population of the sulfur-oxidizing thermophile *Sulfolobus acidocaldarius* has

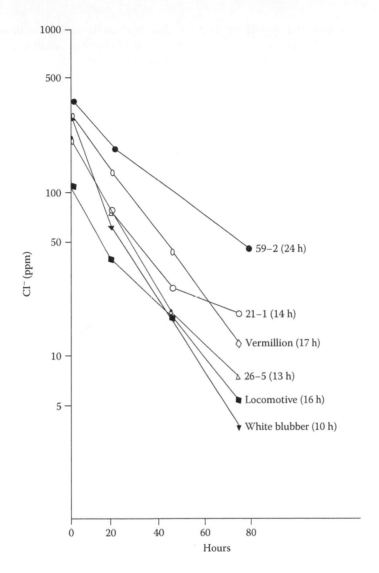

Figure 8.11. Chloride dilution in several small springs in Yellowstone National Park, Montana. Estimated halftimes for chloride dilution are given in parentheses. At site 21–1, chloride concentration had reached the natural background level by the final sampling time, and the dilution rate was estimated from the data from the first three sampling times. (From Mosser, JL et al., *J Bacteriol*, 118, 1075, 1974. With permission.)

been found to be in steady state in hot springs in Yellowstone National Park, implying, as in the case of the algae in the mat, that the growth rate of the organism equals its washout rate from the spring. In this instance, the washout rate was measured by following the water turnover rate in terms of dilution rate of a small measured amount of NaCl added to the spring pool (Figure 8.11). The dilution rate was then translated into the growth rate of *S. acidocaldarius* in the spring (Mosser et al., 1974).

8.9 QUANTITATIVE STUDY OF GROWTH ON SURFACES

The chemostat or its principle of operation is not applicable to all culture situations. In the study of geomicrobial phenomena, the central microbial activity often occurs on the surface of inorganic or organic solids. Indeed, the solid may be the growth-limiting substrate upon which the organism acts. Under these conditions, it may be assumed that the microbial population as it colonizes the surface will increase geometrically once it has settled on it, approximating the relationship

$$\log N = \log N_0 + \frac{\Delta t}{g} \log 2 \qquad (8.4)$$

where
  N is the final cell concentration per unit area
  N_0 is the initial cell concentration per unit area
  Δt is the time required for N_0 to multiply to N cells
  g is the average doubling time (generation time)

It is also assumed that the cell multiplication rate is significantly slower than the rate of attachment of cells in the bulk phase to the solid surface. Once all available space on the surface has been occupied, however, the cell population on it will remain constant (provided that the surface area does not decrease significantly because of solid substrate consumption or dissolution). The cell population in the liquid in contact with the solid will then show an arithmetic increase in cell numbers according to the relationship

$$N_{final,liquid} = N_{initial,liquid} + z N_{solid} \qquad (8.5)$$

where
  z is the number of cell doublings on the solid
  N_{solid} is the N cells on the solid when all attachment sites are occupied

Equation 8.6 states that for every cell doubling on the surface of the cell-saturated solid, one daughter cell will be displaced into the liquid medium for lack of space on the solid. This model assumes that the bulk phase, liquid medium by itself, cannot support growth of the organism. As long as there is no significant change in the surface area of the solid phase, the model applies. An example to which the model can be applied is the growth of autotrophic thiobacteria on water-insoluble metal sulfide, which serves as their sole energy source in a mineral salt solution that satisfies all other nutrient requirements (Chapter 21).

To introduce the concept of time into Equation 8.6, the following relationship should be considered:

$$g = \frac{\Delta t}{z} \qquad (8.6)$$

where
  g is the doubling time
  Δt is the time interval between $N_{initial,liquid}$ and $N_{final,liquid}$ determinations

If Equations 8.6 and 8.7 are combined, we get

$$N_{final,liquid} = N_{initial,liquid} + \frac{\Delta t}{g} N_{solid} \qquad (8.7)$$

when solving for g, we get

$$g = \frac{\Delta t N_{solid}}{N_{final,liquid} - N_{initial,liquid}} \qquad (8.8)$$

Espejo and Romero (1987) took a different mathematical approach to bacterial growth on a solid surface. They developed the following relationship to describe growth before surface saturation was reached:

$$\frac{dN_a}{dt} = \frac{\mu N_a (N_s - N_a)}{N_s} \qquad (8.9)$$

where
  N_a is the number of attached bacteria
  N_s is the limit value of bacteria that can attach to the surface under consideration
  μ is the specific growth rate

After surface saturation by the bacteria, they propose the relationship

$$\frac{dN_f}{dt} = \mu N_s \qquad (8.10)$$

where N_f represents the number of free (unattached) bacteria in the liquid phase with which the surface is in contact. From Equation 8.11, the following relationship for the specific growth rate of the culture after surface saturation is derived:

$$\mu = \frac{\Delta N_f}{\Delta t N_s} \qquad (8.11)$$

These relationships were tested by the researchers in a real experiment growing *Acidithiobacillus ferrooxidans* on elemental sulfur in which the sulfur was the only available energy source for the bacterium. They observed logarithmic population increase before surface saturation and a linear increase thereafter. They concluded, however, that the value for N_s was not constant in their experiment but changed gradually after linear growth had begun.

Another mathematical model was presented by Konishi et al. (1994), in which the change in particle size of the solid substrate with time as well as cell adsorption to the solid substrate was taken into account.

8.10 TEST FOR DISTINGUISHING BETWEEN ENZYMATIC AND NONENZYMATIC GEOMICROBIAL ACTIVITY

Observed changes in geochemistry and/or mineralogy in the presence of microorganisms does not *a priori* indicate if the reaction is due to a direct, enzyme-mediated biochemical process the cell and it's biological machinery dictate or if the reaction in question is due to an indirect process in which products of other enzymatic reactions are linked to the specific reaction in question. To determine if a geomicrobial transformation is an enzymatic or a nonenzymatic process, attempts should be made to reproduce the phenomenon with cell-free extract, especially if a single organism is involved. If catalysis is observed with the cell-free extract, identification of the responsible enzyme system should be undertaken by standard techniques. Spent culture medium from which all cells have been removed, for example, by filtration or by inactivation by heating, should also be tested for activity. If equal levels of activity are observed with untreated and treated spent medium, operation of a nonenzymatic process may be inferred. If the level of activity in treated cell-free extract is lower than in untreated cell-free spent medium,

extracellular enzyme activity may be present in the untreated cell-free spent medium. This can be verified by testing activity in unheated cell-free medium over a range of temperatures. If a temperature optimum is observed, at least part of the activity in the cell-free medium can be attributed to extracellular enzymes.

8.11 *IN SITU* STUDY OF PAST GEOMICROBIAL ACTIVITY

In situ geomicrobial activity that ceased in sedimentary deposits as far back as the early Precambrian or during the Phanerozoic up to modern times can sometimes be inferred through the detection of specific organic biomarkers and/or isotopic biosignatures in samples from such deposits. These biomarkers represent preserved organic derivatives formed from some characteristic cellular constituent, for example, chlorophylls from photosynthetic prokaryotes (bacteria, cyanobacteria), phototrophic eukaryotic microbes (Roselle-Melé and Koç, 1997; Brocks and Pearson, 2005; Brocks et al., 2005), carotenoids of anoxygenic photosynthetic bacteria and the oxygenic cyanobacteria (Hebting et al., 2006), and lipid cell membrane components from archaea or bacteria (Brocks and Pearson, 2005; Hebting et al., 2006). The process by which these cell constituents were preserved appears to have involved mostly or exclusively chemical reduction (abiotic) under anaerobic conditions. In situ, the agents of preservation in the case of preservation of carotenoids such as β- (phytoplankton) and γ-carotene (cyanobacteria and green nonsulfur bacteria, i.e., Chloroflexaceae) and okenone (purple sulfur bacteria, i.e., Chromatiaceae) appear to have been reducing agents in the form of reduced sulfur compounds such as H_2S. The corresponding products of preservation from these pigments were β- and γ-carotane and okenane or perhydrookenone (Brocks et al., 2005; Hebting et al., 2006). The detection of the biomarker okenane, derived naturally from okenone of photosynthetic purple nonsulfur bacteria (Bradyrhizobiaceae) and chlorobactane from green sulfur bacteria (Chlorobiacea) in sediments of the 1.64-Gyr-old Barney Creek Formation (BCF) of the McArthur Group in northern Australia, has been used to argue that oxygen levels at this site at that time were well below modern levels although such levels had been approached or reached elsewhere on Earth (Brocks et al., 2005). This interpretation was supported by the simultaneous finding of extremely low concentrations of 2α-methylhopanes, biomarkers from cyanobacteria at the BCF site, which are normally considered to be oxygenic photosynthesizers (Brocks et al., 2005). However, Rashby et al. (2007) have demonstrated that a contemporary strain of the phototroph *Rhodopseudomonas palustris* (purple nonsulfur bacteria) is capable of synthesizing 2-methylbacteriohopanepolyols from which 2-methylhopane biomarkers could be formed during fossilization, indicating that this type of biomarker may not be a unique indicator for cyanobacteria.

Under certain circumstances, the occurrence of past geomicrobial activity can also be identified in terms of isotopic fractionation. Certain prokaryotic and eukaryotic microbes have been shown to distinguish between stable isotopes of elements such as C, H, O, N, S, Si, Li, and Fe. These microbes prefer to metabolize molecules containing the lighter isotopes of these elements (^{12}C in preference to ^{13}C, H in preference to D, ^{16}O in preference to ^{18}O, ^{14}N in preference to ^{15}N, ^{32}S in preference to ^{34}S, ^{28}Si in preference to ^{30}Si, ^{6}Li in preference to ^{7}Li, ^{54}Fe in preference to ^{56}Fe), especially under conditions of slow growth (see Jones and Starkey, 1957; Wellman et al., 1968; Emiliani et al., 1978; Estep and Hoering, 1980; Mortimer and Coleman, 1997; Beard et al., 1999; Mandernack et al., 1999; De La Rocha et al., 2000; Sakaguchi and Tomita, 2000; Croal et al., 2004; Crosby et al., 2007). They distinguish kinetically between different stable isotopes of the same element in a substrate on the basis of a difference in the rate of reaction the substrate undergoes in a specific biochemical step, the reaction rate involving the substrate with the lighter isotope being faster than that with heavier stable isotope. The isotope fractionation may occur during a specific enzyme-catalyzed chemical reaction into which the isotope-containing substrate enters, or it may occur during membrane transport of the isotope-containing substrate, or both (see Hoefs, 1997). Thus, the products of metabolism will be enriched in the lighter isotope compared to the starting compound or to some reference standard that has not been subjected to isotope fractionation. In practice, isotope fractionation is measured by determining isotope ratios of the heavier isotope of an

element to the lighter, using mass spectrometry, and then calculating the amount of enrichment from the relationship

$$\delta \, \text{Isotope (‰)}$$
$$= \frac{\text{Isotopic ratio of sample} - \text{Isotope ratio of standard}}{\text{Isotopic ratio of standard}}$$
$$\times 1000 \qquad (8.12)$$

If the enrichment value (δ) is negative, the sample tested was enriched in the lighter isotope relative to a reference standard, and if the value is positive, the sample tested was enriched in the heavier isotope relative to a reference standard. Thus, for instance, to determine if a certain metal sulfide mineral deposit is of biogenic origin, various parts of the deposit are sampled, and $\delta^{34}S$ values of the sulfide are determined. If the values are generally negative (although the magnitude of the $\delta^{34}S$ may vary among the samples and fall in the range of -5‰ to -50‰), the deposit can be viewed as of biogenic origin because a chemical explanation for such ^{32}S enrichment under natural conditions is not likely. If the $\delta^{34}S$ values are positive and fall in a narrow range, the deposit is viewed as being abiogenic in origin.

8.12 SUMMARY

Geomicrobial phenomena can be studied in the field and in the laboratory. Direct observation may involve microscopic examination and chemical and physical measurements. Specific and sensitive molecular biological techniques can be applied in tandem to determine the abundance, identity, interactions, and physiological potential of microorganisms in field samples. Laboratory study may involve reconstruction of a geomicrobial process using microcosms or stable isotope amendment experiments. Special methods have been devised for sampling, direct observation, and laboratory manipulation of microorganisms and their physicochemical environment. The latter two categories include the use of fluorescence microscopy, radioactive and stable isotope tracers, electrochemical sensors, and mass spectrometry for observing microorganisms *in situ*, measuring process rates and metabolic activity and measuring microbial isotope fractionation, respectively. The chemostat principle has been applied in the field to measure natural growth rates of geomicrobial agents. It has also been used under simulated conditions for determining limiting substrate concentrations. Extension of these methods for studying the growth and interactions of microorganisms on particle surfaces may require special experimental approaches.

REFERENCES

Amend JP, Rogers KL, Shock EL, Gurrieri S, Inguaggiato S. 2003. Energetics of chemolithoautotrophy in the hydrothermal system of Vulcano Island, southern Italy. *Geobiology* 1:37–58.Barker TW, Worgan JT. 1981. The application of air-lift fermenters to the cultivation of filamentous fungi. *Eur J Appl Microbiol Biotechnol* 13(2):77–83.

Barker WW, Banfield JF. 1998. Zones of chemical and physical interaction at interfaces between microbial communities and minerals: A model. *Geomicrobiol J* 15:223–244.

Beard BL, Johnson CM, Cox L, Sun H, Nealson KH, Aguilar C. 1999. Iron isotope biosignatures. *Science* 285:1889–1892.

Beeman RE, Suflita JM. 1989. Evaluation of deep subsurface sampling procedures using serendipitous microbial contaminants as tracer organisms. *Geomicrobiol J* 7:223–233.

Berg P, Risgaard-Petersen N, Rysgaard S. 1998. Interpretation of measured concentration profiles in sediment pore water. *Limnol Oceanogr* 43(7):1500–1510.

Blazewicz SJ, Barnard RL, Daly RA, Firestone MK. 2013. Evaluating rRNA as an indicator of microbial activity in environmental communities: Limitations and uses. *ISME J* 7:2061–2068.

Boetius A, Ravenschlag K, Schubert CJ, Rickert D, Widdel F, Gieseke A, Amann R, Jørgensen BB, Witte U, Pfannkuche O. 2000. A marine microbial consortium apparently mediating anaerobic oxidation of methane. *Nature (London)* 407:623–626.

Bontognali TR, Sessions AL, Allwood AC, Fischer WW, Grotzinger JP, Summons RE, Eiler JM. 2012. Sulfur isotopes of organic matter preserved in 3.45-billion-year-old stromatolites reveal microbial metabolism. *Proc Natl Acad Sci USA* 109(38):15146–15151.

Boyd ES, Druschel GK. 2013. Involvement of intermediate sulfur species in biological reduction of elemental sulfur under acidic, hydrothermal conditions. *Appl Environ Microbiol* 79(6):2061–2068.

Brendel PJ, Luther GW. 1995. Development of a gold amalgam voltammetric microelectrode for the determination of dissolved Fe, Mn, O-2, and S(-II) in porewaters of marine and fresh-water sediments. *Environ Sci Technol* 29:751–761.

Brock TC, Brock ML. 1968. Measurement of steady-state growth rates of a thermophilic alga directly in nature. *J Bacteriol* 95:811–815.

Brock TD. 1978. *Thermophilic Microorganisms and Life at High Temperatures*. New York: Springer.

Brocks JJ, Love GD, Summons RE, Knoll AH, Logan GA, Bowden SA. 2005. Biomarker evidence for green and purple sulphur bacteria in a stratified Paleoproterozoic sea. *Nature (London)* 437:866–870.

Brocks JJ, Pearson A. 2005. Building the biomarker tree of life. *Rev Mineral Geochem* 59:233–258.

Bromfield SM, Skerman VBD. 1950. Biological oxidation of manganese in soils. *Soil Sci* 69:337–348.

Buffle J, Tercier-Waeber MI. 2000. In situ voltammetry: Concepts and practice for trace analysis and speciation. In: *In situ Monitoring of Aquatic Systems: Chemical Analysis and Speciation* (eds. Buffle J, Horvai G). New York: John Wiley & Sons, pp. 281–405.

Burrows ND, Penn RL. 2013. Cryogenic transmission electron microscopy: Aqueous suspensions of nanoscale objects. *Microsc Microanal* 19(6):1542–1553.

Casida LE. 1971. Microorganisms in unamended soil as observed by various forms of microscopy and staining. *Appl Microbiol* 21:1040–1045.

Chan CS, De Stasio G, Welch SA, Girasole M, Frazer BH, Nesterova MV, Fakra S, Banfield JF. 2004. Microbial polysaccharides template assembly of nanocrystal fibers. *Science* 303:1656–1658.

Chan CS, Fakra SC, Emerson D, Fleming EJ, Edwards KJ. 2011. Lithotrophic iron-oxidizing bacteria produce organic stalks to control mineral growth: Implications for biosignature formation. *ISME J* 5:717–727.

Christner BC et al. 2014. A microbial ecosystem beneath the West Antarctic ice sheet. *Nature* 512:310–315.

Clesceri LS, Greenberg AE, Trussell RR, eds. 1989. *Standard Methods for the Examination of Water and Wastewater*, 17th ed. Washington, DC: American Public Health Association, pp. 9-65–9.66.

Comolli LR, Duarte R, Baum D, Luef B, Downing KH, Larson DM, Csentsits R, Banfield JF. 2012. A portable cryo-plunger for on-site intact cryogenic microscopy sample preparation in natural environments. *Microsc Res Techniq* 75:829–836.

Cowen JP et al. 2012. Advanced instrument system for real-time and time-series microbial geochemical sampling of the deep (basaltic) crustal biosphere. *Deep-Sea Res I* 61:43–56.

Croal LR, Johnson CM, Beard BL, Newman DK. 2004. Iron isotope fractionation by Fe(II)-oxidizing photoautotrophic bacteria. *Geochim Cosmochim Acta* 68:1227–1242.

Crosby HA, Roden EE, Johnson CM, Beard BL. 2007. The mechanism of iron isotope fractionation produced during dissimilatory Fe(III) reduction by Shewanella putrefaciens and Geobacter sulfurreducens. *Geobiology* 5:169–189.

Crusius J, Anderson RF. 1991. Compression and surficial sediment loss of lake sediments of high porosity caused by gravity coring. *Limnol Oceanogr* 36(5):1021–1031.

De Beer D. 2002. Microsensor studies of oxygen, carbon, and nitrogen cycles in lake sediments and microbial mats. In: *Environmental Electrochemistry: Analyses of Trace Element Biogeochemistry.* ACS Symposium Series 811 (eds. Taillefert M, Rozan TF). Washington, DC: American Chemical Society, pp. 227–246.

De Beer D, Kuhl M. 2000. Interfacial processes, gradients and metabolic activity in microbial mats and biofilms. In: *The Benthic Boundary Layer* (eds. Boudreau B, Jørgensen BB). Oxford, UK: Oxford University Press, pp. 374–394.

De Beer D, Sweerts JPRA. 1989. Measurement of nitrate gradients with an ion-selective microelectrode. *Anal Chim Acta* 219:351–356.

DeFlandre B, Duchene J. 2010. PRO2FLUX—A software program for profile quantification and diffusive O_2 flux calculations. *Environ Model Softw* 25:1059–1061.

De La Rocha CL, Brzczinski MA, DeNiro MJ. 2000. A first look at the distribution of the stable isotopes of silicon in natural waters. *Geochim Cosmochim Acta* 64:2467–2477.

DeLong EF, Wickham GS, Pace NR. 1989. Phylogenetic stains: Ribosomal RNA-based probes for the identification of single cells. *Science* 243:1360–1363.

Desai MS, Assig K, Dattagupta S. 2013. Nitrogen fixation in distinct microbial niches within a chemoautotrophy-driven cave ecosystem. *ISME J* 7:2411–2423.

Diaz JM, Hansel CM, Voelker BM, Mendes CM, Andeer PF, Zhang T. 2013. Widespread production of extracellular superoxide by heterotrophic bacteria. *Science* 340:1223–1226.

Druschel GK, Emerson D, Sutka R, Luther GW. 2008. Low oxygen and chemical kinetic constraints on the geochemical niche of neutrophilic iron(II) oxidizing microorganisms. *Geochimica et Cosmochimica Acta* 72:3358–3370.

Dubinina GA. 1970. Untersuchungen über die Morphologie und die Beziehung zu Mycoplasma. *Z Allg Mikrobiol* 10:309–320.

Edwards KJ et al. 1999. Geomicrobiology of pyrite (FeS2) dissolution: Case study at Iron Mountain, CA. *Geomicrobiol J* 16:155–179.

Edwards KJ, Wheat CG, Sylvan JB. 2011. Under the sea: Microbial life in volcanic oceanic crust. *Nat Rev Microbiol* 9(10):703–712.

Elsgaard L, Jorgensen BB. 1992. Anoxic transformations of radiolabeled hydrogen sulfide in marine and freshwater sediments. *Geochim Cosmochim Acta* 56(6):2425–2435.

Emiliani C, Hudson J, Shinn A, George RY. 1978. Oxygen and carbon isotope growth record in a reef coral from the Florida Keys and a deep-sea coral from the Blake Plateau. *Science* 202:627–629.

Eren J, Pramer D. 1966. Application of immunofluorescent staining to studies of the ecology of soil microorganisms. *Soil Sci* 101:39–49.

Espejo RT, Romero P. 1987. Growth of *Thiobacillus ferrooxidans* on elemental sulfur. *Appl Environ Microbiol* 53:1907–1912.

Estep MF, Hoering TC. 1980. Biogeochemistry of the stable hydrogen isotopes. *Geochim Cosmochim Acta* 44:1197–1206.

Ferris MJ, Kuhl M, Wieland A, Ward DM. 2003. Cyanobacterial ecotypes in different optical microenvironments of a 68 C hot spring mat community revealed by 16S-23S rRNA internal transcribed spacer region variation. *Appl Environ Microbiol* 65(5):2893–2898.

Fike DA, Finke N, Zha J, Blake G, Hoehler TM, Orphan VJ. 2009. The effect of sulfate concentration on (sub) millimeter-scale sulfide $\delta^{34}S$ in hypersaline cyanobacterial mats over the diurnal cycle. *Geochimica et Cosmochimica Acta* 73(20):6187–6204.

Fogaren KE, Sansone FJ, De Carlo E. 2013. Porewater temporal variability in a wave impacted permeable nearshore sediment. *Mar Chem* 149:74–84.

Froehlich PN, Klinkhammer GP, Bender ML, Luedtke NA, Heath GR, Cullen D, Dauphin P, Hammond D, Hartman B, Maynard V. 1979. Early oxidation of organic matter in pelagic sediments of the eastern equatorial Atlantic: Suboxic diagenesis. *Geochim Cosmochim Acta* 43:1075–1090.

Gaidos E et al. 2008. An oligarchic microbial assemblage in the anoxic bottom waters of a volcanic subglacial lake. *ISME J* 3(4):486–497.

Gerhardt P, Murray RGE, Costilow RN, Nester EW, Wood WA, Krieg NR, Phillips GB, eds. 1981. *Manual of Methods for General Bacteriology.* Washington, DC: The American Society for Microbiology.

Gerhardt P, Murray RGE, Wood WA, Krieg NR. 1993. *Methods for General and Molecular Bacteriology.* Washington, DC: ASM Press.

Ghiorse WC. 1984. Biology of iron- and manganese-depositing bacteria. *Ann Rev Microbiol* 38:515–550.

Giovannoni SJ, DeLong EF, Schmidt TM, Pace NR. 1990. Tangential flow filtration and preliminary phylogenetic analysis of marine picoplankton. *Appl Environ Microbiol* 56:2572–2575.

Glazer BT, Cary CS, Hohmann L, Luther GW. 2002. In situ sulfur speciation using Au/Hg microelectrodes as an aid to microbial characterization of an intertidal salt marsh microbial mat. In: *Environmental Electrochemistry: Analyses of Trace Element Biogeochemistry.* ACS Symposium Series 811 (eds. Taillefert M, Rozan TF). Washington, DC: American Chemical Society, pp. 283–305.

Glazer BT, Luther III GW, Konovalov SK, Freiderich GE, Nuzzio DB, Trouwborst RE, Tebo BM, Clement BG, Romanov AS. 2006. Documenting the suboxic zone of the Black Sea via high-resolution real time redox profiling. *Deep Sea Res II* 53:1740–1755.

Greef O, Glud RN, Gundersen J, Holby O, Jorgensen BB. 1998. A benthic lander for tracer studies in the sea bed: In situ measurements of sulfate reduction. *Cont Shelf Res* 18:1581–1594.

Griffin WT, Phelps TJ, Colwell FS, Fredrickson JK. 1997. Methods for obtaining deep subsurface microbiological samples by drilling. In: *The Microbiology of the Terrestrial Deep Subsurface* (eds. Amy PS, Haldeman DL). Boca Raton, FL: CRC Lewis, pp. 23–44.

Hebsgaard MB, Philips MJ, Willerslev E. 2005. Geologically ancient DNA: Fact or artefact? *Trends Microbiol* 13(5):212–220.

Hebting Y, Schaeffer P, Behrens A, Adam P, Schmitt G, Schneckenburger P, Bernasconi SM, Albrecht P. 2006. Biomarker evidence for a major preservation pathway of sedimentary organic carbon. *Science* 312:1627–1630.

Hoefs J. 1997. *Stable Isotope Geochemistry,* 4th ed. Berlin, Germany: Springer, pp. 5, 41–42, 59–60.

House CH, Cragg B, Teske A, The Leg 201 Shipboard Scientific Party. 2003. Drilling contamination tests on ODP Leg 201 using chemical and particulate tracers. In: *Proceedings of the Ocean Drilling Program,* Initial Reports, Vol. 201. College Station, TX: Ocean Drilling Program, Chapter 2, pp. 1–19.

House CH, Schopf JW, McKeegan KD, Coath CD, Harrison TM, Stetter KO. 2000. Carbon isotopic composition of individual Precambrian microfossils. *Geology* 28(8):707–710.

Huber H, Huber G, Stetter KO. 1985. A modified DAPI fluorescence staining procedure suitable for the visualization of lithotrophic bacteria. *Syst Appl Microbiol* 6:105–106.

Hubert C et al. .2009. A constant flux of diverse thermophilic bacteria into the cold Arctic seabed. *Science* 325(5947):1541–1544.

Hurst CJ, Knudsen GR, McInerney MJ, Stetzenbach LD, Walter MV. 1997. *Manual of Environmental Microbiology.* Washington, DC: ASM Press.

Ireland TR. 1995. Ion microprobe mass spectrometry: Techniques and applications in cosmochemistry, geochemistry, and geochronology. In: *Advances in Analytical Geochemistry* (eds. Hyman M, Rowe M), Vol. 2. JAI Press, pp. 1–118.

Ivarsson M, Holmström S. 2012. The use of ESEM in geobiology. In: *Scanning Electron Microscopy* (ed. Kazmiruk V). Rijeka, Croatia: InTech. ISBN: 978-953-51-0092-8.

Imachi H et al. 2011. Cultivation of methanogenic community from subseafloor sediments using a continuous-flow bioreactor. *ISME J* 5:1913–1925.

Jannasch HW. 1967. Growth of marine bacteria at limiting concentrations of organic carbon in seawater. *Limnol Oceanogr* 12:264–271.

Jannasch HW. 1969. Estimation of bacterial growth in natural waters. *J Bacteriol* 99:156–160.

Jannasch HW, Mottl MJ. 1985. Geomicrobiology of the deep-sea hydrothermal vents. *Science* 229:717–725.

Jannasch HW, Wirsen CO. 1981. Morphological survey of microbial mats near deep-sea thermal vents. *Appl Environ Microbiol* 41:528–538.

Jensen G, Briegel A. 2007. How electron cryotomography is opening a new window onto prokaryotic ultrastructure. *Curr Opin Struct Biol* 17:260–267.

Jeroschewski P, Steuckart C, Kuhl M. 1996. An amperometric microsensor for the determination of H_2S in aquatic environments. *Anal Chem* 68:4351–4357.

Jin Q, Bethke CM. 2005. Predicting the rate of microbial respiration in geochemical environments. *Geochim Cosmochim Acta* 69(5):1133–1143.

Jones AA, Sessions AL, Campbell BJ, Li C, Valentine DL. 2008. D/H ratios of fatty acids from marine particulate organic matter in the California Borderland Basins. *Org Geochem* 39(5):485–500.

Jones GE, Starkey RL. 1957. Fractionation of stable isotopes of sulfur by microorganisms and their role in deposition of native sulfur. *Applied Microbiol* 5:111–118.

Jorgensen BB. 2006. Bacteria and marine biogeochemistry. In: *Marine Geochemistry,* 2nd ed (eds. Schulz HD, Zabel M). New York: Springer, Chapter 5, pp. 169–206.

Jørgensen BB, Revsbech NP. 1983. Colorless sulfur bacteria, Beggiatoa spp. and Thiovulum spp. in oxygen and hydrogen sulfide microgradients. *Appl Environ Microbiol* 45:1261–1270.

Kallmeyer J, Ferdelman TG, Weber A, Fossing H, Jørgensen BB. 2004. A cold chromium distillation procedure for radiolabeled sulfide applied to sulfate reduction measurements. *Limnol Oceanogr Methods* 2:171–180.

Kallmeyer J, Smith DC, Spivack AJ, D'Hondt S. 2008. New cell extraction procedure applied to deep subsurface sediments. *Limnol Oceanogr Methods* 6:236–245.

Kalyuzhnaya MG, Lidstrom ME, Chistoserdova L. 2008. Real-time detection of actively metabolizing microbes by redox sensing as applied to methylotroph populations in Lake Washington. *ISME J* 2:696–706.

Kemp PF, Lee S, LaRoche J. 1993. Estimating the growth rate of slowly growing marine bacteria from RNA content. *Appl Environ Microbiol* 59(8):2594–2601.

Kepner RL Jr., Pratt JR. 1994. Use of fluorochromes for direct enumeration of total bacteria in environmental samples: Past and present. *Microbiol Rev* 58:603–615.

König B, Kohls O, Holst G, Glud RN, Kühl M. 2005. Fabrication and test of sol–gel based planar oxygen optodes for use in aquatic sediments. *Mar Chem* 97(3):262–276.

Konishi YY, Takasaka Y, Asai S. 1994. Kinetics of growth and elemental sulfur oxidation in batch culture of *Thiobacillus ferrooxidans*. *Biotechnol Bioeng* 44:667–673.

Krepski ST, Emerson D, Hredzak-Showalter PL, Luther GW, Chan CS. 2013. Morphology of biogenic iron oxides records microbial physiology and environmental conditions: Toward interpreting iron microfossils. *Geobiology* 11:457–471.

Kumar K, Das D. 2012. Growth characteristics of *Chlorella sorokiniana* in airlift and bubble column photobioreactors. *Bioresour Technol* 116:307–313.

LaRock PA, Ehrlich HL. 1975. Observations of bacterial microcolonies on the surface of ferromanganese nodules from Blake Plateau by scanning electron microscopy. *Microb Ecol* 2:84–96.

Lassen C, Ploug H, Jørgensen BB. 1992. Microalgal photosynthesis and spectral scalar irradiance in coastal marine sediments of Limfjorden, Denmark. *Limnol Oceanogr* 37:760–772.

Lawrence JR, Korber DR, Wolfaardt GM, Caldwell DE. 1997. Analytical imaging and microscopy techniques. In: *Manual of Environmental Microbiology* (eds. Hurst CJ, Knudsen GR, McInerney MJ, Stetzenbach LD, Walter MV). Washington, DC: ASM Press, pp. 29–51.

Lay CY, Mykytczuk NC, Yergeau É, Lamarche-Gagnon G, Greer CW, Whyte LG. 2013. Defining the functional potential and active community members of a sediment microbial community in a high-arctic hypersaline subzero spring. *Appl Environm Microbiol* 79:3637–3648.

Lepot K, Williford KH, Ushikubo T, Sugitani K, Mimura K, Spicuzza MJ, Valley JW. 2013. Texture-specific isotopic compositions in 3.4 Gyr old organic matter support selective preservation in cell-like structures. *Geochimica et Cosmochimica Acta* 112:66–86.

Lever MA et al. 2013. Evidence for microbial carbon and sulfur cycling in deeply buried ridge flank basalt. *Science* 339(6125):1305–1308.

Lindberg RD, Runnels D. 1984. Ground water redox reactions: An analysis of equilibrium state applied to Eh measurements and geochemical modeling. *Science* 225:925–927.

Luther GW III, Brendel PJ, Lewis BL, Sundby F, Lefrançois L, Silverberg N, Nuzzio DB. 1998. Simultaneous measurement of O_2, Mn, Fe, I and S(-II) in marine pore waters with a solid-state voltammetric microelectrode. *Limnol Oceanogr* 43:325–333.

Luther GW III, Glazer BT, Hohmann L, Popp JI, Taillefert M, Rozan TF, Brendel PJ, Theberge SM, Nuzzio DB. 2001a. Sulfur speciation monitored in situ with solid state gold amalgam voltammetric microelectrodes: Polysulfides as a special case in sediments, microbial mats and hydrothermal vent waters. *J Environ Monit* 3:61–66.

Luther GW III, Rozan TF, Taillefert M, Nuzzio DB, DiMeo C, Shank TM, Lutz RA, Cary SC. 2001b. Chemical speciation drives hydrothermal vent ecology. *Nature* 410:813–816.

Luther GW III et al. 2008: Use of voltammetric solid state (micro)electrodes for studying biogeochemical processes: Laboratory measurements to real time measurements with an in situ electrochemical analyzer (ISEA). *Mar Chem* 108(3–4):221–235.

Madison AS, Tebo BM, Mucci A, Sundby B, Luther GW III. 2013. Abundant porewater Mn(III) is a major component of the sedimentary redox system. *Science* 341:875–878.

Malfatti F, Azam F. 2009. Atomic force microscopy reveals microscale networks and possible symbioses among pelagic marine bacteria. *Aquat Microb Ecol* 58:1–14.

Mandernack KW, Bazilinski DA, Shanks WC III, Bullen TD. 1999. Oxygen isotope studies of magnetite produced by magnetotactic bacteria. *Science* 285:1892–1896.

McInerney MJ, Sieber JR, Gunsalus RP. 2009. Syntrophy in anaerobic global carbon cycles. *Curr Opin Biotechnol* 20:623–632.

Michaelis W et al. 2002. Microbial reefs in the Black Sea fueled by anaerobic oxidation of methane. *Science* 297(5583):1013–1015.

Moore TS, Mullaugh KM, Holyoke RR, Madison AS, Yucel M, Luther GW. 2009. Marine chemical technology and sensors for marine waters: Potentials and limits. *Ann Rev Mar Sci* 1:91–115.

Morris BEL, Henneberger R, Huber H, Moissl-Eichinger C. 2013. Microbial syntrophy: Interaction for the common good. *FEMS Microbiol Rev* 37:384–406.

Morris JJ, Kirkegaard R, Szul MJ, Johnson ZI, Zinser ER. 2008. Facilitation of robust growth of *Prochlorococcus* colonies and dilute liquid cultures by "helper" heterotrophic bacteria. *Appl Environ Microbiol* 74(14):4530–4534.

Mortimer RJG, Coleman ML. 1997. Microbial influence on the oxygen isotopic composition of diagenetic siderite. *Geochim Cosmochim Acta* 61:1705–1711.

Mosser JL, Bohlool BB, Brock TD. 1974. Growth rates of *Sulfolobus acidocaldarius* in nature. *J Bacteriol* 118:1075–1081.

Müller AL, de Rezende JR, Hubert CRJ, Kjeldsen KU, Lagkouvardos I, Berry D, Jørgensen BB, Loy A. 2014. Endospores of thermophilic bacteria as tracers of microbial dispersal by ocean currents. *ISME J* 8(6):1153–1165.

Murray AE et al. 2012. Microbial life at −13°C in the brine of an ice-sealed Antarctic lake. *Proc Natl Acad Sci USA* 109(50):20626–20631.

Muyzer G, de Bruyn AC, Schmedding DJM, Bos P, Westbroek P, Kuenen GJ. 1987. A combined immunofluorescence DNA fluorescence staining technique for enumeration of *Thiobacillus ferrooxidans* in a population of acidophilic bacteria. *Appl Environ Microbiol* 53:660–664.

Nesje A. 1992. A piston corer for lacustrine and marine sediments. *Arctic Alpine Res* 24:257–259.

Orcutt B, Wheat CG, Edwards KJ. 2010. Subseafloor ocean crust microbial observatories: Development of FLOCS (FLow-through Osmo Colonization System) and evaluation of borehole construction materials. *Geomicrobiol J* 27:143–157.

Orcutt BN, Sylvan JB, Knab NJ, Edwards KJ. 2011. Microbial ecology of the dark ocean above, at, and below the seafloor. *Microbiol Mol Biol Rev* 75(2):361–422.

Onstott TC et al. 2003. Indigenous and contaminant microbes in ultradeep mines. *Environ Microbiol* 5:1168–1191.

Orphan VJ, House CH. 2009. Geobiological investigations using secondary ion mass spectrometry: Microanalysis of extant and paleo-microbial processes. *Geobiology* 7:360–372.

Orphan VJ, House CH, Hinrichs KU, McKeegan KD, DeLong EF. 2001. Methane-consuming archaea revealed by directly coupled isotopic and phylogenetic analysis. *Science* 293(5529):484–487.

Orphan VJ, Taylor LT, Hafenbradl D, Delong EF. 2000. Culture-dependent and culture-independent characterization of microbial assemblages associated with high-temperature petroleum reservoirs. *Appl Environ Microbiol* 66:700–711.

Overmann J. 2005. Chemotaxis and behavioral physiology of not-yet-cultivated microbes. *Methods Enzymol* 397:133–147.

Overmann J, Schubert K. 2002. Phototrophic consortia: Model systems for symbiotic interrelations between prokaryotes. *Arch Microbiol* 177:201–208.

Pedersen K. 1993. The deep subterranean biosphere. *Earth-Sci Rev* 34:243–260.

Perfil'ev BV, Gabe DR. 1965. The use of the microbial-landscape method to investigate bacteria which concentrate manganese and iron in bottom deposits. In: *Applied Capillary Microscopy: The Role of Microorganisms in the Formation of Iron and Manganese Deposits* (eds. Perfil'ev BV, Gabe DR, Gal'perina AM, Rabinovich VA, Sapotnitskii AA, Sherman EE, Troshanov EP). New York: Consultants Bureau, pp. 9–54.

Perfil'ev BV, Gabe DR. 1969. *Capillary Methods for Studying Microorganisms*. Toronto, Ontario, Canada: University Toronto Press (English Translation by Toronto SJ).

Pernthaler A et al. 2002. Identification of DNA-synthesizing bacterial cells in coastal North Sea plankton. *Appl Environ Microbiol* 68:5728–5736.

Phelps TJ, Russell BF. 1990. Drilling and coring deep subsurface sediments for microbiological investigations. In: *Proceedings of the First International Symposium on Microbiology of the Deep Subsurface* (eds. Fliermans CB, Hazen TC). Aiken, SC: WSRC Information Services Section Publications Group, pp. 2-35–2-47.

Prange A, Chauvistré R, Modrow H, Hormes J, Trüper HG, Dahl C. 2002. Quantitative speciation of sulfur in bacterial sulfur globules: X-ray absorption spectroscopy reveals at least three different species of sulfur. *Microbiology* 148(1):267–276.

Priscu JC et al. 2013. A microbiologically clean strategy for access to the Whillans Ice Stream subglacial environment. *Antarct Sci* 2:1–11.

Radajewski S, Ineson P, Parekh NR, Murrell JC. 2000. Stable-isotope probing as a tool in microbial ecology. *Nature* 403(6770):646–649.

CHAPTER NINE

Molecular Methods in Geomicrobiology

Maureen L. Coleman and Dianne K. Newman

CONTENTS

9.1 Introduction / 187
9.2 Who Is There? Assessing the Diversity and Abundance of Geomicrobial Organisms / 187
    9.2.1 Taxonomic Diversity In Situ / 188
    9.2.2 Quantifying Abundance of Specific Taxa / 189
    9.2.3 New Culturing Techniques / 190
9.3 What Are They Doing? Deducing Activities of Geomicrobial Organisms / 190
    9.3.1 Single-Cell Isotopic Techniques / 190
    9.3.2 Single-Cell Metabolic Assays / 192
    9.3.3 Community Techniques Involving Isotopes / 193
    9.3.4 Community Genomics and Gene Expression / 194
    9.3.5 Single-Cell Genomic Approaches / 195
9.4 How Are They Doing It? Unraveling the Mechanisms of Geomicrobial Organisms / 196
    9.4.1 Genetic Approaches / 197
    9.4.2 Bioinformatic Approaches / 200
    9.4.3 Follow-Up Studies / 201
9.5 Summary / 201
References / 201

9.1 INTRODUCTION

In addition to the techniques described in Chapter 8, molecular tools have become increasingly important in the study of the presence, activity, and mechanisms of catalysis by geomicrobial organisms. Today, various molecules (deoxyribonucleic acid [DNA], ribonucleic acid [RNA], protein, and lipids) are used to detect specific geomicrobial agents in situ and to make inferences about their metabolic activity. DNA sequencing has become routine, expanding our appreciation of the genetic potential of uncultivated organisms and complex natural communities from the environment. Isotope labeling approaches allow us to measure metabolic activity more directly and to specifically link organisms with geochemical fluxes. Finally, the application of molecular genetic, cell biological, and biochemical

techniques to study the genes and gene products that catalyze geochemically significant reactions is unraveling the mechanisms underlying these processes. Together, these molecular approaches provide a window into the interactions between microorganisms and their geochemical environment and enable predictions about how these geomicrobial processes may be altered in response to environmental perturbations.

9.2 WHO IS THERE? ASSESSING THE DIVERSITY AND ABUNDANCE OF GEOMICROBIAL ORGANISMS

A key first step in understanding the functioning of microbial communities is to identify which organisms are present. This *taxonomic diversity* is most often assessed using the 16S ribosomal RNA transcribed (the metabolic rates of some important geomicrobial organisms may be so slow that insufficient rRNA is present to detect), it can only be used with sequences that are already known, and there is a chance that related but nontarget genes will interfere with meaningful identification. One approach to circumvent the first of these issues is to amplify the signal using catalyzed reporter deposition (CARD-FISH) (Pernthaler et al., 2002); this method has been used to identify organisms in biofilm communities in soil and marine environments (Fazi et al., 2005; Ferrari et al., 2006). In another approach, taxon-specific sequences of the 16S rRNA or any other marker gene can be used to design primers or probes for quantitative PCR (qPCR) (Walker, 2002). This approach has been used to quantify niche partitioning among *Prochlorococcus* ecotypes along ocean-scale environmental gradients (Johnson et al., 2006). Unlike FISH, this method does not provide spatial resolution for imaging structured microbial communities.

9.2.3 New Culturing Techniques

Although molecular approaches have revolutionized our ability to detect organisms in nature, linking sequence information with metabolic function and physiology ultimately depends on studies of pure cultures and model systems. For many years, the great plate count anomaly (i.e., the failure of the majority of cells from nature to form colonies on plates, despite being visualized by microscopy as viable) vexed microbiologists interested in studying dominant organisms from the environment (Eilers et al., 2000). In recent years, new strategies have brought geomicrobially important organisms into cultivation, including the ubiquitous SAR11 group of marine heterotrophs and the mesophilic ammonia-oxidizing Archaea (Rappe et al., 2002; Konneke et al., 2005). These strategies include using diffusion chambers in media that simulate the composition of the natural environment (Kaeberlein et al., 2002), high-throughput procedures for isolating cell cultures through the dilution of natural microbial communities into very low nutrient media (Rappe et al., 2002), and encapsulating cells in gel microdroplets for massively parallel cultivation under low nutrient flux conditions, followed by flow cytometry to detect microdroplets containing microcolonies (Zengler et al., 2002).

Two key concepts underlying the success of these methods are as follows: (1) the growth medium must approximate the natural environment as closely as possible, often with very low nutrient concentrations, and (2) organisms often need to be cultured in combination with others, as cross-feeding between cells may be essential for their growth and not readily simulated in the laboratory (D'Onofrio et al., 2010; Morris et al., 2012). Even with these developments, it may not be possible to culture all organisms detected by molecular methods; there is growing recognition that many cells in natural environments exist in a metabolically dormant state that is poorly understood (Jones and Lennon, 2010).

9.3 WHAT ARE THEY DOING? DEDUCING ACTIVITIES OF GEOMICROBIAL ORGANISMS

Although the use of 16S rRNA molecular signatures to identify specific organisms in the environment has revolutionized microbial ecology, these molecules can only tell us *who is there* but tell us little about their metabolic activities. To infer what geomicrobial organisms are doing in any particular system, the methods described in Chapter 8 have great value. In addition, several techniques have been developed in recent years that permit concomitant molecular identification of geomicrobial organisms and their metabolic activity (Figure 9.1). Some of these techniques are even applicable at the single-cell level.

9.3.1 Single-Cell Isotopic Techniques

The first attempts to combine molecular identification (using 16S rRNA) with isotopic analysis of single cells were made in 1999 by two different groups (Lee et al., 1999; Ouverney and Fuhrman, 1999). Both employed FISH (described in Section 9.2.2) with microautoradiography (MAR) and enabled direct microscopic observation of whether particular cells had consumed a radioactively labeled substrate under specific incubation conditions or not. Subsequently, a quantitative FISH-MAR technique was developed and used to measure the substrate affinity, K_s, of uncultured target organisms in activated sludge (Nielsen et al., 2003). In the past several years, this technique has been perfected and employed in many different contexts, from single cells to

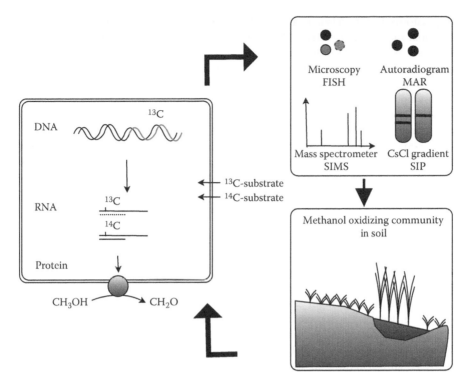

Figure 9.1. An illustration of some of the techniques described in Section 8.3. Starting with an environmental sample (e.g., a methanol oxidizing soil community), a labeled substrate (labeled with either stable isotopes or radioisotopes) is provided to the cells within the sample. Cells that can metabolize this substrate incorporate the label into DNA, RNA, or other cellular constituents. Depending on the method used, cells that incorporate this label are identified using MAR or SIMS, and their phylogenetic identify is confirmed using fluorescent probes that bind to their DNA or RNA (FISH). Before attempting FISH-MAR or FISH-SIMS, stable isotope probing (SIP) can be used to identify the organisms that take up the substrate of interest. See text for details.

biofilms (for a recent review, see Wagner et al., 2006). FISH-MAR only measures assimilation of the radiolabeled substrates into macromolecules; unincorporated labeled compounds are not retained by the FISH fixation process. Nevertheless, FISH-MAR is a very sensitive technique because radiotracers are incorporated into *all* macromolecules, not just nucleic acids. An advantageous consequence of this is that FISH-MAR requires relatively short incubation times (generally a few hours), which minimizes the possibility of substrate breakdown and cross-feeding of secondary organisms.

The main limitations of FISH-MAR for geomicrobial studies are as follows: (1) it depends on *a priori* knowledge about the phylogenetic affiliation of the studied organism(s), (2) the number of populations that can be specifically detected in a single FISH experiment is limited by the number of available fluorophores that can be applied simultaneously, (3) some geomicrobial samples may

not be well suited for FISH analysis, as sediment/soil particles might obscure the resident bacteria, and (4) the radioactive substrates of geomicrobial interest may not be available. The second limitation is partially overcome by a new method called CLASI-FISH (combinatorial labeling and spectral imaging fluorescent in situ hybridization), which enables dozens, potentially hundreds, of taxa to be imaged simultaneously (Valm et al., 2011). The first, second, and third limitations can be circumvented by molecular techniques that cast a broader net, such as stable isotope probing (SIP) or the isotope array (Section 9.3.3); the fourth limitation may be circumvented in some cases by simultaneously incubating the sample with unlabeled complex substrate and labeled $^{14}CO_2$. This technique, called $HetCO_2$-MAR, takes advantage of the fact that most heterotrophs assimilate CO_2 during biosynthesis in various carboxylation reactions, though specific inhibitors must be used to block rapid consumption of the $^{14}CO_2$ by limitations to FISH-MAR, organisms that are metabolically quiescent (or simply growing slowly) at the time of sampling would be missed, and the technique is limited to substrates for which there is a radiolabel.

9.3.4 Community Genomics and Gene Expression

Although isotopic labeling can be a powerful way to link specific metabolic capabilities with particular organisms in the environment, it is limited to metabolisms for which there are isotopic tracers and is typically used to target one metabolic pathway at a time. A complementary approach is to use genomic information to infer potential metabolic capabilities encoded throughout the community, an approach called metagenomics or community genomics (Riesenfeld et al., 2004; DeLong, 2005). Rather than target a single gene like the 16S rRNA (Section 9.2.1), metagenomics captures and sequences genome fragments without a gene-specific PCR amplification step. Rapid advances in DNA sequencing technology continue to lower the cost and increase the sequence output that can be achieved by metagenomics; hence this approach is now both extremely powerful and accessible.

Initial metagenomic studies involved cloning high molecular weight DNA from mixed communities into large-insert libraries using cosmid, fosmid, or bacterial artificial chromosomes as vectors. These libraries were then screened with hybridization probes or by end sequencing to identify particular clones of interest for full sequencing. The large inserts provided substantial genome context to link phylogenetically informative genes (e.g., 16S rRNA) and novel pathways and led to the discovery that the marine gamma-proteobacteria SAR86 encode a light-driven proton pump never previously observed in bacteria (Beja et al., 2000). This large-insert approach remains useful, particularly for functional metagenomics, in which clone libraries are screened to identify novel genes conferring a phenotype of interest (see also Section 9.4.1). A more common approach for reconstructing community metabolism is shotgun metagenomics, where small fragments are randomly sequenced from clone libraries or directly using next-generation sequencing technologies. These short sequences are then bioinformatically matched to databases to infer taxonomic

origin and biochemical function, enabling reconstruction of both community structure and metabolic pathways that may be operating in the entire community (Tyson et al., 2004; DeLong et al., 2006). Annotation of short fragment sequences is challenging, particularly in cases where there are no closely related microbial reference genomes in databases and for the ever-present "hypothetical proteins" whose functions are unknown (see also Section 9.4.2). It is also difficult to infer whether two fragments came from the same organism, since physical linkage is disrupted by the shotgun approach. This problem is partially overcome by improved assembly methods, aided by compositional binning techniques (Dick et al., 2009) and by the increasingly high coverage and longer reads afforded by new sequencing technologies.

Despite its challenges, community metagenomics is extremely powerful for generating a variety of testable hypotheses about the functioning of particular organisms and communities and their interactions with the geochemistry of their environment. For example, genome reconstruction from short fragments has enabled predictions about the metabolism of uncultured organisms, including the widespread sulfur oxidizing SUP05 group (Walsh et al., 2009; Anantharaman et al., 2013) and the photoheterotrophic marine group II Euryarchaeota (Iverson et al., 2012). These metabolic predictions may then enable successful cultivation of previously enigmatic taxa (Tyson et al., 2005; Marshall and Morris, 2012). Shotgun metagenomics can also reveal pathways encoded by a community that may not be readily evident by geochemical measurements, as in the discovery of an active sulfur cycle in ocean oxygen minimum zones (Canfield et al., 2010). The predicted activity of these pathways can then be followed up with targeted geochemical (e.g., isotopic tracers) and biochemical approaches. Beyond describing the metabolism of a single ecosystem, metagenomic comparisons can identify genes or metabolic features that differ across environments, and these patterns of gene presence/absence can provide clues to the ecological function of unknown genes or to biogeochemical differences between ecosystems. For example, a role for 2-methylhopanoid lipids in facilitating interspecies ecological interactions was proposed based on the distribution of the methylase gene *hpnP* and environmental factors (Ricci et al., 2014). Metagenomic comparisons across ocean basins revealed that microbial

EHRLICH'S GEOMICROBIOLOGY

communities in the Atlantic experience greater phosphorus limitation than their counterparts in the Pacific Ocean and identified novel candidate genes involved in P assimilation (Coleman and Chisholm 2010). Ultimately, hypotheses about community metabolism or gene function arising from metagenomics must be tested through classical genetic and biochemical analyses (see Section 9.4 and Martinez et al., 2007). For a more complete discussion of the opportunities and challenges of environmental genomics in the context of geomicrobial problems, see Banfield et al. (2005).

While analysis of environmental DNA is a powerful approach for assessing the functional potential of microbial communities, it offers limited insight into the activity of particular organisms and pathways at a given time and place. Measuring gene expression at the RNA and protein levels can help fill this gap. The expression of specific metabolic genes at the transcript level can be measured by first reverse-transcribing mRNA to the more stable cDNA and then amplifying target genes by PCR (or quantitatively, by qPCR). This approach has been used to detect expression of the *arrA* gene, which encodes one of the subunits of the respiratory arsenate reductase common to most arsenate-respiring bacteria, in arsenic-contaminated sediments (Malasarn et al., 2004). Analogous to metagenomics, it is also possible to analyze community-wide gene expression by metatranscriptomics. RNA is extracted from environmental samples, converted to cDNA, and sequenced, providing insight into the dynamics of gene expression across space and time (Frias-Lopez et al., 2008). This approach has shown that the metabolic activity of deep-sea hydrothermal plumes is dominated by water-column methanotrophs and chemolithoautotrophs and by a poorly characterized group of sulfur-oxidizing gammaproteobacteria (Lesniewski et al., 2012).

An even more direct assessment of activity in a community is by measuring protein-level expression. Proteins are the catalytic machines that actually perform biogeochemical transformations; hence their abundance should allow inferences about which pathways are active in a given sample. Moreover, the abundance of a protein is often not reliably predictable from the abundance of its cognate mRNA (Lee et al., 2003; Nie et al., 2006; Waldbauer et al., 2012). Specific proteins of interest can be detected using antibody-based methods. This has been used to study (per)

chlorate reduction, where immunoprobes were raised against the enzyme chlorite dismutase (CD) (O'Connor and Coates, 2002). CD is highly conserved among the dominant (per)chlorate-reducing bacteria in the environment; thus CD appears to be a good target for a probe specific to this process. Genome-wide detection and quantification of proteins by mass spectrometry is also possible, an approach called proteomics (for review, see Yates et al., 2009). In the most commonly applied proteomics strategy (termed "bottom-up" proteomics), protein is extracted from the sample, purified of interfering substances (salts, lipids, nucleic acids, etc.), and enzymatically cleaved into peptides by a protease (commonly trypsin). These peptides are then separated by liquid chromatography, ionized by electrospray, and analyzed by mass spectrometry. Most often, a sequence database (e.g., an organismal genome) is used to computationally match the observed mass spectra with peptide sequences, so proteomics leverages the rapidly expanding databases of DNA and RNA sequence information. When applied to mixed communities of microbes, the approach is termed metaproteomics (reviewed in Siggins et al., 2012). To date, metaproteomic analyses have been made of a number of microbial ecosystems, including soils, marine plankton, and acid mine drainage communities (e.g., Ram et al., 2005; Morris et al., 2010). Proteomics can also be combined with stable isotope probing (SIP) to track and quantify microbial activity at the protein level (von Bergen et al., 2013; Justice et al., 2014).

We refer those interested in applying metagenomic, metatranscriptomic, and metaproteomic to problems in geomicrobiology to a comprehensive *Methods in Enzymology* volume on these topics (edited by DeLong, 2013).

9.3.5 Single-Cell Genomic Approaches

A major limitation of community genomics is the difficulty in reconstructing which genes occurred together in a single cell. New approaches allow detection of multiple gene targets in a single cell by PCR, or even whole genome sequencing from a single cell. In microfluidic digital PCR, 16S rRNA identification of single, uncultivated cells can be coupled with the simultaneous detection of other genes in these cells. Cells are diluted and separated from a complex mixture using a microfluidic device. This separation step is critical, as

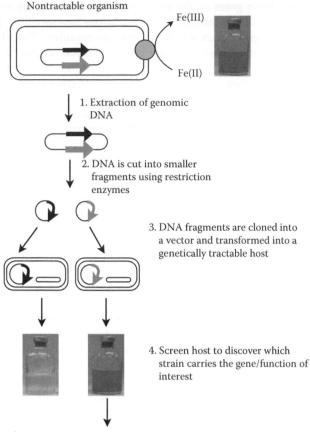

Nontractable organism

Fe(III)

Fe(II)

1. Extraction of genomic DNA

2. DNA is cut into smaller fragments using restriction enzymes

3. DNA fragments are cloned into a vector and transformed into a genetically tractable host

4. Screen host to discover which strain carries the gene/function of interest

5. DNA fragment containing gene(s) of interest is sequenced

Figure 9.2. Steps involved in heterologous complementation (i.e., a *gain-of-function* assay). As an example, Fe(II) oxidation to Fe(III) is provided as the phenotype of interest. The black and gray arrows symbolize different parts of the chromosome. The grey fragment represents the one that contains the gene(s) that encodes the function of interest. Qualitatively, bottles turn a rusty brown color when iron oxidation occurs. See text for details.

control a process of interest (Figure 9.2). If one seeks to learn more about the importance of particular residues within a given gene, however, transposons are not appropriate; in this case, site-directed mutagenesis (a controlled technique for generating mutations, including point mutations), chemical mutagenesis (the addition of chemicals that promote modifications of DNA at the scale of single residues), or PCR mutagenesis (using an error-prone polymerase to encourage mistakes) must be used. For example, these more highly resolved mutagenic techniques would be appropriate if one wanted to understand the function of a specific protein in greater detail (e.g., which sites are necessary for it to bind its substrate, which are required for the binding of cofactors, and which control its ability to associate with other molecules in the cell). Given the plummeting costs of

whole genome sequencing, chemical mutagenesis may resurge as an attractive approach to studying new isolates (Levin et al., 2014).

After mutagenesis is performed, mutants are identified either through a selection or a screen. A selection permits only those mutants that have the desired properties to grow, whereas a screen requires characterizing the behavior of thousands of mutants to identify those that have the phenotype of interest. Depending on the manner in which one has identified candidate mutants, secondary screens may be required to narrow the pool of candidates down to only those that are interesting. For example, if one performs a screen to find genes that control various steps in a biochemical reaction and if the assay for mutant identification involves looking at the rate at which a reaction proceeds, then false mutants could be identified by the screen that are simply slow growers but which do not have a specific defect in the reaction of interest. These mutants could be sorted out by measuring the growth rate of all candidates and only continuing to study those whose growth were normal with respect to the wild-type (i.e., parent) strain. Once interesting mutants are identified, the nature of the mutation must be determined through sequencing and genetic verification. Often this includes a complementation experiment, where a wild-type copy of the gene of interest is put back into the mutagenized strain (often on a plasmid), to demonstrate that it can restore the original phenotype. In the event that the wild-type phenotype is not rescued by this experiment, it means that the mutation responsible for the phenotype of interest lies elsewhere on the chromosome. Sequence analysis can help generate hypotheses to explain why the mutant behaves the way it does (see the discussion on bioinformatics in Section 9.4.2) and thus infer what affects the process of interest. To test these hypotheses, however, physiological, biochemical, or cell biological experiments are required.

However, not every organism is amenable to mutagenesis. For a practical discussion of this, see Newman and Gralnick (2005). An alternative strategy to identify genes from these organisms (or from environmental genomic sequences for which an organismal host may or may not be known) that are necessary and sufficient to catalyze a particular reaction is to clone and express them in a foreign host (i.e., heterologous complementation). This is a gain-of-function strategy, in contrast to the loss-of-function mutagenesis approach described earlier (Figure 9.3). Toward this end, the host must have the metabolic machinery necessary to process the gene product(s) one wishes to express. For example, if one seeks to express genes encoding multiheme cytochromes, such as those found in *Geobacter* and *Shewanella*, in a genetically tractable foreign host such as *E. coli*, *E. coli*

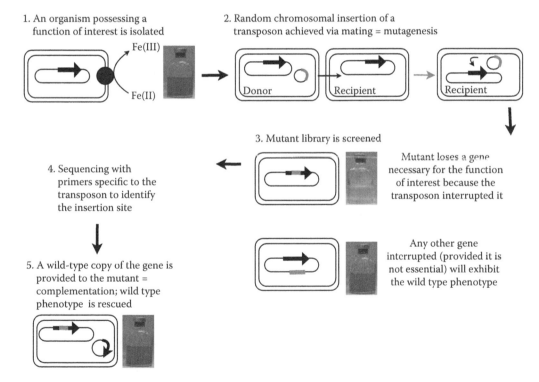

1. An organism possessing a function of interest is isolated

2. Random chromosomal insertion of a transposon achieved via mating = mutagenesis

Donor   Recipient   Recipient

3. Mutant library is screened

4. Sequencing with primers specific to the transposon to identify the insertion site

Mutant loses a gene necessary for the function of interest because the transposon interrupted it

5. A wild-type copy of the gene is provided to the mutant = complementation; wild type phenotype is rescued

Any other gene interrupted (provided it is not essential) will exhibit the wild type phenotype

Figure 9.3. Steps involved in transposon mutagenesis (i.e., a *loss-of-function* assay). As an example, Fe(II) oxidation to Fe(III) is provided as the phenotype of interest. The black arrow in the cell on the top in section 3 represents the part of the chromosome that contains the gene(s) that encodes the function of interest. The gray fragment represents the transposon. Qualitatively, bottles turn a rusty brown color when iron oxidation occurs. While transposon mutagenesis is convenient because it is straightforward to map the site of disruption, as sequencing costs have plummeted, it is now realistic to perform chemical or UV mutagenesis and sequence a strain's entire genome to identify genes responsible for a given function. See text for details.

Amann RI, Krumholz L, Stahl, DA. 1990. Fluorescent-oligonucleotide probing of whole cells for determinative, phylogenetic, and environmental studies. *J Bacteriol* 172:762–770.

Amann RI, Ludwig W, Schleifer K-H. 1995. Phylogenetic identification and in situ detection of individual microbial cells without cultivation. *Microbiol Rev* 59:143–169.

Amann RI, Zarda B, Stahl DA, Schleifer K-H. 1992. Identification of individual prokaryotic cells using enzyme-labeled, rRNA-targeted oligonucleotide probes. *Appl Environ Microbiol* 58:3007–3011.

Anantharaman K, Breier JA, Sheik CS, Dick GJ. 2013. Evidence for hydrogen oxidation and metabolic plasticity in widespread deep-sea sulfur-oxidizing bacteria. *Proc Natl Acad Sci USA* 110(1):330–335.

Banfield JF, Verberkmoes NC, Hettich RL, Thelen MP. 2005. Proteogenomic approaches for the molecular characterization of natural microbial communities. *OMICS—J Integr Biol* 9:301–333.

Behrens S, Lösekann T, Pett-Ridge J, Weber PK, Ng WO, Stevenson BS, Hutcheon ID, Relman DA, Spormann AM. 2008. Linking microbial phylogeny to metabolic activity at the single-cell level by using enhanced element labeling-catalyzed reporter deposition fluorescence in situ hybridization (EL-FISH) and NanoSIMS. *Appl Environ Microbiol* 74(10):3143–3150.

Beja O, Aravind L, Koonin EV, Suzuki MT, Hadd A, Nguyen LP, Jovanovich SB et al. 2000. Bacterial rhodopsin: Evidence for a new type of phototrophy in the sea. *Science* 289:1902–1906.

Bose A, Newman DK. 2011. Regulation of the phototrophic iron oxidation (pio) genes in *Rhodopseudomonas palustris* TIE-1 is mediated by the global regulator, FixK. *Mol Microbiol* 79(1):63–75.

Braun-Howland EB, Danielsen SA, Nierzwicki-Bauer SA. 1992. Development of a rapid method for detecting bacterial cells in situ using 16S rRNA-targeted probes. *Biotechniques* 13:928–934.

Braun-Howland EB, Vescio PA, Nierzwicki-Bauer SA. 1993. Use of a simplified cell blot technique and 16S rRNA-directed probes for identification of common environmental isolates. *Appl Environ Microbiol* 59:3219–3224.

Canfield DE, Stewart FJ, Thamdrup B, De Brabandere L, Dalsgaard T, Delong EF, Revsbech NP, Ulloa O. 2010. A cryptic sulfur cycle in oxygen-minimum–zone waters off the Chilean coast. *Science* 330(6009):1375–1378.

Caporaso JG, Lauber CL, Walters WA, Berg-Lyons D, Lozupone CA, Turnbaugh PJ, Fierer N, Knight R. 2011. Global patterns of 16S rRNA diversity at a depth of millions of sequences per sample. *Proc Natl Acad Sci USA* 108(Suppl 1):4516–4522.

Clement BG, Kehl LE, Debord KL, Kitts CL. 1998. Terminal restriction fragment patterns (TRFPs), a rapid, PCR-based method for the comparison of complex bacterial communities. *J Microbiol Method* 31:135–142.

Cliff JB, Gaspar DJ, Bottomley PJ, Myrold DD. 2002. Exploration of inorganic C and N assimilation by soil microbes with time-of-flight secondary ion mass spectrometry. *Appl Environ Microbiol* 68:4067–4073.

Coleman ML, Chisholm SW. 2010. Ecosystem-specific selection pressures revealed through comparative population genomics. *Proc Natl Acad Sci USA* 107(43):18634–18639.

Coppi MV, Leang C, Sandler SJ, Lovley DR. 2001. Development of a genetic system for *Geobacter sulfurreducens*. *Appl Environ Microbiol* 67:3180–3187.

Croal LR, Jiao YQ, Newman DK. 2007. The fox operon from *Rhodobacter* strain SW2 promotes phototrophic Fe(II) oxidation in *Rhodobacter capsulatus* SB1003. *J Bacteriol* 189:1774–1782.

DeLong EF. 2005. Microbial community genomics in the ocean. *Nat Rev Microbiol* 3:459–469.

DeLong EF (ed.). 2013. Microbial metagenomics, metatranscriptomics, and metaproteomics, *Methods in Enzymology*, Vol. 531. San Diego, CA: Elsevier Academic Press.

DeLong EF, Preston CM, Mincer T, Rich V, Hallam SJ, Frigaard NU, Martinez A et al. 2006. Community genomics among stratified microbial assemblages in the ocean's interior. *Science* 311:496–503.

DeLong EF, Wickham GS, Pace NR. 1989. Phylogenetic stains: Ribosomal RNA-based probes for the identification of single cells. *Science* 243:1360–1363.

Delsuc F, Brinkmann H, Philippe H. 2005. Phylogenomics and the reconstruction of the tree of life. *Nat Rev Genet* 6:361–375.

DiChristina TJ, Moore CM, Haller CA. 2002. Dissimilatory Fe(III) and Mn(IV) reduction by *Shewanella putrefaciens* requires ferE, a homolog of the pulE (gspE) type II protein secretion gene. *J Bacteriol* 184:142–151.

Dick GJ, Andersson AF, Baker BJ, Simmons SL, Thomas BC, Yelton AP, Banfield JF. 2009. Community-wide analysis of microbial genome sequence signatures. *Genome Biol* 10(8):R85.

D'Onofrio A, Crawford JM, Stewart EJ, Witt K, Gavrish E, Epstein S, Clardy J, Lewis K. 2010. Siderophores from neighboring organisms promote the growth of uncultured bacteria. *Chem Biol* 17(3):254–264.

Dumont MG, Murrell JC. 2005. Stable isotope probing—Linking microbial identity to function. *Nat Rev Microbiol* 3:499–504.

Eilers H, Pernthaler J, Glockner FO, Amann R. 2000. Culturability and in situ abundance of pelagic bacteria from the North Sea. *Appl Environ Microbiol* 66:3044–3051.

Fazi S, Amalfitano S, Pernthaler J, Puddu A. 2005. Bacterial communities associated with benthic organic matter in headwater stream microhabitats. *Environ Microbiol* 7:1633–1640.

Ferrari BC, Tujula N, Stoner K, Kjelleberg S. 2006. Catalyzed reporter deposition-fluorescence in situ hybridization for enrichment-independent detection of microcolony-forming soil bacteria. *Appl Environ Microbiol* 72:918–922.

Frias-Lopez J, Shi Y, Tyson GW, Coleman ML, Schuster SC, Chisholm SW, DeLong EF. 2008. Microbial community gene expression in ocean surface waters. *Proc Natl Acad Sci USA* 105(10):3805–3810.

Galperin MY, Koonin EV. 2000. Who's your neighbor? New computational approaches for functional genomics. *Nat Biotechnol* 18(6):609–613.

Giovannoni SJ, DeLong EF, Olson GJ, Pace NR. 1988. Phylogenetic group-specific oligonucleotide probes for identification of single microbial cells. *J Bacteriol* 170:720–726.

Gralnick JA, Vali J, Lies DP, Newman DK. 2006. Extracellular respiration of dimethyl sulfoxide by *Shewanella oneidensis* strain MR-1. *Proc Natl Acad Sci USA* 103:4669–4674.

Guyer RL, Koshland DE. 1989. The molecule of the year. *Science* 246:1543–1546.

Hesselsoe M, Nielsen JL, Roslev P, Nielsen PH. 2005. Isotope labeling and microautoradiography of active heterotrophic bacteria on the basis of assimilation of $^{14}CO_2$. *Appl Environ Microbiol* 71(2):646–655.

Huse SM, Dethlefsen L, Huber JA, Welch DM, Relman DA et al. 2008. Exploring microbial diversity and taxonomy using SSU rRNA hypervariable tag sequencing. *PLoS Genet* 4(11):e1000255.

Huynen, MA, Snel B, Mering CV, Bork P. 2003. Function prediction and protein networks. *Curr Opin Cell Biol* 15(2):191–198.

Iverson V, Morris RM, Frazar CD, Berthiaume CT, Morales RL, Armbrust EV. 2012. Untangling genomes from metagenomes: Revealing an uncultured class of marine Euryarchaeota. *Science* 335(6068):587–590.

Jiao YQ, Kappler A, Croal LR, Newman DK. 2005. Isolation and characterization of a genetically tractable photoautotrophic Fe(II)-oxidizing bacterium, *Rhodopseudomonas palustris* strain TIE-1. *Appl Environ Microbiol* 71:4487–4496.

Jiao YQ, Newman DK. 2007. The *pio* operon is essential for phototrophic Fe(II) oxidation in *Rhodopseudomonas palustris* TIE-1. *J Bacteriol* 189:1765–1773.

Johnson ZI, Zinser ER, Coe A, McNulty NP, Woodward EMS, Chisholm SW. 2006. Niche partitioning among *Prochlorococcus* ecotypes along ocean-scale environmental gradients. *Science* 311:1737–1740.

Jones SE, Lennon JT. 2010. Dormancy contributes to the maintenance of microbial diversity. *Proc Natl Acad Sci USA* 107(13):5881–5886.

Jurtshuk RJ, Blick M, Bresser J, Fox GE, Jurtshuk PJ. 1992. Rapid in situ hybridization technique using 16S rRNA segments for detecting and differentiating the closely related gram-positive organisms *Bacillus polymyxa* and *Bacillus macerans*. *Appl Environ Microbiol* 58:2571–2578.

Justice NB, Li Z, Wang Y, Spaudling SE, Mosier AC, Hettich RL, Pan C, Banfield JF. 2014. 15N- and 2H proteomic stable isotope probing links nitrogen flow to archaeal heterotrophic activity. *Environ Microbiol* 16(10):3224–3237.

Kaeberlein T, Lewis K, Epstein SS. 2002. Isolating uncultivable microorganisms in pure culture in a simulated natural environment. *Science* 296:1127–1129.

Karkhoff-Schweizer RR, Huber DPW, Voordouw G. 1995. Conservation of the genes for dissimilatory sulfite reductase from *Desulfovibrio vulgaris* and *Archaeoglobus fulgidus* allows their detection by PCR. *Appl Environ Microbiol* 61:290–296.

Klappenbach JA, Saxman PR, Cole JR, Schmidt TM. 2001. rrndb: The ribosomal RNA operon copy number database. *Nucleic Acids Res* 29:181–184.

Klindworth A, Pruesse E, Schweer T, Peplies J, Quast C, Horn M, Glöckner FO. 2013. Evaluation of general 16S ribosomal RNA gene PCR primers for classical and next-generation sequencing-based diversity studies. *Nucleic Acids Res* 41(1):e1.

Koch B, Jensen LE, Nybroe O. 2001. A panel of Tn7-based vectors for insertion of the *gfp* marker gene or for delivery of cloned DNA into gram-negative bacteria at a neutral chromosomal site. *J Microbiol Method* 45:187–195.

Komeili A, Vali H, Beveridge TJ, Newman DK. 2004. Magnetosome vesicles are present before magnetite formation, and MamA is required for their activation. *Proc Natl Acad Sci USA* 101:3839–3844.

Siggins A, Gunnigle E, Abram F. 2012. Exploring mixed microbial community functioning: Recent advances in metaproteomics. *FEMS Microbiol Ecol* 80(2):265–280.

Soergel DA, Dey N, Knight R, Brenner SE. 2012. Selection of primers for optimal taxonomic classification of environmental 16S rRNA gene sequences. *ISME J* 6(7):1440–1444.

Stackebrandt E, Goebel BM. 1994. Taxonomic note: A place for DNA–DNA reassociation kinetics and sequence analysis in the present species definition in bacteriology. *Int J Syst Bacteriol* 44:846–849.

Stahl DA, Lane DJ, Olsen GJ, Pace NR. 1984. The analysis of hydrothermal vent-associated symbionts by ribosomal RNA sequences. *Science* 224:409–411.

Stepanauskas R, Sieracki ME. 2007. Matching phylogeny and metabolism in the uncultured marine bacteria, one cell at a time. *Proc Natl Acad Sci USA* 104(21):9052–9057.

Stewart PS, Franklin MJ. 2008. Physiological heterogeneity in biofilms. *Nat Rev Microbiol* 6:199–210.

Sullivan NL, Tzeranis DS, Wang Y, So PT, Newman DK. 2011. Quantifying the dynamics of bacterial secondary metabolites by spectral multiphoton microscopy. *ACS Chem Biol* 6(9):893–899.

Swan BK et al. 2013. Prevalent genome streamlining and latitudinal divergence of planktonic bacteria in the surface ocean. *Proc Natl Acad Sci USA* 110:11463–11468.

Tadmor AD, Ottesen EA, Leadbetter JR, Phillips R. 2011. Probing individual environmental bacteria for viruses by using microfluidic digital PCR. *Science* 333(6038):58–62.

Teal TK, Lies DP, Wold BJ, Newman DK. 2006. Spatiometabolic stratification of *Shewanella oneidensis* biofilms. *Appl Environ Microbiol* 72:7324–7330.

Thiel V, Toporski J, Schumann G, Sjövall P, Lausmaa J. 2007. Analysis of archaeal core ether lipids using Time of Flight—Secondary Ion Mass Spectrometry (ToF-SIMS): Exploring a new prospect for the study of biomarkers in geobiology. *Geobiology* 5:75–83.

Tsien HC, Bratina BJ, Tsuji K, Hanson RS. 1990. Use of oligonucleotide signature probes for identification of physiological groups of methylotrophic bacteria. *Appl Environ Microbiol* 56:2858–2865.

Tyson GW, Chapman J, Hugenholtz P, Allen EE, Ram RJ, Richardwon PM, Solovyev VV, Rubin EM, Rokhsar DS, Banfield JF. 2004. Community structure and metabolism through reconstruction of microbial genomes from the environment. *Nature* 428:37–43.

Tyson GW, Lo I, Baker BJ, Allen EE, Hugenholtz P, Banfield JF. 2005. Genome-directed isolation of the key nitrogen fixer *Leptospirillum ferrodiazotrophum* sp nov from an acidophilic microbial community. *Appl Environ Microbiol* 71:6319–6324.

Valm AM, Welch JLM, Rieken CW, Hasegawa Y, Sogin ML, Oldenbourg R, Dewhirst FE, Borisy GG. 2011. Systems-level analysis of microbial community organization through combinatorial labeling and spectral imaging. *Proc Natl Acad Sci USA* 108(10):4152–4157.

van Waasbergen LG, Hildebrand M, Tebo BM. 1996. Identification and characterization of a gene cluster involved in manganese oxidation by spores of the marine *Bacillus* sp. strain SG-1. *J Bacteriol* 178:3517–3530.

von Bergen M, Jehmlich N, Taubert M, Vogt C, Bastida F, Herbst FA, Schmidt F, Richnow H, Seifert J. 2013. Insights from quantitative metaproteomics and protein-stable isotope probing into microbial ecology. *ISME J* 7(10):1877–1885.

Wagner MP, Nielsen P, Loy A, Nielsen J, Daims H. 2006. Linking microbial community structure with function: Fluorescence in situ hybridization-microautoradiography and isotope arrays. *Curr Opin Biotechnol* 17:83–91.

Waldbauer JR, Rodrigue S, Coleman ML, Chisholm SW. 2012. Transcriptome and proteome dynamics of a light-dark synchronized bacterial cell cycle. *PLoS ONE* 7(8):e43432.

Walker NJ. 2002. A technique whose time has come. *Science* 296:557–559.

Walsh DA, Zaikova E, Howes CG, Song YC, Wright JJ, Tringe SG, Tortell PD, Hallam SJ. 2009. Metagenome of a versatile chemolithoautotroph from expanding oceanic dead zones. *Science* 326(5952):578–582.

Ward DM, Weller R, Bateson MM. 1990. 16S rRNA sequences reveal numerous uncultured microorganisms in a natural community. *Nature* 345:63–65.

Wegener G, Bausch, M, Holler T, Thang NM, Mollar XP, Kellermann MY, Hinrichs KU, Boetius A. 2012. Assessing sub-seafloor microbial activity by combined stable isotope probing with deuterated water and 13C-bicarbonate. *Environ Microbiol* 14:1517–1527.

Welander PV, Coleman ML, Sessions AL, Summons RE, Newman DK. 2010. Identification of a methylase required for 2-methylhopanoid production and implications for the interpretation of sedimentary hopanes. *Proc Natl Acad Sci USA* 107(19):8537–8542.

Welander PV, Doughty DM, Wu CH, Mehay S, Summons RE, Newman DK. 2012. Identification and characterization of *Rhodopseudomonas palustris* TIE-1 hopanoid biosynthesis mutants. *Geobiology* 10(2):163–177.

Whiteley AS, Manefield M, Lueders T. 2006. Unlocking the *microbial black box* using RNA-based stable isotope probing technologies. *Curr Opin Biotechnol* 17:67–71.

Woese CR, Fox GE. 1977. Phylogenetic structure of the prokaryotic domain: The primary kingdoms. *Proc Natl Acad Sci USA* 74:5088–5090.

Yates JR, Ruse CI, Nakorchevsky A. 2009. Proteomics by mass spectrometry: Approaches, advances, and applications. *Ann Rev Biomed Eng* 11:49–79.

Zengler K, Toledo G, Rappé M, Elkins J, Mathur EJ, Short JM, Keller M. 2002. Cultivating the uncultured. *Proc Natl Acad Sci USA* 99:15681–15686.

CHAPTER TEN

Microbial Formation and Degradation of Carbonates

Tanja Bosak, Jaroslav Stolarski, and Anders Meibom

CONTENTS

10.1 Distribution of Carbon in Earth's Crust / 209
10.2 Biological Deposition of Calcium Carbonate / 210
    10.2.1 Principles of Microbial Carbonate Deposition / 211
    10.2.2 Microbially Influenced Precipitation of Calcite, Aragonite, and Dolomite / 212
    10.2.3 Models for the Formation of Ooids and Fine Carbonate Grains in Marine Environments / 218
    10.2.4 Deposition of Calcium Carbonate by Microbial Eukaryotes and Microbially Induced Calcification in Animals / 219
    10.2.5 Microbial Formation of Other Carbonates / 222
        10.2.5.1 Sodium Carbonate / 222
        10.2.5.2 Manganous Carbonate / 222
        10.2.5.3 Ferrous Carbonate / 223
        10.2.5.4 Strontium Carbonate / 223
        10.2.5.5 Magnesium Carbonate / 223
10.3 Biodegradation of Carbonates / 224
    10.3.1 Biodegradation of Limestone / 224
    10.3.2 Cyanobacteria, Algae, and Fungi That Bore into Limestone / 226
10.4 Summary / 228
References / 228

10.1 DISTRIBUTION OF CARBON IN EARTH'S CRUST

Carbon is one of the less abundant elements in the crust (320 ppm) (Weast and Astle, 1982), but it is central to all life on Earth. Organisms contribute to the carbon cycle on Earth's surface principally by fixing inorganic carbon (CO_2) to form organic carbon and remineralizing organic carbon back to inorganic carbon (Figure 10.1). The burial of organic matter removes a small portion of the organic carbon from the carbon cycle. Buried organic matter may be gradually transformed into coal, petroleum, bitumen, kerogen, organic matter in shales, natural gas (see Chapter 23) or, rarely,

graphite or diamonds. The carbon in these substances reenters the carbon cycle through human exploitation of fossil fuels or the weathering of organic-rich rocks.

The largest reservoirs of the carbon on the surface of the Earth are aged organic matter and inorganic carbon in carbonate rocks such as limestone and dolomite (Figure 10.1). Unaged, dead organic matter in soils and sediments contains nearly three orders of magnitude less C (Figure 10.1). Even less organic and inorganic C is present in living matter and the atmosphere (estimates from Bolin, 1970; Bowen, 1979; Fenchel and Blackburn, 1979; Post et al., 1990) (Figure 10.1). The carbon in limestone, dolomite, and aged organic matter

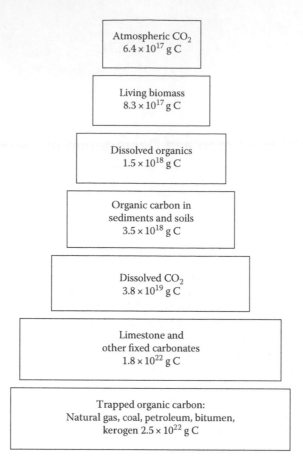

Figure 10.1. Distribution of carbon in the lithosphere of the Earth. (Estimates from Fenchel, T and Blackburn, TH, *Bacteria and Mineral Cycling*, Academic Press, London, U.K., 1979; Bowen, HJM, *Environmental Chemistry of the Elements*, Academic Press, London, U.K., 1979.)

is not readily available for assimilation by living organisms. Therefore, living systems source inorganic carbon from unaged, dead organic matter and atmospheric carbon. To maintain these short-term fluxes between organic matter and the reservoirs of inorganic carbon, organic matter has to be recycled by biological mineralization (see Chapter 7). In the recycling process, some of the carbon released by mineralization of the organic matter enters the atmosphere as CO_2, some is trapped in carbonate deposits, and some is fixed yet again by living organisms. In the absence of human interference, the transfer of carbon from living and dead organic matter to the atmosphere and back is assumed to be in *homeostasis*. However, the use of fossil fuels is increasing the size of the atmospheric reservoir of carbon, trapping more heat on Earth's surface, warming the climate, and changing the cycling of carbon among the reservoirs.

10.2 BIOLOGICAL DEPOSITION OF CALCIUM CARBONATE

Carbonate minerals that precipitate in natural waters remove inorganic carbon from the atmosphere and the hydrosphere. Magmatic and metamorphic processes create a portion of these insoluble minerals (see Berg, 1986; Bonatti, 1966; Skirrow, 1975), but bacteria, archaea, fungi, algae, foraminifera, coccolithophorids, and some animals mediate the deposition of most carbonate minerals on Earth's surface. Bacteria, archaea, and some fungi form calcium carbonate extracellularly (Bavendamm, 1932; Chafetz and Buczynski, 1992; Krumbein, 1974, 1979; Monty, 1972; Morita, 1980; Verrecchia et al., 1990). Only exceptionally, bacteria deposit calcium carbonate or amorphous Ba–Sr–Ca–Mg carbonate intracellularly (Buchanan and Gibbons, 1974; Couradeau et al., 2012; De Boer et al., 1971). Some green, brown,

EHRLICH'S GEOMICROBIOLOGY

and red algae, chrysophytes (coccolithophores), foraminifera, testate amoebae, and other microbial eukaryotes use genetically controlled mechanisms to precipitate calcium carbonate and form scales, tests, and coccoliths. Calcium carbonate also forms the skeletons of calcifying sponges and invertebrates such as coelenterates (corals), echinoderms, bryozoans, brachiopods, and mollusks or is associated with the chitinous exoskeleton of arthropods. In some siliceous sponges, calcium carbonate is present as a small component of the composite spicules (Ehrlich et al., 2010). Structural calcium carbonate in each of these organisms provides support and protection.

10.2.1 Principles of Microbial Carbonate Deposition

Table 10.1 lists the solubility constants (K_{sol}) of several geologically important carbonate minerals. Because carbonates are rather insoluble in water, they readily precipitate from aqueous solutions at relatively low concentrations of carbonate and the divalently charged cations.

In an aqueous solution containing $10^{-4.16}$ M Ca^{2+},* $CaCO_3$ will precipitate if the concentration of CO_3^{2-} exceeds $10^{-4.16}$ M. This is because the product of the concentrations of these two ions will exceed the solubility constant of $CaCO_3$:

$$[Ca^{2+}][CO_3^{2-}] = K_{sol} = 10^{-8.32} \quad (10.1)$$

TABLE 10.1
Solubility products of some carbonates.

Compound	Solubility constant (K_{sol})	Reference
$CaCO_3$	$10^{-8.32}$	Latimer and Hildebrand (1942)[a]
$MgCO_3 \cdot 3H_2O$	10^{-5}	Latimer and Hildebrand (1942)
$MgCO_3$	$10^{-4.59}$	Weast and Astle (1982)
$CaMg(CO_3)_2$	$10^{-16.7}$	Stumm and Morgan (1981)

[a] Note that Stumm and Morgan (1981) give a solubility product in freshwater of $10^{-8.42}$ (25°C) for calcite and $10^{-8.22}$ (25°C) for aragonite.

* The ion concentrations here refer really to ion activities.

In general, Ca^{2+} precipitates as $CaCO_3$ when the carbonate ion concentration exceeds the ratio $10^{-8.32}/[Ca^{2+}]$ (see Equation 10.1).

Carbonate ion in an unbuffered aqueous solution undergoes hydrolysis. The following reactions explain this phenomenon:

$$Na_2CO_3 \Leftrightarrow 2Na^+ + CO_3^{2-} \quad (10.2)$$

$$CO_3^{2-} + H_2O \Leftrightarrow HCO_3^- + OH^- \quad (10.3)$$

$$HCO_3^- + H_2O \Leftrightarrow H_2CO_3 + OH^- \quad (10.4)$$

The third reaction can be neglected at pH values higher than 7.3, but is important in solutions with the lower pH.

Bicarbonate dissociates according to the reaction

$$HCO_3^- \Leftrightarrow CO_3^{2-} + H^+ \quad (10.5)$$

with the equilibrium constant (K_2)

$$[CO_3^{2-}][H^+]/[HCO_3^-] = 10^{-10.33} \quad (10.6)$$

The ionization constant of water (K_w) at 25°C is described by

$$[H^+][OH^-] = 10^{-14} \quad (10.7)$$

and the equilibrium constant ($K_{equilib}$) for the hydrolysis of CO_3^{2-} (Reaction 10.3) is

$$[HCO_3^-][OH^-]/[CO_3^{2-}] = 10^{-14}/10^{-10.33} = 10^{-3.67} \quad (10.8)$$

At pH 7, the concentration of the hydroxyl ion is 10^{-7} (see Equation 10.7), and the ratio of bicarbonate to carbonate is

$$[HCO_3^-]/[CO_3^{2-}] = 10^{-3.67}/10^{-7} = 10^{3.33} \quad (10.9)$$

This means that at pH 7, the concentration of bicarbonate is $10^{3.33}$ times greater than the concentration of carbonate, assuming that an equilibrium exists between the CO_2 in the gas above the solution and CO_2 in the solution.

If a solution contains 0.03 g of Ca^{2+} per liter (i.e., at $10^{-3.14}$ M Ca^{2+}), Equation 10.1 shows that at an excess of $10^{-5.18}$ M carbonate ion is required to precipitate calcium carbonate:

$$[CO_3^{2-}] = 10^{-8.32}/10^{3.14} = 10^{-5.18} \quad (10.10)$$

a suitable buffer system, or the development of alkaline conditions.

Sulfate is the second most abundant anion in modern seawater. Because the chemistry of microbial sulfate reduction (MSR) in anaerobic environments can favor precipitation of calcium carbonate, precipitation of carbonate minerals in modern marine anaerobic sediments often occurs in the zones of sulfate reduction. The work of Abd-el-Malek and Rizk (1963a) described some of the first links between $CaCO_3$ deposition and MSR in culture experiments that used fertile clay loam soil enriched with starch and sulfate or sandy soil enriched with sulfate and plant matter. The works of Ashirov and Sazonova (1962) and Roemer and Schwartz (1965) also offer evidence of microbial carbonate formation during sulfate reduction in siliciclastic and petroliferous environments. Ashirov and Sazonova (1962) described the deposition of secondary calcite in an enrichment of sulfate-reducing bacteria from in quartz sand using hydrogen (H_2), lactate, acetate, or petroleum as electron donors. The petroleum may have first been broken down to usable hydrogen donors for sulfate reduction by other organisms in the mixed culture (see Nazina et al., 1985). These early experiments suggested that the activity of sulfate reducing microbes at the petroleum–water interface could influence the sealing of hydrocarbon reservoirs by $CaCO_3$. Indeed, the accumulation of a number of microbial carbonate deposits is thought to have influenced the formation of restricted basins and, the sealing, preservation and extraction of hydrocarbons (Al-Marjeby and Nash, 1986; Brognon and Verrier, 1966; Carroll and Bohacs, 2001; Collins et al., 2006; Hung Kiang et al., 1992; Mancini et al., 2000; Schröder et al., 2005).

Studies in the last three decades have refined our understanding of the role of MSR in the precipitation of carbonate minerals. Changes in the pH and the concentration of carbonate ion associated with MSR favor the formation of thin carbonate crusts in a modern alkaline lake in Indonesia (Arp et al., 2003; Kempe and Kazmierczak, 1993). Less than 1 mm

thick calcium carbonate laminae also precipitate in the vertically narrow zones of active sulfate reduction within modern marine stromatolites and cyanobacterial mats (Krumbein and Cohen, 1977; Lyons et al., 1984; Visscher et al., 2000). Spatial association also exists between dolomite, $CaMgCO_3$, and MSR in various hypersaline microbialites (e.g., Bontognali et al., 2010; Dupraz et al., 2004; Vasconcelos and McKenzie, 1997; Wright, 1999) and organic-rich sediments in deep sea (see Warren, 2000, for a review). MSR and methanogenesis have also been implicated in the formation of microbial dolomite (Kenward et al., 2009; Roberts et al., 2004; Sanchez Roman et al., 2008). More recent studies of microbial carbonates have focused on the role of organic surfaces and extracellular matrices (Figure 10.3). Organic compounds and surfaces appear to be critical in the nucleation of dolomite and can impact both the nucleation and the morphology of calcite and aragonite (Arp et al., 1999; Bosak and Newman, 2003, 2005; Braissant et al., 2004; 2007; Gautret et al., 2004; Kawaguchi and Decho, 2002; Roberts et al., 2013; Van Lith et al., 2003; Zhang et al., 2012). Particularly promising in the investigations of microbially mediated organomineral processes are synchrotron-based analytical techniques that reveal small-scale interactions among organic ligands and metal cations, the mineralogy (e.g., Lepot et al., 2008; Obst et al., 2009; Wacey et al., 2010) and the composition of trace elements or isotopes in the precipitates (e.g., Kunioka et al., 2006; Nehrke et al., 2013).

The effect of MSR on carbonate precipitation can be ambiguous even in the modern, sulfate-rich environments. When MSR occurs in closed systems and reduces less than 10 mM sulfate, it can lower the pH and cause dissolution of carbonate minerals (Ben-Yaakov, 1973). If MSR removes little sulfate, or large quantities of sulfide diffuse through Fe-poor sediments and are oxidized, the pH is lowered and carbonates dissolve. An active, microbially mediated S cycle in sediments is responsible for ~50% of the dissolution of calcium carbonate

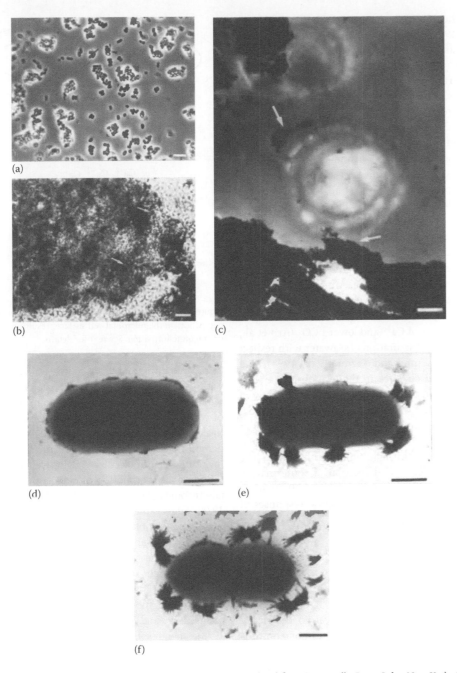

Figure 10.4. Calcite and gypsum precipitation by *Synechococcus* sp. isolated from Fayetteville Green Lake, New York. (a) Phase contrast photomicrograph of *Synechococcus* laboratory culture. (b) Petrographic thin-section photomicrograph of calcite crystal from Green Lake showing evidence of occlusion of numerous small bacterial cells within calcite grain (arrows). Note similar size of *Synechococcus* in (a) and occluded bacterial cells in (b). Scale bars in (a) and (b) equal 5 μm. (c) Thin-section transmission electron micrograph of two *Synechococcus* cells and calcite from a 72 h culture (cell represented by white oval area between arrows). Arrows point to calcite (electron-dense material) on the cell surface. Cells are unstained to avoid dissolution by heavy metal stains that are used to provide contrast to biological specimens. Scale bar equals 200 nm. (d through f) Series of transmission electron micrographs showing progression of gypsum precipitation on the cell wall of *Synechococcus* (whole mounts). (d) Initiation of numerous nucleation sites on the cell surface. (e) Gypsum precipitation spreading away from the cell. Gypsum still appears to be covered by a thin layer of bacterial slime. (f) Dividing *Synechococcus* cell shedding some of the precipitated gypsum. Scale bars equal 500 nm. (Courtesy of Thompson JB, Ferris FG, *Geology*, 18, 995, 1990. With permission.)

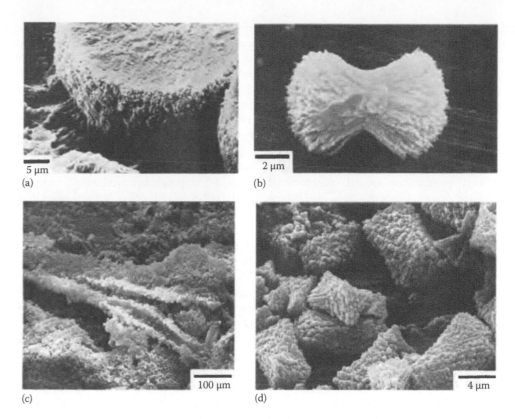

Figure 10.3. Bacterial precipitation of calcium carbonate as aragonite and calcite. Scanning electron micrographs of: (a) A crust of a hemisphere of aragonite precipitate formed by bacteria in liquid culture in the laboratory. (b) An aragonite dumbbell precipitated by bacteria in liquid medium in the laboratory. (c) Crystal bundles of calcite encrusting dead cyanobacterial filaments that were placed in gelatinous medium inoculated with bacteria from Baffin Bay. The specimen was treated with 30% H_2O_2 to remove organic matter. It should be noted that the crystal aggregates have been cemented to form a rigid crust that does not depend on the organic matter for support. (d) Higher magnification view of some of the crystal aggregates from (c) (area close-up not from field of view in [c]). These aggregates sometimes resemble rhombohedra, tetragonal disphenoids, or tetragonal pyramids. ([a] and [b]: From Buczynski, C and Chafetz, HS, *J Sedim Petrol*, 61, 226, 1991. With permission; [c] and [d]: From Chafetz, HS and Buczynski, C, *Palaios*, 7, 277, 1992. With permission.)

and high-Mg calcite in shallow carbonate sediments of Florida Bay (Ku et al., 1999; Walter and Burton, 1990). Sulfate reduction is thought to have had a smaller impact on the precipitation of calcium carbonate during the first 80% of Earth history (Bosak and Newman, 2003), when seawater contained less sulfate (Kah et al., 2002) and oxygen. At the same time, the exquisitely preserved microbial textures in some microbial carbonates from the early Earth (e.g., Bosak et al., 2009, 2013a; Knoll et al., 2013; Mata et al., 2012; Pruss et al., 2010; Sumner, 1997) suggest that carbonate minerals precipitated around organic surfaces before extensive organic degradation. In fact, the preservation of original microbial textures may have been promoted by the presence of lower concentrations of both sulfate and oxygen, which would have reduced the impact of pH-lowering processes such as sulfide oxidation and aerobic degradation of organic matter (Bosak et al., 2013a; Higgins et al., 2009).

4. *Removal of CO_2 from a bicarbonate-containing solution.* Removal of CO_2, a weak acid, causes an increase in the concentration of carbonate ion according to the following relationship:

$$2HCO_3^- \Leftrightarrow CO_2\uparrow + H_2O + CO_3^{2-} \quad (10.18)$$

If sufficient Ca^{2+} is present, $CaCO_3$ will precipitate.

Carbonate minerals precipitate around photo-synthetic organisms in various freshwater environments, where the waters are poorly buffered by bicarbonate (Arp et al., 1999; Golubic, 1973; Merz-Preiß and Riding, 1999; Pentecost, 1978; Pentecost and Bauld, 1988; Thompson and Ferris, 1990), forming travertine and lacustrine carbonate crusts and nodules. Travertine, porous limestone, forms by rapid calcium carbonate precipitation in waterfalls and streambeds of fast-flowing rivers. Mineral precipitation around cyanobacterial cells creates porous layers (Golubic, 1973), but cyanobacteria contribute to less than 10% of calcium carbonate precipitation in some travertine deposits (Pentecost, 1978, 1995). The rest of $CaCO_3$ forms abiotically because the stream water loses CO_2 to the atmosphere by degassing. At Fell Beck in Yorkshire, England, Spiro and Pentecost (1991) observed that calcite deposited in the presence of cyanobacteria contained more ^{13}C than nearby travertine, suggesting a photosynthetic contribution to travertine formation during the summer (Pentecost and Spiro, 1990). In general, textures and density of travertines depend on the relative rates of flow, microbial growth, and water chemistry. Less porous layers can be expected where the flow rates are large, solutions are more saturated with respect to calcium carbonate, and dense microbial growth is absent from the precipitating mineral surfaces (Okumura et al., 2012, 2013).

Benthic cyanobacteria attached to rocks or sediment in shallow portions of lakes with carbonate-saturated waters can stimulate the deposition of calcium carbonate by removing CO_2 (Golubic, 1973). When this process occurs in the presence of currents, it creates calcareous nodules. High-magnesian calcite found in the sheaths of certain cyanobacteria such as *Scytonema* may be related to the ability of the sheaths to concentrate magnesium three to five times over the concentrations of magnesium in seawater (Monty, 1967; see also discussion by Golubic, 1973). Photosynthetic removal of CO_2 also may be responsible for the formation of structural carbonate minerals in eukaryotes. Examples are found in some green, brown, and red algae and chrysophytes that deposit calcium carbonate in their walls (see Friedmann et al., 1972; Lewin, 1965).

Thompson and Ferris (1990) demonstrated the precipitation of calcite, gypsum, and probably magnesite around the cells of *Synechococcus* from Green Lake, Lafayette, New York, grown in filter-sterilized water from the lake (Figure 10.4). This lake is meromictic with a distinct, permanent chemocline at a depth of 18 m (Brunskill and Ludlam, 1969). Its water is naturally alkaline (pH 7.95, alkalinity 3.24 meq/L) and contains 11 mM Ca^{2+}, 2.8 mM Mg^{2+}, and 10 mM SO_4^{2-} (Thompson and Ferris, 1990). Gypsum crystals on the surface of *Synechococcus* cells developed before calcite crystals, but calcite crystals became larger and remained associated with the cells after cell division. Calcite deposition coincided with the photosynthetically induced rise in the pH at the cell surface and took place through interactions between calcium ions bound at the cell surface and carbonate ions. Gypsum deposition occurred in the dark as well as the light and, hence, was not driven by photosynthesis. This led Thompson and Ferris (1990) to suggest that calcite may replace gypsum in the deposits of microbial carbonates (bioherms) and fine carbonate muds in Green Lake (Thompson et al., 1990).

Where biological sources of silica are present, the deposition of calcium carbonate can occur together with the deposition of silica. In Flathead Lake delta, Montana, cyanobacteria and algae deposit calcareous nodules and crusts on sub-aqueous levees (Moore, 1983). Calcium carbonate forms in the outer portion of the nodules and concretions. Dissolved silica, probably derived by the dissolution of diatom frustules, precipitates together with calcium carbonate in the interior zones of the concretions during periods of high productivity (Moore, 1983). An amorphous Mg silicate phase also precipitates in association with aragonite in the alkaline waters of Lake Satonda, Indonesia (Arp et al., 2003). Silica is also present in the interior regions of micrometer-scale calcite precipitates that form in cultures of cyanobacterium *Synechococcus* PCC 7942 (Matsko et al., 2011). Silica thus may help nucleate calcite or stabilize amorphous calcium carbonate in the early stages of mineral precipitation (Gal et al., 2010).

The record of $CaCO_3$ precipitation on cyanobacterial sheaths (calcified cyanobacteria) has varied through time and in space (Arp et al., 2001; Kah and Riding, 2007; Riding, 1991, 1992, 2006). As noted previously, this mode of preservation is common in modern freshwater environments, but rare in the marine ones. In modern marine settings, calcification around cyanobacterial sheaths does occur (e.g., Sprachta et al., 2001), but is most commonly reported in periodically exposed environments (Planavsky et al., 2009)

and other highly evaporative settings (Défarge et al., 1994). Textures formed by these calcified filaments are rarely preserved during subsequent organic decay and calcification on extracellular polymeric matrices. In contrast, carbonate grains (oncoids, layered, rounded carbonate grains) or carbonate crusts from a number of Phanerozoic marine deposits preserve calcified cyanobacteria. Also confounding are temporal variations in the abundance of calcified filaments through time: the oldest textures formed by encrusted filaments are Neoproterozoic (Knoll et al., 1993; Swett and Knoll, 1985; Turner et al., 1993, 2000), although well-preserved microbial textures are present in much older carbonates (e.g., Bosak et al., 2013a; Mata et al., 2012; Sumner, 1997). Suggested explanations for the enhanced preservation of calcified filaments during some time intervals (e.g., the Cambrian) include lower buffering of the seawater by dissolved inorganic carbon, a higher concentration of Ca^{2+} and lower pCO_2 (Arp et al., 2001), elevated saturation of seawater with respect to $CaCO_3$ (Kempe and Kazmierczak, 1994), and physiological changes in the cyanobacterial carbon concentration mechanisms induced by a low CO_2/O_2 ratio (Kah and Riding, 2007; Riding, 2006). These explanations do not take into account the influence of changing sulfate and oxygen concentrations through time. Furthermore, the seemingly selective calcification of some marine cyanobacteria (e.g., Planavsky et al., 2009) hints at specific ecology and physiological responses of calcifying marine cyanobacteria.

10.2.3 Models for the Formation of Ooids and Fine Carbonate Grains in Marine Environments

Calcitic or aragonitic ooids, spherical, concentrically laminated carbonate grains, are common constituents of modern, carbonate sands from wave-agitated environments (Figure 10.5). These rounded grains have been present in the shallow water environments for more than 2.5 billion years (e.g., Beukes, 1983). The darker and lighter concentric laminae of ooids precipitate around other carbonate grains and skeletal fragments or around nonlaminated and sometimes nonspherical nuclei (e.g., Bathurst, 1967; Plee et al., 2008). The precipitation of calcium carbonate in the ooid laminae is typically thought to require solutions saturated with respect to calcium carbonate (Weyl, 1967), the presence of biofilms

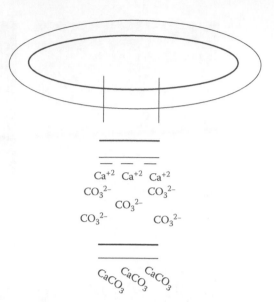

Figure 10.5. Schematic diagram of calcium carbonate deposition on the microbial cell surface, resulting in calcite or aragonite formation. See text for details.

or organic films, and wave action. However, ooids with laminae composed of carbonate crystals with radially oriented long axes also form in the absence of waves (Freeman, 1962). Experiments that investigated ooid formation in the absence of microbes demonstrated the ability of organic compounds, including high-molecular-weight humic acids to influence the morphology of radially oriented crystals in ooids (Davies et al., 1978; Ferguson et al., 1978; Suess and Fütterer, 1972). According to these studies, ooids may owe their lamination to the episodic inhibition of carbonate precipitation by organic compounds and the regeneration of organic-rich surfaces that facilitate carbonate precipitation (Davies et al., 1978; Ferguson et al., 1978). Indeed, modern marine ooids contain humic compounds and aspartate (Carter and Mitterer, 1978), spend some time within microbial mats (Bathurst, 1967), and contain a rather high content of organic carbon (1%-3%) (e.g., Summons et al., 2013). Microbes may mediate the formation of ooid fabrics even more directly, if some ooid carbonate nucleates within microbial coatings. For example, calcium carbonate crystals nucleate around the concentrically arranged cyanobacterial filaments in some spherical microbial aggregates in the laboratory (Brehm et al., 2006). Most marine ooids occur in sediments that are stirred by waves and currents, but neither the importance of agitation nor the microbial contribution to ooid growth is well constrained. Observations of ooids in high-energy areas led Bathurst (1967) to suggest that waves expose ooids to supersaturated waters for a longer time. However, the same study reported more ooids in an area of the Bimini Lagoon, Bahamas, that experienced less agitation than some neighboring areas. Furthermore, continuous agitation and transport of ooids prevents the growth of microbial mats and biofilms (Bosak et al., 2013b; Neumann et al., 1970) and abrades ooids (Van Ee and Wanless, 2008) at rates which exceed the estimated rates of carbonate precipitation in oolitic environments (Weyl, 1967). Therefore, the absence of quiet times that allow for microbial colonization of grains may actually impede the accretion of carbonate laminae in the organic-rich matrix around ooids. These times should last for more than 2 weeks and up to multiple months at a time (Bosak et al., 2013b).

Fine grained aragonite is abundant in the seas around the Bahamas and is occasionally suspended in the water column, causing the whiting events. The origin of this fine grained carbonate is debated, although at least a portion precipitates directly from the seawater (Shinn et al., 1989). Lowenstam and Epstein (1957) favored algal involvement in the formation of the fine-grained Bahamian aragonite based on a study of oxygen and carbon isotope ratios of various aragonitic carbonates, including algal precipitates and carbonate grains. Other geochemical indicators and morphological investigations suggested a different origin for tropical carbonate muds, including the production by fish (Macintyre and Aronson, 2006; Milliman et al., 1993; Perry et al., 2011).

Mechanisms behind the whiting events are also unclear. Some geochemists attribute whiting events entirely (Broecker et al., 2000) or partially (Morse et al., 2003) to physical resuspension of preexisting fine carbonate sediment and the precipitation of a rather soluble calcium carbonate phase around resuspended aragonite (Morse et al., 2003). Proposed ideas for the stirring mechanism include both sharks (Broecker et al., 2000) and wind-induced circulation (Dierssen et al., 2009). Other investigations suggested an episodic contribution of phytoplankton to the whiting events (Robbins and Blackwelder, 1992; Sondi and Juracic, 2010). On the Great Bahama Bank, this contribution may require the coupling among microbial communities, physical circulation, and water chemistry (Thompson, 2000).

10.2.4 Deposition of Calcium Carbonate by Microbial Eukaryotes and Microbially Induced Calcification in Animals

Morphological and physiological studies have shown that bacteria, including cyanobacteria, and some algae cause precipitation of $CaCO_3$ close to or at the cell surface. Other algae and unicellular eukaryotes form $CaCO_3$ intracellularly and then export it to the cell surface. Examples of algae that form $CaCO_3$ outside of the cell include green algae such as *Chara* and *Halimeda* (de Vrind-de Jong and de Vrind, 1997), *Acetabularia*, *Nitella*, *Penicillus*, and *Padina* and coralline red algae such as *Arthrocardia silvae* and *Amphiroa beauvoisii* (Figure 10.6). Organisms that form $CaCO_3$ intracellularly and then export

(a) (b)

Figure 10.6. Articulated coralline (calcareous) algae. (a) *Arthrocardia silvae*, Johansen, California. Scale bar equals 1.5 cm. (b) *Amphiroa beauvoisii*, Lamouroux, Gulf of California. Scale bar equals 1.5 cm. (Courtesy of Johansen HW, Department of Biology, Clark University, Worcester, MA.)

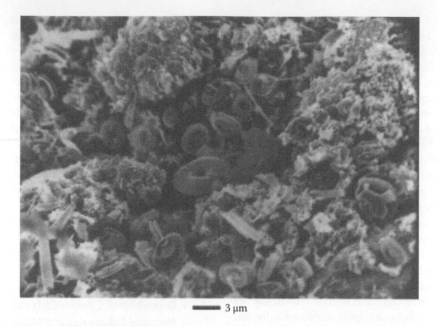

Figure 10.7. Coccoliths, the calcareous skeletons of an alga belonging to the class Chrysophyceae (see sketch in Figure 5.9). These specimens were found residing on the surface of a ferromanganese nodule from Blake Plateau of the Atlantic coast of the United States. (With kind permission from Springer Science+Business Media: *Microb Ecol*, 2, 1975, 84, LaRock, PA and Ehrlich, HL.)

it to the cell surface include coccolithophores (Chromalveolates) (Figure 10.7) and foraminifera (Rhizaria) (Figure 10.8). Calcifying algae may incorporate a significant portion of the total assimilated carbon into carbonate (Jensen et al., 1985). Wefer (1980) reported in situ $CaCO_3$ production by *Halimeda incrassata*, *Penicillus capitatus*, and *Padina sanctae-crucis*, respectively, in Harrington Sound, Bermuda, which amounted to approximately 50, 30, and 240 g m^{-2} per year, respectively.

Chara and *Halimeda* induce calcification by increasing the saturation at specific sites at the cell surface. These sites involve internodal cells in *Chara* and intercellular spaces in *Halimeda* (de Vrind-de Jong and de Vrind, 1997). Gelatinous or mucilaginous substances associated with the cell walls of these algae may be involved in the deposition and organization of the crystalline $CaCO_3$. If the algae incorporate the inorganic carbon from seawater into carbonate, these minerals will be

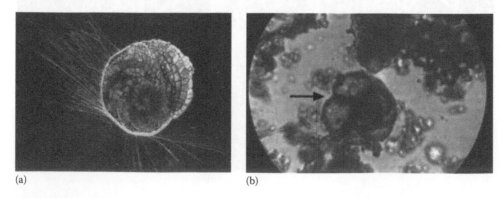

(a)   (b)

Figure 10.8. Foraminifera. (a) A living foraminiferan specimen, *Heterostegina depressa*, from laboratory cultivation. Note the multichambered test and the projection of fine protoplasmic threads from the test. Test diameter is 3 mm. (b) Foraminiferan test (arrow) in Pacific sediment: *Globigerina* (?) (2430×). ([a] Courtesy of Röttger R, *Marine Biol* (Berlin), 26, 5, 1974.)

enriched in ^{13}C relative to the carbon in seawater CO_2. Conversely, if algae form carbonate by using inorganic carbon derived from respiration, this carbonate will be enriched in ^{12}C relative to the carbon in seawater CO_2 (Lewin, 1965).

The basis for the morphogenesis of intricate morphologies of calcium carbonate minerals in eukaryotes remains to be elucidated. Studies of calcification in echinoderms, mollusks, and arthropods indicate that amorphous calcium carbonate, a much more soluble, water-rich, and disordered phase, is a critical precursor that enables mineral growth within confined spaces (e.g., Politi et al., 2004; Weiner et al., 2009). This precursor loses water and matures into the stable, much less soluble calcite or aragonite, but the maturation pathways and the importance of amorphous calcium carbonates in other carbonate-precipitating eukaryotes remain to be determined (Weiner et al., 2009).

McConnaughey (1991) proposed that *Chara corallina*, a calcareous alga, forms $CaCO_3$ at the cell surface by a process involving an ATP-driven H^+/Ca^{2+} exchange. In this process, protons produced in the reaction

$$Ca^{2+} + CO_2 = H_2O \rightarrow CaCO_3 + 2H^+ \qquad (10.19)$$

which is assumed to occur on the $CaCO_3$ surface facing the cell, are exchanged for intracellular Ca^{2+}. The protons then react with HCO_3^- in the cell to generate CO_2 for photosynthesis:

$$HCO_3^- + H^+ \rightarrow CO_2 + H_2O \qquad (10.20)$$

Organisms that form intracellular $CaCO_3$ first have to localize and concentrated Ca^{2+} at the site of calcium deposition (e.g., the plasma membrane, Golgi apparatus) (see de Vrind-de Jong and de Vrind, 1997) and then convert it into $CaCO_3$ in the presence of carbonic anhydrase. This enzyme promotes the conversion of dissolved metabolic CO_2 to bicarbonate and carbonate in a reversible reaction and is present in some microalgae and cyanobacteria (Aizawa and Miyachi, 1986). Although Aizawa and Miyachi focused on the role of carbonic anhydrase in CO_2 assimilation, this enzyme may very well play a role in CO_2 conversion to carbonate under different physiological conditions.

The coccolithophore *Pleurochrysis carterae* forms coccoliths in coccolithosome-containing vesicles and cisternae-containing scales in the Golgi apparatus in the cytoplasm of the cells. Once formed, the coccoliths are exported to the cell surface (de Jong et al., 1983; de Vrind-de Jong and de Vrind, 1997; van der Wal et al., 1983). In contrast, *Emiliania huxleyi* forms coccoliths in special vesicles called the coccolith-production compartments, which may be derived by the fusion of Golgi vesicles (de Vrind-de Jong and de Vrind, 1997). Acidic polysaccharides act as a template for $CaCO_3$ deposition (Isenberg and Lavine, 1973). Acidic Ca^{2+}-binding polysaccharides from some strains A92, L92, and 92D of *E. huxleyi* inhibit precipitation of $CaCO_3$ in vitro (Borman et al., 1982, 1987) and are thought to regulate $CaCO_3$ precipitation in the intracellular coccolith-forming vesicles (see de Jong et al., 1983). *E. huxleyi* derives its carbonate mainly from photosynthesis, whereas *P. carterae* derives it from photosynthesis in the light and respiration in the dark (de Vrind-de Jong and de Vrind, 1997). A more extensive discussion of calcification in coccolithophorids can be found in Young and Henriksen (2003).

Among foraminifera that form calcareous tests, the mineralized components may be in the form of low-Mg calcite, high-Mg calcite, and aragonite (other foraminifera construct tests out of organic material, agglutinated sediment grains or silica). Foraminiferal carbonate biomineralization involves two different pathways: (1) Golgi-mediated secretion of membrane-bound bundles of crystals and (2) extracellular precipitation on an organic template (see Berthold, 1976; Hemleben et al., 1986). Test formation in the imperforate (porcelaneous) foraminifera occurs through the first pathway that produces needle-shaped calcite crystals (usually high-Mg calcite) within intracellular vacuoles (Angell, 1980; Hemleben et al., 1986). Next, the crystals are deposited within an organic matrix, which forms the shape of the chamber. Typically, the crystals do not have any preferred orientation and the tests are opaque. In contrast, perforate foraminifera form tests in a space that is delineated by ectoplasmic pseudopods and use the cytoplasmic bulge as a mold for the organic matrix. Next, calcium carbonate crystals are deposited on both sides of a thin organic layer (primary organic membrane; see Hemleben et al., 1986).

Ion transport to calcification sites is thought to occur principally through the endocytosis of seawater (Bentov et al., 2009; Erez, 2003). This model emerged from the staining of the newly formed chambers of previously decalcified foraminifera, when these organisms were grown in seawater labeled with the membrane-impermeable fluorescent marker fluorescein isothiocyanate–dextran CE2. Challenges to this model include the large volume of seawater needed to supply the required Ca to the site of calcification (de Nooijer et al., 2009) and the requirement for a mechanism that modifies elemental composition of the vacuolized seawater and removes Mg during the transport to the site of calcification. A more recent model, based on stable isotope labeling of benthic foraminifera (Nehrke et al., 2013), suggests that calcium is supplied primarily by transmembrane transport and that a small component of passively transported (vacuolized) seawater is brought to the site of calcification. This small component plays a key role in defining the trace-element composition of the test.

Eukaryotic, multicellular organisms mostly form calcium carbonates through physiologically controlled processes (controlled biomineralization). Nonetheless, at least in some invertebrates, microbes may contribute to formation of calcareous skeletal material. Morris and Soule (2005) suggested that microbial biofilms living on the skeleton of microporellid bryozoans could contribute to the formation of calcareous deposits that become incorporated into the bryozoan skeleton. Uriz et al. (2012) documented calcification in nonsclerosponge siliceous demosponges (*Hemimycale*) that form bacteria-like vesicles that resemble bacteria that are present within archeocyte-like sponge cells. These bacteria lack cell walls and divide within sponge cells and may become surrounded by a calcitic sheet, and calcitic spherules are extruded to the subepithelial zone of the sponge to cover the surface of the host sponges and form cortex-like structures. Some shallow-burrowing venerid bivalves (*Granicorium* and *Samarangia*) have a thick layer of sand cemented to the shell. This cementation is facilitated by secretion of the mucus that harbors microbial biofilms (Braithwaite et al., 2000; Taylor et al., 1999). Vermeij (2014) summarized potential cases of microbial involvement in metazoan calcification and suggested that it can be much more widespread than commonly thought. Particularly in mollusks, the remote calcification, i.e., biomineralization, which is not strongly controlled by either the mantle edge or the inner mantle surface (see Chinzei, 2013; Chinzei and Seilacher, 1993), may depend on microbial activity.

10.2.5 Microbial Formation of Other Carbonates

10.2.5.1 Sodium Carbonate

Natron, hydrated sodium carbonate ($Na_2CO_3 \cdot 10H_2O$), is present in Wadi El Natrun in the Libyan Desert. The wadi (the channel of a watercourse that is dry, except during periods of rainfall, an arroyo) contains a chain of small lakes 23 m below sea level. The smallest of the lakes dries up almost completely during summer, and the larger ones dry out partially at that time. Natron crystals are present at the bottom of some of the lakes. Springs and streamlets feed the water to the lakes, which probably derive their water from the nearby Rosetta branch of the Nile. The surface water passes through cypress swamps before it reaches the lakes. Abd-el-Malek and Rizk (1963b) reported 189–204 meq of carbonate per liter and 324–1107 meq of sulfate per liter of lake water. In contrast, sulfate and carbonate concentrations are low in the cypress swamps (0 meq to traces of carbonate per liter and 2–13 meq of sulfate per liter). Bicarbonate is present in the lakes and swamps (11–294 and 11–16 meq L^{-1}, respectively). Soluble sulfides are present in the lakes and swamps (7–13 and 1–4 meq L^{-1}, respectively), and sulfate reducing microbes are present in the lakes and in the soil at a distance of 150 m from the lakes at concentrations from 1 to 8×10^6 mL^{-1}, but not elsewhere. Sulfate reduction is thought to occur chiefly in the cypress swamps because of these cell densities and the presence of readily available organic matter. Bicarbonate and sulfide generated by sulfate reduction, described by Equations 9.15 through 9.17, are washed into the lakes, concentrated by evaporation and partially precipitated as carbonate, including natron, and sulfide salts. Some sulfide also combines with iron to form ferrous sulfide, imparting a characteristically black color to the swamps. The carbonate in the lakes is produced by the loss of CO_2 from the water to the atmosphere due to warm water temperatures, especially in the summer (CO_2 solubility in water decreases with increase in temperature).

10.2.5.2 Manganous Carbonate

Carbonate may combine with Mn(II) to form rhodochrosite ($MnCO_3$). An example is the occurrence of rhodochrosite together with siderite ($FeCO_3$) in Punnus-Ioki Bay of Lake Punnus-Yarvi on the Karelian Peninsula in Russia (Sokolova-Dubinina and Deryugina, 1967). Lake Punnus-Yarvi is 7 km long, 1.5 km wide at its broadest part, 14 m deep, and only slightly thermally stratified. The oxygen concentration in its waters ranges from 11.8 to 12.1 mg L^{-1} at the surface to 5.7 to 6.6 mg L^{-1} in the bottom waters. The pH is slightly acidic (6.3–6.6), and the Mn^{2+} concentration ranges from 0.09 to 0.02–0.2 mg L^{-1} at the bottom (1.4 mg L^{-1} in winter). The Suantaka-Ioki and Rennel Rivers and 24 small streams that drain surrounding swampland supply the water to the lake. Surface and underground drainage of Cambrian and Quaternary glacial deposits delivers 0.2–0.8 mg of manganese per liter and 0.4–2 mg of iron per liter of drainage to the lake. Oxidized manganese and iron are incorporated into silt and reduced and concentrated by upward migration to the sediment–water interface and reoxidation into lake ore. This occurs mostly in Punnus-Ioki Bay, which has 5–7 cm thick oxide deposits at water depth down to 5–7 m.

All of the sediment and ore samples taken from the lake (mainly Punnus-Ioki Bay) contain manganese-reducing bacteria in the upper sediment layer. In a limited area near the center of Punnus-Ioki Bay, ore contained as much as 4.7% calcite, 5.96% siderite, and 4.99% rhodochrosite, together with 15.8% hydrogoethite and 38.9% barium psilomelanes and wads (complex oxides of manganese) (Sokolova-Dubinina and Deryugina, 1967). These investigators related the relatively localized concentration of carbonates to the localized availability of organic matter and its attack by heterotrophs: extensive carbonate ores in the Punnus-Ioki Bay area are found only in areas with abundant decaying plants.

The anoxic sediments of Landsort Deep in the central Baltic Sea also contain manganous and ferrous carbonates, manganous and ferrous sulfides, and manganous and calcium-iron phosphates. The minerals are thought to have formed authigenically, as a result of microbial mineralization of organic matter (Suess, 1979).

10.2.5.3 Ferrous Carbonate

Observations in rapidly accreting tidal marsh sediments at very shallow depths on the Norfolk Coast, England, support the role of microbial iron reduction in the formation of siderite. Extensive microbial decay of organic matter there occurs in the presence of low interstitial sulfate and

sulfide concentrations (Pye, 1984; Pye et al., 1990). Scanning electron microscopic examination and x-ray powder diffraction analysis of siderite concretions from this site reveal siderite as a void-filling cement and coating around quartz grains. Traces of greigite (Fe_3S_4), iron monosulfide, and calcite are also present (Pye, 1984). Carbon isotope fractionation studies support a microbial role in the formation of siderite (Pye et al., 1990). Microbial reduction of Fe_2O_3 and the subsequent reaction of Fe(II) with HCO^-_3 also may have produced siderite ($FeCO_3$) beds in the Yorkshire Lias in England.

Ehrlich and Wickert (1997) observed siderite formation by microbes colonizing the crushed bauxite in a column experiment in the laboratory. The columns containing bauxite ore were fed with a sucrose–mineral salts solution. Siderite and iron sulfide were detected by x-ray diffraction analysis of ore residue taken from the column at the end of the experiment.

10.2.5.4 Strontium Carbonate

Strontium carbonate is more insoluble ($K_{sol} = 10^{-8.8}$) than calcium carbonate ($K_{sol} = 10^{-8.07}$) and precipitates under appropriate conditions. In nature, Ferris et al. (1995) observed precipitation of strontium-rich calcite on a serpentine outcrop in a groundwater discharge zone near Rock Creek, British Columbia, Canada. This mineral precipitates nucleate around epilithic cyanobacteria, including *Calothrix*, *Synechococcus*, and *Gloeocapsa*. The concentration of Ca in water samples from the study site is 32–36 ppm and the Sr concentration is 5.8–6.6 ppm. The pH of the water 2 m above the outcrop is 8.5 and at its base (0.2 m level) 8.8. This pH difference was attributed to cyanobacterial photosynthesis (Ferris et al., 1995). The strontium content of calcite is as high as 1 wt%, and strontium carbonate forms a homogenous solid solution in calcite. Intracellular Sr–Mg–Ca carbonate also precipitates in the cells of some cyanobacteria from an alkaline lake in Mexico (Couradeau et al., 2012).

10.2.5.5 Magnesium Carbonate

Microbial communities dominated by actinomycetes from the genus *Streptomyces* may play a role in the formation of hydromagnesite ($Mg_5(CO_3)_4(OH)_2 \cdot 4H_2O$) (Cañaveras et al., 1999).

These organisms were found associated with moonmilk deposits containing hydromagnesite and needle-fiber aragonite in Altamira Cave in northern Spain.

10.3 BIODEGRADATION OF CARBONATES

Carbonates in nature can be dissolved as a direct or indirect result of biological activity, especially microbial activity that generates acid (Golubic and Schneider, 1979). For instance,

$$CaCO_3 + H^+ \rightarrow Ca^{2+} + HCO_3^- \quad (10.21)$$

$CaCO_3$ begins to dissolve even in weakly acidic solutions and dissolves more rapidly in more acidic solutions. Therefore, from a biochemical standpoint, any microorganism that generates acidic metabolic wastes is capable of dissolving insoluble carbonates. Even the mere metabolic generation of CO_2 during respiration can have this effect, because

$$CO_2 + H_2O \rightarrow H_2CO_3 \quad (10.22)$$

and

$$H_2CO_3 + CaCO_3 \rightarrow Ca^{2+} + 2HCO_3^- \quad (10.23)$$

Thus, it is not surprising that various kinds of CO_2- and acid-producing microbes can mediate the breakdown of natural carbonates.

10.3.1 Biodegradation of Limestone

Microbes were first implicated in the breakdown of lime in the cement of reservoir walls and docksides as long ago as 1899 (Stutzer and Hartleb, 1899). However, extensive investigation into microbial decay of limestone and the distribution of microbes in limestone was first undertaken three decades later by Paine et al. (1933). These authors estimated the rate of limestone decay through microbial action under controlled conditions, using a special apparatus that trapped CO_2 evolved during the dissolution of $CaCO_3$. The data indicated that as much as 28 g of CO_2 could emanate from 1 kg of stone in 1 year. As expected, the rate of CO_2 evolution from decaying stone was greater than from sound stone. Organic acids and CO_2 from heterotrophic metabolism of organic matter were major causes of the observed limestone decay in these experiments, but autotrophic nitrifying and sulfur-oxidizing microbes that produce nitric and sulfuric acids, respectively, were also able to promote limestone decay (Paine et al., 1933).

In a much later study, Krumbein (1968) reported variable numbers of fungi, algae, and ammonifying, nitrifying, and sulfur-oxidizing bacteria on the surface of some limestones in Germany. The number of detectable organisms depended on the type of stone, the length of elapsed time since the collection of the stone from a natural site, the surface structure of the stone (i.e., the degree of weathering), the cleanliness of the stone, and the climate and microclimatic conditions prevailing at the collection site. In the case of strongly weathered stone, the bacteria had sometimes penetrated the stone to a depth of 10 cm. Ammonifying microbes were generally most numerous. Nitrogen-fixing microbes were few, and denitrifiers were absent. The number of microbes did not correlate with the presence of lichens or algae. The ammonifiers were most abundant on freshly collected and weathered stone, perhaps because of the pH of the stone surface (pH 8.1–8.3 in water extract). Sulfur oxidizers were more numerous in cities, where the atmosphere contains more oxidizable sulfur compounds. The concentration of nitrifiers on limestone did not depend on the provenance of the samples, but the ammonifiers were also more abundant in stones exposed to city air than in stones exposed to country air. This may be due to the greater abundance of organic pollutants, which provide nutrients to microbes, in the city air. Laboratory experiments by Krumbein (1968) confirmed the weathering of limestone by its natural flora. This study identified a role for ammonifiers as sources of ammonia for nitrifying microbes. The latter corroded the limestone.

In yet another important study of the decomposition of limestone, numerous bacteria and fungi were isolated from a number of samples (Wagner and Schwartz, 1965). The active organisms appeared to weather the stone through the production of oxalic and gluconic acids. The investigators also noted the presence of nitrifying bacteria and thiobacilli in their samples.

Microbes can also attack marble, metamorphosed limestone, and dolostone. Figure 10.9 shows the corrosive effect of microcolonial black yeast on marble from the Dionysus Theater of the Acropolis in Athens, Greece (Anagnostidis et al., 1992). *Micrococcus halobius* colonizes the surfaces of Carrara

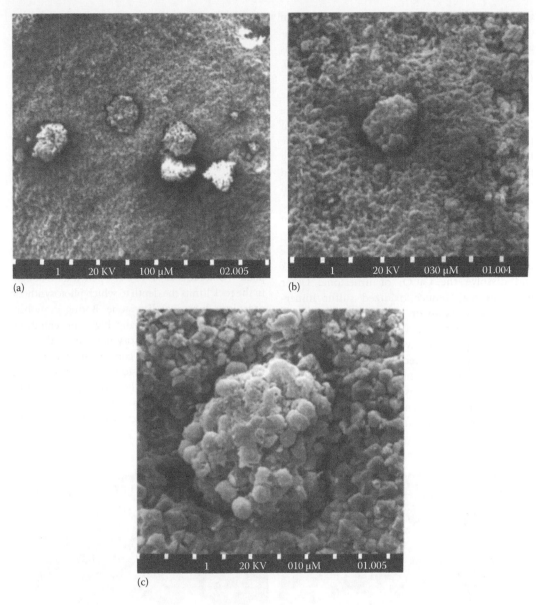

(a)

(b)

(c)

Figure 10.9. (a through c) SEM photomicrographs showing a section of marble from the Dionysus Theater, Acropolis, Athens, Greece, at three different magnifications (note scale and scale marks on the bottom of each photograph). The marble has been extensively corroded by *biopitting*. Deep holes of different sizes (between 2 and 5 mm in diameter and depth) were incised chemically (etched by metabolically produced substances) and physically (mechanical action) by black yeasts and meristematically growing yeast-like fungi. The microcolonial fungi can be confused with algae in SEM micrographs. The fungi have a cell size similar to that of the marble grain. The chemical and physical corrosive actions of these fungi have been demonstrated in laboratory experiments. (Courtesy of Krumbein WE, from Anagnostidis, K et al., Biodeterioration of marbles of the Parthenon and Propylaea, Acropolis, Athens—Associated organisms, decay, and treatment suggestions, in: Decrouez, D et al. (eds.), *2nd International Symposium for the Conservation of Monuments in Mediterranean Basin*, Musée d'Histoire Naturelle, Genève, Switzerland, 1992.)

marble slabs, forming biofilm and producing acids that etched the marble surfaces (Urzì et al., 1991). This organism also caused a discoloration of the marble surface, suggesting that natural patina on marble structures may have a microbial origin.

The microbial weathering of rock surfaces such as marble may not only dissolve limestone but can also lead to the formation of new minerals such as calcite, apatite, and wilkeite (Urzì et al., 1999). Such secondary $CaCO_3$ precipitation caused by microbes may be exploitable in the preservation of monuments and statuary made from carbonate rock, as studied under laboratory conditions with *Myxococcus xanthus* as the inducer of $CaCO_3$ (Rodriguez-Navarro et al., 2003).

Black fungi may contribute to the destruction of some marble and limestone (Gorbushina et al., 1994). However, instead of producing acids that dissolve the marble, the fungi appear to attack the marble by exerting physical pressure in pores and crevices and by changing the water activity in superficial polymers around the cells. The fungal melanin pigment also may blacken the surfaces of marble structures.

Various microbes are present in limestone caves and generate corrosive agents that attack and dissolve the $CaCO_3$ of limestone. Other microbes may reduce oxidized sulfur minerals such as gypsum or oxides of iron or manganese (see Herman, 1994, and other papers cited therein). Some caves, such as Movile Cave in Romania, which receives little input from the surface environment, have developed light-independent ecosystems in which the primary producers are chemosynthetic instead of photosynthetic autotrophs (Sarbu et al., 1994, 1996).

10.3.2 Cyanobacteria, Algae, and Fungi That Bore into Limestone

Endolithic cyanobacteria; green, brown, and red algae; and fungi can dissolve limestone and form tubular passages (Figure 10.10; Golubic, 1969; Golubic et al., 1975). These organisms attack coral reefs, beach rock, and other types of limestone. Mechanisms by which cyanobacteria bore into limestone are not well understood (Garcia-Pichel, 2006), but some filamentous boring cyanobacteria possess a terminal cell that is directly responsible for the boring action (Golubic, 1969). Different boring microorganisms form tunnels with a characteristic morphology (Golubic et al., 1975). In a pure mineral such as Iceland spar, boring follows the planes of crystal twinning, diagonal to the main cleavage planes (Golubic et al., 1975). The penetration of light in the rock limits the depth to which photosynthetic organisms bore into limestone. Boring cyanobacteria may contain unusually high concentrations of phycocyanin, an accessory pigment of the photosynthetic apparatus, to compensate for the low light intensity in the limestone. In contrast, boring fungi are not limited by light penetration. Boring cyanobacteria may affect as much as 60% of newly formed skeleton of scleractinian coral (*Porites*) and contaminate the original aragonite skeleton with

(a)

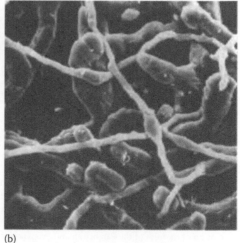
(b)

Figure 10.10. Microorganisms that bore into limestone. (a) Limestone sample experimentally recolonized by the cyanobacterium *Hyella balani* Lehman (234×). The exposed tunnels are the result of boring by the cyanobacterium. (b) Casts of the boring of the green alga *Eugamantia sacculata* Kormann (larger filaments) and the fungus *Ostracoblabe implexa* Bornet and Flahault in calcite spar (2000×). The casts were made by infiltrating a sample of fixed and dehydrated bore mineral with synthetic resin and the etching sections of the embedded material (e.g., with dilute mineral acid) to expose the casts of the organism. (Courtesy of Golubic S.)

EHRLICH'S GEOMICROBIOLOGY

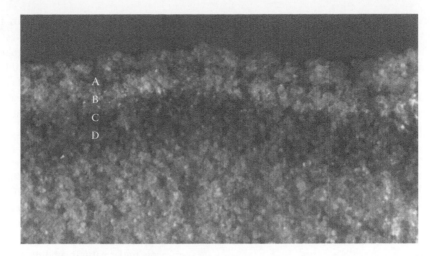

Figure 10.11. Cryptoendolithic microorganisms in vertically fractured Beacon sandstone. (A) Lichen (small black bodies between rock particles). (B) Zone of fungus filaments. (C) Yellowish-green zone of unicellular cyanobacterium. (D) Blue-green zone of unicellular cyanobacterium. The color difference between (C) and (D) is not apparent in this black-and-white photograph. Sample A76–77/36, north of Mount Dido, elevation 1750 m, magnification 4.5×. (Courtesy of Friedmann EI, *Antarctic J* U.S., 12, 26–30, 1977.)

secondary calcite characterized by entirely different trace element ratios (Nothdurft et al., 2007).

Endolithic fungi, cyanobacteria, or algae may coexist with fungi in lichens (Figure 10.11). Cyanobacteria and algae in these associations fix carbon and share it with their fungal partner, whereas the fungi may enhance the mobilization of nutrients or have other, less well-defined functions. Lichens grow within limestone and may serve as food to some snails (Shachak et al., 1987). The snails have a tonguelike organ, the *radula*, with embedded toothlike structures that abrade the lichens. The toothlike structures frequently consist of the iron mineral hematite. The snails use the radula to scrape the surface of the limestone and feed on the lichens, but also ingest some of the pulverized limestone and leave a trail of limestone powder as they move over the limestone surface. This biological weathering in the Negev Desert is estimated to affect 0.7–1.1 metric tons of rock per hectare per year (Shachak et al., 1987). A similar phenomenon was previously noted on some reef structures in Bermuda (see Golubic and Schneider, 1979).

Cyanobacteria and fungi inhabit not only limestone but also other porous rock. To distinguish among rock-inhabiting microbes, the term *euendoliths* is used to refer to true boring microbes in limestone. Opportunists that invade preexistent pores of rock and alter the rock are called *cryptoendoliths*, whereas those that invade

preexistent cracks and fissures without altering the rock structure are classified as *chasmoendoliths* (Golubic et al., 1981). In some environments, such as Antarctic dry valleys, cryptoendolithic cyanobacteria (Friedmann and Ocampo, 1976) and lichens (in this instance, symbiotic associations of a green alga and a filamentous fungus) (Friedmann, 1982; Friedmann and Ocampo-Friedmann, 1984) inhabit superficial cavities in sandstone, 1–2 mm below the surface (Figure 10.11). The near-surface locale in the sandstone (orthoquartzite) is the major habitat for microorganisms in this inhospitable environment (Friedmann, 1982). These cryptoendolithic microorganisms appear to differ from the boring microbes in that they do not form cavities in the sandstone but instead invade preexistent pores. However, they promote exfoliation of the sandstone surface—a physical process facilitated by the solubilization of a cementing substance that holds the quartz grains together (Friedmann and Weed, 1987). The lichens are visible because they mobilize iron. Lichens and coccoidal cyanobacteria (Friedmann, 1982) also inhabit some fissures in granite and granodiorite in the Antarctic dry valleys (Friedmann, 1982).

Cryptoendoliths also live in rock outcroppings (sandstone) in the Canadian High Arctic, that is, Ellesmere Island, Nunavut, Canada, where the temperatures are warmer and the precipitation is more extensive than in the Antarctic dry valleys (Omelon et al., 2006a,b). The cryptoendoliths reside in small spaces, 0.5–5 mm below the sandstone surface and include cyanobacterial, algal, fungal, and heterotrophic bacteria (Omelon et al., 2006a,b). Cryptoendoliths are not unique to the Antarctic dry valleys but are also found in hot desert environments (e.g., the Mojave Desert, California; Sonoran Desert, Mexico; Negev Desert, Israel) (Friedmann, 1980). Because moisture limits microbial growth in both hot deserts and the cold Antarctic deserts, rock interiors afford protection and permit life to persist in both types of environments (Figure 10.11).

10.4 SUMMARY

Calcium and calcium–magnesium carbonates are the major forms of inorganic carbon on Earth's surface. In many cases, these carbonates are of biogenic origin. Bacteria, archaea, and fungi mediate the precipitation of extracellular carbonate minerals by increasing the pH and alkalinity or facilitating the nucleation through interactions between organic surfaces and ions.

Microbial surfaces and organic matrices in microbial mats may act as nuclei for carbonate minerals and may play an indispensable role in the nucleation of dolomite at low temperatures. Carbonate-precipitating algae, microbial eukaryotes, plants, and animals form structural calcium carbonate either extracellularly or intracellularly. Structural $CaCO_3$ formation typically involves active transport of calcium and inorganic carbon prior to the precipitation of carbonate minerals. An important precursor to calcite and aragonite in a number of animal calcified structures is hydrated, relatively soluble and disordered amorphous calcium carbonate.

Microbial calcium carbonates precipitate in soils and in freshwater and marine environments, giving rise to crusts, fine-grained carbonates, larger carbonate grains such as ooids and oncoids, and macroscopic structures such as thrombolites or stromatolites. Temporal trends in microbial calcification and the preservation of microbial textures in carbonates may reflect evolutionary changes as well as major changes in the seawater chemistry.

Microbes can also dissolve carbonate minerals. This occurs when bacteria, fungi, and algae produce organic and inorganic acids as metabolic by-products or create habitats within the rock and may also involve physical processes. This activity is evident on limestones and marble used in building construction and in natural limestone formations such as coral reefs. Bacteria and fungi contribute to the discoloration of structural limestone and marble and may be the cause of patina.

REFERENCES

Abd-el-Malek Y, Rizk SG. 1963a. Bacterial sulfate reduction and the development of alkalinity. II. Laboratory experiments with soils. J Appl Bacteriol 26:14–19.

Abd-el-Malek Y, Rizk SG. 1963b. Bacterial sulfate reduction and the development of alkalinity. III. Experiments under natural conditions in the Wadi Natrun. J Appl Microbiol 26:20–26.

Aizawa K, Miyachi S. 1986. Carbonic anhydrase and CO_2 concentration mechanism in microalgae and cyanobacteria. FEMS Microbiol Rev 39:215–233.

Al-Marjeby, A, Nash, D. 1986. A summary of the geology and oil habitat of the Eastern Flank Hydrocarbon Province of South Oman. Mar Petrol Geol 3:306–314.

Anagnostidis K, Gehrmann K, Gross M, Krumbein WE, Lisi S, Panasidou A, Urzì C, Zagari M. 1992. Biodeterioration of marbles of the Parthenon and Propylaea, Acropolis, Athens-associated organisms, decay, and treatment suggestions. In: Decrouez D, Chamay J, Zezza F, eds., 2nd International Symposium for the Conservation of Monuments in Mediterranean Basin. Genève, Switzerland: Musée d'Histoire Naturelle, pp. 305–325.

Angell RB. 1980. Test morphogenesis (chamber formation) in the foraminifer Spiroloculina hyaline schulze. J Foram Res 10(2):89–101.

Arp G, Reimer A, Reitner J. 2001. Photosynthesis-induced biofilm calcification and calcium concentrations in Phanerozoic oceans. Science 292:1701–1704.

Arp G, Reimer A, Reitner J. 2003. Microbialite formation in seawater of increased alkalinity, Satonda Crater Lake, Indonesia. J Sed Res 73:105–127.

Arp G, Thiel V, Reimer A, Michaelis W, Reitner J. 1999. Biofilm exopolymers control microbialite formation at thermal springs discharging into the alkaline Pyramid Lake, Nevada, USA. Sed Geol 126:159–176.

Ashirov KB, Sazonova IV. 1962. Biogenic sealing of oil deposits in carbonate reservoirs. Mikrobiologiya 31:680–683 (English transl., pp. 555–557).

Bathurst RGC. 1967. Oolitic films on low energy carbonate sand grains, Bimini lagoon, Bahamas. Mar Geol 5:89–109.

Bavendamm W. 1932. Die mikrobiologische Kalkfällung in der tropischen See. *Arch Microbiol* 3:205–276.

Ben-Yaakov S. 1973. pH buffering of pore water of recent anoxic marine sediments. *Limnol Oceanogr* 18:86–94.

Bentov S, Brownlee C, Erez J. 2009. The role of seawater endocytosis in the biomineralization process in calcareous foraminifera. *Proc Natl Acad Sci* 106(21):21500–21504.

Berg GW. 1986. Evidence for carbonate in the mantle. *Nature (Lond)* 324:50–51.

Berthold WU. 1976. Biomineralisation bei miliolid en Foraminiferen und die Matrizen-Hypothese. *Naturwischenschaften* 63:196.

Beukes NJ. 1983. Ooids and oolites of the Proterophytic Boomplaas Formation, Transvaal Supergroup, Griqualand West, South Africa. In: Peryt TM ed., *Coated Grains*, pp. 199–214.

Bolin R. 1970. The carbon cycle. *Sci Am* 223:124–132.

Bonatti E. 1966. Deep-sea authigenic calcite and dolomite. *Science* 152:534–537.

Bontognali TRR, Vasconcelos C, Warthmann RJ, Bernasconi SM, Dupraz C, Strohmenger CJ, McKenzie JA. 2010. Dolomite formation within microbial mats in the coastal sabkha of Abu Dhabi (United Arab Emirates). *Sedimentology* 57:824–844.

Borman AH, de Jong EW, Huizinga M, Kok DJ, Bosch L. 1982. The role of $CaCO_3$ crystallization of an acid Ca^{2+}-binding polysaccharide associated with coccoliths of *Emiliania huxleyi*. *Eur J Biochem* 129:179–183.

Borman AH, de Jong EW, Thierry R, Westbroek P, Bosch L. 1987. Coccolith-associated polysaccharides from cells of *Emiliania huxleyi* (Haptophyceae). *J Phycol* 23:118–123.

Bosak T, Greene SE, Newman DK. 2007. A likely role for anoxygenic photosynthetic microbes in the formation of ancient stromatolites. *Geobiology* 5:119–126.

Bosak T, Knoll AH, Petroff APP. 2013a. The meaning of stromatolites. *Annu Rev Earth Planet Sci* 41:21–44.

Bosak T, Liang B, Sim MS, Petroff AP. 2009. Morphological record of oxygenic photosynthesis in conical stromatolites, *Proc Natl Acad Sci USA* 106:10939–10943.

Bosak T, Mariotti G, Perron JT, Macdonald FA, Pruss SB. 2013b. Microbial sedimentology of stromatolites in the Neoproterozoic cap carbonates, In: Ecosystems Paleobiology and Geobiology, *Paleontological Spec Papers*, v. 19, Paleontological Society, 51–75.

Bosak T, Newman DK. 2003. Microbial nucleation of calcium carbonate in the Precambrian. *Geology* 31:577–580.

Bosak T, Newman DK. 2005. Microbial kinetic controls on calcite morphology in supersaturated solutions. *J Sed Res* 75:190–199.

Bowen HJM. 1979. *Environmental Chemistry of the Elements*. London, U.K.: Academic Press.

Braissant O, Cailleau G, Dupraz C, Verrechia E. 2004. Bacterially induced mineralization of calcium carbonate in terrestrial environments: The role of exopolysaccharides and amino acids. *J Sed Res* 73:485–490.

Braissant O, Decho AW, Dupraz C, Glunk C, Przekop KM, Visscher P. 2007. Exopolymeric substances of sulfate-reducing bacteria: Interactions with calcium at alkaline pH and implication for formation of carbonate minerals. *Geobiology* 5:401–411.

Braithwaite CJRJD, Taylor PU, Glover, EA. 2000. Marine carbonate cements, biofilms, biomineralization, and skeletogenesis: Some bivalves do it all. *Journal of Sedimentary Research* 70:1129–1138.

Brehm U, Krumbein WE, Palinska K. 2006. Biomicrospheres generate ooids in the laboratory. *Geomicrobiol J* 23:545–550.

Broecker WS, Sanyal A, Takahashi T. 2000. The origin of Bahamian whitings revisited. *Geophys Res Lett* 27:3759–3760.

Brognon GP, Verrier GR. 1966. Oil and geology in Cuanza Basin of Angola. *AAPG Bull* 50:108–158.

Brunskill GJ, Ludlam SD. 1969. Fayetteville Green Lake, New York. I. Physical and chemical limnology. *Limnol Oceanogr* 14:817–829.

Buchanan RE, Gibbons NE, eds. 1974. *Bergey's Manual of Determinative Bacteriology*, 8th ed. Baltimore, MD: Williams and Wilkins.

Buczynski C, Chafetz HS. 1991. Habit of bacterially induced precipitates of calcium carbonate and the influence of medium viscosity on mineralogy. *Journal of Sedimentary Petrology* 61:226–233.

Cañaveras JC, Hoyos M, Sanchez-Moral S, Sanz-Rubino E, Bedoya J, Soller V, Groth I, Schumann P, Laiz L, Gonzalez I, Sais-Jimenez C. 1999. Microbial communities associated with hydromagnesite and needle-fiber aragonite deposits in a karstic cave (Altamira, northern Spain). *Geomicrobiol J* 16:9–25.

Carroll AR, Bohacs KM. 2001. Lake-type controls on petroleum source rock potential in nonmarine basins. *American Association of Petroleum Geologists Bulletin* 85:1033–1053.

Carter PW, Mitterer PM. 1978. Amino acid composition of organic matter associated with carbonate and non-carbonate sediments. *Geochim Cosmochim Acta* 42:1231–1238.

Chafetz HS, Buczynski C. 1992. Bacterially induced lithification of microbial mats. *Palaios* 7:277–293.

Chinzei K. 2013. Adaptation of oysters to life on soft substrates. *Historical Biology* 25:223–231.

Chinzei K, Seilacher A. 1993. Remote biomineralization I: Fill skeletons in vesicular oyster shells. *Neues Jahrbuch fur Geologie und Palaontologie Abhandlungen* 190:349–361.

Collins JF, Kenter JAM, Harris PM, Kuanysheva G, Fischer DJ, Steffen KL. 2006. Facies and reservoir-quality variations in the late Visean to Bashkirian outer platform, rim, and flank of the Tengiz buildup, Precaspian Basin, Kazakhstan. In: Harris PM, Weber LJ, eds., *Giant Hydrocarbon Reservoirs of the World: From Rocks to Reservoir Characterization and Modeling*. American Association of Petroleum Geologists Memoir 88, Tulsa, OK, pp. 55–95.

Coureadeau E, Benzerara K, Gerard E, Moreira D, Bernard S, Brown GE Jr, Lopez-Garcia P. 2012. An early-branching microbialite cyanobacterium forms intracellular carbonates. *Science* 336:459–462.

Davies PJ, Bubela B, Ferguson J. 1978. The formation of ooids. *Sedimentology* 25:703–730.

De Boer W, la Rivière J, Schmidt K. 1971. Some properties of *Achromobacter oxaliferum*. *Leeuwenhoek* 37:553–556.

Défarge C, Trichet J, Maurin A, Hucher M. 1994. Kopara in Polynesian atolls: Early stages of formation of calcareous stromatolites. *Sedimentary Geology* 89:9–23.

de Jong EW, van der Wal P, Borman AH, de Vrind JPM, Van Emburg P, Westbroek P, Bosch L. 1983. Calcification in coccolithophorids. In: Westbroek P, de Jong WE, eds., *Biomineralization and Biological Metal Accumulation*. Dodrecht, Netherlands: Reidel, pp. 291–301.

de Nooijer LJ, Langer, G, Nehrke, G, and Bijma, J. 2009. Physiological controls on seawater uptake and calcification in the benthic foraminifer Ammonia tepida. *Biogeosciences* 6:2669–2675.

de Vrind-de Jong EW, de Vrind JPM. 1997. Algal deposition of carbonates and silicates. In: Banfield JG, Nealson KH, eds., *Geomicrobiology: Interactions Between Microbes and Minerals*. Rev Mineral, Vol. 35. Washington, DC: Mineralological Society of America, pp. 267–307.

del Moral A, Roldan E, Navarro J, Monteoliva-Sanchez M, Ramos-Cormenzana A. 1987. Formation of calcium carbonate crystals by moderately halophilic bacteria. *Geomicrobiol J* 5:79–87.

Dierssen HM, Zimmerman RC, Burdige DJ. 2009. Optics and remote sensing of Bahamian carbonate sediment whitings and potential relationship to wind-driven Langmuir circulation. *Biogeosciences* 6:487–500.

Dupraz C, Visscher PT, Baumgartner LK, Reid RP. 2004. Microbe–mineral interactions: Early carbonate precipitation in a hypersaline lake (Eleuthera Island, Bahamas). *Sedimentology* 51:745–765.

Ehrlich H, Simon P, Carillo-Cabrera V et al. 2010. Insights into chemistry of biological materials: Newly discovered silica-aragonite-chitin biocomposites in demosponges. *Chem Mater* 22(4):1462–1471.

Ehrlich HL, Wickert LM. 1997. Bacterial action on bauxites in columns fed with full-strength and dilute sucrose-mineral salts medium. In: Lortie L, Bédard P, Gould WD, eds., *Biotechnology and the Mining Environment. Proceedings of the 13th Annual General Meeting of BIOMINET Sp 97–1*. Ottawa, Canada: CANMET, Natural Resources Canada, pp. 74–89.

Erez J. 2003. The source of ions for biomineralization in Foraminifera and their implications for paleoceanographic proxies. In: Dove PM, De Yoreo JJ, Weiner S, eds., *Biomineralization, Rev Mineralog Geochem*. Washington, DC: Mineralogical Society of America.

Fenchel T, Blackburn TH. 1979. *Bacteria and Mineral Cycling*. London, U.K.: Academic Press, pp. 115–149.

Ferguson J, Bubela B, Davies PJ. 1978. Synthesis and possible mechanism of formation of radial carbonate ooids. *Chem Geol* 22:285–308.

Ferrer MR, Quevedo-Sarmiento J, Bejar V, Delgado R, Ramos-Cormenzana A, Rivadeneyra MA. 1988a. Calcium carbonate formation by *Deleya halophila*: Effect of salt concentration and incubation temperature. *Geomicrobiol J* 6:49–57.

Ferris FG, Fratton CM, Gerits JP, Schultze-Lam S, Lollar BS. 1995. Microbial precipitation of a strontium calcite phase at a groundwater discharge zone near Rock Creek, British Columbia, Canada. *Geomicrobiol J* 13:57–67.

Freeman T. 1962. Quiet water oolites from Laguna Madre, Texas. *J Sed Petrol* 32:475–483.

Friedmann EI. 1977. Microorganisms in Antarctic desert rocks from dry valleys and Dufek Massif. *Antarctic J US* 12:26–30.

Friedmann EI. 1980. Endolithic microbial life in hot and cold deserts. *Orig Life* 10:223–235.

Friedmann EI. 1982. Endolithic microorganisms in the Antarctic Cold Desert. *Science* 215:1045–1053.

Friedmann EI, Ocampo R. 1976. Endolithic blue-green algae in the dry valleys: Producers in the Antarctic desert ecosystem. *Science* 193:1247–1249.

Friedmann EI, Ocampo-Friedmann R. 1984. Endolithic microorganisms in extreme dry environments: Analysis of a lithobiontic microbial habitat. In:

Klug MJ, Reddy CA, eds., *Current Perspectives in Microbial Ecology. Proc 3rd Int Symp Microb Ecol.* Washington, DC. American Society of Microbiology, pp. 177–185.

Friedmann EI, Roth WC, Turner JB, McEwin RS. 1972. Calcium oxalate crystals in the aragonite-producing green alga *Penicillus* and related genera. *Science* 177:891–893.

Friedmann EI, Weed R. 1987. Microbial trace-fossil formation, biogenous and abiotic weathering in the Antarctic Cold Desert. *Science* 236:703–705.

Gal A, Weiner S, Addadi L. 2010. The stabilizing effect of silicate on biogenic and synthetic amorphous calcium carbonate. *Journal of the American Chemical Society Communications* 132:13208–13211.

Garcia-Pichel F. 2006. Plausible mechanisms for the boring on carbonates by microbial phototrophs. *Sed Geol* 185:205–213.

Gautret P, Camoin G, Golubic S, Sprachta S. 2004. Biochemical control of calcium carbonate precipitation in modern lagoonal microbialites, Tikehau Atoll, French Polynesia. *J Sed Res* 74:462–478.

Golubic S. 1969. Distribution, taxonomy, and boring patterns of marine endolithic algae. *Am Zool* 9:747–751.

Golubic S. 1973. The relationship between blue-green algae and carbonate deposits. In: Carr NG, Whitton BA, eds., *The Biology of Blue Green Algae.* Oxford, U.K.: Blackwell Scientific, pp. 434–472.

Golubic S, Friedmann I, Schneider J. 1981. The lithobiontic ecological niche, with special reference to microorganisms. *J Sediment Petrol* 51:475–478.

Golubic S, Perkins RD, Lukas KJ. 1975. Boring microorganisms and microborings in carbonate substrates. In: Frey RW, ed., *The Study of True Fossils.* New York: Springer, pp. 229–259.

Golubic S, Schneider J. 1979. Carbonate dissolution. In: Trudinger PA, Swaine DJ, eds., *Biogeochemical Cycling of Mineral-Forming Elements.* Amsterdam, Netherlands: Elsevier, pp. 107–129.

Gorbushina AA, Krumbein WE, Hamman CH, Panina L. Solukharjevski S, Wollenzien U. 1994. Role of black fungi in color change and biodeterioration of antique marbles. *Geomicrobiol J* 11:205–211.

Herman JS. 1994. Karst geomicrobiology and redox geochemistry: State of the science. *Geomicrobiol J* 12:137–140.

Hemleben C, Anderson RO, Berthold W, Spindler M. 1986. Chamber formation in Foraminifera—A brief overview. In: Leadbeater BS, Riding R, eds., *Biomineralization in Lower Plants and Animals,* Vol. 30. The Systematics Association Spec, Oxford, U.K., pp. 237–249.

Higgins JA, Fischer WW, Schrag DP. 2009. Oxygenation of the ocean and sediments: Consequences for the seafloor carbonate factory. *Earth Planet Sci Lett* 284:25–33.

Hung Kiang C, Kowsmann RO, Figueiredo AMF, Bender AA. 1992. Tectonics and stratigraphy of the East Brazil Rift system: An overview. *Tectonophysics* 213:97–138.

Isenberg HD, Lavine LS. 1973. Protozoan calcification. In: Zipkin I, ed., *Biological Mineralization.* New York: Wiley, pp. 649–686.

Jensen PR, Gibson RA, Littler MM, Littler DS. 1985. Photosynthesis and calcification in four deep-water *Halimeda* species (Chlorophyceae, Caulerpales). *Deep Sea Res* 32:451–464.

Kah LC, Lyons TW, Frank TD. 2002. Low marine sulphate and protracted oxygenation of the Proterozoic biosphere. *Nature* 431:434–438.

Kah LC, Riding R. 2007. Mesoproterozoic carbon dioxide levels inferred from calcified cyanobacteria. *Geology* 35:799–802.

Kawaguchi T, Decho A. 2002. A laboratory investigation of cyanobacterial extracellular polymeric secretions (EPS) in influencing $CaCO_3$ polymorphism. *J Crystal Growth* 240:230–235.

Kempe S, Kaźmierczak J. 1993. Satonda Crater Lake: Hydrogeochemistry and carbonates. *Facies* 28:1–31.

Kempe S, Kazmierczak J. 1994. The role of alkalinity in the evolution of ocean chemistry, organization of living systems, and biocalcification processes. *Bulletin de la Institut Océanographique (Monaco)* 13:61–117.

Kenward PA, Goldstein RH, Gonzalez LA, Roberts JA. 2009. Precipitation of low-temperature dolomite from an anaerobic microbial consortium: The role of methanogenic Archaea. *Geobiology* 7:556–565.

Knoll AH, Fairchild IJ, Swett K. 1993. Calcified microbes in Neoproterozoic carbonates; implications for our understanding of the Proterozoic/Cambrian transition. *Palaios* 8:512–525.

Knoll AH, Wörndle S, Kah LC. 2013. Covariance of microfossil assemblages and microbialite fabrics across an Upper Mesoproterozoic carbonate platform. *Palaios* 28:453–470.

Krumbein WE. 1968. Zur Frage der biologischen Verwitterung: Einfluss der Mikroflora auf die Bausteinverwitterung und ihre Abhängigkeit von adaphischen Faktoren. *Z Allg Mikrobiol* 8:107–117.

Krumbein WE. 1974. On the precipitation of aragonite on the surface of marine bacteria. *Naturwissenschaften* 61:167.

Krumbein WE. 1979. Photolithotrophic and chemoorganotrophic activity of bacteria and algae as related to beachrock formation and degradation (Gulf of Aqaba, Sinai). Geomicrobiol J 1:139–203.

Krumbein WE, Cohen Y. 1977. Primary production, mat formation and lithification: Contribution of oxygenic and facultative oxygenic cyanobacteria. In: Flügel E, ed., Fossil Algae: Recent Results and Developments. Berlin, Germany: Springer, pp. 37–56.

Ku TCW, Walter LM, Coleman ML, Blake RE, Martini AM. 1999. Coupling between sulfur recycling and syndepositional carbonate dissolution: Evidence from oxygen and sulfur isotope composition of pore water sulfate, South Florida Platform, U.S.A. Geochim Cosmochim Acta 63:2529–2546.

Kunioka D, Shirai K, Takahata N, Sano Y, Toyofuku T, Ujiie Y. 2006. Microdistribution of Mg/Ca, Sr/Ca, and Ba/Ca ratios in Pulleniatina obliquiloculata test by using a NanoSIMS: Implication for the vital effect mechanism. Geochem Geophys Geosyst 7:Q12P20.

LaRock PA, Ehrlich HL. 1975. Observations of bacterial microcolonies on the surface of ferromanganese nodules from Blake Plateau by scanning electron microscopy. Microb Ecol 2:84.

Lepot K, Benzerara K, Brown GE, Phillipot P. 2008. Microbially influenced formation of 2,724-million-year-old-stromatolites. Nature Geosci 1:118–127.

Lewin J. 1965. Calcification. In: Lewin RA, ed., Physiology and Biochemistry of Algae. New York: Academic Press, pp. 457–465.

Lowenstam HA, Epstein S. 1957. On the origin of sedimentary aragonite needles of the Great Bahama Bank. J Geol 65:364–375.

Luff R, Wallmann K. 2003. Fluid flow, methane fluxes, carbonate precipitation and biogeochemical turnover in gas hydrate-bearing sediments at Hydrate Ridge, Cascadia Margin: Numerical modeling and mass balances. Geochim Cosmochim Acta 67:3403–3421.

Luff R, Wallmann K, Aloisi G. 2004. Numerical modeling of carbonate crust formation at cold vent sites: Significance for fluid and methane budgets and chemosynthetic biological communities. Earth Planet Sci Lett 221:337–353.

Lyons WB, Long DT, Hines ME, Gaudette HE, Armstrong PB. 1984. Calcification of cyanobacterial mats in Solar Lake, Sinai. Geology 12:623–626.

Macintyre IG, Aronson RB. 2006. Lithified and unlithified Mg-calcite precipitates in tropical reef environments. J Sed Res 76:81–90.

Mancini, EA, Benson, DJ, Hart, BS, Balch, RS, Parcell, WC, Panetta, BJ, 2000. Appleton field case study

(eastern Gulf Coast plain): Field development model for upper Jurassic microbial reef reservoirs associated with paleotopographic basement structures. Am Assoc Petrol Geol Bull 84:1699–1717.

Marine Chemistry. 1971. A report of the Marine Chemistry Panel of the Committee of Oceanography, National Academy of Sciences, Washington, DC.

Mata SA, Harwood CL, Corsetti FA et al. 2012. Influence of gas production and filament orientation on stromatolite microfabric. Palaios 27:206–219.

Matsko NB, Žnidaršič N, Letofsky-Papst I, Dittrich M, Grogger W, Štrus J, Hofer F. 2011. Silicon: The key element in early stages of biocalcification. J Struct Biol 174:180–186.

McConnaughey T. 1991. Calcification in Chara corallina: CO_2 hydroxylation generates protons for bicarbonate assimilation. Limnol Oceanogr 36:619–628.

Merz-Preiß M, Riding R. 1999. Cyanobacterial tufa calcification in two freshwater streams: Ambient environment, chemical thresholds and biological processes. Sed Geol 126:103–124.

Milliman JD, Freile D, Steinen RP, Wilber RJ. 1993. Great Bahama Bank aragonitic muds: Mostly inorganically precipitated, mostly exported. J Sediment Petrol 63:589–595.

Monty CLV. 1967. Distribution and structure of recent stromatolitic algal mats, Eastern Andros Island, Bahamas. Ann Soc Geol Belgi 90:55–99.

Monty CLV. 1972. Recent algal stromatolitic deposits, Andros Island, Bahamas. Preliminary report. Geol Rundschau 61:742–783.

Moore JN. 1983. The origin of calcium carbonate nodules forming on Flathead Lake delta, northwestern Montana. Limnol Oceanogr 28:646–654.

Morita RY. 1980. Calcite precipitation by marine bacteria. Geomicrobiol J 2:63–82.

Morris, PA, Soule, DF. 2004. The potential role of microbial activity and mineralization in exoskeletal development of Microporellidae. In: Moyano H, Cancino, Wyse PN, Jackson J eds. Bryozoan Studies. Leiden, London, New York, Philadelphia, Singapore: Balkima Publishers, pp. 181–186.

Morse JW, Arvidson RS, Luttge A. 2007. Calcium carbonate formation and dissolution. Chem Rev 107:342–381.

Morse JW, Gledhill DK, Millero FJ. 2003. $CaCO_3$ precipitation kinetics in waters from the great Bahama bank: Implications for the relationship between bank hydrochemistry and whitings. Geochim Cosmochim Acta 67:2819–2826.

Nazina TN, Rozanova EP, Kuznetsov SI. 1985. Microbial oil transformation processes accompanied by methane and hydrogen sulfide formation. *Geomicrobiol J* 4:103–130.

Nehrke G, Keul N, Langer G, de Nooijer L, Bijma J, Meibom A. 2013. A new model for biomineralization and trace-element signatures of Foraminifera tests. *Biogeosciences* 10:6759–6767.

Neumann AC, Gebelein CD, Scoffin TP. 1970. The composition, structure and erodibility of subtidal mats, Abaco, Bahamas. *J Sed Petrol* 40:274–297.

Nothdurft LD, Webb GE, Bostrom TE, Rintoul L. 2007. Calcite-filled borings in the most recently deposited skeleton in live-collected Porites (Scleractinia): Implications for trace element archives. *Geochim Cosmochim Acta* 71:5423–5438.

Obst M, Wang J, Hitchcock AP. 2009. Soft X-ray spectro-tomography study of cyanobacterial biomineral nucleation. *Geobiology* 7:577–591.

Okumura T, Takashima C, Shiraishi F, Akmaluddin, Kano A. 2012. Textural transition in an aragonite travertine formed under various flow conditions at Pancuran Pitu, Central Java, Indonesia. *Sed Geol* 265–266:195–209.

Okumura T, Takashima C, Shiraishi F, Nishida S, Kano A. 2013. Processes forming daily lamination in a microbe-rich travertine under low flow condition at the Nagano-yu hot spring, Southwestern Japan. *Geomicrobiol J* 30:910–927.

Omelon CR, Pollard WH, Ferris FG. 2006a. Chemical and ultrastructural characterization of high Arctic cryptoendolithic habitats. *Geomicrobiol J* 23:189–200.

Omelon CR, Pollard WH, Ferris FG. 2006b. Environmental controls on microbial colonization of high Arctic cryptoendolithic habitats. *Polar Biol* 30:19–29.

Paine SG, Lingood FV, Schimmer F, Thrupp TC. 1933. IV. The relationship of microorganisms to the decay of stone. *Roy Soc Phil Trans* 222B:97–127.

Pentecost A. 1978. Blue-green algae and freshwater carbonate deposits. *Proc Roy Soc Lond B* 200:43–61.

Pentecost A. 1995. Significance of the biomineralizing microniche in a *Lyngbya* (Cyanobacterium) travertine. *Geomicrobiol J* 13:213–222.

Pentecost A, Bauld J. 1988. Nucleation of calcite on the sheaths of cyanobacteria using a simple diffusion cell. *Geomicrobiol J* 6:129–135.

Pentecost A, Spiro B. 1990. Stable carbon and oxygen isotope composition of calcites associated with modern freshwater cyanobacteria and algae. *Geomicrobiol J* 8:17–26.

Perry CT, Salter MA, Harborne AR et al. 2011. Fish as major carbonate mud producers and missing components of the tropical carbonate factory. *Proc Natl Acad Sci USA* 108:3865–3869.

Planavsky N, Reid RP, Lyons TW, Myshrall KL, Visscher PT. 2009. Formation and diagenesis of modern calcified cyanobacteria. *Geobiology* 7:566–576.

Plee K, Ariztegui D, Martini R, Davaud E. 2008. Unravelling the microbial role in ooid formation—Results of an in situ experiment in modern freshwater Lake Geneva in Switzerland. *Geobiology* 6:341–350.

Politi Y, Arad T, Klein E, Weiner S, Addadi L. 2004. Sea urchin spine calcite forms via a transient amorphous calcium carbonate phase. *Science* 306:1161–1164.

Post WH, Peng T-H, Emanuael WE, King AW, Dale H, DeAngellis DL. 1990. The global carbon cycle. *Am Sci* 78:310–326.

Pruss SB, Bosak T, Macdonald FA, McLane M, Hoffman PF. 2010. Microbial facies in a Sturtian cap carbonate, the Rasthof Formation, Otavi Group, northern Namibia. *Precambrian Res* 181:187–198.

Pye K. 1984. SEM analysis of siderite cements in intertidal marsh sediments, Norfolk, England. *Mar Geol* 56:1–12.

Pye K, Dickson AD, Schiavon N, Coleman ML, Cox M. 1990. Formation of siderite-Mg-calcite-iron sulfide concentrations in intertidal marsh and sand-flat sediments, north Norfolk, England. *Sedimentology* 37:325–343.

Riding R. 1991. Calcified cyanobacteria. In: Riding R, ed., *Calcareous Algae and Stromatolites*. Berlin, Germany: Springer-Verlag, pp. 55–87.

Riding R. 1992. Temporal variation in calcification in marine cyanobacteria. *J Geol Soc (Lond)* 149:979–989.

Riding R. 2006. Cyanobacterial calcification, carbon dioxide concentrating mechanisms, and Proterozoic–Cambrian changes in atmospheric composition. *Geobiology* 4:299–316.

Rivadeneyra MA, Delgado R, Quesada E, Ramos-Cormenzana A. 1991. Precipitation of calcium carbonate by *Deleya halophila* in media containing NaCl as sole salt. *Curr Microbiol* 22:185–190.

Rivadeneyra MA, Martín-Algarra A, Sánchez-Navas A, Martín-Ramos D. 2006. Carbonate and phosphate precipitation by *Chromohalobacter marismortui*. *Geomicrobiol J* 23:1–13.

Rivadeneyra MA, Martín-Algarra A, Sánchez-Román M, Sánchez-Navas A, Martín-Ramos JD. 2010. Amorphous Ca-phosphate precursors for Ca-carbonate biominerals mediated by *Chromohalobacter marismortui*. *ISME J* 4:922–932.

Rivadeneyra MA, Perez-Garcia I, Salmeron V, Ramos-Cormenzana A. 1985. Bacterial precipitation of calcium carbonate in the presence of phosphate. *Soil Biol Biochem* 17:171–172.

Robbins LL, Blackwelder PL. 1992. Biochemical and ultrastructural evidence for the origin of whitings: A biologically induced calcium carbonate precipitation mechanism. *Geology* 20:464–468.

Roberts JA, Bennett PC, Gonzalez LA, Macpherson GL, Milliken KL. 2004. Microbial precipitation of dolomite in methanogenic groundwater. *Geology* 32:277–280.

Roberts JA, Kenward PA, Fowle DA, Goldstein RH, Gonzalez LA, Moore DS. 2013. Surface chemistry allows for abiotic precipitation of dolomite at low temperature. *Proc Natl Acad Sci USA* 110:14540–14545.

Rodriguez-Navarro C, Rodriguez-Gallego M, Chekroun KB, Gonzalez-Muñoz MT. 2003. Conservation of ornamental stone by *Myxococcus xanthus*-induced carbonate mineralization. *Appl Environ Microbiol* 69:2182–2193.

Roemer R, Schwartz E. 1965. Geomikrobiologische Untersuchungen. V. Verwertung von Sulfatmineralien and Schwermetallen. Tolleranz bei Desulfizierern. *Z Allg Mikrobiol* 5:122–135.

Röttger R. 1974. Larger foraminifera: Reproduction and early stages of development in Heterostegina depressa. *Marine Biol (Berlin)*, 26, 5–12.

Sánchez-Román M, Rivadeneyra MA, Vasconcelos C, McKenzie JA. 2007. Biomineralization of carbonate and phosphate by moderately halophilic bacteria. *FEMS Microbiol Ecol* 61:273–284.

Sánchez-Román M, Vasconcelos C, Schmid T, Dittrich M, McKenzie JA, Zenobi R, Rivadeneyra MA. 2008. Aerobic microbial dolomite at the nanometer scale: Implications for the geologic record. *Geology* 36:879–882.

Sarbu SM, Kane TC, Kinkle BK. 1996. A chemoautotrophically based cave ecosystem. *Science* 272:1953–1955.

Sarbu SM, Kinkle BK, Vlasceanu L, Kane TC, Popa R. 1994. Microbiological characterization of a sulfide-rich groundwater ecosystem. *Geomicrobiol J* 12:175–182.

Schmalz RF. 1972. Calcium carbonate: Geochemistry. In: Fairbanks RW, ed., *The Encyclopedia of Geochemistry and Environmental Sciences*. Encycl Earth Sci Ser, Vol. IVA. New York: Van Nostrand Reinhold, p. 110.

Schröder, S, Grotzinger, JP, Amthor, JE, Matter, A, 2005. Carbonate deposition and hydrocarbon reservoir development at the Precambrian–Cambrian boundary: The Ara Group in South Oman. *Sediment Geol* 180:1–28.

Shachak M, Jones CG, Granot Y. 1987. Herbivory in rocks in the weathering of a desert. *Science* 236:1098–1099.

Shinano H. 1972a. Studies on marine microorganisms taking part in the precipitation of calcium carbonate. II. Detection and grouping of the microorganisms taking part in the precipitation of calcium carbonate. *Bull Jpn Soc Sci Fisheries* 38:717–725.

Shinano H. 1972b. Microorganisms taking part in the precipitation of calcium carbonate. III. A taxonomic study of marine bacteria taking part in the precipitation of calcium carbonate. *Bull Jpn Soc Sci Fisheries* 38:727–732.

Shinano H, Sakai M. 1969. Studies on marine bacteria taking part in the precipitation of calcium carbonate. I. Calcium carbonate deposited in peptone medium prepared with natural seawater and artificial seawater. *Bull Jpn Soc Sci Fisheries* 35:1001–1005.

Shinn E., Steinen RP, Lidz BH, Swart PK. 1989. Whitings, a sedimentologic dilemma. *J Sed Petrol* 59:147–161.

Skirrow G. 1975. The dissolved gases—Carbon dioxide. In: Riley JR, Skirrow G, eds., *Chemical Oceanography*, Vol. 2, 2nd ed. London, U.K.: Academic Press, pp. 144–152.

Sokolova-Dubinina GA, Deryugina ZP. 1967. On the role of microorganisms in the formation of rhodochrosite in Punnus-Yarvi Lake. *Mikrobiologiya* 36:535–542.

Sondi I, Juracic M. 2010. Whiting events and the formation of aragonite in Mediterranean Karstic Marine Lakes: New evidence on its biologically induced inorganic origin. *Sedimentology* 57:85–95.

Spiro B, Pentecost A. 1991. One day in the life of a stream—A diurnal inorganic carbon mass balance for the travertine depositing stream (Waterfall Beck, Yorkshire). *Geomicrobiol J* 9:1–11.

Sprachta S, Camoin G, Golubic S, Le Campion Th. 2001. Microbialites in a modern lagoonal environment: Nature and distribution, Tikehau atoll (French Polynesia). *Palaeogeogr Palaeoclimatol Palaeoecol* 175:103–124.

Stutzer A, Hartleb R. 1899. Die Zersetzung von Cement unter dem Einfluss von Bakterien. *Z Angew Chem* 12:402 (cited by Paine et al., 1933).

Suess E. 1979. Mineral phases formed in anoxic sediments by microbial decomposition of organic matter. *Geochim Cosmochim Acta* 43:339–352.

Suess E, Fütterer D. 1972. Aragonite ooids: Experimental precipitation from seawater in the presence of humic acid. *Sedimentology* 19:129–139.

Sumner DY. 1997. Late Archean calcite-microbe interactions; Two morphologically distinct microbial communities that affected calcite nucleation differently. *Palaios* 12:302–318.

Swett K, Knoll AH. 1985. Stromatolitic bioherms and microphytolites from the Late Proterozoic Draken Conglomerate Formation, Spitsbergen. *Precambr Res* 28:327–347.

Taylor JD, Glover EA, Braithwaite CJR. 1999. Bivalves with "concrete overcoats": *Granicorium* and *Samarangia. Acta Zoologica* 80:285–300.

Thompson JB. 2000. Microbial whitings. In: Riding RE, Awaramik SM, eds., *Microbial Sediments*. Berlin, Germany: Springer, pp. 250–260.

Thompson JB, Ferris FG. 1990. Cyanobacterial precipitation of gypsum, calcite and magnesite from natural alkaline lake waters. *Geology* 18:995–998.

Thompson JB, Ferris FG, Smith DA. 1990. Geomicrobiology and sedimentology of the mixolimnion and chemocline in Fayetteville Green Lake, New York. *Palaios* 5:52–75.

Turner EC, Narbonne GM, James NP. 1993. Neoproterozoic reef microstructures from the Little Dal Group, northwestern Canada. *Geology* 21:259–262.

Turner EC, Narbonne GM, James NP. 2000. Taphonomic control on microstructure in early Neoproterozoic reefal stromatolites and thrombolites. *Palaios* 15:87–111.

Uriz MJ, Agell G, Blanquer A, Turon X, Casamayor EO. 2012. Endosymbiotic calcifying bacteria: A new cue to the origin of calcification in metazoa? *Evolution* 66(10):2993–2999.

Urzì C, Garcia-Valles M, Vendrell M, Pernice A. 1999. Biomineralization processes on rock and monument surfaces observed in field and laboratory conditions. *Geomicrobiol J* 16:39–54.

Urzì C, Lisi S, Criseo G, Penrice A. 1991. Adhesion to and decomposition of marble by a *Micrococcus* strain isolated from it. *Geomicrobiol J* 9:81–90.

van der Wal P, de Jong EW, Westbroek P, de Bruijn WC, Mulder-Stapel AA. 1983. Polysaccharide localization, coccolith formation, and Golgi dynamics in the coccolithophorid *Hymenomonas carterae. J Ultrastruct Res* 85:139–158.

Van Ee NJ, Wanless HR 2008. Ooids and grapestone—A significant source of mud on Caicos platform, Proc. Dev. *Models Analogs Isol. Carbonate Platf.—Holocene Pleistocene Carbonates Caicos Platf. Br. West Indies Soc. Sediment. Geol.*, pp. 121–125.

Van Lith Y, Warthmann R, Vasconcelos C, McKenzie JA. 2003. Microbial fossilization in carbonate sediments: A result of the bacterial surface involvement in dolomite precipitation. *Sedimentology* 50:237–245.

Vasconcelos C, McKenzie JA. 1997. Microbial mediation of modern dolomite precipitation and diagenesis under anoxic conditions (Lagoa Vermelha, Rio de Janeiro, Brazil). *J Sed Res* 67:378–390.

Vermeij GJ. 2014. The oyster enigma variations: A hypothesis of microbial calcification. *Paleobiology* 40(1):1–13.

Verrecchia EP, Dumant J-L, Collins KE. 1990. Do fungi building limestone exist in semi-arid regions? *Naturwissenschaften* 77:584–586.

Visscher PT, Reid RP, Bebout BM. 2000. Microscale observations of sulfate reduction: Correlation of microbial activity with lithified micritic laminae in modern stromatolites. *Geology* 28:919–922.

Wacey D, Gleeson D, Kilburn MR. 2010. Microbialite taphonomy and biogenicity: New insights from NanoSIMS. *Geobiology* 8:403–416.

Wagner E, Schwartz W. 1965. Geomikrobiologische Untersuchungen. VIII. Über das Verhalten von Bakterien auf der Oberfläche von Gesteinen und Mineralien und ihre Rolle bei der Verwitterung. *Z Allg Mikrobiol* 7:33–52.

Walter LM, Burton EA. 1990. Dissolution of recent platform carbonate sediments in marine pore fluids. *Am J Sci* 290:601–643.

Warren J. 2000. Dolomite: Occurrence, evolution and economically important associations. *Earth Sci Rev* 52:1–81.

Warthmann R, van Lith Y, Vasconcelos C, McKenzie JA, Karpoff AM. 2000. Bacterially induced dolomite precipitation in anoxic culture experiments. *Geology* 28:1091–1094.

Weast RC, Astle MJ, eds. 1982. *CRC Handbook of Chemistry and Physics*, 63rd ed. Boca Raton, FL: CRC Press.

Wefer G. 1980. Carbonate production by algae *Halimeda, Penicillus*, and *Padina. Nature (Lond)* 285:323–324.

Weiner S, Mahamid J, Politi Y, Ma Y, Addadi L. 2009. Overview of the amorphous precursor phase strategy in biomineralization. *Front Mater Sci Chin* 3:104–108.

Weyl PK. 1967. The solution behavior of carbonate minerals in sea water. *Stud Trop Oceanogr U Miami* 5:178–228.

Wright DT. 1999. The role of sulfate-reducing bacteria and cyanobacteria in dolomite formation in distal ephemeral lakes of the Coorong region, South Australia. *Sed Geol* 126:147–157.

Wright DT, Wacey D. 2005. Precipitation of dolomite using sulphate-reducing bacteria from the Coorong Region, South Australia: Significance and implications. *Sedimentology* 52:987–1008.

Young JR, Henriksen K. 2003. Biomineralization within vesicles: The calcite of coccoliths. In: Dove PM, De Yoreo JJ, Weiner S, eds., *Biomineralization, Rev Mineral Geochem*. Washington, DC: Mineralogical Society of America.

Zhang F, Xu H, Konishi H, Shelobolina ES, Roden EE. 2012. Polysaccharide-catalyzed nucleation and growth of disordered dolomite: A potential precursor of sedimentary dolomite. *Am Mineral* 97:556–567.

CHAPTER ELEVEN

Geomicrobial Interactions with Silicon

Kurt Konhauser

CONTENTS

11.1 Distribution and Some Chemical Properties of Silicon / 237
11.2 Biologically Important Properties of Silicon and Its Compounds / 238
11.3 Bioconcentration of Silicon / 239
    11.3.1 Diatoms / 239
    11.3.2 Radiolarians / 242
    11.3.3 Silicoflaggelates / 242
    11.3.4 Sponges / 242
11.4 Passive Silica Biomineralization / 242
11.5 Biomobilization of Silicon and Other Constituents of Silicates (Bioweathering) / 246
    11.5.1 Solubilization by Ligands / 246
    11.5.2 Solubilization by Acids / 247
    11.5.3 Solubilization by Alkali / 249
    11.5.4 Solubilization by Extracellular Polysaccharide / 249
11.6 Role of Microbes in the Silica Cycle / 249
11.7 Summary / 250
References / 250

11.1 DISTRIBUTION AND SOME CHEMICAL PROPERTIES OF SILICON

The element silicon is one of the most abundant in the Earth's crust, ranking second only to oxygen. Its estimated crustal abundance is 27.7% (w/w), whereas that of oxygen is 46.6% (Mitchell, 1955). The concentration of silicon in various components of the Earth's surface is listed in Table 11.1.

In nature, silicon occurs generally in the form of silicates, including ferromagnesian silicates (e.g., olivine, pyroxenes, and amphiboles), aluminosilicates (e.g., feldspar, mica, and clays), and silicon dioxide (e.g., amorphous silica, quartz). In general, the silicon in silicate minerals is surrounded by four oxygen atoms in tetrahedral fashion (Kretz, 1972). In most ferromagnesian minerals, iron, magnesium, and/or calcium links single or double chains of silica tetrahedral together, while in aluminosilicates, the aluminum is coordinated with oxygen in sheets or 3D frameworks, depending on the mineral (see Tan, 1986). In clays, which form during the weathering of primary aluminosilicates, silica tetrahedral sheets and aluminum hydroxide octahedral sheets are layered in different ways depending on the clay type. In montmorillonite-type clays, structural units consisting of single aluminum hydroxide octahedral sheets are sandwiched between silica tetrahedral sheets. The units are interspaced with layers of water molecules of variable thickness into which other polar molecules, including some organic ones, and ions can enter. This variable water layer allows montmorillonite-type clays to swell. The structural units of illite-type clays resemble those of montmorillonite-type clays but differ from them in that Al replaces some of the Si in the silica tetrahedral sheets. These substituting Al atoms impart extra charges, which are neutralized by

TABLE 11.1
Abundances of silicon on the Earth's surface.

Phases	Concentration	Reference
Granite	336,000 ppm	Bowen (1979)
Basalt	240,000 ppm	Bowen (1979)
Shale	275,000 ppm	Bowen (1979)
Limestone	32,000 ppm	Bowen (1979)
Sandstone	327,000 ppm	Bowen (1979)
Soils	330,000 ppm	Bowen (1979)
Seawater	$3 \cdot 10^3$ µg L^{-1}	Marine Chemistry (1971)
Freshwater	7 ppm	Bowen (1979)

potassium ions between successive silica sheets. The potassium ions act as bridges that prevent the swelling exhibited by montmorillonite in water. In *kaolinite* clays, structural units consist of silica tetrahedral sheets alternating with aluminum hydroxide octahedral sheets are joined to one another by oxygen bridges (see Toth, 1955).

Silicon–oxygen bonds of *siloxane linkages* (Si–O–Si) in silicate minerals are very strong (their energy of formation ranges from 3110 to 3142 kcal mol^{-1} or 13,031 to 13,165 kJ mol^{-1}), whereas Al–O bonds are weaker (their energy of formation ranges from 1793 to 1878 kcal mol^{-1} or 7531 to 7869 kJ mol^{-1}). Bonds between nonframework cations and oxygen are the weakest (energy of formation ranges from 299 to 919 kcal mol^{-1} or 1252 to 3851 kJ mol^{-1}) (values cited by Tan, 1986). The strength of these bonds determines their susceptibility to weathering. Thus, Si–O bonds are relatively resistant to acid hydrolysis (Karavaiko et al., 1984) unlike Al–O bonds. Bonds between cations and oxygen are readily broken by protonation or cation exchange.

Silicon in solution at pH 2–9 exists in the form of undissociated monosilicic acid (H_4SiO_4), whereas at pH 9 and above, it transforms into silicate anions, such as $H_3SiO_4^-$ (see Hall, 1972). Monosilicic acid polymerizes in solutions supersaturated with respect to amorphous silica, forming oligomers of polysilicic acids (Iler, 1979). This polymerization reaction is favored around neutral pH where silica solubility is lowest (Avakyan et al., 1985). Polymerization of monosilicic acid can be viewed as a removal of water from between adjacent monomers to form a siloxane linkage. *Silica* can be viewed as an anhydride of *silicic acid*:

$$H_4SiO_4 \rightarrow SiO_2 + 2H_2O \quad (11.1)$$

Dissociation constants for silicic acid are as follows (see Anderson, 1972):

$$H_4SiO_4 \rightarrow H + H_3SiO_4^- \quad (K_1 = 10^{-9.5}) \quad (11.2)$$

$$H_3SiO_4^- \rightarrow H + H_2SiO_4^{2-} \quad (K_2 = 10^{-12.7}) \quad (11.3)$$

Silica can exist in partially hydrated form called *metasilicic acid* (H_2SiO_3) or in a fully hydrated form called orthosilicic acid (H_4SiO_4). Each of these forms can be polymerized, the ortho acid forming, for instance, $H_3SiO_4 \cdot H_2SiO_3 \cdot H_3SiO_3$ (Latimer and Hildebrand, 1940; Liebau, 1985). The polymers may exhibit colloidal properties, depending on size and other factors. Colloidal particles of silica tend to form at silica supersaturation conditions (Tobler et al., 2009) and are favored by acid conditions (Hall, 1972).

Silica and silicates form an important buffer system in the oceans (Garrels, 1965), together with the $CO_2/HCO_3^-/CO_3^{2-}$ system. The latter is a rapidly reacting system, whereas the system based on reaction with silica and silicates is slow (Garrels, 1965; Sillén, 1967).

Aluminosilicates in the form of clay perform a buffering function in mineral soils. This is because of their ion-exchange capacity, net electronegative charge, and adsorption powers that make them important reservoirs of cations and organic molecules. Montmorillonite exhibits the greatest ion-exchange capacity, illite less, and kaolinite the least (Dommergues and Mangenot, 1970).

11.2 BIOLOGICALLY IMPORTANT PROPERTIES OF SILICON AND ITS COMPOUNDS

Silicon is taken up and concentrated in significant quantities by certain forms of life. These include microbial forms such as diatoms and other chrysophytes; silicoflagellates and some xanthophytes; radiolarians and actinopods; some plants such as horsetails, ferns, grasses, and some flowers and trees; and also some animals such as sponges, insects, and even vertebrates. Some bacteria (Heinen, 1960) and fungi (Holzapfel and Engel, 1954a,b; Heinen, 1960) have also been reported to take up silicon to a limited extent. According to Bowen (1966), diatoms may contain from 1500 to 20,000 ppm silicon, land plants from 200 to 5000 ppm, and marine animals from 120 to 6000 ppm.

Although the function of silicon in higher forms of life, animals and plants, is not presently apparent, it is clearly structural in some microorganisms such as diatoms, actinopods, and radiolarians. In diatoms, silicon also seems to play a metabolic role in the synthesis of chlorophyll (Werner, 1966, 1967), DNA (Darley and Volcani, 1969; Reeves and Volcani, 1984), and DNA polymerase and thymidylate kinase (Sullivan, 1971; Sullivan and Volcani, 1973).

Silicon compounds in the form of clays (aluminosilicates) may exert an effect on microbes in soil. They may stimulate or inhibit microbial metabolism, depending on the conditions (Weaver and Dugan, 1972; Marshman and Marshall, 1981a,b; see also discussion by Marshall, 1971). These effects of clays are mostly indirect, that is, clays tend to modify the microbial habitat physicochemically, thereby eliciting a physiological response by the microbes (Stotzky, 1986). For beneficial effect, clays may buffer the soil environment and help maintain a favorable pH, thereby promoting growth and metabolism of some microorganisms that might otherwise be slowed or stopped if the pH became unfavorable (Stotzky, 1986). They also help with water retention with the pore spaces behaving as nanoreservoirs for water (Phoenix and Konhauser, 2008). Certain clays have been found to enable some bacteria that were isolated from marine ferromanganese nodules or associated sediments to oxidize Mn^{2+}. Intact cells of these organisms can only oxidize Mn^{2+} if it is bound to bentonite (montmorillonite-type clay) or kaolinite (but not illite) that were first pretreated with ferric iron. They cannot oxidize Mn^{2+} that is free in solution (Ehrlich, 1982). Cell-free preparations of these bacteria can oxidize Mn^{2+} bound to bentonite and kaolinite without ferric iron pretreatment, although manganese-oxidizing activity of the cell extracts is greater when clays were pretreated (Ehrlich, 1982). Like intact cells, the cell-free extract cannot oxidize dissolved Mn^{2+} (Ehrlich, 1982). Clays may also enhance the activity of some enzymes such as catalase when the enzymes are bound to their surface (see Stotzky, 1986, p. 404).

By contrast, clays may suppress microbial growth and metabolism by adsorbing organic nutrients, thereby making them less available to microbes. Clays may also adsorb microbial antibiotics and thereby lower the inhibitory activity of these agents (see Stotzky, 1986). In soils, the results may be that an antibiotic producer is outgrown by organisms that in vitro it keeps in check with the help of the antibiotic it excretes. These effects of clay can be explained, at least in part, by the strength of binding to a negatively charged clay surface and the inability of many microbes to attack adsorbed nutrients, or by the inability of adsorbed antibiotics to inhibit growth of susceptible microbes (see Dashman and Stotzky, 1986). High concentrations of clay may interfere with diffusion of oxygen by increasing the viscosity of a solution, which can have a negative effect on aerobic microbial respiration (see Stotzky, 1986). Clays may also modulate other interactions between different microbes and between microbes and viruses in soil, and they may affect the pathogenicity of these disease-causing soil microbes (see Stotzky, 1986).

Although clay-bound organic molecules may be less available or unavailable to organisms in the bulk phase or even attached to the mineral surface, this cannot be a universal phenomenon. Portions of attached large organic polymers may be attacked by appropriate extracellular enzymes, producing smaller unattached units that can be taken up by microbes. Electrostatically bound organic molecules that are potential nutrients may be dislodged by exchange with protons excreted as acids in the catabolism of some microbes. These processes of remobilization must also apply to mineral sorbents other than clays.

11.3 BIOCONCENTRATION OF SILICON

11.3.1 Diatoms

Diatoms are unicellular eukaryotic microorganisms that take up dissolved silica from both freshwater and seawater. Most live planktonically in the photic zone, and in the oceans, where they contribute nearly 50% of the total primary production (Mann, 1999). It is believed that diatoms first evolved in the early Jurassic (around 185 Ma), based on the oldest fossil evidence, although molecular clock studies suggest that they may be as old as 250 Ma (Kooistra and Medlin, 1996). Today, there are more than 100,000 extant species of living diatoms (e.g., Round and Crawford, 1990).

Diatoms have been most extensively studied with respect to their silica uptake ability (Figure 11.1) (Lewin, 1965; de Vrind-de Jong and de Vrind, 1997). They are unicellular algae enclosed in a wall of silica consisting of two

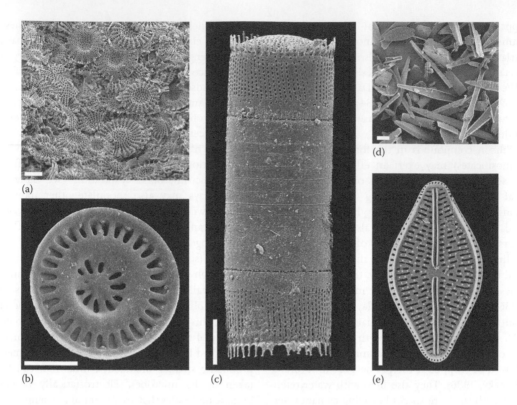

Figure 11.1. An overview of morphological variability in the diatom frustule as viewed in scanning electron microscopy (SEM): Modern and fossil representatives of centric (a–c) and pennate (d, e) genera. (a) Neogene fossil *Cyclostephanos* from British Columbia, Canada, (b) modern *Discostella stelligera* from the Ecuadorian Andes, (c) *Aulacoseira* from the Eocene of northwestern Canada, (d) modern periphyton dominated by the araphid genera *Fragilaria* and *Tabellaria*, and (e) the raphid pennate genus *Brachysira*. All scale bars are 5 μm. (Images courtesy of Alex Wolfe.)

valves, an epivalve and a hypovalve, in pillbox arrangement. One or more girdle bands are loosely connected to the epivalve. The valves are usually perforated plates, which may have thickened ribs. Their shape may be pennate or centric. The pores serve as sites of gas and nutrient exchange (see de Vrind-de Jong and de Vrind, 1997). In cell division, each daughter cell receives either the epivalve or hypovalve of the mother cell and synthesizes the other valve de novo to fit into the one already present. To prevent excessive reduction in size of the daughter diatoms that receive the hypovalve upon each cell division, a special reproductive step called *auxospore formation* returns these daughter cells to maximum size. It occurs when a progeny cell that has received a hypovalve has reached minimum size after repeated divisions. Auxospore formation is a sexual reproductive process in which the cells escape from their frustules and increase in size in their zygote membrane, which may become weakly silicified. After a time, the protoplast in

the zygote membrane contracts and forms the typical frustules of the parent cell (Lewin, 1965).

The silica walls of the diatoms consist of hydrated amorphous silica, a polymerized silicic acid (Lewin, 1965). The walls of marine diatoms may contain as much as 96.5% SiO_2, but only 1.5% Al_2O_3 or Fe_2O_3 and 1.9% water (Rogall, 1939). In clean, dried frustules of freshwater *Navicula pelliculosa*, 9.6% water has been found (Lewin, 1957). Thin parts of diatom frustules reveal a foamlike substructure when viewed under the electron microscope, suggesting silica gel (Helmcke, 1954), which may account for the adsorptive power of such frustules. The silica gel may be viewed as arranged in small spherical particles about 22 μm in diameter (Lewin, 1965). Because of the low solubility of amorphous silica at the pH of most natural waters, frustules of living diatoms do not dissolve readily (Lewin, 1965). At pH 8, however, it has been found that 5% of the silica in the walls of *Thalassiosira nana* and *Nitzschia linearis* dissolves. Moreover, at pH 10, 20% of the silica in the frustules of N. *linearis* and all of the silica in the frustules of T. *nana* dissolve (Jorgensen, 1955). This silica dissolution may reflect the state of integration of newly assimilated silica in the diatom frustule. Some bacteria naturally associated with diatoms have been shown to accelerate dissolution of silica in frustules (e.g., Bidle and Azam, 2001). Frustules of living diatoms are to some extent protected against dissolution by an organic film, when present, and their rate of dissolution has been shown to exhibit temperature dependence (Katami, 1982). After the death of diatoms, their frustules may dehydrate to form more crystalline SiO_2 that is much less soluble in alkali than that in living diatoms. This may account for the accumulation of diatomaceous ooze.

Rates of silica uptake and incorporation by diatoms can be easily measured with radioactive [^{65}Ge]germanic acid as tracer (Azam et al., 1973; Azam, 1974; Chisholm et al., 1978). At low concentration (Ge/Si molar ratio of 0.01), germanium, which is chemically similar to silicon, is apparently incorporated together with silicon into the silicic acid polymer of the frustule. At higher concentrations (Ge/Si molar ratio of 0.1), germanium is toxic to diatoms (Azam et al., 1973). Genetic control of silica transport into diatoms has begun to be studied on a molecular level (see review by Martin-Jézéquel et al., 2000).

Diatoms are able to discriminate between ^{28}Si and ^{30}Si by assimilating the lighter isotope preferentially. The fraction (α) for each of the three diatom species *Skeletonema costatum*, *Thalassiosira weissflogii*, and *Thalassiosira* sp. was 0.9989 ± 0.004. It was independent of temperature between 12°C and 22°C and thus, independent of growth rate (De La Rocha et al., 1997). This fractionation ability appears to be usable as a signature in identifying biogenic silica (De La Rocha et al., 2000).

Diatoms take up silica in the form of orthosilicate. More highly polymerized forms of silicate are not taken up unless first depolymerized, as by some bacteria (Lauwers and Heinen, 1974). Organic silicates are also not available to them. Ge, C, Sn, Pb, P, As, B, Al, Mg, and Fe do not replace silicon extensively if at all (Lewin, 1965). The concentration of silicon accumulated by a diatom depends to some extent on its concentration in the growth medium and on the rate of cell division (the faster the cells divide, the thinner their frustules). Silicon is essential for cell division, but resting cells in a medium in which silica

is not at a limiting concentration continue to take up silica (Lewin, 1965). Synchronously growing cells of N. *pelliculosa* take up silica at a constant rate during the cell division cycle (Lewin, 1965). Silica uptake appears dependent on energy-yielding processes (Lewin, 1965; Azam and Volcani, 1974; Azam et al., 1974; review by Martin-Jézéquel et al., 2000) and seems to involve intracellular receptor sites (Blank and Sullivan, 1979). Uncoupling of oxidative phosphorylation stops silica uptake by N. *pelliculosa* and *Nitzschia angularis*. Starved cells of N. *pelliculosa* show an enhanced silica uptake rate when fed glucose or lactate in the dark or when returned to the light, where they can photosynthesize (Healy et al., 1967). Respiratory inhibitors prevent Si and Ge uptake by *Nitzschia alba* (Azam and Volcani, 1974; Azam et al., 1974).

Total uptake of phosphorus and carbon is decreased during silica starvation of N. *pelliculosa*. Upon restoration of silica to the medium, the total uptake of phosphorus is again increased (Coombs and Volcani, 1968). Sulfhydryl groups (–SH) appear to be involved in silica uptake (Lewin, 1965).

Some progress has been made in understanding how diatoms form their siliceous cell walls (see de Vrind-de Jong and de Vrind, 1997). Valve and girdle-band assembly takes place inside the cell and happens late in the cell cycle during the last part of mitosis. For this assembly, silica is taken into the cell and polymerized in special membrane-bound *silica deposition vesicles* (SDVs), leading to the formation of the girdle bands and valves. The SDV seems to arise from the Golgi apparatus, a special membrane system within the cell. The endoplasmic reticulum, a membrane network within the cell that is connected to the plasma membrane and the nuclear membrane, may participate in SDV development. The active SDVs are located adjacent to the plasma membrane. The shape of the SDV may be determined by interaction with various cell components such as the plasma membrane, actin filaments, microtubules, and cell organelles. The SDV is believed to help determine the morphology of the valves. Frustule buildup in the SDV appears to start along the future raphe, which appears as a longitudinal slot in each mature pennate frustule. The raphe has a central thickening called a nodule. Completed valves are exocytosed by the cell, that is, they are exported to the cell surface. When the valves are in place at the cell surface of the diatom, the raphe plays a role in its motility. Mature frustules have glycoprotein associated with them, which may have played a role in silica assembly during valve formation. It may help in determining valve morphology and in the export of assembled valves to the cell surface. For additional information, the reader is referred to de Vrind-de Jong and de Vrind (1997) and references cited therein.

11.3.2 Radiolarians

Radiolarians are silica-secreting, protozoan zooplankton that mostly live planktonically in the surface ocean. The earliest fossil record for radiolaria is at the beginning of the Cambrian period (543 Ma) (Braun et al., 2013), and it is possible that they may even have evolved by the Neoproterozoic (Blair Hedges et al., 2001). Radiolarians have complex morphologies, most are spherical, but other body shapes, such as cones and tetrahedrons, also exist. Their skeletons tend to have spiny protrusions from the central body, which is used to increase their surface area for buoyancy and to capture prey (see Afanasieva et al., 2005). Due to their ornate skeletons, and ancient lineage, radiolarians have been widely used in biostratigraphy and interpreting the sedimentary record (De Wever et al., 2001).

11.3.3 Silicoflagellates

Silicoflagellates are a small group of unicellular algae found throughout the marine photic zone. Their origins extend back to the early Cretaceous (McCartney, 1993). Today, the number of species is believed to vary between only 1 and 3, despite many recognizable forms in the rock record. Their internal silica skeleton is composed of an outer basal ring, with or without radial spines, and a central apical ring to which several supporting struts attach the two (e.g., McCartney and Loper, 1989).

11.3.4 Sponges

Sponges are sessile aquatic animals of the phylum Porifera. They have bodies perforated with channels that allow water to freely circulate through their central cavity. Most sponges have internal skeletons of a material known as spongin (a type of collagen) and/or spicules of calcium carbonate or silica that provide structural support to

the animal and protection against predation (Bergquist, 1978). The type of mineral spicule is dependent upon the environment that the sponge inhabits; calcium carbonate is predominant in warm marine settings, while silica is restricted to colder waters, such as the poles (Barnes, 1982). The *glass* sponges (or hexactinellid sponges) contain silica spicules that form a scaffolding-like framework. It is believed that the spicule begins with the formation of an organic filament to which silica is secreted (Imsiecke et al., 1995). The earliest known hexactinellid sponges date to the Cambrian (Sperling et al., 2010).

11.4 PASSIVE SILICA BIOMINERALIZATION

A wide variety of ancient and modern hot spring systems are characterized by authigenic silica precipitation. In these systems, the chemical disequilibrium of venting hydrothermal fluids leads to the nucleation and growth of amorphous silica masses and simultaneously the mineralization, and potential fossilization, of many different types of microorganisms. Source waters originating from deep, hot reservoirs commonly contain dissolved silica concentrations significantly higher than the solubility of amorphous silica at 100°C (approximately 400 ppm) (Gunnarsson and Arnórsson, 2000). Upon discharge of these fluids at the surface, decompressional degassing and boiling, rapid cooling to ambient temperatures, evaporation, and changes in solution pH all work together to cause the solution to rapidly and progressively exceed amorphous silica solubility (Fournier, 1985). Under these conditions, silicic acid dissolved in the monomer form $Si(OH)_4$ spontaneously polymerizes initially to oligomers (e.g., dimmers, trimers, and tetramers) and then to polymeric species with spherical diameters of 1–5 nm, as the silanol groups (–Si–OH–) of each oligomer condense and dehydrate to produce the siloxane (–Si–O–Si–) cores of larger polymers. The polymers quickly grow in size such that a bimodal composition of monomers and particles of colloidal dimensions (>5 nm) are generated (Crerar et al., 1981, Tobler and Benning, 2013). Depending on the degree of supersaturation, these either remain in suspension, due to the external silanol groups exhibiting a residual negative surface charge due to a low zero point of charge (around pH 2), they coagulate via cation bridging and nucleate homogenously, or they nucleate heterogeneously on a solid substrate (Iler, 1979). In the latter case, once the first silica monolayer is deposited, silica itself becomes the reactive substrate.

The precipitation of amorphous silica frequently leads to the formation of hard, siliceous crusts, known as sinters. Individual sinter deposits are architecturally complex because of the intricate lateral and vertical variations in texture, mineralogy, and chemical composition, i.e., geyserite, spicules, columnar and stratiform microstromatolites, oncoids, and coccoid microbial mats (see Konhauser et al., 2004). Moreover, most siliceous sinters have been constructed, to some degree, around microorganisms. Indeed, even geyserite, the microbanded, amorphous silica sinter that forms in the proximity of spring vents and fissures, where temperatures in excess of 73°C were supposedly deemed sterile except for scattered thermophilic microorganisms (Walter, 1976), has now been shown to have surfaces commonly covered with biofilms, while their laminae generally contain silicified microorganisms (e.g., Jones et al., 1997).

Much of the textural variation in siliceous sinters can be attributed to the different types of microorganisms that inhabit different discharge aprons or different parts of an individual apron, primarily in response to differences in water temperature, composition, and acidity (Konhauser et al., 2004). In waters from 75°C to 100°C, life consists of a few dominant chemolithoautotrophic and heterotrophic bacteria and archaea—the more extreme an environment, the fewer the number of taxa and the microbial community may be shaped by the biological properties of just its dominant member. In shallow channels, where flow rates are high, streamers of Aquificales and Thermus become mineral encrusted, forming unique fabrics that preserve the original flow directions. On the discharge aprons, where water temperatures have just cooled below the 73°C threshold, thick photosynthetic mats arise, containing species of Synechococcus and Chloroflexus. In waters cooler than 65°C, filamentous cyanobacteria appear, including species of Oscillatoria and Phormidium, as well as some purple and green sulfur bacteria. In the most distal parts of the drainage system, where temperatures are <40°C, other cyanobacteria, such as Spirulina, Fischerella, and Calothrix, begin to predominate. Green algae, diatoms, and fungi also become important mat constituents at the lowermost temperature regimes. At similar temperatures on an individual apron, the determining factor governing species distribution is the composition of the geothermal fluids and gases. For instance, Calothrix prefers to grow in alkaline water; fungi dominate areas with acidic waters (pH of <5.0), while Fischerella appears to be restricted to waters that contain little or no sulfur. Clearly, there are many more such examples of microbial niches in hot spring settings, and hence, it should not come as a surprise that variations in microbial populations give rise to textural variations in siliceous sinters simply because different microorganisms interact uniquely with the waters in which they are bathed.

When sinters are viewed in detail under the electron microscope, they almost always show microbial cells encrusted in spheroidal grains both extracellularly, on the sheaths or walls of living cells (Figure 11.2a), and intracellularly, within the cytoplasm, presumably after the cells have lysed (Figure 11.2b) (Phoenix et al., 2000). The silicification appears to begin with the attachment of silica colloids, on the order of tens of nanometers in diameter (Schultze-Lam et al., 1995; Konhauser and Ferris, 1996; Phoenix et al., 2000; Konhauser et al., 2001). Those silica particles then grow in size on the cell surface, and eventually coalesce until the individual precipitates are no longer distinguishable. Frequently, entire colonies are cemented together in a siliceous matrix several micrometers thick. The timing and rate of silicification relative to death of the microorganisms governs their preservation as intact cells. When silicification is rapid, cell components can rapidly become encased in a silica matrix, thereby retaining intact morphologies. However, experimental studies have shown that unmineralized cells begin to degrade only a few days after death (Bartley, 1996). As a result, their remnants may become progressively obscured, and just a silicified matrix containing sheath and cell wall material may be left of the original organic framework. These observations suggest that silicification begins when microbial communities are living and continues for some time after their death due to the high reactivity of the newly formed silica.

The role that microorganisms play in their own silicification has been the subject of some debate. Some research has suggested that microorganisms play a passive role (Walter et al., 1972), whereas many other studies suggest that microbial surface ligands serve as favorable nucleation sites for silica

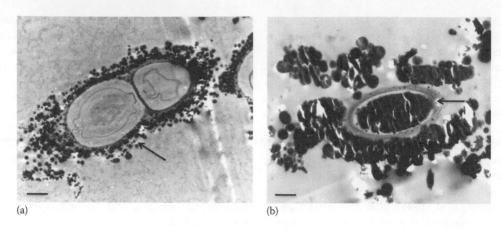

Figure 11.2. Transmission electron micrographs of silicified bacteria from a siliceous microstromatolite at Krisuvik, Iceland. (a) Filamentous cyanobacterium with epicellular, silica spheroids on outer sheath (arrow). Scale bar = 3 μm. (b) Lysed cyanobacterium with epicellular and intracellular silica precipitation. Note selective preservation of cell wall and sheath material (arrow). Scale bar = 1 μm.

precipitation (e.g., Ferris et al., 1986). These latter studies suggested that by reducing the activation energy barriers to nucleation, the microorganisms function as reactive interfaces, or templates, for heterogeneous nucleation. Because a sufficient supply of silica is generally available in hot spring effluent, often in excess of 400 ppm SiO_2 (e.g., Mountain et al., 2003), continued adsorption results in the surface sites becoming saturated, allowing particle nucleation to take place. After silica precipitation is initiated upon the bacterial surface, continued growth presumably occurs autocatalytically due to the increased surface area generated by the small silica phases. This was also used to explain the formation of substantial sinter deposits (silica growth rate ~20 kg year^{-1} m^{-2}) in a highly silica undersaturated geothermal outflow channel in Iceland (Tobler et al., 2008).

In the past few decades, a number of experimental studies have been designed to elucidate the actual mechanisms by which different microorganisms initiate the silicification process. Oehler (1976) was one of the first to experimentally subject various species of cyanobacteria to colloidal silica solutions over different lengths of time. At temperatures of ~100°C, several months were required for complete mineralization, while at higher temperatures (165°C), the cells mineralized quickly, but the filaments fragmented, the trichomes coalesced, intracellular components were destroyed, and there was a preferential preservation of the sheath and wall material. Francis et al. (1978) experimentally silicified 30 different species of algae, bacteria, and fungi for 2–24 weeks at temperatures of 55°C–60°C and found that some microorganisms silicified readily, whereas others were less susceptible to silicification. Westall et al. (1995) later showed that Gram-positive bacteria and eukaryotes silicified with ease, while Gram-negative bacteria did not readily silicify. Toporski et al. (2002) showed that at low silica concentrations, there were species-specific patterns of silicification, but under high silica concentrations, most bacteria suffered significant loss of shape and cellular detail.

Although these studies clearly indicate that silicification is an inevitable outcome of exposing microorganisms to silica supersaturated solutions, recent studies have demonstrated that the actual mechanisms of silicification rely, in part, on the microorganisms providing reactive surface ligands that adsorb silica from solution and, accordingly, reduce the activation energy barriers to heterogeneous nucleation. This means that cell surface charge may have a fundamental control on the initial silicification process. In an early attempt to describe the mechanisms of silicification, Leo and Barghoorn (1976) suggested that monomeric or low-molecular-weight polymeric silica was bound to the bacterial surface through hydrogen bonding as shown in the following reaction:

$$B\text{-}OH + Si(OH)_4 \rightarrow B\text{-}O\text{-}Si(OH)_3 + H_2O \qquad (11.4)$$

where B-OH represents a hydroxy group on the bacterial surface. This hypothesis appears best

EHRLICH'S GEOMICROBIOLOGY

corroborated by electron microscopic observations of microbial silicification, where the mineralization process appears to begin with the immobilization of preformed silica colloids onto the cell's surface (Schultze-Lam et al., 1995; Konhauser and Ferris, 1996; Phoenix et al., 2000). Similarly, recent experimental work with the cyanobacterium *Calothrix* sp. has demonstrated that silicification takes place on the cell's outer sheath (Yee et al., 2003). This structure is electrically neutral at pH 7, consisting predominantly of neutral sugars, along with smaller amounts of negatively charged carboxyl groups and positively charged amine groups, in approximately equal proportions. On the one hand, the low reactivity of *Calothrix*'s sheath gives its surface hydrophobic characteristics that facilitate their attachment to solid submerged substrates, i.e., siliceous sinters. On the other hand, the sheath's electroneutrality makes it less repulsive toward the polymeric silica fraction in solution, and hence, it may actually aid in the silicification process (Phoenix et al., 2002). Furthermore, Benning et al. (2004) used high-resolution synchrotron radiation Fourier transform infrared spectroscopy to show an increase in the integrated area of the absorbance spectra for the silica/polysaccharide region (Si–O/C–O; 1150–950 cm^{-1}), followed later by the occurrence and growth of a Si–O band at 800 cm^{-1}. From this, a two-step process was determined, where in the first stage the polysaccharide sheath thickens (in response to incubation in a silica-supersaturated medium), followed by the abiotic accumulation of amorphous silica precipitates upon the cell surface through the condensation of silanol groups.

In contrast to *Calothrix*, the highly anionic nature of *Bacillus subtilis* may limit silicification from occurring on its cell wall as a result of electrostatic charge repulsion between the organic ligands and the negatively charged silica species. For such anionically charged cell wall surfaces, silicification necessitates some form of cation bridging (e.g., Fe(III), Al(III)). In this regard, Fein et al. (2002) showed that *B. subtilis* precoated with Fe(III) and Al(III) hydroxides could act as templates for silica deposition (in undersaturated solutions) over a wide range of pH conditions. Compared to the negligible silica adsorption directly onto the cell surfaces in undersaturated solutions, virtually all of the monomeric silica was removed from solution by the presence of either Fe or Al. Moreover, increasing the concentrations of Fe or Al increased the extent of silica adsorption over the entire concentration range studied. Following that study, Phoenix et al. (2003) measured the effects of iron bridging in mixed Fe–Si solutions. However, the solutions used were supersaturated with respect to amorphous silica and the cells were not precoated with iron. Their results demonstrate that *B. subtilis* cells immobilize more Fe than bacteria-free systems in solutions with iron concentrations <50 ppm Fe, yet as iron concentrations increase, the difference between iron immobilization in bacterial and bacteria-free systems decreases as abiotic precipitation processes become increasingly dominant. Correspondingly, the bacterial and bacteria-free systems remove nearly identical amounts of silica from solution, whatever the iron concentration, again due to the dominance of abiotic precipitation processes. These results suggest that in natural hot spring systems, where the concentration of soluble silica far exceeds that of iron, not only will the amount of iron partitioned onto the microbial mats be insignificant compared to the abiotic reactions of silica with Fe(OH)$_3$ or various clay phases, but the vast majority of silica precipitated will occur without the aid of a cation bridge.

A third mechanism of silicification was recently discovered with the experimental silicification of the biofilm-forming thermophilic *Aquificales*. That work revealed that not only did the cells remain viable during silicification, but they also produced more rapidly and higher protein concentrations in batch cultures with increasing silica concentration, possibly as a stress response (Lalonde et al., 2005). Biofilm production was visually noted to be greater in cultures with increasing silica concentrations. Acid–base titrations indicated that amine functional groups are highly prominent on their biofilm surfaces and likely serve as positively charged ligands promoting silica colloid adsorption by electrostatic interaction within the protein-rich biofilm matrix. Because silicification was observed to be restricted to the extracellular polymeric substance (EPS), it was proposed that *Aquificales* may prevent cellular silicification to some degree by producing abundant reactive sites in the biofilm matrix and regulating EPS production appropriately in order to restrict sites of silicification away from the cell surfaces (Lalonde et al., 2005).

11.5 BIOMOBILIZATION OF SILICON AND OTHER CONSTITUENTS OF SILICATES (BIOWEATHERING)

Some bacteria and fungi play an important role in mobilization of silica and silicates in nature. The solubilizing action may involve the cleavage of Si–O–Si (siloxane) or Al–O framework bonds or the removal of cations from the crystal lattice of silicate, causing subsequent collapse of the silicate lattice structure. The mode of attack may be microbially produced by (1) ligands of cations; (2) organic or inorganic acids, which are a source of protons; (3) alkali (ammonia or amines); or (4) extracellular polysaccharides acting at acidic pH. The source of the polysaccharides may be the EPS of some bacteria.

Bioweathering action of silica or silicates seems not only to be restricted to corrosive agents that have been excreted by appropriate microorganisms into the bulk phase but also to involve microbes attached to the surface of silica or silicates (Bennett et al., 1996, 2001). Because they are attached, their excreted metabolic products can attack the mineral surface in more concentrated form. Such attack may be manifested in etch marks. Some of the polysaccharides by which the microbes attach to the mineral surface may themselves be corrosive.

Bioweathering, like abiotic weathering, can lead to the formation of new minerals. This is the result of reprecipitation and crystallization of some of the mobilized constituents from the mineral that is weathered (Barker and Banfield, 1996; Adamo and Violante, 2000, Bonneville et al., 2009). New, secondary minerals may form on the surface of the weathered mineral. Microbes attached to the surface of minerals that are weathered may serve as nucleating agents in mineral neoformation (Macaskie et al., 1992; Schultze-Lam et al., 1996). In certain cases, attached microbes can exert substantial mechanical forces thereby inducing physical distortion of the mineral lattice structure that facilitates later chemical weathering (Bonneville et al., 2009).

11.5.1 Solubilization by Ligands

Microbially produced ligands of divalent cations have been shown to cause dissolution of calcium-containing silicates. For instance, a soil strain of *Pseudomonas* that produced 2-ketogluconic acid from glucose dissolved synthetic silicates of calcium, zinc, and magnesium and the minerals wollastonite ($CaSiO_3$), apophyllite ($KCa_4Si_8O_{20}(F,OH) \cdot 8H_2O$), and olivine (($Mg,Fe)_2SiO_4$) (Webley et al., 1960). The demonstration consisted of culturing the organism for 4 days at 25°C on separate agar media, each containing 0.25% (w/v) of one of the synthetic or natural silicates, which rendered the medium turbid. A clear zone was observed around the bacterial colonies when silicate was dissolved (Figure 11.3). A similar silicate-dissolving action was also shown with a Gram-negative bacterium, strain D_{11}, which resembled *Erwinia* spp., and with *Erwinia herbicola* or some *Pseudomonas* strains, all of which were able to produce 2-ketogluconate from glucose (Duff et al., 1963). The action of these bacteria was tested in glucose-containing basal medium: KH_2PO_4, 0.54 g; $MgSO_4 \cdot 7H_2O$, 0.25 g; $(NH_4)_2SO_4$, 0.75 g; $FeCl_3$, trace; Difoc yeast extract, 2 g; glucose 40 g; distilled water, 1 L; and 5–500 mg pulverized mineral per 5–10 mL of medium. It was found that dissolution of silicates in these cases resulted from the complexation of the cationic components of the silicates by 2-ketogluconate. The complexes were apparently more stable than the silicate. For example,

$$CaSiO_3 \Leftrightarrow Ca^{2+} + SiO_3^{2-} \qquad (11.5)$$

$$Ca^{2+} + 2\text{-Ketogluconate} \rightarrow Ca(2\text{-Ketogluconate}) \qquad (11.6)$$

The silicon that was liberated or released in these experiments and subsequently transformed took three forms: (1) low-molecular-weight or ammonium molybdate-reactive silicate (possibly monomeric); (2) a colloidal polymeric silicate of higher molecular weight, which did not react with dilute hydrofluoric acid; and (3) an amorphous form that could be removed from solution by centrifugation and dissolved in cold 5% aqueous carbonate (Duff et al., 1963). Polymerized silicate can be transformed by bacteria into monomeric silicate, as has been shown in studies with *Proteus mirabilis* and *Bacillus caldolyticus* (Lauwers and Heinen, 1974). The *Proteus* culture was able to assimilate some of the monomeric silicate. The mechanism of depolymerization has not been elucidated. It may involve an extracellular enzyme.

Gluconic acid produced from glucose by several different types of bacteria has been shown to

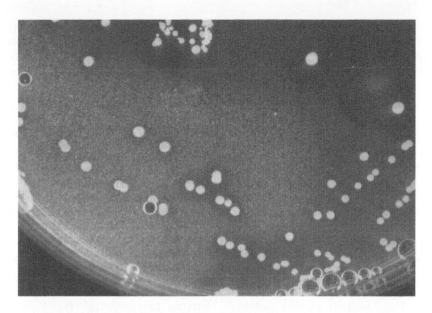

Figure 11.3. Colonies of bacterial isolate C-2 from a sample of weathered rock on synthetic calcium silicate selection medium showing evidence of calcium silicate dissolution around colonies. Basal medium was prepared by aseptically mixing 10 mL of sterile solution A (3 g dextrose or 3 g levulose in 100 mL distilled water), 10 mL of sterile solution B (0.5 g $(NH_4)_2SO_4$, 0.5 g $MgSO_4 \cdot 7H_2O$, 0.5 g Na_2HPO_4, 0.5 g KH_2PO_4, 2 g yeast extract, 0.05 g $MnSO_4 \cdot 2H_2O$ in 500 mL distilled water), and 20 mL of sterile 3% agar and distributing the mixture in Petri plates. Capping agar was prepared by mixing 10 mL of sterile synthetic $CaSiO_3$ suspension with 7.5 mL of sterile solution A, 7.5 mL of sterile solution B, and 15 mL of sterile 3% agar and distributing 3 mL of this mixture aseptically over the surface of the solidified basal agar in the plates.

solubilize bytownite, albite, kaolinite, and quartz at near-neutral pH (Vandevivere et al., 1994). The activity around neutral pH suggests that the mechanism of action of gluconate involves chelation.

Quartz (SiO_2) has been shown to be subject to slow dissolution by organic acids such as citric, oxalic, pyruvic, and humic acids (Bennett et al., 1988), all of which can be formed by fungi or bacteria. In a pH range of 3–7, the effect was greatest at pH 7, indicating that the mechanism of action was not protonation but chelation. Bennett et al. (1988) suggested that the chelation involved an electron donor–acceptor system. Acetate, fumarate, and tartrate were ineffective in dissolving silica as a complex.

11.5.2 Solubilization by Acids

The effect of acids in solubilizing silicates has been noted in various studies. Waksman and Starkey (1931) cited the action of CO_2 on orthoclase

$$2KAlSi_3O_8 + 2H_2O + CO_2 \rightarrow H_4Al_2Si_2O_9 + K_2CO_3 + 4SiO_2 \quad (11.7)$$

The CO_2 is, of course, very likely to be a product of respiration or fermentation.

Its weathering action is best viewed as based on the formation of the weak acid H_2CO_3 through hydration of CO_2. Another example of a silicate attack by acid is that involving spodumene (LiAlSi$_2$O$_6$) (Karavaiko et al., 1979). In this instance, an in situ correlation was observed between the extent of alteration of a spodumene sample and the acidity it produced when in aqueous suspension. Unweathered spodumene generated a pH in the range of 5.1–7.5, whereas altered spodumene generated a pH in the range of 4.2–6.4. Non-spore-forming heterotrophs were found to predominate in weathered spodumene. They included bacteria such as *Arthrobacter pascens*, *A. globiformis*, and *A. simplex* as well as *Nocardia globerula*, *Pseudomonas fluorescens*, *P. putida*, and *P. testeronii* and fungi such as *Trichoderma lignorum*, *Cephalosporium atrium*, and *Penicillium decumbens*.

Acid decomposition of spodumene may be formulated as follows (Karavaiko et al., 1979):

$$4LiAlSi_2O_6 + 6H_2O + 2H_2SO_4 \rightarrow 2Li_2SO_4 + Al_4(Si_4O_{10})(OH)_8 + 4H_2SiO_3 \quad (11.8)$$

The aluminosilicate product in this reaction is kaolinite.

Further investigation into microbial spodumene degradation revealed that among the most active organisms are the fungi *Penicillium notatum* and *Aspergillus niger*, thionic bacteria like *Thiobacillus thiooxidans*, and the slime-producing bacterium *Bacillus mucilaginosus* var. *siliceous* (Karavaiko et al., 1980; Avakyan et al., 1986). The fungi and *T. thiooxidans*, which produce acid, were most effective in solubilizing Li and Al. *B. mucilaginosus* was effective in solubilizing Si in addition to Li and Al by reaction of its extracellular polysaccharide with the silicate in spodumene.

Solubilization of silica along with other constituents in the primary minerals amphibolite, biotite, and orthoclase by acids (presumably citric and oxalic acids) formed by several fungi and yeasts at the expense of glucose has also been demonstrated (Eckhardt, 1980; see also Barker et al., 1997). These findings of silica mobilization are similar to those in earlier studies on the action of the fungi *Botrytis*, *Mucor*, *Penicillium*, and *Trichoderma* isolated from rock surfaces and weathered stone. In these experiments, citric and oxalic acids produced by the fungi solubilized Ca, Mg, and Zn silicates (Webley et al., 1963). In studies by Henderson and Duff (1963), *A. niger* has been shown to release Si from apophyllite, olivine, saponite, vermiculite, and wollastonite, but not augite, garnet, heulandite, hornblende, hypersthene, illite, kaolinite, labradorite, orthoclase, or talc. However, Berner et al. (1980) found in laboratory experiments that augite, hypersthene, hornblende, and diopside in soil samples were subject to weathering by soil acids, presumably of biological origin. Organisms different from those used by Henderson and Duff (1963) and, as a result, different metabolic products were probably involved. *Penicillium simplicissimus* released Si from basalt, granite, granodiorite, rhyolite, andesite, peridotite, dunite, and quartzite with metabolically produced citric acid (Silverman and Munoz, 1970). Acid formed by *P. notatum* and *Pseudomonas* sp. releases Si from plagioclase and nepheline (Aristovskaya and Kutuzova, 1968; Kutuzova, 1969).

In a study of weathering by organic and inorganic acids of three different plagioclase specimens (Ca–Na feldspars), it was found that steady-state dissolution rates were highest at approximately pH 3 and decreased as the pH was increased toward neutrality (Welch and Ullman, 1993). The organic and inorganic acids whose weathering action was studied are representative of some end products of microbial metabolism. Polyfunctional acids, including oxalate, citrate, succinate, pyruvate, and 2-ketoglutarate, were the most effective, whereas acetate and propionate were less effective. However, all organic acids were more effective than the inorganic acids HCl and HNO_3. The polyfunctional acids acted mostly as acidulants at very acidic pH and mainly as chelators at near-neutral pH. Ullman et al. (1996) found that in some instances, the combined effect of protonation and chelation enhanced the solubilizing action of some polyfunctional acids on feldspars by a factor of 10 above the expected proton-promoted rate. Ca and Na were rapidly released in these experiments. The chelate attack appeared to be at the Al sites. Those organic acids that preferentially chelated Al were the most effective in enhancing plagioclase dissolution. Although the products of dissolution of feldspars are usually considered to include separate aluminum and silicate species, soluble aluminosilicate complexes may be intermediates (Browne and Driscoll, 1992).

The practical effect of acid attack of aluminosilicates can be seen in the corrosion of concrete sewer pipes. Concrete is formed from a mixture of cement (heated limestone, clay, and gypsum) and sand. On setting, the cement includes the compounds Ca_2SiO_4, Ca_3SiO_5, and $Ca(AlO_3)_2$, which hold the sand in their matrix. H_2S produced by microbial mineralization of organic sulfur compounds and by bacterial sulfate reduction of sulfate in sewage can itself corrode concrete. But corrosion is enhanced if the H_2S is first oxidized to sulfuric acid by thiobacilli *Thiobacillus neapolitanus* (now renamed *Halothiobacillus neapolitanus*), *T. intermedius*, *T. novellus*, and *T. thiooxidans* (now renamed *Acidithiobacillus thiooxidans*) (Parker, 1947; Milde et al., 1983; Sand and Bock, 1984; Okabe et al., 2007).

Groundwater pollution with biodegradable substances has been found to result in silicate weathering of aquifer rock. Products of microbial degradation of the substances cause the weathering. This was observed in an oil-polluted aquifer near Bemidji, Minnesota (Hiebert and Bennett, 1992). Microcosm experiments of 14 months' duration in the aquifer with a mixture of crystals such as albite, anorthite, anorthoclase, and microcline (all feldspar minerals) and quartz revealed microbial colonization of the mineral surfaces by individual cells and small clusters. Intense etching of the feldspar minerals and light etching of the quartz occurred at or near where the bacteria were seen. Such aquifer rock weathering can affect water quality.

11.5.3 Solubilization by Alkali

Alkaline conditions are very conducive for mobilizing silica, whether from silicates, aluminosilicates, or even quartz (Karavaiko et al., 1984). This is attributable to the significant lability of the Al–O and Si–O bonds under these conditions, because both types of bonds are susceptible to nucleophilic attack (see discussion by Karavaiko et al., 1984). *Sarcina ureae* growing in peptone–urea broth released silica readily from nepheline, plagioclase, and quartz (Aristovskaya and Kutuzova, 1968; Kutuzova, 1969). In this instance, ammonia resulting from the hydrolysis of urea was the source of the alkali. In microbial spodumene degradation, alkaline pH also favors silica release (Karavaiko et al., 1980).

Pseudomonas mendocina was able to enhance mobilization of Al, Si, and Fe impurities from kaolinite in a succinate-mineral salts medium in which the pH rose from 7.7 to 9.2 in 4 days of growth under aerobic conditions (Maurice et al., 2001).

11.5.4 Solubilization by Extracellular Polysaccharide

Extracellular polysaccharide has been claimed to play an important role in silica release, especially in the case of quartz. Such polysaccharide is able to react with siloxanes to form organic siloxanes. It can be of bacterial origin (e.g., from B. *mucilaginosus* var. *siliceous*; Avakyan et al., 1986) or fungal origin (e.g., from A. *niger*; Holzapfel and Engel, 1954a). The reaction appears not to be enzyme-catalyzed because polysaccharide from which the cells have been removed is reactive. Indeed, such organic silica-containing compounds can be formed with reagent-grade organics (Holzapfel, 1951; Weiss et al., 1961) and have been isolated from various biological sources other than microbes (Schwarz, 1973). With polysaccharide from B. *mucilaginosus*, the reaction appears to be favored by acid metabolites (Malinovskaya et al., 1990). It should be noted that Welch and Vandevivere (1995) found that polysaccharides from different sources had either no effect or interfered with solubilization

of plagioclase by gluconate at a pH between 6.5 and 7.

Barker and Banfield (1996) described the weathering by bacteria and lichens of amphibole syenite associated with the Stettin complex near Wausau, Wisconsin. The process involved penetration of grain boundaries, cleavages, and cracks. Mineral surfaces were coated with EPS. During the weathering process, dissolution by metabolically produced corrosive agents and selective transport of mobilized constituents, probably mediated by acid mucopolysaccharides, occurred. Some mobilized constituents reprecipitated, leading to the formation of clay minerals.

A more detailed study revealed that the site of bioweathering by lichens (in this instance a symbiotic consortium of a fungus and an alga) could be divided into four zones (Barker and Banfield, 1998). The authors concluded that in the uppermost zone (zone 1), represented by the upper lichen thallus, no weathering occurs. This is the photosynthetic zone. In zone 2, involving the lower lichen thallus, active weathering due to interaction with lichen products occurs. Mineral fragments coated with organic polymers of incipient secondary minerals that resulted from the weathering may appear in the thallus. In zone 3, reactions occur, which are an indirect consequence of lichen action. In zone 4, any weathering, if it occurs, is due to abiotic processes.

11.6 ROLE OF MICROBES IN THE SILICA CYCLE

As the foregoing discussion shows, some microbes, and even some plants, have a significant influence on the distribution and form of silicon in the biosphere. Those organisms that assimilate silicon clearly act as concentrators of it. Those that degrade silica and silicate minerals act as agents of silicon dispersion. They are an important source of dissolved silica on which the concentrators depend, and they are also important agents of rock weathering. Comparative electron microscopic studies provide clues to the extent of microbial weathering action (see Berner et al., 1980; Callot et al., 1987; Barker and Banfield, 1998).

It has been argued that silicate-mobilizing reactions by microbially produced acids and complexing agents under laboratory conditions occurred at glucose concentrations that may not be encountered in nature and may therefore, be laboratory artifacts. A counterargument can be made, however, that microenvironments exist in soil and sediment that have appropriately high concentrations of utilizable carbohydrates, nitrogenous compounds, and other needed nutrients. They originate from the excretory products and dead remains of organisms from which appropriate bacteria and fungi can form the compounds that can promote breakdown of silica and silicate minerals. Indeed, fungal hyphae in the litter zone and A horizon of several different soils have been shown by SEM to carry calcium oxalate crystals attached to them. This is evidence for extensive in situ production of oxalate by the fungus (Graustein et al., 1977). The basidiomycete *Hysterangium crassum* was shown to weather clay in situ with the oxalic acid it produced (Cromack et al., 1979). Lichens show evidence, observable in situ, of extensive rock weathering activity (Jones et al., 1981). Although Ahmadjian (1967) and Hale (1967) cast doubt on this ability of lichens, current evidence strongly supports the rock weathering activity of these organisms. Biodegradation of silica and silicate minerals is usually a slower process in nature than in the laboratory because the conditions in natural environments are usually less favorable. If this were not so, rock in the biosphere would be very unstable.

At the opposite end of the Si cycle, several planktonic organisms precipitate skeletons comprised of amorphous silica. Today, the oceans are undersaturated with regard to amorphous silica (seawater <10 ppm) predominantly because of the activity of diatoms, but to a lesser extent radiolarians and silicoflagellates as well (Tréguer et al., 1995; Yool and Tyrrell, 2003). Some estimates suggest that diatoms produce annually 240 Tmol silica, which is approximately 40 times greater than estimated silica inputs to the oceans (Tréguer et al., 1995). This discrepancy is balanced by an extremely efficient silica recycling mechanisms by which 97% of the biogenic silica is dissolved in the surface waters and seafloor upon cell lysis. The dissolved silica is then reused by other silica-secreting organisms. Estimates suggest that the Si delivered to the oceans passes through a biological uptake and dissolution cycle on average about 39 times before ultimately being removed to marine sediments (Tréguer et al., 1995).

Thus, silicon in nature may follow a series of cyclic biogeochemical transformations (Harriss, 1972; Lauwers and Heinen, 1974; Kuznetsov, 1975).

Silica and silicate minerals in rocks are subject to the weathering action of biological, chemical, and physical agents. The extent of the contribution of each of these agents must depend on the particular environmental circumstances. Silica liberated in these processes may be leached away by surface water or groundwater, and it may be removed from these waters by abiotic and biologically induced precipitation at new sites, or it may be swept into bodies of freshwater or the sea. There, silicon will tend to be removed by biological agents. Upon their death, other biological agents will release this silicon back into the solution or the siliceous remains will be incorporated into the sediment (Weaver and Wise, 1974; Allison, 1981; Patrick and Holding, 1985), where some or all of the silicon may later be returned to solution by weathering. The sediments of the ocean appear to be a sink for excess silica swept into the oceans because the silica concentrations of seawater tend to remain relatively constant. But over geologic time, this silicon in the sediments is not permanently immobilized. Plate tectonics will ultimately cause it to be recycled.

11.7 SUMMARY

The environmental distribution of silicon is significantly influenced by microbial activity. Certain microorganisms assimilate it and build it into cell-support structures. They include diatoms, some chrysophytes, silicoflagellates, some xanthophytes, radiolarians, and actinopods. Silicon uptake rates by diatoms have been measured, but the mechanism by which silicon is assimilated is still only partially understood. These silicon-assimilating microorganisms are important in the formation of siliceous oozes in the oceans and in lakes. By contrast, some bacteria, fungi, and lichens are able to solubilize silicates and silica. They accomplish this by forming chelators, acids, bases, and exopolysaccharides that react with silica and silicates. These reactions are important in weathering of rock and cycling silicon in nature.

REFERENCES

Adamo P and Violante P. 2000. Weathering of rocks and neogenesis of minerals associated with lichen activity. *Applied Clay Science* 16:229–256.

Afanasieva MS, Amon EO, Agarkov YV, Boltovskoy DS. 2005. Radiolarians in the geological annals. *Paleontol J* 39:135–392.

Ahmadjian V. 1967. *The Lichen Symbiosis.* Waltham, MA: Blaidsell.

Allison CW. 1981. Siliceous microfossils from the Lower Cambrian of northwest Canada: Possible source for biogenic chert. *Science* 211:53–55.

Anderson GM. 1972. Silica solubility. In: Fairbridge RW (ed.) *The Encyclopedia of Geochemistry and Environmental Sciences.* Encyclopedia of Earth Science Series, Vol. IVA. New York: Van Nostrand Reinhold, pp. 1085–1088.

Aristovskaya TV, Kutuzova RS. 1968. Microbiological factors in the mobilization of silicon from poorly soluble natural compounds. *Pochvovedenie* 12:59–66.

Avakyan ZA, Belkanova NP, Karavaiko GI, Piskunov VP. 1985. Silicon compounds in solution during bacterial quartz degradation. *Mikrobiologiya* 54:301–307 (English translation pp. 250–256).

Avakyan ZA, Pivovarova TA, Karavaiko GI. 1986. Properties of a new species, *Bacillus mucilaginosus. Mikrobiologiya* 55:477–482 (English translation pp. 477–482).

Azam F. 1974. Silicic-acid uptake in diatoms studied with [^{68}Ge]germanic acid as tracer. *Planta* 121:205–212.

Azam F, Hemmingsen B, Volcani BE. 1973. Germanium incorporation into the silica of diatom cell walls. *Arch Microbiol* 92:11–20.

Azam F, Hemmingsen B, Volcani BE. 1974. Role of silicon in diatom metabolism. V. Silicic acid transport and metabolism in the heterotrophic diatom *Nitzschia alba. Arch Mikrobiol* 97:103–114.

Azam F, Volcani B. 1974. Role of silicon in diatom metabolism. VI. Active transport of germanic acid in the heterotrophic diatom *Nitzschia alba. Arch Microbiol* 101:1–8.

Barker WW, Banfield JF. 1996. Biologically versus inorganically mediated weathering reactions: Relationships between minerals and extracellular microbial polymers in lithiobiontic communities. *Chem Geol* 132:55–69.

Barker WW, Banfield JF. 1998. Zones of chemical and physical interactions at interfaces between microbial communities and minerals. A model. *Geomicrobiol J* 15:223–244.

Barker WW, Welch SA, Banfield JF. 1997. Biogeochemical weathering of silicate minerals. In: Banfield JF, Nealson KH (eds.) *Geomicrobiology: Interactions Between Microbes and Minerals.* Reviews in Mineralogy, Vol. 35. Washington, DC: Mineralogical Society of America, pp. 391–428.

Barnes RD. 1982. *Invertebrate Zoology.* Philadelphia, PA: Holt-Saunders International.

Bartley JK. 1996. Actualistic taphonomy of cyanobacteria: Implications for the Precambrian fossil record. *Palaios* 11:571–586.

Bennett PC, Hiebert FK, Choi WJ. 1996. Microbial colonization and weathering of silicates in a petroleum-contaminated aquifer. *Chem Geol* 132:45–53.

Bennett PC, Melcer ME, Siegel DI, Hassett JP. 1988. The dissolution of quartz in dilute aqueous solutions of organic acids at 25°C. *Geochim Cosmochim Acta* 52:1521–1530.

Bennett PC, Rogers JR, Choi WJ, Hiebert FK. 2001. Silicates, silicate weathering, and microbial ecology. *Geomicrobiol J* 18:3–19.

Benning LG, Phoenix V, Yee N, Konhauser KO. 2004. The dynamics of cyanobacterial silicification: An infrared micro-spectroscopic investigation. *Geochim Cosmochim Acta* 68:743–757.

Bergquist PR. 1978. *Sponges.* London, U.K.: Hutchinson.

Berner RA, Sjoeberg EL, Velbel MA, Krom MD. 1980. Dissolution of pyroxenes and amphiboles during weathering. *Science* 207:1205–1206.

Bidle KD, Azam F. 2001. Bacterial control of silicon regeneration from diatom detritus: Significance of bacterial ectohydrolases and species identity. *Limnol Oceanogr* 46:1606–1623.

Blair Hedges S, Chen H, Kumar S, Wang DYC, Thompson AS, Watanabe H. 2001. A genomic timescale for the origin of eukaryotes. *BMC Evol Biol* 1:4.

Blank GS, Sullivan CW. 1979. Diatom mineralization of silicic acid. III. Si(OH)$_4$ binding and light dependent transport in *Nitzschia angularis. Arch Microbiol* 123:17–164.

Bonneville S, Smits MM, Brown A, Harrington J, Leake JR, Brydson R, Benning LG. 2009. Plant-driven fungal weathering: Early stages of mineral alteration at the nanometer scale. *Geology* 37:615–618.

Bowen HJM. 1966. *Trace Elements in Biochemistry.* London, U.K.: Academic Press.

Bowen HJM. 1979. *Environmental Chemistry of the Elements.* London, U.K.: Academic Press.

Browne BA, Driscoll CT. 1992. Soluble aluminum silicates: Stoichiometry, stability, and implications for environmental geochemistry. *Science* 256:1667–1670.

Braun A, Chen J, Waloszek D, Maas A. 2013. First early Cambrian radiolaria. *Geol Soc Lond* 286:143–149.

Callot B, Maurette M, Pottier L, Dubois A. 1987. Biogenic etching of microfractures in amorphous and crystalline silicates. *Nature (London)* 328:147–149.

Chisholm SW, Azam F, Epply RW. 1978. Silicic acid incorporation in marine diatoms on light-dark cycles: Use as an assay for phased cell division. *Limnol Oceanogr* 23:518–529.

Coombs J, Volcani BE. 1968. The biochemistry and fine structure of silica shell formation in diatoms. Silicon induced metabolic transients in *Navicula pelliculosa*. *Planta* 80:264–279.

Crerar DA, Axtmann EV, Axtmann RC. 1981. Growth and ripening of silica polymers in aqueous solutions. *Geochim Cosmochim Acta* 45:1259–1266.

Cromack K, Jr., Sollins P, Graustein WC, Speidel K, Todd AW, Spycher G, Li CY, Todd RL. 1979. Calcium oxalate accumulation and soil weathering in mats of the hypogenous fungus *Hysterangium crassum*. *Soil Biol Biochem* 11:463–468.

Darley WM, Volcani BE. 1969. Role of silicon in diatom metabolism. A silicon requirement for deoxyribonucleic acid synthesis in the diatom *Cylindrotheca fusiformis*, Reiman and Lewin. *Exp Cell Res* 58:334–343.

Dashman T, Stotzky G. 1986. Microbial utilization of amino acids and a peptide bond on homoionic montmorillonite and kaolinite. *Soil Biol Biochem* 18:5–14.

De La Rocha CL, Brzezinski MA, DeNiro MJ. 1997. Fractionation of silicon isotopes by marine diatoms during biogenic silica formation. *Geochim Cosmochim Acta* 61:5051–5056.

De La Rocha CL, Brzezinski MA, DeNiro MJ. 2000. A first look at the distribution of the stable isotopes of silicon in natural waters. *Geochim Cosmochim Acta* 64:2467–2477.

de Vrind-de Jong EW, de Vrind JPM. 1997. Algal deposition of carbonates and silicates. In: Banfield JF, Nealson KH (eds.) *Geomicrobiology: Interactions Between Microbes and Minerals. Reviews in Mineralogy*, Vol. 35. Washington, DC: Mineralogical Society of America, pp. 267–307.

De Wever P, Dumitrica P, Caulet J-P, Nigrini C, Caridroit M. 2001. *Radiolarians in the Sedimentary Record*. London, U.K.: Gordon and Breach.

Dommergues Y, Mangenot F. 1970. *Écologie Microbienne du Sol*. Paris, France: Masson.

Duff RB, Webley DM, Scott RO. 1963. Solubilization of minerals and related minerals by 2-ketogluconic acid-producing bacteria. *Soil Sci* 95:105–114.

Eckhardt FEW. 1980. Microbial degradation of silicates. Release of cations from aluminosilicate minerals by yeast and filamentous fungi. In: Oxley TA, Becker G, Allsopp D (eds.) *Biodeterioration. Proceedings of the 4th International Biodeterioration Symposium*. London, U.K.: Pitman, pp. 107–116.

Ehrlich HL. 1982. Enhanced removal of Mn^{2+} from seawater by marine sediment and clay minerals in the presence of bacteria. *Can J Microbiol* 28:1389–1395.

Fein JB, Scott S, Rivera N. 2002. The effect of Fe on Si adsorption by *Bacillus subtilis* cell walls: Insights into non-metabolic bacterial precipitation of silicate minerals. *Chem Geol* 182:265–273.

Ferris FG, Beveridge TJ, Fyfe WS. 1986. Iron-silica crystallite nucleation by bacteria in a geothermal sediment. *Nature* 320:609–611.

Fournier RO. 1985. The behaviour of silica in hydrothermal solutions. In: Berger BR, Bethke PM (eds.) *Geology and Geochemistry of Epithermal Systems. Reviews in Economic Geology*, Vol. 2. El Paso, TX: The Economic Geology Publishing Co., pp. 45–61.

Francis S, Margulis L, Barghoorn ES. 1978. On the experimental silicification of microorganisms. II. On the time of appearance of eukaryotic organisms in the fossil record. *Precamb Res* 6:65–100.

Garrels RM. 1965. Silica: Role in the buffering of natural waters. *Science* 148:69.

Graustein WC, Cromack K, Jr., Sollins P. 1977. Calcium oxalate: Occurrence in soils and effect on nutrient and geochemical cycles. *Science* 198:1252–1254.

Gunnarsson I, Arnórsson S. 2000. Amorphous silica solubility and the thermodynamic properties of H_4SiO_4 in the range of 0° to 350°C at P_{sat}. *Geochim Cosmochim Acta* 64:2295–2307.

Hale ME, Jr. 1967. *The Biology of Lichens*. London, U.K.: Edward Arnold.

Hall FR. 1972. Silica cycle. In: Fairbridge RW (ed.) *The Encyclopedia of Geochemistry and Environmental Sciences*. Encyclopedia of Earth Science Series, Vol. IVA. New York: Van Nostrand Reinhold, pp. 1082–1085.

Harriss RC. 1972. Silica–biogeochemical cycle. In: Fairbridge RW (ed.) *The Encyclopedia of Geochemistry and Environmental Sciences*. Encyclopedia of Earth Science Series, Vol. IVA. New York: Van Nostrand Reinhold, pp. 1080–1082.

Healy FP, Combs J, Volcani BE. 1967. Changes in pigment content of the diatom *Navicula pelliculosa* in silicon starvation synchrony. *Arch Mikrobiol* 59:131–142.

Heinen W. 1960. Silicium-Stoffwechsel bei Mikroorganismen. I. Mitteilung. Aufnahme von Silicium durch Bakterien. *Arch Mikrobiol* 37:199–210.

Helmcke J-G. 1954. Die Feinstruktur der Kieselsäure und ihre physiologische Bedeutung in Diatomschalen. *Naturwissenschaften* 11:254–255.

Henderson MEK, Duff RB. 1963. The release of metallic and silicate ions from minerals, rocks and soils by fungal activity. *J Soil Sci* 14:236–246.

Hiebert FK, Bennett PC. 1992. Microbial control of silicate weathering in organic-rich ground water. *Science* 258:278–281.

Holzapfel L. 1951. Siliziumverbindung in biologischen Systemen. Organ. Kieselsäureverbindungen. XX. Mitteilung. *Z Elektrochem* 55:577–580.

Holzapfel L, Engel W. 1954a. Der Einfluss organischer Kieselsäureverbindungen auf das Wachstum von *Aspergillus niger* und *Triticum*. *Z Naturforsch* 9b:602–606.

Holzapfel L, Engel W. 1954b. Über die Abhängigkeit der Wachstumsgeschwindigkeit von *Aspergillus niger* in Kieselsäurelösungen bei O_2-Belüftung. *Naturwissenschaften* 41:191–192.

Iler R. 1979. *Chemistry of Silica*. New York: Wiley.

Imsiecke G, Steffen R, Custodio M, Borojevic R, Muller WEG. 1995. Formation of spicules by sclerocytes from the freshwater sponge *Ephydatia muelleri* in short-term cultures in vitro. *In Vitro Cell Dev Biol Anim* 31:528–535.

Jones B, Renaut RW, Rosen MR. 1997. Biogenicity of silica precipitation around geysers and hot-spring vents, North Island, New Zealand. *J Sed Res* 67:88–104.

Jones D, Wilson MJ, McHardy WJ. 1981. Lichen weathering of rock-forming minerals: Application of scanning electron microscopy and microprobe analysis. *J Microsc* 124(1):95–104.

Jorgensen EG. 1955. Solubility of silica in diatoms. *Physiol Plant* 8:864–851.

Karavaiko GI, Avakyan ZA, Krutsko VS, Mel'nikova EO, Zhdanov V, Piskunov VP. 1979. Microbiological investigations on a spodumene deposit. *Mikrobiologiya* 48:502–508 (English translation pp. 383–398).

Karavaiko GI, Belkanova NP, Eroshchev-Shak VA, Avakyan ZA. 1984. Role of microorganisms and some physicochemical factors of the medium in quartz destruction. *Mikrobiologiya* 53:976–981 (English translation pp. 795–800).

Karavaiko GI, Krutsko VS, Mel'nikova EO, Avakyan ZA, Ostroushko YI. 1980. Role of microorganisms in spodumene degradation. *Mikrobiologiya* 49:547–551 (English translation pp. 402–406).

Katami A. 1982. Dissolution rates of silica from diatoms decomposing at various temperatures. *Marine Biol (Berlin)* 68:91–96.

Konhauser KO, Ferris FG. 1996. Diversity of iron and silica precipitation by microbial mats in hydrothermal waters, Iceland: Implications for Precambrian iron formations. *Geology* 24:323–326.

Konhauser KO, Jones B, Phoenix VR, Ferris G, Renaut RW. 2004. The microbial role in hot spring silicification. *Ambio* 33:552–558.

Konhauser KO, Phoenix VR, Bottrell SH, Adams DG, Head IM. 2001. Microbial-silica interactions in modern hot spring sinter: Possible analogues for Precambrian siliceous stromatolites. *Sedimentology* 48:415–435.

Kooistra WHCF, Medlin LK. 1996. Evolution of the diatoms (Bacillariophyta): IV. A reconstruction of their age from small subunit rRNA coding regions and the fossil record. *Mol Phylogenet Evol* 6:391–407.

Kretz R. 1972. Silicon: Element and geochemistry. In: Fairbridge RW (ed.) *The Encyclopedia of Geochemistry and Environmental Sciences*. Encyclopedia of Earth Science Series, Vol. IVA. New York: Van Nostrand Reinhold, pp. 1091–1092.

Kutuzova RS. 1969. Release of silica from minerals as result of microbial activity. *Mikrobiologiya* 38:714–721 (English translation pp. 596–602).

Kuznetsov SI. 1975. The role of microorganisms in the formation of lake bottom deposits and their diagenesis. *Soil Sci* 119:81–88.

Lalonde SV, Konhauser KO, Reysenbach A-L, Ferris FG. 2005. The experimental silicification of Aquificales and their role in hot spring sinter formation. *Geobiology* 3:41–52.

Latimer WM, Hildebrand JH. 1940. *Reference Book of Inorganic Chemistry*. Revised edition. New York: Macmillan.

Lauwers AM, Heinen W. 1974. Bio-degradation and utilization of silica and quartz. *Arch Microbiol* 95:67–78.

Leo RF, Barghoorn ES. 1976. Silicification of wood. *Bot Mus Leafl, Harvard Univ* 25:1–29.

Lewin JC. 1957. Silicon metabolism in diatoms. IV. Growth and frustule formation in *Navicula pelliculosa*. *J Gen Physiol* 39:1–10.

Lewin JC. 1965. Silicification. In: Lewin RA (ed.) *Physiology and Biochemistry of Algae*. New York: Academic Press, pp. 445–455.

Liebau F. 1985. *Structural Chemistry of Silicates. Structure, Bonding and Classification*. Berlin, Germany: Springer.

Macaskie LE, Empson RM, Cheetham AK, Gey CP, Skarnulis AJ. 1992. Uranium bioaccumulation by a *Citrobacter* sp. as a result of enzymatically mediated growth of polycrystalline HUO_2PO_4. *Science* 257:782–784.

Malinovskaya IM, Kosenko LV, Votselko SK, Podgorskii VS. 1990. Role of *Bacillus mucilaginosus* polysaccharide in degradation of silicate minerals. *Mikrobiologiya* 59:70–78 (English translation pp. 49–55).

Mann DG. 1999. The species concept of diatoms. *Phycologia* 38:437–494.

Marine Chemistry. 1971. *A Report on the Marine Chemistry Panel of the Committee of Oceanography*. Washington, DC: National Academy of Sciences.

Marshall KC. 1971. Sorption interactions between soil particles and microorganisms. In: McLaren AD, Skujins J (eds.) *Soil Biochemistry*, Vol. 2. New York: Marcel Dekker, pp. 409–445.

Marshman NA, Marshall KC. 1981a. Bacterial growth on proteins in the presence of clay minerals. *Soil Biol Biochem* 13:127–134.

Marshman NA, Marshall KC. 1981b. Some effect of montmorillonite on the growth of mixed microbial cultures. *Soil Biol Biochem* 13:135–141.

Martin-Jézéquel V, Hildebrand M, Brezinski MA. 2000. Silicon metabolism in diatoms: Implications for growth. *J Phycol* 36:821–824.

Maurice PA, Vierkoen MA, Hersman LE, Fulghum JE, Ferryman A. 2001. Enhancement of kaolinite dissolution by an aerobic *Pseudomonas mendocina* bacterium. *Geomicrobiol J* 18:21–35.

McCartney K. 1993. Silicoflagellates. In: Lipps, JH (ed.) *Fossil Prokaryotes and Protists*. Boston, MA: Blackwell Science, pp. 143–154.

McCartney K, Loper DE. 1989. Optimized skeletal morphologies of silicoflagellate genera *Dictyocha* and *Distephanus*. *Palaeobiology* 15(3):283–298.

Milde K, Sand W, Wolff W, Bock E. 1983. Thiobacilli of the corroded concrete walls of the Hamburg sewer system. *J Gen Microbiol* 129:1327–1333.

Mitchell RL. 1955. Trace elements. In: Bear FE (ed.) *Chemistry of the Soil*. New York: Van Nostrand Reinhold, pp. 235–285.

Mountain BW, Benning LG, Boerema JA. 2003. Experimental studies on New Zealand hot spring sinters: Rates of growth and textural development. *Can J Earth Sci* 40:1643–1667.

Oehler JH. 1976. Experimental studies in Precambrian paleontology: Structural and chemical changes in blue-green algae during simulated fossilization in synthetic chert. *Geol Soc Am Bull* 87:117–129.

Okabe S, Odagiri M, Ito T, Satoh H. 2007. Succession of sulfur-oxidizing bacteria in the microbial community on corroding concrete in sewer systems. *App Environ Microbiol* 73:971–978.

Parker CD. 1947. Species of sulfur bacteria associated with corrosion of concrete. *Nature (London)* 159:439.

Patrick S, Holding AJ. 1985. The effect of bacteria on the solubilization of silica in diatom frustules. *J Appl Microbiol* 59:7–16.

Phoenix VR, Adams DG, Konhauser KO. 2000. Cyanobacterial viability during hydrothermal biomineralization. *Chem Geol* 169:329–338.

Phoenix VR, Konhauser KO. 2008. Benefits of bacterial biomineralization. *Geobiology* 6:303–308.

Phoenix VR, Konhauser KO, Ferris FG. 2003. Experimental study of iron and silica immobilization by bacteria in mixed Fe-Si systems: Implications for microbial silicification in hot springs. *Can J Earth Sci* 40(11):1669–1678.

Phoenix VR, Martinez RE, Konhauser KO, Ferris FG. 2002. Characterization and implications of the cell surface reactivity of the cyanobacteria *Calothrix* sp. *App Environ Microbiol* 68:4827–4834.

Reeves CD, Volcani BE. 1984. Role of silicon in diatom metabolism. Patterns of protein phosphorylation in *Cylindrotheca fusiformis* during recovery from silicon starvation. *Arch Microbiol* 137:291–294.

Rogall E. 1939. Über den Feinbau der Kieselmembran der Diatomeen. *Planta* 29:279–291.

Round FE, Crawford RM. 1990. *The Diatoms. Biology and Morphology of the Genera*. Cambridge, U.K.: Cambridge University Press.

Sand W, Bock E. 1984. Concrete corrosion in the Hamburg sewer system. *Environ Technol Lett* 5:517–528.

Schultze-Lam S, Fortin D, Davis BS, Beveridge TJ. 1996. Mineralization of bacterial surfaces. *Chem Geol* 132:171–181.

Schwarz K. 1973. A bound form of silicon in glycosaminoglycans and polyuronides. *Proc Natl Acad Sci USA* 70:1608–1612.

Schultze-Lam S, Ferris FG, Konhauser KO, Wiese RG. 1995. In situ silicification of an Icelandic hot spring microbial mat: Implications for microfossil formation. *Can J Earth Sci* 32:2021–2026.

Sillén LG. 1967. The ocean as chemical system. *Science* 156:1189–1197.

Silverman MP, Munoz EF. 1970. Fungal attack on rock: Solubilization and altered infrared spectra. *Science* 169:985–987.

Sperling EA, Robinson JM, Pisani D, Peterson KJ. 2010. Where's the glass? Biomarkers, molecular clocks, and microRNAs suggest a 200-Myr missing Precambrian record of siliceous sponge spicules. *Geobiology* 8:24–36.

Stotzky G. 1986. Influences of soil mineral colloids on metabolic processes, growth, adhesion, and ecology of microbes and viruses. In: Huang PM, Schnitzer M (eds.) *Interaction of Soil Minerals with Organics and Microbes*. SSSA Special Publication No 17. Madison, WI: Soil Science Society of America, pp. 305–428.

Sullivan CW. 1971. A silicic acid requirement for DNA polymerase, thymidylate kinase, and DNA synthesis in the marine diatom *Cylindrica fusiformis*. PhD Thesis, University of California, San Diego, CA.

Sullivan CW, Volcani EB. 1973. Role of silicon in diatom metabolism. The effects of silicic acid on DNA polymerase, TMP kinase and DNA synthesis in *Cyclotheca fusiformis*. *Biochim Biophys Acta* 308:212–229.

Tan KH. 1986. Degradation of soil minerals by organic acids. In: Huang PM, Schnitzer M (eds.) *Interaction of Soil Minerals with Organics and Microbes*. SSSA Special Publication No 17. Madison, WI: Soil Science Society of America, pp. 1–27.

Tobler DJ, Benning LG. 2013. In situ and time resolved nucleation and growth of silica nanoparticles forming under simulated geothermal conditions. *Geochim Cosmochim Acta* 114:156–168.

Tobler DJ, Stefánsson A, Benning LG. 2008. In-situ grown silica sinters in Icelandic geothermal areas. *Geobiology* 6:481–502.

Tobler DJ, Shaw S, Benning LG. 2009. Quantification of initial steps of nucleation and growth of silica nanoparticles: An in-situ SAXS and DLS study. *Geochim Cosmochim Acta* 73:5377–5393.

Toporski JKW, Steele A, Westall F, Thomas-Keprta KL, McKay DS. 2002. The simulated silicification of bacteria—New clues to the modes and timing of bacterial preservation and implications for the search for extraterrestrial microfossils. *Astrobiology* 2:1–26.

Toth SJ. 1955. Colloidal chemistry of soils. In: Bear FE (ed.) *Chemistry of the Soil*. New York: Van Nostrand Reinhold, pp. 85–106.

Tréguer P, Nelson DM, Van Bennekom AJ, DeMaster DJ, Leynaert A, Quéguiner B. 1995. The silica balance in the world ocean: A reestimate. *Science* 268:375–379.

Ullman WJ, Kirchman DL, Welch SA, Vandevivere P. 1996. Laboratory evidence for microbially mediated silicate mineral dissolution in nature. *Chem Geol* 132:11–17.

Vandevivere P, Welch SA, Ullman WJ, Kirchman DL. 1994. Enhanced dissolution of silicate minerals by bacteria at near-neutral pH. *Microb Ecol* 27:241–251.

Waksman SA, Starkey RL. 1931. *The Soil and the Microbe*. New York: Wiley.

Walter MR. 1976. Geyserites of Yellowstone National Park: An example of abiogenic "stromatolites". In: Walter MR (ed.) *Stromatolites*. Developments in Sedimentology, Vol. 20. Amsterdam, the Netherlands: Elsevier, pp. 88–112.

Walter MR, Bauld J, Brock TD. 1972. Siliceous algal and bacterial stromatolites in hot spring and geyser effluents of Yellowstone National Park. *Science* 178:402–405.

Weaver FM, Wise SW, Jr. 1974. Opaline sediments of the southeastern coastal plain and horizon A: Biogenic origin. *Science* 184:899–901.

Weaver TL, Dugan PR. 1972. Enhancement of bacteria methane oxidation by clay minerals. *Nature (London)* 237:518.

Webley DM, Duff RB, Mitchell WA. 1960. A plate method for studying the breakdown of synthetic and natural silicates by soil bacteria. *Nature (London)* 188:766–767.

Webley DM, Henderson MEF, Taylor IF. 1963. The microbiology of rocks and weathered stones. *J Soil Sci* 14:102–112.

Weiss A, Reiff G, Weiss A. 1961. Zur Kenntnis wasserbeständiger Kieselsäureesster. *Z Anorg Chem* 311:151–179.

Welch SA, Ullman WJ. 1993. The effect of organic acids on plagioclase dissolution rates and stoichiometry. *Geochim Cosmochim Acta* 57:2725–2736.

Welch SA, Vandevivere P. 1995. Effect of microbial and other naturally occurring polymers on mineral dissolution. *Geomicrobiol J* 12:227–238.

Werner D. 1966. Die Kieselsäure im Stoffwechsel von *Cyclotella cryptica* Reimann, Lewin and Guillard. *Arch Mikrobiol* 55:278–308.

Werner D. 1967. Hemmung der Chlorophyllsynthese und des NADP-abhängigen Glyzeraldehyd-3-phosphat-dehydrogenase durch Germaniumsäure bei *Cyclotella cryptica*. *Arch Mikrobiol* 57:51–60.

Westall F, Boni L, Guerzoni E. 1995. The experimental silicification of microorganisms. *Palaeontology* 38:495–528.

Yee N, Phoenix VR, Konhauser KO, Benning LG, Ferris FG. 2003. The effect of cyanobacteria on silica precipitation at neutral pH: Implications for bacterial silicification in geothermal hot springs. *Chem Geol* 199:83–90.

Yool A, Tyrrell T. 2003. Role of diatoms in regulating the ocean's silicon cycle. *Global Biogeochem Cycles* 17:14.1–14.21.

including hydroxamate siderophores, sugar acids, phenols, phenolic acids, and polyphenols (Vance et al., 1995). It can also form sulfato and phosphato complexes (Nordstrom and May, 1995).

The Al^{3+} ion is toxic to most forms of life because it can react with negatively charged groups on proteins, including enzymes, and with other vital polymers (Harris, 1972; Garcidueñas Piña and Cervantes, 1996). However, a few plants not only tolerate it but may have a requirement for it. The tree *Orites excelsa* contains basic aluminum succinate in its woody tissue. The ashes of the club moss *Lycopodium alponium* (Pteridophyta) may contain as much as 33% Al_2O_3 (Gornitz, 1972). The bacterium *Pseudomonas fluorescens* can detoxify Al^{3+} in the bulk phase by producing an extracellular phospholipid that sequesters the aluminum (Appanna and St. Pierre, 1994; Appanna et al., 1994; Appanna and Hamel, 1996). The cyanobacterium *Anabaena cylindrica* can detoxify limited amounts of aluminum taken into the cell with the help of intracellularly occurring inorganic polyphosphate granules, which sequester Al^{3+} (Petterson et al., 1985).

A number of bacteria, fungi, and lichens are known to participate in the formation of secondary aluminum-containing minerals through their ability to weather aluminum-containing rock minerals (see Chapters 4 and 11).

12.2 MICROBIAL ROLE IN BAUXITE FORMATION

12.2.1 Nature of Bauxite

Bauxite is an ore that contains 45%–50% Al_2O_3 in the form of gibbsite, boehmite, and/or diaspore and not more that 20% Fe_2O_3 as hematite, goethite, and/or aluminian goethite. Such ore also contains 3%–5% silica and silicates combined (Valeton, 1972). The silicates in the ore are chiefly in the form of kaolinite.

Bauxite is a product of surficial weathering of aluminosilicate minerals in rock (Butty and Chapalaz, 1984). Warm and humid climatic conditions with wet and dry seasons favor the weathering process. The parent material from which bauxite arises may be volcanic and other aluminosilicate rocks, limestone associated with karsts, and alluvium (Butty and Chapalaz, 1984). The weathering that leads to bauxite formation begins at the surface of an appropriate exposed or buried rock formation and in cracks and fissures. The process

includes breakdown of the aluminosilicates in the parent substance with gradual solubilization of Al, Si, Fe, and other constituents, starting at the mineral surface. The biological contributions to the weathering are favored by warm temperatures and humid conditions (see Section 12.2.2). In the case of limestone as parent substance, an important part of the weathering process is the solubilization of the $CaCO_3$. The solubilized products reprecipitate when their concentration and the environmental pH and E_h are favorable. Groundwater flow may transport some of the solubilized constituents away from the site of weathering, contributing to an enrichment of the constituents left behind. The initial stages of weathering produce material that could serve as precursor in the formation of laterite or bauxite. A key difference between laterite and bauxite is that the former is richer in iron relative to aluminum, whereas, the latter is richer in aluminum. Biotic and abiotic environmental conditions determine whether laterite or bauxite will accumulate (Schellman, 1994).

Bauxite formation in nature is a slow process and impacted by vegetation at the site of formation and by tectonic movement in addition to climate. Vegetation provides cover that protects against erosion of weathered rocks, limits water evaporation, and may generate weathering agents (Butty and Chapalaz, 1984). It is also the source of nutrients for microbiota that participate in rock weathering and some other aspects of bauxite formation. Tectonic movement contributes to topography and geomorphology in the area of bauxite formation. Alterations in topography as well as variation in climate can affect the groundwater level. Alternating moist and dry conditions are needed during weathering of host rock for the formation and buildup of the secondary minerals that make up bauxite.

12.2.2 Biological Role in Weathering of the Parent Rock Material

Biological participation in bauxite formation has been suggested in the past. Butty and Chapalaz (1984) invoked microbial activity in controlling pH and E_h. They viewed rock weathering as being promoted by microbes through generation of acids and/or ligands for mobilizing rock components and in direct participation in redox reactions affecting iron, manganese, and sulfur compounds.

A more detailed proposal of biogenic bauxite formation is that of Taylor and Hughes (1975).

They concluded that bauxite deposits on Rennell Island in the South Solomon Sea near Guadalcanal was the result of biodegradation of volcanic ash that originated in eruptions on Guadalcanal, 180 km distant, and was deposited in pockets of karstic limestone, lagoons, on Rennell Island in the Plio-Pleistocene. The authors established that the bauxite deposit, enclosed in dolomitic limestone from a reef, was not derived from residues left after the dolomite had weathered away. They speculated that sulfate-reducing bacteria generated CO_2, which caused weathering of aluminosilicates and ferromagnesian minerals in the volcanic ash, giving rise to transient kaolin that would dissolve at low pH to yield Al^{3+} and silicic acid. Bacterial pyrite formation by sulfate-reducing bacteria, according to the authors, would create pH and E_h conditions that favor weathering of the minerals in the volcanic ash. As initial microbial activity subsided owing to nutrient depletion, pH was predicted to rise, resulting in the formation of iron and aluminum oxides, the chief constituents of bauxite. The authors speculated that bacteria played a role in the formation of a gel of oxides. Uplift in the northwestern part of the island, groundwater flow, and oxidation of the pyrite were seen to play a role in the maturation of the bauxite.

Natarajan et al. (1997) inferred from the presence of members of the bacterial genera *Thiobacillus*, *Bacillus*, and *Pseudomonas* in the Jamnagar bauxite mines in Gujarat, India, that these microorganisms are involved in bauxite formation. They based their inference on the known ability of these organisms to weather aluminosilicates, to precipitate oxyhydroxides of iron, to dissolve and transform alkaline metal species, and to form alumina, silica, and calcite minerals. On the same basis, they also implicated the fungi of the genus *Cladosporium*, which they suggest can reduce ferric iron and dissolve aluminosilicates.

On the basis of published discussions of bauxite formation (bauxitization) (e.g., Valeton, 1972; Butty and Chapalaz, 1984), the process can be divided into two stages, which may overlap to some extent. The first stage is the weathering of the parent rock or alluvium that leads to the liberation of Al, Fe, and Si from primary and secondary minerals that contain aluminum. The second stage consists of the formation of bauxite from the weathering products. Each of the two stages is thought to be aided by microorganisms.

12.2.3 Weathering Phase

Extensive evidence exists that bacteria, fungi, and lichens have the ability to weather rock minerals (see Chapter 11). The evidence was amassed in laboratory experiments and by in situ observation. The rock weathering resulted from the excretion of corrosive metabolic products by various microbes. These products included inorganic and organic acids, bases, and/or organic ligands. In instances where oxidizable or reducible rock components are present, enzymatic redox reactions may also have come into play. Most studies of microbial weathering have involved aerobic bacteria and fungi. However, anaerobic bacteria must also be considered in some cases of weathering. Many of them are actually a better source of corrosive organic acids needed in rock weathering than are aerobic bacteria. Moreover, bacterial reduction of ferric iron whether in solution or in minerals requires anaerobic conditions.

The products of rock weathering may be soluble or insoluble. In the latter case, they may accumulate as secondary minerals. In bauxitization, Al, Si, and Fe will be mobilized. Control of pH, mostly by microorganisms, helps to segregate these products to some extent by affecting their respective solubilities. Vegetative cover over the site of bauxitization is a source of nutrients required by the microorganisms to grow and form the weathering agents. Warm temperature enhances microbial growth and activity. Infiltrating surface water (rain) and groundwater help to separate the soluble from the insoluble weathering products derived from the source material, leaving behind a mineral mixture that will include aluminum and iron oxides, silica, and silicates (especially kaolinite, formed secondarily from interaction of Al^{3+} and silicate ions). The mineral conglomerate is *protobauxite*.

12.2.4 Bauxite Maturation Phase

In this stage, the protobauxite becomes enriched in aluminum oxides (gibbsite, diaspore, and/or boehmite) by selective removal of iron oxides, silica, and silicates. Such enrichment has been shown to occur in laboratory experiments under anaerobic conditions (Ehrlich et al., 1995; Ehrlich and Wickert, 1997). Unsterilized Australian ore was placed in presterilized glass or Lexan™ columns. The ore in the columns was then completely immersed in a sterile sucrose–mineral salts solution. After outgrowth of bacteria resident on the ore had taken place over 3–5 days at 37°C, the columns were fed daily from the bottom with fresh sterile medium over a time interval of 20–30 min, depending on the size of the columns. The medium was not de-aerated before it was introduced into the columns. Control columns in which the outgrowth of bacteria was suppressed by 0.1% or 0.05% thymol added to the nutrient solution fed to these columns were run in parallel. The effluent of spent medium displaced by each addition of fresh medium was collected and analyzed by measuring pH; determining the concentration of solubilized ferrous and total iron, Si, and Al; and, in the case of the experimental columns, examining the morphology of the bacteria displaced in the effluents.

The content of the experimental columns quickly turned anaerobic as bacteria grew out from the ore. This was indicated by strong foaming and outgassing in the headspace of the columns and by detection of significant numbers of *Clostridia* among the bacteria in the displaced medium in successive effluents from the columns. The evolved gas probably consisted of CO_2 and H_2, both of which are known to be produced from sucrose by clostridial fermentation. Analyses of successive effluents showed that the bacteria solubilized iron in the bauxite, which was in the form of hematite, goethite, and/or aluminian goethite, by reducing it to ferrous iron. As expected, solubilization from unground Australian pisolitic bauxite was slower and less extensive than from the same ore preground to a mesh size of –10 (particle size of 2 mm or less). In one experiment with the unground pisolitic bauxite* (Ehrlich et al., 1995), 25% of the iron was mobilized in 106 days. In the same time, the bacteria also solubilized 2.2% of the SiO_2/kaolinite in the ore. Al was solubilized over this time interval to the extent of 5.9%, but whereas Fe and Si were solubilized at a fairly constant rate once the bacteria had grown out from the ore, Al was not solubilized until the pH in the bulk phase in the column had dropped gradually from about 6.5 initially to about 4.5 after about 20 days. At the start of the experiments, the ore contained 50% Al (calculated as Al_2O_3), 20% Fe (calculated as Fe_2O_3), and 6.5% Si (calculated as SiO_2) by weight. No measurable Fe, Si, or Al solubilization took place in the control columns.

Results from column experiments with bauxite samples from different geographical locations (Ehrlich et al., 1995; Ehrlich and Wickert, 1997) support the notion that for optimal aluminum enrichment of protobauxite, the bulk-phase pH should remain above 4.5. In most cases in the field, the pH probably rarely if ever drops below 4.5. This is because bauxite maturation in nature occurs in an open system where the bacterial activity will be much slower and where acidic metabolites are more readily diluted and carried away by moving groundwater than in the column experiments. However, an exception seems to be a deposit in northern Brazil in which bauxite weathering has given rise to a kaolin deposit as a result of iron mobilization (deferritization) and apparently some aluminum mobilization (Kotschoubey et al., 1999). The experimental results described earlier clearly showed that the action of the bacteria that grew from the ore in the columns enriched the ore in aluminum.

The column effluents contained fermentation acids such as acetic and butyric acids and, sometimes, neutral solvents such as butanol, acetone or isopropanol, and ethanol, detectable by spot confirmation using high performance liquid chromatography (HPLC) analysis and by odor. With unground pisolitic ore, Fe and Si but not Al solubilization leveled off after about 60 days. With the same ore ground to –2 mm particle size, Fe, Si, and Al solubilization continued at a steady rate over the entire experimental period, which in some cases exceeded 100 days.

Some pisolites taken from the active and control columns at the end of the experiment with Australian pisolitic ore described previously were cross sectioned, surface polished, and examined microscopically by reflected light and by scanning electron microscopy (SEM) coupled with energy-dispersive x-ray (EDX) analysis (Ehrlich et al., 1995). Color images of cross sections of pisolites that had been subjected to bacterial action in the active column showed a distinct bleached rim surrounding a reddish-brown core. Comparable sections of pisolites from the control column showed only a faintly bleached zone surrounding a reddish-brown core. A SEM-EDX image of a cross section of a pisolite that had been acted upon by bacteria showed a significant depletion of iron in the bleached zone around the core, whereas a similar cross section of a pisolite from a control column showed no significant iron depletion. No depletion in Si or Al was visible in either of the cross sections, probably because the percentages of these elements that were mobilized were too small (see earlier). Interestingly, cross sections of pisolites collected at the Weipa bauxite deposit in Queensland, Australia (Rintoul and Fredericks, 1995), resemble the cross section of the microbially attacked pisolites in the previously described experiments. The finding of Rintoul and Fredericks supports the idea that what happened to the pisolites in the experimental columns is representative of a natural process.

The iron-depleted bleached zone around the undepleted core in the cross sections from pisolites acted upon by bacteria presents an enigma. The pisolites are not significantly porous, as has been shown by placing untreated pisolites in boiling water and noting a lack of effervescence originating from the pisolitic surface, indicating an absence of air entrapped in pores in the pisolites. Thus, bacteria cannot penetrate the pisolites to effect iron mobilization by Fe(III) reduction below the pisolite surface. It is proposed that the bacteria bring about iron mobilization by enzymatic reduction of Fe(III) because of daily replacement of a major portion of the medium in the columns, which would dilute any chemical reductant, Fe(III)-complexing agent, or extracellular Fe(III)-reducing enzyme in the bulk phase, did not significantly change the rate of iron reduction. Reduction of Fe(III) below the pisolite surface must therefore, depend on a nonenzymatic redox mechanism. Such a mechanism may involve Fe^{2+} produced microbially at the surface by enzymatic reduction of Fe_2O_3 with a suitable electron donor (the CH_2O in Reaction 12.1 representing an unspecified organic electron donor):

$$2Fe_2O_{3\ surface} + (CH_2O) + 8H^+ \rightarrow 4Fe^{2+}_{surface} + CO_2 + 5H_2O \quad (12.1)$$

This Fe^{2+} reacts somehow with Fe_2O_3 below the surface:

$$2Fe^{2+}_{surface} + Fe_2O_{3\ interior} + 6H^+ \rightarrow$$
$$2Fe^{3+}_{surface} + 2Fe^{2+}_{interior} + 3H_2O \quad (12.2)$$

The Fe^{3+} produced at the pisolite surface in Reaction 12.2 is also reduced microbially to Fe^{2+}:

$$4Fe^{3+}_{surface} + (CH_2O) + H_2O \rightarrow 4Fe^{2+}_{surface} + CO_2 + 4H^+ \quad (12.3)$$

Reaction 12.2 is best visualized as involving the conduction of an electron from an $Fe^{2+}_{surface}$

(Reaction 12.4) to the interior, where it reacts with Fe(III) of $Fe_2O_{3\ interior}$ (Reaction 12.5):

$$2Fe^{2+}_{surface} \rightarrow 2Fe^{3+}_{surface} + 2e \quad (12.4)$$

$$Fe_2O_{3\ interior} + 2e + 6H^+ \rightarrow 2Fe^{2+}_{interior} + 3H_2O \quad (12.5)$$

The Fe^{3+} generated at the pisolite surface in Reaction 12.2 is immediately reduced to Fe^{2+} by the bacteria (Equation 12.3). The Fe^{2+} generated in the interior in Reaction 12.5 of the pisolite escapes to the exterior through passages created by the solubilization of Fe and Si, and later Al, if the pH drops below 4.5 in the interior. Because Reaction 12.2 is thermodynamically unfavorable ($\Delta G_r^\circ = +1.99$ kcal or $+8.32$ kJ), it is the rapid, bacterially catalyzed reduction of Fe(III) at the surface of the pisolites (Reactions 12.1 and 12.3) that provide the energy that drives Reaction 12.2. If, for instance, H_2 instead of CH_2O were the reductant in Reaction 12.1, the value of ΔG_r° would be -35.56 kcal (-112.4 kJ), and for Reaction 12.3 it would be -71.12 kcal (-297.3 kJ). If acetate were the electron donor, the value of ΔG_r° would be -26.9 kcal (-112.4 kJ) for Reaction 12.1, and for Reaction 12.3 it would be -62.6 kcal (-261.7 kJ) for reduction of 4 mol Fe^{3+}. Yan et al. (2004) obtained some evidence using x-ray fluorescence and Xanes image analysis that is consistent with this microbe-dependent mechanism of Fe(III) reduction below the surface of pisolites.

The conduction of electrons to the interior of the pisolite to reduce $Fe_2O_{3\ interior}$ to $Fe^{2+}_{interior}$ may be similar to that in a reaction of anaerobic microbial reduction of structural iron(III) within a ferruginous smectite by *Shewanella putrefaciens* MR-1 with formate or lactate as electron donor (Kostka et al., 1996). Stucki et al. (1987) previously showed that a bacterium indigenous to the clay reduced the structural iron in ferruginous smectites including the smectite used by Kostka et al. (1996). The difference between the reaction with smectites and the one postulated for bauxitic pisolites is that the reduced iron formed in the smectite was not mobilized by the bacteria unlike that formed in the reaction with the pisolites. Structural iron(II) produced in smectite remained in place and could be reoxidized, thus, making ferruginous smectite a renewable terminal electron acceptor (Ernstsen et al., 1998). A simulation of electron transfer on hematite surfaces by computer may also have a direct bearing on iron mobilization from pisolites (Kerisit and Rosso, 2006). The recent observations of Fe(III) reduction in kaolinite by *Geobacter pickeringii* sp. nov., *Geobacter argillaceus* sp. nov., and *Pelosinus fermentans* gen. nov., sp. nov. may be a related phenomenon (Shelobolina et al., 2007). Interestingly, the first two of these organisms are Fe(III) respirers, whereas the third is a fermentor, which, though able to reduce Fe(III), does not respire it but uses it as an electron sink.

Bacterial reduction of hematite in oxidized samples from the Central Plateau of Brazil incubated in the presence of sucrose at 25°C was reported by Macedo and Bryant (1989). They observed preferential attack of hematite over aluminian goethite. Initial outgrowth of the bacteria from the soil in these experiments required 3–9 weeks. Microbial activity was correlated with a decrease in redness of the soil.

12.2.5 Bacterial Reduction of Fe(III) in Bauxites from Different Locations

Physiologically similar bacterial cultures grew from bauxite from the pisolitic deposit in Australia, from deposits in the Amazon in Brazil, and from the island of Jamaica in the Caribbean Sea (Ehrlich and Wickert, 1997). When the Amazonian and Jamaican ores in columns were fed sucrose–mineral salts medium, a behavior similar to that observed with the Australian bauxite was noted with respect to Fe, Si, and Al solubilization and pH changes in successive effluents. *Clostridia* were among the bacteria that were first noted in column effluents. The *Clostridia* from these two ores showed a close phylogenetic relationship to *Clostridia* from the Australian ore (B. Methé, unpublished results). These similarities suggest that the natural flora associated with the bauxites may play a role in bauxite maturation, i.e., in its enrichment in Al over time. A caveat is, however, that none of the ore samples were collected under controlled conditions that would have minimized or prevented contamination of the ore in the collection process or in subsequent storage. On the other hand, the probability that all ore samples were so heavily contaminated during collection or storage that very similar mixed anaerobic bacterial assemblages would arise after only 3–4 days of incubation in experimental columns seems small.

12.2.6 Other Observations of Bacterial Interaction with Bauxite

Others have demonstrated bacterial interaction with bauxite, mainly for the purpose of testing whether the ore could be made industrially more attractive (biobeneficiation). These interactions occurred generally under aerobic conditions. Anand et al. (1996) found that *Bacillus polymyxa* strain NCIM 2539 was able to mobilize in shake culture all the calcium and about 45% of the iron from a bauxite ground to 53–74 μm particle size. The bacterial treatment occurred at 30°C in Bromfield medium containing 2% sucrose. The change in composition of the bauxite was attributed to direct action of the cells and to the action of cellular products such as exopolysaccharides and organic acids. The oxidation state of the mobilized iron was not determined.

Groudev (1987) reported silicon removal by *Bacillus circulans* and B. *mucilaginosus* from low-grade bauxites. The Si mobilization was attributed to the action of exopolysaccharides that were elaborated by the bacteria. Some Al was also mobilized in these experiments.

Orgutsova et al. (1989) reported variations in the ability of strains of the fungi *Aspergillus niger* and *Penicillium chrysogenum* and various yeasts and pseudomonads to mobilize Al, Fe, and Si from a ground bauxite of which 70% had a −74 μm particle size. The oxidation state of the mobilized iron was not reported. The mobilization of Al, Fe, and Si was attributed to the action of metabolic products formed by the test organisms.

In another study, Karavaiko et al. (1989) found that a strain of *Bacillus mucilaginosus* removed Si from bauxite ground to −0.074 mm particle size and incubated in a sucrose–mineral salts medium with a 10% inoculum in shake-flask culture at 30°C. The Si removal was attributed to the selective adherence of fine particles of ore rich in Si to the exopolysaccharide at the surface of the bacterial cells and not to dissolution. The mycelial fungi *Aspergillus niger* and *A. pullulans*, on the other hand, were able to mobilize varying amounts of Fe, Al, and Si from the same bauxites by dissolution with acids, which are formed by them metabolically in a sucrose–mineral salts medium.

Bandyopadhay and Banik (1995) were able to mobilize 39.9% silica and 46.4% iron from a bauxite ore with a mutated strain of *Aspergillus niger* in a laboratory experiment in which the fungus was

EHRLICH'S GEOMICROBIOLOGY

allowed to grow at the surface of 80 mL of culture liquid at an initial pH of 4.0 in a flask at 30°C. The culture medium contained glucose as the energy source and $NaNO_3$ as the nitrogen source. The ore was ground to a mesh size of −170 to +200 and then added to the medium at a concentration of 0.3%. The mobilization of Si and Fe was attributed to the action of the organic acids, probably citric and oxalic, produced by the fungus.

12.3 SUMMARY

Aluminum is the third most abundant element in the Earth's crust, silicon and oxygen being more abundant. Of these three elements, aluminum is the only one for which a physiological function has not been found, although a very small number of higher organisms are known to accumulate it. Al^{3+} is generally toxic. At least one known cyanobacterium and a strain of E. coli have each developed a different mechanism of resistance to it. Various microbes are known to participate through weathering action in the formation of some aluminum-containing minerals.

The formation of bauxite (bauxitization), whose major constituents are Al_2O_3 in the form of gibbsite, boehmite, and/or diaspore; Fe_2O_3 in the form of hematite, goethite, and/or aluminian goethite; and SiO_2/aluminosilicate in the form of silica/kaolinite, can be visualized as involving two stages that may overlap to some extent. Evidence to date suggests that microbes are involved in both stages. The source material in bauxitization may be volcanic and other aluminosilicate rocks, limestone associated with karsts, and alluvium. The first stage of bauxitization involves weathering of source rock and the formation of protobauxite and the second stage the maturation of protobauxite to bauxite. The first stage, if aerobic, may be promoted by bacteria and fungi, and if anaerobic, by facultative and anaerobic bacteria. The second stage is promoted by iron-reducing and fermentative bacteria under anaerobic conditions. The first stage involves the mobilization of Al, Fe, and Si from host rock and the subsequent precipitation of these rock constituents as oxides, silica, and silicate minerals. The second stage involves the selective mobilization of iron oxides and silica/ silicate, enriching the solid residue in aluminum. The process is favored in warm, humid climates with alternate wet and dry seasons. The site of formation must be associated with vegetation that can serve as a source of nutrients to the microorganisms and may yield weathering agents as a result of microbial attack of plant residues. In situ, microbes are expected to play a significant role in the control of pH during bauxite maturation, which ensures that little of the aluminum oxide is mobilized.

REFERENCES

Anand P, Modak JM, Natarajan KA. 1996. Biobeneficiation of bauxite using Bacillus polymyxa: Calcium and iron removal. Int J Mineral Process 48:51–60.

Appanna VD, Hamel R. 1996. Aluminum detoxification mechanism in Pseudomonas fluorescens is dependent on iron. FEMS Microbiol Lett 143:223–228.

Appanna VD, St Pierre M. 1994. Influence of phosphate on aluminum tolerance in Pseudomonas fluorescens. FEMS Microbiol Lett 124:327–332.

Appanna VE, Kepes M, Rochon P. 1994. Aluminum tolerance in Pseudomonas fluorescens ATCC 13525; involvement of a gelatinous lipid-rich residue. FEMS Microbiol Lett 119:295–302.

Bandyopadhay N, Banik AK. 1995. Optimization of physical factors for bioleaching of silica and iron from bauxite ore by a mutant strain of Aspergillus niger. Res Ind 40:14–17.

Bowen HJM. 1966. Trace Elements in Biochemistry. London, U.K.: Academic Press.

Butty DL, Chapalaz CA. 1984. Bauxite genesis. In: Jacob LE Jr, ed., Bauxite: Proceedings of the 1984 Bauxite Symposium, Los Angeles, CA. New York: Society of Mining Engineers of American Institute of Mining, Metallurgical, and Petroleum Engineers, Inc., pp. 11–151.

Ehrlich HL, Wickert LM. 1997. Bacterial action on bauxites in columns fed with full-strength and dilute sucrose-mineral salts medium. In: Lortie L, Bédard P, Gould WD, eds., Biotechnology and the Mining Environment. SP 97–1. Ottawa, Ontario, Canada: CANMET Natural Resources Canada, pp. 74–89.

Ehrlich HL, Wickert LM, Noteboom D, Doucet J. 1995. Weathering pisolitic bauxite by heterotrophic bacteria. In: Vargas T, Jerez CA, Wiertz JV, Toledo H, eds., Biohydrometallurgical Processing, Vol. I. Santiago, Chile: University of Chile, pp. 395–403.

Ernstsen V, Gates WP, Stucki JW. 1998. Microbial reduction of structural iron in clays: A renewable resource of reduction capacity. J Environ Qual 27:761–766.

Garcidueñas Piña R, Cervantes C. 1996. Microbial interaction with aluminum. BioMetals 9:311–316.

the negative charges on the phosphates at neutral pH (Westheimer, 1987). It is probably the reason why ATP got selected in the evolution of life as a repository and universal transfer agent of chemical energy in biological systems. The instability of phosphoric acid anhydride linkages at high temperatures is one of the key factors that limit life to temperatures below 130°C (White, 1984).

13.2 OCCURRENCE IN THE EARTH'S CRUST

Phosphorus is found in all parts of the biosphere. Its gross abundance at the surface of the Earth has been cited by Fuller (1972) to be 0.10%–0.12% (w/w). It occurs mostly in the form of inorganic phosphates and organic phosphate derivatives. The organic derivatives in soil are mostly phytins (Paul and Clark, 1996). Total phosphorus concentrations in mineral soil range from 35 to 5300 mg kg^{-1} (average 800 mg kg^{-1}; Bowen, 1979). The ratio of organic to inorganic phosphorus (P_{org}/P_i) varies widely in these environments. In mineral soil, P_{org}/P_i may range from 1:1 to 2:1 (Cosgrove, 1967, 1977). In lake water, as much as 50% of the organic fraction may be phytin and releasable as inorganic phosphorus through hydrolysis catalyzed by phytase (Herbes et al., 1975). The organic phosphorus in lake water may constitute 80%–99% of the total soluble phosphorus. In the particular examples cited by Herbes et al. (1975), the total organic phosphorus rarely exceeded 40 µg phosphate per liter. The authors speculated that hydrolyzable phosphate compounds other than phytins were largely absent because they are much more labile than phytins. Readily measurable phosphatase activity was detected in Sagima and Suruga Bays, Tokyo (Taga and Kobori, 1978; Kobori and Taga, 1979).

Because the water solubility of phosphate minerals (including bones and teeth!) is very low, the concentration of phosphate in surface waters is very low as well. Freshwater lakes and rivers carry contents of total phosphate in the range of 3–150 mg P m^{-3} (= 0.1–5 µM), and the same is true for seawater (ca. 60 mg P m^{-3} = 2 µM; Stumm and Morgan, 1981). In freshwater lakes, phosphorus usually limits primary production, and excess phosphate availability (at concentrations higher than ca. 30 mg P m^{-3}) may cause lake eutrophication, causing unpleasant mass developments of algae and of toxic cyanobacteria (Wetzel, 2001; p. 273 ff.). Also, terrestrial plant production may be phosphate limited, and efficient agriculture depends typically on fertilization with P and N fertilizers. On the medium-term time perspective, our terrestrial phosphorus supply for agriculture may become seriously limiting for food production (Gilbert, 2009).

13.3 CONVERSION OF ORGANIC INTO INORGANIC PHOSPHORUS AND SYNTHESIS OF PHOSPHATE ESTERS

An important source of free organic phosphorus compounds in the biosphere is the breakdown of animal and vegetable matter. However, living microbes such as *Escherichia coli* and organisms from activated sludge have been found to excrete aerobically assimilated phosphorus as inorganic phosphate when incubated under oxygen exclusion (Shapiro, 1967). Organically bound phosphorus is for the most part not directly available to living organisms because it cannot be taken up into the cell in this form. To be taken up, it must be freed from its organic residues through hydrolytic cleavage catalyzed by phosphatases. In soil, a major part of the microbial community is able to contribute to this process (Dommergues and Mangenot, 1970; Tian et al., 2012). Active organisms include bacteria such as *Bacillus megaterium*, *B. subtilis*, *Serratia* sp., *Proteus* sp., *Arthrobacter* sp., *Streptomyces* sp., *Pseudomonas* sp., and *Alcaligenes* sp., and fungi such as *Aspergillus* sp., *Penicillium* sp., *Rhizopus* sp., and *Cunninghamella* sp. (Dommergues and Mangenot, 1970; Paul and Clark, 1996). These organisms secrete, or liberate upon their death, phosphatases with greater or lesser substrate specificity (Skujins, 1967; Nannipieri et al., 2012). Such activity has also been noted in the marine environment (Ayyakkannu and Chandramohan, 1971; Paytan and McLaughlin, 2007; Boge et al., 2012).

Phosphate liberation from phytin requires the enzyme phytase (Menezes-Blackburn et al., 2013):

$$\text{Phytin} + 6H_2O \rightarrow \text{Inositol} + 6P_i \quad (13.1)$$

Phosphate is liberated from nucleic acid by the action of nucleases, which yield nucleotides, followed by the action of nucleotidases, which yield nucleosides and inorganic phosphate:

$$\text{Nucleic acid} \xrightarrow[+H_2O]{\text{Nucleases}} \text{Nucleotides}$$
$$\xrightarrow[+H_2O]{\text{Nucleotidase}} \text{Nucleosides} + P_i \quad (13.2)$$

Phosphate liberation from phosphoproteins, phospholipids, ribitol, and glycerol phosphates requires phosphomono- and phosphodiesterases. Phosphodiesterases attack phosphodiesters at either the 3′ or 5′ carbon linkage, whereas phosphomonoesterases (phosphatases) attack monoester linkages (Lehninger, 1975, pp. 184, 323–325).

Synthesis of organic phosphates (monomeric phosphate esters) is an intracellular process and normally proceeds through reaction between the hydroxyl (OH) of a carbinol group, as for instance in glucose, and ATP in the presence of an appropriate kinase. For example,

$$\text{Glucose} + \text{ATP} \xrightarrow{\text{Glucokinase}} \text{Glucose 6-phosphate} + \text{ADP} \quad (13.3)$$

13.4 ASSIMILATION OF PHOSPHORUS

Whereas animals cover their phosphorus needs with their organic food, primary producers such as algae and cyanobacteria have to take up phosphate from their environment. Also, heterotrophic bacteria require an additional phosphate supply because their relative phosphate content is substantially higher than that of their food. Only little is known about the mechanisms how phosphate is taken up by algae. The K_S values (half-saturation concentration, analogous to the Michaelis constant K_M) for phosphate uptake by diatoms and bacteria are in the range of 0.5–10 μM (Wetzel, 2001). These communities cover their phosphate supply at total phosphate concentrations of less than 1 μM (<30 mg P m^{-3}) phosphate, thus, proving their high affinity for this essential nutrient.

With the green alga *Chlamydomonas reinhardtii*, active uptake of inorganic phosphate has been measured, and the uptake rates increased under conditions of phosphate deprivation (Shimogawara et al., 1999). *E. coli* has two different phosphate uptake systems, one for low and one for high-affinity uptake (van Veen, 1997). At high phosphate supply, phosphate enters the periplasm by diffusion through a porin in the outer membrane and crosses the inner membrane through a constitutively expressed low-affinity transport protein. At phosphate limitation, orthophosphate is bound in the periplasmic space by a specific phosphate-binding protein that transfers the phosphate to an ATP-dependent, phosphate-specific transport system in the cytoplasmic membrane. In addition, also an alkaline phosphatase, a nucleotidase, and a cyclic phosphodiesterase are induced that are localized in the periplasmic space and thus, allow to tap organic phosphate resources (Wanner, 1993). Similar regulatory systems have been identified in *B. subtilis* (Hulett, 1996), *Corynebacterium glutamicum* (Ishige et al., 2003), and the cyanobacterium *Synechococcus* sp. (Aiba et al., 1993).

Since phosphate is available in aquatic environments only at very low concentration and has to be accumulated in the cell by energy-dependent transport systems to intracellular concentrations of about 10 mM (Thauer et al., 1977), exposure of such organisms to growth media of high phosphate content will cause severe disorders in cell energetics. Therefore, many environmental bacteria from freshwater ecosystems are sensitive to elevated phosphate concentrations in growth media.

Also in soil, the majority of phosphate is bound in water-insoluble minerals, and the soil water phase contains bioavailable phosphate only at very low concentrations (<10 μM) (Marschner, 1995).Vascular plants have an active transport system in their fine roots, which takes up primary orthophosphate ($H_2PO_4^-$) in a cotransport with protons (Rausch and Bucher, 2002). Most plants enhance their nutrient uptake capacity (mainly phosphate and nitrogen compounds) by association with fungi in various forms of Mycorrhiza or with symbiotic bacteria in the rhizosphere. Several types of *phosphate-solubilizing bacteria* have been described, such as *Pseudomonas cepacia*, *Serratia marcescens*, *Erwinia herbicola*, *Rhizobium spp.*, and *Bacillus spp.* (Liu et al., 1992; Rodriguez and Fraga, 1999; Krishnaraj and Goldstein, 2001), which are supposed to increase the phosphorus supply in the rhizosphere. The phosphate-mobilizing effect of these bacteria is mainly due to the release of gluconic acid that is formed from glucose via specifically induced, pyrroloquinoline quinone–dependent glucose dehydrogenases; the produced gluconic acid helps to solubilize mineral phosphates especially in alkaline, carbonate-rich soils. Other bacteria may contribute to orthophosphate supply by excretion of extracellular phosphatases to tap the pool of organic phosphates.

Inside the cell, ATP as the most important phosphate carrier may be generated by *substrate level phosphorylation*, as in the following reaction sequence:

$$3\text{-Phosphoglyceraldehyde} + NAD^+ + P_i \xrightarrow[\text{dehydrogenase}]{\text{Triosephosphate}}$$

$$1,3\text{-Bisphosphoglycerate} + NADH + H^+ \quad (13.4)$$

$$1,3\text{-Bisphosphoglycerate} + ADP \xrightarrow{\text{ADP kinase}}$$

$$3\text{-Phosphoglycerate} + ATP \quad (13.5)$$

Alternatively, ATP can be generated through a membrane-bound ATP synthase at the expense of proton or sodium ion fluxes across the cytoplasmic membrane that are established in the context of, e.g., *electron transport phosphorylation* or *photophosphorylation*:

$$ADP + P_i \xrightarrow[\text{ATPase}]{\text{Proton flux}} ATP \quad (13.6)$$

Inorganic pyrophosphate (PP_i) is formed as a side product in several energy-expensive reactions. Its hydrolysis to $2P_i$

$$PP \xrightarrow{\text{Pyrophosphatase}} 2P_i \quad (13.7)$$

can either be catalyzed by membrane-bound pyrophosphatases, which conserve part of the hydrolysis energy in a transmembrane ion gradient (Schöcke and Schink, 1998), or cytoplasmic enzymes do the same without energy conservation. Reports (Liu et al., 1982; Varma et al., 1983) on bacteria that were claimed to be able to grow with inorganic pyrophosphate as an energy source could never be substantiated; probably pyrophosphate cannot enter bacterial cells.

Like pyrophosphate, intracellular inorganic polyphosphate granules formed by some microbes (Friedberg and Avigad, 1968) are a form of metaphosphate and can represent both a phosphorus and an energy storage compound (Van Groenestijn et al., 1987). In the case of the cyanobacterium *Anabaena cylindrica*, it may also play a role as detoxifying agent by combining with aluminum ions that are taken into the cell (Pettersson et al., 1985). The ability of several facultatively anaerobic bacteria to accumulate phosphate inside the cell as polyphosphate, and to release it again under conditions of oxygen (and energy) limitation is

being applied for phosphate removal in wastewater treatment plants (Bonting et al., 1991, 1992; Kortstee et al., 2000).

13.5 MICROBIAL SOLUBILIZATION OF PHOSPHATE MINERALS

Inorganic phosphorus may occur in soluble and insoluble forms in nature. The most common inorganic form is orthophosphate (e.g., $HPO_4^-/H_2PO_4^{2-}$). As an ionic species, the concentration of phosphate is controlled by its solubility in the presence of alkaline earth cations such as Ca^{2+} or Mg^{2+} or in the presence of metal cations such as Fe^{2+}, Fe^{3+}, or Al^{3+} at appropriate pH values (see Table 13.1). In seawater, for instance, the soluble phosphate concentration (3×10^{-6} M, maximum) is controlled by Ca^{2+} ions (4.1×10^2 mg L^{-1}), which form hydroxyapatites with phosphate in a prevailing pH range of 7.4–8.1.Similarly, Mg^{2+} ions precipitate phosphate as apatite at pH 7.0–7.4.

Insoluble phosphate can be found as amorphous minerals such as the carbonated hydroxyapatite (the main mineral of invertebrate bones and teeth) or crystalline fluorapatite or apatite. Some primary phosphates are apatites ($Ca_5(F, Cl, OH)(PO_4)_3$), triphylite ($LiFe(Mn)PO_4$), or lithiophilite ($LiMn(Fe)PO_4$). These two minerals differ by their relative Mn and Fe content. In the apatites, the (F, Cl, OH) ions may be present as one type only or in any combination of these. Apatites are the most common phosphorus mineral and have also industrial importance. In soil, insoluble secondary phosphate minerals exist as aluminum salts (e.g., variscite, $AlPO_4 \times 2H_2O$), as iron salts vivianite ($Fe_3[PO_4]_2 \times 8H_2O$) and strengite ($FePO_4 \times 2H_2O$). The latter one shows low solubility also at low pH.

Insoluble forms of inorganic phosphorus (calcium, aluminum, and iron phosphates) may be solubilized through microbial action. The mechanism by which the microbes accomplish this

TABLE 13.1
Solubility products of some phosphate compounds.

Compound	K^2	References
$CaHPO_4 \cdot 2H_2O$	2.18×10^{-7}	Kardos (1955, p. 185)
$Ca_{10}(PO_4)_6(OH)_2$	1.53×10^{-112}	Kardos (1955, p. 188)
$Al(OH)_2HPO_4$	2.8×10^{-29}	Kardos (1955, p. 184)
$FePO_4$	1.35×10^{-18}	From ΔG of formation

solubilization varies. The first mechanism may be the production of inorganic or organic acids that attack the insoluble phosphates. A second mechanism may be the production of chelators such as gluconate and 2-ketogluconate (Duff and Webley, 1959; Banik and Dey, 1983), citrate, oxalate, and lactate. All these chelators complex the cationic portion of the insoluble phosphate salts and thus, force the dissociation of the salts. A third mechanism of phosphate solubilization may be the reduction of iron in ferric phosphate (e.g., strengite) to ferrous iron by any kind of extracellular electron transfer (Lovley et al., 2004). A fourth mechanism may be the production of hydrogen sulfide (H_2S), which reduces the iron and precipitates it as iron sulfide, thereby mobilizing phosphate:

$$2FePO_4 + 3H_2S \rightarrow 2FeS + 2HPO_4^{2-} + 4H^+ + S^0$$
(13.8)

Solubilization of phosphate minerals has been noted directly in soil (Chatterjee and Nandi, 1964; Dommergues and Mangenot, 1970; Patrick et al., 1973; Alexander, 1977; Babenko et al., 1984). Indeed, soil containing significant amounts of immobilized calcium, aluminum, or iron phosphates has been thought to benefit from inoculation with phosphate-mobilizing bacteria (see discussion by Dommergues and Mangenot, 1970). Important microbial phosphate-solubilizing activity in soil occurs in rhizospheres (Alexander, 1977), probably because root secretions allow phosphate-solubilizing bacteria to generate sufficient acid or ligands to effect dissolution of calcium and other insoluble phosphates. Phosphate-deficient soil may be beneficially fertilized with insoluble inorganic phosphate rather than soluble phosphate salts because the former will be solubilized slowly and thus, will be better conserved than soluble phosphate salts which can be readily leached. Soluble phosphate in soil may consist not only of orthophosphate but also of pyrophosphate (metaphosphate). The latter is readily hydrolyzed by pyrophosphatases, especially in flooded soil (Racz and Savant, 1972).

Experiments with B. megaterium found that phosphate mobilization was 3–10 times slower when the organism was in direct contact with apatite than when it was not in direct contact with the mineral (Hutchens et al., 2006). The authors suggest that the bacterium may block reactive sites at the mineral surface.

13.6 MICROBIAL PHOSPHATE IMMOBILIZATION

Microorganisms can cause fixation or immobilization of phosphate, either by promoting the formation of inorganic precipitates or by assimilating the phosphate into organic cell constituents or intracellular polyphosphate granules. The latter two processes have been called transitory phosphate immobilization by Dommergues and Mangenot (1970) because of the ready solubilization of phosphate through mineralization upon death of the cell. In soil and freshwater environments, transitory phosphate immobilization is often more important, although precipitation of phosphate by Ca^{2+}, Al^{3+}, and Fe^{3+} is recognized. In a few marine environments (coastal waters or shallow seas) where phosphorite deposits occur, the precipitation mechanism may be more important (McConnell, 1965).

13.6.1 Phosphorite Deposition

Phosphorite (Figure 13.1) in nature may form authigenically or diagenetically. In authigenesis, phosphorite forms as a result of a reaction of soluble phosphate with calcium ions forming insoluble calcium phosphate compounds. In diagenesis, phosphate may replace carbonate in calcareous concretions. The role of microbes in these processes may be one or more of the following: (1) making reactive phosphate available, (2) making reactive calcium available, or (3) accumulating both and generating the pH and redox conditions that favor phosphate precipitation.

13.6.1.1 Authigenic Formations

Models of authigenic phosphorite genesis assume that organic phosphorus is mineralized in biologically productive waters, such as at ocean margins, at shallow depths on continental slopes, shelf areas, or plateaus. Here detrital accumulations may be mineralized at the sediment–water interface and in interstitial pore waters, liberating phosphate, some of which may then interact chemically with calcium in seawater to form phosphorite grains. These grains may subsequently be redistributed within the sediment units (Riggs, 1984; Mullins and Rasch, 1985). Dissolution of fish debris (bones) has also been considered an important source of phosphate in authigenic phosphorite

Figure 13.1. Micronodules of phosphorite (phosphatic pellets) from the Peru shelf. The average diameter of such pellets is 0.25 mm. According to Burnett (personal communication), such pellets are more representative of what is found in the geologic record than the larger phosphorite nodules. (Courtesy of Burnett, W.C., Institute for International Cooperative Environmental Research, Florida State University, Tallahassee, FL.)

genesis (Suess, 1981). Upwelling probably plays an important role in many cases of authigenic phosphorite formation on western continental margins at latitudes in both the northern and southern hemispheres, where prevailing winds (e.g., trade winds) cause upwelling (see, e.g., discussion by Burnett et al., 1982; Jahnke et al., 1983; Riggs, 1984). Nathan et al. (1993) cite evidence that in the southern Benguela upwelling system (Cape Peninsula, western coast of South Africa) during nonupwelling periods in winter, phosphate-sequestering bacteria become dominant in the water column. These bacteria sequester phosphate as polyphosphate under oxic conditions and hydrolyze the polyphosphate under anoxic conditions to obtain energy and may thus, contribute to authigenic phosphorite formation. Locally elevated, excreted orthophosphate becomes available for precipitation as phosphorite by reacting with seawater calcium.

Authigenic phosphorite at some eastern continental margins, where upwelling, if it occurs at all, is a weak and intermittent process, may have been formed more directly as a result of intracellular bacterial phosphate accumulation, which became transformed into carbonate fluorapatite upon death of the cells and accumulated in sediments in areas where the sedimentation rate was very low (O'Brien and Veeh, 1980; O'Brien

et al., 1981). Ruttenberg and Berner (1993) concluded that carbonate fluorapatite accumulations in Long Island Sound and Mississippi Delta sediments are the result of mineralization of organic phosphorus. These accumulations increased as organic phosphorus concentrations decreased with depth.

Youssef (1965) proposed that phosphorite could be formed in a marine setting through mineralization of phytoplankton remains that have settled into a depression on the seafloor, leading to localized accumulation of dissolved phosphate. This phosphate could then precipitate as a result of reaction with calcium in seawater. Piper and Codispoti (1975) proposed that carbonate fluorapatite ($Ca_{10}(PO_4,CO_3)_6F_{2-3}$) precipitation in the marine environment may be dependent on bacterial denitrification in the oxygen minimum layer of the ocean as it intersects with the ocean floor. A loss of nitrogen due to denitrification causes lowered biological activity and can lead to excess accumulation of phosphate in this zone. The lower pH (7.4–7.9) in the deeper waters compared to the surface waters keeps phosphate dissolved and allows for its transport by upwelling to regions where phosphate precipitation is favored (pH > 8). This model takes into account the conditions of marine apatite formation described by Gulbrandsen (1969) and helps to explain the occurrence of probable contemporary formation of phosphorite in regions of upwelling such as the continental margin off Peru (Veeh, 1973; see also, however, Suess, 1981) and on the continental shelves of southwestern Africa (Baturin, 1969; Baturin et al., 1972). Evidence has been provided more recently that large sulfur bacteria similar to *Thiomargarita spp.* in the top sediments of marine upwelling regions accumulate large amounts of phosphorite and may also release substantial amounts of dissolved phosphate (Schulz and Schulz, 2005). The sequestration of phosphate in sulfidic marine sediments as phosphorite and apatite by sulfide-oxidizing bacteria was confirmed later by tracer experiments with radioactive phosphate (Goldhammer et al., 2010).

To explain the more extensive ancient phosphorite deposits, a periodic warming of the ocean can be invoked, which would reduce oxygen solubility and favor more intense denitrification in deeper waters, resulting in temporarily lessened biological activity and a consequent increase in dissolved phosphate concentration that would lead to phosphate precipitation (Piper and Codispoti, 1975). Mullins and Rasch (1985) proposed an oxygen-depleted sedimentary environment for biogenic apatite formation along the continental margin of central California during the Miocene. They view oolitic phosphorite as having resulted from organic matter mineralization by sulfate reducers in sediments in which dolomite was concurrently precipitated. The phosphate, according to their model, then tended to precipitate interstitially as phosphorite, in part around bacterial nuclei. O'Brien et al. (1981) had previously reported the discovery of fossilized bacteria in a phosphorite deposit on the East Australian margin.

13.6.1.2 Diagenetic Formation

Models of phosphorite formation through diagenesis generally assume an exchange of phosphate for carbonate in accretions that have the shape of calcite and aragonite. The role of bacteria in this process is to mobilize phosphate by mineralizing detrital organic matter. This has been demonstrated in marine and freshwater environments under laboratory conditions (Lucas and Prevot, 1984; Hirschler et al., 1990a,b). Diagenesis of calcite to form apatite may explain the origin of some

deposits in the North Atlantic. Phosphorite deposits off Baja California and in a core from the eastern Pacific Ocean seem to have formed as a result of partial diagenesis (d'Anglejan, 1967, 1968).

13.6.2 Occurrences of Phosphorite Deposits

Sizable phosphorite deposits are associated with only six brief geological intervals: the Cambrian, Ordovician, Devonian-Mississippian, Permian, Cretaceous, and Cenozoic eras. Because in many instances these phosphorite deposits are associated with black shales and contain uranium in reduced form Altschuler et al. (1958), they are presumed to have accumulated under reducing conditions. Nowadays, apatite appears to be forming in the sediments at the Mexican continental margin (Jahnke et al., 1983) and in the deposits off the coast of Peru (Veeh, 1973; Suess, 1981; Burnett et al., 1982).

13.6.3 Deposition of Other Phosphate Minerals

Microbes can also play a role in the authigenic or diagenetic formation of other phosphate minerals such as vivianite, strengite, and variscite. In these instances, bacteria may contribute orthophosphate to the mineral formation by degrading organic phosphate in detrital matter and may contribute iron or aluminum by mobilizing these metals from other minerals. Authigenic formation of such phosphate minerals is probably most common in soil. Microbial control of pH and E_h can influence the stability of these phosphate minerals (Williams and Patrick, 1971; Patrick et al., 1973).

Citrobacter sp. has been reported to form metal phosphate precipitates, for example, cadmium phosphate ($CdHPO_4$) and uranium phosphate (UO_2HPO_4), which encrust the cells (Macaskie et al., 1987, 1992).

Some strains of *Arthrobacter* sp., *Chromohalobacter marismortui*, *Flavobacterium* sp., *Listeria* sp., *Pseudomonas* sp., and *Desulfotignum* can cause struvite ($MgNH_4PO_4 \times 6H_2O$) precipitation, at least under laboratory conditions (Figure 13.2; Rivadeneyra et al., 1983, 1992b, 2006; Schink et al., 2002). A major, but not necessarily exclusive, microbial contribution to this process appears to be ammonium formation (Rivadeneyra et al., 1992a). Struvite is formed when orthophosphate is added at pH 8 to seawater solutions in which the NH_4^+ concentration is 0.01 M (Handschuh and Orgel, 1973).

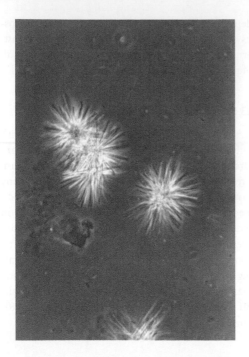

Figure 13.2. Crystals of struvite ($MgNH_4PO_4 \times 6H_2O$) precipitated by *Desulfotignum phosphitoxidans* during sulfate-dependent oxidation of phosphite in an anoxic medium containing magnesium and ammonium salts (for details, see Schink et al., 2002).

The presence of calcium ions in sufficient quantity can suppress struvite formation and promote apatite formation instead (Rivadeneyra et al., 1983). Although struvite formation is probably of little significance in nature today, it may have been significant in the primitive world of Precambrian times if NH_4^+ was present in concentrations as high as 10^{-2} M (Handschuh and Orgel, 1973).

13.7 REDUCTION OF PHOSPHATE

The dominant form of phosphorus on today's earth is phosphate in which phosphorus is in its redox state +V. Nonetheless, also phosphite (P(+III)) and hypophosphite (P(+I)) are found, e.g., in carbonate rocks from the early Archean of Australia (Pasek et al., 2013) and phosphide (P(−III)) minerals like the mineral schreibersite (Fe, Ni)$_3$P (Pasek et al., 2007; Pasek, 2008) in meteoritic rocks. Phosphate can be reduced by lightning strikes (Pasek and Block, 2009), by geothermal reduction in springs or other sources (McDowell et al., 2004; Pech et al., 2009, 2011), and also by microbially catalyzed reduction reactions. The reduced P forms phosphite (HPO_3^{2-})

and hypophosphite ($H_2PO_2^-$) are significantly more soluble than phosphate in waters of neutral pH that are rich in the divalent cations Ca^{2+} and Mg^{2+}, such as modern seawater (Schwartz, 2006).

Phosphate reduction has also been claimed to be a biological process. Reduction of phosphate to phosphine by bacteria has been reported repeatedly in the past (Barrenscheen and Beckh-Widmanstetter, 1923; Rudakov, 1927; Tsubota, 1959; Devai et al., 1988). Phosphite and hypophosphite were claimed to be intermediates in the reduction process (Rudakov, 1927; Tsubota, 1959). Devai et al. (1988) reported phosphine formation in anaerobic sludge of a sewage treatment plant in Hungary. In no case the reduced products were reliably identified; usually a *garlic-like* odor was taken as indication of phosphine formation. This smell may as well have been caused by mercaptanes released from decomposing proteinaceous material. Sniffing of (highly toxic) phosphine could have easily killed the researchers involved! A look at the redox potentials of phosphate reduction to phosphine (Table 13.2) should clearly prove that these reductions at an average redox potential of −713 mV cannot be achieved with biomass as electron donor (average redox potential of CO_2/ glucose = −434 mV at pH 7.0).

The first unequivocal demonstration of phosphine in nature was provided by Gassmann and Schorn (1993) with sediments in Hamburg harbor, and similar findings were reported later. Phosphine in these samples was in the nanomolar range and was probably formed by chemical hydrolysis or thiolysis (with bacterially produced $H_2S!$) of iron phosphides that are present in steels as a consequence of the Thomas conversion

TABLE 13.2
Standard potentials of redox transitions of phosphorus compounds.

	E_0' (mV)[a]
HPO_4^{2-}/HPO_3^{2-}	−690
$HPO_3^{2-}/H_2PO_2^-$	−913
$H_2PO_2^-/P_{white}$	−922
P_{white}/PH_3	−525
Average	−713

[a] Calculated for 1 M concentrations at pH 7.0. (Calculated after Weast, R.C. et al., *CRC Handbook of Chemistry and Physics*, CRC Press, Boca Raton, FL, 1988–1989.)

process. Thus, scrap metals under reducing conditions would be the source of this phosphine, rather than microbial phosphate reduction.

13.8 MICROBIAL METABOLISM OF ORGANOPHOSPHORUS COMPOUNDS

Nonetheless, phosphate can be reduced biochemically to organophosphonates in which the phosphorus is at the (+III) redox state and forms chemically stable C–P linkages. Organophosphonates (or simply phosphonates) are produced by a wide range of marine algae and invertebrates in which phosphonates can constitute up to 50% of the total P-content (Quin, 1965). The most common naturally occurring phosphonate is 2-aminoethylphosphonate (ciliatine), which is a constituent of phosphonolipids, exopolysaccharides, or glycoproteins in various prokaryotes and eukaryotes (Kittredge and Roberts, 1969; White and Metcalf, 2007). Up to 10% of the total P in the cells of the marine cyanobacterium *Trichodesmium erythraeum* strain IMS 101 are phosphonates (Dyhrman et al., 2009). Also, soil actinobacteria are known to produce a wide variety of phosphonates and phosphinates (organic compounds with C–P–C bonds, e.g., phosphinothricin where the central P atom is in the P(+I) state) (White and Metcalf, 2007). These are mainly antibiotics and other secondary metabolites, some of which are used in agriculture as strong herbicides, and some are highly effective against malaria and various bacterial infections.

The biosynthesis of organophosphonates starts with phosphoenolpyruvate, which is isomerized through a mutase reaction to form phosphonopyruvate (Seidel et al., 1988). The other organophosphonates can be derived from phosphonopyruvate through various derivatization reactions.

The microbial degradation of organophosphonates proceeds alternatively through four different key enzymes, namely, (1) phosphonopyruvate hydrolase, (2) phosphonoacetate hydrolase, (3) phosphonoacetaldehyde hydrolase, and (4) carbon–phosphorus (C–P) lyase. In all cases, the phosphorus is released as phosphate and the organic residue as partly reduced derivative, i.e., pyruvate, acetate, or acetaldehyde (McGrath et al., 1995; Ternan and Quinn, 1998; Agarwal et al., 2011). Bacterial degradation of 2-aminoethylphosphonate can proceed either through phosphonoacetaldehyde

dehydrogenase or phosphonoacetate hydrolase (Dumora et al., 1989; Borisova et al., 2011). Although these enzymes are called hydrolases their reaction mechanisms are rather complex.

$$R\text{-}CH_2PO_3^{2-} + H_2O \rightarrow R\text{-}CH_3 + HPO_4^{2-} \quad (13.9)$$

C–P lyase is an enzyme of low specificity, which reacts with nearly all organophosphonates (Equation 13.9). The reaction mechanism has been worked out only recently (Jochimsen et al., 2011; Kamat et al., 2011). Numerous organophosphonates are degraded via phosphonoacetate as key intermediate. Its further metabolism via methylphosphonate to methane plus phosphate (Kamat et al., 2013) by the C–P lyase of the numerically prevalent *Nitrosococcus pumilus* has been identified as an important source of methane formation in oxic water layers of the oceans (Metcalf et al., 2012).

13.9 MICROBIAL OXIDATION OF INORGANIC REDUCED FORMS OF PHOSPHORUS

Reduced inorganic forms of phosphorus, such as phosphite and hypophosphite can be oxidized by aerobic and anaerobic bacteria to be used as phosphorus source under conditions of phosphate starvation. Thus, *Bacillus caldolyticus*, a moderate thermophile, can oxidize hypophosphite to phosphate (Heinen and Lauwers, 1974). The active enzyme reduces nicotinamide adenine dinucleotide (NAD) and oxidizes hypophosphite, but does not oxidize phosphite. Two decades later, two distinct pathways for the oxidation of hypophosphite and phosphite were described in *Pseudomonas stutzeri* WM88 and *Alcaligenes faecalis* WM2072 (White and Metcalf, 2004a,b; Wilson and Metcalf, 2005). Aerobic phosphite oxidation to phosphate for assimilation purposes was first reported in a study with numerous bacteria and fungi (Adams and Conrad, 1953). All phosphite that was oxidized was assimilated, and no phosphate was released into the medium before the organisms died. Phosphate added to the medium inhibited phosphite oxidation. A culture of *P. fluorescens* formed orthophosphate aerobically from orthophosphite in excess of its needs and released it into the medium (Casida, 1960). The phosphite-oxidizing activity was an inducible phosphite-NAD oxidoreductase, which was inactive toward arsenite, hypophosphite, nitrite, selenite, and tellurite and was inhibited by sulfite (Malacinski and Konetzka, 1965, 1966). The enzyme was later purified and characterized (Garcia-Costas et al., 2001).

Oxidation of reduced phosphorus compounds can also occur anaerobically, either as a P source or as an electron donor in energy metabolism. A soil bacillus was isolated that anaerobically oxidizes hypophosphite and phosphite to phosphate (Foster et al., 1978), with preferential oxidation of phosphite over hypophosphite. Phosphate, inhibited phosphite or hypophosphite oxidation, and no phosphate was released into the medium.

An obligate anaerobic marine sulfate-reducing bacterium named *D. phosphitoxidans* was isolated, which uses phosphite not only as a phosphorus source but also as an electron source to support its energy metabolism with sulfate as electron acceptor (Schink and Friedrich, 2000; Schink et al., 2002). This bacterium grows and oxidizes phosphite in the presence of phosphate and releases the excess phosphate into the medium where it may be precipitated as struvite (Figure 13.2) (Simeonova et al., 2010).

Because until recently (see Section 13.7) phosphite and hypophosphite have not been reported in detectable quantities in natural environments, it has been suggested that the microbial ability to utilize these compounds, especially anaerobically, may be a vestigial property that originated at a time when the Earth had a reducing atmosphere that favored the occurrence of phosphite (Foster et al., 1978; De Graaf et al., 1995; Schwartz, 1997).

13.10 SUMMARY

Phosphorus is a very important element for all forms of life. It is important both structural elements and in the metabolic activities of cells. It plays an essential role in transducing biochemically useful energy. When free in the environment, it occurs primarily as inorganic phosphates and as organic phosphate esters. The calcium, aluminum, and iron phosphates are nearly insoluble at neutral or alkaline pH. To become nutritionally available, organic phosphates have to be enzymatically hydrolyzed to liberate orthophosphate. Microbes play a central role in this process. Microbes may also free orthophosphate from insoluble inorganic phosphates, by producing organic or mineral acids or chelators, or in the case of iron phosphates by producing H_2S. Microbes may also promote the formation of

insoluble inorganic phosphates such as those of calcium, magnesium, aluminum, and iron. They have also been implicated in phosphorite formation in the marine environment.

Microbes have been implicated in the reduction of pentavalent phosphorus to lower valence states supported by several pieces of experimental evidence, especially in formation of organophosphonates. Microbes play a major role in the aerobic and anaerobic oxidation of reduced inorganic forms of phosphorus to phosphate. The experimental evidence in this case is strong. It includes demonstration of enzymatic involvement. The geomicrobial significance of these redox reactions in nature is not clearly understood.

REFERENCES

Adams, F. and J.P. Conrad. 1953. Transition of phosphite to phosphate in soils. *Soil Sci.* 75 (5):361–371.

Agarwal, V., S.A. Borisova, W.W. Metcalf, W.A. van der Donk, and S.K. Nair. 2011. Structural and mechanistic insights into C–P bond hydrolysis by phosphonoacetate hydrolase. *Chem. Biol.* 18 (10):1230–1240.

Aiba, H., M. Nagaya, and T. Mizuno. 1993. Sensor and regulation proteins from the *Corynebacterium synechococcus* species PCC7942 that belong to the bacterial signal-transduction protein families: Implication in the adaptive response to phosphate limitation. *Mol. Microbiol.* 8 (1):81–91.

Alexander, M. 1977. *Introduction to Soil Microbiology*, 2nd edn. New York: Wiley.

Altschuler, S., R.S. Clarke, Jr, and E.J. Young, eds. 1958. *Geochemistry of Uranium in Apatite and Phosphorite*. U.S. Geological Survey Professional Paper 314D, pp. 45–90.

Ayyakkannu, K. and D. Chandramohan. 1971. Occurrence and distribution of phosphate solubilizing bacteria and phosphatase in marine sediments at Porto Novo. *Mar. Biol.* 11 (3):201–205.

Babenko, Y.S., G.I. Tyrygina, E.F. Grigorev, L.M. Dolgikh, and T.I. Borisova. 1984. Biological-activity and physiological biochemical-properties of phosphate-dissolving bacteria. *Microbiology* 53 (4):427–433.

Banik, S. and B.K. Dey. 1983. Phosphate-solubilizing potentiality of the microorganisms capable of utilizing aluminium phosphate as a sole phosphate source. *Z. Mikrobiol.* 138 (1):17–23.

Barrenscheen, H.K. and H.A. Beckh-Widmanstetter. 1923. Bacterial reduction of organic bound phosphorus acid. *Biochem. Z.* 140:279–283.

Baturin, G.N. 1969. Autigenic phosphorite nodules in recent sediments of South-West Africa shelf. *Dokl. Akad. Nauk SSSR* 189 (6):1359.

Baturin, G.N., K.I. Merkulova, and P.I. Chalov. 1972. Radiometric evidence for recent formation of phosphatic nodules in marine shelf sediments. *Mar. Geol.* 13 (3):37–41.

Boge, G., M. Lespilette, D. Jamet, and J.-L. Jamet. 2012. Role of sea water DIP and DOP in controlling bulk alkaline phosphatase activity in NW Mediterranean Sea (Toulon, France). *Mar. Pollut. Bull.* 64 (10):1989–1996.

Bonting, C.F.C., G.J.J. Kortstee, and A.J.B. Zehnder. 1991. Properties of polyphosphate—AMP phosphotransferase of *Acinetobacter* strain 210A. *J. Bacteriol.* 173 (20):6484–6488.

Bonting, C.F.C., H.W. van Veen, A. Taverne, G.J.J. Kortstee, and A.J.B. Zehnder. 1992. Regulation of polyphosphate metabolism in *Acinetobacter* strain 210A grown in carbon limited and phosphate limited continuous cultures. *Arch. Microbiol.* 158 (2):139–144.

Borisova, S.A., H.D. Christman, M.E.M. Metcalf, N.A. Zulkepli, J.K. Zhang et al. 2011. Genetic and biochemical characterization of a pathway for the degradation of 2-aminoethylphosphonate in *Sinorhizobium meliloti* 1021. *J. Biol. Chem.* 286 (25):22283–22290.

Bowen, H.J.M. 1979. *Environmental Chemistry of the Elements.* London, U.K.: Academic Press.

Burnett, W.C., M.J. Beers, and K.K. Roe. 1982. Growthrates of phosphate nodules from the continental-margin off Peru. *Science* 215 (4540):1616–1618.

Casida, L.E., Jr. 1960. Microbial oxidation and utilization of orthophosphite during growth. *J. Bacteriol.* 80 (2):237–241.

Chatterjee, A.K. and P. Nandi. 1964. Solubilization of insoluble phosphates by rhizosphere fungi of legumes *Arachis, Cyamopsis, Desmodium. Trans. Bose. Res. Inst. (Calcutta)* 27 (4):115–120.

Cosgrove, D.J. 1967. *Metabolism of organic phosphates in soil microorganisms, plant remains, animal remains, review.* Eds. A.D. Mclaren and G.H. Peterson, *Soil Biochemistry,* Vol. 1. New York: Marcel Dekker.

Cosgrove, D.J. 1977. *Microbial Transformations in the Phosphorus Cycle.* Ed. M. Alexander, Advances in Microbial Ecology, Vol. 1. New York: Plenum Press/Springer.

d'Anglejan, B.F. 1967. Origin of marine phosphorites off Baja California, Mexico. *Mar. Geol.* 5 (1):15–44.

d'Anglejan, B.F. 1968. Phosphate diagenesis of carbonate sediments as a mode of in situ formation of marine phosphorites: Observations in a core from the eastern Pacific. *Can. J. Earth Sci.* 5 (1):81–87.

De Graaf, R.M., J. Visscher, and A.W. Schwartz. 1995. A plausibly prebiotic synthesis of phosphonic acids. *Nature* 378 (6556):474–477.

Devai, I., L. Felfoldy, I. Wittner, and S. Plosz. 1988. Detection of phosphine-new aspects of the phosphorus cycle in the hydrosphere. *Nature* 333 (6171):343–345.

Dommergues, Y.R. and F. Mangenot. 1970. *Écologie Microbienne du sol.* Paris, France: Masson et Cie.

Duff, R.B. and D.M. Webley. 1959. 2-ketogluconic acid as a natural chelator produced by soil bacteria. *Chem. Ind.* 44:1376–1377.

Dumora, C., A.-M. Lacoste, and A. Cassaigne. 1989. Phosphonoacetaldehyde hydrolase from *Pseudomonas aeruginosa*: Purification properties and comparison with *Bacillus cereus* enzyme. *Biochim. Biophys. Acta* 997 (3):193–198.

Dyhrman, S.T., C.R. Benitez-Nelson, E.D. Orchard, S.T. Haley, and P.J. Pellechia. 2009. A microbial source of phosphonates in oligotrophic marine systems. *Nat. Geosci.* 2 (10):696–699.

Emsley, J. 2001. *The Shocking History of Phosphorus: A Biography of the Devil's Element.* London, U.K.: Pan Macmillan Ltd.

Foster, T.L., L. Winans, Jr., and S.J. Helms. 1978. Anaerobic utilization of phosphite and hypophosphite by *Bacillus* sp. *Appl. Environ. Microbiol.* 35 (5):937–944.

Friedberg, I. and G. Avigad. 1968. Structures containing polyphosphate in *Micrococcus lysodeikticus. J. Bacteriol.* 96 (2):544.

Fuller, W.H. 1972. Phosphorus: Element and geochemistry. Ed. W.R. Fairbridge, *The Encyclopedia of Geochemistry and Environmental Sciences.* Encyclopedia of Earth Science Series, Vol. IVA. New York: Van Nostrand Reinhold, pp. 942–946.

Garcia-Costas, A.M., A.K. White, and W.W. Metcalf. 2001. Purification and characterization of a novel phosphorus-oxidizing enzyme from *Pseudomonas stutzeri* WM88. *J. Biol. Chem.* 276 (20):17429–17436.

Gassmann, G. and F. Schorn. 1993. Phosphine from harbor surface sediments. *Naturwissenschaften* 80 (2):78–80.

Gilbert, N. 2009. The disappearing nutrient. *Nature* 461 (7267):716.

Goldhammer, T., V. Bruchert, T.G. Ferdelman, and M. Zabel. 2010. Microbial sequestration of phosphorus in anoxic upwelling sediments. *Nat. Geosci.* 3 (8):557–561.

Gulbrandsen, R.A. 1969. Physical and chemical factors in formation of marine apatite. *Econ. Geol.* 64 (4):365–382.

Handschuh, G.J. and L.E. Orgel. 1973. Struvite and pre-biotic phosphorylation. *Science* 179 (4072):483–484.

Hanrahan, G., T.M. Salmassi, C.S. Khachikian, and K.L. Foster. 2005. Reduced inorganic phosphorus in the natural environment: Significance, speciation and determination. *Talanta* 66 (2):435–444.

Heinen, W. and A.M. Lauwers. 1974. Hypophosphite oxidase from *Bacillus caldolyticus*. *Arch. Microbiol.* 95 (3):267–274.

Herbes, S.E., H.E. Allen, and K.H. Mancy. 1975. Enzymatic characterization of soluble organic phosphorus in lake water. *Science* 187 (4175):432–434.

Hirschler, A, J. Lucas, and J.-C. Hubert. 1990b. Bacterial involvement in apatite genesis. *FEMS Microbiol. Ecol.* 73 (3):211–220.

Hirschler, A., J. Lucas, and J.-C. Hubert. 1990a. Apatite genesis a biologically induced or biologically controlled mineral formation process. *Geomicrobiol. J.* 8 (1):47–56.

Hulett, F.M. 1996. The signal-transduction network for Pho regulation in *Bacillus subtilis*. *Mol. Microbiol.* 19 (5):933–939.

Hutchens, E., E. Valsami-Jones, N. Harouiya, C. Chairat, E.H. Oelkers, and S. McEldoney. 2006. An experimental investigation of the effect of *Bacillus megaterium* on apatite dissolution. *Geomicrobiol. J.* 23 (3–4):177–182.

Ishige, T., M. Krause, M. Bott, V.F. Wendisch, and H. Sahm. 2003. The phosphate starvation stimulon of *Corynebacterium glutamicum* determined by DNA microarray analyses. *J. Bacteriol.* 185 (15):4519–4529.

Jahnke, R.A., S.R. Emerson, K.K. Roe, and W.C. Burnett. 1983. The present day formation of apatite in mexican continental margin sediments. *Geochim. Cosmochim. Acta* 47 (2):259–266.

Jochimsen, B., S. Lolle, F.R. McSorley, M. Nabi, J. Stougaard et al. 2011. Five phosphonate operon gene products as components of a multi-subunit complex of the carbon-phosphorus lyase pathway. *Proc. Nat. Acad. Sci. USA* 108 (28):11393–11398.

Kamat, S.S., H.J. Williams, L.J. Dangott, M. Chakrabarti, and F.M. Raushel. 2013. The catalytic mechanism for aerobic formation of methane by bacteria. *Nature* 497 (7447):132–136.

Kamat, S.S., H.J. Williams, and F.M. Raushel. 2011. Intermediates in the transformation of phosphonates to phosphate by bacteria. *Nature* 480 (7378):570–573.

Kardos, L.T. 1955. Soil fixation of plant nutrients. Ed. F.E. Bear, *Chemistry of the Soil*. New York: Van Nostrand Reinhold, pp. 177–199.

Kittredge, J.S. and E. Roberts. 1969. A carbon-phosphorus bond in nature. *Science* 164 (3875):37–42.

Kobori, H. and N. Taga. 1979. Phosphatase activity and its role in the mineralization of organic phosphorus in coastal sea water. *J. Exp. Mar. Biol. Ecol.* 36 (1):23–39.

Kortstee, G.J.J., K.J. Appeldoorn, C.F.C. Bonting, E.W.J. van Niel, and H.J. van Veen. 2000. Ed. B. Schink, *Advances in Microbial Ecology*. New York: Kluwer Academic/Plenum Publishers.

Krishnaraj, P.U. and A.H. Goldstein. 2001. Cloning of a *Serratia marcescens* DNA fragment that induces quino-protein glucose dehydrogenase-mediated gluconic acid production in *Escherichia coli* in the presence of stationary phase *Serratia marcescens*. *FEMS Microbiol. Lett.* 205 (2):215–220.

Lehninger, A.L. 1975. *Biochemistry: The Molecular Basis of Cell Structure and Function*, 2nd edn. New York: Worth Publishing Inc.

Liu, C.L., N. Hart, and H.D. Peck. 1982. Inorganic pyrophosphate—Energy source for sulfate reducing bacteria of the genus *Desulfotomaculum*. *Science* 217 (4557):363–364.

Liu, S.T., L.Y. Lee, C.Y. Tai, C.H. Hung, Y.S. Chang et al. 1992. Cloning of an *Erwinia herbicola* gene necessary for glucuronic acid production and enhanced mineral phosphate solubilization in *Escherichia coli* HB101—Nucleotide sequence and probable involvement in biosynthesis of the coenzyme pyrroloquinoline quinone. *J. Bacteriol.* 174 (18):5814–5819.

Lovley, D.R., D.E. Holmes, and K.P. Nevin. 2004. Dissimilatory Fe(III) and Mn(IV) reduction. Ed. R.K. Poole, *Advances in Microbial Physiology*, Vol. 49, London, U.K.: Elsevier, Academic Press Ltd, pp. 219–286.

Lucas, J. and L. Prevot. 1984. Apatite synthesis by bacterial activity from phosphatic organic matter and several calcium carbonates in natural fresh water and sea water. *Chem. Geol.* 42 (1–4):101–118.

Macaskie, L.E., A.C.R. Dean, A.K. Cheetham, R.J.B. Jakeman, and A.J. Skarnulis. 1987. Cadmium accumulation by a *Citrobacter* sp—The chemical nature of the accumulated metal precipitate and its location on the bacterial cells. *J. Gen. Microbiol.* 133:539–544.

Macaskie, L.E., R.M. Empson, A.K. Cheetham, C.P. Grey, and A.J. Skarnulis. 1992. Uranium bioaccumulation by a *Citrobacter* sp. as a result of enzymatically mediated growth of polycrystalline HUO_2PO_4. *Science* 257 (5071):782–784.

Malacinski, G. and W.A. Konetzka. 1965. Bacterial oxidation of orthophosphite. *J. Bacteriol.* 91:578–582.

Malacinski, G. and W.A. Konetzka. 1966. Orthophosphite-nicotinamide adenine dinucleotide oxidoreductase from *Pseudomonas fluorescens*. *J. Bacteriol.* 93 (2):578–582.

Marschner, H. 1995. *Mineral Nutrition of Higher Plants*, 2nd edn. London, U.K.: Academic Press.

McConnell, D. 1965. Precipitation of phosphates in sea water. *Econ. Geol.* 60 (5):1059–1062.

McDowell, M.M., M.M. Ivey, M.E. Lee, V.V.V.D. Firpo, T.M. Salmassi et al. 2004. Detection of hypophosphite, phosphite, and orthophosphate in natural geothermal water by ion chromatography. *J. Chromatogr. A* 1039 (1–2):105–111.

McGrath, J.W., G.B. Wisdom, G. McMullan, M.J. Larkin, and J.P. Quinn. 1995. The purification and properties of phosphonoacetate hydrolase, a novel carbon-phosphorus bond-cleavage enzyme from *Pseudomonas Fluorescens* 23F. *Eur. J. Biochem.* 234 (1):225–230.

Menezes-Blackburn, D., M.A. Jorquera, R. Greiner, L. Gianfreda, and M. de la Luz Mora. 2013. Phytases and phytase-labile organic phosphorus in manures and soils. *Crit. Rev. Environ. Sci. Technol.* 43 (9):916–954.

Metcalf, W.W., B.M. Griffin, R.M. Cicchillo, J. Gao, S.C. Janga et al. 2012. Synthesis of methylphosphonic acid by marine microbes: A source for methane in the aerobic ocean. *Science* 337 (6098):1104–1107.

Mullins, H.T. and R.F. Rasch. 1985. Sea-floor phosphorites along the Central California continental margin. *Econ. Geol.* 80 (3):696–715.

Nannipieri, P., L. Giagnoni, G. Renella, E. Puglisi, B. Ceccanti et al. 2012. Soil enzymology: Classical and molecular approaches. *Biol. Fertil. Soils* 48 (7):743–762.

Nathan, Y., J.M. Bremner, R.E. Lowenthal, and P. Monteiro. 1993. Role of bacteria in phosphorite genesis. *Geomicrobiol. J.* 11:69–76.

O'Brien, G.W., J.R. Harris, A.R. Milnes, and H.H. Veeh. 1981. Bacterial origin of East Australian continental margin phosphorites. *Nature* 294 (5840):442–444.

O'Brien, G.W. and H.H. Veeh. 1980. Holocene phosphorite on the East Australian continental margin. *Nature* 288 (5792):690–692.

Pasek, M. and K. Block. 2009. Lightning-induced reduction of phosphorus oxidation state. *Nat. Geosci.* 2 (8):553–556.

Pasek, M.A. 2008. Rethinking early Earth phosphorus geochemistry. *Proc. Nat. Acad. Sci. USA* 105 (3):853–858.

Pasek, M.A., J.P. Dworkin, and D.S. Lauretta. 2007. A radical pathway for organic phosphorylation during schreibersite corrosion with implications for the origin of life. *Geochim. Cosmochim. Acta* 71 (7):1721–1736.

Pasek, M.A., J.P. Harnmeijer, R. Buick, M. Gull, and Z. Atlas. 2013. Evidence for reactive reduced phosphorus species in the early Archean ocean. *Proc. Nat. Acad. Sci. USA* 110 (25):10089–10094.

Patrick, W.H., S. Gotoh, and B.G. Williams. 1973. Strengite dissolution in flooded soils and sediments. *Science* 179 (4073):564–565.

Paul, E.A. and F.E. Clark. 1996. *Soil Microbiology and Biochemistry*, 2nd edn. San Diego, CA: Academic Press.

Paytan, A. and K. McLaughlin. 2007. The oceanic phosphorus cycle. *Chem. Rev.* 107 (2):563–576.

Pech, H., A. Henry, C.S. Khachikian, T.M. Salmassi, G. Hanrahan, and K.L. Foster. 2009. Detection of geothermal phosphite using high-performance liquid chromatography. *Environ. Sci. Technol.* 43 (20):7671–7675.

Pech, H., M.G. Vazquez, J. Van Buren, K.L. Foster, L. Shi et al. 2011. Elucidating the redox cycle of environmental phosphorus using ion chromatography. *J. Chromatogr. Sci.* 49 (8):573–581.

Petterson A, L. Kunst, B. Bergman, and G.M. Roomans. 1985. Accumulations of aluminum by *Anabaena cylindrica* into polyphosphate granules and cell walls: An x-ray energy dispersive microanalysis study. *J. Gen. Microbiol.* 131:2545–2548.

Piper, D.Z. and L.A. Codispoti. 1975. Marine phosphorite deposits and nitrogen cycle. *Science* 188 (4183):15–18.

Quin, L.D. 1965. The presence of compounds with a carbon-phosphorus bond in some marine invertebrates. *Biochemistry* 4 (2):324–330.

Racz, G.J. and N.K. Savant. 1972. Pyrophosphate hydrolysis in soil as influenced by flooding and fixation. *Soil Sci. Soc. Am. Proc.* 36 (4):678.

Rausch, C. and M. Bucher. 2002. Molecular mechanisms of phosphate transport in plants. *Planta* 216 (1):23–37.

Riggs, S.R. 1984. Paleoceanographic model of neogene phosphorite deposition, United States Atlantic continental margin. *Science* 223 (4632):123–131.

Rivadeneyra, M.A., A. Martin-Algarra, A. Sanchez-Navas, and D. Martin-Ramos. 2006. Carbonate and phosphate precipitation by *Chromohalobacter marismortui*. *Geomicrobiol. J.* 23 (2):89–101.

Rivadeneyra, M.A., I. Perez-Garcia, and A. Ramos-Cormenzana. 1992a. Influence of ammonium ion on bacterial struvite production. *Geomicrobiol. J.* 10 (2):125–137.

Rivadeneyra, M.A., I. Perez-Garcia, and A. Ramos-Cormenzana. 1992b. Struvite precipitation by soil and fresh water bacteria. *Curr. Microbiol.* 24 (6):343–347.

Rivadeneyra, M.A., A. Ramos-Cormenzana, and A. Garcia-Cervigon. 1983. Bacterial formation of struvite. *Geomicrobiol. J.* 3 (2):151–163.

Rodriguez, H. and R. Fraga. 1999. Phosphate solubilizing bacteria and their role in plant growth promotion. *Biotechnol. Adv.* 17 (4–5):319–339.

Rudakov KJ. 1927. Die Reduktion der mineralischen Phosphate auf biologischem Wege. II. Mitteilung. Z Bakt Parasitenk Infektionskr Hyg Abt II 79:229–245.

Ruttenberg, K.C. and R.A. Berner. 1993. Authigenic apatite formation and burial in sediments from non-upwelling, continental margin environments. Geochim. Cosmochim. Acta 57 (5):991–1007.

Schink, B. and M. Friedrich. 2000. Phosphite oxidation by sulphate reduction. Nature 406:37.

Schink, B., V. Thiemann, H. Laue, and M.W. Friedrich. 2002. Desulfotignum phosphitoxidans sp. nov., a new marine sulfate reducer that oxidizes phosphite to phosphate. Arch. Microbiol. 177:381–391.

Schöcke, L. and B. Schink. 1998. Membrane-bound proton-translocating pyrophosphatase of Syntrophus gentianae, a syntrophically benzoate-degrading fermenting bacterium. Eur. J. Biochem. 256 (3):589–594.

Schulz, H.N. and H.D. Schulz. 2005. Large sulfur bacteria and the formation of phosphorite. Science 307 (5708):416–418.

Schwartz, A.W. 1997. Speculation on the RNA precursor problem. J. Theoret. Biol. 187 (4):523–527.

Schwartz, A.W. 2006. Phosphorus in prebiotic chemistry. Philos. Trans. R. Soc. B: Biol. Sci. 361 (1474):1743–1749.

Seidel, H.M., S. Freeman, H. Seto, and J.R. Knowles. 1988. Phosphonate biosynthesis—Isolation of the enzyme responsible for the formation of a carbon-phosphorus bond. Nature 335 (6189):457–458.

Shapiro, J. 1967. Induced rapid release and uptake of phosphate by microorganisms. Science 155 (3767):1269.

Shimogawara, K., D.D. Wykoff, H. Usuda, and A.R. Grossman. 1999. Chlamydomonas reinhardtii mutants abnormal in their responses to phosphorus deprivation. Plant Physiol. 120 (3):685–693.

Simeonova, D.D., M.M. Wilson, W.W. Metcalf, and B. Schink. 2010. Identification and heterologous expression of genes involved in anaerobic dissimilatory phosphite oxidation by Desulfotignum phosphitoxidans. J. Bacteriol. 192 (19):5237–5244.

Skujins, J.J. 1967. Enzymes in Soil (review). Eds. A.D. Mclaren and G.H. Peterson, Soil Biochemistry, Vol. 1. New York: Marcel Dekker.

Stumm, W. and J.J. Morgan. 1981. Aquatic Chemistry: An Introduction Emphasizing Chemical Equilibria in Natural Waters, 2nd edn. New York: Wiley & Sons.

Suess, E. 1981. Phosphate regeneration from sediments of the Peru continental margin by dissolution of fish debris. Geochim. Cosmochim. Acta 45 (4):577–588.

Taga, N. and H. Kobori. 1978. Phosphatase activity in eutrophic Tokyo Bay. Mar. Biol. 49 (3):223–229.

Ternan, N.G. and J.P. Quinn. 1998. In vitro cleavage of the carbon–phosphorus bond of phosphonopyruvate by cell extracts of an environmental Burkholderia cepacia isolate. Biochem. Biophys. Res. Commun. 248 (2):378–381.

Thauer, R.K., K. Jungermann, and K. Decker. 1977. Energy conservation in chemotrophic anaerobic bacteria. Bacteriol. Rev. 41 (1):100–180.

Tian, J., X. Wang, Y. Tong, X. Chen, and H. Liao. 2012. Bioengineering and management for efficient phosphorus utilization in crops and pastures. Curr. Opin. Biotech. 23 (6):866–871.

Tsubota, G. 1959. Phosphate reduction in the paddy field I. Soil Sci. Plant Nutr. 5 (1):10–15.

Van Groenestijn, J.W., M.H. Deinema, and A.J.B. Zehnder. 1987. ATP production from polyphosphate in Acinetobacter strain 210A. Arch. Microbiol. 148 (1):14–19.

van Veen, H.W. 1997. Phosphate transport in prokaryotes: Molecules, mediators and mechanisms. Antonie van Leeuwenhoek. 72:299–315.

Varma, A.K., W. Rigsby, and D.C. Jordan. 1983. A new inorganic pyrophosphate utilizing bacterium from a stagnant lake. Can. J. Microbiol. 29 (10):1470–1474.

Veeh, H.H. 1973. Contemporary phosphorites on continental margin of Peru. Science 181 (4102):844–845.

Wanner, B.L. 1993. Gene regulation by phosphate in enteric bacteria. J. Cell. Biochem. 51 (1):47–54.

Weast, R.C., M.J. Astle, and W.H. Beyer, eds. 1988–1989. CRC Handbook of Chemistry and Physics, Boca Raton, FL: CRC Press.

Westheimer, F.H. 1987. Why nature chose phosphates. Science 235 (4793):1173–1178.

Wetzel, R.G. 2001. Limnology: Lake and River Ecosystems, 3rd edn. San Diego, CA: Academic Press.

White, A.K. and W.W. Metcalf. 2004a. The htx and ptx operons of Pseudomonas stutzeri WM88 are new members of the Pho regulon. J. Bacteriol. 186 (17):5876–5882.

White, A.K. and W.W. Metcalf. 2004b. Two C-P lyase operons in Pseudomonas stutzeri and their roles in the oxidation of phosphonates, phosphite, and hypophosphite. J. Bacteriol. 186 (14):4730–4739.

White, A.K. and W.W. Metcalf. 2007. Microbial metabolism of reduced phosphorus compounds. Annu. Rev. Microbiol. 61:379–400.

White, R.H. 1984. Hydrolytic stability of biomolecules at high temperatures and its implication for life at 250°C. *Nature* 310 (5976):430–432.

Williams, B.G. and W.H. Patrick. 1971. Effect of Eh and pH on dissolution of strengite. *Nat. Phys. Sci.* 234 (44):16.

Wilson, M.M. and W.W. Metcalf. 2005. Genetic diversity and horizontal transfer of genes involved in oxidation of reduced phosphorus compounds by *Alcaligenes faecalis* WM2027. *Appl. Environ. Microbiol.* 71 (1):290–296.

Youssef, M.I. 1965. Genesis of bedded phosphates. *Econ. Geol.* 60 (3):590–600.

CHAPTER FOURTEEN

Geomicrobiology of Nitrogen

Christopher A. Francis and Karen L. Casciotti

CONTENTS

14.1 Nitrogen in the Biosphere / 281
14.2 Microbial Interactions with Nitrogen / 282
    14.2.1 Ammonification / 283
    14.2.2 Nitrification / 283
    14.2.3 Ammonia Oxidation / 283
    14.2.4 Nitrite Oxidation / 284
    14.2.5 Heterotrophic Nitrification / 284
    14.2.6 Anaerobic Ammonium Oxidation / 285
    14.2.7 Denitrification / 285
    14.2.8 Dissimilatory Nitrate Reduction to Ammonium / 286
    14.2.9 Nitrogen Fixation / 286
14.3 Stable Isotopes of Nitrogen / 288
    14.3.1 Natural Abundance Stable Nitrogen Isotopes / 288
    14.3.2 Stable Nitrogen Isotope Tracer Experiments / 289
14.4 Summary / 290
References / 290

14.1 NITROGEN IN THE BIOSPHERE

Nitrogen (N) is an element essential to life. It is abundant in the atmosphere, mostly as dinitrogen (N_2), representing roughly four-fifths by volume or three-fourths by weight of the total gas. Trace amounts of oxides of nitrogen (nitrous oxide [N_2O] and nitric oxide [NO]) are also present, which play a role in the greenhouse effect and stratospheric ozone chemistry. In soil, sediment, freshwater, and ocean water, nitrogen exists in both inorganic and organic forms. Inorganic forms of N include gaseous NO and N_2O, as well as dissolved ammonia (NH_3), ammonium (NH_4^+), nitrate (NO_3^-), and nitrite (NO_2^-). Interconversions between these forms of N can be used microbially to produce energy under aerobic conditions (oxidation of ammonia to nitrite and nitrate via nitrification) and anaerobic conditions (reduction of nitrate and nitrite to gaseous compounds nitric oxide, nitrous oxide, and dinitrogen via denitrification). Figure 14.1 illustrates the complexity of the microbial N cycle in the environment. While research on organic nitrogen is limited compared to inorganic nitrogen, this pool is believed to be important in both aquatic and terrestrial ecosystems (Neff et al., 2003; Aluwihare and Meador, 2008; Letscher et al., 2013). Organic nitrogen compounds include humic and fulvic acids, proteins, peptides, amino acids, purines, pyrimidines, pyridines, other amines, and amides.

Generally, inorganic nitrogen compounds exist in nature either as gases in the atmosphere and dissolved in water or as ionic compounds in aqueous solution. While small deposits of other mineral forms of nitrogen exist, principally due to microbial transformations of organic nitrogen deposits (Ericksen, 1983), they are largely irrelevant in regard to the global nitrogen budget and

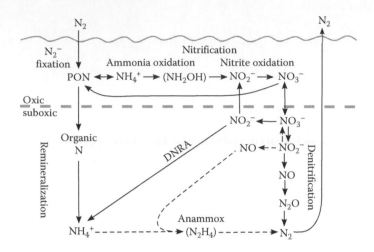

Figure 14.1. Microbial nitrogen transformations above, below, and across an oxic/anoxic interface (based in part on diagram from Francis et al. [2007]). Dashed arrows highlight steps of the anaerobic nitrogen cycle specifically associated with the anammox pathway. PON = particulate organic nitrogen. Parentheses highlight key biochemical pathway intermediates that rarely accumulate in the environment.

the N cycle in most ecosystems. An important geomicrobiological consequence of the N cycle, however, is that the nitric acid formed through nitrification can accelerate the weathering of rocks and minerals (Chapters 4, 7, and 10 through 12).

The availability of fixed nitrogen is seen as a limiting factor to primary productivity in many ecosystems, and primary producers such as plants, algae, and cyanobacteria depend largely on the microbial transformations of nitrogen species. The release of ammonium from organic matter via ammonification and the production of ammonium from atmospheric nitrogen via nitrogen fixation are carried out exclusively by microorganisms. In addition, different species of plants, algae, and cyanobacteria have different preferences for nitrogen species for uptake (for instance, some plants preferentially take up ammonium whereas others have a higher affinity for nitrate); the availability of different nitrogen species depends on the chemical environment (for instance, in temperate clay soils, which are negatively charged, nitrate is immediately available to plants but also prone to leaching, whereas ammonium is retained via attraction to the charged soil particles, and therefore, forms a longer-term but physically less-accessible form of N for plants; the reverse is true in positively charged tropical soils). Therefore, the transformations of inorganic nitrogen compounds, largely carried out by microorganisms as a means of gaining energy, have important implications for primary production in all ecosystems.

Perhaps more than any other biogeochemical cycle, the nitrogen cycle has been perturbed by recent anthropogenic activities (Rockström et al., 2009; Howarth et al., 2012). Due to the ability of nitrogen to control or significantly affect primary productivity discussed earlier, the massive nitrogen inputs to terrestrial and aquatic ecosystems from agricultural runoff, wastewater and (sub)urban discharge, and fossil fuel combustion can lead to a range of negative ecological impacts, including severe eutrophication, oxygen depletion, and nitrous oxide emissions (Erisman et al., 2013).

14.2 MICROBIAL INTERACTIONS WITH NITROGEN

Owing to their unique capacities of transforming inorganic compounds, microbes, including bacteria, archaea, and fungi, play a central role in the nitrogen cycle (Figure 14.1). Many reactions of the cycle, such as nitrogen fixation, nitrification, and ammonification, are entirely dependent on them. The direction of transformations in the cycle is determined by environmental conditions, especially the availability of oxygen and the supply of particular nitrogen compounds. Anaerobic conditions may encourage denitrification and anammox and thus, cause nitrogen limitation, unless these N-removal processes are counteracted by nitrogen fixation or other external inputs of nitrogen. Availability of fixed nitrogen is often viewed as a growth-limiting factor in the marine environment but not in unpolluted freshwater, in which phosphate is more likely to limit productivity. Fixed nitrogen can be a limiting factor in soil, especially in agriculturally exploited soils.

14.2.1 Ammonification

One of the most important sources of biologically available N in all ecosystems is the decomposition of nitrogenous organic matter (plant, animal, and microbial excretions and remains). The first step in the recycling process is *ammonification*, in which the organic nitrogen is transformed into ammonia. An example of ammonification is the deamination of amino acids:

$$RCHCOOH + NAD^+ \rightarrow RCCOOH + NADH + H^+$$
$$| \qquad\qquad\qquad ||$$
$$NH_2 \qquad\qquad\qquad NH$$

$$RCCOOH + H_2O \rightarrow RCCOOH + NH_3$$
$$|| \qquad\qquad\qquad ||$$
$$NH \qquad\qquad\qquad O$$

$$(14.1)$$

The NH_3 reacts with water to form ammonium hydroxide, which dissociates to form ammonium and hydroxyl as follows:

$$NH_3 + H_2O \rightarrow NH_4OH \rightarrow NH_4^+ + OH^- \quad (14.2)$$

In the laboratory, it is commonly observed that when heterotrophic bacteria grow in proteinaceous medium, such as nutrient broth consisting of peptone or beef extract, in which the organic nitrogen serves as the source of energy, carbon, and nitrogen, the pH rises over time owing to the liberation of ammonia and its hydrolysis to ammonium ion. Indeed, ammonification is always an essential first step when an amino compound serves as an energy source.

Ammonia is also formed as a result of urea hydrolysis catalyzed by the enzyme urease:

$$NH_2CONH_2 + 2H_2O \rightarrow 2NH_4^+ + CO_3^{2-} \quad (14.3)$$

Urea is a nitrogen waste product excreted in the urine of many mammals. Although urease is produced by a variety of prokaryotic and eukaryotic microbes, some soil bacteria (e.g., *Bacillus pasteurii* and *Bacillus freudenreichii*) seem to be specialists in degrading urea. They prefer to grow at an alkaline pH such as that generated when urea is hydrolyzed (Alexander, 1977).

14.2.2 Nitrification

Nitrification—the microbial oxidation of ammonia to nitrate via a nitrite intermediate—plays a critical role in linking the decomposition of nitrogenous organic matter to anaerobic N-removal processes. The two steps of nitrification are carried out by distinct functional groups, namely those involved in ammonia oxidation (typically the rate-limiting step of nitrification) and those involved in nitrite oxidation. Most known nitrifying organisms are chemolithoautotrophic, coupling energy gained from the oxidation of NH_3 or NO_2^- to CO_2 fixation, but some also appear to be facultative heterotrophs (e.g., *Nitrobacter winogradskyi*). Because ammonia oxidation was long believed to be catalyzed by only a few genera within the domain *bacteria*, the recent discovery (~10 years ago) of ammonia-oxidizing archaea (AOA) has significantly altered our understanding of the ecology and biogeochemistry of microbial ammonia oxidation.

14.2.3 Ammonia Oxidation

Ammonia-oxidizing bacteria (AOB) are phylogenetically restricted to two different lineages within the *Proteobacteria*: the betaproteobacterial genera *Nitrosomonas* and *Nitrosospira* and the gammaproteobacterial genus *Nitrosococcus*. *Nitrosopumilus maritimus* (Könneke et al., 2005) was the first cultivated representative of the AOA, but several different AOA candidate genera have recently emerged, including *Nitrosoarchaeum*, *Nitrosocaldus*, *Nitrososphaera*, and *Nitrosotalea* (Stahl and de la Torre, 2012). Since AOA were only recently discovered, most of what is known about the biochemistry of ammonia oxidation is based on studies of AOB.

Bacterial oxidation of ammonia is catalyzed by the enzyme ammonia monooxygenase (AMO), which produces hydroxylamine (NH_2OH):

$$NH_3 + 0.5O_2 \rightarrow NH_2OH \qquad (14.4)$$

This reaction does not yield biochemically useful energy. Indeed, it is slightly endothermic and proceeds in the direction of hydroxylamine because it is coupled to the subsequent oxidation of hydroxylamine to nitrous acid, which is strongly exothermic. The overall oxidation of hydroxylamine to nitric acid (HNO_2) can be summarized as follows:

$$NH_2OH + O_2 \rightarrow HNO_2 + H_2O \qquad (14.5)$$

Reaction 14.5 is where chemoautotrophic ammonia oxidizers obtain their energy, by using chemiosmotic coupling (i.e., oxidative phosphorylation). The conversion of hydroxylamine to nitrous acid involves some intermediate steps (Hooper, 1984). While NH_2OH appears to be an important intermediate for ammonia oxidation, it is rarely found as a free pool in the environment.

In addition to NH_3, bacterial AMO can also catalyze the oxygenation of methane (CH_4) to methanol (CH_3OH) (Jones and Morita, 1983):

$$CH_4 + 0.5O_2 \rightarrow CH_3OH \qquad (14.6)$$

Under standard conditions, this reaction is thermodynamically more favorable than Reaction 14.4. This does not mean, however, that ammonia oxidizers can grow on methane, as they lack the ability to oxidize methanol.

Although AOA possess genes encoding an AMO enzyme, the exact mechanism by which they oxidize ammonia to nitrite appears to be different than the AOB. In particular, there is no evidence for a hydroxylamine oxidoreductase gene (or analogue) in any AOA genomes examined to date (Hallam et al., 2006a,b; Walker et al., 2010; Blainey et al., 2011; Stahl and de la Torre, 2012). However, hydroxylamine does appear to be a central intermediate in the ammonia oxidation pathway of N. maritimus (Vajrala et al., 2013), presumably oxidized by a novel archaeal enzyme complex. Clearly, further physiological studies are needed to elucidate the unique biochemistry of ammonia oxidation by AOA.

Ammonia oxidizers can also produce some nitric oxide and nitrous oxide in side reactions of ammonia oxidation, and via nitrite reduction, under both oxic conditions and when oxygen limited (Knowles, 1985; Bock et al., 1991; Davidson, 1993; Stein and Yung, 2003; Kozlowski et al., 2014). This is important, because it means that biogenic nitric and nitrous oxides are not solely the products of denitrification (see Section 14.2.7).

14.2.4 Nitrite Oxidation

Nitrite-oxidizing bacteria (NOB) convert nitrite to nitrate:

$$NO_2^- + 0.5O_2 \rightarrow NO_3^- \qquad (14.7)$$

They obtain useful energy from this process by coupling it chemiosmotically to ATP generation (Aleem and Sewell, 1984; Wood, 1988). Nitrite oxidation is carried out by bacteria within the genera Nitrobacter, Nitrococcus, Nitrospira, and Nitrospina. However, many recent discoveries have placed NOB more broadly in the tree of life, including the genera Nitrotoga and Nitrolancetus from the phylum Chloroflexi (Sorokin et al., 2012). There is considerable evolutionary and metabolic diversity among nitrite oxidizers, although many are chemoautotrophic and use nitrite oxidoreductase (NXR) to catalyze nitrite oxidation for energy generation. NXR, however, is quite diverse in nature. There are two families of NXR: a cytoplasmic form (possessed by Nitrobacter, Nitrococcus, Nitrolancetus) that is most closely related to nitrate reductase and a periplasmic form (possessed by Nitrospira, Nitrospina, Nitrotoga) similar to the DMSO reductase found in anammox bacteria (see Section 14.2.6) used for nitrate reduction (Lucker et al., 2010). The two forms of NXR may allow for a variety of metabolic capabilities in the organisms that posses them and most likely derive from multiple lines of evolution of this nitrogen cycle process (Lucker et al., 2010).

NOB are widespread in terrestrial and aquatic environments, as well as engineered environments such as wastewater treatment plants. Despite their low energy yield, they are often abundant (up to 9% of the microbial community) in oxygen deficient zones (Füssel et al., 2012). The reason for their persistence in oxygen-deficient environments is currently unknown, but there is independent evidence through nitrogen isotope studies (Casciotti et al., 2013) that nitrite oxidation has affected the distribution of nitrate and nitrite isotopes in oxygen-deficient waters.

14.2.5 Heterotrophic Nitrification

Ammonia can also be converted to nitrate by some heterotrophic microorganisms, although the process is of uncertain importance in nature. Rates of heterotrophic nitrification measured under laboratory conditions so far are significantly slower than those of autotrophic nitrification. However, most measured nitrification rates in the environment would include both autotrophic and heterotrophic nitrification. The organisms capable of heterotrophic nitrification include both bacteria, such as Arthrobacter sp., and fungi, such as Aspergillus flavus. The pathway from ammonia to nitrate may involve intermediates such as hydroxylamine, nitrite, and 1-nitrosoethanol in the case of bacteria and 3-nitropropionic acid in the case of fungi (see Alexander, 1977; Paul and Clark, 1996). These organisms apparently gain no energy from this conversion.

14.2.6 Anaerobic Ammonium Oxidation

Within the last decade, strictly anaerobic, chemoautotrophic bacteria within the phylum Planctomycetes have been found to be involved in the process of *anammox*, an abbreviation for ANaerobic AMMonium OXidation with nitrite to form dinitrogen (reviewed by Kartal et al., 2012). An equation summarizing the overall reaction is as follows:

$$NH_4^+ + NO_2^- \rightarrow N_2 + 2H_2O \qquad (14.8)$$

Anammox was first discovered in an anaerobic sewage treatment process (van de Graaf et al., 1995) but has since been demonstrated to occur in nature in both freshwater and marine environments (Jetten et al., 2003; Trimmer et al., 2003; Tal et al., 2005). In the oceans, it may account for 50% of the loss of fixed nitrogen (Dalsgaard et al., 2005), although more recent estimates place the contribution of anammox closer to 30% of fixed N loss (Babbin et al., 2014). The free energy from this reaction (Equation 14.8) supports CO_2 fixation. The anammox reaction appears to occur within a special intracytoplasmic organelle, called an *anammoxosome*, which is bounded by a membrane that contains unique ladderane lipids (van Niftrik et al., 2004). Hydrazine (N_2H_4) and hydroxylamine are intermediates but are rarely found outside the cell. Acetylene and methanol have been found to inhibit different reactions in anammox (Jensen et al., 2007). The bacteria capable of the anammox reaction belong to five genera within the phylum Planctomycetes: *Kuenenia* (Strous et al., 2006), *Brocadia* (Strous et al., 1999; Kartal et al., 2008; Oshiki et al., 2011), *Anammoxoglobus* (Kartal et al., 2007), *Jettenia* (Quan et al., 2008; Hu et al., 2011), and *Scalindua* (Schmid et al., 2003; Woebken et al., 2008; Van de Vossenberg et al., 2012). All genera have been found in wastewater treatment plants; *Scalindua* is also present in marine and freshwater environments. None of the described species has been isolated as a pure culture, so all bear the status *Candidatus* (Kartal et al., 2013).

14.2.7 Denitrification

Nitrate, nitrite, nitrous oxide, and nitric oxide can serve as electron acceptors in microbial respiration, usually under anaerobic conditions. The transformation of nitrate to nitrite as an anaerobic respiratory process is called *dissimilatory nitrate reduction*, and the further reduction of nitrite to gaseous products (NO, N_2O, and N_2) is called *denitrification*.

The ability to denitrify is widespread among many phylogenetically unrelated microbial groups, including all three domains of life and over 50 different genera (Zumft, 1997). Most known denitrifiers are heterotrophic, but autotrophic denitrifiers that use inorganic sulfur compounds, hydrogen, or reduced iron as electron donors may be prevalent in some environments (e.g., oxic/anoxic interfaces). Some nitrate-respiring bacteria are only capable of nitrate reduction, lacking the enzymes for further reduction of nitrite to dinitrogen, whereas others are capable of reducing nitrite to ammonia instead of dinitrogen (a process known as dissimilatory nitrate reduction to ammonium [DNRA]; see Section 14.2.8). All nitrate respiratory processes have been found to operate to varying degrees in terrestrial, freshwater, and marine environments and represent an important part of the nitrogen cycle under anaerobic conditions.

Nitrate reduction is described by the half reaction

$$NO_3^- + 2H^+ + 2e^- \rightarrow NO_2^- + H_2O \qquad (14.9)$$

The enzyme catalyzing reaction (Equation 14.9) is an iron molybdopterin oxidoreductase called *nitrate reductase*. A denitrifying organism may carry a periplasmic nitrate reductase, a phylogenetically widespread enzyme that may also be involved in fermentation or phototrophy; or a membrane-bound, respiratory nitrate reductase, which is found primarily in denitrifiers; or both.

The reduction of nitrite to nitric oxide is described by the following reaction:

$$NO_2^- + 2H^+ + e^- \rightarrow NO + H_2O \qquad (14.10)$$

This process is carried out by the periplasmic enzyme nitrite reductase (NIR), which in some organisms is a copper-containing enzyme (Cu-NIR) and in others is a cytochrome-cd_1 enzyme (Fe-NIR). Most denitrifying organisms carry only one form of this enzyme, and it is unclear whether either confers a specific advantage in certain environments (Kraft et al., 2011, Chen and Straus, 2013). Nitrite reduction is a critical step in the denitrification pathway, because it represents the first committed step to a gaseous product.

The remaining two steps in the full denitrification pathway are nitric oxide reduction, carried out by the enzyme nitric oxide reductase (Reaction 14.11), and nitrous oxide reduction, carried out by the enzyme nitrous oxide reductase (Reaction 14.12):

$$2NO + 2H^+ + 2e^- \rightarrow N_2O + H_2O \quad (14.11)$$

$$N_2O + 2H^+ + 2e^- \rightarrow N_2 + H_2O \quad (14.12)$$

Although it was previously believed that these reactions could only occur at low oxygen tension or in the absence of oxygen, evidence now indicates that in some cases organisms can perform the reactions at near-normal bulk oxygen tension, likely as a means of disposing of excess reducing power. An analysis of the current knowledge on aerobic denitrification may be found in Chen and Strous (2013). For a more detailed discussion of classical (anaerobic) denitrification, the reader is referred to review articles by Zumft (1997), Shapleigh (2006), and Seitzinger et al. (2006).

14.2.8 Dissimilatory Nitrate Reduction to Ammonium

Although denitrification has long been known to be a major nitrate reduction process in low-oxygen environments, an alternative pathway, the transformation of nitrate to ammonium, has been recognized only for the last few decades. Known as nitrate ammonification or DNRA, it is also a dissimilatory process and may be linked either with fermentation or respiration:

$$NO_3^- + 9H^+ + 8e^- \rightarrow NH_3 + 3H_2O \quad (14.13)$$

Fermentative DNRA, which couples the transfer of electrons from complex organic compounds to nitrate, is carried out by many genera of bacteria, such as *Pseudomonas*, *Clostridia*, and *Desulfovibrio*. Environments conducive to denitrification—low in oxygen and rich in organic carbon and nitrate—are also conducive to fermentative DNRA. There is some evidence that

high-carbon, low-nitrate environments may favor DNRA over denitrification, as DNRA transfers more electrons per nitrate molecule than does denitrification (Tiedje, 1988; Bonin, 1996; Nijburg et al., 1997). However, the relative importance of DNRA versus denitrification in nitrate consumption in most ecosystems is still poorly understood.

Chemolithoautotrophic DNRA couples the reduction of nitrate to the oxidation of reduced sulfur forms such as free sulfide and elemental sulfur. The organisms involved, including *Thiobacillus*, *Thiomicrospira*, and *Thioploca*, may reduce nitrate either to N_2 or to NH_4^+ (Brunet and Garcia-Gil, 1996; Otte et al., 1999). For further reading on nitrate removal pathways, the reader is referred to Burgin and Hamilton (2007).

While *assimilatory nitrate reduction* is also a form of microbial nitrate reduction, it differs from dissimilatory nitrate reduction in that nitrate is reduced to ammonia for the purpose of assimilation—a process that consumes only as much nitrate as the cell requires for growth. It is not a form of respiration and in fact is performed by many organisms that cannot use nitrate for respiration.

14.2.9 Nitrogen Fixation

If nature had not provided for microbial nitrogen fixation to reverse the effect of microbial depletion of fixed nitrogen from soil or water as a result of denitrification and anammox, life on Earth would not have long continued after these anaerobic N-removal processes first evolved. Nitrogen fixation is dependent on a special enzyme, *nitrogenase*, which is found exclusively in prokaryotes, including aerobic and anaerobic, as well as photosynthetic and nonphotosynthetic bacteria and archaea. Nitrogenase (encoded by the *nif* genes) is extremely oxygen sensitive. It is usually a combination of an iron protein and a molybdoprotein (Eady and Postgate, 1974; Orme-Johnson, 1992), but in some cases (e.g., *Azotobacter chroococcum*) may also be a combination of an iron protein and a vanadoprotein (Robson et al., 1986; Eady et al., 1987), and in another case (*Azotobacter vinelandii*) may be a combination of two iron proteins (Chiswell et al., 1988). In each case, nitrogenase catalyzes the reaction

$$N_2 + 8H^+ + 8e^- \rightarrow 2NH_3 + H_2 \quad (14.14)$$

The enzyme is not specific for dinitrogen, and it can catalyze the reduction of acetylene (C_2H_2), as well as of hydrogen cyanide (HCN), cyanogen (NCCN), hydrogen azide (HN_3), hydrogen thiocyanate (HCNS), protons (H^+), carbon monoxide (CO), and some other compounds (Smith, 1983).

The reducing power (the term "$8e^-$" in Equation 14.14) needed for dinitrogen reduction is provided by reduced ferredoxin. Reduced ferredoxin can be formed in a reaction in which pyruvate is oxidatively decarboxylated (Lehninger, 1975):

$$CH_3COCOOH + NAD^+ + CoASH \rightarrow$$
$$CH_3CO \sim SCoA + CO_2 + NADH + H^+ \quad (14.15)$$

$$NADH + (ferredoxin)ox \rightarrow$$
$$NAD^+ + (ferredoxin)red + H^+ \quad (14.16)$$

In phototrophs, the reduced ferredoxin is produced as part of the photophosphorylation mechanism (see Chapter 7).

Nitrogen fixation is a very energy-intensive reaction, consuming as many as 16 mol of ATP in the reduction of 1 mol of dinitrogen to ammonia (Newton and Burgess, 1983), but is extremely beneficial to organisms growing in low-nutrient environments.

Nitrogen fixation may proceed symbiotically or asymbiotically. Symbiotic nitrogen fixation requires that the nitrogen-fixing bacterium associates with a specific host to carry out nitrogen fixation. Even then, dinitrogen will be fixed only if the fixed nitrogen level in the surrounding environment of the host plant or alga is low or the diet of the animal host is nitrogen deficient. In some plants (legumes or alder), the diazotroph may be localized in the cells of the cortical root tissue that are transformed into nodules. Invasion of the plant tissue may occur via root hairs. In some other plants, the nitrogen fixers may be localized in special leaf structures (e.g., in Azolla). In yet other plants (e.g., certain cereal grasses such as maize), nitrogen-fixing spirilla such as Azospirillum lipoferum may not invade host plant tissue but rather live in close association with the roots in the rhizosphere (Day et al., 1975; Von Bülow and Döbereiner, 1975; Smith et al., 1976). In animals, diazotrophs may be found to inhabit the digestive tract (Knowles, 1978).

The plant host in symbiotic nitrogen fixation provides the energy source required by the diazotroph for generating ATP. The energy source may take the form of compounds such as succinate, malate, and fumarate (Paul and Clark, 1996). The plant host can also provide an environment in which access to oxygen is controlled so that nitrogenase is not inactivated. In root nodules of legumes, leghemoglobin is involved in the control of oxygen. In return for protection from oxygen and a stable environment, the diazotroph shares the ammonia that it forms from dinitrogen with its plant or animal host.

Symbiotic nitrogen-fixing bacteria include members of the genera Rhizobium, Bradyrhizobium, Frankia, and Anabaena. Some strains of Bradyrhizobium japonicum have been found to grow autotrophically on hydrogen as an energy source. They can couple hydrogen oxidation to ATP synthesis, which is then used in carbon fixation via the ribulose bisphosphate carboxylase/oxidase system. In nitrogen fixation, the ability to couple hydrogen oxidation to ATP synthesis may represent an energy recovery system because a significant amount of hydrogen is generated during nitrogen fixation (see Equation 14.14), the energy content of this would otherwise be lost to the system.

In nonsymbiotic nitrogen fixation, the active organisms are free-living in soil or water and fix nitrogen if other nitrogen forms are limiting. Their nitrogenase is not distinctly different from that of symbiotic nitrogen fixers. However, unlike the symbiotic nitrogen fixers, aerobic nonsymbiotic nitrogen fixers appear to be able to maintain an intracellular environment in which nitrogenase is active in the presence of oxygen.

The capacity for nonsymbiotic nitrogen fixation is widespread among prokaryotes. Some well-known examples include the aerobes Azotobacter and Beijerinckia and the anaerobe Clostridium pasteurianum, but many other aerobic and anaerobic genera include species with nitrogen-fixing capacity, including some photo- and chemolithotrophs. Most of the nitrogen fixers are active only at environmental pH values between 5 and 9, but some strains of the acidophile Acidithiobacillus ferrooxidans have been shown to fix nitrogen at a pH as low as 2.5.

Nitrogen fixation in the marine environment roughly balances the loss of fixed N from denitrification and anammox (DeVries et al., 2012). It is carried out by colonial nitrogen-fixing cyanobacteria such as Trichodesmium, as well as free-living unicellular cyanobacteria and a variety of

initially developed to distinguish the nitrate source of denitrification (Nielsen, 1992), isotopic measurements of dinitrogen gas produced following parallel spikes of labeled ammonium or nitrate can distinguish whether nitrogen gas is formed through a one-to-one combination of ammonium and nitrate (anammox) or different ratios of substrates (denitrification) (Thamdrup and Dalsgaard, 2002; Risgaard-Petersen et al., 2003). This technique has gained widespread use to measure the contributions of anammox and denitrification to nitrogen loss in a variety of ecosystems, including the ocean (Lam et al., 2007; Ward et al., 2009), estuaries (Rich et al., 2008), and soils (Long et al., 2013). Advances in isotope tracer techniques have allowed for estimations of the importance of anammox compared to denitrification in the environment, and the impacts of these two processes to both global and localized nitrogen loss is an area of intense research.

14.4 SUMMARY

Nitrogen is essential to all forms of life. While N is assimilated by cells in the form of ammonia or amino acids, it primarily occurs as dinitrogen and nitrate in the environment. These forms must be converted to ammonia before they can be assimilated. N is released from organic compounds in the form of ammonia, by a process called ammonification, which occurs both aerobically and anaerobically. Ammonia is an energy-rich compound and can be oxidized to nitrate by way of nitrite by certain aerobic, autotrophic bacteria and archaea (nitrifiers). It can also be converted to nitrate by some heterotrophic bacteria and some fungi in a non-energy-yielding process, but this is much less common. Under reducing conditions, nitrate can be respired to nitrite, nitric and nitrous oxide, and dinitrogen, or to ammonium by a variety of microbial species. The reduction of nitrate to dinitrogen can have the effect of lowering fertility of soil and aquatic ecosystems, as can the anaerobic oxidation of ammonium to dinitrogen (anammox). Depletion of fixed nitrogen availability through dinitrogen evolution can, however, be reversed by symbiotic and nonsymbiotic nitrogen fixation. These biological transformations are part of a complex cycle (Figure 14.1) that is essential to the sustenance of life on Earth, through production and degradation of living biomass as well as the generation of energy for microbial metabolism. Records of these geomicrobiological processes can ultimately be traced through modern and ancient systems through their impact on N (and O) stable isotopes.

REFERENCES

Aleem MIH, Sewell DL. 1984. Oxidoreductase system in *Nitrobacter agilis*. In: Strohl WR, Tuovinen OH, eds. *Microbial Chemoautotrophy*. Columbus, OH: Ohio State University Press, pp. 185–210.

Alexander M. 1977. *Introduction to Soil Microbiology*. New York: Wiley.

Alexander M. 1984. *Biological Nitrogen Fixation: Ecology, Technology, and Physiology*. New York: Plenum Press.

Altabet MA. 1988. Variations in nitrogen isotopic composition between sinking and suspended particles: Implications for nitrogen cycling and particle transformation in the open ocean. *Deep Sea Res A* 35(4):535–554.

Altabet MA, Francois R. 1994. Sedimentary nitrogen isotopic ratio as a recorder for surface ocean nitrate utilization. *Global Biogeochem Cy* 8(1):103–116.

Altabet MA, Francois R, Murray DW, Prell WL. 1995. Climate-related variations in denitrification in the Arabian Sea from sediment $^{15}N/^{14}N$ ratios. *Nature* 373:506–509.

Aluwihare LI, Meador T. 2008. Chemical composition of marine dissolved organic nitrogen. In: Capone DG, Bronk DA, Mulholland MR, Carpenter EJ, eds. *Nitrogen in the Marine Environment*. Oxford, U.K.: Elsevier, pp. 95–140.

Babbin AR, Keil RG, Devol AH, Ward BB. 2014. Organic matter stoichiometry, flux, and oxygen control nitrogen loss in the ocean. *Science* 344(6182):406–408.

Balows A. 1992. *The Prokaryotes: A Handbook on the Biology of Bacteria: Ecophysiology, Isolation, Identification, Applications*. New York: Springer.

Beman JM, Popp BN, Francis CA. 2008. Molecular and biogeochemical evidence for ammonia oxidation by marine Crenarchaeota in the Gulf of California. *ISME J* 2(4):429–441.

Binnerup SJ, Jensen K, Revsbech NP, Jensen MH, Sørensen J. 1992. Denitrification, dissimilatory reduction of nitrate to ammonium, and nitrification in a bioturbated estuarine sediment as measured with ^{15}N and microsensor techniques. *Appl Environ Microbiol* 58:303–313.

Blainey PC, Mosier AC, Potanina A, Francis CA, Quake SR. 2011. Genome of a low-salinity ammonia oxidizing archaeon determined by single-cell and metagenomic analysis. *PLoS ONE* 6:e16626.

Bock E, Koops HP, Harms H, Ahlers B. 1991. The biochemistry of nitrifying organisms. In: Shively JM, Barton LL, eds. *Variations in Autotrophic Life*. London, U.K.: Academic Press, pp. 171–200.

Böhlke JK, Harvey JW, Voytek MA. 2004. Reach-scale isotope tracer experiment to quantify denitrification and related processes in a nitrate-rich stream, mid-continent United States. *Limnol Oceanogr* 49(3):821–838.

Bonin P. 1996. Anaerobic nitrate reduction to ammonium in two strains isolated from coastal marine sediment: A dissimilatory pathway. *FEMS Microbiol Ecol* 19:27–38.

Bonin P, Gilewicz M, Bertrand JC. 1989. Effects of oxygen on each step of denitrification on *Pseudomonas nautica*. *Can J Microbiol* 35:1061–1064.

Brandes JA, Devol AH. 2002. A global marine-fixed nitrogen isotopic budget: Implications for Holocene nitrogen cycling. *Global Biogeochem Cy* 16(4):1120.

Brandes JA, Devol AH, Yoshinari T, Jayakumar A, Naqvi SWA. 1998. Isotopic composition of nitrate in the central Arabian Sea and eastern tropical North Pacific: A tracer for mixing and nitrogen cycles. *Limnol Oceanogr* 43(7):1680–1689.

Brunet RC, Garcia-Gil LJ. 1996. Sulfide-induced dissimilatory nitrate reduction to ammonia in anaerobic freshwater sediments. *FEMS Microbiol Ecol* 21:131–38.

Burgin AJ, Hamilton SK. 2007. Have we overemphasized the role of denitrification in aquatic ecosystems? A review of nitrate removal pathways. *Front Ecol Environ* 5(2):89–96.

Casciotti KL. 2009. Inverse kinetic isotope fractionation during bacterial nitrite oxidation. *Geochim Cosmochim Acta* 73:2061–2076.

Casciotti KL, Buchwald C, McIlvin MR, 2013. Implications of nitrate and nitrite isotopic measurements for the mechanisms of nitrogen cycling in the Peru oxygen deficient zone. *Deep Sea Res I* 80:78–93.

Casciotti KL, McIlvin MR. 2007. Isotopic analyses of nitrate and nitrite from reference mixtures and application to Eastern Tropical North Pacific waters. *Mar Chem* 107(2):184–201.

Casciotti KL, Sigman DM, Hastings MG, Böhlke JK, Hilkert A. 2002. Measurement of the oxygen isotopic composition of nitrate in seawater and freshwater using the denitrifier method. *Anal Chem* 74(19):4905–4912.

Casciotti KL, Trull TW, Glover DM, Davies D. 2008. Constraints on nitrogen cycling at the subtropical North Pacific Station ALOHA from isotopic measurements of nitrate and particulate nitrogen. *Deep Sea Res II* 55(14–15):1661–1672.

Chen J, Strous M. 2013. Denitrification and aerobic respiration, hybrid electron transport chains and co-evolution. *BBA—Bioenerget* 1827(2):136–144.

Chiswell JR, Premarkumar R, Bishop PE. 1988. Purification of a second alternative nitrogenase from a *nif* HDK deletion strain of *Azotobacter vinelandii*. *J Bacteriol* 170:27–33.

Dalsgaard T, Thamdrup B, Canfield DE. 2005. Anaerobic ammonium oxidation (anammox) in the marine environment. *Res Microbiol* 156:457–464.

Davidson EA. 1993. Soil water content and the ratio of nitrous to nitric oxide emitted from soil. In: Oremland RS, ed. *Biogeochemistry of Global Change: Radiatively Active Trace Gases*. New York: Chapman & Hall, pp. 369–386.

Day JM, Neves MCP, Döbereiner J. 1975. Nitrogenase activity on the roots of tropical forage grasses. *Soil Biol Biochem* 7:107–112.

Deutsch C, Sarmiento JL, Sigman DM, Gruber N, Dunne JP. 2007. Spatial coupling of nitrogen inputs and losses in the ocean. *Nature* 445:163–167.

Deutsch C, Sigman DM, Thunell RC, Meckler AN, Haug GH. 2004. Isotopic constraints on glacial/interglacial changes in the oceanic nitrogen budget. *Global Biogeochem Cy* 18(4):GB4012.

DeVries T, Deutsch C et al. 2012. Global rates of water-column denitrification derived from nitrogen gas measurements. *Nat Geosci* 5(8):547–550.

Eady RR, Postgate JR. 1974. Nitrogenase. *Nature* 249:805–810.

Eady RR, Robson RL, Richardson TH, Miller RW, Hawkins M. 1987. The vanadium nitrogenase of *Azotobacter chroococcum*. Purification and properties of the vanadium iron protein. *Biochem J* 244:197–207.

Ericksen GE. 1983. The Chilean nitrate deposits. *Am Sci* 71:366–374.

Erisman JW, Galloway JN, Seitzinger S, Bleeker A, Dise NB, Petrescu AMR, Leach AM, de Vries W. 2013. Consequences of human modification of the global nitrogen cycle. *Philos Trans Roy Soc B* 368:20130116.

Foster RA, Zehr JP. 2006. Characterization of diatom-cyanobacteria symbioses on the basis of nifH, hetR and 16S rRNA sequences. *Environ Microbiol* 8:1913–1925.

Francis CA, Beman JM, Kuypers MMM. 2007. New processes and players in the nitrogen cycle: the microbial ecology of anaerobic and archaeal ammonia oxidation. *ISME J* 1:19–27.

Fulweiler RW, Nixon SW et al. 2007. Reversal of the net dinitrogen gas flux in coastal marine sediments. *Nature* 448(7150):180–182.

Füssel J, Lam P, Lavik G, Jensen MM, Holtappels M, Gunter M, Kuypers MMM. 2012. Nitrite oxidation in the Namibian oxygen minimum zone. *ISME J* 6:1200–1209.

Ganeshram RS, Pedersen TF, Calvert SE, Murray JW. 1995. Large changes in oceanic nutrient inventories from glacial to interglacial periods. *Nature* 376(6543):755–758.

Gribsholt B, Boschker HTS et al. 2005. Nitrogen processing in a tidal freshwater marsh: A whole-ecosystem ^{15}N labeling study. *Limnol Oceanogr* 50(6):1945–1959.

Hallam SJ, Konstantinidis KT, Putnam N, Schleper C, Watanabe YI, Sugahara J, Preston C, de la Torre J, Richardson PM, DeLong EF. 2006a. Genomic analysis of the uncultivated marine crenarchaeote *Cenarchaeum symbiosum*. *Proc Natl Acad Sci USA* 103:18296–18301.

Hallam SJ, Mincer TJ, Schleper C, Preston CM, Roberts K, Richardson PM, DeLong EF. 2006b. Pathways of carbon assimilation and ammonia oxidation suggested by environmental genomic analyses of marine Crenarchaeota. *PLoS Biol* 4:e95.

Hastings MG, Casciotti KL, Elliot EM. 2013. Stable isotopes as tracers of anthropogenic nitrogen sources, deposition, and impacts. *Elements* 9(5):339–344.

Higgins MB, Robinson RS, Husson JM, Carter SJ, Pearson A. 2012. Dominant eukaryotic export production during ocean anoxic events reflects the importance of recycled NH_4^+. *Proc Natl Acad Sci USA* 109(7):2269–2274.

Hobbie EA, Högberg P. 2012. Nitrogen isotopes link mycorrhizal fungi and plants to nitrogen dynamics. *New Phytol* 196(2):367–382.

Holmes RM, McClelland JW, Sigman DM, Fry B, Peterson BJ. 1998. Measuring ^{15}N-NH_4^+ in marine, estuarine and fresh waters: An adaptation of the ammonia diffusion method for samples with low ammonium concentrations. *Mar Chem* 60(3–4):235–243.

Hooper AB. 1984. Ammonia oxidation and energy transduction in the nitrifying bacteria. In: Strohl WR, Tuovinen OH, eds. *Microbial Chemoautotrophy*. Columbus, OH: Ohio State University Press, pp. 133–167.

Howarth R, Swaney D, Billen G, Garnier J, Hong B, Humborg C, Johnes P, Mörth CM, Marino R. 2012. Nitrogen fluxes from the landscape are controlled by net anthropogenic nitrogen inputs and by climate. *Front Ecol Environ* 10(1):37–43.

Hu BL, Rush D, van der Biezen E, Zheng P, van Mullekom M, Schouten S, Sinninghe Damsté JS, Smolders AJ, Jetten MSM, Kartal B. 2011. New anaerobic, ammonium-oxidizing community enriched from peat soil. *Appl Environ Microbiol* 77:966–971.

Jensen MM, Thamdrup B, Dalsgaard T. 2007. Effects of specific inhibitors on anammox and denitrification in marine sediments. *Appl Environ Microbiol* 73:3151–3158.

Jetten MSM, Slickers O et al. 2003. Anaerobic ammonium oxidation by marine and freshwater planctomycete-like bacteria. *Appl Microbiol Biotechnol* 63:107–114.

Jones RD, Morita RY. 1983. Methane oxidation by *Nitrosococcus oceanus* and *Nitrosomonas europaea*. *Appl Environ Microbiol* 45:401–410.

Kartal B, de Almeida NM, Maalcke WJ, Op den Camp HJM, Jetten MSM, Keltjens JT. 2013. How to make a living from anaerobic ammonium oxidation. *FEMS Microbiol Rev* 37:428–461.

Kartal B, Rattray J et al. 2007. Candidatus "*Anammoxoglobus propioncus*" a new propionate oxidizing species of anaerobic ammonium oxidizing bacteria. *Syst Appl Microbiol* 30:39–49.

Kartal B, van Niftrik L, Keltjens JT, Op den Camp HJ, Jetten MSM. 2012. Anammox–growth physiology, cell biology, and metabolism. *Adv Microb Physiol* 60:211–262.

Kartal B, van Niftrik L, Rattray J, van de Vossenberg JL, Schmid MC, Sinninghe Damsté J, Jetten MSM, Strous M. 2008. Candidatus 'Brocadia fulgida': An autofluorescent anaerobic ammonium oxidizing bacterium. *FEMS Microbiol Ecol* 63:46–55.

Kendall C, Elliot EM, Wankel SD. 2007. Tracing anthropogenic inputs of nitrogen to ecosystems. In: Michener RH, Lajtha K, eds. *Stable Isotopes in Ecology and Environmental Science*. Malden, MA: Blackwell Publishing, pp. 375–449.

Knapp AN, Dekaezemacker J et al. 2012. Sensitivity of *Trichodesmium erythraeum* and *Crocosphaera watsonii* abundance and N_2 fixation rates to varying NO_3^- and PO_4^{3-} concentrations in batch cultures. *Aquat Microb Ecol* 66(3):223–236.

Knapp AN, DiFiore PJ, Deutsch C, Sigman DM, Lipschultz F. 2008. Nitrate isotopic composition between Bermuda and Puerto Rico: Implications for N_2 fixation in the Atlantic Ocean. *Global Biogeochem Cy* 22:GB3014.

Knapp AN, Sigman DM, Lipschultz F. 2005. N isotopic composition of dissolved organic nitrogen and nitrate at the Bermuda Atlantic Time-series Study site. *Global Biogeochem Cy* 19:GB1018.

Knowles R. 1978. Free-living bacteria. In: Döbereiner R, Burris H, Hollaender A, Franco AA, Neyra CA, Scott DB, eds. *Limitations and Potentials for Biological Nitrogen Fixation in the Tropics*. New York: Plenum Press, pp. 25–40.

Knowles R. 1985. Microbial transformations as sources and sinks of nitrogen oxides. In: Caldwell DE, Brierley JA, Brierley CL, eds. *Planetary Ecology*. New York: Van Nostrand Reinhold, pp. 411–426.

EHRLICH'S GEOMICROBIOLOGY

Könneke M, Bernhard AE, de la Torre JR, Walker CB, Waterbury JB, Stahl DA. 2005. Isolation of an autotrophic ammonia-oxidizing marine archaeon. *Nature* 437(7058):543–546.

Kozlowski, JA, Price, J, Stein, LY. 2014. Revision of N₂O-producing pathways in the ammonia-oxidizing bacterium *Nitrosomonas europaea* ATCC 19718. *Appl Environ Microbiol* 80(16):4930.

Kraft B, Strous M, Tegetmeyer HE. 2011. Microbial nitrate respiration-genes, enzymes and environmental distribution. *J Biotechnol* 155:104–117.

Lam P, Jensen MM et al. 2007. Linking crenarchaeal and bacterial nitrification to anammox in the Black Sea. *Proc Natl Acad Sci USA* 104(17):7104–7109.

Lehninger AL. 1975. *Biochemistry*, 2nd edn. New York: Worth.

Letscher RT, Hansell DA, Carlson CA, Lumpkin R, Knapp AN. 2013. Dissolved organic nitrogen in the global surface ocean: Distribution and fate. *Global Biogeochem Cy* 27:141–153.

Lipschultz F. 2008. Isotope tracer methods for studies of the marine nitrogen cycle. In: Capone DG, Bronk DA, Mulholland MR, Carpenter EJ, eds. *Nitrogen in the Marine Environment*, 2nd edn. Oxford, U.K.: Elsevier, pp. 1345–1384.

Lomas MW, Glibert PM. 1999. Interactions between NH₄⁺ and NO₃⁻ uptake and assimilation: Comparison of diatoms and dinoflagellates at several growth temperatures. *Mar Biol* 133:541–551.

Long A, Heitman J, Tobias C, Philips R, Song B. 2013. Co-occurring anammox, denitrification, and codenitrification in agricultural soils. *Appl Environ Microbiol* 79(1):168–176.

Lucker S, Wagner M et al. 2010. A *Nitrospira* metagenome illuminates the physiology and evolution of globally important nitrite-oxidizing bacteria. *Proc Natl Acad Sci USA* 107(30):13479–13484.

Mariotti A, Germon JC et al. 1981. Experimental determination of nitrogen kinetic isotope fractionation: Some principles; illustration for denitrification and nitrification processes. *Plant Soil* 62(3):413–430.

McCarthy JJ, Taylor WR, Taft JL. 1977. Nitrogenous nutrition of the plankton in the Chesapeake Bay. 1. Nutrient availability and phytoplankton preferences. *Limnol Oceanogr* 22(6):996–1011.

McIlvin MR, Altabet MA. 2005. Chemical conversion of nitrate and nitrite to nitrous oxide for nitrogen and oxygen isotopic analysis in freshwater and seawater. *Anal Chem* 77(17):5589–5595.

McIlvin MR, Casciotti KL. 2011. Technical updates to the bacterial method for nitrate isotopic analyses. *Anal Chem* 83(5):1850–1856.

Montoya JP. 2008. Nitrogen stable isotopes in marine environments. In: Capone DG, Bronk DA, Mulholland MR, Carpenter EJ, eds. *Nitrogen in the Marine Environment*, 2nd edn. Oxford, U.K.: Elsevier, pp. 1277–1302.

Montoya JP, Carpenter EJ, Capone DG. 2002. Nitrogen fixation and nitrogen isotope abundances in zooplankton of the oligotrophic North Atlantic. *Limnol Oceanogr* 47(6):1617–1628. Needoba JA, Waser NAD, Harrison PJ, Calvert SE. 2003. Nitrogen isotope fractionation by 12 species of marine phytoplankton during growth on nitrate. *Mar Ecol Prog Ser* 255:81–91.

Neff JC, Chapin FSC, Vitousek PM. 2003. Breaks in the cycle: Dissolved organic nitrogen in terrestrial ecosystems. *Front Ecol Environ* 1:205–211.

Newell SE, Fawcett SE, Ward BB. 2013. Depth distribution of ammonia oxidation rates and ammonia-oxidizer community composition in the Sargasso Sea. *Limnol Oceanogr* 58(4):1491–1500.

Newton WE, Burgess BK. 1983. Nitrogen fixation: Its scope and importance. In: Mueller A, Newtan WE, eds. *Nitrogen Fixation. The Chemical–Biochemical–Genetic Interface*. New York: Plenum Press, pp. 1–19.

Nielsen LP. 1992. Denitrification in sediment determined from nitrogen isotope pairing. *FEMS Microbiol Ecol* 86:357–362.

Nier AO. 1950. A redetermination of the relative abundances of the isotopes of carbon, nitrogen, oxygen, argon, and potassium. *Phys Rev* 77(6):789–793.

Nijburg JW, Coolen MJL et al. 1997. Effects of nitrate availability and the presence of Glyceria maxima on the composition and activity of the dissimilatory nitrate-reducing bacterial community. *Appl Environ Microbiol* 63(3):931–937.

Orme-Johnson WH. 1992. Nitrogenase structure: Where to now? *Science* 257:1639–1640.

Oshiki M, Shimokawa M, Fujii N, Satoh H, Okabe S. 2011. Physiological characteristics of the anaerobic ammonium-oxidizing bacterium *Candidatus* "Brocadia sinica." *Microbiology* 157:1706–1713.

Otte S, Kuenen JG, Nielsen LP, et al 1999. Nitrogen, carbon, and sulfur metabolism in natural *Thioploca* samples. *Appl Environ Microbiol* 65:3148–57.

Paul EA, Clark PF. 1996. *Soil Microbiology and Biochemistry*. San Diego, CA: Academic Press.

Pennock JR, Velinsky DJ, Ludlam JM, Sharp JH, Fogel ML. 1996. Isotopic fractionation of ammonium and nitrate during uptake by *Skeletonema costatum*: Implications for δ¹⁵N dynamics under bloom conditions. *Limnol Oceanogr* 41(3):451–459.

Quan ZX, Rhee SK, Zuo JE, Yang Y, Bae JW, Park JR, Lee ST, Park YH. 2008. Diversity of ammonium-oxidizing bacteria in a granular sludge anaerobic ammonium-oxidizing (anammox) reactor. *Environ Microbiol* 10:3130–3139.

Rafter PA, Sigman DM, Charles CD, Kaiser J, Haug GH. 2012. Subsurface tropical Pacific nitrogen isotopic composition of nitrate: Biogeochemical signals and their transport. *Global Biogeochem Cy* 26:GB1003.

Reed SC, Cleveland CC, and Townsend AR. 2011. Functional ecology of free-living nitrogen fixation: A contemporary perspective. *Annu Rev Ecol Syst* 42(1):489–512.

Ren H, Sigman DM et al. 2009. Foraminiferal isotope evidence of reduced nitrogen fixation in the ice age Atlantic Ocean. *Science* 323(5911):244–248.

Rich JJ, Dale OR, Song B, Ward BB. 2008. Anaerobic ammonium oxidation (anammox) in Chesapeake Bay sediments. *Microb Ecol* 55(2):311–320.

Risgaard-Petersen N, Nielsen LP, Rysgaard S, Dalsgaard T, Meyer RL. 2003. Application of the isotope pairing technique in sediments where anammox and denitrification coexist. *Limnol Oceanogr Method* 1:63–73.

Robinson RS, Brunelle BG, Sigman DM. 2004. Revisiting nutrient utilization in the glacial Antarctic: Evidence from a new method for ditom-bound N isotopic analysis. *Paleoceanography* 19:PA3001.

Robinson RS, Kienast M et al. 2012. A review of nitrogen isotopic alteration in marine sediments. *Paleoceanography* 27(4):PA4203.

Robson RL, Eadky RR, Richardson TH, Miller RW, Hawkins M, Postgate JR. 1986. The alternative nitrogenase of *Azotobacter chroococcum* is a vanadium enzyme. *Nature* 322:388–390.

Rockström J, Steffen W et al. 2009. A safe operating space for humanity. *Nature* 461:472–475.

Santoro AE, Casciotti KL, Francis CA. 2010. Activity, abundance and diversity of nitrifying archaea and bacteria in the central California Current. *Environ Microbiol* 12(7):1989–2006.

Santoro AE, Sakamoto CM et al. 2013. Measurements of nitrite production in and around the primary nitrite maximum in the central California Current. *Biogeosciences* 10(11):7395–7410.

Schmid M, Walsh K et al. 2003. *Candidatus* "*Scalindua brodae*", sp. nov., *Candidatus* "*Scalindua wagneri*", sp. nov., two new species of anaerobic ammonium oxidizing bacteria. *Syst Appl Microbiol* 26:529–538.

Seitzinger S, Harrison JA, Böhlke JK, Bouwman AF, Lowrance R, Peterson B, Tobias C, Van Drecht G. 2006. Denitrification across landscapes and waterscapes: A synthesis. *Ecol Appl* 16(6):2064–2090.

Shapleigh JP. 2006. The denitrifying prokaryotes. In Dworkin M, Falkow S, Rosenberg E, Schliefer KH, eds. *The Prokaryotes*. Vol. 2: *Ecophysiology and Biochemistry*. New York: Springer, pp. 769–792.

Sigman DM, Altabet MA, Francois R, McCorkle DC, Gaillard JF. 1999. The isotopic composition of diatom-bound nitrogen in Southern Ocean sediments. *Paleoceanography* 14(2):118–134.

Sigman DM, Altabet MA, Michener R, McCorkle DC, Fry B, Holmes RM. 1997. Natural abundance-level measurements of the nitrogen isotopic composition of oceanic nitrate: An adaptation of the ammonia diffusion method. *Mar Chem* 57(3–4):227–242.

Sigman DM, Casciotti KL, Andreani M, Barford C, Galanter M, Böhlke JK. 2001. A bacterial method for the nitrogen isotopic analysis of nitrate is seawater and freshwater. *Anal Chem* 73(17):4145–4153.

Sigman DM, DiFiore PJ et al. 2009. The dual isotopes of deep nitrate as a constraint on the cycle and budget of oceanic fixed nitrogen. *Deep Sea Res I* 56(9):1419–1439.

Sigman DM, Granger J et al. 2005. Coupled nitrogen and oxygen isotope measurements of nitrate along the eastern North Pacific margin. *Global Biogeochem Cy* 19(4):GB4022.

Sigman DM, Robinson R, Knapp AN, van Geen A, McCorkle DC, Brandes JA, Thunell RC. 2003. Distinguishing between water column and sedimentary denitrification in the Santa Barbara Basin using the stable isotopes of nitrate. *Geochem Geophy Geosy* 4(5):1040.

Silva SR, Kendall C, Wilkison DH, Ziegler AC, Chang CCY, Avanzino RJ. 2000. A new method for collection of nitrate from fresh water and the analysis of nitrogen and oxygen isotope ratios. *J Hydrol* 228(1–2):22–36.

Smith BE. 1983. Reactions and physicochemical properties of the nitrogenase MoFe proteins. In: Mueller A, Newton WE, eds. *Nitrogen Fixation. The Chemical-Biochemical-Genetic Interface*. New York: Plenum Press, pp. 23–62.

Smith JM, Casciotti KL, Chavez FP, Francis CA. 2014. Differential contributions of archaeal ammonia oxidizer ecotypes to nitrification in coastal surface waters. *ISME J* 8:1704–1714.

Smith RL, Bouton JH, Schank SC, Queensberry KH, Tyler ME, Milam JR, Gaskin MH, Littell RC. 1976. Nitrogen fixation in grasses inoculated with *Spirillum lipoferum*. *Science* 193:1003–1005.

Sohm JA, Webb EA, Capone DG. 2011. Emerging patterns of marine nitrogen fixation. *Nat Rev Microbiol* 9(7):499–508.

Sorokin DY, Lucker S et al. 2012. Nitrification expanded: Discovery, physiology and genomics of a nitrite-oxidizing bacterium from the phylum *Chloroflexi*. *ISME J* 6(12):2245–2256.

Stahl DA, de la Torre JR. 2012. Physiology and diversity of ammonia-oxidizing archaea. *Annu Rev Microbiol* 66:83–101.

Stein LY, Yung YL. 2003. Production, isotopic composition, and atmospheric fate of biologically produced nitrous oxide. *Annu Rev Earth Planet Sci* 31(1):329–356.

Strous, M, Fuerst JA, Kramer EHM, Logemann S, Muyzer G, Van De Pas-Schoonen KT, Webb R, Kuenen JG, Jetten MSM. 1999. Missing lithotroph identified as new planctomycete. *Nature* 400:446–449.

Strous M, Pelletier E et al. 2006. Deciphering the evolution and metabolism of an anammox bacterium from a community genome. *Nature* 440:790–794.

Tal Y, Watts JEM, Schreier HJ. 2005. Anaerobic ammonia-oxidizing bacteria and related activity in Baltimore Inner Harbor sediment. *Appl Environ Microbiol* 71:1816–1821.

Thamdrup B, Dalsgaard T. 2002. Production of N_2 through anaerobic ammonium oxidation coupled to nitrate reduction in marine sediments. *Appl Environ Microbiol* 68(3):1312–1318.

Thompson AW, Foster RA, Krupke A, Carter BJ, Musat N, Vaulot D. 2012. Unicellular cyanobacterium symbiotic with a single-celled eukaryotic alga. *Science* 337:1546–1550.

Tiedje JM. 1988. Ecology of denitrification and dissimilatory nitrate reduction to ammonium. In: Zehnder AJB, ed. *Biology of Anaerobic Microorganisms*. New York: John Wiley & Sons, pp. 179–244.

Trimmer M, Nicholls JC, Deflandre B. 2003. Anaerobic ammonium oxidation measured in sediments along the Thames Estuary, United Kingdom. *Appl Environ Microbiol* 69:6447–6454.

Twomey LJ, Piehler MF, Paerl HW. 2005. Phytoplankton uptake of ammonium, nitrate and urea in the Neuse River Estuary, NC, USA. *Hydrobiologia* 533:123–134.

van de Graaf AA, Mulder A, de Bruijn P, Jetten MSM, Robertson LA, Kuenen JG. 1995. Anaerobic oxidation of ammonium is a biologically mediated process. *Appl Environ Microbiol* 61:1246–1251.

van de Vossenberg J, Woebken D et al. 2012. The metagenome of the marine anammox bacterium 'Candidatus Scalindua profunda' illustrates the versatility of this globally important nitrogen cycle bacterium. *Environ Microbiol* 15(5):1275–1289.

van Niftrik LA, Fuerst JA, Sinninghe Damsté JS, Kuenen JG, Jetten MSM, Strous M. 2004. The anammoxosome: An intracytoplasmic compartment in anammox bacteria. *FEMS Microbiol Lett* 233:7–13.

Vajrala N, Martens-Habbena W et al. 2013. Hydroxylamine as an intermediate in ammonia oxidation by globally abundant marine archaea. *Proc Natl Acad Sci USA* 110(3):1006–1011.

Villareal TA. 1991. Nitrogen-fixation by the cyanobacterial symbiont of the diatom genus Hemialus. *Mar Ecol Prog Ser* 76:201–204.

Vitousek PM, Menge DNL, Reed SC, Cleveland CC. 2013. Biological nitrogen fixation: Rates, patterns and ecological controls in terrestrial ecosystems. *Philos Trans Roy Soc B* 368(1621):20130119.

Von Bülow JFW, Döbereiner J. 1975. Potential for nitrogen fixation in maize genotypes in Brazil. *Proc Natl Acad Sci USA* 72:2389–2393.

Voss M, Dippner JW, Montoya JP. 2001. Nitrogen isotope patterns in the oxygen-deficient waters of the Eastern Tropical North Pacific Ocean. *Deep Sea Res I* 48:1905–1921.

Walker CB, de la Torre JR et al. 2010. Nitrosopumilus maritimus genome reveals unique mechanisms for nitrification and autotrophy in globally distributed marine crenarchaea. *Proc Natl Acad Sci USA* 107(19):8818–8823.

Ward BB. 2005. Temporal variability in nitrification rates and related biogeochemical factors in Monterey Bay, California, USA. *Mar Ecol Prog Ser* 292:97–109.

Ward BB. 2012. The global nitrogen cycle. In: Knoll AH, Canfield DE, Konhauser KO, eds. *Fundamentals of Geobiology*. West Sussex, U.K.: Wiley-Blackwell, pp. 36–48.

Ward BB, Devol AH et al. 2009. Denitrification as the dominant nitrogen loss process in the Arabian Sea. *Nature* 461:78–81.

Ward BB, Olson RJ, Perry MJ. 1982. Microbial nitrification rates in the primary nitrite maximum off southern California. *Deep Sea Res* 29(2A):247–255.

Ward BB, Zafiriou OC. 1988. Nitrification and nitric oxide in the oxygen minimum of the eastern tropical North Pacific. *Deep Sea Res A* 35:1127–1142.

Waser NAD, Harrison PJ, Nielsen B, Calvert SE, Turpin DH. 1998. Nitrogen isotope fractionation during the uptake and assimilation of nitrate, nitrite, ammonium, and urea by a marine diatom. *Limnol Oceanogr* 43(2):215–224.

Weber, T., Deutsch, C. 2014. Local versus basin-scale limitation of marine nitrogen fixation. *Proc Natl Acad Sci USA* 111(24):8741–8746.

Wilkerson FP, Dugdale RC. 1992. Measurements of nitrogen productivity in the equatorial Pacific. *J Geophys Res* 97(C1):669–679.

arsenopyrite (FeAsS); with selenium, as in As_2Se_3; with tellurium, as in As_2Te; or with sulfosalt, as in enargite (Cu_3AsS_4). It is also found in arsenides of heavy metals such as iron (loellingite, $FeAs_2$), copper (domeykite, Cu_3As), nickel (nicolite, $NiAs$), and cobalt (Co_2As). Sometimes, it occurs in the form of arsenite minerals (arsenolite or claudetite, As_2O_3) or arsenate minerals (erythrite, $Co_3(AsO_4)_2 \cdot 8H_2O$; scorodite, $FeAsO_4 \cdot 2H_2O$; olivenite, $Cu_2(AsO_4)$ (OH)). Arsenopyrite is the most common and widespread mineral form of arsenic, but orpiment and realgar are also fairly common. The ultimate source of arsenic on the Earth's surface is igneous activity. On weathering of arsenic-containing rocks, which may harbor as much as 1.8 ppm of the element, the arsenic is dispersed through the upper lithosphere and the hydrosphere.

Arsenic concentration in soil may range from 0.1 to more than 1000 ppm. The average concentration in seawater has been reported to be 3.7 $\mu g\ L^{-1}$ and in freshwater 1.5–53 $ng\ m^{-3}$ (Bowen, 1979). However, in groundwater in the Munshiganj District of Bangladesh, the As concentration approaches a maximum of 640 $mg\ m^{-3}$ at a depth of 30–40 m but decreases to 58 $mg\ m^{-3}$ at 107 m (Swartz et al., 2004). Some living organisms may concentrate arsenic manyfold over its level in the environment. For example, some algae have been found to accumulate arsenic 200–3000 times in excess of its concentration in the growth medium (Lunde, 1973). Humans may artificially raise the arsenic concentrations in soil and water through the introduction of sodium arsenite $(NaAsO_3)$ or cacodylic acid $((CH_3)_2AsO \cdot OH)$ as herbicides.

15.2.2 Some Chemical Characteristics

In nature, arsenic is usually encountered in the oxidation states of −3, 0, +3, and +5. Its coordination numbers are in the range of 3–6 (Cullen and Reimer, 1989). Except for the oxidation states of As in arsenate and arsenite, the oxidation state of other compounds, whether organic or inorganic, is often unclear and depends on its definition (Cullen and Reimer, 1989, p. 715). According to Cullen and Reimer (1989, p. 715), arsenious acid and its salts in aqueous solution exist in the ortho form (H_3AsO_3) but not in the meta $(HAsO_2)$ form. Environmentally, arsenite is more mobile than arsenate, but it can be significantly adsorbed under certain conditions. This is because of the tendency of arsenate to become strongly adsorbed to mineral

surfaces such as those of ferrihydrite. However, an ability of ferrihydrite and schwertmannite to adsorb both As(V) and As(III) has been observed by Raven et al. (1998) and Carlson et al. (2002). The ability of disordered mackinawite (FeS), which can be formed in sediments where bacterial sulfate reduction occurs, to adsorb As(V) and As(III) has been reported by Wolthers et al. (2005). As(III) adsorption by amorphous iron oxide and goethite has also been studied by Dixit and Hering (2003, 2006), who noted preferential uptake of As(V) in a pH range of 5–6 and As(III) in a pH range of 7–8. They also noted that with goethite, current single-sorbate models were able to predict the adsorption of Fe(II) and As(III) more satisfactorily than double-sorbate models. Overall, because of the lesser mobility of arsenate compared with arsenite under common environmental conditions, reduction of arsenate to arsenite in the environment, whether chemical or biological, leads to an increase in arsenic toxicity (Cullen and Reimer, 1989). The environmental chemistry of arsenic is further complicated in anoxic, sulfide-rich environments where various thioarsenate complexes are commonly observed (Planer-Friederich and Wallschläger, 2009; Planer-Friedrich et al., 2009).

15.2.3 Toxicity

Arsenic compounds are toxic for most living organisms. Arsenite (AsO_3^{3-}) has been shown to inhibit dehydrogenases such as pyruvate, α-ketoglutarate-, and dihydrolipoate-dehydrogenases (Mahler and Cordes, 1966). Arsenate (AsO_4^{3-}) uncouples oxidative phosphorylation, that is, it inhibits chemiosmotic ATP synthesis (Da Costa, 1972).

Both the uptake and the inhibitory effect of arsenate on metabolism can be modified by phosphate (Da Costa, 1971, 1972; Button et al., 1973). This is because of the existence of a common transport mechanism for phosphate and arsenate in the membranes of some organisms. However, a separate transport mechanism for phosphate may also exist (Bennett and Malamy, 1970). In the latter case, phosphate uptake does not affect arsenate uptake, nor does arsenate uptake affect phosphate uptake. In one reported case of a fungus, Cladosporium herbarum, arsenite toxicity could also be ameliorated by the presence of phosphate. In that instance, prior oxidation of the arsenite to arsenate by the fungus appeared to be the basis for the effect (Da Costa, 1972).

In growing cultures of *Candida humicola*, phosphate can inhibit the formation of trimethylarsine from arsenate, arsenite, and monomethylarsonate, but not from dimethylarsinate (Cox and Alexander, 1973). In similar cultures, phosphite can suppress the formation of trimethylarsine from monomethylarsonate, but not from arsenate or dimethylarsinate. Hypophosphite can cause temporary inhibition of the conversion of arsenate, monomethylarsonate, and dimethylarsinate (Cox and Alexander, 1973). High antimonite concentrations lower the rate of conversion of arsenate to trimethylarsine by resting cells of *C. humicola* (Cox and Alexander, 1973).

Phosphate transport in *Escherichia coli* is achieved by an inorganic phosphate transporter ("Pit") that can be inhibited by its arsenate analog. However, cells also contain a phosphate-specific transporter ("Pst") that that was still able to bring phosphate into the cells when arsenate was in 10-fold excess of phosphate in the surrounding medium (Willsky and Malamy, 1980a,b). A considerable amount of work has been done subsequently with this and other microorganisms examining this arsenate/phosphate interplay, and for more details, the reader is referred to the reviews by Rosen (1999) and Silver and Phung (2005). The close proximity and expression of genes involved in phosphorus acquisition with those of arsenite oxidation during phosphate starvation was examined in *Agrobacterium tumefaciens* (Kang et al., 2012). The authors concluded that there was a strong degree of coregulation of these two metabolic functions.

A biological curiosity that attracted considerable controversy was the report that a halomonad, strain GFAJ-1 isolated from arsenic-rich Mono Lake in California, had the ability to grow by using arsenate in lieu of phosphate (Wolfe-Simon et al., 2011). The results implied that some arsenate could replace phosphate in the 3′–5′ deoxyribose-phosphate diester bonds of this microbe's DNA, a finding once reported in mouse tumor cells (Kay, 1965). Considerable published commentary and opinions immediately ensued (e.g., Rosen et al., 2011; Kim and Rensing, 2012; Knodle et al, 2012) as well as theoretical calculations of the similar conformational properties of arsenodiester vs. phosphodiester linkages in DNA (Denning and MacKerell, 2011). However, independent attempts to reproduce the growth results of the initial report failed, as did the search for the presence of significant arsenic in DNA extracted from the GFAJ-1 organism (Erb et al., 2012; Reaves et al., 2012). However,

the GFAJ-1 bacterium was found to possess a unique phosphate transporter that allowed for it to selectively bind and uptake phosphate in the presence of a 4500-fold excess of arsenate (Elias et al., 2012). An arsenate-induced ribosomal phosphate scavenging was observed in *E. coli* that was thought to explain the reported growth of strain GFAJ-1 at low background phosphate when cultivated with an abundance of arsenate (Basturea et al., 2012).

Bacteria can develop genetically determined resistance to arsenic (Ji and Silver, 1992; Ji et al., 1993). The gene locus for this resistance may reside on a plasmid as, for example, in *Staphylococcus aureus* (Dyke et al., 1970) and *E. coli* (Hedges and Baumberg, 1973). The mechanism of resistance in these bacterial species is a special pumping mechanism that expels the arsenic taken up as arsenate by the cells (Silver and Keach, 1982). It involves intracellular reduction of arsenate to arsenite followed by efflux of the arsenite promoted by an oxyanion-translocating ATPase (Broeer et al., 1993; Ji et al., 1993). Some of the resistant organisms have the capacity to oxidize reduced forms of arsenic to arsenate and others to reduce oxidized forms (see Sections 15.2.4 and 15.2.6). In *Shewanella* sp. strain ANA-3, the traits of arsenate resistance and arsenate respiration are encoded in two distinct genetic loci, operons *ars* and *arrs*, respectively (Saltikov and Newman, 2003; Saltikov et al., 2003; see also Section 15.2.7). There are also examples of arsenic resistance in the Archaea, including halophiles like *Halobacterium* sp. strain NRC-1 (Wang et al., 2004) and the acidophile "*Ferroplasma acidarmanus*" Fer1 (Baker-Austin et al., 2007). While these organisms contain homologs of arsenite resistance genes found in Bacteria, it is also suggested that they have unique mechanisms of arsenate resistance that do not involve its reduction to arsenite.

15.2.4 Microbial Oxidation of Reduced Forms of Arsenic

15.2.4.1 Aerobic Oxidation of Dissolved Arsenic

Bacterial oxidation of arsenite to arsenate was first reported by Green (1918). He discovered bacteria with this ability in arsenical cattle-dipping solution used for protection against insect bites. He isolated an organism with this trait, which he named *Bacillus arsenoxydans*. Quastel and Scholefield (1953) observed arsenite oxidation in perfusion experiments in which they passed 2.5–10 mM sodium arsenite solution through columns charged with Cardiff soil. They did not isolate the organism or organisms responsible for the oxidation but showed that a 0.1% solution of NaN_3 inhibited oxidation. The onset of arsenite oxidation in their experiments occurred after a lag. The length of this lag was increased when sulfanilamide was added with the arsenite. A control of pH was found to be important for sustained bacterial activity. They observed an almost stoichiometric O_2 consumption during arsenite oxidation.

Further investigation of arsenical cattle-dipping solutions led to the isolation of 15 aerobic arsenite-oxidizing strains of bacteria (Turner, 1949, 1954). These organisms were assigned to the genera *Pseudomonas*, *Xanthomonas*, and *Achromobacter*. *Achromobacter arsenoxydans* was later considered synonymous with *Alcaligenes faecalis* (Hendrie et al., 1974). This organism was described in the eighth edition of *Bergey's Manual of Determinative Bacteriology* (Buchanan and Gibbons, 1974) as frequently having the capacity of arsenite oxidation.

Of Turner's 15 isolates, *Pseudomonas arsenoxydans* was studied in detail with respect to arsenite oxidation. Resting cells of this culture oxidized arsenite at an optimum pH of 6.4 and an optimum temperature of 40°C (Turner and Legge, 1954). Cyanide, azide, fluoride, and pyrophosphate inhibited the activity. Under anaerobic conditions, 2,6-dichlorophenol indophenol, m-carboxyphenolindo-2,6-dibromophenol, and ferricyanide could act as electron acceptors. Pretreating cells with toluene and acetone or desiccating them rendered them incapable of oxidizing arsenite in air. The arsenite-oxidizing enzyme was described as adaptable. Examination of cell-free extracts of *P. arsenoxydans* suggested the presence of soluble dehydrogenase activity, which under anaerobic conditions conveyed electrons from arsenite to 2,6-dichlorophenol indophenol (Legge and Turner, 1954). This activity was partly inhibited by 1 mM p-chloromercuribenzoate. The entire arsenite-oxidizing system was believed to consist of dehydrogenase and an oxidase (Legge, 1954).

A soil strain of *A. faecalis* was isolated by Osborne (1973) whose arsenite-oxidizing ability was found to be inducible by arsenite and arsenate (Osborne and Ehrlich, 1976). Very recent phylogenetic study of this strain by 16S rDNA analysis showed that it should be reassigned to the genus *Achromobacter* (Santini, 2001, personal communication). It was

TABLE 15.1

Stoichiometry of oxygen uptake by Alcaligenes faecalis *on arsenite based on the reaction of* $AsO_2^- + H_2O + 1/2O_2 \rightarrow AsO_4^{3-} + 2H^+$.

NaAsO$_2$ Added (μmol)	Theoretical	Experimental	Percent of theoretical
19.25	9.63	8.79	91.3
38.50	19.25	18.48	96.0
57.75	28.88	27.05	93.7
77.00	38.50	37.05	96.2

SOURCE: Osborne, FH, Arsenite oxidation by a soil isolate of *Alcaligenes*, PhD thesis, Rensselaer Polytechnic Institute, Troy, NY, 1973. With permission.

shown to oxidize arsenite stoichiometrically to arsenate (Table 15.1)*:

$$AsO_2^- + H_2O + 0.5O_2 \rightarrow AsO_4^{3-} + 2H^+ \quad (15.1)$$

Inhibitor and spectrophotometric studies suggested that the oxidation involved an oxidoreductase with a bound flavin that passed electrons from arsenite to oxygen by way of a c-type cytochrome and cytochrome oxidase (Osborne, 1973; Osborne and Ehrlich, 1976).

Anderson et al. (1992) isolated an inducible arsenite-oxidizing enzyme that was located on the outer surface of the plasma membrane of *A. faecalis* strain NCIB 8687. The enzyme location suggests that in intact cells of this organism, arsenite oxidation occurs in the periplasmic space. Biochemical characterization showed that this enzyme is a molybdenum-containing hydroxylase consisting of a monomeric 85 kDa peptide with a pterin cofactor and one atom of molybdenum, five or six atoms of iron, and inorganic sulfide. Both azurin and cytochrome c from *A. faecalis* served as electron acceptors in arsenite oxidation catalyzed by this enzyme. A more detailed discussion of arsenite oxidase and its phylogenetic distribution can be found in Silver and Phung (2005), and a larger compendium devoted to the broad subject of microbial

---

* The formula for meta-arsenite is used in Equation 15.1 although ortho-arsenite is the form of As(III) in aqueous solution according to Cullen and Reimer (1989). This equation as shown conveys that, if unbuffered, the reaction mixture turns acid as the oxidation progresses. The increase in acidity results from the fact that arsenious acid is a much weaker acid ($K = 10^{-9.2}$) than arsenic acid ($K_1 = 10^{-2.25}$; $K_2 = 10^{-6.77}$; $K_3 = 10^{-11.4}$) (Weast and Astle, 1982).

arsenite oxidation recently appeared in book edited by Santini and Ward (2012).

A strain of *A. faecalis* similar to the one isolated by Osborne (1973) (see also Osborne and Ehrlich, 1976) was independently isolated and characterized by Philips and Taylor (1976). Neither their strain nor that of Osborne oxidized arsenite strongly until late exponential or stationary phase of growth was reached in batch culture (Philips and Taylor, 1976; Ehrlich, 1978). Other heterotrophic arsenite-oxidizing bacteria that have been identified more recently include *Pseudomonas putida* strain 18 and *Alcaligenes eutrophus* strain 280, both of which were isolated from gold–arsenic deposits (Abdrashitova et al., 1981) and a strain that appears to belong to the genus *Zoogloea* (Weeger et al., 1999).

The observation reported by Osborne and Ehrlich (1976) that their strain passes electrons from arsenite to oxygen via an electron transport system that involves c-type cytochrome and cytochrome oxidase suggested that their organism is able to conserve energy from this oxidation. Indeed, indirect evidence indicated that the organism may be able to derive maintenance energy from arsenite oxidation (Ehrlich, 1978). Donahoe-Christiansen et al. (2004) found that a strain of *Hydrogenobaculum*, an obligate chemolithoautotroph using H_2 as sole energy source, was able to oxidize arsenite but not use it as a sole energy source. They had isolated their culture from a geothermal spring in Yellowstone National Park. Arsenite-oxidizing bacteria were also found in Hot Creek, California, by Salmassi et al. (2002, 2006). Salmassi et al. (2002) concluded that arsenite oxidation by their organism, *Agrobacterium albertimagni*, from Hot Creek was a detoxification process, whereas Donahoe-Christiansen et al. (2004) concluded that arsenite oxidation by *Hydrogenobaculum* was not by means of detoxification because this organism was more sensitive to As(V) than As(III).

Strong evidence that arsenite can be used as a sole energy source by some arsenite-oxidizing bacteria was presented by Ilyaletdinov and Abdrashitova (1981), who isolated a culture from a gold–arsenic ore deposit and named it *Pseudomonas arsenitoxidans*. This culture was an obligate autotroph that grew on arsenite as sole energy source. It also oxidized arsenic in arsenopyrite. Santini et al. (2000) reported the isolation of two new strains of autotrophic arsenite oxidizers, N-25 and N-26, from a gold mine in the Northern Territory of Australia. Both strains belong to the *Agrobacterium–Rhizobium*

branch of the α-Proteobacteria. Growth of strain N-26 was accelerated by the addition of a trace of yeast extract to an arsenite–mineral salts growth medium. In addition to growing on arsenite as energy source, strain N-26 was also able to grow heterotrophically on a wide range of organic carbon or energy sources including glucose, fructose, succinate, fumarate, and pyruvate. This strain thus, appears to be facultatively autotrophic, being also able to grow mixotrophically and heterotrophically. The arsenite oxidase of strain N-26 appears to be periplasmic. The oxidase was shown to have a molecular mass of 219 kDa and consist of two heterologous subunits, AroA (98 kDa) and AroB (14 kDa) in an $\alpha_2\beta_2$ configuration containing 2 Mo and 9 or 10 Fe atoms per $\alpha_2\beta_2$ unit. Gene sequence analysis revealed similarities when compared with the arsenite oxidase of *A. faecalis* (Santini and vanden Hoven, 2004). To avoid confusion with the multitude of terms to designate arsenite oxidase (e.g., Aro, Aso, and Aox), a new nomenclature designate of Aio was agreed upon, with AioB and AioA referring to the small and large (molybdenum containing) subunits, respectively (Lett et al., 2012).

More recently, an arsenite-oxidizing chemolithotroph, phylogenetically related to *Acidicaldus*, was isolated from a microbial mat in an acid-sulfate-chloride spring in Yellowstone National Park, United States (D'Imperio et al., 2007). The ability of this organism to oxidize arsenite was inhibited by H_2S, which affected its distribution in the spring. The inhibition appeared to be noncompetitive. Updated tabulations of all the currently known species of aerobic arsenite-oxidizing bacteria, both autotrophic and heterotrophic, have been assembled by Osborne and Santini (2012) and Cavalca et al. (2013). The former authors listed some 41 individual species for which arsenite oxidation was observed in physiological tests as well as by the presence of *aio* genes.

Arsenite (As(III)) may under some conditions also be subject to abiotic oxidation by Mn(IV), but apparently to a much lesser extent or not at all by Fe(III) (Oscarson et al., 1981).

15.2.4.2 Anaerobic Oxidation of Dissolved Arsenic

Oremland et al. (2002) isolated a facultative chemoautotrophic bacterium, strain MLHE-1, from arsenite-enriched bottom water from Mono Lake, California, United States, that oxidized arsenite anaerobically to arsenate using nitrate as terminal electron acceptor. In laboratory culture experiments, H_2 and sulfide were able to replace arsenite as electron donor in a nitrate–mineral salts medium. This organism was shown to possess the gene for ribulose-1,5-bisphosphate carboxylase/oxidase consistent with the observation that it was able to fix CO_2 in the dark. It was also able to grow heterotrophically with acetate as carbon and energy sources and oxygen (aerobic growth) or nitrate (anaerobic growth) as terminal electron acceptors. Phylogenetic analysis based on its 16S rDNA places this organism with the haloalkaliphilic Ectothiorhodospira of the γ-Proteobacteria. It was formally named Alkalilimnicola ehrlichii to honor Professor H.L. Ehrlich's many contributions to the field of geomicrobiology (Hoeft et al., 2007).

Subsequent genomic analysis of strain MLHE-1 provided an unexpected surprise. It was discovered that this organism lacked homologs of aio genes for arsenite oxidation, but instead harbored distant homologs of respiratory arsenate reductase (arr). This was despite the fact that it lacked the ability to grow via arsenate respiration. A reverse functionality for arsenate reductase in vivo for strain MLHE-1 was postulated and subsequently proven to be the case (Richey et al., 2009). These authors also reported an in vitro reverse functionality for the arsenate reductases of Shewanella sp. strain ANA-3 and Alkaliphilus oremlandii. Further investigations with insertion mutations of strain MLHE-1 demonstrated the necessity of one of the annotated arrA gene homologs for chemoautotrophic growth on arsenite and nitrate (Zargar et al., 2010). Its close proximity to the family of respiratory arsenate reductases (arrAB) as opposed to that of aerobic arsenite oxidases (aio) prompted a relabeling of this novel gene as "arx." ArxA was proposed as a new clade in the family of molybdenum oxidoreductases (Zargar et al., 2012). This claim was underscored by the discovery of arsenite oxidation in photosynthetic bacteria (Buddinhoff and Hollibaugh, 2008; Kulp et al., 2008). The isolate taken from Mono Lake, Ectothiorhodospira sp. strain PHS-1, was able to employ arsenite as well as sulfide as electron donors for anoxygenic photosynthesis. Its genome lacked aio genes but arxA was inducible upon photosynthetic growth with arsenite. These microorganisms were found to dominate the microbial population of red-colored biofilms lining hot spring seepages on Paoha Island in Mono Lake (Hoeft et al., 2010).

The phenomenon of anaerobic arsenite oxidation linked to nitrate has been shown to also occur in soils (Rhine et al., 2006; Rodríguez-Freire et al., 2012) as well as in sludge and sediments from sewerage treatment systems (Sun et al., 2008, 2010a). When combined with ferrous iron oxidation, the phenomenon has been proposed as a means of removing both nitrate and arsenic from treated wastewaters (Sun et al., 2009). Several novel species of facultative chemoautotrophic arsenite-oxidizing denitrifiers were isolated from soils (Rhine et al., 2006; Garcia-Dominguez et al., 2008). The amino acid sequences derived from the arsenite oxidase genes (aioBA) of these autotrophs were substantially different than those of many heterotrophic arsenite oxidizers so as to form as distinct phylogenetic group (Rhine et al., 2007). When compared with the previously characterized α-Proteobacterium strain NT-26 (Santini et al., 2000), only a 72%–74% sequence homology was found, suggesting a greater diversity in the realm of arsenite oxidation, aerobic or anaerobic, than had been thought possible. Indeed, a number of enrichment and pure cultures (α-, β-, and γ-Proteobacteria) were established that could link their growth to the chemoautotrophic oxidation of arsenite with chlorate as the electron acceptor and that possess the aio gene (Sun et al., 2010b). A selenate-dependent oxidation of arsenite was observed for strain ML-SRAO isolated from Mono Lake, although the oxidation did not couple with energy conservation (Fisher and Hollibaugh, 2008).

15.2.5 Interaction with Arsenic-Containing Minerals

Arsenic in arsenic-containing minerals that also contain iron, copper, and sulfur may be mobilized by bacteria. This mobilization may or may not involve bacterial oxidation of the arsenic if it exists in reduced form in the mineral. The simplest of these compounds, orpiment (As_2O_3), was found to be oxidized by Thiobacillus (now renamed Acidithiobacillus) ferrooxidans TM in a mineral salts solution (9K medium without iron; for formulation, see Silverman and Lundgren, 1959) to which the mineral had been added in pulverized form (Ehrlich, 1963) (Figure 15.1). Arsenite and arsenate accumulated in the medium over time. Because T. ferrooxidans does not oxidize arsenite to

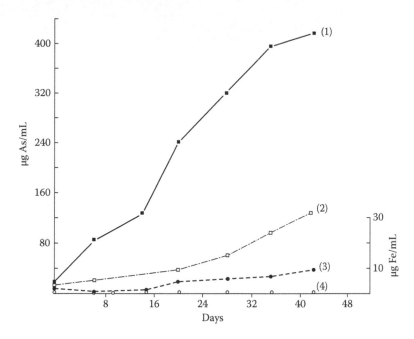

Figure 15.1. Bacterial solubilization of orpiment. (1) Total arsenic mobilized with bacteria (T. *ferrooxidans*), (2) total arsenic mobilized without bacteria, (3) total iron mobilized with bacteria (T. *ferrooxidans*), (4) total iron mobilized without bacteria. (From Ehrlich, HL, *Econ Geol*, 58, 991, 1963, Fig. 1. With permission.)

arsenate, the arsenate that appeared in the inoculated flasks probably resulted from the autoxidation of arsenite produced by this organism. Chemical oxidation by traces of ferric iron from the mineral may also have contributed to the formation of arsenate. The initial pH in the inoculated flasks was 3.5, which was dropped to 2 in 35 days. In contrast, in an uninoculated control in which orpiment autoxidized, but only slowly, the pH rose from 3.5 to 5, suggesting differences between the reactions in the experimental and control flasks. Realgar (As_2S_2) was not attacked by T. *ferrooxidans* TM.

Arsenopyrite (FeAsS) and enargite (Cu_3AsS_4) were also oxidized by an iron-oxidizing acidophilic *Thiobacillus* culture under the same test conditions as those used with orpiment (Ehrlich, 1964). During growth on arsenopyrite, the arsenic in the ore was transformed to arsenite and arsenate. The iron in the ore appeared ultimately as mobilized ferric iron. It precipitated extensively as ferric arsenite and arsenate. Because acidophilic T. *ferrooxidans* did not oxidize arsenite to arsenate, the arsenate in the culture was probably formed through the oxidation by ferric iron, although some autoxidation cannot be ruled out (Braddock et al., 1984; Wakao et al., 1988; Monroy-Fernandez et al., 1995). The pH dropped from 3.5 to 2.5 in both

the inoculated and uninoculated flasks in the first 21 days. However, in the uninoculated controls, it rose to 4.3 by the 40th day, whereas it remained at 2.5 in the inoculated flasks. Oxidation of arsenopyrite in the absence of bacteria was significantly slower than that in their presence (Figure 15.2). In a later study, Carlson et al. (1992) identified the minerals jarosite and scorodite among the solid products formed in the oxidation of arsenic-containing pyrite by a mixed culture of moderately thermophilic acidophiles. In another study, with T. *ferrooxidans* at 22°C and with a moderately thermophilic mixed culture at 45°C on arsenopyrite, Tuovinen et al. (1994) found that the mixed culture oxidized the mineral nearly completely, whereas acidophilic T. *ferrooxidans* did not. Jarosite was a major sink for ferric iron in both cultures, but some amorphous ferric arsenate and S^0 were also formed. The absence of scorodite formation in this case was attributed to an insufficient drop in pH in the experiments. Groudeva et al. (1986) and Ngubane and Baecker (1990) reported that the extremely thermophilic archaeon *Sulfolobus* sp. oxidizes arsenopyrite.

Although acidophilic T. *ferrooxidans* appears not to be able to oxidize arsenite to arsenate, the thermophilic archaeon *Sulfolobus acidocaldarius* strain BC is able to do so (Sehlin and Lindström, 1992).

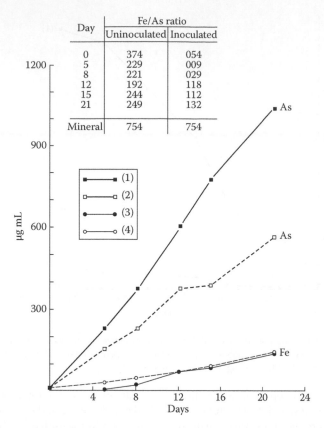

Figure 15.2. Oxidation of arsenopyrite by *T. ferrooxidans*. Curves 1 and 3 represent changes in inoculated reaction mixture in flasks. Curves 2 and 4 represent changes in uninoculated reaction mixture in flasks. (From Ehrlich, HL, *Econ Geol*, 1308, 1964, Fig. 1. With permission.)

Exposure of sterilized arsenopyrite particles for ~2 months in an underground area of the Richmond Mine at Iron Mountain, Northern California (see Figure 20.5), revealed a very extensive surface attachment of rod-shaped bacteria, probably *Sulfobacillus* sp., accompanied by extensive pitting of the mineral particles and the deposition of elemental sulfur on the particle surface (Figure 15.3). The S^0 represented more than 50% of the sulfur in the arsenopyrite.

Ehrlich (1964) found that a strain of *T. ferrooxidans*, when attacking enargite (Cu_3AsS_4), released cupric copper and arsenate into the bulk phase together with some iron that was present as an impurity in the mineral. The culture medium in which he performed the experiments contained (in g L^{-1}) ($NH_4)_2SO_4$, 3.0; KCl, 0.1; K_2HPO_4, 0.5; $MgSO_4 \cdot 7H_2O$, 0.5; and $Ca(NO_3)_2$, 0.01 and was acidified to pH 3.5. To this salt solution, 0.5 g enargite ground to a mesh size of +63 μm was added. He found that in the presence of active bacteria, the pH of the reaction mixture dropped from 3.5

to between 2 and 2.5. In the absence of bacteria, the pH usually rose to 4.5. Some mobilized copper and arsenic precipitated in the experiments. The rate of enargite oxidation without bacteria was significantly slower and may have followed a different course of reaction on the basis of slower rates of Cu and As solubilization and the difference in pH changes.

Escobar et al. (1997) reported on the action of *T.* (now *Acidithiobacillus*) *ferrooxidans* strain ATCC 19859 on enargite, ground to a mesh size of −147 + 104 μm and added to a salt solution containing (in g L^{-1}) ($NH_4)_2SO_4$, 0.4; $MgSO_4 \cdot 7H_2O$, 0.4; and KH_2PO_4, 0.056, acidified to pH 1.6. In the absence of added ferric iron, the mobilization of copper in the enargite was more rapid and sustained in the presence of the bacteria than in the sterile medium. In the presence of ferric iron (3.0 g L^{-1}), the mobilization of copper in the enargite by the bacteria was more rapid and sustained than that without added ferric iron. It was also more rapid than that in sterile medium with ferric iron. Copper mobilization

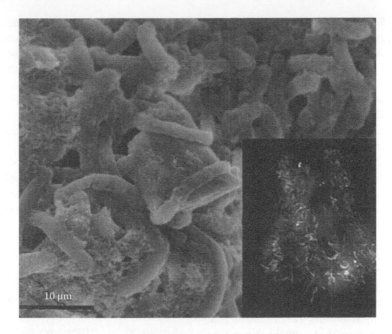

Figure 15.3. Arsenopyrite particle with bacteria colonizing its surface. The particle was sterilized and exposed in an underground stream in the Richmond Mine at Iron Mountain, Northern California, for approximately 2 months (see Figure 20.5) prior to examination by scanning electron microscopy (SEM). The image reveals a highly pitted surface covered with elongate cells and some secondary minerals. Based on microbial ecological studies at the site, the cells are probably *Sulfobacillus* sp. Insert shows an epifluorescence optical microscopic image in which the cells on an arsenopyrite grain are labeled with a fluorescent dye that binds to DNA. Parallel experimental studies by Molly McGuire (see McGuire et al., 2001) demonstrated that ~50% of the sulfur in the arsenopyrite ended up as elemental sulfur on the surface. (Images courtesy of Jill Banfield, University of California at Berkeley.)

without bacteria but with added ferric iron was significantly faster than that without ferric iron and the bacteria, but in both of these instances, much less sustained than with the bacteria.

The ability of T. (*Acidithiobacillus*) *ferrooxidans* and other acidophilic, iron-oxidizing bacteria to oxidize arsenopyrite is now being industrially exploited in beneficiation of sulfidic ores containing precious metals (Pol'kin and Tanzhnyanskaya, 1968; Pol'kin et al., 1973; Livesey-Goldblatt et al., 1983; Olson, 1994; Dew et al., 1997; Brierley and Brierley, 1999). This process involves the selective removal of arsenopyrite as well as pyrite by bacterial oxidation (bioleaching) before gold or silver is chemically (nonbiologically) extracted by cyanidation (complexing with cyanide). Arsenopyrite and pyrite interfere with cyanidation for several reasons. They encapsulate finely disseminated gold and silver, blocking access of cyanide reagent used to solubilize gold and silver. They consume oxygen that is needed to convert metallic gold to aurous (Au^{2+}) and auric (Au^{3+}) gold before it can be complexed by cyanide. Finally, the oxidation products of arsenopyrite and pyrite consume cyanide by forming thiocyanate and iron–cyanide complexes from which cyanide cannot be regenerated (Livesey-Goldblatt et al., 1983; Karavaiko et al., 1986). This increases the amount of cyanide reagent required for precious metal recovery.

As of 1997, Chile began to consider microbial removal of enargite from sulfidic gold ores (Acevedo et al., 1997; Escobar et al., 1997).

Beneficiation of carbonaceous gold ores containing as much as 6% arsenic by the action of T. *ferrooxidans* was reported by Kamalov et al. (1973). They succeeded in removing as much as 90% of the arsenic in the ore in 10 days under some circumstances. In another study, it was found to be possible to accelerate leaching of arsenic from a copper–tin–arsenic concentrate by six- to sevenfold, using an adapted strain of T. *ferrooxidans* (Kulibakin and Laptev, 1971).

The previously cited bioleaching processes can result in the release of the mineral-associated, solid-phase arsenite (i.e., As(III)) that is then oxidized to water-soluble arsenate. This causes an arsenic contamination problem if the processing fluids are to be released untreated into the environment. Sequestration of this arsenic into an insoluble precipitate, followed by its collection and responsible removal that would prevent pollution, would therefore be desirable. It would be beneficial to the overall liability of the mining/processing operation as well as to the human health of the people inhabiting the mining region. Scorodite ($FeAs_4 \cdot 2H_2O$) is an insoluble arsenic mineral that can be generated by biological/chemical treatments conducted at low pH. Biological treatment process employing acidophilic, iron-oxidizing archaea like *Acidianus sulfidivorans* (Gonzales-Contreras et al., 2010) and *Sulfolobales*-dominated consortia (Gonzales-Contreras et al., 2012) has demonstrated efficient crystallization of arsenate into scorodite as a promising means to eliminate arsenic release by the biotreatment. It should be borne in mind, however, that the As(V) sequestered in scorodite can be accessed by anaerobes like *Desulfotomaculum auripigmentum* (Newman et al., 1997b) that makes the precise mode of disposal, such as using sealed canisters in landfills, an important consideration.

15.2.6 Microbial Reduction of Oxidized Arsenic Species

Some bacteria, fungi, and algae are able to reduce arsenic compounds. One of the first reports on arsenite reduction involves fungi (Gosio, 1897). Although originally the product of this reduction was thought to be diethylarsine (Bignelli, 1901), it was later shown to be trimethylarsine (Challenger et al., 1933, 1951). A bacterium from cattle-dipping tanks that reduced arsenate to arsenite was also described early in the twentieth century (Green, 1918). Much more recently, it was reported that a strain of *Chlorella* was able to reduce a part of the arsenate it absorbs from the medium to arsenite (Blasco et al., 1971). Woolfolk and Whiteley (1962) demonstrated arsenate reduction to arsenite by H_2 with cell extracts of *Micrococcus* (*Veillonella*) *lactilyticus* and intact cells of *M. aerogenes*. The enzyme hydrogenase was involved. Arsine (AsH_3) was not formed in these

reactions. *Pseudomonas* sp. and *Alcaligenes* sp. have been reported to be able to reduce arsenate and arsenite anaerobically to arsine (Cheng and Focht, 1979).

Intact cells and cell extracts of the strict anaerobe *Methanobacterium* M.o.H. were shown to produce dimethylarsine from arsenate with H_2 (McBride and Wolfe, 1971). The extracts of this organism used methylcobalamin (CH_3B_{12}) as methyl donor. The reaction required consumption of ATP. The compounds 5-CH_3-tetrahydrofolate and serine could not replace CH_3B_{12}, although CO_2 could when tested by isotopic tracer technique. The reaction sequence as described by McBride and Wolfe (1971) was

$$AsO_4^{3-} \xrightarrow{2e} AsO_3^{3-} \xrightarrow{B_{12}CH_3}$$

$$\underset{\substack{\text{Methylarsonic} \\ \text{acid}}}{CH_3-\overset{\overset{\displaystyle O}{\|}}{\underset{\underset{\displaystyle OH}{|}}{As}}-OH} \xrightarrow{B_{12}CH_3} \underset{\substack{\text{Dimethylarsinic} \\ \text{acid}}}{CH_3-\overset{\overset{\displaystyle O}{\|}}{\underset{\underset{\displaystyle CH_3}{|}}{As}}-OH} \longrightarrow \underset{\substack{\text{Dimethylarsine}}}{CH_3-\overset{}{\underset{\underset{\displaystyle CH_3}{|}}{As}}-H}$$

$$(15.2)$$

In an excess of arsenite, methylarsonic acid was the final product, the supply of CH_3B_{12} being limiting. In an excess of CH_3B_{12}, a second methylation step yielding dimethylarsinic acid (cacodylic acid) followed the first. The last step was shown to occur in the absence of CH_3B_{12}. All of these steps were enzymatic. Reaction sequence (Equation 15.2) was later found to be more complex (see later), and the natural methyl donor in *Methanobacterium* M.o.H (now *M. bryantii*) was shown to be methyl coenzyme M (see discussion by Cullen and Reimer, 1989). Cell extracts of *Desulfovibrio vulgaris* were also found to produce a volatile arsenic derivative—presumably an arsine (McBride and Wolfe, 1971).

The fungus *Scopulariopsis brevicaulis* is able to convert arsenite to trimethylarsine. The reaction sequence originally proposed by Challenger (1951) has been modified and includes the following steps (Cullen and Reimer, 1989, p. 720):

$$As^{III}O_3^{3-} \xrightarrow{RCH_3} \underset{\substack{\text{Methylarsonic} \\ \text{acid}}}{CH_3As^VO(OH)_2} \xrightarrow{2e^-} \underset{\substack{\text{Methylarsinous} \\ \text{acid}}}{CH_3As^{III}(OH)_2} \xrightarrow{RCH_3} \underset{\substack{\text{Dimethylarsinic acid} \\ \text{or cacodylic acid}}}{(CH_3)_2As^VO(OH)}$$

$$\xrightarrow{2e} \underset{\substack{\text{Unknown} \\ \text{intermediate}}}{\{(CH_3)_2As^{III}OH\}} \xrightarrow{RCH_3} \underset{\substack{\text{Trimethylarsine} \\ \text{oxide}}}{(CH_3)_3As^VO} \xrightarrow{2e} \underset{\substack{\text{Trimethylarsine}}}{(CH_3)_3As^{III}} \quad (15.3)$$

It should be noted that in this reaction sequence, the oxidation state of the arsenic changes successively from +3 to +5 to +3. The reaction sequence leading to dimethylarsinic acid is similar to that observed in arsenic detoxification by methylation in human beings (Aposhian et al., 2000; Petrick et al., 2000). The methyl donors (RCH_3) in Reaction 15.3 can be a form of methionine (see Cullen and Reimer, 1989). Besides S. brevicaulis, other fungi such as Aspergillus, Mucor, Fusarium, Paecilomyces, and C. humicola have also been found to be active in such reductions (Cox and Alexander, 1973; Alexander, 1977; Pickett et al., 1981). Trimethylarsine oxide was found to be an intermediate in trimethylarsine formation by C. humicola (Pickett et al., 1981). A review of biomethylation of arsenic can be found in Bentley and Chasteen (2002). The reduction of arsenate and arsenite to volatile arsines is a means of detoxification. It has also been observed in phototrophs, such as Rhodopseudomonas palustris, which can form trimethylarsine from As(III) via the arsenic resistance gene arsM (Qin et al., 2006). The eukaryotic, thermophilic alga Cyanidioschyzon sp. isolate 5508 from a Yellowstone hot spring mat is surprisingly flexible in its metabolism of As(III) (Qin et al., 2009). Depending on how cultivated, from added As(III), it can form trimethylarsine oxide, dimethylarsenate, oxidize As(III) to As(V), and form trimethylarsine gas. While these reactions are involved in arsenic resistance rather than energy conservation, they do imply the importance of this abundant phototroph in the biogeochemical cycling of arsenic in acidic hot spring outflows of Yellowstone National Park.

15.2.7 Arsenic Respiration

Some bacteria reduce arsenate only as far as arsenite. These organisms use arsenate as terminal electron acceptor in anaerobic respiration. A number of them have been characterized (Table 15.2) (Stolz and Oremland, 1999; Oremland and Stolz, 2000), and they include the members of the domains Bacteria and Archaea. Ahmann et al. (1994) isolated a comma-shaped motile rod, strain MIT-13, from arsenic-contaminated sediment from the Aberjona watershed in Massachusetts, United States. This organism was later named Sulfurospirillum arsenophilum. It belongs to the domain Bacteria (Stolz et al., 1999). Anaerobically growing cultures of S. arsenophilum MIT-13 respired on arsenate using lactate but not acetate as electron donor and produced stoichiometric amounts of arsenite.

Chrysiogenes arsenatis, isolated from gold mine wastewater in Australia, is the only arsenate respirer isolated so far that is capable of using acetate as electron donor (Macy et al., 1996). Its arsenate reductase, a periplasmic enzyme, has been analyzed and found to consist of two subunits. One of these subunits has a molecular mass of 87 kDa, and the other

TABLE 15.2
Arsenate-respiring bacteria and archaea.

Organism	References
Bacillus arsenicoselenatis	Switzer Blum et al. (1998)
Bacillus selenitireducens	Switzer Blum et al. (1998) and Oremland et al. (2000)
Chrysiogenes arsenatis	Macy et al. (1996) and Krafft and Macy (1998)
D. auripigmentum	Newman et al. (1997a,b)
Desulfomicrobium strain Ben-RB	Macy et al. (2000)
Pyrobaculum aerophilum[a]	Huber et al. (2000)
Pyrobaculum arsenaticum[a]	Huber et al. (2000)
Shewanella species strain ANA-3	Saltikov et al. (2003)
Sulfurihydrogenibium subterraneum strain HGMK1[T]	Takai et al. (2003)
Sulfurospirillum arsenophilum	Stolz et al. (1999)
Sulfurospirillum barnesii	Laverman et al. (1995) and Zobrist et al. (2000)
Strain, MLMS-1	Hoeft et al. (2004)

[a] Pyrobaculum belongs to the domain archaea.

29 kDa. The enzyme structure includes Zn, Fe, Mo, and acid-labile S (Krafft and Macy, 1998).

Sulfurospirillum barnesii strain SES-3 is another member of the domain Bacteria that can respire anaerobically on lactate using arsenate to arsenite reduction for terminal electron disposal (Laverman et al., 1995). Besides arsenate, it can also use selenate (see Chapter 22), Fe(III), thiosulfate, elemental sulfur, Mn(IV), nitrate, nitrite, trimethylamine N-oxide, and fumarate as terminal electron acceptors (Oremland et al., 1994). It can also grow microaerophilically, albeit sluggishly (15% O_2, optimum) (Laverman et al., 1995).

Newman et al. (1997a,b) isolated *D. auripigmentum* strain OREX-4 from sediment of Upper Mystic Lake, Massachusetts, United States. It is a grampositive rod belonging to the domain Bacteria. It anaerobically reduces arsenate to arsenite and sulfate to sulfide with lactate as electron donor. When both arsenate and sulfate are present together, this organism precipitates orpiment (As_2S_3). The reactions can be summarized as follows (Newman et al., 1997b):

$$CH_3CHOHCOO^- + 2HAsO_4^- + 4H^+ \rightarrow$$

$$CH_3COO^- + 2HAsO_2 + CO_2 + 3H_2O$$

$$\Delta G^{0'} = -41.1 \, \text{kcal mol}^{-1} \, \text{or} -172 \, \text{kJ mol}^{-1} \, \text{lactate}$$
$$(15.4)$$

$$CH_3CHOHCOO^- + 0.5SO_4^{2-} + 0.5H^+ \rightarrow$$

$$CH_3COO^- + 0.5HS^- + CO_2 + H_2O$$

$$(\Delta G^{0'} = -21.3 \, \text{kcal mol}^{-1} \, \text{or} -89 \, \text{kJ mol}^{-1} \, \text{lactate})$$
$$(15.5)$$

$$2HAsO_2 + 3HS^- + 3H^+ \rightarrow As_2S_3 + 4H_2O \quad (15.6)$$

Reactions 15.4 and 15.5 are catalyzed by the organism, whereas Reaction 15.6 is not.

The microbial reduction of arsenate and sulfate when both are present together is sequential at rates that avoid the accumulation of excess sulfide. Orpiment did not form when arsenate and sulfide were added to the culture medium in the absence of the bacteria. Maintenance of a pH of ~6.8 was essential for orpiment stability.

Shewanella sp. strain HN-41 was recently shown to form arsenic-sulfide nanotubes (20–100 nm × 30 µm) when growing anaerobically in a lactate medium with a mixture of thiosulfate and As(V) as electron acceptors at circumneutral pH. The nanotubes initially consisted of amorphous As_2S_3 but with continued incubation developed polycrystalline phases of realgar (AsS) and duranusite (As_4S). Mature nanotubes exhibited metallic properties in terms of electrical and photoconductive properties (Lee et al., 2007).

Macy et al. (2000) isolated two sulfate reducers belonging to the domain Bacteria that had the capacity to reduce arsenate to arsenite when lactate was used as the source of reductant. The lactate was oxidized to acetate. One of these organisms was *Desulfomicrobium* strain Ben-RB, which was able to reduce arsenate and sulfate concurrently and respired on arsenate as sole terminal electron acceptor in the absence of sulfate. The other strain was *Desulfovibrio* strain Ben-RA, which was also able to reduce arsenate and sulfate concurrently. However, it did not grow with arsenate as sole terminal electron acceptor. In this organism, arsenate reduction may be part of an arsenate-resistance mechanism (see Silver, 1998).

Switzer Blum et al. (1998) discovered two species of the genus *Bacillus* in Mono Lake, California, United States, which were able to reduce not only arsenate but also selenium oxyanions (see Chapter 22). They are *Bacillus arsenicoselenatis* (now renamed *B. arseniciselenatis*) and *B. selenitireducens* belonging to the domain Bacteria. Both organisms are moderate halophiles and alkaliphiles, growing maximally at pH between 9 and 11. Each reduces arsenate to arsenite.

Saltikov et al. (2003) isolated *Shewanella* sp. strain ANA-3 from an arsenic-treated wooden pier located in Eel Pond, Woods Hole, Massachusetts, which was found to respire anaerobically on lactate using arsenate as terminal electron acceptor. The arsenate could be replaced by soluble ferric iron, oxides of iron and manganese, nitrate, fumarate, thiosulfate, and the humic acid functional analog 2,6-anthraquinone disulfonate as terminal electron acceptors. This organism is also able to respire using oxygen. Like other arsenate respirers, it produces arsenite from the arsenate it consumes. Besides reducing arsenate anaerobically, it is also able to reduce it aerobically to arsenite. It is resistant to arsenite concentrations as high as 10 mM aerobically and anaerobically, enabled by an *ars* operon—*arsDABC*. Although the *ars* operon is not required for arsenate respiration, its presence in the genome enhances the strain's growth rate and cell yield on lactate. The arsenate reductase of the strain is encoded in two gene clusters—arrAB. The arrA cluster codes for a 95.2 kDa molybdoprotein, and the arrB cluster codes for a 25.7 kDa iron–sulfur protein (Saltikov and Newman, 2003). Malasarn et al. (2008) determined kinetic properties of this enzyme.

The arrA gene has been shown to be a reliable marker for As(V) respiration in bacteria (Malasarn et al., 2004). The arsenate-resistance operon, ars, is expressed both aerobically and anaerobically, whereas the arsenate reductase gene cluster, arrAB, is expressed only anaerobically and is repressed in the presence of oxygen and nitrate. Arsenate respiration and resistance are both induced specifically by arsenite. Arsenate respiration is induced at an arsenite concentration 1000 times lower than arsenate resistance. Arsenate respiration but not arsenite resistance is inducible by low micromolar concentrations of arsenate (Saltikov et al., 2005).

Takai et al. (2003a) described a new member of the Aquificales in the domain Bacteria, isolated from the subsurface hot aquifer water of the Hishikari gold mine, Kagoshima Prefecture, Japan, which was capable of reducing arsenate to arsenite. They named the organism *Sulfurihydrogenibium subterraneum* strain HGMK1^T. The organism, which is a motile, straight to slightly curved rod ($1.5–2.5 \times 0.3–0.5$ μm) is a thermophile with optimum growth at 65°C in a pH range of 6.4–8.8. It grows chemolithoautotrophically using H_2, S^0, or $S_2O_3^{2-}$ as an energy source and O_2, NO_3^-, Fe(III), SeO_4^{2-}, and SeO_3^{2-} besides As(V) as alternative electron acceptors. Nakagawa et al. (2005) subsequently found that this organism is facultatively heterotrophic, being able to use acetate as an organic carbon source. A comparison of S. subterraneum HGMK1^T with S. azorense Az-Fu1^T and S. yellowstonense SS-S^T, of which S. yellowstonense cannot use As(V) as terminal electron acceptor, has been presented by Nakagawa et al. (2005).

Hoeft et al. (2004) isolated a bacterial strain, MLMS-1, from Mono Lake bottom water that was able to reduce arsenate to arsenite using sulfide as electron donor. This organism is a chemoautotrophic, gram-negative, motile curved rod belonging to the δ-Proteobacteria distantly related to Desulfobulbus. It assimilated CO_2 by an enzyme system that did not seem to involve ribulose-1,5-bisphosphate carboxylase/oxidase (Hoeft et al., 2004). Strain MLMS-1 is the first example of an obligate arsenate respirer, being incapable of using other electron acceptors to sustain its growth. It is also an obligate chemoautotroph, restricted in its metabolism to employing only sulfide or thiosulfate as electron donors to sustain arsenate respiration.

Huber et al. (2000) isolated a member of the domain Archaea, *Pyrobaculum arsenaticum*, from a hot spring at Pisciarelli Solfataras at Naples, Italy, capable of reducing arsenate as terminal electron acceptor using H_2 as electron donor. This organism is a thermophilic, facultative autotroph that reduces arsenate to arsenite in its anaerobic respiration. In addition to arsenate, it can also use thiosulfate and elemental sulfur as alternative electron acceptors for autotrophic growth. Organotrophically, this organism produces realgar (As_2S_2) when using arsenate and thiosulfate or arsenate and cysteine as terminal electron acceptors. It can also reduce selenate but not selenite organotrophically to elemental selenium in a respiratory process.

Another species of *Pyrobaculum*, P. aerophilum, was found to reduce arsenate and selenate as terminal electron acceptors in autotrophic growth on H_2 as electron donor (Huber et al., 2000). Organotrophically, this organism can use arsenate as terminal electron acceptor when growing anaerobically. However, it does not form a visible precipitate of realgar on arsenate and thiosulfate or arsenate and cysteine. Using a transcriptomic approach, Cozen et al. (2009) found evidence for the presence of an arsenate reductase that was a distant homolog either of that found in bacteria or of a bacterial polysulfide reductase. P. aerophilum can also use selenate and selenite as alternative electron acceptors. The selenium compounds are reduced to elemental selenium. A further discussion of respiratory arsenate reductase can be found in Silver and Phung (2005).

Despite the fact that deep-sea hydrothermal vents are sites of intense chemoautotrophic processes, little work has been done on such systems from the perspective of arsenic geomicrobiology. Nonetheless, a heterotrophic thermophilic anaerobe, *Deferribacter desulfuricans* (Takai et al., 2003b), was isolated that was capable of arsenate, nitrate, and sulfur respiration. Similarly, shallow-water hydrothermal marine vents can also release arsenic (Price et al., 2007) and a dissimilatory arsenate reducer, *Marinobacter santoriniensis*, was isolated from shallow hydrothermal systems in the seas lying off Greece (Handley et al., 2009).

15.2.8 Direct Observations of Arsenite Oxidation and Arsenate Reduction In Situ

Newman et al. (1998) reviewed environmental microbial arsenate respiration with emphasis on nonmarine systems, that is, freshwater as well as hypersaline systems. They compared and contrasted relevant physiological traits of four different respirers.

Oremland et al. (2000) developed extensive evidence of arsenate reduction in meromictic, alkaline, hypersaline Mono Lake, California, United States, which contains ~200 μM total arsenic. The oxidation state of the dissolved arsenic changed predominantly from +5 in the oxic surface waters (mixolimnion) to +3 in the anoxic bottom waters (monimolimnion). No significant bacterial reduction was noted in the oxic waters, but it was significant in the anoxic waters. A rate of ~5.9 μmol L^{-1} day^{-1} was measured at depths between 18 and 21 m. Sulfate reduction occurred at depths below 21 m; the highest rate of ~2.3 μmol L^{-1} day^{-1} was measured at a depth of 28 m. Oremland et al. (2000) estimated that arsenate respiration is second in importance after sulfate respiration in Mono Lake and may account for mineralization of 14.2% of the photosynthetically fixed carbon annually in the lake. Using an entirely different approach consisting of seasonal mass balances and water column arsenic speciation data, Hollibaugh et al. (2005) derived a similar conclusion. Oremland et al. (2002) subsequently demonstrated anaerobic oxidation of arsenite by a facultative, chemoautotrophic isolate, strain MLHE-1 (recently named *A. ehrlichii* by Hoeft et al., 2007), from Mono Lake using nitrate as terminal electron acceptor. The nitrate could be replaced by Fe(III) and Mn(IV), which also occur in Mono Lake at low concentrations (10 and 0.4 μM, respectively). Thus, in the anaerobic hypolimnion of Mono Lake, a bacterially promoted arsenic cycle seems to occur (Oremland et al., 2002, 2004). The microbial diversity with respect to the domain Bacteria at different depths in Mono Lake has been examined by Humayoun et al. (2003). They found greater diversity in the anoxic bottom waters of the lake than in the oxic surface waters, despite the presence of toxic chemical species including sulfide, arsenate, and arsenite. Microbial precipitation of arsenic sulfides was observed during incubation of sediments taken from salt flats (playas) of the arid regions of Chile (Demergasso et al., 2007). Further research found that arsenate-respiring bacteria could be successfully cultivated from two saline, arsenic-rich desert lakes located in Chile, Salar de Ascotán and Salar de Atacama (Lara et al., 2012).

Searles Lake in the Mojave Desert in California, ~270 km south–southeast of Mono Lake, featuring a salt-saturated, alkaline brine with an unusually elevated arsenic content (total As ~ 3.9 mM), is another lake studied by Oremland et al. (2005). They found that the brine contained traces of sulfide, methane, and ammonia, but they did not detect any nitrate. The sulfide and methane concentrations in the brine of Searles Lake were 0.1–0.01 of those in anoxic bottom waters of Mono Lake, and methane concentration in the sediment of Searles Lake was 0.1–0.01 of its concentration in sediments of Mono Lake. They found that in the sediment of Searles Lake, the oxidation state of arsenic changed from As(V) to As(III) with depth. In sediment slurries, As(V) reduction was stimulated by H_2 or sulfide under anaerobic conditions, whereas As(III) was rapidly oxidized under aerobic conditions. Some of the As(III) formed in the As(V) reduction formed thioarsenites, especially when sulfide was the electron donor. As(V)-reducing and As(III)-oxidizing activities were abolished when the sediment slurries were autoclaved. From the lake sediment, they isolated an anaerobic, extremely alkalihalophilic, facultatively lithotrophic As(V)-respiring bacterium, strain SLAS-1, which is a curved, motile rod. This organism oxidized lactate to acetate while reducing As(V) to As(III). It also could use H_2 or sulfide as electron donor for As(V) reduction. It fixed CO_2 when growing lithoautotrophically with sulfide as electron donor. Phylogenetically, strain SLAS-1 is affiliated with the Halanaerobacteriales and appears to represent a new species, named *Halarsenatibacter silvermanii* (Switzer Blum et al., 2009). The enrichment culture from which strain SLAS-1 was isolated also contained a sulfate reducer, designated strain SLSR-1 from the delta-Proteobacteria (Switzer Blum et al., 2012). However, due to the high salinity and borate content of the Searles Lake brine, sulfate reduction does not operate in situ therein as it does in Mono Lake (Kulp et al., 2006, 2007). The findings of the authors clearly show that a complete arsenic cycle is operative in the sediments of the extreme environment of Searles Lake.

Islam et al. (2004), using a microcosm approach, showed that metal-reducing bacteria played a key role in mobilizing arsenic in sediments from a contaminated aquifer in West Bengal. In this instance, soluble As(III) was generated only after Fe(III) was reduced to Fe(II) by the members of the Geobacteraceae in acetate-fed microcosms, indicating that As(V) bound to iron oxyhydroxides was reduced after its release due to bacterial Fe(III) reduction, but the agent responsible for the reduction of As(V) was not identified. A subsequent study by Islam et al. (2005) with *Geobacter sulfurreducens* showed that this organism was unable to reduce As(V) to As(III) with or without energy conservation. When respiring Fe(III) in a suitable culture medium in the presence of As(V), *G. sulfurreducens* generated vivianite ($Fe_3(PO_4)_2 \cdot 8H_2O$) accompanied by arsenic immobilization, 85% as As(V) and 15% as As(III). Similar arsenic removal occurred when the organism formed siderite ($FeCO_3$) under different culture conditions, with 50% of the immobilized arsenic as As(V) and 50% as As(III), or magnetite (Fe_3O_4) with 100% of the immobilized arsenic as As(V).

Johnson (1972) made direct observations relating to microbial reduction of arsenic in the ocean. He found that bacteria from phytoplankton samples in Narragansett Bay and from Sargasso Sea water were able to reduce added arsenate. An arsenate reduction rate of 10^{-11} µmol cell^{-1} min^{-1} was measured over a 12 h period of incubation at 20°C–22°C. Arsenic was not accumulated by the bacteria, and none was lost from the medium through volatilization. Scudlark and Johnson (1982) presented evidence of bacterial oxidation in aged seawater from Narragansett Bay and some other marine sites. Depending on the geographic location of the collected samples, initial oxidation rates ranged from 3 to 1280 nmol L^{-1} day^{-1}. These observations may help to explain why the observed As(V)/As(III) ratio in the seawater was 10^{-1}:10 instead of 10^{-26}:1, as predicted under equilibrium conditions in oxygenated seawater at pH 8.1 (Johnson, 1972; Andreae, 1979).

Cheng and Focht (1979) observed that bacteria are able to reduce arsenate and arsenite to arsine under anaerobic conditions in soils. They indicated that, unlike fungi, the bacteria produce mono- and dimethylarsine only when methylarsonate or dimethylarsinate is available.

Oxidation of arsenite to arsenate in activated sludge under aerobic conditions was reported by Myers et al. (1973). Under anaerobic conditions,

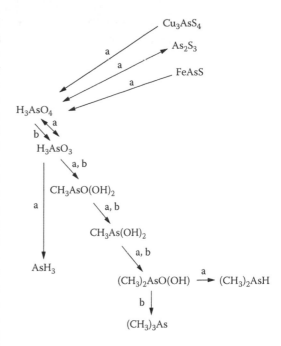

Figure 15.4. Summary of observed microbial interactions with arsenic compounds (a) performed by bacteria and (b) performed by fungi.

arsenate was found to be reduced stepwise to lower oxidation states over an extended period in the activated sludge. *Pseudomonas fluorescens* was an active reducer in this system under overall aerobic conditions (Myers et al., 1973). Reduction may actually have occurred in anaerobic microenvironments in the floc.

Figure 15.4 summarizes the reactions involving arsenic compounds in nature that are catalyzed by microorganisms. The oxidation of methylated arsine, although not indicated in this figure, has been suggested by Cheng and Focht (1979). The arsenic cycle in natural freshwaters has been discussed by Ferguson and Gavis (1972) and Scudlark and Johnson (1982).

15.3 ANTIMONY

15.3.1 Antimony Distribution in Earth's Crust

Antimony is a rare element. Its abundance in the Earth's crust when normalized to its abundances in chondritic stone meteorites and adjusted to Si equal to 1.00 has been given as 0.1 (Winchester, 1972). Its average concentration in igneous rocks is 0.2 ppm, in shales 1.5 ppm, in limestone 0.3 ppm, in sandstone 0.05 ppm, and in soil 1 ppm. Its average concentration in seawater is 0.124 ppb and in freshwater 0.2 ppb (Bowen, 1979). As a mineral, it may occur as stibnite (Sb_2S_3), kermesite (Sb_2S_2O), senarmontite (Sb_2O_3), jamesonite ($2PbS \cdot Sb_2S_3$), boulangerite ($5PbS \cdot 2Sb_2S_3$), and tetrahedrite ($Cu_{12}Sb_4S_{13}$). It may also occur as sulfantimonides of silver and nickel, and sometimes as elemental antimony (Gornitz, 1972). Stibnite is the most common antimony mineral.

Antimony can exist in the oxidation states -3, 0, $+3$, and $+5$. Like arsenic compounds, antimony compounds tend to be toxic for most living organisms. The basis for this toxicity has not been established.

Elevated antimony concentrations have been measured in soils surrounding mining sites and smelters dealing with antimony-containing ores, which fall in the range of 100–1489 mg kg^{-1} (Ainsworth et al., 1990; Li and Thornton, 1993). By using bacterial biosensors, the polluting antimony at five former British sites with levels up to 700 mg kg^{-1} was shown not to be bioavailable (Flynn et al., 2003).

15.3.2 Microbial Oxidation of Antimony Compounds

The report by Bryner et al. (1954) is among the earliest reports of the biooxidation of antimony-containing minerals in which the oxidation of tetrahedrite ($Cu_{12}Sb_4S_{13}$) by T. (now *Acidithiobacillus*) *ferrooxidans* is recorded. Lyalikova (1961) reported the oxidation of antimony trisulfide (Sb_2S_3) by T. *ferrooxidans*. In both of these instances, the oxidation proceeded under acidic conditions (pH 2.45) (Kuznetsov et al., 1963). The oxidation state of the solubilized antimony was not determined in either case. More recently, Silver and Torma (1974) reported on the oxidation of synthetic antimony sulfides by T. *ferrooxidans*. Torma and Gabra (1977) examined the oxidation of stibnite (Sb_2S_3) by the same organism. The latter authors suspected that T. *ferrooxidans* oxidized trivalent antimony (Sb(III)) to pentavalent antimony (Sb(V)) but offered no proof. Lyalikova et al. (1972) reported on the oxidation of Sb–Pb sulfides, Sb–Pb–Te sulfides, and Sb–Pb–As sulfides, resulting in the formation of minerals such as anglesite ($PbSO_4$) and valentinite (Sb_2O_3). The formation of valentinite suggests that antimony in the minerals was solubilized but not oxidized. Ehrlich (1986, 1988) reported on the oxidation of a mixed sulfide ore that included tetrahedrite by T. *ferrooxidans* strain 19759. Although

he followed mobilization of silver, copper, and zinc from the ore in these experiments, he did not follow the mobilization of antimony. In all the foregoing studies, most or all of the oxidation apparently involved sulfide in the minerals.

Trivalent antimony is susceptible to direct oxidation by *Stibiobacter senarmontii* (Lyalikova, 1972, 1974; Lyalikova et al., 1976). This organism, which was isolated from a sample of mine drainage in the former Yugoslavia, oxidizes senarmontite (Sb_2O_3) or Sb_2O_4 to Sb_2O_5, deriving useful energy from this process. It is a gram-positive rod (0.5–1.8 × 0.5 μm) with a single polar flagellum and has the ability to form rudimentary mycelia in certain stages of development. It grows at neutral pH and generates acid when oxidizing Sb_2O_3 (the pH can drop from 7.5 to 5.5). When grown on reduced antimony oxide (senarmontite), this organism possesses the enzyme ribulose bisphosphate carboxylase, indicating that it has chemolithotrophic propensity (Lyalikova et al., 1976). Antimony sulfide ores can thus, be oxidized to antimony pentoxide in the following two steps:

$$Sb_2S_3 \xrightarrow[(1)]{O_2} Sb_2O_3 \xrightarrow[(2)]{O_2} Sb_2O_5 \qquad (15.7)$$

The first step is catalyzed by an organism such as *Thiobacillus* Y or T. *ferrooxidans* (see [1] in Reaction 15.7) and the second by S. *senarmontii* (see [2] in Reaction 15.7) (Lyalikova et al., 1974).

Lehr et al. (2007) studied the oxidation of Sb(III) and As(III) by *A. tumefaciens* and by using knockout mutants reported that the pathways were not analogous, suggesting that a separate Sb-specific metabolic pathway exists in this bacterium. They also reported Sb(III) oxidation by a eukaryotic alga of the order *Cyanidiales*, although neither of these two organisms could conserve energy from these oxidations.

15.3.3 Microbial Reduction of Oxidized Antimony Minerals

In the previous edition (volume 5) of this book, there was a clear dearth of information about this topic, as reflected by the following prescient sentences: "No strong evidence exists to date that microbes are able to reduce oxidized antimony species as terminal electron acceptor in anaerobic respiration, as is the case for oxidized arsenic species" (see Iverson and Brinckman, 1978). However, bioreduction of Sb(V) to Sb(III) is suggested from antimony methylation studies (Bentley and Chasteen, 2002). This limited evidence should not be taken to mean that microbial antimony reduction is of rare occurrence, but rather that it has so far not been thoroughly studied. At the time of this writing, two preliminary reports have appeared that indicate that a substantive change is in the wind in this regard. First, Kulp et al. (2014) reported that anoxic stream sediments in Idaho polluted with arsenic and antimony mine runoff were able to reduce Sb(V) to Sb(III) and that this activity was abolished with autoclaving. Sediments that were amended with 5 mM Sb(V) were able to oxidize the methyl group of acetate to CO_2, while unamended sediments produced CH_4 via acetoclastic methanogenesis, thereby indicating that the former process was a dissimilatory reduction. Meanwhile, from a freshwater spring located on the shoreline of Mono Lake, California, Abin and Hollibaugh (2014) isolated an anaerobic bacillus that demonstrated clear, antimony-dependent growth that coupled lactate oxidation to the reduction of Sb(V) to Sb(III). Undoubtedly, this area of research will expand exponentially in the years to come.

15.4 SUMMARY

Although arsenic and antimony compounds are toxic to most forms of life, some microbes metabolize them. Arsenite has been found to be enzymatically oxidized by several bacteria. The enzyme system has been shown to be inducible in at least some microbes. In laboratory experiments, A. faecalis oxidized arsenite most intensely only after having gone through active growth. This organism probably can derive maintenance energy from arsenite oxidation. P. arsenitoxidans and two as yet unnamed isolates from Australia can grow autotrophically with arsenite as their sole source of energy. Simple and complex arsenic sulfides are oxidized by T. (now Acidithiobacillus) ferrooxidans. However, no evidence has been obtained that this organism can oxidize trivalent to pentavalent arsenic enzymatically.

Arsenate and arsenite have been shown to be reduced by certain bacteria and fungi. When used as terminal electron acceptor, arsenate is reduced to arsenite. In some forms of arsenic detoxification, bacteria can reduce arsenate or arsenite to

arsine or dimethylarsine, whereas fungi produce trimethylarsine. All of these arsines are volatile.

Antimony compounds have also been shown to be microbially oxidized. T. (now Acidithiobacillus) ferrooxidans has been shown to attack various antimony-containing sulfides. Although enzymatic oxidation by T. ferrooxidans of Sb(III) to Sb(V) has been claimed in the case of Sb_2S_3, clear proof is lacking. Generally, only the sulfide moiety and ferrous iron, if present, in an antimony mineral are oxidized by this organism. S. senarmontii, an autotroph that was isolated from an ore deposit in the former Yugoslavia, oxidizes trivalent antimony in Sb_2O_3 or Sb_2O_4 to pentavalent antimony (Sb_2O_5). Microbial reduction of oxidized antimony compounds was originally reported only in connection with antimony methylation, but there are now examples of antimonate serving as an electron acceptor for energy conservation by anaerobes.

REFERENCES

Abdrashitova SA, Mynbaeva BN, Ilyaletdinov AN. 1981. Oxidation of arsenic by the heterotrophic bacteria Pseudomonas putida and Alcaligenes eutrophus. Mikrobiologiya 50:41–45 (Engl transl pp. 28–31).

Abin CA, Hollibaugh, JT. 2014. Biosynthesis of antimony trioxide microcrystals by a novel dissimilatory antimonate-reducing bacterium. Environ Sci Technol 48:681–688.

Acevedo F, Gentina JC, Alegre C, Arévalo P. 1997. Biooxidation of a gold bearing enargite concentrate. In: International Biohydrometallurgy Symposium IBS97 Biomine 97, "Biotechnology Comes of Age." Conference Proceedings, Glenside, South Australia, Australia: Australian Mineral Foundation, pp. M3.2.1–M3.2.9.

Ahmann D, Roberts AL, Krumholz LR, Morel FMM. 1994. Microbe grows by reducing arsenic. Nature (London) 371:750.

Ainsworth N, Cooke JA, Johnson MS. 1990. Distribution of antimony in contaminated grassland: 1—Vegetation and soils. Environ Pollut 65:65–77.

Alexander M. 1977. Introduction to Soil Microbiology. New York: Wiley.

Anderson GL, Williams J, Hille R. 1992. The purification and characterization of arsenite oxidase from Alcaligenes faecalis, a molybdenum-containing hydroxylase. J Biol Chem 267:23674–23682.

Andreae MO. 1979. Arsenic speciation in seawater and interstitial waters: The influence of biological–chemical interactions on the chemistry of a trace element. Limnol Oceanogr 24:440–452.

Aposhian HV, Zheng B, Aposhian MM, Le XC, Cebrian ME, Cullen W, Zakharian RA et al. 2000. DMPS: Arsenic challenge test. II. Modulation of arsenic species, including monomethylarsonous acid (MMA^{III}), excreted in human urine. *Toxicol Appl Pharm* 165:74–83.

Baker-Austin C, Dopson M, Wexler M, Sawers RG, Stemmler A, Rosen BP, Bond PL. 2007. Extreme arsenic resistance by the acidophilic archaeon "*Ferroplasma acidarmanus*" Fer1. *Extremophiles* 11:425–434.

Basturea GN, Harris TK, Deutscher MP. 2012. Growth of a bacterium that apparently uses arsenic instead of phosphorus is a consequence of massive ribosome breakdown. *J Biol Chem* 287:28816–28819.

Bennett RL, Malamy MH. 1970. Arsenate-resistant mutants of *Escherichia coli* and phosphate transport. *Biochem Biophys Res Commun* 40:496–503.

Bentley R, Chasteen TG. 2002. Microbial methylation of metalloids: Arsenic, antimony, and bismuth. *Microbiol Molec Biol Rev* 66:250–271.

Bignelli P. 1901. *Gazz Chim Ital* 31:58 (as cited by Challenger, 1951).

Blasco F, Gaudin C, Jeanjean R. 1971. Absorption of arsenate ions by *Chlorella*. Partial reduction of arsenate to arsenite. *CR Acad Sci (Paris) Ser D* 273:812–815.

Bowen HJM. 1979. *Environmental Chemistry of the Elements*. London, U.K.: Academic Press.

Braddock JG, Luong HV, Brown EJ. 1984. Growth kinetics of *Thiobacillus ferrooxidans* isolated from mine drainage. *Appl Environ Microbiol* 48:48–55.

Brierley JA, Brierley CL. 1999. Present and future commercial applications of biohydrometallurgy. In: Amils R, Ballester A (eds.), *Biohydrometallurgy and the Environment toward the Mining of the 21st Century. Part A*. Amsterdam, the Netherlands: Elsevier, pp. 81–89.

Broeer S, Ji G, Broeer A, Silver S. 1993. Arsenic efflux governed by the arsenic resistance plasmid pI258. *J Bacteriol* 175:3480–3485.

Bryner LC, Beck JV, Davis BB, Wilson DG. 1954. Microorganisms in leaching sulfide minerals. *Ind Eng Chem* 46:2587–2592.

Buchanan RE, Gibbons NE. 1974. *Bergey's Manual of Determinative Bacteriology*, 8th edn. Baltimore, MD: Williams & Wilkins.

Budinhoff CR, Hollibaugh JT. 2008. Arsenite-dependent photoautotrophy by an *Ectothiorhodospira* dominated consortium. *ISME J* 2:340–344.

Button DK, Dunker SS, Moore ML. 1973. Continuous culture of *Rhodotorula rubra*: Kinetics of phosphate–arsenate uptake, inhibition and phosphate limited growth. *J Bacteriol* 113:599–611.

Carapella SC Jr. 1972. Arsenic: Element and geochemistry. In: Fairbridge RW (ed.), *The Encyclopedia of Geochemistry and Environmental Sciences. Encyclopedia of Earth Science, Series IVA*. New York: Van Nostrand Reinhold, pp. 41–42.

Carlson L, Bigham JM, Schwertmann U, Kyek A, Wagner F. 2002. Scavenging of As from acid mine drainage by schwertmannite and ferrihydrite: A comparison with synthetic analogues. *Environ Sci Technol* 36:1712–1719.

Carlson L, Lindström EB, Hallberg KB, Tuovinen OH. 1992. Solid-phase production of bacterial oxidation of arsenic/pyrite. *Appl Environ Microbiol* 58:1046–1049.

Cavalca L, Corsini A, Zaccheo P, Andreoni V, Muyzer G. 2013. Microbial transformations of arsenic: Perspectives for biological removal of arsenic from water. *Future Med* 8:753–768.

Challenger F. 1933. The formation of organo-metalloid compounds by microorganisms. Part II. Trimethylarsine and dimethylarsine. *J Chem Soc* 95:101.

Challenger F. 1951. Biological methylation. *Adv Enzymol* 12:429–491.

Cheng C-N, Focht DD. 1979. Production of arsine and methylarsines in soil and in culture. *Appl Environ Microbiol* 38:494–498.

Cox DP, Alexander M. 1973. Effect of phosphate and other anions on trimethylarsine formation by *Candida humicola*. *Appl Environ Microbiol* 25:408–413.

Cozen AE, Weirauch MT, Pollard KS, Bernick DL, Stuart JM, Lowe TM. 2009. Transcriptional map of respiratory versatility in the hyperthermophilic crenarchaeon *Pyrobaculum aerophilum*. *J Bacteriol* 191:782–794.

Cullen WR, Reimer KJ. 1989. Arsenic speciation in the environment. *Chem Rev* 89:713–784.

D'Imperio S, Lehr CR, Breary M, McDermott TR. 2007. Autecology of an arsenite chemolithotroph: Sulfide constraints on function and distribution in a geothermal spring. *Appl Environ Microbiol* 73:7067–7074.

Da Costa EWB. 1971. Suppression of the inhibitory effects of arsenic compounds by phosphate. *Nature (New Biol) (London)* 231:32.

Da Costa EWB. 1972. Variation in toxicity of arsenic compounds to microorganisms and the suppression of the inhibitory effects of phosphate. *Appl Environ Microbiol* 23:46–53.

Demergasso C, Chong G, Escudero L, Pueyo J, Pedrós-Alió C. 2007. Microbial precipitation of arsenic sulfides In Andean salt flats. *Geomicrobiol J* 24:111–123.

Denning EJ, MacKerell Jr AD. 2011. Impact of the arsenic/phosphorus substitution on the intrinsic conformational properties of the phosphodiester backbone of DNA investigated using ab initio quantum mechanical calculations. *J Am Chem Soc* 133:5770–5772.

Dew DW, Lawson EN, Broadhurst JL. 1997. The BIOX® process for biooxidation of gold-bearing ores and concentrates. In: Rawlings DE (ed.), *Biomining. Theory, Microbes and Industrial Processes*. Berlin, Germany: Springer, pp. 45–80.

Dixit S, Hering JC. 2003. Effects of arsenate reduction and iron oxide transformation on arsenic mobility. *Environ Sci Technol* 37:4182–4189.

Dixit S, Hering JC. 2006. Sorption of Fe(II) and As(III) on goethite in single- and dual-sorbate systems. *Chem Geol* 228:6–15.

Donahoe-Christiansen J, D'Imperio S, Jackson CR, Innskeep WP, McDermott TR. 2004. Arsenite-oxidizing *Hydrogenobaculum* strain isolated from an acid-sulfate-chloride geothermal spring in Yellowstone National Park. *Appl Environ Microbiol* 70:1865–1868.

Drewniak H, Sklodowska A. 2013. Arsenic-transforming microbes and their role in biomining processes. *Environ Sci Pollut Res* 20:7728–7739.

Dyke KGH, Parker MT, Richmond MH. 1970. Penicillinase production and metal-ion resistance in *Staphylococcus aureus* cultures isolated from hospital patients. *J Med Microbiol* 3:125–136.

Ehrlich HL. 1963. Bacterial action on orpiment. *Econ Geol* 58:991–994.

Ehrlich HL. 1964. Bacterial oxidation of arsenopyrite and enargite. *Econ Geol* 59:1306–1312.

Ehrlich HL. 1978. Inorganic energy sources for chemolithotrophic and mixotrophic bacteria. *Geomicrobiol J* 1:65–83.

Ehrlich HL. 1986. Bacterial leaching of silver from silver-containing mixed sulfide ore by a continuous process. In: Lawrence RW, Branion RMR, Ebneer HG (eds.), *Fundamental and Applied Biohydrometallurgy*. Amsterdam, the Netherlands: Elsevier, pp. 77–88.

Ehrlich HL. 1988. Bioleaching of silver from a mixed sulfide ore in a stirred reactor. In: Norris PR, Kelly DP (eds.), *Biohydrometallurgy*. Kew Surrey, U.K.: Science & Technology Letters, pp. 223–231.

Elias M, Wellner A, Goldin-Azulay K, Chabriere E, Vorholt JA, Erb TJ, Tawfik DS. 2012. The molecular basis of phosphate discrimination in arsenate-rich environments. *Nature* 491:134–137.

Erb TJ, Kiefer P, Hattendorf B, Gunther D, Vorholt JA. 2012. GFAJ-1 is an arsenate-resistant, phosphate-dependent organism. *Science* 337:467–470.

Escobar B, Huenupi E, Wirtz J. 1997. Chemical and biological leaching of enargite. *Biotechnol Lett* 19:719–722.

Ferguson JF, Gavis J. 1972. A review of the arsenic cycle in natural waters. *Water Res* 6:1259–1274.

Filella M, Belzile N, Chen Y-W. 2002a. Antimony In the environment: A review focused on natural waters. I. Occurrence. *Earth-Sci Rev* 57:125–176.

Filella M, Belzile N, Chen Y-W. 2002b. Antimony in the environment: A review focused on natural waters. II. Relevant solution chemistry. *Earth-Sci Rev* 59:265–285.

Fillela M, Belzile N, Lett M-C. 2007. Antimony in the environment: A review focused on natural waters. III. Microbiota relevant interactions. *Earth Sci Rev* 80:195–217.

Fisher JC, Hollibaugh JT. 2008. Selenate-dependent anaerobic arsenite oxidation by a bacterium from Mono Lake, California. *Appl Environ Microbiol* 74:2588–2594.

Flynn HC, Meharg AA, Bowyer PK, Paton GI. 2003. Antimony bioavailability in mine soils. *Environ Pollut* 124:93–100.

Garcia-Dominguez E, Mumford A, Rhine ED, Paschal A, Young LY. 2008. Novel autotrophic arsenite-oxidizing bacteria isolated from soil and sediments. *FEMS Microbiol Ecol* 66:401–410.

Gonzales-Contreras P, Weija J, van der Weijden R, Buisman CJN. 2010. Biogenic scorodite crystallization by *Acidianus sulfidivorans* for arsenic removal. *Environ Sci Technol* 44:675–680.

Gonzales-Contreras P, Weijma J, Buisman CJN. 2012. Continuous bioscorodite crystallization in CSTRs for arsenic removal and disposal. *Water Res* 46:5883–5892.

Gornitz V. 1972. Antimony: Element and geochemistry. In: Fairbridge RW (ed.), *The Encyclopedia of Geochemistry and Environmental Sciences. Encyclopedia Earth Science Series IVA*. New York: Van Nostrand Reinhold, pp. 33–36.

Gosio B. 1897. Zur Frage, Wodurch die Giftigkeit arsenhaltiger Tapeten bedingt wird. *Ber Deut Chem Ges* 30:1024–1027 (as cited by Challenger, 1951).

Green HH. 1918. Description of a bacterium which oxidizes arsenite to arsenate, and of one which reduces arsenate to arsenite, isolated from a cattle-dipping tank. *S Afr J Sci* 14:465–467.

Groudeva VI, Groudev SN, Markov KI. 1986. A comparison between mesophilic and thermophilic bacteria with respect to their ability to leach sulfide minerals. In: Lawrence RW, Branion RMR, Ebner HG (eds.), *Fundamental and Applied Biohydrometallurgy*. Amsterdam, the Netherlands: Elsevier, pp. 484–485.

Handley KM, Héry M, Lloyd JR. 2009. Redox cycling of arsenic by the hydrothermal marine bacterium *Marinobacter santoriniensis*. Environ Microbiol 11:1601–1611.

Hedges RW, Baumberg S. 1973. Resistance to arsenic compounds conferred by a plasmid transmissible between strains of *Escherichia coli*. J Bacteriol 115:459–460.

Hendrie MS, Holding AJ, Shewan JM. 1974. Emended description of the genus *Alcaligenes* and of *Alcaligenes faecalis* and a proposal that the generic name of *Achromobacter* be rejected: Status of the named species *Alcaligenes* and *Achromobacter*. Int J Syst Bacteriol 24:534–550.

Hoeft SE, Kulp TR, Han S, Lanoil B, Oremland RS. 2010. Coupled arsenotrophy in a hot spring photosynthetic biofilm at Mono Lake, California. Appl Environ Microbiol 76:4633–4639.

Hoeft SE, Kulp TR, Stolz JF, Hollibaugh JT, Oremland RS. 2004. Dissimilatory arsenate reduction with sulfide as electron donor: Experiments with Mono Lake water and isolation of strain MLMS-1, a chemoautotrophic arsenate respirer. Appl Environ Microbiol 70:2741–2747.

Hoeft SE, Switzer Blum J, Stolz JF, Tabita R, Witte B, King GM, Santini JM, Oremland RS. 2007. *Alkalilimnicola ehrlichii* sp. nov., a novel, arsenite oxidizing haloalkaliphilic gammaproteobacterium capable of chemoautotrophic or heterotrophic growth with nitrate or oxygen as the electron acceptor. Int J Syst Evol Microbiol 57:504–512.

Hollibaugh JT, Carini S, Gürleyük H, Jellison R, Joye S, LeCleir G, Meile C, Vasquez L, Wallschläger D. 2005. Distribution of arsenic species in alkaline, hypersaline Mono Lake, California, and response to seasonal stratification and anoxia. Geochim Cosmochim Acta 69:1925–1937.

Huber R, Sacher M, Vollman A, Huber H, Rose D. 2000. Respiration of arsenate and selenate by hyperthermophilic Archaea. Syst Appl Microbiol 23:305–314.

Humayoun SB, Banno N, Hollibaugh JT. 2003. Depth distribution of microbial diversity in Mono Lake, a meromictic soda lake in California. Appl Environ Microbiol 69:1030–1042.

Ilyaletdinov AN, Abdrashitova SA. 1981. Autotrophic oxidation of arsenic by a culture of *Pseudomonas arsenitoxidans*. Mikrobiologiya 50:197–204 (Engl transl pp. 135–140).

Islam FS, Gault AG, Boothman C, Polya DA, Charnock JM, Chatterjee D, Lloyd JR. 2004. Role of metal-reducing bacteria in arsenic release from Bengal delta sediments. Nature (London) 430:68–71.

Islam FS, Pederick RL, Gault AG, Adams LK, Polya DA, Charnock JM, Lloyd JR. 2005. Interactions between the Fe(III)-reducing bacterium *Geobacter sulfurreducens* and arsenate, and capture of the metalloid by biogenic Fe(II). Appl Environ Microbiol 71:8642–8648.

Iverson WP, Brinckman FE. 1978. Microbial metabolism of heavy elements. In: Mitchell R (ed.), *Water Pollution Microbiology*, Vol. 2. New York: Wiley, pp. 201–232.

Ji G, Silver S. 1992. Regulation and expression of the arsenic resistance operon from *Staphylococcus aureus* plasmid pI258. J Bacteriol 174:3684–3694.

Ji G, Silver S, Garber EAE, Ohtake H, Cervantes C, Corbisier P. 1993. Bacterial molecular genetics and enzymatic transformations of arsenate, arsenite, and chromate. In: Torma AE, Apel ML, Brierley CL (eds.), *Biohydrometallurgical Technologies*, Fossil Energy Materials Bioremediation, Microbial Physiology, Vol. 2. Warrendale, PA: The Minerals, Metals, and Materials Society, pp. 529–539.

Johnson DL. 1972. Bacterial reduction of arsenite in seawater. Nature (London) 240:44–45.

Kamalov MR, Karavaiko GI, Ilyaletdinov AN, Abrashitova SA. 1973. The role of *Thiobacillus ferrooxidans* in leaching arsenic from a concentrate of carbonaceous gold-containing ore. Izv Akad Nauk Kaz SSR Ser Biol 11:37–44.

Kang Y-K, Heinemann J, Bothner B, Rensing C, McDermott TR. 2012. Integrated co-regulation of bacterial arsenic and phosphorus metabolisms. Environ Microbiol 12:3097–3109.

Karavaiko GI, Chuchalin LK, Pivovarova TA, Yemel'yanov BA, Forofeyev AG. 1986. Microbiological leaching of metals from arsenopyrite containing concentrates. In: Lawrence RW, Branion RMR, Ebner HG (eds.), *Fundamental and Applied Biohydrometallurgy*. Amsterdam, the Netherlands: Elsevier, pp. 125–126.

Kay ERM. 1965. Incorporation of radioarsenate into proteins and nucleic acids of the Ehrlich lettré ascites carcinoma in vitro. Nature 206:371–373.

Kim EH, Rensing S. 2012. Genome of *Halomonas* strain GFAJ-1, a blueprint for fame or business as usual. J Bacteriol 194:1643–1645.

Knodle R, Agarwal P, Brown M. 2012. From phosphorus to arsenic: Changing the classic paradigm for the structure of biomolecules. Biomolecules 2:282–287.

Krafft T, Macy JM. 1998. Purification and characterization of the respiratory arsenate reductase of *Chrysiogenes arsenatis*. Eur J Biochem 255:647–653.

Krüger MC, Bertin PN, Heipieper HJ, Arsène-Ploetze F. 2013. Bacterial metabolism of environmental arsenic—Mechanisms and biotechnological applications. *Appl Microbiol Biotechnol* 97:3827–3841.

Kulibakin VG, Laptev SF. 1971. Effect of adaptation of *Thiobacillus ferrooxidans* to a copper–arsenic–tin concentrate on the arsenic leach rate. *Sb Tr Tsent Nauk-Issled Inst Olomyan Prom* 1:75–76.

Kulp TR, Han S, Saltikov CW, Lanoil BD, Zargar K, Oremland RS. 2007. Effects of imposed salinity gradients on dissimilatory arsenate reduction, sulfate reduction, and other microbial processes in sediments from two California soda lakes. *Appl Environ Microbiol* 73:5130–5137.

Kulp TR, Hoeft SE, Asao M, Madigan MT, Hollibaugh JT, Fisher JC, Stolz JF, Culbertson CW, Miller LG, Oremland RS. 2008. Arsenic(III) fuels anoxygenic photosynthesis in hot spring biofilms from Mono Lake, California. *Science* 321:967–970.

Kulp TR, Hoeft SE, Miller LG, Saltikov C, Murphy JN, Han S, Lanoil B, Oremland RS. 2006. Dissimilatory arsenate and sulfate reduction in sediments of two hypersaline, arsenic-rich soda lakes: Mono and Searles Lakes, California. *Appl Environ Microbiol* 72:6514–6526.

Kulp TR, Miller L, Braiotta F, Kocar B, Switzer Blum J, Webb S, Oremland, R. 2014. Microbiological reduction of Sb(V) in anoxic freshwater sediments. *Environ Sci Technol* 48:218–226.

Kuznetsov SI, Ivanov MV, Lyalikova NN. 1963. *Introduction to Geological Microbiology*. Engl Transl. New York: McGraw-Hill.

Lara J, González LE, Ferrero M, Díaz GC, Pedrós-Alió C, Demergasso C. 2012. Enrichment of arsenic transforming and resistant heterotrophic bacteria from sediments of two salt lakes in Northern Chile. *Extremophiles* 16:523–538.

Laverman AM, Switzer Blum J, Schaefer JK, Phillips EJP, Lovley DR, Oremland RS. 1995. Growth of strain SES-3 with arsenate and other diverse electron acceptors. *Appl Environ Microbiol* 61:3556–3561.

Lee J-H, Kim M-G, Yoo B, Myung NV, Maeng J, Lee T, Dohnalkova AC, Fredrickson JK, Sadowsky MJ, Hur H-G. 2007. Biogenic formation of photoactive arsenic-sulfide nanotubes by *Shewanella* sp. strain HN-41. *Proc Natl Acad Sci USA* (Washington, DC) 104:20410–20415.

Legge JW. 1954. Bacterial oxidation of arsenite. IV. Some properties of the bacterial cytochromes. *Aust J Biol Sci* 7:504–514.

Legge JW, Turner AW. 1954. Bacterial oxidation of arsenite. III. Cell-free arsenite dehydrogenase. *Aust J Biol Sci* 7:496–503.Lehr CR, Kashyap DR, McDermott TR. 2007. New insights into microbial oxidation of antimony and arsenic. *Appl Environ Microbiol* 73:2386–2389.

Lett MC, Muller D, Lièvremont D, Silver S, Santini J. 2012. Unified nomenclature for genes in prokaryotic aerobic arsenite oxidation. *J Bacteriol* 194:207–208.

Li X, Thornton I. 1993. Arsenic, antimony, and bismuth in soil and pasture herbage in some old metalliferous mining areas in England. *Environ Geochem Health* 15:135–144.

Livesey-Goldblatt E, Norman P, Livesey-Goldblatt DR. 1983. Gold recovery from arsenopyrite/pyrite ore by bacterial leaching and cyanidation. In: Rossi G, Torma AE (eds.), *Recent Progress in Biohydrometallurgy*. Iglesias, Italy: Assoc Mineraria Sarda, pp. 627–641.

Lunde G. 1973. Synthesis of fat and water soluble arseno organic compounds in marine and limnetic algae. *Acta Chem Scand* 27:1586–1594.

Lyalikova NN. 1961. The role of bacteria in the oxidation of sulfide ores. *Tr In-ta Mikrobiol AN SSR*, No. 9 (as cited by Kuznetsov et al., 1963).

Lyalikova NN. 1972. Oxidation of trivalent antimony to higher oxides as an energy source for the development of a new autotrophic organism *Stibiobacter* gen. n. *Dokl Akad Nauk SSSR Ser Biol* 205:1228–1229.

Lyalikova NN. 1974. *Stibiobacter senarmontii*: A new antimony-oxidizing microorganisms. *Mikrobiologiya* 43:941–948 (Engl transl pp. 799–805).

Lyalikova NN, Shlain LB, Trofimov VG. 1974. Formation of minerals of antimony(V) under the effect of bacteria. *Izv Akad Nauk SSSR Ser Biol* 3:440–444.

Lyalikova NN, Shlain LB, Unanova OG, Anisimova LS. 1972. Transformation of products of compound antimony and lead sulfides under the effect of bacteria. *Izv Akad Nauk SSSR Ser Biol* 4:564–567.

Lyalikova NN, Vedenina IYa, Romanova AK. 1976. Assimilation of carbon dioxide by a culture of *Stibiobacter senarmontii*. *Mikrobiologiya* 45:552–554 (Engl transl pp. 476–477).

Macy JM, Nunan K, Hagen KD, Dixon DR, Harbour PJ, Cahill M, Sly L. 1996. *Chrysiogenes arsenatis* gen. nov., spec. nov., a new arsenate-respiring bacterium isolated from gold mine wastewater. *Int J Syst Bacteriol* 46:1153–1157.

Macy JM, Santini JM, Pauling BV, O'Neill AH, Sly LI. 2000. Two new arsenate/sulfate-reducing bacteria: Mechanism of arsenate reduction. *Arch Microbiol* 173:49–57.

Mahler HR, Cordes EH. 1966. *Biological Chemistry.* New York: Harper & Row.

Malasarn D, Keeffe JR, Newman DK. 2008. Characterization of the arsenate respiratory reductase from *Shewanella* sp. strain ANA-3. *J Bacteriol* 190:135–142.

Malasarn D, Saltikov CW, Campbell KM, Santini JM, Hering JG, Newman DK. 2004. *arrA* is a reliable marker for As(V) respiration. *Science* 306:455.

McBride BS, Wolfe RS. 1971. Biosynthesis of dimethylarsine by *Methanobacterium. Biochemistry* 10:4312–4317.

McGuire MM, Banfield JF, Hamers RJ. 2001. Quantitative determination of elemental sulfur at the arsenopyrite surface after oxidation by ferric iron: Mechanistic implications. *Geochem Trans* 2:25.

Monroy-Fernández MG, Mustin C, de Donato P, Berthelin J, Barion P. 1995. Bacterial behavior and evolution of surface oxidized phases during arsenopyrite oxidation by *Thiobacillus ferrooxidans.* In: Vargas T, Jerez JV, Toledo H (eds.), *Biohydrometallurgical Processing,* Vol. 1. Santiago, Chile: University of Chile, pp. 57–66.

Myers DJ, Heimbrook ME, Osteryoung J, Morrison SM. 1973. Arsenic oxidation state in the presence of microorganisms. Examination by differential pulse polarography. *Environ Lett* 5:53–61.

Nakagawa S, Shtaih Z, Banta A, Beveridge TJ, Sako Y, Reysenbach A-L. 2005. *Sulfurihydrogenibium yellowstonense* sp. nov., an extremely thermophilic, facultative heterotrophic, sulfur-oxidizing bacterium from Yellowstone National Park, and emended descriptions of the genus *Sulfurihydrogenibium, Sulfurihydrogenibium subterraneum* and *Sulfurihydrogenibium azorense. Int J Syst Evol Microbiol* 55:2263–2268.

Newman DK, Ahmann D, Morel FMM. 1998. A brief review of microbial arsenate respiration. *Geomicrobiol J* 15:255–268.

Newman DK, Beveridge TJ, Morel FM. 1997a. Precipitation of arsenic trisulfide by *Desulfotomaculum auripigmentum. Appl Environ Microbiol* 63:2022–2028.

Newman DK, Kennedy EK, Coates JD, Ahmann D, Ellis DJ, Lovley DR, Morel FM. 1997b. Dissimilatory arsenate and sulfate reduction in *Desulfotomaculum auripigmentum,* sp. nov. *Arch Microbiol* 168:380–388.

Ngubane WT, Baecker AAW. 1990. Oxidation of gold-bearing pyrite and arsenopyrite by *Sulfolobus acidocaldarius* and *Sulfolobus* BC in airlift bioreactors. *Biorecovery* 1:225–269.

Olson GJ. 1994. Microbiological oxidation of gold ores and gold bioleaching. *FEMS Microbiol Lett* 119:1–6.

Oremland RS, Dowdle PR, Hoeft S, Sharp JO, Schaefer JK, Miller LG, Switzer Blum J, Smith RL, Bloom NS, Wallschläger D. 2000. Bacterial dissimilatory reduction of arsenate and sulfate in meromictic Mono Lake, California. *Geochim Cosmochim Acta* 64:3073–3084.

Oremland RS, Hoeft SE, Santini JM, Bano N, Hollibaugh RA, Hollibaugh JT. 2002. Anaerobic oxidation of arsenite in Mono Lake water and by a facultative, arsenite-oxidizing chemoautotroph, strain MLHE-1. *Appl Environ Microbiol* 68:4795–4802.

Oremland RS, Kulp TR, Switzer Blum J, Hoeft SE, Baesman S, Miller LG, Stolz JF. 2005. A microbial arsenic cycle in a salt-saturated, extreme environment. *Science* 308:1305–1308.

Oremland RS, Stolz J. 2000. Dissimilatory reduction of selenate and arsenate in nature. In: Lovley DR (ed.), *Environmental Microbial–Metal Interactions.* Washington, DC: ASM Press, pp. 199–224.

Oremland RS, Stolz JF, Hollibaugh JT. 2004. The microbial arsenic cycle in Mono Lake, California. *FEMS Microbiol Ecol* 48:15–27.

Oremland RS, Switzer Blum J, Culbertson CW, Visscher PT, Miller LG, Dowdle P, Strohmaier FE. 1994. Isolation, growth, and metabolism of an obligately anaerobic, selenate-respiring bacterium, strain SES-3. *Appl Environ Microbiol* 60:3011–3019.

Oremland RS, Wolfe-Simon F, Saltikov CW, Stolz JF. 2009. Arsenic in the evolution of earth and extraterrestrial ecosystems. *Geomicrobiol J* 26:522–536.

Osborne FH. 1973. Arsenite oxidation by a soil isolate of *Alcaligenes.* PhD thesis. Rensselaer Polytechnic Institute, Troy, NY.

Osborne FH, Ehrlich HL. 1976. Oxidation of arsenite by a soil isolate of *Alcaligenes. J Appl Bacteriol* 41:295–305.

Osborne TH, Santini JM. 2012. Prokaryotic aerobic oxidation of arsenite. In: Santini JM, Ward SA (eds.), *The Metabolism of Arsenite.* Boca Raton, FL: CRC Press, pp. 61–72.

Oscarson DW, Huang PM, Defosse C, Herbillon A. 1981. Oxidative power of Mn(IV) and Fe(III) oxides with respect to As(III) in terrestrial and aquatic environments. *Nature (London)* 291:50–51.

Petrick JS, Ayala-Fierro F, Cullen WR, Carter DE, Aposhian HV. 2000. Monomethylarsonous acid (MMA^III) is more toxic than arsenite in Chang human erythrocytes. *Toxixol Appl Pharmacol* 163:203–207.

Philips SE, Taylor ML. 1976. Oxidation of arsenite to arsenate by *Alcaligenes faecalis. Appl Environ Microbiol* 32:392–399.

EHRLICH'S GEOMICROBIOLOGY

Pickett AW, McBride BC, Cullen WR, Manji H. 1981. The reduction of trimethylarsine oxide by *Candida humicola*. *Can J Microbiol* 27:773–778.

Planer-Friederich B, Fisher JC, Hollibaugh JT, Süs E, Wallschläger D. 2009. Oxidative transformation of trithioarsenate along alkaline geothermal drainages—Abiotic versus microbially mediated processes. *Geomicrobiol J* 26:339–350.

Planer-Friederich B, Wallschläger D. 2009. A critical investigation of hydride generation-based arsenic speciation in sulfidic waters. *Environ Sci Technol* 43:5007–5013.

Pol'kin SI, Tanzhnyanskaya ZA. 1968. Use of bacterial leaching for ore enrichment. *Izv Vyssch Ucheb Zaved Tsvet Met* 11:115–121.

Pol'kin SI, Yudina N, Nanin VV, Kim DKh. 1973. Bacterial leaching of arsenic from an arsenopyrite gold-containing concentrate in a thick pulp. *Nauchno-Issled Geologorazved Inst Tsvetn Blagorodn Metall* 107:34–41.

Price RE, Amend JP, Pichler T. 2007. Enhanced geochemical gradients in a marine shallow-water hydrothermal system: Unusual arsenic speciation in horizontal and vertical pore water profiles. *Appl Geochem* 22:2595–2605.

Qin J, Lehr CR, Yuan C, Le C, McDermott TR, Rosen BP. 2009. Biotransformation of arsenic by a Yellowstone thermoacidophilic eukaryotic alga. *Proc Natl Acad Sci USA* 106:5213–5217.

Qin J, Rosen BP, Zhang Y, Wang G, Franke S, Rensing C. 2006. Arsenic detoxification and evolution of trimethylarsine gas by a microbial arsenite S-adenosylmethionine methyltransferase, *Proc Natl Acad Sci USA* 103:2075–2080.

Quastel JH, Scholefield PG. 1953. Arsenite oxidation in soil. *Soil* 75:279–285.

Raven KP, Jain A, Loeppert RH. 1998. Arsenite and arsenate adsorption on ferrihydrite: Kinetics, equilibrium, and adsorption envelopes. *Environ Sci Technol* 32:344–349.

Reaves ML, Sinha S, Rabinowitz JD, Kruglyak L, Redfield RJ. 2012. Absence of detectable arsenate In DNA from arsenate-grown GFAJ-1 cells. *Science* 337:470–477.

Rhine ED, Ní Chadhain SM, Zylstra GJ, Young LY. 2007. The arsenite oxidase genes (aroAB) in novel chemoautotrophic arsenite oxidizers. *Biochem Biophys Res Commun* 354:662–667.

Rhine ED, Phelps CD, Young LY. 2006. Anaerobic arsenite oxidation by novel denitrifying isolates. *Environ Microbiol* 8:899–908.

Richey C, Chovanec P, Hoeft SE, Oremland RS, Basu P, Stolz JF. 2009. Respiratory arsenate reductase as a bidirectional enzyme. *Biochem Biophys Res Commun* 382:298–302.

Rodríguez-Freire L, Sun W, Sierra-Alvarez R, Field JA. 2012. Flexible bacterial strains that oxidize arsenite in anoxic or aerobic conditions and utilize hydrogen or acetate as alternative electron acceptors. *Biodegradation* 23:133–143.

Rosen BP. 1999. Families of arsenic transporters. *Trends Microbiol* 7:207–2012.

Rosen BP, Ajees AA, McDermott TR. 2011. Life and death with arsenic. Arsenic life: An analysis of the recent report "A bacterium that can grow by using arsenic Instead of phosphorus." *Bioessays* 33:350–357.

Salmassi TM, Venkateswaren K, Satomi M, Nealson KH, Newman DK, Hering JG. 2002. Oxidation of arsenite by *Agrobacterium albertimagni*, AOL15, sp. nov., isolated from Hot Creek, California. *Geomicrobiol J* 19:53–66.

Salmassi TM, Walker JJ, Newman DK, Leadbetter JR, Pace NR, Hering JG. 2006. Community and cultivation analysis of arsenite oxidizing biofilms at Hot Creek. *Environ Microbiol* 8:50–59.

Saltikov CW, Cifuentes A, Venkateswaran K, Newman DK. 2003. The *ars* detoxification system is advantageous but not required for As(V) respiration by the genetically tractable *Shewanella* species strain ANA-3. *Appl Environ Microbiol* 69:2800–2809.

Saltikov CW, Newman DK. 2003. Genetic identification of a respiratory arsenate reductase. *Proc Natl Acad Sci USA* 100:10983–10988.

Saltikov CW, Wildman RA Jr, Newman DK. 2005. Expression dynamics of arsenic respiration and detoxification in *Shewanella* sp. strain ANA-3. *J Bacteriol* 187:7390–7396.

Santini JM, Sly LI, Schnagl RD, Macy JM. 2000. A new chemolithoautotrophic arsenite-oxidizing bacterium isolated from a gold mine: Phylogenetic, physiological, and preliminary biochemical studies. *Appl Environ Microbiol* 66:92–97.

Santini JM, vanden Hoven RN. 2004. Molybdenum-containing arsenite oxidase of the chemolithoautotrophic arsenite oxidizer NT-26. *J Bacteriol* 186:1614–1619.

Santini JM, Ward SM (eds.). 2012. *The Metabolism of Arsenite*. Boca Raton, FL: CRC Press, 189pp.

Scudlark JR, Johnson DL. 1982. Biological oxidation of arsenite in seawater. *Estuar Coast Shelf Sci* 14:693–706.

Sehlin HM, Lindström EB. 1992. Oxidation and reduction of arsenic by *Sulfolobus acidocaldarius* strain BC. *FEMS Microbiol Lett* 93:87–92.

Silver M, Torma AE. 1974. Oxidation of metal sulfides by *Thiobacillus ferrooxidans* grown on different substrates. *Can J Microbiol* 20:141–147.

Silver S. 1998. Genes for all metals: A bacterial view of the periodic table. *J Ind Microbiol Biotechnol* 20:1–12.

Silver S, Keach D. 1982. Energy-dependent arsenate flux: The mechanism of plasmid-mediated resistance. *Proc Natl Acad Sci USA* 79:6114–6118.

Silver S, Phung LT. 2005. Genes and enzymes involved in bacterial oxidation and reduction of inorganic arsenic. *Appl Environ Microbiol* 71:599–608.

Silverman MP, Lundgren DG. 1959. Studies on the chemoautotrophic iron bacterium *Ferrobacillus ferrooxidans*. 1. An improved medium and a harvesting procedure for securing high cell yields. *J Bacteriol* 77:642–647.

Slyemi D, Bonnefoy V. 2012. How prokaryotes deal with arsenic. *Environ Microbiol Rpts* 4:571–586.

Stolz JF, Ellis DJ, Switzer Blum J, Ahmann D, Lovley DR, Oremland RS. 1999. *Sulfurospirillum barnesii* sp. nov., *Sulfurospirillum arsenophilum* sp. nov., and the *Sulfurospirillum* clade in the Epsilon Proteobacteria. *Int J Syst Bacteriol* 49:1177–1180.

Stolz JF, Oremland RS. 1999. Bacterial respiration of arsenic and selenium. *FEMS Microbiol Rev* 23:615–627.

Sun W, Sierra R, Field JA. 2008. Anoxic oxidation of arsenite linked to denitrification in sludges and sediments. *Water Res* 42:4569–4577.

Sun W, Sierra-Alvarez R, Field JA. 2010a. The role of denitrification on arsenite oxidation and arsenic mobility in an anoxic sediment column model with activated alumina. *Biotechnol Bioeng* 107:786–794.

Sun W, Sierra-Alvarez R, Milner L, Field J. 2010b. Anaerobic oxidation of arsenite linked to chlorate reduction. *Appl Environ Microbiol* 76:6804–6811.

Sun W, Sierra-Alvarez R, Milner L, Oremland R, Field JA. 2009. Arsenite and ferrous iron oxidation linked to chemolithotrophic denitrification for the immobilization of arsenic in anoxic environments. *Environ Sci Technol* 43:6585–6591.

Swartz CH, Blute NK, Badruzzman B, Ali A, Brabander D, Jay J, Besancon J, Islam S, Hemond HF, Harvey CF. 2004. Mobility of arsenic in a Bangladesh aquifer: Inferences from geochemical profiles, leaching data, and mineralogical characterization. *Geochim Cosmochim Acta* 68:4539–4557.

Switzer Blum J, Burns Bindi A, Buzzelli J, Stolz JF, Oremland RS. 1998. *Bacillus arsenicoselenatis*, sp. nov., and *Bacillus selenitireducens*, sp. nov.: Two haloalkaliphiles from Mono Lake, California that respire oxyanions of selenium and arsenic. *Arch Microbiol* 171:19–30.

Switzer Blum J, Han S, Lanoil B, Saltikov C, Whitte B, Tabita FR, Langley S, Beveridge TJ, Jahnke L, Oremland RS. 2009. Ecophysiology of "*Halarsenatibacter silvermanii*" strain SLAS-1^T, gen. nov., sp. nov., a facultative chemoautotrophic arsenate respirer from salt-saturated Searles Lake, California. *Appl Environ Microbiol* 75:1950–1960.

Switzer Blum, J, Kulp TR, Han S, Lanoil B, Saltikov CW, Stolz JF, Miller LG, Oremland RS. 2012. *Desulfohalophilus alkaliarsenatis* gen. nov., sp. nov., an extremely halophilic sulfate- and arsenate-respiring bacterium from Searles Lake, California. *Extremophiles* 16:727–742.

Takai K, Kobayashi H, Nealson KH, Horikoshi K. 2003a. *Sulfurihydrogenibium subterraneum* gen. nov., sp. nov., from a subsurface hot aquifer. *Int J Syst Evol Microbiol* 53:823–827.

Takai K, Kobayashi H, Nealson KH, Horikoshi K. 2003b. *Deferribacter desulfuricans* sp. nov., a novel sulfur-, nitrate- and arsenate-reducing thermophile isolated from a deep-sea hydrothermal vent. *Int J Syst Evol Microbiol* 53:839–846.

Torma AE, Gabra GG. 1977. Oxidation of stibnite by *Thiobacillus ferrooxidans*. *Antonie v Leeuwenhoek* 43:1–6.

Tuovinen OH, Bhatti TM, Bigham JM, Hallberg KB, Garcia O Jr, Lindström EB. 1994. Oxidative dissolution of arsenopyrite by mesophilic and moderately thermophilic acidophiles. *Appl Environ Microbiol* 60:3268–3274.Turner AW. 1949. Bacterial oxidation of arsenite. *Nature (London)* 164:76–77.

Turner AW. 1954. Bacterial oxidation of arsenite. I. Description of bacteria isolated from arsenical cattle-dipping fluids. *Aust J Biol Sci* 7:452–478.

Turner AW, Legge JW. 1954. Bacterial oxidation of arsenite. II. The activity of washed suspensions. *Aust J Biol Sci* 7:479–495.

van Lis R, Nitschke W, Duval S, Schoep-Cothenet B. 2012. Evolution of arsenite oxidation. In: Santini JM, Ward SA (eds.), *The Metabolism of Arsenite*. Boca Raton, FL: CRC Press, pp. 125–144.

Wakao N, Koyatsu H, Komai Y, Shimokawara H, Sakurai Y, Shibota H. 1988. Microbial oxidation of arsenite and occurrence of arsenite-oxidizing bacteria in acid mine water from a sulfur-pyrite mine. *Geomicrobiol J* 6:11–24.

Wang G, Kennedy SP, Fasiludeen S, Rensing C, DasSarma S. 2004. Arsenic resistance in *Halobacterium* sp. strain NRC-1 examined by using an improved gene knockout system. *J Bacteriol* 186:3187–3194.

EHRLICH'S GEOMICROBIOLOGY

Weast RC, Astle MJ. 1982. *CRC Handbook of Chemistry and Physics*. 63rd edn. Boca Raton, FL: CRC Press.

Weeger W, Lièvremont D, Perret M, Lagarde F, Hubert J-C, Leroy M, Lett M-C. 1999. Oxidation of arsenite to arsenate by a bacterium from an aquatic environment. *BioMetals* 12:141–149.

Willsky GR, Malamy MH. 1980a. Characterization of two genetically separable inorganic phosphate transport systems in *Escherichia coli*. *J Bacteriol* 144:356–365.

Willsky GR, Malamy MH. 1980b. Effect of arsenate on inorganic phosphate transport in *Escherichia coli*. *J Bacteriol* 144:366–374.

Winchester JW. 1972. Geochemistry. In: Fairbridge RW (ed.), *The Encyclopedia of Geochemistry and Environmental Sciences*, Encyclopedia Earth Science Series IVA, Vol. IVA. New York: Van Nostrand Reinhold, pp. 402–410.

Wolfe-Simon F, Blum JS, Kulp TR, Gordon GW, Hoeft SE, Stolz JF, Webb SM, Davies PCW, Anbar AD, Oremland RS. 2011. A bacterium that can grow by using arsenic instead of phosphorous. *Science* 332:1163–1166.

Wolthers M, Charlet L, van der Weijden CH, van der Linde PR, Rickard D. 2005. Arsenic mobility in the ambient sulfidic environment: Sorption of arsenic(V) and arsenic(III) onto disordered mackinawite. *Geochim Cosmochim Acta* 69:3483–3492.

Woolfolk CA, Whiteley HR. 1962. Reduction of inorganic compounds with molecular hydrogen by *Micrococcus lactilyticus*. I. Stoichiometry with compounds of arsenic, selenium, tellurium, transition and other elements. *J Bacteriol* 84:647–658.

Zargar K, Hoeft S, Oremland R, Saltikov CW. 2010. Genetic identification of a novel arsenite oxidase, *arxA*, in the haloalkaliphilic, arsenite oxidizing bacterium *Alkalilimnicola ehrlichii* strain MLHE-1. *J Bacteriol* 192:3755–3762.

Zargar K.A, Conrad DL, Bernick TM, Lowe V, Stolc S, Hoeft SE, Oremland RS, Stolz J, Saltikov CW. 2012. ArxA, a new clade of arsenite oxidase within the DMSO reductase family of 3 molybdenum oxidoreductases. *Environ Microbiol* 14:1635–1645.

Zobrist J, Dowdle PR, Davis JA, Oremland RS. 2000. Mobilization of arsenate by dissimilatory reduction of adsorbed arsenate. *Environ Sci Technol* 34:4747–4753.

Figure 3.5. Stromatolite cones from the 3.43 Ga Strelley Pool Formation, Pilbara Craton, Western Australia.

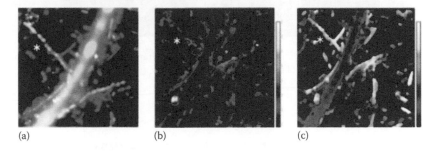

(a)   (b)   (c)

Figure 8.7. Stable isotope probing and single-cell activity measurements of environmental microorganisms using fluorescence in situ hybridization–nano-secondary ion mass spectrometry (nano-SIMS).

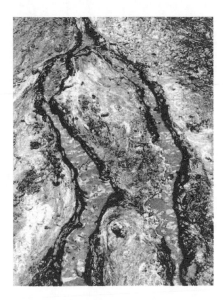

Figure 17.4. Iron-rich spring Fuschna in Switzerland. Iron(II)-oxidizing bacteria precipitate red/orange iron (oxyhydr) oxide minerals in the channels of the outflow of the spring water.

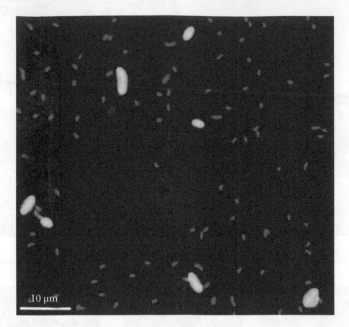

Figure 17.11. Autotrophic nitrate-reducing iron(II)-oxidizing enrichment culture KS in the late exponential phase cultivated with 10 mM iron(II) chloride and 4 mM sodium nitrate.

Figure 17.18. Banded iron formations from the Transvaal Supergroup in South Africa.

Figure 17.19. Gleying of a rice paddy soil. The oxic soil of a rice paddy is orange-brown due to the presence of iron(III) minerals. The paddy soil turns to a grayish-green-blue color when the iron(III) minerals are reduced under anoxic conditions.

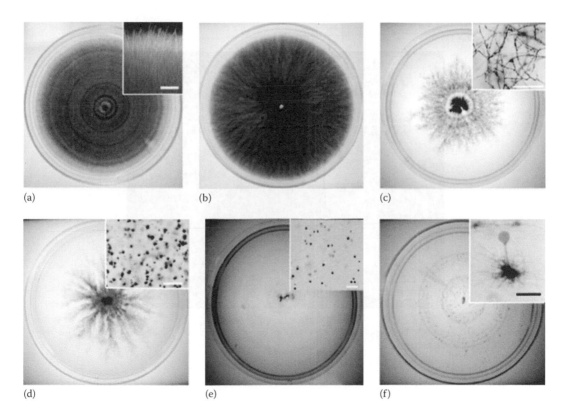

Figure 18.6. Diversity of Mn (II)-oxidizing ascomycete fungi isolated from coal mine drainage treatment systems.

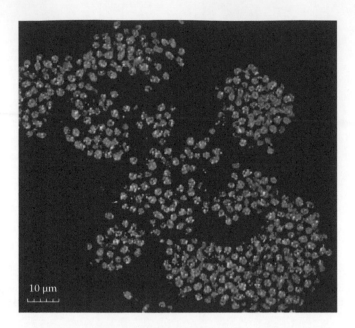

Figure 20.4. Epireflective confocal microscopy of sectioned "pink berry" microbial consortium from Sippewissett Salt Marsh, Cape Cod, MA.

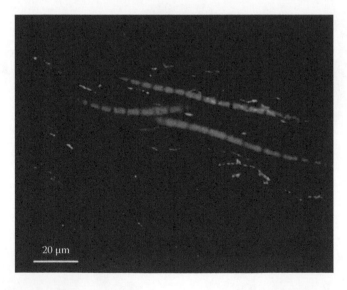

Figure 20.5 CARD-FISH image of the upper portion (ca. 1.5 mm beneath the surface) of a microbial mat from Guerrero Negro, Baja California Sur, Mexico.

CHAPTER SIXTEEN

Geomicrobiology of Mercury*

Robert Mason

CONTENTS

16.1 Introduction / 323
16.2 Distribution of Mercury in the Biosphere and Its Toxicity / 324
16.3 Anthropogenic Mercury / 326
16.4 Microbial Methylation of Mercury / 327
    16.4.1 Early Studies of Microbial Mercury Methylation / 327
    16.4.2 Enzymatic Methylation of Mercury by Microbes / 328
    16.4.3 Genetic Basis for Mercury Methylation / 329
16.5 Other Microbial Interactions with Mercury / 331
    16.5.1 Enzymatic Microbial Reduction of Mercury / 331
    16.5.2 Other Pathways for Mercury Reduction / 332
    16.5.3 Microbial Decomposition of Organomercurials / 333
    16.5.4 Oxidation of Metallic Mercury / 334
    16.5.5 Microbial Precipitation of Mercury / 335
16.6 Summary / 335
References / 335

16.1 INTRODUCTION

The element mercury (Hg) has been known as a specific chemical from around 1500 B.C. The physician Paracelsus (A.D. 1493–1541) attempted to cure syphilis by administering metallic mercury ($Hg^0(l)$) to sufferers (Clarkson and Magos, 2006). His treatment was probably based on intuitive or empirical knowledge that at an appropriate dosage, Hg^0 was more toxic to the cause of the disease than the patient. The true etiology of syphilis was, however, unknown to him. The toxicity of Hg to human beings and other animals varies for the different inorganic (ionic Hg (Hg^{II}, Hg^I) and Hg^0) and organic forms (e.g., monomethylmercury [CH_3Hg^+], dimethylmercury [$(CH_3)_2Hg$], both of which can be formed in the environment, phenylmercury [$C_6H_5Hg^+$], and other mercurials

that have been manufactured as preservatives, antiseptics, and fungicides) (Clarkson and Magos, 2006; Fitzgerald and Clarkson, 1991). Inorganic Hg is found in large-scale deposits in the environment not only as Hg^0 but also in ores, primarily as cinnabar (HgS). It has been known since the Roman era that working in mercury mines was hazardous to the worker's health. More recently, the term "mad as a hatter" refers to the impact on the workers of inorganic Hg used in the hat-making industry in the eighteenth and nineteenth centuries (Clarkson and Magos, 2006).

The extent of the current environmental hazard from Hg compounds became apparent recently as a consequence of their extensive use and through the understanding of the ability of the organic forms, primarily CH_3Hg, to bioaccumulate in the

---

* Throughout the chapter Roman numerals will be used to indicate oxidation state while the use of + or − indicates the charge of the ion or complex. Use of the chemical formula without a charge indicates that all forms are included.

food chain (Clarkson and Magos, 2006; Wiener et al., 2003). The toxicity manifests itself in major physical impairments and death from the acute intake of Hg compounds in food and water or through inhalation of Hg^0. Attention to the problem was heightened by incidents of acute poisoning in Japan (from CH_3Hg in fish, now called "Minamata disease"), Iraq, Pakistan, Guatemala, and the United States (Clarkson and Magos, 2006; Goldwater, 1972 and references therein). In some cases, food was consumed that had been made from seed grain, not intended for consumption, which was treated with Hg compounds, including CH_3Hg, to inhibit fungal damage before planting (the Iraq incident). In other cases, meat was consumed that was tainted because the animals drank Hg contaminated water or had eaten Hg-tainted feed (Goldwater, 1972). More recently, the impact of chronic exposure to CH_3Hg from the consumption of fish and seafood with elevated CH_3Hg levels has focused attention on the sources and cycling of Hg and its compounds in the environment (Mahaffey et al., 2004; Sunderland, 2007). This exposure pathway has highlighted the important and intimate role of microbes in the interconversion of Hg compounds and in the resultant impacts on humans and other wildlife of CH_3Hg in aquatic organisms and other foods.

The maximum permissible level in potable waters was set as 1 µg/L in 1990 by the World Health Organization (WHO) (Cotruvo and Vogt, 1990), and there are now different regulatory levels in countries around the world (e.g., 2 µg/L in the United States; EPA, 2013). More importantly, levels of CH_3Hg in fish and seafood are the basis

of many health advisories and regulations with limits on consumption proposed by some countries for fish with concentrations as low as 0.3 mg/kg (Mahaffey et al., 2004; Sunderland, 2007). These levels are based on estimated consumption levels—the current WHO recommended consumption value is <1.6 µg CH_3Hg/kg body weight/week (WHO, 2013). These advisories are derived for the protection of the developing fetus and for young children.

16.2 DISTRIBUTION OF MERCURY IN THE BIOSPHERE AND ITS TOXICITY

The abundance of Hg in the Earth's crust has been reported as 0.08 mg/kg (Jonasson, 1970). The total dissolved concentration in uncontaminated freshwater ranges up to 10 ng/L, although much higher concentrations have been measured in contaminated waters (Fitzgerald and Lamborg, 2004). Concentrations of Hg in rain and snow are typically 5–10 times higher (5–100 ng/L). The Hg concentration in seawater is typically <1 ng/L (Fitzgerald et al., 2007; Mason et al., 2012). Methylated Hg concentrations in uncontaminated waters are typically <10% of the total Hg content but can be higher in anoxic waters or in sediment porewater and groundwater.

In environmental waters, sediments, and soils, Hg can exist as the liquid metal or in inorganic and organic compounds with Hg mostly in the +II oxidation state (Fitzgerald and Lamborg, 2004) (Figure 16.1). The metal is liquid at ambient temperature and has a significant vapor pressure (1.2×10^{-3} mm Hg at 20°C) and heat

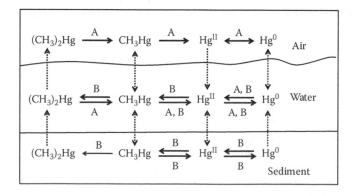

Figure 16.1. Representation of the major transformations (solid arrows) of mercury in the biosphere (sediments/soils, water, and atmosphere) with an indication of whether the primary pathway for the transformation is mediated by microorganisms (B) or is an abiotic reaction (A). Physical transport between reservoirs is indicated by the dotted arrows with an indication of the major direction of transport.

of vaporization of 14.7 cal/mol at 25°C (Vostal, 1972). It is relatively insoluble (~60 µg/L in pure water) compared to $HgCl_2$ (54 g/L). The most prevalent mineral of Hg is cinnabar (HgS). It is found in highest concentration in volcanically active zones such as the circum-Pacific volcanic belt, the mid-ocean ridges and spreading centers, and the global metalliferous belt. The occurrence of Hg in its metallic state is rarer. In water, Hg^{II} may exist as hydroxo and halido complexes (e.g., $HgCl_2$, $Hg(OH)^+$), and the free ion concentration (Hg^{2+}) is typically a small fraction of the total dissolved Hg (Fitzgerald and Lamborg, 2004). Additionally, it is strongly adsorbed to particulate or colloidal materials and also forms strong associations with natural organic matter (NOM) in water, primarily through interaction with thiol ligands (Fitzgerald and Lamborg, 2004). Overall, the concentration of Hg^{II} in natural waters is not controlled by precipitation reactions but rather due to its propensity to bind to solids and biological material, resulting in its rapid scavenging from the water column. In soil, inorganic Hg may exist in the form of Hg^0 vapor adsorbed to soil matter, but it is mostly as Hg^{II}-humate complexes except in the absence of NOM. In addition to ionic Hg, CH_3Hg is also found as complexes or the free ion (CH_3Hg^+) and may be adsorbed to solids in a similar fashion to ionic Hg, although the binding is weaker.

Mercuric ions (Hg^{2+}) and CH_3Hg^+ are toxic because they bind readily to exposed sulfhydryl (–SH) groups of enzyme proteins and are therefore, potent nonspecific enzyme inhibitors (Clarkson and Magos, 2006). Their toxicity can be modulated by the formation of strong complexes that will inhibit such interactions. This is of special significance because it can affect the determination of Hg toxicity for microbes in growth assays (Farrell et al., 1993). Many recent studies have also demonstrated that the rate of uptake of Hg and CH_3Hg into microbes is determined by its complexation in the medium (e.g., Benoit et al., 2003; Hsu-Kim et al., 2013) and that in many instances, this plays a pivotal role in the rate of microbial transformation between the various forms. While passive uptake of neutrally charged complexes has been stated as the most likely uptake mechanism, there is also evidence for facilitated uptake (e.g., Golding et al., 2008; Kelly et al., 2003), and there is evidence for the enhanced uptake into bacteria of complexes such as Hg and CH_3Hg thiol

complexes (Ndu et al., 2012; Schaefer et al., 2011). Additionally, as discussed further in the following text, bacteria with the mer genes have specific membrane transporters for Hg uptake. Daguene et al. (2012) have also shown that the concentration of major cations can change the permeability of the bacterial membrane to Hg complexes and therefore, the uptake rate. In many environments, the bioavailability of the Hg^{II} for methylation is as important a factor controlling CH_3Hg production as the activity of the methylating organisms.

Inorganic Hg compounds are mostly considered less toxic than organic Hg compounds, but as Hg^{II} compounds can be converted in the environment into more toxic organic compounds (e.g., CH_3Hg) (Figure 16.1), focus has shifted to examining the potential for their transformation in the environment (e.g., Benoit et al., 2003). Both methylation and demethylation occur in the environment and both processes are microbially mediated (Benoit et al., 2003; Berman and Bartha, 1986; Marvin-DiPasquale et al., 2000; Schaefer et al., 2004). Living tissue has a high affinity for CH_3Hg, which is bioaccumulated through all levels of the food chain (Wiener et al., 2003). Bioconcentration factors for piscivorous fish are typically 10^5–10^6 times the CH_3Hg concentration found in water. Inorganic Hg is bioconcentrated into microorganisms but is not readily transferred through the food chain, making it similar to other metal cations. The higher bioaccumulation of CH_3Hg relates both to its tendency to bind to thiols in proteins and its enhanced solubilization during food digestion compared to ionic Hg, i.e., its higher assimilation efficiency (e.g., Mason et al., 1996). One suggested reason for its enhanced uptake is that its complex with cysteine (CH_3HgSR; R = remainder of cysteine molecule) is an analog for methionine (CH_3SR) (Clarkson and Magos, 2006).

Dimethylmercury (($CH_3)_2Hg$) is found in ocean waters and some freshwater environments and is volatile (Fitzgerald and Lamborg, 2004; Mason et al., 2012). It can thus enter the atmosphere from soil or water phases where it is photodecomposed by the ultraviolet component of sunlight, forming CH_3Hg or ionic Hg depending on the reaction mechanism (Niki et al., 1983).

While abiotic transformations of Hg occur, microorganisms have been shown to be intimately involved in interconversions of inorganic and organic Hg compounds (Barkay et al., 2003; Benoit et al., 2003; Robinson and Tuovinen, 1984;

Trevors, 1986) (Figure 16.1). The initial discoveries of such microbial environmental methylation has been attributed to Jensen and Jernelöv (1969), who demonstrated the production of CH_3Hg from $HgCl_2$ added to lake sediment samples and incubated for several days. They also noted the production of $(CH_3)_2Hg$ from decomposing fish tissue containing CH_3Hg or supplemented with Hg^{II} after several weeks. Topping and Davies (1981) showed that methylation also occurred in marine systems. Later work has established that environmental methylation is carried out by archaea, bacteria, and fungi (see Section 15.4). Mercury methylation can be influenced by pH (Steffan et al., 1988) and by many other environmental factors, including the presence of sulfide and dissolved organic matter that influence Hg^{II} speciation in the water (Benoit et al., 2003; Graham et al., 2012, 2013; Hsu-Kim et al., 2013). The role of sulfide and NOM is complex, as it has been shown to both enhance and hinder methylation. As noted earlier, these factors impact both the activity of the methylating organisms and the bioavailability of the Hg^{II} to these organisms.

Microbial action on Hg compounds can be a means of detoxification although this is not so for all known microbial reactions. By forming volatile species, Hg^0 (see Section 15.5) or $(CH_3)_2Hg$, microbes can ensure the removal of Hg from their environment into the atmosphere. While not proven, even the microbial formation of CH_3Hg may be a form of Hg detoxification, as it appears in laboratory cultures that most of the CH_3Hg produced is rapidly exported from the cells into the culture medium (e.g., Benoit et al., 2003; Graham et al., 2012). Methylation and rapid export of the products is suggested as a detoxification mechanism for arsenic, and the same may apply for Hg, even though the methylation pathways are different (Mason, 2013). Export of CH_3Hg appears to be more efficient than for Hg^{II}, which binds more strongly to intracellular ligands. The precipitation of HgS by reaction of Hg^{II} with biogenic H_2S could also be a type of Hg detoxification, because the solubility of HgS is very low ($K_{sol} = 10^{-49}$). However, Hg^{II} forms a number of soluble complexes with sulfide and some of these are neutrally charged (e.g., $Hg(SH)_2$, HOHgSH) and are thought to passively diffuse across the membrane of microorganisms (Benoit et al., 2003). Indeed, many of the microbes identified as Hg methylators thrive in reduced sulfidic environments although the accumulated evidence indicates that the methylating ability decreases as sulfide levels increase and that the maximal zone of methylation in environmental sediments is in the oxic/anoxic transition zone (Benoit et al., 2003).

Of all the detoxification mechanisms, formation of volatile Hg^0 has been thought to be the predominant microbial mechanism (Barkay et al., 2003; Robinson and Tuovinen, 1984) with early identification of a significant number of Hg-resistant bacteria (e.g., Baldi et al., 1987, 1989) that could reduce Hg^{II} to Hg^0 but not methylate Hg^{II}. Most of these bacteria were from sites with high Hg content. Some of the strains were also able to degrade CH_3Hg. As discussed further in Section 15.5, reduction and demethylation are under genetic control in environments with elevated Hg and CH_3Hg, but there is also the potential for these reactions to occur and be microbially mediated in more pristine environments, although the pathways are not fully understood (Benoit et al., 2003; Marvin-DiPasquale et al., 2000).

16.3 ANTHROPOGENIC MERCURY

The Hg level in the environment has been largely affected by human activity (Driscoll et al., 2013; Mason et al., 1994). These activities include industrial operations such as the synthesis of vinyl chloride and acetaldehyde that employ Hg^{II} compounds as catalysts and the use of liquid Hg^0 as electrodes in the electrolytic production of chlorine gas and caustic soda (Munthe et al., 2007). Other industrial applications include use in the manufacture of paper pulp, which makes use of phenylmercuric acetate as a slimicide (Jonasson, 1970; Swain et al., 2007). However, because of the resultant environmental pollution, many of these industrial applications are no longer used in many countries. In agriculture, organic compounds used as fungicides to prevent fungal attack of seeds have polluted the soil. Again, these compounds are mostly no longer used. In mining, the exposure of Hg ore deposits and other deposits, typically sulfide ores in which Hg is only a trace component, leads to weathering and resultant solubilization, which introduces the Hg into the environment (Swain et al., 2007). Recently, a global treaty, the Minamata Convention, was signed by 140 countries to reduce the releases of Hg into the environment (UNEP, 2013). This historic treaty will endeavor to limit releases of Hg to the atmosphere and to other environmental compartments.

Currently, the major source of Hg to the environment is due to its release to the atmosphere during the combustion of coal for electricity and other industrial uses (Pirrone et al., 2010; Swain et al., 2007). Mercury is a trace component in coal, being present typically at levels of <1 mg/kg, but its volatility results in its release to the atmosphere during combustion. Other fossil fuels typically have lower levels of Hg than coal. Waste incineration and other combustion processes can also release Hg to the atmosphere.

Mercury has been used for centuries in the extraction of precious metals with which it readily amalgamates. Currently, industrial Hg amalgamation is not the main recovery method for gold (Au) and silver (Ag), but it is used in many countries by artisanal miners (Swain et al., 2007), and the burning of the amalgam to drive off the Hg and recover the Au or Ag is a measurable contributor of Hg^0 to the atmosphere (Pirrone et al., 2010). Artisanal gold mining using Hg is an important source of atmospheric Hg (Pirrone et al., 2010).

16.4 MICROBIAL METHYLATION OF MERCURY

16.4.1 Early Studies of Microbial Mercury Methylation

Early studies of the biochemistry of microbial methylation of Hg using a methanogenic culture cell extract and low concentrations of Hg^{II} found the formation of $(CH_3)_2Hg$ through the interaction of methylcobalamin and Hg^{2+} (Wood et al., 1968). Although this production of methylcobalamin depended on enzymatic catalysis, the production of $(CH_3)_2Hg$ from the reaction of Hg^{2+} with methylcobalamin did not. This apparent nonenzymatic nature of Hg methylation by methylcobalamin was confirmed by others (Bertilsson and Neujahr, 1971: Imura et al., 1971; Schrauzer et al., 1971) and is now a widely used method to synthesize CH_3Hg abiotically in the laboratory. However, as discussed further in the following text, this methylation pathway is related to the pathway now identified in various microorganisms (Choi and Bartha, 1993; Gilmour et al., 2013; Parks et al., 2013). The mechanism of Hg methylation by methylcobalamin was explained as follows (DeSimone et al., 1973):

$$Hg^{2+} \xrightarrow{\text{CH}_3\text{B}_{12}} (CH_3)Hg^+ \xrightarrow{\text{CH}_3\text{B}_{12}} (CH_3)_2Hg$$

(16.1)

According to Wood (1974), the initial methylation of Hg^{2+} in this reaction sequence proceeded 6000 times faster than the second. Some studies of these reactions have indicated that the methylation of CH_3Hg can proceed as fast as the initial methylation of Hg (Baldi, 1997; Filipelli and Baldi, 1993). However, in most environmental systems, and especially in freshwater, the primary product appears to be CH_3Hg (Benoit et al., 2003; Fitzgerald and Lamborg, 2004). In the ocean, $(CH_3)_2Hg$ is found in comparable concentrations to CH_3Hg but the mechanism of its formation is unknown (Fitzgerald et al., 2007; Mason et al., 2012).

Early studies also indicated that $(CH_3)_2Hg$ could arise from the reaction of CH_3Hg^+ with hydrogen sulfide, through the decomposition of the intermediate compound (Baldi, 1997; Craig and Barlett, 1978), although the experiments were done at high concentrations in the laboratory. The transformation of CH_3Hg to $(CH_3)_2Hg$ involved the formation of an intermediate di-monomethylmercurysulfide ($CH_3Hg-S-HgCH_3$), which then decomposed to $(CH_3)_2Hg$ and HgS(s). The final decomposition reaction was shown to be the rate-controlling reaction (Baldi, 1997; Baldi et al., 1993). In nature, the hydrogen sulfide is mostly of biogenic origin, formed anaerobically by sulfate-reducing bacteria, which are also known methylating organisms. There has been little recent study of these reactions and their potential to form at environmental Hg levels. As CH_3Hg^+ can form a number of other sulfide complexes that seem more predominant at lower concentrations, this likely limits the extent of this reaction in nature. In a similar mechanism, conversion of phenylmercuric acetate to diphenylmercury has been reported (Matsumara et al., 1971) in which the diphenylated product was produced as well as an unknown Hg compound and a trace of Hg^{2+}.

Other early studies revealed that Hg can be methylated by microbes other than methanogens, including both aerobes and anaerobes (Robinson and Tuovinen, 1984). Among anaerobes, *Clostridium cochlearium* was shown to methylate various Hg compounds (Yamada and Tonomura, 1972a,b). Among aerobes, *Pseudomonas* spp., *Bacillus megaterium*, *Escherichia coli*, *Enterobacter aerogenes*, and others have been implicated (see summary by Robinson and Tuovinen, 1984). Even fungi such as *Aspergillus niger, Scopulariopsis brevicaulis,* and *Saccharomyces cerevisiae* have been found capable of Hg[II] methylation (Robinson and Tuovinen, 1984). The mechanism of methylation in these instances is not known. Early studies suggested that the fungus *Neurospora crassa* used a different mechanism of methylating Hg[II] (Landner, 1971) through complexation of Hg[II] with homocysteine or cysteine nonenzymatically, and then, with the help of a methyl donor besides methylcobalamin and the enzyme transmethylase, methylmercury is cleaved from this complex. There have been no recent studies to confirm or refute this proposed mechanism.

Bacteria and other microbes have been found to methylate other metals and metalloids (cadmium, lead, tin, thallium, arsenic, selenium, and tellurium; Bentley and Chasteen, 2002; Brinckman et al., 1976; Guard et al., 1981; Schedlbauer and Heumann, 2000; Summers and Silver, 1978; Thayer, 2002; Wong et al., 1975), but in many instances, the reaction pathway is different to that for Hg^{2+} methylation, which requires the addition of a methyl radical (CH$_3$Hg$^{\bullet}$) or a carbanion (CH$_3$Hg$^-$) if the Hg ion is not reduced prior to methylation. There is little evidence for the biotic methylation of Hg^0 in the environment (Benoit et al., 2003), and the recent demonstration that the methylation in the presence of Hg^0 by bacteria in culture requires Hg oxidation prior to methylation supports this contention (Colombo et al., 2013; Hu et al., 2013). In contrast, for As and other metalloids, it has been shown that methylation occurs after reduction of the metal(loid) and is an oxidative methylation mechanism involving a carbocation (CH$_3^+$) (Bentley and Chasteen, 2002; Mason, 2013). There are many more mechanisms for the transfer of a carbocation within a cell than for transfer of a methyl radical and anion, so the pathway of Hg[II] methylation is relatively specific.

Methylation may occur as a result of nonbiological transmethylation by methylated donor compounds of biogenic origin such as trimethyltin and methyl iodide (Brinckman and Olson, 1986). Methyl halides, including methyl iodides (Brinckman and Olson, 1986; White, 1982), are produced by some marine algae and associated microorganisms (Manley and Dastoor, 1988; White, 1982) and by fungi (Harper, 1985). They can react nonenzymatically with Hg[II] (Brinckman and Olson, 1986).

16.4.2 Enzymatic Methylation of Mercury by Microbes

Sulfate reducers, such as *Desulfovibrio desulfuricans,* which has been used as a model organism for Hg methylation studies (Bartha et al., 1994; Gilmour et al., 2011), appear to be the principal methylators of Hg in anoxic estuarine sediments when sulfide is low and fermentable organic energy sources are available (Benoit et al., 2001, 2003; Compeau and Bartha, 1984, 1985). Many studies have shown that various sulfate reducers have this capacity, both in culture and through inhibition experiments with environmental sediments (e.g., Benoit et al., 2001, 2003; Berman and Bartha, 1986; Bridou et al., 2011; King et al., 2000). Environmental methylation has been shown in both saline and freshwater environments. More recently, studies also demonstrated that iron-reducing bacteria (Fleming et al., 2006; Kerin et al., 2006) and methanogens also methylate Hg[II] in culture (Hamelin et al., 2011; Yu et al., 2013). These studies indicated that methylation ability was relatively widespread among bacteria and archaea, although there are both methylators and nonmethylators within a genus, or within closely related microorganisms.

Detailed studies with *D. desulfuricans* showed that the methyl group originated from carbon-3 of the amino acid serine, which was transferred to tetrahydrofolate (THF) to form methylene-THF, a process catalyzed by serine hydroxymethyltransferase (Berman et al., 1990; Choi and Bartha, 1993; Choi et al., 1994a,b). The methylene-THF was then reduced to methyl-THF by reduced ferredoxin. In this organism, the methylation of Hg[II] involved the methyl group of methyl-THF, which was transferred to Hg[II] via a cobalamin–protein complex, the cobalamin being the transfer agent and the protein being the catalyst (enzyme) of the methyl transfer from methyl-THF to Hg^{2+} (Berman et al., 1990; Choi and Bartha, 1993; Choi et al., 1994a,b). It was found that methyl-THF may also arise from other pathways. The normal role of the methylcobalamin–protein complex in *D. desulfuricans* is to provide the methyl group in acetate synthesis from CO$_2$ by the acetyl~SCoA pathway (see Chapters 6 and 19), and Hg^{2+} evidently acts as a competing methyl acceptor in this organism (Choi et al., 1994b). Further studies suggested that the link to the acetyl–CoA pathway was not essential for methylation as some organisms with this pathway did not methylate Hg, while others without the pathway did (Ekstrom et al., 2003).

The studies with dissimilatory iron-reducing bacteria showed that bacteria belonging to the genera *Geobacter* (*G. sulfurreducens*, *G. metallireducens*, and *G. hydrogenophilus*) and *Desulfuromonas palmitatis* had the ability to methylate Hg while reducing Fe(III) and other substrates, whereas *Shewanella* strains were not able to do so (Fleming et al., 2006; Kerin et al., 2006). As Kerin et al. (2006) point out, *Geobacter* and *Desulfuromonas* are closely related to Hg-methylating sulfate reducers in the *Deltaproteobacteria* whereas *Shewanella* is not.

16.4.3 Genetic Basis for Mercury Methylation

The studies by Parks et al. (2013) and Gilmour et al. (2013) have provided a consistent explanation for the methylation of Hg^{II} by these different organisms. Parks et al. demonstrated that specific genes were needed for Hg^{II} methylation in *D. desulfuricans* ND132 and *G. sulfurreducens* PCA by showing that both of the identified two genes (*hgcA* and *hgcB*) were needed for methylation to

occur and removing one stopped methylation. The gene *hgcA* was shown to code for a putative corrinoid protein, while the gene *hgcB* was shown to be a 2[4Fe-4S] ferredoxin. The roles of the proteins are, respectively, methyl donation and electron transfer for the corrinoid cofactor reduction (Figure 16.2). To further demonstrate the necessity of the genes, they showed that methylation could be stimulated in microbes whose genes had been removed by the reinsertion of both genes. Removal of the genes did not impact growth of the bacteria, and therefore, these genes are not essential for cell survival. Overall, the methylation scheme of Parks et al. (2013) is consistent with that of Bartha et al. discussed earlier.

Parks et al. (2013) found the gene pair within the Proteobacteria (24 strains of Deltaproteobacteria), Firmicutes (16 Clostridia strains and one Negativicute) and in the Euryarchaeota (11 Methanomicrobia). Six of these bacteria are known methylators (*D. desulfuricans* ND132, *D. aespoeensis* Aspo-2, *D. africanus* Walvis Bay, *Desulfobulbus propionicus* DSM 2032, *G. sulfurreducens* PCA, and *G. metallireducens* GS-15) (Figure 16.3). Gilmour et al. (2013) further

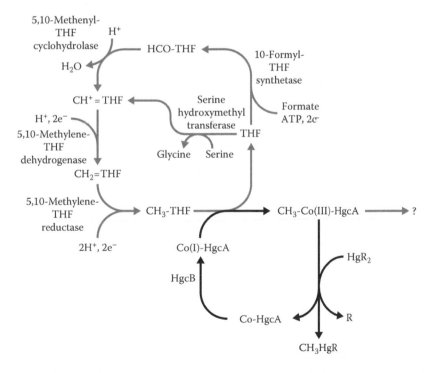

Figure 16.2. Diagrammatic representation of the proposed mercury methylation pathway including the potential sources of the C1 units entering the reductive acetyl–CoA pathway and the mechanisms of transfer from CH$_3$-THF to cobalamin-HgcA. The subsequent methylation pathway is shown including the reduction of cobalamin-HgcA by HgcB. (Figure taken from Parks, JM et al., *Science*, 339, 1332, 2013 and used with permission.)

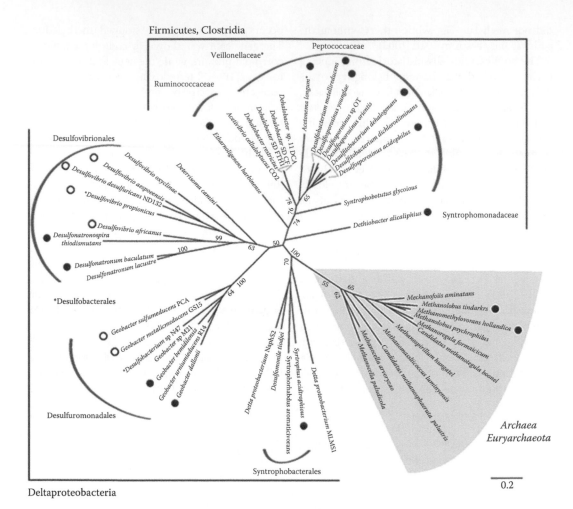

Figure 16.3. Phylogenetic tree using amino acid sequences from all microorganisms with available genome sequences containing hgcAB orthologs. Open circles signify species that were previously known to methylate mercury, while filled circles represent species identified as methylators by Gilmour et al. (2013). (Reproduced with permission from Gilmour, CC. et al., Sulfate-reducing bacterium *Desulfovibrio desulfuricans* ND132 as a model for understanding bacterial mercury methylation, *Appl Environ Microbiol*, 77, 3938–3951. Copyright 2011 American Chemical Society.)

demonstrated that a subset of the organisms identified by Parks et al. (2013) as having the required methylating genes did methylate Hg^{II}, while others without the genes did not. As shown in Figure 16.3, these experiments showed that the following organisms that contained the genes were Hg^{II} methylators (Archaea [*Methanomethylovorans hollandica* and *Methanolobus tindarius*], Firmicutes [*Ethanoligenens harbinense*, *Dethiobacter alkaliphilus*, *Desulfosporosinus acidophilus*, *Desulfitobacterium metallireducens*, *D. dehalogenans*, *Desulfosporosinus youngiae*, and *Acetonema longum*], and Deltaproteobacteria [*Geobacter bemidjiensis*, *G. daltonii*, *Desulfonatronospira thiodismutans*, *Desulfomicrobium baculatum*, and *Syntrophus aciditrophicus*]). Also, Yu et al. (2013) showed that the methanogen *Methanospirillum hungatei* has the two genes and also methylated Hg.

These two studies also showed that related organisms without the genes did not methylate Hg^{II}.

It is worth noting that the experiments of Gilmour et al. (2013) used conditions to enhance the uptake of Hg into the organisms by including cysteine in the medium as it has been shown that methylation is much faster in the presence of cysteine compared to cultures without free cysteine in solution (Schaefer and Morel, 2009; Schaefer et al., 2011). As noted earlier, the speciation of Hg^{II} in the cell's environment may impact the rate of assimilation of Hg^{II} and therefore, its rate of methylation. It has been shown that Hg can be taken up passively by diffusion of neutral complexes across the cell membrane (Mason et al., 1996), and this was thought to be the primary mechanism for uptake in sulfate-reducing bacteria and related organisms as Hg^{II} forms neutral complexes with sulfide (HOHgSH; $Hg(SH)_2$) (Benoit et al., 2003). However, there is also strong evidence of facilitated uptake, especially when Hg is bound to cysteine and other thiols (Golding et al., 2008; Ndu et al., 2012; Schaefer and Morel, 2009; Schaefer et al., 2011). Little is yet known about the uptake mechanisms for Hg^{II} by the recently identified methylating bacteria, but it is apparent that passive diffusion is not the only mechanism of uptake.

Overall, there have been recent dramatic advances in our understanding of the mechanisms of Hg^{II} methylation by microorganisms, and it is clear that there is at least one pathway of methylation that is related to specific genes identified in a broad suite of microbes. Further studies are needed to examine the extent to which these genes are found or expressed in a variety of environmental settings to confirm the overall importance of this pathway in the environment.

16.5 OTHER MICROBIAL INTERACTIONS WITH MERCURY

16.5.1 Enzymatic Microbial Reduction of Mercury

Mercuric ion is reduced to volatile Hg^0 by various members of the domain bacteria. Initially isolated in medical settings and shown to be associated with plasmids that also conferred antibiotic resistance (e.g., Richmond and John, 1964), the genes for Hg^{II} reduction were rapidly shown to be widely distributed in the environment within aerobic and facultative bacteria (Barkay et al., 2003). These genes are mostly associated with the mer operon, discussed further later. It has also been demonstrated that there is a large diversity in the gene structure. Examples of Hg^0-reducing bacteria are strains of Pseudomonas spp., enteric bacteria, Staphylococcus aureus, Acidithiobacillus ferrooxidans, group B Streptococcus, Bacillus, Vibrio, coryneform bacteria, Cytophaga, Flavobacterium, Achromobacter, Alcaligenes, and Acinetobacter and Streptomyces (Barkay et al., 2003; Bogdanova et al., 2001; Komura et al., 1970; Nakahara et al., 1985; Nelson and Colwell, 1974; Nelson et al., 1973; Olson et al., 1981; Osborn et al., 1997; Summers and Lewis, 1973; Yurieva et al., 1997). Many of these bacteria are, as a result, Hg resistant. They possess the mer operon, which may be located on a plasmid, transposon, or bacterial chromosome. Its genetic components include the gene sequence merR, merT, merP, merC, merD, merA, and merB (e.g., Barkay et al.,

2003; Nascimento and Chartone-Souza, 2003) (Figure 16.4). The representation in Figure 16.4 is for the model system in gram-negative bacteria. The gene (merA) codes for the enzyme mercuric reductase (MerA), which is responsible for Hg^{2+} reduction to Hg^0. It is a cytosolic flavin disulfide oxidoreductase using the NAD(P)H system that is active in the presence of an excess of exogenously supplied thiols (RSH). The thiols react with Hg^{2+} to form a thiol complex (RS–Hg–SR) and ensure the reduced state of mercuric reductase and the formation of Hg^0. The reaction catalyzed by mercuric reductase may be summarized as follows (Barkay et al., 2003; Foster, 1987; Fox and Walsh, 1982):

$$NADPH + H^+ + RS-Hg-SR \rightarrow NADP^+ + Hg^0\uparrow + 2RSH \quad (16.2)$$

Gene (merP) codes for a periplasmic protein that scavenges Hg^{2+} from the surrounding medium, and gene (merT) codes for a membrane protein involved in transporting the Hg^{2+} from the periplasm into the cytoplasm where the reduction takes place. Prior to reduction, Hg is bound as a dithiol complex ($HgSR_2$). The expression of the mer operon is controlled by merR as it blocks RNA polymerization until Hg^{II} binds and changes the protein configuration, which results in transcription (see reviews by Barkay et al., 2003; Schelert et al., 2004 for details) (Figure 16.4). The other operons in the mer system (Figure 16.2) are involved in transport (merC) and gene regulation and are not present in all bacteria.

In some reactions of merA, the dimercaptal derivative of Hg^{2+} may be replaced by a monomercaptal or an ethylenediamine derivative. NADPH can be replaced by NADH with enzyme preparations from some organisms, but the preparation is then less active. The kinetics for the purified enzyme is biphasic (see review by Robinson and Tuovinen (1984) for a more detailed discussion). Although the reaction occurs under reducing conditions, it is performed by many obligate and facultative aerobes (e.g., Barkay et al., 2003; Nelson et al., 1973; Spangler et al., 1973a,b).

Mercuric reductase activity is not entirely substrate specific. Besides the mercuric ion, the enzyme can also reduce ionic silver and ionic gold to corresponding metal colloids (Summers, 1986; Summers and Sugarman, 1974). Silver and gold resistance in bacteria is not, however, related to this enzyme activity (Summers, 1986).

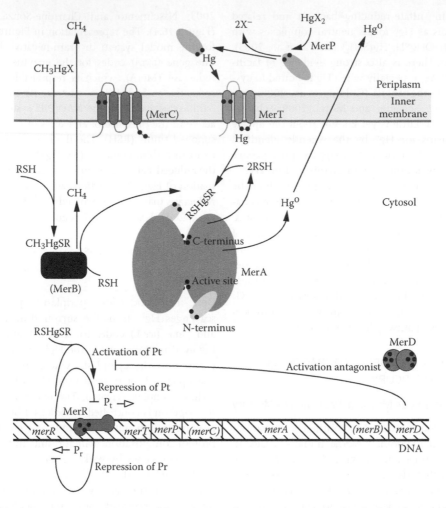

Figure 16.4. A diagrammatic representation of the pathways and major *mer* genes involved in the detoxification of mercury and methylmercury. (Figure reprinted from Barkay, T. et al., *FEMS Microbiol Rev*, 27, 355, 2003 and used with permission.)

Information on the ability of *archaea* to reduce Hg^{2+} to Hg^0 is more limited (Wang et al., 2011). Schelert et al. (2004) first detected the presence of Hg-resistant *Crenarchaeota* in Coso Hot Springs of Yellow Stone National Park, United States, whose DNA included a *merA* gene. Further study of the *merA* gene in *Sulfolobus solfataricus* revealed that it was constitutive, unlike *merA* in *bacteria*. Nevertheless, expression of *merR* was shown to be essential for *S. solfataricus* to be resistant to Hg^{2+} (Simbahan et al., 2004). A number of studies have now confirmed that Hg resistance involving the mer system is widespread among thermophilic archaea and bacteria (Freedman et al., 2012; Simbahan et al., 2005; Vetriani et al., 2005; Wang et al., 2011).

Various species of the yeast and microfungi have been shown to reduce Hg^{II} (Berdicevsky et al., 1993; Kelly et al., 2006). Brunker and Bott (1974)

found that *Cryptococcus albidus* reduced Hg^{2+} to Hg^0. However, the Hg^0 appeared to accumulate in the cell wall, membrane, and vacuoles of the yeast. Hg^0 volatilization was not reported. Two *Cryptococcus* species *fungi*, *Eukaryota*, *C. albidus* and *C. neoformans*, have been shown to contain a *merA* gene (see Tax Blast report: *Cryptococcus*,%20Hg/927499...928899. fa_tax. html), although there is no evidence for this in the published literature. Kelly et al. (2006) examined microfungi and found that volatilization of Hg^0 occurred at high exposure concentrations but that at lower exposures, Hg^{II} was precipitated as HgS(s).

16.5.2 Other Pathways for Mercury Reduction

Not all reduction of Hg observed in nature is biological (Barkay et al., 2003; Nelson and Colwell, 1975; Sciliano et al., 2002) or results from the mer pathway.

EHRLICH'S GEOMICROBIOLOGY

Many field studies in both the ocean and in freshwaters have found a relationship between the levels of Hg^0 and primary productivity suggesting a biological role in its formation. Microbial reduction can occur via various pathways and has been shown to occur in algae cultures (Ben-Bassat and Mayer, 1978; Mason et al., 1995; Poulain et al., 2004; Siciliano et al., 2002). The role of phytoplankton has been implied from the results of photosynthesis inhibition experiments and the lack of reduction in dark samples. Recently, there has been evidence that cyanobacteria are able to reduce Hg using an enzyme that is similar to merA (Marteyn et al., 2013). The bacteria were able to reduce both Hg and uranium, and there were similarities in the pathway to that of the merA reduction system. This result provides evidence that phototrophs also have a dedicated pathway for Hg reduction.

Other pathways for reduction exist. Metal-reducing bacteria appear capable of Hg reduction (Wiatrowski et al., 2006). It has also been shown that an iron-oxidizing bacteria can produce Hg^0 when exposed to either inorganic Hg or organic mercurial (Sugio et al., 2010; Takeuchi et al., 2005). The authors demonstrate that the pathway involves cytochrome c oxidase enzymes. Additionally, it appears that Hg reduction can occur by the direct interaction of the cation with the surface of Fe minerals such as magnetite (Wiatrowski et al., 2009).

Additionally, chemical reduction, especially when photochemically mediated, has been demonstrated, and it has been shown that UV is important in mediating the reduction (e.g., Amyot et al., 1994, 1997; Mason et al., 1995; Whalin and Mason, 2006). It has been shown that the interaction with humic acid can enhance the reduction (Alberts et al., 1974; Gardfelt and Jonsson, 2003). The abiotic formation of Hg^0 tends to dominate in surface waters, while biological production is likely to dominate in deeper waters or in the dark.

16.5.3 Microbial Decomposition of Organomercurials

Phenyl- and CH_3Hg have been shown to be microbially converted to volatile Hg^0 by bacteria in lake and estuarine sediments and in soil (e.g., Barkay et al., 2003; Benoit et al., 2003; Bogdanova et al., 2001; Bruce et al., 1992; Nelson and Colwell, 1975; Nelson et al., 1973; Osborn et al., 1993; Spangler

et al., 1973a,b; Tonomura et al, 1968; Yurieva et al., 1997). The bacteria involved are Hg-resistant strains of Pseudomonas and other genera, such as E. coli and S. aureus (Nelson and Colwell, 1975). The removal of phenyl or methyl groups linked to Hg^{II} is catalyzed by a class of enzymes called mercuric lyases, encoded by the gene merB in the mer operon (see Section 15.5.1). These genes have been shown to be active in the environment (e.g., Schaefer et al., 2004). They catalyze the cleavage of carbon–Hg bonds and in laboratory studies require the presence of an excess of a reducing agent such as L-cysteine. They release Hg^{2+}, which is then reduced to Hg^0 by mercuric reductase (Barkay et al., 2003; Furukawa and Tonomura, 1971, 1972a,b; Robinson and Tuovinen, 1984; Tezuka and Tonomura, 1976, 1978). Phenylmercuric lyase can be inducible (Nelson et al., 1973; Robinson and Tuovinen, 1984). The overall reactions the lyases catalyze can be summarized as follows:

$$C_6H_5Hg^+ + H^+ + 2e^- \rightarrow Hg^0 + C_6H_6 \quad (16.3)$$

$$CH_3Hg^+ + H^+ + 2e^- \rightarrow Hg^0 + CH_4 \quad (16.4)$$

Besides methyl- and phenylmercury compounds, some bacteria are able to decompose one or more of the following organomercurials: ethylmercuric chloride, fluorescein mercuric acetate, para-mercuribenzoate (pHMB), thimerosal, and merbromin (see review by Robinson and Tuovinen, 1984).

In general, resistance to the toxicity of inorganic or organic Hg compounds in bacteria is attributable to the ability to form mercuric reductase and/or mercuric lyase. However, in a strain of E. aerogenes, bacterial resistance to some organomercurials may be due to membrane impermeability (Pan Hou et al., 1981), and in C. cochlearium, it is due to demethylation followed by precipitation with H_2S generated by the organism (Pan Hou and Imura, 1981).

As previously noted, the genes encoding mercuric reductase and mercuric lyase are part of the mer operon, which may occur on a plasmid (Belliveau and Trevors, 1990; Komura and Izaki, 1971; Loutit, 1970; Novick 1967; Richmond and John, 1964; Schottel et al., 1974; Silver, 1997; Silver and Phung, 1996; Smith, 1967; Summers and Silver, 1972), on a transposon, or in a bacterial chromosome (e.g., Barkay et al., 2003; Foster, 1987; Nascimento and Chartone-Souza, 2003). Except in A. ferrooxidans, the Hg resistance genes (mer) in all bacteria so far tested are expressed only in the presence of Hg compounds, i.e., the gene products for which they code are inducible (Robinson and Tuovinen, 1984). Depending on the organism, induction of the two enzymes, mercuric lyase and mercuric reductase, may be coordinated, i.e., the two genes are under common regulatory control. In such an instance, an organomercurial induces both the lyase and the mercuric reductase (see Robinson and Tuovinen, 1984). In *A. ferrooxidans* (formerly *T. ferrooxidans*), the mercuric resistance (Hg^r) trait appears to be constitutive (Olson et al., 1982). In *C. cochlearium*, which lacks the mer operon, the genetic determinants for demethylation of CH_3Hg are also plasmid encoded (Pan Hou and Imura, 1981). The demethylated Hg may be precipitated as HgS with biogenic H_2S to render it nontoxic.

Besides the mer pathway, there is also environmental demethylation of CH_3Hg by mechanisms that are not as well understood (Barkay et al., 2003; Marvin-DiPasquale et al., 2000). In uncontaminated environments, the major products appear to be Hg^{II} and CO_2, and therefore, this pathway has been termed *oxidative demethylation*, in contrast to *reductive demethylation*, where CH_4 and Hg^0 are the major product (Marvin-DiPasquale et al., 2000). The mechanism of oxidative demethylation may be analogous to monomethylamine degradation by methanogens or acetate oxidation by sulfate-reducing bacteria. Reductive demethylation appears to be prevalent in more contaminated environments.

While it has been demonstrated that methane and Hg^0 are the primary products of mer-mediated Hg demethylation, CO_2 has also been observed as a major demethylation product in many studies (Marvin-DiPasquale et al., 2000; Oremland et al., 1991, 1995). A variety of aerobes and anaerobes have been implicated in carrying out oxidative demethylation, which has been observed in freshwater, estuarine, and alkaline-hypersaline sediments (Marvin-DiPasquale and Oremland, 1998; Oremland et al., 1991, 1995). However, the identity of the organisms responsible remains poorly understood and no specific organism has been isolated. One study confirmed the ability of two sulfate-reducing bacterial strains and one methanogen strain to demethylate Hg in pure culture (Pak and Bartha, 1998). The authors however argued that the CO_2 seen in these studies resulted from oxidation of methane released from CH_3Hg after cleavage via organomercurial

lyase and was actually a secondary product and not the primary product of demethylation. However, this view is not universally accepted based on other studies under both aerobic and anaerobic conditions (Marvin-DiPasquale et al., 2000). Overall, the relative importance of mer-mediated versus oxidative demethylation in the environment is poorly understood (Barkay et al., 2003; Benoit et al., 2003). In systems that are not highly contaminated, oxidative demethylation appears to dominate, under both aerobic and anaerobic conditions. The Hg concentrations that would cause a switch from one pathway to the other are only loosely defined. The end product of oxidative demethylation has been presumed to be Hg(II), but that has not been confirmed in most studies.

16.5.4 Oxidation of Metallic Mercury

Elemental mercury (Hg^0) has been reported to be oxidized to Hg^{II} in the presence of certain bacteria (Holm and Cox, 1975). Strains of *P. aeruginosa*, *P. fluorescens*, *E. coli*, and *Citrobacter* oxidized small amounts of Hg^0, while strains of *B. subtilis* and *B. megaterium* oxidized more significant amounts. In none of these cases was CH_3Hg formed. The observed oxidation was reported not to have been enzymatic but due to reaction with metabolic products, which acted as oxidants. Even yeast extract was found to be able to oxidize Hg^0.

By contrast, Smith et al. (1998) demonstrated enzymatic oxidation of metallic Hg in vapor form (monatomic) by hydroperoxidase-catalase, KatG, in growing *E. coli*. The hydroperoxidase-catalase active in stationary cells, Kat E, was found proportionately less active. *B. subtilis* PY79 exhibiting strong catalase activity also promoted Hg^0 oxidation, as did *Streptomyces venezuelae* with weaker catalase activity (Smith et al., 1998). Because of the great toxicity of Hg^{2+}, it is unclear how this activity benefits organisms capable of it unless it occurs in environments where Hg^{2+} is rapidly immobilized. The oxidation of monatomic Hg by catalase has been previously observed in mammals and plants (see refs. cited by Smith et al., 1998).

Recently, it has been shown that the methylating bacteria *D. desulfuricans* ND132 can oxidize and methylate Hg in laboratory culture (Colombo et al., 2013). The authors determined that the oxidation of Hg^0 occurred outside the cell and that the cells were likely taking up Hg^{II} as a thiol complex. Therefore, the methylation pathway was not different from that described earlier but was linked to the organism's ability to oxidize Hg^0. Hu et al. (2013) examined the potential for both iron and sulfate-reducing bacteria to reduce and methylate Hg, finding that some bacteria can do both, while some can reduce Hg but not methylate the ionic Hg formed. A coupled oxidation–methylation pathway could be relatively prevalent in the environment as there are many mechanisms for biotic and abiotic oxidation, and this would result in bioavailable Hg^{II} for methylation.

Abiotic oxidation of Hg^0 has been found in environmental waters, especially in the presence of sunlight, although the exact mechanisms are not well known and likely depend on the presence of various oxidants (e.g., Lalonde et al., 2001). In many studies, there is little difference between the oxidation rate in filtered and unfiltered waters (Quereshi et al., 2010; Whalin et al., 2007) or in dark controls relative to light exposures (Lalonde et al., 2001) indicating that in the environment, oxidation is mostly abiotic. In surface ocean waters, it appears that the rates of photochemical oxidation and reduction are of similar magnitude (e.g., Whalin and Mason, 2007). Further assessment is needed of the relative contributions of biotic and abiotic Hg^0 oxidation in different environments to the overall fate and transport of Hg.

16.5.5 Microbial Precipitation of Mercury

The cyanobacteria Limnothrix planctonica (Lemm.), Synechococcus leopoliensis (Racib.) Komarek, and Phormidium limnetica (Lemm.) were recently shown to form mercuric sulfide ($HgS(s)$) from Hg^{II} and to accumulate it intracellularly (Lefebvre et al., 2007). The sulfide with which Hg^{2+} taken into the cells was combined was formed by the cells from an intracellular thiol pool, as demonstrated by use of the inhibitors dimethylfumarate and iodoacetamide. On initial exposure to Hg(II) in the medium in laboratory experiments, some Hg^0 was formed as well as ß-HgS, but the rate of Hg^0 formation decreased rapidly. Increase in growth temperature enhanced ß-HgS formation and decreased Hg^0 evolution. As noted earlier, microfungi also can precipitate Hg^{II} with the relative amount of precipitation versus Hg^0 formation being a function of the exposure concentration (Kelly et al., 2006).

16.6 SUMMARY

Mercuric ion (Hg^{2+}) may be microbially methylated to CH_3Hg. The enhanced trophic transfer and bioaccumulation of CH_3Hg results in its larger impact on human health and wildlife, except for specific point source contamination locations where exposure to ionic or Hg^0 may be substantial. Some bacteria can further methylate CH_3Hg forming volatile $(CH_3)_2Hg$. Organic Hg compounds such as CH_3Hg and phenylmercury can be enzymatically reduced to volatile Hg^0 by some bacteria using the mer pathway. Most importantly, Hg^{II} can be enzymatically reduced to Hg^0 by bacteria and fungi. Metallic Hg may also be oxidized to Hg^{II} by bacteria although this pathway has not been considered important in the environment. The recent demonstration that the oxidation pathway is relatively prevalent in the environment however counters this notion. Overall, microbes capable of metabolizing the various forms of Hg are generally resistant to its toxic effects. Given these interactions, it is apparent that the Hg cycle in nature is strongly impacted by the activities of microorganisms.

REFERENCES

Alberts JJ, Schindler JE, Miller RW, Nutter DE Jr. 1974. Elementary mercury evolution mediated by humic acid. *Science* 184:895–897.

Amyot M, Gill GA, Morel FMM. 1997. Production and loss of dissolved gaseous mercury in coastal sea water. *Environmental Science & Technology* 31:3606–3611.

Amyot M, Mierle G, Lean DRS, McQueen DJ. 1994. Sunlight-induced formation of dissolved gaseous mercury in lake waters. *Environmental Science & Technology* 28: 2366–2371.

Anderson A. 1967. Mercury in soil. *Grundförbättring* 20:95–105.

Baldi, F. 1997. Microbial transformation of mercury species and their importance in the biogeochemical cycle of mercury. In: Sigel A, Sigel H, eds., *Metal Ions in Biological Systems*. New York: Marcel Dekker, pp. 213–257.

Baldi F, Filippelli M, Olson GJ. 1989. Biotransformation of mercury by bacteria isolated from a river collecting cinnabar mine waters. *Microbial Ecology* 17:263–274.

Baldi F, Olson GJ, Brinckman FE. 1987. Mercury transformation by heterotrophic bacteria isolated from cinnabar and other metal sulfide deposits in Italy. *Geomicrobiology Journal* 5:1–16.

Baldi F, Pepi M, Filippelli M. 1993. Methylmercury resistance in Desulfovibrio desulfuricans strains in relation to methylmercury degradation. Applied and Environmental Microbiology 59:2479–2485.

Barkay T, Miller SM, Summers AO. 2003. Bacterial mercury resistance from atoms to ecosystems. FEMS Microbiology Reviews 27:355–384.

Bartha R, Chase TJ, Choi S-C. 1994. Metabolic pathways leading to mercury methylation in Desulfovibrio desulfuricans LS. Applied and Environmental Microbiology 60:4072–4077.

Belliveau BH, Trevors JT. 1990. Mercury resistance determined by a self–transmissible plasmid in Bacillus cereus 5. Biology of Metals 3:188–196.

Ben-Bassat D, Mayer AM. 1978. Light-induced mercury volatilization and oxygen evolution in Chlorella and the effect of DCMU and methylamine. Physiologia Plantarum 42:33–38.

Benoit JM, Gilmour CC, Heyes A, Mason RP, Miller CL. 2003. Geochemical and biological controls over methylmercury production and degradation in aquatic ecosystems. In: Cai Y, Braids OC, eds., Biogeochemistry of Environmentally Important Trace Elements. ACS Symposium Series, Washington DC: American Chemical Society, pp. 262–297.

Benoit JM, Gilmour CC, Mason RP. 2001. Aspects of bioavailability of mercury for methylation in pure cultures of Desulfobulbus propionicus (1pr3). Journal of Bacteriology 183:51–58.

Bentley R, Chasteen TG. 2002. Microbial methylation of metalloids: Arsenic, antimony, and bismuth. Microbiology and Molecular Biology Reviews 66:250–271.

Berdicevsky I, Duek L, Merzbach D, Yannai S. 1993. Susceptibility of different yeast species to environmental toxic metals. Environmental Pollution 80:41–44.

Berman M, Bartha R. 1986. Levels of chemical versus biological methylation of mercury in sediments. Bulletin of Environmental Contamination and Toxicology 36:401–404.

Berman M, Chase T Jr, Bartha R. 1990. Carbon flow in mercury biomethylation by Desulfovibrio desulfuricans. Applied and Environmental Microbiology 56:298–300.

Betilsson L, Neujahr HY. 1971. Methylation of mercury compounds by methylcobalamin. Biochemistry 10:2805–2808.

Bogdanova E, Minakhin L, Bass I, Volodin A, Hobman JL, Nikiforov V. 2001. Class II broad-spectrum mercury resistance transposons in Gram-positive bacteria from natural environments. Research in Microbiology 152:503–514.

Bogdanova ES, Mindlin SZ, Pakrova E, Kocur M, Rouch DA. 1992. Mercuric reductase in environmental gram-positive bacteria sensitive to mercury. FEMS Microbiology Letters 97:95–100.

Bridou R, Monperrus M, Gonzalez PR, Guyoneaud R, Amouroux D. 2011. Simultaneous determination of mercury methylation and demethylation capacities of various sulfate-reducing bacteria using species-specific isotopic tracers. Environmental Toxicology and Chemistry 30:337–344.

Brinckman FE, Iverson WP, Blair W. 1976. Approaches to the study of microbial transformation of metals. In: Sharpley JM, Kaplan AM, eds., Proceedings of the Third International Biodegradation Symposium. London, U.K.: Applied Science Publishers, pp. 916–936.

Brinckman FE, Olson GJ. 1986. Chemical principles underlying bioleaching of metals from ores and solid wastes, and bioaccumulation of metals from solution. In: Ehrlich HL, Holmes DE, eds., Workshop on Biotechnology for the Mining, Metal-Refining, and Fossil Fuel Processing Industries. Biotechnology and Bioengineering Symposium 16. New York: Wiley, pp. 35–44.

Brown TA, Smith DG. 1977. Cytochemical-localization of mercury in Cryptococcus-albidus grown in presence of mercuric-chloride. Journal of General Microbiology 99:435–439.

Bruce KD, Hiorns WD, Hobman JL, Osborn AM, Strike P, Ritchie DA. 1992. Amplification of DNA from native populations of soil bacteria by using the polymerase chain-reaction. Applied and Environmental Microbiology 58:3413–3416.

Brunker RL, Bott TL. 1974. Reduction of mercury to the elemental state by a yeast. Applied Microbiology 27:870–873.

Choi S-C, Bartha R. 1993. Cobalamin-mediated mercury methylation by Desulfovibrio desulfuricans LS. Applied and Environmental Microbiology 59:290–295.

Choi S-C, Chase T Jr, Bartha R. 1994a. Enzymatic catalysis of mercury methylation by Desulfovibrio desulfuricans LS. Applied and Environmental Microbiology 60:1342–1346.

Choi S-C, Chase T Jr, Bartha R. 1994b. Metabolic pathways leading to mercury methylation in Desulfovibrio desulfuricans LS. Applied and Environmental Microbiology 60:4072–4077.

Clarkson TW, Magos L. 2006. The toxicology of mercury and its chemical compounds. Critical Reviews in Toxicology 36:609–662.

Colombo MJ, Ha JY, Reinfelder JR, Barkay T, Yee N. 2013. Anaerobic oxidation of Hg(0) and methylmercury formation by Desulfovibrio desulfuricans ND132. Geochimica et Cosmochimica Acta 112:166–177.

Compeau G, Bartha R. 1984. Methylation and demethylation of mercury under controlled redox, pH, and salinity conditions. Applied and Environmental Microbiology 48:1203–1207.

Compeau G, Bartha R. 1985. Sulfate-reducing bacteria: Principal methylators of mercury in anoxic estuarine sediments. *Applied and Environmental Microbiology* 50:498–502.

Cotruvo JA, Vogt CD. 1990. Rationale of water quality standards and controls. In: *American Water Works Association. Water Quality and Treatment. A Handbook of Community Water Supplies*, 4th ed. New York: McGraw-Hill, pp. 1–62.

Craig PJ, Bartlett PD. 1978. The role of hydrogen sulfide in the environmental transport of mercury. *Nature (London)* 275:635–637.

Daguene V, McFall E, Yumvihoze E, Xiang SR, Amyot M, Poulain AJ. 2012. Divalent base cations hamper Hg-II uptake. *Environmental Science & Technology* 46:6645–6653.

DeSimone RE, Penley MW, Charbonneau L, Smith SG, Wood JKM, Hill HAO, Pratt JM, Ridsale S, Williams JP. 1973. The kinetics and mechanism of cobalamin-dependent methyl and ethyl transfer to mercuric ion. *Biochimica et Biophysica Acta* 304:851–863.

Driscoll CT, Mason RP, Chan HM, Jacob DJ, Pirrone N. 2013. Mercury as a global pollutant: Sources, pathways, and effects. *Environmental Science & Technology* 47:4967–4983.

Ekstrom EB, Morel FMM, Benoit JM. 2003. Mercury methylation independent of the acetyl-coenzyme a pathway in sulfate-reducing bacteria. *Applied and Environmental Microbiology* 69:5414–5422.

EPA. 2013. US Environmental Protection Agency mercury advisory website, accessed September 23, 2013; http://www.epa.gov/hg/advisories.htm.

Fagerstrom T, Jernelöv A. 1971. Formation of methylmercury from pure mercuric sulfide in aerobic organic sediment. *Water Research* 5:121–122.

Farrell RE, Germida JJ, Huang PM. 1993. Effects of chemical speciation in growth media on the toxicity of mercury(II). *Applied and Environmental Microbiology* 59:1507–1514.

Filippelli M, Baldi F. 1993. Alkylation of ionic mercury to methylmercury and dimethylmercury by methylcobalamin: Simultaneous determination by purge-and-trap GC in line with FTIR. *Applied Organometallic Chemistry* 7:487–493.

Fitzgerald WF, Clarkson TW. 1991. Mercury and monomethylmercury—Present and future concerns. *Environmental Health Perspectives* 96:159–166.

Fitzgerald WF, Lamborg CH. 2004. Geochemistry of mercury in the environment. In: Holland HD, Turekian KK, Exec. eds., *Environmental Geochemistry*, Vol. 9 of *Treatise of Geochemistry*, Chapter 9.04, Amsterdam: Elsevier, pp. 107–148.

Fitzgerald WF, Lamborg CH, Hammerschmidt CR. 2007. Marine biogeochemical cycling of mercury. *Chemical Reviews* 107:641–662.

Fleming EJ, Mack EE, Green PG, Nelson DC. 2006. Mercury methylation from unexpected sources: Molybdate-inhibited freshwater sediments and an iron-reducing bacterium. *Applied and Environmental Microbiology* 72:457–464.

Foster TJ. 1987. The genetics and biochemistry of mercury resistance. *CRC Critical Reviews in Microbiology* 15:117–140.

Fox B, Walsh CT. 1982. Mercuric reductase: Purification and characterization of a transposon-encoded flavoprotein containing an oxidation-reduction-active disulfide. *Journal of Biological Chemistry* 257:2498–2503.

Freedman Z, Zhu CS, Barkay T. 2012. Mercury resistance and mercuric reductase activities and expression among chemotrophic thermophilic aquificae. *Applied and Environmental Microbiology* 78:6568–6575.

Furukawa K, Tonomura K. 1971. Enzyme system involved in decomposition of phenyl mercuric acetate by mercury-resistant *Pseudomonas. Agricultural and Biological Chemistry* 35:604–610.

Furukawa K, Tonomura K. 1972a. Metallic mercury-releasing enzyme in mercury-resistant *Pseudomonas. Agricultural and Biological Chemistry* 36:217–226.

Furukawa K, Tonomura K. 1972b. Induction of metallic-mercury-releasing enzyme in mercury resistant Pseudomonas. *Agricultural and Biological Chemistry* 36(Suppl 13):2441–2448.

Gardfeldt K, Jonsson M. 2003. Is bimolecular reduction of Hg(II) complexes possible in aqueous systems of environmental importance. *Journal of Physical Chemistry A* 107:4478–4482.

Gilmour CC et al. 2011. Sulfate-reducing bacterium *Desulfovibrio desulfuricans* ND132 as a model for understanding bacterial mercury methylation. *Applied and Environmental Microbiology* 77:3938–3951.

Gilmour CC, Podar M, Bullock AL, Graham AM, Brown SD, Somenahally AC, Johs A, Hurt RA, Bailey KC, Elias DA. September 2013. Mercury methylation by novel microorganisms from new environments. *Environmental Science Technology*.

Golding GR, Kelly CA, Sparling R, Loewen PC, Rudd JWM, Barkay T. 2002. Evidence for facilitated uptake of Hg(II) by *Vibrio anguillarum* and *Escherichia coli* under anaerobic and aerobic conditions. *Limnology and Oceanography* 47:967–975.

Golding GR, Sparling R, Kelly CA. 2008. Effect of pH on intracellular accumulation of trace concentrations of Hg(II) in Escherichia coli under anaerobic conditions, as measured using a merlux bioreporter. Applied and Environmental Microbiology 74:667–675.

Goldwater LJ. 1972. Mercury, Baltimore, MD: York Press, 318 pp.

Graham AM, Aiken GR, Gilmour CC. 2013. Effect of dissolved organic matter source and character on microbial Hg methylation in Hg-S-DOM solutions. Environmental Science & Technology 47:5746–5754.

Graham AM, Bullock AL, Maizel AC, Elias DA, Gilmour CC. 2012. Detailed assessment of the kinetics of Hg-cell association, Hg methylation, and methylmercury degradation in several Desulfovibrio species. Applied and Environmental Microbiology 78:7337–7346.

Guard HE, Cobet AB, Coleman WM III. 1981. Methylation of trimethyltin compounds by estuarine sediments. Science 213:770–771.

Gupta A, Phung LT, Chakravarty L, Silver S. 1999. Mercury resistance in Bacillus cereus RC607: Transcriptional organization and two new open reading frames. Journal of Bacteriology 181:7080–7086.

Hamelin S, Amyot M, Barkay T, Wang YP, Planas D. 2011. Methanogens: Principal methylators of mercury in lake periphyton. Environmental Science & Technology 45:7693–7700.

Hamlett NV, Landsdale EC, Davis BH, Summers AO. 1992. Roles of the Tn21 merT, merP, and merC gene products in mercury resistance and mercury binding. Journal of Bacteriology 174:6377–6385.

Harper DB. 1985. Halomethane from halide ion: A highly effective fungal conversion of environmental significance. Nature (London) 315:55.

Holm HW, Cox MF. 1975. Transformation of elemental mercury by bacteria. Applied Microbiology 29:491–494.

Hsu-Kim et al. 2013. Mechanisms regulating mercury bioavailability for methylating microorganisms in the aquatic environment: A critical review. Environmental Science & Technology 47:2441–2456.

Hu, HY, Lin, H, Zheng, W, Tomanicek, SJ, Johs, A, Feng, XB, Elias, DA, Liang, LY, Gu, BH. 2013. Oxidation and methylation of dissolved elemental mercury by anaerobic bacteria. Nature Geoscience 6:751–754.

Imura N, Sukegawa E, Pan S-K, Nagao K, Kim J-Y, Kwan T, Ukita T. 1971. Chemical methylation of inorganic mercury with methyl-cobalamin, a vitamin B_{12} analog. Science 172:1248–1249.

Jensen S, Jernelöv A. 1969. Biological methylation of mercury in aquatic organisms. Nature (London) 223:753–754.

Jonasson IP. 1970. Mercury in the natural environment. A review of recent work. Geological Survey of Canada Paper, p. 57.

Kelly DJA, Budd K, Lefebvre DD. 2006. The biotransformation of mercury in pH-stat cultures of microfungi. Canadian Journal of Botany 84:254–260.

Kelly CA, Rudd JWM, Holoka MH. 2003. Effect of pH on mercury uptake by an aquatic bacterium: Implications for Hg cycling. Environmental Science & Technology 37:2941–2946.

Kerin EJ, Gilmour CC, Roden E, Suzuki MR, Coates JD, Mason RP. 2006. Mercury methylation by dissimilatory iron-reducing bacteria. Applied and Environmental Microbiology 72:7919–7921.

King JK, Kostka JE, Frischer ME, Saunders FM. 2000. Sulfate-reducing bacteria methylate mercury at variable rates in pure culture and in marine sediments. Applied and Environmental Microbiology 66:2430–2437.

Komura I, Izaki K. 1971. Mechanism of mercuric chloride resistance in microorganisms. I. Vaporization of a mercury compound from mercuric chloride by multiple drug resistant strains of Escherichia coli. Journal of Biochemistry (Tokyo) 70:885–893.

Komura I, Izaki K, Takahashi H. 1970. Vaporization of inorganic mercury by cell-free extracts of drug-resistant Escherichia coli. Agricultural and Biological Chemistry 34:480–482.

Korthals ET, Winfrey MR. 1987. Seasonal and spatial variations in mercury methylation and demethylation in an oligotrophic lake. Applied and Environmental Microbiology 53:2397–2404.

Lalonde JD, Amyot M, Kraepiel AML, Morel FMM. 2001. Photooxidation of Hg(0) in artificial and natural waters. Environmental Science & Technology 35:1367–1372.

Landner L. 1971. Biochemical model for the biological methylation of mercury suggested from methylation studies in vivo with Neurospora crassa. Nature (London) 230:452–454.

Lefebvre D, Kelly D, Budd K. 2007. Biotransformation of Hg(II) by cyanobacteria. Applied and Environmental Microbiology 73:243–249.

Loutit JS. 1970. Mating systems of Pseudomonas aeruginosa strain I. VI. Mercury resistance associated with the sex factor (FP). Genetical Research 16:179–184.

Mahaffey KR, Clickner RP, Bodurow CC. 2004. Blood organic mercury and dietary mercury intake: National Health and Nutrition Examination Survey, 1999 and 2000. Environmental Health Perspectives 112:562–570.

Manley SL, Dastoor MN. 1988. Methyl iodide (CH_3I) production by kelp and associated microbes. *Marine Biology (Berlin)* 98:477–482.

Mason RP. 2013. *Trace Metals in Aquatic Systems*, Chichester, U.K.: Wiley-Blackwell, 413 pp.

Marteyn B, Sakr S, Farci S, Bedhomme M, Chardonnet S, Decottignies P, Lemaire SD, Cassier-Chauvat C, Chauvat F. 2013. The Synechocystis PCC6803 mera-like enzyme operates in the reduction of both mercury and uranium under the control of the glutaredoxin 1 enzyme. *Journal of Bacteriology* 195:4138–4145.

Marvin-Dipasquale M, Oremland R. 1998. Bacterial methylmercury degradation in Florida Everglades peat sediment. *Environmental Science & Technology* 32:2556–2563.

Marvin-Dipasquale M et al. 2000. Methyl-mercury degradation pathways: A comparison among three mercury-impacted ecosystems. *Environmental Science & Technology* 34:4908–4916.

Mason RP et al. 2012. Mercury biogeochemical cycling in the ocean and policy implications. *Environmental Research* 119:101–117.

Mason RP, Fitzgerald WF, Morel FMM. 1994. The biogeochemical cycling of elemental mercury—Anthropogenic influences. *Geochimica et Cosmochimica Acta* 58:3191–3198.

Mason RP, Morel FMM, Hemond HF. 1995. The role of microorganisms in elemental mercury formation in natural-waters. *Water, Air, and Soil Pollution* 80:775–787.

Mason RP, Reinfelder JR, Morel FMM. 1996. Uptake, toxicity, and trophic transfer of mercury in a coastal diatom. *Environmental Science & Technology* 30:1835–1845.

Matsumara F, Gotoh Y, Brush GM. 1971. Phenyl mercuric acetate: Metabolic conversion by microorganisms. *Science* 173:49–51.

Munthe J et al. 2007. Recovery of mercury-contaminated fisheries. *Ambio* 36:33–44.

Nakahara H, Schottle JL, Yamada T, Miyagawa Y, Asakawa M, Harville J, Silver S. 1985. Mercuric reductase enzymes from *Streptomyces* species and group B *Streptococcus*. *Journal of General Microbiology* 131:1053–1059.

Nascimento AMA, Chartone-Souza E. 2003. Operon mer: Bacterial resistance to mercury and potential for bioremediation of contaminated environments. *Genetics and Molecular Research* 2:92–101.

Ndu U, Mason RP, Zhang H, Lin SJ, Visscher PT. 2012. Effect of inorganic and organic ligands on the bio-availability of methylmercury as determined by using a mer-lux Bioreporter. *Applied and Environmental Microbiology* 78:7276–7282.

Nelson JD, Blair W, Brinckman FE, Colwell RR, Iverson WP. 1973. Biodegradation of phenylmercuric acetate by mercury-resistant bacteria. *Applied Microbiology* 26:321–326.

Nelson JD, Colwell RR. 1974. Metabolism of mercury compounds by bacteria in Chesapeake Bay. In: Acker RF, Brown BF, De Palma JR, eds., *Proceedings of International Congress on Marine Corrosion and Fouling, 3rd, 1972*. Evanston, IL: Northwestern University Press, pp. 767–777.

Nelson JD, Colwell RR. 1975. The ecology of mercury-resistant bacteria in Chesapeake Bay. *Microbial Ecology* 1:191–218.

Niki H, Maker PD, Savage CM. 1983. A long-path fourier-transform infrared study of the kinetics and mechanism for the ho-radical initiated oxidation of dimethylmercury. *Journal of Physical Chemistry* 87:4978–4981.

Norvick RP. 1967. Penicillinase plasmids of *Staphylococcus aureus*. *Federation Proceedings* 26:29–38.

Olson GJ, Iverson WP, Brinckman FE. 1981. Volatilization of mercury by *Thiobacillus ferrooxidans*. *Current Microbiology* 5:115–118.

Olson GJ, Porter FD, Rubinstein J, Silver S. 1982, Mercuric reductase enzyme from a mercury-volatilizing strain of *Thiobacillus ferrooxidans*. *Journal of Bacteriology* 151:1230–1236.

Oremland RS, Blum JS, Bindi AB, Dowdle PR, Herbel M, Stolz JF. 1999. Simultaneous reduction of nitrate and selenate by cell suspensions of selenium-respiring bacteria. *Applied and Environmental Microbiology* 65:4385–4392.

Oremland RS, Culbertson CW, Winfrey MR. 1991. Methylmercury decomposition in sediments and bacterial cultures: Involvement of methanogens and sulfate reducers in oxidative demthylation. *Applied and Environmental Microbiology* 57:130–137.

Oremland RS, Miller LG, Dowdle P, Connel T, Barkay T. 1995. Methylmercury oxidative degradation potentials in contaminated and pristine sediments of the Carson River, Nevada. *Applied and Environmental Microbiology* 61:2745–2753.

Osborn AM, Bruce KD, Strike P, Ritchie DA. 1993. Polymerase chain reaction-restriction fragment length polymorphism analysis shows divergence among mer determinants from gram-negative soil bacteria indistinguishable by DNA-DNA hybridization. *Applied and Environmental Microbiology* 59:4024–4030.

Osborn AM, Bruce KD, Strike P, Ritchie DA. 1997. Distribution, diversity and evolution of the bacterial mercury resistance (mer) operon. *FEMS Microbiology Reviews* 19:239–262.

Pak K-R, Bartha R. 1998. Mercury methylation and demethylation in Anoxic Lake sediments and by strictly anaerobic bacteria. *Applied and Environmental Microbiology* 64:1013–1017.

Pan Hou HS, Imura N. 1981. Role of hydrogen sulfide in mercury resistance determined by plasmid of *Clostridium cochlearium* T-2. *Archives of Microbiology* 129:49–52.

Pan Hou HS, Nishimoto M, Imura N. 1981. Possible role of membrane proteins in mercury resistance of *Enterobacter aerogenes*. *Archives of Microbiology* 130:93–95.

Parks JM, Johs A, Podar M, Bridou R, Hurt RA, Smith SD, Tomanicek SJ et al. 2013. The genetic basis for bacterial mercury methylation. *Science* 339:1332–1335.

Pirrone N et al. 2010. Global mercury emissions to the atmosphere from anthropogenic and natural sources. *Atmospheric Chemistry and Physics* 10:5951–5964.

Poulain AJ, Amyot M, Findlay D, Telor S, Barkay T, Hintelmann H. 2004. Biological and photochemical production of dissolved gaseous mercury in a boreal lake. *Limnology and Oceanography* 49:2265–2275.

Qureshi A, O'Driscoll NJ, Macleod M, Neuhold YM, Hungerbuhler K. 2010. Photoreactions of Mercury in surface ocean water: Gross reaction kinetics and possible pathways. *Environmental Science & Technology* 44:644–649.

Richmond MH, John M. 1964. Co-transduction by a staphylococcal phage of the genes responsible for penicillinase synthesis and resistance to mercury salts. *Nature (London)* 202:1360–1361.

Robinson JB, Tuovinen OH. 1984. Mechanism of microbial resistance and detoxification of mercury and organomercury compounds: Physiological, biochemical and genetic analyses. *Microbiological Reviews* 48:95–124.

Schaefer JK, Morel FMM. 2009. High methylation rates of mercury bound to cysteine by *Geobacter sulfurreducens*. *Nature Geoscience* 2:123–126.

Schaefer JK, Rocks SS, Zheng W, Liang LY, Gu BH, Morel FMM. 2011. Active transport, substrate specificity, and methylation of Hg(II) in anaerobic bacteria. *Proceedings of the National Academy of Sciences of the United States of America* 108:8714–8719.

Schaefer JK et al. 2004. Role of the bacterial organomercury lyase (MerB) in controlling methylmercury accumulation in mercury-contaminated natural waters. *Environmental Science & Technology* 38:4304–4311.

Schedlbauer OF, Heumann KG. 2000. Biomethylation of thallium by bacteria and first determination of biogenic dimethylthallium in the ocean. *Applied Organometallic Chemistry* 14:330–340.

Schelert J, Dixit V, Hoang V, Simbahan J, Drozda M, Blum P. 2004. Occurrence and characterization of mercury resistance in the hyperthermophilic Archaeon *Sulfolobus solfataricus* by use of gene disruption. *Journal of Bacteriology* 186:427–437.

Schottel JA, Mandel D, Clark S, Silver S, Hodges RW. 1974. Volatilization of mercury and organomercurials determined by inducible R-factor systems in enteric bacteria. *Nature (London)* 251:335–337.

Schrauzer GN, Weber JH, Beckham TM, Ho RKY. 1971. Alkyl group transfer from cobalt to mercury: Reaction of alkyl cobalamins, alkylcobaloximes, and of related compounds with mercury acetate. *Tetrahedron Letters* 3:275–277.

Siciliano SD, O'Driscoll NJ, Lean DRS. 2002. Microbial reduction and oxidation of mercury in freshwater lakes. *Environmental Science & Technology* 36:3064–3068.

Silver S. 1997. The bacterial view of the periodic table: Specific functions for all elements. In: Banfield J, Nealson KH, eds., *Geomicrobiology: Interactions between Microbes and Minerals. Rev Mineral*, Vol. 35. Washington, DC: Mineralogical Society of America, pp. 345–360.

Silver S, Phung LT. 1996. Bacterial heavy metal resistance: New surprises. *Annual Review of Microbiology* 50:753–789.

Simbahan J, Drijber R, Blum P. 2004. *Alicyclobacillus vulcanalis* sp. nov., a new thermophilic, acidophilic bacterium isolated from Coso Hot Springs, California, USA. *International Journal of Systematic and Evolutionary Microbiology* 54:1703–1707.

Simbahan J, Kurth E, Schelert J, Dillman A, Moriyama E, Javanovich S, Blum P. 2005. Community analysis of a mercury hot spring supports occurrence of domain-specific forms of mercuric reductase. *Applied and Environmental Microbiology* 71:8836–8845.

Smith DH. 1967. R factors mediate resistance to mercury, nickel, and cobalt. *Science* 156:1114–1116.

Smith T, Pitts K, McGarvey JA, Summers AO. 1998. Bacterial oxidation of mercury metal vapor, Hg(0). *Applied and Environmental Microbiology* 64:1328–1332.

Spangler WA, Spigarelli JL, Rose JM, Miller HM. 1973a. Methylmercury: Bacterial degradation in lake sediments. *Science* 180:192–193.

Spangler WA, Spigarelli JL, Rose JM, Fillipin RS, Miller HM. 1973b. Degradation of methylmercury by bacteria isolated from environmental samples. *Applied Microbiology* 25:488–493.

Steffan RJ, Korthals ET, Winfrey MR. 1988. Effects of acidification on mercury methylation, demethylation, and volatilization in sediments from an acid-susceptible lake. *Applied and Environmental Microbiology* 54:2003–2009.

Sugio T, Komoda T, Okazaki Y, Takeda Y, Nakamura S, Takeuchi F. 2010. Volatilization of metal mercury from organomercurials by highly mercury-resistant Acidithiobacillus ferrooxidans MON-1. Bioscience Biotechnology and Biochemistry 74:1007–1012.

Summers AO. 1986. Genetic mechanisms of heavy-metal and antibiotic resistances. In: Carlisle D, Berry WI, Kaplan IR, Watterson JR, eds., Mineral Exploration: Biological Systems and Organic Matter. Rubey Vol. 5. Englewood Cliffs, NJ: Prentice-Hall, pp. 265–281.

Summers AO, Lewis E. 1973. Volatilization of mercuric chloride by mercury-resistant plasmid-bearing strains of Escherichia coli, Staphylococcus aureus, and Pseudomonas aeruginosa. Journal of Bacteriology 113:1070–1072.

Summers AO, Silver S. 1972. Mercury resistance in a plasmid-bearing strain of Escherichia coli. Journal of Bacteriology 112:1228–1236.

Summers AO, Silver S. 1978. Microbial transformations of metals. Annual Review of Microbiology 32:637–672.

Summers AO, Sugarman LI. 1974. Cell-free mercury(II)-reducing activity in a plasmid-bearing strain of Escherichia coli. Journal of Bacteriology 119:242–249.

Sunderland EM. 2007. Mercury exposure from domestic and imported estuarine and marine fish in the US seafood market. Environmental Health Perspectives 115:235–242.

Swain EB et al. 2007. Socioeconomic consequences of mercury use and pollution. Ambio 36:45–61.

Takeuchi F, Negishi A, Nakamura S, Kanao T, Kamimura K, Sugio T. 2005. Existence of an iron-oxidizing bacterium Acidithiobacillus ferrooxidans resistant to organomercurial compounds. Journal of Bioscience and Bioengineering 99:586–591.

Tezuka T, Tonomura K. 1976. Purification and properties of an enzyme catalyzing the splitting of carbon-mercury linkages from mercury resistant Pseudomonas K 62-strain. Journal of Biochemistry (Tokyo) 80:779–787.

Tezuka T, Tonomura K. 1978. Purification and properties of a second enzyme catalyzing the splitting of carbon–mercury linkages from mercury-resistant Pseudomonas K-62. Journal of Bacteriology 135:138–143.

Thayer JS. 2002. Biological methylation of less-studied elements. Applied Organometallic Chemistry 16:677–691.

Tonomura K, Makagami T, Futai F, Maeda D. 1968. Studies on the action of mercury-resistant microorganisms on mercurials. I. The isolation of a mercury-resistant bacterium and the binding of mercurials to cells. Journal of Fermentation Technology 46:5506–5512.

Topping G, Davies I. 1981. Methylmercury production in the marine water column. Nature 290:243–244.

Trevors JR. 1986. Mercury methylation by bacteria. Journal of Basic Microbiology 26:499–504.

UNEP. 2013. United Nations Environmental Program, Minamata Convention, accessed November 25, 2013; URL: http://www.unep.org/hazardoussubstances/MinamataConvention/tabid/106191/Default.aspx.

Vetriani C et al. 2005. Mercury adaptation among bacteria from a deep-sea hydrothermal vent. Applied and Environmental Microbiology 71:220–226.

Vostal J. 1972. Transport and transformation of mercury in nature and possible routes of exposure. In: Friberg L, Vostal J, eds., Mercury in the Environment. An Epidemiological and Toxicological Appraisal. Cleveland, OH: CRC Press, pp. 15–27.

Wang YP et al. 2011. Environmental conditions constrain the distribution and diversity of archaeal merA in Yellowstone National Park, Wyoming, U.S.A. Microbial Ecology 62:739–752.

Whalin L, Kim E-H, Mason RP. 2007. Factors influencing the oxidation, reduction, methylation and demethylation of mercury species in coastal waters. Marine Chemistry 107:278–294.

Whalin LM, Mason RP. 2006. A new method for the investigation of mercury redox chemistry in natural waters utilizing deflatable Teflon (R) bags and additions of isotopically labeled mercury. Analytica Chimica Acta 558:211–221.

White RH. 1982. Analysis of dimethyl sulfonium compounds in marine algae. Journal of Marine Research 40:529–536.

WHO. 2013. World Health Organization mercury advisory homepage, accessed September 23, 2013; http://www.who.int/ipcs/assessment/public_health/mercury/en/.

Wiatrowski HA, Das S, Kukkadapu R, Ilton ES, Barkay T, Yee N. 2009. Reduction of Hg(II) to Hg(O) by Magnetite. Environmental Science & Technology 14:5307–5313.

Wiatrowski HA, Ward PM, Barkay T. 2006. Novel reduction of mercury(II) by mercury-sensitive dissimilatory metal reducing bacteria. Environmental Science & Technology 21:6690–6696.

Wiener J, Krabbenhoft D, Heinz G, Scheuhammer AM. 2003. Ecotoxicology of mercury. In: Hoffman DJ, Rattner BA, Burton GA, Cairns J, eds., Handbook of Ecotoxicology. Boca Raton, FL: CRC Press, pp. 409–463.

Wong PTS, Chau YK, Luxon PL. 1975. Methylation of lead in the environment. Nature (London) 253:263–264.

Wood JM. 1974. Biological cycles of toxic elements in the environment. Science 183:1049–1052.

Wood JM, Kennedy FS, Rosen CG. 1968. Synthesis of methylmercury compounds by extracts of a methanogenic bacterium. Nature (London) 220:173–174.

Yamada M, Tonomura K. 1972a. Formation of methylmercury compounds from inorganic mercury by *Clostridium cochlearium*. *Journal of Fermentation Technology* 50:159–166.

Yamada M, Tonomura K. 1972b. Further study of formation of methylmercury from inorganic mercury by *Clostridium cochlearium* T-2. *Journal of Fermentation Technology* 50:893–900.

Yu RQ, Flanders JR, Mack EE, Turner R, Mirza MB, Barkay T. 2013. Contribution of coexisting sulfate and iron reducing bacteria to methylmercury production in freshwater river sediments. *Environmental Science & Technology* 46:2684–2691.

Yurieva O et al. 1997. Intercontinental spread of promiscuous mercury-resistance transposons in environmental bacteria. *Molecular Microbiology* 24:321–329.

CHAPTER SEVENTEEN

Geomicrobiology of Iron

Andreas Kappler, David Emerson, Jeffrey A. Gralnick, Eric E. Roden, and E. Marie Muehe

CONTENTS

17.1 Iron Distribution in the Earth's Crust / 344
17.2 Geochemically Important Properties of Iron / 344
17.3 Biological Importance of Iron / 345
  17.3.1 Function of Iron in Cells / 345
  17.3.2 Iron Assimilation by Microbes / 346
  17.3.3 Exploiting Iron Redox Chemistry for Microbial Life / 347
17.4 Iron(II) as Electron (and Energy) Source for Bacteria / 348
  17.4.1 Neutrophilic, Microaerophilic Iron(II) Oxidizers / 348
    17.4.1.1 Appendaged Bacteria / 349
    17.4.1.2 Unicellular Bacteria / 353
  17.4.2 Acidophilic Iron(II) Oxidizers / 354
    17.4.2.1 Domain Bacteria: Mesophiles / 354
    17.4.2.2 Domain Bacteria: Thermophiles / 358
    17.4.2.3 Domain Archaea: Mesophiles / 358
    17.4.2.4 Domain Archaea: Thermophiles / 358
  17.4.3 Anaerobic Oxidation of Ferrous Iron / 359
    17.4.3.1 Phototrophic Iron(II) Oxidation / 359
    17.4.3.2 Chemotrophic Iron(II) Oxidation: Nitrate-Reducing Iron(II) Oxidizers / 361
  17.4.4 Ferrous Iron Oxidation: Comparison of Energetics, Carbon Dioxide Fixation, and Biochemistry / 364
17.5 Iron(III) as Terminal Electron Acceptor in Bacterial Respiration / 365
  17.5.1 Bacterial Ferric Iron Reduction Accompanying Fermentation / 365
  17.5.2 Ferric Iron Respiration / 366
    17.5.2.1 Microbial Sulfur Oxidation Coupled to Iron(III) Reduction at Acidic pH / 367
    17.5.2.2 Microbial Hydrogen Oxidation Coupled to Iron(III) Reduction at Neutral pH / 368
    17.5.2.3 Microbial Ammonium and Methane Oxidation Coupled to Iron(III) Reduction at Neutral pH / 368
    17.5.2.4 Microbial Organic Matter Oxidation Coupled to Iron(III) Reduction / 369
17.6 Nonenzymatic Oxidation of Ferrous Iron and Reduction of Ferric Iron by Microbes / 374
  17.6.1 Nonenzymatic Iron(II) Oxidation / 374
  17.6.2 Nonenzymatic Iron(III) Reduction / 374
17.7 Microbial Precipitation of Iron / 375
  17.7.1 Enzymatic Iron(II) Oxidation and Iron(III) Precipitation Processes / 375
  17.7.2 Bioaccumulation of Iron / 376
    17.7.2.1 Magnetotactic Bacteria / 376

17.8 Sedimentary Iron Deposits of Putative Biogenic Origin / 378
17.9 Microbial Mobilization of Iron from Minerals in Ore, Soil, and Sediments / 380
17.10 Microbes and the Iron Cycle / 382
Acknowledgments / 384
References / 384

17.1 IRON DISTRIBUTION IN THE EARTH'S CRUST

Iron is the fourth most abundant element in the Earth's crust and the most abundant element in the Earth as a whole. Its average concentration in the crust has been estimated to be approximately 5.0% (Rankama and Sahama, 1950). It is found in a number of minerals in rocks, soil, and sediments. Table 17.1 lists minerals in which iron is a major structural component.

The primary source of iron accumulated at the Earth's surface is volcanic activity. Weathering of iron-containing rocks and minerals is often an important process in the formation of local iron accumulations, including sedimentary ore deposits. The worldwide largest sources of iron are banded iron formations (BIFs) deposited in Canada, South Africa, and Australia. Furthermore, the combination of physical and chemical weathering of iron-containing rocks leads to the hydration, oxidation, and fragmentation of iron-containing minerals and thereby increases the amount of iron, which is easily accessible for biota and thus, supports life.

17.2 GEOCHEMICALLY IMPORTANT PROPERTIES OF IRON

Iron is a very reactive element. Its common oxidation states are 0, 2+, and 3+. However, in the environment, iron occurs mostly in its ferrous (2+) and ferric (3+) oxidation state, which are easily convertible in many different abiotic and biotic reactions and processes (Figure 17.1). In a moist environment exposed to air or in aerated solution at pH values greater than 5, iron's ferrous form (2+) is readily autoxidized by molecular oxygen to the ferric form (3+):

$$4Fe^{2+} + O_2 + 4H^+ \rightarrow 4Fe^{3+} + 2H_2O \quad (17.1)$$

Due to its poor solubility, ferric iron may exist as a hydroxide ($Fe(OH)_3$), oxyhydroxide ($FeO(OH)$), or oxide (e.g., Fe_2O_3) in neutral to slightly alkaline solution. In acidic solutions, it remains as dissolved Fe^{3+}. Ferric hydroxides dissolve in strongly alkaline solution because of their amphoteric nature, which allows them to exist as an oxyanion (e.g., $Fe(OH)_2O^-$). Under reducing conditions, i.e., in the absence of oxygen and in the presence of an appropriate reducing agent, ferric iron is readily reduced to the ferrous state:

$$Fe^{3+} + \text{Reducing agent} \rightarrow$$
$$Fe^{2+} + \text{Oxidized reducing agent} \quad (17.2)$$

Ferrous iron is considered to be more mobile and therefore, more bioavailable than ferric iron, as it is better soluble in many environments and under most geochemical conditions.

TABLE 17.1
Iron minerals found in sediments and soils.

Fe(II) minerals	Fe(III) minerals	Mixed Fe(II)/Fe(III) minerals
Siderite ($FeCO_3$)	Goethite (α-$FeO(OH)$)	Magnetite (Fe_3O_4)
Vivianite ($Fe_2(PO_4)_3$)	Limonite ($FeO(OH) \cdot nH_2O$)	Greigite (Fe_3S_4)
Ferrous sulfide (FeS)	Hematite (α-Fe_2O_3)	Green rust ($Fe_{1-x}^{II}Fe_x^{III}(OH)_2)^{x+*-}$
Pyrite, marcasite (FeS_2)	Maghemite (γ-Fe_2O_3)	$[(x/n)An^{-*}(m/n)H_2O]x^-$
	Lepidocrocite (γ-$FeO(OH)$)	
	Akaganeite (β-$FeO(OH,Cl)$)	
	Ferrihydrite ($Fe(OH)_3$ or $Fe_{8.2}O_{8.5}OH_{7.4} \cdot 3H_2O^a$)	

NOTE: Iron is also present in clay minerals such as montmorillonite, smectite, and illite.

[a] Current formula for the structure of ferrihydrite according to Michel et al. (2010).

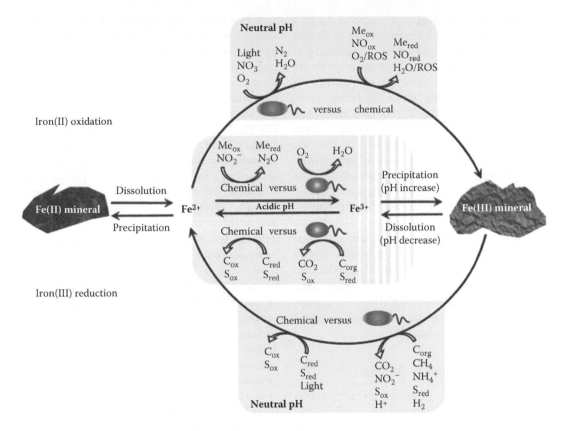

Figure 17.1. Microbial and abiotic cycling of iron C_{org}, organic carbon including humics; C_{red}, reduced organic carbon; C_{ox}, oxidized organic carbon; Me_{ox}, oxidized metal; Me_{red}, reduced metal; ROS, reactive oxygen species; NO_{ox}, oxidized nitrogen oxide species; NO_{red}, reduced nitrogen oxide species; S_{ox}, oxidized sulfur; S_{red}, reduced sulfur. For a full overview of microbial and abiotic iron cycling, see Melton et al. (2014a). (Modified from Kappler, A and Straub, KL, *Rev Mineral Geochem*, 59, 85, 2005).

17.3 BIOLOGICAL IMPORTANCE OF IRON

17.3.1 Function of Iron in Cells

All organisms, prokaryotic and eukaryotic, single-celled and multicellular, require iron nutritionally, except for a small group of homolactic bacteria (e.g., Pandey et al., 1994). Iron-containing proteins are needed in some enzymatic processes that involve the transfer of electrons. Examples of such processes are aerobic and anaerobic respiration in which cytochromes (special proteins that bear iron-containing heme groups) and nonheme iron–sulfur proteins play a role in the transfer of electrons to a terminal electron acceptor, e.g., oxygen. Phototrophic organisms also need iron for ferredoxin (nonheme iron–sulfur protein) and for some cytochromes that are part of the photosynthetic system. Some cells also employ iron in certain superoxide dismutases, which convert superoxide to hydrogen peroxide

by a disproportionation reaction (see Chapter 3). Most aerobically living cells produce the enzymes catalase and peroxidase for catalyzing the decomposition of toxic hydrogen peroxide to water and oxygen. Prokaryotes capable of nitrogen fixation employ ferredoxin and another nonheme iron–sulfur protein (component II of nitrogenase) as well as an iron–molybdo, an iron–vanado, or an iron–iron protein (component I of nitrogenase) (see Chapter 14).

As will be discussed in Section 17.4, ferrous iron may serve as a major energy source and/or source of reducing power for certain bacteria (Konhauser et al., 2011; Melton et al., 2014a). Ferric iron may serve as a terminal electron acceptor for a number of different bacteria, usually under anoxic conditions (Konhauser et al., 2011) (Section 17.5). As discussed in Chapter 3, ferrous iron may have served as an important reductant during the evolutionary emergence of oxygenic photosynthesis by scavenging toxic oxygen produced in the process until the appearance of oxygen-protective superoxide dismutase. Cloud (1973) believed that BIFs that appeared in the sedimentary record from 3.3 to 2 billion years ago are evidence for the oxygen-scavenging action of ferrous iron. However, modifications of this view have been offered more recently as a consequence of the discovery of anoxygenic phototrophic iron(II)-oxidizing bacteria (see discussion in Section 17.8). From a biogeochemical viewpoint, large-scale microbial iron(II) oxidation is important because it may lead to extensive iron precipitation (immobilization of iron and many other elements that are associated with iron minerals), and microbial iron(III) reduction is important because it may lead to an extensive solubilization (mobilization of iron and associated elements). In some anoxic environments, microbial iron(III) reduction plays an important role in the mineralization of organic carbon (iron(III) respiration).

17.3.2 Iron Assimilation by Microbes

In oxic environments of neutral pH, ferrous iron becomes readily oxidized and the resulting ferric iron precipitates from solution as hydroxide, oxyhydroxide, and oxide. At circumneutral pH, these ferric compounds are poorly soluble. This is a prominent problem in seawater, where iron is a growth-limiting nutrient) (e.g., Gelder, 1999; also this book, Chapter 6). As previously stated, iron is an essential trace nutrient for nearly all forms of cellular life. Because iron can be taken into cells only in its dissolved form, a number of microorganisms have acquired the ability to synthesize chelators (ligands). These chelators help to keep ferric iron in solution at circumneutral pH or can return it to solution in sufficient quantities from an insoluble

state. Collectively, such chelators are known as iron siderophores. They have an extremely high affinity for ferric iron but very low affinity for ferrous iron. They exhibit proton-independent binding constants for Fe^{3+} ranging from $10^{22.5}$ to 10^{49}–10^{53} (Reid et al., 1993) compared to around 10^8 for Fe^{2+} (for additional data, see Dhungana and Crumbliss [2005]). Examples of iron siderophores are enterobactin or enterochelin, catechol derivatives from *Salmonella typhimurium* (Pollack and Neilands, 1970); aerobactin, a hydroxamate derivative produced by *Enterobacter* (formerly *Aerobacter*) *aerogenes* (Gibson and Magrath, 1969); alterobactin from *Alteromonas luteoviolacea* (Reid et al., 1993); ferrioxamine E (nocardamine) and ferrioxamines G, D₂, and X₁ from *Pseudomonas stutzeri* (Essén et al., 2007); and rhodotorulic acid, a hydroxamate derivative produced by the yeast *Rhodotorula* (Neilands, 1974) (examples of iron siderophores are depicted in Figure 17.2). Citrate, produced by some fungi, can also serve as a chelator for iron for some bacteria (Rosenberg and Young, 1974). The iron mobilization action of siderophores under circumneutral, oxic conditions may be assisted by an extracellular reductant. Hersman et al. (2000) reported on the mobilization of iron from hematite (Fe_2O_3) by a siderophore secreted by *Pseudomonas mendocina* and by extracellular iron-reducing activity. The reducing activity was maximal under conditions of extreme iron deprivation in the absence of hematite or iron–EDTA. The siderophore mobilized the iron in form of a ferric chelate whereas the iron-reducing activity mobilized the iron in its ferrous form. The authors speculated that the iron-reducing activity was due to an extracellular enzyme produced by *P. mendocina* that was similar to the extracellular iron reductases produced by *Escherichia coli* and *Pseudomonas aeruginosa* (Vartanian and Cowart, 1999). Alternatively, small redox-active metabolites (such as phenazine antibiotics) released

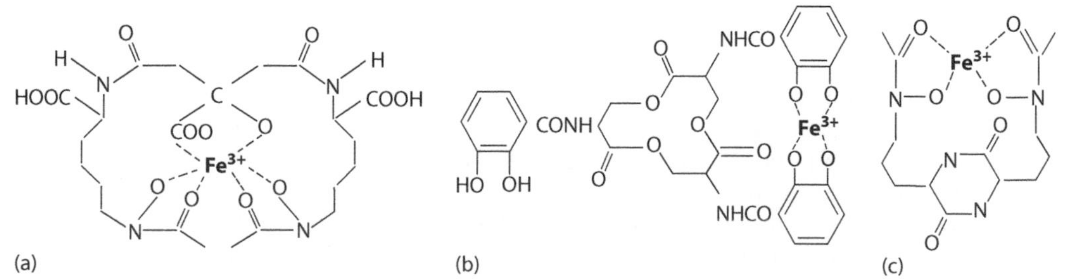

Figure 17.2. Examples of iron siderophores. (a) Aerobactin (bacterial). (b) Enterochelin (bacterial). (c) Rhodotorulic acid (fungal).

by *Pseudomonas* species may also facilitate iron acquisition via electron shuttling leading to iron(III) reduction (Wang and Newman, 2008).

Chelated ferric iron is usually taken up by first binding to ferrisiderophore-specific receptors at the cell surface of the microbial species that produces the siderophores (Figure 17.3). In some cases, it may also bind to surface receptors of certain species of microorganisms incapable of synthesizing the desferrisiderophore (iron-free siderophore). After transport into the cell, the chelated ferric iron is usually enzymatically reduced to ferrous iron and rapidly released from the siderophore because it has only low affinity for Fe^{2+} (Brown and Ratledge, 1975; Tait, 1975; Ernst and Winkelman, 1977; Cox, 1980; Wandersman and Delepelaire, 2004). The desferrisiderophore may then be excreted again for further scavenging of iron. In some instances, the ferrisiderophore taken up by the cell is degraded first before the ferric iron is reduced to ferrous iron. The desferrisiderophore in that instance is not recyclable. In other instances, chelated ferric iron is exchanged with another ligand at the initial binding site at the cell surface before being transported into the

cell and reduced to ferrous iron. The ferrous iron, by whatever mechanism it is generated in the cell, is immediately assimilated into a heme protein or nonheme iron–sulfur protein.

17.3.3 Exploiting Iron Redox Chemistry for Microbial Life

As discussed in Section 17.2, iron is redox active and can be cycled between its 2+ and 3+ redox state. This is used by many physiologically and phylogenetically diverse microorganisms for harvesting electrons and/or energy for growth under acidic, neutral, and alkaline pH conditions. Ferrous iron can serve as electron and/or energy source for oxic, microaerophilic, nitrate-reducing, and phototrophic iron(II)-oxidizing nitrate or carbon dioxide as explained in detail in Section 17.4. *Vice versa*, ferric iron can serve as terminal electron acceptor for iron(III)-reducing microorganisms during anaerobic respiration (see Section 17.5), e.g., coupled to the oxidation of organic compounds, hydrogen, ammonium, methane, or sulfur (see Figure 17.1 or Melton et al. [2014a] for an overview).

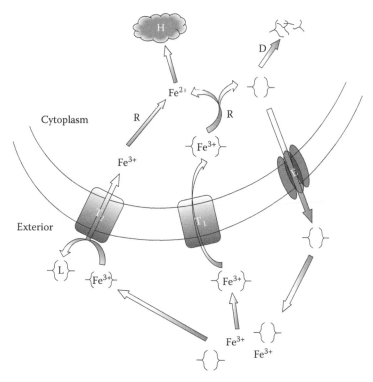

Figure 17.3. Mechanism of iron siderophore binding to the cell surface and subsequent fate of siderophores and iron. Wavy brackets, iron siderophores; T_1, Fe^{3+}-siderophore transporter; T_2, Fe^{3+} transporter; L, ligand; R, iron(III) reductase; D, siderophore degradation; E, siderophore exporter; H, heme or iron–sulfur protein.

17.4 IRON(II) AS ELECTRON (AND ENERGY) SOURCE FOR BACTERIA

Iron(II)-oxidizing bacteria are capable of drawing energy and/or electrons for growth and proliferation by oxidizing ferrous iron (Andrews et al., 2013; Melton et al., 2014a). When a pH-neutral habitat is microoxic, oxygen can serve as terminal electron acceptor for iron(II) oxidation. Microorganisms performing this process can use the iron(II) as electron and energy source and are called microaerophilic iron(II) oxidizers. Once growth conditions turn anoxic, iron(II)-oxidizing microorganisms use energetically less favorable compounds as terminal electron acceptor. Photoautotrophic iron(II) oxidizers use light energy as energy source and carbon dioxide as electron acceptor for biomass synthesis with the electrons stemming from iron(II) oxidation, while nitrate-reducing iron(II) oxidizers were suggested to use iron(II) as an electron and energy source with nitrate as terminal electron acceptor. Under acidic and oxic conditions, the oxidation of ferrous iron is usually performed biotically by acidophilic iron(II) oxidizers that use oxygen as terminal electron acceptor.

For most of the microbially catalyzed iron(II) oxidation processes, no general information is available regarding the proteins involved in iron(II) oxidation and their genetic information. Limited information on iron(II) oxidation enzymatics and genetics is available only for individual strains. Therefore, this information here is not presented in general sections on genetics or biochemistry but is part of the sections where individual strains are introduced.

17.4.1 Neutrophilic, Microaerophilic Iron(II) Oxidizers

Neutrophilic, microaerophilic iron(II)-oxidizing microorganisms oxidize ferrous iron at neutral pH (5–8) by using atmospheric oxygen as terminal electron acceptor:

$$4Fe^{2+} + 10H_2O + O_2 \rightarrow 4Fe(OH)_3 + 8H^+ \quad (17.3)$$

Due to the competition with the spontaneous abiotic oxidation of ferrous iron at circumneutral pH, these organisms are found most consistently in microoxic habitats charged with moderate to high concentrations of iron(II). A pure culture of a microaerophilic iron(II)-oxidizing bacterium was shown to be able to compete with chemical oxidation up to an oxygen concentration of 50 µM (Druschel et al., 2008). Under environmental conditions, microbial iron(II) oxidation has been shown to occur at higher oxygen concentrations, perhaps due to the capacity of organic ligands to stabilize iron(II) and slow the chemical reaction (Fleming et al., 2013). The kinetics of abiotic and biotic iron(II) oxidation, and thus, the balance between these two processes, is governed by the concentration of ferrous iron, the partial pressure of oxygen, the abundance and activity of microaerophilic iron(II) oxidizers, the formation of ferric minerals, and their surface area/properties, temperature, and pH (Neubauer et al., 2002; Druschel et al., 2008; Vollrath et al., 2012, 2013). Overall, the biotic process of iron(II) oxidation becomes more favorable compared to the abiotic oxidation at low oxygen and high Fe^{2+} concentrations, i.e., in opposing gradients of oxygen and aqueous Fe^{2+}. Microelectrode profile measurements have been used to estimate the relative contribution of abiotic and microbial iron(II) oxidation activities at circumneutral pH in the field and in the laboratory (Druschel et al., 2008).

Several iron(II)-oxidizing microorganisms have been isolated from marine and freshwater environments. Table 17.2 provides examples of different known microaerophilic iron(II)-oxidizing bacteria. They are either unicellular or filamentous. Some produce extracellular biomineral filaments, typically either sheaths or stalks, while others form particulate iron oxyhydroxides. The sheath or stalk morphotypes are easily recognized in the light microscope and thus, are diagnostic for the presence of iron(II)-oxidizing bacteria. It has been suggested that these structures help cells to avoid encrustation by directing iron(III) mineral precipitation away from the cell surface toward these organic structures. Freshwater neutrophilic iron(II)-oxidizing bacteria are most commonly found in wetlands and slow-moving streams and ditches (Emerson et al., 2010) and are also present in iron plaque on root systems of wetland plants together with iron(III)-reducing bacteria (Weiss et al., 2003, 2007). Marine iron(II)-oxidizing microorganisms are found in so-called iron flocs or mats associated with hydrothermal vent fluids that are rich in iron(II) (Langley et al., 2009). These iron flocs are a pool of nutrients and trapped metals, i.e., providing good growth

TABLE 17.2
Examples of neutrophilic, microaerophilic iron(II)-oxidizing microorganisms.

Organism	Literature
Unicellular	
Sideroxydans lithotrophicus ES-1	Emerson and Moyer (1997), Weiss et al. (2007)
Gallionella capsiferriformans ES-2	Emerson and Moyer (1997), Emerson et al. (2013)
Dechlorospirillum sp. M1	Picardal et al. (2011)
Bradyrhizobium japonicum 22	Benzine et al. (2013)
Stalk forming	
Gallionella ferruginea	Ehrenberg (1836), Cholodny (1924), Hallbeck and Pedersen (1991)
Mariprofundus ferrooxidans PV-1	Emerson and Moyer (2002), Singer et al. (2011)
Strain R-1	Krepski et al. (2012)
Strain OYT1	Kato et al. (2013)
Ferriphaselus amnicola	Kato et al. (2014)
Sheathed	
Leptothrix spp.	Ghiorse (1984), Ghiorse and Ehrlich (1992), Fleming et al. (2011)
Sphaerotilus	Ghiorse (1984), Ghiorse and Ehrlich (1992)
Crenothrix polyspora	Ghiorse (1984), Ghiorse and Ehrlich (1992)
Calothrix sp.	Ghiorse (1984), Ghiorse and Ehrlich (1992)
Lieskeella bifida	Ghiorse and Ehrlich (1992)
Wall-less	
Acholeplasma laidlawii[a]	Balashova and Zavarzin (1972)

[a] Former name: *Mycoplasma laidlawii*.

habitats for iron(II)-oxidizing and iron(III)-reducing bacteria. Iron flocs are also investigated to better understand the interaction of ferrous iron-oxidizing microorganisms in natural environments (Kato et al., 2013). These studies might also help to identify genes that are involved in the oxidation of ferrous iron (e.g., an iron(II) oxidase) in such neutrophilic iron(II)-oxidizing bacteria. Comparative genome analysis of different known neutrophilic iron(II) oxidizers was undertaken to find homologous genes in these strains that are involved in iron(II) oxidation (Singer et al., 2011; Emerson et al., 2013; Kato et al., 2013).

17.4.1.1 Appendaged Bacteria

17.4.1.1.1 GALLIONELLA FERRUGINEA AND OTHER STALK-FORMING IRON(II)-OXIDIZING BACTERIA

Gallionella ferruginea is a mesophilic microorganism that consists of a bean-shaped cell that excretes a nanofibrillar stalk composed of an organic matrix and iron(III) oxides. It is taxonomically assigned to the domain Bacteria: Proteobacteria; Betaproteobacteria: *Gallionellales; Gallionellaceae*, but its genome has not been sequenced as yet. The organism was first described by Ehrenberg (1836). Cholodny (1924) was the first to recognize that the bean-shaped cell on the lateral stalk was an integral part of the organism. When the cell divides the stalk bifurcates, however, cells are easily dislodged from the stalk, so if care is not taken, cells will be rarely observed on stalks either in culture or in natural samples. The cells do have a motile phase and may detach from their stalk, swim away via polar flagella as swarmers, seek a new site for attachment, and develop a stalked growth habit.

Enrichments using opposing gradients of iron(II) and oxygen of *G. ferruginea* were first reported by Kucera and Wolfe (1957), and isolates have been obtained subsequently, although a pure culture is not available at present. *G. ferruginea* has been described as a gradient organism, i.e., it grows best under low oxygen tension (0.1–1 mg oxygen L^{-1}) and in an E_h range of +200 to +320 mV (Kucera and Wolfe, 1957; Hanert, 1981). This is consistent

Figure 17.4. **(See color insert.)** Iron-rich spring Fuschna in Switzerland. Iron(II)-oxidizing bacteria precipitate red/orange iron (oxyhydr) oxide minerals in the channels of the outflow of the spring water. On the channel rims, cyanobacterial mats form. (Courtesy of Kappler, A., Geomicrobiology, Center for Applied Geosciences, University of Tuebingen, Tuebingen, Germany.)

with the microaerophilic requirements of other oxygen-dependent iron(II)-oxidizing bacteria. In nature, its growth has been observed between 4°C and 47°C (Hanert, 1981). It prefers a pH range of 6.0–7.6 (Hanert, 1981). *Gallionella*-related ferrous iron-oxidizing bacteria have also been observed in high abundance at pH 4.4 in a metal-contaminated creek (Fabisch et al., 2013) and freshwater springs such as the Fuschna spring depicted in Figure 17.4 (Hegler et al., 2012).

It is important to note that stalk formation by itself is not a definitive identifier for *Gallionella*. The closest phylogenetic relative of G. *ferruginea* that is currently in culture, *Gallionella capsiferriformans*, does not produce a stalk, while another recently described genus of iron(II)-oxidizing bacterium, *Ferriphaselus amnicola* that is also a member of the *Gallionellales*, does produce a helical stalk

with similar characteristics to G. *ferruginea* (Kato et al., 2014).

Using scanning transmission x-ray microscopy and high-resolution transmission electron microscopy, Chan et al. (2009) analyzed the nature of iron mineralization associated with *Gallionella* stalks. They found that the predominant mineral associated with these structures was ferrihydrite and that akaganeite was also present. The filaments also contained organic components that were judged to be acidic polysaccharides that presumably were produced by the bacteria and played an important role in at least initiating and perhaps controlling the deposition of iron oxyhydroxides. This work indicated that there was likely a complex interplay between the organic and mineral matrices that ultimately helped control the stability and aging of the minerals and

EHRLICH'S GEOMICROBIOLOGY

concluded that the microbially produced biominerals had fundamentally different properties than chemically synthesized minerals. Suzuki et al. (2011) also showed that the twisted stalks consisted of a backbone of bacterial exopolymers. This central carbon core with oxygen-containing functional groups at the surface interacted with aqueous Fe^{2+} leading to the precipitation of iron oxides at the surface of the stalks. So far, two different types of stalk morphology have been identified, which were similar in chemical composition but were attached at different spots (Suzuki et al., 2012).

G. ferruginea is reported to grow both autotrophically and mixotrophically. Hallbeck and Pedersen (1991) demonstrated that a pure culture obtained its carbon from carbon dioxide when growing in a mineral salts medium in a gradient of oxygen and ferrous sulfide as energy source. The same investigators showed that glucose, fructose, and sucrose could meet part of the energy requirement and part or the entire carbon requirement of the organism, depending on the concentrations of the sugars. It may well be that strictly autotrophic as well as facultatively autotrophic strains exist.

In a natural environment, the rate of growth of *Gallionella* was measured in terms of the rate of attachment to a solid surface such as a submerged microscope slide and stalk elongation and shown in the relationship (Hanert, 1973):

$$V_t = (b_v \times l_v) \times \left(\frac{t^2}{2} \right) \qquad (17.4)$$

Here

b_v is the average rate of attachment

l_v is the average rate of stalk elongation

t is the length of the growth period, which should not be longer than 10 h if this relationship is to hold

V_t is a measure of the amount of growth at time t

Among marine neutrophilic iron(II)-oxidizing bacteria, *Mariprofundus ferrooxydans* also produces a helical stalk with many of the same characteristics as G. ferruginea. It was isolated from iron(II)-rich hydrothermal vent fluids at a depth of 1200 m at Loihi Seamount, a submarine volcano located near Hawaii (Figure 17.5) (Emerson and Moyer, 2002). The genome of M. ferrooxydans has

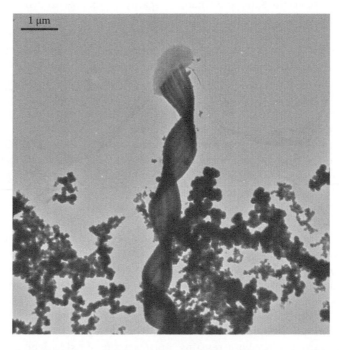

Figure 17.5. Bean-shaped cell of the microaerophilic iron(II) oxidizer *Mariprofundus ferrooxydans* PV-1 with twisted stalks made of organics and iron oxides. Transmission electron micrograph taken with a Zeiss 10CA TEM operated at 100 kV accelerating voltage. (Courtesy of Chan, CS, Department of Geological Sciences, University of Delaware, Newark, DE.)

been sequenced (Singer et al., 2011), and this information helped to confirm that it represents a novel class of Proteobacteria, called the Zetaproteobacteria. Field et al. (2014) identified two types of molybdopterin oxidoreductase genes in single amplified genomes, which could be involved in iron(II) oxidation. Several other molecular studies of marine iron mats have shown that the Zetaproteobacteria are dominant members of marine iron(II)-oxidizing communities (Jannasch and Mottl, 1985; Emerson et al., 2007; Kato et al., 2009; Forget et al., 2010; McBeth et al., 2011) and that the common freshwater iron(II) oxidizers are not present. M. ferrooxydans is a microaerophile and an obligate iron(II) oxidizer with the capacity to fix carbon dioxide. Its optimal growth temperature is around 30°C. Thus, it shares many morphological and physiological similarities with Gallionella, yet the overall genetic homology between these two organisms is very low. It has been suggested that Mariprofundus PV-1 cells excrete oxidized iron, i.e., iron(III), bound to organic polymers (Chan et al., 2011). These organic molecules retard mineral growth, thus, preventing cell encrustation, probably in combination with a near-neutral surface charge of the cells (Saini and Chan, 2013). This model describes an essential role for stalk formation in iron(II) oxidizer growth and even implies that these structures could function as a biosignature in the rock record.

17.4.1.1.2 LEPTOTHRIX OCHRACEA AND OTHER SHEATHED IRON(II)-OXIDIZING BACTERIA

In many freshwater habitats where large accumulations of biogenic iron oxides occur as flocs or mats, the dominant filamentous morphotype belongs to the tubular sheathed bacterium Leptothrix ochracea. L. ochracea remains uncultivated in the laboratory, so its true metabolism is something of a mystery. Winogradsky (1888) first reported that L. ochracea (probably Leptothrix discophora, according to Cholodny (1926)) could oxidize ferrous iron. He found that he could grow the organism in hay infusion only if he added ferrous carbonate. The iron(II) was oxidized and deposited on the sheath of the organism. He inferred from this observation that the organism was an autotroph although it remains unclear how this was determined in the presence of hay infusion. Molisch (1910) and Pringsheim (1949) disagreed with Winogradsky's conclusion

about iron(II) oxidation by L. ochracea, believing that the organism merely deposited autoxidized iron on its sheath. However, Lieske (1919) confirmed Winogradsky's observations of growth on ferrous carbonate in very dilute organic solution and suggested that the organism might be mixotrophic. Subsequently, numerous observations of L. ochracea in natural environments have been made, and there is common agreement that this organism only grows in habitats with high iron(II) concentrations and produces large quantities of sheaths that are mostly empty. This is consistent with a lithotrophic lifestyle where cells must oxidize large quantities of iron(II) to produce biomass. Furthermore, attempts to isolate L. ochracea using methods typically used for the isolation of other Leptothrix and Sphaerotilus spp. have not been successful, providing more circumstantial evidence for a lithoautotrophic or mixotrophic metabolism.

It was recently confirmed that phylogenetically L. ochracea is quite closely related to the other iron(II)- and manganese(II)-oxidizing sheath formers, including Leptothrix cholodnii, L. discophora, and Sphaerotilus natans (Fleming et al., 2011). These latter organisms are heterotrophs that have the capacity to oxidize iron(II) and manganese(II), but do not gain energy from this oxidation. As a group, these bacteria are currently classified in the family Comamonadaceae in the Betaproteobacteria. Ecologically, L. ochracea inhabits similar habitats as the stalk formers G. ferruginea and Ferriphaselus. While it is not uncommon to see morphological evidence for both groups of iron(II) oxidizers in a sample, more often than not one morphotype is dominant over the other. The specific reasons for this are unresolved and may have to do with susceptibilities to different oxygen concentrations. A study of an iron-rich stream followed a seasonal succession where stalks were dominant in early spring and then gave way to near total domination of sheaths for the remaining season. One of the parameters that corresponded most strongly to the presence of sheaths was an increase in dissolved organic matter in the water, suggesting this could be another factor in promoting the growth of L. ochracea (Fleming et al., 2013).

In all these sheath formers, the sheath is the primary site of oxide deposition. Ultrastructural examination of the sheath (Figure 17.6) of L. cholodnii (formerly discophora) SP-6 showed it to be a tubular structure of condensed fibrils

EHRLICH'S GEOMICROBIOLOGY

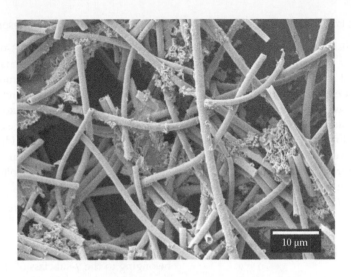

Figure 17.6. Sheaths present in an iron(II)-oxidizing mat from a creek outlet in front of the Black Forest mine Segen Gottes, Germany. Scanning electron micrograph taken with a Leo1450VP SEM operated at 5 kV accelerating voltage. (Courtesy of Obst, M, Environmental Analytical Microscopy, Center for Applied Geosciences, University of Tuebingen, Tuebingen, Germany.)

(6.5 nm in diameter) overlain by a somewhat diffuse capsular layer (Emerson and Ghiorse, 1993a). The fibrils were held together via extensive disulfide bonding that produces a very resistant extracellular structure (Emerson and Ghiorse, 1993a). The primary building block of the sheath is 2-(cysteinyl)amido-2-deoxy-D-galacturonic acid that produces a complex glycoconjugate (Makita et al., 2006). Several studies have looked at the iron or manganese activity associated with the sheath of L. discophora SS-1. De Vrind-de Jong et al. (1990) reported iron(II)-oxidizing activity in spent medium from a culture of L. discophora SS-1. Corstjens et al. (1992) related this iron(II)-oxidizing activity to a 150 kDa protein. It behaved like an enzyme and was not produced by a spontaneous mutant strain that lacked iron(II)-oxidizing activity. The factor was distinct from the manganese-oxidizing protein excreted by the wild-type SS-1 strain described by Adams and Ghiorse (1987).

17.4.1.2 Unicellular Bacteria

Unicellular, microaerophilic, lithoautotrophic, and neutrophilic iron(II) oxidizers were discovered only relatively recently. Sideroxydans lithotrophicus strain ES-1 and G. capsiferriformans ES-1 were both isolated from iron floc-containing groundwater in Michigan (Emerson and Moyer, 1997; Emerson et al., 2013). G. capsiferriformans is an obligate iron(II) oxidizer and is a close relative of G. ferruginea; however, it does not form a stalk but instead precipitates amorphous iron(III) oxyhydroxides around the cell. S. lithotrophicus also produces amorphous iron precipitates. In addition to iron(II), it can grow on reduced sulfur compounds such as thiosulfate. Other novel isolates of freshwater iron(II)-oxidizing bacteria have been isolated from the rhizosphere of wetland plants and include Sideroxydans paludicola and Ferritrophicum radicicola (Weiss et al., 2007). Both of these are obligate iron(II) oxidizers. Bradyrhizobium japonicum strain 22 was described by Benzine et al. (2013) as a novel aerobic, chemolithoautotrophic iron(II)-oxidizing taxa, which is of particular interest in light of its ability to grow via oxidation of the insoluble iron(II)-bearing mineral biotite. Diverse marine, neutrophilic iron(II)-oxidizing microorganisms have been described by Edwards et al. (2004).

Recently, a potential iron(II) oxidation mechanism in neutral-pH organisms such as S. lithotrophicus strain ES-1 was discovered. Liu et al. (2012) described a three-gene cluster homologous to the MtrAB system in Shewanella. The protein MtoA is a decaheme c-type cytochrome, oxidizing aqueous, and complexed iron(II). In combination with MtoB and $CymA_{ES-1}$, MtoA was suggested to transfer electrons obtained during extracellular iron(II) oxidation at the outer membrane to the inner membrane-based quinone pool.

17.4.2 Acidophilic Iron(II) Oxidizers

Many microorganisms with the capacity to oxidize ferrous iron at acidic pH are nowadays used in the extraction of economically important metals and rare earth elements from mine material and even from industrial (electrical) waste during a process called bioleaching. Such bacteria include members of the domains Bacteria and Archaea, which can be active at ambient (mesophiles) or elevated temperatures (thermophiles). Currently, at least 34 species in 14 genera are known capable of acidophilic iron(II) oxidation and are spread throughout gram-negatives, gram-positives, and Archaea. Table 17.3 depicts examples of microorganisms known to perform iron(II) oxidation at acidic pH.

17.4.2.1 Domain Bacteria: Mesophiles

A variety of acidophilic iron(II)-oxidizing bacteria have been recognized and characterized in various acidic environments at moderate temperatures of 20°C–45°C (Pronk and Johnson, 1992). They include autotrophs, mixotrophs, and heterotrophs. See Table 17.3a for examples of mesophilic acidophilic iron(II)-oxidizing bacteria. A more detailed characterization of the two most studied organisms, *Acidithiobacillus ferrooxidans* and *Leptospirillum ferrooxidans*, follows.

17.4.2.1.1 ACIDITHIOBACILLUS (FORMERLY THIOBACILLUS) FERROOXIDANS

General traits: The most widely studied acidophilic, ferrous iron-oxidizing bacterium is *A. ferrooxidans* belonging to the γ-subclass of the Proteobacteria.

TABLE 17.3

Examples of acidophilic aerobic iron(II)-oxidizing microorganisms.

Organism	Metabolisms	Literature
a. Bacteria: Mesophiles		
Acidithiobacillus ferrooxidans[a]	Chemolithoautotroph	Temple and Colmer (1951)
Acidithiobacillus ferridurans	Chemolithoautotroph	Hedrich and Johnson (2013)
Acidithiobacillus prosperus[b]	Chemolithoautotroph	Huber and Stetter (1989)
Leptospirillum ferrooxidans	Chemolithoautotroph	Markosyan (1972)
Metallogenium	Chemoorganotroph, possibly heterotroph	Walsh and Mitchell (1972)
Ferromicrobium acidophilum	Obligate heterotroph	Johnson and Roberto (1997)
Ferrovum myxofaciens	Chemolithoautotroph	Johnson et al. (2014)
b. Bacteria: Thermophiles		
Sulfobacillus thermosulfidooxidans	Auto- and heterotroph	Golovacheva and Karavaiko (1978)
Sulfobacillus acidophilus	Auto- and heterotroph	Norris et al. (1988, 1996)
Acidimicrobium ferrooxidans	Auto- and heterotroph	Norris (1997), Clark and Norris (1996)
Acidiferrobacter thiooxydans	Chemolithoautotroph	Hallberg et al. (2011)
c. Archaea: Mesophiles		
Ferroplasma acidiphilum	Chemolithoautotroph, facultative chemoorganotroph for some	Golyshina et al. (2000)
Ferroplasma acidarmanus	Chemolithotroph, facultative chemoorganotroph for some	Barker et al. (1998)
Ferroplasma thermophilum	Chemomixotroph	Zhou et al. (2008)
d. Archaea: Thermophiles		
Acidianus brierleyi	Auto- and heterotroph	Segerer et al. (1986)
Sulfolobus acidocaldarius[c]	Heterotroph	Brock et al. (1972)

[a] Former name: *Thiobacillus ferrooxidans*.
[b] Former name: *Thiobacillus prosperus*.
[c] Former name: *Sulfolobus brierleyi*.

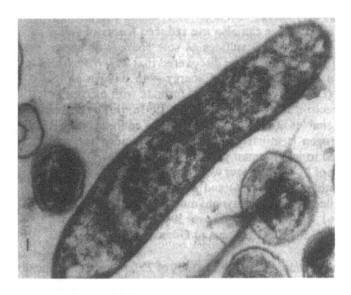

Figure 17.7. *Acidithiobacillus ferrooxidans* (×30,000). Transmission electron micrograph of a thin section. (Courtesy of Lundgren, DG, Department of Biology, Syracuse University, Syracuse, NY.)

Some strains of *A. ferrooxidans* were originally named *Ferrobacillus ferrooxidans* (Leathen et al., 1956) and *Ferrobacillus sulfooxidans* (Kinsel, 1960). These were subsequently considered synonymous with *A. ferrooxidans* (Unz and Lundgren, 1961; Ivanov and Lyalikova, 1962; Hutchinson et al., 1966; Kelly and Tuovinen, 1972; Buchanan and Gibbons, 1974).

A. ferrooxidans is a gram-negative, motile rod (0.5 × 1.0 μm) (Figure 17.7). Morphologically, the cells of *A. ferrooxidans* exhibit a multilayered cell wall (outer membrane, periplasm, thin peptidoglycan layer) typical of gram-negative bacteria (Remsen and Lundgren, 1966; Avakyan and Karavaiko, 1970). Cell division is mostly by constriction but occasionally also by partitioning (Karavaiko and Avakyan, 1979).

A. ferrooxidans is mesophilic, i.e., it grows within a temperature range of 15°C–42°C with an optimum in the range of 30°C–35°C (Silverman and Lundgren, 1959a; Ahonen and Tuovinen, 1989). In general, all strains of *A. ferrooxidans* grow and oxidize iron(II) at around pH 2. Growth can be observed over a range of initial pH of 1.5–6.0. Ferric iron produced by the organism precipitates above pH 1.9 (e.g., Buchanan and Gibbons, 1974; Starr et al., 1981).

Autotrophy: Ferrous iron is not the only energy source and reducing power for *A. ferrooxidans*. It can also use reduced forms of sulfur such as hydrogen sulfide, elemental sulfur, thiosulfate, or tetrathionate and some metal sulfides as sole sources of energy

(see Chapters 20 and 21). In addition, it has been shown that *A. ferrooxidans* can use hydrogen as sole source of energy with oxygen as terminal electron acceptor (Drobner et al., 1990). Formate can be used with oxygen as terminal electron acceptor or ferric iron in anoxic environments (Pronk et al., 1991a,b). The bacterium gets its carbon from carbon dioxide, and it derives its nitrogen preferentially from ammonium and nitrate (Temple and Colmer, 1951; Lundgren et al., 1964). Some strains can fix nitrogen (Mackintosh, 1978; Stevens et al., 1986).

Facultative heterotrophy: A number of claims have appeared in the literature that some strains of *A. ferrooxidans* can be adapted to grow heterotrophically if glucose is used instead of ferrous iron as the sole source of energy and carbon (Lundgren et al., 1964; Shafia and Wilkinson, 1969; Tabita and Lundgren, 1971a; Shafia et al., 1972; Tuovinen and Nicholas, 1977; Sugio et al., 1981). However, the discovery that some cultures of *A. ferrooxidans* contained heterotrophic satellite organisms (Guay and Silver, 1975; Mackintosh, 1978; Harrison, 1981, 1984; Lobos et al., 1986; Wichlacz et al., 1986) casts serious doubt on the existence of any *A. ferrooxidans* strains that can grow heterotrophically.

Consortia with A. ferrooxidans: The existence of satellite organisms that appear to live in close association with *A. ferrooxidans* was first reported by Zavarzin (1972), and confirmation of their existence soon followed. Indeed, taxonomically different organisms were isolated from different *A. ferrooxidans*

L. ferrooxidans oxidizes ferrous iron also differs from that of A. ferrooxidans. Its iron(II)-oxidizing system lacks rusticyanin. Strain DSM 2705 of L. ferrooxidans contains a novel red cytochrome that is soluble, acid stable, and reducible by ferrous iron and seems to lack a-type cytochrome (Hart et al., 1991). The reduced cytochrome 579 was shown to be absolutely necessary for aerobic oxidation of ferrous iron (Blake and Griff, 2012). The red cytochrome consists of a single polypeptide having a molecular mass of 17,900 Da and a standard reduction potential of +680 mV at pH 3.5. It contains 1 equivalent each of iron and zinc (Hart et al., 1991). The cells of L. ferrooxidans have been shown to contain an active ribulose 1,5-bisphosophate carboxylase, typical of many chemolithotrophs (autotrophs) that use the Calvin–Benson cycle for carbon dioxide assimilation (Balashova et al., 1974).

17.4.2.1.3 GENETICS OF IRON(II)-OXIDIZING ACIDOPHILES

Understanding the genetics of acidophilic iron(II) oxidizers lagged behind that of many other bacterial groups. One reason for this was the past difficulty of culturing the organisms on solid media to obtain pure cultures when selecting wild-type and mutant strains from a mixed culture. Another reason was that the acidophilic nature of the organisms made the application of standard molecular biological techniques difficult. With the development of new culture techniques, significant advances in unraveling the genetics of A. ferrooxidans have been made (Croal et al., 2004a; Valdes et al., 2008; Quattrini et al., 2009). Like most bacteria, A. ferrooxidans contains most of its genetic information on a chromosome and some nonessential genetic information on extrachromosomal DNA called plasmids (Mao et al., 1980). Some of the chromosomal and plasmid genes have been mapped and characterized (Rawlings and Kusano, 1994). These include genes of nitrogen metabolism, such as glutamine synthetase (see Rawlings and Kusano, 1994) and nitrogenase (e.g., Pretorius et al., 1986, 1987; Rawlings and Kusano, 1994), genes involved in carbon dioxide fixation and in energy conservation (see Rawlings and Kusano, 1994), genes responsible for resistance to mercury, and other genes (see Rawlings and Kusano, 1994).

The genomes of different strains of A. ferrooxidans and Acidithiobacillus thiooxidans have been compared (Harrison, 1982, 1986). The 13 strains of A. ferrooxidans that were examined fell into different DNA homology groups. Though genetically related,

they were not identical. None showed significant DNA homology with six strains of A. thiooxidans, indicating little genetic relationship. The A. thiooxidans strains fell into two different DNA homology groups. On the basis of 16S rRNA relationships, A. ferrooxidans belongs to the Proteobacteria, most to the β-subgroup, but at least one strain to the γ-subgroup (see Rawlings and Kusano, 1994). A gapped genome sequence has now been determined for A. ferrooxidans ATCC 23270 (Selkov et al., 2000).

17.4.2.2 Domain Bacteria: Thermophiles

A number of thermophilic acidophilic bacteria able to oxidize ferrous iron at up to 70°C have been isolated from iron(II)-rich hot springs. Most of these microorganisms are able to grow autotrophically on inorganic electron donors such as iron(II), elemental sulfur, or metal sulfides and need enriched atmospheric carbon dioxide concentrations to fix carbon dioxide (Golovacheva and Karavaiko, 1978; Norris et al., 1988; Clark and Norris, 1996, Norris, 1997). In the absence of a suitable inorganic energy source, they grow heterotrophically (Golovacheva and Karavaiko, 1978). Often autotrophic growth is stimulated by a trace of yeast extract (Golovacheva and Karavaiko, 1978; Norris, 1997). See Table 17.3b for examples of microorganisms.

17.4.2.3 Domain Archaea: Mesophiles

Archean mesophilic acidophiles oxidize ferrous iron optimally at a pH below 2 (Edwards et al., 2000; Golyshina et al., 2000). Most described acidophilic, mesophilic Archaea capable of iron(II) oxidation are chemolithoautotrophic though they require a trace of yeast extract in their medium. Some strains are also able of chemoorganotrophy (Edwards et al, 2000; Dopson et al., 2004). See Table 17.3c for examples of microorganisms.

17.4.2.4 Domain Archaea: Thermophiles

Archean acidophiles are often hyperthermophilic (Brock et al., 1972, 1976; Segerer et al., 1986). For example, Acidianus brierleyi (formerly Sulfolobus brierleyi) (Segerer et al., 1986) grows in a temperature range of 55°C–90°C (optimum 70°C–75°C) (Brierley and Brierley, 1973; McClure and Wyckoff, 1982; Segerer et al., 1986). Many of

EHRLICH'S GEOMICROBIOLOGY

TABLE 17.4
Examples of anaerobic iron(II)-oxidizing microorganisms.

Organism	Literature
a. Nitrate-reducing	
Acidovorax sp. (BoFeN1)	Kappler et al. (2005b), Muehe et al. (2009), Klueglein et al. (2014)
Acidovorax sp. (2AN)	Chakraborty et al. (2011)
Thiobacillus denitrificans	Straub et al. (1996)
Enrichment culture KS	Straub et al. (1996), Blöthe and Roden (2009b)
Pseudogulbenkiania sp. (2002)	Weber et al. (2009)
Bradyrhizobium spp.	Shelobolina et al. (2012), Benzine et al. (2013)
Cupriavidus necator	Shelobolina et al. (2012), Benzine et al. (2013)
Ralstonia solanacearum	Shelobolina et al. (2012), Benzine et al. (2013)
Desulfitobacterium frappieri G2	Shelobolina et al. (2003)
Dechloromonas agitata strain is5	Benzine et al. (2013)
Nocardioides sp. strain in31	Benzine et al. (2013)
b. Photoferrotrophic	
Chlorobium ferrooxidans (KoFox)	Heising et al. (1999), Kappler and Newman (2004), Hegler et al. (2008)
Thiodictyon sp. (F4)	Kappler and Newman (2004), Croal et al. (2004b), Hegler et al. (2008)
Rhodobacter ferrooxidans (SW2)	Ehrenreich and Widdel (1994); Kappler and Newman (2004), Hegler et al. (2008)
Rhodopseudomonas palustris (TIE-1)	Jiao et al. (2005)
Rhodovulum iodosum	Straub et al. (1999), Wu et al. (2014)
Rhodovulum rubiginosum	Straub et al. (1999)

these strains oxidize ferrous iron or elemental sulfur autotrophically. See Table 17.3d for examples of microorganisms. Thermophilic iron(II)-oxidizing acidophiles seem to fix carbon via the 3-hydroxypropionate/4-hydroxybutyrate and reductive tricarboxylic acid cycle pathway (Jennings et al., 2014). Microbial mats present in hot springs seem to depend on the activity of autotrophic microorganisms to fix carbon as at least 42% of the biomass carbon originates from dissolved inorganic carbon sources (Jennings et al., 2014).

17.4.3 Anaerobic Oxidation of Ferrous Iron

Until the early 1990s, bacterial oxidation of ferrous iron at neutral pH was generally assumed to require oxygen as terminal electron acceptor before the first enzymatically catalyzed processes of anaerobic iron(II) oxidation were discovered. These new metabolisms use either light energy and bicarbonate for iron(II) oxidation and biomass production or nitrate as electron acceptor. Nitrate-reducing and phototrophic iron(II) oxidizers were shown to prevail in the same geochemical habitat, e.g., in surface-near freshwater sediments exposed to light (Melton et al., 2012). Therefore, these two physiological types of iron(II)-oxidizing microorganisms compete for the same electron donor, i.e., iron(II), in such habitats. Microcosm studies with sediments incubated under different conditions suggested that photoferrotrophic iron(II) oxidizers outcompete nitrate-reducing iron(II) oxidizers in light conditions, while in the dark, nitrate-reducing iron(II) oxidizers play the dominant role for iron(II) oxidation (Melton et al., 2012). To which extent these two metabolisms compete with microaerophilic iron(II) oxidizers for iron(II) at the oxic–anoxic interface is currently unknown. A list of examples of phototrophic and nitrate-reducing iron(II) oxidizers is given in Table 17.4.

17.4.3.1 Phototrophic Iron(II) Oxidation

Phototrophic iron(II) oxidation proceeds according to the following equation:

$$4Fe^{2+} + HCO_3^- + 10H_2O + h\nu \rightarrow$$
$$4Fe(OH)_3 + CH_2O + 7H^+ \quad (17.5)$$

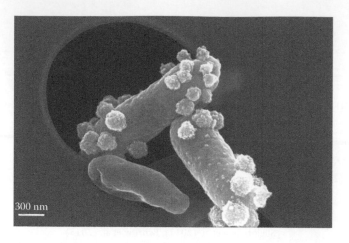

Figure 17.9. Scanning electron micrograph of critical point dried cells of the phototrophic iron(II)-oxidizing bacterium *Chlorobium ferrooxidans* sp. strain KoFox. The particles attached to the surface of the cell are iron minerals precipitated during iron(II) oxidation. Image taken at the Natural and Medical Science Institute at the University of Tuebingen (NMI), using the InLens Detector of a Zeiss Crossbeam Leo 1540 XB microscope at 15 kV acceleration voltage. (Courtesy of Schädler, S, Burkhardt, C, and Kappler, A, Geomicrobiology, Center for Applied Geosciences, University of Tuebingen, Tuebingen, Germany.)

Diverse anoxygenic photosynthetic bacteria, including the freshwater strains R. *palustris* TIE-1 and *Rhodobacter ferrooxidans* SW2 (purple-nonsulfur bacteria), *Chlorobium ferrooxidans* KoFox (a green sulfur bacterium, Figure 17.9), *Thiodictyon* sp. F4 (purple-sulfur bacterium), and two marine *Rhodovulum* strains (Table 17.4b), are able to oxidize iron(II) to iron(III) oxyhydroxides in the light (Hegler et al., 2008) producing a variety of minerals including ferrihydrite, goethite, lepidocrocite, and magnetite (Posth et al., 2014) (see Section 17.7). It has been shown that these organisms can oxidize dissolved iron(II), i.e., Fe^{2+}, but also minerals that are soluble to some degree such as siderite and ferrous sulfide (Kappler and Newman, 2004).

Jiao et al. (2005) studied a genetically tractable variant of R. *palustris*, strain TIE-1, in which they demonstrated the need for two genetically determined components in a functioning phototrophic iron(II)-oxidizing system, using transposon mutagenesis (Figure 17.10): (1) a membrane protein that is part of an ABC transport system and (2) a gene that is a homologue of CobS, an enzyme involved in cobalamin biosynthesis. In a later study, Jiao and Newman (2007) demonstrated that the three-gene *pio* operon (*pioABC*) in this organism is essential for growth by phototrophic iron(II) oxidation. The expression of *pioABC* was shown to be induced by anoxia and is regulated by two fixK boxes

upstream of *pioABC* (Bose and Newman, 2011). One of the genes in this operon, *pioA* encodes a decaheme c-type cytochrome that is upregulated when the strain grows with iron(II). PioB encodes an outer-membrane beta-barrel protein, which is likely to be involved in the transport of iron(II) into the cytoplasm or iron(III) out of the periplasm. PioC encodes a putative high potential iron–sulfur protein (HiPIP) that shows similarity to the putative iron(II) oxidoreductase from A. *ferrooxidans*. Both PioA and PioC are speculated to transfer electrons from iron(II) to the cytoplasmic electron transport chain in a linear manner. The cytochrome c_2, by contrast, is assumed to be responsible for a cyclic electron flow to generate energy for photosynthetic growth (Bird et al., 2014). In contrast to TIE-1, R. *ferrooxidans* strain SW2 contains a foxEYZ operon, which is not homologous to *pioABC*, but also enhances phototrophic iron(II) oxidation in *Rhodobacter capsulatus* SB1003 (Croal et al., 2007). The first gene in this operon, foxE, encodes a c-type cytochrome that acts as an iron-oxidoreductase (Saraiva et al., 2011, 2012). FoxY encodes a pyrroloquinoline quinone-binding protein, while foxZ transcribes into an inner membrane transport protein.

All of these anoxygenic phototrophic iron(II)-oxidizing bacteria, except R. *capsulatus* SB1003, are photoautotrophic and can use the reducing power generated in the oxidation of iron(II) to

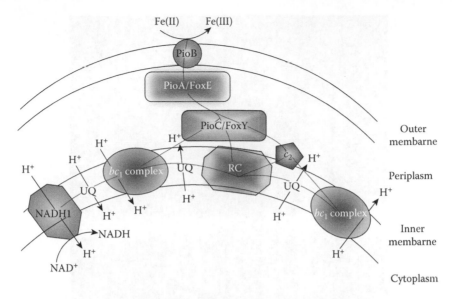

Figure 17.10. Model of biochemical mechanisms of iron(II) oxidation in the phototrophic iron(II) oxidizer *Rhodopseudomonas palustris* TIE1 and *Rhodobacter* sp. SW2. Model according to Bird et al. (2011), considering Bose and Newman (2011), Pereira et al. (2012), Saraiva et al. (2012), and Bird et al. (2014). The dotted line indicates the direction of electron flow. PioB, TIE-1 outer membrane protein potentially involved in iron transfer; PioA, TIE-1 decaheme *c*-type cytochrome; FoxE, SW2 *c*-type cytochrome; PioC, TIE-1 high-potential iron protein; FoxY, SW-2 quinoprotein; RC, phototrophic reaction center; c_2, cytochrome c_2; bc_1, cytochrome bc_1 complex; UQ, ubiquinones; NADH1, NADH dehydrogenase complex.

iron(III) for carbon dioxide assimilation (fixation) (Widdel et al., 1993; Ehrenreich and Widdel, 1994; Caiazza et al., 2007; Melton et al., 2012). Caiazza et al. (2007) have shown that *R. capsulatus* SB1003 can grow photoheterotrophically by oxidizing iron(II) citrate to iron(III) citrate under anoxic conditions followed by a photochemical breakdown of the citrate to acetoacetate, whose carbon is then assimilated. Iron(II)-NTA can also be used in a similar fashion with organic NTA degradation products serving as carbon and electron source for the microorganisms (Kopf and Newman, 2012).

Anoxygenic phototrophic iron(II) oxidizers received a lot of attention in the last years in particular since Widdel et al. (1993) and several other researchers afterwards suggested that anaerobic oxidation of ferrous to ferric iron by phototrophic bacteria could have contributed to the early deposition of BIFs in Archean times, which is discussed in Section 17.8.

17.4.3.2 Chemotrophic Iron(II) Oxidation: Nitrate-Reducing Iron(II) Oxidizers

In addition to anaerobic iron(II)-oxidizing phototrophs, respirers that oxidize iron(II) anaerobically have been described. Straub et al. (1996) found that enrichment cultures from town ditches in Bremen, Germany, and from brackish water lagoons, as well as some denitrifying isolates were able to oxidize ferrous iron anaerobically using nitrate as terminal electron acceptor. According to this publication, some were able to use ferrous iron as exclusive electron donor and grew lithotrophically, and some used acetate concurrently with ferrous iron as electron donors and thus, grew mixotrophically. However, only for the lithoautotrophic iron(II)-oxidizing enrichment culture KS (Figure 17.11) (Blöthe and Roden, 2009a) has it been demonstrated unequivocally that it can be transferred continuously over several generations to fresh growth medium without the addition of any organic compounds. Interestingly, the dominant organism in the KS culture is a phylotype related to the autotrophic iron(II) oxidizer *S. lithotrophicus*. Analysis of the enrichment culture by a 16S rRNA gene library also revealed phylotypes related to known heterotrophic nitrate reducers *Comamonas badia*, *Parvibaculum lavamentivorans*, and *Rhodobacter thiooxidans* (Blöthe and Roden, 2009a). However, exactly how this enrichment culture functions and which organism(s) contribute to iron(II) oxidation and which are responsible for the observed nitrate reduction remains unclear.

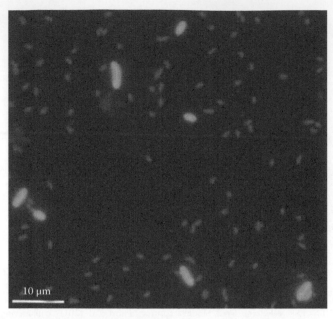

10 μm

Figure 17.11. **(See color insert.)** Autotrophic nitrate-reducing iron(II)-oxidizing enrichment culture KS in the late exponential phase cultivated with 10 mM iron(II) chloride and 4 mM sodium nitrate. Overlay of CARD-FISH and DAPI image. Green cells hybridized with a *Bradyrhizobium*-specific HRP (horse radish peroxidase)-oligonucleotide probe. Signal amplification with a tyramide Alexa 517 conjugate. (Courtesy of Tominski, C, Lösekann-Behrens, T, and Behrens, S, Geomicrobiology, Center for Applied Geosciences, University of Tuebingen, Tuebingen, Germany.)

The following equation summarizes the microbial oxidation of ferrous iron with nitrate as electron acceptor in these observations (Straub et al., 1996):

$$10FeCO_3 + 2NO_3^- + 24H_2O \rightarrow$$
$$10Fe(OH)_3 + N_2 + 10HCO_3^- + 8H^+ \quad (17.6)$$

However, for most of the microbial cultures that are described to perform nitrate-dependent iron(II) oxidation, an organic cosubstrate is necessary in addition to iron(II) to sustain the cultures over several culture transfers. This has cast doubt on the ability of these strains to oxidize iron(II) enzymatically suggesting that a large part if not all of the iron(II) oxidation observed in these cultures comes from abiotic iron(II) oxidation by nitrite formed during heterotrophic iron(II) oxidation (Klueglein et al., 2013, 2014; Melton et al., 2014a). The nitrate was reduced to dinitrogen in all instances. Ammonia production from nitrate was not detected with the exception of *Geobacter* strains that were shown to be able to not only reduce iron(III) to iron(II) using acetate as electron donor but which are also able to reduce nitrate to ammonium with iron(II) as electron donor (Finneran et al., 2002; Weber et al., 2006c). *Thiobacillus denitrificans* was also found

to be able to reduce nitrate with ferrous iron (Straub et al., 1996). However, genetic, transcriptional, and physiological studies with *T. denitrificans* did not show evidence for c-type cytochromes involved in electron transfer from iron(II) to nitrate, nor did the coupling of iron(II) oxidation to nitrate reduction result in energy conservation or growth (Beller et al., 2013). This suggests that iron(II) oxidation with this microorganism is mainly an abiotic process caused by the reactive nitrite. This in turn contradicts a recently postulated mechanism for iron(II) oxidation coupled to nitrate reduction (Carlson et al., 2012), where electron transfer from iron(II) to the quinone pool was suggested to be mediated by cytochromes. Therefore, although it has been described that the addition of iron(II) to cultures growing with nitrate and acetate increased cell growth (Muehe et al., 2009; Chakraborty et al., 2011), and although in some cases, iron(II) oxidation was induced by precultivation in the presence of iron(II) (Chakraborty and Picardal, 2013), the proteins that couple iron(II) oxidation to the reduction of nitrate are still unknown. Nevertheless, even if most of the observed oxidation of iron(II) coupled to the reduction of nitrate in the environment goes via a reactive nitrite

EHRLICH'S GEOMICROBIOLOGY

intermediate, this process can still be of significance contributing to iron cycling in nature.

In particular, the formation of minerals by these organisms could be of interest and relevant for the cycling of trace metals and nutrients such as phosphate. For the mixotrophic nitrate-reducing iron(II) oxidizer *Acidovorax* sp. strain BoFeN1, it has been shown that depending on the geochemical conditions (pH, substrate concentration, presence of humic substances, presence of preexisting minerals), it can produce ferric oxyhydroxides such as ferrihydrite, goethite, and lepidocrocite but also mixed iron(II)–iron(III) minerals such as magnetite and even green rust (Kappler et al., 2005b; Larese-Casanova et al., 2010; Dippon et al., 2012; Pantke et al., 2012). The formation of these iron(III) minerals causes encrustation of strain BoFeN1 cells. Whether this encrustation is useful or harmful to the bacterial cells is discussed in Section 17.7.

Benz et al. (1998) detected anaerobic, nitrate-coupled ferrous iron oxidation in culture enrichments with sediments from freshwater, brackish water, and marine water. They isolated a strain labeled HidR2, which was a motile, nonspore-forming, gram-negative rod that oxidized ferrous iron in the presence of 0.2–1.1 mM acetate as cosubstrate and nitrate as terminal electron acceptor at pH 7.2 and 30°C. The organism is capable of anaerobic growth on acetate in the absence of ferrous iron but in the presence of nitrate. It is also capable of aerobic growth in the presence of low concentrations of acetate with oxygen as terminal electron acceptor. In a more recent study, Straub et al. (2004) reported that three earlier iron(II)-oxidizing, nitrate-reducing isolates, BrG1, BrG2, and BrG3, obtained from freshwater sediments, affiliated with *Acidovorax*, *Aquabacterium*, and *Thermomonas*, respectively, which belong to the β- and γ-subgroups of the Proteobacteria.

Weber et al. (2006b) isolated an iron(II)-oxidizing microorganism, strain 2002, from an anoxic, highly reduced sediment core taken from freshwater Campus Lake at Southern Illinois University in Carbondale, IL. The study suggests that this isolate grew autotrophically while oxidizing iron(II) with nitrate as terminal electron acceptor. However, although strain 2002 is capable of growing with iron(II) as the sole electron donor and fixing carbon dioxide into organic carbon under nitrate-reducing conditions, it cannot be indefinitely cultivated under lithoautotrophic conditions with iron(II), and the exact metabolic pathway remains unknown (Karrie Weber, personal communication). Its autotrophy was facultative, because it was able to grow with some simple organic carbon compounds, like acetate and propionate, as sole carbon and energy source. On acetate, it was able to grow with nitrate, nitrite, nitrous oxide, or oxygen as terminal electron acceptor. The organism was detected in significant numbers throughout the sediment core in a range of approximately 10^3–10^4 g^{-1} of sediment. The organism was closely related to the common soil bacterium *Chromobacterium violaceum* in the Betaproteobacteria.

A number of other iron(II)-oxidizing organisms (*Bradyrhizobium* spp., strains of *Cupriavidus necator* and *Ralstonia solanacearum*, *Dechloromonas agitata* strain is 5) have been isolated recently by Shelobolina et al. (2012) and Benzine et al. (2013). These strains are able to grow via oxidation of iron(II) in smectite with nitrate as the electron acceptor. Also, the first iron(II)-oxidizing *Nocardioides* sp. (strain in31) was isolated (Benzine et al., 2013). Shelobolina et al. (2003) obtained a *Desulfitobacterium frappieri* isolate (strain G2) from subsurface smectite beds in Georgia. This strain oxidizes iron(II) in reduced smectite with nitrate as the electron acceptor. Like the other nitrate-reducing iron(II) oxidizers, it is not autotrophic, i.e., a small amount of fixed carbon (0.05% yeast extract) was required to sustain growth. Additionally, this strain has the capacity to also reduce iron(III) in oxidized smectite (e.g., with hydrogen as the electron donor).

A hyperthermophilic archeon, *Ferroglobus placidus*, has also been found to oxidize ferrous iron anaerobically with nitrate as electron acceptor (Hafenbradl et al., 1996). It can grow lithoautotrophically or heterotrophically between 65°C and 95°C (optimum 85°C) and at pH 7.0. The cells require 0.5%–4.5% NaCl (~2% optimum) in their growth medium to prevent their lysis. The organism reduces the nitrate chiefly to nitrite, but during prolonged incubation, the nitrite is converted to nitric oxide and nitrogen dioxide. No nitrogen, nitrous oxide, or ammonium is formed, however. The following reaction best describes the overall oxidation of ferrous iron by this organism (Hafenbradl et al., 1996, amended):

$$2FeCO_3 + NO_3^- + 5H_2O \rightarrow$$
$$2Fe(OH)_3 + NO_2^- + 2HCO_3^- + 2H^+ \quad (17.7)$$

The bacterial processes described in this section can be considered to be part of an anoxic iron cycle. For more information, we refer the reader to recent reviews (Kappler and Straub, 2005; Weber et al., 2006a, Konhauser et al., 2011; Melton et al., 2014a).

17.4.4 Ferrous Iron Oxidation: Comparison of Energetics, Carbon Dioxide Fixation, and Biochemistry

The energy theoretically obtained in a certain metabolic process can be calculated as Gibbs free energy ΔG. ΔG^0 values are defined for standard conditions at pH 0, i.e., all reactants are assumed to be present at 1 M concentrations or as $\Delta G^{0'}$ for standard conditions at pH 7, i.e., all reactants are given in 1 M concentrations with the exception of protons (concentration of 10^{-7} M). When considering relevant environmental conditions, ΔG values are calculated from ΔG^0 values using the Nernst equation considering the relevant environmental concentrations of all reactants.

Oxidation of ferrous iron does not furnish much energy on a molar basis, compared, for instance, to the oxidation of glucose, which can generate up to 38 molecules of ATP and -2870 kJ mol^{-1} glucose upon complete oxidation:

$$C_6H_{12}O_6 + 6O_2 \rightarrow 6CO_2 + 6H_2O \quad (17.8)$$

As the speciation of iron(II) and iron(III) is strongly dependent on the pH and ambient geochemical conditions (i.e., the presence of phosphate, carbonate, sulfide), the resulting gain in free energy ΔG differs depending on the geochemical conditions.

The oxidation of ferrous iron at low pH deals with mainly dissolved iron species according to the following equation:

$$4Fe^{2+} + 4H^+ + O_2 \rightarrow 4Fe^{3+} + 2H_2O \quad (17.9)$$

This reaction yields (calculated for standard conditions at pH 2) a ΔG^0 of -6 kcal mol^{-1} of oxidized iron(II).

Acidophilic iron(II) oxidizers live under harsh environmental conditions with high extracellular proton concentrations. These protons enter cells by passive and active processes, of which the latter can be used to generate the energy equivalent ATP. To maintain the cell-internal pH neutral, cytosolic protons are exported from the cell. Due to the neutral pH of cells' interior, iron(II) is oxidized extracellularly. In order to also produce reduced NADP$^+$, i.e., NADPH, acidophilic iron(II) oxidizers drive the electrons obtained from iron(II) against the redox gradient of a number of cytochrome complexes including bc_1 (see Figure 17.8). This reverse transport of electrons is fueled by the proton motive force across the plasma membrane (Bird et al., 2011 and references therein).

At neutral pH, depending on the presence of bicarbonate or phosphate, iron(II) can be oxidized using oxygen as electron acceptor from dissolved iron(II), i.e., Fe^{2+}:

$$4Fe^{2+} + O_2 + 10H_2O \rightarrow 4Fe(OH)_3 + 8H^+ \quad (17.10)$$

with a $\Delta G^{0'}$ of -24 kcal mol^{-1} iron(II) oxidized (pH 7) or from siderite:

$$4FeCO_3 + O_2 + 10H_2O \rightarrow 4Fe(OH)_3 + 4HCO_3^- + 4H^+ \quad (17.11)$$

with a $\Delta G^{0'}$ of -14 kcal mol^{-1} iron(II) carbonate oxidized (pH 7). Additionally, even microaerophilic oxidation of ferrous sulfides or iron(II)-containing clays or mixed iron(II)–iron(III) minerals such as magnetite or green rust have been observed.

Providing free energy ΔG estimates for nitrate-dependent iron(II) oxidation under anoxic conditions is more difficult since most of these organisms are mixotrophic, and it is unclear to which extent the iron(II) oxidation in these organisms is enzymatically catalyzed or caused by chemical oxidation of iron(II) by nitrite that is formed during heterotrophic nitrate reduction. Nevertheless, it is possible to calculate the ΔG value for iron(II) oxidation coupled to nitrate reduction:

$$10Fe^{2+} + 2NO_3^- + 24H_2O \rightarrow 10Fe(OH)_3 + N_2 + 18H^+ \quad (17.12)$$

with a $\Delta G^{0'}$ of -23 kcal mol^{-1} iron(II) oxidized (pH 7).

While quite a lot is known about the biochemistry of iron(II) oxidation in *A. ferrooxidans* at acidic pH, almost nothing is known for anaerobic and aerobic iron(II) oxidizers at circumneutral pH. A small number of iron(II) oxidation genes have been identified in phototrophic and microaerophilic iron(II) oxidizers, but not in nitrate-reducing iron(II) oxidizers. For all of these types of iron(II) oxidizers, substantial knowledge is still lacking in order to answer questions such as how much energy these organisms obtain from iron(II) oxidation, how these bacteria access different kinds of iron(II) (soluble, complexed, or mineral bound), whether they transport iron into the cell, and how they deal with the product Fe^{3+}, which rapidly forms iron(III) hydroxide minerals that potentially cover and encrust the cells.

17.5 IRON(III) AS TERMINAL ELECTRON ACCEPTOR IN BACTERIAL RESPIRATION

Ferric iron in nature can be microbiologically reduced to ferrous iron. The ferric iron that is reduced by microbes may be in solution or present as poorly soluble ferric iron minerals. Examples of iron(III) minerals are amorphous oxides or hydroxides, i.e., minerals like ferrihydrite ($Fe(OH)_3$), goethite (α-FeO(OH)), hematite (Fe_2O_3), and so forth (Table 17.1). Also mixed iron(II)/iron(III) minerals such as magnetite ($Fe^{II}Fe_2^{III}O_4$) and Fe^{3+} present in clay minerals can be reduced microbially. As it is the case for iron(II) oxidation, iron(III) reduction may be enzymatic or nonenzymatic (Melton et al., 2014a). Enzymatic ferric iron reduction may manifest itself as a form of respiration, mostly anaerobic, in which ferric iron serves as *dominant* or *exclusive* terminal electron acceptor, or it may accompany fermentation, in which ferric iron serves as a supplementary terminal electron acceptor. When ferric iron reduction is coupled to energy generation, this process is called *dissimilatory iron(III) reduction*.

When ferric iron is reduced during uptake or incorporation of iron into specific cellular components, the process is called *assimilatory iron(III) reduction*. Whereas relatively large quantities of iron(III) are consumed in dissimilatory reduction, only very small quantities are consumed in assimilatory reduction. The ferric iron in assimilatory reduction when acquired at circumneutral pH is usually complexed by iron siderophores and may be reduced in this complexed form or after release from the ligand in the cell envelope (discussed in Section 17.3). Even some fungi have been implicated in iron(III) reduction (e.g., Ottow and von Klopotek, 1969). However, their ability to reduce ferric iron is not likely to involve anaerobic respiration but instead either assimilatory iron(III) reduction or the production of one or more metabolic products that act as a chemical reductant of the ferric iron. Iron(III) dissimilatory and assimilatory reductases have been reviewed by Schröder et al. (2003). The emphasis in the following sections will be on dissimilatory iron(III) reduction.

17.5.1 Bacterial Ferric Iron Reduction Accompanying Fermentation

For some time, ferric iron has been known to influence the fermentative metabolism of bacteria as a result of its ability to act as terminal electron acceptor. Roberts (1947) showed a change in fermentation balance when comparing the anaerobic growth of *Bacillus polymyxa* on glucose in the presence and absence of ferric iron (Table 17.5). The ferric iron seemed to act as a supplementary electron acceptor during fermentation and thereby changed the relative quantities of certain products formed from glucose. Thus, in the presence of ferric iron, less hydrogen, carbon dioxide, and 2,3-butylene glycol but more ethanol were formed in either "organic" or "synthetic" medium than in the absence of iron. Also, approximately twice as much glucose was consumed in the presence of iron than in its absence in either medium.

The iron(III)-reducing activity during fermentation was viewed as a sink for excess reducing power from which the organism could not derive useful energy in its organic carbon oxidation (see reviews by Lovley, 1987, 1991). However, this explanation only holds if it can be demonstrated that iron(III) reduction in these instances is not accompanied by energy conservation, as was reported to be the case with *Clostridium beijerinckii* (Dobbin et al., 1999). Since Roberts (1947) only detected the formation of 42 or 61 mol of iron(III) from 100 mol of glucose, it is believed that ferric iron reduction is only a negligible sink for electrons during fermentative growth. It was often found that less than 5% of electrons obtained from organic carbon ended up on ferric iron, allowing the assumption that iron(III) reduction during fermentation does not significantly support growth. This was supported by thermodynamic calculations by Lovley and Phillips (1989). When considering a mole of glucose and assuming complete transformation to bicarbonate, 24 mol of ferric iron should be reduced:

$$C_6H_{12}O_6 + 24Fe(OH)_3 + 42H^+ \rightarrow$$
$$6HCO_3^- + 24Fe^{2+} + 60H_2O \qquad (17.13)$$

TABLE 17.5

Fermentation balances for Bacillus polymyxa *growing in two different media in the presence and absence of ferric hydroxide.*

Products	Synthetic medium[a] (mol/100 mol glucose)		Organic medium[b] (mol/100 mol glucose)	
	$-Fe(OH)_3$	$+Fe(OH)_3$	$-Fe(OH)_3$	$+Fe(OH)_3$
CO_2	199	170	186	178
H_2	51	31	53	33
HCOOH	11	12	9	12
Lactic acid	17	19	14	7
Ethanol	72	82	78	94
Acetoin	0.5	1	1	2
2,3-Butylene glycol	64	51	49	44
Acetic acid	0	0	0	0
Iron(III) reduced	—	42	—	61
Glucose fermented (mg/100 mL)	1029	2333	1334	2380
C recovery (%)	112.1	101.8	98.8	97.2
O/R index	1.06	1	1.06	1.03

SOURCE: Roberts, JL, *Soil Sci*, 63, 135, 1947. With permission.

NOTE: Incubation was for 7 days at 35°C.

[a] Glucose, 2.4%; asparagine, 0.5%; K_2HPO_4, 0.08%; KH_2PO_4, 0.02%; KCl, 0.02%; $MgSO_4 \cdot 7H_2O$, 0.5%.

[b] Glucose, 2.4%; peptone, 1%; K_2HPO_4, 0.08%; KH_2PO_4, 0.02%; KCl, 0.02%; $MgSO_4 \cdot 7H_2O$, 0.5%.

with a $\Delta G^{0'}$ of -30 kJ electron^{-1} transferred (Lovley and Phillips, 1989). The fermentation of glucose along the following equation:

$$C_6H_{12}O_6 + 2H_2O \rightarrow 2C_2H_5OH + 2HCO_3^- + 2H^+ \quad (17.14)$$

is energetically more favorable with a $\Delta G^{0'}$ of -73 kJ electron^{-1} transferred (Lovley and Phillips, 1989) than complete glucose oxidation during ferric iron reduction. Nonetheless, an increase in growth of fermenters has repeatedly been observed when iron(III) is added to the medium, though this could not be attributed to sufficiently high extents of iron(III) reduction. Yet again, thermodynamic calculations showed that a slightly greater energy yield could be achieved in fermentation, when a marginal amount of electrons were taken up by ferric iron.

17.5.2 Ferric Iron Respiration

The most direct evidence for bacterial iron(III) respiration has come from the studies with *Geobacter metallireducens* (strain GS-15), *G. sulfurreducens*, *S. oneidensis* MR-1, two cultures of *S. putrefaciens*, strains 200 and ATCC 8071, and from *Desulfuromonas acetoxidans* (Obuekwe et al., 1981; Lovley and Phillips, 1988; Myers and Nealson, 1988; Lovley et al., 1989a; Roden and Lovley, 1993; Caccavo et al., 1994). *G. metallireducens* was isolated from freshwater sediment. *G. sulfurreducens* was isolated from a hydrocarbon-contaminated ditch. *S. oneidensis*, a facultative anaerobe, was isolated from sediment of Lake Oneida, NY, whereas *S. putrefaciens* strain 200, also a facultative anaerobe, was isolated from a Canadian oil pipeline. All these isolates are able to use iron(III) and manganese(IV) anaerobically as terminal electron acceptors for growth.

Most of the early evidence for dissimilatory ferric iron reduction rested on observations with growing cultures. Troshanov (1968, 1969) demonstrated that a number of microorganisms isolated from sediments from lakes in the Karelian peninsula in the former USSR could reduce ferric iron to varying degrees. Troshanov noted that the form in which the iron was available to his cultures affected the rate of reduction. Poorly soluble ferric iron minerals in bog ore were reduced more slowly than soluble $FeCl_3$. Cultures also varied in their ability to reduce ferric iron minerals. He found that *Bacillus circulans* actively reduced bog iron ore

TABLE 17.6
Examples of iron(III)-respiring microorganisms.

Organism	Literature
Desulfobulbus propionicus	Lonergan et al. (1996)
Geobacter chapellii	Lonergan et al. (1996)
Geobacter metallireducens GS-15	Lovley (1993)
Geobacter sulfurreducens	Caccavo et al. (1994)
Geothrix fermentans	Coates et al. (1999)
Pelobacter propionicus	Lonergan et al. (1996)
Sulfurospirillum barnesii[a]	Laverman et al. (1995), Lonergan et al. (1996)
Shewanella algae	Caccavo et al. (1992)
Shewanella oneidensis MR-1	Myers and Nealson (1988), Lovley et al. (1989a), Venkateswaran et al. (1999)
Shewanella putrefaciens	Obuekwe and Westlake (1982), Arnold et al. (1986a,b)
Acidimicrobium ferrooxidans	Bridge and Johnson (1998)

[a] Former name: *Geospirillum barnesii*.

whereas *B. polymyxa* did not. Similar findings were made independently with soil bacteria by Ottow and collaborators (Ottow, 1969a, 1971; Ottow and Glathe, 1971; Hammann and Ottow, 1974; Munch et al., 1978; Munch and Ottow, 1980).

Some typical heterotrophic and autotrophic ferric iron(III) respirers are listed in Table 17.6. The ability to use ferric iron as terminal electron acceptor is spread among a variety of members of the domains *Bacteria* and the *Archaea*. These include strictly anaerobic and facultative organisms, the latter of which can grow aerobically as well as anaerobically. In general, all of them reduce iron(III) only anaerobically, but a few exceptions are known (e.g., De Castro and Ehrlich, 1970; Brock and Gustafson, 1976; Short and Blakemore, 1986). The electron donors used by the heterotrophic iron(III) respirers include a wide range of organic compounds as well as hydrogen at circumneutral and sulfur at acidic pH (Lovley et al., 1989a,b; Lovley and Lonergan, 1990; Coates et al., 1999). The organic compounds include substances as simple as acetate and lactate and as complex as palmitate and some aromatic compounds. Different iron(III) reducers utilize different types of these compounds. Some of the organisms are unable to degrade their organic electron donor completely, usually accumulating acetate (e.g., Lovley et al., 1989b). Acetate accumulation is suggestive of ATP generation via substrate level phosphorylation under anoxic conditions, as has been shown for *S. oneidensis* (Hunt et al., 2010). For iron(III) to serve as terminal electron acceptor in anaerobic respiration by iron(III) reducers at circumneutral pH, it may be in the form of a soluble complex formed with citrate, nitrilotriacetate, or some other ligands (Lovley and Woodward, 1996). However, various iron(III) reducers have also been shown to attack poorly soluble forms of iron(III) such as crystalline goethite (e.g., Roden and Zachara, 1996; Nevin and Lovley, 2000), suggesting the presence of an enzymatic mechanism to traffic electrons from the cell membrane to the outside of the cell. Ferrous iron, which is the product of iron(III) reduction, when adsorbed at the surface of an iron(III) oxide, interferes with the microbial attack of the iron(III) oxide. Its removal from the iron(III) oxide surface promotes the reduction of the oxide (Roden and Urrutia, 1999; Roden et al., 2000).

17.5.2.1 Microbial Sulfur Oxidation Coupled to Iron(III) Reduction at Acidic pH

Some autotrophs, such as *A. thiooxidans*, *A. ferrooxidans*, *L. ferrooxidans*, *Sulfolobus* spp., *Sulfobacillus acidophilus*, *Sulfobacillus thermosulfidooxidans*, and *Acidimicrobium ferrooxidans*, can also respire with iron(III) as terminal electron acceptor using elemental sulfur as electron donor (Brock and Gustafson, 1976; Pronk et al., 1992; Sugio et al., 1992; Bridge and Johnson, 1998). The reaction proceeds generally for iron(III) as follows:

$$S^0 + 6Fe^{3+} + 4H_2O \rightarrow HSO_4^- + 6Fe^{2+} + 7H^+$$
$$(17.15)$$

and for solid-phase iron(III) as

$$S^0 + 6Fe(OH)_3 + 11H^+ \rightarrow HSO_4^- + 6Fe^{2+} + 14H_2O \quad (17.16)$$

showing that sulfuric acid is produced during the oxidation of sulfur. As the pH is acidic, ferrous iron is more stable and less likely to be chemically reoxidized by atmospheric oxygen. Thus, the microbial reduction of ferrous iron may not only proceed anaerobically (*A. thiooxidans*) but also aerobically (*A. ferrooxidans*). Under oxic conditions, some growing iron(III) reducers (e.g., *A. ferrooxidans*) appear to use a branched pathway when oxidizing sulfur (Corbett and Ingledew, 1987) meaning that some electrons from sulfur pass to iron(III) and others to oxygen. The reaction can be summarized as follows:

$$S^0 + 2Fe^{3+} + O_2 + 2H_2O \rightarrow HSO_4^- + 2Fe^{2+} + 3H^+ \quad (17.17)$$

Interestingly, strains such as *A. ferrooxidans* are able to oxidize ferrous iron with atmospheric oxygen (see Section 17.4.2) after they have produced iron(II) during iron(III) reduction. Thus, Fe^{2+} accumulates in cultures of *A. ferrooxidans* only, when the strain is grown under anoxic conditions. At higher temperatures of 70°C, oxygen solubility decreases leading to microoxic conditions in solution. Under these conditions, thermophilic iron(III) reducers (e.g., *Sulfolobus acidocaldarius*) can accumulate ferrous iron during ferric iron reduction as they do not reoxidize the ferrous iron produced. Some acidophilic iron(III)-reducing strains (e.g., *A. ferrooxidans*, *S. acidophilus*, and *S. thermosulfidooxidans*) also reduce iron(III) to Fe^{2+} under oxygen limitation, but they perform this reduction best with organic electron donors (Bridge and Johnson, 1998).

Some strains of sulfur-oxidizing iron(III) reducers (e.g., *A. ferrooxidans* and *L. ferrooxidans*) transform elemental sulfur to sulfide in the presence of reduced glutathione (GSH) (Sugio et al., 1992), which does not necessarily lead to growth. The reaction can be summarized as follows:

$$S^0 + 2GSH \rightarrow H_2S + GSSG \quad (17.18)$$

$$H_2S + 6Fe^{3+} + 3H_2O \rightarrow SO_3^{2-} + 6Fe^{2+} + 8H^+ \quad (17.19)$$

$$SO_3^{2-} + 2Fe^{3+} + H_2O \rightarrow SO_4^{2-} + 2Fe^{2+} + 2H^+ \quad (17.20)$$

As sulfite is toxic to many cells when accumulated, the oxidation of this compound by ferric iron is viewed as a mechanism of detoxification (Sugio et al., 1988).

17.5.2.2 Microbial Hydrogen Oxidation Coupled to Iron(III) Reduction at Neutral pH

Evidence that some ferric iron-reducing bacteria can use hydrogen as electron donor was first reported by Balashova and Zavarzin (1979). Since then, a number of different bacteria (including *Pseudomonas*, *Shewanella algae* BrY, *G. sulfurreducens*, *Thiobacillus*, *Alteromonas*) have been shown to use hydrogen as the sole electron source for the reduction of iron(III) according to

$$2Fe(III) + H_2 \rightarrow 2Fe(II) + 2H^+ \quad (17.21)$$

S. oneidensis (MR1), *S. putrefaciens*, and *G. sulfurreducens* can grow with hydrogen as the main energy source but need a source of fixed carbon (e.g., acetate) to grow. By contrast, *G. metallireducens* can grow autotrophically. *G. sulfurreducens* strain PCA is able to use hydrogen as sole electron donor in iron(III) reduction, whereas *G. metallireducens* cannot.

Additionally, there are several hyperthermophilic hydrogen-oxidizing autotrophic iron(III) reducers, e.g., *Geothermobacterium ferrireducens* (Kashefi et al., 2002a), *Geoglobus ahangari* (Kashefi et al., 2002b), and strain 121 (Kashefi and Lovley, 2003).

17.5.2.3 Microbial Ammonium and Methane Oxidation Coupled to Iron(III) Reduction at Neutral pH

Most recently, a biological link between the microbial reduction of ferric iron and the microbial oxidation of nitrogen species, namely, ammonium, has been identified (Shrestha et al., 2009; Yang et al., 2012; Qu et al., 2014). This process has been observed in anoxic environments such as wetlands, but no microorganisms have been identified yet that are catalyzing this process:

$$NH_4^+ + 6FeOOH + 10H^+ \rightarrow NO_2^- + 6Fe^{2+} + 10H_2O \quad (17.22)$$

How this process relates to the production of gaseous nitrogen species and the loss of nitrogen in soils remains to be elucidated in the future.

Similarly, some evidence from lab and field experiments has been found for the existence of anaerobic methane oxidation coupled to ferric iron reduction (Beal et al., 2009; Segarra et al., 2013; Riedinger et al., 2014) although no microorganisms have been identified yet that are able to catalyze and grow by this process.

17.5.2.4 Microbial Organic Matter Oxidation Coupled to Iron(III) Reduction

Dissimilatory iron(III) reduction in the form of anaerobic respiration has now been recognized as an important means of mineralization of organic matter in environments in which sulfate or nitrate occurs in amounts insufficient to sustain sulfate or nitrate respiration, respectively (Figure 17.12) (Lovley, 1987, 1991; Nealson and Saffarini, 1994). The process can operate with various organic acids, including volatile fatty acids, and with aromatic compounds as electron donors. It can displace methanogenesis by outcompeting for hydrogen and acetate (Lovley, 1991).

S. *oneidensis* MR-1 as well as S. *putrefaciens* can use formate, lactate, and pyruvate as electron donors in the anaerobic reduction of iron(III), though lactate and pyruvate are only incompletely oxidized to acetate and carbon dioxide (Myers and Nealson, 1988, 1990; Lovley et al., 1989a) because of the ATP-generating strategy of these bacteria (Hunt et al., 2010). G. *sulfurreducens* strain PCA is able to use acetate as sole electron donor in iron(III) reduction, whereas G. *metallireducens* can use acetate, lactate, and several other organic compounds, such as propionate, butyrate, benzoate, or phenol, as sole electron donors. It oxidizes these compounds to carbon dioxide in the process (Lovley et al., 1993). Elemental sulfur, cobalt(III)–EDTA, fumarate, and malate were found to be able to serve as alternative electron acceptors for G. *sulfurreducens*, but not manganese(IV) or uranium(VI). G. *sulfurreducens* cannot reduce nitrate, sulfate, sulfite, or thiosulfate with acetate as electron donor (Caccavo et al., 1994).

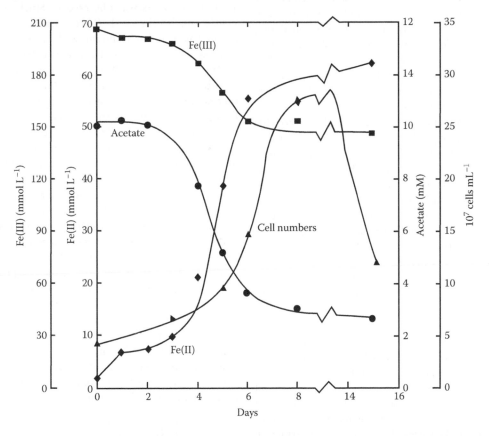

Figure 17.12. Reduction of ferric iron by acetate through mediation of anaerobic bacterial strain GS-15 (*Geobacter metallireducens*). These data illustrate the reduction of oxalate-extractable iron(III) to iron(II) at the expense of acetate consumption during growth of the culture in FWA medium containing poorly crystalline iron(III) hydroxides. (Reproduced from Lovley, DR and Phillips, EJP, *Appl Environ Microbiol*, 54, 1472, 1988. With permission.)

G. metallireducens has been shown to reduce amorphous iron(III) hydroxide to magnetite (Fe_3O_4), but it does not readily reduce crystalline iron oxides (Lovley and Phillips, 1986b). Roden and Urrutia (1999) showed later that the removal of ferrous iron from the surface of a crystalline iron(III) oxide was necessary for ready attack of the iron(III) oxide by the organism. The organism appears to use chemotaxis to locate poorly soluble iron(III) oxide substrates (Childers et al., 2002).

17.5.2.4.1 TYPES OF FERRIC COMPOUNDS ATTACKED BY DISSIMILATORY IRON(III) REDUCTION

The ease with which bacteria reduce ferric iron depends in part on the form in which they encounter it. In the case of less soluble forms, the order of decreasing solubilization in one study (Ottow, 1969b) was $FePO_4 \cdot 4H_2O$ > $Fe(OH)_3$ > lepidocrocite (γ-FeOOH) > goethite (α-FeOOH) > hematite (α-Fe_2O_3) when *B. polymyxa*, *Bacillus sphaericus*, and *P. aeruginosa* were tested with glucose as electron donor. In another study (De Castro and Ehrlich, 1970), using glucose as electron donor, marine *Bacillus* 29A was found to solubilize larger amounts of iron from limonite and goethite than from hematite. Generally, amorphous iron(III) hydroxides were more readily attacked than crystalline iron(III) oxides.

In laboratory studies with synthetic iron(III) oxide and oxyhydroxide minerals that aimed at determining the influence of iron(III) mineral surface area on enzymatic reduction by *Shewanella algae* strain BrY and *S. putrefaciens* strain CN32, it was found that initial rates of iron(III) oxide reduction, and in some cases the long-term extent of reduction, are directly correlated with the specific surface of the oxide phase (Roden and Zachara, 1996; Roden, 2003, 2006).

Various species of *Acidiphilium*, including *A. acidophilus* and strain SJH, are able to reduce iron(III) anaerobically at acidic pH (Johnson and Bridge, 2002). In a study with *Acidiphilium* SJH, Bridge and Johnson (2000) found that the order of decreasing iron(III) reduction was dissolved Fe^{3+} > amorphous $Fe(OH)_3$ > magnetite > goethite = natrojarosite > akaganeite > jarosite > hematite (no significant dissolution). In this case, the minerals were not attacked directly by the organism. The authors found that the bacterial reduction of the ferric iron in the minerals depended on the abiotic mobilization of iron(III) by the acidity of the medium, which was pH 2.0 initially. The abiotic iron(III) mobilization was enhanced by bacterial reduction of the solubilized iron(III), which resulted in product removal in the abiotic dissolution step.

Although magnetite had long been thought to be resistant to microbial reductive attack, its anaerobic reductive dissolution to aqueous Fe^{2+} by two strains of *S. putrefaciens* using glucose and lactate as electron donors has been observed (Kostka and Nealson, 1995). Magnetite, as well as ferrihydrite, goethite, and hematite, has also been shown to be reduced during growth of *Desulfovibrio desulfuricans* strain G-20 in laboratory culture (Li et al., 2006). In the presence of added sulfate, the rate of iron(III) reduction of these minerals was accelerated owing to an interplay of respiratory and chemical iron(III)-reducing reactions and the timing of iron(III) respiration and sulfate reduction in the culture.

The anaerobic reduction of ferric iron replacing some of the Al in the octahedral sheets of the clay smectite by *S. oneidensis* MR-1 and by some unidentified bacteria using lactate and formate as electron donors was observed by Stucki et al. (1987) and Kostka et al. (1996). This microbial action has also been shown with three strains of *Pseudomonas fluorescens* and a strain of *Pseudomonas putida* (Ernstsen et al., 1998). Kashefi et al. (2008) reported growth of some thermophilic and hyperthermophilic iron(III) reducers using iron(III) in ferruginous smectite as the sole electron acceptor. Interestingly, mobilization of reduced iron (Fe^{2+}) in the clay was either very limited or not detected, depending on the study. Stucki and Roth (1977) proposed that the reduction proceeds in two steps. In the first step, an increase in layer charge without structural change occurs during initial reduction of some iron(III), and in the second step, constant layer charge is maintained as a result of elimination of structural OH accompanied by a decrease of the iron in the octahedral sheet during subsequent reduction of additional iron(III). The reduction of Fe^{3+} in the octahedral sheet of smectite results in an increase in net surface charge of the mineral (Stucki et al., 1984) and a decrease in its swellability (Stucki et al., 2000). The mechanism whereby the bacteria inject electrons into the smectite to reduce the structural iron remains to be elucidated.

17.5.2.4.2 ENZYMATICS OF FERRIC IRON RESPIRATION

Our most complete understanding of the enzymatic mechanisms of anaerobic respiration with iron(III) and manganese(IV) come from

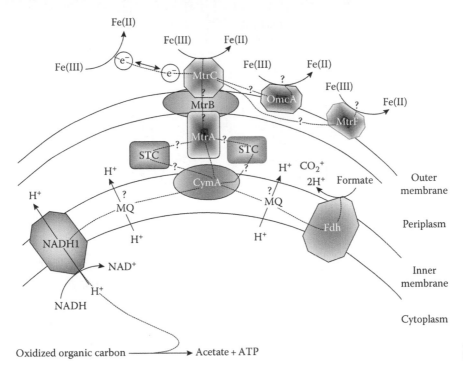

Figure 17.13. Model of biochemical mechanisms of iron(III) reduction in *Shewanella* sp. Model according to Bird et al. (2011), including the cytochrome–porin model described in Gralnick (2012) and Richardson et al. (2012). The dotted line indicates the direction of electron flow. NADH1, NADH dehydrogenase; Fdh, formate dehydrogenase; MQ, menaquinone; CymA, MtrA/C/F, STC, and OmcA, c-type cytochromes; MtrB, outer membrane protein; e⁻, electron shuttle.

studies with *S. oneidensis* MR-1 (Figure 17.13) and *G. sulfurreducens* (Figure 17.14).

When grown anoxically, *S. oneidensis* MR-1 produces a wide variety of c-type cytochromes localized in the cytoplasmic membrane, the periplasm, and the outer membrane. Slightly more than 50% of the formate-dependent ferric reductase activity of anoxically grown *S. oneidensis* MR-1 was found associated with the outer membrane, the rest with the plasma membrane. The membranes of oxically grown cells were devoid of this activity (Myers and Myers, 1993). Most of the formate dehydrogenase activity was soluble (Myers and Myers, 1993) suggesting a possible periplasmic location, as with fumarate reductase in anoxically grown cells of this organism (Myers and Myers, 1992). Addition of nitrate, nitrite, fumarate, or trimethylamine N-oxide as alternative electron acceptor did not inhibit ferric reductase activity. NADH was able to replace formate as electron donor in experiments with a membrane fraction containing ferric reductases (Myers and Myers, 1993). Expression of many of the anaerobic respiratory pathways is controlled by the global regulator cyclic AMP receptor protein (CRP) (Saffarini et al., 2003).

Outer membrane decaheme c-type cytochromes encoded by genes *omcA* and *mtrC* (in early literature also called *omcB*) in *S. oneidensis* MR-1 are required for reduction of a wide range of both poorly soluble oxide minerals, chelated metals, and organic compounds (Figure 17.13). They are required for the reduction of electron-shuttling compound 2,6-anthraquinone disulfonate (Lies et al., 2005) and flavins (riboflavin and flavin mononucleotide) (Coursolle et al., 2010). Lower et al. (2001) obtained direct evidence for the involvement of surface contact between *S. oneidensis* MR-1 and goethite by atomic force electron microscopy during attachment to the mineral under anoxic conditions. Their results support a model in which the contact enables electron transfer at the cell/mineral interface by means of a 150 kDa iron(III) reductase. It is also clear that *S. oneidensis* does not require direct contact to reduce insoluble substrates (Lies et al., 2005), using flavin electron shuttles to access electron acceptors not in direct contact and accelerate the rate of reduction for insoluble accepters close by (reviewed by Brutinel and Gralnick, 2012). Flavin molecules may also play a role as cofactors for MtrC and OmcA (Okamoto et al., 2013),

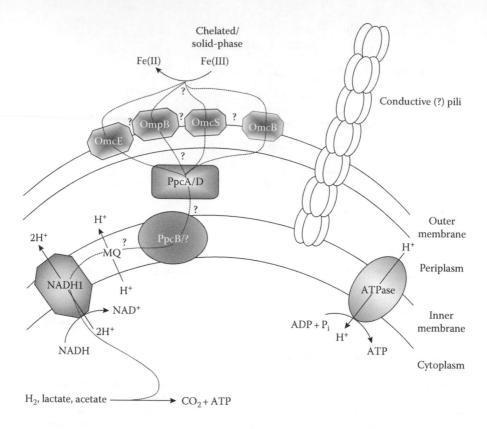

Figure 17.14. Model of biochemical mechanisms of iron(III) reduction in *Geobacter sulfurreducens*. Model according to Bird et al. (2011), considering Lovley et al. (2004). The dotted line indicates the direction of electron flow. NADH1, NADH dehydrogenase; MQ, menaquinone; PpcA/B/D and OmcB/E/S, cytochrome c-type proteins; OmpB, multicopper protein; ATPase, ATP synthase.

though biochemical evidence for binding of flavins to these proteins has not been found (Paquete et al., 2014). In *S. putrefaciens* strain 200, a 91 kDa heme-containing protein in the outer membrane, likely MtrC, is involved in the anaerobic electron transfer to iron(III) and manganese(IV) oxide (DiChristina et al., 2002). Two proteins believed to be involved in direct electron transfer to iron(III) are the outer membrane lipoproteins MtrC and OmcA (reviewed by Bird et al., 2011) located on the outside of the cell (Myers and Myers, 2003). Electrons from the menaquinone pool move through CymA, a cytochrome located in the inner membrane (Myers and Myers, 2000). To transduce electrons across the periplasm, electrons move from CymA to the decaheme c-type cytochrome MtrA (Beliaev and Saffarini, 1998), which associates with the integral outer membrane protein MtrB (Hartshorne et al., 2009). MtrB is believed to form a pore in the outer membrane, binding MtrC from the outside and MtrA from the periplasm

side, facilitating an interaction between these cytochromes such that electrons can pass across the outer membrane (Hartshorne et al., 2009). The biochemical interrogation of the Mtr system by a variety of researchers led to the development of the "porin–cytochrome" model for extracellular electron transfer (Figure 17.13) (Gralnick, 2012; Richardson et al., 2012).

The pathway for iron(III) reduction in *Geobacter* has recently been reviewed by Bird et al. (2011). Electrons are believed to be transferred from the inner membrane through the periplasm to the outer membrane through a series of cytochromes. Specifically, Lovley et al. (2004) proposed a model of an iron(III) reductase system in *G. sulfurreducens* that consists of a 87 kDa c-type cytochrome located in the outer membrane (OmcB) and two cytochromes (OmcD, OmcE) at the outer membrane, which convey electrons to solid-phase and chelated iron(III) at the exterior of the outer membrane. The cytochromes receive electrons from

EHRLICH'S GEOMICROBIOLOGY

NADH in a sequence of redox reactions involving a NADH-dehydrogenase complex, the quinone pool and a 36 kDa cytochrome in the periplasm (PpcB, possible associated with the plasma membrane), and a 9.6 kDa c-type cytochrome in the periplasm (PpcA) (Figure 17.14). The electron transfer from the outer surface to the iron(III) minerals is discussed in the next section.

17.5.2.4.3 ELECTRON TRANSFER FROM THE CELL SURFACE OF A DISSIMILATORY IRON(III) REDUCER TO A FERRIC OXIDE SURFACE

Dissimilatory iron(III) reducers may transfer electrons from a c-type cytochrome-containing enzyme complex in the outer membrane to the surface of a ferric oxide in one of several ways. In S. oneidensis MR-1, this electron transfer process is complex and includes the involvement of several key cytochromes to different extents (Shi et al., 2007). One way may involve physical contact between the cell and the surface of the ferric oxide. Direct electron transfer has been shown in vitro between OmcA and hematite (Xiong et al., 2006). Lower et al. (2007) have obtained direct evidence of the formation of specific bonds between hematite and cytochromes MtrC and OmcA in the outer membrane of S. oneidensis MR-1 that could be involved in electron transfer during anaerobic respiration with hematite as terminal electron acceptor.

Another way may involve contact with the ferric oxide surface via special fimbriae or pili that act as conductors of electrons or nanowires. They have been shown to be formed by G. sulfurreducens (Reguera et al., 2005, 2006) and some evidence presented for their formation in S. oneidensis MR-1 (Gorby et al., 2006). However, these structures in S. oneidensis have recently been shown to be outer membrane extensions (Pirbadian et al., 2014). Electrochemical measurements of active G. sulfurreducens biofilms suggest an electron hopping mechanism where electrons are passed between cytochromes along these extensions (Bond et al., 2012; Snider et al., 2012). In contrast, metal-like conductivity has been proposed for the pili of G. sulfurreducens due to overlapping pi–pi orbitals of aromatic amino acids (Lovley and Malvankar, 2015). The mechanism of long distance electron transfer to insoluble substrates by G. sulfurreducens remains controversial.

If direct physical contact between the iron(III)-reducing cell and a ferric oxide is not essential or is not involved in the reduction of iron(III)

of an iron(III) oxide, an iron(III) chelator or an electron shuttle may be involved (Lovley, 2000; Newman and Kolter, 2000; Royer et al., 2002; von Canstein et al., 2008; Melton et al., 2014a). Nevin and Lovley (2002) found that Geothrix fermentans produced a quinone type of compound that served as an electron shuttle in the reduction of iron(III) oxide when the organism was not in direct contact with iron(III) oxide. In contrast, although an artificial chelator, nitrilotriacetic acid, was found to stimulate iron(III) oxide reduction by G. metallireducens (Lovley and Woodward, 1996), no chelator or electron shuttle was found to be produced by this organisms (Nevin and Lovley, 2000). Feinberg and Holden (2006) found that Pyrobaculum aerophilum, a hyperthermophilic archaeon, appeared to employ an as yet unidentified electron shuttle instead of a chelator when reducing poorly crystalline iron(III) oxide in the absence of direct contact with the oxide. Von Canstein et al. (2008) found that growing cultures of S. oneidensis MR-1 and some other strains of Shewanella released flavin mononucleotide (FMN) into the medium, which could function as an electron shuttle in the reduction of poorly crystalline iron(III) oxide. Nongrowing cells of Shewanella in the stationary phase released riboflavin, which also could serve as an electron shuttle in the reduction of the iron(III) oxide. FMN did not serve as a shuttle in the reduction of the soluble compounds ferric citrate or fumarate. Independently, Marsili et al. (2008) detected riboflavin and FMN in biofilms of S. oneidensis MR-1 and Shewanella sp. MR-4 using electrochemical techniques and demonstrated that these molecules promoted electron transfer to electrodes as well as iron(III) and manganese(IV) oxyhydroxides.

However, not only microbially synthesized and released but also naturally present redox-active organic molecules such as dissolved and solid-phase humic compounds can serve as electron shuttles between microorganisms and iron(III) minerals (Lovley et al., 1996; Kappler et al., 2004; Jiang and Kappler, 2008; Roden et al., 2010). Recently, it was even shown that biochar, redox-active organic remains formed by pyrolysis in the absence of oxygen, can also function as electron shuttle (Kappler et al., 2014). An ecological study in sediments from anoxic aquifers has shown that humics- and iron(III)-reducing microorganisms in anoxic aquifers are rather versatile and able to reduce different extracellular electron acceptors (Piepenbrock et al., 2014).

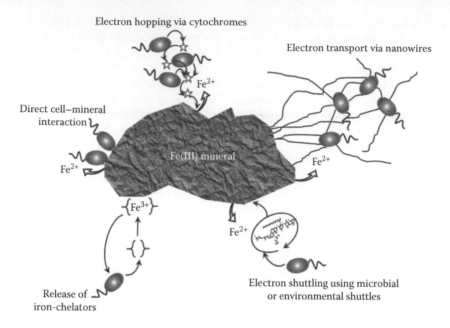

Electron hopping via cytochromes

Electron transport via nanowires

Direct cell–mineral interaction

Fe(III) mineral

Fe^{2+}

Fe^{2+}

Fe^{2+}

Fe^{3+}

Fe^{2+}

Release of iron-chelators

Electron shuttling using microbial or environmental shuttles

Figure 17.15. Microbial iron(III) mineral reduction pathways. For electron hopping, bacteria load electrons on cytochromes (stars) present in the biofilm matrix. These electrons move from cytochrome to cytochrome to the iron(III) mineral. For electron shuttling, the organic compounds transferring the electron from the cell to the iron mineral surface can be used over and over again, which is indicated by an arrowed circle around the organic structure.

The different ways of how electrons move from cells to solid-phase iron(III) is schematically depicted in Figure 17.15.

17.6 NONENZYMATIC OXIDATION OF FERROUS IRON AND REDUCTION OF FERRIC IRON BY MICROBES

17.6.1 Nonenzymatic Iron(II) Oxidation

Many different kinds of microorganisms can promote the oxidation of ferrous iron indirectly (i.e., nonenzymatically). They can accomplish this by affecting the redox potential of the environment. This means they release metabolically formed oxidants into their surroundings that have the ability to oxidize ferrous iron (Melton et al., 2014a). They can also accomplish this by generating a pH above 5 at which Fe^{2+} is oxidized by oxygen in the air (*autoxidation*). Starkey and Halvorson (1927) demonstrated the indirect oxidation of ferrous iron by bacteria in laboratory experiments and inferred that any organism that raises the pH of a medium can promote ferrous iron oxidation in aerated medium.

A more specialized case of indirect microbial iron(II) oxidation is associated with cyanobacterial and algal photosynthesis. The photosynthetic process may promote ferrous iron oxidation by creating conditions that favor autoxidation in two ways: (1) by raising the pH of the water in which they grow (bulk phase) and (2) by raising the oxygen level in the waters around them. Other processes where microbial formation of oxidants promote iron(II) oxidation are denitrification, where the reactive intermediate leads to iron(II) oxidation (Klueglein et al., 2014) (see Section 17.4.3.2) or microbial manganese(II) oxidation leading to the production of manganese(IV) oxides that can function as oxidants for iron(II) (Weber et al., 2006a) (see Chapter 18).

Ferrous iron may be protected from chemical oxidation at elevated pH and E_h by chelation with gluconate, lactate, or oxalate. In that case, bacterial breakdown of the ligand will free the complexed ferrous iron, which then autoxidizes to ferric iron. Microorganisms performing this process do not derive energy from iron(II) oxidation but rather from the oxidation of the ligand.

17.6.2 Nonenzymatic Iron(III) Reduction

Iron(III) oxide and oxyhydroxide minerals are relatively stable in anoxic conditions when a strong reducing agent is not present. However, in the presence of reducing agents (e.g., sulfide, reduced humic substances, or the superoxide radical) and/or under the influence of light, chemical reduction of iron(III) minerals does occur (Melton et al., 2014a). For instance, hydrogen sulfide produced by sulfate-reducing bacteria may reduce ferric to ferrous iron before precipitating ferrous sulfide (Berner, 1962):

$$2FeOOH + 3H_2S \xrightarrow{pH\,7-9} 2FeS + S^0 + 4H_2O \tag{17.23}$$

$$2FeOOH + 3H_2S \xrightarrow{pH\,4} FeS + FeS_2 + 4H_2O \tag{17.24}$$

Marine bacteria that disproportionate elemental sulfur into hydrogen sulfide and sulfate according to the reaction

$$4S^0 + 4H_2O \rightarrow 3H_2S + SO_4^{2-} + 2H^+ \tag{17.25}$$

have been shown to reduce iron(III) and manganese(IV) chemically under anoxic conditions (Thamdrup et al., 1993).

Formate produced by a number of bacteria can reduce iron(III):

$$2Fe^{3+} + HCOOH \rightarrow 2Fe^{2+} + 2H^+ + CO_2 \tag{17.26}$$

Some other metabolic products can also act as reductant of ferric iron (see discussion by Ghiorse, 1988). In general, reduction is favored by acid pH.

With the discovery that some iron(III)-reducing organisms can reduce humic substances (Lovley et al., 1996; Scott et al., 1998), it has been proposed that the microbially reduced humic substances may reduce iron(III) in soils and sediments abiologically, with the hydroquinone groups of the humics acting as electron shuttle between cells and extracellular iron(III) (Lovley, 2000; Kappler et al., 2004; Bauer and Kappler, 2009; Roden et al., 2010).

17.7 MICROBIAL PRECIPITATION OF IRON

17.7.1 Enzymatic Iron(II) Oxidation and Iron(III) Precipitation Processes

The clearest example of enzymatic iron(III) precipitation is obvious with microaerophilic iron(II)-oxidizers such as M. ferrooxydans or G. ferruginea. The ferric iron (FeOOH) produced by the bacterium by oxidizing ferrous iron is deposited on its stalk fibrils (see Figure 17.5 and Section 17.4.1.1,

discussion by Ghiorse (1984), Ghiorse and Ehrlich (1992), Hallbeck (1993), and Chan et al. (2011)). James and Ferris (2004) showed that G. ferruginea and L. ochracea played a significant role in the oxidation of iron(II) and the subsequent precipitation of the resulting iron(III) as ochreous bacteriogenic iron oxide in the water at the mouth of a neutral groundwater spring draining into Ogilvie Creek near Deep River, Ontario, Canada. This resulted in a progressive decrease in ferrous and total iron concentration and increase in redox potential and dissolved oxygen with distance from the spring.

Also anaerobic nitrate-reducing and phototrophic iron(II)-oxidizing microorganisms are known to form minerals in proximity to their cells after ferrous iron oxidation (Kappler and Newman, 2004; Kappler et al., 2005b; Miot et al., 2009a,b). One of these strains, the nitrate-reducing iron(II) oxidizer Acidovorax sp. strain BoFeN1 (Kappler et al., 2005b), encrusts itself entirely during iron(II) oxidation (in contrast to the phototrophic iron(II) oxidizers, Schädler et al. (2009)) and forms a variety of minerals such as goethite, ferrihydrite, magnetite, and green rust (Muehe et al., 2009; Larese-Casanova et al., 2010; Dippon et al., 2012; Pantke et al., 2012). The advantages or disadvantages of cell encrustation are not fully understood yet. The organisms potentially have an advantage by protecting themselves from toxic compounds that bind to the mineral crust and therefore, are not taken up (Muehe et al., 2009). In contrast, this mineral crust could also immobilize nutrients and thus, limit microbial growth.

The biogenic minerals that are formed during microbial ferrous iron oxidation are geochemically different compared to their chemically produced counterparts (Konhauser, 1997; Posth et al., 2014). Their formation is highly dependent on the geochemical environment, i.e., the presence of other substances interacting with the crystal formation. Biogenic minerals tend to be smaller in size, less crystalline, and bear a number of impurities such as organics, metals, and nutrients that change the surface and bulk properties of the minerals (Posth et al., 2010; Dippon et al., 2012). In marine environments, silicification of ferrihydrite-dominated iron(III) oxyhydroxides through their chemical acquisition of the silica naturally present in seawater also affects the mineralogy and can effectively stabilize the mineral from further diagenesis to more crystalline forms (Toner et al., 2012).

Enzymatic iron(III) reduction can also result in the precipitation of magnetite (Fe_3O_4) (Lovley et al., 1987) and iron(II) as siderite ($FeCO_3$) (Coleman et al., 1993). This process is iron(III) reduction rate dependent and does not necessarily lead to iron(II) mineral formation (Zachara et al., 2002). Only if the reduction of ferric iron proceeds at a fast enough rate, iron(II)-bearing minerals such as siderite and vivianite (in the presence of phosphate) are formed. If the ferric iron reduction proceeds at a medium rate, magnetite is formed, and at a slow rate iron(III), minerals such as goethite, lepidocrocite, akaganeite, and hematite are formed due to restructuring and surface-bulk electron transfer processes (Zachara et al., 2002; Hansel et al., 2003; Latta et al., 2012). In the presence of other metal ions, green rust, a poorly crystalline iron(II)/iron(III) mixed mineral, is formed.

An example of a cell and mineral aggregate is presented in Figure 17.16.

17.7.2 Bioaccumulation of Iron

Most or all of the bacteria listed in the previous section may also accumulate ferric iron produced either enzymatically or nonenzymatically by other organisms in the bulk phase (e.g., Ferris, 2005). Usually, the iron is passively collected on the cell surface by reacting with acidic exopolymer, which exposes negative charges. The exopolymer may be organized in the form of a sheath, a capsule, or slime. Some protozoans such as *Anthophysa*, *Euglena* (Mann et al., 1987), *Bikosoeca*, and *Siphomonas* are also known to deposit iron(III) on their cells.

17.7.2.1 Magnetotactic Bacteria

Magnetotactic bacteria occupy a special position among iron(II)-oxidizing bacteria (Blakemore, 1982; Bazylinski et al., 2013) (Figure 17.17a and b). They comprise some motile rods, spirilla, and cocci with crystalline inclusions with magnetic properties. The inclusions consist of magnetite or greigite or rarely both. The inclusions are formed by uptake of either ferric or ferrous iron, which is then transformed into magnetite (Fe_3O_4) or greigite (Fe_3S_4) by mechanisms that in the case of magnetite seem to involve reduction of ferric iron and/or partial reoxidation of ferrous iron (Frankel et al., 1983; Bazylinski and Frankel, 2000). The magnetite or greigite is deposited as crystals in membrane-bound structures called magnetosomes (Figure 17.17b) (Blakemore, 1982; Bazylinski, 1999). Magnetosome synthesis in *Magnetospirillum gryphiswaldense* and other magnetotactic bacteria is controlled by more than 30 gene functions governing the uptake of iron and its deposition and controlled precipitation within vesicles of the magnetosome membrane

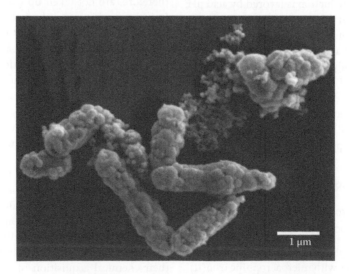

Figure 17.16. Cell-mineral aggregates of the nitrate-reducing, iron(II)-oxidizing *Acidovorax* sp. strain BoFeN1 grown on 10 mM iron(II) chloride, 5 mM sodium nitrate, and 10 mM sodium acetate. The cells are entirely encrusted in iron(III) minerals. Scanning electron micrograph taken with a Leo1450VP SEM operated at 5 kV accelerating voltage. (Courtesy of Stuhr, M and Obst, M, Environmental Analytical Microscopy, Center for Applied Geosciences, University of Tuebingen, Tuebingen, Germany.)

EHRLICH'S GEOMICROBIOLOGY

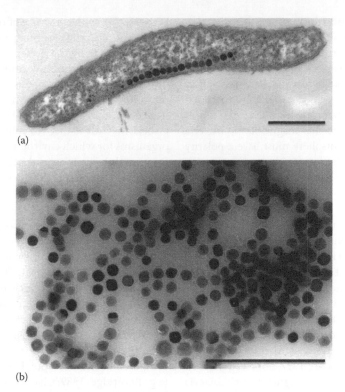

(a)

(b)

Figure 17.17. (a) Transmission electron micrograph of a thin-sectioned cell of the Alphaproteobacterium *Magnetospirillum gryphiswaldense* showing the intracellular chain of magnetosomes (TEM by Wanner, G, LMU Munich, Germany). (b) Transmission electron micrograph of isolated magnetosomes from *M. gryphiswaldense*. The particles consist of membrane enveloped crystals of the mixed valence iron oxide magnetite (Fe_3O_4) (TEM by Schüler, D, University of Bayreuth, Bayreuth, Germany). Bars represent 0.5 μm. (Courtesy of Schüler, D, Department of Microbiology, University of Bayreuth, Bayreuth, Germany.)

(Jogler and Schüler, 2009; Lohße et al., 2014). This is followed by the assembly of nascent magnetite crystals into linear chains along dedicated cytoskeletal structures (Komeili et al., 2006; Scheffel et al., 2006). Biomineralization of the mixed valence oxide magnetite requires proper redox conditions poised by enzymes of oxygen and nitrate/nitrite respiration as well as specific "magnetochromes" residing in the magnetosome membrane (Fukumori, 2004; Li et al., 2012, 2013, 2014, Siponen et al., 2013).

Whereas magnetite-containing magnetotactic bacteria have been found in both freshwater and marine environments, greigite-containing bacteria have been found only in marine environments (see Simmons et al., 2004). The magnetite and greigite crystals are single-domain magnets. The distribution of magnetotactic bacteria with magnetite- or greigite-containing magnetosomes, which are both present in the water column of seasonally stratified Salt Pond in Woods Hole, MA, differed and was found to be affected by depth

and oxygen and sulfide concentration (Simmons et al., 2004). The greigite-containing organisms were most numerous in the region where low concentrations of sulfide were present near the oxic/anoxic interface, whereas the magnetite-containing organisms were most numerous just above this zone.

It is believed that the magnetite- and greigite-containing magnetosomes enable the respective bacteria containing them to align with the Earth's magnetic field, which is inclined downward in the northern hemisphere and upward in the southern hemisphere.

Magnetotactic bacteria are highly motile by means of flagella, and most, if not all of them, display a clear swimming polarity along the geomagnetic field under environmental conditions (Frankel et al., 1997; Popp et al., 2014). It was recently shown for microaerophilic magnetospirilla that their swimming direction in oxygen gradients along magnetic fields is controlled by an aerotoxic sensory pathway (Popp et al., 2014).

Because by convention the magnetic field direction is defined in terms of the north-seeking end of a compass needle, north-seeking magnetic polarity in magnetotactic bacteria in the northern hemisphere has been thought to assist them passively in their downward movement from an oxygen-rich to a more oxygen-poor environment. As a corollary, it was predicted that magnetotactic bacteria in the southern hemisphere must have a polarity opposite to that of bacteria in the northern hemisphere. Indeed, the existence of such bacteria in the southern hemisphere has been demonstrated (Blakemore, 1982; Bazylinski, 1999). Although previous observations of both north-seeking and south-seeking magnetotactic bacteria in the water column of Salt Pond in Woods Hole, MA (Simmons et al., 2006) raised questions about the general adaptive value of magnetosome biosynthesis, it is now widely accepted that magnetotaxis assists the bacteria in locating their preferred habitat, which are partially reduced microenvironments within redox gradients of stratified marine and aquatic sediments (Frankel et al., 1997; Popp et al., 2014).

Upon the death of magnetotactic bacteria with magnetite magnetosomes, the magnetite crystals in them are liberated and in nature may become incorporated in sediment. It has been suggested that remanent magnetism detected in some rocks may be due to magnetite residues from magnetotactic bacteria (Blakemore, 1982; Karlin et al., 1987). However, magnetite is also formed extracellularly by some nonmagnetotactic iron(III)-reducing and phototrophic and nitrate-reducing iron(II)-oxidizing bacteria (Lovley et al., 1987; Jiao et al., 2005; Dippon et al., 2012) (see Section 17.10).

17.8 SEDIMENTARY IRON DEPOSITS OF PUTATIVE BIOGENIC ORIGIN

Sedimentary iron deposits, many quite extensive, may feature iron in the form of oxides, sulfides, or carbonates. Microbes appear to have participated in the formation of many of these deposits (Konhauser, 1997; Koehler et al., 2010; Posth et al., 2014). Their participation can be inferred (1) by the presence in some deposits of fossilized microbes with imputed iron(II)-oxidizing or iron-accumulating potential when they were alive, (2) by the presence of living iron(II)-oxidizing or iron-accumulating bacteria in currently forming deposits of ferric oxide (Fe_2O_3 or

FeOOH), (3) by the presence of iron(III)-reducing bacteria in currently forming magnetite (Fe_3O_4) deposits, or (4) by the presence of sulfate-reducing bacteria in currently forming iron(II) sulfide deposits. The probable environmental conditions prevailing at the time of formation of a biogenic iron deposit may be inferred from microfossils associated with it that resemble modern microorganisms for which environmental requirements for growth in their natural habitat are known. In this section, only iron(III) oxide deposits are considered. Biogenic iron(II) sulfide formation will be discussed in Chapter 21. Biogenic iron(II) carbonate (siderite) formation was briefly examined in Chapter 10.

It has been thought that some mineral deposits may have been initiated by passive adsorption of a particular metal ion by one or more components of the cell surface of certain bacteria acting as a template and subsequent chemical reaction of the adsorbed iron with an appropriate counter ion to form the nucleus in the formation of a mineral composed of the corresponding metal compound (e.g., Beveridge, 1989; Chan et al., 2004). Rancourt et al. (2005), by contrast, obtained some evidence from a comparative study of hydrous ferric oxide formation in the presence and absence of nonmetabolizing *Bacillus subtilis* or *Bacillus licheniformis*. He showed that the bacterial surface did not act as a template in the formation of hydrous ferric oxides by passively adsorbing iron(III) or iron(II). Instead, he found evidence that the hydrous ferric oxide formed by coprecipitation, in which redox functional groups of bacterial cell wall components may have inhibited some reactions that lead to an observable difference in biotically and abiotically formed hydrous ferric oxide.

Among the most ancient iron deposits in the formation of which microbes may have played a central role are the BIFs that arose mostly in the Precambrian, in the time interval roughly between 3.3 and 1.8 billion years ago. The major deposits were formed between 2.5 and 1.8 billion years ago, some are up to 3.8 billion years old (Klein, 2005). These formations have been found in various parts of the world and in many places are extensive enough to be economically exploitable as iron ore. They are characterized by alternating layers rich in chert, a form of silica (SiO_2), and layers rich in iron minerals such as hematite (Fe_2O_3), magnetite (Fe_3O_4), the iron silicate chamosite, and even siderite ($FeCO_3$) (Figure 17.18).

EHRLICH'S GEOMICROBIOLOGY

Figure 17.18. **(See color insert.)** Banded iron formations from the Transvaal Supergroup in South Africa. BIFs are made up of repeated layers of black iron oxides (magnetite, hematite) and red-colored shales and cherts (iron poor). (Courtesy of Kappler, A, Geomicrobiology, Center for Applied Geosciences, University of Tuebingen, Tuebingen, Germany; Posth, N, Nordisk Center for Jordens Udvikling [NordCEE], Odense, Denmark; Beukes, N, Department of Geology, University of Johannesburg, Johannesburg, South Africa.)

Average thickness of the layers is 1–2 cm, but they may be thinner or thicker. Because the most important BIFs were formed during that part of the Precambrian when the atmosphere around the Earth changed from reducing or nonoxidizing to oxidizing as a result of the emergence and incremental dominance of oxygenic photosynthesis, Cloud (1973) argued that the alternating layers in BIFs reflected episodic deposition of iron oxides, involving seasonal, annual, or longer-period cycles. Recently, it has been suggested that ocean water temperature fluctuation has modulated microbial iron(II)-oxidizing activity leading to the alternating precipitation of the silica and iron minerals (Posth et al., 2008). In the reducing or nonoxidizing atmosphere of the Archean, iron in the Earth's crust was mostly in the ferrous form and thus, could act as a scavenger of oxygen initially produced by oxygenic photosynthesis. The scavenging reaction involved autoxidation of the ferrous iron. Because autoxidation of iron(II) is a very rapid reaction at circumneutral pH and higher, the supply of ferrous iron could have been periodically depleted. Further, oxygen scavenging then must have had to wait until the supply of ferrous iron was replenished by leaching from rock or hydrothermal emissions (Holm, 1987). Alternatively, if much of the ferrous iron for scavenging oxygen was formed by ferric iron-respiring bacteria while they consumed organic

carbon produced by seasonally growing photosynthesizers, periodic depletion of organic carbon rather than ferrous iron could have been the basis for the iron-poor layers (Nealson and Myers, 1990). Bacterial or abiotic, diagenetic reduction of oxidized iron could also explain the origin of magnetite-rich layers of BIFs (Koehler et al., 2013; Posth et al., 2013). Lovely et al. (1987) and Lovley and Phillips (1988) found in their laboratory experiments that G. *metallireducens* precipitated magnetite when reducing Fe^{3+}. Previously, it had been proposed that the magnetite in the iron-rich layers could have resulted from partial reduction of hematitic iron(III) by organic carbon from biological activity (Perry et al., 1973). For a recent discussion of the potential significance of microbial iron(III) reduction in BIF deposition see Konhauser et al. (2005).

An ongoing process of iron deposition involving microbial mats consisting of cyanobacteria in a hot spring may provide another model for the origin of some BIFs (Pierson and Parenteau, 2000). Pierson and Parenteau observed three kinds of mats in Chocolate Pots Hot Springs in Yellowstone National Park. One kind consisted of *Synechocystis* and *Chloroflexus* and another of *Pseudanabaena* and *Mastigocladus*. Both of these were present at 48°C–54°C. A third kind of mat consisted mostly of *Oscillatoria* and occurred at 36°C–45°C. The iron minerals occurred below the upper 0.5 mm of the mats, having become entrapped in the biofilm formed by the cyanobacteria. Gliding motility seemed to facilitate the encrustation process of the cyanobacteria. The iron in the water bathing the mats was mostly ferrous. Oxygenic photosynthetic activity of the mats appeared to promote ferrous iron oxidation (Pierson et al., 1999), leading to the formation of amorphous iron(III) oxides that are the source of the iron deposit. The ultimate development of the mineral deposit appeared to involve the death and decay of iron(III) encrusted cells.

If iron was indeed the scavenger of the oxygen of early oxygenic photosynthesis, it means that the sediments became increasingly more oxidizing relative to the atmosphere owing to the accumulation of the oxidized iron (Walker, 1987). Only when free oxygen began to accumulate would the atmosphere have become oxidizing relative to the sediment. Chert was deposited in BIFs because no silicate-depositing microorganisms (e.g., diatoms, radiolarians) had yet evolved. As increasing amounts of oxygen were generated by a growing population of oxygenic photosynthesizers, ferrous iron reserves became largely depleted, permitting the oxygen content of the atmosphere to increase to levels we associate with an oxidizing atmosphere. This in turn would have restricted iron(III) reduction to the remaining and ever-shrinking oxygen-depleted (anoxic) environments. Although an extensive array of microfossils has been found in the cherty layers of various BIFs (e.g., Gruner, 1923; Cloud and Licari, 1968), not all sedimentologists agree that the origin of BIFs depended on biological activity (see discussion by Walker et al., 1983).

Some microbiologists, who have found that some purple photosynthetic bacteria can oxidize iron(II) anaerobically by using it in place of reduced sulfur in the reduction of carbon dioxide have suggested that BIF formation may have been initiated before oxygenic photosynthesis had evolved (Hartman, 1984; Widdel et al., 1993; Kappler et al., 2005a). Anbar and Holland (1992) suggested that BIFs may have begun to form in anoxic Precambrian oceans as a result of abiotic photochemical oxidation of manganese^{2+} to a manganese(IV) oxide like birnessite, which then served as oxidant in the oxidation of ferrous hydroxide complexes to lepidocrocite (γ-FeOOH) or amorphous ferric hydroxide. Because a variety of biological and abiotic pathways can generate the same ferric hydroxide products, distinguishing which process was involved in BIF formation for any given interval in Earth history is a challenge. Multiple approaches must be used to address this question, but one that has gained traction in recent years is measuring iron isotopic fractionation in the iron minerals in BIFs (Johnson et al., 2008). Although it is not yet clear whether iron isotopes can be used to distinguish different pathways from one another (Anbar, 2004), as more is understood about the mechanisms of the reactions (both biological and abiotic), iron isotope fractionation determinations applied to BIFs may become more informative.

Another alternative is that as small reservoirs of oxygen built up in marine environments as a result of photosynthesis, aerobic iron(II)-oxidizing bacteria evolved to begin utilizing the abundant iron(II) source available in the ocean. They could have then precipitated large amounts of iron oxides. In modern environments where oxygen and iron(II) counter gradients are present, microbial iron mats containing members of the iron-oxidizing Zetaproteobacteria of greater than a meter in thickness have been shown to form on the seafloor (Edwards et al., 2011). These marine iron mats are not restricted to the deep sea but can also be found in shallow vent systems such as Santorini (Puchelt et al., 1973; Holm, 1987; Handley et al., 2010) and off the coast of New Guinea (Meyer-Dombard et al., 2013), where the presence of microaerobic iron(II)-oxidizing bacteria is also noted. There is well-preserved microfossil evidence for the presence of aerobic iron(II) oxidizers in 400 million year old iron deposits (Little et al., 2004) and in the 1.8 billion year old Gunflint deposits. More mineralized microfossils are also present that are indicative of aerobic iron(II) oxidizers (Planavsky et al., 2009). Future work, should shed light on the potential of aerobic iron(II)-oxidizing bacteria to have contributed to some of the major BIF formations in the Precambrian.

17.9 MICROBIAL MOBILIZATION OF IRON FROM MINERALS IN ORE, SOIL, AND SEDIMENTS

Bacteria and fungi are able to mobilize significant quantities of iron from minerals in ore, soil, and sediment in which the iron is a major constituent. The minerals include carbonates, oxides, and sulfides. Attack of iron(II) sulfides will be discussed in Chapter 21. Attack of carbonates was briefly discussed in Chapter 10. In this chapter, the focus is on iron(III) oxides.

As indicated in Section 17.5, iron(III) reduction has been observed for some time with bacteria from soil and aquatic sources. In recent years, a more systematic study of the importance of this activity in soils and sediments has been undertaken (see, for instance, review of activity in marine sediments by Burdige (1993)). Ferric iron reduction (iron(III) respiration) can be an important form of anaerobic respiration in environments in which nitrate or sulfate is present in insufficient quantities as terminal electron acceptor and in which methanogenesis is not occurring (Lovley, 1987, 1991, 2000; Sørensen and Jørgensen, 1987; Lowe et al., 2000). Under these conditions, it can make an important contribution to the anaerobic mineralization of organic carbon. Indeed, the presence of bioreducible ferric iron can inhibit sulfate reduction and methanogenesis when electron donors like acetate or hydrogen are limiting (Lovley and Phillips, 1986a, 1987).

The form in which ferric iron occurs can determine whether bacterial iron(III) respiration will occur. In sediments from a freshwater site in the Potomac River estuary in Maryland, United States, only amorphous FeOOH present in the upper 4 cm was bacterially reducible. Mixed iron(II)–iron(III) compounds that were present in the deeper layer were not attacked (Lovley and Phillips, 1986b). Local E_h and pH conditions can determine the form in which the iron produced in bioreduction of FeOOH will occur. A mixed culture from sediment of Contrary Creek in central Virginia, United States, produced magnetite (Fe_3O_4) in the laboratory when the culture medium was allowed to go alkaline (pH 8.5 in the absence of glucose), but it produced Fe^{2+} when the medium was allowed to go acid (to pH 5.5 in the presence of glucose) (Bell et al., 1987). E_h values dropped from a range between 0 and −100 mV to less than −200 mV at the end of the experiments. A growing pure culture of G. metallireducens strain GS-15 produced magnetite from amorphous iron(III) oxide directly with acetate as reductant at pH 6.7–7.0 and 30°C–35°C (Lovley et al., 1987; Lovley and Phillips, 1988) (see also Section 17.5.2).

The phenomenon of gleying in soil has come to be associated with bacterial reduction of iron(III) oxides. It is a process that occurs under anoxic conditions that may result from waterlogging. The affected soil becomes sticky and takes on a gray or light greenish-blue coloration (Alexander, 1977, p. 377). The gleying of rice paddies when they turn anoxic is depicted in Figure 17.19. Although once attributed to microbial sulfate reduction, this was later considered not to be the primary cause of gleying (Bloomfield, 1950). Waterlogged soil has been observed to bleach before sulfate reduction is detectable (Takai and Kamura, 1966). Bloomfield

Figure 17.19. **(See color insert.)** Gleying of a rice paddy soil. The oxic soil of a rice paddy is orange-brown due to the presence of iron(III) minerals. The paddy soil turns to a grayish-green-blue color when the iron(III) minerals are reduced under anoxic conditions. (Courtesy of Winkler, E, Hegler, F, and Kappler, A, Geomicrobiology, Center for Applied Geosciences, University of Tuebingen, Tuebingen, Germany; Zahid, A, Department of Geology, University of Dhaka, Dhaka, Bangladesh.)

(1951) suggested that gleying was due at least in part to the action of products of plant decomposition, although he had earlier demonstrated gleying under artificial conditions in a sugar-containing medium that was allowed to ferment. Presently, microbial reduction of ferric iron and structural iron(III) in iron-bearing phyllosilicates (e.g., smectites) is favored as an explanation of gleying (Ottow, 1969a, 1970, 1971; Benzine et al., 2013). The gray green minerals that are formed are amorphous iron(II)–iron(III) containing minerals with a general formula of $Fe_3(OH)_8$.

17.10 MICROBES AND THE IRON CYCLE

Understanding the microbial iron cycle in different environments is important for evaluating the availability of the essential nutrient iron to the biota. For example, oceans are depleted in bioavailable iron thus, constraining phytoplankton development and hereby affecting the whole ecosystem. Current research is directed at understanding the complex interplay of abiotic reactions and microbial activities in the cycling of iron, i.e., the oxidation of iron(II) to iron(III) that is followed by a rereduction of the iron(III) to iron(II) (Melton et al., 2014a) (Figure 17.1). Weathering of iron-containing minerals in rocks, soils, and sediments mobilizes iron in the cycle. This weathering action is promoted partly by bacterial and fungal action and partly by chemical activity (e.g., Bloomfield, 1953a,b; Lovley, 2000; Melton et al., 2014a). The microbial action often involves the interaction of minerals with metabolic end products from microorganisms (Berthelin and Dommergues, 1972; Berthelin and Koblevi, 1972, 1974; Berthelin et al., 1974). The mobilized iron, if ferrous, may be biologically or nonbiologically oxidized to ferric iron at a pH above 5 under anoxic or partially or fully oxic conditions by phototrophic, nitrate-reducing, or microaerophilic microorganisms (see Section 17.4.3.1). At a pH below 4, ferrous iron is oxidized mainly biologically. The oxidation of ferrous iron may be followed by immediate precipitation of the ferric iron as a hydroxide, oxide, phosphate, or sulfate. If natural, soluble organic ligands abound, the ferric iron may be converted to soluble complexes and be dispersed from its site of formation. It may also be complexed by humic substances. In podzolic soils (spodosols) in temperate climates, complexed iron may be transported by groundwater from the upper A horizon to the B horizon. In hot, humid climates, ferric iron more likely precipitates at the site of its formation after release from iron-containing soil minerals, owing to intense microbial activity that rapidly and fairly completely mineralizes available organic matter including ligands, which prevents extensive formation of soluble ferric iron complexes. The iron precipitates tend to cement soil particles together in a process known as laterization. Aluminum hydroxide liberated in the weathering process may also be precipitated and contribute to laterization (Merkle, 1955, p. 294; see also Chapter 12).

A bacterially promoted iron cycle appears to operate in the rhizosphere of wetland plants (Weiss et al., 2003). Weiss et al. (2003) detected microaerophilic, neutrophilic, lithotrophic iron(II) oxidizers as well as acetate-utilizing iron(III) reducers associated with roots covered by iron plaque. The concentration of iron(II) oxidizers and iron(III) reducers in the rhizosphere was significantly higher than in the bulk soil. The iron(II) oxidizers are thought to be the major contributor to iron plaque formation. Because the iron(II) oxidizers live microaerobically whereas the iron(III) reducers live anaerobically, Weiss et al. (2003) suggest that these two kinds of organisms inhabit different niches in the rhizosphere. Microoxic conditions prevail closest to the roots, which release some oxygen into the rhizosphere in a process called radial oxygen loss (e.g., Christensen et al., 1994). Thus, microaerophilic iron(II) oxidizers prevail close to the roots, where oxygen is present, and aid in forming iron plaque. In a laboratory batch experiment, a lithotrophic iron(II) oxidizer, strain BrT, isolated from the rhizosphere of Typha latifolia, accounted for 18%–53% of total iron(II) oxidation, whereas in a continuous flow experiment it accounted for 62% of total iron(II) oxidation (Neubauer et al., 2002). Iron(III) reducers, by contrast, locate on the other side of the iron plaque, where no oxygen is present, but sufficient amounts of organics are found to reduce and dissolve the iron plaque again.

The microbial cycling of iron also plays an important role in groundwater seeps and freshwater lake sediments, where the downward redox gradient leads to different niches that are occupied by functionally different microorganisms (Sobolev and Roden, 2002; Roden et al., 2004; Weber et al., 2006c; Blöthe and Roden, 2009b; Haaijer et al., 2012; Melton et al., 2012; Roden et al., 2012). The interplay of geochemical processes with functional microbial consortia (exerting competition, beneficiation, and niche-transitional interactions) in these niches is investigated on a conceptual (Blöthe and Roden, 2009; Schmidt et al., 2010) and laboratory basis (Roden et al., 2004; Melton et al., 2014b,c).

Based on experimental results with a lithotrophic iron(II)-oxidizing bacterium and an iron(III) reducer in microcosms that contained sand coated with an iron(III) oxyhydroxide, a conceptual model for how such coupled metabolism could take place on a microscale was developed by Sobolev and Roden (2002) and Roden et al. (2004) (Figure 17.20). A few years later, Blöthe and Roden (2009b) quantified, isolated and identified iron(III)-reducing and iron(II)-oxidizing microorganisms in solids collected from a groundwater seep. They

quantified iron(III) reduction rates and established a simple kinetic model to explore quantitatively the potential coupling of iron(II) oxidation and iron(III) reduction in the solid material collected at this groundwater seep. In principle, such models apply to virtually any redox interfacial environment, from commonplace environments such as surface soils/sediments, the rhizosphere, and redox niches in subsoil/subsurface environments to more specialized environments such as groundwater iron seeps and iron-rich hot springs. All such iron-containing environments with permanent or fluctuating redox interfaces have the potential for coupled iron(III) reduction and iron(II) oxidation leading to iron cycling on various temporal and spatial scales.

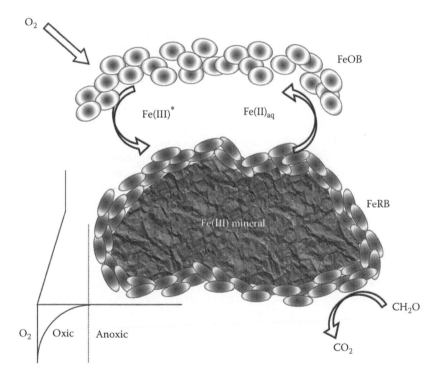

Figure 17.20. Conceptual model of microscale microbial iron redox cycling within a redox interfacial environment (i.e., at an oxic–anoxic interface). Fe(II)$_{aq}$ refers to aqueous (soluble) iron(II), and Fe(III)* refers to ligand-bound or colloidal iron(III) that has the potential to move via diffusion/dispersion. FeOB and FeRB correspond to iron(II)-oxidizing and iron(III)-reducing bacteria, respectively. Close juxtapositioning of FeOB and FeRB and rapid microscale iron redox cycling is hypothesized to occur within the interfacial zone. Clumping of FeOB and FeRB around iron(III) oxide aggregates permits coexistence of iron(II) oxidation and iron(III) reduction within the same horizon, with the FeRB taking advantage of anoxic microzones within the aggregates. These microzones are postulated to be generated by oxygen scavenging by FeOB on the periphery of the aggregates, leading to the development of "ultramicrogradients" of oxygen at the surfaces of aggregates, with oxygen declining from the bulk aqueous phase concentration to essentially zero at some distance from the surface (lower left). As a result of these interactions, anoxic conditions are maintained at the aggregate surface, allowing iron(III) reduction to proceed within a bulk oxic environment. Production and inward flux of soluble iron(III) would presumably facilitate rapid iron cycling at the aggregate surface. (Courtesy of Roden, EE; Modified from Sobolev, D and Roden, EE, *Antonie v Leeuwenhoek Int J Gen Mol Microbiol*, 81, 587, 2002; Roden, EE et al., *Geomicrobiol. J.*, 21, 379, 2004.)

ACKNOWLEDGMENTS

We thank Caroline Schmidt and Dirk Schüler for their invaluable input of revising this chapter.

REFERENCES

Adams LF, Ghiorse WC. 1987. Characterization of extracellular Mn^{2+}-oxidizing activity and isolation of Mn^{2+}-oxidizing protein from *Leptothrix discophora* SS-1. *J Bacteriol* 169:1279–1285.

Ahonen L, Tuovinen OH. 1989. Microbial oxidation of ferrous iron at low temperature. *Appl Environ Microbiol* 55:312–316.

Aleem MIH, Lees H, Nicholas DJD. 1963. Adenosine triphosphate-dependent reduction of nicotinamide adenine dinucleotide by ferro-cytochrome *c* in chemoautotrophic bacteria. *Nature (London)* 200:759–761.

Alexander M. 1977. *Introduction to Soil Microbiology*, 2nd edn. New York: Wiley.

Amouric A, Brochier-Armanet C, Johnson DB, Bonnefoy V, Hallberg KB. 2011. Phylogenetic and genetic variation among Fe(II)-oxidizing acidithiobacilli supports the view that these comprise multiple species with different ferrous iron oxidation pathways. *Microbiology* 157:111–122.

Anbar AD. 2004. Iron stable isotopes: Beyond biosignatures. *Earth Planet Sci Lett* 217:223–236.

Anbar AD, Holland HD. 1992. The photochemistry of manganese and the origin of banded iron formations. *Geochim Cosmochim Acta* 56:2595–2603.

Andrews S, Norton I, Salunkhe AS, Goodluck H, Aly WSM, Mourad-Agha H, Cornelis P. 2013. Control of iron metabolism in bacteria. In: Banci L, ed. *Metallomics and the Cell*. the Netherlands: Springer, pp. 203–239.

Appia-Ayme C, Guiliani N, Ratouchniak J, Bonnefoy V. 1999. Characterization of an operon encoding two *c*-type cytochromes, an aa3-type cytochrome oxidase, and rusticyanin in *Thiobacillus ferrooxidans* ATCC 33020. *Appl Environ Microbiol* 65:4781–4787.

Arkesteyn GJMW, DeBont JAM. 1980. *Thiobacillus acidophilus*: A study of its presence in *Thiobacillus ferrooxidans* cultures. *Can J Microbiol* 26:1057–1065.

Arnold RG, DiChristina TJ, Hoffmann MR. 1986a. Inhibitor studies of dissimilative Fe(III) reduction by *Pseudomonas* sp. 200 ("*Pseudomonas ferrireducens*"). *Appl Environ Microbiol* 52:281–289.

Arnold RG, Olson TM, Hoffmann MR. 1986b. Kinetics and mechanism of dissimilative Fe(III) reduction by *Pseudomonas* sp. 200. *Biotech Bioeng* 28:1657–1671.

Avakyan AA, Karavaiko GI. 1970. Submicroscopic organization of *Thiobacillus ferrooxidans*. *Mikrobiologiya* 39:855–861 (Engl transl, pp. 744–751).

Balashova VV, Vedina IYa, Markosyan GE, Zavazin GA. 1974. *Leptospirillum ferrooxidans* and aspects of its autotrophic growth. *Mikrobiologiya* 43:581–585 (Engl transl, pp. 491–494).

Balashova VV, Zavarzin GA. 1972. Iron oxidation by *Mycoplasma laidlawii*. *Mikrobiologiia* 41:909.

Balashova VV, Zavarzin GA. 1979. Anaerobic reduction of ferric iron by hydrogen bacteria. *Mikrobiologiya* 48:773–778 (Engl transl, pp. 635–639).

Barker WW, Welch SA, Chu S, Banfield JF. 1998. Experimental observations of the effects of bacteria on aluminosilicate weathering. *Am Mineral* 83:1551–1563.

Bauer I, Kappler A. 2009. Rates and extent of reduction of Fe (III) compounds and O$_2$ by humic substances. *Environ Sci Technol* 43:4902–4908.

Bazylinski DA, Frankel RB. 2000. Biologically controlled mineralization of magnetic iron minerals by magnetotactic bacteria. In: Lovley DR, ed. *Environmental Microbe-Metal Interactions*. Washington, DC: ASM Press, pp. 109–144.

Bazylinski DA, Lefèvre CT, Schüler D. 2013. Magnetotactic bacteria. In: Dworkin M, Falkow S, Rosenberg S, Schleifer KH, Stackebrandt E, eds. *The Procaryotes—Proteobacteria: Alpha and Beta Subclasses*. Berlin, Germany: Springer, pp. 453–493.

Beal EJ, House CH, Orphan VJ. 2009. Manganese- and iron dependent marine methane oxidation. *Science* 325:184–187.

Beliaev AS, Saffarini DA. 1998. *Shewanella putrefaciens* mtrB encodes an outer membrane protein required for Fe(III) and Mn(IV) reduction. *J Bacteriol* 180:6292–6297.

Bell PE, Mills AL, Herman JS. 1987. Biogeochemical conditions favoring magnetite formation during anaerobic iron reduction. *Appl Environ Microbiol* 53:2610–2616.

Beller HR, Zhou P, Legler TC, Chakicherla A, Kane S, Letain TE, O'Day PA. 2013. Genome-enabled studies of anaerobic, nitrate-dependent iron oxidation in the chemolithoautotrophic bacterium *Thiobacillus denitrificans*. *Front Microbiol* 4:1–16.

Benz M, Brune A, Schink B. 1998. Anaerobic and aerobic oxidation of ferrous iron at neutral pH by chemoheterotrophic nitrate-reducing bacteria. *Arch Microbiol* 169:159–165.

Benzine J, Shelobolina E, Xiong MY, Kennedy DW, McKinley JP, Lin XJ, Roden EE. 2013. Fe-phyllosilicate redox cycling organisms from a redox transition zone in Hanford 300 Area sediments. *Front Microbiol* 4:1–13.

Berner RA. 1962. Experimental studies of the formation of sedimentary iron sulfides. In: Jensen ML, ed. *Biogeochemistry of Sulfur Isotopes*. New Haven, CT: Yale University Press, pp. 107–120.

Berthelin J, Dommergues Y. 1972. Rôle de produits du métabolisme microbien dans la solubilization de minéraux d'une arène granitique. *Rev Ecol Biol Sol* 9:937–1406.

Berthelin J, Kogblevi A. 1972. Influence de la sterilization partielle sur la solubilization microbienne des minéraux dans les sols. *Rev Ecol Biol* 9:407–419.

Berthelin J, Kogblevi A. 1974. Influence de l'engorgement sur l'altération microbienne des minéraux dans les sols. *Rev Ecol Biol* 11:499–509.

Berthelin J, Kogblevi A, Dommergues Y. 1974. Microbial weathering of a brown forest soil: Influence of partial sterilization. *Soil Biol Biochem* 6:393–399.

Beveridge TJ. 1989. Metal ion and bacteria. In: Beveridge TJ, Doyle RJ, eds. *Metal Ions and Bacteria*. New York: Wiley, pp. 1–29.

Bird LJ, Bonnefoy V, Newman DK. 2011. Bioenergetic challenges of microbial iron metabolisms. *Trend Microbiol* 19:330–340.

Bird LJ, Saraiva IH, Park S, Calcada EO, Salgueiro CA, Nitschke W, Louro RO, Newman DK. 2014. Nonredundant roles for cytochrome c_2 and two high-potential iron-sulfur proteins in the photoferrotroph *Rhodopseudomonas palustris* TIE-1. *J Bacteriol* 96(4):850–858.

Blake RC, Griff MN. 2012. In situ spectroscopy on intact *Leptospirillum ferrooxidans* reveals that reduced cytochrome 579 is an obligatory intermediate in the aerobic iron respiratory chain. *Front Microbiol* 3:1–10.

Blakemore RP. 1982. Magnetotactic bacteria. *Annu Rev Microbiol* 36:217–238.

Bloomfield C. 1950. Some observations on gleying. *J Soil Sci* 1:205–211.

Bloomfield C. 1951. Experiments on the mechanism of gley formation. *J Soil Sci* 2:196–211.

Bloomfield C. 1953a. A study of podzolization. Part I. The mobilization of iron and aluminum by Scots pine needles. *J Soil Sci* 4:5–16.

Bloomfield C. 1953b. A study of podzolization. Part II. The mobilization of iron and aluminum by the leaves and bark of *Agathis australis* (Kauri). *J Soil Sci* 4:17–23.

Blöthe M, Roden EE. 2009a. Composition and activity of an autotrophic Fe (II)-oxidizing, nitrate-reducing enrichment culture. *Appl Environ Microbiol* 75:6937–6940.

Blöthe M, Roden EE. 2009b. Microbial iron redox cycling in a circumneutral-pH groundwater seep. *Appl Environ Microbiol* 75:468–473.

Bond DR, Strycharz-Glaven SM, Tender LM, Torres CI. 2012. On electron transport through *Geobacter* biofilms. *ChemSusChem* 5:1099–1105.

Bonnefoy V, Holmes DS. 2012. Genomic insights into microbial iron oxidation and iron uptake strategies in extremely acidic environments. *Environ Microbiol* 14:1597–1611.

Bose A, Gardel EJ, Vidoudez C, Parra EA, Girguis PR. 2014. Electron uptake by iron-oxidizing phototrophic bacteria. *Nat Commun* 5:1–7.

Bose A, Newman DK. 2011. Regulation of the phototrophic iron oxidation (pio) genes in *Rhodopseudomonas palustris* TIE-1 is mediated by the global regulator, FixK. *Mol Microbiol* 79:63–75.

Bridge TAM, Johnson DB. 1998. Reduction of soluble iron and reductive dissolution of ferric iron containing minerals by moderately thermophilic iron-oxidizing bacteria. *Appl Environ Microbiol* 64:2181–2186.

Bridge TAM, Johnson DB. 2000. Reductive dissolution of ferric iron minerals by *Acidiphilium* SJH. *Geomicrobiol J* 17:193–206.

Brierley CL, Brierley JA. 1973. A chemoautotrophic microorganism isolated from an acid hot spring. *Can J Microbiol* 19:183–188.

Brock TD, Brock KM, Belly RT, Weiss RL. 1972. *Sulfolobus*: A new genus of sulfur-oxidizing bacteria living at low pH and high temperature. *Arch Microbiol* 84:54–68.

Brock TD, Cook S, Peterson S, Mosser JL. 1976. Biogeochemistry and bacteriology of ferrous iron oxidation in geothermal habitats. *Geochim Cosmochim Acta* 40:493–500.

Brock TD, Gustafson J. 1976. Ferric iron reduction by sulfur- and iron-oxidizing bacteria. *Appl Environ Microbiol* 32:567–571.

Brown KA, Ratledge C. 1975. Iron transport in *Mycobacterium smegmatis*: Ferrimycobactin reductase (NAD(P)H:ferrimycobactin oxidoreductase, the enzyme releasing iron from its carrier. *FEBS Lett* 53:262–266.

Bruscella P, Appia-Ayme C, Levcán G, Ratouchniak J, Jedlicki E, Holmes DS, Bonnefoy V. 2007. Differential expression of two bc_1 complexes in the strict acidophilic chemolithoautotrophic bacterium *Acidithiobacillus ferrooxidans* suggests a model for their respective roles in iron and sulfur oxidation. *Microbiology (Reading)* 153:102–110.

Brutinel ED, Gralnick JA. 2012. Shuttling happens: Soluble flavin mediators of extracellular electron transfer in *Shewanella*. *Appl Microbiol Biotechnol* 93:41–48.

Buchanan RE, Gibbons NE, eds. 1974. *Bergey's Manual of Determinative Bacteriology*, 8th edn. FEBS Lett 53:262–266.

Burdige DJ. 1993. The biogeochemistry of manganese and iron reduction in marine sediments. *Earth Sci Rev* 35:249–284.

Caccavo F, Blakemore RP, Lovley DR. 1992. A hydrogen-oxidizing, Fe(III)-reducing microorganism from the Great Bay Estuary, New-Hampshire. *Appl Environ Microbiol* 58:3211–3216.

Caccavo F Jr, Lonergan DJ, Lovley DR, Davis M, Stolz JF, McInerney MJ. 1994. *Geobacter sulfurreducens* sp. nov., hydrogen-, and acetate-oxidizing dissimilatory metal-reducing microorganism. *Appl Environ Microbiol* 60:3752–3759.

Caiazza NC, Lies DP, Newman DK. 2007. Phototrophic Fe(II) oxidation promotes organic carbon acquisition by *Rhodobacter capsulatus* SB1003. *Appl Environ Microbiol* 73:6150–6158.

Carlson HK, Clark IC, Melnyk RA, Coates JD. 2012. Toward a mechanistic understanding of anaerobic nitrate-dependent iron oxidation: Balancing electron uptake and detoxification. *Front Microbiol* 3:1–18.

Chakraborty A, Roden EE, Schieber J, Picardal F. 2011. Enhanced growth of *Acidovorax* sp strain 2AN during nitrate-dependent Fe(II) oxidation in batch and continuous-flow systems. *Appl Environ Microbiol* 77:8548–8556.

Chakraborty AK, Picardal F. 2013. Induction of nitrate-dependent Fe(II) oxidation by Fe(II) in *Dechloromonas* sp. strain UWNR4 and *Acidovorax* sp. strain 2AN. *Appl Environ Microbiol* 79:748–752.

Chan CS, De Stasio G, Welch SA, Girasole M, Frazer BH, Nesterova MV, Fakra S, Banfield JF. 2004. Microbial polysaccharides template assembly of nanocrystal fibers. *Science* 303:1656–1658.

Chan CS, Fakra SC, Edwards DC, Emerson D, Banfield JF. 2009. Iron oxyhydroxide mineralization on microbial extracellular polysaccharides. *Geochim Cosmochim Acta* 73:3807–3818.

Chan CS, Fakra SC, Emerson D, Fleming EJ, Edwards KJ. 2011. Lithotrophic iron-oxidizing bacteria produce organic stalks to control mineral growth: Implications for biosignature formation. *ISME J* 5:717–727.

Childers SE, Ciufo S, Lovley DR. 2002. *Geobacter metallireducens* accesses insoluble Fe(III) oxide by chemotaxis. *Nature (London)* 416:767–769.

Cholodny N. 1924. Zur Morphologie der Eisenbakterien *Gallionella* and *Spirophyllum*. *Ber deutsch Bot Ges* 42:35–44.

Cholodny N. 1926. *Die Eisenbakterien. Beiträge zu einer Monographie.* Jena, Germany: Verlag von Gustav Fischer.

Christensen PB, Revsbech NP, Sand-Jensen K. 1994. Microsensor analysis of oxygen in the rhizosphere of the aquatic macrophyte *Littorella uniflora* (L.) Ascherson. *Plant Physiol* 105:847–852.

Clark DA, Norris PR. 1996. *Acidimicrobium ferrooxidans* gen. nov., sp. nov.: Mixed culture ferrous iron oxidation by *Sulfolobus* species. *Microbiology (Reading)* 142:785–790.

Cloud PE Jr. 1973. Paleoecological significance of the banded-iron formations. *Econ Geol* 68:1135–1143.

Cloud PE Jr, Licari GR. 1968. Microbiotas of the banded iron formations. *Proc Natl Acad Sci USA* 61:779–786.

Coates JD, Ellis DJ, Gaw CV, Lovley DR. 1999. *Geothrix fermentans*, gen. nov., sp. nov., a novel Fe(III)-reducing bacterium from a hydrocarbon contaminated aquifer. *Int J Syst Bacteriol* 49:1615–1622.

Coleman ML, Hedrick DB, Lovley DR, White DC, Pye K. 1993. Reduction of Fe(III) in sediments by sulfate-reducing bacteria. *Nature (London)* 361:436–438.

Corbett CM, Ingledew WJ. 1987. Is Fe^{3+}/Fe^{2+} cycling an intermediate in sulfur oxidation by Fe^{2+}-grown *Thiobacillus ferrooxidans*? *FEMS Microbiol Lett* 41:1–6.

Corstjens PLAM, de Vrind JPM, Westbroek P, de Vrind-de Jong EW. 1992. Enzymatic iron oxidation by *Leptothrix discophora*: Identification of an iron-oxidizing protein. *Appl Environ Microbiol* 58:450–454.

Coursolle D, Baron DB, Bond DR, Gralnick JA. 2010. The Mtr respiratory pathway is essential for reducing flavins and electrodes in *Shewanella oneidensis*. *J Bacteriol* 192:467–474.

Cox CD. 1980. Iron reductases from *Pseudomonas aeruginosa*. *J Bacteriol* 141:199–204.

Croal LR, Gralnick JA, Malasarn D, Newman DK. 2004a. The genetics of geochemistry. *Ann Rev Genet* 38:175–202.

Croal LR, Jiao Y, Newman DK. 2007. The fox operon from *Rhodobacter* strain SW2 promotes phototrophic Fe(II) oxidation in *Rhodobacter capsulatus* SB1003. *J Bacteriol* 189:1774–1782.

Croal LR, Johnson CM, Beard BL, Newman DK. 2004b. Iron isotope fractionation by Fe(II)-oxidizing photoautotrophic bacteria. *Geochim Cosmochim Acta* 68:1227–1242.

De Castro AF, Ehrlich HL. 1970. Reduction of iron oxide minerals by a marine *Bacillus*. *Antonie v Leeuwenhoek* 36:317–327.

De Vrind-de Jong EW, Corstjens PLAM, Kempers ES, Westbroek P, de Vrind JPM. 1990. Oxidation of manganese and iron by *Leptothrix discophora*: Use of N,N,N',N'-tetramethyl-*p*-phenylenediamine as an indicator of metal oxidation. *Appl Environ Microbiol* 56:3458–3462.

Dhungana S, Crumbliss AL. 2005. Coordination chemistry and redox processes in siderophore-mediated transport. *Geomicrobiol J* 22:87–89.

DiChristina TJ, Moore CM, Haller CA. 2002. Dissimilatory Fe(III) and Mn(IV) reduction by *Shewanella putrefaciens* requires *ferE*, a homolog of *pulE* (*gspE*) type II protein secretion gene. *J Bacteriol* 184:142–151.

Dippon U, Pantke C, Porsch K, Larese-Casanova P, Kappler A. 2012. Potential function of added minerals as nucleation sites and effect of humic substances on mineral formation by the nitrate-reducing Fe(II)-oxidizer *Acidovorax* sp. BoFeN1. *Environ Sci Technol* 46:6556–6565.

Dobbin PS, Carter JP, García-Salamanca San Juan C, Hobe M, Powell AK, Richardson DJ. 1999. Dissimilatory Fe(III) reduction by *Clostridium beijerinckii* isolated from freshwater sediment using Fe(III) maltol enrichment. *FEMS Microbiol Lett* 176:131–138.

Dopson M, Baker-Austin C, Hind M, Bowman JP, Bond PL. 2004. Characterization of *Ferroplasma* isolates and *Ferroplasma acidarmanus* sp. nov., extreme acidophiles from acid mine drainage and industrial bioleaching environments. *Appl Environ Microbiol* 70:2079–2088.

Drobner E, Huber H, Stetter KO. 1990. *Thiobacillus ferrooxidans*, a facultative hydrogen oxidizer. *Appl Environ Microbiol* 56:2911–2923.

Druschel GK, Emerson D, Sutka R, Suchecki P, Luther GW. 2008. Low-oxygen and chemical kinetic constraints on the geochemical niche of neutrophilic iron(II) oxidizing microorganisms. *Geochim Cosmochim Acta* 72:3358–3370.

Edwards KJ, Bach W, McCollum TM, Rogers DR. 2004. Neutrophilic iron-oxidizing bacteria in the oceans: Their habitats, diversity, and roles in mineral deposition, rock alteration, and biomass production in the deep-sea. *Geomicrobiol J* 21:393–404.

Edwards KJ, Bond PL, Gihrig TM, Banfield JF. 2000. An archaeal iron-oxidizing extreme acidophile important in acid mine drainage. *Science* 287:1796–1799.

Edwards KJ, Glazer BT, Rouxel OJ, Bach W, Emerson D, Davis RE, Toner BM, Chan CS, Tebo BM, Staudigel H, Moyer CL. 2011. Ultra-diffuse hydrothermal venting supports Fe-oxidizing bacteria and massive umber deposition at 5000 m off Hawaii. *ISME J* 5:1748–1758.

Ehrenberg CG. 1836. Vorläufige Mitteilungen über das wirkliche Vorkommen fossiler Infusorien und ihre gross Verbreitung. *Poggendorf's Ann Phys Chem* 38:213–227.

Ehrenreich A, Widdel F. 1994. Anaerobic oxidation of ferrous iron by purple bacteria, a new type of phototrophic metabolism. *Appl Environ Microbiol* 60:4517–4526.

Elbehti A, Brasseur G, Lemesle-Meunier D. 2000. First evidence for existence of an uphill electron transfer through the bc_1 and NADH-Q oxidoreductase complexes of the acidophilic obligate chemolithotrophic ferrous iron-oxidizing bacterium *Thiobacillus ferrooxidans*. *J Bacteriol* 182:3602–3606.

Emerson D, Field EK, Chertkov O, Davenport KW, Goodwin L, Munk C, Nolan M, Woyke T. 2013. Comparative genomics of freshwater Fe-oxidizing bacteria: Implications for physiology, ecology, and systematics. *Front Microbiol* 4:254.

Emerson D, Fleming EJ, McBeth JM. 2010. Iron-oxidizing bacteria: An environmental and genomic perspective. *Annu Rev Microbiol* 64:561–583.

Emerson D, Ghiorse WC. 1993a. Ultrastructure and chemical composition of the sheath of *Leptothrix discophora* SP-6. *J Bacteriol* 175:7808–7818.

Emerson D, Moyer C. 1997. Isolation and characterization of novel iron-oxidizing bacteria that grow at circumneutral pH. *Appl Environ Microbiol* 63:4784–4792.

Emerson D, Moyer CL. 2002. Neutrophilic Fe-oxidizing bacteria are abundant at the Loihi Seamount hydrothermal vents and play a major role in Fe oxide deposition. *Appl Environ Microbiol* 68:3085–3093.

Emerson D, Rentz JA, Lilburn TG, Davis RE, Aldrich H, Chan C, Moyer CL. 2007. A novel lineage of *Proteobacteria* involved in formation of marine Fe-oxidizing microbial mat communities. *PLoS One* 2:8.

Essén SA, Johnsson A, Bylund D, Pedersen K, Lundström US. 2007. Siderophore production by *Pseudomonas stutzeri* under aerobic and anaerobic conditions. *Appl Environ Microbiol* 73:5857–5864.

Ernst JG, Winkelmann G. 1977. Enzymatic release of iron from sideramines in fungi. NADH:sideramine oxidoreductase in *Neurospora crassa*. *Biochem Biophys Acta* 500:27–41.

Ernstsen V, Gates WP, Stucki JW. 1998. Microbial reduction of structural iron in clays: A renewable resource of reduction capacity. *J Environ Qual* 27:761–766.

Fabisch M, Beulig F, Akob DM, Kusel K. 2013. Surprising abundance of *Gallionella*-related iron oxidizers in creek sediments at pH 4.4 or at high heavy metal concentrations. *Front Microbiol* 4:1–12.

Feinberg LF, Holden JF. 2006. Characterization of dissimilatory Fe(III) versus NO_3^- reduction in the hyperthermophilic archaeon Pyrobaculum aerophilum. J Bacteriol 188:525–531.

Ferris FG. 2005. Biogeochemical properties of bacteriogenic iron oxides. Geomicrobiol J 22:79–85.

Field EK, Sczyrba A, Lyman AE, Harris CC, Woyke T, Stepanauskas R, Emerson D. 2014. Genomic insights into the uncultivated marine Zetaproteobacteria at Loihi Seamount. ISME J. 9:857–870.

Finneran KT, Housewright ME, Lovley DR. 2002. Multiple influences of nitrate on uranium solubility during bioremediation of uranium-contaminated subsurface sediments. Environ Microbiol 4:510–516.

Fleming EJ, Cetinić I, Chan CS, King DW, Emerson D. 2013. Ecological succession among iron-oxidizing bacteria. ISME J 8:804–815.

Fleming EJ, Langdon AE, Martinez-Garcia M, Stepanauskas R, Poulton NJ, Masland EDP, Emerson D. 2011. What's new is old: Resolving the identity of Leptothrix ochracea using single cell genomics, pyrosequencing and FISH. PLoS One 6:e17769.

Forget NL, Murdock SA, Juniper SK. 2010. Bacterial diversity in Fe-rich hydrothermal sediments at two South Tonga Arc submarine volcanoes. Geobiology 8:417–432.

Frankel RB, Bazylinski DA, Johnson MS, Taylor BL. 1997. Magneto-aerotaxis in marine coccoid bacteria. Biophys J 73:994–1000.

Frankel RB, Papaefthymiou GC, Blakemore RP, O'Brien W. 1983. Fe_3O_4 precipitation in magnetotactic bacteria. Biochim Biophys Acta 763:147–159.

Fukumori Y. 2004. Enzymes for magnetite synthesis in Magnetospirillum magnetotacticum. In: Bäuerlein E, ed. Biomineralization, 2nd edn. Weinheim, Germany: Wiley-VCH Verlag, pp. 75–90.

Gelder RJ. 1999. Complex lessons of iron uptake. Nature (London) 400:815–816.

Ghiorse WC. 1984. Biology of iron- and manganese-depositing bacteria. Annu Rev Microbiol 38:515–550.

Ghiorse WC. 1988. Microbial reduction of manganese and iron. In: Zehnder AJB, ed. Biology of Anaerobic Microorganisms. New York: Wiley, pp. 305–331.

Ghiorse WC, Ehrlich HL. 1992. Microbial biomineralization of iron and manganese. In: Skinner HCW, Fitzpatrick RW, eds. Biomineralization, Process of Iron and Manganese: Modern and Ancient Environments. Catena Suppl 21. Cremlingen, Germany: Catena Verlag, pp. 75–99.

Gibson F, Magrath DI. 1969. The isolation and characterization of a hydroxamic acid (aerobactin) formed by Aerobacter aerogenes 62–1. Biochim Biophys Acta 192:175–184.

Golovacheva RS, Karavaiko GI. 1978. A new genus of thermophilic spore-forming bacteria, Sulfobacillus. Mikrobiologiya 47:815–822 (Engl transl, pp. 658–665).

Golyshina OV, Pivovarova TA, Karavaiko GI, Kondrat'eva TF, Moore ERB, Abraham WR, Lunsdorf H, Timmis KN, Yakimov MM, Golyshin PM. 2000. Ferroplasma acidiphilium gen, nov., spec. nov., an acidophilic member of the Ferroplasmaceae fam. nov., comprising a distinct lineage of Archaea. Int J Syst Evol Microbiol 50:997–1006.

Gómez JM, Cantero D, Johnson DB. 1999. Comparison of the effects of temperature and pH on iron oxidation and survival by Thiobacillus ferrooxidans (type strain) and a "Leptospirillum ferrooxidans"-like isolate. In: Amils R, Ballester A, eds. Biohydrometallurgy and the Environment toward the Mining of the 21st Century: Part A. Amsterdam, the Netherlands: Elsevier, pp. 689–696.

Gorby YA, Vanina S, McLean JS, Rosso KM, Moyles D, Dohnalkova A, Beveridge TJ et al. 2006. Electrically conductive bacterial nanowires produced by Shewanella oneidensis strain MR-1 and other microorganisms. Proc Natl Acad Sci USA 103:11358–11363.

Gralnick JA. 2012. On conducting electron traffic across the periplasm. Biochem Soc Trans 40:1178–1180.

Gruner JW. 1923. Algae, believed to be Archean. J Geol 31:146–148.

Guay R, Silver M. 1975. Thiobacillus acidophilus sp. nov. Isolation and some physical characteristics. Can J Microbiol 21:281–288.

Haaijer SCM, Crienen G, Jetten MSM, den Camp HJMO. 2012. Anoxic iron cycling bacteria from an iron sulfide- and nitrate-rich freshwater environment. Front Microbiol 3:1–8.

Hafenbradl D, Keller M, Dirmeier R, Rachel R, Rossnagel P, Burggrag S, Huber H, Stetter KO. 1996. Ferroglobus placidus gen nov., spec. nov., a novel hyperthermophilic archeum that oxidizes Fe^{2+} at neutral pH under anoxic conditions. Arch Microbiol 166:308–314.

Hallbeck L. 1993. On the biology of the iron-oxidizing and stalk-forming bacterium Gallionella ferruginea, Ph.D. thesis. Göteborg, Sweden: University of Göteborg.

Hallbeck L, Pedersen K. 1991. Autotrophic and mixotrophic growth of Gallionella ferruginea. J Gen Microbiol 137:2657–2661.

Hallberg KB, Hedrich S, Johnson DB. 2011. Acidiferrobacter thiooxydans, gen. nov. sp. nov.; an acidophilic, thermo-tolerant, facultatively anaerobic iron- and sulfur-oxidizer of the family Ectothiorhodospiraceae. Extremophiles 15(2):271–279.

Hammann R, Ottow JCG. 1974. Reductive dissolution of Fe₂O₃ by saccharolytic Clostridia and Bacillus polymyxa under anaerobic conditions. *Z Pflanzenernähr Bodenk* 137:108–115.

Handley KM, Boothman C, Mills RA, Pancost RD, Lloyd JR. 2010. Functional diversity of bacteria in a ferruginous hydrothermal sediment. *ISME J* 4:1193–1205.

Hanert H. 1973. Quantifizierung der Massenentwiclung des Eisenbakteriums *Gallionella ferruginea* unter natürlichen Bedignungen. *Arch Mikrobiol* 88:225–243.

Hanert HH. 1981. The genus *Gallionella*. In: Starr MP, Stolp H, Trüper HG, Balows A, Schlegel HG, eds. *The Prokaryotes. A Handbook on Habitats, Isolation, and Identification of Bacteria.* Berlin, Germany: Springer-Verlag, pp. 509–515.

Hansel CM, Benner SG, Neiss J, Dohnalkova A, Kukkadapu RK, Fendorf S. 2003. Secondary mineralization pathways induced by dissimilatory iron reduction of ferrihydrite under advective flow. *Geochim Cosmochim Acta* 67:2977–2992.

Harrison AP Jr. 1981. *Acidiphilium cryptum* gen. nov., spec. nov., heterotrophic bacterium from acidic mineral environments. *Int J Syst Bacteriol* 31:327–332.

Harrison AP Jr. 1982. Genomic and physiological diversity amongst strains of *Thiobacillus ferrooxidans* and genomic comparison with *Thiobacillus thiooxidans. Arch Microbiol* 131:68–76.

Harrison AP Jr. 1983. Genomic and physiological comparisons between heterotrophic thiobacilli and *Acidiphilium cryptum, Thiobacillus versutus* sp. nov., and *Thiobacillus acidophilus* nom. rev. *Int J Syst Bacteriol* 33, 211–217.

Harrison AP Jr. 1984. The acidophilic thiobacilli and other acidophilic bacteria that share their habitat. *Annu Rev Microbiol* 38:265–292.

Harrison AP Jr. 1986. The phylogeny of iron-oxidizing bacteria. In: Ehrlich HL, Holmes DS, eds. *Workshop on Biotechnology for the Mining, Metal-Refining and Fossil Fuel Processing Industries. Biotech Bioeng Symp* 16. New York: Wiley, pp. 311–317.

Hart A, Murrell JC, Poole RK, Norris PR. 1991. An acid-stable cytochrome in iron oxidizing *Leptospirillum ferrooxidans. FEMS Microbiol Lett* 81:89–94.

Hartmann H. 1984. The evolution of photosynthesis and microbial mats. A speculation on banded iron formations. In: Cohen Y, Castenholz RW, Halvorson HO, eds. *Microbial Mats: Stromatolites.* New York: Alan R. Liss, pp. 451–453.

Hartshorne RS, Reardon CL, Ross D, Nuester J, Clarke TA, Gates AJ, Mills PC et al. 2009. Characterization of an electron conduit between bacteria and the extracellular environment. *Proc Natl Acad Sci USA* 106:22169–22174.

Hedrich S, Johnson DB. 2013. *Acidithiobacillus ferridurans* sp. nov., an acidophilic iron-, sulfur-and hydrogen-metabolizing chemolithotrophic gammaproteobacterium. *Int J Syst Evol Microbiol* 63:4018–4025.

Hegler F, Losekann-Behrens T, Hanselmann K, Behrens S, Kappler A. 2012. Influence of seasonal and geochemical changes on the geomicrobiology of an iron carbonate mineral water spring. *Appl Environ Microbiol* 78:7185–7196.

Hegler F, Posth NR, Jiang J, Kappler A. 2008. Physiology of phototrophic iron(II)-oxidizing bacteria: Implications for modern and ancient environments. *FEMS Microbiol Ecol* 66:250–260.

Heising S, Richter L, Ludwig W, Schink B. 1999. *Chlorobium ferrooxidans* sp. nov., a phototrophic green sulfur bacterium that oxidizes ferrous iron in coculture with a "*Geospirillum*" sp. strain. *Arch Microbiol* 172:116–124.

Hersman LE, hung A, Maurice PA, Forsythe JH. 2000. Siderophore production and iron reduction by *Pseudomonas mendocina* in response to iron deprivation. *Geomicrobiol J* 17:261–273.

Holm NG. 1987. Biogenic influence on the geochemistry of certain ferruginous sediments of hydrothermal origin. *Chem Geol* 63:45–57.

Huber H, Stetter KO. 1989. *Thiobacillus prosperus* sp. nov., represents a new group of halotolerant metal-mobilizing bacteria isolated from a marine geothermal field. *Arch Microbiol* 151:479–485.

Hunt KA, Flynn JM, Naranjo B, Shikhare ID, Gralnick JA. 2010. Substrate-level phosphorylation is the primary source of energy conservation during anaerobic respiration of *Shewanella oneidensis* strain MR-1. *J Bacteriol* 192:3345–3351.

Hutchinson M, Johnstone KI, White D. 1966. Taxonomy of the acidophilic Thiobacilli. *J Gen Microbiol* 44:373–381.

Ivanov MV, Lyalikova NN. 1962. Taxonomy of iron-oxidizing bacilli. *Mikrobiologiya* 31:468–469 (Engl transl, pp. 382–383).

James RE, Ferris FG. 2004. Evidence for microbial-mediated iron oxidation at a neutrophilic groundwater spring. *Chem Geol* 212:301–311.

Jannasch HW, Mottl MJ. 1985. Geomicrobiology of deep-sea hydrothermal vents. *Science* 229:717–725.

Jennings RM, Whitmore LM, Moran JJ, Kreuzer HW, Inskeep WP. 2014. Carbon dioxide fixation by Metallosphaera yellowstonensis and acidothermophilic iron-oxidizing microbial communities from Yellowstone National Park. *Appl Environ Microbiol* 80:2665–2671.

Jiang J, Kappler A. 2008. Kinetics of microbial and chemical reduction of humic substances: Implications for electron shuttling. *Environ Sci Technol* 42:3563–3569.

Jiao Y, Kappler A, Croal LR, Newman DK. 2005. Isolation and characterization of a genetically tractable photoautotrophic Fe(II)-oxidizing bacterium, *Rhodopseudomonas palustris* strain TIE-1. *Appl Environ Microbiol* 71:4487–4496.

Jiao Y, Newman DK. 2007. The *pio* operon is essential for phototrophic Fe(II) oxidation in *Rhodopseudomonas palustris* TIE-1. *J Bacteriol* 189:1765–1773.

Jogler C, Schüler D. 2009. Genomics, genetics, and cell biology of magnetosome formation. *Annu Rev Microbiol* 63:501–521.

Johnson CM, Beard BL, Klein C, Beukes NJ, Roden EE. 2008. Iron isotopes constrain biologic and abiologic processes in banded iron formation genesis. *Geochim Cosmochim Acta* 72:151–169.

Johnson DB, Bridge TAM. 2002. Reduction of ferric iron by acidophilic heterotrophic bacteria: Evidence for constitutive and inducible enzyme systems in *Acidiphilium* spp. *J Appl Microbiol* 92:315–321.

Johnson DB, Hallberg KB, Hedrich S. 2014. Uncovering a microbial enigma: Isolation and characterization of the streamer-generating, iron-oxidizing, acidophilic bacterium "*Ferrovum myxofaciens*". *Appl Environ Microbiol* 80:672–680.

Johnson DB, Roberto FF. 1997. Heterotrophic acidophiles and their roles in the bioleaching of sulfide minerals. In: Rawlings ED, ed. *Biomining*. Berlin Heidelberg: Springer, pp. 259–279.

Kappler A, Benz M, Schink B, Brune A. 2004. Electron shuttling via humic acids in microbial iron (III) reduction in a freshwater sediment. *FEMS Microbiol Ecol* 47:85–92.

Kappler A, Newman DK. 2004. Formation of Fe(III)-minerals by Fe(II)-oxidizing photoautotrophic bacteria. *Geochim Cosmochim Acta* 68:1217–1226.

Kappler A, Pasquero C, Konhauser KO, Newman DK. 2005a. Deposition of banded iron formations by photoautotrophic Fe(II)-oxidizing bacteria. *Geology* 33:865–868.

Kappler A, Schink B, Newman DK. 2005b. Fe(III) mineral formation and cell encrustation by the nitrate-dependent Fe(II)-oxidizer strain BoFeN1. *Geobiology* 3:235–245.

Kappler A, Straub KL. 2005. Geomicrobiological cycling of iron. *Rev Mineral Geochem* 59:85–108.

Kappler A, Wuestner ML, Ruecker A, Harter J, Halama M, Behrens S. 2014. Biochar as an electron shuttle between bacteria and Fe (III) minerals. *Environ Sci Technol Let* 1:339–344.

Karavaiko GI, Avakyan AA. 1979. Mechanism of *Thiobacillus ferrooxidans* multiplication. *Mikrobiologiya* 39:950–952.

Karlin R, Lyle M, Heath GR. 1987. Authigenic magnetite formation in suboxic marine sediments. *Nature (London)* 326:490–493.

Kashefi K, Holmes DE, Reysenbach AL, Lovley DR. 2002a. Use of Fe(III) as an electron acceptor to recover previously uncultured hyperthermophiles: Isolation and characterization of *Geothermobacterium ferrireducens* gen. nov., sp nov. *Environ Microbiol* 68:1735–1742.

Kashefi K, Lovley DR. 2003. Extending the upper temperature limit for life. *Science* 301:934.

Kashefi K, Shelobolina·ES, Elliorr WC, Lovley DR. 2008. Growth of thermophilic and hyperthermophilic Fe(III)-reducing microorganisms on a ferruginous smectite as the sole electron acceptor. *Appl Environ Microbiol* 74:251–258.

Kashefi K, Tor JM, Holmes DE, Van Praagh CVG, Reysenbach AL, Lovley DR. 2002b. *Geoglobus ahangari* gen. nov., sp nov., a novel hyperthermophilic archaeon capable of oxidizing organic acids and growing autotrophically on hydrogen with Fe(III) serving as the sole electron acceptor. *Int J Syst Evol Microbiol* 52:719–728.

Kato S, Chan C, Itoh T, Ohkuma M. 2013. Functional gene analysis of freshwater iron-rich flocs at circumneutral pH and isolation of a stalk-forming microaerophilic iron-oxidizing bacterium. *Appl Environ Microbiol* 79:5283–5290.

Kato S, Krepski S, Chan C, Itoh T, Ohkuma M. 2014. *Ferriphaselus amnicola* gen. nov., sp. nov., a neutrophilic, stalk-forming, iron-oxidizing bacterium isolated from an iron-rich groundwater seep. *Int J Syst Evol Microbiol* 64:921–925.

Kato S, Yanagawa K, Sunamura M, Takano Y, Ishibashi J, Kakegawa T, Utsumi M et al. 2009. Abundance of *Zetaproteobacteria* within crustal fluids in back-arc hydrothermal fields of the Southern Mariana Trough. *Environ Microbiol* 11:3210–3222.

Kelly DP, Jones CA. 1978. Factors affecting metabolism and ferrous iron oxidation in suspensions and batch cultures of *Thiobacillus ferrooxidans*: Relevance to ferric iron leach solution generation. In: Murr LE, Torma AE, Brierley JA, eds. *Metallurgical Applications of Bacterial Leaching and Related Microbiological Phenomena*. New York: Academic Press, pp. 19–44.

Kelly DP, Tuovinen OH. 1972. Recommendation that the names *Ferrobacillus ferrooxidans* Leathen and Braley and *Ferrobacillus sulfooxidans* Kinsel be recognized as synonymous of *Thiobacillus ferrooxidans* Temple and Colmer. *Int J Syst Bacteriol* 22:170–172.

Kinsel N. 1960. New sulfur oxidizing iron bacterium: *Ferrobacillus sulfooxidans* sp. n. *J Bacteriol* 80:628–632.

Klein C. 2005. Some Precambrian banded iron-formations (BIFs) from around the world: Their age, geologic setting, mineralogy, metamorphism, geochemistry, and origin. *Am Mineral* 90:1473–1499.

Klueglein N, Lösekann-Behrens T, Obst M, Behrens S, Appel E, Kappler A. 2013. Magnetite formation by the novel Fe(III)-reducing *Geothrix fermentans* strain HradG1 isolated from a hydrocarbon-contaminated sediment with increased magnetic susceptibility. *Geomicrobiol J* 30:863–873.

Klueglein N, Zeitvogel F, Stierhof Y-D, Floetenmeyer M, Konhauser KO, Kappler A, Obst M. 2014. Potential role of nitrite for abiotic Fe(II) oxidation and cell encrustation during nitrate reduction by denitrifying bacteria. *Appl Environ Microbiol* 80:1051–1061.

Koehler I, Konhauser KO, Kappler A. 2010. Role of microorganisms in banded iron formations. In: Barton LL, Mandl M, Loy A, eds. *Geomicrobiology: Molecular and Environmental Perspective*, the Netherlands: Springer.

Koehler I, Konhauser KO, Papineau D, Bekker A, Kappler A. 2013. Biological carbon precursor to diagenetic siderite with spherical structures in iron formations. *Nat Commun* 4:1741.

Komeili A, Li Z, Newman DK, Jensen GJ. 2006. Magnetosomes are cell membrane invaginations organized by the actin-like protein MamK. *Science* 311:242–245.

Konhauser KO. 1997. Bacterial iron biomineralization in nature. *FEMS Microbiol Rev* 20:315–326.

Konhauser KO, Kappler A, Roden EE. 2011. Iron in microbial metabolisms. *Elements* 7:89–93.

Konhauser KO, Newman DK, Kappler A. 2005. The potential significance of microbial Fe(III) reduction during deposition of Precambrian banded iron formations. *Geobiology* 3:167–177.

Kopf SH, Newman DK. 2012. Photomixotrophic growth of *Rhodobacter capsulatus* SB1003 on ferrous iron. *Geobiology* 10:216–222.

Kostka JE, Nealson KH. 1995. Dissolution and reduction of magnetite by bacteria. *Environ Sci Technol* 29:2535–2540.

Kostka JE, Stucki JW, Nealson KH, Wu J. 1996. Reduction of structural Fe(III) in smectite by a pure culture of *Shewanella putrefaciens* strain MR-1. *Clays Clay Miner* 44:522–529.

Kovalenko TV, Karavaiko GI, Piskunov VP. 1982. Effect of Fe^{2+} ions on the oxidation of ferrous iron by *Thiobacillus ferrooxidans* at various temperatures. *Mikrobiologiya* 51:156–160 (Engl transl, pp. 142–146).

Krepski ST, Hanson TE, Chan CS. 2012. Isolation and characterization of a novel biomineral stalk-forming iron-oxidizing bacterium from a circumneutral groundwater seep. *Environ Microbiol* 14:1671–1680.

Kucera K-H, Wolfe RS. 1957. A selective enrichment method for *Gallionella ferruginea*. *J Bacteriol* 74:344–349.

Langley S, Gault A, Ibrahim A, Renaud R, Fortin D, Clark ID, Ferris FG. 2009. A comparison of the rates of Fe(III) reduction in synthetic and bacteriogenic iron oxides by *Shewanella putrefaciens* CN32. *Geomicrobiol J* 26:57–70.

Larese-Casanova P, Haderlein SB, Kappler A. 2010. Biomineralization of lepidocrocite and goethite by nitrate-reducing Fe(II)-oxidizing bacteria: Effect of pH, bicarbonate, phosphate, and humic acids. *Geochim Cosmochim Acta* 74:3721–3734.

Latta DE, Gorski CA, Scherer MM. 2012. Influence of Fe(2+)-catalysed iron oxide recrystallization on metal cycling. *Biochem Soc Trans* 40:1191–1197.

Laverman AM, Switzer Blum J, Schaefer JK, Phillips EJP, Lovley DR, Oremland RS. 1995. Growth of strain SES-3 with arsenate and other diverse electron acceptors. *Appl Environ Microbiol* 61:3556–3561.

Lazaroff N. 1963. Sulfate requirement for iron oxidation by *Thiobacillus ferrooxidans*. *J Bacteriol* 85:78–83.

Leathen WW, Kinsel NA, Braley SA Jr. 1956. *Ferrobacillus ferrooxidans*: A synthetic autotrophic bacterium. *J Bacteriol* 72:700–704.

Li Y, Bali S, Borg S, Katzmann E, Ferguson SJ, Schüler D. 2013. Cytochrome cd_1 nitrite reductase NirS is involved in anaerobic magnetite biomineralization in *Magnetospirillum gryphiswaldense* and requires NirN for proper d_1 heme assembly. *J Bacteriol* 195:4297–4309.

Li Y, Katzmann E, Borg S, Schüler D. 2012. The periplasmic nitrate reductase Nap is required for anaerobic growth and involved in redox control of magnetite biomineralization in *Magnetospirillum gryphiswaldense*. *J Bacteriol* 194:4847–4856.

Li Y, Raschdorf O, Silva KT, Schüler D. 2014. The terminal oxidase cbb_3 functions in redox control of magnetite biomineralization in *Magnetospirillum gryphiswaldense*. *J Bacteriol*. 196(14):2552–2562.

Li Y-L, Vali H, Yang J, Phelps TJ, Zhang CL. 2006. Reduction of iron oxides enhanced by a sulfate-reducing bacterium and biogenic H$_2$S. *Geomicrobiol J* 23:103–117.

Lies DP, Hernandez ME, Kappler A, Mielke RE, Gralnick JA, Newman DK. 2005. *Shewanella oneidensis* MR-1 uses overlapping pathways for iron reduction at a distance and by direct contact under conditions relevant for biofilms. *Appl Environ Microbiol* 71:4414–4426.

Lieske R. 1919. Zur Ernährungsphysiologie der Eisenbakterien. *Z Bakt Parasitenk Infektsionskr Hyg Abt II* 49:413–425.

Little CTS, Glynn SEJ, Mills RA. 2004. Four-hundred-and-ninety-million-year record of bacteriogenic iron oxide precipitation at sea-floor hydrothermal vents. *Geomicrobiol J* 21:415–429.

Liu J, Wang ZM, Belchik SM, Edwards MJ, Liu CX, Kennedy DW, Merkley ED et al. 2012. Identification and characterization of MtoA: A decaheme *c*-type cytochrome of the neutrophilic Fe(II)-oxidizing bacterium *Sideroxydans lithotrophicus* ES-1. *Front Microbiol* 3:1–11.

Lobos JH, Chisholm TE, Bopp LH, Holmes DS. 1986. *Acidiphilium organovorum* sp. nov., an acidophilic heterotroph isolated from a *Thiobacillus ferrooxidans* culture. *Int J Syst Bacteriol* 36:139–144.

Lohße A, Borg S, Raschdorf O, Kolinko I, Tompa E, Posfai M, Faivre D, Baumgartner J, Schüler D. 2014. Genetic dissection of the *mamAB* and *mms6* operons reveals a gene set essential for magnetosome biogenesis in *Magnetospirillum gryphiswaldense*. *J Bacteriol* 196(14):2658–2669.

Lonergan DJ, Jenter HL, Coates JD, Phillips EJ, Schmidt TM, Lovley DR. 1996. Phylogenetic analysis of dissimilatory Fe(III)-reducing bacteria. *J Bacteriol* 178:2402–2408.

Lovley DR. 1987. Organic matter mineralization with the reduction of ferric iron: A review. *Geomicrobiol J* 5(3–4):375–399.

Lovley DR. 1991. Dissimilatory Fe(III) and Mn(IV) reductions. *Microbiol Rev* 55:259–287.

Lovley DR. 1993. Dissimilatory metal reduction. *Annu Rev Microbiol* 47:263–290.

Lovley DR. 2000. Fe(III) and Mn (IV) reduction. In: Lovley DR, ed. *Environmental Microbe-Metal Interactions*. Washington, DC: ASM Press, pp. 3–30.

Lovley DR, Badaeker MJ, Lonergan DJ, Cozzarelli IM, Phillips EJP, Siegel DI. 1989b. Oxidation of aromatic contaminants coupled to microbial iron reduction. *Nature (London)* 339:297–299.

Lovley DR, Coates JD, Blunt-Harris EL, Phillips EJP, Woodword JC. 1996. Humic substances as electron acceptors for microbial respiration. *Nature (London)* 382:445–448.

Lovley DR, Giovannoni SJ, White DC, Champine JE, Phillips EJP, Gorby YA, Goodwin S. 1993. *Geobacter metallireducens* gen. nov., sp. nov., a microorganism capable of coupling the complete oxidation of organic compounds to the reduction of iron and other metals. *Arch Microbiol* 159:336–344.

Lovley DR, Holmes DE, Nevin KP. 2004. Dissimilatory Fe(III) and Mn(IV) reduction. *Adv Microb Phys* 49:219–286.

Lovley DR, Lonergan DJ. 1990. Anaerobic oxidation of toluene, phenol, and p-cresol by the dissimilatory iron-reducing organism, GS-15. *Appl Environ Microbiol* 56:1858–1864.

Lovley DR, Malvankar NS. 2015. Seeing is believing: Novel imaging techniques help clarify microbial nanowire structure and function. *Environ Microbiol*, Jul;17(7):2209–2215.

Lovley DR, Phillips EJP. 1986a. Organic matter mineralization with the reduction of ferric iron in anaerobic sediments. *Appl Environ Microbiol* 51:683–689.

Lovely DR, Phillips EJP. 1986b. Availability of ferric iron for microbial reduction in bottom sediments of the freshwater tidal Potomac River. *Appl Environ Microbiol* 52:751–757.

Lovley DR, Phillips EJP. 1987. Competitive mechanisms for inhibition of sulfate reduction in sediments. *Appl Environ Microbiol* 53:2636–2641.

Lovley DR, Phillips EJP. 1988. Novel mode of microbial energy metabolism: Organic carbon oxidation coupled to dissimilatory reduction of iron or manganese. *Appl Environ Microbiol* 54:1472–1480.

Lovley DR, Phillips EJP. 1989. Requirement for a microbial consortium to completely oxidize glucose in Fe (III)-reducing sediments. *Appl Environ Microbiol* 55:3234–3236.

Lovley DR, Phillips EJP, Lonergan DJ. 1989a. Hydrogen and formate oxidation coupled to dissimilatory reduction of iron and manganese by *Alteromonas putrefaciens*. *Appl Environ Microbiol* 55:700–706.

Lovley DR, Stolz JF, Nord GL, Phillips EJP. 1987. Anaerobic production of magnetite by a dissimilatory iron-reducing microorganism. *Nature (London)* 330:252–254.

Lovley DR, Woodward JC. 1996. Mechanism for chelator stimulation of microbial Fe(III) oxide reduction. *Chem Geol* 132:19–24.

Lowe KL, DiChristina TJ, Toychoudhury AN, van Cappellen P. 2000. Microbiological and geochemical characterization of microbial Fe(III) reduction in salt marsh sediments. *Geomicrobiol J* 17:163–178.

Lower SK, Hochella MR Jr, Beveridge TJ. 2001. Bacterial recognition of mineral surfaces: Nanoscale interactions between *Shewanella* and α-FeOOH. *Science* 292:1360–1363.

Lower BH, Shi L, Yongsunthon R, Doubray TC, McCready DE, Lower SK. 2007. Specific bonds between an iron oxide surface and outer membrane cytochromes MtrC and OmcA from *Shewanella oneidensis* MR-1. *J Bacteriol* 189:4944–4952.

Lundgren DG, Andersen KJ, Remsen CC, Mahoney RP. 1964. Culture, structure, and physiology of the chemoautotroph *Ferrobacillus ferrooxidans. Dev Indust Microbiol* 6:250–259.

Mackintosh ME. 1978. Nitrogen fixation by *Thiobacillus ferrooxidans. J Gen Microbiol* 105:215–218.

Makita H, Nakahara Y, Fukui H, Miyanoiri Y, Katahira M, Seki H, Takeda M, Koizumi JI. 2006. Identification of 2-(cysteinyl)amido-2-deoxy-D-galacturonic acid residue from the sheath of *Leptothrix cholodnii. Biosci Biotechnol Biochem* 70:1265–1268.

Mann H, Tazaki T, Fyfe WS, Beveridge TJ, Humphrey R. 1987. Cellular lepidocrocite precipitation and heavy-metal sorption in *Euglena* sp. (unicellular alga): Implications for biomineralization. *Chem Geol* 63:39–43.

Mao MWH, Dugan PR, Martin PAW, Tuovinen OH. 1980. Plasmid DNA in chemoorganotrophic *Thiobacillus ferrooxidans* and *T. acidophilus. FEMS Microbiol Lett* 8:121–125.

Markosyan GE. 1972. A new iron-oxidizing bacterium *Leptospirillum ferrooxidans* gen. et sp. nov. *Bio Zh Arm* 25:26.

Marsili E, Baron DB, Shikhare ID, Coursolle D, Gralnick JA, Bond DR. 2008. *Shewanella* secretes flavins that mediate extracellular electron transfer. *Proc Natl Acad Sci USA* 105:3968–3973.

Mason J, Kelly DP. 1988. Mixotrophic and autotrophic growth of *Thiobacillus acidophilus. Arch Microbiol* 149:317–323.

Mason J, Kelly DP, Wood AP. 1987. Chemolithotrophic and autotrophic growth of *Thermothrix thiopara* and some thiobacilli on thiosulfate and polythionates, and a reassessment of the growth yields of Thx. thiopara in chemostat culture. *J Gen Microbiol* 133:1249–1256.

McBeth JM, Little BJ, Ray RI, Farrar KM, Emerson D. 2011. Neutrophilic iron-oxidizing "*Zetaproteobacteria*" and mild steel corrosion in nearshore marine environments. *Appl Environ Microbiol* 77:1405–1412.

McClure MA, Wyckoff RWG. 1982. Ultrastructural characteristics of *Sulfolobus acidocaldarius. J Gen Microbiol* 128:433–437.

Melton ED, Schmidt C, Behrens S, Schink B, Kappler A. 2014b. Metabolic flexibility and substrate preference by the Fe(II)-oxidizing purple non-sulphur bacterium *Rhodopseudomonas palustris* strain TIE-1. *Geomicrobiol J* 31:835–843.

Melton ED, Schmidt C, Kappler A. 2012. Microbial iron(II) oxidation in littoral freshwater lake sediment: The potential for competition between phototrophic vs. nitrate-reducing iron(II)-oxidizers. *Front Microbiol* 3:1–12.

Melton ED, Stief P, Behrens S, Kappler A, Schmidt C. 2014c. High spatial resolution of distribution and interconnections between Fe- and N-redox processes in profundal lake sediments. *Environ Microbiol* 16:3287–303.

Melton ED, Swanner ED, Roden EE, Behrens S, Schmidt C, Kappler A. 2014a. The interplay of microbially mediated and abiotic reactions in the biogeochemical Fe cycle. *Nat Rev Microbiol* 12(12):797–808.

Merkle FG. 1955. Oxidation-reduction processes in soils. In: Bear FE, ed. *Chemistry of the Soil*. New York: Reinhold, pp. 200–218.

Meyer-Dombard DR, Amend JP, Osburn MR. 2013. Microbial diversity and potential for arsenic and iron biogeochemical cycling at an arsenic rich, shallow-sea hydrothermal vent (Tutum Bay, Papua New Guinea). *Chem Geol* 348:37–47.

Michel FM, Barron V, Torrent J, Morales MP, Serna CJ, Boily JF, Liu Q, Ambrosini A, Cismasu AC, Brown GE Jr. 2010. Ordered ferrimagnetic form of ferrihydrite reveals links among structure, composition, and magnetism. *Proc Natl Aca Sci USA* 107:2787–2792.

Miot J, Benzerara K, Morin G, Kappler A, Bernard S, Obst M, Ferard C et al. 2009a. Iron biomineralization by anaerobic neutrophilic iron-oxidizing bacteria. *Geochim Cosmochim Acta* 73:696–711.

Miot J, Benzerara K, Obst M, Kappler A, Hegler F, Schadler S, Bouchez C, Guyot F, Morin G. 2009b. Extracellular iron biomineralization by photoautotrophic iron-oxidizing bacteria. *Appl Environ Microbiol* 75:5586–5591.

Molisch H. 1910. *Die Eisenbacterien*. Jena, Germany: Gustav Fischer Verlag.

Muehe EM, Gerhardt S, Schink B, Kappler A. 2009. Ecophysiology and the energetic benefit of mixotrophic Fe(II) oxidation by various strains of nitrate-reducing bacteria. *FEMS Microbiol Ecol* 70:335–343.

Munch JC, Hillbrand Th, Ottow JCG. 1978. Transformation in the Fe_o/Fe_d ratio of pedogenic iron oxides affected by iron-reducing bacteria. *Can J Soil Sci* 58:475–486.

Munch JC, Ottow JCG. 1980. Preferential reduction of amorphous crystalline iron oxides by bacteria activity. *Soil Sci* 129:1–21.

Myers CR, Myers JM. 1992. Fumarate reductase is a soluble enzyme in anaerobically grown *Shewanella putrefaciens* MR-1. *FEMS Microbiol Lett* 98:13–20.

Myers CR, Myers JM. 1993. Ferric reductase is associated with the membranes of anaerobically grown *Shewanella putrefaciens* MR-1. *FEMS Microbiol Lett* 108:15–22.

Myers JM, Myers CR. 2000. Role of the tetraheme cytochrome CymA in anaerobic electron transport in cells of *Shewanella putrefaciens* MR-1 with normal levels of menaquinone. *J Bacteriol* 182:67–75.

Myers CR, Myers JM. 2003. Cell surface exposure of the outer membrane cytochromes of *Shewanella oneidensis* MR-1. *Lett Appl Microbiol* 37:254–258.

Myers CR, Nealson KH. 1988. Bacterial manganese reduction and growth with manganese oxide as sole electron acceptor. *Science* 240:1319–1321.

Myers CR, Nealson KH. 1990. Respiration-linked proton translocation coupled to anaerobic reduction of manganese(IV) and iron(III) in *Shewanella putrefaciens* MR-1. *J Bacteriol* 172:6236–6238.

Nealson KH, Myers CR. 1990. Iron reduction by bacteria: A potential role in the genesis of banded iron formations. *Am J Sci* 290A:35–40.

Nealson KH, Saffarini D. 1994. Iron and manganese in anaerobic respiration: Environmental significance, physiology, and regulation. *Annu Rev Microbiol* 48:311–343.

Neilands JB, ed. 1974. *Microbial Iron Metabolism*. New York: Academic Press.

Neubauer SC, Emerson D, Megonigal JP. 2002. Life at the energetic edge: Kinetics of circumneutral iron oxidation by lithotrophic iron-oxidizing bacteria isolated from the wetland-plant rhizosphere. *Appl Environ Microbiol* 68:3988–3995.

Nevin KP, Lovley DR. 2000. Lack of production of electron-shuttling compounds or solubilization of Fe(III) during reduction of insoluble Fe(III) oxide by *Geobacter metallireducens*. *Appl Environ Microbiol* 66:2248–2251.

Nevin KP, Lovley DR. 2002. Mechanisms of accessing insoluble Fe(III) oxide during dissimilatory Fe(III) reduction by *Geothrix fermentans*. *Appl Environ Microbiol* 68:2294–2299.

Newman DK, Kolter R. 2000. A role for excreted quinones in extracellular electron transfer. *Nature (London)* 405:94–97.

Norris PR. 1997. Thermophiles and bioleaching. In: Rawlings DE, ed. *Biomining: Theory, Microbes and Industrial Processes*. Berlin, Germany: Springer, pp. 247–258.

Norris PR, Barr DW, Hinson D. 1988. Iron and mineral oxidation by acidophilic bacteria: Affinities for iron and attachment to pyrite. In: Norris PR, Kelly DP, eds. *Biohydrometallurgy*. Kew Surrey, U.K.: Science & Technology Letters, pp. 43–59.

Norris PR, Clark DA, Owen JP, Waterhouse S. 1996. Characteristics of *Sulfobacillus acidophilus* sp. nov. and other moderately thermophilic mineral-sulphide-oxidizing bacteria. *Microbiology* 142:775–783.

Norris PR, Marsh RM, Lindstrom EB. 1986. Growth of mesophilic and thermophilic bacteria on sulfur and tetrathionate. *Biotechnol Appl Biochem* 8:313–329.

Obuekwe CO, Westlake DWS, Cook FD. 1981. Effect of nitrate on reduction of ferric iron by a bacterium isolated from crude oil. *Can J Microbiol* 27:692–697.

Obuekwe CO, Westlake DWS. 1982. Effect of reducible compounds (potential electron acceptors) on reduction of ferric iron by *Pseudomonas* sp. *Microbios Lett* 19:57–62.

Okamoto A, Hashimoto K, Nealson KH, Nakamura R. 2013. Rate enhancement of bacterial extracellular electron transport involves bound flavin semiquinones. *Proc Natl Acad Sci USA* 110:7856–7861.

Ottow JCG. 1969a. The distribution and differentiation of iron-reducing bacteria in gley soils. *Zentralbl Baketeriol Parasitenk Infektionskr Hyg Abt 2* 123:600–615.

Ottow JCG. 1969b. Der Einfluss von Nitrat, Chlorat, Sulfat, Eisenoxydform und Wachsbedingungen auf das Ausmass der bakteriellen Eisenreduktion. *Z Pflanzenernär Bodenk* 124:238–253.

Ottow JCG. 1970. Bacterial mechanism of gley formation in artificially submerged soil. *Nature (London)* 225:103.

Ottow JCG. 1971. Iron reduction and gley formation by nitrogen-fixing *Clostridia*. *Oecologia* 6:164–175.

Ottow JCG, Glathe H. 1971. Isolation and identification of iron-reducing bacteria from gley soils. *Soil Biol Biochem* 3:43–55.

Ottow JCG, von Klopotek A. 1969. Enzymatic reduction of iron oxide by fungi. *Appl Microbiol* 18:41–43.

Pandey A, Bringel F, Meyer J-M. 1994. Iron requirement and search for siderophores in lactic acid bacteria. *Appl Microb Cell Physiol* 40:735–739.

Pantke C, Obst M, Benzerara K, Morin G, Ona-Nguema G, Dippon U, Kappler A. 2012. Green rust formation during Fe(II) oxidation by the nitrate-reducing *Acidovorax* sp strain BoFeN1. *Environ Sci Technol* 46:1439–1446.

Paquete CM, Fonseca BM, Cruz DR, Pereira TM, Pacheco I, Soares CM, Louro RO. 2014. Exploring the molecular mechanisms of electron shuttling across the microbe/metal space. *Front Microbiol* 5:1–12.

Pereira L, Saraiva IH, Coelho R, Newman DK, Louro RO, Frazao C. 2012. Crystallization and preliminary crystallographic studies of FoxE from *Rhodobacter ferrooxidans* SW2, an FeII oxidoreductase involved in photoferrotrophy. *Acta Crystallogr F* 68:1106–1108.

Perry EC Jr, Tan FC, Morey GB. 1973. Geology and stable isotope geochemistry of Biwabic Iron Formation, northern Minnesota. *Econ Geol* 68:1110–1125.

Picardal FW, Zaybak Z, Chakraborty A, Schieber J, Szewzyk U. 2011. Microaerophilic, Fe (II)-dependent growth and Fe (II) oxidation by a *Dechlorospirillum* species. *FEMS Microbiol Lett* 319:51–57.

Piepenbrock A, Behrens S, Kappler A. 2014. Comparison of humic substance- and Fe(III)-reducing microbial communities in anoxic aquifers. *Geomicrobiol J* 31:917–928.

Pierson BK, Parenteau MN. 2000. Phototrophs in high iron microbial mats: Microstructure of mats in iron-depositing hot springs. *FEMS Microbiol Ecol* 32:181–196.

Pierson BK, Parenteau MN, Griffin BM. 1999. Phototrophs in high-iron concentration microbial mats: Physiological ecology of phototrophs in an iron-depositing hot spring. *Appl Environ Microbiol* 65:5474–5483.

Pirbadian S, Barchinger SE, Leung KM, Byun HS, Jangir Y, Bouhenni RA, Reed SB et al. 2014. *Shewanella oneidensis* MR-1 nanowires are outer membrane and periplasmic extensions of the extracellular electron transport components. *Proc Natl Acad Sci USA* 111:12883–12888.

Planavsky N, Rouxel O, Bekker A, Shapiro R, Fralick P, Knudsen A. 2009. Iron-oxidizing microbial ecosystems thrived in late Paleoproterozoic redox-stratified oceans. *Earth Planet Sci Lett* 286:230–242.

Pollack JR, Neilands JB. 1970. Enterobactin, an iron transport compound. *Biochem Biophys Res Commun* 38:989–992.

Popp F, Armitage JP, Schüler D. 2014. Polarity of bacterial magnetotaxis is controlled by aerotaxis through a common sensory pathway. *Nat Commun* 5:5398.

Posth N, Canfield DE, Kappler A. 2014. Biogenic Fe(III) minerals: From formation to diagenesis and preservation in the rock record. *Earth Sci Rev* 135:103–121.

Posth NR, Hegler F, Konhauser KO, Kappler A. 2008. Alternating Si and Fe deposition caused by temperature fluctuations in Precambrian oceans. *Nat Geosci* 1:703–708.

Posth NR, Huelin S, Konhauser KO, Kappler A. 2010. Size, density and composition of cell-mineral aggregates formed during anoxygenic phototrophic Fe(II) oxidation: Impact on modern and ancient environments. *Geochim Cosmochim Acta* 74:3476–3493.

Posth NR, Kohler I, Swanner ED, Schroder C, Wellmann E, Binder B, Konhauser KO, Neumann U, Berthold C, Nowak M, Kappler A. 2013. Simulating Precambrian banded iron formation diagenesis. *Chem Geol* 362:66–73.

Pretorius IM, Rawlings DE, O'Neill EF, Jones WA, Kirby R, Woods DR. 1987. Nucleotide sequence of the gene encoding the nitrogenase iron protein of *Thiobacillus ferrooxidans*. *J Bacteriol* 169:367–370.

Pretorius IM, Rawlings DE, Woods DR. 1986. Identification and cloning of *Thiobacillus ferrooxidans* structural nif-genes in *Escherichia coli*. *Gene* 45:59–65.

Pringsheim EG. 1949. The filamentous bacteria *Sphaerotilus*, *Leptothrix* and *Cladothrix* and their relation to iron and manganese. *Philos Trans Roy Soc Lond Ser B: Biol Sci* 233:453–482.

Pronk JT, de Bruyn JC, Bos P, Kuenen JG. 1992. Anaerobic growth of *Thiobacillus ferrooxidans*. *Appl Environ Microbiol* 58:2227–2230.

Pronk JT, Johnson DB. 1992. Oxidation and reduction of iron by acidophilic bacteria. *Geomicrobiol J* 10:149–171.

Pronk JT, Liem K, Bos P, Kuenen JG. 1991b. Energy transduction by anaerobic ferric iron respiration in *Thiobacillus ferrooxidans*. *Appl Environ Microbiol* 57:2063–2068.

Pronk JT, Meijer WM, Hazeu W, van Dijken JP, Bos P, Kuenen JG. 1991a. Growth of *Thiobacillus ferrooxidans* on formic acid. *Appl Environ Microbiol* 57:2057–2062.

Puchelt H, Schock HH, Schroll E, Hanert H. 1973. Rezente marine Eisenerze aus Santorin, Griechenland. *Geol Rundschau* 62:786–812.

Qu B, Fan B, Zhu SK, Zheng YL. 2014. Anaerobic ammonium oxidation with an anode as the electron acceptor. *Environ Microbiol Rep* 6:100–105.

Quatrini R, Appia-Ayme C, Denis Y, Jedlicki E, Holmes DS, Bonnefoy V. 2009. Extending the models for iron and sulfur oxidation in the extreme acidophile *Acidithiobacillus ferrooxidans*. *BMC Genom* 10:394.

Rancourt DG, Thibault P-J, Mavrocordatos D, Lamarche G. 2005. Hydrous ferric oxide precipitation in the presence of nonmetabolizing bacteria: Constraints on the mechanism of a biotic effect. *Geochim Cosmochim Acta* 69:553–577.

Rankama K, Sahama TG. 1950. *Geochemistry*. Chicago, IL: University of Chicago Press, pp. 657–676.

Rawlings DE, Kusano T. 1994. Molecular genetics of *Thiobacillus ferrooxidans*. *Microbiol Rev* 58:39–55.

Reguera G, McCarthy KD, Mehta T, Nicoll JS, Tuominen MR, Lovley DR. 2005. Extracelluar electron transfer via microbial nanowires. *Nature (London)* 435:1098–1101.

Reguera G, Nevin KP, Nicoll JS, Covalla SF, Woodward TL, Lovley DR. 2006. Biofilm and nanowire production lead to increased current in microbial fuel cells. *Appl Environ Microbiol* 72:7345–7348.

Reid RT, Live DH, Faulkner DJ, Butler A. 1993. A siderophore from a marine bacterium with an exceptional ferric ion affinity constant. *Nature (London)* 366:455–458.

Remsen C, Lundgren DG. 1966. Electron microscopy of the cell envelope of *Ferrobacillus ferrooxidans* prepared by freeze-etching and chemical fixation techniques. *J Bacteriol* 92:1765–1771.

Richardson DJ, Butt JN, Fredrickson JK, Zachara JM, Shi L, Edwards MJ, White G, Baiden N, Gates AJ, Marritt SJ, Clarke TA. 2012. The porin-cytochrome' model for microbe-to-mineral electron transfer. *Mol Microbiol* 85:201–212.

Riedinger N, Formolo MJ, Lyons TW, Henkel S, Beck A, Kasten S. 2014. An inorganic geochemical argument for coupled anaerobic oxidation of methane and iron reduction in marine sediments. *Geobiology* 12:172–181.

Roberts JL. 1947. Reduction of ferric hydroxide by strains of *Bacillus polymyxa*. *Soil Sci* 63:135–140.

Roden EE. 2003. Fe(III) oxide reactivity toward biological versus chemical reduction. *Environ Sci Technol* 37:1319–1324.

Roden EE. 2006. Geochemical and microbiological controls on dissimilatory iron reduction. *CR Geosci* 338:456–467.

Roden EE, Kappler A, Bauer I, Jiang J, Paul A, Stoesser R, Konishi H, Xu HF. 2010. Extracellular electron transfer through microbial reduction of solid-phase humic substances. *Nat Geosci* 3:417–421.

Roden EE, Lovley DR. 1993. Dissimilatory Fe(III) reduction by the marine microorganism *Desulfuromonas acetoxidans*. *Geomicrobiol J* 11:734–742.

Roden EE, McBeth JM, Blöthe M, Percak-Dennett EM, Fleming EJ, Holyoke RR, Luther GW, Emerson D, Schieber J. 2012. The microbial ferrous wheel in a neutral pH groundwater seep. *Front Microbiol* 3:1–18.

Roden EE, Sobolev D, Glazer B, Luther GW. 2004. Potential for microscale bacterial Fe redox cycling at the aerobic-anaerobic interface. *Geomicrobiol J* 21:379–391.

Roden EE, Urrutia MM. 1999. Ferrous iron removal promotes microbial reduction of crystalline iron(III) oxides. *Environ Sci Technol* 33:1847–1853.

Roden EE, Urrutia MM, Mann CJ. 2000. Bacterial reductive dissolution of crystalline Fe(III) oxide in continuous-flow column reactors. *Appl Environ Microbiol* 66:1062–1065.

Roden EE, Zachara JM. 1996. Microbial reduction of crystalline iron(III) oxides: Influence of oxide surface area and potential for cell growth. *Environ Sci Technol* 30:1618–1628.

Rosenberg H, Young IG. 1974. Iron transport in the enteric bacteria. In: Neilands JB, ed. *Microbial Iron Metabolism: A Comprehensive Treatise.* New York: Academic Press, pp. 67–82.

Royer RA, Burgos WD, Fisher AS, Unz RF, Dempsey BA. 2002. Enhancement of biological reduction of hematite by electron shuttling and Fe(II) complexation. *Environ Sci Technol* 36:1939–1946.

Saffarini DA, Schultz R, Beliaev A. 2003. Involvement of cyclic AMP (CAMP) and cAMP receptor protein in anaerobic respiration of *Shewanella oneidensis*. *J Bacteriol* 185:3668–3671.

Saini G, Chan CS. 2013. Near-neutral surface charge and hydrophilicity prevent mineral encrustation of Fe-oxidizing micro-organisms. *Geobiology* 11:191–200.

Saraiva IH, Bird L, Calcada E, Salgueiro CA, Newman DK, Louro RO. 2011. Molecular characterization of photosynthetic iron oxidation. *Eur Biophys J* 40:185.

Saraiva IH, Newman DK, Louro RO. 2012. Functional characterization of the FoxE iron oxidoreductase from the photoferrotroph *Rhodobacter ferrooxidans* SW2. *J Biol Chem* 287:25541–25548.

Schädler S, Burkhardt C, Hegler F, Straub KL, Miot J, Benzerara K, Kappler A. 2009. Formation of cell-iron-mineral aggregates by phototrophic and nitrate-reducing anaerobic Fe(II)-oxidizing bacteria. *Geomicrobiol J* 26:93–103.

Scheffel A, Gruska M, Faivre D, Linaroudis A, Plitzko JM, Schüler D. 2006. An acidic protein aligns magnetosomes along a filamentous structure in magnetotactic bacteria. *Nature* 440(7080):110–114.

Schmidt C, Behrens S, Kappler A. 2010. Ecosystem functioning from a geomicrobiological perspective—A conceptual framework for biogeochemical iron cycling. *Environ Chem* 7:399–405.

Schröder I, Johnson E, de Vries S. 2003. Microbial ferric iron reductases. *FEMS Microbiol Rev* 27:427–447.

Scott DT, McKnight DM, Blunt-Harris EL, Kolesar SE, Lovley DR. 1998. Quinone moieties act as electron acceptors in the reduction of humic substances by humics-reducing microorganisms. *Environ Sci Technol* 32:2984–2989.

Segarra KEA, Comerford C, Slaughter J, Joye SB. 2013. Impact of electron acceptor availability on the anaerobic oxidation of methane in coastal freshwater and brackish wetland sediments. *Geochim Cosmochim Acta* 115:15−30.

Segerer A, Neuner A, Kristjansson JK, Stetter KO. 1986. *Acidianus infernus*, gen. nov., spec. nov.: Facultatively aerobic, extremely acidophilic thermophilic sulfur-metabolizing archaeabacteria. *Int J Syst Bacteriol* 36:559−564.

Selkov E, Overbeek R, Kogan Y, Chu L, Vonstein V, Holmes D, Silver S, Haselkorn R, Fonstein M. 2000. Functional analysis of gapped microbial genomes: Amino acid metabolism of *Thiobacillus ferrooxidans*. *Proc Natl Acad Sci USA* 97:3509−3514.

Shafia F, Brinson KR, Heinzman MW, Brady JM. 1972. Transition of chemolithotroph *Ferrobacillus ferrooxidans* to obligate organotrophy and metabolic capabilities of glucose-grown cells. *J Bacteriol* 111:56−65.

Shafia F, Wilkinson RF. 1969. Growth of *Ferrobacillus ferrooxidans* on organic matter. *J Bacteriol* 97:256−260.

Shelobolina ES, Konishi H, Xu H, Benzine J, Xiong MY, Wu T, Blöthe M, Roden EE. 2012. Isolation of phyllosilicate-iron redox cycling microorganisms from an illite-smectite rich hydromorphic soil. *Front Microbiol* 3:1−10.

Shelobolina ES, Vanpraagh CG, Lovley DR. 2003. Use of ferric and ferrous iron containing minerals for respiration by *Desulfitobacterium frappieri*. *Geomicrobiol J* 20:143−156.

Shi L, Squier TC, Zachara JM, Fredrickson JK. 2007. Respiration of metal (hydr)oxides by *Shewanella* and *Geobacter*: A key role for multihaem c-type cytochromes. *Mol Microbiol* 65:12−20.

Short KA, Blakemore RP. 1986. Iron respiration-driven proton translocation in aerobic bacteria. *J Bacteriol* 167:729−731.

Shrestha J, Rich JJ, Ehrenfeld JG, Jaffe PR. 2009. Oxidation of ammonium to nitrite under iron-reducing conditions in wetland soils laboratory, field demonstrations, and push-pull rate determination. *Soil Sci* 174:156−164.

Silverman MP, Lundgren DG. 1959a. Studies on the chemoautotrophic iron bacterium *Ferrobacillus ferrooxidans*. I. An improved medium and a harvesting procedure for securing high cell yields. *J Bacteriol* 77:642−647.

Simmons SL, Bazylinski DA, Edwards KJ. 2006. South-seeking magnetotactic bacteria in the Northern Hemisphere. *Science* 311:371−374.

Simmons SL, Sievert SM, Frankel RB, Bazylinski DA, Edwards KJ. 2004. Spatiotemporal distribution of marine magnetotactic bacteria in a seasonally stratified coastal salt pond. *Appl Environ Microbiol* 70:6230−6239.

Singer E, Emerson D, Webb EA, Barco RA, Kuenen JG, Nelson WC, Chan CS et al. 2011. *Mariprofundus ferrooxydans* PV-1 the first genome of a marine Fe(II) oxidizing Zetaproteobacterium. *PLoS One* 6:9.

Siponen MI, Legrand P, Widdrat M, Jones SR, Zhang WJ, Chang MCY, Faivre D, Arnoux P, Pignol1 D. 2013. Structural insight into magnetochrome-mediated magnetite biomineralization. *Nature* 502:681−684.

Snider RM, Strycharz-Glaven SM, Tsoi SD, Erickson JS, Tender LM. 2012. Long-range electron transport in *Geobacter sulfurreducens* biofilms is redox gradient-driven. *Proc Natl Acad Sci USA* 109:15467−15472.

Sobolev D, Roden EE. 2002. Evidence for rapid microscale bacterial redox cycling of iron in circumneutral environments. *Antonie v Leeuwenhoek Int J Gen Mol Microbiol* 81:587−597.

Sørensen J, Jørgensen BB. 1987. Early diagenesis in sediments from Danish coastal waters: Microbial activity and Mn-Fe-S geochemistry. *Geochim Cosmochim Acta* 51:1583−1590.

Starkey RL, Halvorson HO. 1927. Studies on the transformations of iron in nature. II. Concerning the importance of microorganisms in the solution and precipitation of iron. *Soil Sci* 24:381−402.

Starr MP, Stolp H, Trüper HG, Balows A, Schlegel HG, eds. 1981. *The Prokaryotes: A Handbook on Habitats, Isolation, and Identification of Bacteria*. Berlin, Germany: Springer-Verlag.

Stevens CJ, Dugan PR, Tuovinen OH. 1986. Acetylene reduction (nitrogen fixation) by *Thiobacillus ferrooxidans*. *Biotechnol Appl Biochem* 8:351−359.

Straub KL, Benz M, Schink B, Widdel F. 1996. Anaerobic, nitrate-dependent microbial oxidation of ferrous iron. *Appl Environ Microbiol* 62:1458−1460.

Straub KL, Rainey FA, Widdel F. 1999. *Rhodovulum iodosum* sp. nov. and *Rhodovulum robiginosum* sp. nov., two new marine phototrophic ferrous-iron-oxidizing purple bacteria. *Int J Syst Bacteriol* 49:729−735.

Straub KL, Schönhuber WA, Buchholz-Cleven BEE, Schink B. 2004. Diversity of ferrous iron-oxidizing, nitrate-reducing bacteria and their involvement in oxygen-dependent iron cycling. *Geomicrobiol J* 21:371−378.

Stucki JW, Golden DC, Roth CB. 1984. Effects of reduction and reoxidation of structural iron(III) in smectites. *Clays Clay Min* 32:350−356.

Stucki JW, Komadel P, Wilkinson HT. 1987. Microbial reduction of structural iron(III) in smectites. *Soil Sci Soc Am J* 51:1663−1665.

Stucki JW, Roth CB. 1977. Oxidation-reduction mechanism for structural iron in nontronite. *Soil Sci Soc Am J* 41:808–814.

Stucki JW, Wu J, Gan H, Komadel P, Banin A. 2000. Effects of iron oxidation state and organic cations on dioctahedral smectite hydration. *Clays Clay Min* 48:290–298.

Sugio T, Anzai Y, Tano T, Imai K. 1981. Isolation and some properties of an obligate and a facultative iron-oxidizing bacterium. *Agric Biol Chem* 45:1141–1151.

Sugio T, Katagiri T, Moriyama M, Zhen YL, Inagaki K, Tano T. 1988. Existence of a new type of sulfite oxidase which utilizes ferric ions as an electron acceptor in *Thiobacillus ferrooxidans*. *Appl Environ Microbiol* 54:153–157.

Sugio T, White KJ, Shute E, Choate D, Blake RC II. 1992. Existence of a hydrogen sulfide:ferric ion oxidoreductase in iron-oxidizing bacteria. *Appl Environ Microbiol* 58:431–433.

Suzuki T, Hashimoto H, Ishihara H, Matsumoto N, Kunoh H, Takada J. 2012. Two types of morphologically distinct fibers comprising *Gallionella ferruginea* twisted stalks. *Microbes Environ* 27:338–341.

Suzuki T, Hashimoto H, Matsumoto N, Furutani M, Kunoh H, Takada J. 2011. Nanometer-scale visualization and structural analysis of the inorganic/organic hybrid structure of *Gallionella ferruginea* twisted stalks. *Appl Environ Microbiol* 77:2877–2881.

Tabita R, Lundgren DG. 1971a. Utilization of glucose and the effect of organic compounds on the chemolithotroph *Thiobacillus ferrooxidans*. *J Bacteriol* 108:328–333.

Tait GH. 1975. The identification and biosynthesis of siderochromes formed by *Micrococcus denitrificans*. *Biochem J* 146:191–204.

Takai Y, Takamura T. 1966. The mechanism of reduction in water-logged paddy soil. *Folia Microbiol* 11:304–315.

Temple KL, Colmer AR. 1951. The autotrophic oxidation of iron by a new bacterium: *Thiobacillus ferrooxidans*. *J Bacteriol* 62:605–611.

Thamdrup B, Finster K, Wuergler Hansen J, Bak F. 1993. Bacterial disproportionation of elemental sulfur coupled to chemical reduction of iron or manganese. *Appl Environ Microbiol* 99:101–108.

Tikhonova GV, Lisenkova LL, Doman NG, Skulachev VP. 1967. Electron transport pathways in *Thiobacillus ferrooxidans*. *Mikrobiologiya* 32:725–734 (Engl transl, pp. 599–606).

Toner BM, Berquo TS, Michel FM, Sorensen JV, Templeton AS, Edwards KJ. 2012. Mineralogy of iron microbial mats from Loihi Seamount. *Front Microbiol* 3:1–18.

Troshanov EP. 1968. Iron- and manganese-reducing microorganisms in ore-containing lakes of the Karelian isthmus. *Mikrobiologiya* 37:934–940 (Engl transl, pp. 786–791).

Troshanov EP. 1969. Conditions affecting the reduction of iron and manganese by bacteria in the ore-bearing lakes of the Karelian isthmus. *Mikrobiologiya* 38:634–643 (Engl transl, pp. 528–535).

Tuovinen OH, Nicholas DJD. 1977. Transition of *Thiobacillus ferrooxidans* KG-4 from heterotrophic growth on glucose to autotrophic growth on ferrous-iron. *Arch Microbiol* 114:193–195.

Unz RF, Lundgren DG. 1961. A comparative nutritional study of three chemoautotrophic bacteria: *Ferrobacillus ferrooxidans*, *Thiobacillus ferrooxidans* and *Thiobacillus thiooxidans*. *Soil Sci* 92:302–313.

Valdes J, Pedroso I, Quatrini R, Dodson RJ, Tettelin H, Blake R, Eisen JA, Holmes DS. 2008. *Acidithiobacillus ferrooxidans* metabolism: From genome sequence to industrial applications. *BMC Genom* 9:597.

Vartanian SE, Cowart RE. 1999. Extracellular iron reductases: Identification of a new class of enzymes by siderophore-producing microorganisms. *Arch Biochem Biophys* 364:75–82.

Venkateswaran K, Moser DP, Dollhopf ME, Lies DP, Saffarini DA, MacGregor BJ, Ringelberg DB et al. 1999. Polyphasic taxonomy of the genus *Shewanella* and description of *Shewanella oneidensis* sp. nov. *Int J Syst Bacteriol* 49:705–724.

Vollrath S, Behrends T, Koch CB, Cappellen PV. 2013. Effects of temperature on rates and mineral products of microbial Fe(II) oxidation by *Leptothrix cholodnii* at microaerobic conditions. *Geochim Cosmochim Acta* 108:107–124.

Vollrath S, Behrends T, Van Cappellen P. 2012. Oxygen dependency of neutrophilic Fe(II) oxidation by *Leptothrix* differs from abiotic reaction. *Geomicrobiol J* 29:550–560.

von Canstein H, Ogawa J, Shimizu S, Lloyd JR. 2008. Secretion of flavins by *Shewanella* species and their role in extracellular electron transfer. *Appl Environ Microbiol* 74:615–623.

Walker JCG. 1987. Was the Archean biosphere upside down? *Nature* 329:710–712.

Walker JCG, Klein C, Schidlowski M, Schopf JW, Stevenson DJ, Walter MR. 1983. Environmental evolution of the Archean-Early Proterozoic Earth. In: Schopf JW, ed. *Earth's Earliest Biosphere: Its Origin and Evolution*. Princeton, NJ: Princeton University Press, pp. 260–290.

Walsh F, Mitchell R. 1972. An acid-tolerant iron-oxidizing *Metallogenium*. *J Gen Microbiol* 72:369–376.

Wandersman C, Delepelaire P. 2004. Bacterial iron sources from siderophores to hemophores. *Annu Rev Microbiol* 88:611–647.

Wang Y, Newman DK. 2008. Redox reactions of phenazine antibiotics with ferric (hydr)oxides and molecular oxygen. *Environ Sci Technol* 42:2380–2386.

Weber KA, Achenbach LA, Coates JD. 2006a. Microorganisms pumping iron: Anaerobic microbial iron oxidation and reduction. *Nat Rev Microbiol* 4:L752–764.

Weber KA, Hedrick DB, Peacock AD, Thrash JC, White DC, Achenbach LA, Coates JD. 2009. Physiological and taxonomic description of the novel autotrophic, metal oxidizing bacterium, *Pseudogulbenkiania* sp. strain 2002. *Appl Microbiol Biotechnol* 83:555–565.

Weber KA, Pollock J, Cole KA, O'Connor SM, Achenbach LA, Coates JD. 2006b. Anaerobic nitrate-dependent iron(II) bio-oxidation by a novel lithoautotrophic Betaproteobacterium, strain 2002. *Appl Environ Microbiol* 72:686–694.

Weber KA, Urrutia MM, Churchill PF, Kukkadapu RK, Roden EE. 2006c. Anaerobic redox cycling of iron by freshwater sediment microorganisms. *Environ Microbiol* 8:100–113.

Weiss JV, Emerson D, Backer SM, Megonigal JP. 2003. Enumeration of Fe(II)-oxidizing and Fe(III)-reducing bacteria in the root zone of wetland plants: Implications for a rhizosphere iron cycle. *Biogeochemistry* 64:77–96.

Weiss JV, Rentz JA, Plaia T, Neubauer SC, Merrill-Floyd M, Lilburn T, Bradburne C, Megonigal JP, Emerson D. 2007. Characterization of neutrophilic Fe(II)-oxidizing bacteria isolated from the rhizosphere of wetland plants and description of *Ferritrophicum radicicola* gen. nov. sp. nov., and *Sideroxydans paludicola* sp. nov. *Geomicrobiol J* 24:559–570.

Wichlacz PL, Unz RF, Langworthy TA. 1986. *Acidiphilium angustum* sp. nov., *Acidiphilium facilis* sp. nov., and *Acidiphilium rubrum* sp. nov.: Acidophilic heterotrophic bacteria isolated from acid coal mine drainage. *Int J Syst Bacteriol* 36:197–201.

Widdel F, Schnell S, Heising S, Ehrenreich A, Assmus B, Schink B. 1993. Ferrous oxidation by anoxygenic phototrophic bacteria. *Nature (London)* 362:834–836.

Winogradsky S. 1888. Über Eisenbakterien. *Bot Ztg* 46:261–276.

Wu W, Swanner ED, Hao L, Zeitvogel F, Obst M, Pan Y, Kappler A. 2014. Characterization of the physiology and cell–mineral interactions of the marine anoxygenic phototrophic Fe (II) oxidizer *Rhodovulum iodosum*–implications for Precambrian Fe (II) oxidation. *FEMS Microbiol Ecol* 88:503–515.

Xiong Y, Shi L, Chen B, Mayer MJ, Lower BH, Londer Y, Saumyaditya B, Hochella MF, Fredrickson JK, Squier TC. 2006. High-affinity binding and direct electron transfer to solid metals by the *Shewanella oneidensis* MR-1 outer membrane c-type cytochrome OmcA. *J Am Chem Soc* 128:13978–13979.

Yang WH, Weber KA, Silver WL. 2012. Nitrogen loss from soil through anaerobic ammonium oxidation coupled to iron reduction. *Nat Geosci* 5:538–541.

Yarzábal A, Appia-Ayme C, Ratouchniak J, Bonnefoy V. 2004. Regulation of the expression of the *Acidithiobacillus ferrooxidans* rus operon encoding two cytochromes c, a cytochrome oxidase and rusticyanin. *Microbiology (Reading, U.K.)* 150:2113–2123.

Yarzábal A, Brasseur G, Ratouchniak J, Lund K, Lemesle-Meunier D, DeMoss JA, Bonnefoy V. 2002a. The high-molecular-weight cytochrome c Cyc2 of *Acidithiobacillus ferrooxidans* is an outer membrane protein. *J Bacteriol* 184:313–317.

Yarzábal A, Brasseur G, Bonnefoy V. 2002b. Cytochromes c of *Acidithiobacillus ferrooxidans*. *FEMS Microbiol Lett* 209:189–195.

Zachara JM, Kukkadapu RK, Fredrickson JK, Gorby YA, Smith SC. 2002. Biomineralization of poorly crystalline Fe(III) oxides by dissimilatory metal reducing bacteria (DMRB). *Geomicrobiol J* 19:179–207.

Zavarzin GA. 1972. Heterotrophic satellite of *Thiobacillus ferrooxidans*. *Mikrobiologiya* 41:369–370 (Engl transl, pp. 323–324).

Zhou H, Zhang R, Hu P, Zeng W, Xie Y, Wu C, Qiu G. 2008. Isolation and characterization of *Ferroplasma thermophilum* sp. nov., a novel extremely acidophilic, moderately thermophilic archaeon and its role in bioleaching of chalcopyrite. *J Appl Microbiol* 105:591–601.

CHAPTER EIGHTEEN

Geomicrobiology of Manganese

Colleen M. Hansel and Deric R. Learman

CONTENTS

18.1   Important Properties of Manganese / 403
18.2   Chemistry of Manganese / 404
    18.2.1   Physicochemical Properties of Mn / 404
    18.2.2   Thermodynamics and Kinetics / 404
        18.2.2.1   Oxidation / 404
        18.2.2.2   Reduction / 407
18.3   Microbial Oxidation of Mn / 408
    18.3.1   Taxonomic and Environmental Diversity / 408
    18.3.2   Physiological Reasons for Oxidation / 409
    18.3.3   Mechanisms of Biological Oxidation / 411
        18.3.3.1   Direct Enzymatic: MCOs and Laccases / 412
        18.3.3.2   Direct Enzymatic: Peroxidases / 414
        18.3.3.3   Indirect Biological Mn(II) Oxidation / 415
18.4   Microbial Reduction of Mn / 416
    18.4.1   Taxonomic and Environmental Diversity / 416
    18.4.2   Physiological Reasons for Reduction / 416
    18.4.3   Mechanisms of Biological Reduction / 416
        18.4.3.1   Enzymatic Reduction via C-Type Cytochromes / 417
        18.4.3.2   Indirect Biological Reduction / 418
18.5   Products of Microbial Mn Cycling / 420
    18.5.1   Mn(III) Complexes / 420
    18.5.2   Mn Mineral Phases / 420
    18.5.3   Association of Mn Oxides with Microbes and Microbial Exudates / 423
18.6   Manganese in the Earth's Crust / 425
18.7   Linking Microbial Activity to Manganese Deposits / 425
18.8   Terrestrial Manganese Oxide Deposits / 427
    18.8.1   Soils / 427
    18.8.2   Desert Varnish / 427
    18.8.3   Caves / 428
    18.8.4   Ores / 428
18.9   Freshwater Manganese Oxide Deposits / 428
    18.9.1   Rivers, Creeks, and Ponds / 428
    18.9.2   Ferromanganese Nodules / 429
    18.9.3   Engineered Systems / 429

18.10 Marine Manganese Oxide Deposits / 430
    18.10.1 Water Column/Coastal Waters / 430
    18.10.2 Hydrothermal Systems / 431
    18.10.3 Deep-Sea Concretions, Crusts, and Nodules / 431
18.11 Concluding Comments / 433
References / 433

It is difficult to discuss the manganese (Mn) cycle in isolation. Manganese is inherently and indiscriminately interactive due to the unique geochemical characteristics of the various Mn species. The redox cycling of Mn involves a Mn(II) antioxidant and both a Mn(III) reactive intermediate and Mn(III,IV) oxide mineral, two of the strongest oxidants on our planet. Further, the immense reactivity of Mn oxides is responsible for their role and utility as environmental sponges (Figure 18.1). Indeed, the cycling of Mn comes to bear on nearly all other elemental cycles (spanning nutrients to radionuclides), touches every possible ecological niche (from the soils to the deep ocean and even below), and involves all domains of life (from bacteria to fungi). It must be inferred from the widespread occurrence of manganese-oxidizing and manganese-reducing microorganisms in terrestrial, freshwater, and marine systems that they play an essential role in the cycling of manganese, and therefore, by

extension, environmental biogeochemistry in general. Though in no way comprehensive (we refer the reader to many informative reviews— e.g., Nealson and Saffarini, 1994; Tebo et al., 1997; Post, 1999; Morgan, 2000; Tebo et al., 2004; Canfield et al., 2005; Lovley, 2006; Gadd, 2007; Spiro et al., 2010), this chapter attempts to tell the story of Mn, a story that spans the biological, chemical, and geological sciences.

It's a story of redemption, a coming of age if you will. Despite the historical perception that Mn concentrations are too low to significantly impact global cycles, particularly in terrestrial systems (in comparison to its neighbor Fe), recent discoveries of high Mn concentrations on the "red" planet (Mars) (Lanza et al., 2014), of Mn-dependent leaf litter decomposition on the forest floor (a major source of atmospheric CO_2) (Berg et al., 2007), of the importance of Mn in scrubbing the ocean of toxic oxygen radicals (Wuttig et al., 2013), and of evidence of ancient Mn oxide deposits that

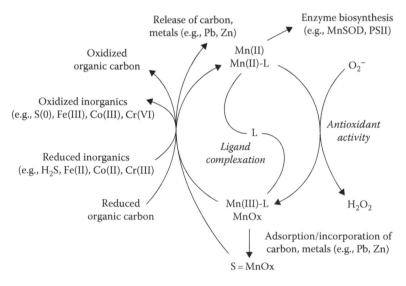

Figure 18.1. Simplified schematic of the impact of Mn on other biogeochemical processes, including Mn(III)- and/or Mn oxide–mediated oxidation of (in)organics, Mn oxide adsorption/incorporation of metals and carbon, scavenging of the reactive oxygen species superoxide (O_2^-) by Mn(II), and Mn(II) incorporation into essential microbial enzymes such as Mn superoxide dismutase and photosystem II (PSII). (Reprinted with permission from Tebo BM and He LM, Microbially mediated oxidative precipitation reactions, in Sparks DL and Grundl TJ, eds., *Mineral-Water Interfacial Reactions Kinetics and Mechanisms*, Washington, DC, pp. 393–414. Copyright 1999 American Chemical Society.)

predate the rise of O_2 with a potential link to the evolution of photosynthesis (Johnson et al., 2013) have brought the Mn cycle to the forefront in ancient and modern biogeochemistry. This story remains incomplete. Despite decades of study, we still have much to learn about the players, mechanisms, and consequences of the Mn cycle, and new and exciting discoveries are being made at a rapid rate. What is clear is the dynamic and ever-inspiring complexity of reactions involving Mn and the acknowledgment that microbial processes are the catalytic engine driving the Mn cycle.

18.1 IMPORTANT PROPERTIES OF MANGANESE

The transition metal Mn is an essential trace nutrient for all forms of life, serving as a cofactor in a broad range of enzymes in eukaryotes, bacteria, and archaea that are important in antioxidant and metabolic functions. For instance, the Mn-containing superoxide dismutase is essential for most oxygen-based life by converting superoxide radicals to hydrogen peroxide during ATP synthesis (Leach and Harris, 1997). It is also required in oxygenic photosynthesis, where it functions in the production of oxygen from water by photosystem II (e.g., Klimov, 1984). Mn also has antioxidant properties, whereby it scavenges toxic reactive oxygen species (ROS) such as superoxide (O_2^-) and hydrogen peroxide (H_2O_2). In fact, the ability for Mn to effectively scavenge ROS has been linked to Mn-facilitated ionizing radiation resistance in the bacterium *Deinococcus radiodurans*, where this organism has been shown to accumulate exceedingly high (mM) levels of intracellular Mn (Daly et al., 2004, 2007). In humans, however, excessive levels of Mn may be toxic and lead to neurological disorders, including Parkinson's disease (Keen et al., 1999).

Microbes may respire Mn(III, IV) when coupled to oxidation of organic carbon or hydrogen. The organic carbon may be completely degraded to CO_2 either by a single species of organism (Lovley et al., 1993a) or the successive action of two or more species (see, e.g., Lovley et al., 1989a). Mn(III, IV) oxide respiration can therefore be viewed environmentally as a form of mineralization of organic compounds (Ehrlich, 1993a,b; Nealson and Saffarini, 1994) and thereby have an impact on the carbon cycle. Indeed, Thamdrup et al. (2000) found that in shelf sediments of the Black Sea, dissimilatory

Mn reduction was the most important means of organic carbon mineralization in the surface layer down to ~1 cm depth, whereas dissimilatory sulfate reduction was the main carbon mineralization process below this depth. Similarly, microbial reduction of Mn oxides was found to be the only important anaerobic carbon oxidation process within coastal sediments enriched in manganese (>3%) (Canfield et al., 1993a,b).

Oxidized Mn species may also directly influence the cycling of carbon. Mn(III) and Mn(IV) oxides are among the strongest natural oxidants on our planet possessing the rare ability to oxidize a wide range of organic compounds and contaminants (e.g., tetracycline) (Rubert and Pedersen, 2006). In fact, Mn(III) and Mn oxides have the capacity to degrade lignin and complex organics (e.g., humic acids) (Sunda and Kieber, 1994; Banerjee and Nesbit, 2001). Further, a highly evolved and widespread strategy employed by fungi ("wood-rot fungi") to degrade lignin involves the regulated production of highly oxidizing Mn(III) chelates (Perez and Jeffries, 1992; Schlosser and Höfer, 2002). In addition to carbon, Mn oxides are one of the only known naturally occurring oxidants of chromium (Cr), oxidizing the benign form Cr(III) to the toxic and carcinogenic form, Cr(VI) (Bartlett and James, 1979; Oze et al., 2007). Further, both Mn oxides and soluble Mn(III) complexes oxidize the low solubility uranium (U) mineral uraninite (UO_2) to the soluble and toxic U(VI) species uranyl (Wang et al., 2012, 2014). In this way, Mn(II)-oxidizing microbes may indirectly influence the cycling of metals and radionuclides. For instance, Mn oxides formed by the spores of *Bacillus* SG-1 oxidize both cobalt (Co) and Cr, taking Co(II) to Co(III) (Murray et al., 2007) and Cr(III) to Cr(VI) (Murray and Tebo, 2007). These biogenic Mn oxides even oxidize solid-phase U(IV) (as UO_2) to soluble U(VI) (Chinni et al., 2008).

Owing to their high surface area and reactivity, Mn oxides are also extremely powerful sorbents having demonstrated large capacities for metal adsorption, including lead (Pb), nickel (Ni), Co, copper (Cu), and zinc (Zn), among others (Jenne, 1968; Loganathan and Burau, 1973; Nelson et al., 1999a,b; Kay et al., 2001; Matocha et al., 2001; O'Reilly and Hochella, 2003; Toner et al., 2006; Takahashi et al., 2007; Peña et al., 2010). Metal ions may also be structurally incorporated within the Mn oxide structure, either isomorphically by replacing Mn(III) or Mn(IV) ions or within interstitial vacancies or tunnels (Manceau et al., 1992; Kay et al., 2001; O'Reilly and Hochella, 2003). In fact, Mn(III, IV) oxides have been coined the "scavengers of the sea" due to their immense adsorptive capacity and subsequent disproportionately high trace metal content relative to the surrounding seawater (Goldberg, 1954).

Due to their physicochemical properties, Mn oxides can have important roles in water treatment, environmental remediation, and biotechnology. For instance, owing to their nanosheet architecture (Villalobos et al., 2003; Villalobos et al., 2006), Mn oxides have emerged as a potential new material in a wide range of technological applications, including energy storage, electrochromism, and catalysis (Sakai et al., 2005; Izawa et al., 2006). Their strong oxidative and sorptive capacities have made Mn oxides an attractive substrate to target in remediation system design. For instance, stimulating the formation of Mn oxides in passive treatment systems has become a common worldwide approach to treating contaminated waters, such as metal-laden coal mine drainage (Cravotta and Trahan, 1999; Vali and Riley, 2000).

Lastly, Mn(III, VI) oxides are receiving increasing attention as a particularly useful paleoredox indicator since oxidation of Mn(II) requires oxidants with a high redox potential. The distribution of Mn oxide deposits is variable throughout the geological record. Rocks and sediments enriched in Mn oxides have been observed on Mars and in early Earth sedimentary strata (Roy, 2006; Maynard, 2010; Lanza et al., 2014). Their distribution in the rock record provides a window to the past and predictions about the presence of oxidants, including the rise of oxygen.

18.2 CHEMISTRY OF MANGANESE

18.2.1 Physicochemical Properties of Mn

Manganese is a transition element that exists in oxidation states ranging from 0 to +7. In nature, however, only the +2, +3, and +4 oxidation states are commonly found. Of the three naturally occurring oxidation states, only Mn in the +2 oxidation state can occur as a free ion in aqueous solution, where it exists primarily as the hydrolyzed ion $Mn(H_2O)_6^{2+}$. Mn(II) may complex with a number of organic (e.g., citrate) and inorganic (e.g., chloride) ligands. Mn(II) can also exist as a trace element in primary and secondary silicates and oxides and also precipitate as the Mn(II)-bearing mineral rhodochrosite ($MnCO_3$) under saturated carbonate conditions.

The intermediate Mn species, Mn(III), is unstable in aqueous solution unless it is complexed to stabilizing ligands (Kostka et al, 1995; Klewicki and Morgan, 1998, 1999; Luther et al., 1998). In the absence of these ligands, Mn(III) tends to disproportionate into the +2 and +4 oxidation states:

$$2Mn^{3+} + 2H_2O \Leftrightarrow Mn^{2+} + MnO_2 + 4H^+ \quad (18.1)$$

Citrate, pyrophosphate, and siderophores are examples of effective ligands for Mn(III). Of these, siderophores form the strongest complexes. For instance, the siderophores pyoverdine (PVD) and desferrioxamine B (DFOB) have stability constants (log K) of 44.6 ± 0.5 and 28.6 ± 0.5, respectively (Parker et al., 2004, 2007; Duckworth and Sposito, 2005). Mn(III) may also precipitate as lone or mixed valence sparingly soluble hydroxide minerals.

Mn(IV) is not stable in solution and exists primarily as Mn hydroxides, oxyhydroxides or oxides (hereinafter referred to as oxides). The solubility of Mn(III, IV) oxides in water is low but varies with structure, crystal size, composition, and structural disorder. Mn oxides have a high affinity for various cations, especially for metal ions such as Co(II), Ni(II), and Cu(II) (Goldberg, 1954; Geloso, 1927), and are frequently associated with ferric iron.

18.2.2 Thermodynamics and Kinetics

18.2.2.1 Oxidation

Despite thermodynamic favorability starting at a pH of approximately 8 (Luther, 2005, 2010), the homogenous oxidation of Mn(II) by molecular oxygen is exceedingly slow in seawater (Von Langen et al., 1997; Morgan, 2005). Indeed, oxidation of free Mn(II) ions do not homogeneously oxidize in aqueous solution (pH 8.4) even after seven years (Diem and Stumm, 1984b). Mn(II) will homogeneously oxidize within a few weeks to months only under more extreme conditions (pH > 8.5; pO$_2$ ~ 1; [Mn$_T$] > 450 µM) (Morgan and Stumm, 1964; Hem and Lind, 1983; Murray et al., 1985). One explanation that has been proposed for this resistance of Mn(II) ions to homogenous oxidation by O$_2$ is the high energy of activation requirement for the reaction (Crerar and Barnes, 1974). Another explanation proposed is that the Mn(II) may be extensively complexed and thereby stabilized by such inorganic ions as chloride, sulfate, and bicarbonate (Goldberg and Arrhenius, 1958; Hem, 1963) or by organic compounds such as amino acids and humic acids (Graham, 1959; Hood, 1963; Hood and Slowey, 1964).

More recently, thermodynamic modeling confirmed that homogeneous oxidation of Mn(II) to Mn(III) by O_2 is thermodynamically prohibited owing to a reactivity barrier imposed by this first electron transfer step (Luther, 2010) (Figure 18.2). In fact, below a pH of about 8, the oxidation of Mn(II) to various Mn(III) forms is endergonic. In contrast, a two-electron transfer process between

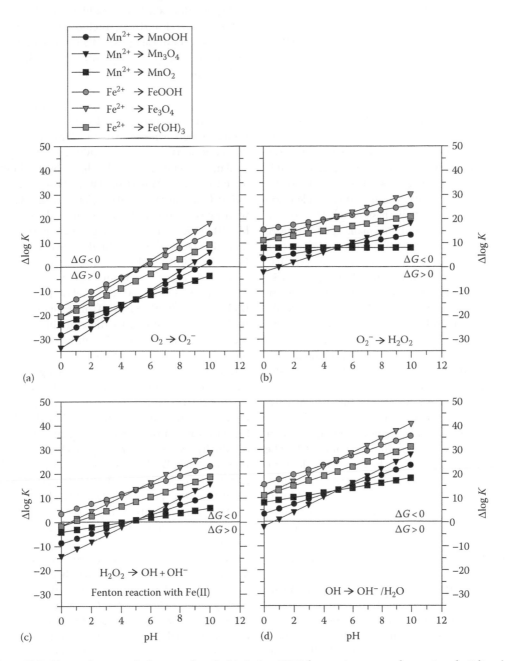

Figure 18.2. Thermodynamic calculations as described in Luther (2010) for one-electron transfer reactions for Fe^{2+} and Mn^{2+} with oxygen and reactive oxygen species (a) O_2; (b) superoxide, O_2^-; (c) hydrogen peroxide, H_2O_2; and (d) hydroxyl, OH. The $+ \Delta\log K$ on the y-axis indicates a favorable complete reaction, and $-\Delta\log K$ indicates an unfavorable reaction as $\Delta G° = -RT \ln K = -2.303RT \log K$. (Modified from Luther, GW III, *Aquat Geochem*, 16, 395, 2010.)

Mn(II) and molecular oxygen would be exergonic at pH values greater than 4 (Figure 18.2). The overall oxidation of Mn(II) by O_2 (Reaction 18.2)

$$Mn^{2+} + \tfrac{1}{2}O_2 + H_2O \rightarrow MnO_2 + 2H^+ \quad (18.2)$$

involves two successive electron transfer steps (Reactions 18.3 and 18.4) as

$$Mn^{2+} \rightarrow \{Mn^{3+}\} + e^- \quad (E^{\circ\prime} = 0.84 \text{ V}) \quad (18.3)$$

$$\{Mn^{3+}\} + 2H_2O \rightarrow MnO_2 + 4H^+ + e^-$$
$$(E^{\circ\prime} = +0.08 \text{ V}) \quad (18.4)$$

where $\{Mn^{3+}\}$ represents Mn(III) bound within an enzyme complex. Although oxidation of Mn^{2+} to MnO_2 in a single step (Reaction 18.2) by removal of two electrons is thermodynamically viable ($\Delta G^{\circ\prime} = \sim 68$ kJ) (Figure 18.2; Luther, 2005, 2010), the $E^{\circ\prime}$ for the Mn(II)/Mn(III) redox couple is too high to allow for ATP synthesis coupled to electron transport (Ehrlich and Newman, 2008). Instead, it has been proposed that energy conservation could only occur in the second step, the Mn(III)/Mn(IV) couple (Ehrlich and Newman, 2008). Many Mn redox reactions proceed, however, in discrete one-electron steps (Luther, 2005, 2010), including microbial Mn(II) oxidation (Glenn and Gold, 1985; Schlosser and Höfer, 2002; Webb et al., 2005a; Anderson et al., 2009; Learman et al., 2011a), and therefore, the first electron transfer will control the reaction.

Nevertheless, oxidation of Mn(II) to Mn(III) by the ROS superoxide (O_2^-) is thermodynamically viable over all environmentally relevant pH conditions (Luther, 2005) (Figure 18.2). The ROS superoxide and hydrogen peroxide have almost no thermodynamic barrier to reaction with Mn(II). The kinetics of homogenous Mn(II) oxidation by superoxide are rapid and can be faster than many heterogeneous oxidation rates and those induced by biological processes (Nico et al., 2002; Hansard et al., 2011; Learman et al., 2011b).

Based on thermodynamics, microbial Mn(II) oxidation could be coupled to electron acceptors other than oxygen, in particular nitrate (Luther et al., 1997), yet empirical evidence for these reactions is lacking, and instead Mn(II) oxidation observed within suboxic conditions has been linked to low levels of molecular oxygen (Schippers et al., 2005; Clement et al., 2009).

Although homogenous O_2-mediated Mn(II) oxidation rates are slow, a wide spectrum of Mn(II) oxidation rates have been observed for heterogeneous reactions, with Mn(II) half-lives ranging from 5 to 2800 days for surface-catalyzed oxidation (Diem and Stumm, 1984; Davies and Morgan, 1989; Junta and Hochella, 1994; Wehrli et al., 1995; Madden and Hochella, 2005). Mineral surfaces, including Mn oxides themselves (autocatalysis), also accelerate the oxidation of Mn(II) by O_2 (Morgan and Stumm, 1965; Brewer, 1975; Coughlin and Matsui, 1976; Sung and Morgan, 1981; Davies and Morgan, 1989), presumably by destabilizing the Mn(II) ion symmetry through H_2O ligand exchange between the hydrated Mn(II) ion and the oxide surface (Morgan, 2000, 2005). In particular, biogenic Mn oxides have been shown to induce rapid Mn(II) oxidation (Bargar et al., 2005; Learman et al., 2011b; Tang et al., 2014) with rates equivalent to other mineral-catalyzed Mn(II) oxidation rates (e.g., nanoparticulate hematite rates of $\sim 10^{-4}$ M d^{-1}) (Madden and Hochella, 2005) and faster than some reported biological (i.e., enzymatic) rates in natural waters (e.g., ~ 3–12 nM h^{-1}) (Tebo et al., 1997; Dick et al., 2009). Rates of microbially mediated Mn(II) oxidation may therefore include also these superimposed biooxide-induced reactions once the mineral has nucleated.

Complexation of Mn to (in)organic ligands can also increase the rate of Mn(II) oxidation by O_2. In particular, organic carbon in Mn(III)–ligand complexes can donate electrons to Mn(II) allowing electron transfer and subsequent oxidation of Mn(II) by molecular oxygen (Klewicki and Morgan, 1998; Duckworth and Sposito, 2005). In fact, the formation constant of Mn(III)HDFOB$^+$ is 22 orders of magnitude higher than the Mn(II)–DFOB complex, illustrating a large thermodynamic driving force for the oxidation of the complex.

Rates of bacterially mediated oxidation of Mn(II) in the presence of O_2 has been reported as ~ 5 orders of magnitude higher than abiotic homogenous oxidation by O_2 (Nealson et al., 1988; Tebo, 1991). Rates of aerobic Mn(II) oxidation vary widely in different environments and as a function of a number of geochemical variables (see Tebo et al., 1997, and references therein, Zhang et al., 2002 and Section 18.7). For instance, environmental Mn(II) oxidation rates have been shown to range from around 2 pM h⁻¹ in deep-sea hydrothermal vent plumes (Cowen et al., 1990; Mandernack and Tebo, 1993) to greater than 65 nM h⁻¹ at the oxic/anoxic interface of the Black Sea (Tebo, 1991). More recently, rates as high as 2 nM h⁻¹ were measured in deep-sea hydrothermal plumes in Guaymas Basin in the Gulf of California (Dick et al., 2009). $Mn(II)$ oxidation at submicromolar oxygen concentrations in the Black Sea ranged from 4 to 50 nM h⁻¹ (Clement et al., 2009). Residence times in various marine settings have ranged from hours to years (Emerson et al., 1982; Landing and Bruland, 1987; Tebo 1991; Mandernack and Tebo, 1993; Delgadillo-Hinojosa et al., 2006; Clement et al., 2009; Dick et al., 2009). Rates in freshwaters can exceed those in marine settings, approaching high nM h⁻¹ rates (see, for instance, Tipping et al., 1984).

18.2.2.2 Reduction

The reduction potential of $Mn(II)/Mn(III, IV)$ couples depends on the speciation of both the reduced and oxidized metal species (Figure 18.3). From a thermodynamic perspective, the possible oxidants of $Mn(II)$ are situated above the Mn redox couples in Figure 18.3, while the reductants of $Mn(III, IV)$ sit below the couples.

Mn oxides are abiotically reduced by a variety of organic and inorganic compounds, particularly $Fe(II)$ and sulfide. These reactions are energetically favorable under most natural conditions with rates that vary widely with environmental conditions (Stone and Morgan, 1984a,b; Burdige and Nealson, 1986; Stone, 1987a,b). In the presence of light, for instance, in the photic zone, Mn oxides rapidly undergo photodissolution, where

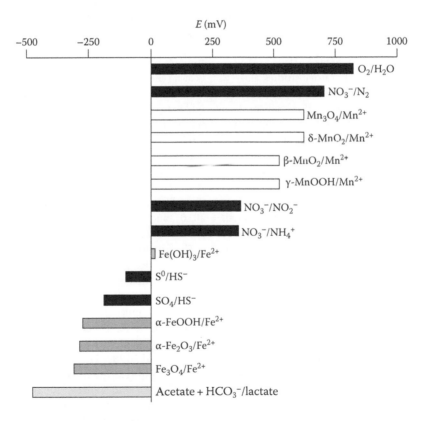

Figure 18.3. Electrode potentials versus standard hydrogen electrode of selected iron and manganese species, compared to other important redox couples. Potentials were calculated for pH 7 and environmentally relevant concentrations of solutes. As in temperature = 25°C, pH = 7, $[Fe^{2+}] = [Mn^{2+}] = [NO_3^-] = [NH_4^+] = 10$ μM, $[SO_4^{2-}] = 10$ mM, $[HS^-] = [NO_2^-] = [acetate] = [lactate] = 1$ μM, $[HCO_3^-] = 1$ mM, $P_{N2} = 1$ atm. (Modified from Thamdrup, B, *Adv Microb Ecol*, 16, 41, 2000.)

the intensity and wavelength of light have a profound impact on the reduction rate (Sunda et al., 1983; Sunda and Huntsman, 1988; Waite and Szymczak, 1993).

Microbial respiration of Mn(III, IV) oxides coupled to the oxidation of H_2 and various carbon sources is thermodynamically viable under most environmental conditions. Initial rates of microbial Mn oxide reduction vary with cell density, Mn species, and type and concentration of electron donor. Specifically, Mn oxide reduction rates by pure and mixed Mn-reducing communities span a wide range, with initial reduction rates as low as 20 nM h^{-1} and as high as 1 mM h^{-1} (Burdige and Nealson, 1985; Lovley and Phillips, 1988a; Myers and Nealson, 1988a,b; Rusin et al., 1991a; Burdige et al., 1992; Kostka et al., 1995; Greene et al., 1997; Bratina et al., 1998; Kieft et al., 1999; Fredrickson et al., 2002; Fredrickson et al., 2004). A recent detailed compilation of microbial reduction rates under a wide variety of conditions indicated that the zero- and first-order rate constants appear to follow a log-normal distribution (Bandstra et al., 2011). When using the zero-order rate data, the authors found that the average surface area normalized rate for these laboratory studies (1.47 × 10^{-4} mmol m^{-2} h^{-1}) are close to the upper end of the range of measured reduction rates in deep subseafloor sediments (2.5 × 10^{-8} to 6.96 × 10^{-5} mmol m^{-2} h^{-2}) (D'Hondt et al., 2004).

Within soils, the rate of Mn(III, IV) reduction has been found to vary with a number of factors, including temperature and water potential (Sparrow and Uren, 2014). In fact, the authors found that even small changes in soil temperature, water potential, and pH, shifted the rates of Mn(II) oxidation and Mn(III, IV) reduction that would result in a shift in the balance to favor one process over the other.

18.3 MICROBIAL OXIDATION OF MN

18.3.1 Taxonomic and Environmental Diversity

The ability of microorganisms to catalyze the oxidation of Mn(II) has been known for over a century (Jackson, 1901a,b; Beijerinck, 1913). Since this discovery, a significant number of organisms have been reported to oxidize manganese (Figure 18.4). To date, only Mn(II)-oxidizing bacteria and fungi have been identified in the literature, and it is unknown if organisms within the archaea domain can oxidize Mn(II). Biological oxidation has been found in a taxonomically diverse group of bacterial phyla that include Firmicutes, Actinobacteria, and Bacteroidetes and class Alpha-, Beta-, and Gammaproteobacteria (e.g., Tebo et al., 2005; Santelli et al., 2010; Yang et al., 2013; Akob et al., 2014). Characterized isolates within these phyla include *Aurantimonas manganoxydans* SI85-9A1 (Caspi et al., 1996; Dick et al., 2008a), *Erythrobacter* sp. SD-21 (Francis et al., 2001), *Leptothrix discophora* SS-1 (Corstjens et al., 1997; Brouwers et al., 2000a; Takeda et al., 2012), *Pedomicrobium* sp. ACM 3067 (Larsen et al., 1999), *Escherichia coli* (Su et al., 2013), *Pseudomonas putida* strain MnB1 and GB-1 (Okazaki et al., 1997; Caspi et al., 1998; de Vrind et al., 1998; Brouwers et al., 1999; Francis et al., 2001; Geszvain et al., 2013), *Roseobacter* sp. AzwK-3b (Hansel and Francis, 2006), *Bacillus* sp. SG-1 (Van Waasbergen et al., 1993, 1996; Francis et al., 2002), *Ralstonia* sp. (Yang et al., 2013), *Variovorax* sp. (Stein et al., 2001; Yang et al., 2013), *Marinobacter* sp. (Templeton et al., 2005), and *Albidiferax* sp. TB-2 (Akob et al., 2014) (Figure 18.5).

In contrast, fungal oxidation of Mn(II) has only been found in two different phyla, Basidiomycota and Ascomycota, which include *Phanerochaete chrysosporium* (Glenn and Gold, 1985; Glenn et al., 1986; Lackner et al., 1991), *Trametes versicolor* (Höfer and Schlosser, 1999), *Stropharia rugosoannulata* (Schlosser and Höfer, 2002), *Stilbella aciculosa* (Hansel et al., 2012), *Stagonospora* sp. (Santelli et al., 2010), *Acremonium* sp. (Miyata et al., 2004; Santelli et al., 2014), *Pleosporales* sp. (Miyata et al., 2006a; Mariner et al., 2008; Santelli et al., 2014), *Plectosphaerella* sp. (Santelli et al., 2010), *Pyrenochaeta* sp. (Santelli et al., 2010), and *Paraconiothyrium* sp. (Takano et al., 2006; Santelli et al., 2014) (Figure 18.6).

The diversity of Mn(II)-oxidizing organisms is also reflected in the various environments that they inhabit. Mn(II)-oxidizing bacteria are ubiquitously distributed in the environment and are readily cultured from soils (Drosdoff and Nikiforoff, 1940; Douka, 1977; Letunova et al., 1978; Yang et al., 2013), desert varnish (Taylor-George et al., 1983; Hungate et al., 1987), freshwater springs (Hariya and Kikuchi, 1964; Mustoe, 1981; Ehrlich and Zapkin, 1985), rivers (Brauer et al., 2011), lakes (Sokolova-Dubinina and Deryugina, 1967a,b; Dubinina and Deryugina, 1971; Jaquet et al., 1982; Stabel and Kleiner, 1983), and various marine environments including ocean water (Moffett, 1997; Cowen and Silver, 1984), ocean floor (Ehrlich et al., 1972; Ehrlich,

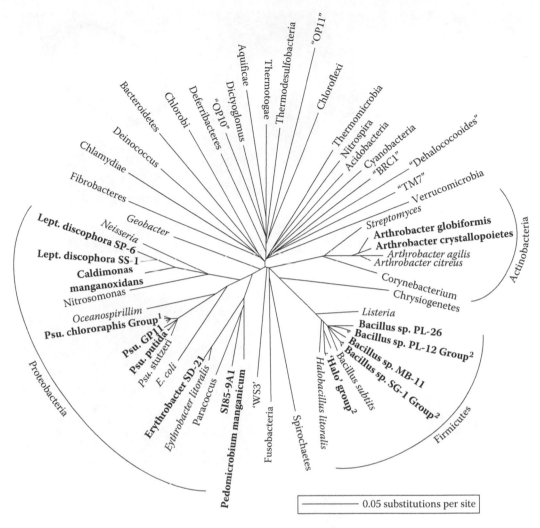

Figure 18.4. Phylogenetic tree illustrating the diversity of bacteria that oxidize Mn(II). *Note:* Mn(II) oxidizing bacterial strains are shown in bold. (From Tebo, BM et al., *Trends Microbiol*, 13, 421, 2005.)

2000; Templeton et al., 2005), estuaries/bays (Thiel, 1925; Krumbein, 1971; Tebo et al., 1984; Tebo and Emerson, 1985; Hansel and Francis, 2006), and hydrothermal vents (Cowen et al., 1990, 1998; Dick and Tebo, 2010).

Although less studied, Mn(II)-oxidizing fungi are also frequently observed in association with Mn oxides within soils (Thompson et al., 2005), desert varnish (Krumbein and Jens, 1981), decaying wood (Blanchette, 1984), and deteriorating monuments (de la Torre and Gomez-Alarcon, 1994). They have also been isolated from a diverse range of environments, including stream pebbles (Tani et al., 2003; Miyata et al., 2006b), hot springs (Sasaki et al., 2004), ponds (Santelli et al., 2014), deep-sea iron oxide mats (Templeton et al., 2005; Connell et al., 2009), soil

nodules (Golden et al., 1992), and coal mine drainage treatment systems (Santelli et al., 2010).

18.3.2 Physiological Reasons for Oxidation

The physiological reasons for microbial Mn(II) oxidation remains an enigma. Early reports suggested either autotrophic (Beijerinck, 1913) or mixotrophic growth (Lieske, 1919; Sartory and Meyer, 1947). It has been proposed that Mn(II)-oxidizing bacteria could gain energy from this reaction. The gram-negative marine bacterium, strain SSW_{22}, is thought to couple ATP synthesis to Mn(II) oxidation (energy conservation) (Ehrlich, 1983; Ehrlich and Salerno, 1988, 1990). Further, a bacterium isolated from a marine ferromanganese

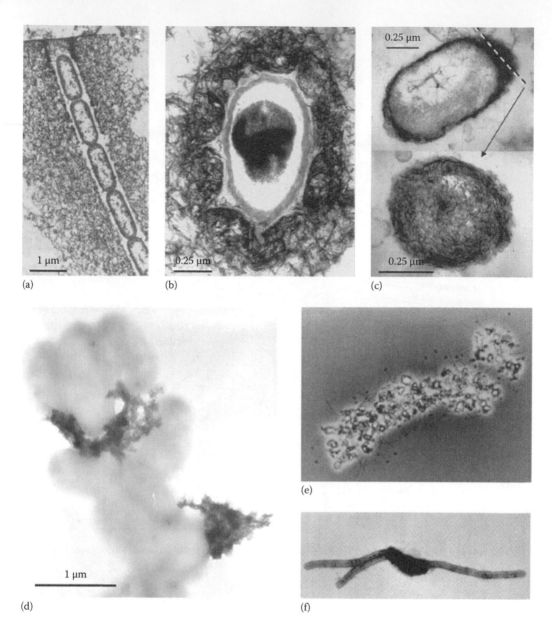

Figure 18.5. Transmission electron micrograph of representatives of Mn(II)-oxidizing bacteria. (a) An unidentified *Leptothrix* sp. from a forest pond near Kiel, Germany. (b) Spores of the marine *Bacillus* sp. strain SG-1 with Mn oxides on the outermost layer of the spores. (c) *Pseudomonas putida* strain MnB1 showing Mn oxides on the extracellular organic matrix. (d) *Roseobacter* sp. AzwK-3b isolated from Elkhorn Slough, CA, showing a heterogeneous distribution of Mn oxides outside the cell. (e) *Pedomicrobium* in association with manganese oxide particles (×2800). (f) Manganese-oxidizing *Hyphomicrobium* sp. isolated from a Baltic Sea iron–manganese crust (×15,600). ([a–c] Reproduced from Tebo, BM et al., *Annu Rev Earth Planet Sci*, 32, 287, 2004; [d] Courtesy of Hansel, CM, Woods Hole Oceanographic Institution, Woods Hole, MA; [e,f] Courtesy of Ghiorse WC, Department of Microbiology, Cornell University, Ithaca, NY.)

nodule (BIII 45) is believed to gain energy during mixotrophic growth through the oxidation of Mn(II), but only if the Mn(II) is adsorbed first to Mn oxides (Ehrlich, 1976). As discussed in Section 18.2.2, energy conservation is believed to occur only during the second electron transfer step, Mn(III) to Mn(IV). In contrast, most recent investigations of Mn(II)-oxidizing bacteria have indicated that they are strict heterotrophs and Mn(II) oxidation is not involved in energy generation.

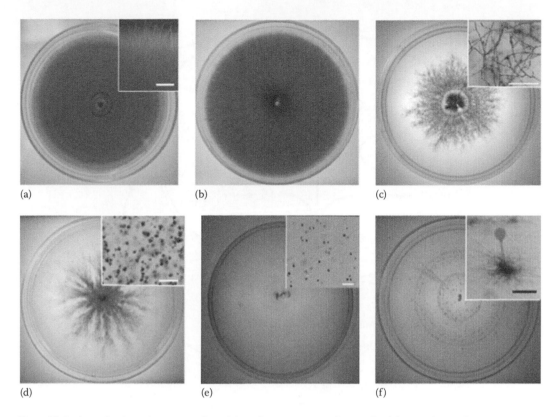

Figure 18.6. **(See color insert.)** Diversity of Mn (II)-oxidizing ascomycete fungi isolated from coal mine drainage treatment systems. Fungi were inoculated in the center of the petri dish, and mycelia grew radially outward from the inoculation point. Mycelia are visible due to the precipitation of Mn(III, IV) oxides (brown/black coloring). (a) *Plectosphaerella cucumerina* isolate DS2psM2a2, (b) *Microdochium bolleyi* isolate SRC1dJ1a, and (c) *Pithomyces chartarum* isolate DS1bioJ1b have oxide products that are closely associated with hyphae (insets). (d) *Stagonospora* sp. SRC1lsM3a, (e) *Acremonium strictum* isolate DS1bioAY4a, and (f) *Stilbella aciculosa* isolate DS2rAY2a have Mn oxides precipitated away from the hyphae and at the base of conidiophores in the case of *S. aciculosa* (see insets).

Beyond energy conservation, other reasons have been proposed to describe the physiological benefits microbes might gain for Mn(II) oxidation. Whether intentional or not, organisms may benefit from the Mn(III, IV) oxides produced via Mn(II) oxidation. Mn oxides are highly reactive minerals that can break down complex recalcitrant carbon (Sunda and Keiber, 1994). Thus, Mn oxide formation could generate a pool of more labile carbon that could be used in metabolism. Further, Mn oxides have a unique ability to sorb heavy metals and also scavenge ROS: thus Mn oxides could serve as protection from these harmful chemicals (Ghiorse, 1984; Emerson et al., 1989). In addition, Mn oxides can be used as terminal electron acceptors in anaerobic respiration, and therefore, bacteria could generate Mn oxides to store electron acceptors to use when conditions require anaerobic metabolism to survive (Tebo, 1983). At present, however, there is actually no direct evidence for a physiological benefit for most bacterial Mn(II)-oxidizing activity, and instead, in at least some organisms, this activity may instead be a fortuitous side reaction (Learman and Hansel, 2014).

18.3.3 Mechanisms of Biological Oxidation

The mechanisms used by microorganisms to oxidize Mn(II) are more diverse than originally thought and include a number of microbially mediated processes (Figure 18.7). Some organisms use an enzyme to directly oxidize Mn(II), while others oxidize Mn(II) via indirect pathways (Brouwers et al., 2000a; Tebo et al., 2004, 2005; Geszvain et al., 2012). There is experimental evidence for a Mn(III) intermediate formed during the oxidation of Mn(II) to Mn(IV) by a broad group of microbes that oxidize Mn(II) via direct or indirect means (Webb et al., 2005a; Miyata et al.,

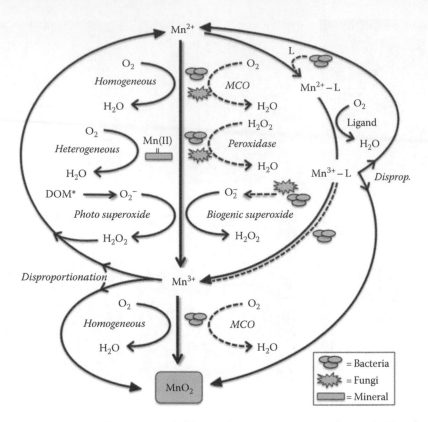

Figure 18.7. Schematic representing the major pathways for Mn(II) oxidation, including direct enzymatic activity by multicopper oxidases and peroxidases, abiotic reaction with superoxide produced biogenically and via photochemically excited dissolved organic matter (DOM*), and ligand (L)-mediated oxidation. Abiotic oxidation pathways also include direct homogeneous oxidation by molecular oxygen (O_2) (at pH > 8) and heterogeneous oxidation by O_2 catalyzed by Mn(II) adsorption onto mineral surfaces. Abiotic chemical reactions are curved solid black lines and biologically mediated are curved dashed lines. Mn(III) formed upon Mn(II) oxidation can undergo further oxidation to Mn oxides (MnO_2) or disproportionate (Disprop.) to both Mn(II) and Mn oxides.

2006a; Dick et al., 2008; Johnson and Tebo, 2008; Anderson et al., 2009; Learman et al., 2011a; Su et al., 2013). The fate of this Mn(III) intermediate is less clear and may involve either disproportionation to Mn(II) and Mn(IV) or further oxidation to Mn oxides either through a second electron transfer reaction within the enzyme (Butterfield et al., 2013) or by another oxidant.

18.3.3.1 Direct Enzymatic: MCOs and Laccases

Bacterial Mn(II) oxidation by multicopper oxidases (MCOs) is a direct enzymatic pathway that has been found in numerous bacteria. MCOs are a structurally and functionally diverse family of enzymes with the ability to oxidize a number of (in)organic substrates (e.g., Fe(II), diphenolics) (Solomon et al., 1996; Brouwers et al., 2000b). MCOs have been implicated in Mn(II) oxidation

in various bacteria: *Leptothrix* species (Corstjens et al., 1997; Brouwers et al., 2000a; Takeda et al., 2012), *Pedomicrobium* species (Larsen et al., 1999; Ridge et al., 2007), *E. coli* (Su et al., 2014), *P. putida* species (de Vrind et al, 1998; Brouwers et al., 1999; Francis and Tebo, 2001, 2002; Okazaki et al., 1997; Geszvain et al., 2013), and *Bacillus* species (van Waasbergen et al., 1996; Francis et al., 2002; Dick et al., 2008; Butterfield et al., 2013; Su et al., 2013). Another common feature shared by these organisms is that they deposit the Mn oxides on their cells, sheath, or spore (Mulder, 1972; Rosson and Nealson, 1982; Okazaki et al., 1997).

Bacillus species have served as a model organism to define the molecular mechanism of MCO-directed Mn(II) oxidation. The dormant spores, rather than the vegetative cells, of *Bacillus* sp. SG-1 oxidize Mn(II) (Neaslson and Ford, 1980; Rosson and Nealson, 1982; de Vrind et al., 1986b;

Francis and Tebo, 1999). Initial work showed that the spores bind and oxidize Mn(II) via a protein component in the exosporium (Rosson and Nealson, 1982; de Vrind et al., 1986a; Francis and Tebo, 1999). Mn(II)-oxidizing activity in *Bacillus* is linked to an operon labeled *mnx* (Van Waasbergen et al., 1993, 1996). The *mnxG* gene codes for a protein that belongs to the MCO family. Addition of Cu(II) to the growth medium stimulates Mn(II) oxidation by the spores (Van Waasbergen et al., 1996), consistent with the notion that the oxidation involves an MCO; however, difficulties in purifying the MnxG protein (Dick et al., 2008) once limited the understanding of this system. This difficulty was related to the fact that Mn(II) oxidation activity required a protein complex. Butterfield et al. (2013), however, recently purified this complex, MnxEFG, and showed that it has Mn(II)-oxidizing activity. As previously thought, Mn(II) is oxidized to Mn(III) (Webb et al., 2005a), but the complex MnxEFG is also able to oxidize Mn(III) to Mn(IV) (Butterfield et al., 2013). Thus, MCOs may mediate oxidation of both Mn(II) and Mn(III). The ability to measure Mn(III) using exogenous ligands may indicate that the Mn(III)–enzyme complex is somewhat labile

and/or exchangeable. A theoretical model involving intermediate polynuclear Mn(IV) complexes has been proposed to provide a potential pathway by which the manganese oxidase active site could allow for the coupled successive electron transfer steps (Soldatova et al., 2012) (Figure 18.8).

The Mn(II)-oxidizing activity of two *P. putida* strains is also connected to MCOs. *P. putida* strains MnB1 and GB-1 generate a Mn(II)-oxidizing protein complex that includes an MCO that appears to reside in the outer membrane (Okazaki et al., 1997; Brouwers et al., 2000a). A recent study of *P. putida* GB-1 found that oxidation in this bacterium is related to two different MCOs: MnxG and McoA (Geszvain et al., 2013). Interestingly, each protein has the ability to oxidize Mn(II) independently. The reason for this genetic overlap in function is unknown, but may be related to growth state (planktonic vs. biofilm) or that one protein is specific for Mn(II) oxidation and the other more specific to other metals or organic compounds (Geszvain et al., 2013).

While MCOs have been implicated in Mn(II)-oxidizing activity in *L. discophora* SS-1, this role is unresolved. Corstjens et al. (1997) identified the *mofA* gene in *L. discophora* SS-1, which codes for a

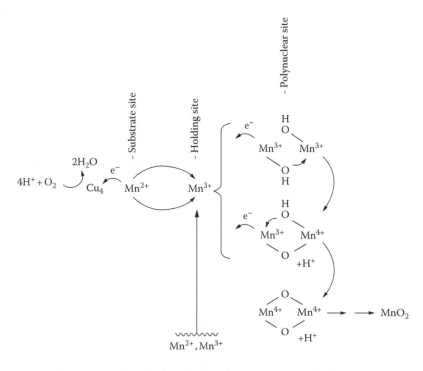

Figure 18.8. Proposed mechanism of Mn(II) oxidation and Mn oxide formation catalyzed by the multicopper manganese oxidase. (From Soldatova, A et al., *J Biol Inorg Chem*, 17, 1151, 2012.)

Mn(II)-oxidizing protein that includes copper in its structure and is related to MCOs. Addition of aqueous Cu(II) to the culture medium stimulates production of the Mn(II)-oxidizing protein in growing cultures (Brouwers et al., 2000a). Iron limitation significantly decreases (~75%) Mn(II) oxidation by L. discophora SS-1, which is not linked to siderophore production (El Gheriany et al., 2009). Nevertheless, while the Mn(II)-oxidizing protein is thought to be MofA, mofA transcript levels are not impacted by the level of Fe or Cu and also do not correspond with Mn(II) oxidation (El Gheriany et al., 2009; El Gheriany et al., 2011). It is possible that MofA might require multiple proteins or a protein complex like the Mnx system found in Bacillus. A genetic system has been recently developed for strain SS-1 (Bocioaga et al., 2014) that is predicted to be amenable for systematic genetic interrogation of manganese oxidation genes, which should greatly aid in resolving the pathways involved in Mn(II) oxidation by this organism.

An MCO, specifically laccase, has also been implicated in Mn(II) oxidation by fungi. Several studies have shown both basidiomycete and ascomycete fungi utilize a laccase to oxidize Mn(II) (Miyata et al., 2004; Miyata et al., 2006a). Fungal laccases, similar to bacterial MCOs, have an ability to oxidize a remarkably wide variety of phenolic compounds and also oxidize Mn(II) via a one electron transfer to Mn(III) (Hofer and Schlosser, 1999; Schlosser and Höfer, 2002). Various ascomycetes exhibit laccase-like activity, and Mn(II) oxidation to Mn(IV) is thought to occur in two sequential one-electron transfer reactions (Miyata et al., 2004; Miyata et al., 2006a,b; Takano et al., 2006), but this mechanism remains unclear.

18.3.3.2 Direct Enzymatic: Peroxidases

Peroxidase-driven Mn(II) oxidation is a well-described enzymatic pathway used by fungi. Manganese peroxidase (MnP) is a common lignin-degrading protein that is important for recalcitrant carbon degradation by basidiomycetes (Gold and Alic, 1993; Hofrichter, 2002). For instance, the white rot fungus P. chrysosporium has an extracellular MnP that oxidizes Mn(II) to Mn(III) coupled with the consumption of H_2O_2 (Glenn and Gold, 1985; Glenn et al., 1983). Mn(III) is stabilized by an organic complex (e.g.,

lactate), which may then react with an organic compound, such as a lignin component, that reduces the Mn(III) back to Mn(II) (Glenn et al., 1986; Paszczynski et al., 1986). In this system, the reduction of Mn(III) to Mn(II) makes Mn merely an electron shuttle in the peroxidase system and has no geochemical significance insofar as Mn redistribution in the environment is concerned. However, Kenten and Mann (1949) proposed that at a very low H_2O_2 concentration, oxidized Mn might accumulate because reduction of the oxidized Mn would be negligible. In the absence of organic compounds, Glenn et al. (1986) observed the formation of a brown precipitate identified as a Mn oxide. Mn oxide formation via this mechanism therefore appears to involve a balance between Mn(II) oxidation and hydrogen peroxide scavenging.

Recent evidence also suggests that a laccase–MnP cooperation may be relevant to Mn(II) oxidation in some basidiomycetes. Specifically, Schlosser and Höfer (2002) found that Mn(II) oxidation by MnP in the fungus S. rugosoannulata only occurred in the presence of laccase. The authors proposed that laccase oxidized Mn(II) to Mn(III), which in the presence of the organic chelators oxalate and malonate resulted in the formation of superoxide. Reduction of the superoxide to hydrogen peroxide provides a substrate for the MnP to initiate Mn(II)-oxidizing activity (Figure 18.9). The authors propose that Mn(II) oxidation may therefore involve an enzyme cooperation based on a hydrogen peroxide intermediate.

A heme peroxidase has also been connected to bacterial Mn(II) oxidation and Mn oxide formation in bacteria. Heme peroxidases are involved in Mn(II) oxidation in two alphaproteobacteria, Erythrobacter sp. SD-21 and A. manganoxydans SI85-9A1 (Anderson et al., 2009). As seen in other bacteria, difficulties in genetically manipulating these organisms have limited the understanding of the mechanisms involved in Mn(II) oxidation. Protein fractionation and chromatography lead to the discovery of a large Mn(II)-oxidizing protein (>250 kDa) that was identified as a heme peroxidase (Anderson et al., 2009). The involvement of a heme peroxidase in Mn(II) oxidation was recently further demonstrated by heterologous expression of the implicated heme proteins in E. coli (Nakama et al, 2014). Similar to fungal MnP oxidation, Mn(III) is also a product of Mn(II) oxidation, but

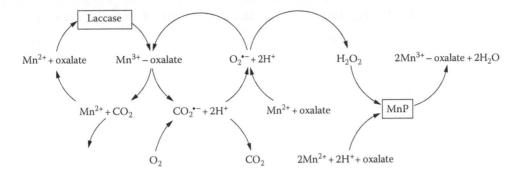

Figure 18.9. Proposed scheme for Mn(II) oxidation via the linked enzymatic activity of laccase and manganese peroxidase (MnP) and extracellular generation of hydrogen peroxide (H_2O_2). Initially, laccase-catalyzed Mn(II) oxidation in the presence of oxalate leads to the formation of H_2O_2 via various intermediate radical reactions. The resulting H_2O_2 is then used by MnP to oxidize Mn(II). (From Schlosser, D and Höfer, C, *Appl Environ Microbiol*, 68, 3514, 2002.)

bacterial oxidation with heme peroxidases do ultimately generate Mn oxides.

18.3.3.3 Indirect Biological Mn(II) Oxidation

Bacteria and fungi both also oxidize Mn(II) via indirect pathways. While these pathways still involve enzymatic processes, these enzymes do not act directly on Mn(II). Indirect Mn(II) oxidation may be promoted through production of one or more metabolic end products, which cause chemical oxidation of Mn(II). According to Soehngen (1914), a large number of microorganisms can cause such Mn(II) oxidation in the presence of hydroxycarboxylic acids such as citrate, lactate, malate, gluconate, or tartrate. In apparent agreement, Van Veen (1973) found that with *Arthrobacter* 216, hydroxycarboxylic acids are required for Mn(II) oxidation. Bacteria may also utilize polysaccharides to oxidize Mn(II) (van Veen et al., 1978; Ghiorse and Hirsch, 1979; Ehrlich, 1983; Adams and Ghiorse, 1987; Boogerd and de Vrind, 1987; Beveridge, 1989). Further, phytoplankton have been shown to change local pH conditions that allow for Mn(II) oxidation (Richardson et al., 1988).

The mechanisms of Mn(II) oxidation by bacteria and fungi were recently expanded to include an indirect oxidation pathway involving extracellular superoxide (O_2^-). Specifically, a bacterium with the common *Roseobacter* clade (*Roseobacter* sp. AzwK-3b) and several ascomycete fungi rapidly oxidize Mn(II) to Mn(III) via the enzymatic production of superoxide (Learman et al., 2011a; Hansel et al., 2012; Tang et al., 2013) (Figure 18.7).

Although the production of superoxide by bacteria was once thought to be limited to a small number of bacterial species, it is now considered widespread in heterotrophic bacteria (Diaz et al., 2013). Production of extracellular superoxide has already been shown to be ubiquitous in fungi (Aguirre et al., 2005), where it is involved in host defense, posttranslational modification of proteins, cell signaling, and cell differentiation (Bedard et al., 2007). Fungal production of extracellular superoxide is primarily carried out by the NOX family of NADPH oxidases (Bedard et al., 2007). In both bacteria and fungi, Mn(III) is generated from superoxide-mediated Mn(II) oxidation, which is either further oxidized or disproportionated to Mn oxides. However, it appears that reaction between superoxide and Mn(II) alone is not enough to produce Mn oxides. In detail, Mn oxide formation is inhibited by hydrogen peroxide (H_2O_2), a product of the reaction between Mn(II) and superoxide (Reaction 18.5). Specifically, hydrogen peroxide reacts with Mn(III), reducing it back to Mn(II) (Learman et al., 2013) (Reaction 18.6). The precipitation of Mn oxides therefore requires also the scavenging of hydrogen peroxide that is formed upon the reaction of superoxide and Mn(II). It is also proposed that ligands may help stabilize Mn(III) during Mn(II) oxidation by biogenic superoxide (Learman et al., 2013; Tang et al., 2013) (Figure 18.7):

$$Mn(II) + O_2^- + 2H^+ \rightarrow Mn(III) + H_2O_2 \quad (18.5)$$

$$Mn(III) + \tfrac{1}{2}H_2O_2 \rightarrow Mn(II) + \tfrac{1}{2}O_2 + H^+ \quad (18.6)$$

18.4 MICROBIAL REDUCTION OF MN

18.4.1 Taxonomic and Environmental Diversity

A wide taxonomic diversity is also found in bacteria that can reduce Mn, which includes *Geobacter metallireducens* (Lovley and Phillips, 1988a); *G. sulfurreducens* (Caccavo et al., 1994); *Geothrix fermentans* (Coates et al., 1999); *Pyrobaculum islandicum* (Kashefi and Lovley, 2000); *Sulfurospirillum barnesii* SES-3 (Laverman et al., 1995); *Desulfovibrio desulfuricans, Desulfomicrobium baculatum, Desulfobacterium autotrophicum,* and *Desulfuromonas acetoxidans* (Lovley and Phillips, 1994); *Shewanella oneidensis* MR-1 (Myers and Nealson, 1988a) and MR-4 (Larsen et al., 1998); *Bacillus polymyxa* D1 and *Bacillus* MBX1 (Rusin et al., 1991a,b); and *Pantoea agglomerans* SP1 (Francis et al., 2000). Even sparingly soluble forms of Mn(III) oxides (MnOOH) can be anaerobically reduced by bacteria (Larsen et al., 1998). In addition, some fungi have been found to reduce Mn oxides under laboratory conditions (Gadd, 2010; Wei et al., 2012). Only one member of the archaea has so far been found to reduce Mn(IV) oxide, namely, *Pyrobaculum islandicum* (Kashefi and Lovley, 2000).

Like Mn(II) oxidizers, organisms that reduce Mn(IV) are also found in a variety of different environments. The occurrence of microbial reduction of Mn oxide was suggested as far back as the 1890s (Adeny, 1894). Mann and Quastel (1946) reported the microbial reduction of biogenically formed Mn oxide in a soil column upon the addition of glucose. Further, Troshanov (1968) isolated a number of Mn(IV)-reducing bacteria from reduced horizons of several Karelian lakes (former USSR). Some of these strains were able to reduce both Mn and Fe oxides, but strains that reduced only Mn oxides were encountered most frequently. In addition, Fe(III) and Mn(IV) reducers have been found in marine and freshwater sediments (Lovley, 1991; Thamdrup, 2000; Vandieken et al., 2014), contaminated sediments (Lovley et al., 1989a; Lovley and Phillips, 1992; Lovley and Anderson, 2000), and hydrothermal environments (Brock et al., 1976; Tor et al., 2003; Lovley et al., 2004). Although a significant part of the Mn(IV) seems to be reduced enzymatically, some of it may also be reduced chemically (Troshanov, 1968). These nonenzymatic reactions are dependent on the form of the oxide (see Ehrlich, 1984; Nealson and Saffarini, 1994).

18.4.2 Physiological Reasons for Reduction

The majority of the bacteria studied to date reduce Mn(IV) oxides as a terminal electron acceptor. This reduction is coupled to carbon oxidation (electron donor), but a few types can also use H_2 anaerobically. Since Mn oxide minerals are the dominant form of oxidized Mn in the environment, bacteria have developed unique pathways to donate electrons from the cytoplasm into a solid-phase mineral outside the cell. Some bacteria can reduce the oxidized Mn both aerobically or anaerobically (e.g., Ehrlich, 1966; Troshanov, 1968; Di-Ruggiero and Gounot, 1990), whereas others can reduce it only anaerobically (e.g., Burdige and Nealson, 1985; Lovley and Phillips, 1988a; Myers and Nealson, 1988a). Among these latter bacteria, some, like *S. oneidensis*, are facultative but nevertheless reduce Mn(IV) oxide only anaerobically (Myers and Nealson, 1988a), whereas others, like *G. metallireducens*, are strict anaerobes and reduce Mn oxides only anaerobically (Lovley and Phillips, 1988a). In some cases, Mn may be reduced to satisfy a nutritional need for soluble Mn(II) (see de Vrind et al., 1986b; also discussion by Ehrlich, 1987) or to scavenge excess reducing power, as in some cases of nitrate (Robertson et al, 1988) and ferric reduction (Ghiorse, 1988; Lovley, 1991).

18.4.3 Mechanisms of Biological Reduction

Organisms that are strict or facultative anaerobes are able to reduce Mn(III, IV) oxides in an anaerobic respiratory process. All are quite versatile in regard to the electron donors that they can couple to Mn reduction. For instance, *G. metallireducens* can use butyrate, propionate, lactate, succinate, acetate, and several other compounds, which are completely oxidized to CO_2 (Lovley and Phillips, 1988a; Lovley, 1991; Myers et al, 1994; Langenhoff et al., 1997). *S. oneidensis*, on the other hand, can utilize lactate and pyruvate as electron donors but only incompletely oxidizes these to acetate. It can, however, use H_2 and formate as electron donors, whereas *G. metallireducens* cannot (Lovley et al., 1989a,b). Unlike *G. metallireducens, G. sulfurreducens* can use H_2 as electron donor (Caccavo et al., 1994).

The reduction of Mn oxides has also been linked to the anaerobic oxidation of methane and ammonium. Incubations of sediments from the Eel River basin (California) have shown Mn reduction coupled to anaerobic methane oxidation

(Beal et al., 2009). In this coupled anaerobic reaction, Mn reduction was faster than Fe reduction; however, both were slower than sulfate reduction. This is likely due to Mn and Fe being solid-phases, whereas sulfate is soluble. A consortium of microorganisms oxidize methane coupled to sulfate reduction with the archaeal member driving methane oxidation (Orphan et al., 2001a; Knittel et al., 2005) and the bacterial member conducting sulfate reduction (Orphan et al., 2001b; Orphan et al., 2002; Niemann et al., 2006). Mn reduction and methane oxidation could be performed with a similar consortia or possibly by only bacteria but with a different enzymatic mechanism used for methane oxidation than what is known for archaea (Beal et al., 2009). The coupling of methane oxidation to Mn reduction has now been implicated in various environments (Crowe et al., 2011; Jones et al., 2011; Segarra et al., 2013; Wang et al., 2014).

Further, Mn reduction has also been linked to the oxidation of nitrogen. Specifically, Mn reduction has been proposed to play a role in the oxidation of ammonium to N_2 (Luther et al., 1997; Hulth et al., 1999). There has been difficulty in demonstrating the coupling of these reactions (Thamdrup and Dalsgaard, 2000, 2002) until recently (Lin and Taillefert, 2014). Surface area and the Mn oxide structure play a role in this reaction coupling (Lin and Taillefert, 2014).

Another unique aspect of certain Mn reducers (e.g., *Geobacter* and *Shewanella*) is that they are not restricted to Mn oxides as terminal electron acceptors. Depending on the organism, ferric iron, uranyl ion, chromate, nitrate, iodate, elemental sulfur, sulfite, sulfate, thiosulfate, fumarate, and glycine may also serve as electron acceptors. In the case of *S. oneidensis*, oxygen, nitrate, and fumarate inhibits Mn oxide reduction, but sulfate, sulfite, molybdate, nitrite, or tungstate do not (Myers and Nealson, 1988a).

In contrast to bacteria, fungi are not known to gain energy through the reduction of Mn. Instead, fungal-induced reduction of Mn oxides is indirect (nonenzymatic), where fungal metabolites, such as oxalate, induce Mn oxide dissolution and reduction (Gadd, 2010; Wei et al., 2012).

18.4.3.1 Enzymatic Reduction via C-Type Cytochromes

Anaerobic bacterial reduction of Mn oxides has been well documented (Burdige and Nealson, 1985; Lovley and Goodwin, 1988; Lovley and Phillips, 1988a; Myers and Nealson, 1988a; Lovley et al., 1989a, 1993a). Two of the more detailed enzymatic pathways described are found in *S. oneidensis* MR-1 and *Geobacter sulfurreducens*. Both bacteria are able to transport electrons generated in the cytoplasm to extracellular electron acceptors via multiple c-type cytochromes. In *G. sulfurreducens*, two c-type cytochromes, MacA and PpcA, transport electrons from the cytoplasmic membrane to the periplasm (Lloyd et al., 2003; Butler et al., 2004; Lovley, 2006) (Figure 18.10). These electrons are transported to the outer membrane via OmcB (Leang et al., 2003) to two additional outer membrane bound c-type cytochromes, OmcE and OmcS (Lovley et al., 2004; Mehta et al., 2005; Holmes et al., 2006; Lovley, 2006). It was once thought that these outer membrane cytochromes would transfer electrons to insoluble electron acceptors like Mn oxides; however, electrons are now thought to be transferred to a conductive pilus, where the final electron transfer reaction takes place (Reguera et al., 2005).

In *S. oneidensis* MR-1, various soluble and membrane bound cytochromes also form a complex pathway that transports electrons to extracellular Mn oxides (Figure 18.11). The outer membrane c-type cytochrome involved in Mn oxide reduction also plays a role in dissimilatory Fe(III) reduction (Myers and Myers, 1992, 1993, 2000). Electrons generated from central metabolism are passed through the electron transport chain to a tetraheme c-type cytochrome, CymA (Myers and Myers, 1997). While CymA is also involved in the reduction of other electron acceptors (e.g., nitrate and fumarate), for solid-phase reduction, electrons are transferred to a soluble periplasmic c-type cytochrome, MtrA (Schuetz et al., 2009). Electrons are then transferred through the outer membrane via MtrB (transmembrane protein) to two different outer membrane decaheme cytochromes, MtrC and OmcA (Shi et al., 2006; Ross et al., 2007; Coursolle and Gralnick, 2010; Richardson et al., 2012a,b; Wee et al., 2014). A type II secretion protein, GspD, located in the outer membrane has also been shown to be important in the export of exoproteins important for Mn reduction (DiChristina et al., 2002; Shi et al., 2008). Figure 18.11 shows a general model for the electron transport pathway to MnO_2 in *S. oneidensis* MR-1 (Szeinbaum et al., 2014).

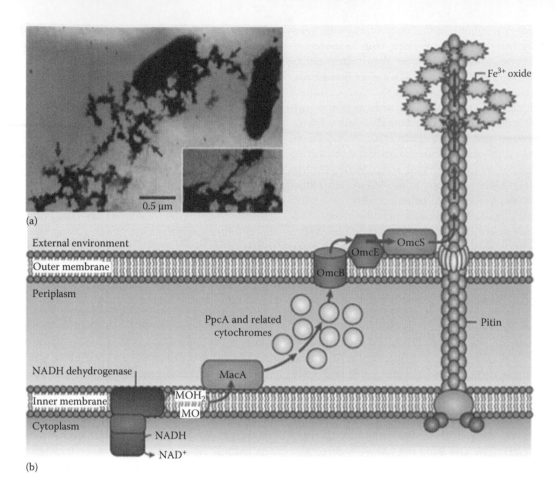

Figure 18.10. Mn and Fe oxide reduction pathway used by *Geobacter metallireducens*. The illustration is depicted for Fe oxide reduction, but Mn oxides are believed to occur via similar processes. (a) Transmission electron microscopy image shows direct contact between the cell and presumed pilin with the Fe oxide substrate. (b) Enzymatic pathway proposed for Fe and Mn oxide reduction. MacA and PpcA transport electrons from the cytoplasmic membrane to the periplasm where they are transferred to outer membrane proteins OmcB, OmcE, and OmcS. Electrons are then passed from these outer membrane proteins to a conductive pilus where the electrons reduce Mn and Fe oxides. (From Lovley, DR, *Nat Rev Microbiol*, 4, 497, 2006.)

Mn(IV) was once thought to be reduced directly to Mn(II) in a single two electron transfer process. However, recent work provides evidence that reduction proceeds via a two-step, single-electron transfer with Mn(III) as an intermediate (Lin et al., 2012; Szeinbaum et al., 2014). Further, it is proposed that reduction of the solid-phase Mn(IV) requires an initial solubilization step. This solubilization step is similar to the reaction proposed for solid-phase iron reduction, which involves an endogenous organic ligand (Taillefert et al., 2007; Fennessey et al., 2010; Jones et al., 2010). While MtrC and OmcA are still involved in the reductive dissolution of Mn(IV), it is proposed that only the reduction of Mn(III) is

coupled to carbon oxidation. The reduction of Mn(III) is thought to be reduced solely by OmcA (Szeinbaum et al., 2014).

18.4.3.2 Indirect Biological Reduction

Some bacteria and most fungi reduce Mn(IV) oxides via an indirect (nonenzymatic) reaction. A likely mechanism of reaction is the production of metabolic products that are strong enough reductants for Mn(IV) oxides. *E. coli*, for instance, produces formic acid from glucose, which is capable of abiotically reducing Mn oxides. Similarly, pyruvate (a metabolic byproduct) can also nonenzymatically reduce manganese oxides at acidic pH

EHRLICH'S GEOMICROBIOLOGY

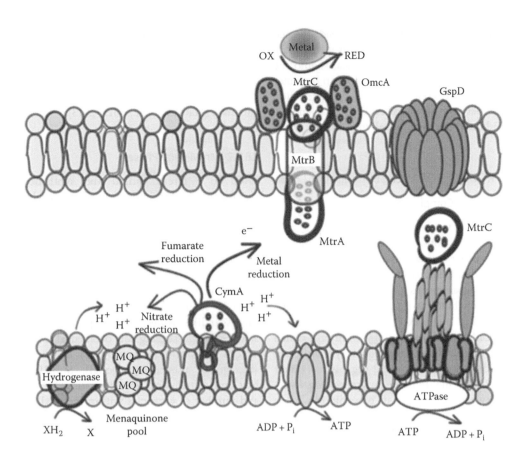

Figure 18.11. Schematic of the proposed mechanism employed by *Shewanella oneidensis* MR-1 to reduce Mn and Fe oxides. Electrons are transported from the cytoplasm to the c-type cytochrome, CymA. Electrons move from the periplasm to the outer membrane via transfer from CymA to MtrA and MtrB. Electrons can be transferred to Mn oxides (metal) by a reaction with two outer membrane cytochromes, MtrC and OmcA. The enzyme GspD is also depicted, as it is used to export extracellular proteins involved in metal reduction. (From Szeinbaum, N et al., *Environ Microbiol Rep*, 6, 490, 2014.)

(Stone, 1987a). In addition, many fungi produce oxalic acid, which also reduces Mn oxides (e.g., Stone, 1987; Wei et al., 2012). Inorganic compounds like H_2S, sulfite, and Fe^{2+} can also readily reduce some Mn(IV) oxides nonenzymatically (Mulder, 1972; Burdige and Nealson, 1986; Lovley and Phillips, 1988b; Myers and Nealson, 1988b; Nealson and Saffarini, 1994; Aller and Rude, 1988).

Growing cultures of marine *Bacillus* 29 are able to reduce Mn oxides aerobically and anaerobically using metabolites formed from glucose as electron donors. Both Mn(II) and Mn oxides appear to serve as inducers for this reductive reaction (Ehrlich, 1966; Trimble and Ehrlich, 1968; Trimble and Ehrlich, 1970; Ghiorse and Ehrlich, 1976). The proposed pathway for this reaction involves a component of the electron transport system (Ehrlich, 1966; Trimble and Ehrlich, 1968, 1970) and indirect reduction by hydrogen peroxide

(Ghiorse, 1988). This reduction by metabolically produced hydrogen peroxide is similar to that previously reported by Dubinina (1978a,b) with *Leptothrix pseudoochracea*, *Arthrobacter siderocapsulatus*, and *Metallogenium*. Ghiorse (1988) suggested that in *Bacillus* 29, Mn oxide reduction may serve as a means of disposing of excess reducing power without energy coupling. The organization of the Mn oxide-reducing system in *Bacillus* 29 is far from universal in Mn(IV)-reducing bacteria.

The gram-negative marine *Pseudomonad* strain BIII 88 is able to reduce insoluble Mn oxide either aerobically or anaerobically using acetate as electron donor Ehrlich (1993a,b). Reduction requires that the bacteria be in direct contact with Mn oxide particles (Ehrlich, 1993a). The reaction involves the oxidation of Mn(II) bound to the outer membrane. The bound Mn is oxidized to Mn(III) via a reaction with the Mn(IV) oxide. Electrons are then

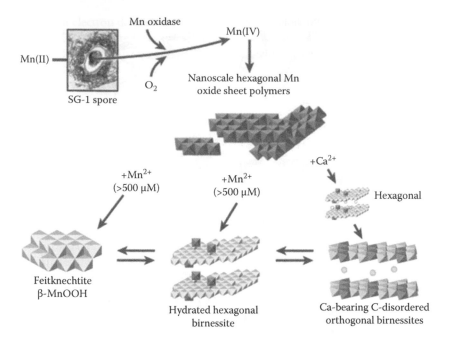

Figure 18.13. Oxidation of Mn(II), for instance, by spores of *Bacillus* SG-1, leads to the formation of nanoparticulate hexagonal Mn(IV) oxide minerals. These biogenic oxides will then transform to a variety of other Mn oxide structures depending on the environmental conditions such as aqueous Ca^{2+} and Mn^{2+} levels. (Modified from Templeton, AS and Knowles, E, *Annu Rev Earth Planet Sci*, 37, 367, 2009; Based on ideas from Bargar, JR et al., *Am Min*, 90, 143, 2005; Webb, SM et al., *Am Min*, 90, 1342, 2005b.)

a more crystalline pseudo-orthogonal birnessite (Bargar et al., 2005; Webb et al., 2005b; Learman et al., 2011b; Tang et al., 2014). Structural evolution of birnessite is accelerated by redox transfer reactions at the mineral surface, as observed for surface reactions with Mn(II) and organic (Banerjee and Nesbitt, 2001; Bargar et al., 2005; Learman et al., 2011b) or inorganic compounds (e.g., Cr(III), Tang et al., 2014). Aqueous Mn(II), for instance, leads to size coarsening and structural ripening of biogenic nanohexagonal birnessite to particulate triclinic birnessite (Bargar et al., 2005; Webb et al., 2005b; Learman et al., 2011b).

Considerably, fewer studies have characterized the composition and structure of fungal biooxides. Similar to bacteria, the initial Mn oxide formed by a wide diversity of ascomycete fungi is a poorly ordered, nanocrystalline phyllomanganate similar to δ-MnO_2 (Miyata et al., 2006b; Grangeon et al., 2010; Santelli et al., 2011; Tang et al., 2013). However, structural and morphological differences of these oxides and the secondary products upon ageing (e.g., triclinic birnessite, and todorokite) are largely species dependent (Santelli et al., 2011). For instance, while most

fungi form a nanoparticulate hexagonal birnessite phase that is stable after long periods of growth, some species appear to accelerate ageing to triclinic birnessite (e.g., *Stagonospora* sp.) relative to others (e.g., *Plectosphaerella cucumerina*). The products formed by *Acremonium strictum* have been identified as either solely todorokite (Petkov et al., 2009; Saratovsky et al., 2009) or layered Mn oxides (Grangeon et al., 2010). The observed difference in biomineral structure appears to be due to growth conditions and lifestyle, as it was recently found that different Mn oxides formed when the organism was grown planktonically (δ-MnO_2) versus surface attached (δ-MnO_2 and todorokite) (Santelli et al., 2011).

Biogenic Mn oxides tend to be more highly disordered and turbostratic and contain a higher degree of layer site vacancies (Tebo et al., 2004). These characteristics have been attributed to higher sorption and oxidation capacities of biogenic versus abiotic Mn oxides. For instance, it has been observed that both bacterial (Nelson et al., 1999b) and fungal (Tani et al., 2004) Mn biooxides are more reactive than their synthetic (i.e., abiotic) analogs. Also, it has been theoretically demonstrated that the higher degree of layer site vacancies typified by biogenic oxides will enhance their photochemical reactivity by narrowing the band gap (Kwon et al., 2009).

Microbial activities can lead to the formation of Mn-bearing mineral phases other than Mn oxides. For instance, microbial respiration of Mn oxides can lead to the formation of the manganese carbonate mineral rhodochrosite ($MnCO_3$). Thamdrup et al. (2000) found that in Black Sea shelf sediments, Mn(III, IV) respiration was accompanied by $MnCO_3$ formation. Reduction of MnO_2 by axenic cultures of the metal-reducing bacterium *G. metallireducens* in anaerobic incubations leads to the production of $MnCO_3$ under controlled laboratory conditions (Lovley and Phillips, 1988a). Indeed, detection of Mn-reducing bacteria has previously lead to the inference that microbial respiration of Mn oxides was responsible for observed $MnCO_3$ deposits within a lake (Sokolova-Dubinina and Deryugina, 1967a; Troshanov, 1968). Further, growth of two soil fungi (*Aspergillus niger* and *Serpula himantioides*) in the presence of various Mn oxides results in the formation of Mn oxalate minerals (Wei et al., 2012).

18.5.3 Association of Mn Oxides with Microbes and Microbial Exudates

In all fungal and bacterial systems, Mn oxide products are formed outside of the cell. As with Fe(II)-oxidizing organisms, Mn(II) oxidation leads to Mn oxide deposits on the surface of the cell or on some structure surrounding it, such as a slime layer or a sheath (see reviews by Ghiorse, 1984; Ghiorse and Ehrlich, 1992; Tebo et al., 1997; Ehrlich, 1999) (Figure 18.5). In some instances, the accumulation may represent the product of Mn(II) oxidation by the organism; in others, it may represent electrostatic attraction and subsequent accumulation of Mn oxides that were formed by other organisms or abiotically. Microorganisms that accumulate manganese oxides include sheathed bacteria like *Leptothrix*; budding and appendaged bacteria like *Pedomicrobium* (Ghiorse and Hirsch, 1979; Ghiorse and Ehrlich, 1992), *Hyphomicrobium*, *Planctomyces* (Schmidt et al., 1981, 1982), *Caulococcus*, and *Kuznetsova*; and bacteria with capsules like *Siderocapsa* and *Naumanniella*. Ghiorse and colleagues also described polysaccharide and/or protein polymers underlying Mn and Fe oxides produced by a *Pedomicrobium*-like bacterium (Ghiorse and Hirsch, 1979).

Further, bacteria without appendages such as *Pseudomonas* and *Bacillus* species also accumulate Mn oxides on the cell surface or in the extracellular matrix surrounding the cells (Figure 18.5). In some cases, the Mn oxides encrust portions of the microbial cells (Francis et al., 2002; Tebo et al., 2004, 2005) and/or are embedded within an extracellular polymeric substance (EPS) (Toner et al., 2005). For instance, Mn oxides formed by the common Mn(II)-oxidizing bacterium *P. putida* are completely enveloped by a chemically heterogeneous biofilm matrix (Toner et al., 2005). Also, Mn oxides formed by spore-forming *Bacillus* spp. are located in the outermost layer of the proteinaceous spore coat (Francis et al., 2002). Conversely, in the case of a *Roseobacter* bacterium, only a minor proportion of the produced Mn oxides are located near the organism, and no evidence for nucleation of the oxide on the cell surface is observed, suggesting that the oxides are simply electrostatically attracted to the cell either in solution or during sample processing (Learman et al., 2011a) (Figure 18.5).

Fungal cellular structures, such as hyphae, also serve as a substrate for Mn oxide accumulation (Miyata et al., 2006b, 2007; Santelli et al., 2011; Hansel et al., 2012; Tang et al., 2013). Oftentimes, however, Mn oxides produced via fungal Mn(II) oxidation are not associated with any cellular structure (Figure 18.14). More recently, filamentous ascomycete fungi grown in both aqueous suspension and on mineral surfaces (both quartz and rhodochrosite) form Mn oxides at a considerable distance (averaging 10–100 μm) away from any visible cellular structure (such as hyphae, conidiophores) (Santelli et al., 2011; Tang et al., 2013). These Mn oxides are unique in character and illustrate an intriguing ordered morphology suggesting that the oxides are in fact templated on an unknown organic molecule (Figure 18.14). Similarly, fungal Mn oxides have been observed to encrust "thin, narrow, delicate" filaments (termed "arai") that are less electron dense than the oxides themselves (Zavarzin, 1964), pointing to an organic carbon core. Thin, electron-poor filaments have also been observed in transmission electron microscopy cross sections of mycogenic Mn oxide filaments, speculated to be composed of mucous or polysaccharides (Klaveness, 1977). Mycogenic Mn oxides have also been observed to be connected by "slime threads" that were "too thin for real cellular structures" (Krumbein and Jens, 1981). The spatial decoupling of fungal cells and Mn oxides hints to

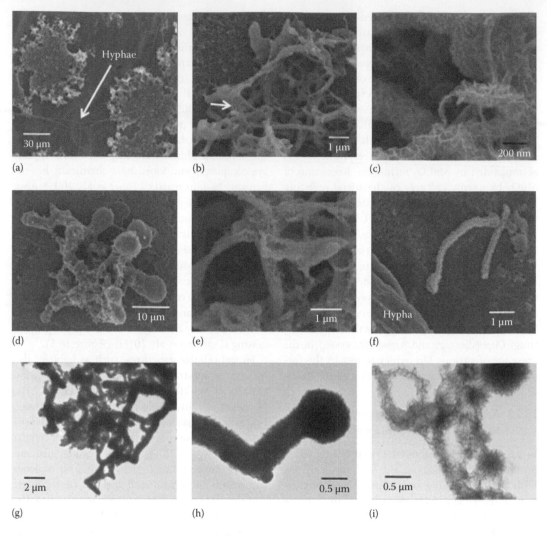

Figure 18.14. Scanning electron and transmission electron microscopy images of fungal Mn oxides made by the ascomycete fungi (a–c) *Pyrenochaeta* sp. DS3sAY3a and (D-I) *Stagonospora* sp. SRClsM3a grown on (a–f) rhodochrosite or on (g–i) Mn(II)-supplemental agar plates. Hyphae are not in the field of view for images (b–i). (Modified from Tang, Y et al., *Environ Microbiol*, 15, 1063, 2013.)

a role for either soluble oxidants or Mn(III) complexes that allow for Mn oxide nucleation away from the cell surface. Observed differences in cellular-oxide associations by Mn(II)-oxidizing fungi likely reflect different oxidation pathways including outer membrane proteins, soluble secreted exoproteins, and/or soluble, extracellular reactive metabolites (e.g., superoxide).

Organic polymer-mediated Mn oxide formation is now considered the likely explanation for *Metallogenium*-like structures—dendritic Mn oxide–encrusted filaments, primarily observed in association with fungi in Mn oxide–rich environments. *Metallogenium sp.* were originally thought to be the

bacteria responsible for the oxidation of Mn(II) to Mn(IV) oxides in various aquatic and geologic environments, including lake waters and sediments (Klaveness, 1977; Jaquet et al., 1982; Maki et al., 1987) and desert rock varnishes (Krumbein and Jens, 1981b). In fact, fossilized *Metallogenium*-like structures have been found in Precambrian sedimentary formations and in Cretaceous–Paleogene cherts (Crerar et al., 1980). However, it was ultimately found that *Metallogenium*-like structures contained no viable cells or evidence of cellular structures (Emerson et al., 1989). In fact, *Metallogenium*-like structures associated with Mn(II)-oxidizing fungi contain filamentous protein

EHRLICH'S GEOMICROBIOLOGY

templates at the core of the Mn oxides (Emerson et al., 1989). The authors suggested that protein cross-linking might be required to maintain the structural integrity of the polymer matrix underlying the oxides. They also suggested that the matrix likely consists of anionic polymers such as proteins or acidic polysaccharides that is exuded by the fungus (an unidentified Basidiomycete) and serves to facilitate mineral nucleation. Since *Metallogenium*-like structures have been observed in association with phylogenetically diverse fungal species (Mirchink et al., 1970; Santelli et al., 2011; Tang et al., 2013) and are found frequently in modern and ancient environments (Crerar et al., 1980; Neretin et al., 2003), organic templation may be a prevalent yet unrealized mechanism of Mn oxide formation in ancient terrestrial and extraterrestrial environments.

Finally, protein-rich coatings on foraminiferal tests have been proposed as a likely substrate for Mn oxide precipitation. Specifically, foraminiferal tests from the Blake Plateau were coated with a veneer of manganese-rich material containing Cu, Ni, Co, and Fe (Graham and Cooper, 1959).

It is clear that organics are intricately involved in the precipitation and likely nucleation of biogenic Mn oxides and possibly even secondary abiotic oxides. For instance, Mn oxides produced by chemically generated superoxide are dispersed, highly disordered birnessite colloids that do not aggregate to form particles unless microbial cells or organic metabolites are present (Learman et al., 2013). The importance of organic polymers in nucleating and directing mineral precipitation is widely appreciated for other biominerals, such as iron oxides, magnetite, and metal sulfides (Chan et al., 2004; Komeili et al., 2006; Komeili, 2007; Moreau et al., 2007; Chan et al., 2011). At present, it is assumed that the microbial cell surfaces, EPS in biofilms, and/or organic metabolites are merely involved in the deposition of the Mn oxides. Whether these extracellular organics or proteins are directly involved either in electron transfer to Mn(II) or Mn(III) or in the precipitation and structural evolution of the Mn oxides is not clear. For instance, even the presence of sterile organic-rich medium has been shown to aid the auto-oxidation of biogenic Mn oxides to more crystalline phase (Learman et al, 2011b). Furthermore, Mn oxides formed by ascomycete fungi often are precipitated at the tips of hyphae (Miyata et al., 2006a; Santelli et al., 2011), where high concentrations of organic molecules and extracellular proteins are exuded (Gadd, 1999).

18.6 MANGANESE IN THE EARTH'S CRUST

Manganese is the second most abundant redox-active metal in the Earth's crust (crustal concentration is 0.1%) (Alexandrov, 1972). Although Mn has a 50-fold lower crustal abundance than Fe, it concentrates to high levels in certain environments. Indeed, the distribution of Mn in the crust is by no means uniform (Table 18.1). In soils, for instance, its concentration range from 0.002% to 10% (Goldschmidt, 1954). An average concentration in freshwater has been reported to be 8 $\mu g\ kg^{-1}$ (Bowen, 1979). Concentrations slightly in excess of 1 $mg\ kg^{-1}$ can be encountered in anoxic hypolimnia of some lakes. In seawater, an average concentration has been reported to be 0.2 $\mu g\ kg^{-1}$ (Bowen, 1979), but concentrations more than three orders of magnitude greater can be encountered near active hydrothermal vents at mid-ocean spreading centers.

Manganese is found as a major or minor component in more than 100 naturally occurring minerals (Bureau of Mines, 1965; Post, 1999). Major accumulations of manganese occur in the form of oxides, carbonates, and silicates. Among the oxides, psilomelane $(Ba,Mn^{2+},(Mn^{4+})_8O_{16}(OH)_4)$, birnessite $((Na,Ca)Mn_7O_{14} * 2.8H_2O)$, pyrolusite (MnO_2), vernadite $(MnO_2 * nH_2O)$, manganite $(MnOOH)$, hausmannite (Mn_3O_4), and todorokite $((Ca,Na,K)_x(Mn^{4+}, Mn^{3+})_6O_{12} * 3.5H_2O)$ are relatively abundant examples (Post, 1999). Among the carbonates, rhodochrosite $(MnCO_3)$ is prominent, and among the silicates, rhodonite $(MnSiO_3)$ and braunite $((Mn,Si)_2O_3)$ are frequently observed. The oxides, carbonates, and silicates of manganese originated mostly as secondary authigenic minerals formed by reprecipitation of dissolved manganese. Minerals that contain manganese as a minor constituent include ferromagnesian minerals such as pyroxenes and amphiboles (Trost, 1958) and micas such as biotite (Lawton, 1955), all of which are igneous in origin.

18.7 LINKING MICROBIAL ACTIVITY TO MANGANESE DEPOSITS

Microbes have been implicated in the formation of Mn deposits within the environment by several lines of evidence, including (1) observations of Mn oxide–encrusted cells; (2) isolation and/or identification of Mn-depositing, including Mn(II)-oxidizing, organisms within samples; (3) enzymatic-like Mn removal and/or Mn(II) oxidation kinetics; and (4) inhibition of Mn oxide formation

TABLE 18.1
Approximate ranges of manganese concentrations and global fluxes (F).

Location	$[Mn_T]$ (mol kg^{-1})	Reference	Source	FMn (mol yr^{-1})	Reference
Crust	0.015−0.028	1	River		2, 3
River		2	Dissolved	$5 \times 10^9 - 2 \times 10^{10}$	
Dissolved	2×10^{-7}		Particles	$2 \times 10^{11} - 4 \times 10^{11}$	
Particles	8×10^{-6}		Aeolian	2×10^9	3
Ocean		3	Vents	$5 \times 10^{10} - 2 \times 10^{11}$	7, 8
Mixed	$5 \times 10^{-10} - 3 \times 10^{-16}$				
Deep	$8 \times 10^{-11} - 5 \times 10^{-16}$		Sedimentation		9
Pelagic sediments		4	Pelagic	5×10^{10}	
Solids	0.1		Marine	$1.5 \times 10^{11} - 3.5 \times 10^{11}$	2
Opore waters	$10^{-8} - 10^{-4}$				
Nodules	0.5−6				
Deep basins	$3 \times 10^{-6} - 1 \times 10^{-5}$	5			
Fjords	$5 \times 10^{-6} - 2\ 10^{-5}$	6			

SOURCE: Morgan, JJ, in Sigel, A, Sigel, H, Eds., *Metal Ions in Biological Systems, Vol. 37: Manganese and Its Role in Biological Processes*, Marcel Dekker, New York, pp. 1–34.

NOTES: (1) Garrels and Machenzie (1971), (2) Martin and Maybeck (1979), (3) Donat and Bruland (1995), (4) Cronan (1974), (5) Lewis and Landing (1991), (6) Grill (1982), (7) Von Damm et al. (1985a), (8) Von Damm et al. (1985b), (9) Bender et al. (1977).

in the presence of microbial poisons. These observations have spanned many diverse habitats in the terrestrial, freshwater, and marine environment. Through these approaches, microbial activity has been implicated in the majority of Mn(II) retention and precipitation within soils, sediments, lakes, marine environments, and engineered systems.

A number of approaches have been used to explore the role that microbes play in the formation of Mn deposits in soil (Gerretsen, 1937; Mann and Quastel, 1946; Bromfield and Sherman, 1950; Aristovskaya, 1961; Perfil'ev and Gabe, 1965; Schweisfurth, 1971; Timonin et al., 1972; Van Veen, 1973; Bromfield, 1978). These primarily include soil percolation experiments (Mann and Quastel, 1946) and different versions of incubations within soil–agar mixtures (Gerretsen, 1937; Leeper and Swaby, 1940; Ten Khak-mun, 1967; Uren and Leeper, 1978). A link to microbial activity has been made by observing the degree of inhibition after sterilizing the soil, for instance, by adding poisons such as sodium azide (Mann and Quastel, 1946). The link to microbial activity has also been made by observing growth of bacteria and fungi on plates (Gerretsen, 1937; Leeper and Swaby, 1940) or via direct microscopic observations using pedoscopes (Perfil'ev and Gabe, 1965).

Within aqueous systems that include lakes, estuaries, and ocean waters, microbial Mn(II) oxidation has been demonstrated by Mn removal in the presence and absence of cell poisons (such as sodium azide, penicillin, and tetracycline) from water samples collected at various depths and spiked with $^{54}Mn^{2+}$ (Emerson et al., 1982; Rosson et al., 1984; Tebo et al., 1984; Jacobs et al., 1985; Moffett, 1997; Dick et al., 2009). The removal of manganese reflected both the binding of Mn(II) to particulates (bacterial cells and [in]organic aggregates) and oxidation of Mn(II). Binding has been distinguished partially from oxidation by the use of formaldehyde. Mn(II) oxidation rates have also been obtained by measuring Mn(II) removal from the solution in the presence and absence of oxygen (e.g., Tebo and Emerson, 1985). In the absence of oxygen, Mn(II) removal should be due only to surface binding. The difference between total Mn(II) removal in air and Mn(II) removal in the absence of oxygen should therefore represent the amount of Mn that was oxidized.

EHRLICH'S GEOMICROBIOLOGY

Mn removal oftentimes follows a Michaelis–Menten-like response to O_2, suggesting an underpinning enzymatic mechanism (Tebo and Emerson, 1986; Sunda and Huntsman, 1987; Moffett, 1994a; Clement et al., 2009; Dick et al., 2009). Mn(II) oxidation rates within both terrestrial and marine systems oftentimes also have a distinct temperature, pH, and optimum Mn(II) activity (Leeper and Swaby, 1940; Bromfield and David, 1976; Uren and Leeper, 1978; Tebo and Emerson, 1985; Sunda and Huntsman, 1987; Thompson et al., 2005; Clement et al., 2009; Uren, 2013; Sparrow and Uren, 2014), further implicating microbial, and likely enzymatic, activity in the Mn deposition (Thamdrup et al., 1994; Tebo et al., 1997).

Another approach to link microbial activity and environmental Mn oxide deposits has been via the visualization and culturing of Mn(II)-oxidizing organisms from the deposits. Various microscopic and spectroscopic techniques have been used to define relationships between microbes and Mn oxides. A number of model organisms have been isolated from these deposits, as discussed in Section 18.3. Also, culture-independent approaches have gained substantial momentum over the past decade, allowing for the direct identification of microbial species and communities within Mn oxide deposits. This allows for inferences to be made on other uncultured members that may be key players in the formation of Mn oxides.

Based on investigations using these approaches, examples of microbial activities involved in the formation and deposition of Mn oxides, primarily through the oxidation of Mn(II), are discussed for terrestrial (Section 18.8), freshwater (Section 18.9), and marine (Section 18.10) systems.

18.8 TERRESTRIAL MANGANESE OXIDE DEPOSITS

18.8.1 Soils

Manganese oxides are frequently observed in soils, in sediments, and at the sediment–water interface within freshwater springs (Mustoe, 1981) and lakes (Sokolova-Dubinina and Deryugina, 1967a,b), where the Mn oxide content in some sediments is as high as 50% (weight %). The mechanisms of Mn oxidation within soils have only been minimally investigated (see, for instance, Uren, 2013; Sparrow and Uren, 2014) however.

In these studies, a number of factors have been shown to influence the formation of Mn oxide deposits in terrestrial systems, including pH, O_2 levels, and water tension, for instance (Sparrow and Uren, 1987; Uren, 2013). Mn oxide formation has also been only observed in soil incubations in the presence of CO_2 in the headspace ($\sim$1%), the activity of which was inhibited by azide (Leeper and Swaby, 1940).

Using a unique approach, Thompson et al. (2005) conducted soil incubation experiments in thin transparent sample cells to allow for direct observations of fungal Mn oxide formation under near in situ soil conditions. The authors found that during incubation, Mn was redistributed at the air–water interface within distinct dendritic and circular shapes reminiscent of precipitation on fungal hyphae that was subsequently confirmed using the soil-derived fungal isolates.

18.8.2 Desert Varnish

Manganese oxides are sometimes found as thin, brown to black veneers (up to 100 μm thick) covering rock surfaces in some semiarid and arid regions of the world. In these black coatings, manganese and/or iron oxides are major components (20%–30%) along with clay (about 60%) and various trace elements (Potter and Rossman, 1977; Dorn, 1991). These manganese- and iron-rich coatings are known as desert varnish or rock varnish. Manganese-rich coatings have been detected on some rocks in the Sonoran and Mojave Deserts (North America), Negev Desert (Middle East), the Gibson and Victoria Deserts (Western Australia), and the Gobi Desert (Asia). The mechanisms of desert varnish formation have been widely studied and debated. Although a number of mechanisms have been proposed, microbial activity is often attributed to their formation (Krumbein, 1969; Dorn and Oberlander, 1981; Taylor-George et al., 1983; Hungate et al., 1987). Bacteria, fungi, cyanobacteria, and algae, either alone or as part of a symbiotic community, have all been implicated in Mn(II) oxidation leading to desert varnish (Krumbein, 1969; Dorn and Oberlander, 1981; Krumbein and Jens, 1981). Prevalence of Mn(II)-oxidizing organisms, such as *Bacillus*, *Geodermatophilus*, *Arthrobacter*, and *Micrococcus*, within desert varnish samples from geographically distinct desert sites suggests that they play a role in its formation (Hungate et al., 1987). Deposits resembling desert varnish have been recreated in the laboratory by refining the conditions that favored their formation, which include moist rock surfaces, low nutrient content, and circumneutral pH, in the presence of desert varnish–derived bacterial isolates (Dorn and Oberlander, 1981; Dorn, 1991).

18.8.3 Caves

Mn oxide deposits are widespread through caves worldwide (Northup and Lavoie, 2001). Manganese deposits in caves are found in clastic deposits (Cílek and Fábry, 1989), as coatings on walls or speleothems (Moore and Sullivan, 1978, 1997; Gascoine, 1982; Hill, 1982; Rogers and Williams, 1982; Kashima, 1983) or as consolidated crusts (Moore, 1981; Hill, 1982; Peck, 1986; Jones, 1992; Carmichael et al., 2013). The mineralogy of these deposits can be complex and include an assemblage of oxide phases (Post, 1999; White et al., 2009; Onac and Forti, 2011). Historically, cave deposits of Mn oxides were considered abiotic in origin stimulated by microsite changes favorable for abiotic Mn(II) oxidation (e.g., Northup and Lavoie, 2001; Barton and Northup, 2007). More recently, both direct and indirect microbial activity has been implicated in the formation of Mn oxides in caves (Northup et al., 1997; Jones, 2001; Melin et al., 2001; Spilde et al., 2005; Cânaveras et al., 2006; Taboroši, 2006; de los Ríos et al., 2011; Carmichael et al., 2013). Cultivation and 16S rRNA sequencing identified known Mn(II)-oxidizing bacteria, including Leptothrix, Pedomicrobium, and Pseudomonas species, yet only minimal taxonomic similarity was observed within Mn deposits in shallow versus deep caves.

18.8.4 Ores

Some sedimentary manganese ore deposits are believed to be of biogenic origin, including Precambrian, Cretaceous–Paleocene, and Paleozoic deposits, based in part on the observation of structures in the ore that have been identified as microfossils (Shternberg, 1967; Crerar et al., 1979). For example, examination of manganese ore from the Groote Eylandt deposits in Australia revealed the presence of stromatolite structures, cyanobacterial oncolites, coccoid microfossils, and microfossils enclosed in Mn oxides (Ostwald, 1981). In this deposit, the microbes are viewed as being the main cause of Mn oxide formation and accretion, with subsequent nonbiological diagenetic changes leading to the ultimate form of the deposit.

18.9 FRESHWATER MANGANESE OXIDE DEPOSITS

18.9.1 Rivers, Creeks, and Ponds

As early as the start of the twentieth century, Mn(II)-oxidizing organisms have been observed in sediments, organic debris, and manganiferous crusts (Neufeld, 1904; Molisch, 1910; Lieske, 1919; Thiel, 1925; von Wolzogen-Kühr, 1927; Zappfe, 1931; Sartory and Meyer, 1947; Moese and Brantner, 1966). Mn oxides observed in freshwater springs have also been attributed to the activity of Mn(II)-oxidizing bacteria, including sheathed bacteria (Hariya and Kikuchi, 1964), pseudomonads (Mustoe, 1981), and a diversity of gram-positive spore-forming and non-spore-forming rods and gram-negative rods (Ehrlich and Zapkin, 1985). Mn oxides observed in lakes and within lab incubations of lake sediments have resembled *Metallogenium*-like structures (Sokolova-Dubinina and Deryugina, 1967a,b; Gregory et al., 1980; Jaquet et al., 1982; Stabel and Kleiner, 1983).

Temperature has been shown to influence Mn oxide formation. For instance, Ghiorse and Chapnik (1983) noted that a *Leptothrix* strain isolated from water from a swamp (Sapsucker Woods, Ithaca, NY) grew optimally between 20°C and 30°C in the laboratory. This correlated with observations that *Leptothrix* and particulate Fe and Mn oxides were most abundant in the surface water of the swamp when the temperature in the surface water was in the range of 20°C–30°C.

Mn oxide deposits have been observed in a number of pond, creek, and river sediments (Lind and Hem, 1993; Bilinski et al., 2002; Tani et al., 2003). Mn oxides in these environments have been attributed in part to the activity of fungi, particularly ascomycete fungi (Tani et al., 2003; Miyata et al., 2006b; Santelli et al., 2014). In fact, in an extensive cultivation effort in a freshwater pond, Santelli et al. (2014) isolated a large diversity of Mn(II)-oxidizing ascomycete fungi. Despite extensive Mn oxide deposits in the pond, the Mn(II)-oxidizing bacteria and fungi, however, only represent a minor proportion of the microbial communities in the sediments. Extensive Mn oxide deposits (up to 35 μm thick) in sediments of Pinal Creek in Arizona (Harvey and Fuller, 1998; Marbel et al., 1999; Robbins and Corley, 2005; Bargar et al., 2009) have been attributed to microbial activity, including the known Mn oxidizer L. discophora (Robbins and Corley, 2005), due to substantially less Mn oxide formation within sediment incubations upon the addition of microbial poisons (Harvey and Fuller, 1998; Marble et al., 1999).

18.9.2 Ferromanganese Nodules

Ferromanganese concretions and nodules have been observed in lakes throughout the world, including Oneida Lake and Lake George in New York, Lake Charlotte in Nova Scotia, Mosque Lake in Ontario, and the Great Lakes in the United States (Kindle, 1932; Rossman and Callender, 1968; Cronan and Thomas, 1970; Dean, 1970; Harriss and Troup, 1970; Schöttle and Friedman, 1971; Chapnick et al., 1982; Kepkay, 1985a–c). Many concretions exhibit coarse concentric banding of alternating zones rich in manganese and iron (Moore, 1981). Growth rates of concretions in Oneida Lake, as determined by natural-radio-isotope analysis, have been estimated to vary between > 1 mm per 100 years during some periods and no growth during other periods (Moore et al., 1980). Analysis of the manganese budget for the lake supports the notion that most of the dissolved manganese (95%) becomes incorporated into the nodules (Chapnick et al., 1982). Growth rates may be very slow, as in Lake Ontario (0.015 mm yr^{-1}), or more rapid, as in Mosque Lake (1.5 mm yr^{-1}). In general, lake concretions form in areas where the sedimentation rate is low.

Microbial activity has also been implicated in the formation of these ferromanganese nodules and concretions (e.g., Gillette, 1961; Dean, 1970; Dean and Ghosh, 1978; Dean and Greeson, 1979; Chapnick et al., 1982). In some cases, this role has been proposed to be direct microbial activity (Gillette, 1961; Burdige and Kepkay, 1983; Kepkay, 1985a,b; Stein et al., 2001), due in part to the presence of metal- and Mn(II)-oxidizing bacteria within the nodules. For instance, Stein et al. (2001) identified several Mn(II)-oxidizing organisms within ferromanganese micronodules that accounted for a substantial proportion (22%) of the 16S rRNA sequences obtained. Conversely, an indirect role of phytoplankton has been proposed, in which their photosynthetic activity induces alkaline conditions (pH > 9) (Kindle, 1932; Richardson

et al., 1988) that lead to abiotic Mn(II) oxidation by O_2. In these cases, the oxidized Fe and Mn is believed to be transported with dead cyanobacterial and algal cells to the lake bottom that is then released upon decay of the biomass and somehow incorporated into concretions. Alternatively, Chapnick et al. (1982) suggested that cyanobacteria and algae transport Mn(II) from surface waters to nodule-forming regions at the sediment–water interface where bacteria participate in the oxidation of Mn(II).

18.9.3 Engineered Systems

Deposits of Mn oxides are observed in various engineered systems, including water pipelines (Schweisfurth and Mertes, 1962; Tyler and Marshall, 1967a). Mn oxides formed in water pipelines are suspected to form primarily through microbial activity (Schweisfurth and Mertes, 1962). Mn oxide–encrusted microbes have been frequently observed within pipelines, where Hyphomicrobium sp. are dominant species in these Mn oxide pipeline deposits at various sites around the world (Tyler and Marshall, 1967a; Tyler, 1970).

The low maintenance and operating costs for treating acid and coal mine drainage using passive limestone systems has resulted in extensive Mn oxide deposits within engineered systems throughout mine-impacted regions in the United States (e.g., Appalachia) and worldwide (Cravotta and Trahan, 1999; Vali and Riley, 2000; Hallberg and Johnson, 2005; Johnson and Younger, 2005). The Mn oxides formed include both hexagonal and triclinic birnessite and todorokite, with elevated levels of associated metals such as zinc, nickel, and cobalt (Tan et al., 2010). Both abiotic and biotic processes have been implicated in the formation of Mn oxide deposits in mine drainage treatment systems (Cravotta and Trahan, 1999; Haack and Warren, 2003; Halberg and Johnson, 2005; Johnson et al., 2005; Bamforth et al., 2006; Luan et al., 2012). More recently, an extensive diversity of Mn(II)-oxidizing fungi and bacteria was isolated from coal mine drainage treatment systems in Appalachia (Santelli et al., 2010). Incubation of sediments from a subset of these sites reveals both abiotic and biotic contributions to Mn oxide formation that vary depending on the site and the rate of overall Mn(II) oxidation (Luan et al., 2012).

18.10 MARINE MANGANESE OXIDE DEPOSITS

Manganese is unevenly distributed in the marine environment (Table 18.1). The concentration of manganese in surface seawater from the Pacific Ocean has been reported to fall in the range of 0.3–3.0 nmol kg^{-1} (16.4–164.8 ng kg^{-1}) (Klinkhammer and Bender, 1980; Landing and Bruland, 1980). The concentration in bottom water is generally less than that in surface water. Klinkhammer and Bender (1980) found it to be one-fourth or less at some stations in the Pacific Ocean. Manganese concentrations in surface waters over the continental slope near the mouth of major rivers such as the Columbia River on the coast of the state of Washington, USA, is as high as 5.24 nmol kg^{-1} (164.8 ng kg^{-1}) (Jones and Murray, 1985). The dominant oxidation state of manganese in seawater is +2 despite the alkaline pH of seawater (7.5–8.3) (Park, 1968) and its E_h of +430 mV (ZoBell, 1946). Divalent manganese in seawater is complexed by ions such as chloride (Goldberg and Arrhenius, 1958), sulfate, and bicarbonate (Hem, 1963) and by organic substances like amino acids (Graham, 1959). In water from suboxic zones of the Black Sea and Chesapeake Bay, complexed Mn(III) was recently reported to be the prevailing manganese species (up to 100% of total manganese; 5 μmol max.), as previously discussed in Section 18.5 (Trouwborst et al., 2006). Mn occurs in greater quantities on and in sediments than in the seawater (Table 18.1). At the sediment interface at abyssal depths, a significant portion of manganese is concentrated in ferromanganese concretions (nodules) and crusts.

18.10.1 Water Column/Coastal Waters

Microbially mediated Mn(II) oxidation and biogenic Mn oxides have been observed from a number of coastal waters (Sunda and Huntsman, 1987; Moffett, 1994a,b), bays (Krumbein, 1971; Krumbein and Altmann, 1973; Moffett and Ho, 1996), fjords (Tebo et al., 1984; Jacobs et al., 1985), inlets (Emerson et al., 1982; Tebo and Emerson, 1985), and estuaries (Hansel and Francis, 2006; Vojak et al., 1985a,b). A number of factors have been found to impact the extent and rate Mn(II) oxidation in these environments, including O$_2$ levels, temperature, light, salinity, and amount of suspended matter (Tebo et al., 1984; Tebo and Emerson, 1985; Sunda and Huntsman, 1987; Richardson et al., 1988; Sunda and

Huntsman, 1988). Fluctuations in these factors, for instance, due to seasonal and diel variation (Sunda and Huntsman, 1990; Miyuajima, 1992; Thamdrup et al., 1994), lead to changes in Mn(II) oxidation and Mn oxide formation. Excess concentrations of aqueous Mn(II) is also found to inhibit oxidation (Emerson et al., 1982; Tebo and Emerson, 1985). Although Mn oxides undergo rapid photoreduction (Sunda et al., 1983) and light has been observed to inhibit microbial Mn(II) oxidation (Sunda and Huntsman, 1988), Mn oxidation and accumulation has at times been observed in the euphotic zone (Jacobs et al., 1985).

In the suboxic zone of the Black Sea, microbially catalyzed Mn(II) oxidation has been observed (Tebo, 1991; Schippers et al., 2005). Thermodynamic calculations predict that Mn(II) oxidation could be coupled to denitrification (Luther et al., 1997; Luther, 2010). Field observations have hinted at the presence of Mn(II)-based denitrification (Vandenabeele et al., 1995; Luther et al., 1997), but to date strong evidence for this reaction mechanisms is still lacking. Attempts to enrich and isolate O$_2$-independent (anaerobic) Mn(II)-oxidizing bacteria have not been successful, and, instead, Mn(II) oxidation in the suboxic waters of the Black Sea has been attributed to lateral intrusions of O$_2$ into this zone (Schippers et al., 2005) and low levels of O$_2$ (Clement et al., 2009).

The contribution of biogenic Mn(II) oxidation on Mn removal in ocean waters has been found to vary with site. For instance, contrasting Mn(II)-oxidizing activity has been observed in the mixed layer (waters above the thermocline) of the Sargasso Sea (Atlantic Ocean) at the Bermuda Atlantic Time Series Station and the equatorial Pacific (0°N, 140°W, and 9°N, 147°W) (Moffett, 1997). In particular, the processes responsible for Mn removal are dominantly Mn(II) oxidation in the Sargasso Sea but nonoxidative Mn uptake in the Pacific. In the Pacific, light stimulates ^{54}Mn^{2+} uptake suggesting phytoplankton involvement, whereas light inhibits Mn oxidation in the Sargasso Sea (Sunda and Huntsman, 1988). The proposed role for microbes in precipitating Mn oxides in the ocean water column has been suggested due in part to observations of Mn oxide encapsulated bacteria that are often associated with flocculent amorphous aggregates (marine snow) and occasionally in fecal pellets (Cowen and Silver, 1984). In the eastern subtropical North

Pacific, manganese deposits on bacterial capsules are absent in a depth range from 100 to 700 m but become increasingly noticeable below 700 m (Cowen and Bruland, 1985).

18.10.2 Hydrothermal Systems

In the deep ocean, manganese exists predominantly as Mn(II) in basaltic rocks and hydrothermal fluids and as Mn oxides in sediments and ferromanganese nodules and crusts (Kalhorn and Emerson, 1984; Murray et al., 1984; Pattan and Mudholkar, 1990).

Hydrothermal solutions are loaded with H_2S, Fe(II), Mn(II), Cu(II), Zn(II), etc. (Jannasch and Mottle, 1985). Exceptionally high Mn concentrations occur around active hydrothermal vents on mid-ocean spreading centers. Manganese in hydrothermal fluids is oftentimes enriched over a million times above ambient seawater (Edmond et al., 1982) and can be as high as 1002 μmol kg^{-1} (55.05 mg kg^{-1}) (Von Damm et al., 1985a). Although diluted as much as 8500 times within tens of meters from a vent (via hydrothermal solution mixing with bottom water), elevated manganese concentrations may be encountered in plumes extending 1 km (Baker and Massoth, 1986) and in some instances for hundreds of kilometers or more from the vent source. Mixing of hydrothermal fluids with oxygenated deep-sea water leads to the scavenging of dissolved Mn(II) onto particles and/or oxidation to Mn oxide minerals. These minerals sink to the seafloor leading to enriched Mn oxide sediments adjacent to hydrothermal sources. Bacteria have been implicated in the formation of Mn deposits in deep-sea sediments, including those surrounding hydrothermal vents (Cowen et al., 1990; Mandernack and Tebo, 1993; Juniper and Tebo, 1995; Fortin et al., 1998).

Mn-scavenging activity by encapsulated bacteria is very prominent in hydrothermal vent plumes (Cowen et al., 1986; Campbell et al., 1988; Cowen et al., 1998). Indeed, microbial processes, including enzymatic activity, have been linked to Mn(II) oxidation and Mn oxide formation within hydrothermal plumes and surrounding waters (Cowen et al., 1990; Tambiev and Demina, 1992; Mandernack and Tebo, 1993; Dick et al., 2009). The relative role of biotic versus abiotic mechanisms in Mn(II) oxidation, however, appears to vary among different hydrothermal sites and locations along the ridge axis (Cowen et al., 1990; Mandernack

and Tebo, 1993; Dick et al., 2009). Mn(II) oxidation activity and Mn(II)-oxidizing bacteria have also been isolated from water samples collected around white smokers (Mussel Bed Vent) on the Galapagos Rift (Ehrlich, 1983) and black smokers on the East Pacific Rise at 21° north (Ehrlich, 1985) and 10° north (Ehrlich, unpublished results). Heterotrophic, gram-negative Mn(II)-oxidizing bacteria, particularly *Aeromonas* and *Pseudomonas* species, have also been found associated with the epidermis of polychaete worms and their tubes from vent sites (Durand et al., 1990).

Basalt-derived Mn(II) is often oxidized and immobilized upon weathering leading to Mn oxides rinds and ferromanganese crusts on deep-sea basalt surfaces (Hekinian and Hoffert, 1975; Hein et al., 1996; Giorgetti et al., 2000; Templeton et al., 2005; Templeton et al., 2009). The main Mn phases are todorokite, birnessite, and δ-MnO_2 (Burns et al., 1974). Endolithic microbial communities have been frequently documented in fractures on the surfaces of seafloor basalts, where Mn oxide–encrusted cells are typically present (Thorseth et al., 2003, 2001; McLoughlin et al., 2011). A diverse assemblage of Mn(II)-oxidizing bacteria (Templeton, 2005) and fungi (Connell et al., 2009) have been isolated from deep-sea basalts. All of the Mn(II)-oxidizing bacteria are obligate heterotrophs and primarily α- and γ-proteobacteria. Along with yeast-like fungal species, deep-sea fungi include known Mn(II) oxidizers *Acremonium* and *Aspergillus* species.

18.10.3 Deep-Sea Concretions, Crusts, and Nodules

Manganese oxides can be found in large quantities in concretions (nodules) or crusts on the ocean floor at great distances from hydrothermal discharges, where the rate of sedimentation is low (Figure 18.15) (Margolis and Burns, 1976). Such concretions have been found in all the oceans of the world (Horn et al., 1972). They may cover vast areas of the ocean floor, as on some parts of the Pacific Ocean floor, or be distributed in patches. The chemical components of a nodule are not evenly distributed throughout its mass (Sorem and Foster, 1972). When examined in cross section, nodules seem to develop around a nucleus, which may be a foraminiferal test, a piece of pumice, a clay particle, an older nodule fragment, etc. The growth rate of manganese

Figure 18.15. A bed of ferromanganese nodules on the ocean floor at a depth of 5,292 m in the southwest Pacific Ocean at 43°01′S and 139°37′W. Nodules may range in size from <1 to 25 cm in diameter. Average size has been given as 3 cm. (Courtesy of Woods Hole Oceanographic Institute, Woods Hole, MA.)

nodules in the deep sea is reportedly very slow. Ku and Broecker (1969) and Kadko and Burckle (1980), for instance, reported rates ranging from 1 to 10 mm per 10^6 years. Assuming a constant rate of growth, a nodule in the Peru Basin was estimated radiometrically to have grown 168 ± 24 mm per 10^6 years (Reyss et al., 1982). Heye and Beyersdorf (1973) found variable growth rates varying from 0 to 15.1 mm per 10^6 years, suggesting that conditions must not be continually favorable for nodule growth. Bender et al. (1970) estimated manganese accumulation rates from five nodule specimens as ranging from 0.2 to 1.0 mg cm^{-2} per 1000 years.

While the Mn mineralogy of ferromanganese crusts is dominated by hexagonal birnessite (Post, 1999; Manceau et al., 2007a; Takahashi et al., 2007; Peacock and Moon, 2012), both triclinic birnessite and todorokite are common in diagenetic Fe–Mn nodules and hydrothermal deposits (Burns and Burns, 1977). Furthermore, enriched Ba within hydrothermal waters promotes the formation of either romanechite (lower temperatures) or hollandite (higher temperatures) within

ferromanganese deposits (Feng et al., 1998; Manceau et al., 2007b). The Mn oxide mineralogy within Fe–Mn crusts is oftentimes heterogeneous with layered and/or zoned mineralogy and chemistry, involving both phyllo- and tectomanganates with variable trace metal enrichment (Post, 1999). For instance, recent investigations of ferromanganese nodules revealed a biphasic mineral association with alternating layers of Ni-rich vernadite (a phyllomanganate whose synthetic analog is δ-MnO_2) and Ba-rich romanechite (a tectomanganate) (Manceau et al., 2007a). The authors concluded that this binary Mn oxide banding is due to a two-mode accretionary model, with romanechite formation during episodic hydrothermal activity superimposed on continuous diagenetic formation of vernadite (Manceau et al., 2007a). The Mn(IV) oxides in the nodule have a strong capacity to scavenge cations (e.g., Ehrlich et al., 1973; Loganathan and Burau, 1973; Varentsov and Pronina, 1973; Crerar and Barnes, 1974; Bodei et al., 2007), particularly Mn(II) and other transition metals.

Like freshwater nodules, microbial activity has been implicated in the formation of marine ferromanganese nodules. A high diversity of bacteria has been observed in marine ferromanganese concretions (Ehrlich, 1963; Ehrlich et al., 1972; Wu et al., 2013; Yli-Hemminki et al., 2014), where known Mn(II)-oxidizing bacteria typically represent only a small fraction of the total community (Yli-Hemminki et al., 2014). These concretions seem to universally select for *Alteromonas* and *Pseudoalteromonas* species (Wu et al., 2013) and genera that have a number of Mn(II)-oxidizing representatives (Caspi et al., 1996; Templeton et al., 2005; Dick et al., 2008b). The mechanisms of nodule formation vary widely and are a matter of debate. Proposed mechanisms include both direct and indirect microbial Mn(II) oxidation. For instance, indirect microbial activity (Kalinenko et al., 1962) has been suggested whereby bacterial destruction of organic complexes of Mn(II) in seawater liberates aqueous Mn(II) ions, which then abiotically oxidize and precipitate (Graham, 1959). Iron oxides have also been implicated in Mn(II) oxidation (Butkevich, 1928).

Microorganisms, including those resembling sheathed bacteria and fungal hyphae, have been observed on the surface and within nodules (Ehrlich et al., 1972; LaRock and Ehrlich, 1975; Ghiorse, 1980; Burnett and Nealson, 1981, 1983).

Greenslate (1974a) observed manganese deposition in microcavities of planktonic debris, especially from diatoms, and proposed that such deposition was the beginning of nodule growth. He also found remains of shelter-building organisms such as benthic foraminifera on nodules, which became encrusted and ultimately buried in the nodule structure. He proposed that the skeletal remains provided a framework on which manganese and other components were deposited, perhaps with the help of bacterial action (Greenslate, 1974b). Others have since reported evidence of traces of such organisms on nodules (Fredericks-Jantzen et al., 1975; Bignot and Dangeard, 1976; Dugolinsky et al., 1977; Harada, 1978; Riemann, 1983).

18.11 CONCLUDING COMMENTS

Through decades of interrogation, the intricate web of reactions involved in and influenced by the redox cycling of manganese has continued to grow. This research now provides the foundation of the Mn cycle, highlighting not only the importance of this cycle in biogeochemistry but also its immense complexity. Through this research, many questions remain, and many more have been introduced.

In particular, the root of biological oxidation of Mn(II) remains an enigma. Specifically, the question of why an organism would produce enzymes to oxidize Mn(II) if it is not specifically linked to energy conservation is perplexing and remains unanswered. It is clear now that a number of direct and indirect pathways are involved in biological oxidation, but how and why these pathways are invoked still needs to be resolved. On the flip side, the ecological and geochemical constraints on a lifestyle reliant on "breathing" (reducing) solid-phase Mn oxides with only limited solubility are unknown. The evolution of such a unique biochemical pathway, while awe-inspiring, is not understood.

The recent unveiling of Mn(III) as a dominant Mn species in aqueous and sedimentary systems presents an exciting new research frontier in the Mn cycle and biogeochemistry, in general. The consequences of overlooking such a strong oxidant has no doubt led to a substantial underestimation of the oxidation capacity of Mn-bearing environmental systems, with direct implications on the carbon cycle in particular. Biogenic Mn oxides, also, are receiving increasing acknowledgment as an important component of ancient and modern systems. The mechanisms and controls on Mn oxide nucleation and structural evolution need further resolution, however, to allow for better understanding of their utility as paleoredox indicators, remediation agents, and controls on the carbon cycle.

Over the coming decades, the scientific community will continue to be called upon to develop new strategies and methodologies to overcome the challenges associated with investigating the Mn cycle. These challenges include a paucity of genetic systems in Mn(II)-oxidizing organisms to allow for the identification of Mn-oxidizing enzymes, the lack of a Mn-specific oxidase or reductase to link environmental Mn dynamics directly to specific microbial activities, a need for emergent methods for identifying Mn(III) intermediate complexes, and the looming limitations in spectral resolution for identifying organic–oxide interfaces involved in Mn oxide nucleation and precipitation. Despite and perhaps because of, these challenges, it is an exciting time to be immersed in manganese geomicrobiology and to witness the emergence of this element as a key player in the biogeochemistry of Earth and beyond.

REFERENCES

Adams LF, Ghiorse WC. 1987. Characterization of an extracellular Mn^{2+}-oxidizing activity and isolation of Mn^{2+}-oxidizing protein from *Leptothrix discophora* SS-1. *J Bacteriol* 169:1279–1285.

Adams LF, Ghiorse WC. 1988. Oxidation state of Mn in the Mn oxide produced by *Leptothrix discophora* SS-1. *Geochim Cosmochim Acta* 52:2073–2076.

Adeny WE. 1894. On the reduction of manganese peroxide in sewage. *Proc Roy Dublin Soc*, Chap. 27:247–251.

Aguirre J, Rios-Momberg M, Hewitt D, Hansberg W. 2005. Reactive oxygen species and development in microbial eukaryotes. *Trends Microb* 13:111–118.

Akob, DM, Bohu T, Beyer A, Schaffner F, Handel M, Johnson CA, Merten D et al. 2014. Identification of Mn(II)-oxidizing bacteria from a low-pH contaminated former uranium mine. *Appl Environ Microbiol* 80:5086–5097.

Alexandrov EA. 1972. Manganese: Element and geochemistry. In: Fairbridge RW, ed. *Encyclopedia of Geochemistry and Environmental Sciences*. Encyclopedia of Earth Sciences Series, Vol. IVA. New York: Van Nostrand Reinhold, pp. 670–671.

Aller RC, Rude PD. 1988. Complete oxidation of solid phase sulfides by manganese and bacteria in anoxic marine sediments. *Geochim Cosmochim Acta* 52:751–765.

Anderson CR, Johnson HA, Caputo N, Davis RE, Torpey JW, Tebo BM. 2009. Mn(II) oxidation is catalyzed by heme peroxidase in *Aurantimonas manganoxydans* strain SI85–9A1 and *Erythrobacter* sp. strain SD-21. *Appl Environ Microbiol* 75:4130–4138.

Aristovskaya TV. 1961. Accumulation of iron in break-down of organo-mineral humus complexes by microorganisms. *Dokl Akad Nauk SSSR* 136:954–957.

Baker ET, Massoth GJ. 1986. Hydrothermal plume measurements. A regional perspective. *Science* 234:980–982.

Bamforth SM, Manning DAC, Singleton I, Younger PL, Johnson KL. 2006. Manganese removal from mine waters—Investigating the occurrence and importance of manganese carbonates. *Appl Geochem* 21:1274–1287.

Bandstra JZ, Ross DE, Brantley SL, Burgos WD. 2011. Compendium and synthesis of bacterial manganese reduction rates. *Geochim Cosmochim Acta* 75:337–351.

Banerjee D, Nesbitt HW. 2001. XPS study of disso-lution of birnessite by humate with constraints on reaction mechanism. *Geochim Cosmochim Acta* 65:1703–1714.

Bargar JR, Fuller CC, Marcus MA, Brearley AJ, De la Rosa MP, Webb SM, Caldwell WA. 2009. Structural char-acterization of terrestrial microbial Mn oxides from Pinal Creek, AZ. *Geochim Cosmochim Acta* 73:889–910.

Bargar JR, Tebo BM, Bergmann U, Webb SM, Glatzel P, Chiu VQ, Villalobos M. 2005. Biotic and abi-otic products of Mn(II) oxidation by spores of the marine *Bacillus* sp. strain SG-1. *Am Min* 90:143–154.

Bargar JR, Tebo BM, Villinski JE. 2000. In situ char-acterization of Mn(II) oxidation by spores of the marine *Bacillus* sp. strain SG-1. *Geochim Cosmochim Acta* 64:2775–2778.

Bartlett RJ, James B. 1979. Behavior of chromium in soils: III. Oxidation. *J Environ Qual* 8:31–35.

Barton HA, Northup DE. 2007. Geomicrobiology in cave environments: Past, current and future per-spectives. *J Cave Karst Stud* 69:163–178.

Beal EJ, House CH, Orphan VJ. 2009. Manganese- and iron-dependent marine methane oxidation. *Science* 325:184–187.

Bedard K, Lardy B, Krause KH. 2007. NOX family NADPH oxidases: Not just in mammals. *Biochimie* 89:1107–1112.

Beijerinck MW. 1913. Oxydation des Mangancarbonates durch Bakterien und Schimmelpilze. *Fol Microbiol Holländ Beitr Gesamt Mikrobiol Delft* 2:1–12.

Bender ML, Klinkhammer GP, Spencer DW. 1977. Manganese in seawater and marine manganese balance. *Deep-Sea Res* 24:799–812.

Bender ML, Ku T-L, Broecker WS. 1970. Accumulation rates of manganese in pelagic sediments and nod-ules. *Earth Planet Sci Lett* 8:143–148.

Berg B, Steffen KT, McClaugherty C. 2007. Litter decomposition rate is dependent on litter Mn con-centration. *Biogeochem* 82:29–39.

Beveridge TJ. 1989. Role of cellular design in bacterial metal accumulation and mineralization. *Annu Rev Microbiol* 43:147–171.

Bignot G, Dangeard L. 1976. Contribution à l'étude de la fraction biogène des nodules polymétalliques des fonds océanique actuels. *CR Somm Soc Geol Fr* 3:96–99.

Bilinski H, Giovanoli R, Usui A, Hanzel D. 2002. Characterization of Mn oxides in cemented stream-bed crusts from Pinal Creek, Arizona, U.S.A., and in hot-spring deposits from Yuno-Taki Falls, Hokkaido, Japan. *Amer Mineral* 87:580–591.

Blanchette RA. 1984. Manganese accumulation in wood decayed by white rot fungi. *Ecol Epidem* 74:725–730.

Bocioaga D, El Gheriany IA, Lion L, Ghiorse WC, Schuler ML, Hay AG. 2014. Development of a genetic system for a model Mn-oxidizing proteo-bacterium, *Leptothrix discophora* SS1. *Microbiol.*

Bodei S, Manceau A, Geoffroy N, Baronnet A, Buatier, M. 2007. Formation of todorokite from vernadite in Ni-rich hemipelagic sediments. *Geochim Cosmochim Acta* 71:5698–5716.

Boogerd FC, de Vrind JPM. 1987. Manganese oxida-tion by *Leptothrix discophora*. *J Bacteriol* 169:489–494.

Bowen HJM. 1979. *Environmental Chemistry of the Elements.* London, U.K.: Academic Press.

Bratina BJ, Stevenson BS, Green WJ, Schmidt TM. 1998. Manganese reduction by microbes from oxic regions of the lake vanda (Antarctica) water col-umn. *Appl Environ Microbiol* 64:3791–3797.

Bräuer SL, Adams C, Kranzler K, Murphy D, Xu M, Zuber P, Simon HM, Baptista AM, Tebo BM. 2011. Culturable *Rhodobacter* and *Shewanella* species are abundant in estuarine turbidity maxima of the Columbia River. *Environ Microbiol* 13:589–603.

Brewer PG. 1975. Minor elements in sea water. In: Riley JP, Skirrow G, eds. *Chemical Oceanography, I*, 2nd edn. London, U.K.: Academic Press, pp. 415–496.

Brock TD, Cook S, Petersen S, Mosser JL. 1976. Biogeochemistry and bacteriology of ferrous iron oxidation in geothermal habitats. *Geochim Cosmochim Acta* 40:493–500.

Bromfield SM. 1978. The oxidation of manganous ion under acid conditions by an acidophilous actinomycete from acid soil. *Aust J Biol Soil Res* 16:91–100.

Bromfield SM, David DJ. 1976. Sorption and oxidation of manganous ions and reduction of manganese oxide by cell suspensions of a manganese oxidizing bacterium. *Soil Biol Biochem* 8:37–43.

Bromfield SM, Sherman VDB. 1950. Biological oxidation of manganese in soils. *Soil Sci* 69:337–348.

Brouwers GJ, Corstjens PLAM, de Vrind JPM, Verkammen A, De Kuyper M, de Vrind-de Jong EW. 2000b. Stimulation of Mn^{2+} oxidation in *Leptothrix discophora* SS-1 by Cu^{2+} and sequence analysis of the region flanking the gene encoding putative multicopper oxidase MofA. *Geomicrobiol J* 17:25–33.

Brouwers, GJ., de Vrind JPM, Corstjens PLAM, Cornelis P, Baysse C, de Vrind-de Jong EW. 1999. cumA, a gene encoding a multicopper oxidase, is involved in Mn^{2+} oxidation in *Pseudomonas putida* GB-1. *Appl Environ Microb* 65:1762–1768.

Brouwers GJ, Vijgenboom E, Corstjens PLAM, de Vrind JPM, de Vrind-de Jong EW. 2000a. Bacterial Mn^{2+} oxidizing systems and multicopper oxidases: An overview of mechanisms and functions. *Geomicrobiol J* 17:1–24.

Burdige DJ, Dhakar SP, Nealson KH. 1992. Effects of manganese oxide mineralogy on microbial and chemical manganese reduction. *Geomicrobiol J* 10:27–48.

Burdige DJ, Kepkay PE. 1983. Determination of bacterial manganese oxidation rates in sediments using an in-situ dialysis technique. I. Laboratory studies. *Geochim Cosmochim Acta* 47:1907–1916.

Burdige DJ, Nealson KH. 1985. Microbiological manganese reduction by enrichment cultures from coastal marine sediments. *Appl Environ Microbiol* 50:491–497.

Burdige DJ, Nealson KH. 1986. Chemical and microbiological studies of sulfide-mediated manganese reduction. *Geomicrobiol J* 4:361–378.

Bureau of Mines. 1965. *Mineral Facts and Problems*. Bull. 630. Washington, DC: Bureau of Mines, U.S. Department of the Interior.

Burnett BR, Nealson KH. 1981. Organic films and microorganisms associated with manganese nodules. *Deep-Sea Res* 28A:637–645.

Burnett BR, Nealson KH. 1983. Energy dispersive X-ray analysis of the surface of a deep-sea ferromanganese nodule. *Mar Geol* 53:313–329.

Burns RG, Burns VM. 1977. Mineralogy of ferromanganese nodules. In: Glasby GP, ed. *Marine Manganese Deposits*. Amsterdam, the Netherlands: Elsevier.

Burns RG, Burns VM, Smig W. 1974. Ferromanganese mineralogy: Suggested terminology of the principal manganese oxide phases. *Abstr Progr. Ann Meet Geol Soc Am* 6(7):1029–1031.

Butkevich VS. 1928. The formation of marine manganese deposits and the role of microorganisms in the latter. *Wissenschaft Meeresinst Ber* 3:7–80.

Butler JE, Kaufmann F, Coppi MV, Nunez C, Lovley DR. 2004. MacA, a diheme c-type cytochrome involved in Fe(III) reduction by *Geobacter sulfurreducens*. *J Bacteriol* 186:4042–4045.

Butterfield CN, Soldatova AV, Lee SW, Spiro TG, Tebo BM. 2013. Mn(II,III) oxidation and MnO_2 mineralization by an expressed bacterial multicopper oxidase. *Proc Natl Acad Sci USA* 110:11731–11735.

Caccavo F Jr, Lonergan DJ, Lovley DR, Davis M, Stolz SF, McInerney MJ. 1994. *Geobacter sulfurreducens* sp. nov., a new hydrogen- and acetate-oxidizing dissimilatory metal-reducing microorganism. *Appl Environ Microbiol* 60:3752–3759.

Campbell AC, Gieskes JM, Lupton JE, Lonsdale PF. 1988. Manganese geochemistry in the Guaymas Basin, Gulf of California. *Geochim Cosmochim Acta* 52:345–357.

Cañaveras JC, Cuezva S, Sanchez-Moral S, Lario J, Laiz L, Gonzalez JM, Saiz-Jimenez C. 2006. On the origin of fiber calcite crystals in moonmilk deposits. *Naturwissenschaften* 93:27–32.

Canfield DE, Jorgensen BB, Fossing H, Glud R, Gundersen J, Ramsing NB, Thamdrup B, Hansen JW, Nielsen LP, Hall POJ. 1993a. Pathways of organic carbon oxidation in three continental margin sediments. *Mar Geol* 113:27–40.

Canfield DE, Thamdrup B, Hansen JW. 1993b. The anaerobic degradation of organic matter in Danish coastal sediments: Iron reduction, manganese reduction, and sulfate reduction. *Geochim Cosmochim Acta* 57:3867–3885.

Canfield DE, Thamdrup B, Kristensen E. 2005. The iron and manganese cycle. In: *Aquatic Geomicrobiology, Advances in Marine Biology*. New York: Academic Press, pp. 269–312, Chapter 8.

Carmichael MJ, Carmichael SK, Santelli C, Strom A, and Bräuer SL. 2013. Environmentally relevant members of ferromanganese deposits in caves of the upper Tennessee River Basin. *Geomicrobiol J* 30:779–800.

Caspi R, Haygood MG, Tebo BM. 1996. Unusual ribulose-1,5- bisphosphate carboxylase/oxygenase genes from a marine manganese-oxidizing bacterium. *Microbiol* 142:2549–2559.

Caspi R, Tebo BM, Haygood MG. 1998. c-type cytochromes and manganese oxidation in *Pseudomonas putida* strain MnB1. *Appl Environ Microbiol* 64:3549–3555.

Chan CS, De Stasio G, Welch S, Girasole M, Frazer BH, Nesterova MV, Fakra S, Banfield J. 2004. Microbial polysaccharides template assembly of nanocrystal fibers. *Science* 303:1656–1658.

Chan CS, Fakra SC, Emerson D, Fleming EJ, Edwards KJ. 2011. Lithotrophic iron-oxidizing bacteria produce organic stalks to control mineral growth: Implications for biosignature formation. *ISME J* 5:717–727.

Chapnik SD, Moore WS, Nealson KH. 1982. Microbially mediated manganese oxidation in a freshwater lake. *Limnol Oceanogr* 27:1004–1014.

Chinni S, Anderson C, Ulrich KU, Giammar DE, Tebo BM. 2008. Indirect UO_2 oxidation by Mn(II)-oxidizing spores of *Bacillus* sp. strain SG-1 and the effect of U and Mn concentrations. *Environ Sci Technol* 42:8709–8714.

Cílek V, Fábry J. 1989. Epigenetické, manganembohaté polohy v krasových v'yplních Zlatého koně v Českém krasu. *Českonolov Kras* 40:37–55.

Clement BG, Luther III GW, Tebo BM. 2009. Rapid, oxygen-dependent microbial Mn(II) oxidation kinetics at sub-micromolar oxygen concentrations in the Black Sea suboxic zone. *Geochim Cosmochim Acta* 73:1878–1889.

Coates JD, Ellis DJ, Gaw CV, Lovley DR. 1999. *Geothrix fermentans* gen nov., sp. nov., a novel Fe(III)-reducing bacterium from a hydrocarbon contaminated aquifer. *Int J Syst Bacteriol* 49:1615–1622.

Connell L, Barret A, Templeton A, Staudigel H. 2009. Fungal diversity associated with Active Deep Sea Volcano: Vailulu'u Seamount, Samoa. *Geomicrobiol J* 26:597–605.

Corstjens PLAM, de Vrind JPM, Goosen T, de Vrind-de Jong EW. 1997. Identification and molecular analysis of the *Leptothrix discophora* SS-1 *mofA* gene, a gene putatively encoding a manganese oxidizing protein with copper domains. *Geomicrobiol J* 14:91–108.

Coughlin RW, Matsui I. 1976. Catalytic-oxidation of aqueous Mn(II). *J Catalysis* 41:108–123.

Coursolle D, Gralnick JA. 2010. Modularity of the Mtr respiratory pathway of *Shewanella oneidensis* strain MR-1. *Mol Microbiol* 77:995–1008.

Cowen JP, Bertram MA, Baker ET, Feely RA, Massoth GJ, Summit M. 1998. Geomicrobial transformation of manganese in Gorda Ridge event plumes. *Deep-Sea Res II* 45:2713–2737.

Cowen JP, Bruland KW. 1985. Metal deposits associated with bacteria: Implications for Fe and Mn marine biogeochemistry. *Deep-Sea Res* 32:253–272.

Cowen JP, Massoth GJ, Baker ET. 1986. Bacterial scavenging of Mn and Fe in a mid- to far-field hydrothermal particle plume. *Nature* 322:169–171.

Cowen JP, Massoth GJ, Feely RA. 1990. Scavenging rates of dissolved manganese in a hydrothermal vent plume. *Deep-Sea Res* 37:1619–1637.

Cowen JP, Silver MW. 1984. The association of iron and manganese with bacteria on marine microparticulate material. *Science* 224:1340–1342.

Cravotta III CA, Trahan MK. 1999. Limestone drains to increase pH and remove dissolved metals from acidic mine drainage. *Appl Geochem* 14:581–606.

Crerar DA, Barnes HL. 1974. Deposition of deep-sea manganese nodules. *Geochim Cosmochim Acta* 38:279–300.

Crerar DA, Fischer AG, Plaza CL. 1979. *Metallogenium* and biogenic deposition of manganese from Precambrian to recent time. In: *Geology and Geochemistry of Manganese*, Vol III. Budapest, Hungary: Publ House Hungarian Acad Sci, pp. 285–303.

Crerar DA, Fischer AG, Plaza CL. 1980. *Metallogenium* and biogenic deposition of manganese from Precambrian to recent time. In: Vanentsov IM, Grasselly G, eds. *Geology and Geochemistry of Manganese. Volume III. Manganese on the Bottom of Recent Basins.* Stuttgart, Germany: Verlagsbuchhandlung, pp. 285–303.

Cronan DS. 1974. Authigenic minerals in deep-sea sediments. In: Goldberg, ED, ed. *The Sea*, Vol. 5. New York: Wiley Interscience, pp. 491–525.

Cronan DS, Thomas RL. 1970. Ferromanganese concretions in Lake Ontario. *Can J Earth Sci* 7:1346–1349.

Crowe SA, Katsev S, Leslie K, Sturm A, Magen C, Nomosatryo S, Pack MA et al. 2011. The methane cycle in ferruginous Lake Matano. *Geobiol* 9:61–78.

Daly MJ, Gaidamakova EK, Matrosova VY, Vasilenko A, Zhai M, Leapman RD, Lai B, Ravel B, Li SMW, Kemner KM, Fredrickson JK. 2007. Protein oxidation implicated as the primary determinant of bacterial radioresistance. *PLoS Biol* 5:0769–0779.

Daly MJ, Gaidamakova EK, Matrosova VY, Vasilenko A, Zhai M, Venkateswaran A, Hess M et al. 2004. Accumulation of Mn(II) in *Deinococcus radiodurans* facilitates gamma-radiation resistance. *Science* 306:1025–1028.

Davies SHR, Morgan JJ. 1989. Manganese(II) oxidation kinetics on metal oxide surfaces. *J Colloid Interface Sci* 129:63–77.

de la Torre MA, Gomez-Alarcon G. 1994. Manganese and iron oxidation by fungi isolated from building stone. *Microb Ecol* 27:177–188.

de los Ríos A, Bustillo M, Ascaso C, Carvalho MR. 2011. Bioconstructions in ochreous speleothems from lava tubes on Terceira Island (Azores). *Sediment Geol* 236:117–128.

de Vrind JPM, Boogerd FC, de Vrind-de Jong EW. 1986b. Manganese reduction by a marine *Bacillus* species. *J Bacteriol* 167:30–34.

de Vrind JPM, Brouwers GJ, Corstjens PLAM, den Dulk J, de Vrind-de Jong EW. 1998. The cytochrome c maturation operon is involved in manganese oxidation in *Pseudomonas putida* GB-1. *Appl Environ Microb* 64:3556–3562.

de Vrind JPM, de Vrind-de Jong EW, de Vogt J-WH, Westbroek P, Boogerd FC, Rosson R. 1986a. Manganese oxidation by spore coats of marine *Bacillus* sp. *Appl Environ Microbiol* 52:1096–1100.

Dean WE. 1970. Fe-Mn oxidate crusts in Oneida Lake, New York. *Proc Conf Great Lakes Res Int Assoc Great Lakes Res* 13:217–226.

Dean WE, Ghosh SK. 1978. Factors contributing to the formation of ferromanganese nodules in Oneida Lake, New York, *J Res US Geol Surv* 6:231–240.

Dean WE, Greeson PE. 1979. Influences of algae on the formation of freshwater ferromanganese nodules, Oneida Lake, New York. *Arch Hydrobiol* 86:181–192.

Delgadillo-Hinojosa F, Segovia-Zavala JA, Huerta-Diaz MA. 2006. Influence of geochemical and physical processes on the vertical distribution of manganese in Gulf of California waters. *Deep Sea Res I* 53:1301–1319.

Dellwig O, Schnetger B, Brumsack HJ, Grossart HP, Umlauf L. 2012. Dissolved reactive manganese at pelagic redoxclines (part II): Hydrodynamic conditions for accumulation. *J Mar Syst* 90:31–41.

D'Hondt S, Jørgensen BB, Miller DJ, Batzke A, Blake R, Cragg BA, Cypionka H et al. 2004. Distributions of microbial activities in deep subseafloor sediments. *Science* 306:2216–2221.

Diaz JM, Hansel CM, Voelker BM, Mendes CM, Andeer PF, Zhang T. 2013. Widespread production of extracellular superoxide by heterotrophic bacteria. *Science* 340:1223–1226.

DiChristina TJ, Moore CM, Haller CA. 2002. Dissimilatory Fe(III) and Mn(IV) reduction by *Shewanella putrefaciens* requires *ferE*, a homolog of the *pulE* (*gspE*) type II protein secretion gene. *J Bacteriol* 184:142–151.

Dick GJ, Clement BG, Webb SM, Fodrie FJ, Bargar JR, Tebo BM. 2009. Enzymatic microbial Mn(II) oxidation and Mn biooxide production in the Guaymas Basin deep-sea hydrothermal plume. *Geochim Cosmochim Acta* 73:6517–6530.

Dick GJ, Podell S, Johnson HA, Rivera-Espinoza Y, Bernier-Latmani R, McCarthy JK, Torpey JW, Clement BG, Gaasterland T, Tebo BM. 2008b. Genomic insights into Mn(II) oxidation by the marine alphaproteobacterium *Aurantimonas* sp. strain SI85-9A1. *Appl Environ Microbiol* 74:2646–2658.

Dick GJ, Tebo BM. 2010. Microbial diversity and biogeochemistry of the Guaymas Basin deep-sea hydrothermal plume. *Environ Microbiol* 12:1334–1347.

Dick GJ, Torpey JW, Beveridge TJ, Tebo BM. 2008a. Direct identification of a bacterial manganese(II) oxidase, the multicopper oxidase MnxG, from spores of several different marine *Bacillus* species. *Appl Environ Microb* 74:1527–1534.

Diem D, Stumm W. 1984. Is dissolved Mn^{2+} being oxidized by O_2 in the absence of Mn-bacteria or surface catalysts? *Geochim Cosmochim Acta* 48:1571–1573.

Di-Ruggiero J, Gounot AM. 1990. Microbial manganese reduction by bacterial strains isolated from aquifer sediments. *Microb Ecol* 20:53–63.

Donat JR, Bruland KW. 1995. Trace elements in the ocean. In: Salbu B, Steinnes E, eds. *Trace Elements in Natural Waters*. Boca Raton, FL: CRC Press, Inc., pp. 247–281.

Dorn RI. 1991. *Rock varnish. Am Sci* 79:542–553.

Dorn RI, Oberlander TM. 1981. Microbial origin of desert varnish. *Science* 213:1245–1247.

Douka CE. 1977. Study of bacteria from manganese concretions. Precipitation of manganese by whole cells and cell-free extracts of isolated bacteria. *Soil Biol Biochem* 9:89–97.

Drosdoff M, Nikiforoff CC. 1940. Iron-manganese concretions in Dayton soils. *Soil Sci* 49:333–345.

Dubinina GA. 1978a. Functional role of bivalent iron and manganese oxidation in *Leptothrix pseudoochracea*. *Mikrobiologiya* 47:783–789 (Engl transl., pp. 631–636).

Dubinina GA. 1978b. Mechanism of oxidation of divalent iron and manganese by iron bacteria growing at neutral pH of the medium. *Mikrobiologiya* 47:591–599 (Engl transl., pp. 471–478).

Dubinina GA, Deryugina ZP. 1971. Electron microscope study of iron-manganese concretions from Lake Punnus-Yarvi. *Dokl Akad Nauk SSSR* 201:714–716 (Engl transl., pp. 738–740).

Duckworth OW, Sposito G. 2005. Siderophore-manganese(III) interactions. I. Air-oxidation of manganese(II) promoted by desferrioxamine B. *Environ Sci Technol* 39:6037–6044.

Dugolinsky BK, Margolis SV, Dudley WC. 1977. Biogenic influence on growth of manganese nodules. *J Sediment Petrol* 47:428–445.

Durand P, Prieur D, Jeanthon C, Jacq E. 1990. Occurrence and activity of heterotrophic manganese oxidizing bacteria associated with Alvinellids (polychaetous annelids) from a deep hydrothermal vent site on the East Pacific Rise. *CR Acad Sci (Paris)* 310 (ser III):273–278.

Edmond JM, Von Damm KL, McDuff RE, Measure CI. 1982. Chemistry of hot springs on the East Pacific Rise and their effluent dispersal. *Nature* 297:187–191.

Ehrlich HL. 1963. Bacteriology of manganese nodules. I. Bacterial action on manganese in nodule enrichments. *Appl Microbiol* 11:15–19.

Ehrlich HL. 1966. Reactions with manganese by bacteria from ferromanganese nodules. *Dev Ind Microbiol* 7:43–60.

Ehrlich HL. 1973. The biology of ferromanganese nodules. Determination of the effect of storage by freezing on the viable flora, and a check on the reliability of the results from a test to identify MnO_2-reducing cultures. In: *Interuniversity Program of Research of Ferromanganese Deposits on the Ocean Floor. Phase I Rep, Seabed Assessment Program, International Decade of Ocean Exploration.* Washington, DC: Nat Sci Foundation, pp. 217–219.

Ehrlich HL. 1976. Manganese as an energy source for bacteria. In: Nriagu JO, ed. *Environmental Biogeochemistry*, Vol. 2. Ann Arbor, MI: Ann Arbor Sci Pub, pp. 633–644.

Ehrlich HL. 1983. Manganese oxidizing bacteria from a hydrothermally active area on the Galapagos Rift. *Ecol Bull* 35:357–366.

Ehrlich HL. 1984. Different forms of bacterial manganese oxidation. In: Strohl WR, Tuovinen OH, eds. *Microbial Chemoautotrophy*. Columbus, OH: Ohio State Univ Press, pp. 47–56.

Ehrlich HL. 1985. Mesophilic manganese-oxidizing bacteria from hydrothermal discharge areas at 21° north on the East Pacific Rise. In: Caldwell DE, Brierley JA, Brierley CL, eds. *Planetary Ecology*. New York: Van Nostrand Reinhold, pp. 186–194.

Ehrlich HL. 1987. Manganese oxide reduction as a form of anaerobic respiration. *Geomicrobiol J* 5:423–431.

Ehrlich HL. 1993a. Electron transfer from acetate to the surface of MnO_2 particles by a marine bacterium. *J Ind Microbiol* 12:121–128.

Ehrlich HL. 1993b. A possible mechanism for the transfer of reducing power to insoluble mineral oxide in bacterial respiration. In: Torma AE, Apel ML, Brierley CL, eds. *Biohydrometallurgical Technologies*, Vol. II. Warrendale, PA: Min, Metals Mater Soc, pp. 415–422.

Ehrlich HL. 1999. Microbes as geologic agents: Their role in mineral formation. *Geomicrobiol J* 16:135–153.

Ehrlich HL. 2000. Ocean manganese nodules: Biogenesis and Bioleaching possibilities. *Mineral Metallurg Process* 17:121–128.

Ehrlich HL, Ghiorse WC, Johnson GL II. 1972. Distribution of microbes in manganese nodules from the Atlantic and Pacific Oceans. *Dev Ind Microbiol* 13:57–65.

Ehrlich HL, Newman DK. 2008. *Geomicrobiology*, 5nd rev ed. New York: Marcel Dekker.

Ehrlich HL, Salerno JC. 1988. Stimulation by ADP and phosphate of Mn^{2+} oxidation by cell extracts and membrane vesicles of induced Mn-oxidizing bacteria. *Abstr Annu Meet Am Soc Microbiol* K-151:231.

Ehrlich HL, Salerno JC. 1990. Energy coupling in Mn^{2+} oxidation by a marine bacterium. *Arch Microbiol* 154:12–17.

Ehrlich HL, Zapkin MA. 1985. Manganese-rich layers in calcareous deposits along the western shore of the Dead Sea may have a bacterial origin. *Geomicrobiol J* 4:207–221.

El Gheriany IA, Bocioaga D, Hay AG, Ghiorse WC, Shuler ML, Lion LW. 2009. Iron requirement for Mn(II) oxidation by *Leptothrix discophora* SS-1. *Appl Environ Microbiol* 75:1229–1235.

El Gheriany IA, Bocioaga D, Hay AG, Ghiorse WC, Shuler ML, Lion LW. 2011. An uncertain role for Cu(II) in stimulating Mn(II) oxidation by *Leptothrix discophora* SS-1. *Arch Microbiol* 193:89–93.

Emerson D, Garen RE, Ghiorse W. 1989. Formation of *Metallogenium*-like structures by a manganese-oxidizing fungus. *Arch Microbiol* 151:223–231.

Emerson S, Kalhorn S, Jacobs L, Tebo BM, Nealson KH, Rosson RA. 1982. Environmental oxidation rate of manganese(II): Bacterial catalysis. *Geochim Cosmochim Acta* 46:1073–1079.

Feng Q, Yanagisawa K, Yamasaki N. 1998. Hydrothermal soft chemical process for synthesis of manganese oxides with tunnel structures. *J Porous Mater* 5:153–161.

Feng XH, Zhu MQ, Ginder-Vogel M, Ni CY, Parikh SJ, Sparks DL. 2010. Formation of nano-crystalline todorokite from biogenic Mn oxides. *Geochim Cosmochim Acta* 74:3232–3245.

Fennessey CM, Jones ME, Taillefert M, DiChristina TJ. 2010. Siderophores are not involved in Fe(III) solubilization during anaerobic Fe(III) respiration by *Shewanella oneidensis* MR-1. *Appl Environ Microbiol* 76:2425–2432.

Fortin D, Ferris FG, Scott SD. 1998. Formation of Fe-silicate and Fe oxides on bacterial surfaces in samples collected near hydrothermal vents on the Southern Explorer Ridge in the northeast Pacific Ocean. *Am Mineral* 83:1399–1408.

Francis CA, Casciotti KL, Tebo BM. 2002. Localization of Mn(II)-oxidizing activity and the putative multicopper oxidase, MnxG, to the exosporium of the marine *Bacillus* sp. strain SG-1. *Arch Microbiol* 178:450–456.

Francis CA, Co E-M, Tebo MB. 2001. Enzymatic manganese(II) oxidation by a marine α-Proteobacterium. *Appl Environ Microbiol* 67:4024–4029.

Francis CA, Obraztsova AY, Tebo BM. 2000. Dissimilatory metal reduction by the facultative anaerobe *Pantoea agglomerans* SP1. *Appl Environ Microbiol* 66:543–548.

Francis CA, Tebo BM. 1999. Marine *Bacillus* spores as catalysts for oxidative precipitation and sorption of metals. *J Mol Microbiol Biotechnol* 1:71–78.

Francis CA, Tebo BM. 2001. *cumA* multicopper oxidase genes from diverse Mn(II)-oxidizing and non-Mn(II)-oxidizing *Pseudomonas* strains. *Appl Environ Microbiol* 67:4272–4278.

Francis CA, Tebo BM. 2002. Enzymatic manganese(II) oxidation by metabolically dormant spores of diverse *Bacillus* species. *Appl Environ Microbiol* 68:874–880.

Fredricks-Jantzen CM, Herman H, Herley P. 1975. Microorganisms on manganese nodules. *Nature (Lond)* 285:270.

Fredrickson JK, Zachara JM, Kennedy DW, Kukkadapu RK, McKinley JP, Heald SM, Liu C, Plymale AE. 2004. Reduction of TcO$_4^-$ by sediment-associated biogenic Fe(II). *Geochim Cosmochim Acta* 68:3171–3187.

Fredrickson JK, Zachara JM, Kennedy DW, Liu C, Duff MC, Hunter DB, Dohnalkova A. 2002. Influence of Mn oxides on the reduction of uranium(VI) by the metal-reducing bacterium *Shewanella putrefaciens*. *Geochim Cosmochim Acta* 66:3247–3262.

Gadd GM. 1999. Fungal production of citric and oxalic acid: Importance in metal speciation, physiology and biogeochemical processes. *Adv Microb Physiol* 41:47–92.

Gadd GM. 2007. Geomycology: Biogeochemical transformations of rocks, minerals, metals and radionuclides by fungi, bioweathering and bioremediation. *Mycological Res* 111:3–49.

Gadd GM. 2010. Metals, minerals and microbes: Geomicrobiology and bioremediation. *Microbiol* 156:609–643.

Garrels RM, Mackenzie FT. 1971. Gregors denudation of continents. *Nature* 231:382–383.

Gascoine W. 1982. The formation of black deposits in some caves of south east Wales. *Cave Sci* 9:165–175.

Geloso M. 1927. Adsorption au sein des solutions salines par le bioxide de manganese colloidal. *Ann Chim* 7:113–150.

Gerretsen FC. 1937. Manganese deficiency of oats and its relation to soil bacteria. *Ann Bot* 1:207–230.

Geszvain K, Butterfield C, Davis R, Madison A, Lee S, Parker D, Soldatova A, Spiro T, Luther G, Tebo B. 2012. The molecular biogeochemistry of manganese(II) oxidation. *Biochem Soc Trans* 40:1244–1248.

Geszvain K, McCarthy JK, Tebo BM. 2013. Elimination of manganese(II,III) oxidation in *Pseudomonas putida* GB-1 by a double knockout of two putative multicopper oxidase genes. *Appl Environ Microbiol* 79:357–366.

Ghiorse WC. 1980. Electron microscopic analysis of metal-depositing microorganisms in surface layers of Baltic Sea ferromanganese concretions. In: Trudinger PA, Walter MR, Ralph BJ, eds. *Biogeochemistry of Ancient and Modern Environments.* Canberra, Australia: Aust Acad Sci, pp. 345–354.

Ghiorse WC, 1984. Biology of iron- and manganese-depositing bacteria. *Annu Rev Microbiol* 38:515–550.

Ghiorse WC. 1988. Microbial reduction of manganese and iron. In: Zehnder JB, ed. *Biology of Anaerobic Microorganisms.* New York: Wiley, pp. 305–331.

Ghiorse WC, Chapnik SD. 1983. Metal-depositing bacteria and the distribution of manganese and iron in swamp waters. *Ecol Bull (Stockh)* 35:367–376.

Ghiorse WC, Ehrlich HL. 1976. Electron transport components of the MnO$_2$ reductase system and the location of the terminal reductase in a marine bacillus. *Appl Environ Microbiol* 31:977–985.

Ghiorse WC, Ehrlich HL. 1992. Microbial biomineralization of iron and manganese. In: Skinner HCW, Fitzpatrick RW, eds. *Biomineralization. Process of Iron and Manganese. Modern and Ancient Environments*, Catena Suppl 21. Cremlingen, Germany: Catena Verlag, pp. 75–99.

Ghiorse WC, Hirsch P. 1979. An ultrastructural study of iron and manganese deposition associated with extracellular polymers of *Pedomicrobium*-like budding bacteria. *Arch Microbiol* 123:213–226.

Gillette NJ. 1961. Oneida Lake Pancakes. *NY State Conservationist* 41(April–May):21.

Giorgetti G, Marescotti P, Cabella R, Lucchetti G. 2000. Clay mineral mixtures as alteration products in pillow basalts from the eastern flank of Juan de Fuca Ridge: A TEM-AEM study. *Clay Miner* 36:75–91.

Giovanoli R. 1980. Vernadite is random-stacked birnessite. *Mineral. Deposita* 15:251–285.

Glenn JK, Akileswaram L, Gold MH. 1986. Mn(II) oxidation is the principal function of the extracellular Mn-peroxidase from *Phanerochaete chrysosporium*. *Arch Biochem Biophys* 251:688–696.

Glenn JK, Gold MH. 1985. Purification and characterization of an extracellular Mn(II)-dependent peroxidase from the lignin degrading basidiomycete *Phanerochaete chrysosporium*. *Arch Biochem Biophys* 242:329–341.

Glenn JK, Morgan MA, Mayfield MB, Kuwahara M, Gold MH. 1983. An extracellular H_2O_2-requiring enzyme preparation involved in lignin biodegradation by the white rot basidiomycete *Phanerochaete chrysosporium*. *Biochem Biophys Res Commun* 114:1077–1083.

Gold MH, Alic M. 1993. Molecular biology of the lignin-degrading basidiomycete *Phanerochaete chrysosporium*. *Microbiol Rev* 57:605–622.

Goldberg ED. 1954. Marine geochemistry. I. Chemical scavengers of the sea. *J Geol* 62:249–265.

Goldberg ED, Arrhenius GOS. 1958. Chemistry of Pacific pelagic sediments. *Geochim Cosmochim Acta* 13:153–212.

Golden DC, Zuberer DA, Dixon JB. 1992. Manganese oxides produced by fungal oxidation of manganese from siderite and rhodochrosite. In: Skinner HCW, Fitzpatrick RW, eds. *Biomineralization Processes of Iron and Manganese: Modern and Ancient Environments*. Destedt, Germany: Catena.

Goldschmidt VM. 1954. Geochemistry. Oxford, U.K.: Clarendon Press, pp. 621–642.

Graham JW. 1959. Metabolically induced precipitation of trace elements from sea water. *Science* 129:1428–1429.

Graham JW, Cooper S. 1959. Biological origin of manganese-rich deposits on the sea floor. *Nature* 183:1050–1051.

Grangeon S, Lanson B, Miyata N, Tani Y, Manceau A. 2010. Structure of nanocrystalline phyllomanganates produced by freshwater fungi. *Am Mineral* 95:1608–1616.

Greene AC, Patel BKC, Sheehy AJ. 1997. *Deferribacter thermophilus* gen. nov., sp. nov., a novel thermophilic manganese-and iron-reducing bacterium isolated from a petroleum reservoir. *Int J Syst Bacterio* 47:505–509.

Greenslate J. 1974a. Manganese and biotic debris associations in some deep-sea sediments. *Science* 186:529–531.

Greenslate J. 1974b. Microorganisms participate in the construction of manganese nodules. *Nature* 249:181–183.

Gregory E, Perry RS, Staley JT. 1980. Characterization, distribution, and significance of *Metallogenium* in Lake Washington. *Microb Ecol* 6:125–140.

Grill EV. 1982. Kinetic and thermodynamic factors controlling manganese concentrations in oceanic waters. *Geochim Cosmochim Acta* 46:2435–2446.

Haack EA, Warren LA. 2003. Biofilm hydrous manganese oxyhydroxides and metal dynamics in acid rock drainage. *Environ Sci Technol* 37:4138–4147.

Hallberg KB, Johnson DB. 2005. Biological manganese removal from acid mine drainage in constructed wetlands and prototype bioreactors. *Sci Total Environ* 338:115–124.

Hansard SP, Easter HD, Voelker BM. 2011. Rapid reaction of nanomolar Mn(II) with superoxide radical in seawater and simulated freshwater. *Environ Sci Technol* 45:2811–2817.

Hansel CM, Francis CA. 2006. Coupled photochemical and enzymatic Mn(II) oxidation pathways of a planktonic *Roseobacter*-like bacterium. *Appl Environ Microb* 72:3543–3549.

Hansel CM, Zeiner CA, Santelli CM, Webb SM. 2012. Mn(II) oxidation by an ascomycete fungus is linked to superoxide production during asexual reproduction. *Proc Natl Acad Sci USA* 109:12621–12625.

Harada K. 1978. Micropaleontologic investigation of Pacific manganese nodules. *Mem Fac Sci Kyoto Univ Ser Geol Mineral* 45:111–132.

Hariya T, Kikuchi T. 1964. Precipitation of manganese by bacteria in mineral springs. *Nature* 202:416–417.

Harriss RC, Troup AG. 1970. Chemistry and origin of freshwater ferromanganese concretions. *Limnol Oceanogr* 15:702–712.

Harvey JW, Fuller CC. 1998. Effect of enhanced manganese oxidation in the hyporheic zone on basin-scale geochemical mass balance. *Water Resources Res* 34:623–636.

Hein J, Gibbs A, Clague D, Torresan M. 1996. Hydrothermal mineralization along submarine rift zones in Hawaii. *Mar Geores Geotech* 14:177–203.

Hekinian R, Hoffert M. 1975. Rate of palagonitization and manganese coating on basaltic rocks from the rift valley in the Atlantic ocean near 36 deg 50 N. *Mar Geol* 19:91–109.

Hem JD. 1963. Chemical equilibria and rate of manganese oxidation. US Geol Surv Water Supply Paper 1667-A.

Hem JD, Lind CJ. 1983. Nonequilibrium models for predicting forms of precipitated manganese oxides. *Geochim Cosmochim Acta* 47:2037–2046.

Heye D, Beiersdorf H. 1973. Radioaktive und magnetische Untersuchungen an Manganknollen zur Ermittlung der Wachstumsgeschwindigkeit bzw. Zur Altersbestimmung. *Z Geophysik* 39:703–726.

Hill CA. 1982. Origin of black deposits in caves. *NSS Bull* 44:15–19.

Hochella Jr. MF, Lower S, Maurice P, Penn R, Sahai N, Sparks DL, Twinning B. 2008. Nanominerals, mineral nanoparticles, and earth systems. *Science* 319:1631.

Höfer C, Schlosser D. 1999. Novel enzymatic oxidation of Mn^{2+} to Mn^{3+} catalyzed by a fungal laccase. *FEBS Lett* 451:186–190.

Hofrichter M. 2002. Review: Lignin conversion by manganese peroxidase (MnP). *Enzyme Microbial Technol* 30:454–466.

Holmes DE, Chaudhuri SK, Nevin KP, Mehta T, Methé BA, Liu A, Ward JE, Woodard TL, Webster J, Lovley DR. 2006. Microarray and genetic analysis of electron transfer to electrodes in *Geobacter sulfurreducens*. *Environ Microbiol* 8:1805–1815.

Hood DW. 1963. Chemical oceanography. *Oceanogr Mar Biol Annu Rev* 1:129–155.

Hood DW, Slowey JF. 1964. Texas A and M Univ Progr Rept, Proj 276, AEC Contract No. AT-(40–1)-2799.

Horn DR, Horn BM, Delach MN. 1972. Distribution of ferromanganese deposits in the world ocean. In: Horn DR, ed. *Ferromanganese Deposits on the Ocean Floor*. The Office of the International Decade of Ocean Exploration. Washington, DC: Natl Sci Foundation, pp. 9–17.

Hulth S., Aller R. C. and Gilbert F. (1999) Coupled anoxic nitrification manganese reduction in marine sediments. *Geochim Cosmochim Acta* 63:49–66.

Hungate B, Danin A, Pellerin NB, Stemmler J, Kjellander P, Adams BJ, Staley JT. 1987. Characterization of manganese oxidizing ($Mn^{II} \rightarrow Mn^{IV}$) bacteria from Negev Desert rock varnish: Implications in desert varnish formation. *Can J Microbiol* 33:939–943.

Izawa K, Yamada T, Unal U, Ida S, Altuntasoglu O, Koinuma M, Matsumoto Y. 2006. Photoelectrochemical oxidation of methanol on oxide nanosheets. *J Phys Chem B* 110:4645–4650.

Jackson DD. 1901a. A new species of *Crenothrix* (*C. manganifera*). *Trans Am Microsc Soc* 23:31–39.

Jackson DD. 1901b. The precipitation of iron, manganese, and aluminum by bacterial action. *J Soc Chem Ind* 21:681–684.

Jacobs L, Emerson S, Skei J. 1985. Partitioning and transport of metals across an O_2/H_2S interface in a permanently anoxic basin: Framvaren Fjord, Norway. *Geochim Cosmochim Acta* 49:1433–1444.

Jannasch HW, Mottle MJ. 1985. Geomicrobiology of deep-sea hydrothermal vents. *Science* 229:717–725.

Jaquet JM, Nembrini G, Garcia J, Vernet JP. 1982. The manganese cycle in Lac Léman, Switzerland: The role of *Metallogenium*. *Hydrobiology* 91:323–340.

Jenne EA. 1968. Controls on Mn, Co, Ni, Cu, and Zn concentrations in soil and water: The significant role of hydrous Mn and Fe oxides. In: Gould RF, ed. *Trace Inorganics in Water*. Washington, DC: American Chemical Society, pp. 337–387.

Johnson HA, Tebo BM. 2008. In vitro studies indicate a quinone is involved in bacterial Mn(II) oxidation. *Arch Microbiol* 189:59–69.

Johnson JE, Webb SM, Thomas K, Ono S, Kirschvink JL, Fischer WW. 2013. Manganese-oxidizing photosynthesis before the rise of cyanobacteria. *Proc Natl Acad Sci USA* 110:11238–11243.

Johnson KL, Baker A, Manning DAC. 2005. Passive treatment of Mn-rich mine water: Using fluorescence to observe microbiological activity. *Geomicrobiol J* 22:141–149.

Johnson KL, Younger PL. 2005. Rapid manganese removal from mine waters using an aerated packed-bed bioreactor. *J Environ Qual* 34:987–993.

Jones B. 1992. Manganese precipitates in the karst terrain of Grand Cayman, BritishWest Indies. *Can J Earth Sci* 29:1125–1139.

Jones B. 2001. Microbial activity in caves— A geological perspective. *Geomicrobiol J* 18:345–357.

Jones C, Crowe SA, Sturm A, Leslie KL, MacLean LCW, Katsev S, Henny C, Fowle DA, Canfield DE. 2011, Biogeochemistry of manganese in ferruginous Lake Matano, Indonesia. *Biogeosci* 8:2977–2991.

Jones CJ, Murray JW. 1985. The geochemistry of manganese in the northeast Pacific Ocean off Washington. *Limnol Oceanogr* 30:81–92.

Jones ME, Fennessey CM, DiChristina TJ, Taillefert M. 2010. *Shewanella oneidensis* MR-1 mutants selected for their inability to produce soluble organic-Fe(III) complexes are unable to respire Fe(III) as anaerobic electron acceptor. *Environ Microbiol* 12:938–950.

Juniper SK, Tebo BM. 1995. Microbe-mineral interactions and mineral deposition at hydrothermal vents. In: Karl DM, ed. *The Mineralogy of Deep-Sea Hydrothermal Vents*. Boca Raton, FL: CRC Press, pp. 219–253.

Junta J, Hochella MF. 1994. Manganese(II) oxidation at mineral surfaces—A microscopic and spectroscopic study. *Geochim Cosmochim Acta* 58:4985–4999.

Kadko D, Burckle LH. 1980. Manganese nodule growth rates determined by fossil diatom dating. *Nature* 287:725–726.

Kalhorn S, Emerson S. 1984. The oxidation state of manganese in surface sediments of the deep sea. *Geochim Cosmochim Acta* 48:897–902.

Kalinenko VO, Belokopytova OV, Nikolaeva GG. 1962. Bacteriogenic formation of iron-manganese concretions in the Indian Ocean. *Okeanologiya* 11:1050–1059 (Engl transl).

Kashefi K, Lovley DR. 2000. Reduction of Fe(III), Mn(IV), and toxic metals at 100°C by *Pyrobaculum islandicum*. *Appl Environ Microbiol* 66:1050–1056.

Kashima N. 1983. On the wad minerals from the cavern environment. *Int J Speleol* 13:67–72.

Kay JT, Conklin MH, Fuller CC, O'Day PA. 2001. Processes of nickel and cobalt uptake by a manganese oxide forming sediment in Pinal Creek, globe mining district, Arizona. *Environ Sci Technol* 35:4719–4725.

Keen CL, Ensunsa JL, Watson MH, Baly DL, Donovan SM, Monaco MH, Clegg MS. 1999. Nutritional aspects of manganese from experimental studies. *Neurotoxicology* 20:213–223.

Kenten RH, Mann PJG. 1949. The oxidation of manganese by plant extracts in the presence of hydrogen peroxide. *Biochem J* 45:225–263.

Kepkay PE. 1985a. Kinetics of microbial manganese oxidation and trace metal binding in sediments: Results from an in situ dialysis technique. *Limnol Oceanogra* 30:713–726.

Kepkay PE. 1985b. Microbial manganese oxidation and trace metal binding in sediments: Results from an in situ dialysis technique. In: Caldwell DE, Brierley JA, Brierley CL, eds. *Planetary Ecology*. New York: Van Nostrand Reinhold, pp. 195–209.

Kepkay PE. 1985c. Microbial manganese oxidation and nitrification in relation to the occurrence of macrophyte roots in lacustrine sediments. *Hydrobiology* 128:135–142.

Kieft TL, Fredrickson JK, Onstott TC, Gorby YA, Kostandarithes HM, Bailey TJ, Kennedy DW, Li SW, Plymale AE, Spandoni CM, Gray MS. 1999. Dissimilatory reduction of Fe(III) and other electron acceptors by Thermus isolate. *Appl Environ Microbiol* 65:1214–1221.

Kindle EM. 1932. Lacustrine concretions of manganese. *Am J Sci* 24:496–504.

Klaveness D. 1977. Morphology, distribution and significance of the manganese-accumulating microorganisms Metallogenium in lakes. *Hydrobiologia* 56:25–33.

Klewicki JK, Morgan JJ. 1998. Kinetic behavior of Mn(III) complexes of pyrophosphate, EDTA and citrate. *Environ Sci Technol* 32:2916–2922.

Klewicki JK, Morgan JJ. 1999. Dissolution of α-MnOOH particles by ligand: Pyrophosphate, ethylenediaminetetraacetate, and citrate. *Geochim Cosmochim Acta* 63:2017–3024.

Klimov VV. 1984. Charge separation in photosystem II reaction centers. The role of phaeophytin and manganese. In Sybesina C, ed. *Advances in Photosynthetic Research, Vol 1. Proc Int Congr Photosynth 6th*. The Hague, the Netherlands: Nijhoff, pp. 131–138.

Klinkhammer GP, Bender ML. 1980. The distribution of manganese in the Pacific Ocean. *Earth Planet Sci Lett* 46:361–384.

Knittel K, Losekann T, Boetius A, Kort R, Amann, R. 2005. Diversity and distribution of methanotrophic archaea at cold seeps. *Appl Environ Microbiol* 71:467–479.

Komeili A. 2007. Molecular mechanisms of magnetosome formation. *Annu Rev Biochem* 76:351–366.

Komeili A, Newman DK, Jensen GJ. 2006. Magnetosomes are cell membrane invaginations organized by the actin-like protein MamK. *Science* 311:242–245.

Kostka JE, Luther GW III, Nealson KH. 1995. Chemical and biological reduction of Mn(III) pyrophosphate complexes: Potential importance of dissolved Mn(III) as an environmental oxidant. *Geochim Cosmochim Acta* 59:885–894.

Krumbein WE. 1969. Ueber den Einfluss der Mikroflora auf die exogene Dynamik (Verwitterung und Krustenbildung). *Geol Rundschau* 58:333–365.

Krumbein WE. 1971. Manganese oxidizing fungi and bacteria in recent shelf sediments of the Bay of Biscay and the North Sea. *Naturwissenschaften* 58:56–57.

Krumbein WE, Altmann HJ. 1973. A new method for the detection and enumeration of manganese-oxidizing and reducing microorganisms. *Helgoländer Wissenschaftliche Meeresuntersuchungen* 25:347–356.

Krumbein WE, Jens K. 1981. Biogenic rock varnishes of the Negev Desert (Israel): An ecological study of iron and manganese transformation by cyanobacteria and fungi. *Oecologia (Berl)* 50:25–38.

Ku TL, Broecker WS. 1969. Radiochemical studies on manganese nodules of deep-sea origin. *Deep-Sea Res Oceanogr Abstr* 16:625–635.

Kwon KD, Refson K, Sposito G. 2009. On the role of Mn(IV) vacancies in the photoreductive dissolution of hexagonal birnessite. *Geochim Cosmochim Acta* 73:4142–4150.

Lackner R, Srebotnik E, Messner K. 1991. Oxidative degradation of high molecular weight chlorolignin by manganese peroxidase of Phanerochaete chrysosporium. *Biochem Biophys Res Commun* 178:1092–1098.

Landing WM, Bruland KW. 1987. The contrasting biogeochemistry of iron and manganese in the Pacific Ocean. *Geochim Cosmochim Acta* 51:29–43.

Landing WM, Bruland K. 1980. Manganese in the North Pacific. *Earth Planet Sci Lett* 49:45–56.

Langenhoff AAM, Brouwers-Ceiler DL, Engelberting JHL, Quist JJ, Wolkenfeldt JGPN, Zehnder AJB, Schraa G. 1997. Microbial reduction of manganese coupled to toluene oxidation. *FEMS Microbiol Ecol* 22:119–127.

Lanza NL, Fischer WW, Wiens RC, Grotzinger J, Ollila AM, Cousin A, Anderson RB et al. 2014. High manganese concentrations in rocks at Gale crater, Mars. *Geophys Res Lett* 41:5755–5763.

LaRock PA, Ehrlich HL. 1975. Observations of bacterial microcolonies on the surface of ferromanganese nodules from Blake Plateau by scanning electron microscopy. *Microb Ecol* 2:84–96.

Larsen I, Little B, Nealson KH, Ray R, Stone A, Tian J. 1998. Manganite reduction by *Shewanella putrefaciens* MR-4. *Am Min* 83:1564–1572.

Larsen EI, Sly LI, McEwan AG. 1999. Manganese(II) adsorption and oxidation by whole cells and a membrane fraction of *Pedomicrobium* sp. ACM 3067. *Arch Microbiol* 171:257–264.

Laverman AM, Switzer Blum J, Schaefer JK, Philips EJP, Lovley DR, Oremland RS. 1995. Growth of strain SES-3 with arsenate and other diverse electron acceptors. *Appl Environ Microbiol* 61:3556–3561.

Lawton K. 1955. Chemical composition of soils. In: Bear FE, ed. *Chemistry of the Soil*. New York: Reinhold, pp. 53–54.

Leach RM, Harris ED. 1997. Manganese. In: O'Dell BL, Sunde RA, eds. *Handbook of Nutritionally Essential Minerals*. New York: Marcel Dekker, Inc., pp. 335–355.

Leang C, Coppi MV, Lovley DR. 2003. OmcB, a c-type polyheme cytochrome, involved in Fe(III) reduction in *Geobacter sulfurreducens*. *J Bacteriol* 185:2096–2103.

Learman DR, Hansel CM. 2014. Comparative proteomics of Mn(II)-oxidizing and non-oxidizing Roseobacter clade bacteria reveal an operative manganese transport system but minimal Mn(II)-induced expression of manganese oxidation and antioxidant enzymes. *Environ Microbiol Rep*.

Learman DR, Voelker BM, Madden AS, Hansel CM. 2013. Constraints on superoxide mediated formation of manganese oxides. *Front Microbiol* 4:1–11.

Learman DR, Voelker BM, Vazquez-Rodriguez AI, Hansel CM (2011a) Formation of manganese oxides by bacterially generated superoxide. *Nat Geosci* 4:95–98.

Learman DR, Wankel SD, Webb SM, Martinez N, Madden AS, Hansel CM. 2011b. Coupled biotic-abiotic Mn(II) oxidation pathway mediates the formation and structural evolution of biogenic Mn oxides. *Geochim Cosmochim Acta* 75:6048–6063.

Leeper GW, Swaby RJ. 1940. The oxidation of manganous compounds by microorganisms in the soil. *Soil Sci* 49:163–169.

Letunova SV, Ulubekova MV, Shcherbakov VI. 1978. Manganese concentration by microorganisms inhabiting soils of the manganese biogeochemical province of the Georgian SSR. *Mikrobiologiya* 47:332–337 (Engl transl., pp. 273–278).

Lewis BL, Landing WM. 1991. The biogeochemistry of manganese and iron in the Black-Sea. *Deep-Sea Res A Ocean Res Papers* 38:S773-S803.

Lieske R. 1919. Zur Ernährungsphysiologie der Eisenbakterien. *Zentralbl Bakteriol Parasitenk Infektionskr Hyg Abt II* 49:413–425.

Lin H, Szeinbaum NH, DiChristina TJ, Taillefert M. 2012. Microbial Mn(IV) reduction requires an initial one-electron reductive solubilization step. *Geochim Cosmochim Acta* 99:179–192.

Lin H, Taillefert M. 2014. Key geochemical factors regulating Mn(IV)-catalyzed anaerobic nitrification in coastal marine sediments. *Geochim Cosmochim Acta* 133:17–33.

Lind CJ, Hem JD. 1993. Manganese minerals and associated fine particulates in the streambed of Pinal Creek, Arizona, U.S.A.: A mining-related acid drainage problem. *Appl Geochem* 8:67–80.

Lloyd JR, Leang C, Hodges Myerson AL, Coppi MV, Cuifo S, Methe B, Sandler SJ, Lovley DR. 2003. Biochemical and genetic characterization of PpcA, a periplasmic c-type cytochrome in *Geobacter sulfurreducens*. *Biochem J* 369:153–161.

Loganathan P, Burau RG. 1973. Sorption of heavy metals by hydrous manganese oxides. *Geochim Cosmochim Acta* 37:1277–1293.

Lovley DR. 1991. Dissimilatory Fe(III) and Mn(IV) reduction. *Microbiol Rev* 55:259–287.

Lovley DR. 2006. Bug juice: Harvesting electricity with microorganisms. *Nat Rev Microbiol* 4:497–508.

Lovley DR, Anderson RT. 2000. The influence of dissimilatory metal reduction on the fate of organic and metal contaminants in the subsurface. *J Hydrol* 8:77–88.

Lovley DR, Baedecker MJ, Lonergan DJ, Cozzarelli IM, Phillips EJP, Siegel DI. 1989b. Oxidation of aromatic contaminants coupled to microbial iron reduction. *Nature* 339:297–299.

Lovley DR, Giovannoni SJ, White DC, Champine JE, Phillips EJP, Gorby YA, Goodwin S. 1993a. *Geobacter metallireducens* gen nov. sp. nov., a microorganism capable of coupling the complete oxidation of organic compounds to the reduction of iron and other metals. *Arch Microbiol* 159:336–344.

Lovley DR, Goodwin S. 1988. Hydrogen concentrations as an indicator of the predominant terminal electron-accepting reactions in aquatic sediments. *Geochim Cosmochim Acta* 52:2993–3003.

Lovley DR, Holmes DE, Nevin KP. 2004. Dissimilatory Fe(III) and Mn(IV) reduction. *Adv Microb Physiol* 49:219–286.

Lovley DR, Phillips EJP. 1988a. Novel mode of microbial energy metabolism: Organic carbon oxidation coupled to dissimilatory reduction of iron and manganese. *Appl Environ Microbiol* 54:1472–1480.

Lovley DR, Phillips EJP. 1992. Reduction of uranium by *Desulfovibrio desulfuricans*. *Appl Environ Microbiol* 58:850–856.

Lovley DR, Phillips EJP. 1994. Novel process for anaerobic sulfate production from elemental sulfur by sulfate-reducing bacteria. *Appl Environ Microbiol* 60:2394–2399.

Lovley DR, Phillips EJP, Lonergan DJ. 1989a. Hydrogen and formate oxidation coupled to dissimilatory reduction of iron and manganese by *Alteromonas putrefaciens*. *Appl Environ Microbiol* 55:700–706.

Luan F, Santelli CM, Hansel CM, Burgos WD. 2012. Defining Mn(II) removal processes in coal mine drainage treatment systems through laboratory incubation experiments. *Appl Geochem* 27:1567–1578.

Luther GW III. 2005. Manganese(II) oxidation and Mn(IV) reduction in the environment—Two one-electron transfer steps versus a single two-electron step. *Geomicrobiol J* 22:195–203.

Luther GW III. 2010. The role of one- and two-electron transfer reactions in forming thermodynamically unstable intermediates as barriers in multi-electron redox reactions. *Aquat Geochem* 16:395–420.

Luther GW III, Ruppel DT, Burkhard C. 1998. Reactivity of dissolved Mn(III) complexes and Mn(IV) species with reductants: Mn redox chemistry without a dissolution step? In: Sparks DL, Grundl T, eds. *Mineral-Water Interfacial Reactions: Kinetics and Mechanisms*. ACS Symposium Volume, pp. 265–280.

Luther GW III, Sunby B, Lewis BL, Brendel PJ, Silverberg N. 1997. Interactions of manganese with the nitrogen cycle: Alternative pathways to dinitrogen. *Geochim Cosmochim Acta* 61:4043–4052.

Madden AS, Hochella Jr. MF. 2005. A test of geochemical reactivity as a function of mineral size: Manganese oxidation promoted by hematite nanoparticles. *Geochim Cosmochim Acta* 69:389–398.

Madison AS, Tebo BM, Luther III GW. 2011. Simultaneous determination of soluble manganese(III), manganese(II) and total manganese in natural (pore) waters. *Talanta* 84:374–81.

Madison AS, Tebo BM, Mucci A, Sundby B, Luther III GW. 2013. Abundant porewater Mn(III) is a major component of the sedimentary redox system. *Science* 341:875–878.

Maki JS, Tebo BM, Palmer FE, Nealson KH, Staley JR. 1987. The abundance and biological activity of manganese-oxidizing bacteria and *Metallogenium*-morphotypes in Lake Washington. *FEMS Microbiol Ecol* 45:21–29.

Manceau A, Gorshkov AI, Drits VA. 1992. Structural chemistry of Mn, Fe, Co, and Ni in manganese hydrous oxides: Part II. Information from EXAFS spectroscopy and electron and X-ray diffraction. *Am Min* 77:1144–1157.

Manceau A, Kersten M, Marcus MA, Geoffroy N., Granina L. 2007a. Ba and Ni speciation in a nodule of binary Mn oxide phase composition from Lake Baikal. *Geochim Cosmochim Acta* 71:1967–1981.

Manceau A, Lanson M, Geoffroy N. 2007b. Natural speciation of Ni, Zn, Ba, and As in ferromanganese coatings on quartz using X-ray fluorescence, absorption, and diffraction. *Geochim Cosmochim Acta* 71:95–128.

Manceau A, Marcus MA, Grangeon S. 2012. Determination of Mn valence states in mixed-valent manganates by XANES spectroscopy. *Amer Mineral* 97:816–827.

Mandernack KW, Post J, Tebo BM. 1995. Manganese mineral formation by bacterial spores of the marine *Bacillus* strain SG-1: Evidence for the direct oxidation of Mn(II) to Mn(IV). *Geochim Cosmochim Acta* 59:4393–4408.

Mandernack KW, Tebo BM. 1993. Manganese scavenging and oxidation at hydrothermal vents and in vent plumes. *Geochim Cosmochim Acta* 57:3907–3923.

Mann PJG, Quastel JH. 1946. Manganese metabolism in soils. *Nature* 158:154–156.

Marble JC, Corley TL, Conklin MH, Fuller CC. 1999. Environmental factors affecting oxidation of manganese in Pinal Creek, Arizona. In: Morganwalp DW, Buxton HT, eds. *Water-Resources Investigation Report, 99-4018A*. West Trenton, NJ: U.S. Geological Survey, pp. 173–183.

Margolis JV, Burns RG. 1976. Pacific deep-sea manganese nodules: Their distribution, composition, and origin. *Annu Rev Earth Planet Sci* 4:229–263.

Mariner R, Johnson D, Hallberg K. 2008. Characterisation of an attenuation system for the remediation of Mn(II) contaminated waters. *Hydrometallurgy* 94:100–104.

Martin JM, Meybeck M. 1979. Elemental mass-balance of material carried by major world rivers. *Mar Chem* 7:173–206.

Matocha CJ, Elzinga EJ, Sparks DL. 2001. Reactivity of Pb(II) at the Mn(III,IV) (oxyhydr)oxide-water interface. *Environ Sci Technol* 35:2967–2972.

Maynard J.B. 2010. The chemistry of manganese ores through time: A signal of increasing diversity of Earth-surface environments. *Econ Geol* 105:535–552.

McLoughlin N, Wacey D, Kruber C, Kilburn MR, Thorseth IH, Pedersen RB. 2011. A combined TEM and NanoSIMS study of endolithic microfossils in altered seafloor basalt. *Chem Geol* 289:154–162.

Mehta T, Coppi MV, Childers SE, Lovley DR. 2005. Outer membrane c-type cytochromes required for Fe(III) and Mn(IV) oxide reduction in *Geobacter sulfurreducens*. *Appl Environ Microbiol* 71:8634–8641.

Melim LA, Shinglman KM, Boston PJ, Northup DE, Spilde MN, Queen JM. 2001. Evidence for microbial involvement in pool finger precipitation, Hidden Cave, New Mexico. *Geomicrobiol J* 18:311–329.

Mirchink TG, Zaprometova KM, Zvyagintsev DG. 1970. Satellite fungi of manganese oxidizing bacteria. *Mikrobiologiya* 39:379–383 (Engl transl., pp. 327–330).

Miyajima T. 1992. Biological manganese oxidation in a lake I: Occurrence and distribution of *Metallogenium* sp. and its kinetic properties. *Arch Hydrobiol* 124:317–335.

Miyata N, Maruo K, Tani Y, Tsuno H, Seyama H, Soma M, Iwahori K. 2006b. Production of biogenic manganese oxides by anamorphic ascomycete fungi isolated from streambed pebbles. *Geomicrobiol J* 23:63–73.

Miyata N, Tani Y, Iwahori K, Soma M. 2004. Enzymatic formation of manganese oxides by an *Acremonium*-like hyphomycete fungus, strain KR21-2. *FEMS Microb Ecol* 47:101–109.

Miyata N, Tani Y, Maruo K, Tsuno H, Sakata M, Iwahori K. 2006a. Manganese(IV) oxide production by *Acremonium* sp. strain KR21-2 and extracellular Mn(II) oxidase activity. *Appl Environ Microbiol* 72:6467–6473.

Miyata N, Tani Y, Sakata M, Iwahori K. 2007. Microbial manganese oxide formation and interaction with toxic metal ions. *J Biosci Bioeng* 104:1–8.

Moese JR, Brantner H. 1966. Mikrobiologische Studien an manganoxydierenden Bakterien. *Zentralbl Bakt Parasitenk Infektionskr Hyg Abt II* 120:480–495.

Moffett JW. 1994a. The relationship between Ce and Mn oxidation in seawater. *Limnol Oceanogr* 39:1309–1318.

Moffett JW. 1994b. A radiotracer study of Ce and Mn uptake onto suspended particles in Chesapeake Bay. *Geochim Cosmochim Acta* 58:695–703.

Moffett JW. 1997. The importance of microbial Mn oxidation in the upper ocean: A comparison of the Sargasso Sea and equatorial Pacific. *Deep-Sea Res I* 44:1277–1291.

Moffett JW, Ho J. 1996. Oxidation of cobalt and manganese in seawater is a common microbially catalyzed pathway. *Geochim Cosmochim Acta* 60:3415–3424.

Mollisch H. 1910. *Die Eisenbakterien*. Jena, Germany: Gustav Fischer Verlag.Moore GW. 1981. Manganese deposition in limestone caves. In: Beck BF, ed. *Proc 8th Intl Congr Speleol II*, pp. 642–645.

Moore WS. 1981. Iron-manganese banding in Oneida Lake ferromanganese nodules. *Nature* 292:233–235.

Moore WS, Dean WE, Krishnaswami S, Borole DV. 1980. Growth rates of manganese nodules in Oneida Lake, New York. *Earth Planet Sci Lett* 46:191–200.

Moore GW, Sullivan GN. 1978. *Speleology: The Study of Caves*, rev 2nd ed. St. Louis, MO: Cave Books, Inc. 150pp.

Moore GW, Sullivan GN. 1997. *Speleology: Caves and the Cave Environment*, rev 3rd ed. St. Louis, MO: Cave Books, Inc. 176pp.

Moreau JW, Weber PK, Martin MC, Gilbert B, Hutcheon ID, Banfield JF. 2007. Extracellular proteins limit the dispersal of biogenic nanoparticles. *Science* 316:1600–1603.

Morgan JJ. 2000. Manganese in natural waters and earth's crust: Its availability to organisms. In: Sigel A, Sigel H, eds. *Metal Ions in Biological Systems, Vol 37: Manganese and Its Role in Biological Processes*. New York: Marcel Dekker, pp. 1–34.

Morgan JJ. 2005. Kinetics of reaction between O_2 and Mn(II) species in aqueous solutions. *Geochim Cosmochim Acta* 69:35–48.

Morgan JJ, Stumm W. 1965. The role of multivalent metal oxides in limnological transformations as exemplified by iron and manganese. In: Jaag O, ed. *Second Water Pollution Research Conference 1*. New York: Pergamon.

Mulder GE. 1972. Le cycle biologique tellurique et aquatique du fer et du manganèse. *Rev Ecol Biol Sol* 9:321–348.

Murray JW, Balistrieri LS, Paul B. 1984. The oxidation state of manganese in marine sediments and ferromanganese nodules. *Geochim Cosmochim Acta* 48:1237–1247.

Murray JW, Dillard JG, Giovanoli R, Moers H, Stumm W. 1985. Oxidation of Mn(II): Initial mineralogy, oxidation state and ageing. *Geochim Cosmochim Acta* 49:463–470.

Murray KJ, Tebo BM. 2007. Cr(III) is indirectly oxidized by the Mn(II)-oxidizing bacterium *Bacillus* sp. strain SG-1. *Environ Sci Technol* 41:528–533.

Murray KJ, Webb SM, Bargar JR, Tebo BM. 2007. Indirect oxidation of Co(II) in the presence of marine Mn(II)-oxidizing bacterium *Bacillus* sp. strain SG-1. *Appl Environ Microbiol* 73:6905–6909.

Mustoe GE. 1981. Bacterial oxidation of manganese and iron in a modern cold spring. *Geol Soc Am Bull Part 1* 92:147–153.

Myers CR, Alatalo LJ, Myer JM. 1994. Microbial potential for the anaerobic degradation of simple aromatic compounds in sediments of the Milwaukee harbor, Green Bay, and Lake Erie. *Environ Toxicol Chem* 13:461–471.

Myers CR, Myers JM. 1992. Localization of cytochromes to the outer membrane of anaerobically grown *Shewanella putrefaciens* MR-1. *J Bacteriol* 174:3429–3438.

Myers CR, Myers JM. 1993. Ferric reductase is associated with the membranes of *Shewanella putrefaciens* MR-1. *Microbiol Lett* 108:15–22.

Myers CR, Myers JM. 1997. Outer membrane cytochromes of *Shewanella putrefaciens* MR-1: Spectral analyses, and purification of the 83-kDa c-type cytochrome. *Biochim Biophys Acta* 1326:307–318.

Myers JM, Myers CR. 2000. Role of the tetraheme cytochrome Cym A in anaerobic electron transport in cells of *Shewanella putrefaciens* MR-1 in reduction of manganese dioxide. *J Bacteriol* 182:67–75.

Myers CR, Nealson KH. 1988a. Bacterial manganese reduction and growth with manganese oxide as the sole electron acceptor. *Science* 240:1319–1321.

Myers CR, Nealson KH. 1988b. Microbial reduction of manganese oxides: Interaction with iron and sulfur. *Geochim Cosmochim Acta* 52:2727–2732.

Nakama K, Medina M, Lien A, Ruggieri J, Collins K, Johnson HA. 2014. Heterologous expression and characterization of the manganese-oxidizing protein from *Erythrobacter* sp. strain SD21. *Appl Environ Microbiol* 80:6837–6842.

Nealson KH, Ford J. 1980. Surface enhancement of bacterial manganese oxidation: Implications for aquatic environments. *Geomicrobiol J* 2:21–37.

Nealson KH, Saffarini D. 1994. Iron and manganese in anaerobic respiration: Environmental significance, physiology, regulation. *Annu Rev Microbiol* 48:311–343.

Nealson KH, Tebo BM, Rosson RA. 1988. Occurrence and mechanism of microbial oxidation of manganese. *Adv Appl Microbiol* 33:279–318.

Nelson YM, Lion LW, Ghiorse WC, Shuler ML. 1999b. Production of biogenic Mn oxides by *Leptothrix discophora* SS-1 in a chemically defined growth medium and evaluation of their Pb adsorption characteristics. *Appl Environ Microbiol* 65:175–180.

Nelson YM, Lion LW, Shuler ML, Ghiorse WC. 1999a. Lead binding to metal oxide and organic phases of natural aquatic biofilms. *Limnol* 44:1715–1729.

Neretin LN, Pohl C, Jost G, Leipe T, Pollehne F. 2003. Manganese cycling in the Gotland Deep, Baltic Sea. *Mar Chem* 82:125–143.

Neufeld CA. 1904. *Z. Uners. Nahrungs-Genussmitt* 7:478 (as cited by Moese and Brantner, 1966).

Nico PS, Anastasio C, Zasoski RJ. 2002. Rapid photo-oxidation of Mn(II) mediated by humic substances. *Geochim Cosmochim Acta* 66:4047–4056.

Niemann H, Losekann T, de Beer D, Elvert M, Nadalig T, Knittel K, Amann R, Sauter EJ, Schluter M, Klages M, Foucher JP, Boetius A. 2006. Novel microbial communities of the Haakon Mosby mud volcano and their role as a methane sink. *Nature* 443:854–858.

Northup DE, Lavoie KH. 2001. Geomicrobiology of caves: A review. *Geomicrobiol J* 18:199–222.

Northup DE, Reysenbach AL, Pace NR. 1997. Microorganisms and speleothems. In: Hill C, Forti P, eds. *Cave Minerals of the World*. Huntsville, AL: National Speleological Society, pp. 261–266.

Okazaki M, Sugita T, Shimizu M, Ohode Y, Iwamoto K, de Vrind-de Jong EW, de Vrind JP, Corstjens PL. 1997. Partial purification and characterization of manganese-oxidizing factors of *Pseudomonas fluorescens* GB-1. *Appl Environ Microbiol* 63:4793–4799.

Onac BP, Forti P. 2011. State of the art and challenges in cave mineral studies. *Studia UBB Geologia* 56:29–38.

O'Reilly SE, Hochella MF. 2003. Lead sorption efficiencies of natural and synthetic Mn and Fe-oxides. *Geochim Cosmochim Acta* 67:4471–4487.

Orphan VJ, Hinrichs KU, Ussler W, Paull CK, Taylor LT, Sylva SP, Hayes JM, Delong EF. 2001b. Comparative analysis of methane-oxidizing archaea and sulfate-reducing bacteria in anoxic marine sediments. *Appl Environ Microbiol* 67:1922–1934.

Orphan VJ, House CH, Hinrichs KU, McKeegan KD, DeLong EF. 2001a. Methane-consuming archaea revealed by directly coupled isotopic and phylogenetic analysis. *Science* 293:484–487.

Orphan VJ, House CH, Hinrichs KU, McKeegan KD, DeLong EF. 2002. Multiple archaeal groups mediate methane oxidation in anoxic cold seep sediments. *Proc Natl Acad Sci USA* 99:7663–7668.

Ostwald J. 1981. Evidence for a biogeochemical origin of the Groote Eylandt manganese ores. *Econ Geol* 76:556–567.

Oze C, Bird DK, Fendorf S. 2007. Genesis of hexavalent chromium from natural sources in soil and groundwater. *Proc Natl Acad Sci USA* 104:6544–6549.

Park PK. 1968. Seawater hydrogen-ion concentration: Vertical distribution. *Science* 162:357–358.

Parker DL, Morita T, Mozafarzadeh ML, Verity R, Mccarthy JK, Tebo BM. 2007. Inter-relationships of MnO₂ precipitation, siderophore-Mn-(III) complex formation, siderophore degradation, and iron limitation in Mn-(II)-oxidizing bacterial cultures. *Geochim Cosmochim Acta* 71:5672–5683.

Parker DL, Sposito G, Tebo BM. 2004. Manganese(III) binding to a pyoverdine siderophore produced by a manganese(II)-oxidizing bacterium. *Geochim Cosmochim Acta* 68:4809–4820.

Paszczynski A, Huynh VB, Crawford R. 1986. Comparison of lignase-1 and peroxidase-M2 from the white rot fungus *Phanerochaete chrysosporium*. *Arch Biochem Biophys* 244:750–765.

Pattan JN, Mudholkar AV. 1990. The oxidation state of manganese in ferromanganese nodules and deep-sea sediments from the Central Indian Ocean. *Chem Geol* 85:171–181.

Peacock CL, Moon EM. 2012. Oxidative scavenging of thallium by birnessite: Controls on thallium sorption and stable isotope fractionation in marine ferromanganese precipitates. *Geochim Cosmochim Acta* 84:297–313.

Peck SB. 1986. Bacterial deposition of iron and manganese oxides in North American caves. *NSS Bull* 48:26–30.

Peña J, Kwon KD, Refson K, Bargar JR, Sposito G. 2010. Mechanisms of nickel sorption by a bacteriogenic birnessite. *Geochim Cosmochim Acta* 74:3076–3089.

Perez J, Jeffries TW. 1992. Role of manganese and organic acid chelators in regulating lignin degradation and biosynthesis of peroxidases by *Phanerochaete chrysosporium*. *Appl Environ Microb* 58:2402–2409.

Perfil'ev BV, Gabe DR. 1965. The use of the microbial-landscape method to investigate bacteria which concentrate manganese and iron in bottom deposits. In: Perfil'ev BV, Gabe DR, Gal'perina AM, Rabinovich VA, Saponitskii AA, Sherman EE, Troshanov EP, eds. *Applied Capillary Microscopy. The Role of Microorganisms in the Formation of Iron Manganese Deposits*. New York: Consultants Bureau, pp. 9–54.

Petkov V, Ren Y, Saratovsky I, Pasten P, Gurr SJ, Hayward MA, Poeppelmeier KR, Gaillard JF. 2009. Atomic-scale structure of biogenic materials by total X-ray diffraction: A study of bacterial and fungal MnOx. *ACS Nano* 3:441–445.

Post J. 1999. Manganese oxide minerals: Crystal structures and economic and environmental significance. *Proc Natl Acad Sci USA* 96:3447–3454.

Potter RM, Rossman GR. 1977. Desert varnish: The importance of clay minerals. *Science* 196:1446–1448.

Reguera G, Mccarthy KD, Mehta T, Nicoll JS, Tuominen MT, Lovley DR. 2005. Extracellular electron transfer via microbial nanowires. *Nature* 435:1098–1101.

Reyss JL, Marchig V, Ku TL. 1982. Rapid growth of a deep-sea manganese nodule. *Nature* 295:401–403.

Richardson LL, Aguilar C, Nealson KH. 1988. Manganese oxidation in pH and O₂ microenvironments produced by phytoplankton. *Limnol Oceanogr* 33:352–363.

Richardson DJ, Butt JN, Fredrickson JK, Zachara JM, Shi L, Edwards MJ, White G, Baiden N, Gates AJ, Marritt SJ, Clarke TA. 2012b. The 'porin-cytochrome' model for microbe-to-mineral electron transfer. *Mol Microbiol* 85:201–212.

Richardson DJ, Edwards MJ, White GF, Baiden N, Hartshorne RS, Fredrickson J, Shi L, Zachara J, Gates AJ, Butt JN, Clark TA. 2012a. Exploring the biochemistry at the extracellular redox frontier of bacterial mineral Fe(III) respiration. *Biochem Soc Trans* 40:493–500.

Ridge JP, Lin M, Larsen EI, Fegan M, McEwan AG, Sly LI. 2007. A multicopper oxidase is essential for manganese oxidation and laccase-like activity in *Pedomicrobium* sp. ACM 3067. *Environ Microbiol* 9:944–953.

Riemann F. 1983. Biological aspects of deep-sea manganese nodule formation. *Oceanol Acta* 6.303–311.

Robbins EI, Corley TL. 2005. Microdynamics and seasonal changes in manganese oxide epiprecipitation in Pinal Creek, Arizona. *Hydrobiologia* 534:165–180.

Robertson LA, Van Niel EWJ, Torremans RAM, Kuenen GJ. 1988. Simultaneous nitrification and denitrification in aerobic chemostat cultures of *Thiosphaera pantotropha*. *Appl Environ Microbiol* 54:2821–2818.

Rogers BW, Williams KM. 1982. Mineralogy of Lilburn Cave, Kings Canyon National Park, California. *NSS Bull* 44:23–31.

Ross DE, Ruebush SS, Brantley SL, Hartshorne RS, Clarke TA, Richardson DJ, Tien M. 2007. Characterization of protein-protein interactions involved in iron reduction by *Shewanella oneidensis* MR-1. *Appl Environ Microbiol* 73:5797–5808.

Rossman R, Callender E. 1968. Manganese nodules in Lake Michigan. *Science* 162:1123–1124.

Rosson RA, Nealson KH. 1982. Manganese binding and oxidation by spores of a marine bacillus. *J Bacteriol* 151:1027–1034.

Rosson RA, Tebo BM, Nealson KH. 1984. Use of poisons in determination of microbial manganese binding rates in seawater. *Appl Environ Microbiol* 47:740–745.

Roy S. 2006. Sedimentary manganese metallogenesis in response to the evolution of the Earth system. *Earth-Science Rev* 77:273–305.

Rubert KF, Pedersen JA. 2006. Kinetics of oxytetracycline reaction with a hydrous manganese oxide. *Environ Sci Technol* 40:7216–7221.

Rusin PA, Quintana L, Sinclair NA, Arnold RG, Oden KL. 1991a. Physiology and kinetics of manganese-reducing *Bacillus polymyxa* strain D1 isolated from manganiferous silver ore. *Geomicrobiol J* 9:13–25.

Rusin PA, Sharp JE, Arnold RG, Sinclair NA. 1991b. Enhanced recovery of manganese and silver from refractory ores. In: Smith RW, Mishra M, eds. *Minerals Bioprocessing*. Warrendale, PA: Mineral Met Maters Soc, pp. 207–218.

Sakai N, Ebina Y, Takada K, Sasaki, T. 2005. Electrochromic films composed of MnO_2 nanosheets with controlled optical density and high coloration efficiency. *J Electrochem Soc* 152:E384–E389.

Santelli CM, Chaput DL, Hansel CM. 2014. Microbial communities promoting Mn(II) oxidation in Ashumet Pond, a historically polluted freshwater pond undergoing remediation. *Geomicrobiol J* 31:605–616.

Santelli CM, Pfister DH, Lazarus D, Sun L, Burgos WD, Hansel CM. 2010. Promotion of Mn(II) oxidation and remediation of coal mine drainage in passive treatment systems by diverse fungal and bacterial communities. *Appl Environ Microb* 76:4871–4875.

Santelli CM, Webb SM, Dohnalkova AC, Hansel CM. 2011. Diversity of Mn oxides produced by Mn(II)-oxidizing fungi. *Geochim Cosmochim Acta* 75:2762–2776.

Saratovsky I, Gurr S, Hayward M. 2009. The structure of manganese oxide formed by the fungus *Acremonium* sp. strain KR21–2. *Geochim Cosmochim Acta* 73:3291–3300.

Sartory A, Meyer J. 1947. Contribution à l'étude du métabolisme hydrocarboné des bactéries ferrugineuses. *CR Acad Sci (Paris)* 225:541–542.

Sasaki K Konno H, Endo M, Takano K. 2004. Removal of Mn(II) ions from aqueous neutral media by manganese-oxidizing fungus in the presence of carbon fiber. *Biotechnol Bioeng* 85:489–496.

Schippers A, Neretin LN, Lavik G, Leipe T, Pollehne F. 2005. Manganese(II) oxidation driven by lateral oxygen intrusions in the western Black Sea. *Geochim Cosmochim Acta* 69:2241–2252.

Schlosser D, Höfer C. 2002. Laccase-catalyzed oxidation of Mn^{2+} in the presence of natural Mn^{3+} chelators as a novel source of extracellular H_2O_2 production and its impact on manganese peroxidase. *Appl Environ Microbiol* 68:3514–3521.

Schmidt JM, Sharp WP, Starr MP. 1981. Manganese and iron encrustations and other features of *Planctomyces crassus* Hortobagyi 1965, morphotype Ib of the *Blastocaulis-Planctomyces* group of budding and appendaged bacteria, examined by electron microscopy and X-ray micro-analysis. *Curr Microbiol* 5:241–246.

Schmidt JM, Sharp WP, Starr MP. 1982. Metallic oxide encrustations of the nonprothecate stalks of naturally occurring populations of *Planctomyces belfie*. *Curr Microbiol* 7:389–394.

Schöttle M, Friedman GM. 1971. Fresh water iron-manganese nodules in Lake George, New York. *Geol Soc Am Bull* 82:101–110.

Schuetz B, Schicklberger M, Kuermann J, Spormann AM, Gescher J. 2009. Periplasmic electron transfer via the c-type cytochromes MtrA and FccA of *Shewanella oneidensis* MR-1. *Appl Environ Microbiol* 75:7789–7796.

Schweisfurth R. 1971. Manganoxidierende Pilze. I. Vorkommen, Isolierung und mikroskopische Untersuchungen. *Z Allg Mikrobiol* 11:415–430.

Schweisfurth R, Mertes R. 1962. Mikrobiologische und chemische Untersuchungen über Bildung and Bekämpfung von Manganschlammablagerung einer Druckleitung für Talsperrenwasser. *Arch Hyg Bakteriol* 146:401–417.

Segarra KEA, Comerford C, Slaughter J, Joye SB. 2013. Impact of electron acceptor availability on the anaerobic oxidation of methane in coastal freshwater and brackish wetland sediments. *Geochim Cosmochim Acta* 115:15–30.

Shi L, Chen BW, Wang ZM, Elias DA, Mayer MU, Gorby YA, Ni S et al. 2006. Isolation of a high-affinity functional protein complex between OmcA and MtrC: Two outer membrane decaheme c-type cytochromes of *Shewanella oneidensis* MR-1. *J Bacteriol* 188:4705–4714.

Shi L, Deng S, Marshall MJ, Wang ZM, Kennedy DW, Dohnalkova AC, Mottaz HM et al. 2008. Direct involvement of type II secretion system in extracellular translocation of *Shewanella oneidensis* outer membrane cytochromes MtrC and OmcA. *J Bacteriol* 190:5512–5516.

Shternberg LE. 1967. Biogenic structures in manganese ores. *Mikrobiologiya* 36:710–712 (Engl transl., pp. 595–597).

Soehngen NL. 1914. Umwandlung von Manganverbindungen unter dem Einfluss mikrobiologischer Prozesse. *Zentralbl Bakt Parasitenk Infektionskr Hyg II Abt* 40:545–554.

Sokolova-Dubinina GA, Deryugina ZP. 1967a. On the role of microorganisms in the formation of rhodochrosite in Punnus-Yarvi lake. *Mikrobiologiya* 36:535–542 (Engl transl., pp. 445–451).

Sokolova-Dubinina GA, Deryugina ZP. 1967b. Process of iron-manganese concretion formation in Lake Punnus-Yarvi. *Mikrobiologiya* 36:1066–1076 (Engl transl., pp. 892–900).

Soldatova A, Butterfield C, Davis R, Oyerinde OF, Tebo BM, Spiro T. 2012. Multicopper oxidase involvement in both Mn(II) and Mn(III) oxidation during bacterial formation of MnO_2. *J Biol Inorg Chem* 17:1151–1158.

Solomon EI, Sundaram UM, Machonkin TE. 1996. Multicopper oxidases and oxygenases. *Chem Rev* 96:2563–605.

Sorem RK, Foster AR. 1972. Internal structure of manganese nodules and implications in beneficiation. In: Horn DR, ed. *Ferromanganese Deposits on the Ocean Floor*. Washington, DC: Office of the International Decade of Ocean Exploration, Natl Sci Foundation, pp. 167–181.

Sparrow LA, Uren NC. 1987. Oxidation and reduction of Mn in acid soils: Effect of temperature and soil pH. *Soil Biol Biochem* 19:143–148.

Sparrow LA, Uren NC. 2014. Manganese oxidation and reduction in soils. Effects of temperature, water potential, pH and their interactions. *Soil Res* 52:483–494.

Spilde MN, Northup DE, Boston PJ, Schelble RT, Dano KE, Crossey LJ, Dahm CN. 2005. Geomicrobiology of cave ferromanganese deposits: A field and laboratory investigation. *Geomicrobiol J* 22:99–116.

Spiro TG, Bargar JR, Sposito G, Tebo BM. 2010. Bacteriogenic manganese oxides. *Acc Chem Res* 43:2–9

Stabel HH, Kleiner J. 1983. Endogenic flux of manganese to the bottom of Lake Constance. *Arch Hydrobiol* 98:307–316.

Stein LY, La Duc MT, Grundl TJ, Nealson KH. 2001. Bacterial and archaeal populations associated with freshwater ferromanganous micronodules and sediments. *Environ Microbiol* 3:10–18.

Stone AT. 1987a. Microbial metabolites and the reductive dissolution of manganese oxides: Oxalate and pyruvate. *Geochim Cosmochim Acta* 51:919–925.

Stone AT. 1987b. Reductive dissolution of manganese(III/IV) oxides by substituted phenols. *Environ Sci Technol* 21:979–988.

Stone AT, Morgan JJ. 1984a. Reduction and dissolution of manganese(III) and manganese(IV) oxides by organics. 1. Reaction with hydroquinone. *Environ Sci Technol* 18:450–456.

Stone AT, Morgan JJ. 1984b. Reduction and dissolution of manganese(III) and manganese(IV) oxides by organics. 2. Survey of the reactivity of organics. *Environ Sci Technol* 18:617–624.

Su J, Bao P, Bai T, Deng L, Wu H, Liu F, He J. 2013. CotA, a multicopper oxidase from *Bacillus pumilus* WH4, exhibits manganese-oxidase activity. *PLoS ONE* 8:e60573.

Sunda WG, Huntsman SA. 1987. Microbial oxidation of manganese in a North Carolina estuary. *Limnol Oceanogr* 32:552–564.

Sunda WG, Huntsman SA. 1988. Effect of sunlight on redox cycles of manganese in the southwestern Sargasso Sea. *Deep-Sea Res* 35:1297–1317.

Sunda WG, Huntsman SA. 1990. Diel cycles in microbial manganese oxidation and manganese redox speciation in coastal waters of the Bahama Islands. *Limnol Oceanogr* 35:325–338.

Sunda WG, Huntsman SA, Harvey GR. 1983. Photoreduction of manganese oxides in seawater and its geochemical and biological implications. *Nature* 301:234–236.

Sunda WG, Kieber DJ. 1994. Oxidation of humic substances by manganese oxides yields low-molecular-weight organic substrates. *Nature* 367:62–64.

Sung W, Morgan JJ. 1981. Oxidative removal of Mn(II) from solution catalyzed by the g-FeOOH (lepidocrocite) surface. *Geochim Cosmochim Acta* 45:2377–2383.

Szeinbaum N, Burns JL, DiChristina TJ. 2014. Electron transport and protein secretion pathways involved in Mn(III) reduction by *Shewanella oneidensis*. *Environ Microbiol Rep* 6:490–500.

Taboroši D. 2006. Biologically influenced carbonate speleothems. In: Harmon RS, Wicks C, eds. *Perspectives on Karst Geomorphology, Hydrology, and Geochemistry*. Boulder, CO: Geological Society of America, pp. 307–317.

Taillefert M, Beckler JS, Carey E, Burns JL, Fennessey CM, DiChristina TJ. 2007. *Shewanella putrefaciens* produces an Fe(III)-solubilizing organic ligand during anaerobic respiration on insoluble Fe(III) oxides. *J Inorg Biochem* 101:1760–1767.

Takahashi Y, Manceau A, Geoffroy N, Marcus MA, Usui A. 2007. Chemical and structural control of the partitioning of Co, Ce, and Pb in marine ferromanganese oxides. *Geochim Cosmochim Acta* 71:984–1008.

Takano K, Itoh Y, Ogino T, Kurosawa K, Sasaki K. 2006. Phylogenetic analysis of manganese-oxidizing fungi isolated from manganese-rich aquatic environments in Hokkaido, Japan. *Limnology* 7:219–223.

Takeda M, Kawasaki Y, Umezu T, Shimura S, Hasegawa M, Koizumi JI. 2012 Patterns of sheath elongation, cell proliferation, and manganese(II) oxidation in *Leptothrix cholodnii*. *Arch Microbiol* 94:667–673.

Tambiev SB, Demina LS. 1992. Biogeochemistry and fluxes of manganese and some other metals in regions of hydrothermal activities (Axial Mountain, Juan de Fuca Ridge and Guaymas Basin, Gulf of California). *Deep-Sea Res* 39:687–703.

Tan H, Zhang G, Heaney PJ, Webb SM, Burgos WD. 2010. Characterization of manganese oxide precipitates from Appalachian coal mine drainage treatment systems. *Appl Geochem* 25:389–399.

Tang Y, Estes E, Webb S, Hansel CM. 2014. Chromium(III) oxidation by biogenic manganese oxides of varying structural ripening. *Environ Sci Process Impacts*.

Tang Y, Zeiner CA, Santelli CM, Hansel CM. 2013. Fungal oxidative dissolution of the Mn(II)-bearing mineral rhodochrosite and the role of metabolites in manganese oxide formation. *Environ Microbiol* 15:1063–1077.

Tani Y, Miyata N, Iwahori K, Soma M, Tokuda SI, Seyama H, Theng BKG. 2003. Biogeochemistry of manganese oxide coatings on pebble surfaces in the Kikukawa River System, Shizuoka, Japan. *Appl Geochem* 18:1541–1554.

Tani Y, Ohashi M, Miyata N, Seyama H, Iwahori K, Soma M. 2004. Sorption of Co(II), Ni(II), and Zn(II) on biogenic manganese oxides produced by a Mn-oxidizing fungus, strain KR21-2. *J Environ Sci Health* A39:2641–2660.

Taylor-George S, Palmer F, Staley JT, Borns DJ, Curtiss B, Adams JB. 1983. Fungi and bacteria involved in desert varnish formation. *Microb Ecol* 9:227–245.

Tebo BM. 1983. The ecology and ultrastructure of marine manganese oxidizing bacteria. PhD dissertation thesis. Univ. Calif., San Diego, San Diego, CA, 220pp.

Tebo BM. 1991. Manganese(II) oxidation in the suboxic zone of the Black Sea. *Deep-Sea Res* 38:S883–S905.

Tebo BM, Bargar JR, Clement BG, Dick GJ, Murray KJ, Parker DL, Verity R, Webb SM. 2004. Biogenic manganese oxides: Properties and mechanisms of formation. *Annu Rev Earth Planet Sci* 32:287–328.

Tebo BM, Emerson S. 1985. Effect of oxygen tension, Mn(II) concentration, and temperature on the microbially catalyzed Mn(II) oxidation rate in a marine fjord. *Appl Environ Microbiol* 50:1268–1273.

Tebo BM, Emerson S. 1986. Microbial manganese(II) oxidation in the marine environment: A quantitative study. *Biogeochem* 2:149–161.

Tebo BM, Ghiorse WC, van Waasbergen LG, Siering PL, Caspi R. 1997. Bacterially mediated mineral formation: Insights into genetic and biochemical studies. In: Banfield JF, Nealson KH, eds. *Geomicrobiology: Interactions between Microbes and Minerals*. Washington, DC: Mineral. Soc. Am, pp. 225–266.

Tebo BM, He LM. 1999. Microbially mediated oxidative precipitation reactions. In Sparks DL, Grundl TJ, eds. *Mineral-Water Interfacial Reactions Kinetics and Mechanisms*. Washington, DC: Am Chem Soc, pp. 393–414.

Tebo BM, Johnson HA, McCarthy JK, Templeton AS. 2005. Geomicrobiology of manganese(II) oxidation. *Trends Microbiol* 13:421–428.

Tebo BM, Nealson KH, Emerson S, Jacobs L. 1984. Microbial mediation of Mn(II) and Co(II) precipitation at the O_2/H_2S interfaces in two anoxic fjords. *Limnol Oceanogr* 29:1247–1258.

Templeton AS, Knowles E. 2009. Microbial transformations of minerals and metals: Recent advances in geomicrobiology derived from synchrotron-based X-ray spectroscopy and X-ray microscopy. *Annu Rev Earth Planet Sci* 37:367–391.

Templeton AS, Knowles EJ, Eldridge DL, Arey BW, Dohnalkova AC, Webb SM, Bailey BE, Tebo BM, Staudigel H. 2009. A seafloor microbial biome hosted within incipient ferromanganese crust. *Nature Geosci* 2:872–876.

Templeton AS, Staudigel H, Tebo BM. 2005. Diverse Mn(II)-oxidizing bacteria isolated from submarine basalts at Loihi Seamount. *Geomicrobiol J* 22:127–139.

Ten Khak-mun. 1967. Iron- and manganese-oxidizing microorganisms in soils of South Sakhalin. *Mikrobiologiya* 36:337–344 (Engl transl., pp. 276–281).

Thamdrup B. 2000. Microbial manganese and iron reduction in aquatic sediments. *Adv Microb Ecol* 16: 41–84. Thamdrup B, Dalsgaard T. 2000. The fate of ammonium in anoxic manganese oxide-rich marine sediment. *Geochim Cosmochim Acta* 64:4157–4164.

Thamdrup B, Dalsgaard T. 2002. Production of N_2 through anaerobic ammonium oxidation coupled to nitrate reduction in marine sediments. *Appl Environ Microbiol* 68:1312–1318.

Thamdrup B, Glud RN, Hansen JW. 1994. Manganese oxidation and in situ fluxes from coastal sediment. *Geochim Cosmochim Acta* 58:2563–2570.

Thamdrup B, Rosselló-Mora R, Amann R. 2000. Microbial manganese and sulfate reduction in Black Sea shelf sediments. *Appl Environ Microbiol* 66:2888–2897.

Thiel GA. 1925. Manganese precipitated by microorganisms. *Econ Geol* 20:301–310.

Thompson IA, Huber DM, Guest CA, Schulze DG. 2005. Fungal manganese oxidation in a reduced soil. *Environ Microbiol* 7:1480–1487.

Thorseth HI, Pedersen RB, Christie DM. 2003. Microbial alteration of 0–30 Ma seafloor and subseafloor basaltic glasses from the Australian-Antarctic Discordance. *Earth Planet Sci Letts* 215:237–247.

Thorseth HI, Torsvik T, Torsvik V, Daae FL, Pedersen RB, Keldysh-98 Scientific Party. 2001. Diversity of life in ocean floor basalt. *Earth Planet Sci Lett* 194:31–37.

Timonin MI, Illman WI, Hartgering T. 1972. Oxidation of manganous salts of manganese by soil fungi. *Can J Microbiol* 18:793–799.

Tipping E, Thompson DW, Davidson W. 1984. Oxidation products of Mn(II) oxidation in lake waters. *Chem Geol* 44:359–383.

Toner B, Fakra S, Villalobos M, Warwick T, Sposito G. 2005. Spatially resolved characterization of biogenic manganese oxide production within a bacterial biofilm. *Appl Environ Microbiol* 71:1300–1310.

Toner B, Manceau A, Webb SM, Sposito G. 2006. Zinc sorption to biogenic hexagonal-birnessite particles within a hydrated bacterial biofilm. *Geochim Cosmochim Acta* 70:27–43.

Tor JM, Kashefi K, Lovley DR. 2003. Potential importance of dissimilatory Fe(III)-reducing microorganisms in hot sedimentary environments. In: Wilcock WS, De Long EF, eds. *The Subseafloor Biosphere at Mid-Ocean Ridges*. Washington, DC: American Geophysical Union.

Toshanov EP. 1968. Iron- and manganese-reducing microorganisms in ore-containing lakes of the Karelian Isthmus. *Mikrobiologiya* 37:934–940 (Engl transl., pp. 786–791).

Trimble RB, Ehrlich HL. 1968. Bacteriology of manganese nodules. III. Reduction of MnO_2 by two strains of nodule bacterial. *Appl Microbiol* 16:695–702.

Trimble RB, Ehrlich HL. 1970. Bacteriology of manganese nodules. IV. Induction of an MnO_2-reductase system in Bacillus. *Appl Microbiol* 19:966–972.

Trost WR. 1958. The chemistry of manganese deposits. Mines Branch Res Rep R8. Ottawa, Ontario, Canada: Dept Mines Tech Surveys.

Trouwborst RE, Clement BG, Tebo BM, Glazer BT, Luther GW III. 2006. Soluble Mn(III) in suboxic zones. *Science* 313:1955–1957.

Tyler PA. 1970. Hyphomicrobia and the oxidation of manganese in aquatic ecosystems. *Antonie v Leeuwenhoek* 36:567–578.

Tyler PA, Marshall KC. 1967a. Microbial oxidation of manganese in hydro-electric pipelines. *Antonie v Leeuwenhoek* 33:171–183.

Uren NC. 2013. Cobalt and manganese. In: Alloway BJ, ed. *Heavy Metals in Soils: Trace Metals and Metalloids in Soils and Their Bioavailability (Environmental Pollution)*, 3rd edn. Dordrecht, the Netherlands: Springer.

Uren NC, Leeper GW. 1978. Microbial oxidation of divalent manganese. *Soil Biol Biochem* 10:85–87. Vail W, Riley R. 2000. The pyrolusite process: A bioremediation process for the abatement of acid mine drainage. *Green Lands* 30:40–46.

Van Veen WL. 1973. Biological oxidation of manganese in soils. *Antonie v Leeuwenhoek* 39:657–662.

van Veen WL, Mulder EG, Deinema MH. 1978. The Sphaerotilus-Leptothrix group of bacteria. *Microbiol Rev* 42:329–356.

Van Waasbergen LG, Hildebrand M, Tebo BM. 1996. Identification and characterization of a gene cluster involved in manganese oxidation by spores of the marine Bacillus sp. SG-1. *J Bacteriol* 178:3517–3530.

Van Waasbergen LG, Hoch JA, Tebo BM. 1993. Genetic analysis of the marine manganese-oxidizing Bacillus sp. SG-1: Protoplast transformation, Tn917 mutagenesis, and identification of chromosomal loci involved in manganese oxidation. *J Bacteriol* 175:7594–7603.

Vandenabeele J, De Beer D, Germonpré R, Van de Sande R, Verstraete W. 1995. Influence of nitrate on manganese removing microbial consortia from sand filters. *Water Res* 29:579–587.

Vandieken V, Finke N, Thamdrup B. 2014. Hydrogen, acetate, and lactate as electron donors for microbial manganese reduction in a manganese-rich coastal marine sediment. *FEMS Microbiol Ecol* 87:733–745.

Varentsov IM, Pronina NV. 1973. On the study of mechanisms of iron-manganese ore formation in recent basins: The experimental data on nickel and cobalt. *Mineral Depos (Berl)* 8:161–178.

Villalobos M, Lanson B, Manceau A, Toner B, Sposito G. 2006. Structural model for the biogenic Mn oxide produced by Pseudomonas putida. *Am Min* 91:489–502.

Villalobos M, Toner B, Bargar J, Sposito G. 2003. Characterization of the manganese oxide produced by Pseudomonas putida strain MnB1. *Geochim Cosmochim Acta* 67:2649–2662.

Vojak PWL, Edwards C, Jones MV. 1985a. A note on the enumeration of manganese-oxidizing bacteria in estuarine water and sediment samples. *J Appl Microbiol* 59:375–379.

Vojak PWL, Edwards C, Jones MV. 1985b. Evidence for microbiological manganese oxidation in the River Tamar estuary, South West England. *Estuarine Coastal Shelf Sci* 20:661–671.

Von Damm KL, Edmond JM, Grant B, Measures CI, Walden B, Weiss RF. 1985a. Chemistry of submarine hydrothermal solutions at 21°N, East Pacific Rise. *Geochim Cosmochim Acta* 49:2197–2220.

Von Damm KL, Edmond JM, Measures CI, Grant B. 1985b. Chemistry of submarine hydrothermal solutions at Guaymas Basin, Gulf of California. *Geochim Cosmochim Acta* 49:2221–2237.

Von Langen PJ, Johnson KS, Coale KH, Elrod VA. 1997. Oxidation kinetics of manganese(II) in seawater at nanomolar concentrations. *Geochim Cosmochim Acta* 61:4945–4954.

Von Wolzogen-Kühr CAH. 1927. Manganese in waterworks. *J Am Water Works Assoc* 18:1–31.

Waite T. D. and Szymczak R. 1993. Manganese dynamics in surface waters of the eastern Caribbean. *J Geophys Res* 98:2361–2369.

Wang PL, Chiu YP, Cheng TW, Chang YH, Tu WX, Lin LH. 2014. Spatial variations of community structures and methane cycling across a transect of Lei-Gong-Hou mud volcanoes in eastern Taiwan. *Front Microbiol* 5:121.

Wang Z, Lee SW, Kapoor P, Tebo BM, Giammar DE. 2012. Uraninite oxidation and dissolution induced by manganese oxide: A redox reaction between two insoluble minerals. *Geochim Cosmochim Acta* 101:24–40.

Wang Z, Xiong W, Tebo BM, Giammar DE. 2014. Oxidative UO$_2$ dissolution induced by soluble Mn(III). *Environ Sci Technol* 48:289–298.

Webb SM, Dick GJ, Bargar JR, Tebo BM. 2005a. Evidence for the presence of Mn(III) intermediates in the bacterial oxidation of Mn(II). *Proc Natl Acad Sci USA* 102:5558–5563.

Webb SM, Fuller CC, Tebo BM, Bargar JR. 2006. Determination of uranyl incorporation into biogenic manganese oxides using X-ray absorption spectroscopy and scattering. *Environ Sci Technol* 40:771–777.

Webb SM, Tebo BM, Bargar JR. 2005b. Structural characterization of biogenic Mn oxides produced in seawater by the marine *Bacillus* sp. strain SG-1. *Am Min* 90:1342–1357.

Wee S, Burns JL, DiChristina TJ. 2014. Identification of a molecular signature unique to metal-reducing Gammaproteobacteria. *FEMS Microbiol Lett* 350:90–99.

Wehrli B, Friedl G, Manceau A. 1995. Reaction rates and products of manganese oxidation at the sediment-water interface. In: Huang CP, O'Melia CR, Morgan JJ, eds. *Aquatic Chemistry Interfacial and Interspecies Processes.* Washington, DC: American Chemical Society, pp. 111–134.

Wei Z, Hillier S, Gadd GM. 2012. Biotransformation of manganese oxides by fungi: Solubilization and production of manganese oxalate minerals. *Environ Microbiol* 14:1744–1752.

White WB, Vito C, Scheetz BE. 2009. The mineralogy and trace element chemistry of black Manganese oxide deposits from caves. *J Cave Karst Stud* 71:136–143.

Wu YH, Liao L, Wang CS, Ma WL, Meng FX, Wu M, Xu XW. 2013. A comparison of microbial communities in deep-sea polymetallic nodules and the surrounding sediments in the Pacific Ocean. *Deep-Sea Res. Part 1 Ocean Res Papers* 79:40–49.

Wuttig K, Heller M, Croot PL. 2013. Pathways of superoxide (O$_2^-$) decay in the Eastern Tropical North Atlantic|. *Environ Sci Technol* 47:10249–10256.

Yakushev E, Pakhomova S, Sørenson K, Skei J. 2009. Importance of the different manganese species in the formation of water column redox zones: Observations and modeling. *Mar Chem* 117:59–70.

Yang W, Zhang Z, Zhang Z, Chen H, Liu J, Ali M, Liu F, Li L. 2013. Population structure of manganese-oxidizing bacteria in stratified soils and properties of manganese oxide aggregates under manganese–complex medium enrichment. *PLoS ONE* 8:e73778.

Yli-Hemminki P, Jorgensen KS, Lehtoranta J. 2014. Iron-manganese concretions sustaining microbial life in the Baltic Sea: The structure of the bacterial community and enrichments in metal-oxidizing conditions. *Geomicrobiol J* 31:263–275.

Zavarzin GA. 1964. Metallogenium symbioticum. *Zeitschrift fur Allg. Mikrobiologie* 4:390–395.

Zhang J, Lion LW, Nelson YM, Shuler ML, Ghiorse W. 2002. Kinetics of Mn(II) oxidation by *Leptothrix discophora* SS1. *Geochim Cosmochim Acta* 65:773–781.

Zapffe C. 1931. Deposition of manganese. *Econ Geol* 26:799–832.

ZoBell CE. 1946. *Marine Microbiology.* Waltham, MA: Chronica Botanica.

CHAPTER NINETEEN

Geomicrobial Interactions with Other Transition Metals (Chromium, Molybdenum, Vanadium, Technetium), Metalloids (Polonium), Actinides (Uranium, Neptunium, and Plutonium), and the Rare Earth Elements

Jonathan R. Lloyd, John D. Coates, Adam J. Williamson, and Matthew P. Watts

CONTENTS

19.1   Microbial Interactions with Chromium / 454
    19.1.1 Occurrence of Chromium / 454
    19.1.2 Chemically and Biologically Important Properties / 454
    19.1.3 Mobilization of Chromium with Microbially Generated Lixiviants / 455
    19.1.4 Biooxidation of Chromium(III) / 455
    19.1.5 Bioreduction of Chromium(VI) / 455
    19.1.6 In Situ Chromate-Reducing Activity / 459
    19.1.7 Applied Aspects of Chromium(VI) Reduction / 459
19.2   Microbial Interactions with Molybdenum / 460
    19.2.1 Occurrence and Properties of Molybdenum / 460
    19.2.2 Microbial Oxidation and Reduction / 460
19.3   Microbial Interactions with Vanadium / 461
    19.3.1 Bacterial Oxidation of Vanadium / 461
19.4   Microbial Interactions with Technetium / 462
19.5   Bacterial Interactions with Polonium / 463
19.6   Microbial Interactions with Uranium / 463
    19.6.1 Occurrence and Properties of Uranium / 463
    19.6.2 Microbial Oxidation of U(IV) / 463
    19.6.3 Microbial Reduction of U(IV) / 464
    19.6.4 Bioremediation of Uranium Pollution / 465
19.7   Microbial Interactions with Neptunium / 466
19.8   Microbial Interactions with Plutonium / 466
19.9   Microbial Interactions with REEs / 467
    19.9.1 Microbial Interactions with Rare Earth Ions / 467
    19.9.2 Biogenic Mineral Interactions / 468
    19.9.3 Microbial Oxidation of Cerium / 468
    19.9.4 REE Summary and Perspectives / 469
19.10 Summary / 470
References / 470

19.1 MICROBIAL INTERACTIONS WITH CHROMIUM

19.1.1 Occurrence of Chromium

Chromium is not a very plentiful element in the Earth's crust, but it is nevertheless fairly widespread. Its average crustal abundance of 122 ppm (Fortescue, 1980) is less than that of manganese. The average concentration in rocks ranges from 4 to 90 mg kg^{-1}, in soil around 70 mg kg^{-1}, in freshwater around 1 µg kg^{-1}, and in seawater around 0.3 µg kg^{-1} (Bowen, 1979). Higher concentrations in the environment are associated with anthropogenic activities such as mining, metallurgy, and leather tanning (Kamaludeen et al., 2003).

Its chief mineral occurrence is as chromite, in which the chromium has an oxidation state of +3. It is a spinel whose end members are $MgCr_2O_4$ and $FeCr_2O_4$. The chromium in this mineral can be partially replaced by Al or Fe. Chromite is of igneous origin. Other chromium minerals of minor occurrence include eskolite (Cr_2O_3), daubréelite ($FeS·Cr_2S_3$), crocoite ($PbCrO_4$), and uvarovite, which is also known as garnet ($Ca_3Cr_2(SiO_4)_3$) (Smith, 1972).

19.1.2 Chemically and Biologically Important Properties

The element chromium is a member of the first transition series of elements in the periodic table together with scandium, titanium, vanadium, manganese, iron, cobalt, nickel, copper, and zinc. The chief oxidation states of chromium are 0, +2, +3, and +6. A +5 oxidation state is also known, which appears to be of significance in at least some biochemical reductions of Cr(VI) (Shi and Dalal, 1990; Suzuki et al., 1992). The geomicrobially important oxidation states are +3 and +6.

Chromium in the hexavalent state (+6) is very toxic, in part because of its high solubility as chromate (CrO_4^{2-}) and dichromate ($Cr_2O_7^{2-}$) in the physiological pH range. At high enough concentrations, Cr(VI) can be mutagenic and carcinogenic. Chromium in the trivalent state is less toxic, in part probably because it is less soluble in this oxidation state at physiological pH. At neutral pH, Cr^{3+} tends to precipitate as a hydroxide ($Cr(OH)_3$) or a hydrated oxide. Chromate and dichromate are strong oxidizing agents.

Chromium has been reported to be nutritionally essential in trace amounts (Miller and Neathery, 1977; Mertz, 1981) for lipid and glucose metabolism (Wang, 2000).

As $Cr_2(SO_4)_3$, K_2CrO_4, or $K_2Cr_2O_7$, chromium is inhibitory to growth of bacteria at appropriate concentrations (e.g., Forsberg, 1978; Wong et al., 1982; Bopp et al., 1983). When taken into the cell, hexavalent chromium can act as a mutagen in prokaryotes and eukaryotes (Venitt and Levy, 1974; Nishioka, 1975; Petrelli and DeFlora, 1977) and as a carcinogen in animals (Gruber and Jennette, 1978; Sittig, 1985). In bacteria, chromate can be taken into the cell via the sulfate uptake system, which involves active transport (Ohtake et al., 1987; Silver and Walderhaug, 1992); see Figure 19.1. Whether dichromate is transported by the same system is not clear. Intracellular Cr(VI) reduction generates reactive oxygen species within the cell, implicated in protein and DNA damage (Cervantes et al., 2007). Bacterial resistance to Cr(VI) has been observed (e.g., Bader et al., 1999). In some instances, at least, the genetic trait is plasmid borne (Summers and Jacoby, 1978; Bopp et al., 1983; Cervantes, 1991; Silver and Walderhaug, 1992). In *P. fluorescens* LB300, resistance to chromate (CrO_4^{2-}) was found to be due to decreased chromate uptake (Bopp, 1980; Ohtake et al., 1987). In *Pseudomonas ambigua*, chromate resistance was attributed to formation of a thickened cell envelope that reduced permeability of Cr(VI) and to an ability to reduce Cr(VI) to Cr(III) (Horitsu et al., 1987). Genes coding for the chromate ion transport superfamily (CHR) are widely acknowledged to encode for proteins capable of efflux of chromate from within the cell, limiting intracellular Cr(VI) reduction (Nies et al., 1998; Juhnke et al., 2002; Viti et al., 2014). Proteomic and genomic studies have noted the upregulation and expression of genes related to detoxification of reactive oxygen species and DNA repair in bacteria exposed to Cr(VI) (Hu et al., 2005; Kilic et al., 2009), implicating these in Cr(VI) resistance.

The basis for resistance to dichromate ($Cr_2O_7^{2-}$) and chromite (Cr^{3+}) has not been clearly established. The resistance mechanism for dichromate need not be the same as for chromate, because *P. fluorescens* LB300, which is resistant to chromate, is much more sensitive to dichromate (Bopp, 1980;

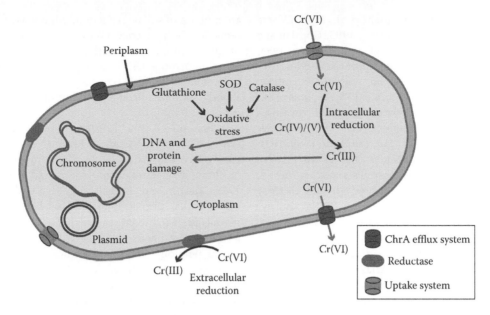

Figure 19.1. Mechanisms of bacterial Cr(VI) uptake, efflux, and reduction in Gram-negative bacteria.

Bopp et al., 1983). Some bacteria have an ability to accumulate chromium. In at least some cases, the accumulation may be due to adsorption (Johnson et al., 1981; Marques et al., 1982; Coleman and Paran, 1983).

19.1.3 Mobilization of Chromium with Microbially Generated Lixiviants

Acidithiobacillus thiooxidans and *Acidithiobacillus ferrooxidans* have been found to solubilize only a limited amount of chromium contained in the mineral chromite (Cr_2O_3) with sulfuric acid generated by the oxidation of sulfur (Ehrlich, 1983). Similarly, the acid produced during iron oxidation by *A. ferrooxidans* was able to solubilize only limited amounts of chromium from chromite (Wang et al., 1982). On the other hand, chromium can be successfully leached from some solid industrial wastes with biologically formed sulfuric acid (Bosecker, 1986).

19.1.4 Biooxidation of Chromium(III)

No observations of enzymatic oxidation of Cr(III) to Cr(VI) have been reported. However, nonenzymatic oxidation of Cr(III) to Cr(VI) may occur in soil environments, where biogenic (or abiogenic) Mn(III) or Mn(IV) oxides may oxidize

Cr(III) to Cr(VI) (Bartlett and James, 1979). These interactions can be summarized as

$$Cr^{3+} + 3MnOOH + H^+ \rightarrow CrO_4^{2-} + 3Mn^{2+} + 2H_2O \tag{19.1}$$

$$2Cr^{3+} + 3MnO_2 + 2H_2O \rightarrow 2CrO_4^{2-} + 3Mn^{2+} + 4H^+ \tag{19.2}$$

These reactions have been implicated in Cr(III) oxidation in low organic matter and high-Mn(IV)-bearing oxide environments (Kozuh et al., 2000; Fandeur et al., 2009). As most Mn(III) or Mn(IV) oxides in the environment are believed to be biogenic in origin (Tebo et al., 2004) and biogenic Mn oxides have been found to be more reactive than synthetic alternatives (Murray and Tebo, 2007), the oxidation of Cr(III) is likely to be indirectly mediated by Mn(II)-oxidizing bacteria.

19.1.5 Bioreduction of Chromium(VI)

A number of bacterial species have been shown to reduce Cr(VI) to Cr(III) (Romanenko and Koren'kov, 1977; Horitsu et al., 1978; Lebedeva and Lyalikova, 1979; Shimada, 1979; Kvasnikov et al., 1985; Gvozdyak et al., 1986;

Wang et al., 1989; Ishibashi et al., 1990; Llovera et al., 1993; Shen and Wang, 1993; Gopalan and Veeramani, 1994; Lovley and Phillips, 1994; Garbisu et al., 1998; Philip et al., 1998). They include *Achromobacter eurydice*, *Aeromonas dechromatica*, *Agrobacterium radiobacter* strain EPS-916, *Arthrobacter* spp., *Bacillus subtilis*, *B. cereus*, *B. coagulans*, *Desulfovibrio vulgaris* (Hildenborough) ATCC 29579, *Escherichia coli* K-12 and ATCC 33456, *Enterobacter cloacae* HO1, *Flavobacterium devorans*, *Sarcina flava*, *Micrococcus roseus*, *Pseudomonas* spp., and *Shewanella putrefaciens* (now *S. oneidensis*) MR-1. It is unclear whether all these strains reduce Cr(VI) enzymatically. Sulfate-reducing bacteria can reduce chromate with the H_2S they produce from sulfate (Bopp, 1980), but *D. vulgaris* can do it enzymatically as well (Lovley and Phillips, 1994). Arias and Tebo (2003) observed, however, that sulfate-reducing bacteria in general are inhibited at elevated Cr(VI) concentrations so that chromate reduction by biogenic H_2S is likely to be significant only at low Cr(VI) concentrations in the environment. *Thiobacillus ferrooxidans* can reduce dichromate with partially reduced sulfur species it forms during the oxidation of elemental sulfur (Sisti et al., 1996). Of those bacteria that reduce Cr(VI) enzymatically, some of the facultative strains reduce it only anaerobically, whereas others will do so aerobically and anaerobically. Many bacterial strains reduce Cr(VI) as a form of respiration, but at least one (*P. ambigua* G-1) reduces it as a means of detoxification (Horitsu et al., 1987). Kwak et al. (2003) showed that the Cr(VI) reductase of *P. ambigua* G-1 is homologous with the nitroreductase of strain KCTC of *Vibrio harveyi* and strain DH5α of *E. coli*. Marsh et al. (2000) explored some of the factors that affect biological chromate reduction in microcosms of sandy aquifer material. They found that biological reduction occurred only in light-colored sediment, abiotic reduction being observed in black, clay-like sediment. The pH optimum for Cr(VI) reduction in this sediment was 6.8. Although they detected two temperature optima, at 22°C and 50°C, the lower optimum probably represented that of the dominant bacterial group in view of the ambient in situ temperature of the sediment. The presence of oxygen in the sediment was inhibitory as was the addition of nitrate, but not of selenate or ferrous iron. The presence of Cr(VI) prevented a loss of sulfate or production of Fe(II). Molybdate,

an inhibitor of sulfate reduction, inhibited Cr(VI) reduction only at concentrations 40 times that of Cr(VI). Bromoethanesulfonic acid, an inhibitor of methanogenesis, strongly inhibited Cr(VI) reduction at 20 mM concentration, but only slightly inhibited it at concentrations between 0.2 and 2.0 mM.

P. fluorescens LB 300, isolated from the northern portion of the Hudson River (New York), can reduce chromate aerobically with glucose or citrate as electron donor (Bopp and Ehrlich, 1988; DeLeo and Ehrlich, 1994). Conditions under which aerobic reduction has been studied include batch culturing with shaking at 200 rpm and continuous culturing with stirring and forced aeration (DeLeo and Ehrlich, 1994). The organism converts chromate to Cr^{3+} in batch culture when growing in glucose–mineral salts solution (Vogel–Bonner medium) and in continuous culture (chemostat) when growing in a citrate–yeast extract–tryptone solution buffered with phosphate (Figure 19.2). *P. fluorescens* strain LB300 was found to reduce chromate anaerobically, only when growing with acetate as an energy source (electron donor). Furthermore, whereas aerobically *P. fluorescens* LB300 will reduce chromate at an initial concentration as high as 314 µg mL^{-1}, anaerobically, it reduces chromate only at a concentration below 50 µg mL^{-1} (Bopp and Ehrlich, 1988; DeLeo and Ehrlich, 1994). Other bacteria that can reduce chromate aerobically and anaerobically include *E. coli* ATCC 33456 (Shen and Wang, 1993), *A. radiobacter* EPS-916 (Llovera et al., 1993), and *Burkholderia cepacia* MCMB-821 (Wani et al., 2007). Reduction of chromate by *E. coli* ATCC 33456 is, however, partially repressed by oxygen through uncompetitive inhibition (Shen and Wang, 1993). Reduction of chromate by resting cells of *A. radiobacter* EPS-916 proceeded initially at similar rates aerobically and anaerobically but subsequently slowed significantly in air (Llovera et al., 1993). *Pseudomonas putida* PRS2000 reduces chromate aerobically more rapidly than anaerobically (Ishibashi et al., 1990). *Pseudomonas* sp. strain C7 has so far been tested only aerobically (Gopalan and Veeramini, 1994).

By contrast, *Pseudomonas dechromaticans*, *Pseudomonas chromatophila*, *E. cloacae* OH1, and *D. vulgaris* reduce Cr(VI) only anaerobically with organic electron donors or H_2 in the case of *D. vulgaris* (Romanenko and Koren'kov, 1977;

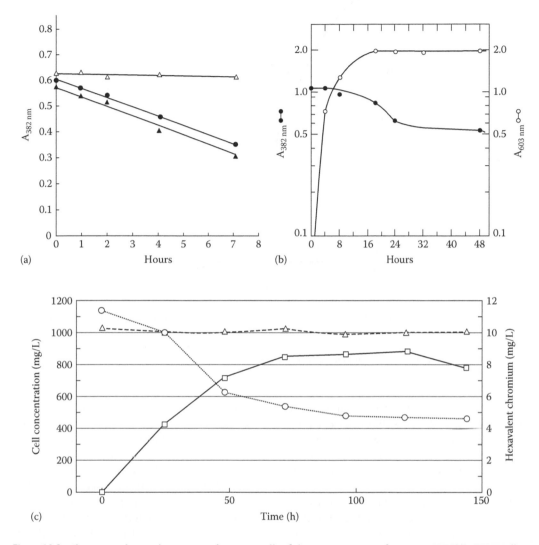

Figure 19.2. Chromate reduction by resting and growing cells of chromate-resistant *P. fluorescens* LB300. (a) Resting cells grown with and without chromate. (Δ) Chromate-grown cells in the absence of electron donor (results were the same for cells grown without chromate and assayed in the absence of electron donor; no chromate reduction was observed; (=) chromate-grown cells with 0.5% (wt vol⁻¹) glucose; (π) cells grown without chromate and assayed with 0.5% (wt vol⁻¹) glucose). Chromate was not reduced by spent medium from either chromate-grown cells or cells grown without chromate or by assay buffer containing either 0.25% or 0.5% glucose. Chromate concentration was measured as absorbance at 328 nm after cell removal. (b) Growing cells in VP broth at an initial K_2CrO_4 concentration of 40 μg mL⁻¹; the growth of the culture was measured photometrically as turbidity at 600 nm; chromate concentration was measured as absorbance at 382 nm after first removing the cells from replicate samples by centrifugation followed by filtration. (c) Chromate reduction by cells growing in citrate–chromate medium in a chemostat at a dilution rate of 1.17 mLh⁻¹; (ρ), chromate concentration in uninoculated reactor; (™), chromate concentration in inoculated reactor; (≤), cell concentration in inoculated reactor. ([a,b] With kind permission from Springer Science+Business Media: *Arch Microbiol*, Chromate resistance and chromate reduction in Pseudomonas fluorescens strain LB300, 150, 1988, 426–431, Bopp, LH and Ehrlich, HL; [c] With kind permission from Springer Science+Business Media: *Appl Microbiol Biotechnol*, Reduction of hexavalent chromium by *Pseudomonas fluorescens* LB300 in batch and continuous culture, 40, 1994, 756–759, DeLeo, PC, Ehrlich, HL.)

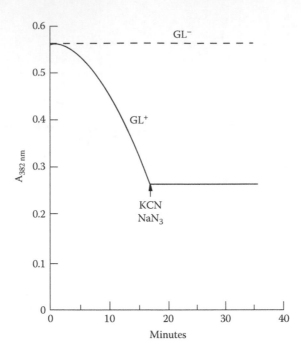

Figure 19.3. Chromate reduction by cell extract from *P. fluorescens* LB300. GL−, without added glucose; GL+, with glucose from time 0 and KCN and NaN₃ at time indicated. Chromate concentration was measured as absorbance at 382 nm after cell removal. (With kind permission from Springer Science+Business Media: *Arch Microbiol*, Chromate resistance and reduction in Pseudomonas fluorescens strain LB300, 150, 1988, 426–431, Bopp, LH and Ehrlich, HL.)

Lebedeva and Lyalikova, 1979; Komori et al., 1989; Lovley and Phillips, 1994). Except for *E. cloacae* OH1, these organisms cannot use glucose as reductant. *P. dechromaticans* and *P. dechromatophila* appear to be able to reduce chromate and dichromate.

Cell extracts of *P. fluorescens* LB300 reduce chromate with added glucose or NADH (Figure 19.3). One or more plasma membrane components appear to be required (Bopp and Ehrlich, 1988). *E. cloacae* HO1 also uses a membrane-bound respiratory system to reduce chromate, but it functions only under anaerobic conditions (Wang et al., 1991). By contrast, most of the chromate-reducing activity in *E. coli* ATCC 33456 appears to be soluble, i.e., it does not involve plasma membrane components, but is mediated by NADH (Shen and Wang, 1993). The chromate-reducing activity of *P. putida* PRS2000 also does not depend on plasma membrane components. This organism mediates reduction via NADH or NADPH (Ishibashi et al., 1990). *D. vulgaris* ATCC 29759 uses its cytochrome c_3 as its Cr(VI) reductase coupled to hydrogenase when using H_2 as

reductant (Lovley and Phillips, 1994). Ackerley et al. (2004) detected soluble Cr(VI)–reducing activity in aerobically growing strains of *P. putida* and *E. coli* associated with soluble bacterial flavoproteins. The flavoprotein from *P. putida*, labeled ChrR, and that from *E. coli*, labeled YieF, were dimers. These authors tested the activity of the two reductases with NADH as electron donor. Unlike the YieF dimer, the ChrR dimer generated a flavin semiquinone transiently and reactive oxygen species diminished over time, which suggested that Cr(V) may be an intermediate in Cr(VI) reduction by this reductase. As the YieF dimer did not generate flavin semiquinone and used only 25% of the electrons from NADH for reduction of reactive oxygen species, Ackerley et al. (2004) suggested that in reducing Cr(VI), this reductase may transfer three electrons to Cr(VI) and one to reactive oxygen species for every two NADH oxidized.

In anaerobically grown cells of *S. oneidensis* MR-1, chromate reductase activity is associated with the cytoplasmic membrane (Myers et al., 2000). Although this organism is facultative, it reduces chromate only anaerobically. Both formate and NADH but not l-lactate or NADPH can serve as electron donors to the Cr(VI) reductase system, which includes a multicomponent electron transport system. Some of the activity is inducible in cells grown anaerobically with lactate as electron donor and nitrate or fumarate as electron acceptor, but this activity is inhibited by nitrite (Viamajala et al., 2002). Although Cr(III) is the usual product of respiratory Cr(VI) reduction by S. oneidensis and other unrelated Cr(VI) reducers, one case in which Cr(II) accompanied Cr(III) in Cr(VI) respiration by S. oneidensis MR-1 has been reported (Daulton et al., 2007). The Cr(II) in this instance was concentrated near the cytoplasmic membrane.

19.1.6 In Situ Chromate-Reducing Activity

In situ rates of microbial Cr(VI)-reducing activity are generally not readily available in the literature, although such measurements would be useful because a number of different bacteria possess the ability to reduce Cr(VI). On the other hand, Wang and Shen (1997) examined rate parameters for a number of pure cultures under laboratory conditions. For example, they reported that the half-maximal Cr(VI)-reducing velocity constant K_s, in mg Cr(VI) L^{-1}, is 5.43 for B. subtilis (aerobic), 19.2 for D. vulgaris ATCC 29579 (anaerobic), 8.64 for E. coli ATCC 33456 (anaerobic), 641.9 for P. ambigua G-1 (aerobic), and 5.55 for P. fluorescens LB300 (aerobic), based on Monod kinetics. Natural levels of Cr(VI) in most environments can be expected to be low. However, anthropogenic pollution can cause very significant elevation in environmental chromium concentrations.

Lebedeva and Lyalikova (1979) isolated a strain of P. chromatophila from an effluent in a chromite mine in Yugoslavia that contained the mineral crocoite (PbCrO4) in its oxidation zone. The isolate was shown to use a range of carbon and energy sources anaerobically for chromate reduction. These included ribose, fructose, benzoate, lactate, acetate, succinate, butyrate, glycerol, and ethylene glycol but not glucose or hydrogen. Crocoite reduction with lactate as electron donor was demonstrated anaerobically in the laboratory. Since this early work, a wide variety of Cr(VI)-reducing bacteria have been isolated from Cr(VI)-contaminated mine sites,

for example, Bacillus sp. (Dhal et al., 2010), E. cloacae (Harish et al., 2012), and Arthrobacter sp. (Dey and Paul, 2012).

19.1.7 Applied Aspects of Chromium(VI) Reduction

The first practical application of microbial Cr(VI) reduction as a bioremediation process was explored by Russian investigators. They presented evidence that indicated that bacterial chromate reduction can be harnessed in wastewater and sewage treatment to remove chromate (Romanenko et al., 1976; Pleshakov et al., 1981; Serpokrylov et al., 1985; Simonova et al. 1985). The process also has potential application in treatment of tannery and, especially, electroplating wastes and in situ bioremediation. In the case of tannery and electroplating waste treatment, prior dilution of the waste may be necessary to bring the Cr(VI) concentration into a range tolerated by the Cr(VI)-reducing bacteria. Cr(VI) contamination is often associated with alkali wastes, such as residues from chromite ore processing, and for this reason, there has been several more recent studies on alkaliphilic Cr(VI)-reducing bacteria, for example, Alkaliphilus metalliredigens (Ye et al., 2004) and B. cepacia (Wani et al., 2007).

Extensive research in the use of E. cloacae HO1 in bioremediation of Cr(VI)-containing wastewaters was performed in Japan (Komori et al., 1990a,b; Ohtake et al., 1990; Yamamoto et al., 1993; Fuji et al., 1994). A variety of more recent studies have also sought to use Cr(VI)-reducing bacteria in bioreactor systems to treat Cr(VI)-contaminated waters (Tripathi, 2002). Several bioreactors have also employed immobilized cells that form a biofilm (Chirwa and Wang, 1997; Smith, 2001), which often exhibit an increased tolerance to Cr(VI) toxicity (Tripathi, 2002). Other studies have also suggested the use of Cr(VI)-reducing bacteria for in situ reduction of Cr(VI) via a bioaugmentation method (Jeyasingh et al., 2010).

Other studies have focused on indirect mechanisms of Cr(VI) reduction (Wielinga et al., 2001), mediated by Fe(II)-bearing biominerals generated by Fe(III)-reducing bacteria, for example, nanoscale magnetite (Cutting et al., 2010); Figure 19.4.

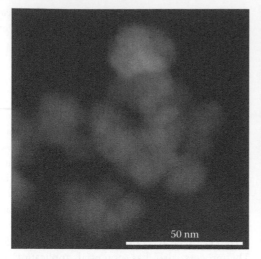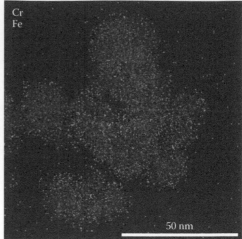

Figure 19.4. STEM-EDX map of Cr(III) associated with magnetite synthesized by Fe(III)-reducing bacteria and used to reductively precipitate soluble Cr(VI). (Courtesy of Mathew Watts and Jon Lloyd.)

19.2 MICROBIAL INTERACTIONS WITH MOLYBDENUM

19.2.1 Occurrence and Properties of Molybdenum

Molybdenum is an element of the second transition series in the periodic table. In mineral form, it occurs extensively as molybdenite (MoS_2). The minerals wulfenite ($PbMoO_4$) and powellite ($CaMoO_4$) are often associated with the oxidation zone of molybdenite deposits (Holliday, 1965). Molybdite (MoO_3) is another molybdenum-containing mineral that may be encountered in nature. The abundance of Mo has been reported to be 2–4 g ton^{-1} in basaltic rock, 2.3 g ton^{-1} in granitic rock, and 0.001–0.005 g ton^{-1} in ocean waters (Enzmann, 1972).

The oxidation states in which molybdenum can exist include 0, +2, +3, +4, +5, and +6. Of these, the +4 and +6 states are the most common, but the +5 state is of biological significance. Molybdenum oxyanions of the +6 oxidation state tend to polymerize, the complexity of the polymers depending on the pH of the solution (Latimer and Hildebrand, 1942).

Molybdenum is a biologically important trace element. A number of enzymes feature it in their structure, e.g., nitrogenase, nitrate reductase (Brock and Madigan, 1991), sulfite reductase, and arsenite oxidase (Anderson et al., 1992). Molybdate is an effective inhibitor of bacterial sulfate reduction (Oremland and Capone, 1988).

19.2.2 Microbial Oxidation and Reduction

Molybdenite (MoS_2) is aerobically oxidizable as an energy source by *Acidianus brierleyi* (Brierley and Murr, 1973) (see also Chapter 20) with the formation of molybdate and sulfate. *A. ferrooxidans* can also oxidize molybdenite, but it is poisoned by the resulting molybdate (Tuovinen et al., 1971) unless the molybdate is rendered insoluble, for instance, by reaction with Fe^{3+}. Sugio et al. (1992) reported that T. (now *Acidithiobacillus*) *ferrooxidans* AP-19–3 contains an enzyme that oxidizes molybdenum blue (Mo^{5+}) to molybdate (Mo^{6+}). They purified the molybdenum oxidase and found it to be an enzyme complex that included cytochrome oxidase as an important component. The function of molybdenum oxidase in the organism remains unclear in view of its sensitivity to molybdate.

Molybdate was first shown to be reduced by *Sulfolobus* sp. by Brierley and Brierley (1982). In a more detailed study, molybdate was shown to be reduced microbially to molybdenum blue (containing Mo^{5+}) by *A. ferrooxidans* using sulfur as electron donor (Sugio et al., 1988). The enzyme that reduced the Mo^{6+} was identified as sulfur/ferric ion oxidoreductase. Molybdate has also been shown to be reduced anaerobically to molybdenum blue by *E. cloacae* strain 48 using glucose as electron donor (Ghani et al., 1993). The reduction appears to be mediated via NAD and *b*-type cytochrome. Other bacteria reported to be able to reduce Mo(VI) include *Pseudomonas guilliermondii*, *Micrococcus* sp., and *Desulfovibrio desulfuricans* (reviewed by Lloyd, 2003).

19.3 MICROBIAL INTERACTIONS WITH VANADIUM

Vanadium belongs to the first transition series of elements in the periodic table. In mineral form, it often occurs in complex forms such as patronite (a complex sulfide), roscoelite (a vanadium mica), vanadinite (a lead vanadate), and carnotite (a hydrous potassium uranium vanadate) (De Huff, 1965). Its average abundance (mg kg^{-1}) in granites is 72, in basalts 270, and in soil 90. Its average concentration (ng m^{-3}) in freshwater is 0.0005 and in seawater 0.0025 (Bowen, 1979).

Vanadium occurs in the oxidation states of 0, +2, +3, +4, and +5. Pentavalent vanadium in solution occurs as the oxyanion VO_3^- (vanadate) and is colorless. Tetravalent vanadium in solution occurs as VO^{2+} and is deep blue. Trivalent vanadium (V^{3+}) forms a green solution and divalent vanadium (V^{2+}) a violet solution (Dickerson et al., 1979).

As a trace element in prokaryotes, vanadium has been found to occur in place of molybdenum in certain nitrogenases (Brock and Madigan, 1991) (see also Chapter 13). It also occurs in oxygen-carrying blood pigment of ascidian worms.

19.3.1 Bacterial Oxidation of Vanadium

Five different bacteria have been reported to be able to reduce vanadate. The first three are *Veillonella* (*Micrococcus*) *lactilyticus*, *D. desulfuricans*, and *Clostridium pasteurianum*, which were shown by Woolfolk and Whiteley (1962) to be able to reduce vanadate to vanadyl with hydrogen using a hydrogenase:

$$VO_3^- + H_2 \rightarrow VO(OH) + OH^- \quad (19.3)$$

The fourth and fifth organisms are isolates assigned to the genus *Pseudomonas* (Yurkova and Lyalikova, 1990; Lyalikova and Yurkova, 1992). One of these, isolated from a waste stream from a ferrovanadium factory, was named *P. vanadiumreductans*, and the other, isolated from seawater in Kraternaya Bay, Kuril Islands, *P. issachenkovii*. Both are Gram-negative, motile,

non-spore-forming rods that can grow as facultative chemolithotrophs and facultative anaerobes. Anaerobically, chemolithotrophic growth was observed with H_2 and CO as alternative energy sources, CO_2 as carbon source, and vanadate as terminal electron acceptor. However, the organisms can also grow organotrophically under anaerobic conditions with glucose, maltose, ribose, galactose, lactose, arabinose, lactate, proline, histidine, threonine, and serine as a carbon and energy sources. *P. issachenkovii* can also use asparagine as carbon and energy source. Vanadate reduction by the organism involved transformation of pentavalent vanadium to tetravalent and trivalent vanadium. The tetravalent oxidation state was identified in the medium by the development of a blue color and the trivalent state by formation of a black precipitate and by its reaction with tairon reagent. An equation describing the overall reduction of vanadium by lactate in these experiments was presented by the authors as

$$2NaVO_3 + NaC_3H_5O_3 \rightarrow V_2O_3 + NaC_2H_3O_2$$
$$+ NaHCO_3 + NaOH \quad (19.4)$$

It accounts for the alkaline pH developed by the medium during growth that started at pH 7.2. Antipov et al. (2000) found that molybdenum- and molybdenum cofactor–free nitrate reductases of *P. issachenkovii* appear to mediate the vanadate reduction. Homogeneous membrane-bound nitrate reductase from the organism reduced vanadate with NADH as electron donor. In a medium containing both nitrate and vanadate, the organism reduced nitrate before vanadate. A vanadium mineral similar to sherwoodite was detected in cultures reducing vanadium, suggesting that these bacteria may play a role in epigenetic vanadium mineral formation (Lyalikova and Yurkova, 1992).

More recently, *S. oneidensis* MR-1 was shown to be able to use vanadate as sole electron acceptor anaerobically, i.e., it was able to respire anaerobically on lactate, pyruvate, formate, fumarate, Fe(III), or citrate using vanadate as the terminal electron acceptor (Carpentier et al., 2003, 2005). It reduces vanadate [V^V] to vanadyl [V^{IV}] (Carpentier et al., 2003). The reduction was inhibited by 2-heptyl-4-hyroxyquinoline N-oxide and antimycin A, and proton translocation associated with vanadate reduction was abolished by the protonophores dinitrophenol and carbonyl cyanide m-chlorophenylhydrazone (Carpentier et al., 2005).

Geobacter metallireducens has also been shown to be able to grow anaerobically with acetate as the electron donor and vanadate as the terminal electron acceptor (Ortiz-Bernad et al., 2004a). It reduced V^V to V^{IV}, which subsequently precipitated. Microprobe analysis of the precipitate suggested that it could have been vanadyl phosphate. In a bioremediation study of groundwater contaminated with uranium(VI) and vanadium(V), acetate stimulation of *Geobacteraceae* caused precipitation of the vanadium as a result of its reduction (Ortiz-Bernad et al., 2004a). More recently, the V(V)-reducing *E. cloacae* EVSA01 was isolated from a South African deep gold mine, suggesting a more widespread distribution of bacteria capable of V(V) reduction (van Marwijk et al., 2009).

Vanadium(V) can also be reduced nonenzymatically by bacteria. An example is the reduction of vanadium(V), at concentrations up to 5 mM, to vanadium(IV) by *A. thiooxidans*, using elemental sulfur as its energy source. The vanadium reduction in this instance is brought about by partially reduced sulfur intermediates produced by the organism during oxidation of the elemental sulfur (Briand et al., 1996).

19.4 MICROBIAL INTERACTIONS WITH TECHNETIUM

Technetium is a long-lived (half-life 2.13×10^5 years) β-emitter, generated in appreciable quantities as a fission product during the production of nuclear energy and the testing of nuclear weapons (Wildung et al., 1979). Within most oxic environments, Tc(VII) dominates as the pertechnetate anion (TcO_4^-), which has weak ligand-complexing capabilities and is therefore highly mobile in the environment. However, under reducing conditions, relatively insoluble Tc(IV) dominates, typically as soluble hydrated TcO_2 or Tc(IV) sorbed to mineral surfaces (Icenhower et al., 2008). Two geomicrobiological mechanisms of Tc(VII) reduction have been characterized: direct enzymatic reduction and indirect mechanisms mediated via biogenic Fe(II) and sulfide (Lloyd, 2003).

Early studies showed that Tc(VII) was removed from solution under anaerobic conditions in the presence of mixed cultures of bacteria (Henrot,

1989) or in microbiologically active sediments (Pignolet et al., 1989). The mechanisms of removal were not identified in these studies; however, later work confirmed direct enzymatic reduction of Tc(VII) was possible in cultures of *G. metallireducens* and *S. oneidensis* (Lloyd and Macaskie, 1996). Subsequently, *S. oneidensis* (Wildung et al., 2000) and *Geobacter sulfurreducens* (Lloyd et al., 2000a) both showed optimal activity against Tc(VII) with H_2 as the electron donor, rather than organic acids such as lactate or acetate. The fate of the reduced Tc(IV) in the former study was an insoluble Tc(IV) oxide that was associated primarily with the periplasm and the outer surface of the cell, although extracellular particulates and carbonate complexes were formed in carbonate buffer. The enzymology of Tc(VII) reduction remains best studied in *E. coli* (Lloyd et al., 1997), in which the hydrogenase 3 component of a formate hydrogenlyase complex has been shown to be responsible for the transfer of electrons from dihydrogen to Tc(VII). More recent studies have confirmed a role for a periplasmic NiFe hydrogenase in Tc(VII) reduction by the sulfate-reducing bacterium *Desulfovibrio fructosovorans* (De Luca et al., 2001). It should be stressed that Tc(VII) reduction has not been shown to support growth in any of these studies and seems to be a fortuitous biochemical side reaction in all organisms studied to date.

Although the ability to enzymatically reduce Tc(VII) may be widespread, indirect mechanisms are thought to be more important in controlling the fate of Tc in geomicrobiological systems, with both biogenic Fe(II) and sulfide (Lloyd, 2003) playing dominant roles. Studies on Fe(II)-mediated reactions have been reported widely, including experiments using aquifer sediments (Wildung et al., 2004; McBeth et al., 2007), estuarine sediments (Burke et al., 2005), and model Fe(II) biomineral phases (McBeth et al., 2011). Recent studies have also shown that Tc(IV) associated with Fe(II)-rich sediments can be very resistant to reoxidation and remobilization under oxic- or nitrate-rich conditions (Burke et al., 2006; McBeth et al., 2007). Although the reasons for this surprising level of recalcitrance remains unclear, a recent study has shown that under alkaline conditions, Tc incorporated in magnetite is stable to oxidative dissolution by an armoring process with reoxidized Fe(III) on the magnetite/maghemite surface (Marshall et al., 2014).

19.5 BACTERIAL INTERACTIONS WITH POLONIUM

Polonium is a radioactive element that occurs naturally in association with uranium and thorium minerals. Different isotopes of polonium are produced in the decay of ^{238}U, ^{235}U, and ^{232}Th. Of these isotopes, ^{210}Po, which originates from decay of ^{238}U, has the longest half-life (138.4 days) (Lietzke, 1972). Its known oxidation states are +2 and +4 (see LaRock et al., 1996).

Polonium-210 can be an environmental pollutant arising by release from uranium-containing phosphorite, which is commercially exploited for its phosphate and phosphogypsum, which is a by-product in the manufacture of phosphoric acid from phosphorite (see LaRock et al., 1996).

Sulfate-reducing bacteria that were found by LaRock et al. (1996) to be associated with phosphogypsum are able to mobilize Po contained in the mineral. In laboratory experiments, this mobilization required that the dissolved sulfate concentration in the bulk phase was below 10 μM. Above this sulfate concentration, enough H_2S was produced to coprecipitate the mobilized Po as a metal sulfide (LaRock et al., 1996). The release of polonium apparently depended on the reduction of the sulfate in the gypsum. Aerobic bacteria were also able to mobilize Po in the phosphogypsum. The mechanism in this instance may involve Po complexation by ligands produced by the active organisms (LaRock et al., 1996).

Immobilization of Po by at least one aerobic bacterial isolate has also been reported (Cherrier et al., 1995). The Po was taken into the cell by a mechanism that appeared to differ from that of sulfate uptake. However, its partitioning after uptake paralleled that of sulfur among cell components that included cell envelope, cytoplasm, and cytoplasmic protein. Most polonium and sulfur were detected in the cytoplasmic fraction. In nature, such immobilization of Po must be considered transient if upon the death of these cells, the Po becomes redissolved in the bulk phase.

19.6 MICROBIAL INTERACTIONS WITH URANIUM

19.6.1 Occurrence and Properties of Uranium

Uranium is one of the naturally occurring radio-active elements. Its abundance in the Earth's crust is only 0.0002%. It is found in more than 150 minerals, but the most important are the igneous minerals pitchblende and coffinite and the secondary mineral carnotite (Baroch, 1965). It is found in small amounts in granitic rocks (4.4 mg kg^{-1}) and in even smaller amounts in basalt (0.43 mg kg^{-1}). In freshwater, it has been reported in concentrations of 0.0004 mg kg^{-1} and in seawater 0.0032 mg kg^{-1} (Bowen, 1979).

Uranium can exist in the oxidation states of 0, +3, +4, +5, and +6 (Weast and Astle, 1982). The +4 and +6 oxidation states are of greatest significance microbiologically. In nature, the +4 oxidation state usually manifests itself in insoluble forms of uranium, e.g., UO_2. The +6 oxidation state predominates in nature in a soluble and hence mobile form, e.g., UO_2^{2+} (Haglund, 1972). In radioactive decay of an isotopic uranium mixture, alpha, beta, and gamma radiation are emitted, but the overall rate of decay is very slow because the dominant isotopes have very long half-lives (Stecher, 1960). This slow rate of decay probably accounts for the ability of bacteria to interact with uranium species without experiencing lethal radiation damage.

19.6.2 Microbial Oxidation of U(IV)

A. ferrooxidans has been shown to oxidize tetravalent U^{4+} to hexavalent UO_2^{2+} in a reaction that yields enough energy to enable the organism to fix CO_2. Nevertheless, experimental demonstration of growth of *A. ferrooxidans* with U^{4+} as the sole energy source has not succeeded to date. *Thiobacillus acidophilus* (now renamed *Acidiphilium acidophilum*) was also found to oxidize U^{4+} but without energy conservation (DiSpirito and Tuovinen, 1981, 1982a,b) (see also Chapter 20).

Anaerobic microbial metabolism can also catalyze U(IV) oxidation with nitrate as the electron acceptor, potentially yielding enough energy to support microbial growth or generation of chemical energy in the form of ATP:

$$2.5UO_2 + NO_3^- + 5HCO_3^- + H^+ \rightarrow$$
$$2.5UO_2(CO_3)_2^{2-} + 0.5N_2 + 3H_2O -$$
$$(\Delta G^{\circ\prime} = -352.9 \text{ kJ mol}^{-1})$$

The autotroph, *Thiobacillus denitrificans*, was shown to promote oxidation of U(IV) oxide, uraninite, anaerobically in the presence of nitrate in a laboratory glovebox experiment (Beller, 2005).

Although U(IV) oxidation was accompanied by significant nitrate reduction, the amount of nitrate consumed significantly exceeded the amount of U(IV) oxidized. The excess nitrate reduction was attributable to biooxidation of H_2 in the atmosphere in the glovebox (Beller, 2005). Subsequent studies have shown that nitrate-dependent U(IV) oxidation is prevalent in both pristine and contaminated soils and sediments and is conserved across a phylogenetically diverse range of microorganisms across several phyla and also subphyla of the Proteobacteria suggesting that it may be environmentally prevalent (Weber et al., 2011). In these studies, H_2 was not a confounding factor, and the U(IV) oxidation was demonstrated to be directly coupled to the electron transport chain. In general, although the energetics is favorable, growth is not associated with this metabolism.

19.6.3 Microbial Reduction of U(IV)

A number of organisms have been shown to be able to reduce hexavalent uranium (UO_2^{2+}) to tetravalent uranium (UO_2). The first demonstration was with *V. (Micrococcus) lactilyticus* using H_2 as electron donor under anaerobic conditions (Woolfolk and Whiteley, 1962). Much more recently, some other bacteria were shown to be able to reduce U(VI) to U(IV) anaerobically. They are the facultative organism *S. putrefaciens* (Lovley et al., 1991), *S. alga* strain BrY (Caccavo et al., 1992), the strict anaerobe *G. metallireducens* strain GS-15 (Lovley et al., 1951, 1993a), *D. desulfuricans* (Lovley and Phillips, 1992), *D. vulgaris* (Lovley et al., 1993c), and *Desulfovibrio* sp. (Pietzsch et al., 1999). In *Desulfovibrio*, cytochrome c_3 appears to be the U(VI) reductase (Figure 19.5) (Lovley et al., 1993c). The electron donors used by these organisms may be organic or, in some instances, H_2. Whereas *G. metallireducens* and *S. putrefaciens* can gain energy from the process of U(VI) reduction (Lovley et al., 1991), *D. desulfuricans* and *D. vulgaris* are unable to do so (Lovley and Phillips, 1992; Lovley et al., 1993b). One other *Desulfovibrio* sp., on the other hand, has been reported to be able to gain energy from the process (Pietzsch et al., 1999). A comprehensive list of the many known uranium-reducing prokaryotes is given by Kostka and Green (2011).

Insofar as the study of electron transport pathways in U(VI) and Cr(VI) reduction under anaerobic conditions by *S. oneidensis* MR-1 is concerned, Bencheikh-Latmani et al. (2005), using whole-genome DNA microarrays, showed that U(VI)-reducing conditions caused upregulation of 121 genes and Cr(VI)-reducing conditions 83 genes. Genes whose products are known to enable the use of the alternative electron acceptors fumarate, dimethyl sulfoxide, Mn(IV), or soluble Fe(III) were upregulated under both U(VI)- and Cr(VI)-reducing conditions. Mutant studies confirmed that several genes, whose products are known to be involved in ferric citrate reduction, are also involved in the reduction of both U(VI) and Cr(VI) by *S. oneidensis* MR-1. The genes are mtrA, mtrB, mtrC, and menC. On the other hand, genes coding for efflux pumps were upregulated under Cr(VI)-reducing conditions but not U(VI)-reducing conditions. These findings by Bencheikh-Latmani et al. (2005) show that the same electron transport pathway or parts thereof may be involved in the reduction of U(VI) and Cr(VI) and several other terminal electron acceptors. Parallel studies of the electron transfer pathway to U(VI) in *Geobacter* species have also shown a role for a diversity of outer membrane c-type cytochromes (Lovley et al., 2011), with recent work suggesting that conductive pili may also be involved (Cologgi et al., 2011). The mechanisms of U(VI) reduction in Gram-positive bacteria remain largely explored, although these organisms are potentially important in radioactive waste geodisposal scenarios, where they dominate under high pH conditions associated with concrete-rich intermediate-level radioactive waste, and can reduce and immobilize U(VI) and other metals (Williamson et al., 2014).

The isolation of some of the organisms from freshwater and marine sediments suggests that this microbial activity may play or may have played a significant role in the immobilization of uranium and its accumulation in sedimentary rock. The first evidence in support of such immobilization was obtained by Gorby and Lovley (1992) in experiments with groundwater amended with 0.4 or 1.0 mM uranyl acetate and 30 mM $NaHCO_3$ and inoculated with *G. metallireducens* GS-15, which produced a black precipitate from the dissolved UO_2^{2+}. The black precipitate was identified as uraninite (UO_2). This nanoscale precipitate is the most studied end product of microbial U(VI) reduction; however, recent studies have reported poorly ordered coordination polymers of U(IV) coordinated to phosphoryl and/or carboxylate groups in cultures of both Gram-positive and Gram-negative bacteria (Bernier-Latmani et al., 2010). Fredrickson et al. (2000) studied U(VI)

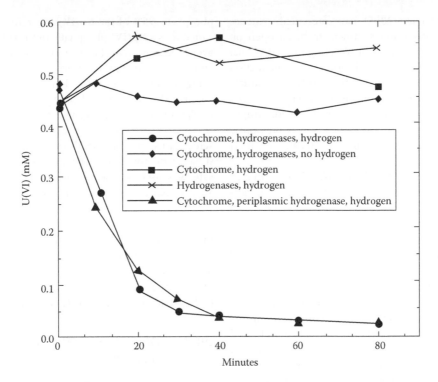

Figure 19.5. Reduction of U(VI) by electron transfer from H_2 to U(VI) via hydrogenase and cytochrome c_3. As noted, pure periplasmic hydrogenase or a protein fraction containing two hydrogenases was used. (From Lovley, DR et al., *Appl Environ Microbiol*, 59, 3572, 1993c. With permission.)

reduction by *S. putrefaciens* CN32 in the presence of goethite ($Fe_2O_3 \cdot H_2O$). They found that besides enzymatic reduction of the U(VI) by *S. putrefaciens* CN32, Fe(II) that had been sorbed to goethite was able to reduce U(VI) abiotically, as did the humic analog anthraquinone-2,6-disulfonate (AQDS).

19.6.4 Bioremediation of Uranium Pollution

Uranium(VI)-contaminated aquifer groundwater can be bioremediated by reduction of the U(VI) to insoluble U(IV) under various conditions. Thus in a field experiment, the uranium level in a contaminated aquifer in Rifle, Colorado, USA, was lowered to less than 0.18 µM from a range of 0.4 to 1.4 µM within 50 days by injection of 1 to 3 mM acetate per day (Anderson et al., 2003). Initial loss of uranium from the groundwater was attributed to the activity of *Geobacter* spp., which can reduce U(VI) enzymatically as well as produce Fe(II) from the reduction of Fe(III). As acetate injection continued, *Geobacter* spp. were gradually replaced by sulfate reducers, stimulated by the injected acetate, which was used as the carbon and energy source, and by using

sulfate in the groundwater as the terminal electron acceptor, then causing the uranium concentration in the groundwater to increase because the particular sulfate reducers that became selectively enriched were incapable of reducing U(VI) or promoting its reduction. It appears that sustained bioremediation of a uranium-contaminated aquifer by *Geobacter* spp. would benefit from the maintenance of conditions that favor the activity of *Geobacter* spp. over sulfate reducers. It also appears that only U(VI) dissolved in the groundwater at the Rifle, Colorado, aquifer site is amenable to bioremediation. U(VI) bound to the sediment in the aquifer was not microbially reducible (Ortiz-Bernad et al., 2004b), although this contrasts with recent studies using sediments from Sellafield in the United Kingdom (Law et al., 2011), where sorbed U(VI) was bioreducible. For a full description of the extensive work done at Rifle, involving sophisticated molecular analyses using microarray technologies, environmental proteomics, and more recently systems analysis and genome-scale modeling, the reader is referred to the comprehensive review of Williams et al. (2012).

Suzuki et al. (2005) have presented evidence from shallow-water sediment from an open pit of a uranium mine in Washington, USA, that showed U(VI) immobilization by bioreduction to U(IV) oxide, in which both Fe(III) and sulfate reducers, members of the Geobacteraceae and Desulfovibrionacea, respectively, were implicated. Both families of organisms were previously known to contain members capable of U(VI) reduction.

Nevin et al. (2003) demonstrated uranium bioremediation in a high-salinity subsurface sediment from an aquifer associated with uranium mine tailings at Shiprock, New Mexico, USA, by stimulation with acetate, which enriched populations of microorganisms related to *Pseudomonas* and *Desulfosporosinus* species.

Shelobolina et al. (2004) obtained evidence for potential U(VI) anaerobic bioreduction at low pH in nitrate- and U(VI)-contaminated subsurface sediment from the Natural and Accelerated Bioremediation Research Field Center in Oak Ridge, TN, USA; *Salmonella subterranean* sp. nov., a Gram-negative, motile rod capable of using O_2, NO_3^-, $S_2O_3^{2-}$, fumarate, and malate as terminal electron acceptors and of reducing U(VI) in cell suspension, was isolated from the sediment.

19.7 MICROBIAL INTERACTIONS WITH NEPTUNIUM

Neptunium has a very long half-life (2.14×10^6 years) and is therefore considered a very important radioactive component of high- and intermediate-level wastes over geological timespans (Sasaki et al., 2001). Np(V) occurs primarily in oxic conditions, as the soluble NpO_2^+ species, which is highly mobile, sorbing poorly to mineral phases (Kaszuba and Runde, 1999). In contrast, Np(IV) species dominates under reducing conditions and is far less soluble (Law et al., 2010). Thus, in common with the microbial reduction of U(VI) to U(IV), the reduction of Np(V) to Np(IV) also has the potential to immobilize Np in contaminated sediments.

Direct enzymatic reduction of soluble Np(V) species has been demonstrated for *S. oneidensis* (Lloyd et al., 2000b; Icopini et al., 2007) but not with *G. sulfurreducens* (Renshaw et al., 2005). In the former study, the majority of the resulting Np(IV) remained in solution, but in combination with a *Citrobacter* sp. that produced extracellular phosphate, 95% of the Np(IV) was removed, presumably as a Np(IV) phosphate. In the latter, however, *S. oneidensis* was able to simultaneously reduce and precipitate Np(IV) from a Np(V) solution, using lactate as the electron donor. A further study confirmed reduction and precipitation of Np(V) to Np(IV), utilizing either pyruvate or H_2 as electron donors (Rittmann et al., 2002). This was achieved using a microbial consortium including sulfate-reducing *Desulfovibrio* species. When pyruvate was used as the electron donor, complexation of Np(IV) with fermentation intermediates was thought to prevent complete precipitation of the radionuclide. Other more complex sediment systems have been shown to support Np(V) reduction and precipitation under anaerobic conditions, but in these experiments, the mechanism of reduction was associated with Mn(IV)- and Fe(III)-reducing conditions (Law et al., 2010).

19.8 MICROBIAL INTERACTIONS WITH PLUTONIUM

Early work suggested bacterial interactions with plutonium in a study with *Bacillus polymyxa* and *B. circulans* (Rusin et al., 1994). However, reduction of plutonium(IV) to plutonium(III) by the bacteria was only inferred but not directly demonstrated. More recently, enzymatic reduction of Pu(IV) to the potentially more soluble Pu(III) under anaerobic conditions was definitively demonstrated with *G. metallireducens* GS-15 and *S. oneidensis* MR-1 (Boukhalfa et al., 2007, Icopini et al., 2009, Renshaw et al., 2009). In the study of Boukhalfa et al. (2007), the organisms reduced the Pu(IV)(EDTA) and amorphous $Pu(IV)(OH)_4$ in the presence of ethylenediaminetetracetic acid (EDTA), forming Pu(III)(EDTA). In the absence of a complexing ligand like EDTA, *S. oneidensis* MR-1 produced only minor amounts of Pu(III), and *G. metallireducens* produced little or none. The Pu(IV) reduction did not support growth of either organism. A recent study demonstrated reduction of Pu(V) (PuO_2^+) via a Pu(IV) intermediate to Pu(III) as a $PuPO_4$ precipitate, using *S. alga* (Deo et al., 2011). Plymale et al. (2012) observed that Pu(IV) reduction by *G. sulfurreducens* and *S. oneidensis* was enhanced in the presence of EDTA and AQDS. Despite the potential to mobilize Pu(IV) via reduction to the potentially more soluble Pu(III), few studies have addressed the impact of stimulating microbial reduction processes on Pu solubility in environmental systems. However, in a recent study using Pu-contaminated soils from the Atomic Weapons Establishment in the United Kingdom, Pu(IV) proved highly refractory under anoxic (Fe(III) reducing) conditions (Kimber et al., 2012). Thus in sediments contaminated with a broad-range of redox active radionuclides, reductive immobilization of uranium, neptunium, and technetium would be expected, without the mobilization of any Pu(IV) present.

19.9 MICROBIAL INTERACTIONS WITH REEs

The 17 rare earth elements (REEs) are comprised of yttrium (Y) and scandium (Sc) in addition to the lanthanide series (lanthanum [La], cerium [Ce], praseodymium [Pr], neodymium [Nd], promethium [Pm], samarium [Sm], europium [Eu], gadolinium [Gd], terbium [Tb], dysprosium [Dy], holmium [Ho], erbium [Er], thulium [Tm], ytterbium [Yb], and lutetium [Lu]). These elements are not rare because of natural abundance, but rather their inability to concentrate in pure ore mineral deposits. Instead, aggregates of REE oxides form as they exhibit similar physiochemical properties such as ionic radii and valence and insensitivity to ligands. For this reason, REE separations are technically and economically challenging. This is important as REEs have broad application in mature technologies including chemical, glass, lighting, and metallurgical industries (accounting for 59% of the world's REE consumption, 80% of which are based on La and Ce). In addition, there is a growing application of REEs in developing advanced technologies, including battery alloys, ceramics, magnets, and electronic components (41% world's REE consumption; 85% Dy, Nd, and Pr) (Goonan, 2011).

There are more than 250 minerals known to contain significant quantities of REEs, varying from 10 to 300 ppm (Jordens et al., 2013); however, only bastnäsite ((Ce, La)CO_3(F,OH)), monazite ((Ce, La, Nd, Th)PO_4, SiO_4)), and xenotime (YPO_4) are exploited commercially (Tasman metals Ltd). REEs can also be found in accessory minerals, amphiboles, pyroxenes, feldspars, micas, garnets, and trace amounts in pyrite, fluorite, and chalcopyrite (Kothe and Varma, 2012). Currently, bastnäsite and monazite contribute ~95% of REEs produced and contain significant quantities of cerium by mass (49% and 45.5%, respectively). Monazite is also the principal ore of thorium

(containing up to 30%), and xenotime is primarily comprised of yttrium orthophosphate, with expressive secondary components dominated by heavier REEs. Ion-adsorbed clays are also a significant source of REEs, with 60% comprised of the yttrium element group (Jordens et al., 2013). Excessive mining of REEs over the last 30 years has also led to significant quantities of metals leaching out into the aqueous environment, with 50–100 million tonnes of rare earth ores entering Chinese agricultural systems each year (Gonzalez et al., 2014). No evidence of REE carcinogenesis has been reported to date; however, as trivalent REEs have similar cationic radii to Ca(II), they may induce cytophysiological effects through blocking calcium channels (Pałasz and Czekaj, 2000) and are becoming an increasing environmental concern.

19.9.1 Microbial Interactions with Rare Earth Ions

REEs, especially scandium, are known to activate secondary metabolism and extracellular enzyme production in certain microorganisms. It has been shown that REEs enhance the production of acid and aromatic nuclei by *Arthrobacter luteolus*, which can influence mineral acid, chelating agent, and polysaccharide production (Emmanuel et al., 2012).

Scandium has been reported to stimulate the production of amylase, bacilysin (a dipeptide antibiotic), and undecaprenyl pyrophosphate (C_{55}-PP, an essential molecule involved in construction of the bacterial peptidoglycan cell wall) in B. *subtilis* (Inaoka and Ochi, 2011; Inaoka and Ochi, 2012). Scandium and lanthanum have also been found to enhance antibiotic production by *Streptomyces* spp. and activate many secondary metabolite biosynthetic genes in *Streptomyces coelicolor* A3 (Kawai et al., 2007; Tanaka et al., 2010).

Lanthanide biosorption has been observed in *E. coli* (Takahashi et al., 2005), *Alcaligenes faecalis*, *S. putrefaciens*, and *P. fluorescens* (Takahashi, 2007). The potential of REE biosorption for bioremediation has led to a number of studies that have focused on microbial interactions with REEs, reviewed in Andrès et al. (2003) and Moriwaki and Yamamoto (2013). For example, Eu(III) coordination to bacterial cells varied not only between Gram-positive and Gram-negative species but also within three types of Gram-negative bacteria (Ozaki et al., 2002b, 2005). To further explore Gram-positive

REE adsorption, a recent study investigated the potential of wild-type B. *subtilis* to adsorb REEs (Tm(III), La(III), and Eu(III)) in comparison to a lipoteichoic acid (LTA)–deficient mutant. They found that lower REE adsorption occurred in the LTA-deficient strain, suggesting that REE preferentially adsorbs to teichoic acids on the cell wall (Moriwaki et al., 2013). Selective accumulation of REEs by a number of microorganisms has also been reported (Tsuruta, 2005, 2006; Emmanuel et al., 2012); however, the reasons for this selectivity were not explored and clearly warrant further work. *Arthrobacter nicotianae* preferentially absorbed Sm(III), Tb(III), and Y(III), while *A. luteolus* accumulated significantly more Sm than other REEs tested (Ce, Eu, and Sc). Alternatively, *Streptomyces albus* accumulated higher amounts of Lu than any other REE (Tsuruta, 2006). In the case of *A. nicotianae*, the extent of biosorption was strongly influenced by REE concentration and pH (Tsuruta, 2006).

Carboxylate ligand complexation with REEs have been reported to affect REE sorption, with iminodiacetic acid, nitrilotriacetic acid, and EDTA all suppressing biosorption (Takahashi et al., 2005). Similar impacts on biosorption were also observed with the biogenic siderophores hydroxamate and catecholate (Yoshida et al., 2004; Cervini-Silva et al., 2008; Challaraj Emmanuel et al., 2011; Emmanuel et al., 2012). The strong affinity of REE adsorption onto extracellular polymeric substances such as polyglutamic acid and polysaccharide has also been observed. Biomass accumulation can lead to flocculation of these biopolymers with a large surface area and high affinity for REEs (Merroun et al., 2003).

To further understand the complex nature of lanthanide–microbe interactions, spectroscopic techniques have been used to elucidate mechanisms of biosorption. Time-resolved laser-induced fluorescence spectroscopy (TRLFS) has been used to predict the coordination states and structure of REEs adsorbed to surfaces in the presence of water (Ozaki et al., 2002a; Markai et al., 2003). TRFLS has also been used to determine interactions of REEs in the presence of organic ligands (Ozaki et al., 2006). It was found that Eu(III) preferentially adsorbs onto bacterial cells by an inner spherical process and in the presence of organic ligands with low chelating ability (Ozaki et al., 2006). In the presence of high chelating organics, biodegradation of the free organic compounds

proceeds until recalcitrant stoichiometric Eu(III)–organic acid complexes form and degradation is inhibited (Ozaki et al., 2006). Extended x-ray absorption fine structure (EXAFS) has additionally been used to determine the coordination number and structure of solid phase REEs in biological systems. One EXAFS study determined that REE–phosphate binding was preferential at low pH, with increasing carboxylate binding at higher pH (Ngwenya et al., 2009). Takahashi and coworkers have shown that REE adsorption to cells is influenced by the structure and number of phosphate sites associated with N-acetylglucosamine phosphates, in the lipid A outer membrane of Gram-negative cells (Takahashi et al., 2010).

19.9.2 Biogenic Mineral Interactions

Interactions with biogenic minerals may also influence REE transport in the environment, and it has been shown that biogenic manganese and iron oxides can adsorb the entire lanthanide inventory (Anderson and Pedersen, 2003; Tanaka et al., 2010). It was also reported that Ce and Yb biosorption followed by the release of phosphorous from the cells of *Saccharomyces cerevisiae* resulted in the biomineralization of REE(III)–phosphate nanocrystallites (Jiang et al., 2010, 2012). Abiotic controls with Yb and P however showed no evidence for REE–P mineral formation, indicating that the cell surface played an integral role as a nucleation site for biomineralisation (Jiang et al., 2012).

19.9.3 Microbial Oxidation of Cerium

Cerium is the most abundant REE and generally is the only one to undergo redox transformations on the Earth's surface. Cerium typically exists as cerous Ce(III) under most natural environments, but in the presence of a powerful oxidizing agent, oxidation to ceric Ce(IV) with the resultant formation of poorly soluble cerianite (CeO_2) under most environmental pHs (4–9) and cerium carbonate Ce_2CO_3 under higher pH and Eh exists (Cotton et al., 1999; Gonzalez et al., 2014). Oxidation of Ce(III) by O_2 in seawater is slow, but has been shown to increase in the presence of microorganisms (Moffett, 1990). It was suggested that enzymatic Ce(III) oxidation was occurring in these systems resulting in bioprecipitation of cerium(IV) oxide (Moffett, 1990, 1994); however, no characterization of the solid cerium mineral phase was conducted leaving this result open to interpretation as to whether the observed bioprecipitation was the result of Ce absorption onto bacterial cell surfaces or was caused by Ce(III) oxidation. Targeted EXAFS experiments on these cerium phases would resolve this ambiguity. The variable redox chemistry of cerium (and europium) has given rise to anomalies in the typical lanthanide pattern (tetrad effect), which describes the relative abundance of cerium (or europium) to other lanthanides in a solid (Equation 19.5). The index n represents the normalized element content. Positive anomalies usually develop in hydrogenic oceanic and soil concentrations as a result of Ce(III) oxidation. Negative anomalies develop in cerium-bearing waters during solid phase interactions such as cerianite sedimentation or oxidizing purification of cerium from seawater, which involves lanthanide adsorption onto Fe or Mn oxides, oxidation at the surface, and subsequent desorption:

$$Ce^1_{an} = \log\left[\frac{3Ce_n}{2La_n + Nd_n}\right] \qquad (19.5)$$

Manganese oxides are often enriched in Ce, suggesting an association of Mn and Ce geochemistry. Recent studies have shown that Ce(III) preferentially associates with the biogenic MnO_2 mineral surface, rather than biomass, where it is subsequently oxidized to Ce(IV) (Ohnuki et al., 2008; Tanaka et al., 2010). REE(III)–humate complexes have been shown to inhibit the tetrad effect and suppress the cerium anomaly. It was suggested that humate shields the REE from the oxidative surface of MnO_2. Furthermore, if oxidation did occur, the REE could not be removed from the surface due to humate complexation (Davranche et al., 2005). Further work showed that competition for humate–REE binding was stronger at lower pH (<6) and under low dissolved organic content/ Mn ratios (Davranche et al., 2008). Ce(III) oxidation has also been observed in systems supplied with Fe hydroxides (Nath et al., 1994; Bau, 1999; Bau and Koschinsky, 2009). Additionally, workers have shown that trihydroxamate siderophores such as desferrioxamine (DFO) can oxidize Ce(III) to Ce(IV), forming strong Ce(IV)–DFO complexes in comparison to Ce(III). Another ubiquitous biogenic compound, catechol ($C_6H_6O_2$), has also been shown to induce Ce(III) oxidation to form CeO_2 (Cervini-Silva et al., 2008). DFO and catechol-mediated Ce(III) oxidation could be judiciously applied to the biooxidation of Ce(III) and provide an alternative approach to selective cerium separation chemistry. Overall, these studies highlight that cerium content in organic-rich soils is not a true representation of the redox environment.

19.9.4 REE Summary and Perspectives

REE applications and demand have increased exponentially over the last few decades and will continue to grow in the future; thus, the need for alternative cheaper environmentally friendly metal extractions is becoming more and more desirable. Investment is needed in the research and development of REE recycling methods, new technologies with greater material efficiency, and deposit exploration. REEs have a high affinity for bacterial surfaces, and selective accumulation of REEs has been reported in a number of microorganisms (Tsuruta, 2005, 2006; Emmanuel et al., 2012), but no further insight into these processes was given. A recent study has shown that it is possible to engineer a sensory system to detect lanthanide ion and regulate a bacterial response using the *Salmonella* PmrA/PmrB 2 component system in *E. coli*, which has a significant potential in future lanthanide remediation (Liang et al., 2013). Understanding REE biosorption mechanisms may also provide an alternative route for REE separations in the future.

Cerium represents a significant proportion of the lanthanide content of natural ores bastnäsite and monazite, so removing this metal from the ore leachate would make further REE separations and isolations much easier. However, few studies have explored cerium biogeochemistry, and the potential for cerium separation through microbial Ce(III) oxidation could be of interest. Furthermore, alternative Ce(III) oxidation pathways could be explored through interactions with biologically generated superoxides that are produced by a variety of heterotrophic bacteria (Learman et al., 2011; Diaz et al., 2013).

Given the extent of agricultural applications in conjunction with the significant REE mine tailings, there is a paucity of studies that have reviewed the ecotoxicity of these metals in environmental systems (Evans, 1990; Gonzalez et al., 2014). A number of microorganisms have been studied for their ability to sorb REEs as a method of bioremediation, but again the mechanisms are poorly understood and clearly warrant further work.

19.10 SUMMARY

Enzymatic oxidation of Cr(III) by bacteria has not been demonstrated. Nonenzymatic oxidation of Cr(III), which is dependent on biogenic (bacterial, fungal) formation of Mn(IV), which then oxidizes Cr(III) to Cr(VI) chemically, may occur in soil.

Aerobic and anaerobic reduction of Cr(VI) by bacteria has been demonstrated. Various organic electron donors may serve, but not all act equally well aerobically and anaerobically. Chromate and dichromate are not necessarily reduced equally well. The ability to reduce chromate does not always correlate with chromate tolerance.

Although chromite is not very susceptible to leaching by acid formed by acidophiles like A. ferrooxidans and A. thiooxidans, chromium in some solid inorganic industrial wastes can by leached by sulfuric acid formed by A. thiooxidans in sulfur oxidation.

Cr(VI) reduction to Cr(III) is beneficial ecologically because it lowers chromium toxicity. At equivalent concentrations, Cr(VI) is more toxic than Cr(III). Cr(III) also tends to precipitate as a hydroxo compound around neutrality, the pH range at which all known Cr(VI) reducers operate. Microbially produced Fe(II) is also important in mediating the reduction of Cr(VI) in sediments.

Bacteria have been discovered that can enzymatically oxidize Mo(IV) and Mo(V) to Mo(VI) in air. Other bacteria have been found that can enzymatically reduce Mo(VI) to Mo(V), some aerobically, others anaerobically.

Vanadate (VO_3^-) has been found to be reduced anaerobically to vanadyl (VO(OH)) by a number of bacteria. At least two of these organisms can use vanadate as the terminal electron acceptor during chemolithotrophic growth with H_2 as electron donor.

Tetravalent uranium can be oxidized to hexavalent uranium, an oxidation that serves as a source of energy to several microorganisms, although it does not support their growth as the sole energy source. Anaerobically, hexavalent uranium can be reduced to tetravalent uranium by a number of bacteria using either H_2 or one of a variety of organic electron donors. Similar reduction processes have been described for Np(V) to Np(IV) and Tc(VII) to Tc(IV). Plutonium biogeochemistry is more complex, but microbial reductions of Pu(V) and Pu(IV) to Pu(III) have been demonstrated.

The bacterial oxidations and reductions of Cr, Mo, V, and U can play an important role in their mobilization and immobilization in soils and sediments.

Polonium can be mobilized or immobilized by bacteria, but whether redox reactions involving Po are involved remains unknown.

REEs such as Sc, La, and Eu have a high affinity for bacterial surfaces, and Ce(III) can be oxidized microbially to Ce(IV).

REFERENCES

Ackerley DF, Gonzalez CF, Park CH, Blake II R, Keyhan M, Matin A. 2004. Chromate-reducing properties of soluble flavoproteins from Pseudomonas putida and Escherichia coli. Appl Environ Microbiol 70:873–882.

Anderson CR and Pedersen K. 2003. In situ growth of Gallionella biofilms and partitioning of lanthanides and actinides between biological material and ferric oxyhydroxides. Geobiology 1:169–178.

Anderson GL, Williams J, Hille R. 1992. The purification and characterization of arsenite oxidase from Alcaligenes faecalis, a molybdenum-containing hydroxylase. J Biol Chem 267:23674–23682.

Anderson RT, Vrionis HA, Ortiz-Bernad I, Resch CT, Long PE, Dayvault R, Karp K et al. 2003. Stimulating the in situ activity of Geobacter species to remove uranium from the groundwater of a uranium-contaminated aquifer. Appl Environ Microbiol 69:5884–5891.

Andrès Y, Texier AC, Le Cloirec P. 2003. Rare earth elements removal by microbial biosorption: A review. Environ Technol 24:1367–1375.

Antipov AN, Lyalikova NN, Khijniak TV, L'vov NP. 2000. Vanadate reduction by molybdenum-free dissimilatory nitrate reductase from vanadate-reducing bacteria. IUBMB Life 50:39–42.

Arias YM, Tebo BM. 2003. Cr(VI) reduction by sulfidogenic and nonsulfidogenic microbial consortia. Appl Environ Microbiol 69:1847–1853.

Bader JL, Gonzalez G, Goodell PC. 1999. Chromium-resistant bacterial populations from a site heavily contaminated with hexavalent chromium. Water Air Soil Pollut 109:263–276.

Baroch CT. 1965. Uranium. In: *Mineral Facts and Problems. Bulletin 630.* Washington, DC: Bureau of Mines, US Department of the Interior, pp. 1007–1037.

Bartlett RJ, James BR. 1979. Behavior of chromium in soils: III. Oxidation. *J Environ Qual* 8:31–25.

Bau M. 1999. Scavenging of dissolved yttrium and rare earths by precipitating iron oxyhydroxide: Experimental evidence for Ce oxidation, Y-Ho fractionation, and lanthanide tetrad effect. *Geochim Cosmochim Acta* 63:67–77.

Bau M, Koschinsky A. 2009. Oxidative scavenging of cerium on hydrous Fe oxide: Evidence from the distribution of rare earth elements and yttrium between Fe oxides and Mn oxides in hydrogenetic ferromanganese crusts. *Geochem J* 43:37–47.

Beller HR. 2005. Anaerobic, nitrate-dependent oxidation of U(IV) oxide minerals by the chemolithoautotrophic bacterium *Thiobacillus denitrificans. Appl Environ Microbiol* 71:2170–2174.

Bencheikh-Latmani R, Williams SM, Haucke L, Criddle CS, Wu L, Zhou J, Tebo BM. 2005. Global transcriptional profiling of *Shewanella oneidensis* MR-1 during Cr(VI) and U(VI) reduction. *Appl Environ Microbiol* 71:7453–7460.

Bernier-Latmani R, Veeramani H, Vecchia ED, Junier P, Lezama- Pacheco JS, Suvorova EI, Sharp JO, Wigginton NS, Bargar JR. 2010. Non-uraninite products of microbial U(VI) reduction. *Environ Sci Technol* 44:9456–9462.

Bopp LH. 1980. Chromate resistance and chromate reduction in bacteria. PhD thesis. Rensselaer Polytechnic Institute, Troy, NY.

Bopp LH, Chakrabarty AM, Ehrlich HL. 1983. Chromate resistance plasmid in *Pseudomonas fluorescens. J Bacteriol* 155:1105–1109.

Bopp LH, Ehrlich HL. 1988. Chromate resistance and reduction in *Pseudomonas fluorescens* strain LB300. *Arch Microbiol* 150:426–431.

Bosecker K. 1986. Bacterial metal recovery and detoxification of industrial waste. In: Ehrlich HL, Holmes DS, eds., *Workshop on Biotechnology for the Mining, Metal-Refining and Fossil Fuel Processing Industries. Biotech Bioeng Symp* 16. New York: Wiley, pp. 105–120.

Boukhalfa H, Icopini GA, Reilly SD, Neu MP. 2007. Plutonium(IV) reduction by the metal-reducing bacteria *Geobacter metallireducens* GS15 and *Shewanella oneidensis* MR1. *Appl Environ Microbiol* 73:5897–5903.

Bowen HJM. 1979. *Environmental Chemistry of the Elements.* London, U.K.: Academic Press, p. 333.

Briand L, Thomas H, Donati E. 1996. Vanadium(V) reduction by *Thiobacillus thiooxidans* cultures on clemental sulfur. *Biotechnol Lett* 18:505–508.

Brierley CL, Brierley JA. 1982. Anaerobic reduction of molybdenum by *Sulfolobus* species. *Zentralbl Bakteriol Hyg I Abt Orig* C3:289–294.

Brierley CL, Murr LE. 1973. A chemoautotrophic and thermophilic chemoautotrophic microbe. *Science* 179:488–490.

Brock TD, Madigan MT. 1991. *Biology of Microorganisms,* 6th ed. Englewood Cliffs, NJ: Prentice-Hall, p. 874

Burke IT, Boothman C, Lloyd JR, Livens FR, Charnock JM, McBeth JM, Mortimer RJG, Morris K. 2006. Redoxidation behaviour of technetium, iron and sulfur in estuarine sediments. *Env Sci Technol* 40:3529–3535.

Burke IT, Boothman C, Lloyd, JR, Mortimer RJG, Morris K. 2005. Technetium solubility during the onset of progressive anoxia. *Env Sci Technol* 39:4109–4116.

Caccavo F Jr, Blakemore RP, Lovley DR. 1992. A hydrogen-oxidizing, Fe(III) reducing microorganism from the Great Bay Estuary, New Hampshire. *Appl Environ Microbiol* 58:3211–3216.

Carpentier W, De Smet L, Van Beeumen J, Brigé A. 2005. Respiration and growth of *Shewanella oneidensis* MR-1 using vanadate as the sole electron acceptor. *J. Bacteriol* 187:3293–3301.

Carpentier W, Sandra K, De Smet I, Brigé A, De Smet L, Van Beeumen J. 2003. Microbial reduction and precipitation of vanadium by *Shewanella oneidensis. Appl Environ Microbiol* 69:3636–3639.

Cervantes C. 1991. Bacterial interactions with chromium. *Antonie v Leeuwenhoek* 59:229–233.

Cervantes C, Campos-García J. et al. 2007. Reduction and efflux of chromate by bacteria. *Mol Microbiol Heavy Metals* 6:407–419.

Cervini-Silva J, Gilbert B, Fakra S, Friedlich S, Banfield J. 2008. Coupled redox transformations of catechol and cerium at the surface of a cerium(III) phosphate mineral. *Geochim Cosmochim Acta* 72:2454–2464.

Challaraj Emmanuel, ES, Vignesh, V, Anandkumar, B, Maruthamuthu, S. 2011. Bioaccumulation of cerium and neodymium by *Bacillus cereus* isolated from rare earth environments of Chavara and Manavalakurichi, India. *Indian J Microbiol* 51:488–495.

Cherrier J, Burnett WC, LaRock PA. 1995. Uptake of polonium and sulfur by bacteria. *Geomicrobiol J* 13:103–115.

Chirwa EMN, Wang Y-T. 1997. Use hexavalent chromium reduction by *bacillus* sp. in a packed-bed bioreactor. *Environ Sci Technol* 31(5):1446–1451.

Coleman RN, Paran JH. 1983. Accumulation of hexavalent chromium by selected bacteria. *Environ Technol Lett* 4:149–156.

Cologgi DL, Lampa-Pastirk S, Speers AM, Kelly SD, Reguera G. 2011. Extracellular reduction of uranium via *Geobacter* conductive pili as a protective cellular mechanism. *Proc Natl Acad Sci USA* 108:15248–15252.

Cotton F, Wilkinson G, Murillo C. 1999. *Advanced Inorganic Chemistry*, 6th ed. Toronto, Ontario, Canada: A. Wiley, p. 1376.

Cutting RS, Coker VS, Telling ND, Kimber RL, Pearce CI, Ellis BL, Lawson RS et al. 2010. Optimizing Cr(VI) and Tc(VII) remediation through nanoscale biomineral engineering. *Environ Sci Technol* 44:2577–2584.

Daulton TL, Little BJ, Jones-Meehan J, Blom DA, Allard LF. 2007. Microbial reduction of chromium from the hexavalent to the divalent state. *Geochim Cosmochim Acta* 71:556–565.

Davranche M, Pourret O, Gruau G, Dia A, Jin D, Gaertner D. 2008. Competitive binding of REE to humic acid and manganese oxide: Impact of reaction kinetics on development of cerium anomaly and REE adsorption. *Chem Geol* 247:154–170.

Davranche M, Pourret O, Gruau G, Dia A, Le Coz-Bouhnik M. 2005. Adsorption of REE(III)-humate complexes onto MnO_2: Experimental evidence for cerium anomaly and lanthanide tetrad effect suppression. *Geochim Cosmochim Acta* 69(20):4825–4835.

De Luca G, Philip P, Dermoun Z, Rousset M, Vermeglio A. 2001. Reduction of technetium(VII) by *Desulfovibrio fructosovorans* is mediated by the nickel-iron hydrogenase. *Appl Environ Microbiol* 67:4583–4587.

DeHuff GL. 1965. Vanadium. In: *Mineral Facts and Problems*. Bull 630. Washington, DC: Bur Mines, US Department of the Interior, pp. 1039–1053.

DeLeo PC, Ehrlich HL. 1994. Reduction of hexavalent chromium by *Pseudomonas fluorescens* LB300 in batch and continuous culture. *Appl Microbiol Biotechnol* 40:756–759.

Deo R., Rittmann B, Reed D. 2011. Bacterial Pu(V) reduction in the absence and presence of Fe(III)–NTA: Modeling and experimental approach. *Biodegrad* 22:921–929.

Dey S, Paul A. 2012. Optimization of cultural conditions for growth associated chromate reduction by *Arthrobacter* sp. SUK 1201 isolated from chromite mine overburden. *J Hazard Materials* 213:200–206.

Dhal B, Thatoi H et al. 2010. Reduction of hexavalent chromium by *Bacillus* sp. isolated from chromite mine soils and characterization of reduced product. *J. Chem. Technol. Biotechnol.* 85:1471–1479.

Diaz JM, Hansel CM, Voelker BM, Mendes CM, Andeer PF, Zhang T. 2013. Widespread production of extracellular superoxide by heterotrophic bacteria. *Science* 340(6137):1223–1226.

Dickerson RE, Gray HB, Haight GP Jr. 1979. *Chemical Principles*, 3rd ed. Menlo Park, CA: Benjamin/Cummings, p. 1037.

DiSpirito AA, Tuovinen OH. 1981. Oxygen uptake couples with uranous sulfate oxidation by *Thiobacillus ferrooxidans* and *T. acidophilus. Geomicrobiol J* 2:275–291.

DiSpirito AA, Tuovinen OH. 1982a. Uranous ion oxidation and carbon dioxide fixation by *Thiobacillus ferrooxidans. Arch Microbiol* 133:28–32.

DiSpirito AA, Tuovinen OH. 1982b. Kinetics of uranium and ferrous iron oxidation by *Thiobacillus ferrooxidans. Arch Microbiol* 133:33–37.

Ehrlich HL. 1983. Leaching chromite ore and sulfide matte with dilute sulfuric acid generated by *Thiobacillus thiooxidans* from sulfur. In: Rossi G, Torma AE, eds., *Recent Progress in Biohydrometallurgy*. Iglesias, Italy: Assoc Min Sarda, pp. 19–42.

Emmanuel EC, Ananthi T, Anandkumar B, Maruthamuthu S. 2012. Accumulation of rare earth elements by siderophore-forming *Arthrobacter luteolus* isolated from rare earth environment of Chavara, India. *J Biosci* 37:25–31.

Enzmann RD. 1972. Molybdenum: Element and geochemistry. In: Fairbridge RW, ed., *The Encyclopedia of Geochemistry and Environmental Sciences. Ecycl Earth Sci Ser, Vol IVA.* New York: Van Nostrand Reinhold, pp. 753–759.

Evans CH. 1990. *Biochemistry of the Lanthanides.* Boston, MA: Springer, p. 460.

Fandeur D, Juillot D et al. 2009. XANES evidence for oxidation of Cr(III) to Cr(VI) by Mn-oxides in a lateritic regolith developed on serpentinized ultramafic rocks of New Caledonia. *Env Sci Technol* 43:7384–7390.

Forsberg CW. 1978. Effect of heavy metals and other trace elements on the fermentative activity of the rumen flora and growth of functionally important rumen bacteria. *Can J Microbiol* 24:298–306.

Fortescue JAC. 1980. *Environmental Geochemistry.* New York: Springer-Verlag.

Frederickson JK, Zachara JM, Kennedy DW, Duff MC, Gorby YA, Li S-MW, Krupka KM. 2000. Reduction of U(VI) in goethite (α-FeOOH) suspensions by a dissimilatory metal-reducing bacterium. *Geochim Cosmochim Acta* 64:3085–3098.

Fuji E, Tsuchida T, Urano K, Ohtake H. 1994. Development of a bioreactor system for the treatment of chromate wastewater using *Enterobacter cloacae* HO1. *Water Sci Technol* 30:235–243.

Garbisu C, Alkorta I, Llama MJ, Serra JL. 1998. Aerobic chromate reduction by *Bacillus subtilis. Biodegradation* 9:133–141.

Ghani B, Takai M, Hisham NZ, Kishimoto N, Ismail AKM, Tano T, Sugio T. 1993. Isolation and characterization of a Mo^{6+}-reducing bacterium. *Appl Environ Microbiol* 59:1176–1180.

Gonzalez V, Vignati DA, Leyval C, Giamberini L. 2014. Environmental fate and ecotoxicity of lanthanides: Are they a uniform group beyond chemistry? *Environ Int* 71:148–157.

Goonan TG. 2011. Rare Earth elements—End use and recyclability. US Geological Survey Scientific Investigations Report 2011–5094, p. 15.

Gopalan R, Veeramani H. 1994. Studies on microbial chromate reduction by *Pseudomonas* sp. in aerobic continuous suspended culture. *Biotechnol Bioeng* 43:471–476.

Gorby YA, Lovley DR. 1992. Enzymatic uranium precipitation. *Environ Sci Technol* 26:205–207.

Gruber JE, Jennette KW. 1978. Metabolism of the carcinogen chromate by rat liver microsomes. *Biochem Biophys Res Commun* 82:700–706.

Gvozdyak PI, Mogilevich NF, Ryl'skit AF, Grishchenko NI. 1986. Reduction of hexavalent chromium by collection strains of bacteria. *Mikrobiologiya* 55:962–965 (English transl., pp. 770–773).

Haglund DS. 1972. Uranium: Element and geochemistry. In: Fairbridge RW, ed., *The Encyclopedia of Geochemistry and Environment Sciences*. Encycl Earth Sci Ser, Vol IVA. New York: Van Nostrand Reinhold, pp. 1215–1222.

Harish R, Samuel J et al. 2012. Bio-reduction of Cr (VI) by exopolysaccharides (EPS) from indigenous bacterial species of Sukinda chromite mine, India. *Biodegradation* 23:487–496.

Henrot J. 1989. Bioaccumulation and chemical modification of Tc by soil bacteria. *Health Phys* 57:239–245.

Holliday RW. 1965. Molybdenum. In: *Mineral Facts and Problems*. Bull 630. Washington, DC: US Department of the Interior, pp. 595–606.

Horitsu H, Futo S, Myazawa Y, Ogai S, Kawai K. 1987. Enzymatic reduction of hexavalent chromium by hexavalent chromium tolerant *Pseudomonas ambigua* G-1. *Biol Chem* 51:2417–2420.

Horitsu H, Nishida H, Kato H, Tomoyeda M. 1978. Isolation of potassium chromate-tolerant bacterium and chromate uptake by the bacterium. *Agric Biol Chem* 42:2037–2043.

Hu P, Brodie EI et al. 2005. Whole-genome transcriptional analysis of heavy metal stresses in *Caulobacter crescentus*. *J Bacteriol* 187:8437–8449.

Icenhower JP, Qafoku NP, Martin MJ, Zachara JM. 2008. The geochemistry of technetium: A summary of the behaviour of an artificial element in the natural environment. PNNL-18139, U.S. Department of Energy, Washington, DC.

Icopini GA, Boukhalfa H, Neu MP. 2007. Biological Reduction of Np(V) and Np(V) Citrate by Metal-Reducing Bacteria. *Environ Sci Technol* 41:2764–2769.

Icopini GA, Lack JG, Hersman LE, Neu MP, Boukhalfa H. 2009. Plutonium(V/VI) reduction by the metal-reducing bacteria *Geobacter metallireducens* GS-15 and *Shewanella oneidensis* MR-1. *Appl Environ Microbiol* 75:3641–3647.

Inaoka T, Ochi K. 2011. Scandium stimulates the production of amylase and bacilysin in *Bacillus subtilis*. *Appl Environ Microbiol* 77:8181–8183.

Inaoka T, Ochi K. 2012. Undecaprenyl pyrophosphate involvement in susceptibility of *Bacillus subtilis* to rare earth elements. *J Bacteriol* 194:5632–5637.

Ishibashi Y, Cervantes C, Silver S. 1990. Chromium reduction by *Pseudomonas putida*. *Appl Environ Microbiol* 56:2268–2270.

Jeyasingh J, Somasundaram V et al. 2010. Pilot scale studies on the remediation of chromium contaminated aquifer using bio-barrier and reactive zone technologies. *Chem Eng J* 167:206–214.

Jiang M, Ohnuki T, Kozai N, Tanaka K, Suzuki Y, Sakamoto F, Kamiishi E, Utsunomiya S. 2010. Biological nano-mineralization of Ce phosphate by *Saccharomyces cerevisiae*. *Chem Geol* 277:61–69.

Jiang M, Ohnuki T, Tanaka K, Kozai N, Kamiishi E, Utsunomiya S. 2012. Post-adsorption process of Yb phosphate nano-particle formation by *Saccharomyces cerevisiae*. *Geochim Cosmochim Acta* 93:30–46.

Johnson I, Flower N, Loutit MW. 1981. Contribution of periphytic bacteria to the concentration of chromium in the crab *Helice crassa*. *Microb Ecol* 7:245–252.

Jordens A, Cheng YP, Waters KE. 2013. A review of the beneficiation of rare earth element bearing minerals. *Min Eng* 41:97–114.

Juhnke S, Peitzsch N et al. 2002. New genes involved in chromate resistance in *Ralstonia metallidurans* strain CH34. *Arch Microbiol* 179:15–25.

Kamaludeen S, Megharaj M et al. 2003. Chromium-microorganism interactions in soils: Remediation implications. *Rev Environ Contam Toxicol* 178:93–164.

Kaszuba JP, Runde WH. 1999. The aqueous geochemistry of neptunium: Dynamic control of soluble concentrations with applications to nuclear waste disposal. *Environ Sci Technol* 33:4427–4433.

Kawai K, Wang G, Okamoto S, Ochi K. 2007. The rare earth, scandium, causes antibiotic overproduction in Streptomyces spp. *FEMS Microbiol Lett* 274:311–315.

Kilic NK, Stensballe A et al. 2009. Proteomic changes in response to chromium(VI) toxicity in *Pseudomonas aeruginosa*. *Biores Technol* 101:2134–2140.

Kimber RL, Boothman C, Purdie P, Livens FR, Lloyd JR. 2012. Biogeochemical behaviour of plutonium during anoxic biostimulation of contaminated sediments. *Mineral Mag* 76:567–578

Komori K, Rivas A, Toda K, Ohtake H. 1990a. Biological removal of toxic chromium using an *Enterobacter cloacae* strain that reduces chromate under anaerobic conditions. *Biotechnol Bioeng* 35:951–954.

Komori K, Rivas A, Toda K, Ohtake H. 1990b. A method for removal of toxic chromium using dialysis-sac cultures of the chromate reducing strain of *Enterobacter cloacae*. *Appl Microbiol Biotechnol* 33:117–119.

Komori K, Wang P-C, Toda K, Ohtake H. 1989. Factors affecting chromate reduction in *Enterobacter cloacae* HO1. *Appl Microbiol Biotechnol* 31:567–570.

Kostka JE, Green SJ. 2011. Microorganisms and processes linked to uranium reduction and immobilization. In: Stolz JF, Oremland RS, eds., *Microbial Metal and Metalloid Metabolism: Advances and Applications*. Washington, DC: ASM Press, pp. 117–138.

Kothe E, Varma A. 2012. *Bio-Geo Interactions in Metal-Contaminated Soils*. Berlin, Germany: Springer, p. 426.

Kozuh N, Stupar J, Gorenc B. 2000. Reduction and oxidation processes of chromium in soils. *Environ Sci Technol* 34:112–119.

Kvasnikov EI, Stepnyuk VV, Klyushnikova TM, Serpokrylov NS, Simonova GA, Kasatkina TP, Panchenko LP. 1985. A new chromium-reducing, gram-variable bacterium with mixed type of flagellation. *Mikrobiologiya* 54:83–88 (English transl., pp. 69–75).

Kwak YH, Lee DS, Kim HB. 2003. *Vibrio harveyi* nitroreductase is also a chromate reductase. *Appl Environ Microbiol* 69:4390–4395.

LaRock PA, Huyn J-H, Boutelle S, Burnett WC, Hull CD. 1996. Bacterial mobilization of polonium. *Geochim Cosmochim Acta* 60:4321–4328.

Latimer WM, Hildebrand JH. 1942. *Reference Book of Inorganic Chemistry*, Rev. ed. New York: Macmillan, pp. 357–360.

Law GTW, Geissler A, Lloyd JR, Burke IT, Livens FR, Boothman B, McBeth J, Morris K. 2011. Uranium redox cycling in sediment and biomineral systems. *Geomicrobiol J* 28:5–6, 497–506.

Law GTW, Geissler A, Lloyd JR, Livens FR, Boothman C, Begg JDC, Denecke MA et al. 2010. Geomicrobiological redox cycling of the transuranic element neptunium. *Environ Sci Technol* 44:8924–8929.

Learman DR, Voelker BM, Vazquez-Rodriguez AI, Hansel CM. 2011. Formation of manganese oxides by bacterially generated superoxide. *Nature Geosci* 4:95–98.

Lebedeva EV, Lyalikova NN. 1979. Reduction of crocoite by *Pseudomonas chromatophila* nov. sp. *Mikrobiologiya* 48:517–522.

Liang H, Deng X, Bosscher M, Ji Q, Jensen MP, He C. 2013. Engineering bacterial two-component system PmrA/PmrB to sense lanthanide ions. *J Am Chem Soc* 135(6):2037–2039.

Lietzke MH. 1972. Polonium: Element and geochemistry. In: Fairbridge RW, ed., *The Encyclopedia of Geochemistry and Environmental Sciences. Encycl Earth Sci Ser*, Vol. IVA. New York: Van Nostrand Reinhold, pp. 962–964.

Llovera S, Bonet R, Simon-Pujol MD, Congregado F. 1993. Chromate reduction by resting cells of *Agrobacterium radiobacter* EPS-916. *Appl Environ Microbiol* 59:3516–3518.

Lloyd JR. 2003. Microbial reduction of metals and radionuclides. *FEMS Microbiol Rev* 27:411–425.

Lloyd JR, Cole JA, Macaskie LE. 1997. Reduction and removal of heptavalent technetium from solution by *Escherichia coli*. *J Bacteriol* 179:2014–2021.

Lloyd JR, Macaskie LE. 1996. A novel phosphorImager based technique for monitoring the microbial reduction of technetium. *Appl Environ Microbiol* 62:578–582.

Lloyd JR, Sole VA, Van Praagh CVG, Lovley DR. 2000a. Direct and Fe(II)-mediated reduction of technetium by Fe(III)-reducing bacteria. *Appl Environ Microbiol* 66:3743–3749.

Lloyd JR, Yong P., Macaskie LE. 2000b. Biological reduction and removal of pentavalent Np by the concerted action of two microorganisms. *Environ Sci Tech* 34:1297–1301.

Lovley DR, Giovannoni SJ, White DC, Champine JE, Phillips EJP, Gorby YA, Goodwin S. 1993a. *Geobacter metallireducens* gen. nov. sp. nov., a microorganism capable of coupling the complete oxidation of organic compounds to the reduction of iron and other metals. *Arch Microbiol* 159:336–344.

Lovley DR, Phillips EJP, Gorby YA, Landa ER. 1991. Microbial reduction of uranium. *Nature* 350:413–416.

Lovley DR, Phillips EJP. 1994. Reduction of chromate by *Desulfovibrio vulgaris* and its c_3 cytochrome. *Appl Environ Microbiol* 60:726–728.

Lovley DR, Phillips EJP. 1992. Reduction of uranium by *Desulfovibrio desulfuricans*. *Appl Environ Microbiol* 58:850–856.

Lovley DR, Roden EE, Phillips EJP, Woodward JC. 1993b. Enzymatic iron and uranium reduction by sulfate-reducing bacteria. *Mar Geol* 113:41–53.

Lovley DR, Ueki T, Zhang T, Malvankar NS, Shrestha PM, Flanagan KA, Aklujkar M et al. 2011. *Geobacter*: The microbe electric's physiology, ecology, and practical applications. *Adv Microb Physiol* 59:1–100.

Lovley DR, Widman PK, Woodward JC, Phillips JEP. 1993c. Reduction of uranium by cytochrome c_3 of *Desulfovibrio vulgaris*. *Appl Environ Microbiol* 59:3572–3576.

Lyalikova NN, Yurkova NA. 1992. Role of microorganisms in vanadium concentration and dispersion. *Geomicrobiol J* 10:15–26.

Markai S, Andrès Y, Montavon G, Grambow B. 2003. Study of the interaction between europium (III) and *Bacillus subtilis*: Fixation sites, biosorption modeling and reversibility. *J Colloid Interface Sci* 262:351–361.

Marques AM, Espuny Tomas MJ, Congregado F, Simon-Pujol MD. 1982. Accumulation of chromium by *Pseudomonas aeruginosa*. *Microbios Lett* 21:143–147.

Marsh TL, Leon NM, McInerney MJ. 2000. Physicochemical factors affecting chromate reduction by aquifer materials. *Geomicrobiol J* 17:291–303.

Marshall TA, Morris K, Law GTW, Mosselmans JFW, Bots P, Parry SA, Shaw S. 2014. Incorporation and retention of 99-Tc(IV) in magnetite under high pH conditions. *Environ Sci Technol* 48:11853–11862.

McBeth JM, Lear G, Morris K, Burke IT, Livens FR, Lloyd JR. 2007. Technetium reduction and reoxidation in aquifer sediments. *Geomicrobiol J* 2:189–197.

McBeth JM, Lloyd JR, Law GTW, Livens FR, Burke IT, Morris K. 2011. Redox interactions of technetium with iron-bearing minerals. *Mineral Mag* 75:2419–2430.

Merroun ML, Ben Chekroun K, Arias JM, González-Muñoz MT. 2003. Lanthanum fixation by *Myxococcus xanthus*: Cellular location and extracellular polysaccharide observation. *Chemosphere* 52:113–120.

Mertz W. 1981. The essential trace elements. *Science* 213:1332–1338.

Miller WJ, Neathery MW. 1977. Newly recognized trace mineral elements and their role in animal nutrition. *BioScience* 27:674–679.

Moffett JW. 1990. Moffett microbial Ce(III) oxidation. *Nature* 345:421–423.

Moffett JW. 1994. A radiotracer study of cerium and manganese uptake onto suspended particles in Chesapeake Bay determination of oxidation state. *Limnol Oceanogr* 58:695–703.

Moriwaki H, Koide R, Yoshikawa R, Warabino Y, Yamamoto H. 2013. Adsorption of rare earth ions onto the cell walls of wild-type and lipoteichoic acid-defective strains of *Bacillus subtilis*. *Appl Microbiol Biotech* 97:3721–3728.

Moriwaki H, Yamamoto H. 2013. Interactions of microorganisms with rare earth ions and their utilization for separation and environmental technology. *Appl Microbiol Biotechnol* 97:1–8.

Murray KJ, Tebo BM. 2007. Cr(III) is indirectly oxidized by the Mn(II)-oxidizing bacterium *Bacillus* sp. strain SG-1. *Environ Sci Technol* 41:528–533.

Myers CR, Carstens BP, Antholine WE, Myers JM. 2000. Chromium(VI) reductase activity is associated with the cytoplasmic membrane of anaerobically grown *Shewanella putrefaciens* MR-1. *J Appl Microbiol* 88:98–106.

Nath BN, Roelandts I, Sudhakar M, Pger WL, Balaram V. 1994. Cerium anomaly variations in ferromanganese nodules and crusts from the Indian Ocean, *Marine Geol* 120:385–400.

Nevin KP, Finnernan KT, Lovley DR. 2003. Microorganisms associated with uranium bioremediation in a high-salinity subsurface sediment. *Appl Environ Microbiol* 69:3672–3675.

Ngwenya BT, Mosselmans JFW, Magennis M, Atkinson KD, Tourney J, Olive V, Ellam RM. 2009. Macroscopic and spectroscopic analysis of lanthanide adsorption to bacterial cells. *Geochim Cosmochim Acta* 73:3134–3147.

Nies DH, Koch S. et al. 1998. CHR, a novel family of prokaryotic proton motive force-driven transporters probably containing chromate/sulfate antiporters. *J Bacteriol* 180:5799–5802.

Nishioka H. 1975. Mutagenic activities of metal compounds in bacteria. *Mutat Res* 31:185–190.

Ohnuki T, Ozaki T, Kozai N, Nankawa T, Sakamoto F, Sakai T, Suzuki Y, Francis AJ. 2008. Concurrent transformation of Ce(III) and formation of biogenic manganese oxides. *Chem Geol* 253:23–29.

Ohtake H, Cervantes C, Silver S. 1987. Decreased chromate uptake in *Pseudomonas fluorescens* carrying a chromate resistance plasmid. *J Bacteriol* 169:3853–3856.

Ohtake H, Fuji E, Toda K. 1990. Reduction of toxic chromate in an industrial effluent by use of a chromate-reducing strain of *Enterobacter cloacae*. *Environ Technol* 11:663–668.

Oremland RS, Capone DG. 1988. Use of "specific" inhibitors in biogeochemistry and microbial ecology. *Adv Microb Ecol* 10:285–383.

Ortiz-Bernad I, Anderson RT, Vrionis HA, Lovley DR. 2004a. Vanadium respiration by *Geobacter metallireducens*: Novel strategy for in situ removal of vanadium from groundwater. *Appl Environ Microbiol* 70:3091–3095.

Ortiz-Bernad I, Anderson RT, Vrionis HA, Lovley DR. 2004b. Resistance of solid-phase U(VI) to microbial reduction during in situ bioremediation of uranium-contaminated groundwater. *Appl Environ Microbiol* 70:7558–7560.

Ozaki T, Kimura T, Ohnuki T. 2002b. Association of Eu(III) and Cm(III) with Bacillus subtilis and *Halobacterium salinarum*. *Proceedings of Actinides 2001 International Conference (Japan)*, November 4–9, 2001, Hayama, Japan, pp. 950–953.

Ozaki T, Kimura T, Ohnuki T, Francis AJ. 2005. Associations of Eu(III) with Gram-negative bacteria, *Alcaligenes faecalis, Shewanella putrefaciens*, and *Paracoccus denitrificans*. *J Nucl Radiochem Sci* 6(1):73–76.

Ozaki T, Suzuki Y, Nankawa T, Yoshida T, Ohnuki T, Kimura T, Francis AJ. 2006. Interactions of rare earth elements with bacteria and organic ligands. *J Alloys Compounds* 408–412:1334–1338.

Ozaki TF, Arisaka T, Kimura T, Francis AJ, Yoshida Z. 2002a. Empirical method for prediction of the coordination environment of Eu(III) by time-resolved laser-induced fluorescence spectroscopy. *Anal Bioanal Chem* 374:1101–1104.

Pałasz A, Czekaj P. 2000. Toxicological and cytophysiological aspects of lanthanides action. *Acta Biochim Pol* 47(4):1107–1114.

Petrelli FL, DeFlora S. 1977. Toxicity and mutagenicity of hexavalent chromium on *Salmonella typhimurium*. *Appl Environ Microbiol* 33:805–809.

Philip L, Iyengar L, Venkobachar C. 1998. Cr(VI) reduction by *Bacillus coagulans* isolated from contaminated soils. *J Env Eng* 124:1165–1170.

Pietzsch K, Hard BC, Babel W. 1999. A *Desulfovibrio* sp. capable of growing by reducing U(VI). *J Basic Microbiol* 39:365–372.

Pignolet L, Fonsny K, Capot F, Moureau Z. 1989. Role of various microorganisms on Tc behaviour in sediments. *Health Phys* 57:791–800.

Pleshakov VD, Koren'kov VN, Zhukov IM, Serpokrylov NS, Lemper IA, Pankrova II. 1981. Biochemical removal of chromium(VI) compounds from wastewaters. USSR Patent SU 835,978 June 7, 1981.

Plymale AE, Bailey VL, Fredrickson JK, Heald SM, Buck EC, Shi L, Wang Z, Resch, CT, Moore M, Bolton H. 2012. Biotic and abiotic reduction and solubilization of Pu(IV)O₂•xH2O(am) as affected by anthraquinone-2,6-disulfonate (AQDS) and ethylenediaminetetraacetate (EDTA). *Environ Sci Tech* 46:2132–2140.

Renshaw JC, Butchins LJC, Livens, FR, May I, Charnock JM, Lloyd JR. 2005. Bioreduction of uranium: Environmental implications of a pentavalent intermediate. *Environ Sci Tech* 39:5657–5660.

Renshaw JC, Law N, Geissler AG, Livens FR, Lloyd JR. 2009. Impact of the Fe(III)-reducing bacteria *Geobacter sulfurreducens* and *Shewanella oneidensis* on the speciation of plutonium. *Biogeochemistry* 94:191–196.

Rittmann BE, Banaszak JE, Reed DT. 2002. Reduction of Np(V) and precipitation of Np(IV) by an anaerobic microbial consortium. *Biodegradation* 13:329–342.

Romanenko VI, Koren'kov VN. 1977. A pure culture of bacteria utilizing chromates and bichromates as hydrogen acceptors in growth under anaerobic conditions. *Mikrobiologiya* 46:414–417 (English transl., pp. 329–332).

Romanenko VI, Kuznetsov SI, Koren'kov VN. 1976. Koren'kov method for biological purification of wastewater. USSR Patent SU 521,234, July 15, 1976.

Rusin PA, Quintana L, Brainard JR, Strietelmeier BA, Tait CD, Ekberg SA, Palmer PD, Newton TW, Clark DI. 1994. Solubilization of plutonium hydrous oxide by iron-reducing bacteria. *Environ Sci Technol* 28:1686–1690.

Sasaki T, Zheng J, Asano H, Kudo A., Kudo A. 2001. Interaction of Pu, Np and Pa with anaerobic microorganisms at geologic repositories. *Rad Environ* 1:221–232.

Serpokrylov NS, Zhukov IM, Simonova GA, Kvasnikov EI, Klyushnikova TM, Kasatkina TP, Kostyukov VP. 1985. Biological removal of chromium(VI) compounds from wastewater by anaerobic microorganisms in the presence of an organic substrate. USSR Patent SU 1,198,020. December 1985.

Shebolina ES, Sullivan SA, O'Neill KR, Nevin KP, Lovley DR. 2004. Isolation, characterization, and U(VI)-reducing potential of a facultatively anaerobic, acid-resistant bacterium from low-pH, nitrate- and U(VI)-contaminated subsurface sediment and description of *Salmonella subterranean* sp. nov. *Appl Environ Microbiol* 70:2959–2965.

Shen H, Wang YT. 1993. Characterization of enzymatic reduction of hexavalent chromium by *Escherichia coli* ATCC 33456. *Appl Environ Microbiol* 59:3771–3777.

Shi X, Dalal NS. 1990. One-electron reduction of chromate by NADPH-dependent glutathione reductase. *J Inorg Biochem* 40:1–12.

Shimada A. 1979. Effect of six-valent chromium on growth and enzyme production of chromium-resistant bacteria. *Abst Annu Meet Am Soc Microbiol* 25:238.

Silver S, Walderhaug M. 1992. Gene regulation of plasmid- and chromosome-determined inorganic ion transport in bacterial. *Microbiol Rev* 56:195–228.

Simonova GA, Serpokrylov NS, Tokareva LL. 1985. Adaptation characteristics of a culture of *Aeromonas dechromaticans* KS-11 in microbial removal of hexavalent chromium from water. *Izv Sev Kavk Nauhn Tsentra Vyssh Shk Estestv Nauki* 4:89–91.

Sisti F, Allegretti P, Donati E. 1996. Reduction of dichromate by *Thiobacillus ferrooxidans*. *Biotechnol Lett* 18:1477–1480.

Sittig M., ed. 1985. *Handbook of Toxic and Hazardous Chemicals and Carcinogens*. Park Ridge, NJ: Noyes, p. 638.

Smith CH. 1972. Chromium: Element and geochemistry. In: Fairbridge RW, ed., *The Encyclopedia of Geochemistry and Environmental Sciences*. *Encycl Earth Sci Ser*, Vol. IVA. New York: Van Nostrand Reihold, pp. 167–170.

Smith WL, Gadd GM. 2000. Reduction and precipitation of chromate by mixed culture sulphate-reducing bacterial biofilms. *J Appl Microbiol* 88(6):983–991.

Stecher PG, ed. 1960. *The Merck Index of Chemicals and Drugs*. 7th ed. Rahway, NJ: Merck, p. 1493.

Sugio T, Hirajama K, Inagaki K, Tanaka H, Tano T. 1992. Molybdenum oxidation by *Thiobacillus ferrooxidans*. *Appl Environ Microbiol* 58:1768–1771.

Sugio T, Tsujita Y, Katagiri T, Inagaki K, Tano T. 1988. Reduction of Mo^{6+} with elemental sulfur by *Thiobacillus ferrooxidans*. *J Bacteriol* 170:5956–5959.

Summers AO, Jacoby GA. 1978. Plasmid-determined resistance to boron and chromium compounds in *Pseudomonas aeruginosa*. *Anitmicrob Agents Chemother* 13:637–640.

Suzuki T, Miyata N, Horitsu H, Kawai K, Takamizawa K, Tai Y, Okazaki M. 1992. NAD(P)H-dependent chromium(VI) reductase of *Pseudomonas ambigua* G-1: A Cr(V) intermediate is formed during the reduction of Cr(VI) to Cr(III). *J Bacteriol* 174:5340–5345.

Suzuki Y, Kelly SD, Kemner KM, Banfield JF. 2005. Direct microbial reduction and subsequent preservation of uranium in natural near-surface sediment. *Appl Environ Microbiol* 71:1790–1797.

Takahashi Y, Châtellier X, Hattori KH, Kato K, Fortin D. 2005. Adsorption of rare earth elements onto bacterial cell walls and its implication for REE sorption onto natural microbial mats. *Cheml Geol* 219:53–67.

Takahashi Y, Hirata T, Shimizu H, Ozaki T, Fortin D. 2007. A rare earth element signature of bacteria in natural waters? *Chem Geol* 244:569–583.

Takahashi Y, Yamamoto M, Yamamoto Y, Tanaka K. 2010. EXAFS study on the cause of enrichment of heavy REEs on bacterial cell surfaces. *Geochim Cosmochim Acta* 74:5443–5462.

Tanaka Y, Hosaka T, Ochi K. 2010. Rare earth elements activate the secondary metabolite-biosynthetic gene clusters in *Streptomyces coelicolor* A3(2). *J Antibiotics* 63:477–481.

Tebo BM, Bargar JR et al. 2004. Biogenic manganese oxides: properties and mechanisms of formation. *Annu Rev Earth Planet Sci* 32:287–328.

Tripathi AGA. 2002. Bioremediation of toxic chromium from electroplating effluent by chromate-reducing *Pseudomonas aeruginosa* A 2 Chr in two bioreactors. *Appl Microbiol Biotechnol* 58:416–420.

Tsuruta T. 2005. Separation of rare earth elements by microorganisms. *J Nucl Radiochem Sci* 6:1–84.

Tsuruta T. 2006. Selective accumulation of light or heavy rare earth elements using Gram-positive bacteria. *Colloids Surf B* 52:117–122.

Tuovinen OH, Niemalä SI, Gyllenberg HG. 1971. Tolerance of *Thiobacillus ferrooxidans* to some metals. *Antonie v Leeuwenhoek* 37:489–496.

van Marwijk J, Opperman D, Piater L, van Heerden E. 2009. Reduction of vanadium(V) by *Enterobacter cloacae* EV-SA01 isolated from a South African deep gold mine. *Biotechnol Lett* 31:845–849.

Venitt S, Levy LS. 1974. Mutagenicity of chromates in bacteria and its relevance to carcinogenesis. *Nature* (*Lond*) 250:493–495.

Viamajala S, Peyton BM, Apel WA, Petersen JN. 2002. Chromate reduction in *Shewanella oneidensis* MR-1 is an inducible process associated with anaerobic growth. *Biotechnol Prog* 18:290–295.

Viti C, Marchi E et al. 2014. Molecular mechanisms of Cr (VI) resistance in bacteria and fungi. *FEMS Microbiol Rev* 38:633–659.

Wang P, Mori T, Komori K, Sasatsu M, Toda K, Ohtake H. 1989. Isolation and characterization of an *Enterobacter cloacae* strain that reduces hexavalent chromium under anaerobic conditions. *Appl Environ Microbiol* 55:1665–1669.

Wang P-C, Toda K, Ohtake H, Kusaka I, Yabe I. 1991. Membrane-bound respiratory system of *Enterobacter cloacae* strain HO1 grown anaerobically with chromate. *FEMS Microbiol Lett* 78:11–16.

Wang Y-T. 2000. Microbial reduction of Cr(VI). In: Lovley DR, ed., *Environmental Microbe-Metal Interactions*. Washington, DC: ASM Press, pp. 225–235.

Wang Y-T, Shen H. 1997. Modeling Cr(VI) reduction by pure bacterial cultures. *Water Res* 31:727–732.

Wani R, Kodam KM, Gawai KR, Dhakephalkar PK. 2007. Chromate reduction by *Burkholderia cepacia* MCMB-821, isolated from the pristine habitat of alkaline crater lake. *Appl Microbiol Biotechnol* 75:627–632.

Weast RC, Astle MJ. 1982. *CRC Handbook of Chemistry and Physics*, 63rd ed. Boca Raton, FL: CRC Press, pp. B44–B45.

Weber KA, Thrash, JC, Van Trump JI, Achenbach LA, Coates JD. 2011. Environmental and taxonomic bacterial diversity of anaerobic uranium(IV) biooxidation. *Appl Environ Microbiol* 77:4693–4696.

Wielinga B, Mizuba MM, Hansel CM, Fendorf S. 2001, Iron promoted reduction of chromate by dissimilatory iron-reducing bacteria. *Environ Sci Technol* 35:522–527.

Wildung RE, Gorby YA, Krupka KM, Hess NJ, Li SW, Plymale AE, McKinley JP, Fredrickson JK. 2000. Effect of electron donor and solution chemistry on products of dissimilatory reduction of technetium by *Shewanella putrefaciens*. *Appl Environ Microbiol* 66:2451–2460.

Wildung RE, Li SW, Murray CJ, Krupka KM, Xie Y, Hess NJ, Roden EE. 2004. Technetium reduction in sediments of a shallow aquifer exhibiting dissimilatory iron reduction potential. *FEMS Microbiol Ecol* 49:151–162.

Wildung RE, McFadden KM, Garland TR. 1979. Technetium sources and behaviour in the environment. *J Environ Qual* 8:156–161.

Williams KH, Bargar JR, Lloyd JR, Lovley DR. 2012. Bioremediation of uranium-contaminated groundwater: A systems approach to subsurface biogeochemistry. *Curr Opin Biotech* 24:1–9.

Williamson AJ, Morris K, Charnock JM, Law GT, Rizoulis A, Lloyd JR. 2014. Microbial reduction of U(VI) under alkaline conditions; implications for radioactive waste geodisposal. *Environ SciTech.* 48:13549–13556.

Wong C, Silver M, Kushner DJ. 1982. Effects of chromium and manganese on *Thiobacillus ferrooxidans*. *Can J Microbiol* 28:536–544.

Woolfolk CA, Whiteley HR. 1962. Reduction of inorganic compounds with molecular hydrogen by *Micrococcus lactilyticus*. *J Bacteriol* 84:647–658.

Yamamoto K, Kato J, Yano I, Ohtake H. 1993. Kinetics and modeling of hexavalent chromium reduction in *Enterobacter cloacae*. *Biotechnol Bioeng* 41:129–133.

Ye Q, Roh Y, Carroll SL, Blair B, Zhou J, Zhang CL, Fields MW. 2004. Alkaline anaerobic respiration: Isolation and characterization of a novel alkaliphilic and metal-reducing bacterium. *Appl Environ Microbiol* 70:5595–5602.

Yoshida T, Ozaki T, Ohnuki T, Francis A. 2004. Adsorption of rare earth elements by γ-Al_2O_3 and cells in the presence of desferrioxamine B: implication of siderophores for the Ce anomaly. *Chemical Geol* 212:239–246.

Yurkova NA, Lyalikova NN. 1990. New vanadate-reducing facultative chemolithotrophic bacteria. *Mikrobiologiya* 59:968–975 (English transl., pp. 672–677).

CHAPTER TWENTY

Geomicrobiology of Sulfur

David A. Fike, Alexander S. Bradley, and William D. Leavitt

CONTENTS

20.1 Introduction / 480
20.2 Microbial Sulfate Reduction / 481
    20.2.1 Microbial Sulfate Reducers / 481
    20.2.2 Coupling to Carbon and Nitrogen Cycles / 484
        20.2.2.1 Autotrophy / 484
        20.2.2.2 Mixotrophy / 484
        20.2.2.3 Heterotrophy / 485
        20.2.2.4 Nitrogen / 485
    20.2.3 Sulfate Reduction Pathways / 485
    20.2.4 Oxygen Tolerance of Sulfate Reducers / 487
20.3 Other Reduction Dissimilatory Metabolisms / 487
    20.3.1 Thiosulfate Reduction / 488
    20.3.2 Elemental Sulfur Reduction / 488
20.4 Oxidative Processes / 489
    20.4.1 Physiology and Biochemistry of Microbial Oxidation of Reduced Forms of Sulfur / 489
        20.4.1.1 Oxidation of Sulfide / 489
        20.4.1.2 Oxidation of Elemental Sulfur / 491
        20.4.1.3 Oxidation of Sulfite / 492
        20.4.1.4 Oxidation of Thiosulfate / 492
        20.4.1.5 Oxidation of Tetrathionate / 494
    20.4.2 Common Mechanism for Oxidizing Reduced Inorganic Sulfur Compounds in Domain Bacteria / 494
    20.4.3 Ecology of Reduced S Oxidizers / 494
20.5 Disproportionation / 497
    20.5.1 Elemental Sulfur Disproportionation / 497
    20.5.2 Thiosulfate Disproportionation / 497
20.6 Reduced Forms of Sulfur as an Electron Donor / 498
    20.6.1 Autotrophs / 499
        20.6.1.1 Chemosynthetic Autotrophs / 499
    20.6.2 Mixotrophy / 500
20.7 Sulfur Assimilation / 500
20.8 Sulfur-Cycling Microbial Consortia / 501
20.9 Evolution of Sulfur Cycling over Earth History / 503
20.10 Summary / 503
References / 503

fumarate, malate, and ethanol. Furthermore, none of these organisms are able to degrade their organic energy sources beyond acetate (Postgate, 1984), i.e., they are "incomplete oxidizers" (Rabus et al., 2013). The importance of the sulfate reducers in the anaerobic mineralization of organic matter in sulfate-rich environments remained unappreciated before the discovery of the "complete oxidizers," capable of the oxidation of organic matter to CO_2. This restricted view of sulfate reducers changed rapidly with the discovery of a sulfate reducer, *Desulfotomaculum acetoxidans* (Widdel and Pfennig, 1977, 1981), which is able to oxidize acetate anaerobically to CO_2 and H_2O with sulfate. Subsequently, a wide variety of other sulfate reducers were discovered that differed in the nature of the energy sources they were capable of using, including a wide range of aliphatic, aromatic, and heterocyclic compounds. In many cases, the organisms are able to completely reminerialize the substrate to bicarbonate/carbon dioxide and each substrate by a specific group of sulfate reducers (e.g., Pfennig et al., 1981; Imhoff-Stuckle and Pfennig, 1983; Braun and Stolp, 1985; Bak and Widdel, 1986a,b; Szewzyk and Pfennig, 1987; Platen et al., 1990; Zellner et al., 1990; Aeckersberg et al., 1991; Boopathy and Daniels, 1991; Qatabi et al., 1991; Schnell and Schink, 1991; Tasaki et al., 1991, 1992; Kuever et al., 1993; Rueter et al., 1994; Janssen and Schink, 1995; Rees et al., 1998; Londry et al., 1999; Meckenstock et al., 2000). Some of these sulfate reducers were also found to use H_2 as an energy source. Most require an organic carbon source, but some can grow autotrophically on hydrogen. Table 20.2 presents a list of some of the different kinds of sulfate reducers in the domain Bacteria. While most sulfate reducers discovered to date are mesophiles, thermophilic types are also now in culture (e.g., Pfennig et al., 1981; Zeikus et al., 1983; Stetter et al., 1987; Burggraf et al., 1990; Itoh et al., 1999). At least one moderate psychrophile, *Desulforhopalus vacuolatus* (optimal growth at 10°C–19°C), has been described (Isaksen and Teske, 1996)—isolated from sediment in Kysing Fjord, Denmark, at 10°C.

Sulfate reducers are morphologically diverse and are known to include cocci, sarcinae, rods, vibrios, spirilla, and filaments (Figure 20.2). The cultured representatives in the domain Bacteria are primarily gram-negative, though some gram-positive relatives of the *Clostridiales* are now known (Pereira et al., 2011 and references therein).

TABLE 20.2
Some sulfate-reducing bacteria.[a]

Heterotrophs	Autotrophs[b]
Desulfovibrio desulfuricans [c,d]	*Desulfovibrio baarsii*
Desulfovibrio vulgaris	*Desulfobacter hydrogenophilus*
Desulfovibrio gigas	*Desulfosarcina variabilis*
Desulfovibrio fructosovorans	*Desulfonema limicola*
Desulfovibrio sulfodismutans	
Desulfomonas pigra	
Desulfotomaculum nigrificans	
Desulfotomaculum acetoxidans	
Desulfotomaculum orientis[d]	
Desulfobacter postgatei	
Desulfobulbus propionicus	
Desulfobacterium phenolicum[e]	
Desulfobacterium indolicum[f]	
Desulfobacterium catecholicum[g]	

[a] For a more detailed description of sulfate reducers, see Pfennig et al. (1981), Postgate (1984), Dworkin (2001), and Rabus et al. (2013).
[b] Autotrophic growth on H_2 and CO_2.
[c] Some strains can grow mixotrophically on H_2 and CO_2 and acetate.
[d] At least one strain can grow autotrophically on H_2 and CO_2.
[e] Bak and Widdel (1986b).
[f] Bak and Widdel (1986a).
[g] Szewzyk and Pfennig (1987).

The first sulfate reducers described from the Archaeal domain were the Euryarchaeota *Archaeoglobus fulgidus* (Stetter et al., 1987; Speich and Trüper, 1988) and *Archaeoglobus profundus* (Burggraf et al., 1990). Following these early discoveries, facultative sulfate reducers from the Crenarchaeota *Caldivirga maquilingensis* and *Thermocladium modestius* were reported (Itoh et al., 1999) and more known from genome sequence in recent years (Pereira et al., 2011). The two Archaeal sulfate reducers that were discovered by Stetter et al. (1987) and Burggraf et al. (1990) are extremely thermophilic, anaerobic, gram-negative, irregularly shaped cocci. *A. fulgidus* strains were found to grow naturally in a hydrothermal system at temperatures between 70°C and 100°C in the vicinities of Vulcano and Stufe di Nerone, Italy. Under laboratory conditions, the cultures grow anaerobically in marine mineral salts medium supplemented with yeast extract. In this medium, they produce a large amount of hydrogen sulfide and some methane. Thiosulfate, but not elemental sulfur, can act as alternative electron acceptor. Energy sources

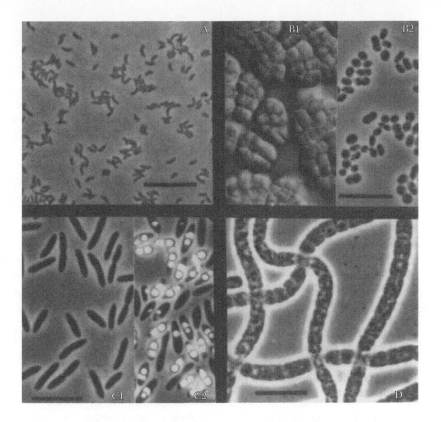

Figure 20.2. Sulfate-reducing bacteria. (A) Desulfovibrio desulfuricans (phase contrast). (B1, B2) Desulfosarcina variabilis: (B1) sarcina packets (interference contrast); (B2) free-living cells (phase contrast). (C1, C2) Desulfotomaculum acetoxidans: (C1) vegetative cells (phase contrast); (C2) cells with spherical spores and gas vacuoles (phase contrast). (D) Desulfonema limicola (phase contrast). (With kind permission from Springer Science+Business Media: *The Prokaryotes: A Handbook on Habitats, Isolation, and Identification of Bacteria*, Vol. I, Copyright 1981, Pfennig, N, Widdel, F, and Trüper, HG, Springer, Berlin, Germany.)

include hydrogen and some simple organic molecules as well as glucose, yeast extract, and other more complex substrates. Cells contain a number of compounds such as 8-OH-5-deazaflavin and methanopterin previously found only in methanogens, which are also members of the domain Archaea, but 2-mercaptoethanesulfonic acid and factor F430, which are found in methanogens, were absent (Stetter et al., 1987).

A. profundus was isolated from the Guaymas hot vent area (Gulf of California, also known as the Sea of Cortez). It grows anaerobically at temperatures between 65°C and 95°C (optimum 82°C) in a pH range of 4.5–7.5 at NaCl concentrations in the range of 0.9%–3.6%. Unlike A. fulgidus, it is an obligate mixotroph that requires H_2 as an energy source. Its organic carbon requirement can be satisfied by acetate, lactate, pyruvate, yeast extract, beef extract, peptone, or acetate-containing crude oil. As for A. fulgidus, sulfate, thiosulfate,

and sulfate can serve as terminal electron acceptors for growth. Although S^0 is reduced by resting cells, it does not support growth (Burggraf et al., 1990). Interestingly, the primary sulfate reduction genes in fully sequenced Archaeal sulfate reducers appear to have derived from Bacterial genomes (Wagner et al., 1998; Pereira et al., 2011).

The presence of as-yet-unidentified, extremely thermophilic sulfate reducers was detected in hot deep-sea sediments at the hydrothermal vents of the tectonic spreading center of Guaymas Basin (Sea of Cortez or Gulf of California). Sulfate-reducing activity was measurable between 100°C and 110°C (optimum 103°C–106°C). The responsible organisms are probably examples of Archaea (Jørgensen et al., 1992).

The Crenarchaeotal strains C. maquilingensis and T. modestius described by Itoh et al. (1999) were also thermophiles, growing over a temperature range of 60°C–92°C with an optimum at 85°C.

These were provided a variety of complex carbon sources such as gelatin, glycogen, beef or yeast extract, peptone, and tryptone. In addition to being able to employ sulfate as a terminal electron acceptor, these strains also were able to utilize sulfur or thiosulfate.

Sulfate reducers play several important roles in geomicrobiology in relation to other elemental cycles. In arsenic-contaminated groundwater, sulfide produced by sulfate reducers can precipitate arsenic phases, removing it from solution (Kirk et al., 2004). Sulfate reducers also play a role in the anaerobic oxidation of methane (AOM; Boetius et al., 2000). This process, which oxidizes methane to CO_2 in marine sediments and methane seeps, is postulated to be catalyzed by a consortium of methane cycling archaea and sulfate-reducing bacteria. The bacterial partner in these consortia typically derives from the Desulfosarcina, Desulfococcus, or Desulfobulbus group within the δ-proteobacteria (Knittel and Boetius, 2009). AOM is thought to operate close to thermodynamic equilibrium, and many details of this process are as yet unclear—including whether a syntrophic relationship between the archaea and sulfate-reducing (or other) bacterial partner is obligate. Furthermore, sulfate reducers play an integral role in the cycling of environmentally relevant iron oxide phases (Flynn et al., 2014).

20.2.2 Coupling to Carbon and Nitrogen Cycles

20.2.2.1 Autotrophy

Autotrophic sulfate reducers use hydrogen as an electron donor. Although the ability of Desulfovibrio desulfuricans to grow autotrophically with hydrogen (H_2) as an energy source had been previously suggested, experiments by Mechalas and Rittenberg (1960) failed to demonstrate it. Seitz and Cypionka (1986), however, obtained autotrophic growth of D. desulfuricans strain Essex 6 with hydrogen, but the growth yield was less when sulfate was the terminal electron acceptor. Better yields were obtained with nitrate or nitrite as terminal electron acceptor, presumably because the latter two acceptors did not need to be activated by ATP, which is a requirement for sulfate reduction (see in the following text), and the oxidation of hydrogen with the N-anions yields greater free energy of reaction than with sulfate (Thauer et al., 1977).

Desulfotomaculum orientis also has the ability to grow autotrophically with hydrogen as energy source using sulfate, thiosulfate, or sulfite as the terminal electron acceptor (Cypionka and Pfennig, 1986). Under optimal conditions, better growth yields were obtained with this organism than had been reported for D. desulfuricans (12.4 versus 9.4 g of dry cell mass per mole of sulfate reduced). This may be because Desulfotomaculum can utilize inorganic pyrophosphate generated in sulfate activation as an energy source whereas Desulfovibrio cannot. D. orientis gave better growth yields when thiosulfate or sulfite was the terminal electron acceptor than when sulfate was; this is also consistent with the ATP required for sulfate activation. The organism excreted acetate that was formed as part of its CO_2 fixation process (Cypionka and Pfennig, 1986). The acetate may have been formed via the activated acetate pathway in which it is formed directly from two molecules of CO_2 as is the case in methanogens and homoacetogens (see Chapters 7 and 23), and as has now been shown to occur in Desulfovibrio baarsii, which can also grow with hydrogen and sulfate (Jansen et al., 1984) and in Desulfobacterium autotrophicum (Schauder et al., 1989). Desulfobacter hydrogenophilus, by contrast, assimilates CO_2 by a reductive tricarboxylic acid cycle when growing autotrophically with H_2 as energy source and sulfate as terminal electron acceptor (Schauder et al., 1987). Other sulfate reducers able to grow autotrophically on hydrogen as energy source and sulfate as terminal electron acceptor include Desulfonema limicola, Desulfonema ishimotoi, Desulfosarcina variabilis (Pfennig et al., 1981; Fukui et al., 1999), and D. autotrophicum (Schauder et al., 1989). The pathways for CO_2-fixation vary among these strains, as reviewed by Rabus et al. (2013), and include the reductive citric acid cycle and the Ljungdahl–Wood pathway.

20.2.2.2 Mixotrophy

D. desulfuricans has been shown to grow mixotrophically with any one of several different compounds as sole energy source, including hydrogen, formate, and isobutanol. The carbon in the organic energy sources was not assimilated. It was derived instead from substances as complex as yeast extract or as simple as acetate and CO_2. Sulfate was the terminal electron acceptor in all instances (Mechalas and Rittenberg, 1960; Sorokin, 1966a–d; Badziong and Thauer,

1978; Badziong et al., 1978; Brandis and Thauer, 1981). A strain of D. *desulfuricans* used by Sorokin (1966a) was able to derive as much as 50% of its carbon from CO_2 when it grew on hydrogen as the energy source and acetate and CO_2 as carbon sources, whereas on lactate and CO_2, it derived only 30% of its carbon from CO_2. Badziong et al. (1978), using a different strain of *Desulfovibrio*, found that 30% of its carbon was derived from CO_2 when it grew on hydrogen, acetate, and CO_2. Members of some other genera of sulfate-reducing bacteria can also grow mixotrophically on hydrogen and acetate and CO_2 (Pfennig et al., 1981). In all the instances, ATP is generated chemiosmotically from hydrogen oxidation in the periplasm.

20.2.2.3 Heterotrophy

The great majority of autotrophic sulfate reducers can grow heterotrophically with sulfate as terminal electron acceptor. In general, sulfate reducers specialize with respect to the carbon or energy source they can utilize (see also Pfennig et al., 1981). When acetate serves as energy source, it may be completely oxidized anaerobically via the tricarboxylic acid cycle, as in the case of *Desulfobacter postgatei* (Brandis-Heep et al., 1983; Gebhardt et al., 1983). More commonly, however, sulfate reducers oxidize acetate by reversal of the active-acetate-synthesis pathway (Schauder et al., 1986). Assimilation of acetate most likely involves carboxylation to pyruvate. ATP synthesis in the heterotrophic mode of sulfate reduction, insofar as it is understood, is mainly by oxidative phosphorylation (chemiosmotically) involving transfer of hydrogen abstracted from an organic substrate into the periplasm followed by its oxidation (Odom and Peck, 1981; but see also Kramer et al., 1987; Odom and Wall, 1987). In the case of lactate, this hydrogen transfer from the cytoplasm to the periplasm across the plasma membrane appears to be energy driven (Pankhania et al., 1988). Some ATP may be formed by substrate-level phosphorylation.

Organic carbon oxidation by many strains is incomplete, producing acetate, although some strains are capable of complete oxidation of organic substrates to CO_2. Those organisms capable of complete versus incomplete oxidation are reviewed by Rabus et al. (2013). The most common substrate for cultivation of substrate reducers

in pure culture is lactate, although depending on the strain a wide variety of substrates can be used including methanol, ethanol, acetate, propionate, succinate, fumarate, fatty acids, or sugars.

20.2.2.4 Nitrogen

Most sulfate reducers derive their nitrogen needs from assimilation of ammonium (Rabus et al., 2013). Some sulfate reducers that are also capable of reducing nitrate derive nitrogen from the ammonium produced through this pathway. A few species of sulfate reducers are capable of nitrogen fixation, including *Desulfovibrio*, *Desulfotomaculum*, and *Desulfobacter* species (Rabus et al., 2013). Sulfate reducing bacteria have been shown to be an important source of nitrogen in benthic hypoxic zones, where nitrogen is lost due to denitrification (Bertics et al., 2013). Despite the large energy requirement imposed during nitrogen fixation, consortia of methane oxidizing archaea and sulfate-reducing bacteria performing AOM (see Section 19.8) have also been demonstrated to be capable of nitrogen fixation (Dekas et al., 2009).

20.2.3 Sulfate Reduction Pathways

Scientific knowledge regarding the biochemistry of sulfate reduction has undergone substantial progress in the last two decades. Much of this progress has come about through the characterization of protein crystal structures, notably that of dissimilatory sulfite reductase, which have improved our mechanistic understanding of the intermediate steps of sulfate reduction (c.f. Oliveira et al., 2008a,b; Schiffer et al., 2008; Parey et al., 2010; Venceslau et al., 2014).

Sulfate is imported to cells through sulfate transporters that operate as symporters transferring sulfate into the cell along with cations (Figure 20.3). The cations are generally protons in freshwater species of sulfate-reducing bacteria and sodium ions in the case of marine species (Cypionka, 1995). Expression of transporters is regulated, and strains are thought to have versions of these transporters adapted for high-affinity accumulation of sulfate (under sulfate limitation) and for low-affinity accumulation (under high sulfate). These transporters are thought to be uncoupled from direct consumption of ATP. This contrasts with sulfate permeases used in sulfate

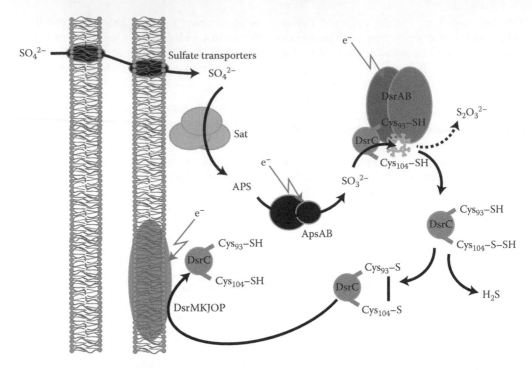

Figure 20.3. Schematic of the processes involved in dissimilatory sulfate reduction. These are as follows: (1) sulfate is transported into the cells through the cell envelope and into the cytoplasm; (2) Sat activates sulfate to APS; (3) APS reductase reduces APS to sulfite; and (4) sulfite interacts with the DsrABC complex. This interaction is complex and in vitro can produce a range of direct products, including trithionate, thiosulfate, and sulfide. Partially reduced sulfur bound to DsrAB can yield thiosulfate. Generally, four electrons are transferred by DsrAB, and zero-valent sulfur is bound to DsrC that undocks from DsrAB and may form a heterodisulfide while releasing H2S. The DsrC acts as an electron acceptor that interacts with the energy-conserving complex in the cell membrane (DsrMKJOP), regenerating reduced DsrC that can redock with DsrAB. Sites of electron donation are indicated by orange arrows.

assimilation by a wide variety of microbes (Piłsyk and Paszewski, 2009), which are ATP dependent.

After import into the cell, sulfate is "activated" by ATP sulfurylase, consuming a molecule of ATP, and generating the "activated" form of sulfate APS (adenosine 5′-phosphosulfate) (Figure 20.3). Direct reduction of sulfate to sulfite is thermodynamically unfavorable at standard state, and the formation of the APS intermediate allows an exergonic reaction from APS to sulfite. However, the formation of APS itself is endergonic, despite the investment of a molecule of ATP to form APS (Cypionka, 1995):

$$SO_4^{2-} + ATP \xrightarrow{\text{ATP sulfurylase}} APS + PP_i \quad (20.1)$$

In members of the genus *Desulfovibrio*, pyrophosphate (PP_i) is hydrolyzed to inorganic phosphate (P_i), which helps to pull Reaction 20.1 in the direction of APS:

$$PP_i + H_2O \xrightarrow{\text{Pyrophosphatase}} 2P_i \quad (20.2)$$

Unlike in assimilatory sulfate reduction (see Section 20.7), APS, once formed, is reduced directly to sulfite and adenylic acid (AMP):

$$APS + 2e \xrightarrow{\text{APS reductase}} SO_3^{2-} + AMP \quad (20.3)$$

The APS reductase, unlike PAPS reductase, does not require NADP as a cofactor but, like PAPS reductase, contains bound flavine adenine dinucleotide and iron (for further discussion see, for instance, Peck, 1993). This is a highly exothermic reaction that is coupled to energy conservation via the QmoABC complex in sulfate-reducing bacteria (Ramos et al., 2012).

The subsequent details on the reduction of sulfite to sulfide were the subject of controversy for many years. Two models competed for several decades: a step-wise reduction of sulfite to sulfide via trithionate and thiosulfate (termed the trithionate pathway), and a direct, six-electron reduction of sulfite (Bandurski et al., 1956; Harrison and Thode, 1958; Peck 1959, 1962; Kobayashi et al., 1969). These proposed pathways are reviewed in detail by Rabus et al. (2013) and Bradley et al. (2011). More recent evidence has come to light suggesting that neither of these two pathways is correct in detail. The reduction of sulfite to sulfide is carried out by the enzyme complex dissimilatory sulfite reductase (Dsr), which consists of three subunits. Crystal structures (Oliveira et al., 2008a; Parey et al., 2010) show that it consists of two subunits, DsrA and DsrB in a symmetrical $\alpha_2\beta_2$ arrangement, with each subunit bound to a third subunit DsrC. Recent work has demonstrated that sulfite is likely to be reduced in two, two-electron transfers (Lui et al., 1993; Parey et al., 2010), after which a zero-valent sulfur bound to DsrC dissociates from the DsrAB complex (Oliveira et al., 2008a) (Figure 20.3). The DsrC subsequently releases sulfide and transfers oxidizing power to the membrane-bound DsrMKJOP complex in the membrane, where energy is conserved.

20.2.4 Oxygen Tolerance of Sulfate Reducers

In general, sulfate reducers are considered strict anaerobes, yet some show limited oxygen tolerance (Abdollahi and Wimpenny, 1990; Wall et al., 1990; Marshall et al., 1993; Minz et al., 1999; Baumgartner et al., 2006). Indeed, *D. desulfuricans*, *Desulfovibrio vulgaris*, *Desulfovibrio desulfodismutans*, *D. autotrophicum*, *Desulfobulbus propionicus*, and *Desulfococcus multivorans* have shown an ability to use oxygen as terminal electron acceptor, that is, to respire microaerophilically (<10 µM dissolved O_2) without being able to grow under these conditions (Dilling and Cypionka, 1990; Baumgartner et al., 2001). These organisms may have several mechanisms for responding to oxidative stress, such as *bd*- and *cox*-oxidases (Ramel et al., 2013).

Some evidence has been presented in support of aerobic growth of sulfate-reducing bacteria (Canfield and Des Marais, 1991; Jørgensen and Bak, 1991; Fründ and Cohen, 1992). Responses of various *Desulfovibrio* species to oxygen exposure have been investigated (Cypionka, 2000; Faraleira et al., 2003). A chemostat study of a coculture of *Desulfovibrio oxyclinae* and *Marinobacter* strain MB isolated from a mat from Solar Lake in the Sinai Peninsula, showed *D. oxyclinae* is able to grow slowly on lactate in the presence of air and the concurrent absence of sulfate or thiosulfate. The lactate is oxidized to acetate by *D. oxyclinae* (Krekeler et al., 1997; Sigalevich and Cohen, 2000; Sigalevich et al., 2000a). *Marinobacter* strain MB is a facultatively aerobic heterotroph. When grown on lactate in the presence of sulfate in a chemostat supplied with oxygen after an initial anaerobic growth phase, a pure culture of *D. oxyclinae* tended to form clumps after ~149 h of exposure to oxygen (Sigalevich et al., 2000b). Such clumps were not formed in coculture with *Marinobacter* strain MB (Sigalevich et al., 2000b). The clumping may represent a defense mechanism against exposure to oxygen for sulfate-reducing bacteria in general because the interior of active clumps >3 µm in size will become anoxic.

20.3 OTHER REDUCTION DISSIMILATORY METABOLISMS

In addition to reduction of sulfate, many microbes are capable of the reduction of other sulfur species such as sulfite, thiosulfate, elemental sulfur, or polysulfides. Most, if not all, sulfate-reducing bacteria such as *Desulfovibrio* are capable of sulfite and thiosulfate reduction. Inducible sulfite reduction has also been observed with obligate and facultative anaerobes *Clostridium pasteurianum* and *Shewanella oneidensis* MR-1 (Burns and DiChristina, 2009), respectively. Neither is capable of dissimilatory sulfate reduction, though both can reduce sulfite or thiosulfate to sulfide for energy conservation. In the absence of added selenite, whole *C. pasteurianum* cells did not release detectable amounts of trithionate or thiosulfate when reducing sulfite, but in the presence of selenite, they do. Selenite was found to inhibit thiosulfate reductase but not trithionate reductase in whole cells, but inhibited both in cell extracts (Harrison et al., 1980). A purified sulfite reductase from *C. pasteurianum* produced sulfide from sulfite. It was also able to reduce NH_2OH, SeO_3^{2-}, and NO_2^- but did not reduce trithionate or thiosulfate (Harrison et al., 1984). Several physical and chemical properties of this enzyme differed from those of bisulfite reductases in sulfate reducers. Its role in *C. pasteurianum* may be in detoxification when excess bisulfite is present (Harrison et al., 1984). Peck (1993) referred to the enzymes involved in the transformation of bisulfite to sulfide collectively as bisulfite reductase. Distinct sulfite, trithionite, and thiosulfate reductases were also identified (reviewed in Peck and LeGall, 1982). However, at the time, they did not visualize a major role for these enzymes in sulfite reduction to sulfide. Furthermore, S. oneidensis MR-1 is not known to contain a Dsr-like sulfite reductase, but does carry an alternative octahaem c cytochrome (SirA) required for sulfite reduction (Shirodkar et al., 2010), while polysulfide reductase is (Psr) is required for thiosulfate reduction (Burns and DiChristina, 2009).

20.3.1 Thiosulfate Reduction

Thiosulfate reducers play an important role in some environments (Jorgensen, 1990a). Liang et al. (2014) demonstrated that *Anaerobaculum* species were abundant in oil pipelines and contributed to biocorrosion, although this was ameliorated during syntrophic growth with methanogens. *Shewanella* has been shown to couple H_2 oxidation to both concurrent reduction of MnO_2 and thiosulfate, indicating a potential coupling with the biogeochemical manganese cycle (Lee et al., 2011).

Growth and growth yield of some members of the anaerobic and thermophilic and hyperthermophilic *Thermotogales* were shown to be stimulated in the presence of thiosulfate (Ravot et al., 1995). The test organisms included *Fervidobacterium islandicum*, *Thermosipho africanus*, *Thermotoga maritima*, *Thermotoga neapolitana*, and *Thermotoga* sp. SERB 2665. The last named was isolated from an oil field. All reduced thiosulfate to sulfide. The *Thermotogales* in this group are able to ferment glucose among various energy-yielding substrates. Thiosulfate, like elemental sulfur (see, e.g., Janssen and Morgan, 1992), appears to serve as an electron sink by suppressing H_2 accumulation in the fermentation of glucose, for instance. This accumulation has an inhibitory effect on the growth of these organisms. The biochemical mechanism by which they reduce thiosulfate remains to be elucidated. *Pyrobaculum islandicum* is able to mineralize peptone by way of the tricarboxylic acid cycle, using thiosulfate as terminal electron acceptor, producing CO_2 and H_2S in a ratio of 1:1 (Selig and Schönheit, 1994). In *Salmonella enterica*, a membrane-bound thiosulfate reductase catalyzes thiosulfate reduction (Stoffels et al., 2012).

20.3.2 Elemental Sulfur Reduction

Elemental sulfur can be used anaerobically as terminal electron acceptor in bacterial respiration or as an electron sink for disposal of excess reducing power. The product of S^0 reduction in either case is sulfide. Polysulfide may be an intermediate in respiration (Fauque et al., 1991; Schauder and Müller, 1993). Some members of both Bacteria and Archaea can respire on sulfur (Schauder and Kröger, 1993; Bonch-Osmolovskaya, 1994; Ma et al., 2000). Examples of Bacteria include *Desulfuromonas acetoxidans*, *Desulfovibrio gigas*, and some other sulfate reducers (Pfennig and Biebl, 1976; Biebl and Pfennig, 1977; Fauque et al., 1991); examples of Archaea include *Pyrococcus furiosus* (Schicho et al., 1993), *Pyrodictium* (Stetter et al., 1983; Stetter 1985), *Pyrobaculum* (Huber et al., 1987), *Acidianus*, *Caldisphaera*, and *Acidilobus* (Boyd et al., 2007).

Organisms that use S^0 reduction as an electron sink include *Thermotoga* spp. in the domain Bacteria and *Thermoproteus*, *Desulfurococcus*, and *Thermofilum* in the domain Archaea (Jannasch et al., 1988a,b). These organisms are fermenters that dispose in this way excess of H_2 they produce, which would otherwise inhibit their growth (Bonch-Osmolovskaya et al., 1990; Janssen and Morgan, 1992; Bonch-Osmolovskaya, 1994). It is possible that these organisms can salvage some energy in the disposal of H_2 (e.g., Schicho et al., 1993). Some fungi, for example, *Rhodotorula* and *Trichosporon* (Ehrlich and Fox, 1967), can also reduce sulfur to H_2S with glucose as electron donor. This is probably not a form of respiration.

The energy source for the sulfur-respiring Archaea is sometimes hydrogen and methane but more often organic molecules such as glucose and small peptides (e.g., Boyd et al., 2007), whereas that for Bacteria may be simple organic compounds (e.g., ethanol, acetate, propanol) or more complex organics. In the case of D. *acetoxidans* (domain Bacteria), an electron transport pathway including cytochromes appears to be involved (Pfennig and Biebl, 1976). When acetate is used as an energy source, oxidation proceeds anaerobically by way of the tricarboxylic acid cycle (see Chapter 7). The oxaloacetate required for initiation of the cycle is formed by carboxylation of pyruvate, which arises from carboxylation of acetate (Gebhardt et al., 1985). Energy is gained in the oxidation of isocitrate and 2-ketoglutarate. Membrane preparations were shown to oxidize succinate using S^0 or NAD as electron acceptor by an ATP-dependent reaction. Similar membrane preparations reduced fumarate to succinate with H_2S as electron donor by an ADP-independent reaction. Menaquinone mediated hydrogen transfer. Protonophores and uncouplers of phosphorylation inhibited reduction of S^0 but not fumarate. The compound 2-n-nonyl-4-hydroxyquinoline N-oxide inhibited electron transport to S^0 and fumarate. Together these observations support the notion that S^0 reduction in *D. acetoxidans* involves a membrane-bound electron transport system and the ATP is formed chemiosmotically, that is, by oxidative phosphorylation, when growing on acetate (Paulsen et al., 1986).

The hyperthermophilic Archaea, *Thermoproteus tenax* and *P. islandicum*, growing on S^0 and glucose or casamino acids in the case of the former and on peptone in the case of the latter, mineralized their carbon substrates completely. They produced CO_2 and H_2S in a ratio of 1:2 using the tricarboxylic acid cycle (Selig and Schönheit, 1994).

Shewanella can reduce S^0 using similar enzymatic machinery to thiosulfate reduction, encoded by the *phs* gene cluster (Burns and DiChristina, 2009). The ability to reduce S^0 may play an important role in alkaline groundwaters, where sulfate- and S^0-reduction produces sulfide that can subsequently reduce iron phases such as goethite, regenerating S^0. Under these conditions, S^0 acts as an iron shuttle promoting iron reduction in conditions under which it would otherwise be thermodynamically unfavorable (Flynn et al., 2014).

20.4 OXIDATIVE PROCESSES

Oxidation of reduced sulfur compounds is a major process in geomicrobiology. Oxidation of sulfide or sulfur, usually coupled to reduction of O_2, is a process that occurs wherever reduced sulfur encounters oxygen-replete environments, such as in marine sediments, oxygen minimum zones (OMZs), and in hydrothermal systems. Bacteria capable of sulfide oxidation are in many cases autotrophs and can even coexist as carbon-fixing symbionts in animals such as mussels or worms (see Section 19.8). Some phototrophic bacteria are also capable of sulfide oxidation and use H_2S or S^0 as an electron donor for CO_2 reduction.

20.4.1 Physiology and Biochemistry of Microbial Oxidation of Reduced Forms of Sulfur

20.4.1.1 Oxidation of Sulfide

20.4.1.1.1 AEROBIC OXIDATION OF SULFIDE Many aerobic bacteria that oxidize sulfide are obligate or facultative chemosynthetic autotrophs (chemolithoautotrophs). When growing in the autotrophic mode, they use sulfide as an energy source to assimilate CO_2. Most of them oxidize the sulfide to sulfate, regardless of the level of oxygen tension (e.g., *Acidithiobacillus thiooxidans* London and Rittenberg, 1964). However, some like *Thiobacillus thioparus* form elemental sulfur (S^0) if the pH of their milieu is initially alkaline and the rH_2* is 12, that is, if the milieu is partially reduced due to an oxygen tension below saturation. Thus, *T. thioparus* T5, isolated from a microbial mat, produces elemental sulfur in continuous culture in a chemostat under conditions of oxygen limitation. In this case, small amounts of thiosulfate together with even smaller amounts of tetrathionate and polysulfide are also formed (van den Ende and van Gemerden, 1993). In batch culture under oxygen limitation, *T. thioparus* has been observed to produce initially a slight increase in pH followed by a drop to 7.5 in 4 days and a rise in rH_2 to 20 (Sokolova and Karavaiko, 1968). The reaction leading to the formation of elemental sulfur can be summarized as

$$HS^- + 0.5O_2 + H^+ \rightarrow S^0 + H_2O \qquad (20.4)$$

Wirsen et al. (2002) described an autotrophic, microaerophilic sulfide-oxidizing organism from a coastal marine environment, *Candidatus Arcobacter sulfidicus* that produced filamentous elemental sulfur. It fixed CO_2 via the reductive tricarboxylic acid cycle rather than the Calvin–Benson–Bassham cycle (Hügler et al., 2005). The organism was able to fix nitrogen (Wirsen et al., 2002). *Thiovulum* sp. is another example of a member of the domain Bacteria that oxidizes sulfide to sulfur under reduced oxygen concentrations (Wirsen and Jannasch, 1978).

---

* $rH_2 = -\log[H_2] = (Eh/0.029) + 2pH$, because $Eh = -0.029$ $\log[H_2] + 0.058 \log[H^+]$.

Under conditions of high oxygen tension (at or near saturation), *T. thioparus* will oxidize soluble sulfide all the way to sulfate (London and Rittenberg, 1964; Sokolova and Karavaiko, 1968; van den Ende and van Gemerden, 1993):

$$HS^- + 2O_2 \rightarrow SO_4^{2-} + H^+ \qquad (20.5)$$

London and Rittenberg (1964) (see also Vishniac and Santer, 1957) suggested that the intermediate steps in the oxidation of sulfide to sulfate involved

$$4S^{2-} \rightarrow 2S_2O_3^{2-} \rightarrow S_4O_6^{2-} \rightarrow SO_3^{2-} + S_3O_6^{2-} \rightarrow$$
$$4SO_3^{2-} \rightarrow 4SO_4^{2-} \qquad (20.6)$$

However, this reaction sequence does not explain the formation of elemental sulfur at reduced oxygen tension. Unless this occurs by way of a specialized pathway, which seems doubtful, a more attractive model of the pathway that explains both the processes, the formation of S^0 and SO_4^{2-}, in a unified way is the one proposed by Roy and Trudinger (1970) (see also Suzuki et al., 1994; Yamanaka, 1996; Suzuki, 1999):

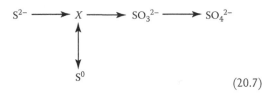

$$\qquad (20.7)$$

Here, X represents a common intermediate in the oxidation of sulfide and elemental sulfur to sulfite. Roy and Trudinger visualized X as a derivative of glutathione or a membrane-bound thiol. It may also be a representative of the *intermediate sulfur* described by Pronk et al. (1990). The scheme of Roy and Trudinger (1970) permits integration of a mechanism for elemental sulfur oxidation into a unified pathway for oxidizing reduced forms of sulfur. Hallberg et al. (1996) found this mechanism consistent with the action of *Acidithiobacillus caldus* on reduced forms of sulfur.

Sorokin (1970) questioned the sulfide-oxidizing ability of Thiobacilli, believing that they oxidize only thiosulfate resulting from chemical oxidation of sulfide by oxygen and that any elemental sulfur formed by Thiobacilli from sulfide is due to the chemical interaction of bacterial oxidation products with S^{2-} and $S_2O_3^{2-}$, as previously proposed by Nathansohn (1902) and Vishniac (1952). This view is not accepted today. Indeed, Vainshtein (1977) and others have presented clear evidence to the contrary. Nübel et al. (2000) showed that hyperthermophilic, microaerophilic, chemolithotrophic *Aquifex aeolicus* VF5 oxidizes sulfide to elemental sulfur using a membrane-bound electron transport pathway that conveys electrons from the oxidation of sulfide to oxygen. The pathway includes a quinone pool, a cytochrome bc_1 complex, and cytochrome oxidase.

20.4.1.1.2 ANAEROBIC OXIDATION OF SULFIDE

Most bacteria that oxidize sulfide anaerobically are photosynthetic autotrophs, but a few, like the facultative anaerobes *Thiobacillus denitrificans* and *Thermothrix thiopara*, are chemosynthetic autotrophs (chemolithoautotrophs). In the presence of nonlimiting concentrations of sulfide, most photosynthetic autotrophs oxidize sulfide to elemental sulfur, using the reducing power from this reaction in the assimilation of CO_2. However, some exceptional organisms exist that never form elemental sulfur. When elemental sulfur is formed, it is usually accumulated intracellularly by purple sulfur bacteria (PSB) and extracellularly by green sulfur bacteria and cyanobacteria. Elemental sulfur accumulated extracellularly by *Chlorobium* appears to be readily available to the cell that formed it, but not to other individuals in the population of the same organism or to other photosynthetic bacteria that can oxidize elemental sulfur. The sulfur is apparently attached to the cell surface (van Gemerden, 1986). Recent study by environmental scanning electron microscopy suggests that the extracellularly deposited sulfur is associated with spinae on the cell surface (Douglas and Douglas, 2000). Spinae are helically arrayed proteins, which form a hollow tube protruding from the cell surface (Easterbrook and Coombs, 1976). Details of the biochemistry of sulfide oxidation by the photosynthetic autotrophs remain to be explored.

The chemosynthetic autotroph *T. denitrificans* can oxidize sulfide to sulfate anaerobically with nitrate as terminal electron acceptor. As sulfide is oxidized, nitrate is reduced via nitrite to nitric oxide (NO), nitrous oxide (N_2O), and dinitrogen (N_2) (Baalsrud and Baalsrud, 1954; Milhaud et al., 1958; Peeters and Aleem, 1970; Adams et al., 1971; Aminuddin and Nicholas, 1973). Acetylene has been found to cause accumulation of sulfur rather than sulfate in gradient culture of a strain of T. denitrificans using nitrous oxide as terminal electron acceptor. In the absence of acetylene, the gradient culture, unlike a batch culture, did not even accumulate sulfur transiently. It was suggested that acetylene prevents the transformation of S^0 to SO_3^{2-} in this culture (Daalsgaard and Bak, 1992). Polysulfide ($S_{n-1}SH^-$), but not free sulfur, appears to be an intermediate in sulfide oxidation to sulfate by this organism (Aminuddin and Nicholas, 1973). The polysulfide appears to be oxidized to sulfite and thence to sulfate (Aminuddin and Nicholas, 1973, 1974a,b).

Oil field brine from the Coleville oil field in Saskatchewan, Canada, yielded two microaerophilic strains of bacteria, one (strain CVO) resembling Thiomicrospira denitrificans and the other (strain FWKO B) resembling Arcobacter. Both of these strains can oxidize sulfide anaerobically with nitrate as terminal electron acceptor (Gevertz et al., 2000). Each can grow autotrophically, but strain CVO can also use acetate in the place of CO_2. Strain CVO produces elemental sulfur or sulfate, depending on the sulfide concentration, while reducing nitrate or nitrite to dinitrogen. Strain FWKO B produces only sulfur and reduces nitrate only to nitrite. Anaerobic sulfide oxidation linked to nitrate reduction to nitrate has also been implicated in OMZs, where a "cryptic" sulfur cycle may exist (Canfield et al., 2010), with sulfate being reduced to sulfide and rapidly reoxidized to sulfate. A similar cryptic cycle may exist in the methane zone of marine sediments, driven by iron (Holmkvist et al., 2011).

20.4.1.1.3 HETEROTROPHIC AND MIXOTROPHIC OXIDATION OF SULFIDE

Hydrogen sulfide oxidation is not limited to autotrophs. Most known strains of Beggiatoa grow mixotrophically or heterotrophically on sulfide. In the former instance, the organisms derive energy from oxidation of the H_2S. In the latter, they apparently use sulfide oxidation to eliminate metabolically formed hydrogen peroxide in the absence of catalase (Burton and Morita, 1964). Beggiatoa deposits sulfur granules resulting from sulfide oxidation in its cells external to the cytoplasmic membrane in invaginated, double-layered membrane pockets (Strohl and Larkin, 1978; see also discussion by Ehrlich, 1999). The sulfur can be further oxidized to sulfate under sulfide limitation (Pringsheim, 1967). At least one strain

of Beggiatoa, isolated from the marine environment, has proven to be autotrophic (Nelson and Jannasch, 1983; see also Jannasch et al., 1989). The heterotrophs Sphaerotilus natans (prokaryote, domain Bacteria), Alternaria, and yeast (eukaryotes, fungi) have also been reported to oxidize H_2S to elemental sulfur (Skerman et al., 1957a,b), using a pathway that includes reverse dissimilatory sulfite reductase (rDsr) coupled to energy metabolism (Belousova et al., 2013).

20.4.1.2 Oxidation of Elemental Sulfur

20.4.1.2.1 AEROBIC OXIDATION OF ELEMENTAL SULFUR

Elemental sulfur may be enzymatically oxidized to sulfuric acid by certain members of the Bacteria and Archaea. The overall reaction may be written as

$$S^0 + 1.5O_2 + H_2O \rightarrow H_2SO_4 \qquad (20.8)$$

Cell extract of A. thiooxidans, to which catalytic amounts of glutathione were added, oxidized sulfur to sulfite (Suzuki and Silver, 1966). Sulfite was also shown to be accumulated when sulfur was oxidized by A. thiooxidans in the presence of 2-n-heptyl-4-hydroxyquinoline N-oxide (HQNO), which has been shown to inhibit sulfite oxidation. The stoichiometry when the availability of sulfur was limited was 1 mol sulfite accumulated per mole each of sulfur and oxygen consumed (Suzuki et al., 1992). A sulfur oxidizing enzyme in T. thioparus used glutathione as a cofactor to produce sulfite (Suzuki and Silver, 1966). The enzyme in both organisms contained nonheme iron and was classed as an oxygenase. The mechanism of sulfur oxidation is consistent with the model described in Reaction 20.7. The glutathione in this instance forms a polysulfide (compound X in Reaction 20.7) with the substrate sulfur, which is then converted to sulfite by the introduction of molecular oxygen. This reaction appears not to yield useful energy to the cell. Sulfur oxidation to sulfite that does not involve oxygenase but an oxidase with a potential for energy conservation has also been considered. Some experimental evidence supports such a mechanism (see Pronk et al., 1990).

20.4.1.2.2 ANAEROBIC OXIDATION OF ELEMENTAL SULFUR

Details are emerging regarding how elemental sulfur is oxidized in anaerobes, especially photosynthetic autotrophs. In these organisms, the sox pathway plays an important role, although the precise details require further elucidation (Friedrich et al., 2001). *T. denitrificans* appears to follow the reaction sequence in Reaction 20.7 except that oxidized forms of nitrogen substitute for oxygen as terminal electron acceptor. *Acidithiobacillus ferrooxidans* has the capacity to oxidize elemental sulfur anaerobically using ferric iron as terminal electron acceptor (Brock and Gustafson, 1976; Corbett and Ingledew, 1987). The anaerobic oxidation yields enough energy to support growth at a doubling time of 24 h (Pronk et al., 1991, 1992).

20.4.1.3 Oxidation of Sulfite

20.4.1.3.1 AEROBIC OXIDATION OF SULFITE
Sulfite may be oxidized by two different mechanisms, one of which includes substrate-level phosphorylation whereas the other does not, although both yield useful energy through oxidative phosphorylation by the intact cell (see, e.g., review by Wood, 1988). In substrate-level phosphorylation, sulfite reacts oxidatively with adenylic acid (AMP) to give APS:

$$SO_3^{2-} + AMP \xrightarrow{\text{APS oxido-reductase}} APS + 2e \quad (20.9)$$

The sulfate of APS is then exchanged for phosphate:

$$APS + P_i \xrightarrow{\text{ADP sulfurylase}} ADP + SO_4^{2-} \quad (20.10)$$

ADP can then be converted to ATP as follows:

$$2ADPS \xrightarrow{\text{ADP sulfurylase}} ATP + AMP \quad (20.11)$$

Hence, the oxidation of 1 mol of sulfite yields 0.5 mol of ATP formed by substrate-level phosphorylation. However, most energy conserved as ATP is gained from shuttling electrons in Reaction 20.9 through the membrane-bound electron transport system to oxygen (Davis and Johnson, 1967).

A number of Thiobacilli appear to use an AMP-independent sulfite oxidase system (Roy and Trudinger, 1970, p. 214). These systems do not all seem to be alike. The AMP-independent sulfite oxidase of autotrophically grown Thiobacillus

novellus may use the following electron transport pathway (Charles and Suzuki, 1966):

$$SO_3^{2-} \rightarrow \text{Cytochrome c} \rightarrow \text{Cytochrome oxidase} \rightarrow O_2 \quad (20.12)$$

The sulfite oxidase of *Thiobacillus neapolitanus* can be pictured as a single enzyme complex that may react either with sulfite and AMP in an oxidation that gives rise to APS and sulfate or with sulfite and water followed by oxidation to sulfate (Roy and Trudinger, 1970). The enzyme complex then transfers the reducing power that is generated to oxygen. Sulfite-oxidizing enzymes that do not require the presence of AMP have also been detected in *A. thiooxidans*, *T. denitrificans*, and *T. thioparus*. *Thiobacillus concretivorus* (now considered a strain of *A. thiooxidans*) was reported to shuttle electrons from SO_3^{2-} oxidation via the following pathway to oxygen (Moriarty and Nicholas, 1970):

$$SO_3^{2-} \rightarrow \text{(Flavin?)} \rightarrow \text{Coenzyme Q} \rightarrow \text{Cytochrome } b$$
$$\rightarrow \text{Cytochrome } c \rightarrow \text{Cytochrome } a_1 \rightarrow O_2 \quad (20.13)$$

The archaeon *Acidianus ambivalens* appears to possess both an ADP-dependent and an ADP-independent pathway. The former occurs in the cytosol, whereas the latter is membrane associated (Zimmermann et al., 1999).

20.4.1.3.2 ANAEROBIC OXIDATION OF SULFITE
T. denitrificans is able to form APS reductase that is not membrane bound (Bowen et al., 1966), as well as a membrane-bound AMP-independent sulfite oxidase (Aminuddin and Nicholas, 1973, 1974a,b). Both enzyme systems appear to be active in anaerobically grown cells (Aminuddin, 1980). The electron transport pathway under anaerobic conditions terminates in cytochrome d, whereas under aerobic conditions it terminates in cytochromes aa_3 and d. Nitrate but not nitrite acts as electron acceptor anaerobically when sulfite is the electron donor (Aminuddin and Nicholas, 1974b).

20.4.1.4 Oxidation of Thiosulfate

Most chemosynthetic autotrophic bacteria that can oxidize elemental sulfur can also oxidize thiosulfate to sulfate. The photosynthetic, autotrophic, and purple and green sulfur bacteria and some purple nonsulfur bacteria oxidize thiosulfate to sulfate as a source of reducing power for CO_2 assimilation (e.g., Trüper 1978; Neutzling et al., 1985). However, the mechanism of thiosulfate oxidation is probably not the same in all these organisms. The chemosynthetic, aerobic autotrophic *T. thioparus* will transiently accumulate elemental sulfur outside its cells when growing in excess thiosulfate in batch culture but only sulfate when growing in limited amounts of thiosulfate. *T. denitrificans* will do the same anaerobically with nitrate as terminal electron acceptor (Schedel and Trüper, 1980). The photosynthetic purple bacteria may also accumulate sulfur transiently, but some green sulfur bacteria (Chlorobiaceae) do not (see discussion by Trüper, 1978). Several of the purple nonsulfur bacteria (Rhodospirillaceae) when growing photoautotrophically with thiosulfate do not accumulate sulfur in their cells (Neutzling et al., 1985). Some mixotrophic bacteria oxidize thiosulfate only to tetrathionate.

Thiosulfate is a reduced sulfur compound with sulfur in a mixed valence state. Current evidence indicates that the two sulfurs are covalently linked, the outer or sulfane sulfur of $S_2O_3^{2-}$ having a valence of −1 and the inner or sulfone sulfur having a valence of +5. An older view was that the sulfane sulfur had a valence of −2 and the sulfone sulfur +6.

Charles and Suzuki (1966) proposed that when thiosulfate is oxidized, it is first cleaved according to the reaction:

$$S_2O_3^{2-} \rightarrow SO_3^{2-} + S^0. \qquad (20.14)$$

The sulfite is then oxidized to sulfate:

$$SO_3^{2-} + H_2O \rightarrow SO_4^{2-} + 2H^+ + 2e, \qquad (20.15)$$

and the sulfur is oxidized to sulfate via sulfite as previously described:

$$S^0 \rightarrow SO_3^{2-} \rightarrow SO_4^{2-} \qquad (20.16)$$

Alternatively, thiosulfate oxidation may be preceded by a reduction reaction, resulting in the formation of sulfite from the sulfone sulfur and sulfide from the sulfane sulfur:

$$S_2O_3^{2-} + 2e \rightarrow S^{2-} + SO_3^{2-} \qquad (20.17)$$

These products are then each oxidized to sulfate (Peck, 1962). In the latter case, it is conceivable that sulfur could accumulate transiently by the mechanisms suggested by Reaction 20.7, but sulfur could also result from asymmetric hydrolysis of tetrathionate resulting from direct oxidation of thiosulfate (see Roy and Trudinger, 1970 for detailed discussion):

$$2S_2O_3^{2-} + 2H^+ + 0.5O_2 \rightarrow S_4O_6^{2-} + H_2O \qquad (20.18)$$

$$S_4O_6^{2-} + OH^- \rightarrow S_2O_3^{2-} + S^0 + HSO_4^- \qquad (20.19)$$

The direct oxidation reaction may involve the enzymes thiosulfate oxidase and thiosulfate cytochrome *c* reductase, a thiosulfate-activating enzyme (Trudinger, 1961; Aleem, 1965). The thiosulfate oxidase may use glutathione as a coenzyme (see summary by Roy and Trudinger, 1970; Wood, 1988).

Thiosulfate may also be cleaved by the enzyme rhodanese, which is found in most sulfur-oxidizing bacteria. For instance, it can transfer sulfane sulfur to acceptor molecules such as cyanide to form thiocyanate. This enzyme may also play a role in thiosulfate oxidation. In anaerobically growing *T. denitrificans* strain RT, for instance, rhodanese initiates thiosulfate oxidation by forming sulfite from the sulfone sulfur, which is then oxidized to sulfate. The sulfane sulfur accumulates transiently as elemental sulfur outside the cells, and when the sulfone sulfur is depleted, the sulfane sulfur is rapidly oxidized to sulfate (Schedel and Trüper, 1980). In another strain of *T. denitrificans*, however, thiosulfate reductase rather than rhodanese catalyzes the initial step of thiosulfate oxidation, and both the sulfane and sulfone sulfur are attacked concurrently (Peeters and Aleem, 1970). *Thiobacillus versutus* (formerly *Thiobacillus* A_2) seems to oxidize thiosulfate to sulfate by a unique pathway (Lu and Kelly, 1983) that involves a thiosulfate multienzyme system that has a periplasmic location (Lu, 1986). No free intermediates appear to be formed from either the reduced ($S-SO_3^{2-}$) or sulfonate ($S-SO_3^{2-}$) sulfurs of thiosulfate.

Pronk et al. (1990) summarized the evidence that supports a model in which *A. ferrooxidans*, *A. thiooxidans*, and *Acidiphilium acidophilum* oxidize thiosulfate by forming tetrathionate in an initial step:

$$2S_2O_3^{2-} \rightarrow S_4O_6^{2-} + 2e^- \qquad (20.20)$$

This is followed in the model by a series of hydrolytic and oxidative steps whereby tetrathionate is transformed into sulfate with transient accumulation of intermediary sulfur from polythionates. Thiosulfate dehydrogenase from *A. acidophilum*, which catalyzes the oxidation of thiosulfate to tetrathionate, was purified and partially characterized by Meulenberg et al. (1993).

20.4.1.5 Oxidation of Tetrathionate

Although bacterial oxidation of tetrathionate has been reported, the mechanism of oxidation is still not certain (see Roy and Trudinger, 1970; Kelly, 1982). It may involve disproportionation (see Section 20.5) and hydrolysis reactions. A more detailed scheme was described by Pronk et al. (1990), which was mentioned earlier in connection with thiosulfate oxidation.

20.4.2 Common Mechanism for Oxidizing Reduced Inorganic Sulfur Compounds in Domain Bacteria

Friedrich et al. (2001) suggested that the mechanism for oxidizing inorganic reduced sulfur compounds by aerobic and anaerobic sulfur-oxidizing bacteria, including anoxygenic phototrophic bacteria, have certain common features. Their suggestion is based on molecular comparisons of the *Sox* genes and the proteins they encode between those in *Paracoccus pantotrophus* and those in other bacteria capable of oxidizing inorganic reduced sulfur compounds. The *Sox* enzyme system in the archeon *Sulfolobus solfataricus* appears to differ from that in Bacteria on the basis of genomic analysis.

20.4.3 Ecology of Reduced S Oxidizers

Bacteria and Archaea comprise most of the geomicrobiologically important microorganisms that oxidize reduced forms of sulfur in relatively large quantities. These include aerobes, facultative organisms, and anaerobes. Most are obligate or facultative autotrophs or mixotrophs. Among aerobes in the domain Archaea, one of the most widely studied groups consists of the genera *Sulfolobus* and *Acidianus* (Table 20.3). Among the aerobes in Bacteria, one of the most important groups in terrestrial environments is that of the *Thiobacillaceae* (Table 20.3). This group includes obligate and facultative autotrophs as well as

TABLE 20.3
Some aerobic sulfur-oxidizing bacteria.[a,b]

Autotrophic	Mixotrophic	Heterotrophic
Acidithiobacillus albertensis[c]	Pseudomonas spp.	Beggiatoa spp.
Acidithiobacillus caldus[c]	Thiobacillus intermedius	Thiobacillus perometabolis
Acidithiobacillus ferrooxidans[c]	Thiobacillus organoparus	
Acidithiobacillus thiooxidans[c]	Thiobacillus versutus[d]	
Acidianus brierleyi[e]		
Alicyclobacillus disulfidooxidans[f,g]		
Alicyclobacillus tolerans[f,g]		
Beggiatoa alba MS-81-6		
Sulfolobus acidocaldarius[e]		
Thermothrix thiopara		
Thiobacillus denitrificans[h]		
Thiobacillus neapolitanus		
Thiobacillus novellus		
Thiobacillus tepidarius		
Thiobacillus thioparus		

[a] A more complete survey of aerobic sulfur-oxidizing bacteria can be found in Balows et al. (1992) and Dworkin (2001).
[b] All members of the domain Bacteria in this table are gram-negative except for *Alicyclobacillus disulfidooxidans* and *A. tolerans*.
[c] Formerly assigned to the genus *Thiobacillus* (see Kelly and Wood, 2000).
[d] Can also grow autotrophically and heterotrophically.
[e] Archeon.
[f] *Alicyclobacillus disulfidooxidans* formerly known as *Sulfobacillus disulfidooxidans* and *Alicyclobacillus tolerans* formerly known as *Sulfobacillus thermosulfidooxidans* subsp. *thermotolerans* (see Karavaiko et al., 2005).
[g] Facultative autotroph.
[h] Facultative anaerobe.

mixotrophs. Another bacterial group that oxidizes sulfide and is important in some freshwater and marine environments is the family *Beggiatoaceae*. Most cultured members of the group use hydrogen sulfide mixotrophically or heterotrophically. In the latter instance, they employ H_2S oxidation as protection against metabolically produced H_2O_2 in the absence of catalase (Nelson and Castenholz, 1981; Kuenen and Beudecker, 1982), but at least one marine strain, *Beggiatoa alba* MS-81-6, can grow autotrophically (Nelson and Jannasch, 1983). Other hydrogen sulfide oxidizers found in aquatic environments include *Thiovulum* (autotrophic) (e.g., Wirsen and Jannasch, 1978), *Achromatium*, *Thiothrix*, *Thiobacterium* (LaRiviere and Schmidt, 1981), and *Thiomicrospira* (Kuenen and Tuovinen, 1981). Of all these groups, only the Thiobacilli produce sulfate directly without accumulating elemental sulfur when oxidizing H_2S in the presence of abundant oxygen. The other groups accumulate sulfur (S^0), which they may oxidize to sulfate when the supply of H_2S is limited or depleted.

Among members of the domain Bacteria, *T. thioparus* oxidizes S^0 slowly to sulfate; this process becomes inhibited as the pH drops below 4.5. *Halothiobacillus halophilus* (formerly *T. halophilus*) is another neutrophilic, but extremely halophilic, obligate chemolithotroph that oxidizes elemental sulfur to sulfate (Wood and Kelly, 1991). By contrast, *A. thiooxidans*, *Acidithiobacillus albertensis* (formerly *T. albertis*) (Bryant et al., 1983), and *A. ferrooxidans* readily oxidize elemental sulfur to sulfate. This reaction produces protons, and being acidophilic, they may lower the pH as low as 1.0 in batch culture. All these organisms are strict autotrophs.

The Archaea *Sulfolobus* spp. and *Acidianus* spp. are also able to oxidize elemental sulfur to sulfate. Both the genera are extremely thermophilic. *S. acidocaldarius* will oxidize sulfur between 55°C and 85°C (70°C–75°C optimum) in a pH range of 0.9–5.8 (pH 2–3 optimum) (Brock et al., 1972; Shivvers and Brock, 1973). The organisms are facultative autotrophs. *Acidianus* (formerly *Sulfolobus*) *brierleyi* has traits similar to those of *S. acidocaldarius* but can also reduce S^0 anaerobically with H_2 and has a different GC (guanine + cytosine) content (31 versus 37 mol.%) (Brierley and Brierley, 1973; Segerer et al., 1986).

Moderately, thermophilic bacteria capable of oxidizing sulfur have also been observed—some were isolated from sulfurous hot springs, others from ore deposits. One of these, *Thiobacillus thermophilica* Imshenetskii, is a motile rod and is a facultative autotroph capable of oxidizing various sulfides and organic compounds besides elemental sulfur (Egorova and Deryugina, 1963). Another is an aerobic, gram-positive, facultative thermophile capable of sporulation, which is able to oxidize not only elemental sulfur but also Fe^{2+} and metal sulfides mixotrophically. It was originally named *Sulfobacillus thermosulfidooxidans* subsp.

thermotolerans strain K1 (Golovacheva and Karavaiko, 1978; Bogdanova et al., 1990), and later renamed *Alicyclobacillus tolerans* (Karavaiko et al., 2005). Still, another is a gram-negative, facultatively autotrophic *Thiobacillus* sp. capable of growth at 50°C and 55°C with a pH optimum of 5.6 (range 4.8–8) (Williams and Hoare, 1972). Other thermophilic *Thiobacillus*-like bacteria have been isolated that can grow on thiosulfate at 60°C and 75°C and a pH range of 4.8–7.5 (LeRoux et al., 1977). A moderately thermophilic acidophile, *A. caldus* (formerly *T. caldus*), with an optimum growth temperature of ·45°C was isolated by Hallberg and Lindström (1994) and found capable of oxidizing S^{2-}, S^0, SO_3^{2-}, $S_2O_3^{2-}$, and $S_4O_6^{2-}$ (Hallberg et al., 1996).

A number of heterotrophs, including bacteria and fungi, have been reported to be able to oxidize reduced sulfur in the form of elemental sulfur, thiosulfate, and tetrathionate. A diversity of heterotrophic thiosulfate-oxidizing bacteria have been detected in marine sediments and around hydrothermal vents (Teske et al., 2000). Many bacteria that oxidize elemental sulfur oxidize it to thiosulfate, whereas others oxidize thiosulfate to sulfuric acid (Guittonneau, 1927; Guittonneau and Keiling, 1927; Grayston and Wainright, 1988; see also Roy and Trudinger, 1970, pp. 248–249). Some marine Pseudomonadaceae can gain useful energy from thiosulfate oxidation by using it as a supplemental energy source (Tuttle et al., 1974; Tuttle and Ehrlich, 1986).

Two examples of facultatively anaerobic sulfur oxidizers in the domain Bacteria are *T. denitrificans* (e.g., Justin and Kelly, 1978) and *T. thiopara* (Caldwell et al., 1976; Brannan and Caldwell, 1980), the former a mesophile and the latter a thermophile. The genome of *T. denitrificans* was sequenced (Beller et al., 2006). Anaerobically, both organisms use nitrate as terminal electron acceptor and reduce it to oxides of nitrogen and dinitrogen, with nitrite being an intermediate product. They can use sulfur in various oxidation states as an energy source. *T. denitrificans* is an obligate autotroph whereas *T. thiopara* is a facultative autotroph.

The strictly anaerobic sulfur oxidizers are represented by photosynthetic purple and green bacteria (Pfennig, 1977) and certain cyanobacteria (Table 20.4). All Cyanobacteria grow aerobically, but not all oxidize reduced sulfur compounds directly. The PSB (Chromatiaceae) (Figure 20.4) are obligate anaerobes that oxidize reduced

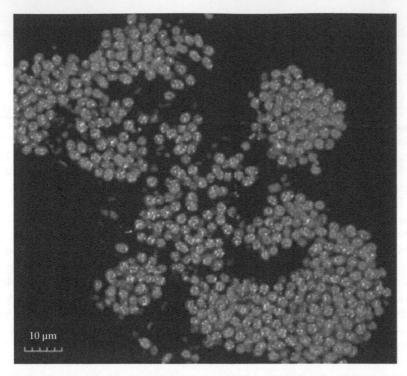

Figure 20.4. **(See color insert.)** Epireflective confocal microscopy of sectioned "pink berry" microbial consortium from Sippewissett Salt Marsh, Cape Cod, MA. Consortia mass is dominated by purple sulfur bacteria (PSB) belonging to the Halochromatium–Thiohalocapsa lineage of the Chromatiaceae. The bright reflective signals are refractile elemental sulfur inclusions within the PSB cells. (Reprinted from Wilbanks, EG et al., *Environ Microbiol*, 2014.)

sulfur, especially H_2S, and use it as a source of reducing power for CO_2 assimilation. Despite the terminology, several purple nonsulfur bacteria (Rhodospirillaceae) can also grow autotrophically on H_2S as a source of reducing power for CO_2 assimilation, but for the most part, they tolerate only low concentrations of sulfide, in contrast to PSB. In the laboratory, purple nonsulfur bacteria can also grow photoheterotrophically, using reduced carbon compounds as a carbon source. Most sulfur-oxidizing phototrophs, when growing on H_2S, oxidize it to S^0, which they deposit intracellularly (Figure 20.3), but *Ectothiorhodospira* spp. deposit it extracellularly. Under conditions of H_2S limitation, these strains oxidize the elemental sulfur they accumulate further to sulfate. Among the purple nonsulfur bacteria, *Rhodopseudomonas palustris* and *Rhodopseudomonas sulfidophila* do not form elemental sulfur as an intermediate from H_2S but oxidize sulfide directly to sulfate (Hansen and van Gemerden, 1972; Hansen and Veldkamp, 1973). In contrast, *Rhodospirillum rubrum*, *Rhodospirillum capsulata*, and *Rhodopseudomonas spheroides* form elemental sulfur from sulfide, which they deposit extracellularly

(Hansen and van Gemerden, 1972). *R. sulfidophila* differs from most purple nonsulfur bacteria in being more tolerant of high concentrations of sulfide.

Green sulfur bacteria (Chlorobiaceae) are strictly anaerobic photoautotrophs that oxidize H_2S by using it as a source of reducing power in CO_2 fixation. They deposit the sulfur (S^0) they produce extracellularly. Under H_2S limitation, they oxidize the sulfur further to sulfate. At least a few strains of *Chlorobium limicola* forma *thiosulfatophilum* do not accumulate sulfur but oxidize H_2S directly to sulfate (Ivanov, 1968, p. 137; Paschinger et al., 1974). Many of these bacteria can also use thiosulfate as electron donor in the place of hydrogen sulfide.

Filamentous gliding green bacteria (Chloroflexacea) grow photoheterotrophically under anaerobic conditions, but at least some can also grow photoautotrophically with H_2S as electron donor under anaerobic conditions (Brock and Madigan, 1988).

A few filamentous Cyanobacteria, including some members of the genera *Oscillatoria*, *Lyngbya*, *Aphanothece*, *Microcoleus*, and *Phormidium*, which are oxygenic

EHRLICH'S GEOMICROBIOLOGY

TABLE 20.4
Some anaerobic sulfur-oxidizing bacteria.[a]

Photolithotrophs	Chemolithotrophs
Chromatinum spp.	Thermothrix thiopara[b,c]
Chlorobium spp.	Thiobacillus denitrificans[c]
Ectothiorhodospira spp.	
Rhodopseudomonas spp.[b]	
Chloroflexus aurantiacus[b]	
Oscillatoria sp.[c]	
Lyngbya spp.[c]	
Aphanothece spp.[c]	
Microcoleus spp.[c]	
Phormidium spp.[c]	

[a] For a more complete description of anaerobic sulfur-oxidizing bacteria, see Holt (1984) and Dworkin (2001).
[b] Facultatively autotrophic.
[c] Facultatively anaerobic.

photoautotrophs, can grow photosynthetically under anaerobic conditions with H_2S as a source of reducing power (Cohen et al., 1975; Garlick et al., 1977). They oxidize H_2S to elemental sulfur and deposit it extracellularly. In the dark, they can re-reduce the sulfur they produce using internal reserves of polyglucose as reductant (Oren and Shilo, 1979). At this time there is no evidence that these organisms can oxidize the sulfur they produce anaerobically further to sulfate under H_2S limitation.

Capability for sulfide-oxidation may also be encoded in viral genomes, some of which have been reported to encode genes for reverse dissimilatory sulfite reductase (rdsr; Anantharamn et al., 2014). This suggests a possibility for the transfer of these capabilities among microorganisms.

20.5 DISPROPORTIONATION

20.5.1 Elemental Sulfur Disproportionation

Anaerobic marine enrichment cultures consisting predominantly of slightly curved bacterial rods have been shown to contain chemolithotrophic bacteria that were able to grow on sulfur by disproportionating it into H_2S and SO_4^{2-}, but only in the presence of sulfide scavengers such as FeOOH, $FeCO_3$, or MnO_2 (Thamdrup et al., 1993; see also Janssen et al., 1996). The disproportionation reaction can be summarized as

$$4S^0 + 4H_2O \rightarrow SO_4^{2-} + 3H_2S + 2H^+ \quad (20.21)$$

Added ferrous iron scavenges the sulfide by forming FeS, whereas added MnO_2 scavenges sulfide in a redox reaction in which MnO_2 is reduced to Mn^{2+} by the sulfide, producing SO_4^{2-}, with S^0 a probable intermediate (Thamdrup et al., 1994). The scavenging action is needed to propel the reaction in the direction of sulfur disproportionation. In the disproportionation reaction, three pairs of electrons from one atom of sulfur are transferred via an as-yet-undefined electron transport pathway to three other atoms of sulfur, generating H_2S in Reaction 20.21. The sulfur atom yielding the electrons is transformed into sulfate. The transfer of the three pairs of electrons is the source of the energy conserved by the organism for growth and reproduction. This sulfur disproportionation reaction is similar to the one that has been observed under laboratory conditions with the photolithotrophic green sulfur bacteria C. *limicola* subspecies *thiosulfaticum* and *Chlorobium vibrioforme* under an inert atmosphere in the light in the absence of CO_2. To keep the reaction going, the H_2S produced had to be removed by continuous flushing with nitrogen (see Trüper, 1984).

A study of sulfur isotope fractionation as a result of sulfur disproportionation by enrichment cultures from Åarhus Bay, Denmark, and other sediment sources revealed that the sulfide produced may be depleted in ^{34}S by as much as 7.3%–8.6‰ and the corresponding sulfate produced may be enriched by as much as 12.6%–15.3‰. Similar fractionation is obtained from laboratory experiments (Habicht et al., 1998).

20.5.2 Thiosulfate Disproportionation

It has been demonstrated experimentally that some bacteria, like *Desulfovibrio sulfodismutans*, can obtain energy anaerobically by disproportionating thiosulfate into sulfate and sulfide (Bak and Cypionka, 1987; Bak and Pfennig, 1987; Jørgensen, 1990a,b):

$$S_2O_3^{2-} + H_2O \rightarrow SO_4^{2-} + HS^- + H^+$$
$$(\Delta G^0 = -21.9 \text{ kJ mol}^{-1}) \quad (20.22)$$

The energy from this reaction enables the organisms to assimilate carbon from a combination of CO_2 and acetate. Energy conservation by thiosulfate disproportionation seems, however, paradoxical if the oxidation state of the sulfane sulfur is −2 and that of the sulfone sulfur is +6, as formerly believed, because no redox reaction would be required to generate a mole of sulfate and sulfide each per mole of thiosulfate. A solution to this paradox has been provided by the report of Vairavamurthy et al. (1993), which demonstrated spectroscopically that the charge density of the sulfane sulfur in thiosulfate is really −1 and that of the sulfone sulfur is +5. Based on this finding, the formation of sulfide and sulfate by disproportionation of thiosulfate requires a redox reaction. Another organism able to disproportionate thiosulfate is *Desulfotomaculum thermobenzoicum* (Jackson and McInerney, 2000). The addition of acetate to the growth medium stimulated thiosulfate disproportionation by this organism. Thiosulfate disproportionation has also been observed with *Desulfocapsa thiozymogenes* (Janssen et al., 1996).

D. *desulfodismutans* can also generate useful energy from the disproportionation of sulfite and dithionite to sulfide and sulfate (Bak and Pfennig, 1987). The overall reaction for sulfite disproportionation is

$$4SO_3^{2-} + H^+ \rightarrow 3SO_4^{2-} + HS^-$$
$$(\Delta G^0 = -58.9 \text{ kJ mol}^{-1}) \qquad (20.23)$$

D. *sulfodismutans* can also grow on lactate, ethanol, propanol, and butanol as energy sources and sulfate as terminal electron acceptor, like typical sulfate reducers, but growth is slower than by disproportionation of partially reduced sulfur compounds. Bak and Pfennig (1987) suggested that from an evolutionary standpoint, D. *sulfodismutans*-type sulfate reducers could be representative of the progenitors of typical sulfate reducers.

Perry et al. (1993) suggested that *Shewanella putrefaciens* MR-4, which they isolated from the Black Sea, disproportionates thiosulfate into either sulfide and sulfite or elemental sulfur and sulfite. They never detected any sulfate among the products in these reactions. These disproportionations are, however, endergonic (+30.98 and +16.10 kJ mol⁻¹ at pH 7, 1 atm, and 25°C, respectively). Perry and coworkers suggested that in S. *putrefaciens* MR-4 these reactions must be coupled to exergonic reactions such as carbon oxidation.

Thiosulfate disproportionation seems to play a significant role in the sulfur cycle in marine environments (Jørgensen, 1990a). In Kysing Fjord (Denmark) sediment, thiosulfate was identified as a major intermediate product of anaerobic sulfide oxidation that was simultaneously reduced to sulfide, oxidized to sulfate, and disproportionated to sulfide and sulfate. This occurred at a rapid rate as reflected by a small thiosulfate pool. The metabolic fate of thiosulfate in these experiments was determined by adding differentially labeled ³⁵S-thiosulfate and following the consumption of the thiosulfate and the isotopic distribution in sulfide and sulfate formed from the sulfane and sulfone sulfur atoms of the labeled thiosulfate over time in separate experiments. According to Jørgensen (1990a), the disproportionation reaction can explain the observed large difference in ³⁴S/³²S in sulfate and sulfide in the sediments. These findings were extended to anoxic sulfur transformations in further experiments with Kysing Fjord sediments and with sediments from Braband Lake, Århus Bay, and Aggersund by Elsgaard and Jørgensen (1992). They showed a significant contribution made by thiosulfate disproportionation in anaerobic production of sulfate from sulfide. Addition of nitrate stimulated anoxic oxidation of sulfide to sulfate. Addition of iron in the form of lepidocrocite (FeOOH) caused partial oxidation of sulfide with the formation of pyrite and sulfur and precipitation of iron sulfides.

20.6 REDUCED FORMS OF SULFUR AS AN ELECTRON DONOR

All evidence to date indicates that to conserve energy, chemosynthetic autotrophic and mixotrophic bacteria that oxidize reduced forms of sulfur feed the reducing power (electrons) into a membrane-bound electron transport system whether oxygen, nitrate, or nitrite is the terminal electron acceptor (Peeters and Aleem, 1970; Sadler and Johnson, 1972; Aminuddin and Nicholas, 1974b; Loya et al., 1982; Lu and Kelly, 1983; Smith and Strohl, 1991; Kelly et al., 1993; also see review by Kelly, 1982). However, the components of the electron transport system in the plasma membrane, that is, the cytochromes, quinones, and nonheme iron proteins, are not identical in all organisms. Whatever the electron transport chain makeup in the plasma membrane, it is the oxidation state of a particular sulfur compound being oxidized, or more exactly the midpoint potential of its redox couple at physiological pH, that determines the entry point into the electron transport chain of the electrons removed during the oxidation of the sulfur compound. Thus, the electrons from elemental sulfur are generally thought to enter the transport chain at the level of a cytochrome bc_1 complex or equivalent. As pointed out earlier, the first step in the oxidation of sulfur to sulfate can be the formation of sulfite by an oxygenation involving direct interaction with oxygen without involvement of the cytochrome system. Only in the subsequent oxidation of sulfite to sulfate is the electron transport system directly involved starting at the level of the cytochrome bc_1 complex or equivalent. Also, as discussed earlier, sulfite may be oxidized by an AMP-dependent or AMP-independent pathway. In either case, electrons are passed into the electron transport system at the level of a cytochrome bc_1 complex. In the AMP-dependent pathway, most of the energy coupling can be assumed to be chemiosmotic, that is, on average 1 or 2 mol of ATP can be formed per electron pair passed to oxygen by the electron transport system, but in addition, 0.5 mL of ATP can be formed via substrate-level phosphorylation (Reactions 20.9 and 20.10). By contrast, only 1 or 2 mol of ATP can be formed on average per electron pair passed to oxygen by the AMP-independent pathway.

Chemiosmosis is best explained if it is assumed that the sulfite oxidation half-reaction occurs at the exterior of the plasma membrane (in the periplasm):

$$SO_3^{2-} + H_2O \rightarrow SO_4^{2-} + 2H^+ + 2e \quad (20.24)$$

and the oxygen reduction half-reaction on the inner surface of the plasma membrane (cytoplasmic side):

$$0.5O_2 + 2H^+ + 2e \rightarrow H_2O \quad (20.25)$$

In T. versutus, a thiosulfate-oxidizing, multienzyme system has been located in the periplasm (Lu, 1986).

The pH gradient resulting from sulfite oxidation and any proton pumping associated with electron transport in the plasma membrane together with any electrochemical gradient provide the proton motive force for ATP generation via F_0F_1 ATP synthase. Proton translocation during thiosulfate oxidation has been observed in T. versutus (Lu and Kelly, 1988). Involvement of energy coupling via chemiosmosis is also indicated for T. neapolitanus using thiosulfate as energy source. The evidence

for this is (1) inhibition of CO_2 uptake by the uncouplers carbonyl cyanide m-chlorophenyl-hydrazone (CCCP) and carbonylcyanide p-trifluoromethoxy-phenylhydrazone (FCCP) and (2) an increase in transmembrane electrochemical potential and CO_2 uptake in response to nigericin (Holthuijzen et al., 1987).

20.6.1 Autotrophs

20.6.1.1 Chemosynthetic Autotrophs

Reduced sulfur is not only an energy source but also a source of reducing power for chemosynthetic autotrophs that oxidize it. Because the midpoint potential for pyridine nucleotides (e.g., NAD(P)H) is lower than that for reduced sulfur compounds that could serve as potential electron donors, reverse electron transport from the electron-donating sulfur substrate to pyridine nucleotide is required (see Chapter 7). Electrons must travel up the electron transport chain, that is, against the redox gradient, to NADP with consumption of ATP providing the needed energy. This applies to both aerobes and anaerobes that use nitrate as terminal electron acceptor (denitrifiers).

Insofar as studied, thiobacilli (domain Bacteria) generally fix CO_2 by the Calvin–Benson–Bassham cycle (see Chapter 7), that is, by means of ribulose 1.5-bisphosphate carboxylase. In at least some Thiobacilli, this enzyme is detected in both the cytosol and the cytoplasmic polyhedral bodies called carboxysomes (Shively et al., 1973). The carboxysomes may represent a means of regulating the level of carboxylase activity in the cytosol (Beudecker et al., 1980, 1981; Holthuijzen et al., 1986a,b). Sulfolobus (domain Archaea) assimilates CO_2 via a reverse, that is, a reductive tricarboxylic acid cycle (see Brock and Madigan, 1988), like green sulfur bacteria (domain Bacteria) (see Chapter 7).

20.6.1.1.1 PHOTOSYNTHETIC AUTOTROPHS

In purple sulfur and nonsulfur bacteria, reverse electron transport, a light-independent sequence, is used to generate reduced pyridine nucleotide (NADPH) using ATP from photophosphorylation to provide the needed energy. In green sulfur bacteria as well as Cyanobacteria, photochemical electron transport is used to generate NADPH (Stanier et al., 1986) (see discussion in Chapter 7).

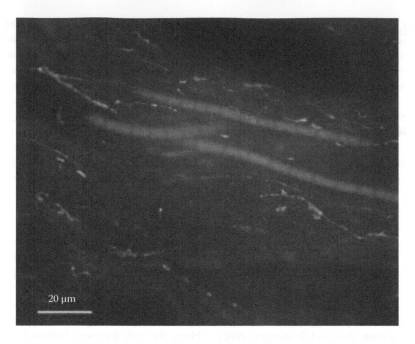

Figure 20.5. **(See color insert.)** CARD-FISH image of the upper portion (ca. 1.5 mm beneath the surface) of a microbial mat from Guerrero Negro, Baja California Sur, Mexico. This highlights the close physical association of cyanobacteria (large filamentous cells in red, chlorophyll autofluorescence) and a group of sulfate-reducing bacteria (thin filamentous cells in green, DSS 658 probe), which morphologically resemble the genus *Desulfonema*. Blue indicates the DNA stain DAPI. (Reprinted from *Geochim et Cosmochim Acta*, 73, Fike, DA, Finke, N, Zha, J, Blake, G, Hoehler, TM, and Orphan, VJ, The effect of sulfate concentration on (sub)millimeter-scale sulfide $\delta^{34}S$ in hypersaline cyanobacterial mats over the diel cycle, 6187–6204, Copyright 2009, with permission from Elsevier.)

body cavity called a trophosome. These organelles when viewed in section under a transmission electron microscope contain tightly packed bacteria (Cavanaugh et al., 1981). Metabolic evidence indicates that these are chemosynthetic, autotrophic bacteria (Felbeck, 1981; Felbeck et al., 1981; Rau, 1981; Williams et al., 1988). The bacteria in the trophosomes appear to be autotrophic sulfur-oxidizing bacteria that share the carbon they fix with the worm. The worm absorbs sulfide (HS) and oxygen from the water through a special organ at its anterior end consisting of a tentacular plume attached to a central supporting obturaculum (Jones, 1981; Goffredi et al., 1997) and transmits these via its circulatory system to the trophosome. The blood of the worm contains hemoglobin for reversible binding of oxygen and another special protein for reversible binding of sulfide. The latter protein prevents reaction of sulfide with the hemoglobin and its consequent destruction (Arp and Childress, 1983; Powell and Somero, 1983). The bound hydrogen sulfide and oxygen are released at the site of the trophosome.

Somewhat less intimate consortia around hydrothermal vents are formed by giant clams and mussels (*Mollusca*) with autotrophic sulfide-oxidizing bacteria. The bacteria in these instances reside not in the gut of the animals but on their gills (see Jannasch and Taylor, 1984, for discussion; also Rau and Hedges, 1979). These looser consortia involving autotrophic sulfide-oxidizing bacteria and mollusks are not restricted to hydrothermal vent communities but also occur in shallow water environments rich in hydrogen sulfide (Cavanaugh, 1983). Another unique symbiosis is found in the scaly foot snail (Waren et al., 2003), which has a symbiotic relationship with gammaproteobacteria within its esophageal gland and a nutritional dependence on chemoautotrophic inputs (Goffredi et al., 2004). Most notably, this organism has a set of mineralized scales (composed of a mix of pyrite and greigite) on its foot. These scales house abundant epsilon- and deltaproteobacteria that may have a role in the precipitation of these scales (Goffredi et al., 2004). Long, filamentous threads of sulfur cycling symbionts are also known to colonize the outer surfaces of the yeti crab (*Kiwa hirsuta*), giving the creature its name

(Goffredi et al., 2008). These organisms are found around deep-sea hydrothermal systems, where sulfide oxidation by microbial symbionts helps detoxify the environment for the crabs (Goffredi et al., 2008).

20.9 EVOLUTION OF SULFUR CYCLING OVER EARTH HISTORY

Microbial sulfur cycling leaves diagnostic chemical fingerprints in the environment, particularly in the stable isotopic composition of metabolites (Szabo et al., 1950; Holland, 1973; Canfield and Teske, 1996; Habicht et al., 1998; Canfield, 2001a,b; Habicht et al., 2002; Farquhar et al., 2003; Johnston et al., 2005a,b; Canfield et al., 2010). The record of this metabolic activity can be preserved in sedimentary rocks up to billions of years old. Of the various S-bearing compounds utilized by microbes, both sulfate salts (e.g., gypsum) and sulfide (particularly pyrite [FeS_2]) are common in the rock record. Building predominantly on the marine sedimentary record, the long-term evolution of sulfur cycling can be reconstructed (see in particular Canfield, 2001b). The patterns have been used to argue for the antiquity of MSR (Shen et al., 2001) and sulfur disproportionation (Canfield and Teske, 1996; Johnston et al., 2005a; Philippot et al., 2007), as well as to reconstruct the redox conditions of the ocean and atmosphere (Farquhar et al., 2000; Farquhar and Wing, 2003; Kah et al., 2004; Kampschulte and Strauss, 2004; Hurtgen et al., 2005; Fike et al., 2006; Riccardi et al., 2006; Fike and Grotzinger, 2008).

20.10 SUMMARY

Sulfur cycling plays a critical role in enabling and regulating a diverse suite of microbial metabolic pathways and has had a profound influence on the evolution of Earth's surface environment. Sulfur, which occurs in myriad organic and inorganic forms in nature, is essential to life and is taken up through assimilatory processes. In addition, myriad microbes make use of dissimilatory sulfur transformations (most notably sulfate reduction) in which oxidation or reduction of sulfur-bearing compounds is coupled to that of another compound in an energy-yielding reaction to drive biochemical processes. These dissimilatory processes are responsible for the overwhelming majority of sulfur transformations in geomicrobiological processes. In these reactions, oxidized forms of sulfur, especially sulfate, but also elemental sulfur and thiosulfate, serve as terminal electron acceptors. Reduced forms of sulfur such as hydrogen sulfide and elemental sulfur can serve as sources of electrons to generate energy and for reducing power. Diverse microorganisms (both Bacteria and Archaea) spanning chemolithoautotrophs, anoxygenic and oxygenic (cyanobacteria) photolithotrophs, mixotrophs, and heterotrophs together play important roles in global sulfur cycling (Figure 20.1). These processes leave behind diagnostic geochemical fingerprints (particularly in the isotopic fractionation between the stable isotopes of sulfur (^{32}S, ^{33}S, ^{34}S, and ^{36}S) that can be used to track the evolution and ecological impact of these sulfur cycling metabolisms over Earth history.

REFERENCES

Abdollahi H, Wimpenny JWT. 1990. Effects of oxygen on the growth of *Desulfovibrio desulfuricans*. *J Gen Microbiol*, 136:1025–1030.

Adams CA, Warnes GM, Nicholas DJD. 1971. A sulfite-dependent nitrate reductase from *Thiobacillus denitrificans*. *Biochim Biophys Acta*, 235:398–406.

Aeckersberg F, Bak F, Widdel F. 1991. Anaerobic oxidation of saturated hydrocarbons to CO_2 by a new type of sulfate-reducing bacterium. *Arch Microbiol*, 156:5–14.

Aleem MIH. 1965. Thiosulfate oxidation and electron transport in *Thiobacillus novellas*. *J Bacteriol*, 90:95–101.

Aminuddin M. 1980. Substrate level versus oxidative phosphorylation in the generation of ATP in *Thiobacillus denitrificans*. *Arch Microbiol*, 128:19–25.

Aminuddin M, Nicholas DJD. 1973. Sulfide oxidation linked to the reduction of nitrate to nitrite in *Thiobacillus denitrificans*. *Biochim Biophys Acta*, 325:81–93.

Aminuddin M, Nicholas DJD. 1974a. An AMP-independent sulfite oxidase from *Thiobacillus denitrificans*. *J Gen Microbiol*, 82:103–113.

Aminuddin M, Nicholas DJD. 1974b. Electron transfer during sulfide to sulfite oxidation in *Thiobacillus denitrificans*. *J Gen Microbiol*, 82:115–123.

Anantharaman K, Duhaime MB, Breier JA, Wendt KA, Toner BM, Dick GJ. 2014. Sulfur oxidation genes in diverse deep-sea viruses. *Science*, 344(6185):757–760.

Arp AJ, Childress JJ. 1983. Sulfide binding by the blood of the hydrothermal vent tube worm *Riftia pachyptila*. *Science*, 219:295–297.

Baalsrud K, Baalsrud KS. 1954. Studies on *Thiobacillus denitrificans*. *Arch Mikrobiol*, 20:34–62.

Cypionka H. 2000. Oxygen respiration by *Desulfovibrio* species. *Annu Rev Microbiol*, 54:827–848.

Cypionka H, Pfennig N. 1986. Growth yields of *Desulfotomaculum orientis* with hydrogen in chemostat culture. *Arch Microbiol*, 143(16):396–399.

Daalsgaard T, Bak F. 1992. Effect of acetylene on nitrous oxide reduction and sulfide oxidation in batch and gradient cultures of *Thiobacillus denitrificans*. *Appl Environ Microbiol*, 58:1601–1608.

Dahl C, Truper HG. 1994. Enzymes of dissimilatory sulfide oxidation in phototrophic sulfur bacteria. *Methods Enzymol*, 243:400–421.

Davis EA, Johnson EJ. 1967. Phosphorylation coupled to the oxidation of sulfide and 2-mercaptoethanol in extracts of *Thiobacillus thioparus*. *Can J Microbiol*, 13:873–884.

Dekas AE, Poretsky RS, Orphan VJ. 2009. Deep-sea Archaea Fix and share nitrogen in methane-consuming microbial Consortia. *Science*, 326(5951):422–426.

Dilling W, Cypionka H. 1990. Aerobic respiration in sulfate reducing bacteria. *FEMS Microbiol Lett*, 71:123–128.

Douglas S, Douglas DD. 2000. Environmental scanning electron microscopy studies of colloidal sulfur deposition in a natural microbial community from a cold sulfide spring near Ancaster, Ontario, Canada. *Geomicrobiol J*, 17:275–289.

Dworkin M, editor-in-chief. 2001. *The Prokaryotes*. Electronic version. New York: Springer.

Easterbrook KB, Coombs RW. 1976. Spinin: The subunit protein of bacterial spinae. *Can J Microbiol*, 23:438–440.

Egorova AA, Deryugina ZP. 1963. The spore forming thermophilic thiobacterium: *Thiobacillus thermophilica* Imschenetskii nov. spec. *Mikrobiologiya*, 32:439–446.

Ehrlich HL. 1999. Microbes as geologic agents: Their role in mineral formation. *Geomicrobiol J*, 16:135–153.

Ehrlich HL, Fox SI. 1967. Copper sulfide precipitation by yeast from acid mine-waters. *Appl Microbiol*, 15:135–139.

Elsgaard L, Jørgensen BB. 1992. Anoxic transformations of radiolabeled hydrogen sulfide in marine and freshwater sediments. *Geochim Cosmochim Acta*, 56:2425–2435.

Fagerbakke KM, Heldal M, Norland S. 1996. Content of carbon, nitrogen, oxygen, sulfur and phosphorus in native aquatic and cultured bacteria. *Aquat Microbial Ecol*, 10:15–27.

Fareleira P. 2003. Response of a strict anaerobe to oxygen: Survival strategies in *Desulfovibrio gigas*. *Microbiology*, 149:1513–1521.

Fareleira P, Santos BS, António C, Moradas-Ferreira P, LeGall J, Xavier AV, Santos H. 2003. Response of a strict anaerobe to oxygen: survival strategies in *Desulfovibrio gigas*. *Microbiology*, 49:1513–1522.

Farquhar J, Bao H, Thiemens MH. 2000. Atmospheric influence of Earth's earliest sulfur cycle. *Science*, 289:756–758.

Farquhar J, Johnston DT, Wing BA, Habicht KS, Canfield DE, Airieau S, Thiemens MH. 2003. Multiple sulphur isotopic interpretations of biosynthetic pathways: Implications for biological signatures in the sulphur isotope record. *Geobiology*, 1(1):27–36.

Farquhar J, Wing B. 2003. Multiple sulfur isotopes and the evolution of the atmosphere. *Earth Planetary Sci Lett*, 213:1–13.

Fauque G, LeGall J, Barton LL. 1991. Sulfate-reducing and sulfur-reducing bacteria. In: Shively JM, Barton LL, eds. *Variations in Autotrophic Life*. London, U.K.: Academic Press, pp. 271–337.

Felbeck H. 1981. Chemoautotrophic potential of the hydrothermal vent tube worm, *Riftia pachyptila* Jones (Vestimentifera). *Science*, 213:336–338.

Felbeck H, Childress JJ, Solmero GN. 1981. Calvin–Benson cycle and sulfide oxidation enzymes in animals from sulfide-rich habitats. *Nature* (Lond), 293:291–293.

Fike DA, Finke N, Zha J, Blake G, Hoehler TM, Orphan VJ. 2009. The effect of sulfate concentration on (sub)millimeter-scale sulfide $\delta^{34}S$ in hypersaline cyanobacterial mats over the diel cycle. *Geochim et Cosmochim Acta*, 73:6187–6204.

Fike DA, Gammon CL, Ziebis W, Orphan VJ. 2008. Micron-scale mapping of sulfur cycling across the oxycline of a cyanobacterial mat: A paired nano-SIMS and CARD-FISH approach. *ISME J*, 2:749–759.

Fike DA, Grotzinger JP. 2008. A paired sulfate-pyrite $\delta^{34}S$ approach to understanding the evolution of the Ediacaran-Cambrian sulfur cycle. *Geochim et Cosmochim Acta*, 72(11):2636–2648.

Fike DA, Grotzinger JP, Pratt LM, Summons RE. 2006. Oxidation of the Ediacaran Ocean. *Nature*, 444:744–747.

Flynn TM, O'Loughlin EJ, Mishra B, DiChristina TJ, Kemner KM. 2014. Sulfur-mediated electron shuttling during bacterial iron reduction. *Science*, 344(6187):1039–1042.

Freney JR. 1967. Sulfur-containing organics. In: McLaren AD, Petersen GH, eds. *Soil Biochemistry*. New York: Marcel Dekker, pp. 229–259.

Friedrich CG, Rother D, Bardischewsky F, Quentmeier A, Fischer J. 2001. Oxidation of reduced inorganic sulfur compounds by bacteria. Emergence of a common mechanism? *Appl Environ Microbiol*, 67:2873–2882.

Fründ C, Cohen Y. 1992. Diurnal cycles of sulfate reduction under oxic conditions in cyanobacterial mats. *Appl Environ Microbiol*, 58:70–77.

Fukui M, Teske A, Assmus F, Muyzer G, Widdel F. 1999. Physiology, phylogenetic relationships, and ecology of filamentous sulfate-reducing bacteria (genus *Desulfonema*). *Arch Microbiol*, 172:193–203.

Garlick S, Oren A, Padan E. 1977. Occurrence of facultative anoxygenic photosynthesis among filamentous and unicellular cyanobacteria. *J Bacteriol*, 129:623–629.

Gebhardt NA, Linder D, Thauer RK. 1983. Anaerobic oxidation of CO_2 by *Desulfobacter postgatei*. 2. Evidence from ^{14}C-labelling studies for the operation of the citric acid cycle. *Arch Microbiol*, 136:230–233.

Gebhardt NA, Thauer RK, Linder D, Kaulfers P-M, Pfennig N. 1985. Mechanism of acetate oxidation to CO_2 with elemental sulfur in *Desulfuromonas acetoxidans*. *Arch Microbiol*, 141:392–398.

Gevertz D, Telang AJ, Voordrouw G, Jenneman GE. 2000. Isolation and characterization of strains CVO and FWKO B, two novel nitrate-reducing, sulfide-oxidizing bacteria isolated from oil field brine. *Appl Environ Microbiol*, 66:2491–2501.

Goffredi SK, Childress JJ, Desaulniers NT, Lallier FH. 1997. Sulfide acquisition by the vent worm *Riftia pachyptila* appears to be via uptake of HS^-, rather than H_2S. *J Exp Biol*, 200 (pt 20):2609–2616.

Goffredi SK, Jones WJ, Ehrlich H, Springer A, Vrijenhoek RC. 2008. Epibiotic bacteria associated with the recently discovered Yeti crab, *Kiwa hirsuta*. *Environ Microbiol*, 10(10):2623–2634.

Goffredi SK, Warén A, Orphan VJ, Van Dover CL, Vrijenhoek RC. 2004. Novel forms of structural integration between microbes and a hydrothermal vent gastropod from the Indian Ocean. *Appl Environ Microbiol*, 70(5):3082–3090.

Golovacheva RS, Karavaiko GI. 1978. *Sulfobacillus*, a new genus of thermophilic sporeforming bacteria. *Mikrobiologiya*, 47:815–822 (Engl transl, pp. 658–665).

Grayston SJ, Wainwright M. 1988. Sulphur oxidation by soil fungi including some species of mycorrhizae and wood-rotting basidiomycetes. *FEMS Microbiol Lett*, 53:1–8.

Guittonneau G. 1927. Sur l'oxidation microbienne du soufre au cours de l'ammonisation. *CR Acad Sci (Paris)*, 184:45–46.

Guittonneau G, Keiling J. 1927. Sur la solubilisation du soufre élémentaire et la formation des hyposulfides dans une terre riche en azote organique. *CR Acad Sci (Paris)*, 184:898–901.

Habicht KS, Canfield DE, Rethmeier J. 1998. Sulfur isotope fractionation during bacterial reduction and disproportionation of thiosulfate and sulfite. *Geochim Et Cosmochim Acta*, 62(15):2585–2595.

Habicht KS, Gade M, Thamdrup B, Berg P, Canfield DE. 2002. Calibration of sulfate levels in the Archean Ocean. *Science*, 298(5602):2372–2374.

Hallberg KB, Dopson M, Lindström EB. 1996. Reduced sulfur compound oxidation by *Thiobacillus caldus*. *J Bacteriol*, 178:6–11.

Hallberg KB, Lindström EB. 1994. Characterization of *Thiobacillus caldus*, sp. nov., a moderately thermophilic acidophile. *Microbiology (Reading)*, 140:3451–3456.

Hansen TA, van Gemerden H. 1972. Sulfide utilization by purple sulfur bacteria. *Arch Microbiol*, 86:49–56.

Hansen TA, Veldkamp H. 1973. *Rhodopseudomonas sulfidophila* nov. spec., a new species of the purple nonsulfur bacteria. *Arch Microbiol*, 92:45–58.

Harrison AG, Thode H. 1958. Mechanism of the bacterial reduction of sulfate from isotope fractionation studies. *Trans Faraday Soc*, 54:84–92.

Harrison G, Curle C, Laishley EJ. 1984. Purification and characterization of an inducible dissimilatory type of sulfite reductase from *Clostridium pasteurianum*. *Arch Microbiol*, 138:172–178.

Harrison GI, Laishley EJ, Krouse HR. 1980. Stable isotope fractionation by *Clostridium pasteurianum*. 3. Effect of SeO_3^- on the physiology of associated sulfur isotope fractionation during SO_3^{2-} and SO_4^{2-} reduction. *Can J Microbiol*, 26:952–958.

Hayes JM, Waldbauer JR. 2006. The carbon cycle and associated redox processes through time. *Philos Trans R Soc B Biol Sci*, 361(1470):931–950.

Holland HD. 1973. Systematics of the isotopic composition of sulfur in the oceans during the Phanerozoic and its implications for atmospheric oxygen. *Geochim Cosmochim Acta*, 37:2605–2616.

Holmkvist L, Ferdelman TG, Jørgensen BB. 2011. A cryptic sulfur cycle driven by iron in the methane zone of marine sediment (Aarhus Bay, Denmark). *Geochim Et Cosmochim Acta*, 75(12):3581–3599.

Holser WT. 1997. Catastrophic chemical events in the history of the ocean. *Nature*, 267:403–408.

Holt JG, ed. 1984. *Bergey's Manual of Systematic Bacteriology*, Vol. 1. Baltimore, MD: Wiliams & Wilkins.

Holthuijzen YA, van Breemen JFL, Konings WN, van Bruggen EFJ. 1986a. Electron microscopic studies of carboxysomes of *Thiobacillus neapolitanus*. *Arch Microbiol*, 144:258–262.

Holthuijzen YA, van Breemen JFL, Kuenen JG, Konings WN. 1986b. Protein composition of the carboxysomes of *Thiobacillus neapolitanus*. *Arch Microbiol*, 144:398–404.

Holthuijzen YA, Van Dissel-Emiliani FFM, Kuenen JG, Konings WN. 1987. Energetic aspects of CO$_2$ uptake in Thiobacillus neapolitanus. Arch Microbiol, 147:285–290.

Huber R, Kristjansson JK, Stetter KO. 1987. Pyrobaculum gen. nov., a new genus of neutrophilic, rod-shaped archaebacteria from continental solfataras growing optimally at 100°C. Arch Microbiol, 149:95–101.

Hügler M, Wirsen CO, Fuchs G, Taylor CD, Sievert SM. 2005. Evidence for autotrophic CO$_2$ fixation via the reductive tricarboxylic acid cycle by members of the ε subdivision of Proteobacteria. J Bacteriol, 187:3020–3027.

Hurtgen MT, Arthur MA, Halverson GP. 2005. Neoproterozoic sulfur isotopes, the evolution of microbial sulfur species, and the burial efficiency of sulfide as sedimentary pyrite. Geology, 33(1):41–44.

Imhoff-Stuckle D, Pfennig N. 1983. Isolation and characterization of a nicotinic acid-degrading sulfate-reducing bacterium, Desulfococcus niacini sp. nov. Arch Microbiol, 136:194–198.

Isaksen MF, Teske A. 1996. Desulforhopalus vacuolatus gen. nov., spec. nov., a new moderately psychrophilic sulfate-reducing bacterium with gas vacuoles isolated from a temperate estuary. Arch Microbiol, 166:160–168.

Itoh T, Suzuki KI, Sanchez PC, Nakase T. 1999. Caldivirga maquilingensis gen. nov., sp. nov., a new genus of rod-shaped crenarchaeote isolated from a hot spring in the Philippines. Int J Syst Bacteriol, 49(3):1157–1163.

Ivanov MV. 1968. Microbiological Processes in the Formation of Sulfur Deposits. Israel Program for Scientific Translations. Washington, DC: US Department of Agriculture and National Science Foundation.

Jackson BE, McInerney MJ. 2000. Thiosulfate disproportionation by Desulfotomaculum thermobenzoicum. Appl Environ Microbiol, 66:3650–3653.

Jannasch HW. 1984. Microbial processes at deep-sea hydrothermal vents. In: Rona PA, Bostrom K, Laubier L, Smith KL Jr, eds. Hydrothermal Processes at Sea Floor Spreading Centers. New York: Plenum Press, pp. 677–709.

Jannasch HW, Huber R, Belkin S, Stetter KO. 1988b. Thermotoga neapolitana sp. nov. of the extremely thermophilic, eubacterial genus Thermotoga. Arch Microbiol, 150:103–104.

Jannasch HW, Mottl MJ. 1985. Geomicrobiology of deep-sea hydrothermal vents. Science, 229:717–725.

Jannasch HW, Nelson DC, Wirsen CO. 1989. Massive natural occurrence of unusually large bacteria (Beggiatoa sp.) at a hydrothermal deep-sea vent site. Nature (Lond), 342:834–836.

Jannasch HW, Taylor CD. 1984. Deep-sea microbiology. Annu Rev Microbiol, 38:487–514.

Jannasch HW, Wirsen CO, Molyneaux SJ, Langworthy TA. 1988a. Extremely thermophilic fermentative archaebacteria of the genus Desulfurococcus from deep-sea hydrothermal vents. Appl Environ Microbiol, 54:1203–1209.

Jansen K, Thauer RK, Widdel F, Fuchs G. 1984. Carbon assimilation pathways in sulfate reducing bacteria. Formate, carbon dioxide, carbon monoxide, and acetate assimilation by Desulfovibrio baarsii. Arch Microbiol, 138:257–262.

Janssen PH, Morgan HW. 1992. Heterotrophic sulfur reduction by Thermotoga sp. strain FjSS3B1. FEMS Microbiol Lett, 96:213–218.

Janssen PH, Schink B. 1995. Metabolic pathways and energetics of the acetone-oxidizing sulfate-reducing bacterium, Desulfobacterium cetonicum. Arch Microbiol, 163:188–194.

Janssen PH, Schuhmann A, Bak F, Liesack W. 1996. Disproportionation of inorganic sulfur compounds by the sulfate-reducing bacterium Desulfocapsa thiozymogenes gen. nov., spec. nov. Arch Microbiol, 166:184–192.

Johnston DT, Farquhar J, Wing BA, Kaufman AJ, Canfield DE, Habicht KS. 2005a. Multiple sulfur isotope fractionations in biological systems: A case study with sulfate reducers and sulfur disproportionators. Am J Sci, 305:645–660.

Johnston DT, Wing BA, Farquhar J, Kaufman AJ, Strauss H, Lyons TW, Kah LC. Canfield, DE. 2005b. Active microbial sulfur disproportionation in the Mesoproterozoic. Science, 310(5753):1477–1479.

Jones ML. 1981. Riftia pachyptila Jones: Observations on the vestimentiferan worm from the Galápagos Rift. Science, 213:333–336.

Jørgensen BB. 1990a. A thiosulfate shunt in the sulfur cycle of marine sediments. Science, 249:152–154.

Jørgensen BB. 1990b. The sulfur cycle of freshwater sediments: Role of thiosulfate. Limnol Oceanogr, 35:1329–1342.

Jørgensen BB, Bak F. 1991. Pathways of microbiology of thiosulfate transformations and sulfate reduction in marine sediment (Kattegat, Denmark). Appl Environ Microbiol, 57:847–856.

Jørgensen BB, Isaksen MF, Jannasch HW. 1992. Bacterial sulfate reduction above 100°C in deep-sea hydrothermal vent sediments. Science, 258:1756–1757.

Justin P, Kelly DP. 1978. Growth kinetics of Thiobacillus denitrificans in anaerobic and aerobic chemostat culture. J Gen Microbiol, 107:123–300.

Kah LC, Lyons TW, Frank TD. 2004. Low marine sulphate and protracted oxygenation of the proterozoic biosphere. *Nature*, 431(7010):834–838.

Kampschulte A, Strauss H. 2004. The sulfur isotopic evolution of Phanerozoic seawater based on the analysis of structurally substituted sulfate in carbonates. *Chem Geol*, 20:255–286.

Karavaiko GI, Bogdanova TI, Tourova TP, Kondrat'eva TF, Tsalpina IA, Egorova MA, Karsil'nikova EN, Zakharchuk LM. 2005. Reclassification of 'Sulfobacillus thermosulfidooxidans subsp. thermotolerans' strain K1 as *Alicyclobacillus tolerans* sp. nov. and *Sulfobacillus disulfidooxidans* Dufresne et al. 1996 as *Alicyclobacillus disulfidooxidans* comb. nov., and emended description of the genus *Alicyclobacillus*. *Int J Syst Evol Microbiol*, 55:941–947.

Kelly DP. 1982. Biochemistry of the chemolithotrophic oxidation of inorganic sulfur. *Phil Trans R Soc Lond B*, 298:499–528.

Kelly DP, Lu W-P, Poole PK. 1993. Cytochromes in *Thiobacillus tepidarius* and the respiratory chain involved in the oxidation of thiosulfate and tetrathionate. *Arch Microbiol*, 160:87–95.

Kelly DP, Wood AP. 2000. Reclassification of some species of *Thiobacillus* to the newly designated genera *Acidithiobacillus* gen. nov., *Halobacillus* gen. nov. and *Thermithiobacillus* gen. nov. *Int J Syst Evol Microbiol*, 50:511–516.

Kirk MF, Holm TR, Park J, Jin Q, Sanford RA, Fouke BW, Bethke CM. 2004. Bacterial sulfate reduction limits natural arsenic contamination in groundwater. *Geology*, 32(11):953–956.

Knittel K, Boetius A. 2009. Anaerobic oxidation of methane: Progress with an unknown process. *Ann Rev Microbiol*, 63(1):311–334.

Kobayashi K, Tachibana S, Ishimoto M. 1969. Intermediary formation of trithionate in sulfite reduction by a sulfate-reducing bacterium. *J Biochem (Tokyo)*, 65:155–157.

Kobiyashi KS, Tashibana S, Ishimoto M. 1969. Intermediary formation of trithionate in sulfite reduction by a sulfate-reducing bacterium. *J Biochem (Tokyo)*, 65:155–157.

Kramer JF, Pope DH, Salerno JC. 1987. Pathways of electron transfer in *Desulfovibrio*. In: Kim CH, Tedeschi H, Diwan JJ, Salerno JC, eds. *Advances in Membrane Biochemistry and Bioenergetics*. New York: Plenum Press, pp. 249–258.

Krekeler D, Sigalevich P, Teske A, Cypionka H, Cohen Y. 1997. A sulfate-reducing bacterium from the oxic layer of a microbial mat from Solar Lake (Sinai), *Desulfovibrio oxyclinae* sp. nov. *Arch Microbiol*, 167:369–375.

Kuenen JG, Beudecker RF. 1982. Microbiology of thiobacilli and other sulfur-oxidizing autotrophs, mixotrophs, and heterotrophs. *Phil Trans R Soc Lond B*, 298:473–497.

Kuenen JG, Tuovinen OH. 1981. The genera *Thiobacillus* and *Thiomicrospira*. In: Starr MP, Stolp H, Trüper HG, Balows A, Schlegel H, eds. *The Prokaryotes: A Handbook of Habitats, Isolation and Identification of Bacteria*. Berlin, Germany: Springer, pp. 1023–1036.

Kuever J, Kulmer J, Jannsen S, Fischer U, Blotevogel K-H. 1993. Isolation and characterization of a new sporeforming sulfate-reducing bacterium growing by complete oxidation of catechol. *Arch Microbiol*, 159:282–288.

LaRiviere JWM, Schmidt K. 1981. Morphologically conspicuous sulfur-oxidizing bacteria. In: Starr MP, Stolp H, Trüper HG, Balows A, Schlegel H, eds. *The Prokaryotes: A Handbook of Habitats, Isolation and Identification of Bacteria*. Berlin, Germany: Springer, pp. 1037–1048.

Lee J-H, Kennedy DW, Dohnalkova A, Moore DA, Nachimuthu P, Reed SB, Fredrickson JK. 2011. Manganese sulfide formation via concomitant microbial manganese oxide and thiosulfate reduction. *Environ Microbiol*, 13(12):3275–3288.

LeRoux N, Wakerley DS, Hunt SD. 1977. Thermophilic thiobacillus-type bacteria from Icelandic thermal areas. *J Gen Microbiol*, 100:197–201.

Liang R, Grizzle RS, Duncan KE, McInerney MJ, Suflita JM. 2014. Roles of thermophilic thiosulfate-reducing bacteria and methanogenic archaea in the biocorrosion of oil pipelines. *Front Microbiol*, 5:89.

London J. 1963. *Thiobacillus intermedius* nov. sp. a novel type of facultative autotroph. *Arch Microbiol*, 46:329–337.

London J, Rittenberg SC. 1964. Path of sulfur in sulfide and thiosulfate oxidation by thiobacilli. *Proc Natl Acad Sci USA*, 52:1183–1190.

London J, Rittenberg SC. 1966. Effects of organic matter on the growth of *Thiobacillus intermedius*. *J Bacteriol*, 91:1062–1069.

London J, Rittenberg SC. 1967. *Thiobacillus perometabolis* nov. sp., a non-autotrophic thiobacillus. *Arch Microbiol*, 59:218–225.

Londry KL, Suflita JM, Tanner RS. 1999. Cresol metabolism by the sulfate-reducing bacterium *Desulfotomaculum* sp. strain Groll. *Can J Microbiol*, 45:458–463.

Loya S, Yanofsky SA, Epel BL. 1982. Characterization of cytochromes in lithotrophically and organotrophically grown cells of *Thiobacillus* A_2. *J Gen Microbiol*, 128:2371–2378.

Lu W-P. 1986. A periplasmic location for the bisulfite-oxidizing multienzyme system from *Thiobacillus versutus*. FEMS Microbiol Lett, 34:313–317.

Lu W-P, Kelly DP. 1983. Purification and some properties of two principal enzymes of the thiosulfate-oxidizing multienzyme system from *Thiobacillus A₂*. J Gen Microbiol, 129:3549–3562.

Lu W-P, Kelly DP. 1988. Respiration-driven proton translocation in *Thiobacillus versutus* and the role of the periplasmic thiosulfate-oxidizing enzyme system. Arch Microbiol, 149:297–302.

Lui S, Soriano A, Cowan J. 1993. Enzymatic reduction of inorganic anions. Pre-steady-state linetic analysis of the dissimilatory sulfite reductase (Desulfoviridin) from *Desulfovibrio vulgaris* (Hildenborough). Mechanistic Implications. J Am Chem Soc, 115(23):10483–10486.

Ma K, Weiss R, Adams MWW. 2000. Characterization of hydrogenase II from the hyperthermophilic archeon *Pyrococcus furiosus* and assessment of its role in sulfur reduction. J Bacteriol, 182:1864–1871.

Markosyan GE. 1973. A new mixotrophic sulfur bacterium developing in acidic media, *Thiobacillus organoparus* sp. n. Dokl Akad Nauk SSSR Ser Biol, 211:1205–1208.

Marshall C, Frenzel P, Cypionka H. 1993. Influence of oxygen on sulfur reduction and growth of sulfate-reducing bacteria. Arch Microbiol, 159:168–173.

Mason J, Kelly DP. 1988. Mixotrophic and autotrophic growth of *Thiobacillus acidophilus* on tetrathionate. Arch Microbiol, 149:317–323.

Mechalas BJ, Rittenberg SC. 1960. Energy coupling in *Desulfovibrio desulfuricans*. J Bacteriol, 80:501–507.

Meckenstock RU, Annweiler E, Michaelis W, Richnow HH, Schink B. 2000. Anaerobic naphthalene degradation by a sulfate-reducing enrichment culture. Appl Environ Microbiol, 66:2743–2747.

Meulenberg R, Pronk JT, Hazeu W, van Dijken JP, Frank J, Bos P, Kuenen JG. 1993. Purification and partial characterization of thiosulfate dehydrogenase from *Thiobacillus acidophilus*. J Gen Microbiol, 139:2033–2039.

Milhaud G, Aubert JP, Millet J. 1958. Role physiologique du cytochrome C de la bactérie chemieautotrophe *Thiobacillus denitrificans*. CR Acad Sci (Paris), 246:1766–1769.

Miller JDA, Wakerley DS. 1966. Growth of sulfate-reducing bacteria by fumarate dismutation. J Gen Microbiol, 43:101–107.

Milucka J, Ferdelman TG, Polerecky L, Franzke D, Wegener G, Schmid M, Lieberwirth I, Wagner M, Widdel F, Kuypers MMM. 2012. Zero-valent sulphur is a key intermediate in marine methane oxidation. Nature, 491:541–546.

Minz D, Flax JL, Green SJ, Muyzer G, Cohen Y, Wagner M, Rittmann BE, Stahl DA. 1999. Diversity of sulfate-reducing bacteria in oxic and anoxic regions of a microbial mat characterized by comparative analysis of dissimilatory sulfite reductase genes. Appl Environ Microbiol, 65(10):4666–4671.

Moran JJ, Beal EJ, Vrentas JM, Orphan VJ, Freeman KH, House CH. 2007. Methyl sulfides as intermediates in the anaerobic oxidation of methane. Environ Microbiol, 10(1):162–173.

Moriarty DJW, Nicholas DJD. 1970. Electron transfer during sulfide and sulfite oxidation by *Thiobacillus concretivorus*. Biochim Biophys Acta, 216:130–138.

Nathansohn A. 1902. Über eine neue Gruppe von Schwefelbakterien und ihren Stoffwechsel. Mitt Zool Sta Neapel, 15:655–680.

Nelson DC, Castenholz RW. 1981. Use of reduced sulfur compounds by *Beggiatoa* sp. J Bacteriol, 147:140–154.

Nelson DC, Jannasch HW. 1983. Chemoautotrophic growth of a marine *Beggiatoa* in sulfide-gradient cultures. Arch Microbiol, 136:262–269.

Neumann S, Wynen A, Trüper H, Dahl C. 2000. Characterization of the cys gene locus from *Allochromatium vinosum* indicates an unusual sulfate assimilation pathway. Mol Biol Rep, 27(1):27–33.

Neutzling O, Pfleiderer C, Trüper HG. 1985. Dissimilatory sulfur metabolism in phototrophic "non-sulfur" bacteria. J Gen Microbiol, 131:791–798.

Nübel T, Klughammer C, Huber R, Hauska G, Schütz M. 2000. Sulfide: Quinone oxidoreductase in membranes of the hyperthermophilic bacterium *Aquifex aeolicus* (VF5). Arch Microbiol, 173:233–244.

Odom JM, Peck HD Jr. 1981. Hydrogen cycling as a general mechanism for energy coupling in the sulfate-reducing bacteria, *Desulfovibrio* sp. FEMS Microbiol Lett, 12:47–50.

Odom JM, Wall JD. 1987. Properties of a hydrogen-inhibited mutant of *Desulfovibrio desulfuricans* ATCC 27774. J Bacteriol, 169:1335–1337.

Oliveira TF, Vonrhein C, Matias PM, Venceslau SS, Pereira IAC, Archer M. 2008a. The crystal structure of *Desulfovibrio vulgaris* dissimilatory sulfite reductase bound to DsrC provides novel insights into the mechanism of sulfate respiration. J Biol Chem, 283(49):34141–34149.

Oliveira TF, Vonrhein C, Matias PM, Venceslau SS, Pereira IAC, Archer M. 2008b. Purification, crystallization and preliminary crystallographic analysis of a dissimilatory DsrAB sulfite reductase in complex with DsrC. J Struct Biol, 164:236–239.

Oren A, Shilo M. 1979. Anaerobic heterotrophic dark metabolism in the cyanobacterium *Oscillatoria limnetica*: Sulfur respiration and lactate fermentation. *Arch Microbiol*, 122:77–84.

Orphan VJ, House CH, Hinrichs KU, McKeegan KD, DeLong EF. 2001. Methane-consuming archaea revealed by directly coupled isotopic and phylogenetic analysis. *Science*, 293(5529):484–487.

Orphan VJ, House CH, Hinrichs KU, McKeegan KD, DeLong EF. 2002. Multiple archaeal groups mediate methane oxidation in anoxic cold seep sediments. *Proc Natl Acad Sci USA*, 99(11):7663–7668.

Pankhania IP, Sporman AM, Hamilton WA, Thauer RK. 1988. Lactate conversion to acetate, CO_2, and H_2 in cell suspensions of *Desulfovibrio vulgaris* (Marburg): Indications for the involvement of an energy driven reaction. *Arch Microbiol*, 150:26–31.

Parey K, Warkentin E, Kroneck PMH, Ermler U. 2010. Reaction cycle of the dissimilatory sulfite reductase from *Archaeoglobus fulgidus*. *Biochemistry*, 49(41):8912–8921.

Paschinger H, Paschinger J, Gaffron H. 1974. Photochemical disproportionation of sulfur into sulfide and sulfate by *Chlorobium limicola* forma thiosulfatophilum. *Arch Microbiol*, 96:341–351.

Paulsen J, Kröger A, Thauer RK. 1986. ATP-driven succinate oxidation in the catabolism of *Desulfuromonas acetoxidans*. *Arch Microbiol*, 144:78–83.

Peck HD Jr. 1959. The ATP-dependent reduction of sulfate with hydrogen in extracts of *Desulfovibrio desulfuricans*. *Proc Natl Acad Sci USA*, 45(5):701–708.

Peck HD Jr. 1962. Symposium on metabolism of inorganic compounds. V. Comparative metabolism of inorganic sulfur compounds in microorganisms. *Bacteriol Rev*, 26:67–94.

Peck HD Jr. 1993. Bioenergetic strategies of the sulfate-reducing bacteria. In: Odom JM, Singleton Jr, eds. *The Sulfate-Reducing Bacteria: Contemporary Perspectives*. New York: Springer, pp. 41–76.

Peck HD Jr., LeGall J. 1982. Biochemistry of dissimilatory sulfate reduction. *Phil Trans R Soc Lond B*, 298:443–466.

Peeters T, Aleem MIH. 1970. Oxidation of sulfur compounds and electron transport in *Thiobacillus denitrificans*. *Arch Microbiol*, 71:319–330.

Pereira IAC, Ramos AR, Grein F, Marques MC, da Silva SM, Venceslau SS. 2011. A comparative genomic analysis of energy metabolism in sulfate reducing bacteria and archaea. *Front Microbiol*, 2:69.

Perry KA, Kostka JE, Luther GW III, Nealson KH. 1993. Mediation of sulfur speciation by a Black Sea facultative anaerobe. *Science*, 259:801–803.

Pfennig N. 1977. Phototrophic green and purple bacteria: A comparative, systematic survey. *Annu Rev Microbiol*, 31:275–290.

Pfennig N, Biebl H. 1976. *Desulfuromonas acetoxidans* gen nov. and sp. nov., a new anaerobic, sulfur-reducing, acetate-oxidizing bacterium. *Arch Microbiol*, 110:3–12.

Pfennig N, Widdel F, Trüper HG. 1981. The dissimilatory sulfate-reducing bacteria. In: Starr MP, Stolp H, Trüper HG, Balows A, Schlegel H, eds. *The Prokaryotes: A Handbook of Habitats, Isolation and Identification of Bacteria*. Vol 1. Berlin, Germany: Springer, pp. 926–940.

Philippot P, Van Zuilen M, Lepot K, Thomazo C, Farquhar J, Van Kranendonk MJ. 2007. Early Archaean microorganisms preferred elemental sulfur, not sulfate. *Science*, 317(5844):1534–1537.

Piłsyk S, Paszewski A. 2009. Sulfate permeases— Phylogenetic diversity of sulfate transport. *Acta Biochim Pol*, 56(3):375–384.

Platen H, Temmes A, Schink B. 1990. Anaerobic degradation of acetone by *Desulfurococcus biacutus* spec. nov. *Arch Microbiol*, 154:355–361.

Postgate JR. 1952. Growth of sulfate reducing bacteria in sulfate-free media. *Research*, 5:189–190.

Postgate JR. 1963. Sulfate-free growth of Cl. nigrificans. *J Bacteriol*, 85:1450–1451.

Postgate JR. 1984. *The Sulfate-Reducing Bacteria*, 2nd edn. Cambridge, U.K.: Cambridge University Press.

Powell MA, Somero GN. 1983. Blood components prevent sulfide poisoning of respiration of the hydrothermal vent tube worm *Riftia pachyptila*. *Science*, 219:297–299.

Pringsheim EG. 1967. Die Mixotrophie von Beggiatoa. *Arch Mikrobiol*, 59:247–254.

Pronk JT, De Bruyn JC, Bos P, Kuenen JG. 1992. Anaerobic growth of *Thiobacillus ferrooxidans*. *Appl Environ Microbiol*, 58:2227–2230.

Pronk JT, Liem K, Bos P, Kuenen JG. 1991. Energy transduction by anaerobic ferric iron respiration in *Thiobacillus ferrooxidans*. *Appl Environ Microbiol*, 57:2063–2068.

Pronk JT, Meulenberg R, Hazeu W, Bos P, Kuenen JG. 1990. Oxidation of reduced inorganic sulfur compounds by acidophilic thiobacilli. *FEMS Microbiol Rev*, 75:293–306.

Qatabi AI, Nivière V, Garcia JL. 1991. *Desulfovibrio alcoholovorans* sp. nov., a sulfate-reducing bacterium able to grow on glycerol, 1,2- and 1,3-propanol. *Arch Microbiol*, 155:143–148.

Rabus R, Hansen T, Widdel F. 2013. Dissimilatory sulfate- and sulfur-reducing prokaryotes. In: Rosenberg E, DeLong E, Lory S, Stackebrandt E, Thompson F, eds. *The Prokaryotes*. Berlin, Heidelberg: Springer, pp. 309–404.

Ramel F, Amrani A, Pieulle L, Lamrabet O, Voordouw G, Seddiki N, Brethes D, Company M, Dolla A, Brasseur G. 2013. Membrane-bound oxygen reductases of the anaerobic sulfate-reducing Desulfovibrio vulgaris Hildenborough: Roles in oxygen defense and electron link with the periplasmic hydrogen oxidation. Microbiology, 159:2663–2673.

Ramos AR, Keller KL, Wall JD, Pereira IAC. 2012. The membrane QmoABC complex interacts directly with the dissimilatory adenosine 5'-phosphosulfate reductase in sulfate reducing bacteria. Front Microbiol, 3:137.

Rau GH. 1981. Hydrothermal vent clam and tube worm $^{13}C/^{12}C$: Further evidence of nonphotosynthetic food sources. Science, 213:338–340.

Rau GH, Hedges JI. 1979. Carbon-13 depletion in a hydrothermal vent mussel: Suggestion of a chemosynthetic food source. Science, 203:648–649.

Ravot G, Magot M, Fardeau ML, Patel BKC, Prensier G, Egan A, Garcia JL, Ollivier B. 1995. Thermotoga elfii sp. nov., a novel thermophilic bacterium from an African oil-producing well. Int J Syst Bacteriol, 45:308–314.

Rees GN, Harfoot CG, Sheehy AJ. 1998. Amino acid degradation by the mesophilic sulfate-reducing bacterium Desulfobacterium vacuolatum. Arch Microbiol, 169:76–80.

Riccardi AL, Arthur MA, Kump LR. 2006. Sulfur isotopic evidence for chemocline upward excursions during the end-Permian mass extinction. Geochim Et Cosmochim Acta, 70(23):5740–5752.

Roy AB, Trudinger PA. 1970. The Biochemistry of Inorganic Compounds of Sulfur. Cambridge, U.K.: Cambridge University Press.

Rueter P, Rabus R, Wilkes H, Aeckersberg F, Rainey FA, Jannasch HW, Widdel F. 1994. Anaerobic oxidation of hydrocarbons in crude oil by new types of sulfate-reducing bacteria. Nature (Lond), 372:455–458.

Sadler MH, Johnson EJ. 1972. A comparison of the NADH oxidase electron transport system of two obligately chemolithotrophic bacteria. Biochim Biophys Acta, 283:167–179.

Schauder R, Eikmanns B, Thauer RK, Widdel F, Fuchs G. 1986. Acetate oxidation to CO_2 in anaerobic bacteria via a novel pathway not involving reactions of ther citric acid cycle. Arch Microbiol, 145:162–172.

Schauder R, Kröger A. 1993. Bacterial sulfur respiration. Arch Microbiol, 159:491–497.

Schauder R, Müller E. 1993. Polysulfide as a possible substrate for sulfur-reducing bacteria. Arch Microbiol, 160:377–382.

Schauder R, Preuss A, Jetten M, Fuchs G. 1989. Oxidative and reductive acetyl CoA/carbon monoxide dehydrogenase pathway in Desulfobacterium autotrophicum. Arch Microbiol, 151:84–89.

Schauder R, Widdel F, Fuchs G. 1987. Carbon assimilation pathways in sulfate-reducing bacteria. II. Enzymes of a reductive citric acid cycle in autotrophic Desulfobacter hydrogenophilus. Arch Microbiol, 148:218–225.

Schedel M, Trüper HG. 1980. Anaerobic oxidation of thiosulfate and elemental sulfur in Thiobacillus denitrificans. Arch Microbiol, 124:205–210.

Schicho RN, Ma K, Adams MWW, Kelly RM. 1993. Bioenergetics of sulfur reduction in the hyperthermophilic archeon Pyrococcus furiosus. J Bacteriol, 175:1823–1830.

Schiffer A, Parey K, Warkentin E, Diederichs K, Huber H, Stetter KO, Kroneck PMH, Ermler U. 2008. Structure of the dissimilatory sulfite reductase from the hyperthermophilic Archaeon Archaeoglobus fulgidus. J Mol Biol, 379:1063–1074.

Schnell S, Schink B. 1991. Anaerobic aniline degradation via reductive deamination of a 4-aminobenzoyl·CoA in Desulfobacterium anilini. Arch Microbiol, 155:183–190.

Segerer A, Neuner A, Kristiansson JK, Stetter KO. 1986. Acidianus infernos gen. nov., sp. nov., and Acidianus brierleyi comb. nov.: Facultative aerobic, extremely acidophilic thermophilic sulfur-metabolizing archaebacteria. Int J Syst Bacteriol, 36:559–564.

Seitz H-J, Cypionka H. 1986. Chemolithotrophic growth of Desulfovibrio desulfuricans with hydrogen coupled to ammonification of nitrate or nitrite. Arch Microbiol, 146:63–67.

Selig M, Schönheit P. 1994. Oxidation of organic compounds to CO_2 with sulfur or thiosulfate as electron acceptor in the anaerobic hyperthermophilic archaea Thermoproteus tenax and Pyrobaculum islandicum proceeds via the citric acid cycle. Arch Microbiol, 162:286–294.

Shen Y, Buick R, Canfield DE. 2001. Isotopic evidence for microbial sulphate reduction in the early Archaean era. Nature, 410:77–81.

Shirodkar S, Reed S, Romine M, Saffarini D. 2010. The octaheme SirA catalyses dissimilatory sulfite reduction in Shewanella oneidensis MR-1. Environ Microbiol, 13:108–115.

Shively JM, Ball F, Brown DH, Saunders RE. 1973. Functional organelles in prokaryotes: Polyhedral inclusions (carboxysomes) of Thiobacillus neapolitanus. Science, 182:584–586.

Shivvers DW, Brock TD. 1973. Oxidation of elemental sulfur by *Sulfolobus acidocaldarius*. *J Bacteriol*, 114:706–710.

Sigalevich P, Baev MV, Teske A, Cohen Y. 2000a. Sulfate reduction and possible aerobic metabolism of the sulfate-reducing bacterium *Desulfovibrio oxyclinae* in a chemostat coculture with *Marinobacter* sp. strain MB under exposure of increasing oxygen concentrations. *Appl Environ Microbiol*, 66:5013–5018.

Sigalevich P, Cohen Y. 2000. Oxygen-dependent growth of sulfur-reducing bacterium *Desulfovibrio oxyclinae* in coculture with *Marinobacter* sp. strain MB in an aerated sulfate-dependent chemostat. *Appl Environ Microbiol*, 66:5019–5023.

Sigalevich P, Meshorer E, Helman Y, Cohen Y. 2000b. Transition from anaerobic to aerobic growth conditions for the sulfate-reducing bacterium *Desulfovibrio oxyclinae* results in flocculation. *Appl Environ Microbiol*, 66:5005–5012.

Skerman VBD, Dementyeva G, Carey B. 1957a. Intracellular deposition of sulfur by *Sphaerotilus natans*. *J Bacteriol*, 73:504–512.

Skerman VBD, Dementyeva G, Skyring GW. 1957b. Deposition of sulfur from hydrogen sulfide by bacteria and yeasts. *Nature (Lond)*, 179:742.

Smith AL, Kelly DP, Wood AP. 1980. Metabolism of *Thiobacillus* A_2 grown under autotrophic, mixotrophic, and heterotrophic conditions in a chemostat culture. *J Gen Microbiol*, 121:127–138.

Smith DW, Rittenberg SC. 1974. On the sulfur-source requirement for growth of *Thiobacillus intermedius*. *Arch Microbiol*, 100:65–71.

Smith DW, Strohl WR. 1991. Sulfur-oxidizing bacteria. In: Shively JM, Barton LL, eds. *Variations in Autotrophic Life*. London, U.K.: Academic Press, pp. 121–146.

Sokolova GA, Karavaiko GI. 1968. *Physiology and Geochemical Activity of Thiobacilli*. Sprinfield, VA: US Department of Commerce/Clearinghouse Fed Tech Info (Engl transl).

Sorokin YI. 1966a. Role of carbon dioxide and acetate in biosynthesis of sulfate-reducing bacteria. *Nature (Lond)*, 210:551–552.

Sorokin YI. 1966b. Sources of energy and carbon for biosynthesis by sulfate-reducing bacteria. *Mikrobiologiya*, 35:761–766 (Engl transl, pp. 643–647).

Sorokin YI. 1966c. Investigation of the structural metabolism of sulfate-reducing bacteria with ^{14}C. *Mikrobiologiya*, 35:967–977 (Engl transl, pp. 806–814).

Sorokin YI. 1966d. The role of carbon dioxide and acetate in biosynthesis in sulfate reducing bacteria. *Dokl Akad Nauk SSSR*, 168:199.

Sorokin YI. 1970. The mechanism of chemical and biological oxidation of sodium, calcium, and iron sulfides. *Mikrobiologiya*, 39:253–258 (Engl transl, pp. 220–224).

Speich N, Trüper HG. 1988. Adenylylsulfate reductase in a dissimilatory sulfate-reducing archaebacterium. *J Gen Microbiol*, 134:1419–1425.

Stanier RY, Ingraham JL, Wheelis ML, Painter PR. 1986. *The Microbial World*, 5th edn. Englewood Cliffs, NJ: Prentice Hall.

Starkey RL. 1934. The production of polythionates from thiosulfate by microorganisms. *J Bacteriol*, 28:387–400.

Stetter KO. 1985. Thermophilic archaebacteria occurring in submarine hydrothermal areas. In: Caldwell DE, Brierley JA, Brierley CL, eds. *Planetary Ecology*. New York: Van Nostrand Reinhold, pp. 320–332.

Stetter KO, Koenig H, Stackebrandt E. 1983. *Pyrodictium* gen. nov., a new genus of submarine disk-shaped sulfur-reducing archaebacteria growing optimally at 105°C. *Syst Appl Microbiol*, 4:535–551.

Stetter KO, Lauerer G, Thomm M, Neuner A. 1987. Isolation of extremely thermophilic sulfate reducers: Evidence for a novel branch of archaebacteria. *Science*, 236:822–824.

Stoffels L, Krehenbrink M, Berks BC, Unden G. 2012. Thiosulfate reduction in *Salmonella enterica* is driven by the proton motive force. *J Bacteriol*, 194(2):475–485.

Strohl WR, Larkin JM. 1978. Enumeration, isolation, and characterization of *Beggiatoa* from freshwater sediments. *Appl Environ Microbiol*, 36:755–770.

Suzuki I. 1965. Oxidation of elemental sulfur by an enzyme system of *Thiobacillus thiooxidans*. *Biochim Biophys Acta*, 104:359–371.

Suzuki I. 1999. Oxidation of inorganic sulfur compounds: Chemical and enzymatic reactions. *Can J Microbiol*, 45:97–105.

Suzuki I, Chan CW, Takeuchi TL. 1992. Oxidation of elemental sulfur to sulfite by *Thiobacillus thiooxidans* cells. *Appl Environ Microbiol*, 58:3767–3769.

Suzuki I, Chan CW, Takeuchi TL. 1994. Oxidation of inorganic sulfur compounds by Thiobacilli. In: Alpers CN, Blowes DW, eds. *Environmental Geochemistry of Sulfide Oxidation*. ACS Symposium 550. Washington, DC: American Chemical Society, pp. 60–67.

Suzuki I, Silver M. 1966. The initial product and properties of the sulfur-oxidizing enzyme of thiobacilli. *Biochim Biophys Acta*, 122:22–33.

Szabo A, Tudge A, Macnamara J, Thode HG. 1950. The distribution of S34 in nature and the sulfur cycle. *Science*, 111:464–465.

Szewzyk R, Pfennig N. 1987. Complete oxidation of catechol by a strictly anaerobic sulfate-reducing *Desulfobacterium catecholicum* sp. nov. *Arch Microbiol*, 147:163–168.

Tasaki M, Kamagata Y, Nakamura K, Mikami E. 1991. Isolation and characterization of a thermophilic benzoatedegrading sulfate-reducing bacterium, *Desulfotomaculum thermobenzoicum* sp. nov. *Arch Microbiol*, 155:348–352.

Tasaki M, Kamagata Y, Nakamura K, Mikami E. 1992. Utilization of methoxylated benzoates and formation of intermediates by *Desulfotomaculum thermobenzoicum* in the presence and absence of sulfate. *Arch Microbiol*, 157:209–212.

Teske A, Brinkhoff T, Muyzer G, Moser DP, Rethmeier J, Jannasch HW. 2000. Diversity of thiosulfate- oxidizing bacteria from marine sediments and hydrothermal vents. *Appl Environ Microbiol*, 66:3125–3133.

Thamdrup B, Finster K, Hansen JW, Bak F. 1993. Bacterial disproportionation of elemental sulfur coupled to chemical reduction of iron and manganese. *Appl Environ Microbiol*, 59:101–108.

Thamdrup B, Fossing H, Jorgensen BB. 1994. Manganese, iron, and sulfur cycling in a coastl marine sediment, Aarhus Bay, Denmark. *Geochim Et Cosmochim Acta*, 58(23):5115–5129.

Thauer RK, Jungermann K, Decker K. 1977. Energy conservation in chemotrophic anaerobic bacteria. *Bacteriol Rev*, 41:100.

Trautwein K. 1921. Beitrag zur Physiologie und Morphologie der Thionsäurebakterien. *Zentralbl Bakteriol Parasitenk Infektionskr Hyg Abt II*, 53:513–548.

Trudinger PA. 1961. Thiosulfate oxidation and cytochromes in *Thiobacillus* X. 2. Thiosulfate oxidizing enzyme. *Biochem J*, 78:680–686.

Trüper HG. 1978. Sulfur metabolism. In: Clayton RK, Sistrom WR, eds. *The Photosynthetic Bacteria*. New York: Plenum Press, pp. 677–690.

Trüper HG. 1984. Phototrophic bacteria and the sulfur metabolism. In: Müller A, Krebs B, eds. *Sulfur, Its Significance for Chemistry, for the Geo-, Bio, and Cosmosphere and Technology*, Vol 5. Amsterdam, the Netherlands: Elsevier, pp. 367–382.

Tuttle JH, Ehrlich HL. 1986. Coexistence of inorganic sulfur metabolism and manganese oxidation in marine bacteria. *Abstr Annu Meet Am Soc Microbiol*, I-21:168.

Tuttle JH, Holmes PE, Jannasch HW. 1974. Growth rate stimulation of marine pseudomonads by thiosulfate. *Arch Microbiol*, 99:1–14.

Vainshtein MB. 1977. Oxidation of hydrogen sulfide by thionic bacteria. *Mikrobiologiya*, 46:1114–1116 (Engl transl, pp 898–899).

Vairavamurthy A, Manowitz B, Luther Iii GW, Jeon Y. 1993. Oxidation state of sulfur in thiosulfate and implications for anaerobic energy metabolism. *Geochim Cosmochim Acta*, 57:1619–1623.

van den Ende FP, van Gemerden H. 1993. Sulfide oxidation under oxygen limitation by a *Thiobacillus thioparus* isolated from a marine microbial mat. *FEMS Microbiol Ecol*, 13:69–78.

van Gemerden H. 1986. Production of elemental sulfur by green and purple sulfur bacteria. *Arch Microbiol*, 146:52–56.

Venceslau SS, Stockdreher Y, Dahl C, Pereira IAC. 2014. The "bacterial heterodisulfide" DsrC is a key protein in dissimilatory sulfur metabolism. *Biochim et Biophys Acta (BBA)—Bioenergetics*, 1837:1148–1164.

Vishniac W. 1952. The metabolism of *Thiobacillus thioparus*. I. The oxidation of thiosulfate. *J Bacteriol*, 64:363–373.

Vishniac W, Santer M. 1957. The thiobacilli. *Bacteriol Rev*, 21:195–213.

Wagner M, Roger A, Flax J, Brusseau G, Stahl D. 1998. Phylogeny of dissimilatory sulfite reductases supports an early origin of sulfate respiration. *J Bacteriol*, 180:2975–2982.

Wall JD, Rapp-Giles BJ, Brown MF, White JA. 1990. Response of *Desulfovibrio desulfuricans* colonies to oxygen stress. *Can J Microbiol*, 36:400–408.

Warén A, Bengtson S, Goffredi SK, Van Dover CL. 2003. A hot-vent gastropod with iron sulfide dermal sclerites. *Science*, 302(5647):1007.

Widdel F, Pfennig N. 1977. A new anaerobic, sporing, acetate-oxidizing, sulfate-reducing bacterium, *Desulfotomaculum* (emend) *acetoxidans*. *Arch Microbiol*, 112:119–122.

Widdel F, Pfennig N. 1981. Sporulation and further nutritional characteristics of *Desulfotomaculum acetoxidans*. *Arch Microbiol*, 129:401–402.

Wilbanks EG, Jaekel U, Salman V, Humphrey PT, Eisen JA, Faccioti MT, Buckley DH et al. 2014. Microscale sulfur cycling in the phototrophic pink berry consortia of the Sippewissett Salt Marsh. *Environ Microbiol*, 16:3398–3415.

Williams CD, Nelson DC, Farah BA, Jannasch HW, Shively JM. 1988. Ribulose bisphosphate carboxylase of the prokaryotic symbiont of a hydrothermal vent tube worm: Kinetics, activity, and gene hybridization. *FEMS Microbiol Lett*, 50:107–112.

Williams RD, Hoare DS. 1972. Physiology of a new facultative autotrophic thermophilic *Thiobacillus*. *J Gen Microbiol*, 70:555–566.

Wirsen CO, Jannasch HW. 1978. Physiological and morphological observations on *Thiovulum* sp. *J Bacteriol*, 136:765–774.

Wirsen CO, Sievert SM, Cavanaugh CM, Molyneaux SJ, Ahmad A, Taylor LT, DeLong EF, Taylor CD. 2002. Characterization of an autotrophic sulfide-oxidizing marine *Arcobacter* sp. that produces filamentous sulfur. *Appl Environ Microbiol*, 68:316–325.

Wood AP, Kelly DP. 1978. Comparative radiorespirometric studies of glucose oxidation in three facultative heterotrophic thiobacilli. *FEMS Microbiol Lett*, 4:283–286.

Wood AP, Kelly DP. 1983. Autotrophic, mixotrophic and heterotrophic growth with denitrification by *Thiobacillus* A$_2$ under anaerobic conditions. *FEMS Microbiol Lett*, 16:363–370.

Wood AP, Kelly DP. 1991. Isolation and characterization of *Thiobacillus halophilus* sp. nov., a sulfur-oxidizing autotrophic eubacterium from a Western Australian hypersaline lake. *Arch Microbiol*, 156:277–280.

Wood P. 1988. Chemolithotrophy. In: Anthony C, ed. *Bacterial Energy Transduction*. London, U.K.: Academic Press, pp. 183–230.

Yamanaka T. 1996. Mechanism of oxidation of inorganic electron donors in autotrophic bacteria. *Plant Cell Physiol*, 37:569–574.

Zehnder AJB, Zinder SH. 1970. The sulfur cycle. In: Hutzinger O, ed. *The Handbook of Environmental Chemistry*. Berlin, Germany: Springer Verlag, pp. 105–145.

Zeikus JG, Swanson MA, Thompson TE, Ingvosen K, Hatchikian EC. 1983. Microbial ecology of volcanic sulfidogenesis: Isolation and characterization of *Thermodesulfobacterium commune* gen. nov. and spec. nov. *J Gen Microbiol*, 129:1159–1169.

Zellner G, Kneifel H, Winter J. 1990. Oxidation of benzaldehydes to benzoic acid derivatives by three *Desulfovibrio* strains. *Appl Environ Microbiol*, 56:2228–2233.

Zerkle AL, Farquhar J, Johnston DT, Cox RP, Canfield DE. 2009. Fractionation of multiple sulfur isotopes during phototrophic oxidation of sulfide and elemental sulfur by a green sulfur bacterium. *Geochim Cosmochim Acta*, 73:291–306.

Zimmermann P, Laska S, Kletzin A. 1999. Two modes of sulfite oxidation in the extremely thermophilic and acidophilic archeon *A. ambivalens*. *Arch Microbiol*, 172:76–82.

CONTE

21.1 In
21.2 Na
    21
    21
21.3 Pr
21.4 La
    21
    21
21.5 Bi
    21
    21
    21
    21
21.6 Bi
    21
    21
    21
    21
    21
21.7 Bi
21.8 Fo
    21
21.9 Su
Referen

21.1 IN

Sulfate-
in some
tion of
pyrite (
pervasi
of meta
expose
origin

such gold ores to remove interfering pyrite and arsenopyrite impurities is now being practiced on a commercial scale. The pyrites in these ores encapsulate the gold, making it inaccessible to a chemical extractant such as aqueous cyanide or thiourea. When cyanide is used as extractant, the pyrites cause excessive consumption of it, resulting in formation of cyanide complexes with the iron and sulfur components of pyrite, i.e., ferro- and ferricyanide and thiocyanate, from none of which the cyanide is readily recoverable. A great potential exists for industrial bioextraction of a variety of other metal sulfide ores.

In the metal industry, a widely used term for metal bioextraction from ores is *bioleaching*. In this chapter, we examine ore biogenesis and biomobilization, including bioleaching, in some detail. Table 21.1 lists metal sulfide minerals of geomicrobial interest.

21.2 NATURAL ORIGIN OF METAL SULFIDES

21.2.1 Hydrothermal Origin (Abiotic)

Most metal sulfides, including those of commercial interest, are of igneous origin. Current theory explaining their formation invokes plate tectonics, which has played and is playing a central role in their formation. Terrestrial deposits of *porphyry copper ore* (small crystals of copper sulfides richly dispersed in host rock) are thought to have originated as a result of subduction of oceanic crust that had become somewhat enriched in copper by hydrothermal activity at mid-ocean spreading centers. Subsequent formation of terrestrial deposits of porphyry sulfide ores from subducted oceanic crust is thought to have involved the following successive steps: (1) remelting of the subducted oceanic crust, (2) rising of the resultant magma, (3) release of water with fracturing of incipient rock and

TABLE 21.1
Metal sulfides of geomicrobial interest.

Mineral or synthetic compound	Formula	Reference
Antimony trisulfide	Sb_2S_3	Silver and Torma (1974); Torma and Gabra (1977)
Argentite	Ag_2S	Baas Becking and Moore (1961)
Arsenopyrite	FeAsS	Ehrlich (1964)
Bornite	Cu_5FeS_4	Cuthbert (1962); Bryner et al. (1954)
Chalcocite	Cu_2S	Bryner et al. (1954); Ivanov (1962); Razzell and Trussell (1963); Sutton and Corrick (1963, 1964); Fox (1967); Nielsen and Beck (1972)
Chalcopyrite	$CuFeS_2$	Bryner and Anderson (1957)
Cobalt sulfide	CoS	Torma (1971)
Covellite	CuS	Bryner et al. (1954); Razzell and Trussell (1963)
Digenite	Cu_9S_5	Baas Becking and Moore (1961); Nielsen and Beck (1972)
Enargite	$3Cu_2S \cdot As_2S_5$	Ehrlich (1964)
Galena	PbS	Silver and Torma (1974)
Gallium sulfide	Ga_2S_3	Torma (1978)
Marcasite, pyrite	FeS_2	Leathen et al. (1953); Silverman et al. (1961)
Millerite	NiS	Razzell and Trussell (1963)
Molybdenite	MoS_2	Bryner and Anderson (1957); Bryner and Jameson (1958); Brierley and Murr (1973)
Orpiment	As_2S_3	Ehrlich (1963a)
Nickel sulfide	NiS	Torma (1971)
Pyrrhotite	Fe_4S_5	Freke and Tate (1961)
Sphalerite	ZnS	Ivanov et al. (1961); Ivanov (1962); Malouf and Prater (1961)
Tetrahedrite	$Cu_8Sb_2S_7$	Bryner et al. (1954)

formation of hydrothermal solution containing hydrogen sulfide during progressive partial cooling of the magma, and finally (4) reformation of copper and other metal sulfides by crystallization of the cooling magma and/or from reaction of H_2S in the hydrothermal solution with metal constituents in the cooled magma in the fractured rock (see Strahler, 1977; Bonatti, 1978; Tittley, 1981).

The enrichment of the surficial deposits of metal sulfide in and on the oceanic crust has occurred and is occurring in hydrothermally active regions at seafloor spreading centers (mid-ocean ridges) at depths of 2500–2699 m. Examples of such sites are the eastern Pacific Ocean at the Galapagos Rift and the East Pacific Rise (Ballard and Grassle, 1979; Corliss et al., 1979) and the Atlantic Ocean at the Mid-Atlantic Ridge (e.g., Klinkhammer et al., 1985). Metal sulfide deposits are evident on the seafloor where some hydrothermal vents ("black smokers"; see Chapters 2 and 18) discharge brine solution that has a temperature near 350°C and is metal laden and charged with H_2S. Metal sulfides such as chalcopyrite ($CuFeS_2$) and sphalerite (ZnS) precipitate around the mouth of these vents as the brine meets cold seawater and are often deposited in the form of hollow tubes (chimneys). The hydrothermal solution discharged by these vents originated from seawater that penetrated deep into porous volcanic rock (basalt) at the mid-ocean spreading centers to depths as great as 10 km below the seafloor (Bonatti, 1978). As this water penetrated ever deeper into the rock, it absorbed heat diffusing away from underlying magma chambers and was subjected to increasing hydrostatic pressure. This caused the seawater to react with the basalt and pick up various metal species and hydrogen sulfide. The reactions responsible for these seawater modifications include, among others, the interaction of magnesium in the seawater with the rock to form new minerals with an accompanying release of acid (H^+) (Seyfried and Mottl, 1982). The acid leaches metals from the basalt (Edmond et al., 1982; Marchig and Grundlach, 1982). H_2S is formed by reduction of the sulfate in seawater and sulfur in basalt by ferrous iron released from the basalt (e.g., Mottl et al., 1979; Shanks et al., 1981; Styrt et al., 1981). As long as the hydrothermal solution is subjected to high temperature and pressure in the basalt, metal sulfides are prevented from precipitating.

A quantitatively more significant deposition of metal sulfides occurs within the upper oceanic crust associated with white smokers. Here, hot,

metal-charged hydrothermal brine rising from the lower crust meets and mixes with cold seawater that penetrated the upper crust. The mixing of the two solutions in the upper crust results in partial cooling of the solution and consequent precipitation of metal sulfides in the upper crust. This contrasts with the precipitation of metal sulfides associated with black smokers, which occurs external to the crust around the mouth of the vents and becomes deposited mostly in the walls of the vent chimneys. The brine emerging from the vents of white smokers is depleted in some base metals but still contains major quantities of iron, manganese, and hydrogen sulfide. It is much cooler than the hydrothermal solution issuing from the vents of black smokers.

A study of bioalteration of sulfur and mineral sulfide samples deployed and incubated under conditions prevailing in the vicinity of a seafloor hydrothermal vent systems (main Endeavor Segment of the Juan de Fuca Ridge axis, Pacific Ocean) revealed ready colonization by bacteria but not archaea (Edwards et al., 2003). Elemental sulfur appeared to be most readily attacked. Extensive Fe-oxide accumulation on Fe-containing minerals suggested the activity of neutrophilic iron-oxidizing bacteria to the investigators.

21.2.2 Sedimentary Metal Sulfides of Biogenic Origin

Among sedimentary metal sulfides of biogenic origin, iron sulfides are the most common. They are usually associated with reducing zones in sedimentary deposits in estuarine environments, which have a plentiful supply of sulfate. The presence of sulfate is important, because the formation of these metal sulfides is usually the result of an interaction of iron compounds with H_2S that originated from bacterial reduction of the sulfate under anaerobic conditions at these sites. The interaction of the H_2S with the iron compounds leads to the formation of iron pyrite (FeS_2). Whether amorphous sulfide (FeS), mackinawite (FeS) and greigite (Fe_3S_4) are intermediates in the formation of the pyrite depends on prevailing environmental conditions (Luther, 1991; Schoonen and Barnes, 1991a,b). In at least one salt marsh (Great Sippewissett Marsh, Massachusetts) where pyrite forms, the pore waters were found to be undersaturated with respect to these compounds (Jørgensen, 1977; Fenchel and Blackburn, 1979; Howarth, 1979; Giblin and Howarth, 1984; Howarth and Merkel, 1984; Berner, 1984). Rapid and extensive microbial pyrite formation has been observed in salt marsh peat on Cape Cod (Massachusetts) (Howarth, 1979). Pyrite formation from biogenic H₂S has also been noted in organic-rich sediments at the Peru Margin of the Pacific Ocean (Mossmann et al., 1991), in Long Island Sound off the Atlantic coast in Connecticut and New York (e.g., Westrich and Berner, 1984), along the Danish coast (Thode-Anderson and Jørgensen, 1989), and in two seepage lakes, Gerritsfles and Kliplo, and two moorland ponds in the Netherlands (Marnette et al., 1993).

In many sedimentary environments, pyrite does not represent a permanent sink for iron because the pyrite may be subject to seasonal reoxidation as conditions in the environment change from reducing to oxidizing (Luther et al., 1982; Giblin and Howarth, 1984; King et al., 1985; Giblin, 1988). Active growth of marsh grass may draw oxygen into the sediment by evapotranspiration (Giblin, 1988). Of all of the biogenic sulfides formed in these environments, only a portion is consumed in the formation of pyrite and other metal sulfides. The rest is reoxidized as it enters the oxidizing zones (Jørgensen, 1977). This oxidation may be biological or abiological (Fenchel and Blackburn, 1979).

Nonferrous sulfide deposits of sedimentary origin, especially biogenic ones, appear to be relatively rare. They are generally thought to have formed syngenetically. The metals in question were precipitated by hydrogen sulfide of hydrothermal origin (abiotic formation) or of microbial origin and then buried in contemporaneously formed sediment. The limiting conditions for sedimentary sulfide formation by bacteria as calculated by Rickard (1973) require a minimum of 0.1% carbon (dry weight) and an enriched source of metals such as a hydrothermal solution if more than 1% metal is to be deposited. More recent studies of microbial sulfate reduction revealed, however, that a significant amount of reducing power for sulfate reduction may be furnished by hydrogen (H₂), which would lower the requirement for organic carbon correspondingly (Nedwell and Banat, 1981; see also Section 20.9).

Examples of nonferrous sedimentary sulfide deposits, which may have been biogenically formed, include the Permian Kupferschiefer of Mansfeld in Germany (Love, 1962; Stanton, 1972, p. 1139), the Black Sea sediments (Bonatti, 1972, p. 51), the Roan Antelope Deposit in Zambia and Katanga (Africa) (Cuthbert, 1962; Stanton, 1972, p. 1139), the

Zechstein deposit in southwestern Poland (Serkies et al., 1967), and the deposits in Pernatty Lagoon (Australia) (Lambert et al., 1971). By contrast, the sulfide deposit in the Pine Point Pb–Zn property in Northwest Territories, Canada, was abiotically formed (Powell and MacQueen, 1984). $\delta^{34}S$ analyses of the metal sulfides in this deposit suggest that the sulfide resulted from a reaction between bitumen and sulfate at elevated temperature and pressure.

As an example of ongoing nonferrous sulfide biodeposition, the following observation at the Piquette Pb–Zn deposit in Tennyson, Wisconsin, must be cited. At this site, investigators examined a flooded tunnel in carbonate rock and found the presence of biofilms in which aerotolerant members of sulfate-reducing bacteria of the family Desulfobacteraceae were precipitating sphalerite (ZnS) at a pH between ~7.2 and 8.6. The sphalerite accumulated in the biofilm in aggregates of particles that had a diameter of 2–5 nm (Labrenz et al., 2000).

Although most instances of metal sulfide biogenesis in nature are associated with bacterial sulfate reduction, at least one case of biogenesis of galena has been attributed to the aerobic mineralization of organic sulfur compounds by *Sarcina flava* Bary (Dévigne, 1968a,b, 1973). The *Sarcina* strain was isolated from earthy concretions between crystals of galena in an accumulation in a karstic pocket located in the lead–zinc deposit of Djebel Azered, Tunisia. In laboratory experiments, the organism was shown to produce PbS from Pb^{2+} bound to sulfhydryl groups of sulfur-containing amino acids in peptone.

21.3 PRINCIPLES OF METAL SULFIDE FORMATION

Metal sulfides in nature result from an interaction between an appropriate metal ion and biogenically or abiogenically formed sulfide ion:

$$M^{2+} + S^{2-} \rightarrow MS \qquad (21.1)$$

The source of the sulfide in the reaction determines whether a metal sulfide is considered to be of biogenic or abiogenic origin. In the case of biogenic sulfide, it does not matter whether the sulfide resulted from bacterial sulfate reduction (Chapter 20) or from bacterial mineralization of organic sulfur-containing compounds (Dévigne, 1968a,b, 1973). Because of their relative insolubility, the metal sulfides form readily at ambient temperatures and pressures. Table 21.2 lists

TABLE 21.2
Solubility products for some metal sulfides.

CdS	1.4×10^{-28}	FeS	1×10^{-19}	NiS	3×10^{-21}
Bi_2S_3	1.6×10^{-72}	PbS	3.4×10^{-28}	Ag_2S	1×10^{-51}
CoS	7×10^{-23}	MnS	5.6×10^{-16}	SnS	8×10^{-29}
Cu_2S	2.5×10^{-50}	Hg_2S	1×10^{-45}	ZnS	1.2×10^{-23}
CuS	8.5×10^{-45}	HgS	3×10^{-53}	H_2S	1.1×10^{-15}
				Hs^-	1×10^{-15}

SOURCES: From Latimer, WM and Hildebrand, JH, *Reference Handbook of Inorganic Chemistry*, Macmillan, New York, 1942; Weast, R.C. and Astle, M.J., *CRC Handbook of Chemistry and Physics*, 63rd edn., CRC Press, Boca Raton, FL, 1982.

solubility products for a few common simple sulfide compounds.

The following calculations will show that relatively low concentrations of metal ions, typical in some lakes, will form metal sulfides by reacting with low concentrations of H_2S. The ionic activities in these calculations are taken as approximately equal to concentration because of the low concentrations involved. The following examines the case of amorphous iron sulfide (FeS) formation.

The ionization constant for FeS is

$$[Fe^{2+}][S^{2-}] = 10^{-19} \qquad (21.2)$$

The ionization constant for H_2S is

$$[S^{2-}] = 10^{-21.96}[H_2S]/[H^+]^2 \qquad (21.3)$$

This relationship is derived from the constant for the dissociation of H_2S into HS^- and H^+,

$$[HS^-][H^+]/[H_2S] = 10^{-6.96} \qquad (21.4),$$

and the constant for the dissociation of HS^- into S^{2-} and H^+,

$$[S^{2-}][H^+]/[HS^-] = 10^{-15} \qquad (21.5)$$

Substituting Equation 20.3 into Equation 20.2, the following relationship is obtained:

$$\begin{aligned}[Fe^{2+}] &= [H^+]^2/[H_2S] \times 10^{-19}/10^{-21.96} \\ &= [H^+]^2/[H_2S] \times 10^{2.96} \qquad (21.6)\end{aligned}$$

Assuming that the bottom water of a lake contains about 34 mg of $H_2S\ L^{-1}$ (i.e., 10^{-3} M) at pH 7, about 5.08 μg $Fe^{2+}\ L^{-1}$ (i.e., $10^{-7.04}$ M) will be precipitated as FeS by 3.4 mg of hydrogen sulfide per

liter (i.e., 10^{-4} M). The remaining H_2S will ensure reducing conditions, which will keep the iron in the ferrous state. Because ferrous sulfide is one of the more soluble sulfides, metals whose sulfides have even smaller solubility products will require even less sulfide for precipitation. In the excess of sulfide, the FeS would probably be transformed into FeS_2, which is more stable than FeS.

21.4 LABORATORY EVIDENCE IN SUPPORT OF BIOGENESIS OF METAL SULFIDES

21.4.1 Batch Cultures

Metal sulfides have been generated in laboratory experiments using H_2S from bacterial sulfate reduction. Miller (1949, 1950) reported that sulfides of Sb, Bi, Co, Cd, Fe, Pb, Ni, and Zn were formed in a lactate-containing broth culture of *Desulfovibrio desulfuricans* to which insoluble salts of selected metals had been added. For instance, he found that bismuth sulfide was formed on addition of $(BiO_2)_2CO_3 \cdot H_2O$, cobalt sulfide on addition of $2CoCO_3 \cdot 3Co(OH)_2$, lead sulfide on addition of $2PbCO_3 \cdot Pb(OH)_2$ or $PbSO_4$, nickel sulfide on addition of $NiCO_3$ or $Ni(OH)_2$, and zinc sulfide on addition of $2ZnCO_3 \cdot 3Zn(OH)_2$. The metal salt reactants were added as insoluble compounds to minimize metal toxicity for *D. desulfuricans*. Metal ion toxicity depends in part on the solubility of the metal compound from which the ion derives. Obviously, for a metal sulfide to be formed from another metal compound that is relatively insoluble, the metal sulfide must be even more insoluble than the source compound of the metal. Miller was not able to demonstrate copper sulfide formation from malachite $[CuCO_3 \cdot Cu(OH)_2]$, probably because malachite was too insoluble relative to copper sulfides in the medium. Miller (1949) also showed that with the addition of Cd or Zn ions to the culture medium, the yield of total sulfide produced from sulfate by the bacteria in batch culture was greater than in the absence of the added metal ions. This was because the uncombined sulfide itself at high enough concentration becomes toxic to sulfate reducers.

Baas Becking and Moore (1961) also undertook a study of biogenesis of sulfide minerals. Like Miller, they worked with batch cultures of sulfate-reducing bacteria. The bacteria they employed were D. desulfuricans and Desulfotomaculum sp. (which they called Clostridium desulfuricans). They grew them in lactate or acetate medium containing steel wool. The steel wool in the medium was meant to serve as a source of hydrogen for the bacterial reduction of sulfate. The hydrogen resulted from corrosion of the steel wool by the following spontaneous reaction:

$$Fe^0 + 2H_2O \rightarrow H_2 + Fe(OH)_2 \quad (21.7)$$

The H_2 was then used by the sulfate reducers in the formation of hydrogen sulfide:

$$4H_2 + SO_4^{2-} + 2H^+ \rightarrow H_2S + 4H_2O \quad (21.8)$$

The media were saline to simulate marine (near shore and estuarine) conditions under which the investigators thought the reactions are likely to occur in nature. They formed ferrous sulfide from strengite ($FePO_4$) and from hematite (Fe_2O_3). They also formed covellite (CuS) from malachite ($CuCO_3 \cdot Cu(OH)_2$), argentite (Ag_2S) from silver chloride (Ag_2Cl_2) and from silver carbonate ($AgCO_3$), galena (PbS) from lead carbonate ($PbCO_3$) and from lead hydroxycarbonate ($PbCO_3 \cdot Pb(OH)_2$), and sphalerite (ZnS) from smithsonite ($ZnCO_3$). All mineral products were identified by x-ray powder diffraction studies. Baas Becking and Moore (1961) were unable to form cinnabar (HgS) from mercuric carbonate ($HgCO_3$), probably owing to the toxicity of the Hg^{2+} ion. They were also unable to form alabandite (MnS) from rhodochrosite ($MnCO_3$) or bornite (Cu_5FeS_4) or chalcopyrite ($CuFeS_2$) from a mixture of cuprous oxide (Cu_2O) or malachite plus hematite and lepidocrocite. They succeeded in forming covellite from malachite where Miller (1950) failed, probably because they performed their experiment in a saline medium (3% NaCl) in which Cl^- could complex Cu^{2+}, thereby increasing the solubility of Cu^{2+}. The starting materials that were the source of metal were all relatively insoluble, as in Miller's experiments. Baas Becking and Moore found that in the formation of covellite and argentite, native copper and silver were respective intermediates that disappeared with continued bacterial H_2S production.

Leleu et al. (1975) synthesized ZnS by passing H_2S produced by unnamed strains of sulfate-reducing bacteria through a solution of $ZnSO_4$. In one experiment, biogenic H_2S formation and ZnS precipitation by the biogenic H_2S occurred in separate vessels. In a second experiment, biogenesis of H_2S and precipitation of ZnS occurred in the same vessel at an initial $ZnSO_4$ concentration in the culture medium of 10^{-2} M. The ZnS formed under either experimental condition was identified as a sphalerite–wurtzite mixture by powder x-ray diffraction. The presence of Zn directly in the culture medium caused a lag in H_2S production, which was not observed when H_2S was generated in a separate vessel.

21.4.2 Column Experiment: A Model for Biogenesis of Sedimentary Metal Sulfides

The relatively high toxicity of many of the heavy metals for sulfate-reducing bacteria has been used as an argument that these organisms could not have been responsible for metal sulfide precipitation in nature (e.g., Davidson, 1962a,b). However, in a sedimentary environment, metal ions will be mostly adsorbed to sediment particles such as clays or complexed by organic matter (Hallberg, 1978), which lessens their toxicity. Such adsorbed or complexed ions are still capable of reacting with sulfide and precipitating as metal sulfides, as was shown experimentally by Temple and LeRoux (1964). They constructed a column in which clay or ferric hydroxide slurry carrying adsorbed Cu^{2+}, Pb^{2+}, and Zn^{2+} ions was separated by an agar plug from an underlying liquid culture of sulfate reducers actively generating hydrogen sulfide in saline medium. They also tested clay that was carrying Fe^{3+} in this setup. They found that, in time, bands of precipitate formed in the agar plug separating the slurry of metal-carrying adsorbent from the culture of sulfate-reducing bacteria (Figure 21.1). The bands formed as upward-diffusing sulfide ion species and downward diffusing, desorbed metal ion species encountered each other in the agar. Differential desorption of metal ions from the adsorbent and differential diffusion in the agar accounted for the discrete banding of the various sulfides. These results demonstrate that biogenesis of relatively large amounts of sulfides in a

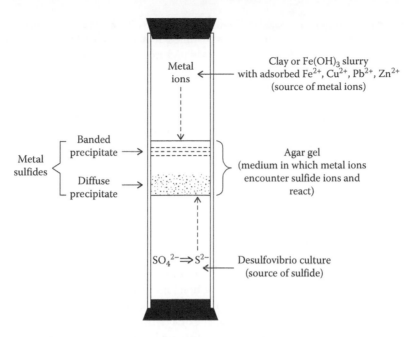

Figure 21.1. Temple and LeRoux column modeling a sedimentary environment in which sulfate reducers can precipitate metal sulfides with sulfide they produce. The adsorbents, clay or $Fe(OH)_3$ slurry, control the concentration of metal ions in solution, and the plug of agar gel prevents physical contact of the sulfate reducers with metal ions. In nature, sediments can act as adsorbents of metal ions. They keep the metal ion concentration in the interstitial water at a level that does not poison the sulfate reducers.

sedimentary environment is possible, even in the presence of relatively large amounts of metal ions. The main requirement is that the metal ions are in a nontoxic form (e.g., adsorbed or complexed) or combined in the form of insoluble mineral oxide, carbonate, or sulfate. As Temple (1964) pointed out, syngenetic microbial production of metal sulfide in nature is possible. Restrictions on the process, according to him, are not metal toxicity but free movement of the bacterially generated sulfide and a need for metal-enriched zones in the sedimentary environment. On a biochemical basis, Temple suspected that microbial sulfate reduction evolved in the Precambrian. Subsequent stable isotope analyses of samples representing the early Precambrian in South Africa indicated that extensive biogenic sulfate reduction occurred at least 2350 million years ago (Cameron, 1982).

21.5 BIOOXIDATION OF METAL SULFIDES

Regardless of whether they are of abiogenic or biogenic origin, metal sulfides in nature may be subject to microbial oxidation. This may take the form of direct or indirect interaction (Silverman and Ehrlich, 1964). In direct interaction, the microbes

oxidize a metal sulfide in physical contact with the mineral surface. In indirect interaction, the microbes usually generate an oxidant (commonly ferric iron from ferrous iron) in the bulk phase. The oxidant then attacks the metal sulfide. In most instances, the metal is solubilized as a metal ion by either direct or indirect mode of oxidation. The biooxidation of galena (PbS) is an exception because the mobilized metal reacts with sulfate ion, which is generated during the oxidation of the lead sulfide and which is also present in the bulk phase from other sources, to form insoluble lead sulfate ($PbSO_4$). Some microbes can mobilize metals in metal sulfides in an indirect mode by generating ligands, which may also be acids. These mobilize the metals by forming soluble metal complexes.

21.5.1 Organisms Involved in Biooxidation of Metal Sulfides

A number of different acidophilic, iron-oxidizing bacteria have been detected at sites where metal sulfide oxidation is occurring (Norris, 1990; Rawlings, 1997b; Okibe et al., 2003; Mousavi et al., 2005). The most important of these have been identified as the mesophiles *Acidithiobacillus ferrooxidans*,

Leptospirillum ferrooxidans, Ferroplasma acidiphilum, and *Ferroplasma acidarmanus,* the moderate thermophiles *Alicyclobacillus tolerans* (formerly *Sulfobacillus thermosulfidooxidans*) and *Acidimicrobium ferrooxidans,* and the extreme thermophiles *Sulfolobus* spp. and *Acidianus brierleyi* (formerly *Sulfolobus brierleyi*). All are autotrophs, and all but *F. acidarmanus* grow best in a pH range of about 1.5–2.5. *F. acidarmanus,* a recent discovery, grows at a pH as low as 0 (optimum pH 1.2) at a temperature of ~40°C. It was isolated from pyrite surfaces of the ore body at Iron Mountain, California, and has been described by Edwards et al. (2000a) as a cell-wall-lacking, iron-oxidizing autotroph. *F. acidiphilum,* also a recent discovery and a close relative of *F. acidarmanus,* was isolated from a bioleaching pilot plant (Golyshina et al., 2000). It grows in a pH range of 1.3–2.2 (optimum pH 1.7) in a temperature range of 15°C–45°C. An organism closely related to *F. acidiphilum* was isolated from a bioleaching operation by Mintek in South Africa.

Acidithiobacillus ferrooxidans, L. ferrooxidans, A. tolerans, and *Acidimicrobium ferrooxidans* are members of the domain bacteria (Norris, 1997). *Sulfolobus* spp., *A. brierleyi, F. acidarmanus,* and *F. acidiphilum* are members of the domain archaea. Whereas *L. ferrooxidans* and *Acidimicrobium ferrooxidans* oxidize Fe^{2+} and pyrite, they do not oxidize reduced forms of sulfur as *Acidithiobacillus ferrooxidans* and *A. tolerans* do. This seems to suggest that *L. ferrooxidans* and *Acidimicrobium ferrooxidans* can promote metal sulfide oxidation only by generating Fe^{3+} from dissolved Fe^{2+}, which then oxidizes metal sulfide abiotically. However, because of a structural feature possessed by both *Acidithiobacillus ferrooxidans* and *L. ferrooxidans,* both organisms may also be able to oxidize metal sulfides by attacking them directly. The common structural feature is exopolymeric substance (EPS) secreted by the cells that contains bound iron (Gehrke et al., 1995, 1998; Sand et al., 1997; Harneit et al., 2006). Barreto et al. (2005) have identified the gene cluster involved in EPS formation in *Acidithiobacillus ferrooxidans.* The exopolymers enable attachment to sulfide mineral surfaces. In addition, as will be explained in Section 21.5.2, the iron in the EPS may serve as an electron shuttle or conductor for conveying electrons from the oxidative half reaction of metal sulfides to the electron transport system in the plasma membrane. This electron transfer occurs via cytochrome in the outer membrane, such as cytochrome Cyc2 in *Acidithiobaccillus ferrooxidans,* and specific electron carriers in the periplasm, such as rusticyanin and cytochrome Cyc1

in *Acidithiobacillus ferrooxidans.* It remains to be determined if *Acidimicrobium ferrooxidans* forms EPS with bound iron and transfers electrons from its outer membrane to its plasma membrane by a mechanism similar to that in *Acidithiobacillus ferrooxidans.*

Whereas *Acidithiobacillus ferrooxidans, Sulfolobus* spp., and *A. brierleyi* are autotrophs, the growth of the two archaea in this group of three is stimulated by a trace of yeast extract in laboratory culture. In the absence of dissolved ferrous iron or reduced forms of sulfur in the medium, all three organisms can use appropriate metal sulfides as energy sources. Depending on the oxidation state of the metal moiety in the metal sulfide, both it and the sulfide may serve as energy sources. For example, in the oxidation of chalcocite (Cu_2S), *Acidithiobacillus ferrooxidans* can use the energy from Cu(I) oxidation for CO_2 fixation (Nielsen and Beck, 1972) (see also further discussion in the next section). Cell extracts from *Acidithiobacillus ferrooxidans* have been prepared that catalyze the oxidation of cuprous copper in Cu_2S but not of elemental sulfur (Imai et al., 1973). The oxidation is not inhibited by quinacrine (Atabrine). It needs the addition of a trace of iron for proper activity. The effect of traces of iron on metal sulfide oxidation had been previously noted in experiments in which the addition of 9 mg of ferrous iron per liter of medium stimulated metal sulfide oxidation by whole cells of *Acidithiobacillus ferrooxidans* (Ehrlich and Fox, 1967).

Acidithiobacillus ferrooxidans can use NH_4^+ and some amino acids as nitrogen sources (see Sugio et al., 1987) (see also Chapter 17). At least some strains are able to fix nitrogen (Mackintosh, 1978; Stevens et al., 1986).

Acidithiobacillus ferrooxidans is very versatile in attacking metal sulfides. It has been reported to oxidize arsenopyrite ($FeS_2.FeAs_2$ or FeAsS), bornite (Cu_5FeS_4), chalcocite (Cu_2S), chalcopyrite ($CuFeS_2$), covellite (CuS) enargite ($3Cu_2S \cdot As_2S_5$), galena (PbS), millerite (NiS), orpiment (As_2S_3), pyrite (FeS_2), marcasite (FeS_2), sphalerite ZnS, stibnite (Sb_2S_3), and tetrahedrite ($Cu_8Sb_2S_7$) (see Silverman and Ehrlich, 1964; Wang et al., 2007). In addition, the oxidation of gallium sulfide, pyrrhotite, and synthetic preparations of CoS, NiS, and ZnS by *Acidithiobacillus ferrooxidans* has been reported (Torma, 1971, 1978; Pinka, 1991; Bhatti et al., 1993). The mode of attack of any of these minerals may be direct, indirect, or both.

Although not as exhaustively tested as *Acidithiobacillus ferrooxidans,* the archaea *A. brierleyi* and

EHRLICH'S GEOMICROBIOLOGY

Sulfolobus sp. can also oxidize a variety of metal sulfides including pyrite, marcasite, arsenopyrite, chalcopyrite, NiS, and probably CoS (Brierley and Murr, 1973; Brierley, 1978a,b; 1982; Dew et al., 1999; Wang et al., 2007). Unlike *Acidithiobacillus ferrooxidans*, *A. brierleyi* can oxidize molybdenite in the absence of added iron (Brierley and Murr, 1973) because molybdate ion is less toxic to *A. brierleyi* than to *Acidithiobacillus ferrooxidans* (Tuovinen et al., 1971).

21.5.2 Direct Oxidation

According to the concept of direct oxidation of susceptible metal sulfides as defined by Silverman and Ehrlich (1964), the crystal lattice of such sulfides is attacked through enzymatic oxidation. To accomplish this, the microbes have to be in intimate contact with the mineral they attack. Evidence for rapid attachment of *Acidithiobacillus ferrooxidans* to mineral surfaces of chalcopyrite particles (CuFeS$_2$) has been presented by McGoran et al. (1969) and Shrihari et al. (1991); to covellite particles by Pogliani et al. (1990); to galena crystals by Tributsch (1976); to pyrite crystals by Bennett and Tributsch (1978), Rodriguez-Leiva and Tributsch (1988), Mustin et al. (1992), Murthy and Natarajan (1992), and Edwards et al. (1998); and to pyrite/arsenopyrite-containing auriferous ore by Norman and Snyman (1988).

Bacterial attachment to mineral sulfide surfaces appears not to be random but to occur at specific sites and even specific crystal faces. Some evidence suggests that direct microbial attack is initiated at sites of crystal imperfections. Selective attachment of *Acidithiobacillus ferrooxidans* or *Sulfolobus acidocaldarius* to newly exposed pyrite crystals in coal is very rapid, i.e., about 90% complete in 2–5 min (Badigian and Myerson, 1986; Chen and Skidmore, 1987, 1988). Although the details of how microbes attack crystal lattices of metal sulfides once they have attached are in most respects not yet understood, a collective model is that bacterial cells possessing this ability act as catalytic conductors in transferring electrons from cathodic areas on crystal surfaces of a metal sulfide via an electron transport system in the cell envelope to oxygen (Figure 21.2). The model assumes the existence of a special electron transport system, which in Gram-negative bacteria involves components of the outer membrane, the periplasm, and the plasma membrane. Indeed, the bacteria can be viewed as cathodic extensions. They benefit from this process by coupling energy conservation (ATP synthesis) to it.

The mere spontaneous dissociation of a mineral to yield oxidizable ion species in solution that *Acidithiobacillus ferrooxidans* can attack is too small in the case of minerals that are very insoluble in acid solution. For example, covellite (CuS), in which the only oxidizable constituent is the sulfide, has a solubility constant of $10^{-44.07}$ (Table 21.2). *Acidithiobacillus ferrooxidans* is able to oxidize this mineral at pH 2.0 (see later in this section). Simple calculations show that at equilibrium at pH 2 in water, the dissociation of CuS will only generate a concentration of HS$^-$ equal to $10^{-15.53}$ M and a concentration of H$_2$S equal to $10^{-13.06}$. This is insufficient for sulfide oxidation by *Acidithiobacillus ferrooxidans* because the most recent K_s value for sulfide oxidase in intact cells of this organism has been reported to be $10^{-5.30}$ (Pronk et al., 1990). K_s values are a measure of the substrate concentration at which a reaction velocity catalyzed by intact cells is half-maximal (Michaelis–Menten kinetics).[*] K_s for a cell is equivalent to K_m for an individual enzyme.) Thus, the mere dissociation of CuS into Cu^{2+}, HS$^-$, and H$_2$S cannot furnish nearly enough sulfide substrate to sustain its oxidation by *Acidithiobacillus ferrooxidans* at a reasonable velocity, regardless of whether HS$^-$ or H$_2$S or both are the actual substrate for sulfide oxidase. Because *Acidithiobacillus ferrooxidans* can oxidize covellite in the absence of added iron, it must be in direct contact with a mineral to attack it under that condition. The need for direct contact in covellite oxidation by *Acidithiobacillus ferrooxidans* in the absence of added Fe^{2+} was demonstrated experimentally by Pogliani et al. (1990). By contrast, *Acidithiobacillus thiooxidans* was shown by Donati et al. (1995) to promote covellite oxidation only with the addition to the medium of Fe(III) or by Fe(II) autoxidized to Fe(III). The oxidation of covellite by Fe(III) generates S^0 (and possibly small amounts of other partially reduced and dissolved sulfur species) as first described by Sullivan (1930):

$$CuS + 2Fe^{3+} \rightarrow Cu^{2+} + S^0 + 2Fe^{2+} \quad (21.9)$$

As pointed out in Chapter 20, *A. thiooxidans* cannot oxidize Fe^{2+}.

---

[*] K_m is defined by the Michaelis–Menten equation $v = V_{max} [S]/([S] + K_m)$, where v is the intial reaction velocity, V_{max} the maximal velocity, $[S]$ the initial substrate concentration, and K_m a constant.

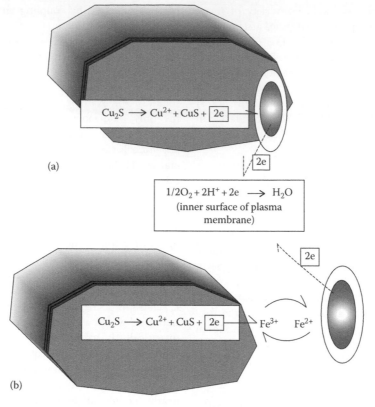

Figure 21.2. Schematic representation of direct and indirect oxidation of a particle of Cu_2S by *Acidithiobacillus ferrooxidans*. (a) Direct oxidation. In this model, the bacterial cell, attached to the surface of a Cu_2S particle, acts essentially as a conductor of electrons it removes in the oxidation of Cu(I) of Cu_2S and transfers to oxygen. Not shown is the mechanism by which the electrons cross the interface between the particle surface and the cell surface. For a possible mechanism for this electron transfer, see discussion in Section 21.5 of this chapter. (b) Indirect oxidation. In this model, planktonic (unattached) bacterial cells generate and regenerate the oxidant (Fe^{3+}) by oxidizing Fe^{2+} in the bulk phase. The iron acts as a shuttle that carries electrons from the oxidation of Cu(I) of Cu_2S to the bacterium, which transfers them to oxygen.

On the other hand, if we consider a more soluble sulfide mineral such as ZnS, which has a solubility constant of $10^{-22.9}$, calculations similar to those for CuS show that at pH 2.0, ZnS dissociation will yield $10^{-4.95}$ M HS$^-$ and $10^{-1.47}$ H$_2$S (Table 21.2). These concentrations of HS$^-$ and H$_2$S are more than sufficient to satisfy the K_s of $10^{-5.30}$ for sulfide oxidase in *Acidithiobacillus ferrooxidans* and to permit its growth without direct attack of ZnS at the mineral surface. Indeed, it has been shown that *A. thiooxidans*, which is unable to oxidize Fe^{2+}, will readily promote the dissolution of ZnS at pH 2.0 (Pistoro et al., 1994). The relative solubility of PbS in acid solution also explains why Garcia et al. (1995) found that *A. thiooxidans* promoted dissolution of PbS (galena). The solubility constant of this metal sulfide is $10^{-27.5}$, a little smaller than that of ZnS ($10^{-22.9}$) but significantly larger than that of CuS

($10^{-44.07}$). At pH 2, PbS dissociates to yield $10^{-7.25}$ M HS$^-$ and $10^{-4.77}$ M H$_2$S (Table 21.2).

The exact nature of the interaction between a sulfide mineral surface and the *Acidithiobacillus ferrooxidans* cell surface, on which the enzyme-catalyzed oxidation of the mineral depends, has become a little clearer in the last few years. Previously, Ingledew proposed that iron bound in the cell envelope of *Acidithiobacillus ferrooxidans* served as an electron shuttle that conveys electron from an external electron donor across the outer membrane to electron carriers in the periplasm of the cell (see Ingledew, 1986; Ehrlich, 2000). Alternatively, Tributsch (1999) proposed that the ferric iron bound in the exopolymer at the cell surface of *Acidithiobacillus ferrooxidans* in contact with a mineral surface generated elemental sulfur according to Reaction 21.9. This sulfur was then supposed to be oxidized by *Acidithiobacillus ferrooxidans* by a known reaction (see Chapter 20). However, this proposal assumes that EPS-bound iron(III) is as strong an oxidant as ferric iron in the bulk phase.

It has now been demonstrated by Yarzábal et al. (2002a,b) that the outer membrane of *Acidithiobacillus ferrooxidans* contains a high-molecular-weight *c*-type cytochrome Cyc2, which has the capacity to promote the oxidation of Fe^{2+} to Fe^{3+} at the outer surface of the outer membrane. Cytochrome Cyc2 in the outer membrane then conveys the electrons it removed from Fe^{2+} to the multicopper oxidase rusticyanin and from it to a low-molecular-weight c-type cytochrome Cyc1, both located in the periplasm. Cytochrome Cyc1 then passes the electrons it receives to aa_3 cytochrome oxidase in the plasma membrane, which passes them to O_2 accompanied by energy conservation via ATP synthase (see also Section 17.4, and Figure 17.4b). Although it remains to be experimentally demonstrated, it seems reasonable to assume that cytochrome Cyc2 or another c-type cytochrome in the outer membrane is involved in conveying electrons from the oxidation of a metal sulfide with which *Acidithiobacillus ferrooxidans* is in physical contact to oxygen either directly or via Fe bound in the EPS of the cells acting as an electron shuttle between the mineral and cytochrome Cyc2 in the outer membrane, as explained below.

Precedents for electron transfer between an outer cell surface and a mineral surface with which it is in contact exist among bacteria that are involved in the reduction of insoluble electron acceptors like ferric oxide or MnO_2 (see Chapters 17 and 18). The electrons in these instances travel in a direction opposite to that when an oxidation of an insoluble electron donor, like metal sulfide or a substrate like Fe^{2+}, at the outer surface of the outer membrane is involved.

Gehrke et al. (1995, 1998) and Sand et al. (1997) have proposed that the iron bound in the EPS around the cells of *Acidithiobacillus ferrooxidans* mediates attack of the sulfur moiety in pyrite. They view bulk-phase iron as inducing EPS formation and enabling attachment to pyrite. However, they do not differentiate between unbound, bulk-phase iron and iron bound (complexed) by EPS of *Acidithiobacillus ferrooxidans*. Bulk-phase ferric iron when interacting with metal sulfide mineral behaves as a chemical reactant, which is consumed in the oxidation of the metal sulfide. EPS-bound iron should be viewed as an *electron shuttle*, which is reversibly reduced and

oxidized during electron transfer from the metal sulfide being oxidized to an acceptor molecule of the cell (e.g., cytochrome Cyc2). This follows from the following consideration. The standard reduction potential for the Fe^{3+}/Fe^{2+} couple is $+777$ mV, whereas that of the $Fe(CN)_6^{3-}/Fe(CN)_6^{4-}$ in 0.01 N NaOH is $+460$ mV (Weast and Astle, 1982) and that for oxidized cytochrome *c*/reduced cytochrome *c* is $+0.245$ mmV (Lehninger, 1975, p. 479). Although no reduction potential for EPS-bound iron is available, its value is likely to lie between that for uncomplexed iron and the porphyrin-bound iron in cytochrome. EPS-bound iron would therefore be a significantly weaker oxidant and function better as an electron shuttle.

Whereas the mechanism by which *Acidithiobacillus ferrooxidans* promotes direct oxidative attack of metal sulfides is becoming clearer, that by which other bacteria known to promote such oxidative attack remains to be clarified. Some may only be able to promote such oxidation by indirect attack, but others are likely to be able to promote it by both direct and indirect attacks. Because *Acidithiobacillus ferrooxidans* is a Gram-negative organism, other Gram-negative organisms, like L. *ferrooxidans*, for instance, may use the same mode of direct attack, but not gram-positive organisms or archaea, like A. *tolerans* and A. *brierleyi*, respectively, which may be capable of it. This is because the envelope structure of Gram-positive bacteria and archaea is very different from that of Gram-negative bacteria (see Chapter 7).

Evidence for enzymatic attack of synthetic covellite (CuS) by noting inhibition of oxygen consumption and Cu^{2+} and SO_4^{2-} ion production by *Acidithiobacillus ferrooxidans* in the presence of the enzyme inhibitor trichloroacetate (8 mM) was obtained by Rickard and Vanselow (1978). In the case of CuS, only the sulfide moiety of the mineral is attacked because the metal moiety is already fully oxidized. The oxidation of the mineral probably proceeds in two steps (Fox, 1967):

$$CuS + 0.5O_2 + 2H^+ \xrightarrow{\text{Bacteria}} Cu^{2+} + S^0 + H_2O \tag{21.10}$$

$$S^0 + 1.5O_2 + H_2O \xrightarrow{\text{Bacteria}} SO_4^{2-} + 2H^+ \tag{21.11}$$

By contrast, *Thiobacillus thioparus* promotes covellite oxidation only after autoxidation of the mineral to $CuSO_4$ and S^0 (similar to Reaction 20.10 but in the absence of bacterial catalysis) (Rickard and Vanselow, 1978). It is the bacterial catalysis of the oxidation of S^0 to sulfate that helps the reaction by removing a product of the autoxidation of CuS.

In some instances, both an oxidizable metal moiety and the sulfide moiety may be attacked by separate enzymes, as, for example, in the case of chalcopyrite ($CuFeS_2$) (assuming the Fe of chalcopyrite to have an oxidation state of +2) (Duncan et al., 1967; Shrihari et al., 1991). Although Duncan and coworkers reported Fe and S to be simultaneously attacked, Shrihari et al. (1991) found that iron-grown *Acidithiobacillus ferrooxidans* oxidized the sulfide sulfur of chalcopyrite by direct attack before oxidizing ferrous iron in solution to ferric iron. When the dissolved ferric iron attained a significant concentration, it promoted chemical oxidation of residual chalcopyrite. The overall reaction by *Acidithiobacillus ferrooxidans* may be written as follows:

$$4CuFeS_2 + 17O_2 + 4H^+ \xrightarrow{\text{Acidithiobacillus ferrooxidans}}$$

$$4Cu^{2+} + 4Fe^{3+} + 8SO_4^{2-} + 2H_2O \quad (21.12)$$

$$4Fe^{3+} + 12H_2O \rightarrow 4Fe(OH)_3 + 12H^+ \quad (21.13)$$

$$4CuFeS_2 + 17O_2 + 10H_2O \xrightarrow{\text{Acidithiobacillus ferrooxidans}}$$

$$4Cu^{2+} + 4Fe(OH)_3 + 8SO_4^{2-} + 8H^+ \quad (21.14)$$

Reaction 21.14 is the sum of Reactions 21.12 and 21.13.

In other cases of direct attack, the oxidizable metal moiety may be oxidized before the sulfide, as in the example of chalcocite (Cu_2S) oxidation (Fox, 1967; Nielsen and Beck, 1972) (Figure 21.2):

$$Cu_2S + 05O_2 + 2H^+ \xrightarrow{\text{Acidithiobacillus ferrooxidans}}$$

$$Cu^{2+} + CuS + H_2O \quad (21.15)$$

$$CuS + 0.5O_2 + 2H^+ \xrightarrow{\text{Acidithiobacillus ferrooxidans}}$$

$$Cu^{2+} + S^0 + H_2O \quad (21.16)$$

$$S^0 + 1.5O_2 + H_2O \xrightarrow{\text{Acidithiobacillus ferrooxidans}}$$

$$SO_4^{2-} + 2H^+ \quad (21.17)$$

Digenite (Cu_9S_5) can be an intermediate in the formation of CuS from Cu_2S (Nielsen and Beck, 1972).

Although in the laboratory it is possible to demonstrate exclusive direct oxidation by *Acidithiobacillus ferrooxidans* of certain nonferrous metal sulfides by using iron-free mineral in an iron-free culture medium, in nature these conditions never occur. This is because nonferrous metal sulfides are always accompanied by pyrites in ore deposits. Thus, in nature, direct and indirect oxidations usually occur together. It is prevailing environmental conditions that determine the extent to which each mode contributes to the overall oxidation of a metal sulfide. Pyrite oxidation presents a special problem in applying the concepts of direct and indirect attacks in bacterial leaching and is treated in a special section to follow.

21.5.3 Indirect Oxidation

In indirect biooxidation of metal sulfides, a major role of the bacteria is the generation of a lixiviant, which chemically oxidizes the sulfide ore. This lixiviant is ferric iron (Fe^{3+}). It is a major *consumable reactant* in the oxidation of the metal sulfide. It may be iron (Fe^{2+}) at pH values of 3.5–5.0 by Metallogenium in a mesophilic temperature range (Walsh and Mitchell, 1972a). At pH values below 3.5, ferric iron may be generated from Fe^{2+} by bacteria such as *Acidithiobacillus ferrooxidans* and L. *ferrooxidans* Markosyan (see Balashova et al., 1974) in a mesophilic temperature range and by *Sulfolobus* spp., *A. brierleyi*, *A. tolerans*, and others in a thermophilic temperature range (Brierley and Brieley, 1973; Brierley and Murr, 1973; Balashova et al., 1974; Brierley and Lockwood, 1977; Brierley et al., 1978; Brierley, 1978b; Golovacheva and Karavaiko, 1979; Pivovarova et al., 1981; Harrison and Norris, 1985; Segerer et al., 1986). In addition, ferric iron can be generated from iron pyrites (e.g., FeS_2) by indirect attack by *Acidithiobacillus ferrooxidans* and other iron-oxidizing acidophiles. At least some of these organisms may also generate Fe^{3+} from pyrites by direct attack. In whichever way it is formed, ferric iron in acid solution acts as an oxidant of the metal sulfides in indirect attack (e.g., Sullivan, 1930; Ehrlich and Fox, 1967):

$$MS + 2Fe^{3+} \rightarrow M^{2+} + S^0 + 2Fe^{2+} \quad (21.18)$$

where M may be any metal in an appropriate oxidation state, which does not always have to be +2 (e.g., Cu(I) in Cu_2S). A central role of *Acidithiobacillus ferrooxidans* in an indirect oxidation process is to regenerate Fe^{3+} from the Fe^{2+} formed in Reaction 20.18. It should be noted that in *chemical* as opposed to *biochemical* oxidation, the sulfide of a mineral is mostly oxidized to elemental sulfur (S^0) (pyrite is an exception). Further oxidation to sulfuric acid (H_2SO_4) is very slow but is likely to be greatly accelerated by microorganisms like *A. thiooxidans*, *Acidithiobacillus ferrooxidans*, *Sulfolobus* spp., and *A. brierleyi*, but not by *L. ferrooxidans*. Elemental sulfur may form a film on the surface of metal sulfide crystals in chemical oxidation and interfere with further chemical oxidation of the residual metal sulfide. The chemical oxidation of metal sulfides must occur in acid solution below pH 5.0 to keep enough ferric iron in solution. In nature, the needed acid may be formed chemically through autoxidation of sulfur and other partially reduced forms of sulfur, but much more likely biologically through bacterial oxidation of the sulfur. The acid may also form as a result of autoxidation or biooxidation of ferrous iron or pyrite. In ferrous iron biooxidation, the acid forms as follows (autoxidation proceeds by the same reaction, but without bacterial participation, and much more slowly):

$$2Fe^{2+} + 0.5O_2 + 2H^+ \xrightarrow{\text{Bacteria}} 2Fe^{3+} + H_2O$$
(21.19)

$$Fe^{3+} + 3H_2O \rightarrow Fe(OH)_3 + 3H^+ \qquad (21.20)$$

Because this reaction normally occurs in the presence of sulfate, the ferric hydroxide may convert to the more insoluble jarosite, especially in the presence of *Acidithiobacillus ferrooxidans* and probably other acidophilic iron oxidizers (Lazaroff et al., 1982, 1985; Lazaroff, 1983; Carlson et al., 1992):

$$A^+ + 3Fe(OH)_3 + 2SO_4^{2-} \rightarrow AFe_3(SO_4)_2(OH)_6 + 3OH^-$$
(21.21)

where A^+ may represent Na^+, K^+, NH_4^+, or H_3O^+ (Duncan and Walden, 1972). The formation of jarosite decreases the ratio of protons produced per iron oxidized from 2:1 to 1:1. In pyrite oxidation, the acid forms as a result of Reactions 21.22 through 21.24 (see Section 21.5.4). Reaction 21.22 proceeds by autoxidation if the process is indirect. Like sulfur, jarosite may also form on the surface of metal sulfide crystals and block further oxidation.

21.5.4 Pyrite Oxidation

Pyrite oxidation by *Acidithiobacillus ferrooxidans* represents a special case in which direct and indirect oxidation of the mineral cannot be readily separated because ferric iron is always a product. Experimentally, Mustin et al. (1992) recognized four phases in the leaching of pyrite by *Acidithiobacillus ferrooxidans* in a stirred reactor. The first phase, which lasted about 5 days under the experimental conditions, featured a measurable decrease in planktonic (*unattached*) bacteria. The small amount of dissolved ferric iron added with the inoculum reacted with some of the pyrite.

The second phase, which also lasted about 5 days, featured the start of pyrite dissolution with oxidation of its iron and sulfur, but with sulfur being preferentially oxidized. Planktonic bacteria multiplied exponentially and the pH began to drop.

The third phase, which lasted about 10 days, featured a significant increase in dissolved ferric iron, the ferrous iron concentration remaining low. Both iron and sulfur in the pyrite were being oxidized at high rates. However, the rate of sulfur oxidation decreased with time relative to iron oxidation, the ratio of sulfate to ferric iron becoming stoichiometric by day 18. Planktonic bacteria continued to increase exponentially, and the pH continued to drop. The surface of the pyrite crystals began to show evidence of corrosion cracks.

In the fourth and last phase, which lasted about 25 days, the dissolved Fe(III)/Fe(II) ratio decreased slightly, iron and sulfur in the pyrite continued to be strongly oxidized, and the planktonic bacteria reached a stationary phase. At the same time, the pH continued to drop to 1.3 by the 45th day. The surface of the pyrite particles now showed easily recognizable square or hexagonal corrosion pits. Mustin and coworkers followed pH and electrochemical (redox) changes during the entire experiment of pyrite oxidation.

Classically, direct bacterial oxidation of iron pyrite (FeS_2) has been described by following the overall reaction:

$$FeS_2 + 3.5O_2 + H_2O \xrightarrow{\text{Attached \textit{Acidithiobacillus ferrooxidans}}}$$
$$Fe^{2+} + 2H^+ + 2SO_4^{2-} \qquad (21.22)$$

The dissolved ferrous iron generated in this reaction is further oxidized by planktonic bacteria according to the overall reaction

$$2Fe^{2+} + 0.5O_2 + 2H^+ \xrightarrow{\text{Planktonic } Acidithiobacillus\ ferrooxidans}$$
$$2Fe^{3+} + H_2O \qquad (21.23)$$

The resultant ferric iron then causes chemical oxidation of residual pyrite according to the following reaction:

$$FeS_2 + 14Fe^{3+} + 8H_2O \rightarrow 15Fe^{2+} + 2SO_4^{2-} + 16\ H^+ \qquad (21.24)$$

whereby Fe^{2+} is regenerated. The oxidation of Fe^{2+} by the planktonic bacteria becomes the rate controlling reaction in this model of pyrite oxidation according to Singer and Stumm (1970), which makes biooxidation of pyrite mostly an indirect process. This is because in the absence of the iron-oxidizing bacteria, the rate of Fe^{2+} oxidation is very slow at acid pH. According to a study by Moses et al. (1987), Fe^{3+} is the preferred chemical oxidant of pyrite over O_2, suggesting that an organism like *Acidithiobacillus ferrooxidans* will facilitate initiation of pyrite oxidation by generating Fe^{3+} from pyrite (Reactions 20.22 and 20.23).

Schippers and Sand (1999) viewed the role of *Acidithiobacillus ferrooxidans* in the oxidation of pyrite to be only indirect, as defined in this book, and they extended this concept to bacterial oxidation of other metal sulfides. In their model, *Acidithiobacillus ferrooxidans* catalyzes the oxidation of the dissolved Fe^{2+} in the bulk phase that results from the chemical oxidation of pyrite by Fe^{3+}:

$$FeS_2 + 6Fe^{3+} + 3H_2O \rightarrow S_2O_3^{2-} + 7Fe^{2+} + 6H^+ \qquad (21.25)$$

in which thiosulfate is formed as a product (see, e.g., Rimstidt and Vaughn [2003] and Descostes et al. [2004] for chemical pyrite oxidation). They proposed furthermore that the thiosulfate is subsequently oxidized by the ferric iron generated by *Acidithiobacillus ferrooxidans*:

$$S_2O_3^{2-} + 8Fe^{3+} + 5H_2O \rightarrow 2SO_4^{2-} + 8Fe^{2+} + 10H^+ \qquad (21.26)$$

ferric iron being the oxidant in this reaction. At the same time, the thiosulfate in their model is oxidized by *Acidithiobacillus ferrooxidans* itself:

$$S_2O_3^{2-} + 2O_2 + H_2O \xrightarrow{Acidithiobacillus\ ferrooxidans}$$
$$2SO_4^{2-} + 2H^+ \qquad (21.27)$$

In the bacterial oxidation, oxygen rather than Fe^{3+} is the terminal electron acceptor (oxidant). It is unclear, however, whether thiosulfate should exist at all at the acid pH (in the range of <2 to 3.5) at which these reactions would occur. Thiosulfate is known to decompose into elemental sulfur and sulfite (SO_3^{2-}) below a pH of 4–5 (see, e.g., Roy and Trudinger, 1970, p. 18; also Rimstidt and Vaughn, 2003, p. 879).

Sand et al. (1995) believe that direct oxidation of pyrite or any other sulfide mineral by *Acidithiobacillus ferrooxidans* does not occur. Thus, their model of bacterial metal sulfide oxidation relies on an abiotic attack by Fe^{3+}, the role of the bacteria being the regeneration of Fe^{3+} from the Fe^{2+} formed in the chemical oxidation of the metal sulfide (Reaction 20.23). Their general model suggests that *Acidithiobacillus ferrooxidans* cannot oxidize any metal sulfide in the absence of iron in the bulk phase. This is, however, contrary to previous observations (see, e.g., experiments by S.I. Fox summarized by Ehrlich, 1978, p. 71; Nielsen and Beck, 1972; Pogliani et al., 1990). In the case of pyrite oxidation, it seems impossible to distinguish between direct and indirect bacterial mechanisms in the absence of any suppression of the chemical action of ferric iron produced in pyrite oxidation. The ferric iron will always be a product of pyrite oxidation, no matter whether pyrite is attacked directly or indirectly. As commented earlier, Sand and collaborators' model does not distinguish between bulk-phase (dissolved) ferric iron and EPS-bound and cell-wall-bound (complexed) iron of *Acidithiobacillus ferrooxidans* in terms of reactivity.

Edwards et al. (2001) approached the question of direct versus indirect action by determining what effect *Acidithiobacillus ferrooxidans* and *F. acidarmanus* had on pitting of the mineral surface of pyrite, marcasite, and arsenopyrite in a mineral salt medium. They used scanning electron microscopy to make their assessments. They found that extensive pitting on the mineral surface occurred in the absence of bacteria with added ferric chloride but not without it. In the presence of bacteria, pitting depended on the kind of bacteria and their attachment, the kind of mineral, and probably other experimental conditions. Thus, *Acidithiobacillus ferrooxidans* produced cell-sized and cell-shaped dissolution pits on pyrite but not marcasite or arsenopyrite. *F. acidarmanus* produced such pits on pyrite and arsenopyrite but not on marcasite. However, individual cells were found in shallow pits on marcasite. The investigators came to the conclusion that overall sulfide dissolution in their experiments was dominated by reaction of a given mineral with bulk-phase Fe^{3+} and not by reaction at the cell/mineral surface interface. Nevertheless, they did not rule out the possibility of a cell/mineral surface interface reaction. If it did occur, they believed it to have been of minor importance in pit formation in these experiments.

If the bound (complexed) iron acts as an electron shuttle in the sense of Ingledew (1986), Moses et al. (1987) and Rawlings et al. (1999) suggest that the bound iron in bacterial cells when attached to a pyrite surface may accept the electrons from the half reactions describing the initial steps in corrosion of pyrite:

$$FeS_2 + 2H_2O \rightarrow Fe(OH)_2S_2 + 2H^+ + 2e \quad (21.28)$$

and

$$FeS_2 + 8H_2O \rightarrow Fe^{3+} + 2SO_4^{2-} + 16H^+ + 15e \quad (21.29)$$

In this way, the bacteria take the place of ferric iron in attacking pyrite. Because *L. ferrooxidans* also features bound iron in its EPS, it may thus also be able to attack pyrite by the direct mechanism when attached to a pyrite surface.

Because iron pyrites usually accompany other metal sulfides in nature, iron pyrite oxidation is an important source of acid for the oxidizing reactions of nonferrous metal sulfides, especially those that consume acid, e.g., the oxidation of chalcocite (Reactions 21.15 through 21.17). In some cases, the host rock in which metal sulfides, including pyrites, are contained may itself consume acid and thus raise the pH of the environment enough to cause extensive precipitation of ferric iron and thereby prevent oxidation of metal sulfide by it (e.g., Ehrlich, 1977, pp. 149–150).

The foregoing discussion of direct and indirect metal sulfide oxidation dealt mainly with *Acidithiobacillus ferrooxidans*, which is capable of the oxidation of both ferrous iron and reduced sulfur. Investigations need to be extended to *L. ferrooxidans*, which oxidizes ferrous iron readily but cannot oxidize reduced forms of sulfur. Yet, *L. ferrooxidans* has been reported to be the dominant member of the leaching microflora in many heap/dump leaching operations (Sand et al., 1992, 1993; Asmah et al. 1999; Bruhn et al., 1999). The ability to oxidize ferrous iron but not reduced forms of sulfur might suggest that *L. ferrooxidans* oxidizes metal sulfides only in an indirect mode. However, like *Acidithiobacillus ferrooxidans*, *L. ferrooxidans* produces exopolymer that has an affinity for iron (Gehrke et al., 1995; Sand et al., 1997). If this bound iron can act as an electron shuttle that transfers the electrons to a cytochrome in the outer membrane of *L. ferrooxidans*, as it may in *Acidithiobacillus ferrooxidans*, the organism should also be capable of oxidizing a metal sulfide in a direct mode. As previously pointed out, investigation of the mode(s) of action in metal sulfide oxidation also needs to be extended to the other microorganisms that have been found active in metal sulfide oxidation. Most of these can oxidize both Fe^{2+} and reduced forms of sulfur.

21.6 BIOLEACHING OF METAL SULFIDE AND URANINITE ORES

21.6.1 Metal Sulfide Ores

When metal sulfide ore bodies are exposed to moisture and air during mining activities, the ore mineral may begin to undergo gradual oxidation, which may be accelerated by native microorganisms, especially acidophilic iron oxidizers. Groundwater passing through a zone of ore oxidation will pick up soluble products of the oxidation and will issue from the site as acid mine drainage (AMD), which contains the metal solubilized in the oxidation. Mining companies harness the microbial oxidizing activity that mobilizes metals in some sulfidic ores on an industrial scale as an economic means of metal extraction. Such metal bioleaching may be applied to rubblized ore in situ (e.g., McCready, 1988), in ore heaps or dumps, or to crushed ore in special reactors. Initially, bioleaching was used commercially only with low-grade portions of an ore and with ore tailings, but with more recent improvements in the process, it is now also used in treating high-grade ore and ore concentrates.

Low-grade sulfide ores generally contain metal values at concentrations below 0.5% (wt/wt). Their extraction by smelting after milling and subsequent ore enrichment (ore beneficiation) by flotation is uneconomic because of an unfavorable gangue/metal ratio. Years of experience have shown that for most efficient bioleaching, ore heaps should be constructed to heights limited to tens of feet to avoid slumping. Ore heaps may consist of waste rock or mine tailings, which are by-products of mining that still contain traces of recoverable metal values, but nowadays may also consist of high-grade ore. The lixiviant can be water, acidified water, or spent acidic leach solution (barren solution) containing ferric sulfate from a previous leaching cycle. It is applied in a fine spray onto ore heaps and dumps (Figure 21.3a). The spraying avoids waterlogging of the heaps and dumps, which would exclude needed oxygen (air). Simultaneous diffusion of oxygen into the ore in a heap or dump being leached is important because the microbial leaching process is aerobic. If oxygen does not reach all parts of a heap or dump because its concentration in the leach solution is insufficient to meet the demand for microbial metal sulfide oxidation, anaerobic conditions will develop in those parts that are not reached by oxygen. In such regions, a microbial flora, mostly heterotrophs, has been shown to develop that includes bacteria that reduce ferric to ferrous iron and others that reduce sulfate to H_2S (Fortin et al., 1995; Fortin and Beveridge, 1997; Johnson and Roberto, 1997). The Fe reducers lower the ferric iron concentration available for chemical oxidation of metal sulfides in the anoxic zone. The sulfate reducers cause metal species mobilized in the oxidized zones to reprecipitate as sulfides. In some heap leaching operations, perforated pipes are placed in strategic positions within the heap during its construction to optimize access of air to deeper portions of the heap.

The lixiviant solution applied to the ore makes possible the growth and multiplication of appropriate acidophilic iron oxidizers and the oxidation of pyrite, chalcopyrite, and nonferrous metal sulfides in the ore. Initially, solutions issuing from the heaps or dumps may be recirculated without any treatment, but ultimately, as microbial and chemical activities continue, the solution in the heaps and dumps becomes charged with dissolved metal values, and after issuing from the heaps or dumps, it is collected as *pregnant solution* in special sumps. Pregnant solution frequently harbors a variety of microorganisms, including autotrophic and heterotrophic bacteria, fungi, and protozoa, despite its very acid pH and high metal content (e.g., Ehrlich, 1963b). Indeed, fungi and protozoa have now been identified in acid mine solution in a pH range from 0.8 to 1.38 in the Richmond Mine at Iron Mountain in California (Baker et al., 2004). Some of the protozoa may harbor prokaryotic endosymbionts affiliated with a Rickettsiales lineage (Baker et al., 2003). When the concentration of desired metal values in the pregnant solution in a sump is high enough, they are stripped from the solution. This may be accomplished in one of several ways. A formerly widely used method for copper separation involved treatment of pregnant solution with sponge iron (Fe^0) in a specially constructed basin called a launder (Figure 21.3b). The sponge iron precipitated the copper by *cementation* in a process involving the following reaction:

$$Cu^{2+} + Fe^0 \rightarrow Cu^0 + Fe^{2+} \qquad (21.30)$$

The copper metal formed in this way was very impure and required further refinement by smelting.

After the metal value is stripped from a pregnant solution, the metal-depleted solution is called "barren solution." It can be recirculated as lixiviant in the heap leaching operation. However, when cementation was used to strip the copper from pregnant solution, the resultant barren solution was significantly enriched in Fe^{2+}. In many instances, it was enriched in ferrous iron to such an extent that some of this iron had to be removed before the solution could be reintroduced in the leach heaps or dumps. Without removal of the excess ferrous iron, there was a danger of excessive jarosite formation upon its oxidation in the ore heaps or dumps being leached. The jarosite could precipitate on the ore mineral surfaces, interfering with further oxidation, and could also clog the drainage channels in the ore heaps and dumps. Plugging and impeding of the leach process would be the result. Excess iron removal from acid barren solution was best accomplished by biooxidation in shallow lagoons called "oxidation ponds." In these ponds, acidophilic iron oxidizers present in the barren solution from previous leaching cycles promoted the oxidation of the

EHRLICH'S GEOMICROBIOLOGY

(a)

(b)

Figure 21.3. Bioleaching of copper from sulfidic copper ores. (a) Top of a leach dump showing corrosion-resistant pipes and hose for watering the dump with barren solution. The dark patches represent moistened areas in which oxidation has occurred. Note the thin streams of solution issuing from the hoses in the distance. (b) Launder used in recovering copper by cementation with sponge iron from pregnant solution from the leach dumps. The copper recovered in this way has to be purified by smelting. Currently, the preferred method of copper recovery from pregnant solution is by electrolysis (electrowinning) because it yields a pure product. (Courtesy of Duval Corporation.)

ferrous iron with concomitant acidification. A significant portion of the oxidized iron precipitated as basic ferric sulfates, including jarosite, in the oxidation ponds. When reintroduced into the heap or dump, the residual iron in the treated barren solution, which was mostly ferric, caused indirect leaching of the metal sulfides in the ore.

The acid in the recirculated barren solution that entered the heaps and dumps caused relatively rapid weathering of the host rock (gangue) of the ore, resulting in liberation of aluminum and aluminosilicates. This weathering is important in exposing occluded metal sulfide crystals to the lixiviant and to the bacteria active in the leaching process. The liberated aluminum could ultimately be separated as $Al(OH)_3$ by neutralizing the pregnant solution (Zimmerley et al., 1958; Moshuyakova et al., 1971), but this has not been done in practice. The recovered $Al(OH)_3$ could be subsequently used in the manufacture of aluminum metal. High acidity of the lixiviant may also play an important role in preventing metal ions formed during leaching from being adsorbed by the host rock (gangue) (Ehrlich and Fox, 1967; Ehrlich, 1977).

A currently preferred method of recovering metal values from pregnant solution involves *electrowinning* when the pregnant solution contains only one major metal value or *solvent extraction* when several different metal values are present, followed by electrowinning of each of the separated metal values. Electrowinning is an electrolytic process in which the metal to be recovered is deposited on a cathode made of the same metal. The anode is usually made of carbon. The metal product of electrowinning is usually of high purity and normally does not need further refining. These metal separation processes have the advantage of not raising the ferrous iron concentration in barren solution. However, recovery of metal values from pregnant solution by solvent extraction can introduce reagents into the resultant barren solution that are inhibitory to the bacteria involved leaching and may have to be removed before recirculation of the barren solution in heaps or dumps.

The acidophilic iron-oxidizing bacteria developing in an ore leaching process play a dual role in solubilizing metal values, as already explained. They generate acid ferric sulfate lixiviant by attacking pyrite and by reoxidizing Fe^{2+}, and they also attack the mineral sulfides directly, as previously discussed. It is usually not possible to assess quantitatively the extent to which these organisms are involved in direct and indirect oxidation if they are capable of both.

The interior temperature of some ore dumps or heaps, especially if they are not well ventilated, can rise as high as 70°C–80°C. Lyalikova (1960) observed that the heating can be accelerated by bacterial action. The heating is due to the fact that metal sulfide oxidation is an exothermic process. Such a temperature rise is unfavorable for the growth of mesophilic bioleaching bacteria, such as *Acidithiobacillus ferrooxidans* and L. *ferrooxidans*, when it occurs in a heap or a dump interior. Microbial activity in oxidation ponds for removal by of excess iron in barren solution is not affected by the heat generated in the heaps and dumps.

The observation of interior heating of ore heaps and dumps during leaching suggested to some in the 1960s that the leaching process in dumps and heaps of metal sulfide ores is mostly abiotic. This view changed after the discovery of thermophiles capable of promoting bioleaching of metal sulfides. *Acidithiobacillus ferrooxidans* and L. *ferrooxidans*, which are unable to live at temperatures in excess of 37°C–40°C, are succeeded by acidophilic, iron-oxidizing thermophiles in the interior of leach dumps and heaps where the temperature has risen to the thermophilic range (e.g., Norris, 1997). These microorganisms, which catalyze reactions similar to those of *Acidithiobacillus ferrooxidans*, operate optimally at the higher temperatures. Thus, contrary to earlier views, leaching in all parts of a heap or dump is most likely biological. It has also been suggested that the thermophiles may be responsible for regulating the internal temperature of active leach heaps and dumps (Murr and Brierley, 1978).

The primary copper mineral in ore bodies of magmatic hydrothermal origin is chalcopyrite ($CuFeS_2$). This mineral tends to be somewhat refractory to chemical (abiotic) leaching by acidic ferric sulfate compared to the secondary copper sulfide minerals chalcocite (Cu_2S) and covellite (CuS). The oxidation of chalcopyrite in nature can thus be significantly enhanced by *Acidithiobacillus ferrooxidans* and *A. brierleyi* (Razzell and Trussell, 1963; Brierley, 1974). Ferric iron, when present in excess of 1000 ppm, has been found to inhibit chalcopyrite oxidation (Duncan and Walden, 1972; Ehrlich, 1977), probably because it precipitates as jarosite or adsorbs to the surface of chalcopyrite crystals and prevents further oxidation. Bacteria themselves may interfere by generating

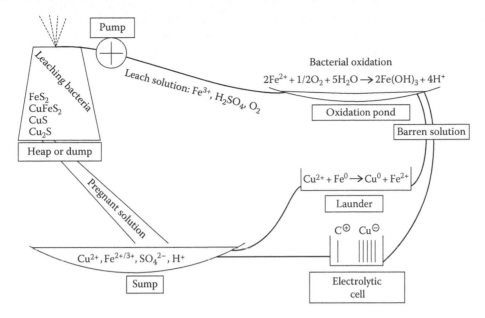

Figure 21.4. Schematic representation of a bioleach circuit for heap or dump leaching of copper sulfide ore. In addition to copper recovery from pregnant solution by cementation with sponge iron in a launder or by electrowinning, copper can also be recovered by solvent extraction in combination with electrowinning when the pregnant solution contains two or more base metals.

excess ferric iron from ferrous iron that precipitates or is adsorbed by residual chalcopyrite.

A typical leach cycle in which copper is recovered either by cementation or by electrowinning is diagrammed in Figure 21.4 (for further discussion of bacterial leaching, see Brierley [1978a, 1982], Lundgren and Malouf [1983], Rawlings [1997a, 2002], Rohweder et al. [2003], and Olson et al. [2003]).

In practice, the leaching of Pb from PbS through oxidation by *Acidithiobacillus ferrooxidans* can present a special problem because the oxidation product, $PbSO_4$, is relatively insoluble. As oxidation of PbS proceeds, $PbSO_4$ is likely to accumulate on the crystal surface and block further access to PbS by the bacteria, dissolved Fe^{3+} if present, and oxygen. In the laboratory, this problem is largely eliminated when *Acidithiobacillus ferrooxidans* oxidizes PbS in batch culture with agitation (Silver and Torma, 1974) or in a large volume of lixiviant in a stirred continuous-flow reactor (Ehrlich, 1988).

Metal sulfide-oxidizing bacteria are naturally associated with metal sulfide-containing deposits, including ore bodies and bituminous coal seams. They usually exist in a consortium when ore from such deposits is bioleached (Bruhn et al., 1999). Therefore, heap, dump, and in situ leaching operations do not require inoculation with

active bacteria, although the leaching process may be improved by it (Brierley et al., 1995). Reactor leaching operations, on the other hand, greatly benefit from inoculation with a strain selected for enhanced activity. The inoculum has to be massive for either leaching process in order to outgrow the organisms naturally present on the ore. The ore cannot be sterilized on an industrial scale.

Under natural conditions, growth and activity of the leaching organisms may be limited by one or more environmental factors. These include limited access to an energy source (metal sulfide crystal), limited nitrogen source, unfavorable temperature, and limited access to air and/or moisture (Ehrlich and Fox, 1967; Brock, 1975; Ahonen and Tuovinen, 1989, 1992). It should be noted, however, that *Acidithiobacillus ferrooxidans* is capable of anaerobic growth with H_2, formate, or S^0 as electron donor and Fe(III) as terminal electron acceptor (see Chapter 17).

21.6.2 Uraninite Leaching

The principles of bioleaching have also been applied on a practical scale to the leaching of uraninite ores, especially if the ores are low grade. The process may involve dump, heap, or in situ leaching (Wadden and Gallant, 1985; McCready and Gould, 1990). *Acidithiobacillus ferrooxidans* is one organism that has been harnessed for this process. Its action in this instance is chiefly indirect by generating an oxidizing lixiviant, acid ferric sulfate, which oxidizes U(IV) in the uraninite ore to soluble U(VI). The overall key reactions leading to uranium mobilization from uraninite can be summarized as follows:

$$2Fe^{2+} + 0.5O_2 + 2H^+ \xrightarrow{\text{\it Acidithiobacillus ferrooxidans}}$$

$$2Fe^{3+} + H_2O \qquad (21.31)$$

Some of the resultant ferric iron hydrolyzes and generates acid:

$$Fe^{3+} + 3H_2O \rightarrow Fe(OH)_3 + 3H^+ \quad (21.32)$$

The ultimate product of Fe^{3+} hydrolysis will more likely be a basic ferric sulfate such as jarosite rather than ferric hydroxide. The consequence of jarosite formation is that the net yield of acid (protons) will be less than in Reaction 21.32 (see Reaction 21.21). The remaining dissolved ferric iron can then react abiologically with the uraninite to form uranyl ions:

$$UO_2 + 2Fe^{3+} \rightarrow UO_2^{2+} + 2Fe^{2+} \quad (21.33)$$

Acidithiobacillus ferrooxidans will reoxidize the ferrous iron from this reaction, thus maintaining the process without the need for continual external resupply of ferric iron to the system. The dissolved uranium may be recovered from solution through concentration by ion exchange.

The acidophilic iron-oxidizing bacteria are often naturally associated with the ore body if it contains pyrite or another form of iron sulfide. In an absence of pyrite in depleted mines, the growth of these bacteria may be stimulated for in situ leaching by intermittent spraying of $FeSO_4$-nutrient-enriched solution onto the floors, walls, and mud of mine stopes (Zajic, 1969). If the leachate becomes anoxic and its pH rises, sulfate-reducing bacteria may develop in it. These bacteria can reprecipitate UO_2 as a result of the reaction of UO_2^{2+} with H_2S:

$$UO_2^{2+} + H_2S \rightarrow UO_2 + 2H^+ + S^0 \quad (21.34)$$

Some observations have indicated that *Acidithiobacillus ferrooxidans* can enzymatically catalyze the oxidation of U(IV) to U(VI), but it has not yet been shown to grow under these conditions in the lab (DiSpirito and Tuovinen, 1981, 1982a,b) (see also Chapter 19). Living cells accumulate significantly less uranium than dead cells and bind it chiefly in their cell envelope (DiSpirito et al., 1983). Despite this capacity of *Acidithiobacillus ferrooxidans* to oxidize (IV) directly, bioleaching of uraninite is believed to involve chiefly the indirect pathway.

Uranium bioleaching is another example of a natural process that is artificially stimulated. Under natural conditions, the reaction described here must occur on a very limited scale and thus only result in slow mobilization of uranium.

UO_2^{2+} in drainage from uranium mines can be microbiologically precipitated by its reduction under anaerobic conditions to insoluble UO_2. Some sulfate-reducing bacteria are capable of this reaction (Lovley et al., 1993). *Geobacter metallireducens* and *Shewanella putrefaciens* are two other bacterial species capable of precipitating U(IV) under anaerobic conditions (Gorby and Lovley, 1992). The electron donors used by the bacteria in these reactions are usually organic compounds but can be H_2 in the case of some sulfate reducers and *Shewanella*. This activity can be useful in remediating uranium-containing mine drainage, although only at a circumneutral pH.

21.6.3 Mobilization of Uranium in Granitic Rocks by Heterotrophs

Acidithiobacillus ferrooxidans is not the only organism capable of uranium mobilization. Heterotrophic microorganisms such as some members of the soil microflora and bacteria from granites or mine waters (*Pseudomonas fluorescens*, *Pseudomonas putida*, *Achromobacter*) can mobilize uranium in granitic rocks, ore, and sand by weathering that results from mineral interaction with organic acids and chelators produced by the microorganisms (Zajic 1969; Magne et al., 1973,1974). Magne et al. found experimentally that the addition of thymol to percolation columns of uraniferous material fed with glucose solution selected a microbial flora whose efficiency in uranium mobilization was improved by greater production of oxalic acid. The authors suggested that in nature, phenolic and quinoid compounds of plant origin can serve the role of thymol. They also reported that microbes can precipitate uranium by digestion of soluble uranium complexes (Magne et al., 1974), i.e., by microbial destruction of the organic moiety that complexes the uranium. These observations may explain how in nature uranium in granitic rock may be mobilized by bacteria and reprecipitated and concentrated elsewhere under the influence of other microbial activities.

21.6.4 Study of Bioleaching Kinetics

A number of studies have been published on the kinetics of bioleaching of metal sulfides under controlled conditions in the presence of ferrous iron. They include studies by Boon et al. (1995), Hansford (1997), Hansford and Vargas (1999), Crundwell (1995, 1997), Nordstrom and Southam (1997), Hansford and Vargas (1999), Driessens et al. (1999), Fowler and Crundwell (1999), and Howard and Crundwell (1999). In these studies, the assumption was made or the inference drawn that bioleaching of metal sulfides proceeds in only one mode. The existence of separate direct and indirect modes that may occur concurrently in the same leaching operation appears to have been rejected. The authors did not consider that the experimental conditions used in their experiments happened to favor strongly an indirect mode of leaching over a direct mode. However, Fowler and Crundwell (1999) do assign oxidation of S^0 that appears at the surface of ZnS during leaching to attached *Acidithiobacillus ferrooxidans* cells.

21.6.5 Industrial versus Natural Bioleaching

Industrial bioleaching of metal sulfide ores harnesses naturally occurring microbiological processes by creating selective and optimized conditions that allow leaching to occur at fast rates. In the absence of human intervention, the same processes occur only at very slow rates in highly localized situations, contributing in the case of sulfide ore to a very slow, gradual change from reduced to oxidized ore. This accounts for the relative stability of undisturbed ore bodies.

21.7 BIOEXTRACTION OF METAL SULFIDE ORES BY COMPLEXATION

Some metal sulfide ores cannot be oxidized by acidophilic iron-oxidizing bacteria because they contain too great an amount of acid-consuming constituents in the host rock (gangue). The metals in such ores may be amenable to extraction by some microorganisms such as fungi (e.g., Burgstaller and Schinner, 1993). Wenberg et al. (1971) reported the isolation of the fungus *Penicillium* sp. from a mine tailing pond of the White Pine Copper Co. in Michigan, United States, which produced unidentified metabolites that could mobilize copper from sedimentary ores of the White Pine deposit in Czapek's broth containing sucrose, $NaNO_3$, and cysteine, methionine, or glutamic acid. *Acidithiobacillus ferrooxidans* could not be employed for leaching of this ore because of the presence of significant quantities of calcium carbonate that would neutralize the required acid. Similar findings were reported by Hartmannova and Kuhr (1974), who found that not only *Penicillium* sp. but also *Aspergillus* sp. (e.g., *Aspergillus niger*) were active in producing complexing agents that leached copper. More recently, Mulligan and Galvez-Cloutier (2000) demonstrated mobilization of copper in an oxidized mining residue by *A. niger* in a sucrose–mineral salt medium. The chief mobilizing agents produced by the fungus were gluconic and citric acids, which can act as acidulants as well as ligands of metal ions.

In some of their experiments, Wenberg et al. (1971) grew their fungus in the presence of copper ore (sulfide or native copper minerals with basic gangue constituents). The addition of some citrate to the medium lowered the toxicity of the extracted copper when the fungus was grown in the presence of the ore. They obtained better results when they grew the fungus in the absence of the ore and then treated the ore with the spent medium from the fungus culture. The principle of action of the fungi in all the earlier cited experiments is similar to that involved in a study by Kee and Bloomfield (1961), who noted the dissolution of the oxides of several trace elements (e.g., ZnO, PbO_2, MnO_2, CoO, Co_2O_3) with anaerobically fermented plant material (Lucerne and Cocksfoot). The principle of action is also similar to that employed in the experiments by Parès (1964a–c), in which *Serratia marcescens* and *Bacillus subtilis*, *B. sphaericus*, and *B. firmus* solubilized copper and some other metals that were associated with laterites and clays. The organisms formed ligands that extracted the metals from the ores by forming complexes with them, which were more stable than the original insoluble form of the metals in the ores. This type of reaction can be formulated as follows:

$$MA + HCh \rightarrow MCh + H^+ + A^- \quad (21.35)$$

where MA is a metal salt (mineral), HCh a ligand (chelating agent), MCh the resultant metal chelate, and A^- the counter ion of the original metal salt, which would be S^{2-} in the case of metal sulfides. The S^{2-} may undergo chemical or bacterial oxidation. The use of carboxylic acids in industrial leaching of ores has been proposed as a general process (Chemical Processing, 1965).

21.8 FORMATION OF ACID COAL-MINE DRAINAGE

When bituminous coal seams that contain pyrite inclusions are exposed to air and moisture during mining, the pyrites undergo oxidation, leading to the formation of AMD. With the onset of pyrite oxidation, acidophilic iron-oxidizing thiobacilli become readily detectable in the drainage (Leathen et al., 1953). *A. thiooxidans* also makes an appearance . When *Acidithiobacillus ferrooxidans* is involved, pyrite biooxidation proceeds by the reactions previously described (Reactions 20.19 through 20.24). *A. thiooxidans*, which cannot oxidize ferrous iron, probably oxidizes elemental sulfur (S^0) and other partially reduced sulfur species, which may form as intermediates in pyrite oxidation, to sulfuric acid (e.g., Mustin et al., 1992, 1993) (Reaction 20.17). The chief products of pyrite oxidation are thus sulfuric acid and basic ferric sulfate (jarosite and amorphous basic ferric sulfates). Streams that receive this mine drainage may exhibit pH values ranging from 2 to 4.5 and sulfate ion concentrations ranging from 1000 to 20,000 mg L^{-1} but a nondetectable ferrous iron concentration (Lundgren et al., 1972). Walsh and Mitchell (1972b) proposed that a Metallogenium-like organism that they isolated from AMD may be the dominant iron-oxidizing organism attacking the pyrite in exposed pyrite-containing coal seams until the pH drops below 3.5. After that, the more acid-tolerant *Acidithiobacillus ferrooxidans* and probably L. ferrooxidans, which was unknown to Walsh and Mitchell, take over. The Metallogenium-like organism does not, however, appear to be essential to lower the pH in a pyritic environment to make it favorable for the acidophilic iron oxidizers. *Acidithiobacillus ferrooxidans* itself may be capable of doing that, at least in pyrite-containing coal or overburden (Kleinmann and Crerar, 1979). It may accomplish this by initial direct attack of pyrite (see Reaction 20.22), creating an acid microenvironment from which the organism and the acid it generates spread.

An early study of microbial succession in coal spoil under laboratory conditions was carried out by Harrison (1978). He constructed an artificial coal spoil by heaping a homogeneous mixture of 1 part crushed, sifted coal plus 2 parts shale and 8 parts subsoil from the overburden of a coal deposit into a mound 50 cm in diameter and 25 cm high on a plastic tray. The mound was inoculated with 20 L of an emulsion of acid soil, drainage water, and mud from a spoil from an old coal strip mine, which was poured on the bottom of the plastic tray. The inoculum was absorbed by the mound and migrated upward, presumably by capillary action. Evaporation losses during the experiment were made up by periodic additions of distilled water to the free liquid on the tray.

Initial samples taken at the base of the mound yielded evidence of the presence of heterotrophic bacteria. These bacteria were dominant and reached a population density of about 10^7 cells g^{-1} within 2 weeks. After 8 weeks, heterotrophs were still dominant, although the pH had dropped from 7 to 5. Between 12 and 20 weeks, the population decreased by about an order of magnitude, coinciding with a slight decrease in pH to just below pH 5 caused by a burst of growth by sulfur-oxidizing bacteria, which then died off progressively. Thereafter, the heterotrophic population increased again to just below 10^7 g^{-1}. In the samples from near the summit of the mound, heterotrophs predominated for the first 15 weeks but then decreased dramatically from 10^6 to 10^2 cells g^{-1}, concomitant with a drop in pH to 2.6. The pH drop was correlated with a marked rise in population density of sulfur- and iron-oxidizing autotrophic bacteria (*A. thiooxidans* and *Acidithiobacillus ferrooxidans*), the former dominating briefly over the latter in the initial weeks. Protozoans, algae, an arthropod, and a moss were also noted, mostly at the higher pH values. The presence of protozoa in coal-mine drainage was earlier reported by Lackey (1938) and Joseph (1953) and later by Johnson and Rang (1993). Metallogenium of the type of Walsh and Mitchell (1972a) was not seen.

The sulfur-oxidizing bacteria were assumed to be making use of elemental sulfur resulting from the oxidation of pyrite by ferric sulfate:

$$FeS_2 + Fe_2(SO_4)_3 \rightarrow 3FeSO_4 + 2S^0 \quad (21.36)$$

More specifically, as Mustin et al. (1992, 1993) indicated, the sulfur may arise as a result of anodic reactions at the surface of pyrite crystals. It is also possible that at least some of the sulfur arises indirectly from the reduction of microbial sulfate in anaerobic zones of the coal spoil. The reduction of sulfate would yield H_2S, which then becomes the energy source for thiobacilli such as T. thioparus that oxidize it to elemental sulfur at the interface of the oxidized and reduced zones, provided the ambient pH is not too far below neutrality. Anaerobic bacteria were not sought in this study.

After 7 weeks of incubation, a mineral efflorescence developed on the surface of the mound. It consisted mainly of sulfates of Mg, Ca, Na, Al, and Fe. The magnesium sulfate was in the form of the hexahydrate rather than epsomite. The leached metals derived from the coal, but magnesium was also leached from the overburden material. Harrison's study, which was reported in 1978 before the introduction of the techniques of molecular biology to microbial ecology, deserves to be repeated using molecular techniques. It could yield new insights into the microbial succession in the development of AMD from coal spoils. Baker and Banfield (2003) have reviewed recent findings from studies of microbial diversity in AMD that relied on DNA analysis rather than on culture techniques. The findings hint at a complex series of microbial successions that must have taken place over time as AMD generation evolved.

Darland et al. (1970) isolated thermophilic, acidophilic Thermoplasma acidophilum from a coal refuse pile that had become self-heated. The organism was described as lacking a true bacterial cell wall and to resemble mycoplasmas; however, it is an archeon. Its growth temperature optimum was 59°C (range 45°C–62°C) and its optimum pH for growth was between 1 and 2 (pH range 0.96–3.5). The organism is a heterotroph growing readily in a medium of 0.02% $(NH_4)_2SO_4$, 0.05% $MgSO_4$, 0.025% $CaCl_2 \cdot 2H_2O$, 0.3% KH_2PO_4, 0.1% yeast extract, and 1.0% glucose at pH 3.0. Its relation to the coal environment and its contribution, if any, to the AMD problem needs to be clarified.

21.8.1 New Discoveries Relating to Acid Mine Drainage

A fairly recent study of abandoned mines at Iron Mountain, California, United States, resulted in a startling discovery insofar as the generation of AMD is concerned, at least at this mine. The ore body at Iron Mountain at the time it was mined contained various metal sulfides and was a source of Fe, Cu, Ag, and Au. A significant part of the iron was in the form of pyrite. The drainage currently coming from the abandoned mine workings contains varying amounts of these metals as well as Cd and is very acidic. A survey of the distribution of Acidithiobacillus ferrooxidans and L. ferrooxidans in solutions from a pyrite deposit in the Richmond Mine, seepage from a tailings pile, and AMD storage tanks outside this mine was undertaken (Schrenk et al., 1998). It revealed that Acidithiobacillus ferrooxidans occurred in slime-based communities at pH >1.3 at temperatures below 30°C, whereas L. ferrooxidans was abundant in subsurface slime-based communities and also occurred in planktonic form at pH values in the range of 0.3–0.7 between 30°C and 50°C (Figure 21.5). Acidithiobacillus ferrooxidans appeared to affect precipitation of ferric iron but seemed to have a minor role in acid generation. Neither Acidithiobacillus ferrooxidans nor L. ferrooxidans was thought to exert a direct catalytic

Figure 21.5. A field site in an abandoned stope within the Richmond Mine at Iron Mountain in northern California, showing streamers of bacterial slime in the foreground. A pH electrode is visible on the right. A fishing line in the center secures a white bottle with a filter lid that contains sulfide mineral samples to study their fate in situ. (Courtesy of Thomas Gihring, Gainesville, FL.)

effect on metal sulfide oxidation, but they were thought to play an active role in generating ferric iron as an oxidizing agent (Schrenk et al., 1998). Microbiological investigation of underground areas (drifts) in the Richmond Mine revealed the presence of archaea as well as *Acidithiobacillus ferrooxidans* and *L. ferrooxidans* (Edwards et al., 1999a), but only the archaea and *L. ferrooxidans* were associated with acid-generating sites (Edwards et al., 1999b). The proportions in which they were detected varied with the site and the season of the year. Members of the Bacteria were most abundant during winter months when archaea were nearly undetectable. The reverse was found in summer and fall months, when archaea represented ~50% of the total population. The authors correlated these population fluctuations with rainfall and conductivity (dissolved solids), pH, and temperature of the mine water. As noted earlier, *Acidithiobacillus ferrooxidans* was the least pH tolerant and was less temperature tolerant than *L. ferrooxidans*. As already mentioned, *Acidithiobacillus ferrooxidans* was absent at acid-generating sites in the mine (Edwards et al., 1999b). Any attachment of iron-oxidizing bacteria was restricted to pyrite phases. Based on dissolution rate measurements, the investigators found that attached and planktonic species contributed to similar extents to acid release. Among the archaea, a newly discovered microbe, *F. acidarmanus*, grew in slime streamers on the pyrite surfaces (Edwards et al., 2000a). It constituted up to 85% of the total communities in the slimes and sediments that were examined. In laboratory study, it was found to be extremely acid tolerant. It was able to grow at pH 0 and exhibited a pH optimum at 1.2 (see also Section 21.5.1). Its cells lack a wall. It belongs to the Archean order Thermoplasmatales (Edwards et al., 2000a) and is a close relative of *F. acidiphilum* (Golyshina et al., 2000). On the basis of studies to date, Edwards et al. (2001) believe that *F. acidarmanus* and *F. acidiphilum* promote pyrite oxidation by generating the oxidant Fe^{3+} from Fe^{2+} (indirect mechanism).

Using molecular phylogenetic techniques, examination of a ~1 cm thick slime on finely disseminated pyrite ore collected in the Richmond Mine revealed the presence of a variety of bacterial types (Bond et al., 2000b). Predominant were *Leptospirillum* species representing 71% of the clones recovered. *Acidimicrobium*-related species, including *Ferromicrobium acidophilus*, were detected. archaea were represented by the family of Ferroplasmaceae. They included organisms closely related to *F. acidiphilum* and *F. acidarmanus* and organisms with an affinity to *T. acidophilum*. Also detected were members of the δ-subdivision of sulfate and metal reducers. These findings indicate the presence of microniches in the slime, some supporting aerobic (oxidizing) activity, while others anaerobic (reducing) activity. The makeups of microbial communities at environmentally different sites in the Richmond Mine were found to differ quantitatively and to some degree qualitatively (Bond et al., 2000a).

The interesting studies at Iron Mountain relegate the position of *Acidithiobacillus ferrooxidans* in AMD formation to the periphery of the remaining ore body at the present time. This position appears to be dictated by prevailing environmental conditions inside and outside the mine. This may not always have been the position of *Acidithiobacillus ferrooxidans*. Its current position may be the result of a species succession in microbial community development that started when the pyrite and other metal sulfides in the ore body first began to be oxidized upon exposure to air and water as mining proceeded. In AMD arising from the weathering of As- and pyrite-rich mine tailings at an abandoned mining site in Carnoulès, France, *Acidithiobacillus ferrooxidans* was readily detectable along with many as yet uncultured organisms (Duquesne et al., 2003; Bruneel et al., 2006). It was also readily detected in addition to *F. acidiphilum* and some other organisms in the Tinto River, southwestern Spain (González-Toril et al. 2003). Thus, microbial succession is likely to have taken place over time in the formation of AMD in the Iron Mountain deposit (Edwards et al., 2000b) and is strongly suggested in a study of a tank bioleaching process at the Mintek plant in South Africa in which Thermoplasmatales appeared (Okibe et al., 2003).

The findings from Iron Mountain raise the question of whether similar changes have occurred or are taking place in the generation of AMD from bituminous coal mines where the pyrite in the coal is more dispersed than in the Iron Mountain ore body. Clearly, AMD formation from bituminous coal mines needs to be reinvestigated. The question of microbial succession in the development of AMD in the mining of metal sulfide ore bodies also needs to be investigated. The evolution of microbial diversity over time in AMD cannot be readily deduced from analysis of a single sample late in the process.

21.9 SUMMARY

Metal sulfides may occur locally at high concentrations, in which case they constitute ores. Although most nonferrous sulfides are formed abiogenically through magmatic and hydrothermal processes, a few sedimentary deposits are of biogenic origin. More importantly, some sedimentary ferrous sulfide accumulations are biogenically formed. The microbial role in biogenesis of any of the sulfide deposits is the generation of H_2S, usually from the bacterial reduction of sulfate, but in a few special cases possibly from the mineralization of organic sulfur compounds. Because metal sulfides are relatively water insoluble, spontaneous reaction of metal ions with the biogenic sulfide proceeds readily. Biogenesis of specific metal sulfide minerals has been demonstrated in the laboratory. These experiments require relatively insoluble metal compounds as starting materials to limit the toxicity of the metal ions to the sulfate-reducing bacteria. In nature, adsorption of the metal ions by sediment components serves a similar function in lowering the concentration below their toxic levels for sulfate reducers.

Metal sulfides are also subject to oxidation by bacteria such as *Acidithiobacillus ferrooxidans, L. ferrooxidans, Sulfolobus* spp, and *A. brierleyi*. The bacterial action may involve direct oxidative attack of the crystal lattice of a metal sulfide or indirect oxidative attack by generation of lixiviant (acid ferric sulfate), which oxidizes the metal sulfide chemically. The indirect mechanism is of primary importance in the solubilization of uraninite (UO_2). Microbial oxidation of metal sulfides is industrially exploited in extracting metal values from low-grade metal sulfide ore and uraninite and has been tested successfully on some high-grade ore and ore concentrates. In bituminous coal seams that are exposed as a result of mining activity, pyrite oxidation by these bacteria is an environmentally deleterious process; it is the source of AMD.

REFERENCES

Ahonen L, Tuovinen OH. 1989. Microbiological oxidation of ferrous iron at low temperatures. *Appl Environ Microbiol* 55:312–316.

Ahonen L, Tuovinen OH. 1992. Bacterial oxidation of sulfide minerals in column leaching experiments at suboptimal temperatures. *Appl Environ Microbiol* 58:600–606.

Asmah RH, Bosompem KM, Osei YD, Rodriguez FK, Addy ME, Clement C, Wilson MD. 1999. Isolation and characterization of mineral oxidizing bacteria from the Obuasi gold mining site, Ghana. In: Amils R, Ballester A, eds. *Biohydrometallurgy and the Environment Toward the Mining of the 21st Century, Part A.* Amsterdam, the Netherlands: Elsevier, pp. 657–662.

Baas Becking LGM, Moore D. 1961. Biogenic sulfides. *Econ Geol* 56:259–272.

Badigian RM, Myerson AS. 1986. The adsorption of *Thiobacillus ferrooxidans* on coal surfaces. *Biotech Bioeng* 28:467–479.

Baker BJ, Banfield JF. 2003. Microbial communities in acid mine drainage. *FEMS Microbiol Ecol* 44:139–152.

Baker BJ, Hugenholtz P, Dawson SC, Banfield JF. 2003. Extremely acidophilic protists from acid miner drainage host *Rickettsiales*-lineage endosymbionts that have intervening sequences in the 16S rRNA genes. *Appl Environ Microbiol* 69:5512–5518.

Baker BJ, Lutz MA, Dawson SC, Bond PL, Banfield JF. 2004. Metabolically active eukaryotic communities in extremely acidic mine drainage. *Appl Environ Microbiol* 70:6264–6271.

Balashova VV, Vedenina IYa, Markosyan GE, Zavarzin GA. 1974. The autotrophic growth of *Leptospirillum ferrooxidans*. *Mikrobiologiya* 43:581–585 (Engl transl, pp. 491–494).

Ballard RD, Grassle JF. 1979. Strange world without sun. Return to oases of the deep. *Natl Geogr Mag* 156:680–703.

Barreto M, Jedlicki E, Holmes DS. 2005. Identification of a gene cluster for the formation of extracellular polysaccharide precursors in the chemolithoautotroph *Acidithiobacillus ferrooxidans*. *Appl Environ Microbiol* 71:2902–2909.

Bennett JC, Tributsch H. 1978. Bacterial leaching patterns of pyrite crystal surfaces. *J Bacteriol* 134:310–317.

Berner RA. 1984, Sedimentary pyrite formation: An update. *Geochim Cosmochim Acta* 48:605–615.

Bhatti TM, Gigham JM, Carlson L, Tuovinen OH. 1993. Mineral products of pyrrhotite oxidation by *Thiobacillus ferrooxidans*. *Appl Environ Microbiol* 59:1984–1990.

Bond PL, Druschel GK, Banfield JF. 2000a. Comparison of acid mine drainage microbial communities in physically and geochemically distinct ecosystems. *Appl Environ Microbiol* 66:4962–4971.

Bond PL, Smriga SP, Banfield JF. 2000b. Phylogeny of microorganisms populating a thick, subaerial, predominantly lithotrophic biofilm at an extreme acid mine drainage site. *Appl Environ Microbiol* 66:3842–3849.

Bonnatti E. 1972. Authigenesis of minerals—Marine. In: Fairbridge RW, ed. *The Encyclopedia of Geochemistry and Environmental Sciences. Encyclopedia Earth sciences series,* Vol. IVA. New York: Van Nostrand Reinhold, pp. 48–56.

Bonnatti E. 1978. The origin of metal deposits in the oceanic lithosphere. *Sci Am* 238:54–61.

Boon M, Hansford GS, Heijnen JJ. 1995. The role of bacterial ferrous iron oxidation in the bio-oxidation of pyrite. In: Vargas T, Jerez CA, Wiertz JV, Toledo H, eds. *Biohydrometallurgical Processing,* Vol. 1. Santiago, Chile: University of Chile, pp. 153–163.

Brierley CL. 1974. Leaching. Use of a high-temperature microbe. In: *Solution Mining Symposium. Proceedings of 103rd AIME Annual Meeting,* Dallas, TX, February 25–27, 1974, pp. 461–469.

Brierley CL. 1978a. Bacterial leaching. *CRC Crit Rev* 6:107–262.

Brierley JA. 1978b. Thermophilic iron-oxidizing bacteria found in copper leaching dumps. *Appl Environ Microbiol* 36:523–525.

Brierley CL. 1982. Microbiological mining. *Sci Am* 247:44–53, 150.

Brierley CL, Brierley JA. 1973. A chemoautotrophic and thermophilic microorganism isolated from an acid hot spring. *Can J Microbiol* 19:183–188.

Brierley CL, Murr LE. 1973. Leaching: Use of a thermophilic chemoautotrophic microbe. *Science* 179:488–490;

Brierley JA, Lockwood SJ. 1977. The occurrence of thermophilic iron-oxidizing bacteria in a copper leaching system. *FEMS Microbiol Lett* 2:163–165.

Brierley JA, Norris PR, Kelly DP, LeRoux NW. 1978. Characteristics of a moderately thermophilic and acidophilic iron-oxidizing *Thiobacillus. Eur J Appl Microbiol Biotechnol* 5:291–299.

Brieley JA, Wan RY, Hill DL, Logan TC. 1995. Biooxidation-heap pretreatment technology for processing lower grade refractory gold ores. In: Vargas T, Jerez CA, Wiertz JV, Toledo H, eds. *Biohydrometallurgical Processing.* Vol. 1. Santiago, Chile: University of Chile, pp. 253–262.

Brock TD. 1975. Effect of water potential on growth and iron oxidation by *Thiobacillus ferrooxidans. Appl Microbiol* 29:495–501.

Bruhn DF, Thompson DN, Noah KS. 1999. Microbial ecology assessment of a mixed copper oxide/sulfide dump leach operation. In: Amils R, Ballester A, eds. *Biohydrometallurgy and the Environment towards the Mining of the 21st Century, Part A.* Amsterdam, the Netherlands: Elsevier, pp. 799–808.

Bruneel O, Duran R, Casiot C, Elbaz-Pouchilet F, Personné J-C. 2006. Diversity of microorganisms in Fe-As-rich acid mine drainage waters at Carnoulès, France. *Appl Environ Microbiol* 72:551–556.

Bryner LC, Anderson R. 1957. Microorganisms in leaching sulfide minerals. *Ind Eng Chem* 49:1721–1724.

Bryner LC, Beck JV, Davis BB, Wilson DG. 1954. Microorganisms in leaching sulfide minerals. *Ind Eng Chem* 46:2587–2592.

Bryner LC, Jameson AK. 1958. Microorganisms in leaching sulfide minerals. *Appl Microbiol* 6:281–287.

Burgstaller W, Schinner F. 1993. Leaching of metals with fungi. *J Biotechnol* 27:91–116.

Cameron EM. 1982. Sulfate and sulfate reduction in early Precambrian oceans. *Nature (Lond)* 296:145–148.

Carlson L, Lindström EB, Hallberg KB, Tuovinen OH. 1992. Solid-phase products of bacterial oxidation of arsenical pyrite. *Appl Environ Microbiol* 58: 1046–1049.

Chemical Processing. 1965. Carboxylic acids in metal extraction. *Chem Proc* 11:24–25.

Chen C-Y, Skidmore DR. 1987. Langmuir adsorption isotherm for *Sulfolobus acidocaldarius* on coal particles. *Biotech Lett* 9:191–194.

Chen C-Y, Skidmore DR. 1988. Attachment of *Sulfolobus acidocaldarius* cells in coal particles. *Biotechnol Progr* 4:25–30.

Corliss JB, Dymond J, Gordon LI, Edmond JM, von Herzen RP, Ballard RD, Green K et al. 1979. Submarine thermal springs on the Galapagos Rift. *Science* 203:1073–1083.

Crundwell FK. 1995. Mathematical modeling and optimization of bacterial leaching plants. In: Vargas T, Jerez CA, Wiertz JV, Toledo H, eds. *Biohydrometallurgical Processing,* Vol 1. Santiago, Chile: Univ Chile, pp 437–446.

Crundwell FK. 1997. Physical chemistry of bacterial leaching. In: Amils R, Ballester A, eds. *Biohydrometallurgy and the Environment toward the Mining of the 21st Century, Part A.* Amsterdam, the Netherlands: Elsevier, pp. 13–26.

Cuthbert ME. 1962. Formation of bornite at atmospheric temperature and pressure. *Econ Geol* 57:38–41.

Darland G, Brock TD, Sansonoff W, Conti SF. 1970. A thermophilic, acidophilic mycoplasma isolated from a coal refuse pile. *Science* 170:1416–1418.

Davidson CF. 1962a. The origin of some strata-bound sulfide ore deposits. *Econ Geol* 57:265–274.

Davidson CF. 1962b. Further remarks on biogenic sulfides. *Econ Geol* 57:1134–1137.

Descostes M, Vitorge P, Beaucaire C. 2004. Pyrite dissolution in acidic media. *Geochim Cosmochim Acta* 68:4559–4569.

Dévigne J-P. 1968a. Précipitation du sulfure de plomb par un micrococcus tellurique. *CR Acad Sci (Paris)* 267:935–937.

Dévigne J-P. 1968b. Une bactérie saturnophile, *Sarcina flava* Bary 1887. *Arch Inst Pasteur (Tunis)* 45:341–358.

Dévigne J-P. 1973. Une métallogenèse microbienne probable en milieux sédimentaires: Celle de la galena. *Cah Geol* 89:35–37.

Dew DW, van Buuren C, McEwan K, Bowker C. 1999. Bioleaching of base metal sulfide concentrates: A comparison of mesophile and thermophile bacterial cultures. In: Amils R, Ballester A, eds. *Biohydrometallurgy and the Environment toward the Mining of the 21st Century*, Part A. Amsterdam, the Netherlands: Elsevier, pp. 229–238.

DiSpirito AA, Talnagi JW Jr, Tuovinen OH. 1983. Accumulation and cellular distribution of uranium in *Thiobacillus ferrooxidans*. *Arch Microbiol* 135:250–253.

DiSpirito AA, Tuovinen OH. 1981. Oxygen uptake coupled with uranous sulfate oxidation by *Thiobacillus ferrooxidans* and *T. acidophilus*. *Geomicrobiol J* 2:275–291.

DiSpirito AA, Tuovinen OH. 1982a. Uranous ion oxidation and carbon dioxide fixation by *Thiobacillus ferrooxidans*. *Arch Microbiol* 133:28–32.

DiSpirito AA, Tuovinen OH. 1982b. Kinetics of uranous and ferrous ion oxidation by *Thiobacillus ferrooxidans*. *Arch Microbiol* 133:33–37.

Donati E, Curutchet G, Pogliani C, Tedesco P. 1995. Bioleaching of covellite by individual or combined cultures of *Thiobacillus ferrooxidans* and *Thiobacillus thiooxidans*. In: Vargas T, Jerez CA, Wjiertz JV, Toledo H, eds. *Biohydrometallurgical Processing*, Vol 1. Santiago, Chile: Univ Chile, pp 91–99.

Driessens YPM, Fowler TA, Crundwell FK. 1999. A comparison of the bacterial and chemical leaching of sphalerite at the same solution conditions. In: Amils R, Ballester A, eds. *Biohydrometallurgy and the Environment toward the Mining of the 21st Century*, Part A. Amsterdam, the Netherlands: Elsevier, pp. 201–208.

Duncan DW, Landesman J, Walden CC. 1967. Role of *Thiobacillus ferrooxidans* in the oxidation of sulfide minerals. *Can J Microbiol* 13:397–403.

Duncan DW, Walden CC. 1972. Microbial leaching in the presence of ferric iron. *Dev Ind Microbiol* 13:66–75.

Duquesne K, Lebrun S, Casiot C, Bruneel O, Personné J-C, Leblanc M, Elbaz-Poulichet F, Morin G, Bonnefoy V. 2003. Immobilization of arsenite and ferric iron by *Acidithiobacillus ferrooxidans* and its relevance to acid mine drainage. *Appl Environ Microbiol* 69:6165–6173.

Edmond JM, Von Damm KL, McDuff RE, Measures CI. 1982. Chemistry of hot springs on the East Pacific Rise and their effluent dispersal. *Nature (Lond)* 297:187–191.

Edwards KJ, Bond PL, Druschel GK, McGuire MM, Hamers RJ, Banfield JF. 2000b. Geochemical and biological aspects of sulfide mineral dissolution: Lessons from Iron Mountain, California. *Chem Geol* 169:383–397.

Edwards KJ, Bond PL, Gihring TM, Banfield JF. 2000a. An archaeal iron-oxidizing extreme acidophile important in acid mine drainage. *Science* 287:1796–1799.

Edwards KJ, Gihring TM, Banfield JF. 1999a. Seasonal variations in microbial populations and environmental conditions in an extreme acid mine drainage environment. *Appl Environ Microbiol* 65:3627–3632.

Edwards KJ, Goebell BM, Rodgers RM, Schrenk MO, Gihring TM, Cardona MM, Hu B, McGuire MM, Hamers RJ, Pace NR. 1999b. Geomicrobiology of pyrite (FeS_2) dissolution: Case study at Iron Mountain, California. *Geomicrobiol J* 16:155–179.

Edwards KJ, Hu B, Hamers RJ, Banfield JF. 2001. A new look at patterns on sulfide minerals. *FEMS Microbiol Ecol* 34:197–206.

Edwards KJ, McCollum TM, Konishi H, Buseck PR. 2003. Seafloor bioalteration of sulfide minerals: Results from in situ incubation studies. *Geochim Cosmochim Acta* 67:2843–2856.

Edwards KJ, Schrenk MO, Hamers R, Banfield JF. 1998. Mineral oxidation of pyrite: Experiments using microorganisms from an extreme acidic environment. *Am Mineral* 83:1444–1453.

Ehrlich HL. 1963a. Bacterial action on orpiment. *Econ Geol* 58:991–994.

Ehrlich HL. 1963b. Microorganisms in acid mine drainage from a copper mine. *J Bacteriol* 86:350–352.

Ehrlich HL. 1964. Bacterial oxidation of arsenopyrite and enargite. *Econ Geol* 59:1306–1312.

Ehrlich HL. 1977. Bacterial leaching of low-grade copper sulfide ore with different lixiviants. In: Schwartz E, ed. *Conference Bacterial Leaching. Gesellschaft für Biotechnologische Forschung mbH, Braunschweig-Stöckheim*. Weinstein, Germany: Verlag Chemie, pp. 145–155.

Ehrlich HL. 1978. Inorganic energy sources for chemolithotrophic and mixotrophic bacteria. *Geomicrobiol J* 1:65–83.

Ehrlich HL. 1988. Bioleaching of silver from a mixed sulfide ore in a stirred reactor. In: Norris PR, Kelly DP, eds. *Biohydrometallurgy*. Kew Surrey, U.K.: *Sci Technol Lett*, pp. 223–231.

Ehrlich HL. 2000. Past, present, and future in biohydrometallurgy. *Hydrometallurgy* 59:127–134.

Ehrlich HL, Fox SI. 1967. Environmental effects on bacterial copper extraction from low-grade copper sulfide ores. *Biotech Bioeng* 9:471–485.

Fenchel T, Blackburn TH. 1979. *Bacteria and Mineral Cycling*. London, U.K.: Academic Press.

Fortin D, Beveridge TJ. 1997. Microbial sulfate reduction within sulfidic mine tailings: Formation of diagenetic Fe sulfides. *Geomicrobiol J* 14:1–21.

Fortin D, Davis B, Southam G, Beveridge TJ.1995. Biogeochemical phenomena induced by bacteria within sulfidic mine tailings. *J Ind Micrbiol* 14:178–185.

Fowler TA, Crundwell FK. 1999. Leaching zinc sulfide by *Thiobacillus ferrooxidans*: Bacterial oxidation of the sulfur product layer increases the rate of zinc sulfide dissolution at high concentrations of ferrous ions. *Appl Environ Microbiol* 65:5285–5292.

Fox SI. 1967. Bacterial oxidation of simple copper sulfides. PhD thesis. Troy, NY: Rensselaer Polytechnic Institute.

Freke AM, Tate D. 1961. The formation of magnetic iron sulfide by bacterial reduction of iron solutions. *J Biochem Microbiol Technol Eng* 3:29–39.

Garcia O Jr, Bigham JM, Tuovinen OH. 1995. Oxidation of galena by *Thiobacillus ferrooxidans* and *Thiobacillus thiooxidans*. *Can J Microbiol* 41:508–516.

Gehrke T, Hallmann R, Sand W. 1995. Importance of exopolymers from *Thiobacillus ferrooxidans* and *Leptospirillum ferrooxidans* for bioleaching. In: Vargas T, Jerez CA, Wiertz KV, Toledo H, eds. *Biohydrometallurgical Processing*, Vol 1. Santiago, Chile: Univ Chile, pp 1–11.

Gehrke T, Telegdi J, Thierry D, Sand W. 1998. Importance of extracellular polymeric substances from *Thiobacillus ferrooxidans* for bioleaching. *Appl Environ Microbiol* 64:2743–2747.

Giblin AE. 1988. Pyrite formation in marshes during early diagenesis. *Geomicrobiol J* 6:77–97.

Giblin AE, Howarth RW. 1984. Porewater evidence for a dynamic sedimentary iron cycle in salt marshes. *Limnol Oceanogr* 29:47–63.

Golovacheva RS, Karavaiko GI. 1979. A new genus of thermophilic spore-forming bacteria, *Sulfobacillus*. *Mikrobiologiya* 47:815–822 (Engl transl, pp. 658–665).

Golyshina OV, Pivovarova TA, Karavaiko GI, Kondrat'eve TF, Moore RB, Abraham W-R, Lunsdorf H, Timmis KN, Yakimov MM, Golyshin PN. 2000. *Ferroplasma acidiphilum* gen nov., sp. nov., an acidophilic, autotrophic, ferrous-iron-oxidizing, cell-wall-lacking, mesophilic member of the Ferroplasmaceae, fam. nov., comprising a distinct lineage of Archaea. *Int J Syst Evol Microbiol* 50: 997–1006.

González-Toril E, Llobet-Brossa E, Casamayor EO, Amann R, Amils R. 2003. Microbial ecology of an extreme acidic environment, the Tinto River. *Appl Environ Microbiol* 69:4853–4865.

Gorby YA, Lovley DR. 1992. Enzymatic uranium precipitation. *Environ Sci Technol* 26:205–207.

Hallberg RO. 1978. Metal-organic interactions at the redoxcline. In: Krumbein WE, ed. *Environmental Biogeochemistry and Geomicrobiology, Vol 3. Methods, Metals, and Assessment*. Ann Arbor, MI: Ann Arbor Sci, pp. 947–953.

Hansford GS. 1997. Recent developments in modeling kinetics of bioleaching. In: Rawlings DE, ed. *Biomining: Theory, Microbes and Industrial Processes*. Berlin, Germany: Springer-Verlag, pp. 153–175.

Hansford GS, Vargas T. 1999. Chemical and electrochemical basis for bioleaching processes. In: Amils R, Ballester A, eds. *Biohydrometallurgy and the Environment Toward the Mining of the 21st Century, Part A*. Amsterdam, the Netherlands: Elsevier, pp. 13–26.

Harneit K, Göksel A, Kock D, Klock J-H, Gehrke T, Sand W. 2006. Adhesion to metal sulfide surfaces by cells of *Acidithiobacillus ferrooxidans, Acidithiobacillus thiooxidans* and *Leptospirillum ferrooxidans*. *Hydrometallurgy* 83:245–254.

Harrison AP Jr. 1978. Microbial succession and mineral leaching in an artificial coal spoil. *Appl Microbiol* 36:861–869.

Harrison AP Jr, Norris PR. 1985. *Leptospirillum ferrooxidans* and similar bacteria: Some characteristics and genomic diversity. *FEMS Microbiol Lett* 30:99–102.

Hartmannova V, Kuhr I. 1974. Copper leaching by lower fungi. *Rudy* 22:234–238.

Howard D, Crundwell FK. 1999. A kinetic study of the leaching of chalcopyrite with *Sulfolobus metallicus*: Amils R, Ballester A, eds. Amsterdam, the Netherlands: Elsevier, pp. 209–217.

Howarth RW. 1979. Pyrite: Its rapid formation in a salt marsh and its importance in ecosystem metabolism. *Science* 203:49–51.

Howarth RW, Merkel S. 1984. Pyrite formation and the measurement of sulfate reduction in salt marsh sediments. *Limnol Oceanogr* 29:598–608.

Imai KH, Sakaguchi H, Sugio T, Tano T. 1973. On the mechanism of chalcocite oxidation by *Thiobacillus ferrooxidans*. *J Ferment Technol* 51:865–870.

Ingledew WJ. 1986. Ferrous iron oxidation by *Thiobacillus ferrooxidans*. In: Ehrlich HL, Holmes DS, eds. *Biotechnology for the Mining, Metal Refining, and Fossil Fuel Processing Industries. Biotechnol Bioeng Symp*, Vol. 16. New York: Wiley, pp. 23–33.

Ivanov VI. 1962. Effect of some factors on iron oxidation by cultures of *Thiobacillus ferrooxidans. Mikrobiologiya* 31:795–799 (Engl transl, pp. 645–648).

Ivanov VI, Nagirnynyak FI, Stepanov BA. 1961. Bacterial oxidation of sulfide ores. I. Role of *Thiobacillus ferrooxidans* in the oxidation of chalcopyrite and sphalerite. *Mikrobiologiya* 30:688–692.

Johnson BS, Rang L. 1993. Effects of acidophilic protozoa on populations of metal-mobilizing bacteria during the leaching of pyritic coal. *J Gen Microbiol* 139:1417–1423.

Johnson BS, Roberto FF. 1997. Heterotrophic acidophiles and their roles in bioleaching of sulfide minerals. In: Rawlings DE, ed. *Biomining: Theory, Microbes and Industrial Processes.* Berlin, Germany: Springer-Verlag, pp. 259–279.

Jørgensen BB. 1977. The sulfur cycle of a coastal marine sediment (Limfjorden, Denmark). *Limnol Oceanogr* 22:814–832.

Joseph JM. 1953. Microbiological study of acid mine water: Preliminary report. *Ohio J. Sci.* 53:123–127.

Kee NS, Bloomfield C. 1961. The solution of some minor element oxides by decomposing plant materials. *Geochim Cosmochim Acta* 24:206–225.

King GM, Howes BL, Dacey JWH. 1985. Short-term end-products of sulfate reduction in a salt marsh: Formation of acid volatile sulfides, elemental sulfur, and pyrite. *Geochim Cosmochim Acta* 49:1561–1566.

Kleinmann RLP, Crerar DA. 1979. *Thiobacillus ferrooxidans* and the formation of acidity in simulated coal mine environments. *Geomicrobiol J* 1:373–388.

Klinkhamer GP, Rona P, Greaves M, Elderfeld, H. 1985. Hydrothermal manganese plumes in the Mid-Atlantic Ridge Rift Valley. *Nature (Lond)* 314:727–731.

Labrenz M, Druschel GK, Thomsen-Ebert T, Gilbert B, Welxh SA, Kemner KM, Logan GA et al. 2000. Formation of sphalerite (ZnS) deposits in natural biofilms of sulfate-reducing bacteria. *Science* 290:1744–1747.

Lackey JB. 1938. The flora and fauna of surface water polluted by acid mine drainage. *Public Health Rep* 53:1499–1507.

Lambert IB, McAndrew J, Jones HE. 1971. Geochemical and bacteriological studies of the cupriferous environment at Pernatty Lagoon, South Australia. *Aust Inst Min Met Proc* 240:15–23.

Latimer WM, Hildebrand JH. 1942. *Reference Book of Inorganic Chemistry*, Rev edn. New York: Macmillan.

Lazaroff N. 1983. The exclusion of D_2O from the hydration sphere of $FeSO_4 \cdot 7H_2O$ oxidized by *Thiobacillus ferrooxidans. Science* 222:1331–1334.

Lazaroff N, Melanson L, Lewis E, Santoro N, Pueschel C. 1985. Scanning electron microscopy and infrared spectroscopy of iron sediments formed by *Thiobacillus ferrooxidans. Geomicrobiol J* 4:231–268.

Lazaroff N, Sigal W, Wasserman A. 1982. Iron oxidation and precipitation of ferric hydroxysulfates by resting *Thiobacillus ferrooxidans* cells. *Appl Environ Microbiol* 43:924–938.

Leathen WW, Braley SA, McIntyre LD. 1953. The role of bacteria in the formation of acid from certain sulfuritic constituents associated with bituminous coal. II. Ferrous iron oxidizing bacteria. *Appl Microbiol* 1:65–68.

Lehninger AL. 1975. Biochemistry, 2nd edn. New York: Worth.

Leleu MT, Gulgalski T, Goni J. 1975. Synthèse de wurtzite par voie bactérienne. *Min Deposita (Berl)* 10:323–329.

Love LG. 1962. Biogenic primary sulfide of the Permian Kupferschiefer and marl slate. *Econ Geol* 57:350–366.

Lovley DR, Roden EE, Phillips EJP, Woodward JC. 1993. Enzymatic iron and uranium reduction by sulfate-reducing bacteria. *Mar Geol* 113:41–53.

Lundgren DG, Malouf EE. 1983. Microbial extraction and concentration of metals. *Adv Biotechnol Proc* 1:223–249.

Lundgren DG, Vestal JR, Tabita FR. 1972. The microbiology of mine drainage pollution. In: Mitchell R, ed. *Water Pollution Microbiology.* New York: Wiley-Interscience, pp. 69–88.

Luther GW III. 1991. Pyrite synthesis via polysulfide compounds. *Geochim Cosmochim Acta* 55:2839–2849.

Luther GW III, Giblin A, Howarth RW, Ryans RA. 1982. Pyrite and oxidized iron mineral phases formed from pyrite oxidation in salt marsh and estuarine sediments. *Geochim Cosmochim Acta* 46:2665–2669.

Lyalikova NN. 1960. Participation of *Thiobacillus ferrooxidans* in the oxidation of ores in pyrite beds of the Middle Ural. *Mikrobiologiya* 29:382–387.

Mackintosh ME. 1978. Nitrogen fixation by *Thiobacillus ferrooxidans. J Gen Microbiol* 105:215–218.

Magne R, Berthelin J, Dommergues Y. 1973. Solubilisation de l'uranium dans les roches par des bactéries n'appartenant pas au genre *Thiobacillus. CR Acad Sci (Paris)* 276:2625–2628.

Magne R, Berthelin JR, Dommergues Y. 1974. Solubilisation et insolubilisation de l'uranium des granites par des bactéries heterotroph. In: *Formation of Uranium Ore Deposits.* Vienna, Austria: International Atomic Energy Commission, pp. 73–88.

Malouf EE, Prater JD. 1961. Role of bacteria in the alteration of sulfide minerals. *J Metals* 13:353–356.

Marichig V, Grundlach H. 1982. Iron-rich metalliferous sediments on the East Pacific Rise: Prototype of undifferentiated metalliferous sediments on divergent plate boundaries. *Earth Planet Sci Lett* 58:361–382.

Marnette ECL, Van Breemen N, Hordijk KA, Cappenberg TE. 1993. Pyrite formation in two freshwater systems in the Netherlands. *Geochim Cosmochim Acta* 57:4165–4177.

McCready RGL, Gould WD. 1990. Bioleaching of uranium. In: Ehrlich HL, Brierley CL, eds. *Microbial Mineral Recovery.* New York: McGraw-Hill, pp. 107–125.

McGoran CJM, Ducan DW, Walden CC. 1969. Growth of *Thiobacillus ferrooxidans* on various substrates. *Can J Microbiol* 15:135–138.

McReady RGL. 1988. Progress in the bacterial leaching of metals in Canada. In: Norris PR, Kelly DP, eds. *Biohydrometallurgy.* Kew, Surrey, U.K.: *Sci Technol Lett*, pp. 177–195.

Miller LP. 1949. Stimulation of hydrogen sulfide production by sulfate-reducing bacteria. *Boyce Thompson Inst Contr* 15:467–474.

Miller LP. 1950. Formation of metal sulfides through the activities of sulfate reducing bacteria. *Boyce Thompson Inst Contr* 16:85–89.

Moses CO, Nordstrom DK, Herman JS, Mills AL. 1987. Aqueous pyrite oxidation by dissolved oxygen and by ferric iron. *Geochim Cosmochim Acta* 51: 1561–1571.

Moshuyakova SA, Karavaiko GI, Shchetinina EV. 1971. Role of *Thiobacillus ferrooxidans* in leaching nickel, copper, cobalt, iron, aluminum, magnesium, and calcium from ores of copper-nickel deposits. *Mikrobiologiya* 40:1100–1107 (Engl transl, pp. 659–969).

Mossman JR, Aplin AC, Curtis CD, Coleman ML. 1991. Geochemistry of inorganic and organic sulfur in organic-rich sediments from the Peru Margin. *Geochim Cosmochim Acta* 55:3581–3595.

Mottl MJ, Holland HD, Corr RF. 1979. Chemical exchange during hydrothermal alteration of basalt by seawater. II. Experimental results for Fe, Mn, and sulfur species. *Geochim Cosmochim Acta* 43:869–884.

Mousavi SM, Yaghmaei S, Vossoughi M, Jafari A, Hoseini SA. 2005. Comparison of bioleaching ability of two native mesophilic and thermophilic bacteria on copper recovery from chalcopyrite concentrate in an airlift bioreactor. *Hydrometallurgy* 80:139–144.

Mulligan CN, Galvez-Coutier R. 2000. Bioleaching of copper mining residues by *Aspergillus niger. Water Sci Technol* 41:255–262.

Murr LE, Brierley JA. 1978. The use of large-scale test facilities in studies of the role of microorganisms in commercial leaching operations. In: Murr LE, Torma AE, Brierely JA, eds. *Metallurgical Applications of Bacteria.* New York: Academic Press, pp. 491–520.

Murthy KSN, Natarajan KA. February 1992. The role of surface attachment of *Thiobacillus ferrooxidans* on the oxidation of pyrite. *Min Metall Process* 9:20–24.

Mustin C, Berthelin J, Marion P, de Donato P. 1992. Corrosion and electrochemical oxidation of a pyrite by *Thiobacillus ferrooxidans. Appl Environ Microbiol* 58:1175–1182.

Mustin C, de Donato P, Berthelin J. 1993. Surface oxidized species, a key factor in the study of bioleaching processes. In: Torma AE, Wey JE, Lakshmanan VI, eds. *Biohydrometallurgical Technologies, Vol 1, Bioleaching Process.* Warrendale, PA: Miner, Metals Mater Soc, pp. 175–184.

Myers JM, Myers CR. 2001. Role of outer membrane cytochromes OmcA and OmcB of *Shewanella putrefaciens* MR-1 in reduction of manganese dioxide. *Appl Environ Microbiol* 67:260–269.

Nedwell DB, Banat IM. 1981. Hydrogen as an electron donor for sulfate-reducing bacteria in slurries of salt marsh sediment. *Microb Ecol* 7:305–313.

Nielsen AM, Beck JV. 1972. Chalcocite oxidation and coupled carbon dioxide fixation by *Thiobacillus ferrooxidans. Science* 175:1124–1126.

Nordstrom DK, Southam G. 1997. Geomicrobiology of sulfide mineral oxidation. *Reviews in Mineralogy* 35:361–390.

Norman PG, Snyman CP. 1988. The biological and chemical leaching of auriferous pyrite/arsenopyrite flotation concentrate: A microscopic examination. *Geomicrobiol J* 6:1–10.

Norris PR. 1990. Acidophilic bacteria and their activity in mineral sulfide oxidation. In: Ehrlich HL, Brierley CL, eds. *Microbial Mineral Recovery.* New York: McGraw-Hill, pp. 3–27.

Norris PR. 1997. Thermophiles and bioleaching. In: Rawlings DE, ed. *Biomining: Theory, Microbes and Industrial Processes.* Berlin, Germany: Springer-Verlag, pp. 3–27.

Okibe N, Gericke M, Hallberg KB, Johnson DB. 2003. Enumeration and characterization of acidophilic microorganisms isolated from a pilot plant stirred-tank bioleaching operation. *Appl Environ Microbiol* 69:1936–1943.

Olson GJ, Brierley JA, Brierley CL. 2003. Bioleaching review part B: Progress in bioleaching: applications of microbial processes by the minerals industries. *Appl Microbiol Biotechnol* 63:249–257.

Parès Y. 1964a. Intervention des bactéries dans la solubilisation du cuivre. *Ann Inst Pasteur (Paris)* 107:132–135.

Parès Y. 1964b. Action de *Serratia marcescens* dans le cycle biologique des métaux. *Ann Inst Pasteur (Paris)* 107:136–141.

Parès Y. 1964c. Action d'*Agrobacterium tumefaciens* dans la mise en solution de l'or. *Ann Inst Pasteur (Paris)* 107:141–143.

Pinka J. 1991. Bacterial oxidation of pyrite and pyrrhotite. *Erzmetall* 44:571–573.

Pistorio M, Curutchet G, Donati E, Tedesco P. 1994. Direct zinc sulfide bioleaching by *Thiobacillus ferrooxidans* and *Thiobacillus thiooxidans*. *Biotechnol Lett* 16:419–424.

Pivovarova TA, Markosyan GE, Karavaiko GI. 1981. *Mikrobiologiya* 50:482–486 (Engl transl, pp. 339–344).

Pogliani C, Curutchet G, Donati E, Tedesco PH. 1990. A need for direct contact with particle surfaces in the bacterial oxidation of covellite in the absence of a chemical lixiviant. *Biotechnol Lett* 12: 515–518.

Powell TG, MacQueen RW. 1984. Precipitation of sulfide ores and organic matter: Sulfate reduction at Pine Point, Canada. *Science* 224:63–66.

Pronk JT, Meulenberg R, Hazeu W, Bos P, Kuenen JG. 1990. Oxidation of reduced inorganic sulfur compounds by acidophilic thiobacilli. *FEMS Microbiol Rev* 75:293–306.

Rawlings DE. 1997a. *Biomining: Theory, Microbes and Industrial Processes*. Berlin, Germany: Springer-Verlag.

Rawlings DE. 1997b. Mesophilic, autotrophic bioleaching bacteria: Description, physiology and role. In: Rawlings DE, ed. *Biomining: Theory, Microbes and Industrial Processes*. Berlin, Germany: Springer-Verlag, pp. 229–245.

Rawlings DE. 2002. Heavy metal mining using microbes. *Annu Rev Microbiol* 56:65–91.

Rawlings DE, Tributsch H, Hansford GS. 1999. Reasons why 'Leptospirillum'-like species rather than *Thiobacillus ferrooxidans* are the dominant iron-oxidizing bacteria in many commercial processes for the biooxidation of pyrite and related ores. *Microbiology (Reading)* 145:5–13.

Razzell WE, Trussell PC. 1963. Microbiological leaching of metallic sulfides. *Appl Microbiol* 11:105–110.

Rickard DT. 1973. Limiting conditions for synsedimentary sulfide ore formation. *Econ Geol* 68:605–617.

Rickard PAD, Vanselow DG. 1978. Investigation into the kinetics and stoichiometry of bacterial oxidation of covellite (CuS) using a polarographic oxygen probe. *Can J Microbiol* 24:998–1003.

Rimstidt JD, Vaughn DJ. 2003. Pyrite oxidation: A state-of-the-art assessment of the reaction mechanism. *Geochim Cosmochim Acta* 67:873–880.

Rodriguez-Leiva M, Tributsch H. 1988. Morphology of bacterial leaching patterns by *Thiobacillus ferrooxidans* on synthetic pyrite. *Arch Microbiol* 149:401–405.

Rohwerder T, Gehrke T, Kinzler K, Sand W. 2003. Bioleaching review part A: Progress in bioleaching: fundamentals and mechanisms of bacterial metal sulfide oxidation. *Appl Microbiol Biotechnol* 63: 239–248.

Roy AB, Trudinger PA. 1970. *The Biochemistry of Inorganic Compounds of Sulphur*. Cambridge, U.K.: Cambridge University Press.

Sand W, Gehrke T, Hallmann R, Rohde K, Sabotke B, Wentzien S. 1993. In-situ bioleaching of metal sulfides: The importance of *Leptospirillum ferrooxidans*. In Torma AE, Wey JE, Lashmanan VL. *Biohydrometallurgical Technologies*, Vol 1. Warrendale, PA: The Minerals, Metals and Materials Society, pp. 15–27.

Sand W, Gehrke T, Hallmann R, Schippers A. 1995. Sulfur chemistry, biofilm, and the (in)direct attack mechanism: A critical evaluation of bacterial leaching. *Appl Microbiol Biotechnol* 43:961–966.

Sand W, Gehrke T, Jozsa P-G, Schippers A. 1997. Novel mechanism for bioleaching of metal sulfides. In: *Biotechnology Comes of Age. International Biohydrometallurgy Symposium IBS97 BIOMINE 97. Conference Proceedings*. Glenside, South Australia, Australia: Australian Mineral Foundation, pp. QP2.1–QP2.10.

Sand W, Rohde K, Sabotke B, Zenneck C. 1992. Evaluation of *Leptospirillum ferrooxidans* for leaching. *Appl Environ Microbiol* 58:85–92.

Schippers A, Sand W. 1999. Bacterial leaching of metal sulfides proceeds by two indirect mechanisms via thiosulfate or via polysulfides and sulfur. *Appl Environ Microbiol* 65:319–321.

Schoonen MAS, Barnes HL. 1991a. Reactions forming pyrite and marcasite from solution. 1. Nucleation of FeS_2 below 100°C. *Geochim Cosmochim Acta* 55:1495–1504.

Schoonen MAS, Barnes HL. 1991b. Reactions forming pyrite and marcasite from solution. II. Via FeS precursors below 100°C. *Geochim Cosmochim Acta* 55:1505–1514.

Schrenk MO, Edwards KJ, Goodman RM, Hamers RJ, Banfield JF. 1998. Distribution of *Thiobacillus ferrooxidans* and *Leptospirillum ferrooxidans*: Implications for generation of acid mine drainage. *Science* 279:1519–1522.

Segerer A, Neuner A, Kristjansson JK, Stetter KO. 1986. *Acidianus infernus* gen nov. sp. nov., and *Acidianus brierleyi* comb. nov.: Facultative aerobic, extremely acidophilic thermophilic sulfur-metabolizing archaebacteria. *Arch Microbiol* 36:559–564.

Serkies J, Oberc J, Idzikowski A. 1967. The geochemical bearings of the genesis of Zechstein copper deposits in southwest Poland as exemplified by the studies on the Zechstein of the Leszczyna syncline. *Chem Geol* 2:217–2232.

Seyfried WE Jr, Mottle MJ. 1982. Hydrothermal alteration of basalt by seawater under seawater-dominated conditions. *Geochim Cosmochim Acta* 46:985–1002.

Shanks WC III, Bischoff JL, Rosenauer RJ. 1981. Seawater sulfate reduction and sulfur isotope fractionation in basaltic systems: Interaction of seawater with fayalite and magnetite at 200–350°C. *Geochim Cosmochim Acta* 45:1977–1981.

Shrihari RK, Ghandi KS, Natarajan KA. 1991. Role of cell attachment in leaching chalcopyrite mineral by *Thiobacillus ferrooxidans*. *Appl Microbiol Biotechnol* 36:278–282.

Silver M, Torma AE. 1974. Oxidation of metal sulfides by *Thiobacillus ferrooxidans* grown on different substrates. *Can J Microbiol* 20:141–147.

Silverman MP, Ehrlich HL. 1964. Microbial formation and degradation of minerals. *Adv Appl Microbiol* 6:153–206.

Silverman MP, Rogoff MH, Wender I. 1961. Bacterial oxidation of pyritic materials in coal. *Appl Microbiol* 9:491–496.

Singer PC, Stumm W. 1970. Acid mine drainage: The rate-determining step. *Science* 167:1121–1123.

Stanton RL. 1972. Sulfides in sediments. In: Fairbridge RW, ed. *The Encyclopedia of Geochemistry and Environmental Sciences. Encyclopedia of Earth Science Series*, Vol. IVA. New York: Van Nostrand Reinhold, pp. 1134–1141.

Stevens CJ, Dugan PR, Tuovinen OH. 1986. Acetylene reduction (nitrogen fixation) by *Thiobacillus ferrooxidans*. *Biotechnol Appl Biochem* 8:351–359.

Strahler AN. 1977. *Principles of Physical Geology*. New York: Harper & Row.

Styrt MM, Brackman AJ, Holland HD, Clark BC, Pisutha-Arnold V, Eldridge CS, Ohmoto H. 1981. The mineralogy and isotopic composition of sulfur in hydrothermal sulfide/sulfate deposits on the East Pacific Rise, 21°N latitude. *Earth Planet Sci Lett* 53:382–390.

Sugio T, Tanijiri S, Fukuda K, Yamargo K, Inagaki K, Tano T. 1987. Utilization of amino acids as sole source of nitrogen by obligate chemoautotroph *Thiobacillus ferrooxidans*. *Agric Biol Chem* 51:2229–2236.

Sullivan JD. 1930. *Chemistry of Leaching Covellite*. Tech Paper 487. Washington, DC: US. Department of Commerce, Bur Mines.

Sutton JA, Corrick JD. 1963. Microbial leaching of copper minerals. *Mining Eng* 15:37–40.

Surron JA, Corrick JD, 1964. Bacteria in mining and metallurgy: Leaching selected ores and minerals; experiments with *Thiobacillus ferrooxidans*. Rept Invest RI 5839. Washington, DC: Bur Mines, US Department of the Interior.

Temple KL. 1964. Syngenesis of sulfide ores. An explanation of biochemical aspects. *Econ Geol* 59:1473–1491.

Temple KL, LeRoux N. 1964. Syngenesis of sulfide ores: Desorption of adsorbed metal ions and the precipitation of sulfides. *Econ Geol* 59:647–655.

Thode-Andersen S, Jørgensen BB. 1989. Sulfate reduction and the formation of ^{35}S-labeled FeS, FeS$_2$, and S^0 in coastal marine sediments. *Limnol Oceanogr* 34:793–806.

Tittley SR. 1981. Porphyry copper. *Am Sci* 69:632–638.

Torma AE. 1971. Microbial oxidation of synthetic cobalt, nickel and zinc sulfides by *Thiobacillus ferrooxidans*. *Rev Can Biol* 30:209–216.

Torma AE. 1978. Oxidation of gallium sulfides by *Thiobacillus ferrooxidans*. *Can J Microbiol* 24:888–891.

Torma AE, Gabra GG. 1977. Oxidation of stibnite by *Thiobacillus ferrooxidans*. *Antonie v Leeuwenhoek* 43:1–6.

Tributsch H. 1976. The oxidative disintegration of sulfide crystals by *Thiobacillus ferrooxidans*. *Naturwissenschaften* 63:88.

Tributsch H. 1999. Direct versus indirect bioleaching. In: Amils R, Ballester A, eds. *Biohydrometallurgy and the Environment toward the Mining of the 21st Century, Part A*. Amsterdam, the Netherlands: Elsevier, pp. 51–60.

Tuovinen OH, Niemelä SI, Gyllenberg HG. 1971. Tolerance of *Thiobacillus ferrooxidans* to some metals. *Antonie v Leeuwenhoek* 37:489–496.

Wadden D, Gallant A. 1985. The in-place leaching of uranium at Denison Mines. *Can J Metall Quart* 24:127–134.

Walsh F, Mitchell R. 1972a. An acid-tolerant iron-oxidizing *Metallogenium*. *J Gen Microbiol* 72:369–376.

Walsh F, Mitchell R. 1972b. The pH-dependent succession of iron bacteria. *Environ Sci Technol* 6:809–812.

Wang H, Bigham JM, Tuovinen OH. 2007. Oxidation of marcasite and pyrite by iron-oxidizing bacteria and archaea. *Hydrometallurgy* 88:127–131.

Weast RC, Astle MJ. 1982. *CRC Handbook of Chemistry and Physics*, 63rd edn. Boca Raton, FL: CRC Press.

Wenberg GM, Erbisch FH, Volin M. 1971. Leaching of copper by fungi. *Trans Soc Min Eng AIME* 250: 207–212.

Westrich JT, Berner RA. 1984. The role of sedimentary organic matter in bacterial sulfur reduction: The G model tested. *Limnol Oceanogr* 29:236–249.

Yarzábal A, Brasseur G, Bonnefoy V. 2002b. Cytochromes *c* of *Acidithiobacillus ferrooxidans*. *FEMS Microbiol Lett* 209:189–195.

Yarzábal A, Brasseur G, Ratouchniak J, Lund K, Lemesle-Meunier D, DeMoss JA, Bonnefoy V. 2002a. The high-molecular-weight cytochrome *c* Cyc2 of *Acidithiobacillus ferrooxidans* is an outer membrane protein. *J Bacteriol* 184:313–317.

Zajic JE. 1969. *Microbial Biogeochemistry*. New York: Academic Press.

Zimmerley SR, Wilson DG, Prater JD. 1958. Cyclic leaching process employing iron oxidizing bacteria. US Patent 2,829,964.

CHAPTER TWENTY-TWO

Geomicrobiology of Selenium and Tellurium

John Stolz

CONTENTS

22.1 Occurrence in the Earth's Crust / 551
22.2 Biological Importance / 551
22.3 Toxicity of Selenium and Tellurium / 552
22.4 Bio-Oxidation of Reduced Forms of Selenium / 553
22.5 Bioreduction of Oxidized Selenium Compounds / 553
    22.5.1 Other Products of Selenate and Selenite Reduction / 555
    22.5.2 Selenium Reduction in the Environment / 556
22.6 Selenium Cycle / 557
22.7 Bio-Oxidation of Reduced Forms of Tellurium / 557
22.8 Bioreduction of Oxidized Forms of Tellurium / 557
22.9 Summary / 558
References / 559

22.1 OCCURRENCE IN THE EARTH'S CRUST

The elements selenium and tellurium, like sulfur, belong to group VI of the periodic table. All three have some properties in common, but selenium and tellurium, especially the latter, have some metallic attributes, unlike sulfur. Selenium and tellurium are much less abundant than sulfur in the Earth's crust. Selenium amounts to only 0.05–0.14 ppm (Rapp, 1972, p. 1080) and tellurium to 10^{-5} to 10^{-2} ppm (Lansche, 1965). Both are associated with metal sulfides in nature and occur in distinct minerals, e.g., ferroselite, ($FeSe_2$), chalomenite ($CuSeO_3 \cdot 2H_2O$), hessite (Ag_2Te), and tetradymite (Bi_2Te_2S). Selenium occurs in small amounts in various soils in concentrations in the range of 0.01–100 ppm. High concentrations are associated with arid, alkaline soils that contain some free $CaCO_3$ (Rosenfeld and Beath, 1964).

22.2 BIOLOGICAL IMPORTANCE

Selenium is required nutritionally as a trace element by many microorganisms and animals, including human beings (Stadtman, 1974; Combs and Scott, 1977; Miller and Neathery, 1977; Patrick, 1978; Mertz, 1981). Some plants, such as *Astragalus* spp. and *Stanleya*, can accumulate large amounts of selenium in the form of organic selenium compounds; however, there is no obligate requirement. Too much or too little can have deleterious effects. Bioaccumulation primarily due to ingestion of selenium-rich foods can cause maladies such as alkali disease and blind staggers (Stolz and Oremland, 1999). Not enough can also have pathological effects such as white muscle disease in ruminants. Lack of selenium results in the compromised functionality of selenium-containing enzymes such as formate dehydrogenase, glycine reductase, and glutathione peroxidase. The latter, which is found in mammalian red blood corpuscles (Rotruck et al., 1973), catalyzes the reaction

$$2GSH + H_2O_2 \rightarrow GSSG + 2H_2O \qquad (22.1)$$

Selenium may also be found in enzymes with molybdenum and tungsten cofactors as it is b subunit (Schroeder et al., 1997). The *ser* operon contains four genes in the order *serABDC* (Krafft et al., 2000). The mature protein is a periplasmic heterotrimer. SerA is the catalytic subunit, SerB the smaller iron sulfur–containing subunit, and SerC the anchoring cytochrome b subunit (Schroeder et al., 1997). Nitrite reductase is found in the periplasm of *T. selenatis* and plays a role in selenite reduction, besides catalyzing nitrite reduction (DeMoll-Decker and Macy, 1993). This helps to explain why *T. selenatis* produces elemental selenium in the presence of nitrate, but selenite in its absence. Selenite does not support growth of *T. selenatis* (DeMoll-Decker and Macy, 1993).

A selenite reductase enzyme has been obtained from the fungus *Candida albicans* (Falcone and Nickerson, 1963; Nickerson and Falcone, 1963). It reduces selenite to Se^0. A characterization of the enzyme has shown that it requires a quinone, a thiol compound (e.g., glutathione), a pyridine nucleotide (NADP), and an electron donor (e.g., glucose 6-phosphate) for activity. Electron transfer between NADP and quinone is probably mediated by flavin mononucleotide in this system. It is possible that this enzyme is part of an assimilatory SeO_4^{2-} and/or SeO_3^{2-} reductase system. How this enzyme compares with that in *T. selenatis* remains to be established.

Sulfurospirillum barnesii (formerly *Geospirillum barnesii*, also called strain SES-3) (Oremland et al., 1994; Stolz et al., 1999) is another bacterium that can reduce selenate to elemental selenium. Cells of this organism grew with lactate as carbon and energy source and selenate as terminal electron acceptor, which was reduced to selenite. As with *T. selenatis*, resting cells of *Ssp. barnesii* but not growing cells were able to reduce selenite to Se^0 (Oremland et al., 1994). One important difference between *Ssp. barnesii* and *T. selenatis* is that *Ssp. barnesii* is able to use a much wider range of reducible anions as terminal electron acceptors than *T. selenatis* (Stolz and Oremland, 1999). *Ssp. barnesii* can reduce selenate and nitrate simultaneously whether pregrown on selenate or nitrate, consistent with the observation that selenate reductase is constitutive in this organism (Oremland et al., 1999).

Another selenate-respiring bacterium that has been well studied is *Enterobacter cloacae* SLD1a-1 (Losi and Frankenberger, 1997; Leaver et al., 2008). A facultative anaerobe capable of selenate respiration, its selenate reductase, is a substrate-specific membrane-bound enzyme (Watts et al., 2003)

that gets its electrons from the menaquinone pool (Ma et al., 2009). Its expression is regulated by the fumarate nitrate reduction (FNR) regulator and is secreted through the twin arginine translocation (TAT) system (Yee et al., 2007; Ma et al., 2007).

Two other reducers of selenium oxyanions, both gram-positive bacteria, are *Bacillus arseniciselenatis* and *Bacillus selenitireducens* (Switzer Blum et al., 1998). The first forms spores but the second does not. Both were isolated from anoxic muds from Mono Lake, California, which is alkaline, hypersaline, and arsenic-rich. *B. arseniciselenatis* reduces selenate to selenite, whereas *B. selenitireducens* reduces selenite to elemental selenium as forms of anaerobic respiration. In coculture, the two strains together can reduce selenate to elemental selenium. Both strains can reduce arsenate as well (Switzer Blum et al., 1998). *Ssp. barnesii* and *B. arseniciselenatis* can reduce selenate and nitrate simultaneously, but unlike the selenate reductase in *Ssp. barnesii*, that in *B. arseniciselenatis* is not constitutive because it does not appear in nitrate-grown cells (Oremland et al., 1999). Therefore, in order for *B. arseniciselenatis* to reduce selenate and nitrate simultaneously, it has to be grown in the presence of a mixture of the two electron acceptors. Another gram-positive selenate reducer is *Bacillus selenatarsenatis*. Interestingly, its selenate reductase, encoded by the *srdBCA* operon, appears to be genetically distinct from the *serABCD* of *T. selenatis* (Kuroda et al., 2011).

A moderately halophilic selenate reducer was isolated from Dead Sea (Israel) sediment. It reduced selenate to selenite and elemental selenium. It is a gram-negative organism and has been named *Selenihalanaerobacter shriftii* (Switzer Blum et al., 2001). When it respires on glycerol or glucose, it forms acetate plus CO_2. Nitrate and trimethylamine N-oxide could serve as alternative electron acceptors, but reduced forms of sulfur, nitrite, arsenate, fumarate, or dimethylsulfoxide could not. More recently, a member of the phylum Chrysiogenetes, *Desulfurispirillum indicus* strain S5, was described (Rauschenbach et al., 2011). Isolated from an estuarine canal in India, it was found to not only respire selenate and selenite but also nitrate, nitrite, and arsenate. Elemental selenium (Se(0)) is formed when either selenate or selenite is used as the terminal electron acceptor.

All previously mentioned selenate- and selenite-reducing bacteria belong in the domain bacteria. They include low and high G + C gram-positive

EHRLICH'S GEOMICROBIOLOGY

bacteria, Halanaerobacter, and Beta, Gamma, and Epsilon proteobacteria (Stolz et al., 2006). An example of a hyperthermophilic member of the domain archaea capable of respiring organotrophically on selenate, Pyrobaculum arsenaticum, was isolated from a hot spring near Naples, Italy (Huber et al., 2000) (see also Chapter 14). It reduces selenate to elemental selenium. Previously isolated Pyrobaculum aerophilum (Völkl et al., 1993) was found capable of respiring organotrophically on selenate and selenite and autotrophically on selenate with H_2 as electron donor (Huber et al., 2000). Elemental selenium was the reduction product.

In most studies of bacterial reduction of selenate and selenite, elemental selenium (red form), when formed, is usually found to be a major, if not the only, product. This is noteworthy because sulfate and sulfite cannot be directly reduced to S^0 but are reduced to H_2S without intermediate formation of S^0. Yet selenium and sulfur are members of the same chemical family. The implication is that enzymatic mechanisms of reduction for oxidized forms of these two elements are different. To date, none of the true selenate respirers have been found capable of sulfate respiration, which could be related to the significantly higher energy yield in selenate respiration ($\Delta G'$, -15.53 kcal mol^{-1} e^{-1}) than in sulfate respiration ($\Delta G'$, -0.10 kcal mol^{-1} e^{-1}) (Newman et al., 1998). It must be noted, however, that D. desulfuricans subsp. aestuarii has been found to reduce nanomolar but not millimolar quantities of selenate to selenite (Zehr and Oremland, 1987). Sulfate inhibited reduction of selenate, suggesting but not proving that the mechanism of sulfate and selenate reduction in this case may be a common one. As Zehr and Oremland (1987) pointed out, when sulfate is being reduced to H_2S in the absence of selenate, some of the H_2S formed may subsequently reduce biogenically formed selenite chemically to Se^0. They found that in nature, the sulfate reducer can reduce selenate only if the ambient sulfate concentration is below 4 mM. Hockin and Gadd (2003) found that in mixed biofilms, Desulfomicrobium norvegicum could reduce selenite that diffused into the biofilm with H_2S it produced anaerobically by reduction of sulfate, resulting in formation of S^0 and Se^0. This reaction was abiotic and can be formulated as follows:

$$3HS^- + SeO_4^{2-} + 5H^+ \rightarrow 3S^0 + Se^0 + 4H_2O \quad (22.5)$$

The sulfur and selenium precipitated within the biofilm as nanometer-sized selenium-sulfur granules. By contrast, Ssp. barnesii, B. selenitireducens, and S. shriftii can form nanospheres consisting exclusively of Se^0 when reducing selenite enzymatically (Oremland et al., 2004).

Whereas selenate and/or selenite reduction by the previously described organisms resulted in extracellular deposition of Se^0, intracellular deposition of Se^0 has been observed with some other organisms. Chromatium vinosum can deposit Se^0 intracellularly as a result of an interaction of H_2Se, which is produced by D. desulfuricans in selenate reduction in coculture with Chr. vinosum. The Se^0 is stored in the form of globules in the Chr. vinosum cells (Nelson et al., 1996). Rhodobacter sphaeroides deposited red Se^0 in or on its cells when it reduced selenate and selenite (Van Fleet-Stadler et al., 2000). Ralstonia metallidurans CH34 can reduce selenite to red Se^0, which it stores in its cytoplasm and occasionally in its periplasm (Roux et al., 2001).

22.5.1 Other Products of Selenate and Selenite Reduction

In E. coli, a significant portion of selenite reduced during glucose metabolism is deposited as Se^0 on its cell membrane but not in its cytoplasm (Gerrard et al., 1974), and another portion is incorporated as selenide in organic compounds such as selenomethionine (Ahluwalia et al., 1968). Some soil microbes reduce selenate or selenite to dimethyl selenide (($CH_3)_2Se$) at elevated selenium concentrations (Kovalskii et al., 1968; Fleming and Alexander, 1972; Alexander, 1977; Doran and Alexander, 1977). Other volatile selenium compounds may also be formed, their relative quantities depending on reaction conditions (e.g., Reamer and Zoller, 1980). The compounds include dimethyl diselenide (($CH_3)_2Se_2$) and dimethyl selenone (($CH_3)_2SeO_2$). Pure cultures of Se-methylating isolates include En. cloacae (Dungan and Frankenberger, 2000), Aeromonas veronii (Rael and Frankenberger, 1996), Pseudomonas sp. (Chasteen and Bentley, 2003; Ranjard et al., 2003), Halomonas sp. (De Souza et al., 2001), and the phototrophic bacterial species Rhodospirillum rubrum and Rhodocyclus tenuis (McCarty et al., 1993).

Some fungi have been found to be effective in forming methylated selenium compounds (Barkes and Fleming, 1974). Alternaria alternata isolated

2006; Baesman et al., 2007). Although genetic evidence indicates that tellurate reduction involves a molybdenum enzyme (Theisen et al., 2013), the enzyme itself has as yet been uncovered. Trutko et al. (2000) presented evidence that in some gram-negative bacteria, the respiratory chain was involved in tellurite reduction. The tellurite was reduced to tellurium crystallites, which appeared in the periplasmic space or on the outer or inner surface of the plasma membrane. The makeup of the respiratory chain differed to some extent among the different bacterial cultures tested. Klonowska et al. (2005) found that although *Shewanella oneidensis* was able to reduce selenite and tellurite anaerobically, the electron transport pathway to the two electron acceptors diverged upstream from tetracytochrome c, CymA.

The biological reduction of Te oxyanions can result in the formation of Te(0) deposits of varying morphologies including granules, rosettes, nanospheres, and nanorods (Pearce et al., 2011). *Veillonella atypica*, an organism isolated from human tonsils, has been shown to precipitate Te(0) nanorods from the reduction of Te(VI), under anaerobic growth conditions (Pearce et al., 2011). Baesman et al. (2007) found that B. *selenitireducens* and *Ssp. barnesii* produced Te⁰ in the form of nanogranules when respiring on tellurate or tellurite. With B. *selenitireducens*, Te⁰ was deposited in the form of nanorods on the surface of the cells, which subsequently formed clusters, called *shards*, and rosettes. Nanorods also appeared in the bulk phase. Some crystals in the form of nanorods that aggregated into shard-like nanocrystals also formed inside the cells. *Ssp. barnesii* deposited Te⁰ as irregularly shaped nanospheres (~20 nm diam.) frequently attached to the cell surface, which coalesced into larger clusters (500–1000 nm diam.). Nanospheres were also observed inside the cells. Another gram-positive bacterium, *Bacillus beveridgei*, was found to not only tolerate high levels of Te(IV) and Te(VI) ~ 5–10 mM but form both nanogranules and rods of Te(0) (Baesman et al., 2009). A question arises whether the difference in morphologies of the Te⁰ nanocrystals is somehow related to a difference in the gram-staining properties of these two organisms, B. *selenitireducens* being gram-positive, whereas *Ssp. barnesii* being gram-negative. This difference in gram reactivity may reflect a difference in organization of their respective electron transport systems, as is probably the case, for instance, in Mn(II) oxidation and

Mn(IV) reduction by gram-positive and gram-negative bacteria (see Chapter 17, Sections 17.5 and 17.6).

Biologically produced tellurium-containing nanomaterials may see greater application in industry and medicine. Tellurium is used in the production of thin-film solar cells. More effective means for tellurium production will be required to meet the need due to the increased emphasis on solar cell technology for alternative energy generation. Recently, the nanorods produced by S. *oneidensis* were shown to have antivirulence properties, as they inhibited the synthesis of pyoverdine in *Pseudomonas aeruginosa* (Mohanty et al., 2014).

The fungus *Penicillium* sp. has been found to produce $(CH_3)_2Te$ from several inorganic tellurium compounds, provided only that reducible selenium compounds were also present (Fleming and Alexander, 1972). The amount of dialkyltelluride formed was related to the relative concentrations of Se and Te in the medium. Microbial reduction of oxidized forms of tellurium may represent detoxification reactions rather than a form of respiration, but this needs further investigation.

22.9 SUMMARY

Selenium, although a very toxic element, is nutritionally required by bacteria, fungi, protists, plants, and animals. In fact, selenocysteine is the 21st amino acid, with its own tRNA[sec]. Microorganisms have been described that can oxidize reduced selenium compounds. At least one, *A. ferrooxidans*, can use selenide in the form of CuSe as a sole source of energy, oxidizing the compound to elemental selenium (Se⁰) and Cu^{2+}. Oxidized forms of inorganic selenium compounds can be reduced by microorganisms, including members of the domains bacteria and archaea. Selenate and selenite may be reduced to one or more of the following: Se⁰, H_2Se, dimethylselenide ($(CH_3)_2Te$), dimethyl diselenide ($(CH_3)_2Se_2$), and dimethyl selenone ($(CH_3)_2SeO_2$). The reductions are enzymatic and in some bacteria represent a form of respiration. The microbial interactions with various forms of selenium contribute to a selenium cycle in nature. Microbial selenate and selenite reduction to elemental selenium in soil and sediment is a form of selenium immobilization that is potentially reversible. Microbial selenate and selenite reduction to volatile forms of selenium in soil, sediment, and water columns of bodies of water is a form of selenium removal that is permanent.

Tellurium occurs in such low concentrations in nature that it does not seem geomicrobially important. Nevertheless, microbial reduction of tellurate and tellurite to elemental tellurium (Te^0) and dimethyltelluride ((CH_3)$_2$Te) has been observed. Microbial oxidation of tellurides has so far not been reported.

REFERENCES

Ahluwalia GS, Saxena YR, Williams HH. 1968. Quantitative studies on selenite metabolism in *Escherichia coli*. *Arch Biochem Biophys* 124:79–84.

Alexander M. 1977. *Introduction to Soil Microbiology*, 2nd edn. New York: Wiley.

Andreesen JR, Ljungdahl LG. 1973. Formate dehydrogenase of *Clostridium thermoaceticum*: Incorporation of selenium-75, and the effects of selenite, molybdate, and tungstate on the enzyme. *J Bacteriol* 116:869–873.

Bacon M, Ingledew WJ. 1989. The reductive reactions of *Thiobacillus ferrooxidans* on sulfur and selenium. *FEMS Microbiol Lett* 58:189–194.

Baesman SM, Bullen TD, Dewald J, Zhang D, Curran S, Islam FS, Beveridge TJ, Oremland RS. 2007. Formation of tellurium nanocrystals during anaerobic growth of bacteria that use Te oxyanions as respiratory electron acceptors. *Appl Environ Microbiol* 73:2135–2143.

Baesman SM, Stolz JF, Kulp TR, Oremland RS. 2009. Enrichment and isolation of *Bacillus beveridgei* sp. nov., a facultative anaerobic haloalkaliphile from Mono Lake, California that respires oxyanions of tellurium, selenium, and arsenic. *Extrophiles* 13:695–705.

Barkes L, Fleming RW. 1974. Production of dimethylselenide gas from inorganic selenium by eleven fungi. *Bull Environ Contam Toxicol* 12:308–311.

Bautista EM, Alexander M. 1972. Reduction of inorganic compounds by soil microorganisms. *Soil Sci Soc Am Proc* 36:918–920.

Breed RS, Murray EGD, Smith NR. 1948. *Bergey's Manual of Determinative Bacteriology*, 6th edn. Baltimore, MD: Williams & Wilkins.

Burton GA Jr, Giddings TH, DeBrine P, Fall R. 1987. High incidence of selenite-resistant bacteria from a site polluted with selenium. *Appl Environ Microbiol* 53:185–188.

Chasteen TG, Bentley R. 2003. Biomethylation of selenium and tellurium: Microorganisms and plants. *Chem Rev* 103:1–26.

Chasteen TG, Fuentes DE, Tantalean JC, Vasquez CC. 2009. Tellurite: History, oxidative stress, and molecular mechanisms of resistance. *FEMS Microbiol Rev* 33:820–832.

Chau YK, Wong PTS, Silverberg BA, Luxon PL, Bengert GA. 1976. Methylation of selenium in the aquatic environment. *Science* 192:1130–1131.

Combs GF Jr, Scott ML. 1977. Nutritional interrelationships of vitamin E and selenium. *BioScience* 27:467–473.

Csotonyi JT, Stackebrandt E, Yurkov V. 2006. Anaerobic respiration on tellurate and other metalloids in bacteria from hydrothermal vent fields in the eastern Pacific Ocean. *Appl Environ Microbiol* 72:4950–4956.

DeMoll-Decker H, Macy JM. 1993. The periplasmic nitrate reductase of *Thauera selenatis* may catalyze the reduction of selenite to elemental selenium. *Arch Microbiol* 160:241–247.

De Souza MP, Amini A, Dojika MA, Pickering IJ, Dawson SC, Pace NR, Terry N. 2001. Identification and characterization of bacteria in a selenium-contaminated hypersaline evaporation pond. *Appl Environ Microbiol* 67:3785–3794.

Doran JW, Alexander M. 1977. Microbial transformation of selenium. *Appl Environ Microbiol* 33:31–37.

Dowdle PR, Oremland RS. 1998. Microbial oxidation of elemental selenium in soil slurries and bacterial cultures. *Environ Sci Technol* 32:3749–3755.

Driscoll DM, Chavatt L. 2004. Finding needles in a haystack: In silico identification of eukaryotic selenoprotein genes. *EMBO Reports* 5:140–141.

Driscoll DM, Copeland PR. 2003. Mechanism and regulation of selenoprotein synthesis. *Annu Rev Nutr* 23:17–40.

Duerre P, Andreesen JR. 1982. Selenium-dependent growth and glycine fermentation by *Clostridium purinolyticum*. *J Gen Microbiol* 128:1457–1466.

Dungan RS, Frankenberger WT Jr. 2000. Factors affecting the volatilization of dimethylselenide by *Enterobacter cloacae* SLD1a-1. *Soil Biol Biochem* 32:1353–1358.

Enoch HG, Lester RL. 1972. Effects of molybdate, tungstate, and selenium compounds on formate dehydrogenase and other enzymes in *Escherichia coli*. *J Bacteriol* 110:1032–1040.

Falcone G, Nickerson WJ. 1963. Reduction of selenite by intact yeast cells and cell-free preparations. *J Bacteriol* 85:754–762.

Fleming RW, Alexander M. 1972. Dimethyl selenide and dimethyl telluride formation by a strain of *Penicillium*. *Appl Microbiol* 24:424–429.

Fourmy D, Guittet E, Yoshizawa S. 2002. Structure of prokaryotic SECIS mRNA hairpin and its interaction with elongation factor SelB. *J Mol Biol* 324:137–150.

Frankenberger WT, Karlson U. 1992. Dissipation of soil selenium by microbial volatilization. In: Adriano DC (ed.), *Biogeochemistry of Trace Metals*. Boca Raton, FL: Lewis Publishers, pp. 365–381.

Frankenberger WT, Karlson U. 1995. Soil management factors affecting volatilization of selenium from dewatered sediments. *Geomicrobiol J* 12:265–277.

Gerrard TL, Telford JN, Williams HH. 1974. Detection of selenium deposits in *Escherichia coli* by electron microscopy. *J Bacteriol* 119:1057–1060.

Gladyshev VN, Hatfield DL. 2010. Selenocysteine biosynthesis, selenoproteins, and selenoproteomes. *Nucleic Acids Mol Biol* 24:3–27.

Guzzo J, Dubow MS. 2000. A novel selenite- and tellurite-inducible gene in *Escherichia coli*. *Appl Environ Microbiol* 66:4972–4978.

Herbel MJ, Johnson TM, Oremland RS, Bullen TD. 2000. Fractionation of selenium isotopes during bacterial respiratory reduction of selenium oxyanions. *Geochim Cosmochim Acta* 64:3701–3709.

Herbel MJ, Johnson TM, Tanji KK, Gao S, Bullen TD. 2002. Selenium stable isotope ratios in California agricultural drainage water management systems. *J Environ Qual* 31:1146–1156.

Herbel MJ, Blum JS, Oremland RS, Borglin SE. 2003. Reduction of elemental selenium to selenide: Experiments with anoxic sediments and bacteria that respire Se-oxyanions. *Geomicrobiology* 20:587–602.

Hille R, Hall J, Basu P. 2014. The mononuclear molybdenum enzymes. *Chem Rev* 114:3963–4038.

Hockin SL, Gadd GM. 2003. Linked redox precipitation of sulfur and selenium under anaerobic conditions by sulfate-reducing bacterial biofilms. *Appl Environ Microbiol* 69:7063–7072.

Huber R, Sacher M, Vollmann A, Huber H, Rose D. 2000. Respiration of arsenate and selenate by hyperthermophilic archaea. *Syst Appl Microbiol* 23:305–314.

Johnson TM, Herbel MJ, Bullen TD, Zawislanski PT. 1999. Selenium isotope ratios as indicators of selenium sources and biogeochemical cycling. *Geochim Cosmochim Acta* 63:2775–2783.

Klonowska A, Heulin T, Vermeglio A. 2005. Selenite and tellurite reduction by *Shewanella oneidensis*. *Appl Microbiol* 71:5607–5609.

Kovalskii VV, Ermakov VV, Letunova SV. 1968. Geochemical ecology of microorganisms in soils with different selenium content. *Mikrobiologiya* 37:122–139.

Krafft T, Bowen A, Theis F, Macy JM. 2000. Cloning and sequencing of the genes encoding the periplasmic-cytochrome *b*-containing selenate reductase of *Thauera selenatis*. *DNA Sequence* 10:365–377.

Kuroda M, Yamashita M, Miwa E, Imao K, Fujimoto N, Ono H, Ike M. 2011. Molecular cloning and characterization of the *srdBCA* operon, encoding the respiratory selenate reductase complex from the selenate-reducing bacterium *Bacillus selenatarsenatis* SF-1. *J Bacteriol* 193:2141–2148.

Lansche AM. 1965. Tellurium. In: *Mineral Facts and Problems*. Washington, DC: Bur Mines, Department of the Interior, pp. 935–939.

Leaver JT, Richardson DJ, Butler CS. 2008. *Enterobacter cloacae* SLD1a-1 gains a selective advantage from selenate reduction when growing in nitrate-depleted anaerobic environment. *J Ind Microbiol Biotechnol* 35:863–873.

Lester RL, DeMoss JA. 1971. Effects of molybdate and selenite on formate and nitrate metabolism in *Escherichia coli*. *J Bacteriol* 105:1006–1014.

Lipman JG, Waksman SA. 1923. The oxidation of selenium by a new group of autotrophic microorganisms. *Science* 57:60.

Lortie L, Gould WD, Rajan S, McCready RGL, Cheng J-J. 1992. Reduction of selenate and selenite to elemental selenium by a *Pseudomonas stutzeri* isolate. *Appl Environ Microbiol* 58:4042–4044.

Losi ME, Frankenberger WT Jr. 1997 Reduction of selenium oxyanions by *Enterobacter cloacae* SLD1a-1: Isolation and growth of the bacterium and its expulsion of selenium particles. *Appl Environ Microbiol* 63:3079–3084.

Ma J, Kobayashi DY, Yee N. 2007. Chemical kinetic and molecular genetic study of selenium oxyanion reduction by *Enterobacter cloacae* SLD1a-1. *Environ Sci Technol* 41:7795–7801.

Ma J, Kobayashi DY, Yee N. 2009. Role of menaquinone biosynthesis genes in selenate reduction by *Enterobacter cloacae* SLD1a-1 and *Escherichia coli* K12. *Environ Microbiol* 11:149–158.

Macy JM, Rech S, Auling G, Dorsch M, Stackebrandt E, Sly L. 1993. *Thauera selenatis* gen. nov. sp. nov., a member of the beta-subclass of Proteobacteria with a novel type of anaerobic respiration. *Int J Syst Bacteriol* 43:135–142.

Maiers DT, Wichlacz PL, Thompson DL, Bruhn DF. 1988. Selenate reduction by bacteria from a selenium-rich environment. *Appl Environ Microbiol* 54:2591–2593.

Masscheleyn PH, Patrick WH Jr. 1993. Biogeochemical processes affecting selenium cycling in wetlands. *Environ Toxicol Chem* 12:2235–2243.

McCarty S, Chasteen T, Marshall M, Fall R, Bachofen R. 1993. Phototrophic bacteria produce volatile, methylated sulfur and selenium compounds. *FEMS Microbiol Lett* 112:93–98.

Mertz W. 1981. The essential trace elements. *Science* 213:1332–1338.

Miller WJ, Neathery MW. 1977. Newly recognized trace mineral elements and their role in animal nutrition. *BioScience* 27:674–679.

Mohanty A, Kathawala MH, Zhang J, Chen WN, Loo JSC, Kjelleberg S, Yang L, Cao B. 2014. Biogenic tellurium nanorods as a novel antivirulence agent inhibiting pyoverdine production in *Pseudomonas aeruginosa*. *Biotechnol Bioeng* 111:858–865.

Nagai S. 1965. Differential reduction of tellurite by growing colonies of normal yeasts and respiration deficient mutants. *J Bacteriol* 90:220–222.

Narasingarao P, Häggblom MM. 2007. Identification of anaerobic selenate-respiring bacteria from aquatic sediments. *Appl Environ Microbiol* 73:3519–3527.

Nelson DC, Casey WH, Sison JD, Mack EE, Ahmad A, Pollack JS. 1996. Selenium uptake by sulfur-accumulating bacteria. *Geochim Cosmochim Acta* 60:3531–3539.

Newman DK, Ahmann D, Morel FMM. 1998. A brief review of microbial arsenate respiration. *Geomicrobiol J* 15:255–268.

Nickerson WJ, Falcone G. 1963. Enzymatic reduction of selenite. *J Bacteriol* 85:763–771.

Niess UM, Klein A. 2004. Dimethylselenide demethylation is an adaptive response to selenium deprivation in the archeon *Methanococcus voltae*. *J Bacteriol* 186:3640–3648.

Oremland RS, Hollibaugh JT, Maest AS, Presser TS, Miller LB, Cuthberson CW. 1989. Selenate reduction to elemental selenium by anaerobic bacteria in sediments and culture: Biogeochemical significance of a novel sulfate-independent respiration. *Appl Environ Microbiol* 55:2333–2343.

Oremland RS, Steinberg NA, Presser TS, Miller LG. 1991. In situ bacterial selenate reduction in the agricultural drainage system of western Nevada. *Appl Environ Microbiol* 57:615–617.

Oremland RS, Herbel MJ, Switzer Blum J, Langley S, Beveridge TJ, Ajayan PM, Sutto T, Ellis AV, Curran S. 2004. Structural and spectral features of selenium nanospheres produced by Se-respiring bacteria. *Appl Environ Microbiol* 70:52–60.

Oremland RS, Switzer Blum J, Burns Bindi A, Dowdle PR, Herbel M, Stolz JF. 1999. Simultaneous reduction of nitrate and selenate by cell suspensions of selenium-respiring bacteria. *Appl Environ Microbiol* 65:4385–4392.

Oremland RS, Switzer Blum J, Culbertson CW, Visscher PT, Miller LG, Dowdle P, Strohmaier FE. 1994. Isolation, growth, and metabolism of an obligately anaerobic, selenate-respiring bacterium, strain SES-3. *Appl Environ Microbiol* 60:3011–3019.

Oremland RS, Zehr JP. 1986. Formation of methane and carbon dioxide from dimethylselenide in anoxic sediments and by a methanogenic bacterium. *Appl Environ Microbiol* 52:1031–1036.

Patrick R. 1978. Effects of trace metals in the aquatic ecosystem. *Am Sci* 66:185–191.

Pearce CI, Baesman SM, Switzer Blum J, Fellowes JW, Oremland RS. 2011. Nanoparticles formed from microbial oxyanion reduction of toxic group 15 and group 16 metalloids. In: Stolz JF, Oremland RS (eds.), *Microbial Metal and Metalloid Metabolism: Advances and Applications*. ASM Press, Washington, DC, pp. 297–319.

Pinsent J. 1954. The need for selenite and molybdate in the coli-aerogenes group of bacteria. *Biochem J* 57:10–16.

Rael RM, Frankenberger WT Jr. 1996. Influence of pH, salinity, and selenium on the growth of *Aeromonas veronii* in evaporation agricultural drainage water. *Water Res* 30:422–430.

Ranjard L, Prigent-Combaret C, Nazaret S, Cournoyer B. 2002. Methylation of inorganic and organic selenium by the bacterial thiopurine methyltransferase. *J Bacteriol* 184:3146–3149.

Ranjard L, Nazaret S, Cournoyer B. 2003. Freshwater bacteria can methylate selenium through the thiopurine methyltransferase pathway. *Appl Environ Microbiol* 69:3784–3790.

Rapp G Jr. 1972. Selenium: Element and geochemistry. In: Fairbridge RW (ed.), *The Encyclopedia of Geochemistry and Environmental Sciences*. Encyclopedia of Earth Sciences Ser, Vol. IVA. New York: Van Nostrand Reinhold, pp. 1079–1080.

Rauschenbach I, Narasingarao P, Häggblom MM. 2011 *Desulfurispirillum indicum* sp. nov., a selenate and selenite respiring bacterium isolated from an estuarine canal in southern India. *Int J Syst Evol Microbiol* 61:654–658.

Reamer DC, Zoller WH. 1980. Selenium biomethylation products from soil and sewage. *Science* 208:500–502.

Rech S, Macy JM. 1992. The terminal reductases from selenate and nitrate respiration in *Thauera selenatis* are two distinct enzymes. *J Bacteriol* 174:7316–7320.

Rosenfeld I, Beath OA. 1964. *Selenium, Geobotany, Biochemistry, Toxicity, and Nutrition*. New York: Academic Press.

Rother M, Mathes I, Lottspeich F, Böck A. 2003. Inactivation of the *selB* gene in *Methanococcus maripaludis*: Effect on synthesis of selenoproteins and their sulfur-containing homologs. *J Bacteriol* 185:107–114.

Rother M, Resch A, Gardner WL, Whitman WB, Böck A. 2001. Heterologous expression of archael seleno-protein genes directed by the SECIS element located in the 3′ non-translated region. *Mol Microbiol* 40:900–908.

Rother M, Wilting R, Commans S, Böck A. 2000. Identification and characterization of the selenocyste-ine-specific translation factor SelB from the archaeon *Methanococcus jannaschii*. *J Mol Biol* 299:351–358.

Rotruck JT, Pope AL, Ganther HE, Swanson AB, Hafman DG, Hoekstra WG. 1973. Selenium: Biochemical role as a component of glutathione peroxidase. *Science* 179:588–590.

Roux M, Sarret G, Pignot-Paintrand I, Fontecave M, Coves J. 2001. Mobilization of selenite by *Ralstonia metallidurans* CH34. *Appl Environ Microbiol* 67:769–773.

Sapozhnikov DI. 1937. The substitution of selenium for sulfur in the photoreduction of carbonic acid by purple sulfur bacteria. *Mikrobiologiya* 6:643–644.

Sarathchandra SU, Watkinson JH. 1981. Oxidation of elemental selenium to selenite by *Bacillus megaterium*. *Science* 211:600–601.

Schilling K, Johnson TM, Wilcke W. 2011. Isotope frac-tionation of selenium during fungal biomethylation by *Alternaria alternate*. *Environ Sci Technol* 45:2670–2676.

Schilling K, Johnson TM, Wilcke W. 2013. Isotope frac-tionation of selenium by biomethylation in micro-cosm incubations of soil. *Chem Geol* 352:101–107.

Schrauzer GN. 2000. Selenomethionine: A review of its nutritional significance, metabolism and toxic-ity. *J Nutr* 130:1653–1656.

Schroeder I, Rech S, Krafft T, Macy JM. 1997. Purification and characterization of the sel-enate reductase from *Thauera selenatis*. *J Biol Chem* 272:23765–23768.

Shrift A. 1964. Selenium in nature. *Nature* (Lond) 201:1304–1305.

Shum AD, Murphy JC. 1972. Effects of selenium com-pounds on formate metabolism and coincidence of selenium-75 incorporation and formic dehydroge-nase activity in cell-free preparations of *Escherichia coli*. *J Bacteriol* 110:447–449.

Silverman MP, Ehrlich HL. 1964. Microbial forma-tion and degradation of minerals. *Adv Appl Microbiol* 6:153–206.

Stadtman RC. 1974. Selenium biochemistry. *Science* 183:915–922.Steinberg NA, Blum JS, Hochstein L, Ormland RS. 1992. Nitrate is a preferred electron acceptor for growth of freshwater selenate-respir-ing bacteria. *Appl Environ Microbiol* 58:426–428.

Steinberg NA, Oremland RS. 1990. Dissimilatory sel-enate reductase potentials in a diversity of sediment types. *Appl Environ Microbiol* 56:3550–3557.

Stolz JF, Basu P, Santini JM, Oremland RS. 2006. Arsenic and selenium in microbial metabolism. *Ann Rev Microbiol* 60:107–130.

Stolz JF, Ellis DJ, Switzer Blum J, Ahmann D, Oremland RS, Lovley DR. 1999. *Sulfurospirillum barnesii* sp. nov. and *Sulfurospirillum arsenophilus* sp. nov., new members of the *Sulfurospirillum* clade of the ε-Proteobacteria. *Int J Syst Bacteriol* 49:1177–1180.

Stolz JF, Oremland RS. 1999. Bacterial respiration of arsenic and selenium. *FEMS Microbiol Rev* 23:615–627.

Switzer Blum J, Burns Bindi A, Buzzelli J, Stolz JG, Oremland RS. 1998. *Bacillus arsenicoselenatis*, sp. nov., and *Bacillus selenitireducens*, sp. nov.: Two haloalkaliphiles from Mono Lake, California that respired oxyanions of selenium and arsenic. *Arch Microbiol* 171:19–30.

Switzer Blum J, Stolz JF, Oren A, Oremland RS. 2001. *Selenihalanaerobacter shriftii* gen nov., spec. nov., a halo-philic anaerobe from Dead Sea sediments that respire selenate. *Arch Microbiol* 175:208–219.

Theisen J, Zylstra GJ, Yee N. 2013. Genetic evidence for a molybdopterin-containing tellurate reductase. *Appl Environ Microbiol* 79:3171–3175.

Thompson-Eagle ET, Frankenberger WT Jr, Karlson U. 1989. Volatilization of selenium by *Alternaria alternata*. *Appl Environ Microbiol* 55:1406–1413.

Tomei FA, Barton LL, Lemansky CL, Zocco TG. 1992. Reduction of selenate and selenite to elemental sele-nium by *Wolinella succinogenes*. *Can J Microbiol* 38:1328–1333.

Torma AE, Habashi F. 1972. Oxidation of copper(II) selenide by *Thiobacillus ferrooxidans*. *Can J Microbiol* 18:1780–1781.

Trutko SM, Akimenko VK, Suzina NE, Anisimova LA, Shlyapnikov MG, Baskunov BP, Duda VI, Boronin AM. 2000. Involvement of the respiratory chain of gram-negative bacteria in the reduction of tellurite. *Arch Microbiol* 173:178–186.

Turner RJ, Hou Y, Weiner JH, Taylor DE. 1992. The arsenical ATPase efflux pump mediates tellurite resistance. *J Bacteriol* 174:3092–3094.

Van Fleet-Stadler V, Chasteen TG, Pickering IJ, George GN, Prince RC. 2000. Fate of selenate and selenite metabolized by *Rhodobacter sphaeroides*. *Appl Environ Microbiol* 66:4849–4853.

Völkl P, Huber R, Drobner E, Rachel R, Burggraf S, Trincone A, Stetter KO. 1993. *Pyrobaculum aerophilum* sp. nov., a novel nitrate-reducing hyperthermo-philic archaeum. *Appl Environ Microbiol* 59:2918–2926.

Watts CA, Ridley H, Condie KL, Leaver JT, Richardson DJ, Butler CS. 2003. Selenate reduction by *Enterobacter cloacae* SLD1a-1 is catalyzed by a molybdenum-dependent membrane-bound nitrate reductase. *FEMS Microbiol Lett* 228:273–279.

Woolfolk CA, Whiteley HR. 1962. Reduction of inorganic compounds with molecular hydrogen by *Micrococcus lactilyticus*. I. Stoichiometry with compounds of arsenic, selenium, tellurium, transition and other elements. *J Bacteriol* 84:647–658.

Yamamoto I, Saiki T, Liu S-M, Ljungdahl LG. 1983. Purification and properties of NADP-dependent formate dehydrogenase from *Clostridium thermoaceticum*, a tungstate-selenium protein. *J Biol Chem* 258:1826–1832.

Yee N, Ma J, Dalia A, Boonfueng T, Kobayashi DY. 2007. Se(VI) reduction and the precipitation of Se(0) by the facultative bacterium *Enterobacter cloacae* SLD1a-1 are regulated by FNR. *Appl Environ Microbiol* 73:1914–1920.

Zalokar M. 1953. Reduction of selenite by *Neurospora*. *Arch Biochem Biophys* 44:330–337.

Zehr JP, Oremland RS. 1987. Reduction of selenate to selenide by sulfate-respiring bacteria: Experiments with cell suspensions and estuarine sediments. *Appl Environ Microbiol* 53:1365–1369.

23.6.8   Economic Significance of the Petroleum Reservoir Deep Biosphere / 597
      23.6.8.1 Biodegradation and the Occurrence of Heavy Oil / 597
      23.6.8.2 Reservoir Souring and Corrosion / 598
      23.6.8.3 Secondary and Tertiary Oil Recovery / 600
      23.6.8.4 Removal of Organic Sulfur from Petroleum / 603
23.6.9   Use of Microbes in Prospecting for Petroleum / 603
23.6.10 Microbes and Shale Oil / 604
23.7 Summary / 604
Acknowledgment / 606
References / 606

23.1 INTRODUCTION

Although much of the organic carbon in the biosphere is continually recycled, a very significant amount has become trapped in special sedimentary formations, where it is inaccessible to mineralization by microbes until it becomes reexposed to water through natural causes or human intervention. Microbial mineralization of such reexposed organic carbon also depends on the access to suitable terminal electron acceptors, that is, oxygen in air in the case of aerobes and inorganic electron acceptors in the case of facultative or anaerobic microbes. The trapped organic carbon exists in various forms. The degree of its chemically reduced state is related to the length of time it has been trapped and any secondary changes that it has undergone during this time. Some of this trapped carbon has value as a fuel, a source of energy for industrial and other human activity, and is exploited for this purpose. Because of the great age of this material, it is known as *fossil fuel*. The remainder of the trapped carbon is chiefly kerogen and bitumen, some of which can be converted to fuel by human intervention. Fossil fuels include methane gas, natural gas (which is largely methane), petroleum, oil shale, coal, and peat. They are generally considered to have had a microbial origin (Ourisson et al., 1984).

23.2 NATURAL ABUNDANCE OF FOSSIL FUELS

A major portion of the total carbon at the Earth's surface is in the form of carbonate (Figure 23.1). It represents a major sink for carbon. The other sink is the trapped organic carbon that is not directly accessible for microbial mineralization. The carbonate carbon is not an absolute sink unless it is deeply buried because it is in a steady-state relationship with dissolved carbonate/bicarbonate and atmospheric CO_2, which in turn are in a steady-state relationship with organic carbon in living and dead biomass. The passage of carbon from one compartment into another is under biological control (Figure 23.1; Fenchel and Blackburn, 1979).

23.3 METHANE

Methane at atmospheric pressure and ambient temperature is a colorless, odorless, and flammable gas. Because its autoignition temperature is 650°C, it does not catch fire spontaneously. It is sparingly soluble in water (3.5 mL/100 mL of water) but readily soluble in organic solvents, including liquid hydrocarbons. It may have an abiotic or biogenic origin. Biogenic accumulations of methane may occur in nature when it is formed in consolidated sediment from which it cannot readily escape. In some deep-ocean sediments under conditions of high pressure and low temperature, methane accumulations are found in the form of methane hydrates (Kvenvolden, 1988; Haq, 1999). A special, mixed community of certain members of the archaea and bacteria living very close to methane hydrate has been detected in the forearc basin of the Nankai Trough off the east coast of Japan by Reed et al. (2002) and in solid gas hydrates in the Gulf of Mexico by Mills et al. (2005), and in some cases, abundant consortia of anaerobic methane-oxidizing archaea and sulfate-reducing bacteria develop (Boetius et al., 2000).

Bioformation of methane comes about when organic matter in the sediment is undergoing anaerobic microbial breakdown in the absence of significant quantities of alternative terminal electron acceptors such as nitrate, Fe(III), Mn(IV), or sulfate. If gas pressure due to methane builds up sufficiently in anaerobic lake or coastal sediment,

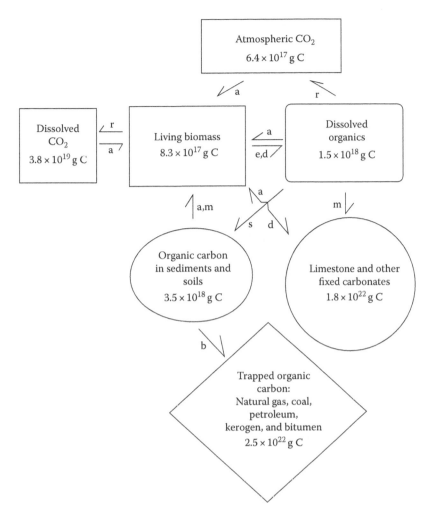

Figure 23.1. Microbial and physical processes contributing to carbon transfer among different compartments in the biosphere. a, Microbial assimilation; b, burial; d, decomposition; e, excretion; m, microbial mineralization; r, respiration; s, sedimentation. (Quantitative estimates from Fenchel, T and Blackburn, TH, *Bacteria and Mineral Cycling*, Academic Press, London, U.K., 1979; Bowen, HJM, *Environmental Chemistry of the Elements*, Academic Press, London, U.K., 1979.)

it may escape in the form of large gas bubbles that break at the water surface to release their methane into the atmosphere (Martens, 1976; Zeikus, 1977). In marshes, escaping methane may be ignited (by biogenic phosphine?) to burn as the so-called will-o'-the-wisps (Roels and Verstraete, 2001).

Many of the methane accumulations on Earth are of biogenic origin. Methane may occur in association with peat, coal, and oil deposits, or independent of them. That, which occurs in association with coal and oil, is most probably a mix of methane microbially formed in the early stages of their formation with the majority generated thermogenically from decomposition of organic matter with increasing burial depth and temperature

(Figure 23.2). In addition, there is growing evidence for the occurrence of so-called secondary biogenic methane that is formed in reservoir from the methanogenic degradation of crude oil components (see Section 23.6.8.1). Methane-associated with coal deposits can be the cause of serious mine explosions when accidentally ignited. Such methane is called *coal damp* by coal miners.

Biogenic methane formation is a unique biochemical process that appears to have arisen very early in the evolution of life. Indeed, the methanogenesis that results from the microbial reduction of CO_2 by H_2 may represent the first process or one of the first processes on Earth that autotrophs have harnessed for energy conservation (see Chapter 3).

dimethyl sulfide as well as methylamines, methanol, and acetate as energy sources. Yang et al. (1992) found that *Methanococcus voltae*, *Methanococcus maripaludis*, and *Methanococcus vannielii* can each use pyruvate as an energy source in the absence of H_2, and it has now been shown that some isolates of the methylotrophic methanogen *Methanococcoides* are capable of growth with choline and N,N-dimethylethanolamine (Watkins et al., 2012).

For methanogens to be able to draw on the wide range of oxidizable carbon compounds that may be available in their environment but that cannot be metabolized by them directly, they associate with heterotrophic fermenters or anaerobic respirers that do not completely mineralize their organic energy sources (e.g., Jain and Zeikus, 1989; Sharak Genthner et al., 1989; Grbic-Galic, 1990). To optimize access to the microbially generated energy sources and the electron acceptor CO_2 that these methanogens need, some of them form intimate consortia (*syntrophic associations*) with other anaerobic bacteria that can furnish them with these energy sources (H_2, acetate) and CO_2 through their metabolic end products (e.g., McInerney et al., 1979; Winter and Wolfe, 1979, 1980; Bochem et al., 1982; Zinder and Koch, 1984; Wolin and Miller, 1987; MacLeod et al., 1990; Stams and Plugge, 2009; Sieber et al., 2012). Frequently, the metabolites that are the basis for these syntrophic associations are not readily detectable when all the members of the consortium are growing together in mixed culture. This is because the metabolites are consumed as quickly as they are formed. When hydrogen is the metabolite, the process is called interspecies hydrogen transfer (Wolin and Miller, 1987; Stams and Plugge, 2009).

Among the most widely recognized genera of methanogens are *Methanobacterium*, *Methanothermobacter*, *Methanobrevibacterium*, *Methanococcus*, *Methanomicrobium*, *Methanogenium*, *Methanospirillum*, *Methanosarcina*, *Methanoculleus*, and *Methanosaeta* (Brock and Madigan, 1988; Bhatnagar et al., 1991; Boone et al., 1993; Atlas, 1997; Madigan et al., 2014). Methanogens may be mesophilic or thermophilic. They are found in diverse anaerobic habitats (Zinder, 1993), including some marine environments such as salt marsh sediments (Oremland et al., 1982; Jones et al., 1983b), coastal sediments (Sansone and Martens, 1981; Gorlatov et al., 1986), anoxic basins (Romesser et al., 1979), geothermally heated seafloor (Huber et al., 1982), hydrothermal vent effluent on the East

Pacific Rise (Jones et al., 1983a), deep subseafloor sediments (Newberry et al., 2004), deep subsurface igneous rocks (Kotelnikova and Pedersen, 1998), sediment effluent channel of the Crystal River Nuclear Power Plant (Florida; Rivard and Smith, 1982), lakes (Deuser et al., 1973; Jones et al., 1982; Giani et al., 1984), soils (Jakobsen et al., 1981), desert environments (Worakit et al., 1986), solfataric fields (Wildgruber et al., 1982; Zabel et al., 1984), oil deposits (Nazina and Rozanova, 1980; Rubinshtein and Oborin, 1986; Stetter et al., 1993), the digestive tract of insects and higher animals, especially ruminants and herbivores (Wolin, 1981; Breznak, 1982; Zimmerman et al., 1982; Brock and Madigan, 1988; Atlas, 1997; Madigan et al., 2014), and as endosymbionts (van Bruggen et al., 1984; Fenchel and Finlay, 1992). Thus, despite their obligately anaerobic nature, methanogens are fairly ubiquitous.

Methanogens play an important but not exclusive role in anaerobic mineralization of organic carbon compounds in soil and aquatic environments, especially freshwater sediments (Wolin and Miller, 1987). In marine sediments, where methanogens have to share hydrogen and acetate as sources of energy with sulfate-reducing bacteria, they tend to be outcompeted by the sulfate reducers because of the latter's higher affinity for hydrogen and acetate (Abrams and Nedwell, 1978; Kristjansson et al., 1982; Schönheit et al., 1982; Robinson and Tiedje, 1984). Thus, in many estuarine or coastal anaerobic muds, sulfate-reducing activity and methanogenesis occur usually in spatially separated zones in the sediment profile, with the zone exhibiting sulfate-reducing activity overlying the zone exhibiting methanogenesis (e.g., Martens and Berner, 1974; Sansone and Martens, 1981). Recent evidence suggested that some sulfate-reducing bacteria can also use methane as electron donor; however, it is now apparent that consortia of methane-oxidizing archaea and sulfate-reducing bacteria are required for this process (Section 23.3.5.2).

Under two special circumstances, methanogenesis and sulfate reduction can be compatible in an anaerobic marine environment. One circumstance is the existence of an excess supply of a shared energy source (H_2 or acetate; Oremland and Taylor, 1978). The other circumstance is one where sulfate reducers and methanogens use different energy sources, namely, products of decaying plant material and methanol or trimethylamine, respectively (Oremland et al., 1982; King, 1984). In anaerobic freshwater sediments and soils where sulfate, nitrate, ferric oxide, and manganese(IV) oxide concentrations are very low, methanogenesis is usually the dominant mechanism of organic carbon mineralization. Yet, even here, certain sulfate-reducing bacteria may grow in the same location as methanogens (e.g., Koizumi et al., 2003). Indeed, they may form a consortium with them. In the absence of sulfate, these sulfate reducers ferment suitable organic carbon with the production of H_2, which the methanogens then use in their energy metabolism to form methane (e.g., Bryant et al., 1977).

A few methanogens, in particular, M. *barkeri* grown with H_2/CO_2 or methanol, have the ability to reduce Fe(III) in place of CO_2 (van Bodegom et al., 2004). This ability may explain in part the inhibition of methanogenesis in soil and sediment by Fe(III).

23.3.2 Methanogenesis and Carbon Assimilation by Methanogens

23.3.2.1 Methanogenesis

One kind of autotrophic methane formation represents a form of anaerobic respiration in which hydrogen (H_2) is the electron donor and CO_2 is the terminal electron acceptor, with the CO_2 being transformed to CH_4 according to the overall reaction

$$4H_2 + CO_2 \rightarrow CH_4 + 2H_2O \qquad (23.1)$$

This reaction is exothermic and yields energy ($\Delta G^0 = -33$ kcal or -137.9 kJ) that can be used by the organism to do metabolic work.

In a few instances, secondary alcohols were found to serve as electron donors, with CO_2 as the terminal electron acceptor. The CO_2 was therefore, the source of the methane formed (Widdel, 1986; Zellner et al., 1989). In these reactions, the alcohols were replacing H_2 as the reductant of CO_2. At least one instance is known in which ethanol served as the electron donor for methane formation from CO_2, the ethanol being oxidized to acetate (Frimmer and Widdel, 1989):

$$2CH_3CH_2OH + HCO_3^- \rightarrow 2CH_3COO^-$$
$$+ CH_4 + H_2O + H^+$$
$$(\Delta G^0 = -27.8 \text{ kcal/mol } CH_4$$
$$\text{or} -116.3 \text{ kJ/mol } CH_4) \qquad (23.2)$$

The organism in this instance was a nonautotrophic methanogen, *Methanogenium organophilum* growing in a medium containing 0.05% trypticase peptone and 0.05% yeast extract as nitrogen sources among other ingredients (Frimmer and Widdel, 1989).

Some methanogens can form methane by a disproportionation reaction, that is, by fermentation, in which a portion of the substrate molecule acts as the electron donor (energy source) and the rest as the electron acceptor. For example, they can produce methane from carbon monoxide, formic acid, methanol, acetate, and methylamines without H_2 as the electron donor (Zeikus, 1977; Mah et al., 1978; Smith and Mah, 1978; Brock and Madigan, 1988; Atlas, 1997; Madigan et al., 2014):

$$4HCOOH \rightarrow CH_4 + 3CO_2 + 2H_2O$$
$$(\Delta G^0 = -35 \text{ kcal or } -146.3 \text{ kJ}) \qquad (23.3)$$

$$4CH_3OH \rightarrow 3CH_4 + CO_2 + 2H_2O$$
$$(\Delta G^0 = -76 \text{ kcal or } 317.7 \text{ kcal}) \qquad (23.4)$$

$$CH_3COOH \rightarrow CH_4 + CO_2 \ (\Delta G^0 = -9 \text{ kcal or } -37.6 \text{ kJ})$$
$$(23.5)$$

$$4CH_3NH_2 + 2H_2O \rightarrow 3CH_4 + CO_2 + 4NH_3$$
$$(\Delta G^0 = -75 \text{ kcal or } -313.5 \text{ kJ}) \qquad (23.6)$$

$$4CO + 2H_2O \rightarrow CH_4 + 3CO_2$$
$$(\Delta G^0 = -44.5 \text{ kcal or } -186 \text{ kJ}) \qquad (23.7)$$

Although methanogenesis from acetate by the disproportionation reaction of *acetoclastic* methanogens (Reaction 23.5) is fairly common during anaerobic degradation of organic matter in the absence of a plentiful external supply of electron acceptors such as Fe(III), Mn(IV), or sulfate, a consortium of anaerobic acetate oxidizers like *Clostridium* spp., which generate H_2 and CO_2 from the acetate, and *hydrogenotrophic* methanogens like *Methanomicrobium* or *Methanobacterium* may form methane in the absence of acetoclastic methanogens (Karakashev et al., 2006). There is increasing evidence that such syntrophic acetate-oxidizing bacteria (Hattori, 2008) may be important in driving methanogenesis in situations where acetoclastic methanogens are inhibited or absent (Westerholm et al., 2011).

Some methanogens can form methane from pyruvate by disproportionation (Yang et al., 1992). Resting cells of *Methanococcus* spp. grown in a pyruvate-containing medium under an N_2 atmosphere were shown to transform pyruvate to acetate, methane, and CO_2 according to the following stoichiometry (Yang et al., 1992):

$$4CH_3COCOOH + 2H_2O \rightarrow 4CH_3COOH + 3CO_2$$
$$+ CH_4 \quad (\Delta G^0 = -74.9 \text{ kcal or } 313.1 \text{ kJ})$$
$$(23.8)$$

This stoichiometry is attained if the organism oxidatively decarboxylates pyruvate:

$$4CH_3COCOOH + 4H_2O \rightarrow 4CH_3COOH$$
$$+ 4CO_2 + 8(H) \quad (23.9)$$

and uses the reducing power [8(H)] to reduce one-fourth of the CO_2 to CH_4:

$$CO_2 + 8(H) \rightarrow CH_4 + 2H_2O \quad (23.10)$$

Bock et al. (1994) found that a spontaneous mutant of *M. barkeri* could grow by fermenting pyruvate to methane and CO_2 with the following stoichiometry:

$$CH_3COCOOH + 0.5H_2O \rightarrow 1.25CH_4 + 1.75CO_2 \quad (23.11)$$

To achieve this stoichiometry, the authors proposed the following mechanism based on known enzyme reactions in methanogens. Pyruvate is oxidatively decarboxylated to acetyl SCoA and CO_2:

$$CH_3COCOOH + CoASH \rightarrow$$
$$CH_3CO \, SCoA + CO_2 + 2(H) \quad (23.12)$$

The available reducing power [2(H)] from this reaction is then used to reduce one-fourth of the CO_2 formed to methane:

$$0.25CO_2 + 2(H) \rightarrow 0.25CH_4 + 0.5H_2O \quad (23.13)$$

and the acetyl·SCoA is decarboxylated to methane and CO_2:

$$CH_3CO·SCoA + H_2O \rightarrow CH_4 + CO_2 + CoASH \quad (23.14)$$

The standard free energy yield at pH 7 (ΔG^0) was calculated to be −22.9 kcal mol^{-1} or −96 kJ mol^{-1} of methane produced (Bock et al., 1994).

Although Reactions 23.1 through 23.7 look very disparate, they share a common metabolic pathway (Figure 23.4). The reason methanogens differ with respect to the methane-forming reactions they can perform is that not all of them possess the same key enzymes that permit entry of particular methanogenic substrates into the common pathway (Vogels and Visser, 1983; Zeikus et al., 1985; Stanier et al., 1986; Brock and Madigan, 1988; Atlas, 1997; Madigan et al., 2014). The pathway involves stepwise reduction of carbon from the +4 to the −4 oxidation state via bound formyl, methylene, and methyl carbon. The operation of the methane-forming pathway requires some unique coenzyme and carrier molecules (Table 23.1). Coenzyme M (2-mercaptoethylsulfonate) is unique to methanogens and may be used to identify them as methane formers. The large majority of methanogens synthesize this molecule de novo.

Methanogenic reactions with hydrogen as the electron donor that utilize formic or acetic acid, methanol, or methylamines as electron acceptors instead of CO_2 such as the following:

$$3H_2 + HCOOH \rightarrow CH_4 + 2H_2O$$
$$(\Delta G^0 = -42 \text{ kcal or } -175.6 \text{ kJ}) \quad (23.15)$$

$$4H_2 + CH_3COOH \rightarrow 2CH_4 + 2H_2O$$
$$(\Delta G^0 = -49 \text{ kcal or } 204.8 \text{ kJ}) \quad (23.16)$$

$$H_2 + CH_3OH \rightarrow CH_4 + H_2O$$
$$(\Delta G^0 = -26.9 \text{ kcal or } -112.4 \text{ kJ}) \quad (23.17)$$

$$H_2 + CH_3NH_2 \rightarrow CH_4 + NH_3$$
$$(\Delta G^0 = -9 \text{ kcal or } -37.6 \text{ kJ}) \quad (23.18)$$

are not known to occur.

New evidence suggests that Reaction 23.1 can occur abiotically in the presence of a nickel–iron alloy under hydrothermal conditions (e.g., 200°C–400°C, 50 MPa), conditions met in parts of the oceanic crust (Horita and Berndt,1999).

23.3.3 Bioenergetics of Methanogenesis

As an anaerobic respiratory process, methane formation is performed to yield useful energy to the cell. Evidence to date indicates that adenosine 5-triphosphate (ATP) is generated by chemiosmotic energy-coupling metabolism (e.g., Mountford, 1978; Doddema et al., 1978, 1979; Blaut and Gottschalk, 1984; Sprott et al., 1985; Blaut et al., 1990, 1992; Gottschalk and Blaut, 1990; Müller

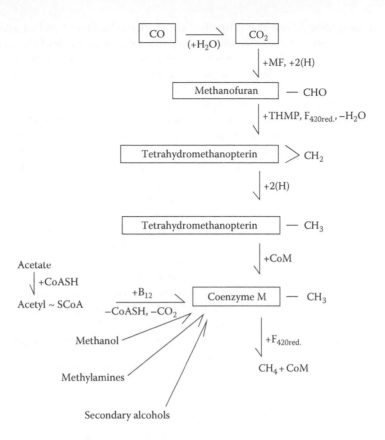

Figure 23.4. Pathways of methanogenesis from CO, CO_2, acetate, methanol, secondary alcohols, and methylamines. (MF, methanofuran; THMP, tetrahydromethanopterin; CoM, coenzyme M.)

TABLE 23.1
Unusual coenzymes in methanogens.

Coenzyme[a]	Function
Methanofuran	CO_2 reduction factor in first step of methanogenesis
Methanopterin (coenzyme F_{342})	Formyl and methene carrier in methanogenesis
Coenzyme M (2-mercaptoethane sulfonate)	Methyl carrier in methanogenesis
Coenzyme F_{430}	Hydrogen carrier for reduction of methyl coenzyme M
Coenzyme F_{420} (nickel-containing tetrapyrrole)	Mediates electron transfer between hydrogenase or formate and NADP, reductive carboxylation of acetyl·CoA, and succinyl·CoA

[a] For structures of these coenzymes, see Brock and Madigan (1988) and Blaut et al. (1992).

et al., 1993; Atlas, 1997; Li et al., 2006). The chemiosmotic coupling mechanism seems to involve pumping of protons or sodium ions across the plasma membrane, depending on the methanogen. Membrane-associated electron transport constituents required in chemiosmotic energy conservation involving proton coupling in *Methanosarcina* strain Göl include reduced factor F_{420} dehydrogenase, an unknown electron carrier, cytochrome b, and heterodisulfide reductase (Blaut et al., 1992). The heterodisulfide consists of coenzyme M covalently linked to 7-mercaptoheptanoyl-threonine by a disulfide bond (Blaut et al., 1992). A proton-translocating ATPase associated with the membrane catalyzes ATP synthesis in this organism.

An example of a methanogen that employs sodium ion coupling is *M. voltae* (Dybas and Konisky, 1992; Chen and Konisky, 1993). It appears to employ Na+-translocating ATPase that is insensitive to proton translocation inhibitors. A scheme for pumping sodium ions from the cytoplasm to the periplasm that depends on membrane-bound methyl transferase was proposed by Blaut et al. (1992). *Methanosarcina acetivorans* is another example of a methanogen that employs sodium ion coupling in its ATP synthesis (Li et al., 2006). Recently, Thauer et al. (2008) have described differences in the energetics of methanogens that do and do not possess cytochromes, which has a major influence on the growth rates and H_2 threshold concentrations for the different methanogens.

23.3.4 Carbon Fixation by Methanogens

When methanogens grow autotrophically, their carbon source is CO_2. The mechanism by which they assimilate CO_2 is different from that of most autotrophs (Simpson and Whitman, 1993). Most autotrophs in the domain bacteria use the pentose diphosphate pathway (Calvin–Benson–Bassham cycle). Among the exceptions are green sulfur bacteria, which use a reverse tricarboxylic acid cycle; *Chloroflexus aurantiacus*, which uses bicyclic CO_2 fixation involving 3-hydroxypropionate as a key intermediate; and the methane-oxidizing bacteria, which use either the hexulose monophosphate or the serine pathway (see Section 23.3.7). A recent review of autotrophic carbon fixation offers a broad perspective on the range of inorganic carbon fixation pathways used by archaea (Berg et al., 2010). In methanogens, as in homoacetogens (see Chapter 7) and some sulfate-reducing bacteria (see Chapter 19),

the chief mechanism of carbon assimilation is by reduction of one molecule of CO_2 to methyl carbon and a second to a formyl carbon followed by the coupling of the formyl carbon to the methyl carbon to form acetyl·SCoA (see Figure 7.12) or form the important metabolic intermediate pyruvate, they next carboxylate the acetyl·SCoA reductively to form pyruvate. All other cellular constituents are then synthesized from pyruvate and may utilize incomplete reductive or oxidative tricarboxylic acid cycles (Simpson and Whitman, 1993).

Examination of the genome sequence of *Methanocaldococcus* (*Methanococcus*) *jannaschii* and *M. acetivorans* has revealed the presence of genes for a ribulose 1,5-bisphosphate carboxylase/oxidase (Finn and Tabita, 2003), but these organisms do not use the Calvin–Benson–Bassham cycle for primary CO_2 fixation (see introduction of the article by Finn and Tabita, 2004, and Sprott et al., 1993).

23.3.5 Microbial Methane Oxidation

23.3.5.1 Aerobic Methanotrophy

Methane can be used as a primary energy source by a number of aerobic bacteria. Some of these are obligate *methanotrophs*; others are facultative (Higgins et al., 1981; see also Theisen and Murrell, 2005). Methane is also oxidized by some yeasts (Higgins et al., 1981). Except for the anaerobic methanotrophic consortia described in Section 23.3.5.2, most, if not all, known methanotrophs are aerobes. Examples of obligate methanotrophs are *Methylomonas*, *Methylococcus*, *Methylobacter*, *Methylosinus*, and *Methylocystis* (Figure 23.5). All are Gram negative

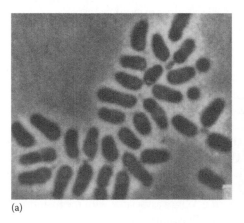 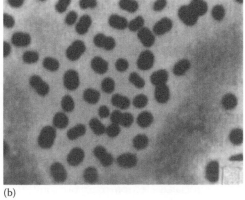

(a)   (b)

Figure 23.5. Aerobic methane-oxidizing bacteria (methanotrophs) (×19,000). (a) *Methylosinus trichosporium* in rosette arrangement. Organisms are anchored by visible holdfast material. (b) *Methylococcus capsulatus*. (From Whittenbury, R et al., *J Gen Microbiol*, 61, 205, 1970. With permission.)

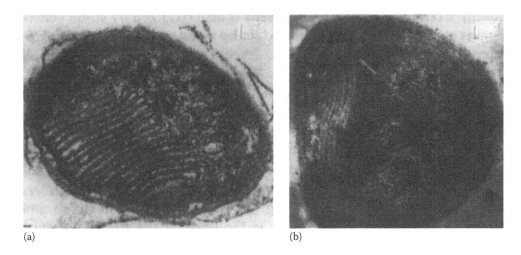

(a)   (b)

Figure 23.6. Fine structure of methane-oxidizing bacteria (×80,000). (a) Section of *Methylococcus* (subgroup *minimus*) showing Type I membrane system. (b) Peripheral arrangement of membranes in *Methylosinus* (subgroup *sporium*) characteristic of Type II membrane systems. (From Davies, SL and Whittenbury, R, *J Gen Microbiol*, 61, 227, 1970. With permission.)

and feature intracytoplasmic membranes. On the basis of the organization of these membranes, each obligate methanotroph can be assigned to one of two types (Davies and Whittenbury, 1970). Members of *Type I* have stacked membranes, whereas members of *Type II* have paired membranes concentric with the plasma membrane and forming vesicle-like or tubular structures (Figure 23.6). Facultative methanotrophs feature internal membranes with an appearance like those of the *Type II* obligate methanotrophs. This terminology has largely been superseded, with *Type I* methanotrophs belonging to the *Gammaproteobacteria* and *Type II* to the *Alphaproteobacteria*. In addition, nonproteobacterial methanotrophs, such as the *Verrucomicrobia*, have now been described (Sharp et al., 2013). All methanotrophs can also use methanol as primary energy source, but not all methanol oxidizers can grow on methane as primary energy source. Methanol oxidizers that cannot oxidize methane are called *methylotrophs*.

Recently, the surprising discovery was made that the filamentous, sheathed bacterium *Crenothrix polyspora* Cohn 1870 and *Clonothrix fusca* Roze 1896 have methanotrophic ability (Stoecker et al., 2006; Vigliotta et al., 2007). *C. polyspora* was found to have an unusual methane monooxygenase (Stoecker et al., 2006). *C. fusca* was shown to possess an elaborate internal membrane system to oxidize and assimilate methane (Vigliotta et al., 2007).

Methanotrophs are important for the carbon cycle in returning the carbon of methane, which is always generated anaerobically, to the reservoir of CO_2 (e.g., Vogels, 1979). Obligate

methanotrophs are found mainly at aerobic/anaerobic interfaces in soils and aquatic environments that are crossed by methane (e.g., Reeburgh, 1976; Alexander, 1977; Ward and Brock, 1978; Sieburth et al., 1987; Hyun et al., 1997; Horz et al., 2002; Inagaki et al., 2004; Sundh et al., 2005) and also in coal and petroleum deposits (Kuznetsov et al., 1963; Ivanov et al., 1978). Some methanotrophs are also important intracellular symbionts in mussels from marine hydrocarbon seeps and in other benthic invertebrates that encounter methane in their habitat. Cavanaugh et al. (1987) found evidence of the presence of such symbiotic methanotrophs in the epithelial cells of the gills of some mussels from reducing sediments at hypersaline seeps at abyssal depths in the Gulf of Mexico at the Florida Escarpment. MacDonald et al. (1990) made similar observations in mussels that occurred in a large bed surrounding a pool of hypersaline water rich in methane at a depth of 650 m on the continental slope south of Louisiana. Transmission electron microscopic examination by Cavanaugh et al. (1987) showed that the symbionts feature typical intracytoplasmic membranes of Type I methanotrophs. They possess the key enzymes associated with methane oxidation in the group (see Section 23.3.6). The basis for the symbiosis between the invertebrate host and the methanotrophs is the sharing of fixed carbon derived from methane taken up by the host and metabolized

23.3.6 Biochemistry of Methane Oxidation in Aerobic Methanotrophs

Obligate aerobic methanotrophs can use methane, methanol, and methylamines as energy sources by oxidizing them to CO_2, H_2O, and for methylamines NH_3. When methane is the energy source, the following steps are involved in its oxidation:

$$CH_4 \xrightarrow{+O_2/+2H/-H_2O} CH_3OH \xrightarrow{-2H} HCHO \xrightarrow{+H_2O/-2H} HCOOH \xrightarrow{-2H} CO_2 \quad (23.21)$$

The first step in this reaction sequence is catalyzed by a monooxygenase that causes the direct introduction of an atom of molecular oxygen into the methane molecule (Anthony, 1986). This step is generally considered not to yield useful energy to the cell. A report by Sokolov (1986), however, suggests the contrary, at least with *Methylomonas alba* BG8 and *Methylosinus trichosporium* OB3b. Because monooxygenase requires reduced pyridine nucleotide ($NADH + H^+$) in its catalytic process to provide electrons for reduction of one of the two oxygen atoms in O_2 to H_2O (the other oxygen atom is introduced into methane to form methanol), a proton motive force is generated in the electron transfer from the reduced pyridine nucleotide, which the cell may be able to couple to ATP synthesis. The enzyme that catalyzes methanol oxidation is methanol dehydrogenase, which in *Methylococcus thermophilus*, as in other methanotrophs, does not use pyridine nucleotide as the cofactor (Sokolov et al., 1981; Anthony, 1986). Instead, the enzyme contains pyrroloquinone (+90 mV) as its prosthetic group, which feeds electrons from methanol into the electron transport chain. The formaldehyde resulting from methanol oxidation is oxidized to formate (Reaction 23.21; see Roitsch and Stolp, 1985). The oxidation of formaldehyde to formate may involve a pyridine nucleotide–linked dehydrogenase or a pyrroloquinone-linked dehydrogenase (Stanier et al., 1986). Whichever the mechanism of formaldehyde oxidation, the reducing power is fed into the electron transport system for energy generation. The pyridine nucleotide–coupled formate dehydrogenase oxidizes formate to CO_2 and H_2O and feeds electrons mobilized by the oxidation into the electron transport system via the pyridine nucleotide to generate energy. ATP synthesis in methanotrophs appears to be mainly or entirely by chemiosmosis (Anthony, 1986).

It should be noted that the methane monooxygenase of methanotrophs is not a very specific enzyme. It can also catalyze NH_3 oxidation (O'Neill and Wilkinson, 1977) and may be important in co-metabolism of some chlorinated solvents (McCarty, 1993). In the case of ammonia, the monooxygenase hydroxylates ammonia to NH_2OH. Ammonia-oxidizing autotrophic bacteria can similarly oxidize methane to methanol (Jones and Morita, 1983). However, just as ammonia oxidizers cannot grow on methane as the energy source, methanotrophs cannot grow on ammonia as the energy source. This is because they lack the enzyme sequences for methanol or hydroxylamine oxidation, respectively.

23.3.7 Carbon Assimilation by Aerobic Methanotrophs

All autotrophically grown methanotrophs of Types I gammaproteobacterial and II alphaproteobacterial methanotrophs assimilate some carbon (up to 30%) in the form of CO_2 (Romanovskaya et al., 1980). The enzyme involved in the fixation appears to be phosphoenolpyruvate (PEP) carboxylase. The mechanism of fixation of the remaining carbon depends on the methanotroph type. Both types assimilate it at the formaldehyde oxidation state of carbon. Gammaproteobacterial Type I methanotrophs fix this carbon via an assimilatory, cyclic ribulose monophosphate pathway (Figure 23.7a), whereas alphaproteobacterial Type II methanotrophs fix it via a cyclic serine pathway (Figure 23.7b). In the ribulose monophosphate pathway, 3-phosphoglyceraldehyde is the key intermediate in carbon assimilation, whereas in the serine pathway it is acetyl·SCoA (Gottschalk, 1986; Stanier et al., 1986; Brock and Madigan, 1988; Atlas, 1997; Madigan et al., 2014). Reducing power for assimilation derives from methane dissimilation and may require reverse electron transport to generate needed $NADPH + H^+$. Methylotrophs generally use the serine pathway for carbon assimilation from C_1 compounds. Genomics and some recent physiological studies have shown that many methanotrophs can also fix carbon via the Calvin–Benson–Bassham cycle (Sharp et al., 2013).

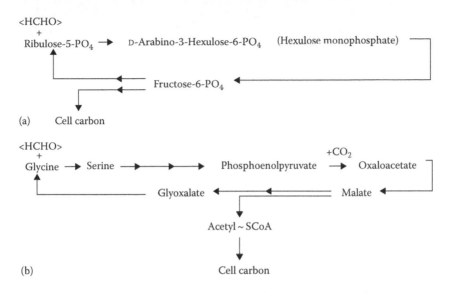

Figure 23.7. Alternative pathways of formaldehyde–carbon assimilation in methanotrophs. (a) The hexulose monophosphate (HuMP) or ribulose monophosphate (RuMP) pathway used by gammaproteobacterial Type I methanotrophs. (b) The serine pathway used by alphaproteobacterial Type II methanotrophs.

Obligate methanotrophs are positioned somewhere between typical autotrophs and heterotrophs in the carbon-assimilating mechanism (Quayle and Ferenci, 1978).

23.3.8 Position of Methane in Carbon Cycle

For a sufficient pool of biologically available carbon to be maintained in the biosphere, carbon has to be continually recycled from organic to inorganic carbon. This is accomplished both aerobically and anaerobically. Quantitatively, the aerobic process has been thought to make the greater contribution, but the anaerobic process is far from negligible; indeed, with newer knowledge about the extent to which the deep subsurface is inhabited by microbes, it is probably quite significant. Henrichs and Reeburgh (1987) have estimated that the rate of anaerobic mineralization in sediment equals approximately the rate of burial of organic carbon.

Anaerobic mineralization involves fermentation processes coupled with anaerobic forms of respiration, including dissimilatory nitrate reduction (nitrate respiration, denitrification, nitrate ammonification) (e.g., Sørensen, 1987), iron and manganese respiration (Lovley, 1987, 1993; Lovley and Phillips, 1988; Myers and Nealson, 1988; Ehrlich, 1993), sulfate respiration (Skyring, 1987), and methanogenesis (Wolin and Miller, 1987; Young

and Frazer, 1987). Each of these forms of respiration is dominant where the respective terminal electron acceptor is dominant and other environmental conditions are optimal. In some instances, more than one form of anaerobic respiration may occur simultaneously in the same general environment, provided there is no competition for the same electron donor or other growth-limiting substance (see Ehrlich, 1993).

In anoxic soils (paddy spoils) or anoxic freshwater or marine sediments where extensive methanogenesis occurs, a small portion of the methane is oxidized to CO_2 without the benefit of oxygen. A larger portion of the methane that is not trapped escapes into an oxidizing environment and is extensively oxidized to CO_2 by aerobic methanotrophs (Higgins et al., 1981). A small amount of methane may be used as energy and carbon source by special marine invertebrates that have formed a symbiotic relationship with specific methanotrophs.

With a growing awareness of anaerobic methane oxidation, in marine systems in particular, it has been estimated that of the 85–300 Tg CH_4 year[−1] methane generated in marine sediments >90% is consumed by AOM (Reeburgh, 2007; Knittel and Boetius, 2009). This equates to around 7%–25% of annual global methane production (Knittel and Boetius, 2009). Any methane that is not bio-oxidized or otherwise combusted or trapped in

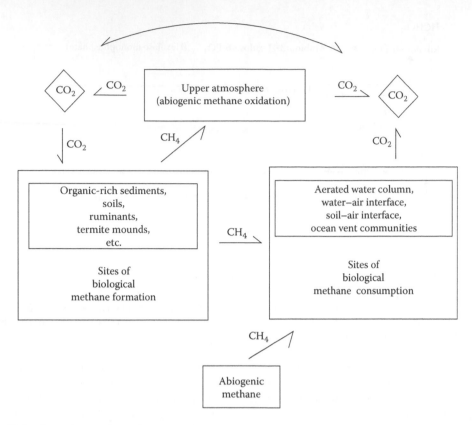

Figure 23.8. The methane cycle, emphasizing methanogenesis and methanotrophy.

natural sedimentary reservoirs escapes this biological attack and enters the atmosphere, where it may be chemically oxidized in the troposphere (Vogels, 1979). The various paths for methanogenesis and methane oxidation are summarized in Figure 23.8.

23.4 PEAT

23.4.1 Nature of Peat

Although peat and coal are two different substances, their modes of origin have included common initial steps. Indeed, the formation of peat may have been an intermediate step in the formation of coal. Peat is a form of organic soil or histosol. It is mostly derived from plant remains that have accumulated in marshes and bogs (Figure 23.9). According to Francis (1954), these remains have come from (1) sphagnum, grasses, and heather, yielding high moor peat; (2) reeds, grasses, sedges, shrubs, and bushes, yielding low moor peat; (3) trees, branches, and debris of large forests in low-lying wet ground, yielding forest peat; or (4) plant debris accumulated in swamps, yielding sedimentary or lake peat. In all these

instances, plant growth outstripped the decay of the plant remains, ensuring a continual supply of new raw material for peat formation.

23.4.2 Roles of Microbes in Peat Formation

Initially, the plant remains may have undergone attack by some of their own enzymes but soon were attacked by fungi, which degraded the relatively stable polymers such as cellulose, hemicellulose, and lignin. Bacteria degraded the more easily oxidizable substances and the breakdown products of fungal activity that were not consumed by the fungi themselves. Fungal activity continued for as long as the organisms had access to air, but as they and their remaining substrate became buried and conditions became anaerobic, bacterial fermentation and anaerobic respiration, including methanogenesis, set in and continued until arrested by accumulation of inhibitory (toxic) wastes, lack of sufficient moisture, depletion of suitable electron acceptors for anaerobic respiration (e.g., nitrate, sulfate, Fe(III), Mn(IV), and carbon dioxide), and other factors (Francis, 1954; Rogoff et al., 1962;

Figure 23.9. Peat. (a) Section of a ditch near Vestburg showing light sphagnum peat over dark peat: (1) living sphagnum, (2) sphagnum peat, (3) shrub remains, (4) sedge rootstock, (5) pond lily rootstocks, (6) laminated peat. (b) View of surface near ditch, showing corresponding vegetative zones: (1) shrub zone, (2) grass zone, (3) sedge zone, (4) pond lily zone. (From Davis, CA, Peat: Essays on the origin, uses and distribution in Michigan, in: *Plate XII from Annual Report, Geological Survey of Michigan*, Wynkoop Hallenbeck Crawford, State Printers, Lansing, MI, 1907, pp. 395. With permission.)

Kuznetsov et al., 1963). Combined with these limiting factors, the overwhelming accumulation of organic debris must also be taken into account. It was more than the system could handle before conditions for continued mineralizing activity became unfavorable.

The uppermost aerobic layers of peat harbor a viable microflora even today, indicating that peat formation may be occurring at the present time (Kuznetsov et al., 1963; Preston et al., 2012). Indeed, viable anaerobic bacteria and actinomycetes have even been detected in the deeper layers of some peats (Rogoff et al., 1962; Zvyagintsev et al., 1993; Metje and Frenzel, 2005). Besides members of the domain bacteria, methanogenic archaea are also found associated with peat (e.g., Chan et al., 2002). Differences in methane production in various peats are a reflection of the differences in their origin with respect to source materials and environmental conditions (e.g., Yavitt and Lang, 1990; Brown and Overend, 1993; Galand et al., 2005).

During its formation, peat becomes enriched in lignin, ulmins, and humic acids. The first of these compounds is a relatively stable polymer of woody tissue, and the second and third compounds are complex materials that resulted from the incomplete breakdown of plant matter, including lignin.

Peat also contains other compounds that are relatively resistant to microbial attack, such as resins and waxes from cuticles, stems, and spore exines of the peat-forming plants. Compared to the C, H, O, N, and S contents of the original undecomposed plant material, peat is slightly enriched in carbon, nitrogen, and sulfur but depleted in oxygen and sometimes hydrogen (Francis, 1954). This enrichment in C, N, and S over oxygen may be explained in part by the volatilization of products formed from the less resistant components by microbial attack and in part by the buildup of residues (resins, waxes, lignin) that have a relatively low-oxygen content owing to their hydrocarbon-like and aromatic properties. The plant origin of peat is still clearly visible on examination of its structure.

23.5 COAL

23.5.1 Nature of Coal

Coal has been defined by Francis (1954) as "a compact, stratified mass of mummified plants, which have been modified chemically in varying degrees, interspersed with smaller amounts of inorganic matter." Peat can be distinguished from coal in chemical terms by its much lower carbon

Figure 23.10. A section of the Pittsburgh coal seam in the Safety Research coal mine of the U.S. Bureau of Mines in Pittsburgh, Pennsylvania (United States). Although not visible in this black-and-white photograph, extensive brown iron stains were present on portions of the face of the coal seam shown here. The stains are evidence of acid mine drainage emanating from the fracture at the upper limit of the coal seam. The acid drainage resulted from microbial oxidation of exposed iron pyrite inclusions in the seam. (Courtesy of the U.S. Bureau of Mines, Pittsburgh, PA.)

content (51%–59% dry weight) and higher hydrogen content (5.6%–6.1% dry weight) compared to coal (carbon, 75%–95% dry weight; hydrogen 2.0%–5.8% dry weight) (Francis, 1954, p. 295). The average carbon content of typical wood has been given as 49.2% (dry weight) and its average hydrogen content as 6.1% (dry weight) (Francis, 1954). Coalification can thus, be seen to have resulted in an enrichment in carbon and a slight depletion in hydrogen of the substance that gives rise to coal. Coal is generally found buried below layers of sedimentary strata (*overburden*). Its geologic age is generally advanced. Significant deposits formed in the upper Paleozoic between 300 and 210 million years ago, in the Mesozoic between 180 and 100 million years ago, and in the Tertiary between 60 and 2.5 million years ago. Peats, however, have generally developed from about 1 million years ago up to the present.

Coal is classified by rank. According to the American Society for Testing and Materials classification system, four major classes are recognized (Bureau of Mines, 1965). They are—starting with the least developed coal—lignitic coal, subbituminous coal, bituminous coal, and anthracite coal. Lignitic coal (lignite) resembles peat structurally and has the highest moisture and lowest carbon content (59.2%–72.3% dry weight) (Francis, 1954, p. 335) as well as has the lowest heat value of any of the coals. This coal formed during the Tertiary times. Subbituminous coal has a slightly lower moisture content and higher

carbon content (72.3%–80.4% dry weight). It also has a higher heat content than lignitic coal. Bituminous coal (Figure 23.10) has a carbon content ranging from 80.4% to 90.9% (dry weight) and a high heat value. Subbituminous and bituminous coals are mostly of the Paleozoic and the Mesozoic age. Anthracite coals have very low moisture content, few volatiles, and high carbon contents (92.9%–94.7% dry weight). They are of Paleozoic age. Cannel coal is a special type of bituminous coal that was derived mainly from wind-blown spores and pollen rather than from woody plant tissue.

23.5.2 Role of Microbes in Coal Formation

As mentioned earlier, coal deposits developed at special periods in geologic time. In these periods, the climate, landscape features, and biological activity were favorable. Large amounts of plant debris accumulated in swamps or shallow lakes, because owing to warm, moist climatic conditions, plant growth was very profuse and provided a continual supply of plant debris. At times, the accumulated debris would become buried, covered by clay and sand under water, before a new layer of plant debris accumulated. As more sediment was deposited, subsidence followed and is believed to have played an integral part in the formation of coal deposits (Francis, 1954).

Bacteria and fungi are generally believed to have had an important role in coalification only

EHRLICH'S GEOMICROBIOLOGY

in the initial stages. Their role was similar to that in peat formation. They destroyed the easily metabolizable substances, such as sugars, amino acids, and volatile acids, in a short time and degraded the more stable polymers, such as cellulose, hemicellulose, lignins, waxes, and resins, more slowly. Many of the latter were degraded only very incompletely before microbial activity ceased for reasons similar to those in peat formation (Section 23.4.2). Hyphal remains, sclerotia, and fungal spores have been identified in some coal remains. Initial microbial attack is believed to have been aerobic and mainly fungal. Later attack, mostly by bacteria, occurred under progressively more anaerobic conditions. Tauson believed, however, that in coal formation, the anaerobic phase was abiological (as cited by Kuznetsov et al., 1963). Conversion of the residue from microbial activity (peat?) to coal is presently believed to have been due to physical and chemical agencies of an unidentified nature, but probably involved heat and pressure, which resulted in the loss of volatile components.

23.5.3 Coal as a Microbial Substrate

Coal is not a very suitable nutrient to support microbial growth. According to early views, this is because coal contains inhibitory substances ("antibiotics") that may suppress it. These "antibiotics" have been thought to be associated with the waxy and resinous part of coal, extractable with methanol (Rogoff et al., 1962). In the first culture experiments with coal slurries, only marginal bacterial growth was obtainable, the limiting factors being the presence of inhibitory substances and lack of assimilable nutrients (Koburger, 1964). Growth of *Escherichia freundii* and *Pseudomonas rathonis* in such slurries improved when coal was first treated with H_2O_2. In much more recent experiments with run-of-the-mine bituminous coal from Pennsylvania, in which coal particles in a size range 0.5–13 mm were wetted in glass columns with air-saturated distilled water at 10- to 14-day intervals, a bacterial community did develop. It consisted mainly of autotrophic bacteria (iron and sulfur oxidizers) and, to a lesser extent, heterotrophic bacteria. Progressive acid production by the autotrophs was thought to limit the development of heterotrophs. Observed changes in the rate of acetate metabolism may be a reflection of microbial succession among the

heterotrophs (Radway et al., 1987, 1989). The bacteria in these experiments lived at the expense of impurities in the coal, such as pyrite.

Two basidiomycete fungi, including *Trametes versicolor* (also known as *Polyporus versicolor* and *Coriolus versicolor*) and *Poria monticola*, have been shown to grow directly on crushed lignite coal as well as in minimal lignite-noble agar medium (Cohen and Gabriele, 1982). With time, the cultures growing on the lignite exuded a black liquid that was a product of lignite attack. Infrared spectra of the exudates gave an indication that conjugated aromatic rings from the lignite had been structurally modified. It must be stressed that the coal in this case was lignite, a low-rank coal, and not bituminous coal. Other fungi, including *Paecilomyces*, *Penicillium* spp., *Phanerochaete chrysosporium*, *Candida* sp., and *Cunninghamella* sp., as well as *Streptomyces* sp. (actinomycete), have also been shown to grow on and degrade lignite and, in some cases, even bituminous coal (see Cohen et al., 1990; Stewart et al., 1990). In the case of *T. versicolor*, Cohen et al. (1987) and Pyne et al. (1987) at first attributed lignite solubilization to a protein secreted by the fungus that had polyphenol oxidase activity (syringaldazine oxidase; Pyne et al., 1987). Subsequently, Cohen et al. (1990) and Fredrickson et al. (1990) reported that a ligand produced by the fungus and identified as ammonium oxalate (Cohen et al., 1990) was the real solubilizing agent. The oxalate acted as a siderophore by removing iron from the test substrate leonardite (oxidized lignin). This later conclusion seems a little puzzling because it implies that iron(III) plays a central role in holding the leonardite structure together and that no modification of the organic skeleton of leonardite takes place. More recently, Hölker et al. (2002) implicated extracellular esterases and phenolic oxidases (probably laccases) in the solubilization of lignite by *Trichoderma atroviride*.

White-rot fungi (e.g., *P. versicolor* and *P. chrysosporium*) are able to produce two kinds of extracellular enzymes: lignin peroxidase and manganese-dependent peroxidase, both of which catalyze lignin attack (e.g., Paszczynski et al., 1986). Indeed, Stewart et al. (1990) reported that *P. chrysosporium* degraded lignite and bituminous coal from Pennsylvania, albeit weakly. Some bacteria also can form such extracellular enzymes that are active on low-rank coals (Crawford and Gupta, 1993). Moreover, dibenzothiophene-degrading aerobic bacteria have been found that are able to break down part of the carbon framework of liquefied bituminous coal (suspension of pulverized coal), and in the process, remove some of the sulfur bound in the framework (Stoner et al., 1990). It would therefore, seem reasonable that such types of enzymes play a role in lignite attack. Depending on the coal, it is probably a combination of enzymatic and nonenzymatic processes that leads to the transformation of low-rank coals. The interested reader is referred to Crawford (1993) for further details on these processes.

Lignin and lignin derivatives can also be biologically attacked under anaerobic conditions (Young and Frazer, 1987). It would be of interest to study this action on lignite.

There is mounting evidence that methane associated with coal measures is biogenic (Strapoc et al., 2011) and the process of methanogenesis from coal is ongoing. Though the components of coal, which are ultimately converted to methane, are currently unclear, it has been suggested that stimulation of methanogenesis in coal beds could be a route to cleaner energy production from coal in the form of methane gas (Strapoc et al., 2011). A number of commercial organizations have explored this technology with several patents filed in this area.

23.5.4 Microbial Desulfurization of Coal

Bituminous coal may contain significant amounts of sulfur in inorganic (pyrite, marcasite, elemental sulfur, sulfate) or organic form. The total sulfur can range from 0.5% to 11% (Finnerty and Robinson, 1986). The proportion of pyritic and organic sulfur in the total sulfur depends on the source of the coal. Iron pyrite or marcasite (FeS_2) originated in some bituminous coal seams during their formation. Some or all of this iron disulfide may well have been formed biogenically, in which case, it is a reflection of the anaerobic biogenic phase of coalification, representative of sulfate respiration (Chapters 20 and 21) and also the mineralization of organic sulfur compounds such as sulfur-containing amino acids. A model system illustrating how pyrite may have formed in coal is provided by the sulfur transformations presently occurring in the formation of Everglades peat (Casagrande and Siefert, 1977; Altschuler et al., 1983). The source of the sulfur in this case is organic.

The presence of pyritic sulfur in coal lowers its commercial value because on combustion of such coal, air pollutants such as SO_2 are generated. Pyritic sulfur in coal can be removed in various ways (Bos and Kuenen, 1990; Blazquez et al., 1993). In all cases, the coal must first be pulverized to expose the pyrite. The pyrite particles can be separated by differential flotation, in which pyrite flotation is suppressed. Flotation suppression can be achieved chemically or biologically. In the latter case, *Acidithiobacillus ferrooxidans*, which attaches rapidly and selectively to pyrite particles, is the suppression agent (Pooley and Atkins, 1983; Bagdigian and Myerson, 1986; Townsley et al., 1987). However, pyrite can also be removed by oxidizing it. This can be accomplished by the action of pyrite-oxidizing bacteria such as *A. ferrooxidans*, *Sulfolobus* spp., *Acidianus* spp., and *Metallosphaera* (Dugan, 1986; Andrews et al., 1988; Merrettig et al., 1989; Larsson et al., 1990; Baldi et al., 1992; Clark et al., 1993).

Bituminous coal may also contain organically bound sulfur. Its presence is undesirable because on combustion, it too contributes to air pollution. Microbiological methods continue to be explored to remove it, but progress has been slow (Dugan, 1986; Finnerty and Robinson, 1986; Mormile and Atlas, 1988; Crawford and Gupta, 1990; Stoner et al., 1990; Van Afferden et al., 1990; Omori et al., 1992; Olson et al., 1993; Izumi et al., 1994). Dibenzothiophene is used as a model structure for organically bound sulfur in coal. Because the sulfur is likely to occur in different types of compounds bound in the structure of bituminous coal, whose degradation may require different enzyme systems, it may be that no one organism in nature can attack the whole range of these substances. Genetic engineering is being applied to find a solution to this problem. In any approach taken to remove organic sulfur from coal, it is important to find a way to remove the sulfur without significant loss of carbon, which would lower the caloric value of the coal.

23.6 PETROLEUM

23.6.1 Nature of Petroleum

Petroleum is a mixture of aromatic and aliphatic hydrocarbons and various heterocyclics including oxygen-, nitrogen-, and sulfur-containing compounds. Crude oils are probably one of the most complex naturally occurring chemical mixtures. Ultrahigh resolution mass spectrometry techniques such as Fourier transform-ion cyclotron resonance-mass spectrometry have shown that crude oils contain tens of thousands of identifiable compounds (Marshall and Rodgers, 2004), leading to a new term to describe oil chemistry, petroleomics. Despite this chemical complexity, and the fact that crude oils from different provinces can have very different compositions, crude oils can be operationally classified into four major fractions: saturated hydrocarbons, aromatic hydrocarbons, and two nonhydrocarbon fractions, the resins and asphaltenes. The saturated or aliphatic hydrocarbons include gaseous compounds of the paraffinic series such as methane, ethane, propane, and butane, besides longer-chain, nongaseous ones. Crude oil also contains a wide variety of aromatic hydrocarbons from the simplest monocyclic compounds (benzene, toluene, and xylene isomers) to complex polycyclic aromatic hydrocarbons like naphthalene, methylnaphthalenes, and phenanthrenes. The resin fraction contains N-, S-, and O-containing heterocyclic compounds and the asphaltene fraction comprises complex macromolecular structures. Some of the heterocyclic compounds, such as porphyrin derivatives, may contain metals such as vanadium and nickel bound in their structure.

Petroleum is formed from the remains of organisms deposited in sediments, which are transformed via a range of diagenetic processes into recalcitrant organic matter that with time is buried to greater and greater depth. With increase pressure and temperature, the preserved organic carbon decomposes to form the oil and gas that constitute petroleum (Tissot and Welte, 1984). This is largely thought to be a physical process of thermal decomposition termed *catagenesis* (Figure 23.2), but there is increasing evidence that at temperatures at the lower end of the "oil window" that biological process may also contribute to the decomposition of sedimentary organic matter (Wellsbury et al. 1997; see Section 23.6.2). Organic-rich sediments that ultimately give rise to petroleum accumulations are known as source rocks, and for petroleum produced by catagenesis in source rocks to result in a petroleum accumulation or reservoir, buoyancy-driven flow of the oil through a porous rock migration pathway typically occurs. If the migrating oil encounters an impermeable "trap," then oil and gas may accumulate as a petroleum reservoir (Tissot and Welte, 1984). More detail of theories that explain the

processes leading to the formation of petroleum are given in the succeeding text.

Petroleum accumulations are found in some folded, porous, sedimentary rock strata, such as limestone or sandstone, or in other fractured rocks such as fissured shale of igneous rock. In petroleum geology, these rock formations are collectively known as reservoir rocks. The age of reservoir rocks may range from late Cambrian (500 million years) to the Pliocene (1–13 million years). Very extensive petroleum reservoirs are found in rocks of the Tertiary age (70 million years; North, 1985).

Most petroleum derived mainly from planktonic debris that was deposited on the floor of depressions of shallow seas and ultimately buried under heavy layers of sediment, deposited perhaps by turbidity currents. Over geologic time, the trapped organic matter became converted to petroleum and natural gas (chiefly methane; North, 1985).

Many theories have been advanced to explain the origin of petroleum and associated natural gas (e.g., Beerstecher, 1954; Robertson, 1966; North, 1985). None of these have been fully accepted. Some theories invoke heat or pressure or both as agents that promoted abiological conversion of planktonic residues to the hydrocarbons and other constituents of petroleum and natural gas. The source of the heat has been viewed as the natural radioactivity of the Earth's interior but was more likely heat diffusing from magma chambers underlying tectonically active areas. Other theories have invoked inorganic catalysis with or without the influence of heat and pressure and with or without prior acid or alkaline hydrolysis. Still other theories have proposed that petroleum represents a residue of naturally occurring hydrocarbons in the planktonic remains after all other components have been biologically destroyed. It has even been proposed that biological agents produced the hydrocarbons by aerobic or anaerobic reduction of fatty acids, proteins or amino acids, carbohydrates, carotenoids, sterols, glycerol, chlorophyll, and lignin–humus complexes, together with appropriate decarboxylations and deaminations. A theory has also been put forward that methane, formed biogenically from planktonic debris, then became polymerized under high temperature and pressure in the possible presence of inorganic catalysts (e.g., Mango, 1992).

Alternatively, it was theorized that bacteria modified the planktonic material to substances closely resembling petroleum components, which were then converted to petroleum and natural gas constituents by heat and pressure. Abiotically formed methane could also have been a source and could have been polymerized as proposed for biogenic methane. Chemical reaction of methane with liquid hydrocarbons has been noted in the laboratory at high temperature and pressure (1000 atm, 150°C–259°C; Gold et al., 1986). At a hydrothermal mound area in the southern rift of the Guaymas Basin in the Gulf of California (Sea of Cortez), organic matter appears to be actively and abiotically transformed into petroliferous substances including gasoline-range aliphatic and aromatic hydrocarbons (Simoneit and Londsdale, 1982; Didyk and Simoneit, 1989). This site may also be a source of methane used by methanotrophic consortia (see Section 23.3).

23.6.2 Roles of Microbes in Petroleum Formation

At present, it is generally thought that bacteria played a role in the initial stages of petroleum formation, but what this role remains obscure, except in the case of methane formation. ZoBell (1952, 1963) suggested that the planktonic debris was fermented, leading to compounds enriched in hydrogen and depleted in oxygen, sulfur, and phosphorus. Davis (1967) visualized microbial processes not unlike those in peat formation, involving aerobic attack of the sedimented planktonic debris followed by anaerobic activity after initial burial. This activity may have included hydrolytic, decarboxylating, deaminating, and sulfate-reducing reactions, resulting in the accumulation of marine humus, that is, stabilized organic matter. Progressively, deeper burial led to the cessation of microbial activity and to compaction, accompanied by evolution of small amounts of hydrocarbon substance plus petroleum precursors. The biotic reactions were followed by an abiotic phase of very long duration during which the microbially produced precursors were transformed under the influence of heat and pressure into the range of hydrocarbons associated with petroleum. This sequence of biotic and abiotic reactions is supported by observations on light hydrocarbon formation in marine sediments (Hunt et al., 1980; Hunt, 1984). Clays could have catalytically promoted further chemical reductions of petroleum precursors. There is now evidence that microorganisms in deep subsurface

sediments may contribute to the decomposition of recalcitrant organic matter that has survived burial. It has been shown that at moderate temperatures, microorganisms can generate hydrogen and short-chain fatty acids such as acetate from sedimentary organic matter (Wellsbury et al., 1997) and such processes may have played a role in early catagenesis.

23.6.3 Roles of Microbes in Petroleum Migration in Reservoir Rock

As hydrocarbons accumulated during petroleum formation, the more volatile compounds generated increasing pressure, which helped force the more liquid components through porous rock (sandstone, limestone, fractured rock) to anticlinal folds. The hydrocarbons were trapped in the apex of these folds below a stratum of impervious rock to form a petroleum reservoir. We tap such reservoirs today for our petroleum supply. The migration of petroleum from the source rock to the reservoir rock was probably helped by groundwater movement and by the action of natural detergents such as fatty acid soaps and other surface-active compounds of microbial origin. ZoBell (1952) suggested that bacteria themselves may help to liberate oil from rock surfaces at the site of its formation and thereby promote its migration by dissolving carbonate and sulfate minerals to which oil may adhere and by generating CO_2, whose gas pressure could help force the migration of petroleum. Bacterially produced methane may lower the viscosity of petroleum liquid by dissolving in it and thus, help migration (ZoBell, 1952). The important microbial contributions to petroleum formation thus, come during the initial action on the source material (planktonic biomass) and the final stages by promoting migration of petroleum from the source rock to the reservoir rock. In addition, sulfate-reducing bacteria may play a role in the sealing of an oil deposit in reservoir rock by deposition of secondary $CaCO_3$ at the interface between the oil-bearing stratum and the stratal waters (Ashirov and Sazanova, 1962; Kuznetsov et al., 1963; Davis, 1967) (see also Chapter 10).

23.6.4 Petroleum Reservoirs as a Microbial Habitat

At first glance, petroleum reservoirs represent an excellent habitat for microorganisms. They are replete with potential organic electron donors, often have temperatures well within the range known to be permissive for microbial life, are usually associated with an aquifer (the formation waters) providing available water, and are found in porous or fractured rocks that improve mass transfer of electron donors and inorganic nutrients for example. Nevertheless, there are also many factors that may predicate against microbial life in such systems. The oil itself is a cocktail of chemicals and solvents that may be toxic to microbial life, the formation of water in a petroleum reservoir may be highly saline, the pressure and temperature may be high, and even though reservoir rocks are relatively porous, the deep aquifers with which they are associated may be hydraulically very inactive.

The existence of the deep biosphere and decades of research showing that a wide range of microorganisms can be isolated from petroleum reservoirs provide compelling evidence for a deep petroleum reservoir biosphere (e.g., Magot, 2005). This is supported by the fact that many petroleum reservoir isolates have physiological properties consistent with life in situ in a petroleum reservoir (i.e., optimum growth temperature and salinity tolerance).

It is often assumed that a petroleum reservoir in its entirety represents a microbial habitat; however, this may not be the case. While oil in a petroleum reservoir does contain a certain amount of water that might harbor microbial cells, this does not represent a continuous water phase, and thus, water in the oil represents an isolated environment that, while rich in electron donors, cannot be resupplied with inorganic nutrients, for example, which are required for growth. Thus, while the oil column in a petroleum reservoir does harbor a microbial population, this will be small and subsisting on very low levels of internally recycled nutrients.

The oil–water transition zone (OWTZ) where the oil column meets the underlying aquifer is thus, a much more hospitable environment for petroleum reservoir microorganisms. In the OWTZ, the condition changes from highly oil saturated sediment with a continuous oil phase and isolated water inclusions to water saturated sediment. Evidence that the OWTZ is a microbial hotspot in petroleum reservoirs comes from biodegraded petroleum reservoirs, which often exhibit compositional gradients from the oil column to the OWTZ (Head et al., 2003; Figure 23.11).

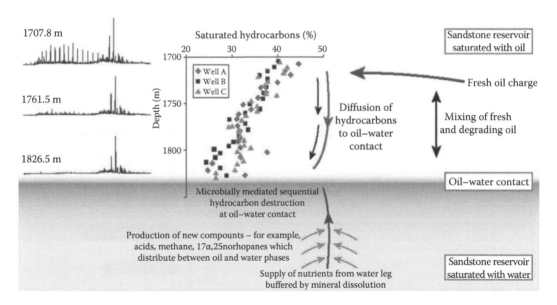

Figure 23.11. A conceptual model of the petroleum reservoir biosphere. The oil column of a petroleum reservoir contains a low microbial population and the oil–water transition zone (OWTZ), which represents the interface between the oil column and the underlying aquifer, is the focus of microbial populations and activity in petroleum reservoirs. The oil provides a plentiful supply of electron donors (both hydrocarbons and nonhydrocarbons) and the aquifer provides water, and potentially inorganic nutrients and/or electron donors. This model is supported by the occurrence of compositional gradients in the oil column of biodegraded petroleum reservoirs and direct measurement of bacterial 16S rRNA genes. (From Head, IM et al., *Nature*, 426, 344, 2003. With permission.)

This has led to the development of a conceptual model of the petroleum reservoir biosphere that proposes that in petroleum reservoirs the microbial population is focused in the OWTZ with electron donor being delivered by diffusion from the oil column, while inorganic nutrients and possibly electron donors are supplied from the underlying aquifer (Figure 23.11). Delivery of nutrients and electron donors may be driven by local dissolution of appropriate mineral phases (e.g., P from apatite inclusions in feldspars; Rogers and Bennett, 2004).

Due to the practical difficulties of sampling depth transects through petroleum reservoirs, it has been difficult to link oil compositional gradients with microbial abundance data. There is currently one report that confirms the coincidence of geochemical gradients and a maximum in bacterial abundance at the OWTZ of a Canadian heavy oil reservoir (Bennett et al., 2013; Figure 23.12).

The first report of the presence of microorganisms in petroleum reservoirs was published in 1926 when Bastin identified sulfate-reducing bacteria in produced fluids from the Illinois Basin (Bastin, 1926). Other early literature reported the

detection of viable bacteria in brines associated with petroleum reservoirs to which access was gained by drilling. They were assigned to three specific groups. One included the sulfate reducers *Desulfovibrio desulfuricans* and *Desulfotomaculum nigrificans* (Kuznetsov et al., 1963; Nazina and Rozanova, 1978). The second group included the methanogens *Methanobacterium mazei*, *Sarcina methanica* (now *Methanosarcina methanica*), and *Methanobacterium omelianskii* (now *Methanobacterium* MOH). The third group included the phototroph *Rhodopseudomonas palustris* (Rozanova, 1971), which is capable of anaerobic respiration in the dark (Yen and Marrs, 1977; Madigan and Gest, 1978). The petroleum-associated brines may be connate seawater whose mineral content was somewhat altered through contact with enclosing rock strata (Chapter 6). Such brines may be low in sulfate but high in chloride and not very conducive to microbial growth. Indeed, one study has demonstrated that microorganisms cannot be isolated from reservoir waters if the salinity exceeds 100 g L^{-1} (Grassia et al., 1996; Röling et al., 2003; Figure 23.13). Sulfate-containing groundwaters and alkaline carbonate waters, on mixing with the brines, can provide

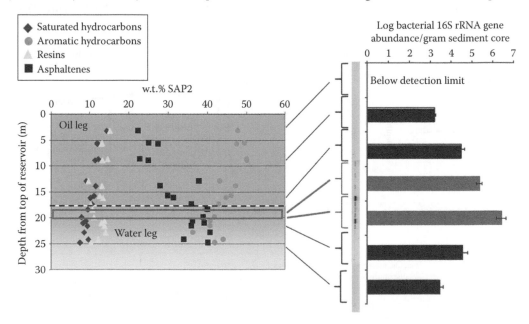

Figure 23.12. Compositional gradients in oil chemistry in relation to the abundance of bacterial 16S rRNA genes measured in core samples across the oil–water transition zone (OWTZ) in a Canadian heavy oil reservoir. This demonstrates that bacterial abundance was highest at the OWTZ implying that these organisms are degrading components of the oil fractions that decrease toward the OWTZ. Note the low percentage of saturated hydrocarbons at the top of the reservoir compared to that shown in Figure 23.12. This is because the oil in this reservoir is a heavy oil resulting from extensive in-reservoir oil biodegradation. (Adapted from Bennett, B et al., *Org Geochem*, 56, 94, 2013; contributions by Barry Bennett and Angela Sherry are gratefully acknowledged.)

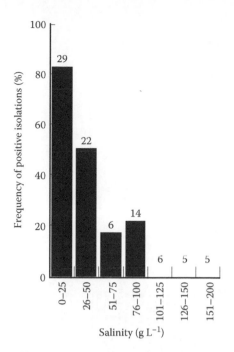

Figure 23.13. Frequency of successful isolation of microbial cultures from petroleum reservoirs using media with different salinity. The numbers above each bar represent the total number of studies considered. (Adapted from Grassia, GS et al., *FEMS Microbiol Ecol*, 21, 47, 1996; Röling, W.F.M. et al., *Res Microbiol*, 154, 321, 2003.)

TABLE 23.2
Composition of petroleum-associated brines (percentage equivalents).

Cl^-	7.4–49.90	Ca^{2+}	0.33–11.02
SO_4^{2-}	0.03–10.06	Mg^{2+}	0.04–4.70
CO_3^{2-}	0.03–42.2	K^+ and Na^+	34.28–49.34

SOURCE: After Kuznetsov, S.I. et al., *Introduction to Geological Microbiology*, McGraw-Hill, New York, 1963, p. 17.

a milieu suitable for the active sulfate-reducing bacteria (Table 23.2). These waters furnish needed moisture and, in the case of sulfate-reducing bacteria, the terminal electron acceptor sulfate for their respiration. By the current understanding of methanogenesis (see Section 23.3), the methanogens in these brines probably rely mainly on H_2 for energy and CO_2 as terminal electron acceptor and as the source of carbon, and in fact, the most salt-tolerant hydrogenotrophic methanogen isolated to date *Methanocalculus halotolerans* was isolated from a petroleum reservoir (Ollivier et al., 1988). The sulfate-reducing bacteria may obtain their energy source directly from petroleum, as has

been claimed by some, but may also depend on other bacteria to produce what they need as carbon and energy sources (Ivanov, 1967; Chapters 20 and 21). If the plutonic waters associated with an oil reservoir are hydraulically connected with infiltrating surface waters, it is also possible that carbon and energy sources for the sulfate reducers derive at least in part from products formed by aerobic bacteria in oxidizing strata (Jobson et al., 1979). However, the likelihood of oxygen being transported to deep petroleum reservoirs in meteoric water is low. Any dissolved oxygen is likely to encounter large quantities of sulfide minerals and organic matter and be consumed by chemical and biological reactions before it reaches a deep petroleum reservoir. The observation of Panganiban and Hanson (1976) that at least one sulfate reducer can use methane for energy and acetate for carbon makes a petroleum reservoir a not impossible direct source of an energy substrate for sulfate reducers and many aromatic and aliphatic hydrocarbon-degrading sulfate-reducing bacteria are now known (Aeckersberg et al., 1991; Widdel and Rabus, 2001; Grossi et al., 2008). A sulfate-reducing archaeon from a hot petroleum reservoir, *Archaeoglobus fulgidis*, has been shown to be capable of growth in the presence of crude oil, and initial reports suggested that it did not use crude oil hydrocarbons as a carbon and energy source (Stetter et al., 1993; Stetter and Huber, 2000). However, analysis of the genome of *A. fulgidis* VC-16 showed that it carried a gene homologous to alkylsuccinate synthase genes that encode the alkane-activating enzyme in a number of known anaerobic hydrocarbon-degrading bacteria, and the ability of *A. fulgidis* to grow on n-alkanes has now been demonstrated (Khelifi et al., 2014).

23.6.5 Microorganisms in Petroleum Reservoirs

Although a relatively wide range of petroleum reservoirs has been subject to microbiological analysis (Figure 23.14), there is as yet no systematic picture of microbial communities associated with reservoirs with particular characteristics such as oil type, salinity of formation waters, or souring from hydrogen sulfide production. Moreover, the vast majority of studies have relied on analysis of well-head waters and often come from reservoirs that have been water-flooded for secondary production. Contamination control is a major issue as

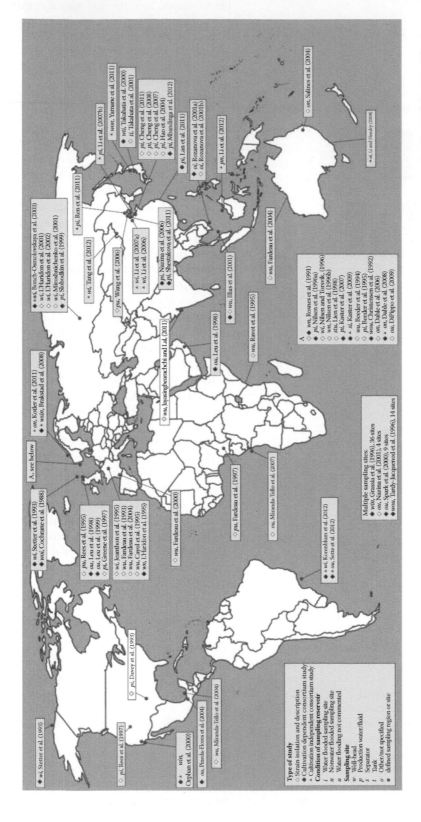

Figure 23.14. Global summary of analysis of microbial communities in petroleum reservoirs. (With kind permission from Springer Science+Business Media: Polyextremophiles: Life under Multiple Forms of Stress, Seckbach J, Oren A, Stan-Lotter H, eds., Deep subsurface oil reservoirs as polyextreme habitats for microbial life. A current review, 2013, pp. 443–466, London, U.K., Wentzel A, Lewin A, Cervantes FJ, Valla S, Kotlar HK.)

EHRLICH'S GEOMICROBIOLOGY

it is rarely possible to suspend production to permit sampling to high microbiological standards. This leads to difficulties in distinguishing organisms that are truly indigenous to a reservoir from those that have been introduced during drilling or with injection water. The changes in the hydraulic regimen in a reservoir during drilling and production also materially affect processes such as mass transfer that may lead to the selection of organisms originally present at low abundance, but which grows preferentially under production conditions. For this reason, it is typical to consider thermophilic organisms from higher temperature reservoirs as genuinely native to petroleum reservoirs. More caution is usually exercised when drawing conclusions from the analysis of low-temperature reservoirs. For a detailed discussion of contaminant versus indigenous reservoir organisms, the reader is directed to Magot (2005).

Bearing in mind these important caveats, a wide range of microbial taxa have been isolated from petroleum reservoirs. The bacteria most frequently isolated from high-temperature reservoirs are anaerobic thermophiles from the Thermotogae (*Petrotoga*, *Thermotoga*, *Geotoga*, *Kosmotoga*, *Oceanotoga*, and *Thermosipho*) followed by Firmicutes (e.g., *Thermoanaerobacter*, *Geobacillus*, *Bacillus*, *Desulfotomaculum*, *Caldanaerobacter*, and *Mahella*) (Wentzel et al., 2013; Figure 23.15). Representatives of *Deferribacter*, *Thermus*, *Anaerobaculum*, and *Thermovirga* have also been isolated from high-temperature petroleum reservoirs as have deltaproteobacterial sulfate reducers (Magot et al., 2000; Magot, 2005). A range of archaea has also been cultured from high-temperature oil reservoirs including fermentative thermophiles such as *Thermococcus*, the sulfate-reducing archaeon *Archaeoglobus* and many methanogens. Interestingly, it is predominantly hydrogenotrophic methanogens that have been isolated including members of the genera *Methanoculleus*, *Methermicoccus*, *Methanothermobacter*, and *Methanococcus* (Wentzel et al., 2013; Figure 23.15).

In addition to cultivation of organisms from petroleum reservoirs, there are a growing number of reports of microorganisms from petroleum reservoirs as analyzed by culture-independent approaches such as 16S rRNA gene analysis (Wentzel et al., 2013; Figure 23.15). A compendium of such studies suggests that the most frequently encountered bacteria in petroleum reservoirs belong to the phylum Firmicutes with Proteobacteria being highly represented

(Gray et al., 2010; Hubert et al., 2012; Figure 23.16). Interestingly, if only reservoirs with temperatures less than 50°C are considered, then the most commonly encountered bacteria are from the *Epsilonproteobacteria* (Hubert et al., 2012).

Although many aerobic bacteria have been isolated from petroleum reservoirs, including hydrocarbon-degrading bacteria (e.g., *Marinobacter* spp.), petroleum reservoirs are generally considered to be anoxic environments, and this is supported by the occurrence of many strictly anaerobic taxa including methanogenic archaea in inventories of reservoir communities (Magot et al., 2000).

23.6.6 Factors Controlling Microbial Activity in Petroleum Reservoirs

A number of environmental factors are known to affect microbial activity; these include temperature, pH, salinity, water activity, radiation, and availability of resources such as carbon and energy sources, electron acceptors, and inorganic nutrients. In the context of in-reservoir crude oil biodegradation, the most important of these are temperature, salinity, and availability of inorganic nutrients.

A relationship between the occurrence of biodegraded oil and reservoir temperature was first proposed by Connan (1984) who suggested that in-reservoir oil biodegradation ceased around a reservoir temperature of 80°C. However, non-degraded reservoirs are also found at lower temperatures and an explanation for the occurrence of these low-temperature, nondegraded reservoirs came with the development of the palaeopasteurization hypothesis, developed at the turn of the twentieth century (Wilhelms et al., 2001). The palaeopasteurization hypothesis proposes that the upper thermal limit for hydrocarbon-degrading microbial life in petroleum reservoirs is typically around 80°C–90°C and that once a reservoir had been heated to temperatures within this range, it was not recolonized by hydrocarbon-degrading microorganisms, even if the reservoir was subsequently uplifted to shallower depths where the in situ temperature is below the 80°C limit proposed for hydrocarbon-degrading microbial life (Wilhelms et al., 2001). Thus, reservoir temperature is considered a primary control on the occurrence of biodegraded petroleum reservoirs.

The apparent upper temperature limit for life in petroleum reservoirs is considerably lower than

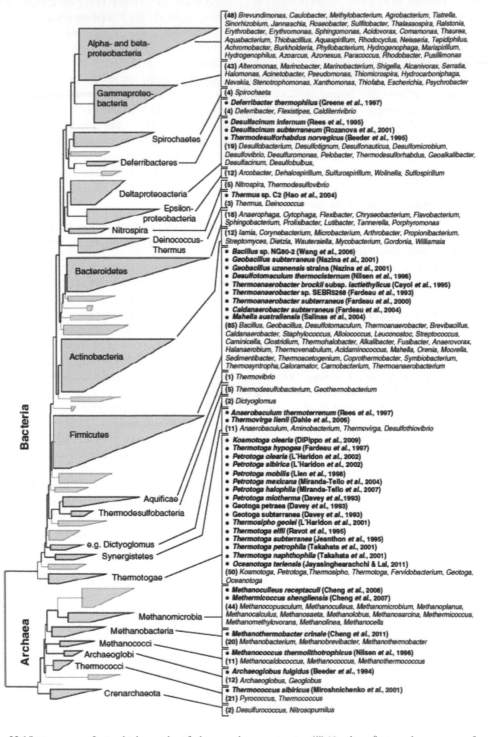

Figure 23.15. Inventory of microbial taxa identified in petroleum reservoirs. (#) Number of reported encounters of genera in the different studies explicitly referred to in Figure 23.14. Bullet points/bold represent detailed strain descriptions. (With kind permission from Springer Science+Business Media: *Polyextremophiles: Life under Multiple Forms of Stress*, Seckbach J, Oren A, Stan-Lotter H, eds., Deep subsurface oil reservoirs as polyextreme habitats for microbial life. A current review, 2013, pp. 443–466, London, U.K., Wentzel A, Lewin A, Cervantes FJ, Valla S, Kotlar HK.)

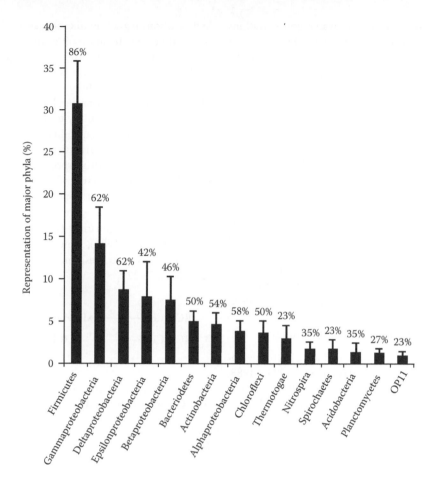

Figure 23.16. Most commonly identified groups of organisms from culture-independent studies of petroleum impacted systems. Bars correspond to average percent representation of major phyla (1× standard error) based on a survey of 26 bacterial clone libraries. Values shown above the columns indicate the percentage of studies in which the phylum was identified. (Modified from Gray, ND et al., *Adv Appl Microbiol*, 72, 137, 2010.)

the maximum temperature for biological activity observed for cultivated organisms isolated from high-temperature hydrothermal systems (ca. 121°C: Kashefi and Lovley, 2003; Takai et al., 2008). Furthermore, there are few reports of deep subsurface hyperthermophiles that have been isolated at temperatures greater than 90°C when oil field waters have been used as an inoculum (Grassia et al., 1996) and, interestingly, methanogenesis and sulfate reduction could be measured only at temperatures between 70°C and 83°C in produced waters from Californian petroleum reservoirs and not at higher temperatures, even when the temperature of the reservoir from which the samples came was up to 120°C (Orphan et al., 2003).

It is well known that salinity affects microbial activity and that petroleum reservoir formation waters can vary from freshwater, to many times seawater salinity. Reservoir water salinity will therefore, affect the occurrence of microorganisms in petroleum reservoirs. The frequency of successful cultivation of microorganisms from petroleum reservoir samples declines with increasing salinity (Figure 23.13). Furthermore, it is likely that there will be an interaction between salinity and other environmental factors such as temperature and, for example, the temperature required for palaeopasteurization of a reservoir may be lower if the salinity is elevated. However, there has been no systematic analysis of salinity–biodegradation relationships in oil fields, though there is anecdotal evidence that some low-temperature, non-uplifted reservoirs that are not biodegraded are associated with high-salinity aquifers. Given that petroleum reservoirs contain plentiful supplies of potential electron donors and carbon sources, formation water concentrations of inorganic N and P are potentially important controls on microbial cell abundance that can be supported in the subsurface (Head et al., 2003).

23.6.7 Microbes in Petroleum Degradation

When natural petroleum reservoirs become industrially exploited, constituents in the oil, whether the oil is still in its reservoir or removed from it, become susceptible to microbial attack. This attack may be aerobic or anaerobic. An extensive literature has accumulated around this subject, largely because such microbial attack can be used in the management of oil pollution. Several reviews on the topic of hydrocarbon biodegradation are available (e.g., Widdel and Rabus, 2001; Van Hamme et al.; 2003). Only the major principles of microbial petroleum degradation will be discussed here.

23.6.7.1 Aerobic Degradation of Alkanes

A variety of bacteria and fungi are able to metabolize hydrocarbons (Atlas, 1981, 1984, 1988; Atlas and Bartha, 1998; So et al., 2003; Johnson and Hyman, 2006). One compilation of known hydrocarbon-degrading microorganisms listed 87 bacterial genera (including 9 cyanobacterial genera), 103 fungal genera, and 14 algal genera that are known to contain organisms that can degrade or transform hydrocarbons (Prince, 2005); this likely represents a considerable underestimate, and it is clear that the ability to degrade hydrocarbons is a widespread capability in the microbial world. Some examples are listed in Table 23.3. The mode of attack of hydrocarbons by microorganisms depends on the kind of organism involved and the environmental conditions. Aerobically, alkanes may be attacked monoterminally to form an alcohol by oxygenation (Doelle, 1975; Atlas, 1981; Gottschalk, 1986), which may subsequently be oxidized to a corresponding carboxylic acid (Reactions 23.22):

$$RCH_2CH_3 \xrightarrow{+0.5O_2} RCH_2CH_2OH \xrightarrow{-2H}$$
$$RCH_2CHO \xrightarrow{+H_2O,-2H} RCH_2COOH$$

$$(23.22)$$

The carboxylic acid may be oxidized to acetate, which may then be oxidized to CO_2 and H_2O.

TABLE 23.3

Examples of microorganisms capable of aerobic hydrocarbon metabolism.

Organism	Substrate	Mode of attack	References[a]
Pseudomonas oleovorans	Octane	Desaturation	Abbott and Hou (1973)
Pseudomonas putida GPo-1	Propane, butane	Oxidation	Johnson and Hyman (2006)
Pseudomonas fluorescens and *P. aeruginosa*	Aromatic hydrocarbons	Oxidation	Van der Linden Thijsse (1965)
Azoarcus evansii	Benzoate	Oxidation	Gescher et al. (2002)
Nocardia salmonicolor	Hexadecane	Desaturation	Abbott and Casida (1968)
Yeasts		NS	Ahearn et al. (1971)
Trichosporon	n-paraffins	NS	Barna et al. (1970)
Arthrobacter	n-alkane	Oxidation	Klein et al. (1968)
	Aromatics	Oxidation	Stevenson (1967)
Mycobacterium	Butane	NS	Nette et al. (1965)
		Oxidation	Phillips and Perry (1974)
Brevibacterium eythrogenes	Alkane	Oxidation	Pirnik et al. (1974)
Nocardia	Mono- and dicyclic hydrocarbons	Oxidation	Raymond et al. (1967)
Cladosporium	n-alkane	NS	Teh and Lee (1973)
		Oxidation	Walker and Cooney (1973)
Graphium	Ethane	Oxidation	Volesky and Zajic (1970)

[a] For more recent reviews, see Assinder and Williams (1990), Cerniglia (1984), and Atlas (1984).
NS, not specified.

EHRLICH'S GEOMICROBIOLOGY

Alkanes may also be monoterminally attacked to form a ketone (Fredricks, 1967) or a hydroperoxide (Stewart et al., 1959). They may also be attacked diterminally (Doelle, 1975). For instance, *Pseudomonas aeruginosa* can attack 2-methylhexane at either end of the carbon chain, forming a mixture of 5-methylhexanoic acid and 2-methylhexanoic acid (Foster, 1962). Furthermore, alkanes may be desaturated terminally or subterminally, forming alkenes (Reaction 23.23; Chouteau et al., 1962; Abbott and Casida, 1968). Subterminal desaturation may proceed as follows:

$$\text{Hexadecane} \rightarrow \begin{matrix} \text{8-Hexadecene} \\ \text{7-Hexadecene} \\ \text{6-Hexadecene} \end{matrix} \quad (23.23)$$

Alkenes may be attacked by forming epoxides, which may then be further metabolized (Abbott and Hou, 1973). Diols may be formed in the process. In all the foregoing processes, atmospheric oxygen acts as a terminal electron acceptor and as the source of oxygen in oxygenation.

Alkane hydroxylases, which fall into different classes in terms of composition and cofactor requirements, location in the cell, and substrate range, play an important role in microbial degradation of petroleum hydrocarbons, chlorinated hydrocarbons, and other related compounds. Their occurrence in different microbes and their mode of action have been reviewed by van Beilen and Funhoff (2007). Interestingly, it appears that a group of specialized so-called obligate hydrocarbonoclastic bacteria has arisen in marine environments, and these are mainly represented by genera from the *Gammaproteobacteria*. These organisms are typically strongly enriched in oil-impacted environments and during brioremediation treatments to deal with spilled oil (Yakimov et al., 2007).

The ability of an organism to attack hydrocarbons aerobically does not necessarily mean that it can use such compounds as the sole source of carbon and energy. Many cases are known in which hydrocarbons are oxidized in a process known as cooxidation, wherein another compound, which may be quite unrelated, is the carbon and energy source but somehow permits the simultaneous oxidation of the hydrocarbon. Examples are the oxidation of ethane to acetic acid, the oxidation of propane to propionic acid and acetone, and the oxidation of butane to butanoic acid and methyl

ethyl ketone by *Pseudomonas methanica* growing on methane, as first shown by Leadbetter and Foster (1959). Methane is the only hydrocarbon on which this organism can grow. Another example is the oxidation of alkylbenzenes by a strain of *Micrococcus cerificans* growing on n-paraffins (Donos and Frankenfeld, 1968). Other examples have also been summarized (Horvath, 1972).

Chain length and branching of aliphatic hydrocarbons can affect microbial attack. In situ observations revealed rapid microbial degradation of pristane and phytane (Atlas and Cerniglia, 1995). Some bacteria that attack alkanes of chain lengths C8–C20 may not be able to attack alkanes of chain lengths C1–C6, whereas others cannot grow on alkanes of chain lengths greater than C10 (Johnson, 1964). Fungi that can grow on alkanes of chain lengths up to C34 are known. It has also been noted that certain placements of methyl or propyl groups in the alkane carbon chain lessen or prevent utilization of the compounds (McKenna and Kallio, 1964).

23.6.7.2 Anaerobic Degradation of Saturated Hydrocarbons

Although hydrocarbon oxidation by microbes was once considered a strictly aerobic process because the initial attack usually involves an oxygenation, clear evidence now exists for anaerobic degradation of some oil constituents as well. Several early reports suggested that anaerobic oil degradation was possible albeit typically slower than aerobic degradation (e.g., Dutova, 1962; Kuznetsova and Gorlenko, 1965; Davis and Yarbrough, 1966; Davis, 1967; Simakova et al., 1968; Kvasnikov et al., 1973; Panganiban and Hanson, 1976; Ward and Brock, 1978).

Until about 1990, anaerobic attack of alkanes and alkenes was believed possible only if these compounds carried one or more substituents, in particular, halogens. For instance, dichloromethane (CH_2Cl_2) was shown to be degraded anaerobically. The initial attack in this case did not involve an oxygenation but instead a dehalogenation by a consortium of two different bacterial strains (Braus-Stromeyer et al., 1993):

$$CH_2Cl_2 + H_2O \rightarrow (HCHO) + 2HCl \quad (23.24)$$

This reaction was followed by an oxidation of the formaldehyde-like intermediate (HCHO) to formic acid. The formic acid was subsequently converted to acetate in an acetogenic reaction (Braus-Stromeyer et al., 1993).

In the case of tetra- and trichloroethylene, anaerobic dechlorination by a reductive process has been observed. In this process, the chlorinated hydrocarbon serves as a terminal electron acceptor. Complete bacterial dechlorination of tetra- and trichloroethylene was observed by Freedman and Gossett (1989), Ensley (1991), and Maymó-Gatell et al. (1999). De Bruin et al. (1992) observed the formation of ethane by reductive transformation of tetrachloroethene.

A clear demonstration that a saturated, unsubstituted alkane can be mineralized by anaerobic bacteria was presented by Aeckersberg et al. (1991) (see also So et al., 2003). These investigators isolated a sulfate-reducing bacterium, HxD3, from the precipitate of an oil–water separator in an oil field near Hamburg, Germany. This organism mineralized hexadecane, using sulfate as the terminal electron acceptor. The nature of the initial attack of the hexadecane was not elucidated, but hydroxylation was probably involved. The overall reaction of hexadecane mineralization was consistent with the following stoichiometry (Aeckersberg et al., 1991):

$$C_{16}H_{34} + 12.25SO_4^{2-} + 1.5H^+ \rightarrow 16HCO_3^- + H_2O$$
$$(23.25)$$

Subsequent studies have suggested that HxD3 (now named *Desulfococcus oleovorans* HxD3) activated alkanes by carboxylation of carbon 3 of the alkane chain to generate a 2-ethyl fatty acid, but it is now believed that there is an initial hydroxylation followed by carboxylation to generate a 2-acetyl carboxylic acid (Callaghan, 2013). However, the best characterized anaerobic alkane activation mechanisms is addition of the alkane to a molecule of fumarate to generate alkyl succinates (Callaghan, 2013), but it is clear that in different organisms a number of other mechanisms have evolved to activate saturated hydrocarbons (Callaghan et al., 2013). Ehrenreich et al. (2000) demonstrated that three distinct types of denitrifying bacteria were able to oxidize alkanes anaerobically, and Kropp et al. (2000) found that a sulfate-reducing enrichment culture was able to mineralize n-dodecane. Zengler et al. (1999) demonstrated methanogenic alkane degradation

in an enrichment culture growing on hexadecane as a sole source of carbon and energy and this has now been shown for a range of pure alkanes and alkanes in different crude oils, but this is a slow process (Townsend et al., 2003; Gieg et al., 2008; Jones et al., 2008; Siddique et al., 2011, Wang et al., 2011).

Methanogenic crude oil alkane degradation involves syntrophic interactions between primary hydrocarbon-degrading bacteria (often *Smithella* spp.) and methanogens. Zehnder and Brock (1979) found that methanogens are able to oxidize small amounts of the methane they formed anaerobically. The methane oxidation mechanism in these organisms seems to differ from the methane-forming mechanism. The slow rate of anaerobic hydrocarbon degradation, when it occurs, helps to explain in part why petroleum has remained preserved over eons of time. However, prolonged periods of an absence of degradative activity of any kind, probably due to high temperatures pasteurizing reservoir sediments (Wilhelms et al. 2001) and possibly high salinity of stratal waters, may be an important factor in petroleum preservation. Moreover, in-reservoir rates of hydrocarbon degradation have been assessed based on oil composition gradients in biodegraded petroleum reservoirs, with first-order rate constants on the order of 10^{-6} to 10^{-7} years^{-1} (Larter et al., 2003) consistent with in-reservoir biodegradation over millions to tens of millions of years, and in situ microbial conversion of petroleum into bitumen has been implicated in the formation of the Alberta (Canada) oil sands, on the basis of laboratory simulation (Rubinstein et al., 1977) (see Section 23.6.8.1).

23.6.7.3 Degradation of Aromatic Hydrocarbons

A wide range of aromatic compounds can be aerobically and anaerobically degraded by microbes (Tables 23.3 and 23.4). *Aerobic scission* of the ring structure of the aromatic compounds involves oxygenation either between adjacent oxygenated carbon atoms (*ortho*-fission) or adjacent to one of them (*meta*-fission) (Dagley, 1975). In the case of benzene, aerobic degradation of the ring structure involves initial hydroxylation catalyzed by a mixed function monooxygenase or a 1,2-dioxygenase to form catechol followed by the action of dioxygenase to cleave the catechol ring to form *cis*, *cis*-muconate by

TABLE 23.4

TABLE 23.4

Bacteria capable of anaerobic metabolism of aromatic and heterocyclic nonhydrocarbons.

Organism	Substrate	Mode of attack	References[a]
R. palustris	Benzoate, hydroxybenzoate	Reductive ring cleavage	Dutton and Evans (1969)
Desulfobacterium phenolicum	Phenol and derivatives	Anaerobic degradation	Bak and Widdel (1986a)
Desulfobacterium indolicum	Indolic compounds	Anaerobic degradation	Bak and Widdel (1986b)
Desulfobacterium catecholicum	Catechol	Anaerobic degradation	Szewzyk and Pfennig (1987)
Desulfococcus niacini	Nicotinic acid	Anaerobic degradation	Imhoff-Stuckle and Pfennig (1983)

[a] For reviews, see Evans and Fuchs (1988), Reineke and Knackmuss (1988), and Higson (1992).

ortho-fission. This product can then be degraded enzymatically in several steps to succinate and acetate (Doelle, 1975). The acetate is then oxidized to CO_2 and H_2O, and the succinate can enter the TCA cycle. Naphthalene, anthracene, and phenanthrene and their derivatives can be degraded by a similar mechanism by attacking each ring in succession (Doelle, 1975). In some instances, benzene derivatives are attacked by meta-fission instead of ortho-fission as in the previous examples (Doelle, 1975).

Anaerobic scission of an aromatic ring structure involves ring saturation, hydration, and dehydrogenation (Evans and Fuchs, 1988; Colberg, 1990; Grbic-Galic, 1990). Substituted aromatic hydrocarbons such as toluene or methylnaphthalenes seem to be activated by fumarate addition reactions leading to benzyl and napthylsuccinates (Safinowski and Meckenstock, 2004), and benzene is activated by a novel carboxylation reaction (Abu Laban et al., 2010). Some aromatic compounds formed by oxidation of hydrocarbons such as benzoate can also be biodegraded anaerobically by photometabolism of certain *Rhodospirillaceae* (purple nonsulfur bacteria; Table 23.4). Ring cleavage is by hydration of pimelate (Dutton and Evans, 1969). Biodegradation of benzene can result in significant carbon and hydrogen isotopic fractionation with enrichment factors ranging from −1.9‰ to −3.6‰ for carbon and −29‰ to −79‰ for hydrogen (Mancini et al., 2003).

Chlorinated aromatic compounds have been shown to be completely degradable anaerobically by a consortium of bacteria (e.g., Sharak Genthner et al., 1989; Colberg, 1990; Grbic-Galic, 1990). In one kind of consortium, *Desulfomonile tiedje* DCB-1 was found to act on chlorobenzoate by using it as a terminal electron acceptor in a respiratory process that includes dechlorination (Shelton and Tiedje, 1984; Dolfing, 1990; Mohn and Tiedje, 1990, 1991). Subsequent mineralization of the dechlorinated aromatic product depends on other anaerobic organisms in the consortium.

23.6.8 Economic Significance of the Petroleum Reservoir Deep Biosphere

23.6.8.1 Biodegradation and the Occurrence of Heavy Oil

One of the most dramatic manifestations of microbial activity in petroleum reservoirs is heavy oil (Figure 23.17). In fact, the bulk of oil on the planet is heavy oil, and it has been estimated that global heavy oil deposits represent a six-trillion barrel resource (Roadifer, 1987). The largest heavy oil deposits occur in Venezuela (the Orinoco heavy oil belt) and Western Canada (e.g., the Athabasca oil sands in Alberta). These deposits are estimated to contain 1200 billion barrels and 900 billion barrels of oil, respectively. This compares with the supergiant conventional oil fields in the middle east, such as Ghawar in Saudi Arabia and Burgan in Kuwait, which contain approximately 190 billion barrels each (Roadifer, 1987).

It is only relatively recently that the capacity for anaerobic hydrocarbon degradation by a large range of microorganisms has been convincingly established and accepted (Widdel and Rabus, 2001). Consequently, the prevailing view of heavy oil formation was that it was the result of aerobic hydrocarbon degradation driven by oxygen carried in meteoric water. However, it is difficult to envisage how oxygen could be transported to kilometer depths and more without encountering reactive organic carbon or mineral phases that, through chemical or biologically mediated

Figure 23.17. An oil sands mine in North East Alberta. In the northeast of the Western Canada Sedimentary Basin, petroleum reservoirs are shallow and cool and contain very heavy, highly biodegraded oil. Toward the southwest, the reservoirs are deeper, hotter, and less biodegraded. When the reservoirs are shallow, they are accessed by mining, as shown earlier. Further to the west, the reservoirs are produced using thermal processes such as steam-assisted gravity drainage (SAGD) for moderately heavy oil or conventional drilling in the deeper hotter reservoirs, which contain lighter oil. These extensive deposits of heavy oil were generated by in-reservoir microbial degradation over periods of tens of millions of years. (Photograph by Ian M Head.)

reactions, would remove the oxygen. This, coupled with the widespread occurrence of anaerobic microorganisms in petroleum reservoirs, suggested that the model of aerobic in-reservoir biodegradation was incorrect, and it was proposed that anaerobic processes must be responsible (Head et al., 2003). Direct evidence that this was the case came from the detection of metabolites characteristic of anaerobic hydrocarbon degradation in samples from a number of biodegraded reservoirs but not in reservoirs that contained light, undegraded oil (Aitken et al., 2004).

The occurrence of high levels of methane associated with biodegraded petroleum reservoirs and the fact that the methane was isotopically enriched in ^{12}C relative to ^{13}C compared to thermogenic methane associated with nondegraded reservoirs suggested that methanogenesis specifically might play an important role in in-reservoir petroleum biodegradation (Pallaser, 2000). This was supported by the similarity of some geochemical markers in oils from biodegraded reservoirs and oils degraded in long-term laboratory incubations under methanogenic conditions and modeling of gas isotope data (Figure 23.18; Jones et al., 2008). Interestingly, methanogens isolated from, or detected in, petroleum reservoirs and measurements of different methanogenic pathways indicate that hydrogen-oxidizing, CO_2-reducing methanogens are predominant with acetoclastic methanogens being encountered more rarely (Gray et al., 2010). This is consistent with the carbon isotope data for carbon dioxide from biodegraded petroleum reservoirs and the composition of methanogen communities

in active hydrocarbon-degrading systems (Jones et al., 2008; Gray et al., 2011). Methanogen communities in active methanogenic oil-degrading systems have a preponderance of hydrogen-oxidizing, CO_2-reducing methanogens, and gases from biodegraded petroleum reservoirs typically contain CO_2 with a very positive $\delta^{13}C$ (i.e., a low proportion of $^{12}CO_2$) consistent with methanogenesis through CO_2 reduction (Figure 23.19; Jones et al., 2008). This has been termed the methanogenic alkane degradation dominated by CO_2 reduction (MADCOR) process of crude oil hydrocarbon degradation.

23.6.8.2 Reservoir Souring and Corrosion

Formation of hydrogen sulfide in petroleum reservoirs (reservoir souring) is a serious problem in petroleum production (Gieg et al., 2011). It represents a health hazard for oil production workers, lowers the value of the produced oil due to higher refining costs, and can also be related to corrosion. Sulfide also plays a role in microbially influenced corrosion (MIC) though other sulfide-independent forms of corrosion also have a microbiological component.

Although there are several sources of sulfide in petroleum, microbial sulfate reduction is undoubtedly an important contributor. It is often associated with seawater injection to repressurize the reservoir for secondary oil recovery. The sulfate-rich seawater (ca. 28 mM) provides the electron acceptor that sulfate-reducing microorganisms use for respiration. Sulfate reduction is thought to be driven by organic compounds

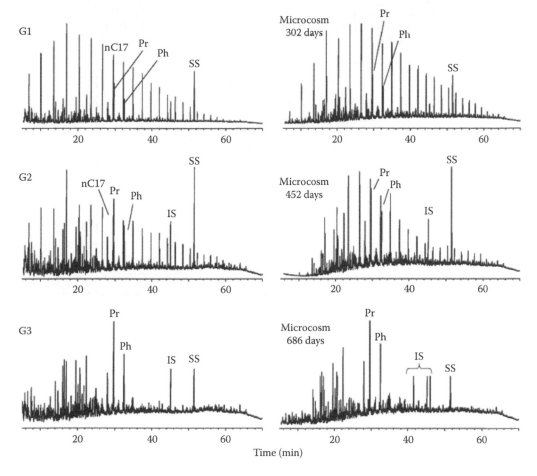

Figure 23.18. Gas chromatograms of oils from a North Sea oil field (left side) and crude oil degraded in laboratory incubations under methanogenic conditions (right side). The North Sea oils are from a field exhibiting a biodegradation gradient. Detailed geochemical analysis of the oils demonstrated that the in situ degraded oils from the reservoirs shared characteristics with oil degraded in the lab under methanogenic conditions, but not oils degraded in the lab under sulfate-reducing conditions. nC17, heptadecane; Pr, pristane; Ph, phytane; IS and SS; standards added prior to analysis. (From Jones, DM et al., *Nature*, 451, 176, 2008. With permission.)

in the crude oil such as organic acids, but it is also possible that crude oil hydrocarbons could act as an electron donor in some circumstances (Figure 23.20; Hubert, 2010; Agrawal et al., 2012). Biocide treatment can be used to control souring, but this typically has limited utility as biocides can have poor activity against biofilms. Alternative control mechanisms have therefore, been devised. One of the most promising of these is treatment with nitrate. This may involve one or more mechanisms acting singly or in consort. Nitrate treatment can stimulate heterotrophic bacteria that outcompete sulfate reducers for electron donor. This can also generate nitrite that can be toxic to sulfate reducers. Alternatively sulfide-oxidizing nitrate reducers can remove sulfide produced by

the sulfate reducers and again may generate toxic nitrite (Figure 23.20; Hubert, 2010).

In 2013, the National Association of Corrosion Engineers estimated that corrosion in all its forms costs the oil and gas industry around $1.4 billion, and a significant proportion of this has been attributed to microbial activity, so-called microbially influenced corrosion (MIC) (Little et al., 2007). The classical mechanism proposed for MIC is through the microbial consumption of hydrogen produced abiotically (electrochemically produced by reduction of protons with electrons from iron oxidation). The microbial consumption of the hydrogen produced drives the reaction forward resulting in extensive corrosion. This has been termed the *cathodic depolarization theory* (Larsen

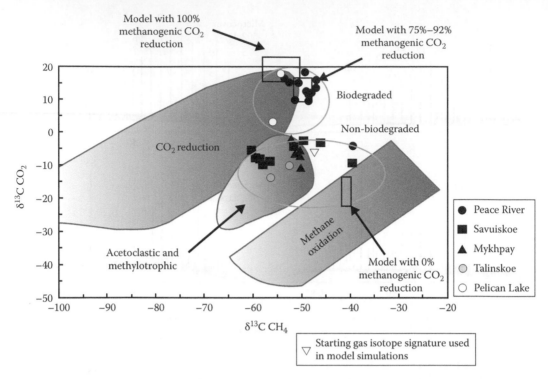

Figure 23.19. Carbon isotopic composition of methane and carbon dioxide from petroleum reservoirs indicating that in-reservoir hydrocarbon degradation is most likely predominantly driven through methanogenic CO_2 reduction with a lesser role for acetoclastic methanogenesis. Measured and modeled methane and carbon dioxide ^{13}C isotopic compositions from degraded and nondegraded oil reservoir gases (Nazina et al., 1995; Rozanova et al., 1995; Grabowski et al., 2005; Jones et al., 2008). These data have been superimposed onto to a plot of $\delta^{13}C$ CH_4 and $\delta^{13}C$ CO_2 showing isotope fractionation ranges (gray shaded areas) for methanogenesis by CO_2 reduction, acetoclastic/methylotrophic methanogenesis, and methane oxidation (adapted from Whiticar, 1999). The measured gas isotope data (see symbol legend on graph) were derived from oil reservoirs from western Canada and western Siberia. Symbols indicate individual gas sample measurements for biodegraded (encompassed by a grey edged circle) and nonbiodegraded oils (encompassed by a grey edged ellipse). Modeled methane and CO_2 compositions were derived from Rayleigh isotope fractionation (Jones et al., 2008) using a defined starting gas isotope signature (inverted open triangle) and simulated for carbon species involved in biodegradation of petroleum via syntrophic alkane oxidation, syntrophic acetate oxidation, acetoclastic methanogenesis, and hydrogenotrophic methanogenesis. Three simulated compositions (black edged boxes) were derived using assuming of a 0%, 75%, or 100% contribution of CO_2 reduction to methanogenesis. (From Gray, ND et al., *Adv Appl Microbiol*, 72, 137, 2010. With permission.)

et al., 2010; Figure 23.21). The validity of microbial hydrogen consumption as a central driver of MIC has been called into question, and in many cases, it seems that prolonged generation of H_2S by SRB and its reaction with metallic iron may be a more significant process in MIC (Figure 23.21; Enning and Garrelfs, 2014). However, it is clear that there are many other abiotic and microbially catalyzed processes that can lead to enhanced corrosion including the direct oxidation of metallic iron as an electron donor for some sulfate-reducing microorganisms (Enning and Garrelfs, 2014; Figure 23.22), and no single theory can fully explain the rates and extent of corrosion observed in practice, reflecting the fact that corrosion

results from a complex interplay of both microbial and abiotic factors (Angell, 1999).

23.6.8.3 Secondary and Tertiary Oil Recovery

When a petroleum reservoir is first tapped for commercial exploitation, the initial oil is recovered by being forced to the surface by gas pressure from the volatile components of the oil and by pumping. This action, however, recovers only part of the total oil in a reservoir. To recover additional oil, a reservoir may be flooded with water by injection to force out additional oil (*secondary oil recovery*). Even secondary oil recovery will yield no more than 30%–40% of the oil (North, 1985).

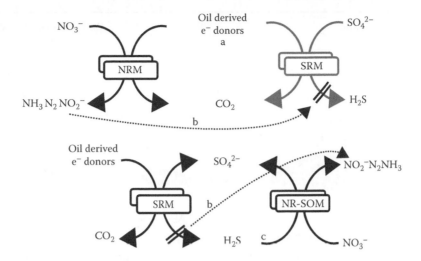

Figure 23.20. Proposed mechanisms of hydrogen sulfide production by sulfate-reducing microorganisms (SRM) in petroleum reservoirs and possible mechanisms whereby nitrate treatment may ameliorate reservoir souring. (a) Heterotrophic nitrate reducers in the reservoir outcompete sulfate reducers for electron donors decreasing sulfide production. (b) Nitrite produced by nitrate-reducing bacteria inhibits sulfate reduction. (c) Nitrate-reducing, sulfide-oxidizing bacteria destroy sulfide and generate nitrite that may inhibit sulfate reducers (b). (With kind permission from Springer Science+Business Media: *Handbook of Hydrocarbon and Lipid Microbiology*, Timmis KN, ed., Microbial ecology of oil reservoir souring control by nitrate injection, 2010, pp. 2753–2766, Berlin, Germany, Hubert, C.)

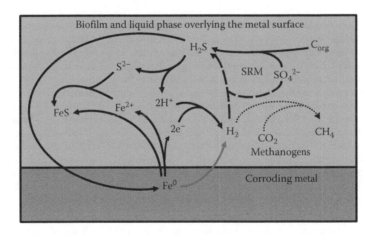

Figure 23.21. The "cathodic depolarization" mechanism of microbially influenced corrosion (dotted and dashed arrows) and chemical corrosion mediated by biogenic hydrogen sulfide production (gray arrows). Metallic iron is oxidized electrochemically and the electrons from this reaction reduce protons to H_2. The H_2 is consumed by sulfate-reducing microorganisms (SRM) and/or methanogens driving the reaction forward. In addition, the SRM generate hydrogen sulfide that reacts chemically with ferrous iron liberated by oxidation of the metallic iron, depositing iron sulfide minerals. Hydrogen sulfide generated by SRM from sulfate, using organic carbon (C_{org}) or hydrogen as an electron donor also reacts chemically with metallic iron resulting in iron oxidation with hydrogen and iron sulfide as the reaction products (light gray arrows). (After Larsen, J et al., Consortia of MIC bacteria and archaea causing pitting corrosion in top side oil production facilities, in: NACE-10252, Corrosion 2010, San Antonio, TX, March 14–18, 2010.)

To recover even part of the remaining oil, which is more viscous than the previously extracted oil, tertiary or enhanced oil recovery treatment is necessary. This may involve a thermal method (e.g., steam injection) to reduce viscosity (e.g., North, 1985) or other chemical or physical methods (e.g., Orr and Taber, 1984). Alternatively, it may involve biological methods such as generation of surface-active agents of microbial origin to facilitate mobility of the oil, or it may involve the generation of

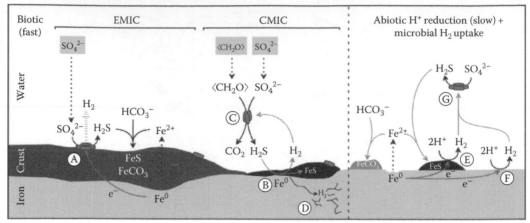

Figure 23.22. Different types of iron corrosion by sulfate-reducing bacteria (SRB) at circumneutral pH. Biotic and abiotic reactions are shown. Depicted biotic reactions tend to be much faster than abiotic corrosion reactions. SRB attack iron via electrical microbially influenced corrosion (EMIC) or chemical microbially influenced corrosion (CMIC). Stoichiometry of the illustrated reactions is given in the lower panel of this figure. Please note that all depicted processes may occur simultaneously on corroding metal surfaces but differ in rates and relative contributions to corrosion. (A) Specially adapted lithotrophic SRB withdraw electrons from iron via electroconductive iron sulfides (EMIC). Excess of accepted electrons may be released as H_2 (via hydrogenase enzyme). Participation of possibly buried (encrusted) SRB in sulfate reduction and hydrogen release is currently unknown. (B) Biogenic, dissolved hydrogen sulfide reacts with metallic iron. (C) Overall representation of CMIC. Organotrophic SRB produce hydrogen sulfide, which reacts with metallic iron. (D) Sulfide stress cracking (SSC) of iron due to biogenic hydrogen sulfide. (E) Catalytic iron sulfides may accelerate reduction of H^+ ions to H_2. (F) Slow, kinetically impeded reduction of H^+ ions to H_2 at iron surfaces. (G) Consumption of H_2 from reaction E or F by SRB does not accelerate the rate of H_2 formation (no "cathodic depolarization"). Note that CMIC quantitatively depends on the availability of biodegradable organic matter (here schematically shown as carbon with the oxidation state of zero, CH_2O). (From Enning, D and Julia Garrelfs, J, *Appl Environ Microbiol,* 80, 1226, 2014. With permission.)

gas pressure by fermentation to force movement of the oil, or it may involve a combination of both processes (McInerney and Westlake, 1990; Tanner et al., 1991; Adkins et al., 1992; Van Hamme et al., 2003; Brown, 2007). Microbially enhanced oil recovery may also involve the promotion of selective plugging by microbes of high-permeability zones in oil reservoirs to increase volumetric sweep efficiency and the microscopic oil displacement efficiency (Raiders et al., 1989; McInerney and Westlake, 1990; Brown, 2007). Volumetric sweep efficiency refers to the ability of injected water to recover oil from less permeable zones. An excellent review summarizing many approaches for using microorganisms to enhance oil recovery has

been provided by Youssef and coauthors (Youssef et al., 2009).

Water injection into oil reservoirs stimulates microbial activity by sulfate reducers as well as methanogens and fermentative bacteria (Rozanova, 1978; Nazina et al., 1985; Belyaev et al., 1990a,b). Because the injected water carries oxygen, hydrocarbon-oxidizing bacteria have also been detected (Belyaev et al., 1990a,b). Indeed, a succession of organisms may occur, with the aerobic hydrocarbon oxidizers and others producing the substrates (e.g., acetate and higher fatty acids) for fermenting and sulfate-reducing bacteria, and the fermenting bacteria producing hydrogen for use by methanogens and sulfate reducers

(Nazina et al., 1985). In the Bondyuzh oil field (former Tatar SSR), maximal numbers of aerobic hydrocarbon-oxidizing bacteria were found at the interface of injection and stratal waters. The destructive effect of these bacteria on petroleum, in one case in which this was studied, appeared to be limited by the salt concentration of the stratal waters (Gorlatov and Belyaev, 1984).

For tertiary oil recovery that involves the use of surface-active agents of microbial origin, xanthan gums of the bacterium *Xanthomonas campestris* or glucan polymers of fungi such as *Sclerotium*, *Stromantinia*, and *Helotium* (Compere and Griffith, 1978) may be generated in separate processes and then injected into an oil well to facilitate oil recovery. As an alternative approach, appropriate organisms may be introduced into the oil well to produce surface-active agents in situ (Finnerty et al., 1984). Promotion of tertiary oil recovery may also involve injection of dilute molasses solution into an oil well and subsequent injection of a gas-producing culture such as *Clostridium acetobutylicum*, which ferments the molasses to the solvents acetone, butanol, and ethanol and large amounts of CO_2 and H_2. The solvents help to lower the viscosity of the oil, and the gases provide pressure to move the oil (Yarbrough and Coty, 1983). With the discovery of methanogenic oil biodegradation as an important process in the formation of heavy oil, it has been proposed that stimulating the conversion of a small fraction of residual oil to methane could be used as an enhanced oil recovery strategy (Parkes 1999; Gieg et al., 2008; Jones et al., 2008). Generation of methane would initially result in higher levels of dissolved gas in the oil reducing oil viscosity and when sufficient gas is produced the reservoir would be repressurized resulting in improved oil recovery (Figure 23.23).

23.6.8.4 Removal of Organic Sulfur from Petroleum

As in the case of coal, petroleum that contains significant amounts of organic sulfur may not be usable as a fuel because of air pollution by the volatile sulfur compounds such as SO_2 that results from the combustion of this fuel. The feasibility of removing this sulfur microbiologically has been actively explored (e.g., Foght et al., 1990; Van Hamme et al., 2003). The experimental approach being taken is similar to that in the investigations to remove organic sulfur from coal. The model

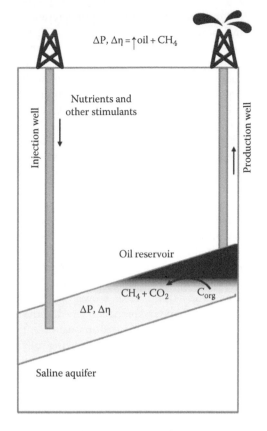

Figure 23.23. Effects of conversion of a portion of residual crude oil to methane on oil properties. Generation of methane would initially result in higher levels of dissolved gas in the oil reducing oil viscosity and when sufficient gas is produced the reservoir would be repressurized resulting improved oil recovery and ultimately recovery of more gas.

substance to evaluate microbial desulfurization of petroleum is also dibenzothiophene. The best microbial agents for any industrially applicable process are those that remove sulfur without significant concomitant oxidation of the carbon to CO_2. *Rhodococcus erythropolis* is an example of such an organism (e.g., Izumi et al., 1994), and *Rhodococcus* sp. strain ECRD-1 (Grossman et al., 2001) is another.

23.6.9 Use of Microbes in Prospecting for Petroleum

Prospecting for petroleum through the detection of hydrocarbon-utilizing microorganisms has been explored (e.g., Brown, 2007). The scientific basis for this method is the detection of microseepage of petroleum or some of its constituents, especially the more volatile components, in the ground overlying a deposit using the presence of any hydrocarbon-utilizing microbes as indicators. One method involves enriching soil, sediment, and water samples from a suspected seepage area for microbes that can metabolize gaseous hydrocarbons and demonstrating hydrocarbon consumption (Davis, 1967). An enrichment medium consisting of a mineral salts solution with added volatile hydrocarbon (ethane, propane, butane, isobutene, but not methane) is satisfactory. Methane-oxidizing bacteria are poor indicators in petroleum prospecting because methane can occur in the absence of petroleum deposits and, moreover, some methane-oxidizing bacteria are unable to oxidize other aliphatic hydrocarbons. Bacteria that can oxidize ethane and longer-chain hydrocarbons, however, provide presumptive evidence for a hydrocarbon seep and an underlying petroleum reservoir (Davis, 1967; Brown, 2007). It is assumed that ethane and propane formed in anaerobic fermentation are produced in quantities too small to select for a hydrocarbon-utilizing microflora. Likely, organisms active in soil enrichments from hydrocarbon seeps may include *Mycobacterium paraffinicum* and *Streptomyces* spp.

Hydrocarbon enrichment cultures may be prepared using ^{14}C-labeled hydrocarbon. This allows for the easy identification of the activity of hydrocarbon-oxidizing bacteria in water and sediments (Caparello and LaRock, 1975). With this method, the hydrocarbon-oxidizing potential of a sample can be correlated with the hydrocarbon burden of the environment from which the sample came.

It has also been suggested that thermophilic spore-forming sulfate-reducing bacteria found in cold arctic sediments may represent organisms released from leaky, hot petroleum reservoirs, and these "thermospores" may offer opportunities in hydrocarbon prospecting (Hubert and Judd, 2010).

23.6.10 Microbes and Shale Oil

North (1985) described *oil shales* as bituminous, nonmarine limestones, or marlstones containing kerogens. *Tar sands* are consolidated or unconsolidated rock coated with bituminous material (North, 1985). *Bitumens* are solid hydrocarbons that are soluble in organic solvents and fusible below 150°C. Kerogens are insoluble in organic solvents. They are intermediate products in the diagenetic transformation of organic matter

in sediments and are considered a precursor in petroleum formation. As in the formation of peat, coal, and petroleum, fungi and bacteria probably played a role in the early stages of transformation of the source material (mostly terrestrial biomass). Later stages involved physicochemical processes. However, in the case of oil sand bitumens of Alberta, the origin appears to be a partial biodegradation of petroleum, leaving behind the high-viscosity compounds (Rubinshtein et al., 1977).

Bitumen and kerogen can be converted to a petroleum-like substance by heat treatment (e.g., retorting; North, 1985). Their separation from host rock, especially if it is limestone, can be facilitated if the limestone is dissolved. This can be achieved, at least on a laboratory scale, by acid-forming microbes (e.g., sulfuric acid from S^0 oxidation by *Acidithiobacillus thiooxidans*) (Meyer and Yen, 1976).

Although raw shale is considered relatively resistant to microbial attack, some reports indicate otherwise. Both aerobic and anaerobic attacks have been observed, but the anaerobic attack proceeded at a much slower rate (Ait-Langomazino et al., 1991; Roffey and Norqvist, 1991; Wolf and Bachofen, 1991). Hydrogenated shale S was metabolized by some Gram-negative organisms (*Alcaligenes* and *Pseudomonas* or *Pseudomonas*-like organisms) (Westlake et al., 1976).

23.7 SUMMARY

Not all carbon in the biosphere is continually being recycled on geologically short timescales. Some is trapped as organic carbon in special sedimentary formations, where microbial attack is much reduced. The forms in which the trapped carbon appears are methane, peat, coal, petroleum, bitumen, and kerogen.

Most methane in sedimentary formations is of biogenic origin. It may occur by itself or in association with peat, coal, or petroleum deposits where a significant proportion may be thermogenic. Its biogenic formation is a strictly anaerobic process involving methanogenic archaea that may reduce CO_2 with H_2 or a few other hydrogen donors, or it may involve disproportionation (fermentation) of formate, methanol, methylamines, secondary alcohols, or acetate in the absence of H_2. The autotrophic methanogens use a mechanism for CO_2 assimilation that involves reduction of one CO_2 to methyl carbon and a second CO_2 to formyl carbon and then coupling the two to form acetate. The acetate is subsequently carboxylated to form pyruvate—the key intermediate for forming building blocks for all the cell constituents. Acetotrophic methanogens obtain their carbon from part of the acetate they consume. Methanogenesis occurs mostly mesophilically and thermophilically.

Methane may be oxidized and assimilated by a special group of microorganisms called *methanotrophs*. This was previously thought to be principally aerobic, although clear evidence of anaerobic methane oxidation now exists and has been shown to be quantitatively important, especially in marine sediments. At least one of the organisms responsible for anaerobic methane oxidation by itself appears to be a sulfate reducer. Limited anaerobic methane oxidation by a methanogen has also been observed. In some anaerobic habitats in the marine environment, significant anaerobic methane oxidation has been shown to occur that is driven by consortia involving an archaeon related to methanogens from the class Methanosarcinales and a sulfate reducer (e.g., *Desulfosarcina*). Carbon assimilation by aerobic methanotrophs may be via a ribulose monophosphate pathway or a serine pathway, in each case involving integration of the carbon at the oxidation level of formaldehyde that is produced as an intermediate in methane oxidation. In addition, both types of methanotrophs derive some of their carbon from CO_2.

Peat is the result of partial biodegradation of plant remains accumulating in marshes and bogs. Aerobic attack by enzymes in the plant debris and by fungi and some bacteria initiates the process. It is followed by anaerobic attack by bacteria during burial resulting from continual sedimentation until inhibited by accumulating wastes, lack of sufficient moisture, and so on. A viable microbial flora can usually be detected in peat even though the peat may have formed over a geologically extended period. Coal is thought to have formed like peat, except that in the advanced stages, as a result of deeper burial, it was subject to physical and chemical influences that converted the peat to coal. Different ranks of coal exist, which differ from one another largely in carbon and moisture content and in heat value, and are a reflection of the maturity of the coal. Coal seams appear to harbor an indigenous microbial flora, and there is evidence that biogenic methane formation from coal occurs.

Bituminous coal has pyrite or marcasite associated with it as inclusions. Upon exposure to air and moisture during mining, this iron disulfide becomes subject to attack by acidophilic, iron-oxidizing thiobacilli, sulfur-oxidizing bacteria, and archaea, and is the source of acid mine drainage.

Whereas peat and coal are derived from terrestrial plant matter, petroleum and associated natural gas (mostly methane) are derived from phytoplankton remains that accumulated in depressions of shallow seas and became gradually buried in sediment. Microbial attack altered these remains biochemically until complete burial by accumulating sediment stopped the organisms. In tectonically active areas, the buried and biochemically altered organic matter became subject to further alteration by heat from magmatic activity and pressure from the weight of overlying sediment. The chemical alterations may have been catalyzed by clay minerals. Thermal decomposition of the preserved organic matter has an important role in the formation of petroleum hydrocarbons a process known as catagenesis. At least some of the natural gas associated with petroleum may represent biogenic methane formed in the initial stages of plankton debris fermentation, but also includes thermogenic gas generated from catagenesis at higher temperature (e.g., via metagenesis). Secondary biogenic gas may also form as a result of in-reservoir oil biodegradation. At its site of formation, petroleum is highly dispersed. As a result of gas (natural gas, CO_2) and hydrostatic pressure as well as lubrication of the surfaces of the sediment matrix by bacteria and some of their metabolic products, matured petroleum may be forced to migrate through pervious sediment strata until it is collected in a trap such as the apex of an anticlinal fold or an impermeable fine-grained rock layer such as a mudstone. These structures are called reservoir seals or caprocks. It is such petroleum-filled traps that constitute commercially exploitable petroleum reservoirs. Sulfate-reducing bacteria may assist in trapping petroleum in reservoirs by laying down impervious calcite layers at the interface between the trapped petroleum and groundwater. This calcite may, however, also interfere with petroleum recovery.

At least some petroleum hydrocarbons can be oxidized in air by many bacteria and fungi. Anaerobic bacterial attack of unsubstituted alkanes as well as chlorinated alkanes and of a wide range of aromatic hydrocarbons has also been demonstrated. Hydrocarbon-utilizing microorganisms may be used as indicators in prospecting for petroleum.

ACKNOWLEDGMENT

Lisa M. Gieg provided invaluable input on an earlier version of the chapter.

REFERENCES

Abbott BJ, Casida LE Jr. 1968. Oxidation of alkanes to internal monoalkenes by a *Nocardia*. *J Bacteriol* 96:925–930.

Abbott BJ, Hou CT. 1973. Oxidation of 1-alkanes to 1,2epoxy-alkanes by *Pseudomonas oleovorans*. *Appl Microbiol* 26:86–91.

Abrams JW, Nedwell DB. 1978. Inhibition of methanogenesis by sulfate reducing bacteria competing for transferred hydrogen. *Arch Microbiol* 117:89–92.

Abu Laban N, Selesi D, Rattei T, Tischler P, Meckenstock RU. 2010. Identification of enzymes involved in anaerobic benzene degradation by a strictly anaerobic iron-reducing enrichment. *Environ Microbiol* 12:2783–2796.

Adkins JP, Tanner RS, Udegbunam EO, McInerney MJ, Knapp RM. 1992. Microbially enhanced oil recovery from unconsolidated limestone cores. *Geomicrobiol J* 10:77–86.

Aeckersberg F, Bak F, Widdel F. 1991. Anaerobic oxidation of saturated hydrocarbons to CO_2 by a new type of sulfate-reducing bacterium. *Arch Microbiol* 156:5–14.

Agrawal A, Park HS, Nathoo S, Gieg LM, Jack TR, Miner K, Ertmoed R, Benko A, Voordouw G. 2012. Toluene depletion in produced oil contributes to souring control in a field subjected to nitrate injection. *Environ Sci Technol* 46:1285–1292.

Ahearn DG, Meyers SP, Standard PG. 1971. The role of yeasts in the decomposition of oils in marine environments. *Dev Ind Microbiol* 12:126–134.

Aitken CM, Jones DM, Larter SR. 2004. Anaerobic hydrocarbon biodegradation in deep subsurface oil reservoirs. *Nature* 431:291–294.

Ait-Langomazino N, Sellier R, Jouquet G, Trescinski M. 1991. Microbial degradation of bitumen. *Experientia* 47:533–539.

Alexander M. 1977. *Introduction to Soil Microbiology*, 2nd ed. New York: Wiley.

Altschuler ZS, Schnepfe MM, Silber CC, Simon FO. 1983. Sulfur diagnosis in Everglades peat and origin of pyrite in coal. *Science* 221:221–227.

Andrews G, Darroch M, Hansson T. 1988. Bacterial removal of pyrite from concentrated coal slurries. *Biotech Bioeng* 32:813–820.

Angell P. 1999. Understanding microbially influenced corrosion as biofilm-mediated changes in surface chemistry *Curr Opin Biotechnol* 10:269–272.

Anthony C. 1986. Bacterial oxidation of methane and methanol. *Adv Microb Physiol* 27:113–210.

Ashirov KB, Sazanova IV. 1962. Biogenic sealing of oil deposits in carbonate reservoirs. *Mikrobiologiya* 31:680–683 (Engl. transl., pp. 555–557).

Assinder SJ, Williams PA. 1990. The TOL plasmids: Determinants of the catabolism of toluene and the xylenes. *Adv Microb Physiol* 31:1–69.

Atlas RM. 1981. Microbial degradation of petroleum hydrocarbons: An environmental perspective. *Microbiol Rev* 45:180–209.

Atlas RM. 1984. *Petroleum Microbiology*. New York: McGraw-Hill.

Atlas RM. 1988. *Microbiology. Fundamentals and Applications*, 2nd ed. New York: Macmillan.

Atlas RM. 1997. *Principles of Microbiology*, 2nd ed. Boston, MA: WCB McGraw-Hill.

Atlas RM, Bartha R. 1998. *Microbial Ecology. Fundamentals and Applications*, 4th ed. Menlo Park, CA: Benjamin Cummings.

Atlas RM, Cerniglia CE. 1995. Bioremediation of petroleum pollutants. *BioScience* 45:332–338.

Bagdigian RM, Myerson AS. 1986. The adsorption of *Thiobacillus ferrooxidans* on coal surfaces. *Biotech Bioeng* 28:467–479.

Bak F, Widdel F. 1986a. Anaerobic degradation of phenol and phenol derivatives by *Desulfobacterium phenolicum* sp. nov. *Arch Microbiol* 146:177–180.

Bak F, Widdel F. 1986b. Anaerobic degradation of indolic compounds by sulfate-reducing enrichment cultures, and description of *Desulfobacterium indolicum* gen. nov. sp. nov. *Arch Microbiol* 146:170–176.

Baldi F, Clark T, Pollack SS, Olson GJ. 1992. Leaching pyrites of various reactivities by *Thiobacillus ferrooxidans*. *Appl Environ Microbiol* 58:1853–1856.

Barna PK, Bhagat SD, Pillai KR, Singh HD, Branah JN, Iyengar MS. 1970. Comparative utilization of paraffins by a *Trichosporon* species. *Appl Microbiol* 20:657–661.

Bastin E. 1926. Microorganisms in oilfields. *Science* 63:21–24.

Beeder J, Nilsen RK, Rosnes JT, Torsvik T, Lien T. 1994. Archaeoglobus fulgidus isolated from hot North Sea oil field waters. *Appl Environ Microbiol* 60:1227–1231.

Beeder J, Torsvik T, Lien T. 1995. Thermodesulforhabdus norvegicus gen. nov., sp. nov., a novel thermophilic sulfate-reducing bacterium from oil field water. *Arch Microbiol* 164:331–336.

Beerstecher E. 1954. *Petroleum Microbiology*. Houston, TX: Elsevier.Belyaev SS, Borzenkov IA, Milekhina EI, Charakhch'yan IA, Ivanov MV. 1990a. Development of microbiological processes in reservoirs of the Romashkino oilfield. *Mikrobiologiya* 59:1118–1126 (Engl. transl., pp. 786–792).

Belyaev SS, Rozanova EP, Borzenkov IA, Charakhch'yan IA, Miller YuM, Sokolov My, Ivanov MV. 1990b. Characteristics of microbiological processes in a water-flooded oilfield in the middle Ob' region. *Mikrobiologiya* 59:1075–1081 (Engl. transl., pp. 754–759).

Bennett B, Adams JJ, Gray ND, Sherry A, Oldenburg TBP, Huang H, Larter SR, Head IM. 2013. The controls on the composition of biodegraded oils in the deep subsurface—Part 3. The Impact of microorganism distribution on petroleum geochemical gradients in biodegraded petroleum reservoirs. *Org Geochem* 56:94–105.

Berg IA, Kockelkorn D, Ramos-Vera WH, Say RF, Zarzycki J, Hügler M, Alber BE, Fuchs G. 2010. Autotrophic carbon fixation in archaea *Nat Rev Microbiol* 8:447–460.

Bhatnagar L, Jain MK, Zeikus JG. 1991. Methanogenic bacteria. In: Shively JM, Barton LL (eds.) *Variations in Autotrophic Life*. London, U.K.: Academic Press, pp. 251–270.

Blaut M, Gottschalk G. 1984. Protonmotive force-driven synthesis of ATP during methane formation from molecular hydrogen and formaldehyde or carbon dioxide in *Methanosarcina barkeri*. *FEMS Microbiol Lett* 24:103–107.

Blaut M, Peinemann S, Deppenmeier U, Gottschalk G. 1990. Energy transduction in vesicles of methanogenic strain Gö 1. *FEMS Microbiol Rev* 87:367–372.

Blaut M, Müller V, Gottschalk G. 1992. Energetics of methanogenesis studied in vesicular systems. *J Bioenerg Biomembr* 24:529–546.

Blazquez ML, Ballester A, Gonzalez F, Mier JL. 1993. Coal biodesulfurization. *Biorecovery* 2:155–157.

Bochem HP, Schoberth SM, Sprey B, Wengler P. 1982. Thermophilic biomethanation of acetic acid: Morphology and ultrastructure of granular consortium. *Can J Microbiol* 28:500–510.

Bock A-K, Prieger-Kraft A, Schönheit P. 1994. Pyruvate: A novel substrate for growth and methane formation in *Methanosarcina barkeri*. *Arch Microbiol* 161:33–46.

Boetius A, Ravenschlag K, Schubert CJ, Rickert D, Widdel F, Gieseke A, Amann R, Jørgensen BB, Witte U, Pfannkuche O. 2000. A marine microbial consortium apparently mediating anaerobic oxidation of methane. *Nature* (London) 407:623–626.

Bonch-Osmolovskaya EA, Miroshnichenko ML, Lebedinsky AV, Chernyh NA, Nazina TN, Ivoilov VS, Belyaev SS, Boulygina ES, Lysov YP, Perov AN, Mirzabekov AD, Hippe H, Stackebrandt E, L'Haridon S, Jeanthon C. 2003. Radioisotopic, culture-based, and oligonucleotide microchip analyses of thermophilic microbial communities in a continental high-temperature petroleum reservoir. *Appl Environ Microbiol* 69:6143–6151.

Boone DR, Whitman WB, Rouvière P. 1993. Diversity and taxonomy of methanogens. In: Ferry JG (ed.) *Methanogenesis*. New York: Chapman & Hall, pp. 35–89.

Bos P, Kuenen JG. 1990. Microbial treatment of coal. In: Ehrlich HL, Brierley CL (eds.) *Microbial Mineral Recovery*. New York: McGraw-Hill, pp. 343–377.

Bowen HJM. 1979. *Environmental Chemistry of the Elements*. London, U.K.: Academic Press.

Brakstad OG, Kotlar HK, Markussen S. 2008. Microbial communities of a complex high-temperature offshore petroleum reservoir. *Int J Oil Gas Coal Technol* 1:211–228.

Braus-Stromeyer SA, Hermann R, Cook AM, Leisinger T. 1993. Dichloromethane as the sole carbon source for an acetogenic mixed culture and isolation of a fermentative dichloromethane-degrading bacterium. *Appl Environ Microbiol* 59:3790–3797.

Breznak JA. 1982. Intestinal microbiota of termites and other xylophagus insects. *Annu Rev Microbiol* 36:323–343.

Brock TD, Madigan MT. 1988. *Biology of Microorganisms*, 5th ed. Englewood Cliffs, NJ: Prentice-Hall.

Brooks JM, Kennicutt MC II, Fisher CR, Macko SA, Cole K, Childress JJ, Bridigare RR, Vetter RD. 1987. Deep-sea hydrocarbon seep communities: Evidence for energy and nutritional carbon sources. *Science* 238:1138–1142.Brown DA, Overend RP. 1993. Methane metabolism in raised bogs of northern wetlands. *Geomicrobiol J* 11:35–48.

Brown LR. 2007. The relationship of microbiology to the petroleum industry. *SIM News* 57:180–190.

Bryant MP, Campbell LL, Reddy CA, Crabill MR. 1977. Growth of *Desulfovibrio* in lactate or methanol media low in sulfate in association with H_2-utilizing methanogenic bacteria. *Appl Environ Microbiol* 33:1162–1169.

Bureau of Mines. 1965. *Mineral Facts and Problems*. Bulletin 630. Washington, DC: Bureau of Mines, U.S. Department of the Interior.

Callaghan AV. 2013. Enzymes involved in the anaerobic oxidation of n-alkanes: From methane to long-chain paraffins *Front Microbiol* 4:89.

Caparello DM, LaRock PA. 1975. A radioisotope assay for quantification of hydrocarbon biodegradation potential in environmental samples. Microb Ecol 2:28–42.

Casagrande D, Siefert K. 1977. Origins and sulfur in coal: Importance of the ester sulfate content in peat. Science 195:675–676.

Cavanaugh CM, Levering PR, Maki JS, Mitchell R, Lidstrom ME. 1987. Symbiosis of methylotrophic bacteria and deep-sea mussels. Nature (London) 325:346–348.

Cayol JL, Ollivier B, Patel BK, Ravot G, Magot M, Ageron E, Grimont PA, Garcia JL. 1995. Description of Thermoanaerobacter brockii subsp lactiethylicus subsp nov., isolated from a deep subsurface French oil well, a proposal to reclassify Thermoanaerobacter finnii as Thermoanaerobacter brockii subsp fi nniicomb. nov., and an emended description of Thermoanaerobacter brockii. Int J Syst Bacteriol 45:783–789.

Cerniglia CE. 1984. Microbial metabolism of polycyclic aromatic hydrocarbons. Adv Appl Microbiol 30:31–71.

Chan OC, Wolf M, Hepperle D, Casper P. 2002. Methanogenic archaeal community in the sediment of an artificially partitioned acidic bog lake. FEMS Microbiol Ecol 42:119–129.

Chen W, Konisky J. 1993. Characterization of a membrane-associated ATPase from Methanococcus voltae, a methanogenic member of the Archaea. J Bacteriol 175:5677–5682.

Cheng L, Qiu TL, Yin XB, Wu XL, Hu GQ, Deng Y, Zhang H. 2007. Methermicoccus shengliensis gen. nov., sp. nov., a thermophilic, methylotrophic methanogen isolated from oil-production water, and proposal of Methermicoccaceae fam. nov. Int J Syst Evol Microbiol 57:2964–2969.

Cheng L, Qiu TL, Li X, Wang WD, Deng Y, Yin XB, Zhang H. 2008. Isolation and characterization of Methanoculleus receptaculi sp. nov. from Shengli oil field, China. FEMS Microbiol Lett 285:65–71.

Cheng L, Dai L, Li X, Zhang H, Lu Y. 2011. Isolation and characterization of Methanothermobacter crinale sp. nov., a novel hydrogenotrophic methanogen from the Shengli oil field. Appl Environ Microbiol 77:5212–5219.

Childress JJ, Fisher CR, Brooks JM, Kennicutt MC II, Bidigare R, Anderson AE. 1986. A methanotrophic marine molluscan (Bivalvia, Mytilidae) symbiosis: Mussel fueled by gas. Science 233:1306–1308.

Chouteau J, Azoulay E, Senez JC. 1962. Dégradation bactérienne des hydrocarbons paraffiniques. IV. Identification par spectrophotométrie infrarouge du hept-1-ene produit à partir du n-heptane par des suspensions nonproliférantes de Pseudomonas aeruginosa. Bull Soc Chim Biol 44:1670–1672.

Christensen B, Torsvik T, Lien T. 1992. Immunomagnetically captured thermophilic sulfate-reducing bacteria from North Sea oil field waters. Appl Environ Microbiol 58:1244–1248.

Clark TR, Baldi F, Olson GJ. 1993. Coal depyritization by the thermophilic archeon Metallosphaera sedula. Appl Environ Microbiol 59:2375–2379.

Cochrane WJ, Jones PS, Sanders PF, Holt DM, Mosley MJ. 1988. Studies on the thermophilic sulfatereducing bacteria from a souring North Sea oil field. SPE European Petroleum Conference, October 16–19, London, U.K., SPE 18368.

Cohen MS, Bowers WC, Aronson H, Gray ET Jr. 1987. Cell-free solubilization of coal by Polyporus versicolor. Appl Environ Microbiol 53:2840–2843.

Cohen MS, Feldman KA, Brown CS, Gray ET Jr. 1990. Isolation and identification of the coal-solubilizing agent produced by Trametes versicolor. Appl Environ Microbiol 56:3285–3291.

Cohen MS, Gabriele PD. 1982. Degradation of coal by the fungi Polyporus versicolor and Poria monticola. Appl Environ Microbiol 44:23–27.

Colberg PJS. 1990. Role of sulfate in microbial transformations of environmental contaminants: Chlorinated aromatic compounds. Geomicrobiol J 8:147–165.

Connan J. 1984. Biodegradation of crude oils in reservoirs. In: Brooks J, Welte DH (eds.) Advances in Petroleum Geochemistry, Vol. 1. London, U.K.: Academic Press, pp. 299–335.

Compere AL, Griffith WL. 1978. Production of high viscosity glucans from hydrolyzed cellulosics. Dev Ind Microbiol 19:601–607.

Crawford DL, ed. 1993. Microbial Transformations of Low Rank Coals. Boca Raton, FL: CRC Press.

Crawford DL, Gupta RK. 1990. Oxidation of dibenzothiophene by Cunninghamella elegans. Curr Microbiol 21:229–231.

Crawford DL, Gupta RK. 1993. Microbial depolymerization of coal. In: Crawford DL (ed.) Microbial Transformations of Low Rank Coals. Boca Raton, FL: CRC Press, pp. 65–92.

Dagley S. 1975. Microbial degradation of organic compounds in the biosphere. Sci Am 63:681–689.

Dahle H, Birkeland NK. 2006. Thermovirga lienii gen. nov., sp. nov., a novel moderately thermophilic, anaerobic, amino-acid-degrading bacterium isolated from a North Sea oil well. Int J Syst Evol Microbiol 56:1539–1545.

Dahle H, Garshol F, Madsen M, Birkeland NK. 2008. Microbial community structure analysis of produced water from a high-temperature North Sea oil-field. *Antonie van Leeuwenhoek* 93:37–49.

Davey ME, Wood WA, Key R, Nakamura K, Stahl DA. 1993. Isolation of three species of Geotoga and Petrotoga: Two new genera, representing a new lineage in the bacterial line of descent distantly related to the "Thermotogales". *Syst Appl Microbiol* 16:191–200.

Davies SL, Whittenbury R. 1970. Fine structure of methane and other hydrocarbon-utilizing bacteria. *J Gen Microbiol* 61:227–232.

Davis CA. 1907. Peat: Essays on the origin, uses and distribution in Michigan. In: *Plate XII from Annual Report, Geological Survey of Michigan*. Lansing, MI: Wynkoop Hallenbeck Crawford, State Printers, pp. 395.

Davis JB. 1967. *Petroleum Microbiology*. Amsterdam, the Netherlands: Elsevier.

Davis JB, Yarbrough HF. 1966. Anaerobic oxidation of hydrocarbons by *Desulfovibrio desulfuricans*. *Chem Geol* 1:137–144.

De Bruin WP, Kotterman MJ, Posthumus MA, Schraa G, Zehnder AJ. 1992. Complete biological reductive transformation of tetrachloroethene to ethane. *Appl Environ Microbiol* 58:1996–2000.

Deuser WG, Degens ET, Harvey GR. 1973. Methane in Lake Kivu: New data bearing on origin. *Science* 181:51–54.

Didyk BM, Simoneit BRT. 1989. Hydrothermal oil of Guaymas Basin and implications for petroleum formation mechanisms. *Nature (London)* 342:65–69.

DiPippo JL, Nesbo CL, Dahle H, Doolittle WF, Birkland NK, Noll KM. 2009. Kosmotoga olearia gen. nov., sp. nov., a thermophilic, anaerobic heterotroph isolated from an oil production fluid. *Int J Syst Evol Microbiol* 59:2991–3000.

Doddema HJ, Hutten TJ, van der Drift C, Vogels GD. 1978. ATP hydrolysis and synthesis by the membrane-bound ATP synthetase complexes of *Methanobacterium thermoautotrophicum*. *J Bacteriol* 136:19–23.

Doddema HJ, van der Drift C, Vogels GD, Veenhuis M. 1979. Chemiosmotic coupling in *Methanobacterium thermoautotrophicum*: Hydrogen-dependent adenosine 5′-triphosphate synthesis by subcellular particles. *J Bacteriol* 140:1081–1089.

Doelle HW. 1975. *Bacterial Metabolism*, 2nd ed. New York: Academic Press.

Dolfing J. 1990. Reductive dechlorination of 3-chlorobenzoate is coupled to ATP production and growth of anaerobic bacterium strain DCB-1. *Arch Microbiol* 153:264–266.

Donos JD, Frankenfeld JW. 1968. Oxidation of alkyl benzenes by a strain of *Micrococcus cerificans* growing on n-paraffins. *Appl Miccrobiol* 16:532–533.

Dugan PR. 1986. Microbiological desulfurization of coal and its increased monetary value. In: Ehrlich HL, Holmes DS (eds.) *Workshop on Biotechnology for the Mining, Metal-Refining and Fossil Fuel Processing Industries. Biotechnology and Bioengineering Symposium*, No 16. New York: Wiley, pp. 185–203.

Duperron S, Nadalig T, Caprais J-C, Sibuet M, Fiala-Médioni A, Amann R, Dubilier N. 2005. Dual symbiosis in a *Bathymodiolus* sp. mussel from a methane seep on the Gabon continental margin (Southeast Atlantic): 16S rRNA phylogeny and distribution of the symbionts in gills. *Appl Environ Microbiol* 71:1694–1700.

Dutova EN. 1962. The significance of sulfate-reducing bacteria in prospecting for oil as exemplified in the study of ground water in Central Asia. In: Kuznetsov SI (ed.) *Geologic Activity of Microorganisms*. New York: Consultants Bureau, pp. 76–78.

Dutton PL, Evans WC. 1969. The metabolism of aromatic compounds by *Rhodopseudomonas palustris*. A new reductive method of aromatic ring metabolism. *Biochem J* 113:525–536.

Dybas M, Konisky J. 1992. Energy transduction in the methanogen *Methanococcus voltae* is based on a sodium current. *J Bacteriol* 174:5575–5583.

Ehrenreich P, Behrends A, Harder J, Widdel F. 2000. Anaerobic oxidation of alkanes by newly isolated denitrifying bacteria. *Arch Microbiol* 173:58–64.

Ehrlich HL. 1993. Bacterial mineralization of organic carbon under anaerobic conditions. In: Bollag J-M, Stotzky G (eds.) *Soil Biochemistry*, Vol 8. New York: Marcel Dekker, pp. 219–247.

Eller G, Känel L, Krüger M. 2005. Cooccurrence of aerobic and anaerobic methane oxidation in the water column of Lake Plußsee. *Appl Environ Microbiol* 71:8925–8928.

Enning D, Garrelfs J. 2014. Corrosion of iron by sulfate-reducing bacteria: New views of an old problem *Appl Environ Microbiol* 80:1226–1236.

Ensley BD. 1991. Biochemical diversity of trichloroethylene metabolism. *Annu Rev Microbiol* 45:283–299.

Ettwig KF, Butler MK, Le Paslier D, Pelletier E, Mangenot S, Kuypers MM, Schreiber F et al. 2010. Nitrite-driven anaerobic methane oxidation by oxygenic bacteria. *Nature* 464:543–548.

Evans WC, Fuchs G. 1988. Anaerobic degradation of aromatic compounds. *Annu Rev Microbiol* 42:289–317.

Fardeau ML, Cayol JL, Magot M, Ollivier B. 1993. H_2 oxidation in the presence of thiosulfate, by a Thermoanaerobacter strain isolated from an oil-producing well. *FEMS Microbiol Lett* 113:327–332.

Fardeau ML, Ollivier B, Patel BK, Magot M, Thomas P, Rimbault A, Rocchiccioli F, Garcia JL. 1997. Thermotoga hypogeasp. nov., a xylanolytic, thermophilic bacterium from an oil-producing well. Int J Syst Bacteriol 47:1013–1019.

Fardeau ML, Magot M, Patel BK, Thomas P, Garcia JL, Ollivier B. 2000. Thermoanaerobacter subterraneussp. nov., a novel thermophile isolated from oil fi eld water. Int J Syst Evol Microbiol 50:2141–2149.

Fardeau ML, Bonilla Salinas M, L'Haridon S, Jeanthon C, Verhe F, Cayol JL, Patel BK, Garcia JL, Ollivier B. 2004. Isolation from oil reservoirs of novel thermophilic anaerobes phylogenetically related to Thermoanaerobacter subterraneus: reassignment of T. subterraneus, Thermoanaerobacter yonseiensis, Thermoanaerobacter tengcongensis and Carboxydibrachium pacificum to Caldanaerobacter subterraneus gen. nov., sp. nov., comb. nov. as four novel subspecies. Int J Syst Evol Microbiol 54:467–474.

Fenchel T, Blackburn TH. 1979. Bacteria and Mineral Cycling. London, U.K.: Academic Press.

Fenchel T, Finlay BJ. 1992. Production of methane and hydrogen by anaerobic ciliates containing symbiotic methanogens. Arch Microbiol 157:475–480.

Finn MW, Tabita FR. 2003. Synthesis of catalytically active form of III ribulose 1,5-bisphosphate carboxylase/oxidase in archaea. J Bacteriol 185:3049–3059.

Finn MW, Tabita FR. 2004. Modified pathway to synthesize ribulose 1,5-bisphosphate in the methanogenic Archaea. J Bacteriol 186:6360–6366.

Finnerty WR, Robinson M. 1986. Microbial desulfurization of fossil fuels: A review. In: Ehrlich HL, Holmes DS (eds.) Workshop on Biotechnology for the Mining, Metal-Refining and Fossil Fuel Processing Industries. Biotechnology and Bioengineering Symposium, No 16. New York: Wiley, pp. 205–221.

Finnerty WR, Singer ME, King AD. 1984. Microbial processes and the recovery of heavy petroleum. In: Meyer RF, Wynne JC, Olson JC (eds.) Future Heavy Crude Tar Sands. 2nd International Conference. New York: McGraw-Hill, pp. 424–429.

Finster K, Tanimoto Y, Bak F. 1992. Fermentation of methanethiol and dimethylsulfide by a newly isolated methanogenic bacterium. Arch Microbiol 157:425–430.

Foght JM, Fedorak PM, Gray MR, Westlake DWS. 1990. Microbial desulfurization of petroleum. In: Ehrlich HL, Brierley CL (eds.) Microbial Mineral Recovery. New York: McGraw-Hill, pp. 379–407.

Foster JW. 1962. Hydrocarbons as substrates for microorganisms. Antonie v Leeuwenhoek 28:241–274.

Francis W. 1954. Coal: Its Formation and Composition. London, U.K.: Edward Arnold.

Fredricks KM. 1967. Products of the oxidation of n-decane by Pseudomonas aeruginosa and Mycobacterium rhodochrous. Antonie v Leeuwenhoek 33:41–48.

Fredrickson JK, Stewart DL, Campbell JA, Powell MA, McMulloch M, Pyne JW, Bean RM. 1990. Biosolubilization of low-rank coal by Trametes versicolor siderophore-like product and complexing agents. J Ind Microbiol 5:401–406.

Freedman DL, Gossett JM. 1989. Biological reductive dechlorination of tetrachloroethylene and trichloroethylene under methanogenic conditions. Appl Environ Microbiol 55:2144–2151.

Frimmer U, Widdel F. 1989. Oxidation of ethanol by methanogenic bacteria. Arch Microbiol 152:479–483.

Galand PE, Fritze H, Conrad R, Yrjälä K. 2005. Pathways for methanogenesis and diversity of methanogenic Archaea in three boreal peatland ecosystems. Appl Environ Microbiol 71:2195–2198.

Gal'chenko VF, Gorlatov SN, Tokarev VG. 1986. Microbial oxidation of methane in Bering Sea sediments. Mikrobiologiya 55:669–673 (Engl. transl., pp. 526–530).

Gescher J, Zaar A, Mohamed M, Schagger H, Fuchs G. 2002. Genes coding for a new pathway of aerobic benzoate metabolism in Azoarcus evansii. J Bacteriol 184:6301–6315.

Giani D, Giani L, Cohen Y, Krumbein WE. 1984. Methanogenesis in the hypersaline Solar Lake (Sinai). FEMS Microbiol Lett 25:219–224.

Gieg LM, Duncan KE, Suflita JM. 2008. Bioenergy production via microbial conversion of residual oil to natural gas. Appl Environ Microbiol 74:3022–3029.

Gieg LM, Jack TR, Foght JM. 2011. Biological souring and mitigation in oil reservoirs. Appl Microbiol Biotechnol 92:263–282.

Gold T, Gordon BE, Streett W, Bilson E, Panaik P. 1986. Experimental study of the reaction of methane with petroleum hydrocarbons in geological conditions. Geochim Cosmochim Acta 50:2411–2418.

Gorlatov SN, Belyaev SS. 1984. The aerobic microflora of an oil field and its ability to destroy petroleum. Mikrobiologiya 53:843–849 (Engl. transl., pp. 701–706).

Gorlatov SN, Gal'chenko VF, Tokarev VG. 1986. Microbiological methane formation in deposits of the Bering Sea. Mikrobiologiya 55:490–495 (Engl. transl., pp. 380–385).

Gottschalk G. 1986. Bacterial Metabolism, 2nd ed. New York: Springer.

Gottschalk G, Blaut M. 1990. Generation of proton and sodium motive forces in methanogenic bacteria. *Biochem Biophys Acta* 1018:263–266.

Grabowski A, Nercessian O, Fayolle F, Blanche D, Jeanthon C. 2005. Microbial diversity in production waters of a low-temperature biodegraded oil reservoir. *FEMS Microbiol Ecol* 54:427–443.

Grassia GS, Mclean KM, Glenat P, Bauld J, Sheehy AJ. 1996. A systematic survey for thermophilic fermentative bacteria and archaea in high-temperature petroleum reservoirs. *FEMS Microbiol Ecol* 21:47–58.

Gray ND, Sherry A, Grant RJ, Rowan AK, Hubert CRJ, Callbeck CM, Aitken CM et al. 2011. The quantitative significance of Syntrophaceae and syntrophic partnerships in methanogenic degradation of crude oil alkanes. *Environ Microbiol* 13:2957–2975.

Gray ND, Sherry A, Hubert C, Dolfing J, Head IM. 2010. Methanogenic degradation of petroleum hydrocarbons in subsurface environments: Remediation, heavy oil formation, and energy recovery. *Adv Appl Microbiol* 72:137–161.

Grbic-Galic D. 1990. Methanogenic transformation of aromatic hydrocarbons and phenols in groundwater aquifers. *Geomicrobiol J* 8:167–200.

Greene AC, Patel BK, Sheehy AJ. 1997. Deferribacter thermophilusgen. nov., sp. nov., a novel thermophilic manganese- and iron-reducing bacterium isolated from a petroleum reservoir. *Int J Syst Bacteriol* 47:505–509.

Grossi V, Cravo-Laureau C, Guyoneaud R, Ranchou-Peyruse A, Hirschler-Réa A. 2008. Metabolism of n-alkanes and n-alkenes by anaerobic bacteria: A summary. *Organic Geochem* 39:1197–1203.

Grossman MJ, Lee MK, Prince RC, Minak-Bernero V, George GN, Pickering IJ. 2001. Deep desulfurization of extensively hydrodesulfurized middle distillate oil by *Rhodococcus* sp. strain ECRD-1. *Appl Environ Microbiol* 67:1949–1952.

Hallam SJ, Putnam N, Preston CM, Detter JC, Rokshar D, Richardson PM, DeLong EF. 2004. Reverse methanogenesis: Testing the hypothesis with environmental genomics. *Science* 305:1457–1462.

Hansen LB, Finster K, Fossing H, Iversen N. 1998. Anaerobic methane oxidation in sulfur depleted sediments: Effects of sulfate and molybdate additions. *Aquat Microb Ecol* 14:195–204.

Hao R, Lu A, Wang G. 2004. Crude-oil-degrading thermophilic bacterium isolated from an oil field. *Can J Microbiol* 50:175–182.

Hattori S. 2008. Syntrophic acetate-oxidizing microbes in methanogenic environments. *Microbes Environ* 23:118–127.

Haq BU. 1999. Methane in the deep blue sea. *Science* 285:543–544.

Head IM, Jones DM, Larter SR. 2003. Biological activity in the deep subsurface and the origin of heavy oil. *Nature* 426:344–352.

Henrichs SM, Reeburgh WS. 1987. Anaerobic mineralization of organic matter: Rates and the role of anaerobic processes in the oceanic carbon economy. *Geomicrobiol J* 5:191–237.

Higgins IJ, Best DJ, Hammond RC, Scott D. 1981. Methane-oxidizing microorganisms. *Microbiol Rev* 45:556–590.

Higson FK. 1992. Microbial degradation of biphenyl and its derivatives. *Adv Appl Microbiol* 37:135–164.

Hinrichs K-U, Hayes JM, Sylva SP, Brewer PG, DeLong EF. 1999. Methane-consuming archaebacteria in marine sediments. *Nature* (*London*) 398:802–805.

Hoehler TM, Alperin MJ, Albert DB, Martens CS. 1994. Field and laboratory studies of methane oxidation in an anoxic marine sediment: Evidence for a methanogen-sulfur reducer consortium. *Global Biogeochem Cycles* 8:451–463.

Hoehler TM, Alperin MJ, Albert DB, Martens CS. 1998. Thermodynamic control on hydrogen concentrations in anoxic sediments. *Geochim Cosmochim Acta* 62:1745–1756.

Hölker U, Schmiers H, Große S, Winkelhöfer M, Polsakiewicz M, Ludwig S, Dohse J, Höfer M. 2002. Solubilization of low-rank coal by *Trichoderma atroviride*: Evidence for the involvement of hydrolytic and oxidative enzymes by using ^{14}C-labelled lignite. *J Ind Microbiol Biotechnol* 28:207–212.

Horita J, Berndt ME. 1999. Abiogenic methane formation and isotope fractionation under hydrothermal conditions. *Science* 285:1055–1057.

Horz H-P, Raghubanshi AS, Heyer J, Kamman C, Conrad R, Dunfield PF. 2002. Activity and community structure of methane-oxidising bacteria in a wet meadow soil. *FEMS Microbiol Ecol* 41:247–257.

Horvath RS. 1972. Microbial catabolism and the degradation of organic compounds in nature. *Bacteriol Rev* 36:146–155.

Huber H, Thomm M, König G, Thies G, Stetter KO. 1982. *Methanococcus thermolithotrophicus*, a novel thermophilic lithotrophic methanogen. *Arch Microbiol* 132:47–50.

Hubert C. 2010. Microbial ecology of oil reservoir souring control by nitrate injection. In: Timmis KN (ed.) *Handbook of Hydrocarbon and Lipid Microbiology*. Berlin, Germany: Springer, pp. 2753–2766.

Hubert C, Judd A. 2010. Using microorganisms as prospecting agents in oil and gas exploration. In: Timmis KN (ed.) *Handbook of Hydrocarbon and Lipid Microbiology*. Berlin, Germany: Springer, pp. 2711–2725.

Hubert CRJ, Oldenburg TBP, Fustic M, Gray ND, Larter SR, Penn K, Rowan AK et al. 2012. Massive dominance of *Epsilonproteobacteria* in formation waters from a Canadian oil sands reservoir containing severely biodegraded oil. *Environ Microbio* 2:387–404.

Hunt JM. 1984. Generation and migration of light hydrocarbons. *Science* 226:1265–1270.

Hunt JM, Whelan JK, Hue AY. 1980. Genesis of petroleum hydrocarbons in marine sediments. *Science* 209:403–404.

Huser BA, Wuhrmann K, Zehnder AJB. 1982. *Methanothrix soehngii* gen. nov. spec. nov. a new acetotrophic non-hydrogen-oxidizing methane bacterium. *Arch Microbiol* 132:1–9.

Hyun J-H, Bennison BW, LaRock PA. 1997. The formation of large aggregates at depth within the Louisiana hydrocarbon seep zone. *Microb Ecol* 33:216–222.

Illias RMD, Wei OS, Idris AK, Rahman WAWA. 2001. Isolation and characterization of halotolerant aerobic bacteria from oil reservoir. *Jurnal Teknologi* 35:1–10.

Imhoff-Stuckle D, Pfennig N. 1983. Isolation and characterization of a nicotinic acid degrading sulfate-reducing bacterium, *Desulfococcus niacini* sp. nov. *Arch Microbiol* 136:194–198.

Inagaki F, Tsunogai U, Suzuki M, Kosaka A, Machiyama H, Takai K, Nunora T, Nealson KH, Horikoshi K. 2004. Characterization of C_1-metabolizing prokaryotic communities in methane seep habitats at the Kuroshima Knoll, Southern Ryukyu Arc, by analyzing *pmoA*, *mmoX*, *mxaF*, *mcrA*, and 16S rRNA genes. *Appl Environ Microbiol* 70:7445.

Ivanov MV. 1967. The development of geological microbiology in the U.S.S.R. *Microbiologiya* 36:849–859 (Engl. transl., pp. 751–722).

Ivanov MV, Nesterov AI, Nasaraev GB, Gal'chenko VF, Nazarenko AV. 1978. Distribution and geochemical activity of methanotrophic bacteria in coal mine waters. *Mikrobiologiya* 47:489–494 (Engl. transl., pp. 396–401).

Iversen N, Jørgensen BB. 1985. Anaerobic methane oxidation rates at the sulfate-methane transition in marine sediments from Kattegat and Skagerrak (Denmark). *Limnol Oceanogr* 30:944–955.

Izumi Y, Ohshiro T, Ogino H, Hine Y, Shimao M. 1994. Selective desulfurization of dibenzothiophene by *Rhodococcus erythropolis* D-1. *Appl Environ Microbiol* 60:223–226.

Jain MK, Zeikus JG. 1989. Bioconversion of gelatin to methane by a coculture of *Clostridium collagenovorans* and *Methanosarcina barkeri*. *Appl Environ Microbiol* 55:366–371.

Jakobsen P, Patrick WH Jr, Williams BG. 1981. Sulfide and methane formation in soils and sediments. *Soil Sci* 132:279–287.

Jayasinghearachchi HS, Lal B. 2011. *Oceanotoga teriensisgen*. nov., sp. nov., a thermophilic bacterium isolated from offshore oil-producing wells. *Int J Syst Evol Microbiol* 61:554–560.

Jeanthon C, Reysenbach AL, l'Haridon S, Gambacorta A, Pace NR, Glenat P, Prieur D. 1995. *Thermotoga subterraneasp*. nov., a new thermophilic bacterium isolated from a continental oil reservoir. *Arch Microbiol* 164:91–97.

Jobson AM, Cook FD, Westlake DWS. 1979. Interaction of aerobic and anaerobic bacteria in petroleum biodegradation. *Chem Geol* 24:355–365.

Johnson EL, Hyman MR. 2006. Propane and n-butane oxidation by *Pseudomonas putida* GPo1. *Appl Environ Microbiol* 72:950–952.

Johnson MJ. 1964. Utilization of hydrocarbons by microorganism. *Chem Ind* 36:1532–1537.

Jones DM, Head IM, Gray ND, Adams JJ, Rowan A, Aitken C, Bennett B et al. 2008. Crude-oil biodegradation via methanogenesis in subsurface petroleum reservoirs. *Nature* 451:176–180.

Jones JG, Simon BM, Gardener S. 1982. Factors affecting methanogenesis and associated anaerobic processes in the sediments of a stratified eutrophic lake. *J Gen Microbiol* 128:1–11.

Jones RD, Morita RY. 1983. Methane oxidation by *Nitrosococcus oceanus* and *Nitrosomonas europaea*. *Appl Environ Microbiol* 45:401–410.

Jones WJ, Leigh JA, Mayer F, Woese CR, Wolfe RS. 1983a. *Methanococcus jannaschii* sp. nov., an extremely thermophilic methanogen from a submarine hydrothermal vent. *Arch Microbiol* 136:254–261.

Jones WJ, Nagle DP Jr, Whitman RB. 1987. Methanogens and the diversity of archaebacteria. *Microbiol Rev* 51:135–177.

Jones WJ, Paynter MJB, Gupta R. 1983b. Characteristics of *Methanococcus maripaludis* sp. nov., a new methanogen from salt marsh sediment. *Arch Microbiol* 135:91–97. Kallmeyer J, Boetius A. 2004. Effects of temperature and pressure on sulfate reduction and anaerobic oxidation of methane in hydrothermal sediment of Guaymas Basin. *Appl Environ Microbiol* 70:1231–1233.

Karakashev D, Batstone DJ, Trably E, Angelidaki I. 2006. Acetate oxidation is the dominant methanogenic pathway from acetate in the absence of *Methanosaetaceae*. *Appl Environ Microbiol* 72:5138–5141.

Kashefi K, Lovley DR. 2003. Extending the upper temperature limit for life. *Science* 301:934.

Kaster KM, Grigoriyan A, Jenneman G, Voordouw G. 2007. Effect of nitrate and nitrite on sul fi de production by two thermophilic, sulfate-reducing enrichments from an oil field in the North Sea. *Appl Microbiol Biotechnol* 75:195–203.

Kaster KM, Bonaunet K, Berland H, Kjeilen-Eilertsen G, Brakstad OG. 2009. Characterisation of culture-independent and dependent microbial communities in a high-temperature offshore chalk petroleum reservoir. *Antonie van Leeuwenhoek* 96:423–439.

Khelifi N, Ali OA, Roche P, Grossi V, Brochier-Armanet C, Valette O, Ollivier B, Dolla A, Hirschler-Réa A. 2014. Anaerobic oxidation of long-chain n-alkanes by the hyperthermophilic sulfate-reducing archaeon, *Archaeoglobus fulgidus*. *ISME J* 8:2153–2166.

King GM. 1984. Metabolism of trimethylamine, choline, and glycine betaine by sulfate-reducing and methanogenic bacteria in marine sediments *Appl Environ Microbiol* 4:719–725.

Klein DA, Davis JA, Casida LE. 1968. Oxidation of n-alkanes to ketones by an *Arthrobacter* species. *Antonie v Leeuwenhoek* 34:495–503.

Knittel K, Boetius A. 2009. Anaerobic oxidation of methane: Progress with an unknown process *Ann Rev Microbiol* 63:311–334.

Koburger JA. 1964. Microbiology of coal: Growth of bacteria in plain and oxidized coal slurries. (39th Annual Session of W Virginia Academic Science Proceedings) *West Virginia Acad Sci* 36:26–30.

Koizumi Y, Takii S, Nishino M, Nakajima T. 2003. Vertical distributions of sulfate-reducing bacteria and methane-producing archaea quantified by oligonucleotide probe hybridization in the profundal sediment of a mesotrophic lake. *FEMS Microbiol Ecol* 44:101–108.

Korenblum E, Souza DB, Penna M, Seldin L. 2012. Molecular analysis of the bacterial communities in crude oil samples from two Brazilian offshore petroleum platforms. *Int J Microbiol* 2012:156537.

Kotelnikova S, Pedersen K. 1998. Distribution and activity of methanogens and homoacetogens in deep granitic aquifers at Äspö Hard Rock Laboratory, Sweden. *FEMS Microbiol Ecol* 26:121–134.

Kotlar HK, Lewin A, Johansen J, Throne-Holst M, Haverkamp T, Markussen S, Winnberg A, Ringrose P, Aakvik T, Ryeng E, Jakobsen K, Drabløs F, Valla S. 2011. High coverage sequencing of DNA from microorganisms living in an oil reservoir 2.5 kilometres subsurface. *Environ Microbiol Rep* 3:674–681.

Kristjansson JK, Schönheit P, Thauer RK. 1982. Different K_s values for hydrogen of methanogenic and sulfate- reducing bacteria: An explanation for the apparent inhibition of methanogenesis by sulfate. *Arch Microbiol* 131:278–282.

Kropp KG, Davidova IA, Suflita JM. 2000. Anaerobic oxidation of n-dodecane by an enrichment reaction in a sulfate-reducing bacterial enrichment culture. *Appl Environ Microbiol* 66:5393–5398.

Kulm LD, Suess E, Moore JC, Carson B, Lewis BT, Ritger SD, Kadko DC et al. 1986. Oregon subduction zone: Venting, fauna, and carbonates. *Science* 231:561–566.

Kuznetsov SI, Ivanov MV, Lyalikova NN. 1963. *Introduction to Geological Microbiology*. New York: McGraw-Hill, pp. 79–81.

Kuznetsova VA, Gorlenko VM. 1965. The growth of hydrocarbon-oxidizing bacteria under anaerobic conditions. *Prikl Biokhim Mikrobiol* 1:623–626.

Kvasnikov EI, Lipshits VV, Zubova NV. 1973. Facultative anaerobic bacteria of producing petroleum wells. *Mikrobiologiya* 42:925–930 (Engl. transl., pp. 823–827).

Kvenvolden KA. 1988. Methane hydrate—A major reservoir of carbon in the shallow geosphere. *Chem Geol* 71:41–51.

Larsson L, Olsson G, Holst O, Karlsson HT. 1990. Pyrite oxidation by thermophilic archaebacteria. *Appl Environ Microbiol* 56:697–701.

Lan G, Li Z, Zhang H, Zou C, Qiao D, Cao Y. 2011. Enrichment and diversity analysis of the thermophilic microbes in a high temperature petroleum reservoir. *Afr J Microbiol Res* 5:1850–1857.

Larsen J, Kim Rasmussen K, Pedersen H, Sørensen K, Lundgaard T, Skovhus TL. 2010. Consortia of MIC bacteria and archaea causing pitting corrosion in top side oil production facilities. In: NACE-10252, Corrosion 2010, 14–18 March, San Antonio, TX.

Larter S, Wilhelms A, Head I, Koopmans M, Aplin A, Di Primio R, Zwach Z, Erdmann M, Telnaes N. 2003. The controls on the composition of biodegraded oils in the deep subsurface—Part 1: Biodegradation rates in petroleum reservoirs. *Org Geochem* 34:601–613.

L'Haridon S, Reysenbacht AL, Glenat P, Prieur D, Jeanthon C. 1995. Hot subterranean biosphere in a continental oil reservoir. *Nature* 377:223–224.

L'Haridon SL, Miroshnichenko ML, Hippe H, Fardeau ML, Bonch-Osmolovskaya E, Stackebrandt E, Jeanthon C. 2001. Thermosipho geoleisp. nov., a thermophilic bacterium isolated from a continental petroleum reservoir in Western Siberia. *Int J Syst Evol Microbiol* 51:1327–1334.

L'Haridon S, Miroshnichenko ML, Hippe H, Fardeau ML, Bonch-Osmolovskaya EA, Stackebrandt E, Jeanthon C. 2002. Petrotoga olearia sp. nov. and Petrotoga sibirica sp. nov., two thermophilic bacteria isolated from a continental petroleum reservoir in Western Siberia. Int J Syst Evol Microbiol 52:1715–1722.

Leadbetter ER, Foster JW. 1959. Oxidation products formed from gaseous alkanes by the bacterium Pseudomonas methanica. Arch Biochem Biophys 82:491–492.

Leu JY, McGovern-Traa CP, Porter AJ, Harris WJ, Hamilton WA. 1998. Identification and phylogenetic analysis of thermophilic sulfate-reducing bacteria in oil fi eld samples by 16S rDNA gene cloning and sequencing. Anaerobe 4:165–174.

Leu JY, McGovern-Traa CP, Porter AJ, Hamilton WA. 1999. The same species of sulphate-reducing Desulfomicrobium occur in different oil field environments in the north sea. Lett Appl Microbiol 29:246–252.

Li D, Hendry P. 2008. Microbial diversity in petroleum reservoirs. Microbiol Aust 29:25–27.

Li Q, Li L, Rejtar T, Lessner DJ, Karger BL, Ferry JG. 2006. Electron transport in the pathway of acetate conversion to methane in the marine archaeon Methanosarcina acetivorans. J Bacteriol 188:702–710.

Li H, Yang SZ, Mu BZ. 2007a. Phylogenetic diversity of the archaeal community in a continental high-temperature, water-flooded petroleum reservoir. Curr Microbiol 55:382–388.

Li H, Yang SZ, Mu BZ, Rong ZF, Zhang J. 2007b. Molecular phylogenetic diversity of the microbial community associated with a high-temperature petroleum reservoir at an offshore oil field. FEMS Microbiol Ecol 60:74–84.

Li D, Midgley DJ, Ross JP, Oytam Y, Abell GC, Volk H, Daud WA, Hendry P. 2012. Microbial biodiversity in a Malaysian oil field and a systematic comparison with oil reservoirs worldwide. Arch Microbiol 194:513–523.

Lien T, Madsen M, Rainey FA, Birkeland NK. 1998. Petrotoga mobilis sp. nov., from a North Sea oil-production well. Int J Syst Bacteriol 48:1007–1013.

Little BJ, Mansfeld FB, Arps PJ, Earthman JC. 2007. Microbiologically influenced corrosion. In Bard AJ, Stratmann M, Frankel GS, eds. Encyclopedia of Electrochemistry, vol 4. Corrosion and Oxide Films. Wiley-VCH Verlag GmbH & Co. KGaA, Weinheim, Germany, pp. 662–685.

Lösekann T, Knittel K, Nadalig T, Fuchs B, Niemann H, Boetius A, Amann R. 2007. Diversity and abundance of aerobic and anaerobic methane oxidizers at the Haakon Mosby Mud Volcano, Barents Sea. Appl Environ Microbiol 73:3348–3362.

Lovley DR. 1987. Organic matter mineralization with the reduction of ferric iron: A review. Geomicrobiol J 5:375–399.

Lovley DR. 1993. Dissimilatory metal reduction. Annu Rev Microbiol 47:263–290.

Lovley DR, Phillips EJP. 1988. Novel mode of microbial energy metabolism: Organic carbon oxidation coupled to dissimilatory reduction of iron and manganese. Appl Environ Microbiol 54:1472–1480.

MacDonald IR, Reilly JF, Guinasso NL Jr, Brooks JM, Carney RS, Bryant WA, Bright TJ. 1990. Chemosynthetic mussels at a brine-filled pockmark in the northern Gulf of Mexico. Science 248:1096–1099.

MacLeod FA, Guiot SR, Costerton JW. 1990. Layered structure of bacterial aggregates produced in an upflow anaerobic sludge bed and filter reactor. Appl Environ Microbiol 56:1598–1607.

Madigan MT, Gest H. 1978. Growth of a photosynthetic bacterium anaerobically in darkness, supported by "oxidant-dependent" sugar fermentation. Arch Microbiol 117:119–122.

Madigan MT, Martinko JM, Bender KS, Buckley DH, Stahl DA. 2014. Brock Biology of Microorganisms, 14th ed. Upper Saddle River, New Jersey: Benjamin Cummings, 1136pp.

Magot M. 2005. Indigenous microbial communities in oil fields. In: B Ollivier, M Magot (eds.) Petroleum Microbiology. Washington, DC: ASM Press

Magot M, Ollivier B, Patel BK. 2000. Microbiology of petroleum reservoirs. Antonie van Leeuwenhoek 77:103–116.

Mah RA, Smith MR, Baresi L. 1978. Studies on an acetate-fermenting strain of Methanosarcina. Appl Environ Microbiol 35:1174–1184.

Mancini SA, Ulrich AC, Lacrampe-Couloume G, Sleep B, Edwards EA, Sherwood Lollar B. 2003. Carbon and hydrogen isotopic fractionation during anaerobic biodegradation of benzene. Appl Environ Microbiol 69:191–198.

Mango FD. 1992. Transition metal catalysis in the generation of petroleum and natural gas. Geochim Cosmochim Acta 56:3851–3854.

Marshall AG, Rodgers RP. 2004. Petroleomics: The next grand challenge for chemical analysis. Acc Chem Res 37:53–59.

Martens CS. 1976. Control of methane sediment-water bubble transport by macroinfaunal irrigation in Cape Lookout Bight, North Carolina. Science 192:998–1000.

Martens CS, Berner RA. 1974. Methane production in the interstitial waters of sulfate-depleted marine sediments. Science 185:1167–1169.

Martens CS, Klump JV. 1984. Biogeochemical cycling in an organic-rich coastal marine basin. 4. An organic carbon budget for sediments dominated by sulfate reduction and methanogenesis. *Geochim Cosmochim Acta* 48:1987–2004.

Maymó-Gatell X, Anguish T, Zinder SH. 1999. Reductive dechlorination of chlorinated ethenes and 1,2-dichloroethane by "*Dehalococcoides ethenogenes*" 195. *Appl Environ Microbiol* 65:3108–3113.

Mbadinga SM, Li KP, Zhou L, Wang LY, Yang SZ, Liu JF, Gu JD, Mu BZ. 2012. Analysis of alkane dependent methanogenic community derived from production water of a high-temperature petroleum reservoir. *Appl Microbiol Biotechnol* 96:531–542.

McCarty PL. 1993. In situ bioremediation of chlorinated solvents. *Curr Opin Biotechnol* 4:323–330.

McInerney MJ, Bryant MP, Pfennig N. 1979. Anaerobic bacterium that degrades fatty acids in syntrophic association with methanogens. *Arch Microbiol* 122:129–135.

McInerney MJ, Westlake DWS. 1990. Microbially enhanced oil recovery. In: Ehrlich HL, Brierley CL (eds.) *Microbial Mineral Recovery*. New York: McGraw-Hill, pp. 409–445.

McKenna EJ, Kallio RE. 1964. Hydrocarbon structure: Its effect on bacterial utilization of alkanes. In: Heukelakian H, Dondero N (eds.) *Principles and Applications in Aquatic Microbiology*. New York: Wiley, pp. 1–14.

Merrettig U, Wlotzta P, Onken U. 1989. The removal of pyritic sulfur from coal by *Leptospirillum*-like bacteria. *Appl Microbiol Biotechnol* 31:626–628.

Meslé M, Dromart G, Oger P (2013). Microbial methanogenesis in subsurface oil and coal. *Res Microbiol* 164:959–972.

Metje M, Frenzel P. 2005. Effect of temperature on anaerobic ethanol oxidation and methanogenesis in acidic peat from a northern wetland. *Appl Environ Microbiol* 71:8191–8200.

Meyer WC, Yen TF. 1976. Enhanced dissolution of oil shale by bioleaching with thiobacilli. *Appl Environ Microbiol* 32:610–616.

Mills HJ, Martinez RJ, Story S, Sopbecky PA. 2005. Characterization of microbial community structure in Gulf of Mexico gas hydrates: Comparative analysis of DNA- and RNA-derived clone libraries. *Appl Environ Microbiol* 71:3235–3247.

Miroshnichenko ML, Hippe H, Stackebrandt E, Kostrikina NA, Chernyh NA, Jeanthon C, Nazina TN, Belyaev SS, Bonch-Osmolovskaya EA (2001) Isolation and characterization of *Thermococcus sibiricus* sp. nov. from a Western Siberia high-temperature oil reservoir. *Extremophiles* 5:85–91.

Miranda-Tello E, Fardeau ML, Thomas P, Ramirez F, Casalot L, Cayol JL, Garcia JL, Ollivier B. 2004. *Petrotoga mexicana* sp. nov., a novel thermophilic, anaerobic and xylanolytic bacterium isolated from an oil-producing well in the Gulf of Mexico. *Int J Syst Evol Microbiol* 54:169–174.

Miranda-Tello E, Fardeau ML, Joulian C, Magot M, Thomas P, Tholozan JL, Ollivier B. 2007. *Petrotoga halophila* sp. nov., a thermophilic, moderately halophilic, fermentative bacterium isolated from an offshore oil well in Congo. *Int J Syst Evol Microbiol* 57:40–44.

Mohn WW, Tiedje JM. 1990. Strain DCB-1 conserves energy for growth from reductive dechlorination coupled to formate oxidation. *Arch Microbiol* 153:267–271.

Mohn WW, Tiedje JM. 1991. Evidence for chemiosmotic coupling of reductive dechlorination and ATP synthesis in *Desulfomonile tiedjei*. *Arch Microbiol* 157:1–6.

Mormile MR, Atlas RM. 1988. Mineralization of dibenzothiophene biodegradation products 3-hydroxy-2-formyl benzothiophene and dibenzothiophene sulfone. *Appl Environ Microbiol* 54:3183–3184.

Mountford DO. 1978. Evidence for ATP synthesis driven by a proton gradient in *Methanosarcina barkeri*. *Biochem Biophys Res Commun* 85:1346–1351.

Müller V, Blaut M, Gottschalk B. 1993. Bioenergetics of methanogenesis. In: Ferry JG (ed.) *Methanogenesis*. New York: Chapman & Hall, pp. 360–406.

Myers CR, Nealson KH. 1988. Bacterial manganese reduction and growth with manganese oxide as the sole electron acceptor. *Science* 240:1319–1321.

Nauhaus K, Boetius A, Krüger M, Widdel F. 2002. In vitro demonstration of anaerobic oxidation of methane coupled to sulphate reduction in sediment from a marine gas hydrate area. *Environ Microbiol* 4:296–305.

Nazina TN, Ivanova AE, Borzenkov IA, Belyaev SS, Ivanov MV. 1995. Occurrence and geochemical activity of microorganisms in high-temperature water-flooded oil fields of Kazakhstan and Western Siberia. *Geomicrobiol J* 13:181–192.

Nazina TN, Rozanova EP. 1978. Thermophilic sulfate-reducing bacteria from oil strata. *Mikrobiologiya* 47:142–148 (Engl. transl., pp. 113–118).

Nazina TN, Rozanova EP. 1980. Ecological conditions for the distribution of methane producing bacteria in oil-containing strata of Apsheron. *Mikrobiologiya* 49:123–129 (Engl. transl., pp. 104–109).

Nazina TN, Rozanova EP, Kuznetsov SI. 1985. Microbial oil transformation processes accompanied by methane and hydrogen-sulfide formation. *Geomicrobiol J* 4:103–130.

Nazina TN, Tourova TP, Poltaraus AB, Novikova EV, Grigoryan AA, Ivanova AE, Lysenko AM, Petrunyaka VV, Osipov GA, Belyaev SS, Ivanov MV. 2001. Taxonomic study of aerobic thermophilic bacilli: Descriptions of *Geobacillus subterraneus* gen. nov., sp. nov. and *Geobacillus uzenensis* sp. nov. from petroleum reservoirs and transfer of *Bacillus stearothermophilus*, *Bacillus thermocatenulatus*, *Bacillus thermoleovorans*, *Bacillus kaustophilus*, *Bacillus thermodenitrificans* to Geobacillusas the new combinations *G. stearothermophilus*, *G. thermocatenulatus*, *G. thermoleovorans*, *G. kaustophilus*, *G. thermoglucosidasius* and *G. thermodenitrificans*. *Int J Syst Evol Microbiol* 51:433–446.

Nazina TN, Shestakova NM, Grigor'ian AA, Mikhailova EM, Turova TP, Poltaraus AB, Feng C, Ni F, Beliaev SS. 2006. Phylogenetic diversity and activity of anaerobic microorganisms of high temperature horizons of the Dagang Oil fileld (China). *Microbiology (Russia)* 75:70–81.

Nette IG, Grechushkina NN, Rabotnova IL. 1965. Growth of certain mycobacteria in petroleum and petroleum products. *Prikl Biokhim Mikrobiol* 1:167–174.

Newberry CJ, Webster G, Cragg BA, Parkes RJ, Weightman AJ, Fry JC. 2004. Diversity of prokaryotes and methanogenesis in deep subsurface sediments from the Nankai Trough, Ocean Drilling Program Leg 190. *Environ Microbiol* 6:274–287.

Niemann H, Lösekann T, de Beer D, Ebert M, Nadalig T, Knittel K, Amann R et al. 2006. Novel microbial communities of the Haakon Mosby mud volcano and their roles of methane sink. *Nature (London)* 443:854–858.

Niewöhner C, Hensen C, Kasten S, Zabel M, Schulz HD. 1998. Deep sulfate reduction completely mediated by anaerobic methane oxidation in sediments of the upwelling area of Namibia. *Geochim Cosmochim Acta* 62:455–464.

Nilsen RK, Torsvik T. 1996. Methanococcus thermolithotrophicus isolated from North Sea oil field reservoir water. *Appl Environ Microbiol* 62:728–731.

Nilsen RK, Torsvik T, Lien T. 1996a. Desulfotomaculum thermocisternum sp. nov., a sulfate reducer isolated from a hot north sea oil reservoir. *Int J Syst Bacteriol* 46:397–402.

Nilsen RK, Beeder J, Thorstenson T, Torsvik T. 1996b. Distribution of thermophilic marine sulfate reducers in North Sea oil field waters and oil reservoirs. *Appl Environ Microbiol* 62:1793–1798.

North FK. 1985. *Petroleum Geology*. Boston, MA: Allen and Unwin.

Ollivier B, Fardeau M-L, Cayol J-L, Magot M, Patel BKC, Prensier G, Garcia J-L. 1988. *Methanocalculus halotolerans* gen. nov., sp. nov., isolated from an oil-producing well. *Int J Syst Evol Microbiol* 48:821–828.

Olson ES, Stanley DC, Gallagher JR. 1993. Characterization of intermediates in the microbial desulfurization of dibenzothiophene. *Energy Fuels* 7:159–164.

Omori T, Monna L, Saiki Y, Kodama T. 1992. Desulfurization of dibenzothiophene by *Corynebacterium* sp. strain SY1. *Appl Environ Microbiol* 58:911–915.

O'Neill JG, Wilkinson JF. 1977. Oxidation of ammonia by methane-oxidizing bacteria and the effects of ammonia on methane oxidation. *J Gen Microbiol* 100:407–412.

Oremland RS, Marsh LM, Polcin S. 1982. Methane production and simultaneous sulfate reduction in anoxic, salt marsh sediments. *Nature (London)* 296:143–145.

Oremland RS, Taylor BF. 1978. Sulfate reduction and methanogenesis in marine sediments. *Geochim Cosmochim Acta* 42:209–214.

Orphan VJ, Boles JR, Goffredi SK, Delong EF. 2003. Geochemical influence on community structure and microbial processes in high temperature oil reservoirs. *Geomicrobiol J* 20:295–311.

Orphan VJ, Hinrichs K-U, Ussler W III, Paull CK, Taylor LlT, Sylva SP, Hayes JM, DeLong EF. 2001b. Comparative analysis of methane-oxidizing Archaea and sulfate-reducing Bacteria in anoxic marine sediments. *Appl Environ Microbiol* 67:1922–1934.

Orphan VJ, House CH, Hinrichs K-U, McKeegan KD, DeLong EF. 2001a. Methane-consuming Archaea revealed by directly coupled isotopic and phylogenetic analysis. *Science* 293:484–487.

Orphan VJ, House CH, Hinrichs K-U, McKeegan KD, DeLong EF. 2002. Multiple archaeal groups mediate methane oxidation in anoxic cold seep sediments. *Proc Natl Acad Sci USA* 99:7663–7668.

Orr FM Jr, Taber JJ. 1984. Use of carbon dioxide in enhanced oil recovery. *Science* 224:563–569.

Ourisson G, Albrecht P, Rohmer M. 1984. Microbial origin of fossil fuels. *Sci Am* 251:44–51.

Pallaser RJ. 2000. Recognising biodegradation in gas/oil accumulations through the $\delta^{13}C$ composition of gas components. *Org Geochem* 31:1363–1373.

Pancost RD, Damsté JSS, de Lint S, van der Maarel MJEC, Gottschal JC, Medinaut Shipboard Scientific Party. 2000. Biomarker evidence for widespread anaerobic methane oxidation in Mediterranean sediments by a consortium of methanogenic Archaea and Bacteria. *Appl Environ Microbiol* 66:1126–1132.

Panganiban A, Hanson RS. 1976. Isolation of a bacterium that oxidizes methane in the absence of oxygen. Abstract. *Annu Meet Am Soc Microbiol* 159:121.

Panganiban AT Jr, Patt TE, Hart W, Hanson RS. 1979. Oxidation of methane in the absence of oxygen in lake water samples. *Appl Environ Microbiol* 37:303–309.

Parkes J. 1999. Cracking anaerobic bacteria *Nature* 401:217.

Paszczynski A, Huynh V-B, Crawford R. 1986. Comparison of lignase-1 and peroxidase-M2 from the white-rot fungus *Phanerochaete chrysosporium*. *Arch Biochem Biophys* 244:750–785.

Patel GB. 1984. Characterization and nutritional properties of *Methanothrix concilii* sp. nov., an acetoclastic methanogen. *Can J Microbiol* 30:1383–1396.

Pernthaler A, Pernthaler J, Amann R. 2002. Fluorescence in situ hybridization and catalyzed reporter deposition for the identification of marine bacteria. *Appl Environ Microbiol* 68:3094–3101.

Phillips WE Jr, Perry JJ. 1974. Metabolism of n-butane and 3-butanone by *Mycobacterium vaccae*. *J Bacteriol* 120:987–989.

Pineda-Flores G, Boll-Arguello G, Lira-Galeana C, Mesta-Howard AM. 2004. A microbial consortium isolated from a crude oil sample that uses asphaltenes as a carbon and energy source. *Biodegradation* 15:145–151.

Pirnik MP, Atlas RM, Bartha R. 1974. Hydrocarbon metabolism by *Brevibacterium erythrogenes*: Normal and branched alkanes. *J Bacteriol* 119:868–878.

Pooley FD, Atkins AS. 1983. Desulfurization of coal using bacteria by both dump and process plant techniques. In: Rossi G, Torma AE (eds.) *Recent Progress in Biohydrometallurgy*. Iglesias, Italy: Assoc Minerar Sarda, pp. 511–526.

Preston MD, Smemo KA, McLaughlin JW, Basiliko N. 2012. Peatland microbial communities and decomposition processes in the James Bay Lowlands, Canada. *Front Microbiol* 3:70.

Prince RC. 2005. The microbiology of marine oil spill bioremediation. In: Ollivier B, Magot M (eds.) *Petroleum Microbiology*. Washington, DC, ASM Press, pp. 317–335.

Pyne JW Jr, Stewart DL, Fredrickson J, Wilson BW. 1987. Solubilization of leonardite by an extracellular fraction from *Coriolus versicolor*. *Appl Environ Microbiol* 53:2844–2848.

Quayle JR, Ferenci T. 1978. Evolutionary aspects of autotrophy. *Microbiol Rev* 42:251–273.

Radway J, Tuttle JH, Fendinger NJ. 1989. Influence of coal source and treatment upon indigenous microbial communities. *J Ind Microbiol* 4:195–208.

Radway J, Tuttle JH, Fendinger NJ, Means JC. 1987. Microbially mediated leaching of low-sulfur coal in experimental coal columns. *Appl Environ Microbiol* 53:1056–1063.

Raghoebarsing AA, Pol A, van de Pas-Schoonen KT, Smolders AJ, Ettwig KF, Rijpstra WI, Schouten S et al. 2006. A microbial consortium couples anaerobic methane oxidation to denitrification. *Nature* 440:918–921.

Raiders RA, Knapp RM, McInerney JM. 1989. Microbial selective plugging and enhanced oil recovery. *J Ind Microbiol* 4:215–230.

Ravot G, Magot M, Fardeau ML, Patel BK, Prensier G, Egan A, Garcia JL, Ollivier B. 1995. Thermotoga elfii sp. nov., a novel thermophilic bacterium from an African oil-producing well. *Int J Syst Bacteriol* 45:308–314.

Raymond RL, Jamison VW, Hudson JO. 1967. Microbial hydrocarbon cooxidation. I. Oxidation of mono- and dicyclic hydrocarbons by soil isolates of the genus *Nocardia*. *Appl Microbiol* 15:857–865.

Reeburgh WS. 1976. Methane consumption in Cariaco Trench waters and sediments. *Earth Planet Sci Lett* 28:337–344.

Reeburgh WS. 1980. Anaerobic methane oxidation: Rate depth distribution in Skan Bay sediments. *Earth Planet Sci Lett* 47:345–352.

Reeburgh WS. 2007. Oceanic methane biogeochemistry. *Chem Rev* 107:486–513.

Reed DW, Fujita Y, Delwiche ME, Blackwelder DB, Sheridan PP, Uchida T, Colwell FS. 2002. Microbial communities from methane hydrate-bearing deep marine sediments in a forearc basin. *Appl Environ Microbiol* 68:3759–3770.

Rees GN, Grassia GS, Sheehy AJ, Dwivedi PP, Patel BKC. 1995. Desulfacinum infernumgen. nov., sp. nov., a thermophilic sulfate-reducing bacterium from petroleum reservoir. *Int J Syst Bacteriol* 45:85–89.

Rees GN, Patel BK, Grassia GS, Sheehy AJ. 1997. *Anaerobaculum hermoterrenumgen*. nov., sp. nov., a novel, thermophilic bacterium which ferments citrate. *Int J Syst Bacteriol* 47:150–154.

Reineke W, Knackmuss H-J. 1988. Microbial degradation of haloaromatics. *Annu Rev Microbiol* 42:263–287.

Rivard CJ, Smith PH. 1982. Isolation and characterization of a thermophilic marine methanogenic bacterium, *Methanogenium thermophilicum* sp. nov. *Int J Syst Bacteriol* 32:430–436.

Roadifer RE. 1987. Size distribution of the world's largest known oil and tar accumulations. In: Meyer RF (ed.) *Exploration for Heavy Crude Oil and Natural Bitumen.* AAPG Studies in Geology, Vol. 25. Tulsa, OH: American Association of Petroleum Geologist, pp. 3–23.

Robertson R. 1966. The origins of petroleum. *Nature (London)* 212:1291–1295.

Robinson JA, Tiedje JM. 1984. Competition between sulfate-reducing and methanogenic bacteria for H_2 under resting and growing conditions. *Arch Microbiol* 137:26–32.

Roels J, Verstraete W. 2001. Biological formation of volatile phosphorus compounds. *Bioresour Technol* 79:243–250.

Roffey R, Norqvist A. 1991. Biodegradation of bitumen used for nuclear waste disposal. *Experientia* 47:539–542.

Rogers JR, Bennett PC. 2004. Mineral stimulation of subsurface microorganisms: Release of limiting nutrients from silicates. *Chem Geol* 203:91–108.

Rogoff MH, Wender I, Anderson RB. 1962. *Microbiology of Coal.* Info Circ 8057. Washington, DC: Bureau of Mines, U.S. Department of the Interior.

Roitsch T, Stolp H. 1985. Distribution of dissimilatory enzymes in methane and methanol oxidizing bacteria. *Arch Microbiol* 143:233–236.

Röling WFM, Larter SR, Head IM. 2003. The microbiology of hydrocarbon degradation in subsurface petroleum reservoirs: Perspectives and prospects. *Res Microbiol* 154:321–328.

Romanovskaya VA, Lyudvichenko ES, Kryshtab TP, Zhukov VG, Sokolov IG, Malashenko YuR. 1980. Role of exogenous carbon dioxide in metabolism of methane-oxidizing bacteria. *Mikrobiologiya* 49:687–693 (Engl. transl., pp. 566–571).

Romesser JA, Wolfe RS, Mayer F, Spiess E, Walther-Mauruschat A. 1979. *Methanogenium,* a new genus of marine methanogenic bacteria, and characterization of *Methanogenium cariaci* sp. nov. and *Methanogenium marisnigri* sp. nov. *Arch Microbiol* 121:147–153.

Rosnes JT, Torsvik T, Lien T. 1991. Spore-forming thermophilic sulfate-reducing bacteria isolated from North Sea oil fi eld waters. *Appl Environ Microbiol* 57:2302–2307.

Rozanova EP. 1971. Morphology and certain physiological properties of purple bacteria from oil-bearing strata. *Mikrobiologiya* 40:152–157 (Engl. transl., pp. 134–138).

Rozanova EP. 1978. Sulfate reduction and water-soluble organic substances in a flooded oil reservoir. *Mikrobiologiya* 47:495–500 (Engl. transl., pp. 401–405).

Rozanova EP, Savvichev AS, Karavaiko SG, Miller YM. 1995. Microbial processes in the Savuiskoe oil-field in the Ob region. *Microbiology* 6:85–90.

Rozanova EP, Borzenkov IA, Tarasov AL, Suntsova LA, Dong CL, Belyaev SS, Ivanov MV. 2001a. Microbiological processes in a high-temperature oil field. *Microbiology (Russia)* 70:102–110.

Rozanova EP, Tourova TP, Kolganova TV, Lysenko AM, Mityushina LL, Yusupov SK, Belyaev SS. 2001b. *Desulfacinum subterraneum* sp. nov., a new thermophilic sulfate-reducing bacterium isolated from a high-temperature oil field. *Microbiology* 70:466–471.

Rubinshtein I, Strausz OP, Spyckerelle C, Crawford RJ, Westlake DW. 1977. The origin of oil and bitumens of Alberta: A chemical and a microbiological study. *Geochim Cosmochim Acta* 41:1341–1353.

Rubinshtein LM, Oborin AA. 1986. Microbial methane production in stratal waters of oilfields in the Perm area of the Cis-Ural region. *Mikrobiologiya* 55:674–678 (Engl. transl., pp. 530–534).

Safinowski M, Meckenstock RU 2004. Enzymatic reactions in anaerobic 2-methylnaphthalene degradation by the sulphate-reducing enrichment culture N47. *FEMS Microbiol Lett* 240:99–104.

Salinas MB, Fardeau ML, Thomas P, Cayol JL, Patel BK, Ollivier B. 2004. Mahella australiensis gen. nov., sp. nov., a moderately thermophilic anaerobic bacterium isolated from an Australian oil well. *Int J Syst Evol Microbiol* 54:2169–2173.

Sansone FJ, Martens CS. 1981. Methane production from acetate and associated methane fluxes from anoxic coastal sediments. *Science* 211:707–709.

Schönheit P, Kristjansson JK, Thauer RK. 1982. Kinetic mechanism for the ability of sulfate reducers to out-compete methanogens for acetate. *Arch Microbiol* 132:285–288.

Schouten S, Wakeham SG, Hopmans EC, Sinninghe Damsté JS. 2003. Biogeochemical evidence that thermophilic Archaea mediate the anaerobic oxidation of methane. *Appl Environ Microbiol* 69:1680–1686.

Sette LD, Simioni KC, Vasconcellos SP, Dussan LJ, Neto EV, Oliveira VM. 2007. Analysis of the composition of bacterial communities in oil reservoirs from a southern offshore Brazilian basin. *Antonie van Leeuwenhoek* 91:253–266.

EHRLICH'S GEOMICROBIOLOGY

Sharak Genthner BR, Price WA II, Pritchard PH. 1989. Anaerobic degradation of chloroaromatic compounds in aquatic sediments under a variety of enrichment conditions. *Appl Environ Microbiol* 55:1466–1477.

Sharp CE, Op den Camp HJM, Tamas I, Dunfield PF. 2013. Unusual members of the PVC superphylum: The methanotrophic *Verrucomicrobia* genus '*Methylacidiphilum*'. In: Fuerst JA (ed.) *Planctomycetes: Cell Structure, Origins, and Biology.* New York: Springer, pp. 221–227.

Shelton DR, Tiedje JM. 1984. Isolation and characterization of bacteria in an anaerobic consortium that mineralizes 3-chlorobenzoic acid. *Appl Environ Microbiol* 48:840–848.

Shestakova N, Korshunova A, Mikhailova E, Sokolova D, Tourova T, Belyaev S, Poltaraus A, Nazina T. 2011. Characterization of the aerobic hydrocarbon-oxidizing enrichments from a high-temperature petroleum reservoir by comparative analysis of DNA- and RNA-derived clone libraries. *Microbiology* 80:60–69.

Siddique T, Penner T, Semple K, Foght JM. 2011. Anaerobic biodegradation of longer-chain n-alkanes coupled to methane production in oil sands tailings. *Environ Sci Technol* 45:5892–5899.

Sieber J, McInerny MJ, Gunsalus RP. 2012. Genomic insights into syntrophy: The paradigm for anaerobic metabolic cooperation. *Annu Rev Microbiol* 66:429–452.

Sieburth J McN, Johson PW, Eberhardt MA, Sieracki ME, Lidstron ME, Laux D. 1987. The first methane-oxidizing bacterium from the upper mixing layer of the deep ocean: *Methylomonas pelagica* sp. nov. *Curr Microbiol* 14:285–293.

Simakova TL, Kolesnik ZA, Strigaleva NV. 1968. Transformation of high-paraffinaceous oil by microorganisms under anaerobic and aerobic conditions. *Mikrobiologiya* 37:233–238 (Engl. transl., pp. 194–198).

Simoneit BRT, Londsdale PG. 1982. Hydrothermal petroleum in mineralized mounds at the seabed of Guaymas Basin. *Nature (London)* 295:198–202.

Simpson PG, Whitman WB. 1993. Anabolic pathways in methanogens. In: Ferry JG (ed.) *Methanogenesis.* New York: Chapman & Hall, pp. 445–472.

Skyring GW. 1987. Sulfate reduction in coastal ecosystems. *Geomicrobiol J* 5:295–374.

Slobodkin AI, Jeanthon C, L'Haridon S, Nazina T, Miroshnichenko M, Bonch-Osmolovskaya E. 1999. Dissimilatory reduction of Fe(III) by thermophilic bacteria and archaea in deep subsurface petroleum reservoirs of Western Siberia. *Curr Microbiol* 39:99–102.

Smith MR, Mah RA. 1978. Gowth and methane genesis by *Methanosarcina* strain 227 on acetate and methanol. *Appl Environ Microbiol* 36:870–879.

So CM, Phelps CD, Young LY. 2003. Anaerobic transformation of alkanes to fatty acids by a sulfate-reducing bacterium, strain Hxd3. *Appl Environ Microbiol* 69:3892–3900.

Sokolov IG. 1986. Coupling of the process of electron transport to methane monooxygenases with the translocation of protons in methane-oxidizing bacteria. *Mikrobiologiya* 55:715–722 (Engl. transl., pp. 559–565).

Sokolov IG, Malashenko YuR, Romanovskaya VA. 1981. Electron transport chain of thermophilic methane-oxidizing culture *Methylococcus thermophilus.* *Mikrobiologiya* 50:13–20 (Engl. transl., pp. 7–13).

Sørensen J. 1987. Nitrate reduction in marine sediment: Pathways and interactions with iron and sulfur cycling. *Geomicrobiol J* 5:401–421.

Spark I, Patey I, Duncan B, Hamilton A, Devine C, McGovern-Traa C. 2000. The effects of indigenous and introduced microbes on deeply buried hydrocarbon reservoirs, North Sea. *Clay Miner* 35:5–12.

Sprott GE, Bird SE, MacDonald IJ. 1985. Proton motive force as a function of the pH at which *Methanobacterium bryantii* is grown. *Can J Microbiol* 31:1031–1034.

Sprott GD, Ekiel I, Patel G. 1993. Metabolic pathways in *Methanococcus jannaschii* and other methanogenic bacteria. *Appl Environ Microbiol* 59:1092–1098.

Stams AJM, Plugge C. 2009. Electron transfer in syntrophic communities of anaerobic bacteria and archaea. *Nature Rev Microbiol* 7:568–577.

Stanier RY, Ingraham JL, Wheelis ML, Painter PR. 1986. *The Microbial World*, 5th ed. Englewood Cliffs, NJ: Prentice-Hall.

Stetter KO, Gaag G, 1983. Reduction of molecular sulfur by methanogenic bacteria. *Nature (London)* 305:309–311.

Stetter KO, Huber R. 2000. The role of hyperthermophilic prokaryotes in oil fields. In: Bell CR, Brylinsky M, Johnson-Green P (eds.) *Microbial Biosystems: New frontiers. Proceedings of the 8th International Symposium on Microbial Ecology,* Halifax, Nova Scotia, Canada, pp. 369–375.

Stetter KO, Huber R, Boechl E, Kurr M, Eden RD, Fielder M, Cash H, Vance I. 1993. Hyperthermophilic archaea are thriving in deep North Sea and Alaskan oil reservoirs. *Nature (London)* 365:743–745.

Stevenson IL. 1967. Utilization of aromatic hydrocarbons by *Arthrobacter* spp. *Can J Microbiol* 13:205–211.

Stewart DL, Thomas BL, Bean RM, Fredrickson JK. 1990. Colonization and degradation of bituminous and lignite coals by fungi. J Ind Microbiol 6:53–59.

Stewart JE, Kallio RE, Stevenson DP, Jones AC, Schissler DO. 1959. Bacterial hydrocarbon oxidation. I. Oxidation of n-hexadecane by a gram-negative coccus. J Bacteriol 78:441–448.

Stoecker K, Bendinger B, Schöning B, Nielsen PH, Nielsen JL, Baranyi C, Toenshoff ER, Daims H, Wagner M. 2006. Cohn's Crenothrix is a filamentous methane oxidizer with an unusual methane monooxygenase. Proc Natl Acad Sci USA 103:2363–2367.

Stoner DL, Wey JW, Barrett KB, Jolley JG, Wright RB, Dugan PR. 1990. Modification of water-soluble coal-derived products by dibenzothiophene-degrading microorganisms. Appl Environ Microbiol 56:2667–2676.

Strapoc D, Mastalerz M, Dawson K, Macalady J, Callaghan AV, Wawrik B, Turich C, Ashby M 2011. Biogeochemistry of microbial coal-bed methane. Ann Rev Earth Planet 39:617–656.

Sundh I, Bastviken D, Tranvik LJ. 2005. Abundance, activity, and community structure of pelagic, methane-oxidizing bacteria in temperate lakes. Appl Environ Microbiol 71:6746–6752.

Szewzyk R, Pfennig N. 1987. Complete oxidation of catechol by the strictly anaerobic sulfate reducing Desulfobacterium catecholicum sp. nov. Arch Microbiol 147:163–168.

Takahata Y, Nishijima M, Hoaki T, Maruyama T. 2000. Distribution and physiological characteristics of hyperthermophiles in the Kubiki oil reservoir in Niigata, Japan. Appl Environ Microbiol 66:73–79.

Takahata Y, Nishijima M, Hoaki T, Maruyama T. 2001. Thermotoga petrophila sp. nov. and Thermotoga naphthophila sp. nov., two hyperthermophilic bacteria from the Kubiki oil reservoir in Niigata, Japan. Int J Syst Evol Microbiol 51:1901–1909.

Takai K, Nakamura K, Tomohiro T, Tsunogai U, Miyazaki M, Miyazaki J, Hirayama H, Nakagawa S, Nunoura T, Horikoshi K. 2008. Cell proliferation at 122°C and isotopically heavy CH$_4$ production by a hyperthermophilic methanogen under high-pressure cultivation. Proc Natl Acad Sci USA 105:10949–10954.

Tang YQ, Li Y, Zhao JY, Chi CQ, Huang LX, Dong HP, Wu XL. 2012. Microbial communities in long-term, water-flooded petroleum reservoirs with different in situ temperatures in the Huabei Oil field, China. PLoS One 7:e33535.

Tanner RS, Udegbunam EO, McInerney MJ, Knapp RM. 1991. Microbially enhanced oil recovery from carbonate reservoirs. J Ind Microbiol 9:169–195.

Tardy-Jacquenod C, Caumette P, Matheron R, Lanau C, Arnauld O, Magot M. 1996. Characterization of sulfate-reducing bacteria isolated from oil- field waters. Can J Microbiol 42:259–266.

Teh JS, Lee KH. 1973. Utilization of n-alkanes by Cladosporium resinae. Appl Microbiol 25:454–457.

Teske A, Hinrichs K-U, Edgcomb V, de Vera Gomez A, Kysela D, Sylva SP, Sogin ML, Jannasch HW. 2002. Microbial diversity of hydrothermal sediments in the Guaymas Basin: Evidence for anaerobic methanotrophic communities. Appl Environ Microbiol 68:1994–2007.

Thauer RK, Kaster AK, Seedorf H, Buckel W, Hedderich R. 2008. Methanogenic archaea: Ecologically relevant differences in energy conservation. Nat Rev Microbiol 6:579–591.

Theisen AR, Murrell JC. 2005. Facultative methanotrophs revisited. J Bacteriol 187:4303–4305.

Thomsen TR, Finster K, Ramsing NB. 2001. Biogeochemical and molecular signatures of anaerobic methane oxidation in a marine sediment. Appl Environ Microbiol 67:1646–1656.

Tissot BP, Welte DH. 1984. Petroleum Formation and Occurrence. New York: Springer-Verlag, 699pp.

Townsend GT, Prince RC, Suflita JM. 2003. Anaerobic oxidation of crude oil hydrocarbons by the resident microorganisms of a contaminated anoxic aquifer. Environ Sci Technol 37:5213–5218.

Townsley CC, Atkins AS, Davis AJ. 1987. Suppression of pyritic sulfur during flotation tests using the bacterium Thiobacillus ferrooxidans. Biotech Bioeng 30:1–8.

Treude T, Niggermann J, Kallmeyer J, Wintersteller P, Schubert CJ, Boetius A, Jøregensen BB. 2005. Anaerobic oxidation of methane and sulfate reduction along the Chilean continental margin. Geochim Cosmochim Acta 69:2767–2779.

Treude T, Orphan V, Knittel K, Gieseke A, House CH, Boetius A. 2007. Consumption of methane and CO$_2$ by methanotrophic microbial mats from gas seeps of the anoxic Black Sea. Appl Environ Microbiol 73:2271–2283.

Valentine DL, Reeburgh WS. 2000. New perspectives on anaerobic methane oxidation. Environ Microbiol 2:477–484.

Van Afferden M, Schacht S, Klein J, Trüper HG. 1990. Degradation of dibenzothiophene by Brevibacterium sp. DO. Arch Microbiol 153:324–328.

van Beilen JB, Funhoff EG. 2007. Alkane hydroxylases involved in microbial alkane degradation. Appl Microbiol Biotechnol 74:13–21.

van Bodegom PM, Scholten JCM, Stams AJM. 2004. Direct inhibition of methanogenesis by ferric iron. FEMS Microbiol Ecol 49:261–268.

Van Bruggen JJA, Zwart KB, van Assema RM, Stumm CK, Vogels GD. 1984. *Methanobacterium formicum*, an endosymbiont of the anaerobic ciliate *Metopus striatus* McMurrich. *Arch Microbiol* 139:1–7.

Van der Linden AC, Thijsse GJE. 1965. The mechanism of microbial oxidations of petroleum hydrocarbons. *Adv Enzymol* 27:469–546.

Van Hamme JD, Singh A, Ward OP. 2003. Recent advances in petroleum microbiology. *Microbiol Mol Biol Rev* 67:503–509.

Vigliotta G, Nurticati E, Carata E, Tredici SM, De Stefano M, Pontieri P, Massardo DR, Prati MV, De Bellis L, Alifano P. 2007. *Clonothrix fusca* Roze 1996, a filamentous, sheathed, methanotrophic γ-Proteobacterium. *Appl Environ Microbiol* 73:3556–3565.

Vogels CD, Visser CM. 1983. Interconnections of methanogenic and acetogenic pathways. *FEMS Microbiol Lett* 20:291–297.

Vogels GD. 1979. The global cycle of methane. *Antonie v Leeuwenhoek* 45:347–352.Volesky B, Zajic JE. 1970. Ethane and natural gas oxidation by fungi. *Dev Ind Microbiol* 11:184–195.

Walker JD, Cooney JJ. 1973. Pathway of n-alkane oxidation in *Cladosporium resinae*. *J Bacteriol* 115:635–639.

Wang L, Tang Y, Wang S, Liu RL, Liu MZ, Zhang Y, Liang FL, Feng L. 2006. Isolation and characterization of a novel thermophilic Bacillus strain degrading long-chain n-alkanes. *Extremophiles* 10:347–356.

Wang L-Y, Gao C-X, Mbadinga SM, Zhou L, Liu J-F, Gu J-D, Mu B-Z. 2011. Characterization of an alkane-degrading methanogenic enrichment culture from production water of an oil reservoir after 274 days of incubation. *Int Biodeter Biodegr* 65:444–450.

Ward BB, Kilpatrick KA, Novelli PC, Scanton MI. 1987. Methane oxidation and methane fluxes in the ocean surface layer and deep anoxic layers. *Nature (London)* 327:226–229.

Ward DM, Brock TD. 1978. Anaerobic metabolism of hexadecane in sediments. *Geomicrobiol J* 1:1–9.

Watkins A, Roussel E, Webster G, Parkes R, Sass H 2012. Choline and N,N-Dimethylethanolamine as direct substrates for methanogens. *Appl Environ Microbiol* 78:8298–8303.

Wellsbury P, Goodman K, Barth T, Cragg BA, Barnes SP, Parkes RJ. 1997. Deep marine biosphere fuelled by increasing organic matter availability during burial and heating *Nature* 388:573–576.

Wentzel A, Lewin A, Cervantes FJ, Valla S, Kotlar HK. 2013. Deep subsurface oil reservoirs as poly-extreme habitats for microbial life. A current review. *Cell Origin Life Ext Hab Astrobiol* 27:443–466.

Wentzel A, Lewin A, Cervantes FJ, Valla S, Kotlar HK. 2013b. Deep subsurface oil reservoirs as poly-extreme habitats for microbial life. A current review. In: Seckbach J, Oren A, Stan-Lotter H (eds.) *Polyextremophiles: Life Under Multiple Forms of Stress.* London, U.K.: Springer, pp. 443–466.

Westerholm M, Dolfing J, Sherry A, Gray ND, Head IM, Schnürer A. 2011. Quantification of syntrophic acetate-oxidizing microbial communities in biogas processes. *Environ Microbiol Rep* 3:500–505.

Westlake DWS, Belicek W, Jobson A, Cook FD. 1976. Microbial utilization of raw and hydrogenated shale oils. *Can J Microbiol* 22:221–227.

Whiticar MJ. 1999. Carbon and hydrogen isotope systematics of bacterial formation and oxidation of methane. *Chem Geol* 161:291–314.

Whittenbury R, Phillips KC, Wilkinson JF. 1970. Enrichment isolation and some properties of methane-utilizing bacteria. *J Gen Microbiol* 61:205–218.

Widdel F. 1986. Growth of methanogenic bacteria in pure culture with 2-propanol and other alcohols as hydrogen donors. *Appl Envirom Microbiol* 51:1056–1062.

Widdel F, Rabus R. 2001. Anaerobic biodegradation of saturated and aromatic hydrocarbons. *Curr Opin Biotechnol* 12:259–276.

Wildgruber G, Thomm M, König H, Ober K, Ricchiuto T, Stetter KO. 1982. *Methanoplanus limicola*, a plate-shaped methanogen representing a novel family, the Methanoplanaceae. *Arch Microbiol* 132:31–36.

Wilhelms A, Larter SR, Head I, Farrimond P, di-Primio R, Zwach C. 2001. Biodegradation of oil in uplifted basins prevented by deep-burial sterilisation. *Nature* 41:1034–1037.

Winter J, Wolfe RS. 1979. Complete degradation of carbohydrate to carbon dioxide and methane by syntrophic cultures of *Acetobacter woodii* and *Methanosarcina barkeri*. *Arch Microbiol* 121:97–102.

Winter JU, Wolfe RS. 1980. Methane formation from fructose by syntrophic associations of *Acetobacterium woodii* and different strains of methanogens. *Arch Microbiol* 124:73–79.

Wolf M, Bachofen R. 1991. Microbial degradation of bitumen. *Experientia* 47:542–548.

Wolin MJ. 1981. Fermentation in the rumen and human large intestine. *Science* 213:1463–1468.

Wolin MJ, Miller TL. 1987. Bioconversion of organic carbon to CH_4 and CO_2. *Geomicrobiol J* 5:239–259.

Worakit S, Boone DR, Mah RA, Abdel-Samie M-E, El-Halwagi MM. 1986. *Methanobacterium alcaliphilum* sp. nov., an H_2-utilizing methanogen that grows at high pH values. *Int J Syst Bacteriol* 36:380–382.

Yakimov MM, Timmis KN, Golyshin PN. 2007. Obligate oil-degrading marine bacteria. *Curr Opin Biotechnol* 18:257–66.

Yamane K, Hattori Y, Ohtagaki H, Fujiwara K. 2011. Microbial diversity with dominance of 16S rRNA gene sequences with high GC contents at 74 and 98°C subsurface crude oil deposits in Japan. *CORD Conf Proc* 76:220–235.

Yang KY-L, Lapado J, Whitman WB. 1992. Pyruvate oxidation by *Methanococcus* spp. *Arch Microbiol* 158:271–275.

Yarbrough HF, Coty VF. 1983. Microbially enhanced oil recovery from the Upper Cretaceous Nacatoch formation, Union County, Arkansas. In: Donaldson EC, Clark JB (eds.) *Proceedings of the International Conference on Microbial Enhancement of Oil Recovery* (Conf 8205140). Springfield, VA: NTIS, pp. 149–153.

Yavitt JB, Lang GE. 1990. Methane production in contrasting wetland sites: Response to organic-chemical components of peat and to sulfate reduction. *Geomicrobiol J* 8:27–46.

Yen HC, Marrs B. 1977. Growth of *Rhodopseudomonas capsulata* under anaerobic dark conditions with dimethy sulfoxide. *Arch Biochem Biophys* 181:411–418.

Young LY, Frazer AC. 1987. The fate of lignin and lignin-derived compounds in anaerobic environments. *Geomicrobiol J* 5:261–293.

Youssef N, Elshahed MS, McInerney MJ. 2009. Microbial processes in oil fields: Culprits, problems, and opportunities. In: Laskin AI, Sariaslani S, Gadd GM (eds.) *Advances in Applied Microbiology*, Vol. 66. Burlington, MA: Academic Press, pp. 141–251.

Zabel HP, König H, Winter J. 1984. Isolation and characterization of a new coccoid methanogen, *Methanogenium tatii* spec. nov. from a solfataric field on Mount Tatio. *Arch Microbiol* 137:308–315.

Zehnder AJB, Brock TD. 1979. Methane formation and methane oxidation by methanogenic bacteria. *J Bacteriol* 137:420–432.

Zehnder AJB, Huser BA, Brock TD, Wuhrmann K. 1980. Characterization of an acetate-carboxylating, non-hydrogen-oxidizing methane bacterium. *Arch Microbiol* 124:1–11.

Zeikus JG. 1977. The biology of methanogenic bacteria. *Bacteriol Rev* 41:514–541.

Zeikus JG, Kerby R, Krzycki JA. 1985. Single-carbon chemistry of acetogenic and methanogenic bacteria. *Science* 127:1167–1173.

Zellner G, Bleicher K, Braun E, Kneifel H, Tindall BJ, Conway de Marcario E, Winter J. 1989. Characterization of a new mesophilic, secondary alcohol-utilizing methanogen, *Methanobacterium palustre* spec nov. from a peat bog. *Arch Microbiol* 151:1–9.

Zellner G, Sleytr UB, Messner P, Kneifel H, Winter J. 1990. *Methanogenium liminatans* spec. nov., a new coccoid, mesophilic methanogen able to oxidize secondary alcohols. *Arch Microbiol* 153:287–293.

Zengler K, Richnow HH, Rossello-Mora R, Michaelis W, Widdel F. 1999. Methane formation from long-chain alkanes by anaerobic microorganisms. *Nature* 401:266–269.

Zimmerman PR, Greenberg JP, Windiga SO, Crutzen PJ. 1982. Termites: A potentially large source of atmospheric methane, carbon dioxide and molecular hydrogen. *Science* 218:563–565.

Zinder SH. 1993. Physiological ecology of methanogens. In: Ferry JG (ed.) *Methanogenesis*. New York: Chapman & Hall, pp. 128–206.

Zinder SH, Koch M. 1984. Non-aceticlastic methanogenesis from acetate: Acetate oxidation by a thermophilic syntrophic coculture. *Arch Microbiol* 138:263–272.

ZoBell CE. 1952. Part played by bacteria in petroleum formation. *J Sediment Petrol* 22:42–49.

ZoBell CE. 1963. The origin of oil. *Int Sci Technol* August:42–48.

Zvyagintsev DG, Zenova GM, Shirokykh IG. 1993. Distribution of actinomycetes with the vertical structure of peat bog ecosystems. *Mikrobiologiya* 62:548–555 (Engl. transl., pp. 339–342).

INDEX

A

Abiotic oxidation
   arsenite, 301
   iron(II), 348, 362
   manganese, 428–429
   mercury, 335
Acid coal-mine drainage (AMD), 531, 538–540
Acidianus ambivalens, 492
Acidianus brierleyi
   iron, 358
   metal sulfides, 524–525
   molybdenite, 460
Acidiphilium acidophilum, 493–494, 500
Acidithiobacillus ferrooxidans
   acid coal mine drainage, 3
   chromium, 455
   iron(II), 354–357
   metal sulfides, 529–531, 536
   selenium, 553
   surface growth study, 177
   water potential requirement, 61
Acidithiobacillus thiooxidans
   chromium, 455
   elemental sulfur, 491, 493
   iron, 358
   metal sulfides, 525–526, 538
   vanadium, 462
Acidophilic iron(II) oxidizers
   mesophilic archaea, 358
   mesophilic bacteria
      Acidithiobacillus ferrooxidans, 354–357
      genetics, 358
      Leptospirillum ferrooxidans, 357–358
   thermophilic archaea, 358–359
   thermophilic bacteria, 358
Acid producing microorganisms, 2, 224
Aerobactin, 346
Aerobic methanotrophy
   biochemistry, 578
   carbon assimilation, 578–579
   methane-oxidizing symbionts, 575–576
   obligate methanotrophs, 574–575

Aerobic mineralization
   definition, 148
   metal sulfide, 520
Aerobic respiration
   chemiosmosis, 142
   energy-yielding processes, 142
   ETS
      components of, 136
      enzyme-catalyzed oxidations, 137
      hydrogen/electrons, 136
      Krebs tricarboxylic acid cycle, 134–135
      oxidative phosphorylation, 134
      plasma membrane, 136
   PMF, 142
   substrate-level phosphorylation, 142
Allochromatium vinosum, 501
Alterobactin, 346
Alteromonas luteoviolacea, 346
Aluminum
   Al^{3+}, 257–258
   bauxite (*see* Bauxite)
   concentration, 257
   crustal abundance, 257
   plants, 258
AMD, *see* Acid coal-mine drainage (AMD)
2-Aminoethylphosphonate, 273
Ammonia oxidation, 283–284
Ammonification, 283
Amorphous iron sulfide, 521
Amphiaerobes, 35
Amplicon/tag sequencing, 188
Anaerobic methanotrophy, 576–577
Anaerobic mineralization
   definition, 148
   iron(III), 381
   methanogens, 570, 579
   sulfate reducers, 482
Anaerobic oxidation
   ammonium, 285
   arsenic compounds, 301–302
   ferrous iron
      chemotrophic oxidation, 361–364
      phototrophic oxidation, 359–361

Anaerobic oxidation of methane (AOM), 484, 501
Anaerobic respiration
    antimony, 312
    arsenic, 307–309
    catabolic reactions, 137–138
    energy generation, 141–142
    iron, 345, 347, 365, 369
    methanogenesis, 148, 571, 579
    Mn oxides, 411
    peat, 580
    petroleum, 588
    selenium, 554
    sulfur, 481
Anoxygenic photosynthesis, 14, 33, 132, 145
    arsenic, 302
    iron(II), 346, 360–361
    sulfur, 494, 500
Anthropogenic mercury, 326–327
Anticyclones, 103
Antimony
    distribution, 311–312
    microbial oxidation, 312
    microbial reduction, 312–313
    partial/sole energy sources, 297
    terminal electron acceptors, 297
AOM, see Anaerobic oxidation of methane (AOM)
Aquiclude, 120
Archaeoglobus
    A. fulgidus, 482–483
    A. profundus, 482–483
Archean acidophilic iron(II) oxidizers, 358–359
Aridisols, 63
Arsenic compounds
    aerobic oxidation, 299–301
    anaerobic oxidation, 301–302
    arsenate reduction in situ, 310–311
    arsenic-containing minerals
        arsenopyrite (FeAsS), 303–305
        bacterial oxidation, 302
        bioleaching process, 305
        carbonaceous gold ores, 305
        copper mobilization, 304
        enargite (Cu$_3$AsS$_4$), 303–305
        orpiment, 302–303
        scorodite (FeAs$_4$·2H$_2$O), 306
        sequestration, 306
        Sulfolobus acidocaldarius, 303–304
    arsenite oxidation, 310–311
    chemical characteristics, 298
    distribution, 297–298
    microbial reduction, 306–307
    partial/sole energy sources, 297
    respiration
        arrA and arrB cluster codes, 309
        arsenate, 307
        ars operon, 308
        Bacillus arsenicoselenatis, 308
        Bacillus selenitireducens, 308
        bacteria and archaea, 307
        Chrysiogenes arsenatis, 307
        D. auripigmentum strain OREX-4, 308
        Deferribacter desulfuricans, 309
        Marinobacter santoriniensis, 309
        Pyrobaculum aerophilum, 309
        Pyrobaculum arsenaticum, 309

Shewanella sp. strain ANA-3, 308
Shewanella sp. strain HN-41, 308
        strain MLMS-1, 309
        Sulfurihydrogenibium subterraneum strain HGMK1, 309
        Sulfurospirillum arsenophilum, 307
        Sulfurospirillum barnesii strain SES-3, 308
    terminal electron acceptors, 297
    toxicity, 298–299
Arsenopyrite (FeAsS), 303–305
Assimilatory reduction
    ammonium, 286
    iron(III), 365
    methane, 578
    sulfur, 481
Asthenosphere, 7–8, 14
Astrobiology, 85–86
Authigenic minerals
    definition, 57
    manganous carbonate, 222–223, 425
    phosphorite, 269–271
    silica, 242
Autotrophic bacteria, 13, 23
    ammonia, 284–285, 289
    arsenite, 301
    coal, 583
    iron(II), 361
    methane, 571, 574, 576
    methanogens, 77–78, 122
    sulfate reducers, 484, 499–500

B

Bacillus arseniciselenatis, 554, 556
Bacillus beveridgei, 558
Bacillus megaterium
    mercury, 334
    phosphate mobilization, 269
    selenium, 553
Bacillus selenitireducens
    arsenic, 308
    selenium, 554–556, 558
Bacterial chromosome, 130, 331, 333
Bacteria sequester phosphate, 270
Bacterioneuston, 110
Banded iron formations (BIFs), 344, 346, 361, 378–380
Barney Creek Formation (BCF), 177
Barophiles, 111, 150
Basalt, 55, 77, 105
Basic local alignment search tool (BLAST), 200
Bauxite
    bacterial interaction with, 262–263
    bacterial reduction of Fe(III), 262
    maturation phase
        column experiments, 259–260
        Fe(III) reduction, 261–262
        hematite reduction, 262
        iron-depleted bleached zone, 261
        pisolites, 260–262
        protobauxite, 259
        unsterilized Australian ore, 259
    nature, 258
    weathering
        parent rock material, 258–259
        phase, 259

BCF, see Barney Creek Formation (BCF)
Beggiatoa
    discovery, 2
    hydrogen sulfide oxidation, 491, 495
    inorganic matter, 132
    methanotrophs, 577
    microelectrodes, 168
BIFs, see Banded iron formations (BIFs)
Bioleaching, metal sulfides
    heterotrophic uranium mobilization, 536–537
    industrial vs. natural, 537
    kinetic study, 537
    metal sulfide ore bodies, 531–535
    uraninite ores, 535–536
Biooxidation
    chromium, 455
    iron(II), 348
    metal sulfides
        direct oxidation, 525–528
        indirect oxidation, 528–529
        organisms involved, 523–525
        pyrite oxidation, 529–531
    selenium, 553
Bioreduction; see also Microbial reduction
    chromium
        bacterial species, 455–456
        P. fluorescens LB300, 456–458
        sulfate-reducing bacteria, 456
    selenium
        environments, 556–557
        methylation, 555–556
    sulfur
        archaeal microorganisms, 482–483
        bacterial microorganisms, 481–482
        chemosynthetic autotrophs, 484, 499–500
        electron transport system, 498–499
        heterotrophy, 485
        mixotrophy, 484–485, 500
        nitrogen fixation, 485
        oxygen tolerance, 487
        reduction pathways, 485–487
    tellurium, 557–558
Bituminous coal, 3, 535, 538, 540, 582–584, 605
BLAST, see Basic local alignment search tool (BLAST)

C

Calcium carbonate (CaCO$_3$), biological deposition
    bacteria deposit, 210
    calcifying sponges and invertebrates, 211
    ferrous carbonate (FeCO$_3$), 223
    magmatic and metamorphic processes, 210
    magnesium carbonate, 223–224
    manganous carbonate (MnCO$_3$), 222–223
    microbial eukaryotes and microbially induced
            calcification
        articulated coralline (calcareous) algae, 219
        cementation, 222
        coccoliths, 220–221
        echinoderms, mollusks, and arthropods
                calcification, 220
        foraminifera, 220–222
        green algae, 219
        ion transport, 221
        nonsclerosponge siliceous demosponges, 222

microbial metabolisms (see Microbial metabolisms,
            calcium carbonate precipitation)
    ooids and fine carbonate grains, 218–219
    principles, 211–212
    sodium carbonate, 222
    strontium carbonate, 223
Calcium–magnesium carbonates, 228
Caldivirga maquilingensis, 482–483
Calvin–Benson cycle, 147
Candida albicans, 554
Candidatus Arcobacter sulfidicus, 489
Carbonaceous gold ores, 305
Carbonates
    biodegradation
        dissolving, 224
        limestone (see Limestone biodegradation)
    calcium carbonate (see Calcium carbonate (CaCO$_3$),
            biological deposition)
    carbon distribution, Earth's crust, 209–210
Carbon–phosphorus (C–P) lyase, 273
Catalyzed reporter deposition (CARD), 190
CD, see Chlorite dismutase (CD)
Cerium, 468–469
Chalcopyrite, 519
Chelated ferric iron, 347
Chemiosmosis, 142
Chemolithoautotrophs, 131–132, 145
Chemostat, 173–174
Chemotrophic iron(II) oxidizers, 361–364
Chlorite dismutase (CD), 195
Chlorobium limicola, 496–497
Chlorobium phaeobacteroides, 109
Chlorobium vibrioforme, 497
Chloroflexus aurantiacus, 500
Chromatium vinosum, 555
Chromite, 454
Chromium
    biooxidation, 455
    bioreduction
        bacterial species, 455–456
        P. fluorescens LB300, 456–458
        S. oneidensis MR-1, 458–459
        sulfate-reducing bacteria, 456
    Earth's crust occurrence, 454
    microbially generated lixiviants, 455
    microbial reducing activity, in situ rates of, 459
    occurrence, 454
    oxidation states, 454
Cinnabar (HgS), 323
Clark-type oxygen electrodes, 167
CLASI-FISH, see Combinatorial labeling and spectral
            imaging fluorescent in situ hybridization
            (CLASI-FISH)
Clostridium pasteurianum, 487
Coal
    bituminous (see Bituminous coal)
    desulfurization, 584
    microbes, 582–583
    microbial substrate, 583–584
    nature, 581–582
Columbia River Basalt, 122
Combinatorial labeling and spectral imaging fluorescent
            in situ hybridization (CLASI-FISH), 191
Confined aquifer, 120
Crustal divergence, 10–11

D

Denaturing gradient gel electrophoresis (DGGE), 188
Denitrification, 285–286
Denitrifying methanotrophic acetogenesis, 24
Desert soils, 63
Desulfobacter hydrogenophilus, 484
Desulfobacterium autotrophicum, 484
Desulfobacter postgatei, 485
Desulfofustis glycolicus, 501
Desulfomicrobium norvegicum, 555
Desulforhopalus vacuolatus, 482
Desulfotomaculum, 163
    D. acetoxidans, 482
    D. thermobenzoicum, 498
Desulfovibrio
    D. aespoeensis, 79
    D. baarsii, 484
    D. desulfuricans, 484–485
    D. sulfodismutans, 497–498
Desulfurispirillum indicus, 554
Devil's element, 265
DGGE, *see* Denaturing gradient gel electrophoresis (DGGE)
Diagenetic phosphorite, 271
Diatoms, silicon
    auxospore formation, 240
    concentration, 241
    frustules, 241
    glycoprotein, 242
    hydrated amorphous silica, 240
    marine, 240
    morphological variability, 239–240
    N. pelliculosa, 241
    rates of silica uptake, 241
    SDVs, 241
    Thalassiosira nana and *Nitzschia linearis*, 240–241
    unicellular algae, 239–240
    unicellular eukaryotic microorganisms, 239
Direct oxidation
    manganese, 414–415
    metal sulfides, 525–528
Dissimilatory reduction
    ammonium, 286
    antimony, 313
    arsenate, 309
    iron(III), 365
    selenium, 557
    sulfur, 481, 488
    thiosulfate, 488
Divergence, 103
Dystrophic lakes, 117

E

Earth's crust occurrence
    chromium, 454
    iron, 344
    manganese, 425
    mercury, 324
    uranium, 463
    zircon evidence, 20
Electron transport system (ETS)
    components of, 136
    enzyme-catalyzed oxidations, 137
    hydrogen/electrons, 136

Krebs tricarboxylic acid cycle, 134–135
    oxidative phosphorylation, 134
    plasma membrane, 136
    sulfur microbial reduction, 498–499
Elemental selenium, 555
Elemental sulfur
    aerobic oxidation, 491
    anaerobic oxidation, 491–492
    disproportionation, 497
    reduction dissimilatory metabolisms, 487
Enargite (Cu_3AsS_4), 303–305
Enterobacter
    E. aerogenes, 346
    E. cloacae, 554
Enterobactin/enterochelin, 346
Entisols, 63
ETS, *see* Electron transport system (ETS)
Eukaryotic microorganisms, 130–131
Euryhaline, 112
Expert Protein Analysis System (ExPASy), 200
Extended x-ray absorption fine structure (EXAFS), 468

F

Facultative barophiles, 111, 150
Ferric iron
    bacterial fermentation, 365–366
    bacterial respiration, 366–367
    chelators, 346–347
    microbial reduction
        acidic pH, 367–368
        assimilatory reduction, 365
        dissimilatory reduction, 365
        fermentation, 365–366
        neutral pH, 368–369
        organic matter oxidation, 369–372
        siderophores, 346
    nonenzymatic reduction, 374–375
    oxidation states, 344
    precipitation process, 375–376
    terminal electron acceptor, 345
Ferriphaselus amnicola, 350
Ferrous carbonate ($FeCO_3$), 223
Ferrous iron
    chelators, 346–347
    magnetotactic bacteria, 376–378
    microbial oxidation, 346
    nonenzymatic oxidation, 374
    oxidation states, 344
    oxidizing bacteria
        acidophilic oxidizers (*see* Acidophilic iron(II) oxidizers)
        anaerobic oxidation, 359–364
        biochemistry, 364–365
        carbon dioxide fixation, 364
        electron and energy source, 348
        free energy, 364
        neutrophilic, microaerophilic oxidizers (*see* Neutrophilic, microaerophilic iron(II) oxidizers)
        precipitation process, 375–376
FISH, *see* Fluorescence in situ hybridization (FISH)
FISH-MAR technique, 190–191
FixK, 200
Florida Escarpment, 14

Fluorescence-activated cell sorting (FACS), 196
Fluorescence in situ hybridization (FISH), 163, 189
    CARD method, 190
    magneto-FISH, 196
    MAR, 190–192
    SIMS, 192
Fossil fuels
    coal (see Coal)
    methane gas (see Methane)
    natural abundance, 566–567
    peat, 580–581
    petroleum (see Petroleum)
    shale oil, 604
    trapped organic carbon, 566
Fougèrite, 26
Freshwater lakes
    evolution, 118
    fertility, 117–118
    lake bottoms, 117
    lentic environments, 114
    microbial populations, 118
    physical and chemical features
        cold monomictic lakes, 115
        dimictic lakes, 115–116
        hydrostatic influence, 114
        meromictic lakes, 115
        morphological influence, 114
        oligomictic lakes, 115
        polymictic lakes, 115
        thermal influence, 114
        thermal stratification, 115
        warm monomictic lakes, 115
    watershed, 114
Freshwater manganese oxide deposits
    engineered systems, 429
    ferromanganese concretions and nodules, 429
    rivers, creeks and ponds, 428–429
Freshwater neutrophilic iron(II)-oxidizing bacteria, 348–349

G

Galena (PbS), 523
Gallionella
    G. capsiferriformans ES-1, 353
    G. ferruginea, 2, 349–351
Geobacter metallireducens, 462, 536
Geological microbiology, 1
Geomicrobial process
    anabolism, 145–147
    carbon assimilation, 147
    catabolism
        aerobic respiration (see Aerobic respiration)
        anaerobic respiration, 137–138
        electron donors/acceptors, 138–140
        fermentation, 140–141
    chemolithoautotrophic bacteria, 143–144
    cyanobacteria, 144
    enzymatic and nonenzymatic, 177
    eukaryotes, 129–131 (see also Eukaryotic
        microorganisms)
    exogenous organisms, 163
    geologic agents of
        concentration, 132
        dispersion, 132–133
        fractionation, 133

green sulfur bacteria, 144
heterotrophic bacteria, 149
indigenous organisms, 163
in situ measurements (see In situ measurements)
isolation and characterization organisms, 171
laboratory reconstruction
    air-lift column, ore leaching, 172
    chemostat, 173–174
    chloride dilution, 175
    downflow hanging sponge reactors,
        172, 174
    percolation columns, 172–173
lithification, 133
magmatic activity/volcanism, 133
mineral diagenesis, 133
mineralization
    aerobic mineralization, 148
    anaerobic mineralization, 148
    definition, 148
    marine humus, 148–149
physical parameters
    hydrostatic pressure, 150
    pH and E_h, 150–151
    temperature, 149–150
prokaryotes (see Prokaryotes)
purple sulfur bacteria, 144
rock weathering, 133
sampling
    aquatic sampling, 159–161
    microbes/microbial biomarkers, 158
    storage, 161–162
    terrestrial surface/subsurface, 159
    water, sediment and mineral/rock, 159
solid reaction product, 169–171
solid substrate, 175–176
surface saturation, 176–177
Gloeobacter, 34
Gold–amalgam (Au–Hg) electrodes, 168
Granite, 55–56
Gravitational water, 60
Greengenes Database, 188
Groundwaters
    aquifers, 120
    autotrophic methanogens, 122
    connate water, 119–120
    definition, 119
    ferrous silicate and water, 122
    homoacetogens, 122
    microbes, 121
    microbial contamination, 121
    surface water, 119–120
    thermodynamic basis, 122
    vadose zone, 120
    water table, 121
Gulf Stream, 100, 103

H

Halomonas, 112
Heliobacteria, 33
$HetCO_2$-MAR technique, 191
Heterotrophic nitrification, 284–285
Horizontal gene transfer (HGT), 189
Hot spots, 10
Hydrogen sulfide (H_2S) amperometric electrodes, 167

I

Indirect oxidation
    metal sulfides, 528–529
    Mn(II), 415
Inorganic minerals, 56–57
In situ chromate-reducing activity, 459
In situ measurements
    geomicrobial agents
        AFM and cryo-EM, 164
        capillary technique, 165
        classical buried slide method, 164
        fluorescent stains, 164
        laser Raman spectroscopy, 166
        SEM, 164, 166
        visual observation, 164
    ongoing geomicrobial activity
        amperometric electrodes, 167
        Au–Hg electrodes, 168
        electrochemical probes, 167
        metabolic compounds, 166
        microelectrode, 167
        Mn cycling, 169
        potentiometric electrodes, 167
        radioisotopes, 166
        redox gradient concentrations, 168
        rRNA, 166
        sulfate-reducing activity, 166–167
        voltammetric electrodes, 167–168
    past geomicrobial activity
        biomarker okenane, 177
        carotenoids preservation, 177
        isotopic fractionation, 178
        sedimentary deposits, 177
Iron
    aerobically living cells, 345
    anoxic environments, 345–346
    BIFs (*see* Banded iron formations (BIFs))
    biological importance
        microorganisms, 346–347
        prokaryotic and eukaryotic cells, 345–346
        redox chemistry, 347
    Earth's crust distribution, 344
    geochemical properties, 344
    microbial and abiotic cycling, 345
    microbial cycle, 382–383
    ore, soil, and sediments, 380–382
    oxidation states, 344
    sedimentary deposits, 378–380
    siderophore, 346–347
Iron(II), *see* Ferrous iron
Iron(III), *see* Ferric iron
Iron-depositing bacteria, 2

K

Kesterson National Wildlife Refuge, California, 557

L

Laccase-driven Mn(II) oxidation, 414
Lakes, *see* Freshwater lakes
Lepidocrocite, 380
Leptospirillum ferrooxidans, 357–358
Leptothrix ochracea, 2

Life on Earth

Life on Earth
    alkaline submarine hydrothermal mound
        hypothesis
        ATP synthetase, 25
        fougèrite, 26
        Hadean Ocean floor, 23–24
        peptides, 27
        redox bifurcating enzymes, 26
    de novo appearance, 21
    evolution
        methanogens, 29–30
        Precambrian (*see* Precambrian period evolution
           of life)
        retrodiction, 29
    organic soup theory, 21–22
    paleontological evidence
        microfossils, 37–38
        molecular biology studies, 38
        organic geochemistry, 38
    panspermia hypothesis, 20–21
    "pyrite-pulled" surface metabolism theory, 22–23
    RNA world theory, 28
    unresolved issues
        endosymbiosis, 39–40
        open-system hydrothermal convection, 39
Limestone biodegradation
    ammonifiers, 224
    black fungi, 225
    chemosynthetic autotrophs, 226
    cyanobacteria, algae, and fungi, 226–228
    decay through microbial action, 224
    microbial weathering of rock surfaces, 225
    microcolonial black yeast, corrosive effect, 224–225
    organic acids and CO_2, 224
    sulfur oxidizers, 224
Lithification, 55

M

Magnesium carbonate, 223
Magneto-FISH method, 196
Magnetospirillum gryphiswaldense, 376
Magnetotactic bacteria, 14
Manganese (Mn)
    chemistry
        kinetics, 406–408
        physicochemical properties, 404
        thermodynamics, 404–406
    cycles
        mineral phases, 420–422
        Mn(III) complexes, 420
        schematic diagram, 402
    Earth's crust, 425
    freshwater oxide deposits
        engineered systems, 429
        ferromanganese concretions and nodules, 429
        rivers, creeks and ponds, 428–429
    lignin degradation, 403
    marine oxide deposits
        crusts, 431–433
        deep-sea concretions, 431–433
        hydrothermal systems, 431
        manganese concentration, 430
        nodules, 431–433
        water column/coastal waters, 430–431

metal adsorption, 403–404
microbes–Mn deposit formation
    implication evidence, 425–426
    Mn removal, 426–427
    soil percolation experiments, 426
    visualization and culturing approach, 427
microbial oxidation
    bacteria, phylogenetic tree of, 408–409
    electron transfer process, 405–406
    fungi, 408–409
    homogenous oxidation, 404–405
    indirect oxidation, 415
    laccases, 414
    MCOs, 412–414
    mechanisms pathways, 411–412
    oxidation rates, 406–407
    peroxidase, 414–415
    physiological reasons for, 409–411
    taxonomic and environmental diversity, 408–409
    transmission electron micrograph, 410–411
microbial reduction
    anaerobic methane oxidation, 416–417
    c-type cytochrome, 417–418
    electrode potentials *vs.* standard hydrogen
      electrode, 407
    environmental diversity, 416
    indirect biological reduction, 418–420
    Metallogenium-like structures, 424–425
    microbial exudates, 423–425
    nitrogen oxidation, 417
    physiological reasons for, 416
    reduction rates, 408
    taxonomic diversity, 416
    terminal electron acceptors, 417
organic carbon mineralization, 403
oxidation states, 404
oxygenic photosynthesis, 403
paleoredox indicator, 404
schematic biogeochemical process impact, 402
technological applications, 404
terrestrial oxide deposits
    caves, 428
    desert varnish, 427–428
    ores, 428
    soils, 427
Manganous carbonate (MnCO$_3$), 222–223
MAR, *see* Microautoradiography (MAR)
Marine humus, 148–149
Marine manganese oxide deposits
    crusts, 431–433
    deep-sea concretions, 431–433
    hydrothermal systems, 431
    manganese concentration, 430
    nodules, 431–433
    water column/coastal waters, 430–431
Marine neutrophilic iron(II)-oxidizing bacteria,
    348–349, 351
Mariprofundus ferrooxydans, 350–351
MCOs, *see* Multicopper oxidases (MCOs)
Mercury (Hg)
    anthropogenic mercury, 326–327
    cinnabar (HgS), 323
    distribution
      abundance, Earth's crust, 324
      bioconcentration factors, piscivorous fish, 325

      concentration, 324–325
      detoxification, 326
      dimethylmercury, 325
      inorganic compounds, 325
      mercuric ions, 325
      microbial environmental methylation, 326
      transformations, 324
    environmental hazard, 323–324
    microbial methylation, 326
      early studies, 327–328
      enzymatic methylation, 328–329
      genetic basis, 329–331
    microbial reduction
      algae cultures, 333
      chemical reduction, 333
      cytochrome c oxidase enzymes, 333
      enzymatic, 331–332
      iron-oxidizing bacteria, 333
    Minamata disease, 324
    organomercurials, 333–334
    oxidation, 334–335
    precipitation, 335
    syphilis treatment, 323
    toxicity, inorganic and organic forms, 323
Mercury methylation
    by archaea, bacteria, and fungi, 326
    detoxification, 326
    early studies, 327–328
    enzymatic methylation, 328–329
    genetic basis
      cysteine, 330
      D. desulfuricans ND132, 329
      Euryarchaeota, 329
      Firmicutes, 329
      G. sulfurreducens PCA, 329
      hgcA and *hgcB* genes, 329
      Hg^{II}, 330–331
      methylators, 329–330
      phylogenetic tree, 330
      Proteobacteria, 329
      sulfate-reducing bacteria, 331
    by pH, 326
    sulfide and NOM, 326
Mesophiles, 111
Mesophilic acidophilic iron(II) oxidizers
    Acidithiobacillus ferrooxidans, 354–357
    genetics, 358
    Leptospirillum ferrooxidans, 357–358
Mesotrophic lakes, 117
Metagenomics, 194
Metal sulfides
    acid coal-mine drainage, 531, 538–540
    biogenesis, laboratory evidence of
      batch cultures, 521–522
      column experiment, 522
    bioleaching
      heterotrophic uranium mobilization,
        536–537
      industrial *vs.* natural, 537
      kinetic study, 537
      metal sulfide ore bodies, 531–535
      uraninite ores, 535–536
    biooxidation
      direct oxidation, 525–528
      indirect oxidation, 528–529

organisms involved, 523–525
pyrite oxidation, 529–531
formation principles, 520
geomicrobial interest, 518
natural origin
biogenic origin, 519–520
hydrothermal origin, 518–519
ore bioextraction, 537
Methane
biogenic accumulations, 566–567
methanogenesis
bioenergetics, 572–574
carbon cycle, 579
carbon fixation, 574
pathways, 573
methanogens
carbon fixation, 574
morphologies, 568–569
sulfate reducers, 570–571
microbial oxidation
aerobic, 574–576
anaerobic, 576–578
physical properties, 566
Methanobacterium palustre, 568
Methanogenium liminatans, 568
Methanogens
carbon fixation, 574
life on Earth, 29–30
morphologies, 568–569
sulfate reducers, 570–571
Methanosaeta concilii, 568
Methylcobalamin–protein complex, 328
Methyl-THF, 328
Microautoradiography (MAR), 76, 190–192
Microbial biogeochemistry, 1–2
Microbial dark matter, 81
Microbial ecology, 1–2
Microbial habitat
oceans (*see* Oceans)
uppermost lithosphere
mineral soil (*see* Mineral soil)
organic soils, 66
rock (*see* Rock)
Microbial habitat, Earth
biosphere
atmosphere, 14
autotrophic bacteria, 13
hydrogen sulfide–oxidizing bacteria, 14
hydrosphere, 12–13
lithosphere, 12–13
living microorganisms, 12
magnetotactic bacteria, 14
seawater–basalt interaction, 13
submarine communities, 14
temperature and hydrostatic pressure, 12
core, 7
crust, 7–9
East Pacific Rise, 10–11
mantle, 7
Mid-Atlantic Ridge, 10–12
origin and movement, continents, 10
Pangaea, 10, 13
volcanic activity, 10

Microbial life, 3
biomolecules, 74
electron acceptors
ferric iron, 75–76
Gibbs free energies, 74–75
respiratory organisms, 74
selenium, 76
shallow and deep subsurface, 74
sulfate reduction, 76–77
electron donors, 74
endospores, 74
H_2-based subsurface biosphere, 77–78
physiological adaptations, 78
temperature and pressure, 73–74
viruses and eukaryotes, 78–79
Microbial metabolisms, calcium carbonate precipitation
aerobic/anaerobic oxidation
carbon compounds, 212–213
organic nitrogen compounds, 213
CO_2 removal
benthic cyanobacteria, 216
calcified cyanobacteria, 216
calcite and gypsum precipitation, 216–217
marine cyanobacteria, 218
photosynthetic organisms, 216
Synechococcus sp., 216–217
travertine and lacustrine carbonate, 216
SO_4^{2-} to H_2S reduction, 213–215
Microbial oxidation
antimony, 312
ferrous iron, 346
manganese
bacteria, phylogenetic tree of, 408–409
electron transfer process, 405–406
fungi, 408–409
homogenous oxidation, 404–405
indirect oxidation, 415
laccases, 414
MCOs, 412–414
mechanisms pathways, 411–412
oxidation rates, 406–407
peroxidase, 414–415
physiological reasons for, 409–411
taxonomic and environmental diversity, 408–409
transmission electron micrograph, 410–411
mercury, 334–335
methane
aerobic, 574–576
anaerobic, 576–578
molybdenum, 460–461
phosphorus, 273–274
sulfur
aerobic oxidation, 489–490
anaerobic oxidation, 490–491
ecology, 494–497
heterotrophic and mixotrophic oxidation, 491
mechanism, 494
sulfite oxidation, 492
tetrathionate oxidation, 494
tellurium, 557
thiosulfate, 492
uranium, 463–464
vanadium, 461–462

Microbial reduction
    antimony, 312–313
    arsenic compounds, 306–307
    ferric iron
        acidic pH, 367–368
        assimilatory reduction, 365
        dissimilatory reduction, 365
        fermentation, 365–366
        neutral pH, 368–369
        organic matter oxidation, 369–372
        siderophores, 346
    manganese
        anaerobic methane oxidation, 416–417
        c-type cytochrome, 417–418
        electrode potentials *vs.* standard hydrogen
            electrode, 407
        environmental diversity, 416
        indirect biological reduction, 418–420
        Metallogenium-like structures, 424–425
        microbial exudates, 423–425
        nitrogen oxidation, 417
        physiological reasons for, 416
        reduction rates, 408
        taxonomic diversity, 416
        terminal electron acceptors, 417
    mercury
        algae cultures, 333
        chemical reduction, 333
        cytochrome c oxidase enzymes, 333
        enzymatic, 331–332
        iron-oxidizing bacteria, 333
    molybdenum, 460–461
    MSR (*see* Microbial sulfate reduction (MSR))
    uranium, 464–465
Microbial sulfate reduction (MSR)
    archaeal microorganisms, 482–483
    bacterial microorganisms, 481–482
    carbonate, 213–214
    chemosynthetic autotrophs, 484, 499–500
    electron transport system, 498–499
    heterotrophy, 485
    mixotrophy, 484–485, 500
    nitrogen fixation, 485
    oxygen tolerance, 487
    reduction pathways, 485–487
Micrococcus selenicus, 553
Microcosms, 158
Microsensors, 167
MICs, *see* Minimum inhibitory concentrations (MICs)
Mineral soil
    evolution
        microbes effect, 59–60
        plants and animals effect, 59
    microbial distribution in mineral
        algae, 65
        bacterial distribution, 65–66
        determination of, 65–66
        fungi, 64–65
        prokaryotes, 64
        protozoa, 65
    nutrient availability, 61–62
    origin, 57–58
    plants and animals effect, 59

soil water distribution, 60–61
structural features of, 58–59
types
    aridisols and entisols, 63
    mollisols, 59, 62
    oxisols, 62
    spodosols, 59, 62
    tundra, 62
water effects, 60
Minimum inhibitory concentrations (MICs), 552
Mitochondria, 130, 138, 140
Molecular methods
    bioinformatic approaches, 200
    community genomics and gene expression,
        194–195
    cultures and model systems, 190
    genomic approaches, 195–196
    isotope array, 193–194
    metabolic assays, 192–193
    mutagenesis
        biochemical reaction, 198
        gain-of-function strategy, 199–200
        loss-of-function approach, 197
        random/directed, 197
        transposons, 197–198
        wild-type phenotype, 199
    putative marker genes, 201
    single-cell isotopic techniques, 190–192
    SIP, 193
    taxonomic diversity (*see* 16S Ribosomal RNA
        (rRNA) molecules)
Molybdenite (MoS_2), 460
Molybdenum
    microbial oxidation and reduction, 460–461
    occurrence and properties, 460
Molybdite (MoO_3), 460
MSR, *see* Microbial sulfate reduction (MSR)
Multicopper oxidases (MCOs), 412–414

N

Natural organic matter (NOM), 325–326
Neptunium, 466
Neutrophilic, microaerophilic iron(II) oxidizers
    abiotic and biotic oxidation, 348
    appendaged bacteria
        Ferriphaselus amnicola, 350
        Gallionella ferruginea, 349–351
        Leptothrix cholodnii, 352–353
        Leptothrix ochracea, 352
        Mariprofundus ferrooxydans, 350–351
    marine and freshwater environments, 348–349
    reaction, 348
    unicellular bacteria
        G. capsiferriformans ES-1, 353
        Sideroxydans lithotrophicus strain ES-1, 353
Nitrate-reducing iron(II) oxidizers, 361–364
Nitrification, 283
Nitrite oxidation, 284
Nitrogen (N)
    ammonia oxidation, 283–284
    ammonification, 283
    anaerobic ammonium oxidation, 285

in biosphere, 281–282
denitrification, 285–286
dissimilatory nitrate reduction, 286
fixation
    asymbiotic/nonsymbiotic, 287
    control, marine environment, 288
    growth-limiting factor, 282–283
    iron protein and molybdoprotein, 286
    iron protein and vanadoprotein, 286
    microbial depletion, 286
    nitrogenase, 286
    reduced ferredoxin, 287
    symbiotic, 287–288
heterotrophic nitrification, 284–285
microbial nitrogen transformations, 282
nitrification, 283
nitrite oxidation, 284
stable isotopes
    ^{14}N and ^{15}N, 288
    natural abundance, 288–289
    tracer experiments, 289–290
Nonenzymatic iron(II) oxidation, 374
Nonferrous sedimentary sulfide deposits, 520

O

Obligate barophiles, 111, 150
Ocean eddies, 103
Oceans
    bacterioplankton, 97
    calcareous oozes, 99
    continental margin, 98
    continental rise, 98–99
    continental slope and shelf, 98
    diatomaceous oozes, 99
    fauna, 97
    flora, 97
    guyots, 99
    microbial distribution
        bacterioneuston, 110
        deep-sea sediments, 111
        euryhaline, 112
        growth and reproduction, 111
        Halomonas, 112
        microbial loop, 110
        nonphotosynthetic microorganisms, 110
        phytoplankton distribution, 109–110
        sedimentary humus, 110
        stenohaline, 112
        temperature and pressure, 111
    microbial flora, 113–114
    microscopic appearance, marine sediments, 99–100
    motion
        anticyclones, 103
        Coriolis force, 100
        ocean eddies, 103
        rings, 103
        surface currents, 100–102
        thermohaline circulation, 103
        upwelling regions, 103
    northern hemisphere, 97
    ocean basin, 98–99
    ocean floor, 99
    oceans cover, 97
    ocean trenches, 98

phytoplankters
    geomicrobial interest, 113
    primary producers, 112–113
phytoplankton, 97
radiolarian oozes, 99
seawater (*see* Seawater)
southern hemisphere, 97
turbidity currents, 98
world, 98
zooplankton, 97, 113
Ocean trenches, 98
Oligotrophic lakes, 117
Operational taxonomic units (OTUs), 188
Oregon Subduction Zone, 14
Organic minerals, 56
Organic soils, 66
Organic soup theory, 21–22
Organomercurials, 333–334
Organophosphorus compounds, 273
Orpiment (As_2O_3), 302–303
OTUs, *see* Operational taxonomic units (OTUs)
Oxidation–reduction potential (ORP)
    electrodes, 167
Oxidative phosphorylation, 134, 141–142
Oxygen-evolving center (OEC), 33–34
Oxygenic photosynthesis
    amphiaerobes, 35
    amphiaerobic prokaryotes, 35
    cyanobacteria, 33–34
    manganese, 34

P

Paleomicrobiology, 3
Paleo pasteurized basins, 73
Passive silica biomineralization
    Aquificales, 245
    authigenic silica precipitation, 242
    Bacillus subtilis, 245
    bacterial and bacteria-free systems, 245
    cyanobacteria, 244–245
    gram-negative bacteria, 244
    gram-positive bacteria and eukaryotes, 244
    nucleation and growth of amorphous silica, 242
    polymers, 242
    siliceous sinters, 243
    silicified bacteria, 243–244
PCR, *see* Polymerase chain reaction (PCR)
Peat
    microbes, 580–581
    nature, 580
Pelagibacter ubique, 113
Penicillium sp., 558
Perched aquifer, 120
Peroxidase-driven Mn(II) oxidation, 414–415
Petroleum
    biodegradation
        aerobic degradation, 594–595
        anaerobic degradation, 595–596
        aromatic compounds, 596–597
        microbes, 603–604
        organic sulfur removal, 603
        shale oil, 604
    environmental factors, 591–594
    microbes, 586

microbial habitat, 586–589
microorganisms, 589–591
nature, 584–586
reservoir rock migration, 586
pH limits, 3
Phosphoenolpyruvate, 273
Phospholipid fatty acids (PLFAs), 193
Phosphonoacetaldehyde hydrolase, 273
Phosphonoacetate hydrolase, 273
Phosphonopyruvate hydrolase, 273
Phosphorite deposition
    authigenic phosphorite, 269–271
    diagenetic phosphorite, 271
    insoluble calcium phosphate compounds, 269
    micronodules, 269–270
    occurrences, 271
    role of microbes, 269
Phosphorus
    anhydrides, 265–266
    assimilation, 267–268
    chemical hydrolysis, 265
    Devil's element, 265
    in Earth's crust, 266
    immobilization
        Citrobacter sp., 271
        inorganic precipitates, 269
        phosphorite deposition, 269–271
        struvite, 271–272
        transitory phosphate, 269
        vivianite, strengite, and variscite, 271
    microbial oxidation, 273–274
    organic–inorganic conversion, 266–267
    organophosphorus compounds, metabolism, 273
    phosphate reduction, 272–273
    phosphodiesterases, 267
    phosphomonoesterases, 267
    solubilization, phosphate minerals, 268–269
Photoautotrophic iron(II) oxidizers, 348, 359–361
Photoheterotrophs, 132, 147
Photolithoautotrophs, 132
Phyllomanganates, 420
Phylogenetic profiling, 200
Pisolites, 260–262
Plasmids, 130
PLFAs, see Phospholipid fatty acids (PLFAs)
Plutonium, 466–467
Polonium, 463
Polymerase chain reaction (PCR)
    amplification, 194–195
    DNA extraction, 188–189
    mutagenesis, 198
    single-cell genomic approaches, 195–196
    16S rRNA gene, 188–189
Porphyry sulfide ores, 518–519
Precambrian period evolution of life
    geologic timescale divisions, 30
    Isua rocks, 31
    metasediments, 31
    ocean floor, 30
    oxygenic photosynthesis
        amphiaerobes, 35
        amphiaerobic prokaryotes, 35
        cyanobacteria, 33–34
        eukaryotic cells, 36
        manganese, 34

stromatolites
    cyanobacteria, 31–32
    Strelley Pool Chert, 31–33
Primary/igneous minerals, 57
Prokaryotes
    anaerobic autotrophs, 29
    bacteria and archaea
        bacterial chromosome, 130
        chemolithoautotrophs, 131–132
        deoxyribonucleic acid, 129–130
        mitochondria, 130
        molecular size, 130
        photoheterotrophs, 132
        photolithoautotrophs, 132
    oxygen-utilizing, 35
    Pangaea, 10
Protobauxite, 259
Proton motive force (PMF), 142
Pseudomonas
    P. aeruginosa, 346, 558
    P. ambigua, 454
    P. fluorescens, 456
    P. mendocina, 346
    P. stutzeri, 346
Psychrotrophs, 111
Pyrite oxidation, 529–531
"Pyrite-pulled" surface metabolism theory, 22–23

R

Radiolarians, 242
Rare earth elements (REEs)
    applications, 469–470
    biogenic mineral interactions, 468
    microbial interactions, 467–468
    occurrence and properties, 467
Rhodobacter sphaeroides, 555
Rhodopseudomonas palustris, 178
Rhodotorula yeast, 346
Ribosomal Database Project (RDP), 188
RNA world theory, 28
Rock
    definition, 55
    igneous origin, 55–56
    metamorphic origin, 56
    sedimentary origin, 55

S

Salmonella typhimurium, 346, 500
Scandium, 467
Scorodite ($FeAs_4 \cdot 2H_2O$), 306
Seawater
    alkaline pH and elevated E_h, 105
    chemical components, 103–104
    chlorinity measurements, 104
    convergence, 107
    density of, 106–107
    euphotic zone, 109
    hydrostatic pressure, 105
    osmotic pressure, 106
    oxygen concentration, 108
    pH-buffering system, 105
    phytoplankton, 108–109
    plagioclase, 105

practical salinity, 104
temperature, 106
titrimetric chlorinity determinations, 104
Secondary ion mass spectrometry (SIMS), 169–170
Secondary mineral, 57
Sedimentary humus, 110
Sedimentary iron deposits, 378–380
Sedimentary metal sulfides, 519–520
Selenate, environmental bacterial reduction, 556–557
Selenihalanaerobacter shriftii, 554
Selenite, environmental bacterial reduction, 556–557
Selenium
    biological importance, 551–552
    bio-oxidation, 553
    bioreduction
        environments, 556–557
        methylation, 555–556
    cycle, 557
    Earth's crust occurrence, 551
    toxicity, 552–553
Shale oil, 604
Shewanella, 488–489
    S. oneidensis MR-1, 458–459, 487–488, 558
    S. putrefaciens, 498, 536
Sideroxydans lithotrophicus strain ES-1, 353
Silica deposition vesicles (SDVs), 241
Silicoflaggelates, 242
Silicon
    biologically important properties, 238–239
    biomobilization
        acids, 247–249
        alkali, 249
        bioweathering, 246
        extracellular polysaccharide, 249
        ligands, 246–247
        mode of attack, 246
    chemical properties, 237–238
    diatoms
        auxospore formation, 240
        concentration, 241
        frustules, 241
        glycoprotein, 242
        hydrated amorphous silica, 240
        marine, 240
        morphological variability, 239–240
        N. pelliculosa, 241
        rates of silica uptake, 241
        SDVs, 241
        Thalassiosira nana and *Nitzschia linearis*, 240–241
        unicellular algae, 239–240
        unicellular eukaryotic microorganisms, 239
    passive silica biomineralization
        Aquificales, 245
        authigenic silica precipitation, 242
        Bacillus subtilis, 245
        bacterial and bacteria-free systems, 245
        cyanobacteria, 244–245
        gram-negative bacteria, 244
        gram-positive bacteria and eukaryotes, 244
        nucleation and growth of amorphous silica, 242
        polymers, 242
        siliceous sinters, 243
        silicified bacteria, 243–244
    radiolarians, 242
    role of microbes, silica cycle, 249–250

silicoflaggelates, 242
sponges, 242
Siljan Ring, central Sweden, 12
16S ribosomal RNA (16S rRNA) molecules
    amplicon/tag sequencing, 188
    clone sequences, 188
    culture-independent approach, 188
    DGGE/t-RFLP, 188
    DNA–DNA hybridization, 188
    fluorescence microscopy technique, 189
    HGT, 189
    Illumina sequencing, 188
    next-generation sequencing, 189
    OTUs, 188
    PCR, 188
    sequence conservation, 188
    sequence variability, 188
    taxon-specific sequences, 190
    V3 and V6 regions, 188
Sodium carbonate, 222
Sphalerite (ZnS), 519, 522
Spodosols, 59, 62
Stable isotope probing (SIP), 193, 195
Stenohaline, 112
Strontium carbonate, 223
Struvite, 271–272
Subduction, 8
Sulfate-reducing bacteria, 517
Sulfolobus acidocaldarius, 174
Sulfur
    assimilation pathway, 500–501
    assimilatory transformations, 481
    biogeochemical cycling
        metabolic activity, 503
        metabolisms and redox transformations, 481
        microbial consortia, 501–503
        roles, 480
    disproportionation, 497–498
    dissimilatory transformations, 481
    Earth's crust occurrence, 480
    geomicrobially important forms, 480
    microbial oxidation
        aerobic oxidation, 489–490
        anaerobic oxidation, 490–491
        ecology, 494–497
        heterotrophic and mixotrophic oxidation, 491
        mechanism, 494
        sulfite oxidation, 492
        tetrathionate oxidation, 494
    microbial reduction
        archaeal microorganisms, 482–483
        bacterial microorganisms, 481–482
        chemosynthetic autotrophs, 484, 499–500
        electron transport system, 498–499
        heterotrophy, 485
        mixotrophy, 484–485, 500
        nitrogen fixation, 485
        oxygen tolerance, 487
        reduction pathways, 485–487
Sulfurospirillum barnesii, 308, 554, 558

T

Technetium, 462
Tellurates, microbial reduction of, 557–558

Tellurium
    biological requirement, 552
    bioreduction, 557–558
    Earth's crust occurrence, 551
    microbial oxidation, 557
    toxicity, 552–553
Temple and LeRoux metal sulfide column modeling, 523
Terminal restriction fragment length polymorphism
            (t-RFLP), 188
Terrestrial manganese oxide deposits
    caves, 428
    desert varnish, 427–428
    ores, 428
    soils, 427
Terrestrial subsurface ecosystem
    assembly-based metagenomic approaches, 81
    biosphere environments
        igneous and metamorphic rocks, 71–72
        sedimentary systems, 71
        shallow and deep microbial habitats, 70
        vadose zone, 72–73
    community metagenomic approaches, 81
    drilling and coring, 79–81
    ecological analyses, 82
    functional microarrays, 82
    genomic inferences, 81
    groundwater wells, 81
    microbial life (see Microbial life)
    subsurface–human interactions
        contamination and remediation, 84
        gas and hydrocarbon production, 82–84
        sequestration of carbon dioxide, 85
        waste repositories, 85
Tetrahydrofolate (THF), 328
Thauera selenatis, 553–554
Thermocladium modestis, 482–483
Thermocline, 106
Thermohaline circulation, 103
Thermoplasma acidophilum, 539
Thermotogales, 488
Thiobacillus denitrificans
    ammonium, 362
    sulfide, 490–491
    uranium, 463
Thiobacillus ferrooxidans, 3, 456
Thiobacillus (Acidithiobacillus) ferrooxidans TM
    antimony
        enzymatic oxidation, 313
        synthetic antimony sulfides oxidation, 312
        tetrahedrite oxidation, 312
        trivalent antimony, 312
    arsenic compounds
        acidophilic, 303
        arsenopyrite, 303–305
        carbonaceous gold ores, 305
        enargite, 304–305
        orpiment, 302–303
Thiobacillus intermedius, 500

Thiobacillus neapolitanus
    silica, 248
    sulfite, 492, 499
Thiobacillus organoparus, 500
Thiobacillus perometabolis, 500
Thiobacillus thioparus
    CuS, 527
    sulfide, 489–493, 495
    sulfur, 539
Thiobacillus versutus, 493, 499–500
Thiomicrospira denitrificans, 491–493
Thiosulfate
    disproportionation, 497
    oxidation, 492
    reduction dissimilatory metabolisms, 487
Time-of-flight SIMS (ToFSIMS), 192
Time-resolved laser-induced fluorescence spectroscopy
            (TRLFS), 468
t-RFLP, see Terminal restriction fragment length
            polymorphism (t-RFLP)
Tundra soil, 62

U

Unconfined aquifer, 81, 84, 120
Upwelling
    AOM, 577
    definition, 103, 122
    phosphorite, 270–271
Uraninite (UO_2)
    leaching, 535–536
    microbial pretreatment, 517–518
Uranium
    bacterial photosynthesis, 31
    bioremediation, 465–466
    microbial oxidation, 463–464
    microbial reduction, 464–465
    occurrence and properties, 463
    terrestrial subsurface, 79

V

Vanadium
    bacterial oxidation, 461–462
    groundwater subsurface contamination, 84
    occurrence, 461
Veillonella atypica, 558

W

Wächtershäuser's surface metabolism, 22–23
Weichert–Gutenberg discontinuity, 7
Whole genome amplification, 194

Y

Yellowstone National Park, Montana, 174–175, 301,
            307, 379